# Science Changed the World



**Anthony N. Stranges**

*Texas A&M University*

Kendall Hunt
publishing company

Cover images are in the public domain.

Kendall Hunt
publishing company

www.kendallhunt.com
*Send all inquiries to*:
4050 Westmark Drive
Dubuque, IA 52004-1840



ISBN 978-1-4652-8448-8



Printed in the United States of America

For Krista and Kara, my daughters and everlasting students

## CONTENTS

# PREFACE

## I. INTRODUCTORY

*Science Changed the World* examines the origin, development, and impact of science from ancient times to the present. It includes subject matter from secondary sources and original research based on primary sources.

Until recently most historians of science considered themselves specialists in a particular branch of science or time period and not generalists. They published books on specific time periods such as the renaissance science, the scientific revolution, the chemical revolution, or modern physical science; on special topics such as natural history and evolution; or biographies of scientists. They wrote for other historians of science and not for a wider audience. As a result, only a few historical surveys of science, such as Stephen Mason *History of the Sciences* (1962), Colin A. Ronan *Science* its *History and Development among the World's Cultures* (1982), John Marks *Science and the Making of the Modern World* (1983), and Anthony Alioto *A History of Western Science* (1992) were available. The situation has begun to change and at least two general surveys have appeared in the last ten years, Andrew Ede and Lesley Cormack, *A History of Science in Society* (2004) and Peter Bowler and Iwan Morus, *Making Modern Science* (2005).

A second reason for the lack of introductory historical surveys of science is the enormity of the project. Writing a chronological and topical survey of science presents a formidable challenge for any historian of science because it requires a comprehensive examination of the contributions of physicists, chemists, biologists, and other scientists over 2,500 years. Nevertheless, writing *Science Changed the World* proved a rewarding experience. Its breadth and depth of coverage provides a wealth of information for those teaching introductory history of science survey courses.

During my years of teaching history of science courses I always have looked for interesting and innovative ways to introduce students to the history of science and at the same time to determine what they know about the history of science, or their understanding of the history of science. Of the many innovations I have tried, *the first day test,* remains the most successful. The 45-50 students who enroll in my introductory survey courses are mainly from the colleges of Liberal Arts, Science, and Engineering and very few, if any, have taken a history of science course. To test their historical knowledge of science I ask them to prepare two lists. The first is a list of ten scientists, engineers, or inventors who have contributed to our understanding of the world in which we live and whom they expect to study in this course. The second is a list of ten scientific discoveries or inventions that have contributed to our understanding of or have changed the world in which we live and which they expect to study in this course, which is *History of Science*. The lists are both surprising and predictable.

Most students have ten names on their first list. Albert Einstein and Isaac Newton appear on every student's list followed in decreasing frequency by Charles Darwin, Thomas Edison, Alexander Graham Bell, and Benjamin Franklin, who appear on one-half or one-third of the lists. J. Robert Oppenheimer, Richard Feynman, Niels Bohr, Werner Heisenberg, Galileo Galilei, Louis Pasteur, Gregor Mendel, James Watson, Francis Crick, Stephen Hawking, Archimedes, and Nicolas Copernicus comprise the next group of scientists mentioned most frequently.

What other information does an examination of the list reveal? First, Newton and Einstein, the two scientists I expected the students to include as important contributors, made all the lists and according to most public surveys really are the best known scientists. Second, for the overwhelming majority of students their remaining choices, with the exception of Newton, Darwin, Galileo, and Mendel, were twentieth-century white males. No women or non-white scientists appear other than Marie Curie whom students mentioned probably ten percent of the time. This is unfortunate but also indicative of the way the sciences actually developed. Women and non-white scientists have not played prominent roles in science until the late nineteenth and twentieth centuries, and even when they did their contributions seldom if ever appeared in most historical accounts of the sciences.

Some of the scientists listed are those who have laws, equations, or scientific units named after them. They include Robert Boyle, James Watt, Alessandro Volta, Charles Coulomb, André Ampere, Michael Faraday, Gregor Mendel, James Clerk Maxwell, J. Willard Gibbs, and Max Planck. Usually students give only the scientists' last name, which suggests that they remember the name from a science class they took either in high school or university. Also surprising but disappointing is the lack of chemists on the lists. Linus Pauling, John Dalton, and Boyle get occasional mention, but few students included Antoine Lavoisier, Humphry Davy, Gilbert (G.N.) Lewis, or even Faraday and Gibbs. Similarly, scientists from the ancient, renaissance, and early modern periods, other than Hippocrates, Archimedes, Aristotle (rarely), and Copernicus, seldom appear on the lists.

For most students, their second list, discoveries and inventions, ranked computers and cell phones, medical developments (mainly drugs for cancer, antibiotics, pain relief), and nuclear related (weapons) at the top, followed in decreasing frequency by electricity, automobiles, space, environment, evolution, genetics, laws of motion, atomic theory, and Sun-centered universe. The list is similar to the first list in that most of the inventions or discoveries mentioned occurred in recent times. The second list also reveals the students' greater awareness of the technological developments resulting from the application of scientific principles than their awareness of the scientific principles that led to the invention. They recognize the impact of science on society more than its origin and development.

I cannot provide evidence that the names and pattern of names and the discoveries-inventions obtained in my classes are the same or different from those likely obtainable from introductory history of science survey courses taught at other universities in the United States, Canada, Europe, and Asia. Nevertheless, I believe that the lists of scientists and of discoveries and inventions would show similarities. The only confirmation of my opinion is the conversations I have had with other historians of science who teach introductory survey courses. None has tested students the way I have, but using their own general questioning of students early in the semester they have reached a similar conclusion.

## II. OBJECTIVES OF THE HISTORY OF SCIENCE

The history of science draws upon the results and methods of the sciences, engineering, and humanities. A culture as widely pervaded by scientific values and as dependent upon the applications of scientific knowledge, as most societies are in the twenty-first century, requires basic scientific literacy for everyone. By providing an integrated view of how the sciences and engineering developed in different times and in diverse cultures, the history of science offers a pathway to understanding how science and engineering have shaped modern society. Its study promotes critical thinking about the origin and development of science and its impact on society, and about the emergence and evolution of modern science and engineering.

With few exceptions if any, students in my history of science courses are learning about the history of science for the first time. History of science courses at the secondary level do not exist, and the history of science is almost entirely ignored in history and science courses at any level, secondary or university. Because of this unique situation, *Science Changed the World* has four objectives.

First, it aims to broaden students' world view and sense of perspective by teaching them about the origin and growth of science, and the impact of science on society. To neglect the history of science in discussing the history of any society, particularly developed or advanced societies, is to neglect the major force that has shaped and continues to shape a society. Students need to understand the connection between science and society's advancement and to answer why some societies have advanced and others have not.

Second, it makes clear that science is a human activity or enterprise. People do science, science does not happen spontaneously. Except for the most notable scientists, humans are missing from or barely mentioned in textbook accounts of most scientific breakthroughs, such as the discoveries in electricity, light, atomic structure and chemical bonding, and genetics. *Science Changed the World* restores the human side of science.

Third, it shows that science is dynamic, not static, and that science changes over time. Scientists are not infallible and neither are their ideas. Beginning in ancient times natural philosophers for centuries debated the composition of matter, whether matter consisted of four interconvertible elements or of indivisible atoms. In the 1500s-1600s they debated the arrangement of the planets, whether the planets encircled Earth or the Sun. For over 2,000 years scientists debated the gravitational theories of Aristotle, Newton, and Einstein and whether gravity was a force or a distortion of geometric space. In the nineteenth century scientists debated theories of heat, whether heat was a fluid or matter in motion, and theories of light, whether light was a particle or a wave. In the behavioral sciences, scientists from the 1600s debated theories of animal behavior, whether animals experienced consciousness or reacted according to their instincts.

Sometimes the change comes slowly, as in the 2,000-year conversion from an Earth-centered universe to a Sun-centered universe, or more rapidly, as with the theories of heat, light, and the atom. But scientific ideas do change. This is the essence of scientific inquiry and research and unfortunately subject to much public misunderstanding even in the twenty-first century.

Fourth, it provides an excellent forum to expose and eliminate misconceptions and erroneous beliefs that have crept into the sciences. These misconceptions and erroneous beliefs, once they become a part of science, are very difficult to remove and leave students with a false sense of historical and scientific truth. All too often students believe, and textbooks usually reinforce the belief, that science progressed logically from one clear idea to the next, that the scientist always advanced in a straight line path leading to some predetermined goal. The path to discovery is never that direct and uncomplicated, and the history of science is littered with examples of how intuition, chance, and even dreams have resulted in scientific breakthroughs. The Cambridge historian Herbert Butterfield (1900-79) once described the inventive or discovery process as "picking up the other end of the stick." The historian of science can put to rest these mistaken ideas of how our modern society emerged.

The flat-Earth misconception is likely the most interesting, longest lasting, and most persistent of scientific mistruths or misstatements, and worthy of discussion. The ancient Babylonians, Egyptians, and Greek pre-Pythagoreans were flat-Earth societies and the misconception arose that medieval and renaissance societies, including Christopher Columbus (1451-1506), continued to believe in a flat Earth. Yet we know that from the time of Pythagoras (572-497 BC) in Crotona in southern Italy, who probably first concluded that Earth was a sphere, widespread acceptance of a spherical Earth prevailed.

Nevertheless, a few flat-Earth believers remained in medieval times and included Lucius Caecelius Fermianus Lactantius and Cosmas Indicopleustus (Indian voyager) as their most prominent proponents. Lactantius (265-345) in Byzantium, an African pagan convert to Christianity who tutored Crispus (d. 326), the son of the Byzantine Emperor Constantine (274-337), claimed the Bible gave no proof of a spherical Earth and was unclear on Earth's shape. He denied Earth's sphericity in order to deny the antipodeans' and the antipodes' existence. No humans lived on the other side of Earth and had their feet above their head. Nor did crops and trees grow upside down and rain and snow fall upward. Lactantius, like many today, failed to recognize that the Bible is not a book of science and does not present a scientific cosmology.

Some of the biblical passages imply a flat Earth, as in the Old Testament's Daniel 4:10-11 and 20: the tree grew and was strong, and the height thereof reached unto heaven, and the sight thereof to the end of the Earth (4:11); the New Testament's Matthew 4:8: the devil took him to a very high mountain and showed him all the kingdoms of the world and their splendor; and the New Testament's Revelation 1:7: behold he is coming with the clouds! Every eye shall see him.

Lactantius had no large following. After his death the Catholic Church condemned his writings, such as *The Divine Institutions (Divinarum Institutionum Libri VII)*, written between 303-11, as heretical for its lack of grasp of Christian principles and ignorance of Scriptures.

Cosmas Indicopleustes (ca. 540), a Greek traveler and geographer in Byzantium, also believed in a flat Earth. He knew that Greek natural philosophers including Aristotle accepted Earth's sphericity, but Cosmas rejected Earth's and the Heaven's sphericity and the antipodes' existence. In *Christian Topography* (547-49), he described the cosmos as a huge rectangular vaulted arch and Earth as its flat floor.

Cosmas's evidence for a flat Earth came from two biblical passages, the New Testament's Epistle to the Hebrews 9:1-5 and the Old Testament's Psalms 104:2-3. Incorrectly interpreting Origen of Alexandria (185-251) who used allegory to interpret the scriptures and whose treatment of Hebrews 9 was metaphor, Cosmas concluded that the enclosed vault of the sky was the tabernacle of Moses and Earth the flat table on which sat the loaves of bread. Psalms 104:2-3 described the Heavens as stretched out like the skin of a tent and as God's palace on the water with the clouds as his chariot. Lactantius and Cosmas were exceptions and not characteristic of their times.

A much later major contributor to the flat Earth misconception was Washington Irving (1783-1859) in New York City. Irving, the author of *Rip Van Winkle* (1820) and *The Legend of Sleepy Hollow* (1820), in 1826 went to Madrid, Spain, as an attaché in the United States embassy and served as the United States minister to Spain in 1842-46. He fabricated the fanciful story in his 1828 biography of Columbus that Spanish religious authorities still believed in a flat Earth and for this reason opposed Columbus's voyage.

Irving wanted to portray the Spanish authorities as doctrinaire and opposed to new knowledge. He was writing at a time when science was challenging authoritarianism, what Irving saw in Spain's Catholic Church. Irving, whose writings revealed his racist views, was wrong. Columbus and his critics disagreed over distance. Columbus incorrectly estimated that the distance between Japan and the Canary Islands was 4,450 kilometers. The actual distance was about 22,000 kilometers. He erred by 80 percent.

Misconceptions about scientists continued to appear in the literature. They include the mistaken reason given for Nicolas Copernicus's delay in publishing his Sun-centered (heliocentric) universe until 1543, the inaccurate accounts of Galileo Galilei's 1633 trial, the erroneous belief that Benjamin Franklin's kite experiment of 1752 was the first test of the lightning-electricity identity, and the incorrect reason for Charles Darwin's delay in publishing his evolutionary theory until 1859.

Another misconception is that the scientific community immediately accepts new scientific ideas or theories, such as its acceptance of Copernicus's Sun-centered universe, John Dalton's early nineteenth-century atomic theory, the nineteenth-century evolutionary theories of Jean Baptiste Lamarck and Charles Darwin, and Albert Einstein's 1905 and 1915-16 theories of relativity. The scientific community moves slowly.

Misconceptions about the scientific method, that it is the only way to discovery or invention and that serendipity is an unlikely alternative pathway to discovery or invention, are common. This ignores the many examples discussed in *Science Changed the World* showing that serendipity provided an alternative pathway to discovery or invention. They include the accidental discoveries of Archimedes, August Kekulé, Louis Pasteur, Wilhelm Röntgen, Alexander Fleming, and a few others that Royston Roberts (1918-96) at the University of Texas described in his book *Serendipity* (1989).

Finally, I need to comment on *Science Changed the World's* content, the topics I have included and excluded. *Science Changed the World* contains eleven chapters. Each chapter has a definitive theme, and when combined the eleven themes show that science moved through three periods of development: from a record keeping period that began with the Babylonians, Egyptians, and Chinese about 2000 BC to a theoretical-philosophical period that arose in Greece in 600-400 BC, and to a mathematical-quantitative period that defined the sixteenth- and seventeenth-centuries (1600-1700) scientific revolution that occurred in Europe.

The transitions did not occur simultaneously in the physical and life sciences and in Eastern and Western societies. The physical sciences completed the three-stage transition in the sixteenth and seventeenth centuries whereas the life sciences moved from record keeping to theoretical-philosophical science in the later eighteenth and nineteenth centuries and to mathematical-quantitative science in the twentieth century. The transitions occurred first in Western societies. As a result of east-west contacts by western missionaries and other travelers to the East, Western science spread to the East beginning in the late 1700s in Japan and in China in the late 1800s, although Jesuit missionaries brought Western astronomy to the Chinese court as early as 1583.

A brief summary of the chapters appears below, chapter one contains a more detailed discussion of the chapter themes.

*Chapter 1*: Introduction to the History of Science: its Language, Objectives, and Scope.

*Chapter 2*: Ancient Science from Earliest Times to the Fall of Rome 476 AD: science undergoes a transition from an oral tradition to record keeping (2000 BC) to the theoretical-philosophical science of the Greeks (600-400 BC).

*Chapter 3*: Science and its Application in the Middle Ages 400-1400: an age of faith, a lack of scientific incentives, technology led science.

*Chapter 4*: Science in the Renaissance 1400-1600: a rebirth of learning, accepting and challenging the authority of the ancients.

*Chapter 5*: Scientific Revolution of the Sixteenth and Seventeenth Centuries 1600-1700: the mechanical philosophy, the study of matter in motion and the search for the mathematical laws to account for the motion dominated the physical sciences; determinism directed the scientific outlook.

*Chapter 6*: Physical and Life Sciences in the Eighteenth Century (the Enlightenment): the study of composition and the classification of matter, theories of heat, and the electrical behavior of matter directed the physical sciences; classification occupied the naturalists; impact of science and mechanics on society.

*Chapter 7*: Darwin's Century: the transition from classification (record keeping) to theories and laws occurs in geology and natural history. Mendel's new science of genetics solved Darwin's problem of inheritance.

*Chapter 8*: Molecular Biology and the Double Helix: a new field of study emerged in the life sciences. It builds on previous advances and the move of physical scientists and their instruments to the life sciences.

*Chapter 9*: Physical Sciences in the Nineteenth Century: the electrodynamic worldview complements the existing mechanical worldview; classical physics and determinism dominate but begin to collapse at the century's end.

*Chapter 10*: Twentieth-Century Physical Science: theoretical science, quantum theory, special and general theories of relativity, quantum mechanics become the foundation of physical science; they replace classical physics and determinism.

*Chapter 11*: Energy and the Environment: conflict or compatibility. Scientists search for new sources of energy, such as nuclear energy and solar energy. Environmental concerns emerge.

Because *Science Changed the World*, focuses on science, the last chapter, Energy and the Environment, does not discuss the history of the space program and the history of computers. They fit better in publications on the history of technology and have received sufficient coverage in history of technology books and articles. I have instead focused on major developments and key scientists in the history of energy and the environment, specifically nuclear and solar energy, water and atmospheric pollution, and climate change. Historians of science have written much on nuclear energy but have not given solar energy and environmental topics as much attention as they deserve. Scientists who contributed significantly to our understanding of solar energy and the environment among them Farrington Daniels, Abe Wolman, Gilbert Plass, Roger Revelle, David Keeling, and Clair Patterson, although well known within their scientific disciplines remain largely unknown outside of their disciplines. *Science Changed the World* will help remedy this deficiency and contribute to a richer discussion of the history of science.

The publication of this manuscript required considerable assistance from several people whom I wish to acknowledge. Rita Walker, a staff member of the Texas A&M History Department, provided steadfast support in finalizing the manuscript, particularly with her computer creations of many diagrams; my wife, Sonya, and granddaughter, Victoria Fazzino, assisted in the compilation of the index and bibliography; my son-in-law, Michael Holmgreen, donated his artistic talents. The Kendall-Hunt team, particularly Abby Davis and Lara McCombie, managed and directed the manuscript's editing and publication. I appreciate everyone's efforts.

*Science Changed the World* is the product of my experiences as a science teacher and my thirty-five-year career at Texas A&M University teaching courses on the history of science. It incorporates the observations and comments of many hundreds of students. Their interest in and appreciation of the influence science has had on life and culture provided the inspiration to take on the challenge of this comprehensive survey of science. My students were and continue to serve as unsuspecting motivators, and I thank them for having participated.

# CHAPTER 1
# INTRODUCTION TO THE HISTORY OF SCIENCE: ITS LANGUAGE, OBJECTIVES, AND SCOPE

The history of science is science itself.
Johann von Wolfgang Goethe (1749-1832)

## I. INTRODUCTION

The history of science provides an integrated examination of the origin, growth, and impact of science in different times and in different cultures from ancient to modern, from East to West. It shows the transition of science from a record keeping period that began with the Babylonians, Egyptians, and Chinese about 2000 BC to a theoretical-philosophical period that arose in Greece in 600-400 BC, and to a mathematical-quantitative period that defined the sixteenth- and seventeenth-centuries (1600-1700) scientific revolution that occurred in Europe.

The transitions did not occur simultaneously in the physical and life sciences and in Eastern and Western societies. The physical sciences completed the three-stage transition in the sixteenth and seventeenth centuries whereas the life sciences moved from record keeping to theoretical-philosophical science in the later eighteenth and nineteenth centuries and to mathematical-quantitative science in the twentieth century. The transitions occurred first in Western societies. As a result of east-west contacts by western missionaries and other travelers to the East, Western science spread to the East beginning in the late 1700s in Japan and in China in the late 1800s, although Jesuit missionaries brought Western astronomy to the Chinese court as early as1583.

A historical study of science requires an understanding of what science is and how science differs from engineering and technology and from the humanities, including religion. Of equal importance to its study is an understanding of how scientists acquire scientific knowledge and develop scientific theories. Is the acquisition of scientific knowledge an inductive or a deductive process? Is this knowledge the product of the scientific method or does serendipity sometimes play a role? Are scientists realists or idealists, and do scientists have an epistemology and an ontology of science? These are important issues and concerns for historians of science and for which a study of the language, objectives, and scope of the history of science has provided appropriate answers.

## II. SCIENCE, ENGINEERING, TECHNOLOGY, AND THE HISTORY OF SCIENCE

***What is science, what is engineering, what is technology?***

Science is a search for natural explanations of observable phenomena. Science describes, explains, and predicts nature's behavior and results in a body of knowledge having a theoretical foundation of laws and principles. Science is knowing, it examines how nature appears to be and accepts only natural causes of observable events.

Engineering is a science that studies the properties of matter and sources of energy and makes them useful to society. Engineering is doing, it tries to transform nature and society into what they ought to be.

Technology is the totality of machines and materials, or the science and engineering, society uses to modify or change the natural environment for its benefit. Technology is a very broad or general investigative field that connects science, engineering, and society. The equation below relates technology's three components.

$$Technology = science + engineering + society$$

Research in nanotechnology serves as an example of how the three components of technology work together. Nanotechnology is an emerging field that deals with manipulating atoms one by one and results in the production of tiny customized molecules having special properties. About 80,000 ball or tube shaped nanomolecules would fit inside the width of a human hair.

In September 2001 researchers at IBM in Yorktown Heights, New York, constructed a computer circuit out of a single stand of carbon atoms known as a nanotube. A nanotube, which resembles a rolled-up tube or cylinder, is about 10 atoms wide, $\frac{1}{15,000}$ of an inch or $\frac{1}{100,000}$ the thickness of a human hair. IBM's computer circuit performed only one simple operation, changing a 0 (false) to a 1 (positive) and vice-versa, but it was the first circuit made of nanotubes of a single kind of atom to perform a logical operation. IBM researchers believe their nanotube circuit will lead to smaller and faster computer chips in which one nanotube transistor occupies the same space as 10,000 of today's silicon transistors.

Other research groups at Rice University in Houston and Penn State University in University Park in June 2001 constructed molecular-size switches and wires so that a computer built on this scale would shrink to the size of a match head and react billions of times faster than current computers. Scientists and engineers imagine joining atoms together to construct an instrument a few atoms wide and a few atoms long that would give them complete control of a nanomaterial's physical properties. They could manufacture chemically-pure materials, super-strong composites, drugs, and even organic tissue; design nanotubes to function as tiny robots sailing through your blood stream cleaning clogged arteries; and redesign genes to give your children mental and physical enhancements. Nanotechnology may soon become a $700 billion industry.

***What is the history of science?***

The history of science studies (1) the origin and development of science, examining whether its origin and development occurred independently of the social-cultural-economic environment or as a consequence of it, and (2) the impact of science on society.

***At least four good reasons justify studying the history of science and make its study relevant***

Science was and remains one of the most important forces shaping the world in which we live. The study of its history is important in understanding how our modern world emerged.

The history of science shows that science is dynamic not static and that scientific ideas change. Science is a human activity. People do science, science does not happen spontaneously.

The history of science provides completeness to our historical understanding. It shows connections between our past and present scientific worldviews and provides an integrated view of how science developed in different times and cultures.

The history of science is interesting.

***When history and science textbooks include history of science, how do they present history of science?***

History textbooks sometimes include a brief mention of the impact or consequence of an important scientific idea or theory but say nothing of the scientific ideas or of the scientists who discovered, invented, or developed the idea or theory. They give an unbalanced presentation of the scientific idea and its application. In textbook discussions of well-known wartime developments, such as gaseous warfare in World War I, the German chemist Fritz Haber (1868-1934), the chief proponent of gaseous warfare, is missing. Haber, a Jew turned Protestant, received the 1918 Nobel Prize in chemistry, despite the controversy over his wartime activities. After the election of Adolf Hitler (1889-1945) in January 1933, Haber had to flee the Nazis and died in exile in Switzerland.

Nor does the role of Haber and the German chemical company BASF in developing the synthetic ammonia process during World War I and BASF's role in developing synthetic liquid fuels in World War II receive any mention. Yet, without an industrial-size synthetic ammonia process that produced fertilizers for food production and explosives for weapons, Germany's war effort probably would have ended two years before it did, in 1918. In World War II, liquid fuels synthesized from coal propelled 95 percent of Germany's air force and provided 50 percent of Germany's total liquid fuel requirements.

Similarly, in relating the World War II development of the atomic bomb and nuclear energy the scientific contributions of the German chemist Otto Hahn, the Italian refugee physicist Enrico Fermi, the Danish physicist Niels Bohr, the American physicist J. Robert Oppenheimer, and others never appear. Just as important an omission is any discussion of why scientists even thought such a process as nuclear fission existed and that they could carry out and control its production.

Another interesting omission is the failure to explain why army General Leslie Groves selected Oppenheimer, whom some government officials suspected of having Communist sympathies, to head the top-secret Manhattan Project that built the first atomic bomb. Groves learned that Oppenheimer was the foremost theoretical physicist in the United States and certainly one of the world's best. This is why he selected Oppenheimer despite the alleged, but untrue, claim of Oppenheimer's disloyalty.

World War II and Cold War textbook discussions on the atomic bomb and the hydrogen bomb say only that a hydrogen bomb is more powerful than an atomic bomb but never say why. The textbooks never say that an atomic bomb is a fission bomb and a hydrogen bomb is a fusion bomb. In fission neutrons split a large atom, such as uranium or plutonium, into two smaller atoms, whereas in fusion small atoms, such as tritium or deuterium, combine to form larger atoms. Fusion releases 1,000 times more energy than fission.

A related issue is Einstein's role in constructing the atomic bomb. He had none. But because most students and teachers have heard of Einstein's $E = mc^2$ equation, even if they do not know its meaning, they usually associate Einstein with the bomb's construction. Einstein, however, was out of mainstream theoretical physics research at that time.

Obviously, the impact of the atomic bomb and nuclear energy requires a place in history texts, but some mention of the science and scientists who made the atomic bomb possible would give a more complete historical account of one of the most tragic, yet fascinating, periods in history.

Science textbooks have the same problem. They omit the historical pathways leading to a scientific discovery and the scientists responsible for the discovery. They present the science without its history, or the end without the beginning. Students of the physical sciences are familiar with the three laws of thermodynamics but seldom know the circumstances of each law's introduction and the uncertainty associated with the meaning of each law. Some historical background to their development would contribute to a better understanding of this difficult subject.

Each law had at least two independent authors: Hermann von Helmholtz, Julius Robert Mayer, and James Joule proposed the first law; Rudolf Clausius and Lord Kelvin proposed the second law; and Walter Nernst, Max Planck, and Gilbert N. Lewis and George E. Gibson proposed the third law. Each of these scientists believed he had given an original or improved version of the laws. The history of the second law is particularly revealing because, for a time, nineteenth-century scientists thought incorrectly that entropy meant the available energy resulting from a chemical process rather than the opposite, or the unavailable energy. The American chemical physicist, Josiah Willard Gibbs, was responsible for clarifying the problem that had confused even James Clerk Maxwell.

Textbook discussions on atomic theory and chemical bonding provide a second example. Students often know that the Bohr atom preceded the quantum atom, but what preceded the Bohr atom and where did these little black dots that chemists put around the symbols for atoms and molecules come from? The same lack of historical context holds true for all of the topics discussed in science textbooks whether it is evolution, electricity and magnetism, the atomic theory, or relativity. The usual justification textbook authors make is lack of space to include the historical context and lack of time to teach it.

***Comparing the methods and objectives of the sciences and the humanities***

*Validity of the distinctions made between science and the humanities*

Positivist and positivist-leaning scientists make clear distinctions between the objectives of science and art, science and the humanities, and science and religion. They believe that scientists observe the phenomena and acquire knowledge only through the senses. Science is discovery, the scientist discovers, whereas art is creativity, the artist creates. They insist that science is objective, it searches for laws and truth, whereas the humanities are subjective and deal with cultural values and taste. They argue that science asks how the universe works, about its mechanics and the mechanics of everything in it, whereas religion asks why the universe exists, who created the universe, and whether the universe has a purpose.

Are these valid distinctions, or does some overlap exist between the sciences and the humanities? Leonardo da Vinci (1452-1519) saw no difference between science and art, both required creativity. Einstein said that scientific theories are creations of the mind, clearly believing that science had progressed beyond merely observing phenomena and acquiring knowledge through the senses, as the positivists maintained.

Do scientists or even anyone outside the scientific community really believe that science is totally objective without any subjectivity? If so, interpretations of scientific observations would never differ. The most a scientist can do is to investigate and account for the universe as it appears to be. Science strives to be objective, to present its results uninfluenced by the environment in which it arose and exists, and to discover the true laws of nature, and so do the humanities. Total objectivity is the ideal, but both science and the humanities sometimes fall short.

Countless instances of subjectivity in history and science exist. They include Trofim Lysenko (1898-1976) and genetics in the Soviet Union, early twentieth-century eugenics and the biased early IQ testing in the United States, and the conflicting historical accounts of the same wars, such as the Japanese and American versions of World War II and the American and North Vietnamese versions of the Vietnam War. More recently the acid rain-global warming debates between scientists and some media personalities and commentators outside the scientific community has intensified considerably, with each side claiming objectivity. These examples expose the misconception of total or absolute historical and scientific objectivity.

The differences that exist between the sciences and the humanities are most pronounced when dealing with issues of testability and of their cumulativity and progressivity. Science is clearly progressive and cumulative, and its legitimacy depends on its testability or reproducibility of results. For some humanities, such as history, this is also true. For others, such as painting, sculpturing, and probably literature, testability and being cumulative and progressive, such as the introduction during the Renaissance of perspective in painting, are certainly important but perhaps not as crucial. No one disagrees that twentieth-century science has surpassed Renaissance science, but not all would conclude the same about the art and sculpture of the two periods. Subjectivity influences judgment more strongly in those two humanistic disciplines than in the sciences. The conclusions of science, such as the laws of motion, of thermodynamics, and of quantum theory, although never final, are the same and will remain the same for all cultures and environments as long as they continue to meet the scientific community's criteria of providing the best explanation of nature.

Finally, in dealing with the how and why of the universe, today's theoretical scientists are asking the same questions as theologian-philosophers. Physicists, such as Alan Guth (b. 1947) at MIT, Stephen Hawking (b. 1942) at Cambridge, George Smoot (b. 1945) at Berkeley, Joel Primack (b. 1945) at the University of California-Santa Cruz, and William Stoeger (b. 1943), a Jesuit priest and astronomer at the Vatican Observatory in Rome and the University of Arizona, want to know why the universe exists, how it originated, and who made the laws of the universe or designed the universe. So did Pythagoras (570-495 BC), Plato (427-347 BC) and Aristotle (384-322 BC) in Athens, Augustine (354-430) in Hippo, Thomas Aquinas (1225-74) at Paris and in Italy, and many other theologians and philosophers.

To answer these questions Guth proposed an inflationary theory of the universe in 1981. He suggested that for a fraction of a second after the universe emerged from a quantum vacuum, it expanded exponentially, propelled by a repulsive gravitational force, doubling in size, and doubling again and again at a rate much faster than the speed of light. The Russian theoretical physicist, Andrei Linde (b. 1948) at Stanford, following Guth's lead, proposed a theory of eternal inflation in 1982. According to Linde, many inflationary events occurred, not just one, creating many universes. Each universe is a minuscule bubble within a giant cauldron of energy that quantum fluctuations have inflated. Primack, who does not believe in a personal God but God as an organizing principle in the universe, also supports a theory of eternal inflation.

Another interpretation of how the universe emerged 12-13 billion years ago in the Big Bang is a revised version of the 1983 *no boundary proposal* of Hawking and James B. Hartle at the University of California-Santa Barbara. The original theory proposed an infinite universe that had no beginning or end either physically or in time. The 1998 revision which Hawking and Neil Turok (b. 1958) at Cambridge derived from quantum mechanics shows how a universe smaller than a proton and having an incredibly high density resulted from a quantum fluctuation.

*Predictive models and laws of nature*

Science seeks to have predictive value, and for this reason scientists propose and develop models to account for their observations. But having predictive value does not guarantee the model's correctness. Compare Earth-centered and Sun-centered models of the universe, both of which have predictive value resulting from the tables of planetary motions and other celestial motions that astronomers compiled over centuries. Scientists abandoned the Earth-centered model after 2,000 years because of the contributions of Nicolaus Copernicus, Galileo Galilei, Johannes Kepler, and Isaac Newton. In 1687, Newton provided a law of universal gravitation that remains valid only for a Sun-centered universe. Experimental proof of Earth's motion came later. The English astronomer James Bradley in 1728 demonstrated stellar aberration and Friedrich Bessel (1784-1846) at Königsberg, Germany, in 1838 demonstrated stellar parallax.

The humanities have focused less on developing predictive models, but throughout history individuals and societies, independently of the scientific community, have searched for laws of nature and laws of history. The Greeks and Romans adopted a cyclical view of time. They believed society went through endlessly recurring cycles. Christianity, however, accepted a day of judgment in the future and a linear and progressive interpretation of time.

Georg W. F. Hegel (1770-1831), a German philosopher who lived in Jena and later in Heidelberg and Berlin, in 1807 proposed a dialectical view of history. His dialectic consisted of a thesis (the dominant consciousness of the age) and its antithesis (a set of opposing propositions) that gave rise to a new synthesis that embodied the best values of both. The new synthesis then became the thesis, and the cycle repeated itself continuously in the historical process. Karl Marx (1818-83), the German economic-political philosopher and social critic in Cologne and from 1849 in London, held a dialectical view of history. He discussed his ideas in the *Communist Manifesto* of 1848 and in his three-volume *Das Kapital* published in 1867, 1885, and 1894.

Lord John Acton (1834-1902), a historian and moralist at Cambridge, had a more realistic view of society. His only law of nature decreed that power tends to corrupt and absolute power corrupts absolutely. Acton mentioned his law in an April 1887 letter to Mandell Creighton (1843-1901), a London bishop and Cambridge historian, who founded the *English Historical Review* in 1886.

***What is a scientific theory and does a scientist discover or create a theory?***

Scientists propose theories to explain or account for an observation or event that has occurred. To a scientist a *theory* is an explanation supported by observations and tests, it is not a guess. Few scientists today question the main tenets of the atomic theory or quantum theory, yet they still refer to them as theories because they know that scientific evidence and interpretations change over time and are not permanent. This differs from adopting a theory that totally lacks supporting evidence.

A hypothesis is a first tentative statement, which if successfully accounts for the natural world, develops into a theory. The difference between a hypothesis and a theory is the degree of confidence. Theories are statements of greater confidence than hypotheses.

Newtonian gravity, in which scientists had a great deal of confidence as a theory, remained unchanged for over 200 years until Einstein replaced it with general relativity and gave a more accurate description of the universe. The atomic theory and Darwin's theory of evolution are other examples of scientific beliefs that have changed over time.

## III. SCIENCE AND RELIGION

***Supernatural or natural causes, the anthropic principle***

The rise of Christian fundamentalism in the 1970s-90s revived interest in the origin and design of the universe. In *A View of the Evidences of Christianity*, his famous 1794 book, William Paley (1743-1805), Anglican archdeacon of Carlisle, England, argued against the skeptics, such as the Scottish philosopher David Hume (1711-76) and the English historian Edward Gibbon (1732-94), in favor of design in the universe and therefore of God, its intelligent designer. Paley's argument today has become the anthropic principle, an expression that the Australian cosmologist Brandon Carter (b. 1942), at the National Center for Scientific Research in Paris, first used in a 1974 lecture before the International Astronomic Union.

In its simplest expression the anthropic principle says that the universe seems destined for the creation of intelligent life. Its supporters include Sir John Polkinghorne (b. 1930), former professor of mathematical physics at Cambridge, retired president of Queen's College (Oxford), and presently an ordained Anglican priest; George Wald (1906-97), an American biologist at Harvard who received the 1967 Nobel Prize in medicine for his work on vitamin A in vision and a 1960s peace activist who held a pantheistic belief in God; and John Barrow (b.1952), a Protestant astronomer and director of the Millennium Mathematics Project at Cambridge. They believe that all the conditions required for intelligent life to exist on Earth could not have happened by chance and instead have resulted because of intelligent design.

Intelligent design implies an intelligent designer, a belief that all proponents of the anthropic principle accept. They disagree on who or what is the intelligent designer. Einstein called his intelligent designer God, but he did not believe in the personal God of Christianity, Judaism, or Islam and to whom you pray. His God existed in nature and was nature. He had a pantheistic view of God. Other physicists, such as the agnostic Hawking, do not accept a personal God but see God as a first force or cause, or organizing principle, whereas Polkinghorne's intelligent designer is the personal God of Christianity.

Wald in an article, "The Cosmology of Life and Mind," published in a 1988 book *Synthesis of Science and Religion* offered four simple conditions that he said confirmed the anthropic principle, the absence of any one of which would have made life impossible. They are: ice floats, the night sky is dark, protons and electrons have identical charges, and Earth has the right kind of Sun.

Considering Wald's first condition, most substances contract when cooled and so does water until it reaches 4° C. But as water continues to cool from 4° C to 0° C it begins to expand as a result of extensive hydrogen bonding. Consequently, the colder water is less dense and rises to the top so that when it freezes ice forms at the surface and not at the bottom. The ice forms a protective layer at the top and enables aquatic life to survive instead of perishing, as would happen if a body of water froze from the bottom up. Wald also pointed out that if the ice formed from the bottom up, as the air temperature rose the thin layer of water that remained on the surface would evaporate. Repetition of this process over a long period of time would result in the entire evaporation and disappearance of the water.

Wald's second condition, the darkness of the night sky despite the presence of billions of stars in the sky, amazed astronomers, such as Kepler, Edmund Halley (1656-1742) in London, and Heinrich Olbers (1758-1840), a German physician and astronomer in Bremen. Olbers in 1779 invented the first successful method for calculating the orbits of comets and published it in 1797. In 1826, he pondered the dilemma that we call Olber's Paradox and carried out a theoretical investigation on the expected brightness of the sky, assuming an infinite static universe of stars. Wald says that the sky is dark only because our expanding universe is still young and light from its most distant stars has not yet reached us. If it did, which would have happened if our universe were much older, we would have received so much radiation in addition to the visible light we receive that life would be impossible.

Wald's third condition, the equality of the proton and electron charges, makes clear that if the two particles had unequal charges, an electromagnetic force rather than a gravitational force (gravity) would have determined the structure of the universe. The electromagnetic force is much stronger than gravity, and if the positive and negative charges are not equal, strong electrical discharges known as lightning result. The universe as we know it, the stars, planets, galaxies, and the life that we know would not exist. Without gravity, our universe might be an expanding shell of unformed matter and energy.

Wald's fourth condition says that Earth has the right kind of Sun. He points out that more than half the stars in our galaxy are double, even triple stars, and if Earth circled such a Sun, moving closer and farther, temperatures would change drastically, probably making life impossible. If the Sun had much more mass it would burn its fuel more quickly and burn itself out so rapidly that simple life forms would have insufficient time to develop.

Barrow's argument in favor of design focuses on the properties of hydrogen, carbon, nitrogen, and oxygen that have resulted in the presence of these elements in every life form. He remarked that because of the balance that exists between the characteristics of carbon and cosmological processes, if carbon's formation required more energy, the universe would contain very little carbon. Barrow also called attention to the special role the Moon plays in Earth's stability and existence. Most planets wobble sharply in their orbits, which produces a climate inhospitable to life, whereas the Moon's gravitational force helps to keep Earth's tilt in line, resulting in a stable climate and an environment suitable for life.

### *Critics of the anthropic principle*

Critics acknowledge the importance of all these conditions, and they argue that for our universe to exist it had to satisfy these conditions. If our universe did not satisfy these conditions, we and the universe would not exist. Millions of other occasions that failed to satisfy all of these conditions may have arisen, and no life began.

Clyde Tombaugh (1906-97), discoverer of the former and now dwarf planet Pluto in 1930 and who spent years at the Lowell Observatory in Arizona, has said that while our Sun is the right size, it emits too much dangerous ultraviolet radiation. A redder Sun would not emit as much ultraviolet radiation and lessen somewhat our concern about the destruction of the ozone layer. Earth also tilts too much, he says. Even though the tilt allows for the seasons, it accounts for the dangerous storms that accompany the seasonal changes such as tornadoes in the spring and hurricanes in late summer. A smaller tilt would result in a more moderate climate, less noticeable season changes, and fewer and milder seasonal storms. Of course, if Earth had a greater tilt the seasonal weather changes would be far worse and Earth possibly uninhabitable.

Steven Weinberg (b. 1933), the 1979 Nobel Prize winning physicist at Texas, remains one of the most severe critics of the anthropic principle. He has argued that the signs of a benevolent designer that some say follow logically from the anthropic principle, seem rather remote. The laws of nature are cold and impersonal and contain no reference to life. Weinberg wants no dialogue between science and religion because that would give religion a kind of legitimacy it should not have. Based on the existing evidence, Weinberg believes that God has no place or role in the universe.

Stephen Gould (1947-2002), the Harvard geologist and zoologist, supported Weinberg's call for no dialogue between religion and science. In his book *Rock of Ages: Science and Religion in the Fullness of Life* (1999) Gould opposed any fusion of science and religion. They should remain wholly different domains of understanding.

### *Naturalist explanations of intelligent life*

Other scientists in the 1960s-90s proposed interesting naturalist explanations of how complex multicellular organisms formed from simple ones, eventually giving rise to intelligent life. The biologist William D. Hamilton (1937-2000) at Cambridge introduced his theory of *kin selection* in a 1964 article "The Genetical Selection of Social Behavior," published in the *Journal of Theoretical Biology*, to account for the formation of complex organisms from simple ones.

Hamilton proposed that nearby cells are related because they have the same kinds of genes. These cells not only code to reproduce themselves, but they have evolved to the point where they recognize that the cells around them have the same genes and that they all are evolving or have evolved for the same purpose. These cells had to cooperate for their survival and development, and their cooperation resulted in simple cells combining to form more complex cellular arrangements.

The English biologist John Maynard Smith (1920-2004) at Sussex suggested an answer to the question of why simple cells have formed complex organisms that have feelings. In a 1982 article "The Evolution of Social Behavior," in *Current Problems in Sociobiology* and in later publications, such as *Evolutionary Genetics* in 1989, Smith wrote that feelings give a moral dimension to life. They make for intelligent and meaningful life which these organisms might not experience if they lacked feelings.

To account for the transition from simple to complex ordered systems Stuart Kauffman (b. 1939), a theoretical biologist at Calgary, formerly associated with the Santa Fe Institute in New Mexico and founder of the BiosGroup (NuTech Solutions later Netezza) in Santa Fe, introduced the idea of *complexity* and the principle of *self-organization*. In his 1993 publications and more thoroughly in his

1995 book *At Home in the Universe: The Search for Laws of Self-Organization and Complexity*, Kauffman suggested the existence of a law beyond Rudolf Clausius's 1850s second law of thermodynamics. Clausius's law says that the entropy or disorderliness of the universe strives to a maximum.

Kauffman's new law would bring order out of chaos in the way that a disorganized mass of air becomes a tornado or a hurricane, or the initially chaotic behavior of water when heated gives rise to convection currents. Out of disorder or chaos, Kauffman says, order may arise. Matter contains within itself an organizing principle that enables it to go from simple to complex, to adapt and survive. Collisions among particles of matter cannot account for intelligent life.

Frank Tipler's *The Physics of Immortality* (1994) also deals with the origin of life, the universe and God. George Ellis (b. 1939), a leading theoretical cosmologist at the University of Cape Town, South Africa, in his review "Pie in the Sky." *Nature* 371 (8 September 1994):115, called Tipler's book "one of the most misleading books ever produced." In comparison, the physicist-philosopher Varadaraja V. Raman (b. 1932) at the Rochester Institute of Technology in his review, "Hope from Physics." *Science* 267 (February 1995):1042-43, called it "a masterpiece for the Age of Aquarius." *Nature*, published in London, and *Science*, published in Washington, DC, are foremost international journals.

According to Tipler (b. 1947), a Platonist and an atheist physicist at Tulane University in New Orleans, life at the end of time will become omnipotent, omnipresent, and omniscient, all of which he notes are attributes of God. At this time, which Tipler calls the *Omega Point/God* and which is the point of ultimate infinite knowledge, Earth's inhabitants will be human computers, that is, beings capable of tremendous information processing. For Tipler, life is information processing. Human beings in Tipler's future will have such vast processing powers that at the end of time on Earth, it will enable them to recreate life all over again somewhere else.

Immortality is therefore only information processing without end. The laws of physics dictate that future beings will recreate everything that has ever lived. Tipler says that to code the entire visible universe will require a limited amount of information and less than $3 \times 10^{45}$ bits of information to code a human. Because the amount of information needed to code the visible universe is finite, a time will come 7 billion years from now, when the beings then existing can easily emulate the universe using the known laws of physics and only a negligible amount of the computer power available to them.

When asked why the future human beings would want to resurrect all of us from the past Tipler says that these future humans, being extremely intelligent and therefore very curious, will have an interest in all the human variations that preceded them, just as today's scientists want to know about and try to recreate the first cells. The future humans will emulate us down to the last atom, giving us a future existence. Sounds fantastic, doesn't it?

## IV. THREE DIFFERENT APPROACHES USED IN STUDYING THE HISTORY OF SCIENCE

The history of science studies the origin and development of science, and the impact of science on society. It does this in several ways.

### *Internal and external history of science*

The *internalist* historian examines the origin and development of science within the scientific community and attaches little significance or influence to the social, cultural, and economic environment in which science originated and developed. External conditions or events contribute little to the origin and development of scientific ideas. Newton arrived at his universal law of gravity by observing the motion of a falling apple in the foreground and wondering about the motion of the Moon in the background. In a daring stroke of intuition he identified the force of gravity as the force acting on the apple and on the Moon. George Sarton (1885-1971), the Belgian-American founder of the history of science who taught at Harvard, emphasized the internalist view.

The *externalist* historian emphasizes the influence of the social, cultural, and economic environment on the origin and development of science and on the applications or impact of science on society. External factors significantly influence the origin and development of scientific ideas. An externalist would argue that the 1887 Michelson-Morley ether-drift experiment, which showed that the speed of light did not depend on the light's direction of motion, played a major role in Einstein's development of special relativity. Einstein, who emphasized the power of scientific ideas or theories in advancing science and aligns with the internalists, claimed that he would have proposed the same theory with the same starting assumptions and reached the same conclusions independently of the Michelson-Morley experiment. The externalist sees science as part of the larger culture surrounding it and subjected to the same forces that act on the rest of society.

#### *Whig history*

The name Whig History refers to England's Glorious Revolution of 1688-89 in which the Whigs and Tories united to depose King James II (1633-1701) and as victors wrote the history to justify their interpretation of events. The Whig historian presents science as moving logically from the unknown to a known objective or goal. No detours or false starts arise in science's straight line path from ignorance to enlightenment. The Whig historian represents science as progressing logically from a four element theory of matter to a theory of indivisible atoms and then to a structural atom; or from an Earth-centered model of the universe to a Sun-centered model and then to a relativistic model of the universe. Science did not develop so neatly and linearly. Scientists over the centuries have wavered between a particle model of light in the 1600s-1700s and a wave model in the 1800s to a duality in the 1900s. Similarly, the dominant model of heat varied between a caloric fluid model in the 1800s to a kinetic or mechanical model in the late 1800s-1900s.

#### *Scientific revolutions*

The American philosopher of science Thomas Kuhn (1922-96) at the Institute for Advanced Study in Princeton, proposed in *The Structure of Scientific Revolutions* (1962) that as science developed over the centuries it passed through a series of normal and revolutionary periods of science. Paradigms (models, theories) compete with one another for the scientific community's acceptance during the revolutionary period, and the model that best fits the scientific community's criteria becomes the adopted model and leads to a period of normal science. As new scientific evidence emerges, paradigms once again compete with one another, and a new period of revolutionary science followed by a new period of normal science arises. The pattern repeats itself again and again. Some historians of science have criticized Kuhn's interpretation, saying that it tends towards an acceptance of relativism in science. It downplays the significance of the objective scientific evidence or criteria that a model has satisfied to gain the scientific community's acceptability.

## V. PHILOSOPHIES OF SCIENCE

The philosophy of science has as its goal to reveal the assumptions a scientist makes in putting forward theories or in interpreting scientific evidence. Its study includes epistemology and ontology, induction and deduction, idealism and realism, positivism and empiricism, rationalism, falsification, and other philosophical-scientific questions and issues.

#### *Epistemology and ontology*

Epistemology investigates theories of knowledge and asks the questions: How do we know? How do we acquire knowledge of the world around us? Are the senses our source of knowledge or does our knowledge come from an intuitive or mental or thought process?

James Frederick Ferrier (1808-64) at Edinburgh introduced the word epistemology (to know, understand) on page 48 of *Institutes of Metaphysic*, second edition published in 1856. The first edition appeared in 1854.

Ontology asks the questions: How do we know what is real? What is reality? Is our universe Sun-centered or Earth-centered? Is it static or evolutionary? Did it have a beginning or did it always exist? How did the laws of nature originate? Where did they come from?

Aristotle discussed ontology in his *Metaphysics*, but the distinction between metaphysics (meaning beyond physics or the study of ultimates, or general concepts) and ontology remained unclear. The word ontology appeared in scholarly literature in the 1600s, first by the philosopher Rudolf Goclenius (1547-1628) at Marburg in *Lexicon Philosophicum* in 1616, and then by the philosopher Johannes Clauberg (1625-65) at Duisburg in *Metaphysica* in 1646. It acquired standard usage by the 1700s.

The relation of ontology to metaphysics is especially unclear because the main interest of metaphysics is also to determine what is most fundamentally real or to distinguish between what appears to be real and what is real. When Galileo and Pope Urban VIII (1568-1644) who reigned from1620 to 1644, met to discuss Galileo's writing a book that debated Earth-centered and Sun-centered models of the universe. Urban agreed to its writing but said that Galileo had to speak of the Sun-centered model as a useful idea that helped us understand the working of the universe. Only the Earth-centered model conformed to reality and was ontologically true. Similarly, in the early nineteenth-century atomic debates John Dalton (1766-1844) in Manchester accepted the reality of atoms, whereas Humphry Davy (1778-1829) at the Royal Institution in London believed that atoms, although useful in understanding the behavior of matter, were not ontologically true.

### *Induction and deduction*

Modern science began when empiricism (induction) and rationalism (deduction) joined to give the scientific method, a new approach to understanding nature. The inductive and deductive methods of acquiring knowledge date from Thales (624-546 BC), Democritus, Plato, Aristotle, and other ancient Greek natural philosophers. The inductivist begins with observations, the deductivist with ideas.

In the inductive method of investigating nature the scientist carries out specific experiments and uses the results obtained to propose a general theory that accounts for or can explain the results. Experiment leads to theory. Mass and volume measurements using different amounts of the same substance (water), and repeating the measurements using different substances (ethyl alcohol, vinegar), will show that the ratio of mass to volume does not change for fixed conditions of temperature and pressure. This leads inductively to the conclusion that the ratio, mass/volume, is a constant called density, $d = m/v$. Aristotle's biological investigations and Antoine Lavoisier's experiments on combustion demonstrate the application of the inductive method.

In the deductive method of investigating nature the scientist begins an investigation with an idea or belief of how nature appears to act and tests the theory with specific experiments to see if the results obtained agree with the theory. Theory guides or directs the experiments. Galileo's experiments on the acceleration of a falling body in the early 1600s and the many experiments physicists have done to test Einstein's special and general theories of relativity after their introduction in 1905 and 1915-16 exemplify the deductive approach.

### *Idealism and realism*

Idealism maintains that objects of knowledge, including both things and thoughts, do not exist independently of the mind or outside of the mind. The universe exists only in our minds, not outside of it, and has no real existence outside of our minds. The mind is the primary source of knowledge, the key to reality. George Berkeley (1685-1753) at Trinity College in Dublin and later Anglican bishop of Cloyne, Ireland, was a strict idealist. He proposed in 1710 that objects of knowledge do not exist independently of the mind. Our perceptions of objects, such as color, taste, temperature, are in us. God is their direct cause; he put the ideas of objects and of the universe in our minds. This is similar to the debate about a tree that falls in a forest. If no one is in the forest to hear the tree fall, how do we know the falling tree made a sound? Physicists will point out, however, that the tree falling in the forest will produce sound waves in the surrounding medium regardless of whether anyone hears the sound. In the same way, if no minds are present to observe the universe, how do we know whether the universe really exists?

An argument based on common sense is the best challenge to Berkeley's idealism in taking the position that all of our collective minds cannot be wrong in thinking that objects of knowledge really do exist outside of our minds. Are we seeing, smelling, touching what really does not exist? *Just because no mind is present to observe the universe does not prove definitively that the universe does not exist.* Berkeley's critics have questioned his meaning of ideas and claimed that his idealism limits the capacity to acquire new knowledge. Berkeley also attacked the logical foundation of Newton's calculus, declaring calculus erroneous because Newton changed its fundamental hypothesis. Berkeley's ideas appear in the philosophies of Johann Fichte (1762-1804) who taught for a time at Jena, Friedrich Schilling (1775-1854) in Munich and later Berlin, and Hegel.

Realism says that objects of knowledge, including both things and thoughts, exist independently of the mind. They exist outside of the mind and have a real existence. Aristotle, Plato, and most philosophers and scientists are realists.

***Positivism, materialism, and empiricism***

Positivism claims that only the phenomena, or knowledge obtained from the senses, are real. Nothing has meaning unless we can verify it empirically. Positivism studies the relations among phenomena. Theories are useful but have no basis in experience, they are creations of the mind and not positive or scientific knowledge. The positivist accepts the existence of atoms, or an atomic theory, as a useful idea or tool but has to verify empirically the presence or existence of atoms before accepting them as real or positive.

Auguste Comte (1798-1857) at the École Polytechnique in Paris was one of the foremost positivists. According to Comte only scientific knowledge is true (positive) knowledge. Positivists, such as the French chemist Marcellin Berthelot (1827-1907) at Paris and the Austrian physicist Ernst Mach (1838-1916) at Vienna, accepted only the observable. God has no place in positivist philosophy.

Positivism flourished in the United States from the 1920s through the 1950s. Well-known positivists included the physicist Percy Bridgman (1882-1961) at Harvard, Alfred North Whitehead (1861-1947) the English mathematician who came to Harvard in 1924, the German emigré physicists Hans Reichenbach (1891-1953) who was at UCLA beginning in 1938, and Phillip Franck (1884-1966) at Harvard in 1939.

Materialism is a natural extension of positivism and the opposite of rationalism or idealism, in which ideas, thoughts, or consciousness, are the most fundamental reality. It maintains that reality consists of matter in motion, forces, and empty space. Natural forces, acting according to natural laws that we can discover, direct the motions of all matter. Thomas Huxley (1825-95), Herbert Spencer (1820-1903) in London, and Ernst Haeckel (1834-1919) at Jena, Germany, were well-known nineteenth-century materialists.

Empiricism (sensationalism) argues that experience, or information obtained from the senses, is the source of all knowledge, including our ideas or theories. This distinguishes a positivist from an empiricist. The empiricist John Locke (1632-1704) in London argued that no difference existed between things as they are, our perceptions or sensations of things such as a book or nature, and our ideas of things. This holds true because all knowledge, including ideas, originates in our experiences.

***Rationalism***

Rationalism (idealism) says that the mind, reason, intuition is the source of knowledge. The idealist Immanuel Kant (1724-1804) at Königsberg, Germany, believed that certain fundamental ideas or categories, such as space, time, and causality exist in the mind *a priori*. God put the ideas or categories in our mind, and we use them to organize and understand the world we experience.

***Popper's falsification***

Karl Popper (1902-94), an Austrian-born philosopher at the London School of Economics, argued in *The Logic of Scientific Discovery* (German edition 1934, English translation 1959) that scientists only provisionally adopt scientific theories. They accept a scientific theory as correct when they can no longer falsify it (falsification). Popper did not accept verification as a reliable test of truth because scientists eventually may falsify an accepted truth.

***Two schools of thought on how scientific discovery or invention occurs: scientific method or non-scientific methods***

A scientist applies the scientific method, or hypothetical-inductive method, when he follows the sequence of observation, induction, deduction, testing, and prediction in order to acquire scientific knowledge.

In general the scientific method has succeeded. The exception occurs at the subatomic level where the act of observing atomic particles influences their behavior. The German physicist Werner Heisenberg, one of the founders of quantum mechanics, gave the first statement of this dilemma in 1927 when he published the indeterminacy principle, later known as the uncertainty principle. Heisenberg noted, at the subatomic level, the impossibility of making accurately, simultaneous measurements of a particle's position and momentum. The more accurate the measurement of momentum, the more uncertain the measurement of position, and vice-versa. This severely undermined the principle of causality, or the law of cause and effect, which in classical deterministic philosophy had remained virtually unquestioned since the ancient Greeks.

The French mathematician Pierre Simon Laplace (1749-1827) in Paris, who like all eighteenth-century mechanical philosophers believed that the universe was a giant machine and its behavior repetitive and perfectly predictable, maintained we could calculate the history and future of the universe if we knew the position and velocity of every particle for any instant of time. Heisenberg's uncertainty principle says that this knowledge is unattainable, and undermines the possibility of certainty at the atomic level. This leads to questioning how we know what we know and how we establish a rational basis for our knowledge. The scientific method is one of the scientist's foremost epistemological tools, and for most scientists, it remains the best method of establishing the criteria for discovering the reality of nature.

Insight, dreams, serendipity, and metaphor in science are events, occasions, or thought processes that enabled scientists to acquire knowledge independent of the scientific method. Many important scientific discoveries or inventions have resulted from these alternative pathways.

| SERENDIPITOUS INVENTIONS OR DISCOVERIES | | |
|---|---|---|
| ***Inventor or Scientist*** | ***Date*** | ***Invention or Discovery*** |
| Archimedes (287-212 BC) | | calculated density and specific gravity |
| Charles Goodyear (1800-60) | 1844 | vulcanized rubber |
| Louis Pasteur (1822-95) | 1848<br>1879 | discovered optical activity<br>prepared vaccine for chicken cholera |
| Henry Bessemer (1813-98) | 1856 | invented a process for the mass production of steel |
| August Kekulé (1829-86) | 1865 | discovered the structure of benzene |
| Wilhelm Röntgen (1845-1923) | 1895 | discovered x-rays |
| Alexander Fleming (1881-1955) | 1928 | discovered penicillin |
| Roy Plunkett (1911-94) | 1939 | invented teflon |

## VI. CHAPTER THEMES AND CHRONOLOGY OF THE HISTORY OF SCIENCE

***Chapter 1: Introduction to the history of science: its language, objectives, and scope***

*i.* Introduction
*ii.* Science, engineering, technology, and the history of science
*iii.* Science and religion
*iv.* Three different approaches used in studying the history of science
*v.* Philosophies of science
*vi.* Chapter themes and chronology of the history of science
*vii.* Other themes
*viii.* Conclusion

***Chapter 2: Ancient science from earliest times to the fall of Rome 476 AD: science undergoes a transition from an oral tradition to record keeping (2000 BC) to the theoretical-philosophical science of the Greeks (600-400 BC)***

*i.* Introduction
*ii.* Communication of ancient thought
*iii.* Middle East and African science and engineering
*iv.* Scientific world of the Greeks: the beginning of theoretical-philosophical science
*v.* Organization of Greek science
*vi.* Technology in ancient Greece
*vii.* Ancients and their environment
*viii.* Science in the East
*ix.* Science in pre-Columbian North America
*x.* Science in MesoAmerica
*xi.* Conclusion

***Chapter 3: Science and its application in the Middle Ages 400-1400: an age of faith, a lack of scientific incentives, technology led science***

*i.* Introduction
*ii.* Rise of the universities
*iii.* Medieval science and medicine
*iv.* Medieval engineering
*v.* Medieval living conditions and practical science
*vi.* Land and water transportation
*vii.* Industrial developments: the environment, textiles, iron, and printing
*viii.* Conclusion

***Chapter 4: Science in the renaissance 1400-1600: a rebirth of learning, accepting and challenging the authority of the ancients***

*i.* Introduction
*ii.* Leonardo da Vinci
*iii.* Nicolas Copernicus and the heliocentric model of the universe
*iv.* Andreas Vesalius and the foundation of anatomy
*v.* Chemistry in medicine: iatrochemistry
*vi.* Witchcraft and vampires
*vii.* Beginning of metallurgy
*viii.* Conclusion

***Chapter 5: Scientific revolution of the sixteenth and seventeenth centuries 1600-1700: the mechanical philosophy, the study of matter in motion and the search for the mathematical laws to account for the motion dominated the physical sciences; determinism directed the scientific outlook.***

*i.* Introduction
*ii.* Experimental method
*iii.* Galileo Galilei
*iv.* Problem of chemical identity in seventeenth-century chemistry
*v.* Studies on air: barometer and air pressure experiments
*vi.* Advances in mathematics
*vii.* Physics of light
*viii.* Establishment of scientific societies and academies
*ix.* Measuring heat
*x.* Sound and its measurement
*xi.* Revolution in physiology: William Harvey and the circulation of the blood
*xii.* René Descartes and the mechanical philosophy
*xiii.* Theory of universal gravitation
*xiv.* Conclusion

***Chapter 6: Physical and life sciences in the eighteenth century: the enlightenment: the study of composition and the classification of matter, theories of heat, and the electrical behavior of matter directed the physical sciences; classification occupied the naturalists; impact of science and mechanics on society***

*i.* Introduction
*ii.* Theories of combustion and the discovery of gases
*iii.* Lavoisier and the eighteenth-century revolution in chemistry: the study of composition
*iv.* Introduction of the metric system
*v.* Theories of heat

*vi.* Applications of heat energy, development of steam power
*vii.* Advances in electrical theory and applications
*viii.* Classification in nature
*ix.* Conclusion

***Chapter 7: Darwin's century: the transition from classification (record keeping) to theories and laws occurs in geology and natural history. Mendel's new science of genetics solved Darwin's problem of inheritance.***

*i.* Introduction
*ii.* How old is Earth
*iii.* How old is humanity
*iv.* Argument for design in the universe
*v.* Early theories of evolution
*vi.* Geological developments: catastrophists and uniformitarians
*vii.* Evolution by natural selection
*viii.* American debate over Darwinism: Louis Agassiz and Asa Gray
*ix.* Mendel: the beginning of genetics: solving the inheritance problem
*x.* A new theory of evolution
*xi.* Evidence supporting evolution
*xii.* How many species are there
*xiii.* Evolution and religion
*xiv.* Conclusion

***Chapter 8: Molecular biology and the double helix: a new field of study emerged in the life sciences. It builds on previous advances and the move of physical scientists and their instruments to the life sciences.***

*i.* Introduction
*ii.* Beginning of genetics
*iii.* Discovery of proteins and nucleic acids
*iv.* Use of x-rays in establishing molecular structure
*v.* Chemical contributions to the structure of DNA and RNA
*vi.* Solving the structure of DNA: x-ray diffraction, models, and the genetic code
*vii.* Human genome project
*viii.* Developments in cloning and stem cells
*ix.* Conclusion

***Chapter 9: Physical sciences in the nineteenth century: the electrodynamic worldview complements the existing mechanical worldview; classical physics and determinism dominate but begin to collapse at the century's end.***

*i.* Introduction
*ii.* Study of light
*iii.* Developments in chemistry
*iv.* Electrical theory and applications
*v.* Thermodynamics: the study of heat
*vi.* Kinetic-molecular theory of gases
*vii.* Scientific studies on atmospheric and water pollution
*viii.* Conclusion

***Chapter 10: Twentieth-century physical science: theoretical science, quantum theory, special and general theories of relativity, quantum mechanics become the foundation of physical science; they replace classical physics and determinism.***

*i.* Introduction
*ii.* Fundamental breakthroughs in physics and chemistry
*iii.* Search for precise atomic weights
*iv.* Electron: the fundamental electrical charge
*v.* Theories of atomic structure
*vi.* Chemical bond
*vii.* Composition of the nucleus
*viii.* Development of quantum mechanics: matrix mechanics and wave mechanics
*ix.* Spectroscopy and chemical bonds
*x.* Quantum mechanics and the structure of the nucleus
*xi.* Search for fundamental particles
*xii.* Conclusion

***Chapter 11: Energy and the environment: conflict or compatibility. Scientists search for new sources of energy, such as nuclear energy and solar energy. Environmental concerns emerge.***

*i.* Introduction
*ii.* Early developments in nuclear fission reactions
*iii.* Manhattan Project: controlling nuclear fission and racing to build the bomb
*iv.* German, Soviet, British, and Japanese atomic bomb projects
*v.* Atomic spy scandals
*vi.* Postwar nuclear developments
*vii.* Nuclear fusion
*viii.* Solar energy as an appropriate alternative technology
*ix.* Environment and pollution
*x.* Conclusion

| *Chronology of the History of Science* | | |
|---|---|---|
| *Time Period* | *Theme* | *Representative Topics* |
| Introduction to the History of Science: Its Language, Objectives, and Scope | The language, objectives, and scope of the history of science | History of science.<br>Science, engineering, and technology<br>Science and the humanities: science and religion.<br>Three approaches used in studying the history of science.<br>Philosophies of science. |
| Ancient Science: from Earliest Times to the Fall of Rome 476 AD | The transition from an oral tradition to record keeping 2000 BC to the theoretical-philosophical science of the Greeks (600-400 BC) | Universe of Plato, Aristotle, Hipparchus and Ptolemy; theories on structure of matter, Democritus, Aristotle and Alchemy; theories of medicine, Hippocrates, Galen; science in Pre-Columbia America: Mayans, Aztecs, Incas; science in the East: Chinese, Japanese, Hindu |
| Science and its Application in the Middle Ages 400-1400 | An age of faith, a lack of scientific incentives, technology led science | Alchemy; impetus theory of motion; Alhazen: light; invention of the compass and gunpowder |
| Science in the Renaissance 1400-1600 | A rebirth of learning, accepting and challenging the authority of the ancients | Leonardo da Vinci; Copernicus: heliocentric model of the universe; Vesalius: anatomy; Paracelsus: alchemy |
| Scientific Revolution of the Sixteenth and Seventeenth Centuries 1600-1700 | The mechanical philosophy, the study of matter in motion and the search for the mathematical laws to account for the motion dominated the physical sciences. Determinism directed the scientific outlook. | Bacon and Empiricism; Gilbert and magnetism; Harvey and the circulation of the blood; Boyle and the gas laws, van Helmont and the composition of matter; Kepler's three laws of planetary motion; Galileo's quantification and laws of falling bodies, Newton's three laws of motion, law of universal gravitation |
| Physical and Life Sciences in the Eighteenth Century: the Enlightenment 1700-1800 | The study of composition and the classification of matter, theories of heat, and the electrical behavior of matter directed the physical sciences; classification occupied the naturalists. Impact of science and mechanics on society. | Lavoisier's revolution in chemistry; law of conservation of mass, definition of an element; theories of heat; development of steam power; advances in electrical theory and applications; classification in nature |

| Chronology of the History of Science | | |
|---|---|---|
| *Time Period* | *Theme* | *Representative Topics* |
| Darwin's Century | A transition from classification (record keeping) to theories and laws occurs in geology and natural history. Mendel's new science of genetics solved Darwin's problem of inheritance. | Age of Earth, theories of evolution: Buffon, Lamarck, Erasmus Darwin; geological developments: catastrophists and uniformitarians; Charles Darwin, natural selection; Mendel and genetics; evidence of evolution; science and religion; Scopes trial, creationists |
| Molecular Biology and the Double Helix | A new field of study emerges in the life sciences, building on previous advances and the moving of physical scientists and their mathematical and experimental methods and instruments to the life sciences. | Earlier developments: cell theory, chromosomes, nucleic acids; Watson and Crick's structure of DNA; Genetic code; genetic engineering; human genome project; ethical issues; cloning and stem cells |
| Physical Sciences in the Nineteenth Century 1800-1900 | The electrodynamic worldview complements the existing mechanical worldview. Classical physics and determinism dominate but begin to collapse at the century's end. | Light: properties, Michelson and speed of light; developments in chemistry; Maxwell: electricity and magnetism; three laws of thermodynamics |
| Twentieth-Century Physical Science 1900-2000 | Theoretical science, quantum theory, special and general theories of relativity, and quantum mechanics become the foundation of physical science. They replace classical physics and determinism. | Fundamental breakthroughs: x-rays, radioactivity, electron; quantum theory: Planck; Einstein's theory of special relativity and general relativity; atomic structure; quantum mechanics: Heisenberg and Schrödinger; elementary particles |
| Energy and the Environment: Conflict or Compatibility | Scientists search for new sources of energy, such as nuclear energy, solar energy. Environmental concerns emerge. | Nuclear fission: Manhattan Project, spy scandals; German, British, Japanese, and Soviet fission research; nuclear fusion; postwar developments in nuclear energy; alternative energies: solar energy; environmental concerns: acid rain, climate change, lead in the environment |

## VII. OTHER THEMES

***External and internal resistance to new scientific ideas***

Examples of resistance to new scientific ideas from outside the scientific community include Galileo and the Catholic Church (1633), smallpox inoculation (1721), use of anesthesia (1846), and the Scopes Trial (1925).

Examples of resistance from within the scientific community include the nineteenth-century debates on the reality of atoms, on whether heat was a fluid (caloric) or a particle (mechanical motion), and on whether light was a wave or particle. In the twentieth century scientists debated relativity and causality.

***Transfer of scientific ideas to other disciplines***

*Pragmatism*

Charles Sanders Peirce (1839-1914), who joined the United States Coast Survey in 1861 and also lectured at Harvard, introduced pragmatism in 1878. William James (1842-1910) at Harvard became a strong advocate of pragmatism beginning in 1898. Pragmatism emerged as a philosophy after Darwin in 1859 published the *Origin of Species* in which he argued that species changed as a consequence of natural selection. Peirce and James extended this line of thinking to claim that ideas or truths also changed. Truth is relative, there is no absolute truth, there are many truths. Just as scientific theories change, so do ideas and truths. We test a truth by its practical consequences.

*Social Darwinism*

Social Darwinism is the application, or better, the misapplication of natural selection to society. Its defenders believe that those individuals who are able to adapt to their environment will survive, those who cannot will perish. Herbert Spencer (1820-1903) in London, England; William Graham Sumner (1840-1910), and the businessmen Andrew Carnegie (1835-1919) and John D. Rockefeller (1839-1932) in the United States, were Social Darwinists. They claimed that Social Darwinism was a law of nature and therefore a law of God and that society including government should not interfere with the enactment and consequences of this law.

Ironically, the most ardent Social Darwinists today remain the most fervent opponents of biological Darwinism. They find biological Darwinism unacceptable as an explanation of the origin of species, but Social Darwinism perfectly acceptable as an explanation for the evolution and behavior of society.

***Science and human values: the impact of science and technology on society***

Vivid accounts describing the impact of the nineteenth-century industrial revolution on western society appeared in several well-known books published at that time. They include Charles Dickens (1812-70) in England, *Hard Times* (1854); and in the United States Henry George (1839-97), *Progress and Poverty* (1879); Edward Bellamy (1850-98), *Looking Backward 2000-1887* (1888); Mark Twain (1835-1910), *A Connecticut Yankee in King Arthur's Court* (1889); and Henry Demarest Lloyd (1847-1903), *Wealth Against Commonwealth* (1894).

The impact of relativity, quantum theory, nuclear energy, space exploration, genetic engineering, robots, and computers on twentieth-century society has received considerable discussion in numerous books and articles on Einstein, Werner Heisenberg, Oppenheimer, and James Watson and Francis Crick. Radio and television broadcasts, lectures, and public forums by noted scientists and popularizers such as Carl Sagan (1934-96) at Cornell, Stephen Gould, and Hawking, have made it impossible for anyone not to have an awareness of the impact of science and technology.

## VIII. CONCLUSION

The historical study of science shows that science has progressed through three time periods, from its record keeping period that began with the Babylonians, Egyptians, and Chinese about 2000 BC, to its theoretical-philosophical period that arose in Greece in 600-400 BC, and to its mathematical-quantitative period that defined the sixteenth- and seventeenth-century (1600-1700) scientific revolution that occurred in Europe.

The history of science makes clear that science is a human activity, that people do science, and that scientific ideas change over time. For centuries scientists and philosophers have debated and continue to debate the origin of scientific ideas, whether they originate in the mind or the senses, and whether they are independent of their environment or a consequence of their environment. The relation of the humanities, including religion, to science and whether the two disciplines are completely separate fields of investigation or sometimes overlap are other topics that the history of science examines.

The history of science also makes clear the methodology and philosophy that scientists use to acquire scientific knowledge and develop scientific theories. Is the acquisition of scientific knowledge an inductive or a deductive process? Is this knowledge the product of the scientific method or does serendipity sometimes play a role? Are scientists realists or idealists, and do scientists have an epistemology and an ontology of science? An understanding of the language, objectives, and scope of the history of science has proved the best way to resolve these longstanding issues.

Today's society must use its centuries of cumulative scientific knowledge to promote progress and to expend its natural resources wisely rather than wasting and exploiting them. It can thereby avoid many of past societies' mistakes and misdirections. A society that does not know its past is likely to repeat the failures of the past and overlook the wisdom of its prophets. This is what the history of science teaches us.

## REVIEW QUESTIONS ON INTRODUCTION TO THE HISTORY OF SCIENCE

1. What is the main theme of this chapter?

2. Distinguish among the three terms: science, engineering, technology.

3. What is the history of science?

4. What are some differences between the sciences and the humanities?

5. What is a scientific theory?

6. What is the anthropic principle? What evidence did George Wald provide to support the anthropic principle? Do you accept or reject the anthropic principle?

7. Distinguish between internalist and externalist history of science.

8. What is the difference between epistemology and ontology?

9. Distinguish between induction and deduction; positivism and empiricism; idealism and realism.

10. What is Immanuel Kant's idealism?

11. Define Karl Popper's falsification method.

12. Is the scientific method the only pathway to new inventions or discoveries? What are other methods of scientific discovery or invention?

13. Identify or define: Frank Tipler, Stuart Kauffman and complexity, pragmatism, Social Darwinism.

# CHAPTER 2
# ANCIENT SCIENCE: FROM EARLIEST TIMES TO THE FALL OF ROME

Where there is matter, there is geometry.
Johann Kepler (1571-1630)

## I. INTRODUCTION

***Beginning of science and its transition from record keeping to theoretical-philosophical science***

Before the invention of writing about 5,000 years ago in 3100 BC, ancient societies for thousands of years used myths and oral communication to pass their knowledge to succeeding generations. These oral traditions included observations of the planets and stars, stone working, and possibly metal working techniques. Observational science began in 2000 BC with Babylonian and Egyptian written record keeping of the planetary and stellar (star) motions, processes for manufacturing dyes, drugs, and pottery, and recipes for cooking and baking. From early record keeping, science moved to the theoretical-philosophical science of the Greek natural philosophers in 600-400 BC, and then to mathematical-quantitative science beginning in the 1600s.

The transition from Babylonian and Egyptian record keeping to the Greek theoretical-philosophical outlook represents the beginning of the *Western intellectual tradition* with its belief that the universe functions according to natural laws that we can discover. Greek natural philosophers beginning with Thales of Miletus (640-546 BC), started to replace belief in supernatural causes, or the actions of the gods, as the cause of natural phenomena with a search for natural explanations of worldly phenomena. The transition from record keeping to theoretical-philosophical science, which is an essential component of modern science, did not occur in the science of contemporary eastern societies, in China, Japan, and India, and in the Americas.

| CHRONOLOGY OF HISTORICAL AGES | |
|---|---|
| ***Event*** | ***Time*** |
| Big Bang formation of the universe | 13.8 billion years ago |
| Earth and planets formed | 4.5 billion years ago |
| Stone Age or Paleolithic Age | began 2.5 million years ago |
| New Stone Age or Neolithic Age | lasted from 8000 to 2000 BC |
| Bronze Age | began in 4000 BC |
| Iron Age | began in 1000 BC |

| CHRONOLOGY OF HUMAN DEVELOPMENT | |
|---|---|
| ***Genus-Species*** | ***Age*** |
| Ardipithecus ramidus | 4.4 million years ago |
| Australopithecus afarensis (Lucy) | 3.4 million years ago |
| Homo habilis, stone age man | 2.2-1.6 million years ago. |
| Homo erectus | 1.6-0.4 million years ago, production and control of fire began 1.5 to 1 million years ago |
| Homo sapiens (early) | 400,000-30,000 years ago |
| Homo sapiens sapiens | emerged 30,000 years ago and continues to occupy Earth to the present time |

| CHRONOLOGY OF SELECTED ANCIENT EVENTS | |
|---|---|
| Hammurabi (1955 BC-1913 BC) | Hammurabi, king of the first dynasty of Babylon, gave a set of laws and codes of governance. It remains the most complete existing collection of Babylonian laws. The best source of the laws and codes is a black Diorite Stela found in 1901 in Susa in Iran. |
| Hellenic Age<br>(1300 BC -323 BC) | Egyptian Empire established from 1500 BC to 1200 BC. The first Greek city states arose in 800-600 BC. The Greek Olympiads began in 776 BC. The Persian Empire in 600 BC extended into the eastern Mediterranean. The Greeks defeated the Persians in 490-479 BC.<br><br>Alexander the Great (356-323 BC) ruled his vast empire from 336 to 323. He defeated the Persian king Darius III (reigned 336-330 BC) in 333 and again in 331 BC. Alexander founded Alexandria in Egypt in 332 BC. |
| Hellenistic Age (323-31 BC) | It dates from Alexander's death to the Roman conquest of Egypt. The spread of Greek-Macedonian culture through the lands Alexander conquered characterized the Hellenistic Age. |
| Rome (753 BC) | Romulus and Remus, descendants of the Trojan prince Aeneas, founded Rome in 753 BC. The Roman Republic lasted from 500 BC to 31 BC. Consuls ruled the Republic, and by 146 BC Rome dominated Greece. Beginning about 100 BC Rome moved toward military dictatorship. Octavian, known as Augustus Caesar (63 BC -14 AD), gained control of all Roman territories in 31 BC. He was the first Roman emperor, ruling from 29 BC to 14 AD. Augustus Caesar returned Rome from military dictatorship to constitutional rule. |

## II. COMMUNICATION OF ANCIENT THOUGHT

### *Origin of speech*

To account for the origin of speech in humans Allan C. Wilson (1935-91), a biochemist at the University of California, Berkeley, in 1987 proposed the controversial theory that the human capacity for language may have come from a genetic mutation in mitochondrial DNA that only mothers could pass on. Wilson based his theory on the research of the Italian medical scientist Luigi Luca Cavalli-Sforza (b. 1922) and his colleagues at Stanford University that suggested that language emerged about the same time as modern humans evolved in Africa at least 30,000-50,000 years ago. Wilson argued that as Africans moved into Europe only the children of African mothers could speak. African men would have prevented African women from mating with European and Asian men, and the mating of African men with European or Asian women would have resulted in mute children. Many anthropologists have rejected Wilson's hypothesis saying that fossil evidence clearly contradicts it.

A more recent study that the neuroscientists Michael Arbib (b. 1940) at the University of Southern California and Giacomo Rizzolatti (b. 1937) at the University of Parma in Italy published in 1998 has offered a different answer to the origin of speech. They have suggested from their analysis of fossil skulls that as the primate brain evolved it gradually developed a language mechanism before primates acquired a larynx capable of articulating speech. They believe that human speech evolved from an ancient gestural communication system, such as a monkey's oral-facial actions of tooth-chattering and lip-smacking, and not from the direct evolution of vocal communication. Using evidence obtained from positron emission and tomography (PET) scans, Arbib and Rizzolatti have concluded that human speech derived from pre-human primates' ancient gestural communication system and that a neural circuit formed in both monkeys and humans facilitated its derivation. Before they had throats capable of speech, primates had brains that could handle communication.

### *Role of myth*

Before there was writing there was myth, a traditional story that unveils part of a society's or culture's worldview. Myths often describe the origin of the world or universe as resulting from the mating of gods or the hatching of a cosmic egg. An ancient Chinese creation myth shows an intertwined double helix, representing an interaction of opposites resulting in creation. In an ancient Egyptian creation myth, Shu, the god of light and air, separates Nut, goddess of the sky, from Geb, god of Earth, reclining below. Animals often enter into creation myths. In ancient China a legendary turtle named Kwei directed the creation of the universe and presided over China's fate. In Hindu mythology four elephants standing on a turtle's back support the universe. In North American Indian myths an island borne on the back of a great turtle represents the universe.

### *Invention of writing*

Archaeologists generally agree that writing first appeared in Mesopotamia, the home of the Babylonians (Sumerians) in the lower Euphrates River region of modern Iraq, around 3100 BC. It began with an elaborate system of symbols that the ancients probably used for keeping temple records. But where did the ancients get the idea for their invention of writing? Denise Schmandt-Basserat (b. 1933), a French-born University of Texas art professor and expert archaeologist on the ancient uses of clay, found evidence that writing evolved from a much older record-keeping system that Middle East herdsmen still use. Her theory could push the roots of writing back at least 5,000 years.

Schmandt-Besserat bases her theory on studies started in 1969. For decades the great numbers of small, geometric clay tokens, some as old as 10,000 years and discovered in digs from Egypt to the Indus Valley, had puzzled archaeologists. Several experts speculated that these tokens were toys or pieces from a still undiscovered prehistoric game. In 1966 Pierre Amiet (b. 1922), curator of Near Eastern art at the Louvre, suggested that the tokens were an ancient recording system. Schmandt-Besserat compared the tokens with samples of early writing discovered in Warka, a community in Mesopotamia, and concluded that they were part of a record keeping system that eventually evolved into writing.

The process, Schmandt-Besserat believes, evolved in four stages. In the first, which ran from about 8500 BC, the date of the oldest tokens found, to around 3500 BC, the small (up to 5 centimeters or 2 inches in diameter) cones, disks, spheres, and pellets represented commodities such as sheep, jugs of oil, bread or clothing that merchants and others in the Middle East used to keep records. In the second stage, merchants shipping goods from one place to another began enclosing the tokens in sealed clay balls known as *bullae*, which the recipients broke open upon delivery to check the shipment against the invoice. The *bullae* represent the first bills of lading. The third stage followed closely and began when merchants realized that they could check cargoes without breaking open the *bullae* if they first impressed each of the tokens on the ball's wet clay surface before enclosing them in the ball. This left a record on the outside of the envelopes as well as inside.

Upon reaching this stage, the fourth step followed quickly. Realizing that impressing the token shapes on the *bullae* made enclosing the tokens unnecessary, merchants abandoned the counters and began keeping their records directly on clay tablets. The technique's efficiency became immediately obvious, says Schmandt-Besserat, and could explain how written record keeping evolved and why writing spread so rapidly along the trade routes and quickly took hold throughout the civilized world of that time. But the development of writing did not lead to the disappearance of the tokens because the written word helped only those who could read. In the twenty-first century, in some regions of the Middle East, illiterate herdsmen use tokens to count the animals in their flocks.

*Materials used in writing*

The Babylonians used clay tablets. They wrote in cuneiform, which consists of wedge-shaped impressions made by a stylus.

**Figure 1**
**Babylonian Cuneiform**
Asmus Koefoed/Shutterstock.com

The Egyptians, beginning around 3100 BC, wrote in hieroglyphics on papyrus, a substance obtained from a reed-like plant that grew along the banks of the Nile River. The discovery of the Rosetta Stone, a dark basaltic stone, on 15 July 1799 by Captain Pierre Bouchard (1772-1832), a French officer of engineers with Napoleon's army in Egypt, while repairing Fort Julien, led to the understanding of hieroglyphics.

**Figure 2**
**Egyptian Hieroglyphics**
Hoika Mikhail//Shutterstock.com

The Rosetta Stone is 3 feet 9 inches long by 2 feet 4½ inches wide. Its erection in 195 BC in the northern town of Rosetta (Rashid) in the Nile Delta 30 miles from Alexandria was to honor Ptolemy V Epiphanes (210-180 BC, reigned 204-180 BC) for remitting the priestly dues. Its writing is in Egyptian demotic, Egyptian hieroglyphics, and Greek. Thomas Young (1773-1829), a London physician, began the translation in 1818 and the French Egyptologist Jean Francois Champollion (1790-1832) at Grenoble and later at the College of France in Paris, completed the translation in 1822. The Rosetta Stone is now in the British Museum in London. It passed into British hands with Napoleon's defeat and the French surrender of Egypt in 1801.

The Chinese used paper and silk. They wrote in more than 7,000 characters.

The Mayans used hieroglyphics, which date from 700 to 500 BC. They wrote on stone and on paper made from bark coated with lime.

***Calendars***

Historical studies require some knowledge of when events occurred, that is, of time. Confusion exists, however because different societies adopted different ways of recording time and developing calendars. The Jews, Mayans, Hindus, Buddhists, Chinese, Romans, Christians, and Muslims have different starting times for their calendars. The Western world beginning in the 500s adopted the Christian calendar, but it, unfortunately, had problems from the beginning because of confusion over the birth date of Jesus Christ and because the mathematics used at that time did not include zero as a number.

The Hebrew 354 day lunar-solar calendar, possibly the oldest calendar, begins in the Christian year 3760 BC. It is their year 1 and identified as the moment of creation. The Christian year 2000 corresponds to the Jewish year 5760. Abraham, the founder of Judaism, lived in the Christian year 1900 BC.

The Mayan solar calendar dates from 3114 BC. It is their year 1 and the beginning of their modern period. The Mayan Great Cycle ended on 23 December 2012. The prophesized world-drenching torrent gushing from the mountain of a stream never occurred. The Christian year 2000 is their year 5113.

The Hindu lunar calendar dates from 3102 BC. It represents their year 1 and the beginning of the Kali Yuga time cycle, one of the four ages of the Hindu world cycle. The Hindus revised their calendar in 79 AD, making it their year 1, and called it 1 Saka. The Christian year 2000 corresponds to 1921 of the Hindu calendar. Hinduism has no known founder or beginning. Belief in Brahman, the one, all, and absolute, is its fundamental doctrine.

Huang Di, China's Yellow emperor, who lived about 5,000 years ago and who allegedly invented the compass, established the Chinese lunar-solar calendar in 2637 BC. It runs in 60-year cycles so that Christian year 2000 corresponds to the Chinese year 4637. Confucianism and Taoism date from the sixth century BC.

The Buddhist year 1 corresponds with the death of Buddha, Siddhartha Gautama (563-483 BC) or (625-543 BC), the founder of Buddhism who lived near Nepal. He developed Buddhism from Hinduism in the sixth century BC. Buddhism's believers follow the noble eightfold path to Nirvana. The Christian year 2000 is the Buddhist year 2483 or 2543.

Julius Caesar (102-44 BC) introduced the Julian solar calendar in 46 BC. January 1 became the beginning of the New Year. He added 90 days to the calendar so that the calendar year now had 365 days divided into 30- and 31-day alternating months, except for February. February had 28 days with an extra day added every fourth February (a leap year) to bring the calendar in agreement with the solar (tropical) year of 365¼ days (365.242199 days). With the correction the Julian calendar exceeded the solar year by 11 minutes. The solar year represents the time for the Sun to make one complete west-to-east revolution around Earth's equator. Without a leap year the calendar would be one day ahead of the solar year. If the calendar indicated January 2, the solar date would be January 1.

The Christian calendar starts with the year of Jesus Christ's birth in approximately 4 BC (Jesus Christ 4 BC-29 AD). Dionysius Exiguus (d. 544), a Scythian monk living in Rome ca. 535, devised for Pope John I (d. 526) the system of counting from the year of Christ's birth. He was working with tables of dates, trying to clarify the date of Easter and calculated Christ's birth date as 753 years after the founding of Rome or 1 AD. The Bible indicates the year of the census that brought Joseph and Mary from Nazareth to Bethlehem was 12 BC. It also says Christ's birth occurred when Herod (73-4 BC) was king of the Jews. Confusion in the chronology results because in the chronologies of Dionysius and Venerable Bede (673-735), a Northumberian (English) monk, the years go from 1 BC to 1 AD (*Anno Domini* meaning in the year of the Lord). This occurred because at the time of Exiguus and Bede, zero as a number was unknown in Europe.

Dionysius's system replaced the Roman numbering *Anno Diocletiani* that counted the years from the beginning of the reign of the Roman emperor Diocletian (245-313), an infamous persecutor of Christians. Later Bede adopted Dionysius's system. He added 1 BC immediately before 1 AD and popularized the chronology in his *Ecclesiastical History of the English People* (731). The chronology spread through much of Europe when Charlemagne (742-814), the Frankish king crowned Emperor of the West in 800, adopted it for dating government facts. The papal chancery adopted it in the tenth century.

The Muslim lunar calendar begins with the year that Mohammed (570-632) fled (Hegira) from Mecca to Medina, 622 AD which is 1 AH (*Anno Hegirae*: in the year of the Hegira or departure).

## III. MIDDLE EAST AND AFRICAN SCIENCE AND ENGINEERING

### *Agriculture and commerce*

The first villages in the Middle East and China sprang up around wild crops of barley, rice, or other grains about 12,000 years ago or in 10,000 BC. The villages predated agriculture, or domestic cultivation of crops, which began around the world, in Egypt, Iran, Iraq, China, and Mesoamerica, about 10,000 years ago (8000 BC).

The rise of city-states along the 1,180-mile Tigris and 2,235-mile Euphrates Rivers in modern Iraq and along the 930-mile Nile River in Egypt occurred in 4000 BC. Upper and lower Egypt united in 3110 BC to form the first dynasty (3110-2890 BC) under Menes, a southern ruler who founded Memphis, a city 19 kilometers south of Cairo, in 3100 BC. The formation of the Egyptian Empire occurred from 1500 to 1200 BC. It fell to Persia (Iran) in 525 BC. Commerce flourished in Egypt because the Nile's current runs south to north, whereas the winds blow north to south, permitting two-way traffic on the Nile. Egypt was a moneyless society. Money was apparently a Greek invention.

Middle East societies used the wheel in 3300 BC.

***Egyptian contributions to engineering and science***

*Agriculture and engineering*

A well-developed agricultural system existed in Egypt. The regular flooding of the Nile River made crops, such as wheat, corn, millet, and cotton plentiful.

The Egyptians constructed about 80 pyramids, mostly on the west side of the Nile River, over a thousand year period from about 2800 BC to 1700 BC. Ten of them are genuine pyramids having a solid masonry core. Imhotep (ca. 2780 BC), the architect, designed the first pyramid, the 60-meter high limestone Step Pyramid of the pharaoh Zoser (ruled 2668 to 2649 BC) at Saqqara near Memphis, during the third (2686 to 2575 BC) and fourth (2613 to 2598 BC) dynasties. It took 20,000 workers about 20 years to build for the pharaoh Khufu (Cheops d. 2566 BC) the Great Pyramid at Giza on the outskirts of Cairo. The Great Pyramid was originally 147 meters (now 137 meters) high with a base 230 meters square. Its construction required over 2 million stone blocks having an average weight of 2.5 tons. There are two other pyramids at Giza, the pyramid of Khufu's son Khafre (ruled 2554-d. 2532 BC), 136 meters high and the pyramid of the pharaoh Menkaure (ruled 2532-d. 2504), 70 meters high. The famous Sphinx, carved during the time of Khafre, is also at Giza.

The Egyptians may have used cradles to move the giant limestone or granite stones, some of which weighed 50 tons, by wrapping four cradles around a rectangular stone to form a cylinder and then pulling the cylinder with a parbuckle. This is a sling for rolling cylindrical objects up or down an inclined plane and consists of a rope looped over a cylinder with its two ends passing around the object to be moved. Models of cradles are extant in Egypt.

Most Egyptologists today believe the Egyptians, after transporting the stones by boat in the Nile River, used sledges pulled by workers to get the stones to the site and ramps to move the stones up the sides of the ever-rising pyramid. Six men could pull a six-ton stone.

*Astronomy and mathematics*

The Egyptians in 2700 BC had a 365-day solar calendar. Egyptian astronomy and mathematics were not particularly advanced and consisted mainly of recording the planetary motions and methods of solving mathematical problems. The Egyptians never developed a physical (astronomical) model of the universe.

The Egyptians used a 10-base system with different numeral hieroglyphs for 1, 10, 100, 1,000, 10,000, 100,000, and 1 million. One was a vertical dash *1*, 10 was an inverted *u* or a heel bone, 100 looked like a comma or a scroll, 100,000 looked like a bird. They also developed a hieratic (a cursive form simpler than hieroglyphic writing) number system of symbols in which 1, 2, and 3 were vertical dashes, 4 a horizontal dash, 5 a symbol that looked like a seven, 6 a symbol that looked like a Z, 100 a symbol that looked like a comma, 2,000 a symbol that looked like a 4, and other symbols for other

numbers. They had no concept of zero.
Ahmose (ca. 1650 BC) made calculations in mathematics and geometry. He calculated pi ($\pi$), the ratio of the circumference of a circle to its diameter, by assuming that the area of a circle is 8/9 the area of the square on a circle's diameter and that $\pi$ represents the ratio of the area of the circle to the area of the square of radius 2r or d. Using $A = \pi r^2$ or $\pi d^2 / 4$ and the relation

$$\frac{\pi d^2}{4} = \left(\frac{8}{9}d\right)^2$$

$$\pi = 4\left(\frac{8}{9}\right)^2$$

$$\pi = 3.1605$$

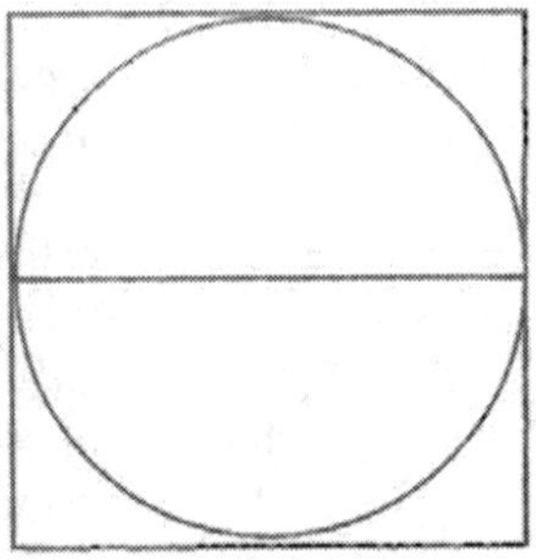

**Figure 3**
**Ahmose's Determination of Pi**
Courtesy of the author

The Rhind Papyrus, 5 meters long by 32 centimeters wide, dating from 1650 BC and written in Hieratic script provided much of our knowledge of Egyptian mathematics. Archaeologists discovered it in Thebes in the ruins of a small building near the Ramesseum. The British Museum acquired the Papyrus in 1864. Egyptologists translated it in 1877.

*Mummification*

The Egyptians practiced mummification by coating the body with natron (sodium carbonate or sodium bicarbonate) to preserve the body for union with the soul after death. They opposed the dissection of the body for this reason. Sodium carbonate is a drying agent and kept moisture from the body. The bicarbonate helped against bacterial decay. Natron also contains sodium chloride and sodium sulfate which protected the body from moisture. The hot dry Egyptian climate provided additional protection.

Probably the most famous mummy ever discovered was that of King Tut (Tutankhamen, 1341-1323 BC), the eighteen-year old boy pharaoh who reigned in 1333-1323 BC. The English Egyptologist Howard Carter (1873-1939), famous for his Valley of the Kings excavations, discovered King Tut's tomb in Thebes in 1922. The tomb's treasures, including Tut's solid gold coffin, gold portrait mask, and other treasures were intact. They are today in the Cairo Museum.

Archaeologists over the decades have discovered thousands of mummies dating from 2,000 to 4,000 years ago in places such as the Valley of the Golden Mummies, in a 2.32 square mile cemetery, about 235 miles southwest of Cairo. The 2,000 year old cemetery contained at least 207 mummies of the wealthy, middle, and poor classes of citizens.

*Egypt and the roots of the western intellectual tradition*

Ptolemy I Soter (367-283 BC) founded the Museum (Library) in Alexandria in 300 BC. It had about 700,000 scrolls. Theophilus (d. 412), patriarch of Alexandria, burned the Museum in 391 AD. Omar, Caliph of Damascus (581-644), and the Muslims supposedly burned it again in 642.

Recently, debate has centered on how much of the Western intellectual tradition, such as Greek rational philosophy, democracy, literature, and science, actually came from Greece and was not wholly or in part borrowed from other societies, particularly from Egypt, that is, from Africa. Thales, Democritus, and Pythagoras, were quite familiar with Egyptian learning. Clearly, the Egyptians made contributions to the growth of learning in engineering, calendar studies, and chemistry, and the issue becomes how much did the Greeks import from these societies, and when did they acquire this information.

One of the totally erroneous claims about Egyptian-borrowed-learning says that Aristotle visited the Museum in Alexandria, which was not built until 300 BC, and used information from the Egyptian *Book of the Dead* in his own book *On the Soul*. Not only could Aristotle not read the Egyptian text, which was totally different from his, but Aristotle had died 22 years earlier.

Nineteenth-century European historians have downplayed the Egyptian achievements as is usually the case when the *victors* write the history of events, and we should ask whether it is really possible for neighboring societies in seemingly close contact with each other not to influence each other. The debate also has led to discussion on whether such well-known figures of that geographic region, such as Socrates and Cleopatra, were white or African. For opposing views on this issue see Martin Bernal (b. 1937), *Black Athena* (1987), and Mary Lefkowitz (b. 1935), *Not Out of Africa* (1997).

**Babylonian contributions to science**

*Mathematics*

The Babylonians had a 60-base system and performed calculations similar to the examples below.

$$8^2 = 1\ 4 \text{ or } (1 \times 60) + 4 = 64$$
$$59^2 = 58\ 1 \text{ or } (58 \times 60) + 1 = 3481$$
$$72^2 = 86\ 24 \text{ or } (86 \times 60) + 24 = 5184$$

They divided the circle into 360 degrees sometime in the period 1000 BC to 1 BC. The Babylonians estimated that in a 24-hour period (one day) an individual could travel 12 of their time-miles (1 time-mile = 7 modern miles). This also represented the time or distance it took the Sun to make one complete revolution, or travel through one circle, and so they divided the circle into 12 parts. But because each division was so large they multiplied 12 by 30 to get the smaller 360 divisions.

The Babylonians used zero for place notation in 500 BC. Without the zero the context alone had to make the meaning clear.

*Astronomy*

The Babylonians developed a star catalogue in 1800 BC and kept a continuous record of the planets' motions from 747 BC. By 400 BC they used a lunar-solar calendar and divided the Zodiac into twelve constellations that make up the ecliptic.

The Babylonians did not develop a physical model of the universe. They compiled excellent planetary tables but never took the next step of devising a model or models to account for the planetary motions.

Kiddinu (ca. 340 BC), also known as Kiddnas or Cidenas, the head of the Babylonian school of astronomy in Sippar, calculated values for the precession of the equinoxes. This is thc slow clockwise, or westward, (when viewed from above, Pisces to Aquarius) movement on the ecliptic of the two dates, March 21 and September 22, when the Sun on its annual eastward journey around the ecliptic crossed the equator. Because of precession the equinoxes, the two dates on which the Sun crosses the equator, occur slightly earlier in consecutive years. Kiddinu also developed methods of calculating the irregular movements of the Moon and other planets.

By about 300 BC the Babylonian astronomer Berosus, who lived sometime between 356 and 261 BC, had constructed a sundial that consisted of a ball-shaped hollow space and a style (stylus) that stuck straight out from it. A typical sundial consisted of a vertical part or style and a base of wood, metal, or stone with marked off divisions over which the style's shadow fell. The shadow moved with the Sun's daily east to west motion and indicated the time. Midday was when the Sun reached due south and was halfway through its journey.

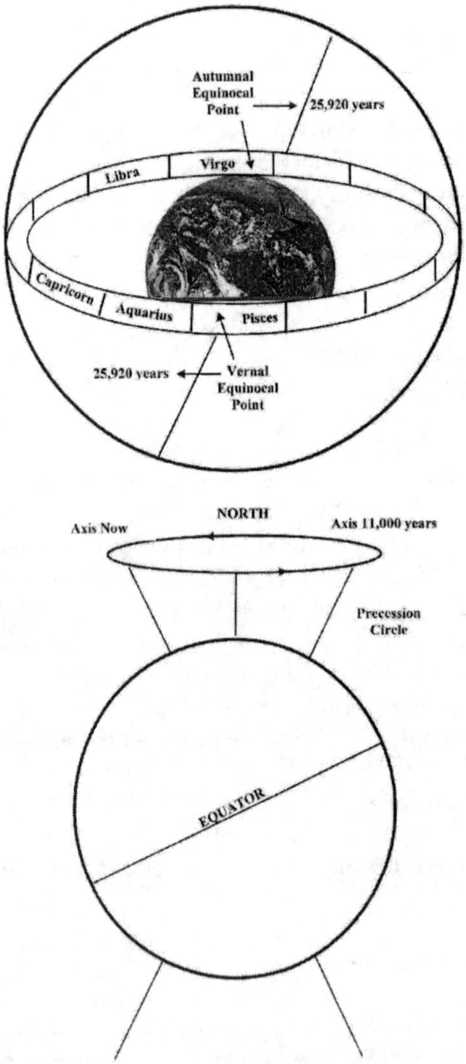


**Figure 4**
**Precession of the Equinoxes**
Courtesy of the author

# IV. SCIENTIFIC WORLD OF THE GREEKS: THE BEGINNING OF THEORETICAL - PHILOSOPHICAL SCIENCE

***Mathematics: number systems***

Greek mathematicians never developed a common number system that all the Greek states adopted. The first Greek number system, the base 10 Attic (Herodianic) or acrophonic system, used in the first millennium BC prior to 300 BC, by definition meant that each symbol except the symbol for one, came from the first letter of the word used for the number.

| ACROPHONIC NUMBER SYSTEM | | |
|---|---|---|
| 1 | | 1 |
| 5 | Pente | Γ |
| 10 | Deka | Δ |
| 100 | Hekaton | H |
| 1,000 | Khilioi | X |
| 10,000 | Murioi | M |

A second ancient number system dating from at least 450 BC, the Ionic learned system or alphabetical numbers, gave numerical values to the letters of the Greek alphabet. It was 10-based and used 27 characters 24 from the Greek alphabet and 3 others. Later Apollonius and Archimedes developed number systems.

| IONIC LEARNED SYSTEM | | | | | | |
|---|---|---|---|---|---|---|
| A | α | 1 | | I | 1 | 10 |
| B | β | 2 | | K | κ | 20 |
| Γ | γ | 3 | | Λ | λ | 30 |
| Δ | δ | 4 | | Π | π | 80 |

***Astronomy***

*Motion of the sun, stars, and planets*

With few exceptions, natural philosophers (astronomers) in Greece and elsewhere believed they were viewing the motions of the heavenly bodies from a stationary Earth. When they observed the Sun it appeared to move in the course of one year counterclockwise from west to east against the background of the distant stars whose relative positions with respect to one another appeared not to change. For this reason the Greek astronomers believed that the stars were attached to a giant sphere. Anaximander (610-546 BC), a pupil of Thales, first introduced spheres to astronomy. He envisioned the sky as a complete sphere. His Earth was a cylinder with an east-west axis and a height 1/3 its diameter. Anaximander very likely invented the gnomon stick, a sundial with a vertical stick.

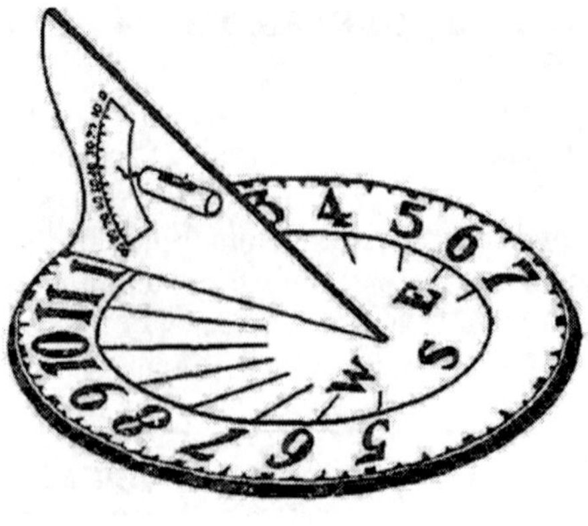

**Figure 5**
**Sundial**
Source: J. Dorman Steele. *Popular Physics.* New York: American Book Company, 1896.

The ancient astronomers also noticed that the planets appeared to move annually from west to east against the background of the fixed stars but not always regularly. Most of the time the planets moved from west to east, but at other times the planets appeared to stop moving and then start moving backwards from east to west. This backward or retrograde motion was really a consequence of Earth's orbital motion.

A different picture emerged when astronomers viewed the Sun, planets, and sphere of the stars daily against the horizon, the extension of which was the celestial equator. The Sun, planets, and sphere of the stars appeared to rotate daily from east to west. They noted that at two times during the year, the equinoxes March 21 and September 22, the Sun rose and set due east and west, making day and night of equal length. At two other times the solstices, June 21 and December 22, they observed that the Sun reached its highest points north and south in the sky giving the longest day-shortest night and the shortest day-longest night in the northern hemisphere, and the shortest day-longest night and longest day-shortest night in the southern hemisphere.

From measurements of the horizontal length of the shadow the Sun, on its annual journey around the ecliptic, cast on the vertical gnomon stick of a sundial, astronomers reached an important conclusion. They found that the shortest length of the gnomon stick's shadow and hence the smallest angle *SGH* of a right angle triangle formed by the vertical gnomon stick *GS*, its horizontal shadow *SH*, and the Sun's ray as the hypotenuse *GH*, occurred at noon on 21 June. The maximum length of the shadow and maximum value of angle *SGH* occurred on 22 December. The calculated maximum value of *SGH*, called the angle of obliquity, was 23 degrees 27 minutes (23° 27' or 23.45°).

For example, an observer measuring from the horizon at 50° latitude north or south sees the Sun 63° 27' above the horizon at noon on 21 June [(90°- 50°) + 23.45° = 63.45°] and 16°33' above the horizon on 22 December [(90°- 50°) - 23.45° = 16.55°]. The difference in the two values, 46° 54', is twice the obliquity, and the angle of obliquity is 23° 27'. At the equator an observer measuring from the equator sees the Sun at 90° + 23.45° = 113.45° above the horizon at noon on 21 June and 90°- 23.45° = 66.55° measured from southern horizon at noon on 22 December. The difference, 46.90°, is twice the obliquity, making the angle of obliquity 23.45°.

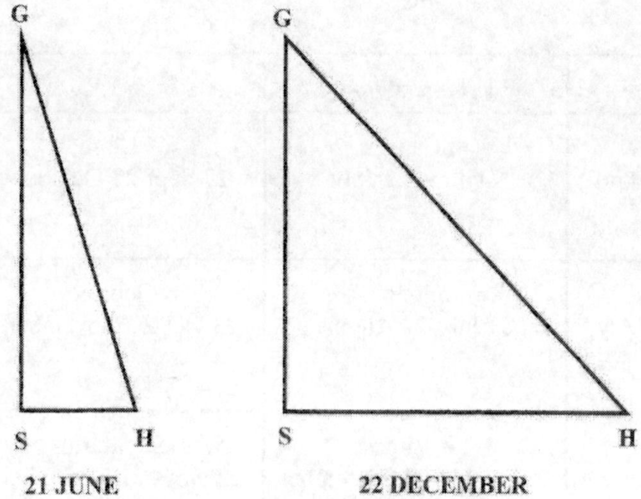


**Figure 6**
**Measuring Earth's Angle of Obliquity (not to scale)**
Courtesy of the author

These observations led ancient astronomers to conclude that the annual path of the Sun around the great circle of the constellations, or the ecliptic, was at an angle 23° 27' to the path of the Sun around the celestial equator. This was the imagined great circle of the celestial sphere or extension of Earth's equator. The ancient astronomers called the Sun's annual path through the constellations the ecliptic because eclipses took place only when the Moon was in or near the ecliptic.

*Spherical Earth and the two-sphere geocentric model of the universe*

From the time of Pythagoras (570-490 BC) Greek natural philosophers believed Earth was a sphere, stationary at the center of the world. Parmenides, Eudoxus, Plato, Aristotle, Aristarchus, Archimedes, and Eratosthenes accepted Earth's sphericity and offered several observational proofs for Earth's sphericity.

A ship's mast appears first when the ship is approaching an observer and disappears last when the ship is departing from the observer.

During an eclipse of the Moon, when the Sun, Earth, and Moon are aligned and Earth is between the Sun and Moon, Earth's curved shadow is visible on the illuminated Moon.

The constellations are different when viewed from different places on Earth's surface, some appear, and others disappear as you move north or south.

Prior to 400 BC Greek natural philosophers accepted a two-sphere model of the world (universe) with a stationary spherical Earth at the center and an outermost sphere containing the Sun, planets, and the fixed stars or twelve constellations. The outermost stellar sphere rotated daily on a fixed axis from east to west (clockwise) once every 24 hours (23 hours and 56 minutes). The ecliptic was a great circle (a diameter) on the sphere and tilted 23½° (23° 27') to the celestial equator which divided the sphere in half. The Sun moved from west to east (counterclockwise) around the ecliptic once every 365¼ days. All the planets including the Sun revolved annually around Earth, through the stars, from west to east.

| ZODIAC | | | |
|---|---|---|---|
| ***Spring Signs*** | ***Summer Signs*** | ***Autumn Signs*** | ***Winter Signs*** |
| 1. Aries<br>21 March-19 April | 4. Cancer<br>21 June-22 July | 7. Libra<br>23 Sept-22 October | 10. Capricorn<br>22 Dec-19 January |
| 2. Taurus<br>20 April-21 May | 5. Leo<br>23 July-22 August | 8. Scorpio<br>23 Oct-21 November | 11. Aquarius<br>20 January-18 February |
| 3. Gemini<br>22 May-20 June | 6. Virgo<br>23 August-22 September | 9. Sagittarius<br>22 Nov- 21 December | 12. Pisces<br>19 February-20 March |

The Sun's annual motion was not simply motion from west to east, otherwise it would always rise and set at the same points on the horizon, and its midday height and time from sunrise to sunset would never change. During the summer period, March to September, when day is longer than night, and more than half of the Sun's daily path is above the horizon, the Sun is north of the equator. During the winter period, September to March, the Sun is south of the equator.

Other evidence that indicated a changing of the Sun's distance from the North Pole (or the South Pole) of the celestial sphere was that in the winter months the Sun was lower in the sky than in the summer, and its midday height was less. All of these observations indicated that the Sun's annual path on the celestial sphere, or ecliptic, was tilted at some angle (oblique) to the equator. Part of its path was on one side of the equator and part was of it on the other side. The angle at which the ecliptic cuts the equator was the obliquity of the ecliptic, 23° 27'. The Chaldeans and Babylonians probably calculated a value for the obliquity, while Hipparchus, Eratosthenes, and Ptolemy later obtained accurate values.

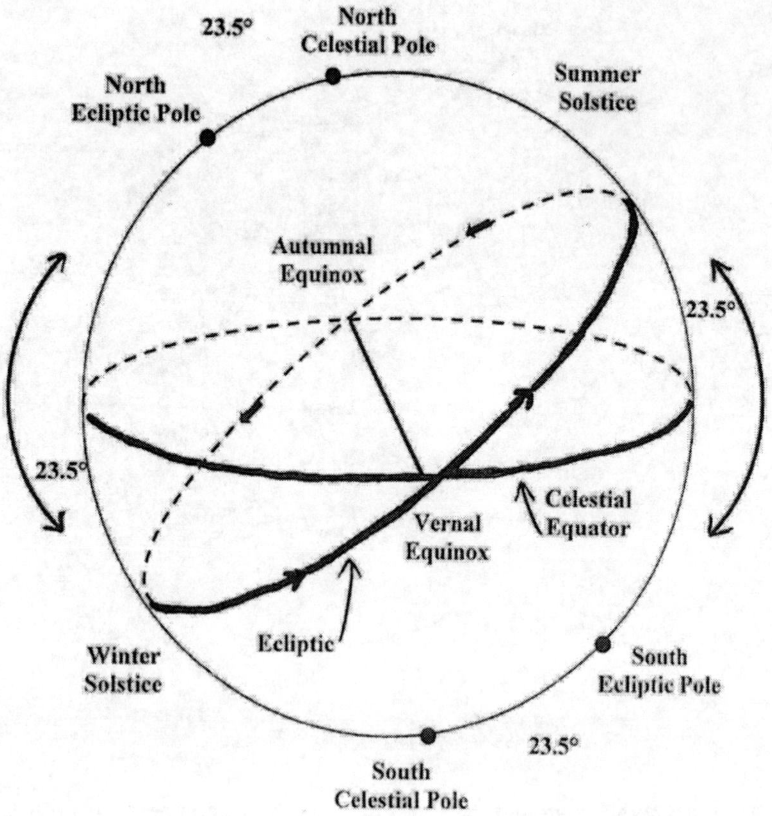


The celestial sphere represents the sky as seen from Earth. It shows the Moon, Sun, planets, and stars having the same distance from Earth, which they do not. The celestial sphere rotates once daily (because of Earth's rotation). The ecliptic is the Sun's apparent path on the celestial sphere during a year. The planets' paths depend on the inclination of their orbits but generally follow the ecliptic. The twelve zodiacal constellations form a band that straddles the ecliptic.

**Figure 7**
**Two-Sphere Model of the Universe**
Courtesy of the author

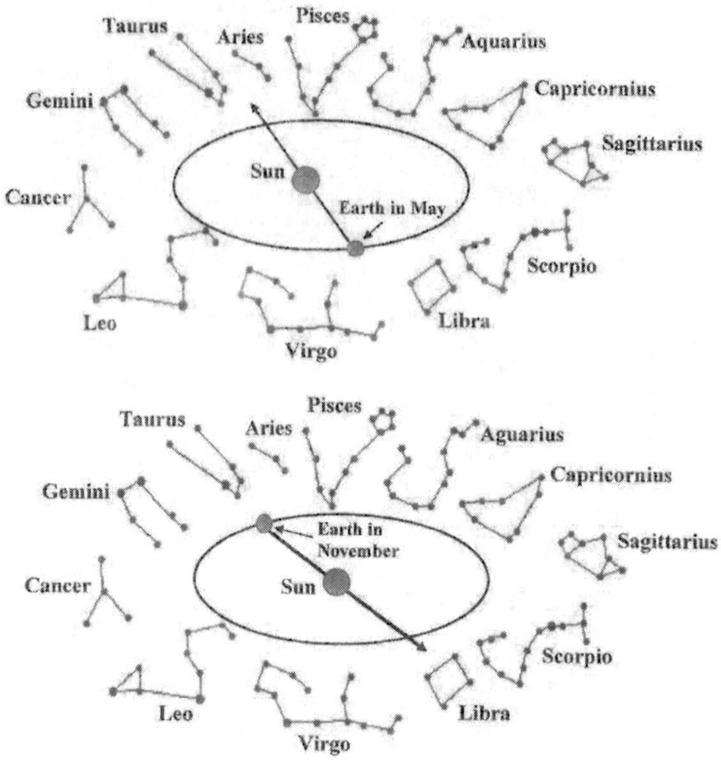


As Earth revolves around the Sun annually, the Sun appears to drift eastward annually among the Zodiac constellations along a circular path called the *ecliptic*. The ecliptic is the projection of Earth's orbit on the sky. This circular path tilts 23.5° with respect to the celestial equator because Earth's rotation axis tilts 23.5° with respect to its orbital plane. The Sun drifts along the ecliptic during the year and rises and sets daily when viewed against the horizon.

**Figure 8**
**Constellations of the Zodiac**
Courtesy of the author

*Order of the planets*

Greek astronomers, who called the planets *wanderers*, established the order of the planets from each planet's time of revolution, or the time it took to move through the constellations. A slower time of revolution around a stationary Earth meant a greater distance from the stationary Earth. The order that resulted was Earth, Moon (29 days), Mercury (88 days), Venus (225 days), Sun (365 days), Mars (687 days), Jupiter (12 years), Saturn (29 years), and the stellar sphere. The increased or decreased brightness of a planet indicated its movement towards or away from Earth.

Greek astronomers believed that circular motion was perfect motion because a circle has no beginning or end. When looking down on the planets from the north pole, they described the motion of all the planets, including the Sun, as revolving counterclockwise, from west to east around a stationary Earth through the sphere of the fixed stars or constellations, during one year.

The Sun, planets, and the fixed stars also revolved daily from east to west when viewed against the horizon (celestial equator). This occurs because of Earth's daily rotation on its axis from west to east, which the ancient astronomers did not accept.

*Solar and sidereal years*

Hipparchus (190-120 BC) in Nicaea in northwestern Asia Minor and later Rhodes appears to be the first astronomer to have a clear understanding and to make a clear distinction between the tropical and the sidereal year.

The solar or tropical year is the time the Sun takes to return to the same place on the equator, from vernal equinox to vernal equinox or from solstice to solstice. It measures the calendar or natural year.

Tropic means belonging to a turn of the Sun at the solstices. It refers to either of the two circles parallel to the equator. The tropics are the limits of the apparent annual north and south journey of the Sun. Tropic of Cancer is 23° 27' north of the equator; it is the northernmost latitude reached by the Sun. Tropic of Capricorn is 23° 27' south of the equator; it is the southernmost latitude reached by the Sun.

Eratosthenes (ca. 100 BC), Hipparchus (ca. 135 BC), and Ptolemy calculated the angle of obliquity (23° 27 minutes), or the tilt of Earth's celestial pole from its equatorial pole, using the method Ptolemy described. Ptolemy calibrated in degrees a brass circle, or an inscribed stone or wood block, having its flat surface oriented north-south, and measured the altitude of the Sun at noon (the shadow the Sun cast) when it was highest in the sky and cast a short shadow (summer solstice, 21 June) and again when the Sun was lowest in the sky at noon and cast a long shadow (winter solstice, 22 December). The difference in the measured values of the angles of altitude on 21 June and 22 December divided by two gave the angle of obliquity.

The celestial or sidereal year is the time the Sun takes to return to the same place in the constellations of the fixed stars. It is longer than the tropical year by about eleven minutes because of Earth's precession which makes the equinoxes occur earlier each year thereby shortening the tropical year. Hipparchus recognized the difference between these two kinds of year and measured the length of each kind of year by comparing the times of the solstices and equinoxes using his measurements and those of earlier astronomers such as Aristarchus.

*Geocentric models of Plato and Aristotle*

Plato (427-347 BC), whose real name was Aristocles but called Plato meaning broad because of his broad shoulders and forehead and also because of his breadth of knowledge, was born into a wealthy and prominent Athenian family. He was a disciple of Socrates (470-399). Plato founded the Academy (380 BC-529 AD) in Athens, which at that time had a population of about 300,000. Plato's existing writings total 260 separate manuscripts and include both fragments and complete works. They are handwritten in Greek and date from 995 to 1600. The manuscripts are copies that professional scribes made for libraries or wealthy patrons. Nothing in Plato's own hand (that Plato himself wrote) exists.

Plato's universe had a beginning, a time of creation. His discussed his planetary model of the universe in *Timaeus*, in which he asked astronomers to consider the motions of the planets as uniform circular motions or combinations of circular motions, and made a crude attempt to develop such a model. The planets were perfect spheres and encircled Earth in perfect circular motion.

Aristotle (384-322 BC), whose father was physician to the Macedonian King Amyntas IV (reigned 359-356 BC), entered Plato's Academy at age 18 and studied there for nearly 20 years, from 367 BC to 347 BC collecting facts and writing on rhetoric, literary criticism, and poetry. After a 12-year period of travel, from 347 to 335 during which Aristotle did most of his biological research and stayed for a year (343-42) in Mieza to tutor Alexander the Great, Aristotle returned to Athens and founded the Lyceum, or Peripatetic school (336 BC-300 BC). He did most of his political research at the Lyceum. Aristotle left Athens for Chalcis in 322 BC where he died. Archeologists in 1997 believed they found the location of the Lyceum at an underground site about a mile from the Acropolis.

Aristotle's writings total about 150 manuscripts, about 50 of which have survived. The Roman General Lucius Sulla (138-78 BC) discovered them in 86 BC in Athens in the library of the famous book collector Apellicon of Teos (d. 84 BC). Sulla took Apellicon's library, including Aristotle's manuscripts to Rome in 84 BC where scribes recopied and published them in 60 BC. Aristotle's works, unlike Plato's, were largely lost to Europe after the fall of Rome in 476 AD. The Muslims in the east preserved Aristotle's writings, and they made their way back into Europe during the time of the Crusades in the twelfth and thirteenth centuries. At that time Aristotle replaced Plato as the source of Greek wisdom.

Aristotle's universe always existed, it had no beginning and no end. He described his planetary model of the universe in *De Caelo*. It was similar to the model of Eudoxus (408-355 BC) of Cnidus, who studied at Plato's Academy and introduced the idea of multiple spheres to describe the motion of each planet. Eudoxus had the pole of one planet embedded in the smaller concentric sphere of another planet which in turn had its pole embedded in the smaller concentric sphere of the next planet, and so on for all the planets. While each planet had regular circular motion, the planet's motion also depended on the motion of the sphere with which its pole was in contact. The net motion was then the combination of the planet's separate motions and, according to Eudoxus, accounted for the irregular motions (retrograde or backward motion) of the planets. Eudoxus had a total of 27 spheres, 1 for the fixed stars, 6 for the Sun and Moon, and 4 for each of the five planets or 20. Callippus (b. 370 BC), who attended both the Academy and the Lyceum, added 7 more spheres, two spheres each for the Sun and Moon, and one each for Venus, Mercury, and Mars, giving a total of 34. Both models gave little agreement with observations. Aristotle's planetary model increased the number of concentric spheres to 55 but gave no more accurate planetary positions.

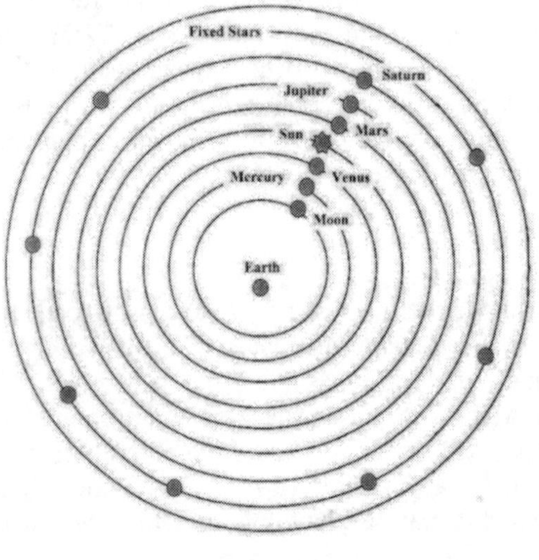


From inner to outer (1) spheres of Earth, water, air and fire that compose the terrestrial region; (2) the planetary spheres carrying Moon, Mercury, Venus, Sun, Mars, Jupiter, Saturn, and the fixed stars; (3) the prime mover, which implies rest.

**Figure 9**
**Aristotle's Universe**
Courtesy of the author

*Archimedes calculates pi*

Archimedes (287-212 BC) in Syracuse provided the Greek astronomers with a valuable mathematical aide when he calculated a value of $\pi$ in 240 BC. He inscribed a circle in a square, circumscribed a circle around the square, and measured the circumference *C* by continuously increasing the number of sides of the square from 4 to 8 to 16 to 32 etc. He measured the diameter *d*, and then calculated the ratio of *C* to *d* which is $\pi$ ($C = \pi d$ or $C = 2 \pi r$). Archimedes discussed his calculation of $\pi$ in *Measurement of a Circle*. A Roman soldier, despite an order not to harm Archimedes, killed Archimedes during Rome's capture of Syracuse in the Second Punic War (218-201 BC) between Rome and Carthage.

**Figure 10**
**Archimedes Planning the Defenses of Syracuse**
Source: André Thévet. *Les Vrais pourtraites et Vies Des Hommes Illustres.* Paris: Blanche Marantin and Guillame Chaudiere, 1584.

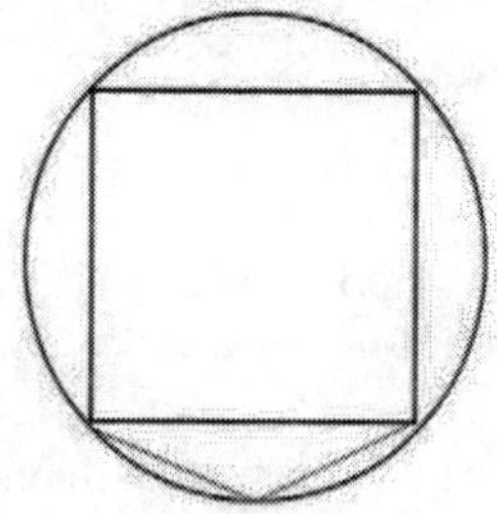

| | | |
|---|---|---|
| **C = Circumference** | **N = 4** | **$\pi$ = 2.8284** |
| **N - Number of sides** | **N = 8** | **$\pi$ = 3.06146** |
| **L - Length of a side** | **N = 16** | **$\pi$ = 3.12144** |
| **C - NL** | **N = 32** | **$\pi$ = 3.13654** |
| | **N = 64** | **$\pi$ = 3.14033** |
| | **N = 128** | **$\pi$ = 3.14127** |
| | **N = 256** | **$\pi$ = 3.14151** |
| | | **$\pi$ = 3.1415926** |

$$\pi = \frac{22}{7} = 3.1415926$$

**Figure 11**
**Archimedes' Calculation of Pi**
Courtesy of the author

*Eratosthenes calculates Earth's circumference*

Eratosthenes (276-195 BC) in Alexandria made another valuable astronomical advance in 215 BC when he determined Earth's circumference. Eratosthenes used a specially designed vertical gnomon stick and measured on successive summer solstices (21 June) the shadow the Sun at noon cast at Alexandria and at Syene (Aswan). The Sun cast no shadow at Syene, and Eratosthenes recognized that this occurred because a line representing a ray of light and extending from Syene into Earth went to the center of Earth. Eratosthenes knew that the two cities were approximately on the same meridian, 5,000 stades (500 Roman miles) apart and that the angle the Sun cast at Alexandria was 1/50 of a full circle's 360° or 7 1/5°. Because the angle, 1/50 of a circle, represented 1/50 of Earth's circumference, the distance from Alexandria to Syene is 1/50 of Earth's circumference, and the complete circumference is 50 x 5,000 stades or 250,000 stades (25,000 Roman miles). From Archimedes' relation, the equation $C = \pi d$, and having a value for π and C, he calculated that Earth's diameter was approximately 8,000 Roman miles.

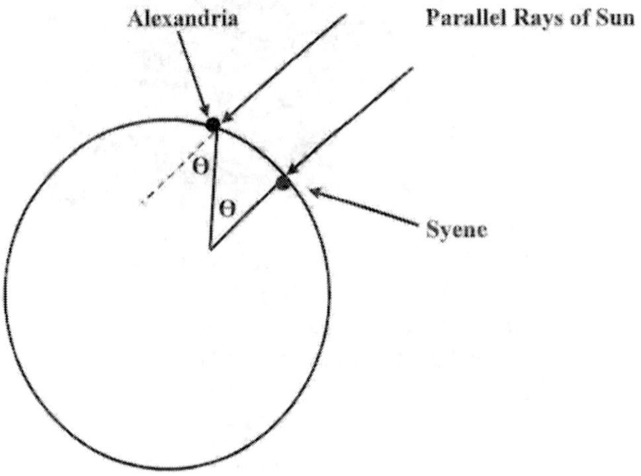


ϴ = 1/50 of a circle (360°)
*d* = distance from Syene to Alexandria, 5,000 stades

Eratosthenes measured Earth's polar circumference in 215 BC, using a base line between Alexandria and Syene, Egypt, as an arc length. Parallel rays of sunlight gave him a measurement of the angle at the center of Earth that subtended his baseline arc. Multiplication provided the value of the circumference.

**Figure 12**
**Eratosthenes' Calculation of Earth's Circumference**
Courtesy of the author

*Hipparchus and Ptolemy: geocentric model of the universe, deferents and epicycles*

Hipparchus, one of the greatest ancient astronomers, made astronomy a quantitative science with his star map, measurements of the solstices, equinoxes, and distances of the Sun and Moon, and discovery of the precession of the equinoxes. All of his important astronomical works are lost and what we know about him came from Ptolemy.

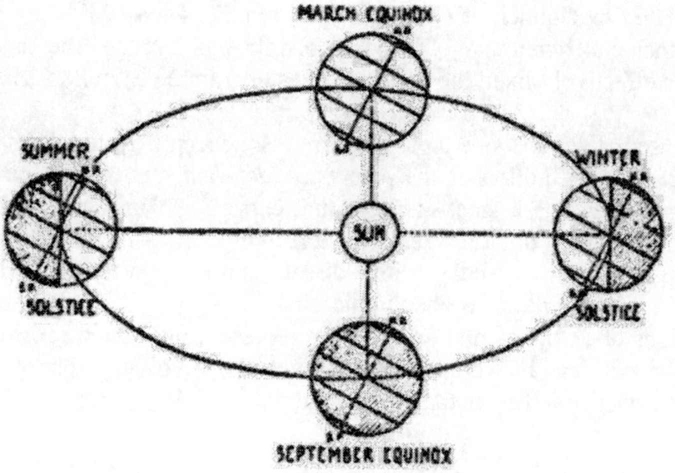


**Figure 13**
**Equinoxes and Solstices**
Source: William H. Snyder. *Everyday Science.* New York: Allen and Bacon, 1919.

The two yearly solstices occur on 21 June (summer solstice when the Sun is farthest from the equator on the north side) and 22 December (winter solstice when the Sun is farthest from the equator on the south side). On these two days the Sun, when viewed against the horizon as it moves through the constellations during one year, rises and sets farthest north of the equator on 21 June and rises and sets farthest south of the equator on 22 December. They are the longest and shortest days of the year in the northern hemisphere and mark the beginning of summer and winter.

The two yearly equinoxes occur on 21 March (vernal equinox when the Sun crosses the equator from south to north) and 22 September (autumnal when the Sun crosses the equator from north to south). On these two dates day and night are equal everywhere on Earth. The Sun rises and sets due east and west when viewed against the horizon as it crosses the equator and moves through the constellations during one year. The equinoxes mark the beginning of spring and autumn or fall. Hipparchus determined the dates of the equinoxes from 162 to 128 BC.

Hipparchus in 125 BC discovered the precession of the equinoxes, upon comparing his values of the celestial coordinates of 850 stars he compiled in his star map with those of earlier Alexandrian astronomers. Precession occurs because the Sun on its annual west to east journey around the ecliptic reaches its equinoctial point on the equator, or crosses the celestial equator, *before* reaching its original position among the stars, instead of at the same time. Successive equinoxes therefore occur slightly earlier each year than they should. Hipparchus calculated its value as 36 seconds of arc per year. Hipparchus, and later Ptolemy and other astronomers, believed that the slow clockwise rotation (one complete rotation or turn in 26,700 years) of the celestial sphere on its axis through the constellations caused precession. Newton discovered the real cause. Earth, because it is not a perfect sphere and bulges at the equator, slowly wobbles or rotates on its axis clockwise through the constellations. Earth makes a complete 360° rotation every 26,700 years or 50 seconds of arc per year. It is presently moving from Pisces (number 12), to Aquarius (number 11), to Capricorn (number 10), and eventually through all the constellations.

Because the equinoxes are occurring earlier and earlier each year, their precession caused a problem by putting the dates of a calendar out of order with the seasons and especially with the date for Easter, which the Catholic Church had fixed according to the vernal equinox. The Catholic Church in Rome and Alexandria from the time of Saints Peter and Paul had celebrated Easter on the first Sunday after the first full Moon after the vernal equinox, which happened on 21 March in Rome and 25 March in

Alexandria. The first Council of Niceae (Turkey) in 325 decreed this practice throughout the Catholic Church. Further clarification by Dionysius Exiguus in 525 set the date of the vernal equinox as 21 March and effectively fixed the period for Easter from 22 March to 25 April.

Hipparchus used deferents and epicycles in his geocentric model to account for the direct and retrograde (backward) motion of the planets. A *deferent* was a large circle rotating around a center whereas an *epicycle* was a smaller circle that carried a planet and rotated on a deferent. Jupiter's retrograde motion lasts for 17 weeks, its direct motion lasts for 39 weeks. Mercury's retrograde motion lasts for 3-4 weeks, its direct motion lasts for 13-14 weeks. Apollonius (262-190 BC) of Perga was probably the first to suggest that astronomers combine the uniform circular motions of deferents and epicycles to represent the observed irregular planetary motions. He introduced the deferent and epicycle to replace the revolving spheres of Eudoxus, Callipus, and Aristotle. No evidence indicates that he worked out a detailed scheme as Hipparchus did.

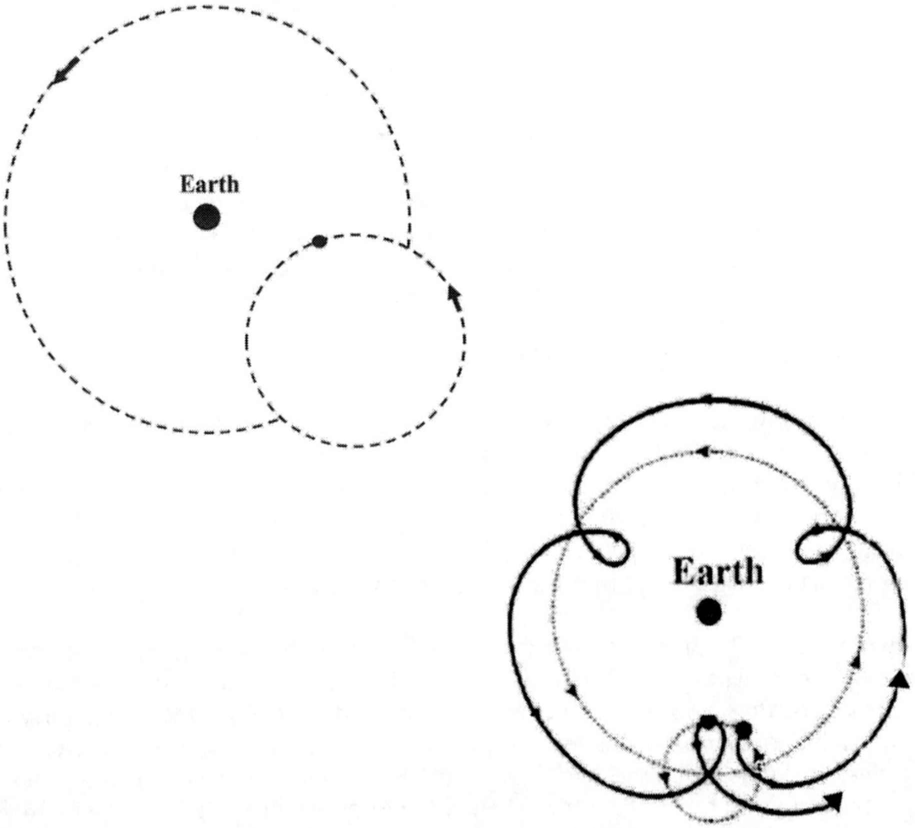


Apollonius showed how circular motion, using epicycles (small circles) and deferents (large circle) accounted for retrograde motion and variations in the distance between a planet and Earth. .

**Figure 14**
**Diagram Illustrating Epicyclic Planetary Motion**
**(Epicycles and Deferents)**
Courtesy of the author

Parallax, the apparent shift in the position of a body, such as a star or planet, when viewed against the background sphere of the fixed stars from different locations on Earth, actually results from Earth's moving in its orbit. The Sun's angle of parallax, 7 seconds, (1 degree = 60 minutes, 1 degree = 60 x 60 or 3,600 seconds) was very small and difficult to measure. Nevertheless, Hipparchus in 129 BC used this value to estimate the distance of the Sun from Earth, obtaining a distance of 4.85 million Roman miles or 1,200 times Earth's 4,000 mile radius. The distance of the Moon calculated from the Moon's angle of parallax, 0.96 degrees, was 240,000 Roman miles from Earth or 59 times Earth's 4,000 mile radius. The Moon's diameter was 2,100 Roman miles.

Hipparchus also used a lunar eclipse method derived from Aristarchus to calculate the Moon's size and distance. He calculated the Moon's size by observing the size of Earth's shadow (the angular diameter or the length measured in degrees) on the Moon using the value of the Moon's distance at the time of a lunar eclipse. Comparing the Moon's distance with the known angular diameters of the Sun and Moon he set up a geometric relation between the distances of the Sun and Moon, so that knowing one he could calculate the other. In his geometric relation, Earth, Moon, and Sun formed a right-angle triangle in which EM was the base, MS the perpendicular side, and ES the hypotenuse. Knowing the distance of a planet enabled Hipparchus to calculate its diameter and then its circumference.

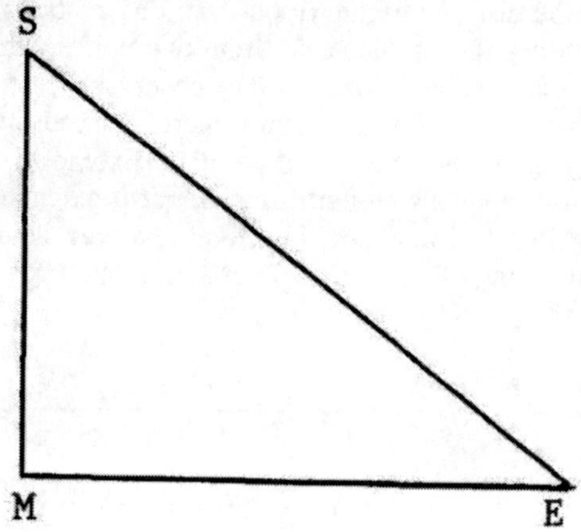


**Figure 15**
**Hipparchus' Calculation of Moon's Distance from Earth**
**Using the Lunar Eclipse Method**
Courtesy of the author

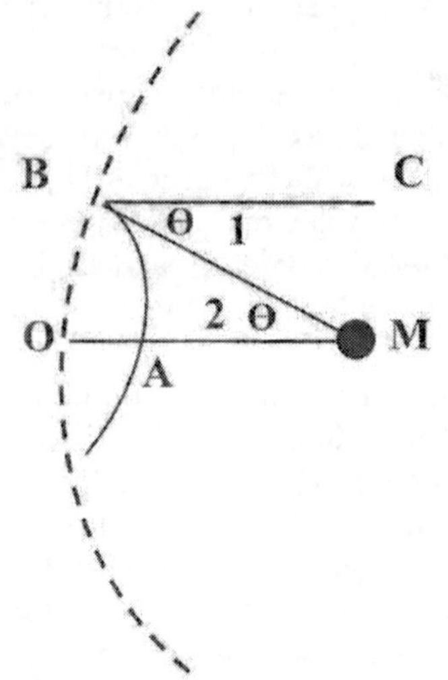


To determine the distance to the Moon from observations made on Earth using parallax, astronomers observed simultaneously from two points: when the Moon is at the zenith or highest point A and on the horizon B. The observer measures accurately angle 1 allowing for the refraction of Earth's atmosphere. Because AM and BC are parallel, angle 1 = angle 2 and is the angle of parallax. Extend MA to Earth's center O and draw OB which is nearly equals an Earth radius, but if extended is actually the arc of a circle with radius OM (dotted line). As the Moon revolves, its distance from Earth varies. The angle of parallax, angle 1 or angle 2, averages about 0.96°. The equations below show the calculation of the Earth to Moon distance.

$$\frac{\textit{Circumference of dotted circle or } 2\,\pi\, OM}{\textit{Radius of Earth or } OB} = \frac{360\ \textit{degrees}}{\textit{angle of parallax}}$$

Let OM = x miles from Earth to Moon

$$\frac{2\,\pi\, x}{4{,}000\ \textit{miles (approximately)}} = \frac{360\ \textit{degrees}}{0.96\ \textit{degrees}}$$

x = 240,000 or OM = 240,000 miles for approximate average distance of Earth to Moon.

When determining the distance to the Moon an observer makes sights from two points that are far apart. By imagining M to move far to the right, angle 2 becomes very small and therefore very difficult to measure. To use parallax on very distant objects, such as stars, requires a very long baseline, much longer than OB, to measure the parallax angle.

**Figure 16**
**Calculating the Distance from Earth to Moon**
Courtesy of the author

Two points on Earth's orbit, diametrically opposite about 186,000 miles apart, serve as observation positions. Most stars are so distant that as we move 186,000,000 miles they show no parallax. These stars form a stationary background against which astronomers measure the observable parallax of the nearer stars. Even the nearest stars exhibit very small parallax. The parallax of Sirius is 0.375 seconds of arc which indicates a distance of 8.6 light years.

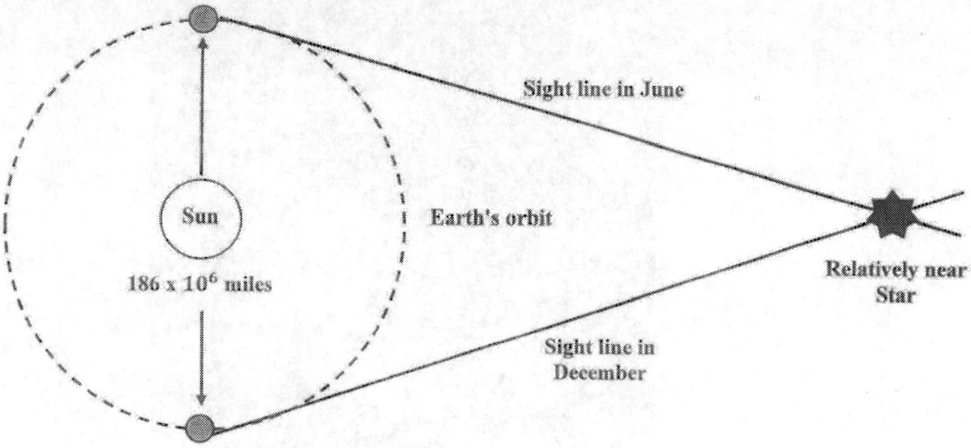


Using parallax and 8.95 seconds of arc as the angle of parallax for the Sun, astronomers have calculated the Earth-Sun distance at perihelion.

$$\frac{Circumference\ of\ Earth's\ orbit}{Radius\ of\ Earth} = \frac{1{,}296{,}000\ seconds\ (60''\ x\ 60'\ x\ 360\ degrees)}{8.95\ seconds}$$

Let x = miles to Sun

$$\frac{2\ \pi\ x\ miles}{3{,}965\ miles} = \frac{1{,}296{,}000\ seconds}{8.95\ seconds}$$

x = approximately 91.5 million miles at perihelion

**Figure 17**
**Calculating the Distance from Earth to Sun**
Courtesy of the author

Hipparchus's model of the planetary motions was the first accurate calculating device or astronomical model. He invented the astrolabe for making measurements of the altitudes, positions, and movement of heavenly bodies. It is essentially a two-dimensional model of the celestial sphere in relation to a central Earth. A simple astrolabe consisted of a wooden or metal disk with its circumference divided into degrees. The disk hung from a metal ring. A moveable pointer pivoted at the disk's center enabled an astronomer to sight a heavenly body and to take readings of its position on the graduated scale. The reading measured the body's angular distance or angle above the horizon, particularly of the Sun and stars. The astrolabe also enabled an astronomer or navigator to determine latitude, longitude, and time.

One fanciful story attributes the astrolabe's invention to Ptolemy who dropped the globe he was studying as he rode on a donkey. The donkey stepped on the globe giving a two-dimensional circle. The astrolabe remained the astronomers' main instrument until the seventeenth and eighteenth centuries when the telescope (1609) and sextant (1780) replaced it. Hipparchus also proposed a workable method of calculating longitude by measuring the times of a lunar eclipse at two different locations, where one of the locations became absolute time or the reference location.

Ptolemy (100-170 AD) in Alexandria provided the most complete account of ancient astronomy in his *Megale Mathematike Syntaxis* or *Almagest* (13 volumes or books). The Muslims translated it into Arabic in 823. A Greek to Latin translation made in 1481 is in the Vatican Library in Rome.

**Figure 18**
**Ptolemy**
Source: Source: André Thévet. *Les Vrais pourtraites et Vies Des Hommes Illustres.* Paris: Blanche Marantin and Guillame Chaudiere, 1584.
John C. Shedd. "The Genesis of the Law of Gravity."
*Popular Science Monthly* 78(April 1911).

In the first and second books Ptolemy discussed the spherical Earth, its immobility, and its circumference, which Posidonius (135-51 BC) in Rhodes had estimated as 180,000 stades (18,000 Roman miles). Eratosthenes earlier in 215 BC had calculated a better value of 250,000 stades (25,000 miles). In the third book Ptolemy discussed the length of the year and the motion of the Sun. In book four he discussed the Moon, in book five the astrolabe and the distance of the Sun and Moon. Ptolemy calculated the distance of the Moon using parallax and like Hipparchus found that the distance was 59 times Earth's radius. He used the Aristarchus-Hipparchus lunar eclipse method to deduce the distance of the Sun from his known distance of the Moon. Knowing the distance of the Moon from parallax measurements, Ptolemy calculated an Earth- Sun distance equal to 1,210 times Earth's radius and nearly identical to Hipparchus's value of 1,200 times, or about 4.8 million Roman miles.

In the sixth, seventh, and eighth books Ptolemy discussed eclipses, his star catalogue, and precession, respectively. He added 170 stars to Hipparchus's star map for a total of 1,028 stars. In the last five books, nine through thirteen, Ptolemy dealt with the motions of the planets. His geocentric model had 80 deferents and epicycles, or 80 spheres total.

In making their astronomical calculations, Ptolemy and other astronomers used zero for place notation, but words represented addition, subtraction, and other mathematical operations. Diophantus (ca. 250) in Alexandria, substituted the letters of the Greek word for *lacking* (*leipis*) to represent subtraction. It looked like an inverted V or Λ with the letter I bisecting the Λ.

After Ptolemy's death astronomers began to estimate the sizes of the planetary spheres and of the universe. Such estimates were quite common after the fifth century. To determine the planetary distances astronomers assumed that each spherical planetary shell had to be thick (wide) enough to contain the planet when it was closest and farthest from Earth. Knowing the relative dimensions of the deferent and epicycle(s) for each planet from observations of each planet's motion through the constellations enabled astronomers to determine the ratio of the inside and outside diameters of each sphere. Because the spheres nested so that they filled the entire celestial region, the outer diameter of one sphere had to equal to the inner diameter of the next, permitting a calculation of the ratio of the distances to each sphere. Knowing that the Moon's distance from Earth was 240,000 miles allowed for a calculation of the absolute distance of each planetary sphere and of the entire universe.

| **EARTH-CENTERED UNIVERSE (GEOCENTRIC)** |
|---|
| PLATO (427-347 BC) |
| ARISTOTLE (384-322 BC) |
| HIPPARCHUS (190-120 BC) |
| PTOLEMY (100-170 AD) |

*Aristarchus's heliocentric model of the universe*

Aristarchus (310-230 BC) in Samos was a pupil of Strato at the Lyceum in Athens. In about 260 BC he proposed a heliocentric model of the universe in which he placed the Sun at the center of an immense sphere of the stars, and Earth and other planets moved in a circle around the Sun. The sphere of the stars was motionless, its apparent motion resulted from Earth's daily rotation on its axis. Proposing a heliocentric model, which clearly did not agree with the picture of the universe the senses provided, Aristarchus placed himself within the rationalist intuitive school of epistemology.

Few astronomers accepted his model because its development lagged compared to the geocentric model, and it violated common sense in putting Earth in motion around the Sun. His model, however, called attention to the idea of relative motion and could deal satisfactorily with the planets' retrograde motion.

Aristarchus, like Hipparchus and Ptolemy after him, also calculated a distance for the Moon and Sun from Earth and described his results in a treatise that has survived entitled *On the Sizes and Distances of the Sun and Moon*. His calculations showed that Earth was larger than the Moon and the Sun larger than Earth. This was another reason Aristarchus thought it more logical to believe Earth revolved around the Sun.

Aristarchus calculated the Moon's size by observing the size of Earth's shadow (its angular diameter or length measured in degrees) on the Moon during an eclipse of the Moon. He found that the Moon's diameter was 1/3 of Earth's diameter, which is slightly larger than it actually is. He calculated the distance of the Sun and Moon from Earth by noting that when the Moon is exactly half light and half dark, Earth, Moon, and Sun each occupy an apex of a right triangle. Using geometry, Aristarchus pointed out that the relative lengths of the sides represented the relative distances. The hypotenuse represented the distance of the Sun from Earth *SE*, the short side (base) represented the distance of the Moon from Earth *ME*, and the vertical side the distance of the Sun to the Moon *SM*. (See Figure 10).

Aristarchus's method was correct, but lacking good instruments he could not measure accurately the angles of his triangle. The estimates he made of the angles led him to conclude that the Sun was twenty times farther than the Moon, 20 x 240,000 Roman miles or 4.8 million Roman miles. We know today the Sun is about 400 times farther, 400 x 240,000 miles or 93 million miles. Aristarchus estimated the Sun's diameter to be twenty times the Moon's diameter, and seven times Earth's diameter, or about 56,000 Roman miles.

*Other cosmological models*

Pythagoras put the stars on an immense moving sphere and placed an immense fire, known as the Altar of Zeus and invisible from Earth, at the center of the sphere. Earth, Sun, all other celestial bodies moved in circles about the central fire.

Heraclides Ponticus (390-339 BC), a student at Plato's Academy, proposed that Earth's daily rotation on its axis and not the rotation of the sphere of the fixed stars accounted for the apparent motion of the stars. He proposed that Venus and Mercury orbited the Sun and that the Sun moved around Earth.

***Structure of matter: atomic theory and four (five) element theory, alchemy***

*Atomists*

Leucippus (ca. 490 BC) in Miletus was a pupil of Zeno of Elea in southern Italy. The Eleatic school denied the value of the senses as a source of truth. This probably led Leucippus about 440 BC to propose an atomic structure of matter. Trusting reason and intuition he deduced that matter consisted of tiny indivisible atoms (Greek *atomos*).

Democritus (470-380 BC) in Abdera was Leucippus's pupil. His friends called him the laughing philosopher. Democritus also adopted a deductive, intuitive philosophy of science. Only atoms moving randomly in empty space, or the void, existed in his mainly empty universe. As a consequence of the atoms' random motion, Democritus believed that other universes composed of atoms existed.

Democritus adopted an atomic theory of matter to answer Parmenides (515-450 BC) in Elea who argued that change was an illusion, permanent change did not occur. Democritus's atoms were eternal and unchanging, but their combining and rearrangement produced change. His atoms had different shapes. The atoms of water were round and smooth, atoms of earth were jagged, atoms of fire were prickly.

Atoms comprised everything including the soul. The soul was therefore material and died with the body. Most Greek natural philosophers did not accept Democritus's materialistic philosophy of nature. The alternative four element theory was much more popular. Democritus wrote 72 books on nature and other topics but none has survived.

Epicurus (341-270 BC) in Samos, who in 306 BC founded the popular Epicurean school in Athens, and Lucretius (95-55 BC) in Rome and the author of *De Natura Rerum* (56 BC), were atomists.

| ATOMISTS |
|---|
| LEUCIPPUS (ca. 490 BC) |
| DEMOCRITUS (460-370 BC) |
| EPICURUS (341-270 BC) |
| LUCRETIUS (95-55 BC) |

*Four (five) element theory*

Empedocles (492-432 BC) in Agrigentum, Sicily, whom his contemporaries considered godlike and honored him whenever he appeared in public, first proposed that all matter consisted of four elements: air, earth, fire, and water. He combined the single element beliefs of four earlier Greek natural philosophers. Thales held that water was the only element because life depended on it. Heraclitus (540-475 BC) in Miletus, known as the weeping philosopher because of his pessimism, said fire was the only element because it was always changing and brought change in other things. Xenophanes (570-480 BC) in Colophon (Ionia) believed earth was the only element, possibly because it gave rise to many other kinds of matter. Anaximenes (570-500 BC) in Miletus, who may have been Anaximander's pupil, held that air was a fundamental element because by compression air could become water and earth, and by rarefaction it became hot and changed into fire.

| **THE ELEMENTS** |
|---|
| THALES (640-546 BC): WATER |
| ANAXIMENES (570-500 BC): AIR |
| XENOPHANES (570-480 BC): EARTH |
| HERACLITUS (540 BC-475 BC): FIRE |

Plato believed that mathematics (geometry) and deduction led to true knowledge. In the *Timaeus*, he associated each element with a geometric shape: the Pythagorean five perfect solids in which all surfaces or sides are equal. He combined geometry, atomism, and the five elements in his geometrical theory of the elements. The geometric solids were convertible one to another and so were the elements. He assigned a shape to each element based on its mobility. Earth had the least mobility of the elements and retained its shape best, so it was a cube. Water was the next least mobile because it was the next heaviest element, and it had the shape of an icosahedron which had the greatest number of unit faces. Air was the next lightest, and it had the shape of an octahedron which had the second smallest number of faces. Fire was the most mobile of the elements, and it had the shape of a tetrahedron which had the smallest number of faces and was most mobile. The heavenly element had the shape of a dodecahedron which Plato associated with the boundary of the universe. Its twelve faces corresponded with the twelve constellations.

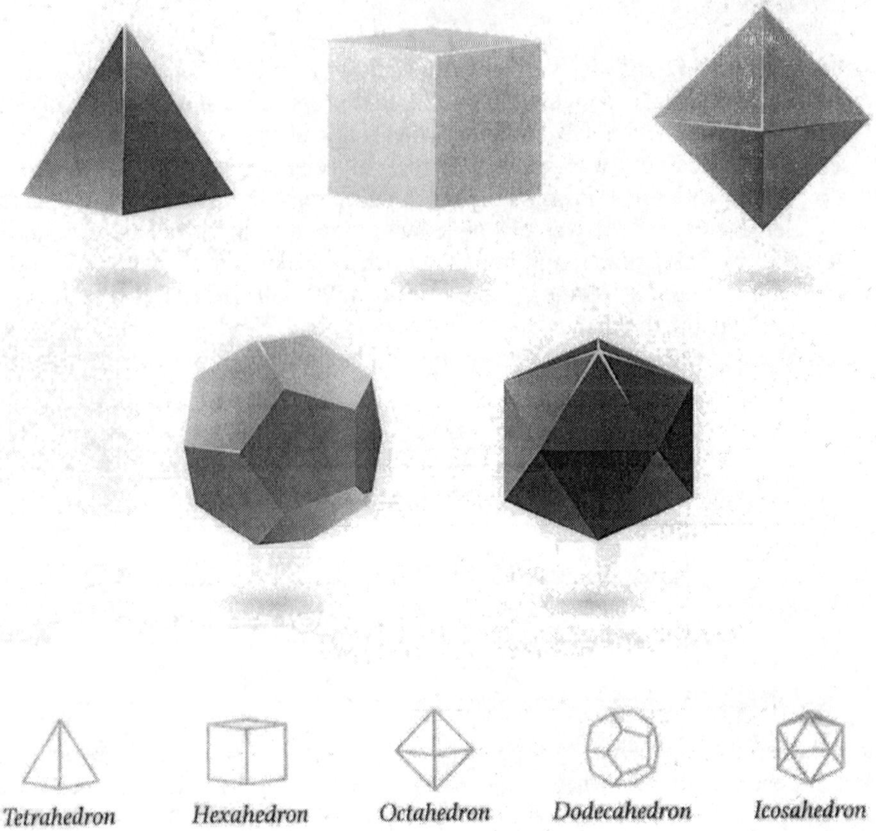


**Figure 19**
**The Five Regular Solids or Platonic Elements**
Courtesy of the author

| MOBILITY OF THE PLATONIC ELEMENTS | |
|---|---|
| ***Element*** | ***Mobility*** |
| Earth: cube, six sides | Least mobile |
| Water: icosahedron, twenty sides | |
| Air: octahedron, eight sides | |
| Fire: tetrahedron, four sides | Most mobile |

In Plato's geometrical atomism when fire changed to air two tetrahedrons became an octahedron. To convert air and fire into water, two octahedrons and a tetrahedron became an icosahedron. Aristotle never accepted Plato's geometrical atoms and deductivism and developed a structure of matter that agreed with his inductive, empirical philosophy of science. Matter and form, the quality impressed on matter, gave rise to the elements. His world was full or a plenum, a vacuum did not exist. The perfect world consisted of concentric spheres of earth, water, air, and fire, but in the real world the elemental spheres intermixed with one another. Surrounding the outermost sphere of fire was the ether or heavenly element.

The terrestrial elements underwent change from one into another and corruption. Each element had specific qualities, or properties, and a natural motion with which it sought its natural place.

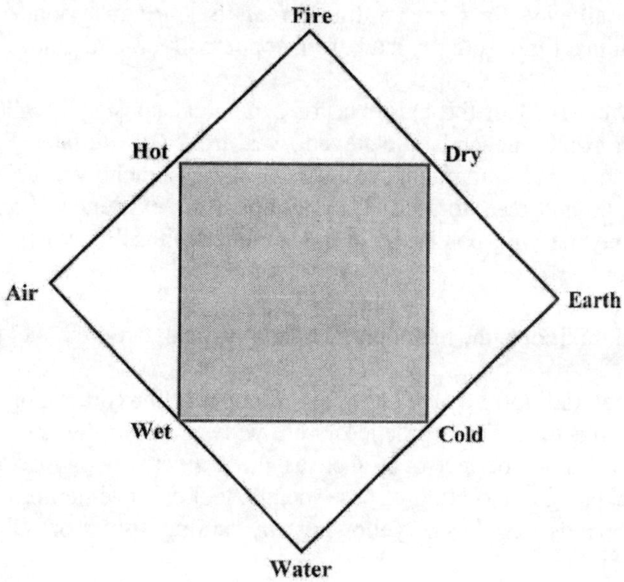


**Figure 20**
**Relation of the Four Elements and the Four Properties**
Courtesy of the author

| PROPERTIES OF ARISTOTELIAN ELEMENTS | |
|---|---|
| ***Elements*** | ***Properties*** |
| Earth | cold and dry and had a downward natural motion |
| Water | cold and wet and had a downward natural motion |
| Air | hot and wet and had an upward natural motion |
| Fire | hot and dry and had an upward natural motion |
| Celestial element or ether | perfect, non-corruptible and had natural circular motion |

*Alchemy*

Alchemy dealt with the transmutation and perfection of metals. The belief that a change in the properties of an element could convert that element into another element provided the theoretical foundation for alchemy.

Hermes Trismegistus (2800 BC), meaning the *thrice greatest* and the Greek name for the Egyptian god Thoth in Egypt's Old Kingdom, was supposedly the first alchemist. Bolos of Mendes or Democritus (ca. 200 BC), another alchemist, practiced in Egypt. Go (Ko) Hung (281-361), a Chinese alchemist, claimed his *lien tan* was the elixir of life.

The alchemists greatly valued Mercury because it was the only known metallic liquid and easily vaporized. This allowed for the easy transfer of its spirit or essence to other elements. The other metals known at that time were tin, lead, iron, copper, silver, and gold.

The alchemists believed in the existence of a magical substance called the philosopher's stone or ferment, which some alchemists believed was iron (II) sulfate, $FeSO_4.7H_2O$. Discovering the philosopher's stone was important because the stone somehow accelerated the conversion of base metals, such as tin and lead, to gold. They accepted the existence of an elixir of life that would give eternal life to anyone who possessed it and anointed himself with it, and the alkahest which was a universal solvent.

Two alchemical traditions, the metallurgical and mystical, evolved over time.

The metallurgical tradition involved alloying of copper-zinc (brass), or copper-tin (bronze), or gilding of metals with gold film. These practices were widespread in the ancient world. The conversion of a base metal into a more perfect one, such as the conversion of lead to gold, required four stages: melanosis (blackening) by heating of base metals; leukosis (whitening) by adding mercury, arsenic or antimony compounds; xanthosis (yellowing) by adding sulfur or sulfides; and iosis (purpling) by adding gold or gilding.

Mystical alchemists believed that metals grew or matured in Earth and that they consisted of a male principle (spirit) of sulfur and a female principle (spirit) of mercury. Transferring the spirit of one metal to another metal transformed one metal into another. Mercury was usually the spirit transferred. The alchemists tried to speed up the transformation with the philosopher's stone.

Alchemists developed some of the apparatus that chemists use today, such as flasks, beakers, crucibles, tongs, stills, and furnaces. They prepared and reacted acids, alloys, salts, and other chemicals in their experiments.

***Theories of medicine***

*Hippocrates in Cos*

Hippocrates (460-370 BC) is significant because he introduced a rational and holistic approach to study illnesses and disease. His rational approach challenged the accepted Greek practice of relying on the gods, cults such as Asclepius (420 BC), diviners, and other irrational treatments for cures. To obtain and to maintain a disease-free or healthy condition, the Hippocratic physician had to balance the four humors, phlegm, black bile, yellow bile, and blood that he believed existed in the body.

The Hippocratic physician emphasized observation and classification of disease and illness. The Hippocratic School believed that an improper diet and climate caused an imbalance of the humors. Cold and moist foods, such as barley, certain kinds of fish, and boiled meat, balanced an excess of the warm and dry humor in the body (choleric). Roasted meats, bread, and fish were warm and dry and balanced an excess of the cold and wet humor (water). *On the Nature of Man and On Airs, Waters, and Places* describe its treatments. Other sources are, *Regimen in Health* and *Of the Epidemics*, which contains 567 case histories, discussions on fevers, and the influence of climate, diet, drugs (herbs) on bodily health. The Hippocratic practice of medicine made little use of anatomy and physiology.

*Physiology and anatomy*

Praxagoras (340 BC) in Cos distinguished between veins and arteries as two kinds of blood vessels, but he believed the arteries carried air because arteries in a corpse contained no air.

Aristotle did not distinguish between arteries and veins and believed they served the same purpose (coextensive) throughout the body. The blood did not circulate in the body because only the heavens had natural circular motion.

Herophilus (335-280 BC) in Alexandria carried out public dissections. At that time no serious opposition to dissections existed, although physicians limited the number of dissections they performed. Erasistratus (304-250 BC) in Alexandria continued the work of Herophilus, but after Erasistratus, anatomy died out.

*Galen*

Galen (130-200 AD) who lived in Alexandria and in Rome was the most influential physician of his time. He wrote his major work *On Medical Experience* in 150. Latin translations of Galen's writings from the Arabic became available in the eleventh century. The original Greek writings became available during the Renaissance (1400-1600).

Galen developed the four humor theory of Hippocrates and added the study of anatomy, physiology, and psychology (personality) to Hippocratic medicine. Galen classified the humors according to the arrangement below. A balance of humors, he said, determined both the body's health and personality.

| **MEDICAL PROPERTIES OF THE ARISTOTELIAN ELEMENTS** | |
|---|---|
| ***Season-Element*** | ***Property of humor*** |
| Winter (water: cold and wet) | phlegmatic (phlegm), cold, indifferent, lazy |
| Spring (air: hot and wet) | sanguine (blood), optimistic |
| Summer (fire: hot and dry) | choleric (yellow bile), fiery, hot-tempered |
| Autumn (earth: cold and dry) | melancholy (black bile), sad |

Galen wrote on anatomy in *On the Usefulness of Parts of the Body*. He dissected dogs, sheep, monkeys, but no humans although for a time he served as physician at the Gladiator School in Pergamum (Bergama) in Turkey and certainly had many opportunities to study human anatomy.

In the study of physiology Galen believed the blood flowed to the body and to the lungs and back again from both sides of the heart and in both the arteries and the veins. It ebbed and flowed in the body. To get from the right side to the left side of the heart, Galen argued that the blood passed through pores (which don't exist) in the septum or inner muscle wall separating the heart's two chambers. He had no understanding of the blood's circulation in the body.

***Contraceptives, abortion, and conception***

Knowledge, use, and social acceptance of effective plant-derived birth control drugs, such as Queen Anne's lace (wild carrot), silphium, artemisia, and pennyroyal, found use in ancient Egypt, China, Greece, and Rome.

Cyrene, on the coast of Libya in North Africa, was a famous center for the growth and export of silphium, a contraceptive and abortifacient used in the period 600-400 BC. The plant became extinct by the fourth century AD.

Hippocrates prescribed the seeds of Queen Anne's lace and silphium as both a contraceptive and a morning-after pill because at that time the belief was that conception, the act of becoming pregnant, did not occur at the time of sexual union but sometime later, such as after a 28-30 day period or at the time of quickening, which is movement of the fetus at the end of the first trimester or beginning of the second.

Most ancient medical practitioners believed, as Aristotle did, that the male provided the form or soul (seed) and the female only the matter (nourishment). The male seed lived in the female for several days before conception occurred.

According to Aristotle if the embryo (the fertilized egg) or the fetus (which exists from three months to birth) developed regularly it resembled the father; if not, the mother; then grandparents, and back to even more remote ancestors. The degree of excitation at the time of sexual union determined the sex of the embryo or fetus. A high degree of excitement produced a male, a lower degree of excitation produced a female.

Galen and Hippocrates, unlike Aristotle, believed male and female contributed equally in procreation and provided both form (soul) and matter (body).

Other ancient cultures considered the time between conception and quickening as one of nonpregnancy. In Israel, under ancient Hebrew law, a woman became pregnant 40 days after conception.

*Ensoulment and the meaning of soul*

Pythagoras and the Theosophists held that all souls existed from the beginning of time and passed from one body to another (metempsychosis). Plato believed that God created souls for all eternity and that they existed before union with the body and after death.

Aristotle in *On the Soul* treated the soul as a separate entity that existed in the body and was the animating spirit of the body. Writing on procreation in *De generatione animalium*, he said that the male semen carries the soul (form), the female provides the passive material or nutrients (matter), and that ensoulment occurred after conception: 40 days later for a male, 80 days later for a female. Aristotle held that, in general, ensoulment occurred when the fetus took on human shape (10-12 weeks). His soul was neither personal nor eternal.

The Stoic school, which Zeno of Citium (335-263 BC) founded in Athens in 300 BC, believed in the potential for life at conception, but that ensoulment occurred at birth.

The Roman Catholic Church in ancient times and the Middle Ages and later taught that God created each human soul from nothing at the moment of the embryo's conception, or at least, as Thomas Aquinas (1225-74) and the scholastics maintained until the seventeenth century, when the body is sufficiently formed, that is, when the fetus took on human shape (10 to 12 weeks).

Later thinkers, such as Gottfried Leibniz (1646-1716) in Germany, said that the creation of all souls occurred at the beginning of time. The souls were enclosed as seeds in matter that evolved and became human.

## *Women physicians and scientists of the ancient world*

### *Middle East*

Women attended the medical schools in Sais and Heliopolis. They diagnosed pregnancies, tested for sterility, performed Caesarean sections, and set bones with splints. Notable women included Merit Ptah (2700 BC) an Egyptian who specialized in gynecology, Zipporah (1300-1200 BC) who attended medical school in Sais, and Tapputi-Belatikallim (1200 BC) in Mesopotamia who produced perfumes. Cleopatra (69-30 BC) in Alexandria may have been an alchemist and the one who introduced and popularized the use of lampblack, ochre, henna, and white lead as cosmetics.

### *Greece*

Theano (500s BC), Pythagoras's wife, in Croton in southern Italy, was a mathematician. Athens at that time had a law that prohibited women from studying medicine, although the Athenians later, during the time Plato's Academy flourished, overturned the law.

Socrates (470-399 BC) and Plato favored women's education. Women studied at the Academy, but both Socrates and Plato had difficulty in dealing with the equality of women in all aspects of Greek society. Aristotle believed women were inferior, even though his first wife, Pythias (362-335 BC), was a zoologist.

### *Roman Empire*

The alchemist, Miriam or Mary or Maria (ca. 100 AD) in Alexandria, prepared alloys and designed apparatus including a still and water bath.

Hypatia (370-415), the daughter of Theon (ca. 360-390) director of the Museum in Alexandria, studied mathematics and astronomy at the museum. She was the last recorded member of the museum and the only noted woman scholar of ancient times. She wrote commentaries on Ptolemy, Diophantes (210-90 BC) the Greek algebraist, and edited Apollonius, author of *Conics*. Christians murdered Hypatia for her pagan beliefs.

## *Natural history: fossils and classification*

Greek natural philosophers proposed a variety of ideas on the origin and evolution of living creatures, on fossils, and on classification of nature.

Anaximander (610-546 BC) in Miletus believed that humans arose from fishlike creatures.

Xenophanes in Colophon, likely the first paleontologist, recognized that fossils were remains of once-living organisms and taught that fossils showed a gradation of animal existence, from primitive to developed.

Herodotus (485-425 BC) of Halicarnassos wrote that fossils present in an area were evidence that a sea had once covered that area. Aristotle and Theophrastus (371-287 BC), his successor at the Lyceum, also recognized fish remains as fossils, but they believed the fossils originated in the rocks.

Strato (340-270 BC), a geographer who studied at Aristotle's Lyceum and served as its third director upon Theophrastus's death in 287 BC, and Eratosthenes claimed that invertebrate fossils provided evidence of the existence of ancient seas.

Aristotle discussed his biological views in *Historia animalium*. He adopted both a *discontinuous* view of nature in which he divided animals into 11 classes that included mammals, birds, fishes, reptiles, and others and a *continuous* (great chain of being) view of nature. He believed in a vegetative soul and an animal soul. Aristotle classified 500 animal species, based on function rather than morphology. He recognized that the whale was a mammal. Theophrastus classified 500 plant species.

***Archimedes: founder of statics and hydrostatics***

In *Equiponderance of Planes*, his major work on statics, Archimedes in Syracuse (Sicily) defined the moment of force as equal to force (weight) x distance from the fulcrum or $T = d \times F$. This is the principle applied in operating a see saw or teeter totter, where the players must balance the weight and distance on each side of the fulcrum in order to move the see saw up and down.

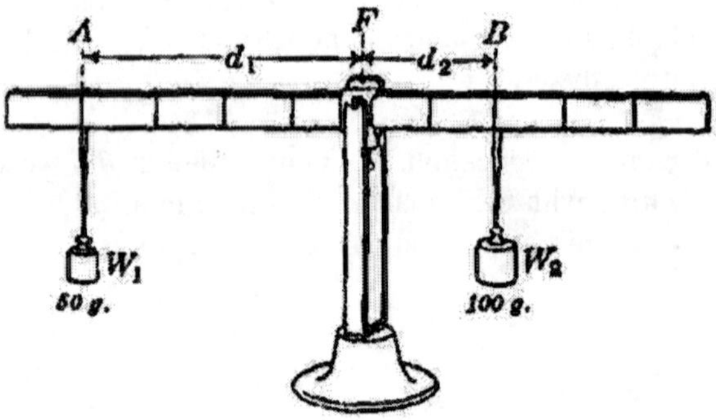


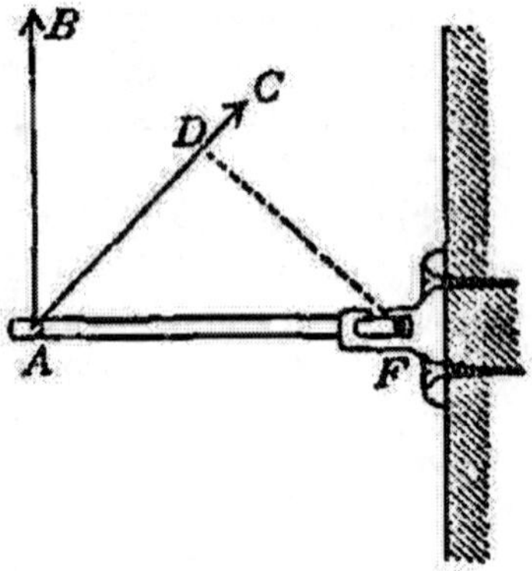


Moment of a force equals force times its perpendicular distance from fulcrum.

**Figure 21**
**Moment of Force**
Source: Newton Henry Black and Harvey Nathaniel Davis. *New Practical Physics.* New York: Macmillan Company, 1913.

Archimedes discussed his results on hydrostatics in *Floating Bodies*, including his famous Archimedes' *principle*. Archimedes demonstrated that the apparent loss in weight of an object immersed in a liquid is equal to the weight of the liquid displaced. He showed that a floating object appears to have lost all its weight. It floats because it has displaced a weight of water equal to its own weight. A piece of lead when flattened sinks when placed in water, whereas the same piece of lead, which has the same weight, when formed in the shape of a hollow dish, will float because it displaces a weight of water equal to its own weight. The combined density of the lead and air in the dish is less than the density of lead, because the total volume has increased. The combined density of lead and air is also less than the density of water. The lead dish therefore floats.

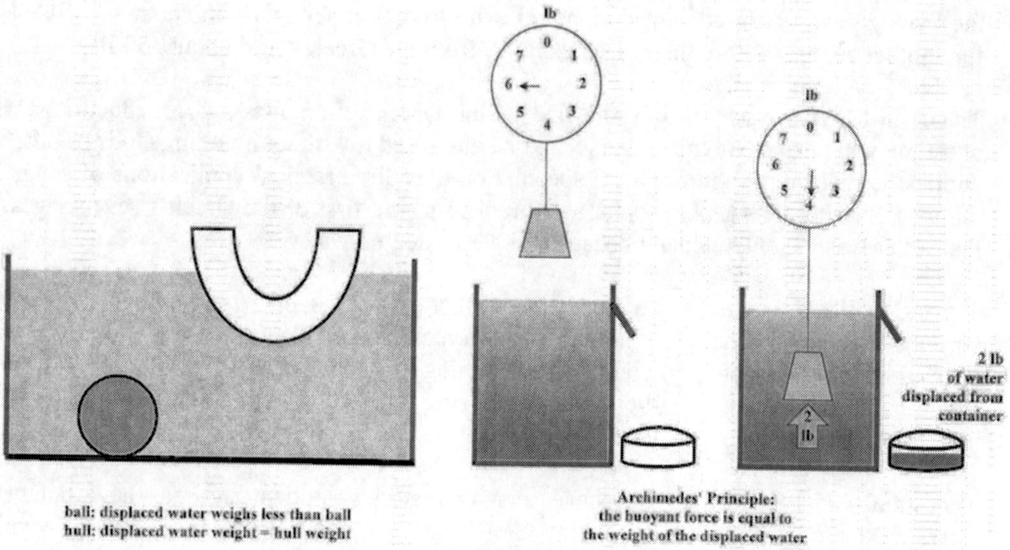


**Figure 22**
**Archimedes' Principle**
Courtesy of the author

Archimedes calculated the volume of an irregularly shaped object by the displacement method. He showed that an object immersed in a liquid displaces a volume of the liquid equal to the volume of the object. This discovery enabled him to define the density of an object as equal to its mass (weight) divided by its volume, $D = M/V$. His discovery of the density relation appeared in the famous story of King Hieron's crown and the dishonest crown maker. The crown maker tried to cheat the king by alloying with silver the crown he was making for the king instead of using pure gold and paid with his life. Archimedes demonstrated experimentally that the crown and a regularly shaped block of gold of weight equal to the crown's weight did not displace the same volume of water, which they should have if the two densities were equal. The crown displaced a greater volume than the gold block of the same weight, giving it a smaller density, and proving that the crown maker had alloyed the crown with silver.

A Latin translation from the Greek of Archimedes' *Works,* completed in 1459 and printed in 1544, was in the library of the Duke Federico Montefeltro Urbino (1422-82) and since about 1884 in the Vatican Library. An earlier Greek to Latin translation of Archimedes *Works* containing most of his writings by the Flemish Dominican, William of Moerbeke (1215-35-86) in Nicea and Thebes, dates from 1269 and was the source of the later translations of Archimedes *Works* including those in the Vatican and other libraries.

***Greek philosophy***

The discussions of the Greek philosophers were often about the senses deceiving us and the reality of change versus lack of change. They accepted the philosophy of rationalism that the universe is orderly, it behaves according to natural laws, and our minds can discover these laws. This belief is the origin and foundation of the Western intellectual tradition.

Pythagoras and the Pythagoreans believed the universe was a book of numbers and geometry. He discovered the Pythagorean Theorem (in a right triangle the square of the hypotenuse is equal to the sum of the squares of the other two sides) and accepted the existence of the five regular solids. The Pythagorean brotherhood, or cult, flourished in Croton, in southern Italy, where it achieved considerable political power as allies of the ruling aristocrats. When power shifted to the democrats the Pythagoreans suffered persecution and exile from southern Italy in about 470 BC. They survived for another century before their disappearance from the Greek world about 350 BC.

Thales in Miletus was probably the first of the Greek rationalists and a deductionist. He asked the question, what makes up the universe, and he answered it without invoking Gods or other supernatural entities. By elevating philosophical speculation over the practical applications of science, Thales set the tone for later Greek philosophers. He ranked as the first of the Greek's seven sages. He believed that water was the only element because life depended on water.

The rationalist Heraclitus in Miletus believed that everything changed, nothing was unchanging. Fire was the element of change. Alternatively, Parmenides in Elea, another rationalist and disciple of Pythagoras and Xenophanes, claimed that nothing ever changed. Only being and non-being existed. Being could not become non-being and non-being could not become being. Change was an illusion, and the senses deceived.

Zeno (490-425 BC) in Elea may have been Parmenides' student and may have taught Socrates the Socratic Method, the method of seeking knowledge by systematic questions and answers. He believed that the senses deceive. Zeno wrote a book of 40 paradoxes concerning the continuum and the infinite of which eight survived in the writings of Aristotle and Simplicius (490-560) of Cilicia, including the paradox of space, the arrow, and of Achilles and the tortoise. His paradox of the 200-yard race between Achilles and the tortoise was to show how the senses can deceive. Zeno argued that mathematically Achilles could never catch the tortoise, whereas the senses showed clearly that he did.

In laying out his paradox, Zeno said that he would give the tortoise a ten yard head start because Achilles could travel ten times as fast as the tortoise. So, if the tortoise moved one yard, Achilles would have moved ten yards in the same time. If Achilles now moved the one yard, the tortoise moved an additional one-tenth of a yard in the same time. If Achilles moved one-tenth of a yard the tortoise moved one-one hundredth of a yard in the same time, and so on. Thus Achilles would never catch the tortoise. Zeno took part in a conspiracy to overthrow Nearchus (440 BC) the tyrant who ruled Elea. It failed and Nearchus had Zeno tortured to death. James Gregory (1638-75), a Scottish mathematician and a reflecting telescope inventor, about 1667 first showed the weakness of Zeno's argument. He demonstrated that space and time are not infinitely divisible and that an infinite series (a series with an infinite number of terms) does converge to a finite sum, proving Zeno incorrect.

Aristotle and Plato were realists. Aristotle used both the deductive and inductive methods in his search for true knowledge. Plato believed deduction and mathematics provided certainty and that mathematical truths are eternal. They and all other eternal truths resided in a perfect world of being. This was a timeless realm of pure thought or ideas in which we can participate through the use of intuition or thought. It was separate from the world of normal experience or becoming, which is subject to change. The physical world was a shadow of the world of being. The lesson taught in Plato's *Allegory of the Cave* is that the senses deceive and that new discoveries will conflict with common sense. Plato's worlds of being and becoming combined the conflicting worldviews of Parmenides and Heraclitus.

## V. ORGANIZATION OF GREEK SCIENCE

***Why did Greek science not advance more than it did, at least four factors may have limited the progress of Greek science***

Greek society was indifferent to science. Its natural philosophers did not agree on what constituted science. Many Greeks still believed an eclipse was a bad omen and did not have a rational explanation. Their interest in literature, drama, theatre, pageantry, and athletic games greatly exceeded their interest in science.

Only a few research centers existed in Greece. Consequently, scientists were few in number and remained largely isolated in Greek society. The most famous center, Athens which had a population of 300,000, was the location of the Academy and Lyceum. Alexandria, which had a population of 1.2 million in 200 BC and where Apollonius, Euclid, Hero, Erasistratus, Eudemus (100s BC), Herophilus, Eratosthenes, and Ptolemy taught at the Museum, was another famous center. Pergamum in northwest Asia Minor, the birthplace of Galen, had a 1.2 million population in 200 BC, and a library second only to Alexandria's library.

**Figure 23**
**World According to Hecateus of Miletus (550-476 BC)**
Source: William Snyder. *Everyday Science.* New York: Allyn and Bacon, 1919.

***Coalescence of Greek science occurred by the second century AD, internal and external factors that contributed to the coalescence***

*Internal factors*

Euclid (325-270 BC), who likely studied at Plato's Academy and later resided at the Museum in Alexandria, had synthesized geometry centuries earlier. A ninth-century Greek manuscript of Euclid's *Elements* (300 BC), 13 volumes, is in the Vatican Library. In the second century AD, Galen synthesized medicine, and Ptolemy synthesized astronomy.

*External factors*

The expansion of the Roman Empire (established 29 BC) led to a codification of law and standardization of education, coinage, and a system of weights and measures.

| A HISTORY OF LIBRARIES | | |
|---|---|---|
| 3000 BC<br>Ancient man started the first libraries with clay tablets. Thousands of these tablets still exist, but scholars have not translated all of them. | 2500 BC-AD 300<br>The Egyptians were the first to use paper-like papyrus made from the papyrus plant. Their libraries held thousands of scrolls. | 200 BC- 1200 AD<br>For many centuries humans wrote on animal skins. Later, they wrote on parchment or specially treated layers of thin animal skins. |
| 105 AD<br>The Chinese invented paper. The art of papermaking reached Europe in the 1100s. | 476-1400<br>(Middle Ages)<br>During the early Middle Ages, monks in monasteries copied and kept books. Hand copying took much skill and time. | Mid-1400s<br>Johannes Gutenberg's invention of the printing press greatly benefitted libraries. |
| 1600-1800<br>This 200-year period represents the golden years of libraries. The number of private and public libraries grew significantly. | 1638<br>John Harvard's gift of 400 books to Harvard College started the first library in the United States. | 1731<br>Benjamin Franklin founded the Library Company of Philadelphia, the first successful library in the United States. Members paid dues but borrowed books for free. |
| 1833<br>Peterborough, NH, had one of the first free public libraries in the United States. | 1876<br>Melvil Dewey (1851-1931) at Amherst and later Columbia invented the Dewey Decimal System. His system made cataloging books easier. | 1900 to today<br>The number of libraries and the services they offer increased. Libraries have become media centers, with many more items than books. |

## VI. TECHNOLOGY IN ANCIENT GREECE

The historian, Benjamin Farrington (1891-1974), suggested four factors that might have contributed to the Greeks not making greater technological application of their scientific knowledge. They are slavery, economics, politics, and philosophy of life. The arguments for each appear below. Greek philosophy of life is the factor most historians have accepted as the best explanation.

***Slavery***

Greek and Roman societies required large numbers of slaves to perform essential mechanical operations. The slaves were not always docile and numerous riots occurred. Most importantly, their use delayed the introduction of labor saving mechanical devices. The tables below show the percentages of slaves in Greece and Rome and accompanying energy equivalencies.

| SLAVERY IN GREECE AND ROME | |
|---|---|
| Athens | about 50 percent slave labor |
| Sparta | about 75 percent slave labor |
| Rome | about 35-40 percent slave labor |

| LABOR SAVING DEVICE AND SLAVE EQUIVALENCY | |
|---|---|
| 1 horse | 10 slaves |
| 1 water mill | 30-50 slaves (3-5 horsepower) |
| 1 windmill | 100 slaves (10 horsepower) |

***Economic factors***

Greek society provided insufficient demand for goods and had little need for inventions. Archimedes constructed his screw, Hero (ca. 60 AD) in Alexandria constructed his steam engine, other Greek craftsmen constructed water wheels, water clocks, windmills and used them for festivals and pageants but not for daily needs.

A bronze box from the first century BC, found in the Mediterranean Sea off the Greek island of Antikythera used gears to predict the motions of the Sun, Moon, planets, and stars. It represents an early form of an analog computer, in which the gears' motions simulated the motions of the Sun, Moon, planets, and stars and provided a means of measuring them. The gear box is a good example of Greek technological advancement.

***Political factors***

Greek society considered science materialistic and atheistic. Greek lawmakers accused Anaximenes's pupil Anaxagoras (500-428 BC) in Athens, Socrates (470-399 BC) in Athens, Aristarchus, and Democritus of holding beliefs, such as atheism and impiety, that were false and contrary to those of Greek society. But they received no punishment because of their beliefs. Aristophanes (448-380 BC), a writer of Greek satire and comedies, nevertheless ridiculed Socrates in the play *The Clouds* for trying to replace the Gods with natural forces and challenging Greek traditional beliefs. Socrates's accusations included corrupting the youth. He chose to drink hemlock rather than accept banishment from Athens.

***Philosophy of life***

Greek natural philosophers mainly acquired knowledge for knowledge's sake. They sought theoretical rather than applied knowledge. They preferred the head over the hand and believed the world was there to live in and not to change. This is the most accepted explanation of why the Greeks did not make more technological applications of their scientific knowledge.

Archimedes' Screw
*Chamber's Encyclopedia,* Philadelphia, J.B. Lippincott Company, 1875.

The screw uses the principle of the inclined plane.
William H. Snyder.
*Everyday Science.*
New York: Allyn and Bacon 1919.

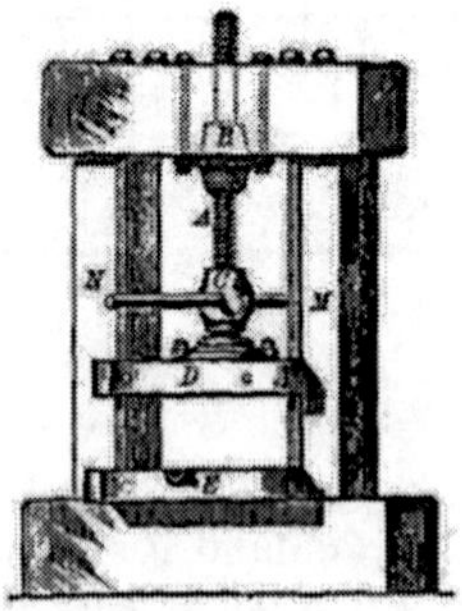


Screw uses the principle of the inclined plane.
J. Dorman Steele. *Popular Physics.*
New York: American Book Company, 1896.

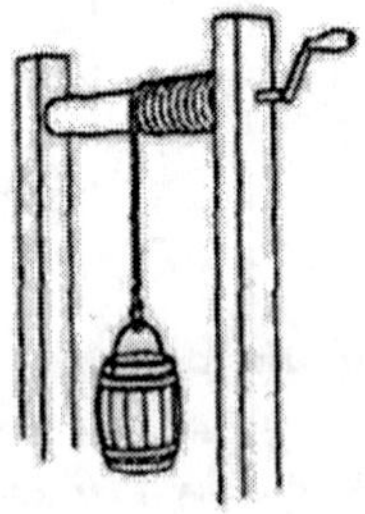

Windlass
William H. Snyder.
*Everyday Science.*
New York: Allyn and Bacon, 1919.

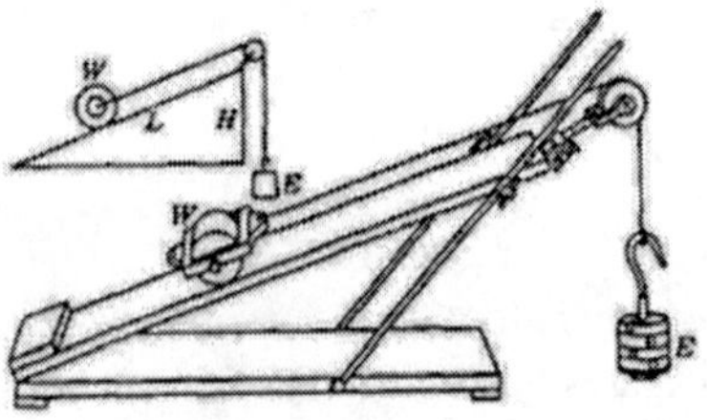


Inclined Plane.
Newton Henry Black and Harvey Nathaniel Davis.
*New Practical Physics.*
New York: Macmillan Company, 1913.

Inclined Plane.
J. Dorman Steele. *Popular Physics.*
New York: American Book Company, 1896.

The machine maker of the ancient world learned to transfer rotary movement into linear motion. The solution lay in a simple machine, the inclined plane: an inclined plane arranged in a continuous spiral formation is a screw. Archimedes' screw, which allowed you to turn a handle to raise water vertically up the shaft, exemplifies the transfer of rotary motion into linear motion. . Archimedes' screw resulted in a worm-gear meshing with a circular cog with the shafts of the two gears at right angles to each other. Mechanics later used the screw as a fixing device.

**Figure 24**
**Archimedes' Screw and Its Later Applications**

| CHRONOLOGY OF ANCIENT GREECE | | | |
|---|---|---|---|
| ***Scientists*** | | | ***Contemporary Events*** |
| Thales of Miletus | 640-546 BC | | |
| Anaximander of Miletus | 610-546 BC | 594 BC | Solon's archonship (magistrate) |
| Anaximenes | 570-500 BC | ca. 545 | Pisistratus in power in Athens |
| Pythagoras of Samos | 570-490 BC | ca. 523 BC | Death of Polycrates of Samos |
| Xenophanes of Colophon (Ionia) | 570-480 BC | 510 BC | War of Sybaris and Croton |
| Heraclitus of Ephesus | 535-475 BC | 508 BC | Cleisthenes reforms constitution of Athens |
| Anaxagoras of Clazomenae | 500-428 BC | 490 BC<br>440-388 BC<br><br>431 BC | Battle of Marathon<br>Thrasybulus, tyrant of Miletus<br><br>Peloponnesian War begins |
| Empedocles of Agrigentum | 492-432 BC | | |
| Leucippus of Miletus | 490 BC | | |
| Zeno of Elea | 490-425 BC | 478 BC | Delian League formed (association of Greek states) |
| Herodotus | 485-425 BC | | |
| Hippocrates of Cos | 460-370 BC | 421 BC<br><br>415 BC<br>404 BC<br>399 BC<br>336 BC | Peace of Nicias (between Athens and Sparta)<br>Athenian expedition to Sicily<br>Peloponnesian War ends<br>Death of Socrates<br>Murder of Philip; Alexander succeeds him |
| Democritus of Abdera | 460-370 BC | | |
| Plato of Athens | 427-347 BC | | |
| Asclepius | 420 BC | | |
| Eudoxus of Cnidus | 408-355 BC | | |
| Aristotle of Stagira | 384-322 BC | 338 BC<br>323 BC | Battle of Chaeronea<br>Death of Alexander |

| CHRONOLOGY OF ANCIENT GREECE | | | |
|---|---|---|---|
| ***Scientists*** | | | ***Contemporary Events*** |
| Theophrastus of Eresus | 371-287 BC | 336 BC<br>323 BC | Philip murdered, Alexander succeeds<br>Death of Alexander |
| Callippus of Cyzicus | 370 BC | | |
| Zeno of Citium | 355-263 BC | 269 BC | Hiero II King of Syracuse |
| Epicurus of Samos | 341-270 BC | | |
| Praxagoras of Cos | 340 BC | | |
| Herophilus of Chalcedon | 335-280 BC | | |
| Euclid | 325-270 BC | 304 BC<br>285 BC | Ptolemy I Soter King of Egypt<br>Ptolemy II Philadelphus joint ruler |
| Aristarchus of Samos | 310-230 BC | | |
| Erasistratus of Ceos | 304-250 BC | | |
| Strato of Lampsacus | 290 BC | Reigned<br>246-222 BC | Ptolemy III Euergetes succeeds |
| Archimedes of Syracuse | 287-212 BC | 221 BC<br>216 BC<br>212 BC | Ptolemy IV Philopator succeeds<br>Battle of Cannae<br>Romans take Syracuse |
| Ctesibius of Alexandria | 285-222 BC | | |
| Eratosthenes of Cyrene | 276-195 BC | | |
| Apollonius of Perga | 262-190 BC | | |
| Hipparchus of Nicaea | 190-120 BC | 168 BC<br>146 BC | Battle of Pydna<br>Rome destroys Carthage and Corinth |
| Posidonius of Apamea (Rhodes) | 135-51 BC | 133 BC | Kingdom of Pergamum bequeathed to Rome |
| Lucretius in Rome | 95-55 BC | 86 BC<br>40 BC<br>31 BC | Sulla sacks Athens<br>*Bellum Alexandrinum*<br>Battle of Actium |
| Vitruvius | 80(70)-15 BC | | |
| Strabo of Amaseia | 64(63)<br>BC-24 AD | | |

| CHRONOLOGY OF ANCIENT GREECE | | | |
|---|---|---|---|
| ***Scientists*** | | | ***Contemporary Events*** |
| Celsus | 25 BC-50 AD | | |
| Hero of Alexandria | 10-70 AD | 14-34 AD<br>69-79 AD | Tiberius emperor<br>Vespasian emperor |
| Ptolemy of Alexandria | 100-170 AD | 98-117 AD<br>117-138 AD | Trajan emperor<br>Hadrian emperor |
| Galen of Pergamum | 130-200 AD | 138-161 AD<br>161-180 AD | Antoninus Pius emperor<br>Marcus Aurelius emperor |

## VII. ANCIENTS AND THEIR ENVIRONMENT

### *Air pollution*

Pollution from natural sources, such as gases from erupting volcanoes, dust from wind, and smoke from lightning-caused fires, occurred long before humans existed. The earliest account of air pollution is probably that of the Greek natural philosopher Theophrastus in the fourth century BC who noted in his *On Stones* the troublesome and disagreeable smell emanating from burning coal. Much later, in the second century AD, Galen, upon observing Greek copper mines, commented on the danger of acid mists arising from the smelting processes.

In Rome, the poet Horace (65-8 BC) lamented the city's smoke-blackened shrines. Rome's citizens clearly recognized their city's considerable air pollution and referred to it as *heavy heaven* or *infamous air*. Vitruvius (first century BC), the Roman engineer, noted the health problems workers in lead and mercury mines experienced. Because of the noxious vapors, mainly sulfur dioxide emitted upon roasting the sulfide ores, and from mercury, the Greek philosopher Plutarch (46-120 AD) recommended that only criminal slaves work in the mines. Non-criminal slaves should not face such hazardous conditions. Pliny the Elder (23-79 AD) wrote that workers in the zinc smelters wore animal bladders as respirators (to protect them from the sulfur dioxide vapors), but the practice did not appear widespread in all the Roman mines and smelters.

### *Water pollution*

For centuries physicians, scientists, engineers, and other concerned citizenry recognized the obvious connection between clean drinking water and good health. They boiled and filtered (strained) their drinking water to improve its appearance and taste and would not drink, or bathe, in dirty or visibly polluted water. The earliest known records on treating water to improve the aesthetic qualities of taste and smell date from 2000-1500 BC. Sanskrit writings record the use of filtration through sand, gravel, or charcoal, boiling, aeration, and exposure to sunlight, as recommended methods. The Egyptians recognized a link between the cloudiness of water and an unpleasant taste and appearance, and as early as 1500 BC they added alum, a known mordant or dye-fixer, to make suspended particles settle out. The ruins of Cretean and Assyrian cities dating from 1500-1400 BC indicate the existence of sanitary sewers. Hippocrates recommended boiling and straining rainwater to prevent it from having a bad smell and causing hoarseness. Aristotle cautioned Alexander the Great not to let his men drink from stagnant pools but to carry boiled water with them.

Roman engineers constructed aqueducts to bring clean water to the gymnasiums and public baths. From 312 BC to 455 AD they constructed throughout their empire extensive aqueduct systems that transported daily 130 million gallons of water. The Romans also built sewage systems and used reservoirs to provide water for cleaning the streets. They and other ancient societies achieved little understanding of water quality and left a sparse record.

## VIII. SCIENCE IN THE EAST

### *Chronology and contacts with the west*

Archeological evidence indicates that Peking (Beijing) man (homo erectus), China's oldest inhabitant, lived in China about 500,000 years ago and that the earliest Chinese farming settlements date from 4000 BC. China's historic period and Bronze Age began about 1500 BC in the Shang dynasty (1751-1027 BC), a united China emerged in the Qin (Ch'in) Dynasty (221-207 BC), and for about 200 years in the Han Dynasty (206 BC-220 AD) the arts and sciences flourished. A period of civil unrest followed from 220 to 589 AD. Beginning with the Tang Dynasty (618-906) Chinese civilization thrived once more.

East-west contacts did not occur until the late Middle Ages, 1200s-1300s, when Catholic missionaries, Dominicans and Franciscans, went to China, although Nestorian missionaries, a heretical sect that originated in Constantinople in 431, visited China in the 600s and remained there until the 1360s. Marco Polo (1254-1324), a Venetian trader and traveler, spent the years 1271-95 on his trip to and from China. He reached China and Emperor Kublai Khan (1216-94) in 1275 during the Sung Dynasty (960-1279). The establishment of the Ming Dynasty (1368-1644) led to the removal of outside influences until 1513 when the Portuguese explorer Jorge Alvares (d. 1521) reached Guangdong in southern China by sea.

The Portuguese in 1555 established a small settlement on the southeast Chinese mainland at Macao (Macau). It became a permanent settlement in1563. During those years the Jesuits began their missions to China. Matteo Ricci (1552-1610), a Jesuit priest who studied mathematics and astronomy under Christopher Clavius (1537-1612) in Rome, went to Chaoking, the administrative capital of Canton, in 1583 and then to Nanking and Peking (Beijing) in 1601. While in China, serving as superior of the 17-member Jesuit mission, Ricci wrote about 20 works, mostly in Chinese, on religious and scientific topics. Ricci never left China, his death and burial occurred in Beijing.

### *Chinese science and engineering*

Confucianism and Taoism were the two main schools of thought. Confucius (551-479 BC), a Chinese philosopher, stressed the moral and practical dimensions of science. Confucianism dominated Chinese thought. Lao-Tse, the Old Master, who lived in the sixth century BC (600-500 BC), founded Taoism, a mystical approach to comprehending nature. His followers described his philosophy in the *Tao-Te-Ching* which dates from the third century BC (300-200 BC). Taoism, or, the way, accepted the yin and yang principles and believed that the world's natural and political orders influenced one another.

The emperor and his court controlled Chinese science and never let science spread beyond the court into Chinese society. No middle class emerged to use and promote science as occurred in the West in the 1600s-1700s, which may explain why a scientific revolution did not occur in China. Alternatively, why should historians assume an inevitable Chinese scientific revolution and then have to explain why no scientific revolution occurred in China?

In practicing science, Chinese natural philosophers abhorred specialization and believed that the same laws governed nature and man. Things naturally belonged in numerical groups of two, three, four, five, or more. The group of two, yin and yang principles of Chinese philosophy, of uncertain origin but possibly dating from 2600-2100 BC or 2100-1600 BC, represents the fundamental and opposite, yet complementary, forces or principles of the universe. Yin is feminine, negative, passive, and represents Earth. Yang is masculine, positive, active, and represents Heaven. In Chinese cosmology yin combined with yang to produce all that comes to be or becomes. The group of four represented the seasons. The group of five had significance in cosmology because it corresponded to earth, wood, metal, water, and fire, the five phases that composed the material world. The five phases first appeared in Chinese writings in 300-200 BC.

The study of numbers allegedly led Fuh Hi (2858?-2738 BC?), the mythical Chinese demigod who founded the Chinese empire in 2852 BC, to the discovery in 2800 BC of magic squares. His *Loh Shu*, or scroll of the River Loh, described magic squares, such as the 3-number squares in which the sums of the three numbers making up each of the four sides |8, 1, 6|6, 7, 2|2, 9, 4|4, 3, 8| and the diagonal |8,5,2| are equal and total 15.

Chinese mathematicians in 2000 BC used a binary system to count by twos. They also developed a base-10 system. The Chinese abacus, suan pan, an early type of digital or counting computer, dates from 190 AD in the Eastern Han Dynasty (25-220 AD) and achieved perfection by 1200 in the Sung Dynasty (960-1279). A biquinary device, like two, five-fingered hands, it consisted of a rectangular wooden frame enclosing vertical wires divided lengthwise into two unequal parts by a horizontal wooden bar. The wires represent the powers of 10. Each of the five beads on a vertical wire on the lower side of the wooden divider counted 1, each of the two beads on a vertical wire on the upper side counted 5.

**Figure 25**
**Chinese Abacus**
sikarin supphatada/Shutterstock.com

Chinese astronomers kept excellent records, and from 1800 BC they compiled comprehensive tables of the planetary movements, sun spots, and earthquakes. They developed lunar calendars from 1400-1200 BC (Shang Dynasty), and by 400 AD they had adopted solar calendars and determined the tropical year to consist of 365.2428 days. Their solar calendar divided the year into 24 parts and specified the type of weather they expected in each part.

No astronomical calculating models of the universe, such as the geocentric heavenly spheres of the Greeks and the heliocentric spheres of later European astronomers, appear in the writings of Chinese astronomers. Ricci and Jesuit missionaries brought the astronomical calculating models of Western astronomy to the Chinese court beginning in 1583. Western science became more widespread in the late 1800s. Chang Heng in the Han dynasty (206 BC-220 AD), a Buddhist monk I Hsing (ca. 725) in the Tang dynasty (618-906), and Guo Shou Ling (ca. 1279) in the Yuan dynasty (1279-1368) made extensive use of the gnomon stick to observe the Sun's motion and to calculate accurately geographical distances.

The Chinese medical practice of acupuncture dates from 2000 BC. Just as Greek physicians tried to balance the body's four humors, Chinese physicians sought to balance the body's positive (yang) and negative (yin) energy or chi. Inserting solid needles into the body's 365 points that connect to form 14 pathways, probably releases endorphins, which are the body's natural pain killers and the most likely reason why acupuncture works.

Shen Kua (1031-95), a civil servant and academician in Yangchow and Jingkou in the Sung dynasty (960-1279), in 1095 gave an excellent summary of Chinese science in *Brush Talks from Dream Brook.* He discussed calendar reform, astronomy, magnetism, and other aspects of natural philosophy.

The construction of the Great Wall remains the most successful engineering accomplishment of the Chinese dynasties. Emperor Tsin Shi Hwang Di in the Zin Dynasty (221-207 BC), the first emperor of a united China, began construction of the 1,500-mile fortification in north China to keep out nomadic wanderers from the north. Completion of the wall, which bends and turns numerous times making its actual length nearly 4,000 miles as it winds through the Mongolian plain's southern border, occurred centuries later in the Ming Dynasty (1368-1644).

An earthen or a rough and irregularly broken stone or mud brick core (rubble) covered or faced with stone or kiln-fired bricks and mortar made up most sections of the wall. Slaves dug the construction materials from quarries, masons dressed the stone used for facing. Wheelbarrows, which the Chinese invented in 200-300 AD, facilitated construction. The wall averages 25 feet in height, 20-23 feet width at the base, and 12 feet at the top which is wide enough for four horsemen to ride abreast. Slabs of stone or brick cover the top roadway which has a 4-foot brick parapet (shielding wall) on the north and south sides. The wall had 25,000 watchtowers along its length, each 46 feet high and 30 feet square. The Chinese constructed another 15,000 watchtowers outside the wall.

***Japanese science and engineering***

Even before Ricci reached China, Portuguese traders in 1543 arrived in Japan. A few years later, on 15 August 1549, St. Francis Xavier (1506-52), a Spanish Jesuit missionary who already reached India about 1541, made the first western mission to Kagoshima, Japan.

Severe persecution and banning of Christianity and westerners followed for about 250 years. During the Tokugawa shogunate (1603-1867), a period of military dictatorship, Japanese natural philosophers deemphasized the naturalistic metaphysical aspects of their science and stressed the Confucian moralistic dimensions of science. The Japanese adopted the yin and yang principles and the five phases of the Chinese philosophers and by the late 1700s began to accept scientific knowledge from the West.

The Shogunate period of no contact with all foreigners, except for a few Dutch and Chinese traders, lasted from 1614 until Commodore Matthew Perry (1794-1858) of the United States Navy in 1853-54 negotiated a diplomatic and trade treaty with the Japanese. After the 1868 restoration of the Emperor Meiji (1852-1912) and throughout his long reign the Japanese were more receptive to western science than the Chinese. Their science became pragmatic, utilitarian, and materialistic, and they began a period of industrialization, particularly in the chemical and electrical industries.

### *Hindu science*

In Hindu thought the individual was an expression of a universe billions of years old and in which an intimate connection existed among all things. The human spirit had infinite flexibility and infinite creativity and shared, with all living things, in one reality called Brahman, the creator of the universe and personification of the absolute.

During the Vedic period, 2000 BC-500 BC, linguistics (Sanskrit) was the model and starting point of study for all sciences. Varuna, the guardian of cosmic law and order, used the Sun to determine the boundaries of air, earth, and sky, and established the four cardinal directions.

Hindu natural philosophers accepted atomism (600-500 BC) and the Greek five elements, air, earth, fire, water, and ether. Atoms formed all the elements except the transmitting ether medium. Disease resulted from an imbalance of three elements, air, fire, and water in the body, and in relation to bodily health Hindu physicians called the three elements the three humors, wind, bile, and phlegm.

From 400 to 800 AD Hindu mathematicians and scientists made important contributions to learning. Ujjain in west-central India became the center of Hindu science. Hindu mathematicians in 458 AD, and independently the Mayans about the same time, invented zero as a number that you could add and subtract. Zero used in place notation appeared earlier in 500 BC in Babylonian mathematics and later in Greek astronomical calculations. The Hindus used a decimal, or base-10 system in mathematics and developed algebra extensively.

Hindu astronomers, like the Chinese and Japanese, never developed astronomical calculating models comparable to the Greek models. Brahmagupta (598-665) at Ujjain, one of the most famous Hindu astronomers, published *Brahma Suphuta Siddhanta*, in 628. In his *Brahma*, Brahmagupta applied algebraic solutions to astronomical problems. He calculated eclipses of the Sun and Moon, and accepted the sphericity of the heavens and Earth but not Earth's motion.

## IX. SCIENCE IN PRE-COLUMBIAN NORTH AMERICA

### *Origin of America's inhabitants*

The inhabitants of the Americas arrived from Asia during the Great Ice Age of 30,000-8000 BC. Glaciers and ice sheets across the globe tied up vast amounts of water, and the resulting low sea levels exposed land across the Bering Strait between Siberia and Alaska. These Asian wanderers that Columbus called Indians crossed the land bridge and spread out across the Americas. They lived by hunting animals, such as bison, horses, and mammoths, and by consuming maize, pepper, cocoa beans, squash, potatoes, tomatoes, and other wild plants that they found and learned to cultivate.

The American Indians mined the copper, gold, and silver deposits they discovered in the regions they settled and designed masks, ornaments, and jewelry, such as earrings, chest and arm bands, from these metals. In Mesoamerica (middle America) Mayan and Inca artisans constructed clay flutes that produced complex musical tones

North American tribes and Mesoamerican societies worshiped the Sun as a deity. They tried to repeat in their lives the order they saw in the universe, which explains why some Indian societies aligned their buildings and laid out their cities as they did. They believed order made the universe sacred and that once they had ordered their universe it acquired meaning.

*North American Indian cosmology and astronomy*

The Chumash Indians, a semi-agricultural 18,000-member tribe who lived in southern California 13,000 years ago and whose descendants still live there, followed the celestial cycles and used a lunar calendar to organize and regulate their society. They and many other Indian tribes constructed medicine wheels, each consisting of a circle of rocks with a central pile of rocks, called a cairn, from which spokes of rock radiated. They aligned the medicine wheels in order to observe the summer and winter solstice, the positions of bright stars, and other celestial events at selected points on the wheel. Ancient Indian tribes, such as the Aztec-Tanoan society that occupied the Great Plains of the United States and Canada, also constructed medicine wheels, about 50 of which remain mainly in Canada on the Rocky Mountains' east flank or on the open plains below. The 3,000-year old Moose Mountain Medicine Wheel in southeastern Saskatchewan, Canada, has a 5-foot high 30-foot diameter cairn that weighs about 80 tons. The longest of its five spokes has a diameter of 123 feet and marks the summer solstice.

North American Plains Indians constructed the Majorville cairn located south of Bassano, Alberta, Canada, on the banks of the Box River, about 4,500 years ago. Its 50-ton, 27-foot diameter center cairn has 28 spokes likely representing the 28 days in a lunar month. They led to an outer rim, or wheel, that had an 81-foot diameter, 60 percent of which archeologists excavated only in 1971. Majorville was once the centerpiece of an enormous Temple of the Sun, Moon, and Venus, the morning star.

The 2,500-year old Bighorn Medicine Wheel that Plains Indians constructed at Medicine Mountain in Bighorn Country in northern Wyoming has a 2-foot high, 12-foot diameter cairn with 28 star-aligned spokes radiating from the center and an outer rim or wheel diameter of 87 feet, only slightly smaller than the largest circle at Stonehenge, England. Two spokes align with the solstices, others with the rising of Sirius, the brightest star in the heavens, and Aldebaran, and Rigel, two other bright stars.

The circle of posts known as the Cahokia Sun Circle in southern Illinois represents a different astronomical monument of uncertain significance but of probable cosmological importance. Cahokia, a large Indian metropolis located about seven miles east of St. Louis, had a population of about 25,000-40,000 people, as many as contemporary London. Its ruins date from about 950 or possibly as late as 1000-1200 and have yielded samples of mica, copper sheets, and marine shell beads.

Other Indians, such as the Omaha of Nebraska's Great Plains and the Sioux in the northern plains, identified their cosmic axis with a tree at the center of the world, or their Tree of Life, around which were the four world directions of north, south, east, and west that ordered their cosmic world. The Pawnees of Nebraska's Great Plains designated four stars as the four pillars of the heavens whose hands supported the stars and whose feet stood firmly on earth. The four stars clearly seen at night, Capella (Auriga), Sirius (Canis Major), Vega (Lyra), and Antares (Taurus), represented their four cardinal directions.

In the southwest, Navajo sand paintings depicted a creation involving Father Sky's constellations including the Big Dipper, and Mother Earth who contained within her the Navajos' four sacred plants: beans, corn, tobacco, and squash. The Pueblos in New Mexico carefully observed the Sun's movements to establish dates for religious and ceremonial events. Their ancestors, the Anasazi, between 900-1100, while farming and building irrigation canals, constructed an extensive network of roads that connected outlying towns with 13 central great houses (apartments) or pueblos at Chaco Canyon.

One of the pueblos, Pueblo Bonito constructed between 919-1080, shows how the Anasazi used the Sun not only for calendrical purposes but for its energy. Bonito's construction and orientation, an

800-room single structure stone village curved in a semicircle facing south, allowed it to function as an efficient solar collector. Sandstone veneer walls filled with heat-trapping adobe absorbed the low winter Sun's rays, while high summer rays fell on an insulated roof constructed of wooden beams covered with brush and dirt. Bonito's walls cast shadows and functioned as a calendar that indicated the summer solstice, 21 June the longest day of the year, and the winter solstice, 22 December the shortest day of the year. A port in the walls of the southeast corner admitted the Sun's rays only on the winter solstice.

## X. SCIENCE IN MESOAMERICA

### *Our changing perception of the Mayan world*

*Mayan society*

Mayan civilization dates from 11 or 13 August 3114 BC on the Gregorian calendar. The Mayans began counting time on this date which represents their zero year and corresponds to 1 January 1 AD on the Gregorian calendar. The date represents a Mayan creation date of the world, the beginning of their last Great Cycle of creation and destruction, although scholars disagree on the exact August date. Nevertheless 11 or 13 August must represent a significant event in Mayan society, possibly the planetary conjunction visible in the constellation of Aries that Hindu astronomers calculated as occurring on the evening of 17-18 February 3100 BC. Each Mayan Great Cycle lasted for 5,125 years, or 13 baktuns each of approximately 394 years, the last one ending on 21 or 23 December 2012 AD, the day of the winter solstice.

The Mayans lived in Honduras, Belize, El Salvador, Guatemala, and Mexico. Their agricultural villages date from 1500 BC. Altun Ha, a city of 8 square kilometers in Belize occupied from 200 BC, had an estimated population of 10,000 in the Mayan classic period (250 to 900 AD). Their main cities, Piedras Negras (Spanish for black stones) occupied from 950 BC to 850 AD; Cerros occupied from 50 BC; Palenque (Spanish for Palisade); Tikal (place of voices, ancient name Mutul) 60 square kilometers, population of 62,000 in 600-750 AD; Calakmul 70 square kilometers estimated population of 50,000; Cobá; and Copán estimated population of 25,000, existed at the beginning of the Christian era. Chichén Itzá (mouth of the well of the Itzás) near the Yucatan capital of Merida, flourished between 800 and 1200 AD. The Mayans occupied 13 city states.

The Mayans adopted a hereditary caste system beginning about 300 BC. Their classic period, from 250 to 900 AD, the high point of their society, declined before the Spanish arrived because of a change from limited warfare to total warfare among the 13 city states, overpopulation, poor agricultural practices such as slash and burn agriculture, overexploitation of the tropical forests, and soil depletion.

The Mayan hieroglyphic writing system which originated in the earlier Olmec culture of 1200 BC, dated from 300 BC, possibly 400 BC. The Mayans developed the only written language in Mesoamerica and provided Mesoamerica's only written historical record, much of which disappeared forever. Diego de Landa (1524-79), a Spanish-born Franciscan priest who arrived in Yucatán in 1549 contributed much to the mystery of the Mayans. He tried in 1562 to eliminate all traces of Mayan idol worshiping and human sacrifice by burning at the Yucatán city of Mani about 5,000 idols and 27 paper codices containing the hieroglyphics that provided descriptive and historical accounts of Mayan society.

Each codex consisted of a paper made from the lime-coated bark and fibers of the maguey plant, and folded accordion style. Quills and brushes of animal hairs served as Mayan writing implements. A thorny flowering tree with hard red-colored wood, belonging in the legume family, *Haematoxylm campechanium* (logwood, bloodwood), and hematite (iron oxide, $Fe_2O_3$) provided a red ink. The Mayans usually wrote their hieroglyphics in black with red highlights, the black ink obtained from soot or charcoal, and other colors such as violet prepared from *campechanium* extracts. The codices, written on both sides, and the hieroglyphics written on stone (stelae) found at temple and city ruins, are the only known written sources on Mayan society.

The Spanish government's Royal and Supreme Council of the Indies in Madrid in 1562 reprimanded Landa for his overzealousness. He returned to Spain for a hearing (trial), and in 1569 the Council absolved him of any crimes. Landa became the second Catholic bishop of Yucatán in Mérida (appointed 1572, ordained 1573).

Despite the destruction Landa caused, he provided a reliable account of Mayan society. He compiled the information he had on Mayan culture, history, architecture, calendars, and language and published it in 1566 in *Relación de las Casas de Yucatán (Relation of the Matters of Yucatán).* Unfortunately, Landa's *Relation of the Matters* lay forgotten until its rediscovery in the 1850s in Madrid's Academy of History and its reprinting in 1864. Although far from a complete account, *Relation of the Matters* remained for a time the only available source on the Mayans and provided a phonetic alphabet that enabled linguists to decipher about one-third of the Mayan hieroglyphics.

Four Mayan codices survived Landa's wrath. The *Codex Dresden*, 3.5 inches wide when folded, 74 pages unfolded, written between 1100 and 1200, identified as Mayan in 1829, and now located in Dresden's state archives, contains tables of the planets' motions especially of Venus, the Moon's motions, eclipses, conjunctions, and constellations, as well as prophecies, medical information, and day counts. The *Codex Tro-Cortesianus* (*Madrid Codex*), 5 inches wide, 112 pages unfolded, written in the 1300s, sent to Spain in the 1500s, rediscovered in the 1860s, and presently located in Madrid's Museum of the Americas, was part of a prophetic calendar that depicted good and bad days. *The Codex Peres (Peresano),* 22 pages unfolded, written between 1224 and 1441, discovered in 1859 in the National Library in Paris and presently located there, contains calendrical cycles and information on the constellations. The *Grolier Codex* written in 1230, discovered in a dry cave in Chiapas, Mexico, in the 1960s, and presently in an unknown location but probably in a private collection, contains half of a 20-page table of Venus's motions. The codices were generally divinatory almanacs containing information on weather, agriculture, disease, hunting, and astronomy.

A different description of the Mayans appeared two hundred years after Landa's history. Francisco Saverio Clavigero (1731-87), a Mexican-born Jesuit priest in Mexico City, whom the Spanish king Charles III (1716-88) in 1767 expelled from Mexico along with all the Jesuits, in 1780 published *Historia antica del Messico (Ancient History of Mexico).* Clavigero's glorified account of Mexican native culture revealed his resentment of Spain's 1767 Jesuit expulsion from Mexico and from the rest of the Spanish empire. The expulsion took place because the Jesuits supported the Vatican against the Spanish crown, and in particular for establishing what Spain considered a Jesuit state within Spain's New World empire. The Spanish also accused the Jesuits of monopolizing Indian laborers. Clavigero's book enjoyed considerable favor, with translations in English in 1787 and German in 1789.

Rediscovery of Mayan ruins began in the 1830s. The American lawyer, diplomat, and explorer John Lloyd Stephens (1805-52) in New York City, beginning in 1839 and joined by the English artist Frederick Catherwood (1799-1854) then in London, in 1839-40 and 1841-42, discovered the Mayan ruins at Palenque, Copán, and Uxmal (thrice built), and those at Chac, Labna, Sayil (place of ants), and other Mayan cities in Yucatán's 2,300 square mile Puuc (hilly) region. Stephens and Catherwood in 1841 and 1843 published two very popular travel books describing their explorations. The second book, *Incidents of Travel in the Yucatan*, sold 9,750 copies in the nine years after its 1843 publication.

Both Stephens and Catherwood died tragically. Stephens died of malaria contracted in Panama while planning the construction of a trans-isthmus railroad for the Panama Railroad Company. Catherwood died in a shipwreck. A French ship rammed his ship *Arctic* as he sailed from Liverpool to the United States. Fifty years after the Stephens and Catherwood explorations, the English archeologist Alfred Maudslay (1850-1931) in 1883-94 at his own expense and support from Harvard's Peabody Museum began archeological investigations that revealed the stone temples, palaces, and recreation areas of Copán, Palenque, Tikal, and Chichén Itzá.

Twentieth-century scholarly research on the Mayans initially proved erroneous. The pioneering 1940s studies of the two Carnegie Institution (Washington) researchers, Sylvanus Morley (1883-1948), who called the Mayans the Greeks of the New World, and the English ethnographer John Eric Thompson (1898-1975), who wrote a detailed commentary on the Mayan *Codex Dresden*, portrayed the Mayans as a peaceful society of calendar constructors and star gazers. A totally different picture of the Mayans emerged in June 1946 when two Lacandon Mayans led the American photographer and chemist Giles Healey (1901-80) to the Bonampak Murals in the southern state of Chiapas. The painted wall murals of three temple ruins in Bonampak dating from 790 showed instead that warfare predominated among the 13 Mayan states and included torture, human sacrifice of captives, and bloodletting in religious, sporting, and other public events.

All the earlier studies, from Landa and Clavigero to Morley and Thompson, contributed to deciphering the Mayan hieroglyphics, but the major breakthroughs occurred in the post-World War II period once linguists recognized that the 800 hieroglyphics represented syllables and not letters or words and were phonetic (the symbol indicated the sound). Yuri V. Knorosov (1922-99) at the N. N. Miklukho-Makiai Institute of Ethnology and Anthropology (History Branch of the Russian Academy of Science) in Moscow in 1952, David H. Kelley (1924-2011) at Calgary and Linda Schele (1942-98) at Texas in the 1960s and 1970s led the renewed deciphering of the Mayan codices and stelae, the first from the temple ruins at Palenque. Their work proved remarkably successful, and today linguists can read about 80 percent of Mayan hieroglyphics.

Yet a significant 60-year gap still remained in Mayan history until a serendipitous event that occurred in October 2001 enabled Arthur Demarest (b. 1953), an anthropologist at Vanderbilt University, to clarify considerably what destroyed the Mayan civilization. As *National Geographic* reported in October 2002 a hurricane that blew through northern Guatemala's jungle uprooted a tree at the base of temple ruins in Dos Pilas, an outpost city established in 629 about 70 miles southwest of Tikal. The hurricane exposed the last 10 steps of inscribed stones that completed the 8 existing steps of a staircase leading up the sides of a pyramid, previously discovered during a 1989-97 expedition. The new discovery filled in the 60-year gap.

The inscribed hieroglyphics record previously unknown accounts of a protracted warfare between the superpower cities of Tikal and its allies in northern Guatemala, and Calakmul, the snake kingdom, and its allies 60 miles farther north in Mexico. Their long war split the Mayan civilization in the 500-600s. The glyphs in graphic detail tell the story of Prince Balaj Chan K'awiil (b. 625) who ascended to the throne of Dos Pilas in 635 and the role he played as Calakmul's ally in the war for complete dominance between Calakmul and Tikal, the city his brother, Nuun Ugol Chaak, ruled.

Tikal recovered after its defeats in the 550s-690s and in 695 crushed Calakmul, although Tikal's rulers abandoned and burned Tikal in 760. Disintegration into regional powers and petty warfare followed and led to the Mayan world's collapse by 900.

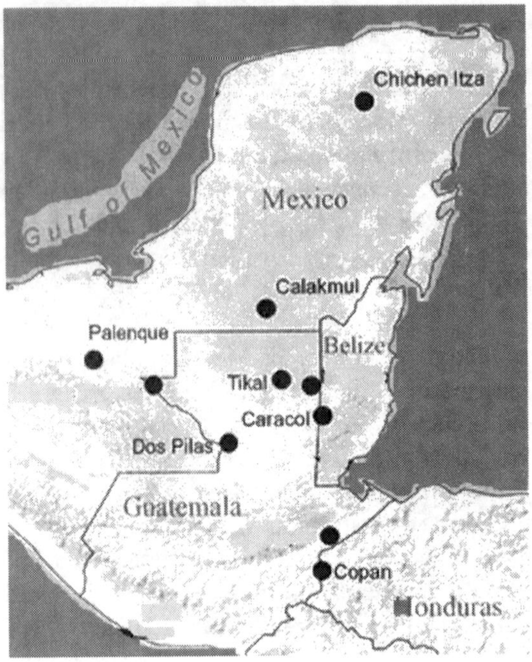


**Figure 26**
**Dos Pilas Region**
Courtesy of the author

*Mayan science, mathematics, and engineering*

Creation myths always have existed in earlier eastern and western cultures. In ancient Mexican, Central and South American societies creation myths often described Earth as an island floating on water. The Warao Indians in Venezuela viewed Earth as an island floating on a sea whose waters met the horizon and mingled with the fluid that made up the sky. Divine sky lifters, or world trees, situated in the key horizontal directions of north, south, east, and west, held up the heavens.

The Mayan creation myth found in the *Popol Vuh* (1554-60) describes the sea and sky as parting. The Maker, the Feathered Serpent, and the Heart of the Sky create the world by calling out Earth, mountains, valleys, and forests. Other creation myths describe a time before creation, a time motionless, still, and suspense, of calm and silence, and the vast expanse of the empty sky. They include a Sun God and the Jaguar God of the Underworld. Their hieroglyphics suggest a universe millions of years, possibly 400 million years, old.

In addition to the creation myths that included descriptions of the heavens, Mayan astronomers observed carefully the heavenly bodies, particularly Venus's motion, and compiled tables that enabled them to predict its motion. But they never developed a geocentric or a heliocentric model of the universe. The Temple of Inscriptions at Palenque in southern Mexico, which dates from 683 and (discovered in 1949), allowed Mayan astronomers to follow the Sun on its annual journey, particularly the winter solstice which represented the Sun's death. They aligned the three temples at Uaxactim in Guatemala with the equinox sunrise and the summer and winter solstice rising Sun.

Similarly, the alignment of the Kukulkan Pyramid, a 75-foot (24 meters) high square-based stepped pyramid constructed about 1050 at Chichén Itzá, on the vernal equinox projected an image of an enormous snake crawling along a stairway. The pyramid has four stairways, each with 91 steps leading to a platform at the top, giving a total of 365 steps.

**Figure 27**
**Pyramid at Chichén Itzá**
Banauke/Shutterstock.com

The Mayans devised three kinds of calendars which, according to Morley, the Mayans introduced between 353 and 235 BC. The first, a *solar calendar,* consisted of 18 months of 20 days known as the *Teolain* or *Long Count*. The *Long Count*, an approximate calendar, coincided with the Mayan 365-day calendar every 52 years or what the Mayans called a bundle. Their bundle corresponded to a century (100 years) in the Gregorian calendar system. The *Long Count* calendar consisted of five time units.

| MAYAN LONG COUNT CALENDAR TIME UNITS | |
|---|---|
| kin | day |
| uinal | 1 uinal = 20 kin or days |
| tun | 1 tun = 18 uinal or 360 days |
| katun | 1 katun = 20 tun = 7,200 days or approximately 20 years |
| baktun | 1 baktun = 20 katun = 144,000 days or approximately 394 years |

The Mayans numbered the kin, tun, and katun from 0 to 19, the uinal from 0 to 17, and the baktun from 1 to 13. Reading from right to left, from kin to baktun, the first date in the Long Count was 13.0.0.0.0., and not 0.0.0.0.0 as expected, because the baktun numbers go from 13 to 1, suggesting an earlier world that lasted 13 baktuns. Mayan scholars disagree whether 13.0.0.0.0 corresponds to 11 or 13 August 3114 BC of the Gregorian calendar and therefore when the Long Count again reached 13.0.0.0.0. Either 21 or 23 December 2012 marked the beginning of a new Long Count, 13 baktuns later.

The second calendar, a *solar civil calendar*, or *haab*, consisted of 18 months of 20 days each, plus an unlucky period of 5 days known as Uayeb. The Uayeb days had no names and were days without souls, or empty nameless days, which the Mayans observed as days of prayer and mourning and on which they extinguished fires and ate no hot food. A miserable future awaited anyone born during Uayeb.

The third calendar, a 260-day *sacred calendar*, or *Tzolkin*, contained 13 numbers, 1 through 13 sequentially, and 20 names of plants and animals, such as serpent, rabbit, deer, flower, grass, and reed, arranged in a definite sequence. Each day received a number 1, 2, 3, up to 13 and the sequence started again, 1, 2, 3. Each numbered day also received a name, starting with the first name in the name sequence to the last, or twentieth, name and the sequence began again. To start the calendar, number 1 paired with the first name in the name sequence, number 13 paired with the 13th name. To continue the pairing, number 1 paired with the 14th name, number 2 with the 15th name, continuing to the 20th name and then the name sequence began again. The cycle took 20 x 13 = 260 days to complete. The Mayan sacred calendar determined important events related to their gods and to humans, such as dates for battles and marriages.

No one has satisfactorily explained why the Mayans and other Mesoamerican Indians adopted the sacred calendar's 260-day cycle. One explanation is that two branches of Mayan society existed, one of which used a 13-day cycle of numbered days possibly related to a set of 13 Mayan gods, and the other a 20-day cycle of 20 named days. At some time in early Mayan history the two branches merged and combined the two calendars so that neither lost its method of counting. Another explanation is that 260 days approximate the nine-month human gestation period and the time of Venus' annual journey through the sky.

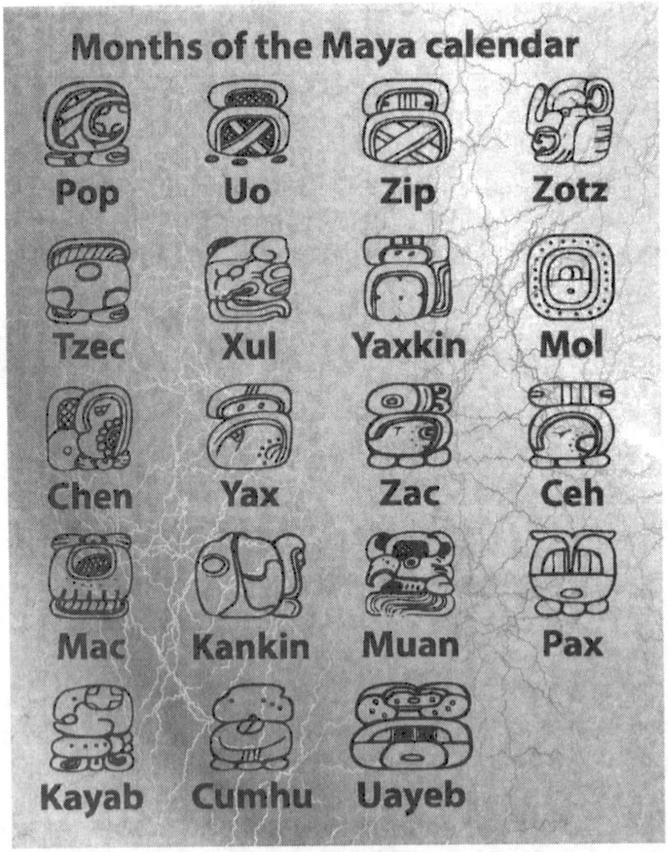


**Figure 28**
**Glyphs of Mayan Solar Civil Calendar or Haab**
Milagli/Shutterstock.com

The Mayans had a 20-base number system and in 500 AD adopted zero as a number. The number 32 on their 20-base system was 1x20+12 equivalent to 3x10+2 on the decimal system. They did not use fractions, dots and bars represented their numbers. A variety of glyphs represented zero, the simplest of which looked like a horizontal oval raindrop with one horizontal line connecting the two farthest points on the oval and four parallel vertical lines dividing the upper half of the oval.

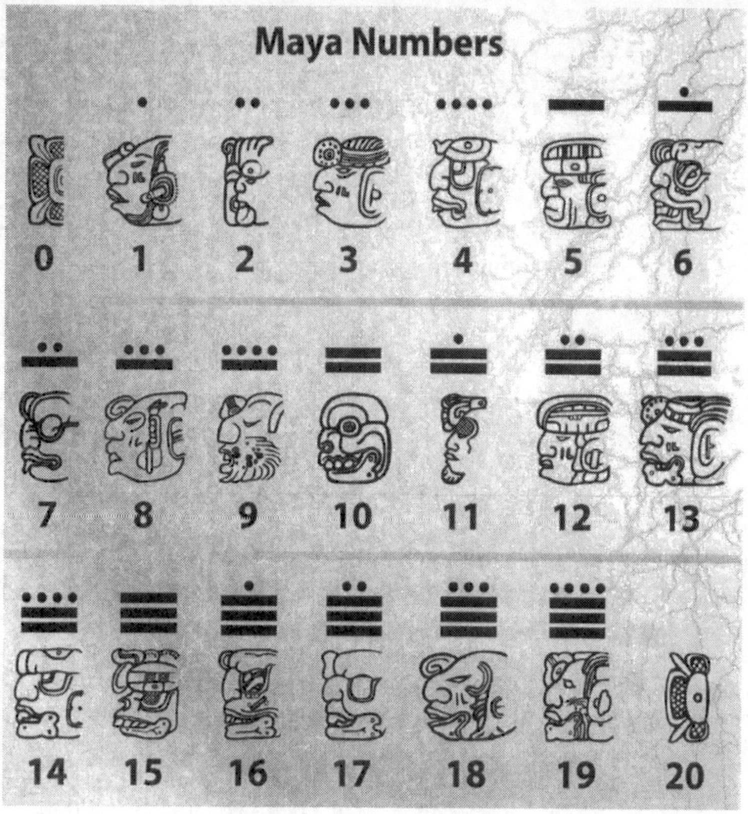


**Figure 29**
**Mayan Glyphs for Numbers**
Milagli/Shutterstock.com

The Mayans made no use of the wheel, which dates from 3300 BC in the Middle East, or of metals for tools, which in Europe and Asia date from 4000 BC for copper and 1000 BC for iron. Mayan metallurgy developed much later, after the end of their classic period in 900 AD. Their metallurgical processes were similar to those of other cultures in and outside of the Americas because the metallic ores they worked with were the same as those found in other cultures. Mayan artists, architects, sculptors, artisans, and metallurgists worked with copper, silver, gold, and jade and applied their metallurgical skills to the production of jewelry and ornaments.

Gold, the most valuable metal, was the easiest metal to obtain because it occurred in the free state, mixed either in fine particles with sand or in veins with quartz and separated by crushing the quartz. Roasting naturally-occurring copper sulfide in air to give copper oxide and reacting the oxide with more copper sulfide produced free copper metal. The same two-step sequence produced free lead from its naturally-occurring sulfide ore.

Tin, which occurred naturally as the oxide, when roasted with a carbon source such as wood charcoal gave the free metal. Heating the free metal and copper produced the copper-tin alloy bronze. Roasting naturally-occurring zinc sulfide gave zinc oxide which when heated with copper produced zinc metal. Silver, the second most valuable metal after gold, occurred naturally as the free metal with gold in the mixed sulfide ores of zinc, copper, and lead. Heating the ores produced a molten mixture of silver, gold, and zinc that rose to the surface. When skimmed off and heated the zinc vaporized leaving behind free silver and free gold. No evidence indicates that the Mayans or any of the Native American or Mesoamerican societies carried out alchemical metallurgical experiments to convert base metals to gold and to discover the philosopher's stone.

The Mayans used a type of volcanic glass (obsidian) for their weapons and constructed roads, bridges, and massive pyramids using stone as their chief building material. Their pyramids, and those of other Mesoamerican societies, date from 200 BC to the 1500s. They were most likely observatories with temples at the top, and unlike the Egyptian pyramids constructed from 2800 BC to 1700 BC, seldom served as tombs. Altun Ha's Temple of the Masonry Altars, a step pyramid, measured 27 meters high. The 45-meter high chalky limestone pyramid at Calakmul (Structure 2) was one of the largest Mayan pyramids constructed during the classic period. Temple-pyramid IV, one of Tikal's six pyramids, completed in 720, measures 72 meters high.

Mesoamerican societies from 1600 BC learned how to process rubber from latex, the sap of *Castilla elastica* and other rubber-producing trees, to produce durable elastic rubber balls for use in a game called chaah. They improved the latex quality by adding juice from pulverized morning glory vines. The juice contains traces of sulfonyl chlorides and sulfonic acids. When added to the latex, chemical analysis using nuclear magnetic resonance spectroscopy and gas chromatography-mass spectroscopy showed that the improved elasticity resulted from purification of latex's polymer component. Purification also increased the number and strength of interchain interactions that the juice's acid components produced. A similar process occurs in the vulcanization of rubber.

Different versions of chaah, a combination of soccer, volleyball, and basketball, which dates from 1600 BC, evolved in Mesoamerica. In most versions two teams, each of six players, tried to hit a rubber ball through a stone ring attached to a wall. The players could use any part of their body except their hands or feet. The balls ranged in size from ping pong to basketball, and the teams competed on a stone court. The Great Ball Court at Chichén Itzá measured 545 feet in length and 225 feet in width. The Mayan version of chaah reenacted parts of a creation myth and sometimes ended in human sacrifice. The Aztecs demanded as tribute thousands of rubber balls from conquered inhabitants of the provinces they ruled.

***Aztec (Mexica or Tenochca) science and society***

The Aztec empire, a confederation of three city states, Tlacopan, Texcoco, and the capital Tenochtitlan (Mexico City), flourished from 1100 to 1520 in south and central Mexico. Tenochtitlan, which the Aztecs established in 1325, had about 300,000 inhabitants in 1512. It sat on a marshy island and had three causeways connecting it to the mainland. Wooden drawbridges controlled access to the city and to several surrounding small towns such as Chalpultepec, Tepeyac, and Itzapalapa. Tehuacan, about 150 miles from Tenochtitlan, dates from 1300. An early Aztec city, Teotihuacan, flourished between 450 and 650 AD and at that time had the largest population in Mesoamerica, 152,000-200,000.

The Aztecs lived in small single-story homes about 15 to 25 square meters that had two doors, walls of adobe or sticks, no windows, and a thatch roof. They constructed wooden bridges, canals, irrigation systems, and rafts of branches that they floated in Lake Texcoco and on which they grew plant foods.

Hernando Cortes (1485-1547), the Spanish conquistador, first invaded Mexico in 1519 and marched on Tenochtitlan where Montezuma II (1466-1520), one of the last and best known Aztec emperors who reigned from 1502 to 1520, welcomed him as a white god. Cortes seized Montezuma, and in the ensuing uprising the Spanish killed Montezuma but had to retreat from the city.

Cortes returned in January 1521, and after the deaths of two other briefly reigning emperors, the Aztec capital and empire that they ruled with their allies and included most of present-day Mexico, collapsed. The Aztecs fell victim to the Spanish and smallpox. An epidemic struck Tenochtitlan a few months earlier, and the disease and its consequences, hunger and dehydration, proved fatal.

The Aztecs had no written language but used pictographs, or glyphs (symbols), painted on deerskin or on paper made from bark to record their events. The National Library in Paris, the Bodlian Library at Oxford, the Vatican Library in Rome, and the Lorenzo Medici Library and National Central Library in Florence hold the twelve existing codices. The *Codex Borgia*, composed about 1490, named for the Italian cardinal Stefano Borgia (1731-1804), and now in the Vatican Library, consists of six illustrated deerskins that describe Aztec astronomy, cosmology, and divination.

Aztec science lagged behind Mayan science. Aztec cosmology described a layered universe consisting of 13 layers of heaven, 9 layers of the underworld, and Earth in the middle layer. The heavenly layers above Earth contained the Moon, stars, Sun, and Venus; followed by a constellation called the Fire Drill; the green heaven of winds and storms; the blue heaven of dust; thunder; three colored zones; and the highest region where Omeyocan, the creator of space, time, and the world, resided.

The Aztec used the same calendars as the Mayans, and like the Mayans they adopted a 20-base number system. They had a 365-day solar calendar and a 260-day sacred calendar and bundled their years into 52-year cycles. A calendar stone represented their five world creations. They counted with dots from 1 to 19 and employed other symbols for larger numbers. A flag represented 20, a feather represented 400, an incense pouch indicated 8,000.

The Aztecs, like the Mayans, were excellent artisans. They worked with gold, silver, and copper to produce fine pottery and mosaics but did not develop alloys such as brass (copper and zinc) and bronze (copper and tin) until the Late Aztec period, the time of the Spanish conquest. During this period alloying processes began to spread through the Aztec empire. The Aztecs constructed stone temples and pyramids. The Great Pyramid at Tenochtitlan reached 60 meters. The Great Pyramid of the Sun at Teotihuacan measured 63 meters high and had a square base 225 meters by 225 meters. They also constructed the Pyramid of the Moon and the Feathered Serpent Pyramid at Teotihuacan.

***Inca science and society***

The Inca civilization began as early as 3,000-6,000 BC in Peru, Ecuador, Chile, and Bolivia, but the Inca Empire lasted only about 100 years, from 1438 to its collapse in 1532, five years after Francisco Pizarro (1475-1541), the Spanish conquistador, landed in Peru. Inca cities included the capital city Cuzco (1100 AD), Tucune (1000 AD), and the stone fortress city of Machu Picchu (1460-70 AD). The Yale archeologist Hiram Bingham (1875-1956) in 1911 discovered the ruins of Machu Picchu, 8,000 feet above sea level in the Andes Mountains. He sent thousands of ceramics, textiles, and bones from Peru to Yale. A century later the Peruvian Justice Ministry has continued to work for their return.

The Incas had no written language and used glyphs (symbols) to depict events. No codices existed. They developed a base-l0 number system and counted and kept records with a quipu. The quipu, of which 550 have survived, consisted of cords of various colors and thicknesses with knots of different sizes and positions tied on the cords to represent numbers.

A simple lunar calendar indicated the seasons for agricultural purposes. It had a 328-day cycle based on the time it took for the Moon to return to its same position among the fixed stars, which was 27⅓ days (the sidereal month) multiplied by 12 months. The Incas invented a sundial called the *Hiwantana* and celebrated the feast of the Sun on the winter solstice (22 December). The star constellations enabled them to navigate the Andes Mountains.

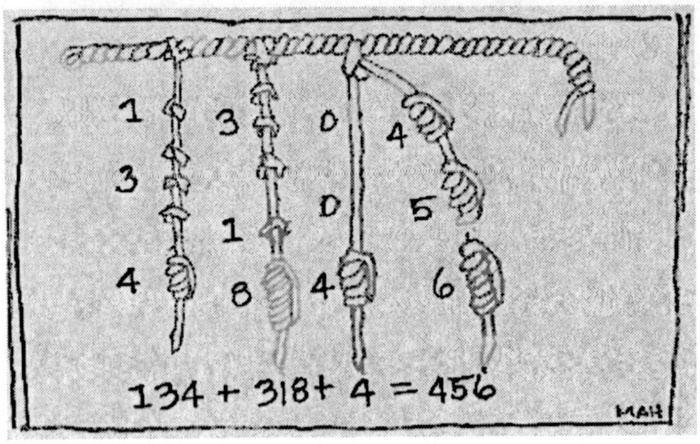


**Figure 30**
**Inca Quipu**
Courtesy of Michael Holmgreen

The Incas mined gold and silver and mastered the alloying of copper and tin to get bronze. Like the Mayans and Aztecs they never invented the wheel. Inca engineering skills enabled them to use stone worked into blocks that fit tightly without mortar to construct buildings, temples, and about 23,000 kilometers of stone roads for trade. They also designed and built irrigation systems, and bridges some of them rope suspension bridges. They planned the layout of their cities and built rectangular-shaped thatch-roofed stone houses. They constructed the 3-square mile city of Machu Picchu entirely of large tightly-fitting stones shaped by hand with heavy stone hammers and sanded to a smooth finish.

The Incas' Nazca lines that lie in a wide arid plain near the southern Peruvian coastal city of Nazca remain a mystery. The Incas from 900 BC to 630 AD constructed the shallow lines that run perfectly straight for 5 miles and cover a 35-mile length by scraping off the top stony rust-colored surface to expose the underlying whitish soil.

The lines form large figures of triangles, rectangles, a humming bird, a condor, animal forms including a monkey, spider, fishes, lizards, and a 900-foot heron and are visible in whole only from the air. The Peruvian archeologists, Toribio Mejia Xesspe (1896-1983) and Julio Tello (1880-1947) in Lima, in 1926 proposed that the Incas related the lines to irrigation systems and religious pilgrimages.

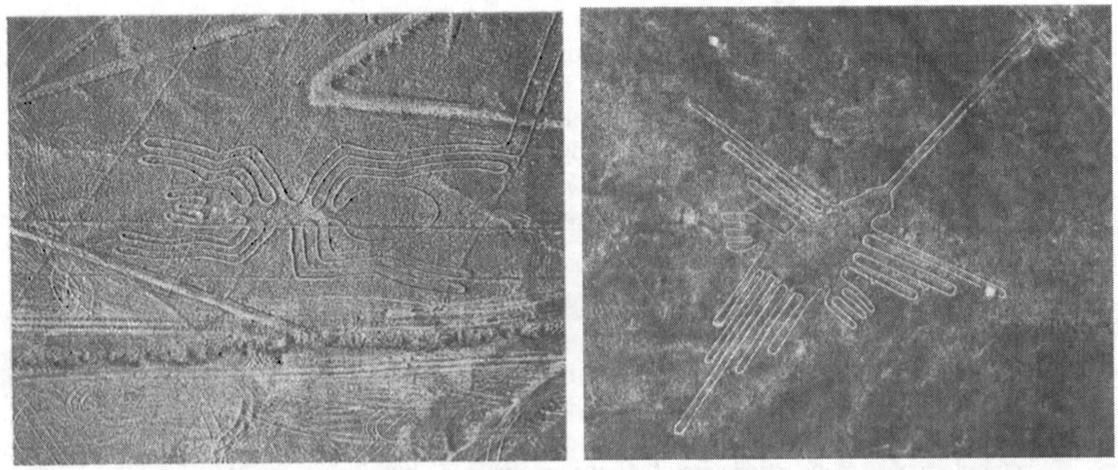

**Figure 31**
**Nazca Lines**
Jarno Gonzalez Zarraonandia/Shutterstock.com (left)
gary yim,/Shutterstock.com. (right)

Convincing evidence exists that as long as 7,000 years ago Indians in Chile practiced the art of mummification because of the respect they had for their ancestors and because they believed the dead remained part of their community. Their mummies predate the earliest known Egyptian mummy of 3500 BC. Archaeologists have found mummies in Peru but they date from 500 BC, a much later period. The Incas also believed in a heaven, a hell, and in the body's resurrection after death.

***Concluding thoughts on Mesoamerica***

When the Spanish arrived in the New World they brought with them horses, dogs, rats, firearms and gunpowder, wheeled vehicles, and smallpox, and introduced alphabetic writing to those societies that had no written language. The Spaniards, particularly Franciscan priests, learned the native languages, and many Mayans learned Spanish and functioned as translators. The best known of them, Malinche, a Mayan woman, accompanied Cortes in Mexico.

Little agreement exists on the population of central and northern Mesoamerica at the time of the Spanish arrival in 1519 with highs ranging from 10 million and lows of 4.5-5 million. In the ensuring 100-150 years, from 1600 to 1650, the population fell to its lowest point, 1.0-1.5 million to 3.4 million, the result of disease, particularly smallpox and measles, exploitation, environmental degradation, and to a lesser extent warfare.

| COMPARISON OF ANCIENT CIVILIZATIONS | | | | |
|---|---|---|---|---|
| ***Civilization*** | ***Science/technology*** | ***Government*** | ***Religion*** | ***Socioeconomics*** |
| Akkad | Metallurgy, walled cities, and irrigation established. | A strong central authority controlled several city-states and kept peace (for a while). Quelled dissent with violence. | Polytheistic tradition centered on city gods, fertility goddess, and harvest deities. Hepatoscopy (examining sheep livers for omens) practiced. | Thriving trade with other city-states led to rise of merchant and artisan classes. |
| China | Grain milling, irrigation, metalworking, wheel-thrown pottery, five-tone music scale calendar existed | Warlords and powerful nobles ruled battling fiefdoms and exacted tribute from peasant farmers. | Ancestors worshipped. Gods created and controlled harmony on Earth and conferred political legitimacy. | Peasant farmers subjugated by wealthy aristocracy. Intense trade networks existed. |
| Egypt | Carried out massive public works projects such as the Great Pyramids. Calendar predicted Nile floods. | Pharaohs had absolute power. Massive hereditary bureaucracy ran the country. | Had a pantheon of gods. Obsessed with life after death. Pharaohs were gods in their own right. | Peasants controlled by the bureaucracy. Economy more isolated than Akkad's, with trade largely within kingdom. |
| Mohonjo-daro (Indus Valley) | Extensive sewer systems constructed. Sea voyages taken. Multi-story dwellings built. | Powerful municipal government organized food storage, city labor forces, and public-works projects. | Possible proto-Hinduism with Shiva-like god depicted as king of the beasts and god of fertility. Cult of mother goddess. | Large middle class, extensive caravan trade throughout Indus Valley and Punjab, and multi-generational households existed. |
| Other | The plow widely used throughout Europe. | Complex Andean civilization flourished. | Proto-Greek religion developed on Minoan Crete. | Native American hunter-gatherers begin to develop settlements. |

## XI. CONCLUSION

We are what we know. In Chapter 1, Introduction to the History of Science, we defined science as a process that involved observation-description, explanation, and prediction. This chapter has made clear that science progressed from the Egyptian and Babylonian record keeping that began around 2000 BC to the theoretical-philosophical science of the Greek natural philosophers in 600-400 BC, and then to mathematical-quantitative science beginning in the 1600s.

For societies, such as Babylonians and ancient Egyptians, the Mayans, the ancient Chinese, and ancient Hindus, their science stalled at the record keeping descriptive stage and the progress of their societies, which was quite advanced for their time, also stalled. For the Mayans their society not only stalled but declined and later disappeared. The Greeks, however, went beyond the descriptive stage of science to the theoretical-philosophical stage and searched for explanations of what they observed. Their explanations resulted in the geocentric astronomical calculating models of the heavens that Hipparchus, Ptolemy, and others developed; the heliocentric model of Aristarchus; the atomic theory of Democritus, and the four (five) element theory of Plato and Aristotle, and the four humor theory of Hippocrates and Galen. Greek astronomers quantified their observations and calculated planetary distances, such as the Earth-Moon and Earth-Sun distance. Eratosthenes calculated Earth's circumference using an accurate value of pi that Archimedes determined. Archimedes laid the foundation for the science of statics and hydrostatics, introducing his famous principle.

The Greeks, beginning with Thales, not only sought explanations of their observations, they sought natural explanations or laws for the behavior of nature. They began to accept that natural laws governed the behavior of the universe and that they were capable of discovering these laws. Our minds had the capability of discovering and understanding these laws, which is the Western intellectual tradition. Once society moved to the second stage of science, natural philosophers expanded their search for models that had predictive value and offered the best understanding of what nature appeared to be.

## REVIEW QUESTIONS ON ANCIENT SCIENCE

1. What is the main theme of this chapter?

2. What are the stages through which science has progressed over the centuries?

3. What is the western intellectual tradition?

4. What appears to be the origin of writing?

5. What base number system did the Babylonians, Egyptians, Greeks, and Mayans use?

6. Give a few examples of the most important advances in ancient science and mathematics?

7. Discuss the evolution of the geocentric model of the universe (Anaximander, Aristotle, Hipparchus, Ptolemy, and others).

8. What motions did the Greek astronomers assign to the Sun and to the sphere of the fixed stars because of their belief in a stationary Earth?

9. What is parallax and what importance does it have?

10. Who invented the gnomon stick (sundial) and why was it important?

11. Who invented the astrolabe? Why was it important?

12. Identify Eratosthenes, Archimedes, Aristarchus.

13. What were Plato's, Aristotle's, and Democritus's beliefs about the origin of the universe?

14. What were the two Greek models of the structure of matter?

15. What is alchemy?

16. Discuss the medical beliefs of Hippocrates and Galen.

17. Discuss the ancients' beliefs including Hippocrates, Aristotle, and Galen on contraceptives, abortion, conception, and ensoulment.

18. Discuss the biological beliefs of the Greek natural philosophers.

19. Discuss the organization of Greek science and technology.

20. What were some of the main differences between western and eastern science?

21. What were the main contributions of the Hindus to science and mathematics?

22. What were the main scientific and mathematical achievements of the Mayans?

# CHAPTER 3
# SCIENCE AND ITS APPLICATION IN THE MIDDLE AGES 400-1400

We know accurately only when we know little;
with knowledge doubt increases.

Johann von Wolfgang Goethe (1749-1832)

## I. INTRODUCTION

Europe's Middle Ages, or medieval years, were an age of faith and characterized to some extent by a lack of scientific incentives. Medieval Europe only gradually underwent a revival of interest in the philosophical and scientific learning of antiquity.

The fall of Rome in 476 AD led to a 500-year Dark Age in the Roman Western Empire, during which time the works of the Greek scholars, Aristotle, Ptolemy, Archimedes, and others were largely unavailable, either lost permanently or temporarily in forgotten locations inside monasteries, courts, or other places where scholars once assembled.

In the Roman Eastern Empire (Byzantium) the primary interests were political and theological, and even when the works of the ancient scholars were available they evoked little interest in natural philosophy.

Two factors contributed to the revival of learning that occurred by 1000-1100. This rise of Islam in the mid-600s led to the compilation of vast encyclopedias in which Muslim scholars collected, translated into Arabic, and added their own commentaries to the writings of Aristotle, Ptolemy, and other Greek scholars. Beginning around 1000, universities emerged in Western Europe, and during the Crusades of 1096-1291 the Arabic translations began to find their way to the West where scholars, such as Gerard of Cremona, translated the Arab writings into Latin.

Compared to the scientific advancements that occurred in ancient Greece, scientific advancement in the Middle Ages lagged significantly. Astronomy and medicine advanced little and remained Ptolemaic and Hippocratic-Galenist. Nevertheless scientific advancement did occur. Alhazen's contributions to optics were of first rank, and the impetus theory of motion offered a challenge to the widely held Aristotelian theory of projectile motion. Gunpowder and the magnetic compass arrived in Europe about 1250. Peter Peregrinus' studies on magnetism expanded understanding of magnetic behavior.

New technology mainly in the form of labor saving machines and mechanical devices emerged in the medieval period, especially after 1000. In every case the *technology led the science*. Artisans and craftsmen improved or invented new mechanical devices but did not know or understand the scientific principles that governed their mechanical devices. The improved or newly invented devices contributed to advances in agriculture, ship construction and navigation, iron making, iron casting, textiles, and printing. While water transportation improved, the quality and quantity of road construction declined with the demise of the Western Roman Empire in 476 and the Byzantine Eastern Empire in 1453.

Feudalism which flourished from 900 to 1200 kept regions of Europe, particularly France, England, and Germany, at a low standard of living. With the rise of trade and economically powerful towns, feudalism began to die out in Western Europe by 1400. The Black Death, or plague, of 1347-61 also had a devastating effect on Europe, wiping out one-third of Europe's 50 million inhabitants.

| CHRONOLOGY OF EVENTS ANCIENT AND MEDIEVAL | |
|---|---|
| ***Date*** | ***Event*** |
| 753 BC-79 AD | Romulus and Remus, descendants of the Trojan prince Aeneas who earlier founded Lavinium the parent city of Rome, founded Rome in 753 BC. The Etruscans controlled Rome until the establishment of the Roman Republic in 500 BC.<br><br>The Roman Republic, which consuls ruled, lasted from 500 BC to 31 BC. By 146 BC Rome dominated Greece.<br><br>Octavian, known as Augustus Caesar (63 BC-14 AD), established the Roman Empire in 29 BC. He ruled as the first Roman emperor from 29 BC to 14 AD. By 117 AD the empire was at its greatest and occupied all of southern Europe, north Africa, and much of the Middle East.<br><br>Pliny the Elder (23-79), the Roman encyclopedist, compiled an encyclopedic but uncritical 37-volume *Naturalis Historia.* His information came from a wide range of sources and included books on astronomy, man, animals, birds, trees, medicine, metals, painting, and other topics. Pliny perished while trying to rescue those caught in the eruption of Vesuvius that destroyed Pompeii and Herculaneum. |
| 79-476 AD | Constantine (280-337 AD) permitted Christianity in the Roman Empire in 311 AD, and it became the official religion of Rome in 324. He called the first Ecumenical Council in Nicaea, in northwest Turkey, in 325 to settle doctrinal disputes. This council established the date of Easter and dealt with the problem of the precession of the equinoxes.<br><br>Constantine in 330 moved the capital of the Roman Empire from Rome whose population had fallen to 30,000-100,000 to Constantinople, which by the 500s had a population of 600,000-800,000.<br><br>The first recorded celebration of Christmas took place in Rome in 336.<br><br>The division of the Roman Empire into a Western and Eastern (Byzantine) Empire occurred in 395 after the death of Theodosius I (346-95). Theodosius's son Honorius (384-423) ruled the West, his son Arcadius (377/378-408) ruled the East.<br><br>Rome fell in 476. The Germanic chief Odoacer (435-93) in 476 overthrew Romulus Augustus (b. 461) the last Roman emperor who ruled in 475-76. A combination of internal and external factors, such as the high cost of maintaining the empire, corruption, and the influx of barbarians, namely the Goths, Visigoths, and Vandals, to the empire, led to the downfall. The Christian Catholic Church and barbarian kingdoms became heirs of the Western Empire. The Eastern or Byzantine Empire existed until 1453. |

| CHRONOLOGY OF EVENTS ANCIENT AND MEDIEVAL | |
|---|---|
| ***Date*** | ***Event*** |
| 529-1453 | The Byzantine emperor Justinian I (483-565) closed Plato's Academy in Athens in 529.<br><br>The Dark Age in Europe lasted from about 500 to 1000-1100. A decline of town life, culture, and learning; the emergence of a rural, subsistence economy, and of feudalism after 732, characterized this age.<br><br>Charlemagne (742-814), king of France in Aachen, founded the Holy Roman Empire in 800. A cultural renaissance spread through central Europe, Germany, France, Austria, and Belgium. The Holy Roman Empire existed in name until 1806 when Napoleon I Bonaparte (1769-1821) dissolved it.<br><br>Ghana in West Africa was a center for traffic in gold and slaves during the period 650-1100.<br><br>Global population in 1000 was 250 million and unchanged for 1,000 years. One-fourth of the population lived in China. Cordova, Spain, was the world's largest city with a population of 450,000.<br><br>Powerful monarchies arose in England and France during the period 1100-1450.<br><br>City states arose in Italy by the late 1300s.<br><br>French clerics sympathetic to the English claimed Joan of Arc (1412-31), French heroine of the Hundred Years War (1337-1453) between the French and English, was a heretic. They executed her, burning her at the stake.<br><br>The rise and decline of Islam occurred in the period 630s-1100s. The Crusades began in 1096. Pope Urban II (1042-99), who reigned 1088-99, began the campaign to retake the Holy Land from Islam. Two centuries of intermittent religious wars ended in 1291 with the Muslim recapture of Acre, in present-day northwest Israel. Increased east-west trade resulted.<br><br>Constantinople and the Byzantine Empire fell to the Islamic Ottoman Turks in 1453. |

| EMERGENCE OF CHRISTIANITY | |
|---|---|
| ***Date*** | ***Event*** |
| 4 BC-1000 AD | Jesus Christ, the founder of Christianity, lived from 4 BC to 30 AD in the Palestine region of the Middle East.<br><br>Writing of the four gospels of Matthew, Mark, Luke, and John took place from 75 to 125 AD. Formation of the New Testament occurred in 125-30 AD.<br><br>Linus (67-79 AD) was the first Roman pope elected, although Tertullian (160-225), the Christian Latin writer, in 199 wrote that Peter (d. 67 AD) ordained Clement I (d. 99 AD) as his successor. |
| 1000-1400 | Roman (Western) church and Eastern (Orthodox) church split over the authority of the Pope in 1054. The four patriarchs of the Eastern church (Constantinople, Alexandria, Antioch, and Jerusalem) became independent of Rome.<br><br>Dominic (1170-1221) in Spain established the Dominican Order of priests in 1205. Francis of Assisi (1181-1226) established the Franciscan Order of priests in 1209.<br><br>Rome established the Inquisition to combat heresy in 1231-42. The Spanish Inquisition lasted from 1483 to 1498. At its height it executed approximately 2,000 persons deemed heretics. |
| 1500-1600 | The Protestant Reformation occurred in 1517-21. Martin Luther (1483-1546), an Augustinian priest in Wittenberg, announced his 95 theses in 1517, challenging errors in the preaching of indulgences. The Catholic Church excommunicated him. This initiated Christianity's second major division.<br><br>The Council of Trent met from 1545 to 1563. It was the Catholic response to the Reformation, responsible for liturgy reform and other church doctrines. |

## II. RISE OF THE UNIVERSITIES

The European universities, derived from the Latin word *universitas* meaning union, were a medieval invention. They usually contained faculties of law, theology, medicine, and the arts (philosophy). Medieval universities had a fixed curriculum, an organized faculty, and awarded degrees.

The first universities grew out of the monastery or cathedral schools that the religious orders of priests had organized in Paris at Notre Dame Cathedral, in Oxford, and in Cambridge, and out of guilds of students and teachers. They were overwhelmingly male institutions, although the University of Bologna in 1236 appointed Bettina Gozzadini to a law chair.

***The three kinds of universities: civic, ecclesiastical (church), and state***

In a *civic* university the students elected the rector and had the power to hire and fire their teachers. Most of the Italian universities, Bologna (1100), Padua (1222), Siena (1241), Rome (1313), Pisa (1343), Florence (1349), Pavia (1361), and Turin (1405) were civic universities.

An *ecclesiastical* university was a guild of teachers under a chancellor. At Paris, founded in 1150, Dominican priests taught 7,000 students. Oxford (1200), which had 2,000 students, and Cambridge

(1209), founded by scholars escaping Oxford after a fight with Oxford townsfolk, were ecclesiastical universities. Franciscan priests taught at Oxford. Franciscans and Dominicans taught at Cambridge.

Salamanca (1218) in Spain, Naples (1224) in Italy, and Heidelberg (1385) in Baden, Germany, were *state* universities founded by monarchs. Alfonso IX (1171-1230) of León founded Salamanca. Frederick II (1194-1250) of Swabia, founded Naples. The elector Rupert I (1352-1410) founded Heidelberg.

***Student life at the universities***

Male students began study at ages 12-15. Students were often rowdy and fought with landlords and merchants over rents and prices. By the thirteenth century students began to rent houses and share expenses instead of living in separate apartments. These *Houses of Scholars* became legally recognized corporate institutions and the colleges that made up the universities of Oxford, Cambridge, and Paris. They included Blackfriars (1221), Ballior (1263-68), Merton (1264), All Souls (1438), and Jesus (1591) at Oxford. At Cambridge the colleges included Peterhouse (1284), Pembrooke (1347), Cauis (1348), and Corpus Christi (1352).

The colleges making up the University of Paris included the Sorbonne (1257), College d' Harcourt (1280), Scoticum (Scots, 1325), and Lombard (1330s). These early colleges were the forerunners of the later European colleges. At Paris, the faculty of arts consisted of four nations. A nation, which included both teachers and students, was a group from a specific region or country who banded together for its safety. In 1249 the four nations were the French nation, English (renamed German) nation, Norman nation, and Picard nation.

***Curriculum, length of study, degrees awarded***

Students studied a two-part curriculum developed from ancient times and known as the trivium and quadrivium. The trivium and quadrivium made up the curriculum of the faculty of arts at the undergraduate level.

Martianus Capella (365-440), a Roman pagan writer in Carthage was an early and a foremost proponent of the seven liberal arts. His textbook *De Nuptiis Philologiae et Mercurii* (*On the Wedding of Philology and Mercury and of the Seven Liberal Arts)* 9 books (410 or 429), was particularly influential in promoting astronomy and music. Anicius Boethius (480-525) in Italy was also influential.

| MEDIEVAL UNIVERSITY CURRICULUM | |
|---|---|
| trivium | grammar, rhetoric, and logic |
| quadrivium | music, arithmetic, geometry, and astronomy |

The teaching of the curriculum was mainly in Latin, the language of the scholar. It was intense, consisting of reading and explaining texts, and required considerable memorization. Examinations were oral disputations. The quadrivium component of the curriculum used Gerard of Cremona's Latin translations of Arabic texts, which were themselves translations of the Greek and Syriac writings.

Mathematics improved in 1145 when Robert (1100s) of Chester, England, introduced Hindu-Arabic numbers including zero to Western Europe, replacing Roman numerals. He was the first to use the term *sine* of an angle (ratio of the side opposite the angle to the hypotenuse in a right angle triangle) to define a trigonometric property of an angle. Robert translated other Arabic works into Latin.
Universities awarded bachelor, master, and doctoral degrees. Professional schools of theology,

medicine, and law existed at Paris, Bologna, Padua, and other universities. Paris excelled in theology, Bologna in cannon and civil law, and Salerno, which dates from the ninth century, in medicine.

| UNIVERSITY DEGREES | | |
|---|---|---|
| ***Degree*** | ***Length of Study*** | ***Course of Study*** |
| Bachelor of Arts (BA) | 4 years of study of the liberal arts | trivium, quadrivium |
| Master of Arts (MA) | 3-4 years of study after the BA | mathematics, science, philosophy, knowledge of Euclid and Aristotle |
| Professional<br><br>Doctor of Medicine (MD)<br>Bachelor of Laws (LLB) | 3-4 years of study after the BA | <br><br>physiology, anatomy<br>cannon law, civil law |
| Doctor of Philosophy (PhD) | 8 years specialized learning after the MA | philosophy and theology |

Plato and Aristotle were the medieval scholars' supreme authorities. Their medieval successors were Augustine and Thomas Aquinas.

Augustine (354-430), who spent his early years in Carthage, North Africa, converted to Christianity in 387 and later became the Bishop of Hippo (Annaba), Algeria. His *City of God* (ca. 413-26), a biblical approach to history and a Christian philosophy of history, was a major work. He was a Platonist and wanted to put scientific knowledge in the service of the Catholic Church.

Thomas Aquinas (1225-74), an Italian Dominican priest at the University of Paris and later in Rome, wrote *Summa Theologia* (1266-74). It became the foundation of Christianity. Aquinas was a rationalist in his scientific outlook.

Augustine and Aquinas made the teachings of Plato and Aristotle compatible or acceptable to Christianity. Aquinas removed any pagan association with science and philosophy, such as Aristotle's eternal universe and a soul that was not personal and eternal, and Plato's teaching that God created souls for all eternity and that the soul existed before union with the body. They both believed that the Bible was not a book of science.

## III. MEDIEVAL SCIENCE AND MEDICINE

During Europe's Dark Age the center of learning in the east moved from Athens and from Alexandria, after the closing of Plato's Academy in 529 and the Alexandria Museum's burning in 642, to Jundishapur. Jundishapur, a city of 100,000 population in Persia's (Iran) southwestern province of Ahwaz, remained the center of learning from 529 to 750. The center of learning then moved west to Baghdad, an even larger city of 250,000 population founded in 762 on the Tigris River in central Iraq, and then back to Europe through southern Italy and Spain at the time of the Crusades (1096-1291). Science was essentially Aristotelian.

In the Byzantine Empire the emphasis was on politics and theology. Byzantines had little interest in science. They associated Greek science with paganism. This was the reason the Byzantine emperor Justinian I (483-565) closed Plato's Academy in Athens in 529.

***Islamic or Muslim science***

The Muslim Arab Empire, which began with the rise of Mohammed (570-632), a merchant in Mecca and founder of Islam, extended from Spain to India.

The Muslims preserved the science of Aristotle, Euclid, Galen, Ptolemy, of Hero who wrote *Mechanics* in 60 AD, and of other Greek natural philosophers. They added their own commentaries to the writings of the Greek natural philosophers making it difficult to know the contents of the original work and what Muslim scholars added later.

They translated writings from the Greek, Syriac, and Aramaic languages into Arabic between 700 and 900. Aramaic was a Semitic language dating from the eighth century BC. It spread throughout Syria and Mesopotamia, was the official language of the Persian Empire, and became the everyday language of Palestine in Christ's time. Parts of the Old Testament are in Aramaic.

*Beliefs and contributions of Islamic science*

Islamic natural philosophers accepted as their starting point that Allah (God) is unknowable and that they must study his works (nature) to learn about Allah. All knowledge is science. The two sources of knowledge were the Koran, which was the source of revealed knowledge, and the study of nature, which gave rational or intellectual knowledge.

Muslim astronomers adopted a lunar calendar that remains in use today. It consists of alternating months of 29 and 30 days and a year having 354 days. The Muslim calendar dates from 622, the year of Mohammed's flight, or Hegira, from Mecca to Medina. The Christian year 2000 is the Muslim year 1420.

Al Farghani (after 861), a Persian astronomer in Baghdad and later Cairo, calculated the obliquity of the ecliptic as 23°35', the circumference of Earth as 20,400 miles, and Earth's diameter as about 6,500 miles. He estimated the size of the universe based on the conception of space-filling spheres just large enough to contain each planet's epicycles and deferent. He calculated Earth's radius as 3,250 miles and that the outer surface of the Moon's sphere was 64 1/6 Earth's radii from Earth's center. The outer surface of Mercury's sphere was 167 times Earth's radius, Venus's 1,020 times Earth's radius, the Sun's 1,220 times Earth's radius, Mars's 8,867 times Earth's radius, Jupiter's 14,405 times Earth's radius, and Saturn's 20,110 times Earth's radius from Earth's center. The sphere of the fixed stars was 75 million miles from Earth.

| PLANET OR HEAVENLY BODY | DISTANCE IN EARTH RADII FROM EARTH |
|---|---|
| Moon | 64 1/6 times Earth's radii |
| Mercury | 167 times Earth's radius |
| Venus | 1,020 times Earth's radius |
| Mars | 8,867 times Earth's radius |
| Jupiter | 14,405 times Earth's radius |
| Saturn | 20,110 times Earth's radius |
| Fixed Sphere of Stars | 75 million miles from Earth |

Muslim astronomers perfected the astrolabe by the ninth century and the instrument became the astronomer's basic tool for the next seven hundred years. In all probability they inherited it directly from the astronomical texts they were translating. Most often made of bronze, the astrolabe was a replica of the stars in the sky, which the astronomer moved to show the positions of the stars at any time in the year. On the reverse side of the circular instrument was a set of gear wheels that the astronomer used to place the Sun and Moon in any position. By setting the stars, Moon, or Sun, in the positions the astronomer saw them, he could then record the date. The astrolabe performed one other vital function. It indicated position, because if used in conjunction with star tables it told the observer where he had to be to see the position of a particular star on a selected date.

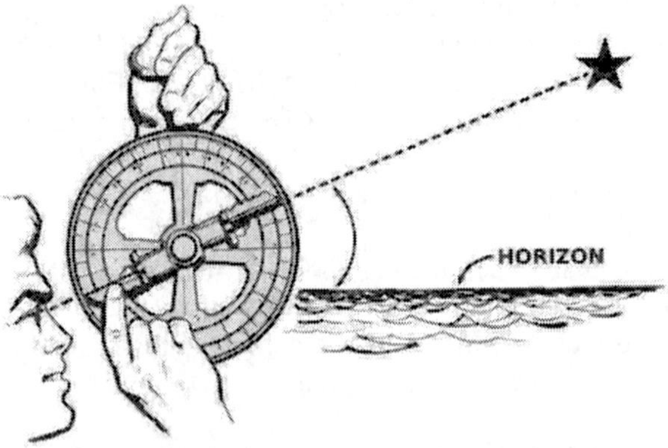


The astrolabe measured the altitude of the Sun, Moon, and stars. The astrolabe hung from a ring at the top. The pivoted limb has sights at each end. The observer moved the pivoted limb until the beam of sunlight entering the hole in the upper sight coincided with the hole in the lower sight. He then read from the graduated scale around the circumference, the angle at which the limb now pointed. This angle represented the Sun's altitude above the horizon. At night the observer made a sighting directly through the two holes.

**Figure 1**
**Astrolabe**
Released into the public domain by Pearson Scott Foresman

Rhazes (854-925) in Tehran, Persia, and Avicenna (980-1037) in Turkestan, central Asia, who was probably the greatest of the medieval Muslim scientists, prepared encyclopedias of science and medicine. Their medical theories were those of Hippocrates and Galen, although Muslim doctors made significant advances in surgery and in the structure and physiology of the eye.

Abu Abdallah Mohammed al-Idrisi (1100-65), a geographer in Sicily, studied climate. As a result of his studies he divided the world into seven climatic zones.

Muslims established libraries in Baghdad and Cairo; hospitals in Jundishapur (Gundeshapur, 705-15), Baghdad (805, 982), Cairo (872, 1284), Damascus (1154), Mecca, Medina, Divrigi (1156) and Beyazid II at Edirne (1471) in Turkey; and observatories in Samarkand, Istanbul (Constantinople), and Baghdad. El Azhar, founded in the 900s in Cairo, was and remains a famed Muslim university. Wealthy Muslims, particularly rulers, financed these institutions.

*Decline of Islam*

A religious dispute occurred after Mohammed's death in 632. It resulted in the emergence of the Shi'ite sect (Shia, followers) who rejected the first three caliphs (rulers) and followed Ali ibn Abi Talib (600-61), Mohammed's cousin and son-in-law and his chosen successor. Ali opponents assassinated Ali in 661, five years after he became the fourth caliph in 656. His descendants and followers who formed the Shi'ite sect and reside today mainly in Iran and Iraq, claimed succession as Mohammed's descendants.

The majority of Muslims at the time of Mohammed's death quickly selected Abu Bakr (578-634), Mohammed's father-in-law (father of his second wife), as the first caliph (632-34). They composed the much more numerous orthodox Sunni sect, meaning the Traditional Way (Sunna) of the Prophet Mohammed. Cultural stagnation and intellectual isolation began in Islam in the 1100s. The center of commerce and trade shifted west to the Americas after Columbus.

*Transmission of science from Islam to the west*

The Italian scholar Gerard of Cremona (1114-87) was the major translator of Arabic works into Latin. He worked in Toledo, Spain, which the Spaniards had reconquered from the Muslims in 1095 and where many Muslim scholars remained. Gerard translated 71 works from Arabic to Latin including Ptolemy's *Almagest* in 1175, which the Muslims had translated from Greek into Arabic in 823; Euclid's *Elements*; and works of Galen, Rhazes, Avicenna, Archimedes, and Hippocrates.

***Chemical science***

*Composition of matter*

Medieval natural philosophers accepted both the Greeks' four elements, air, earth, fire, and water, and the chemical principles of the Muslims Geber, or Jābir Ibyn Hayyan (721-815), in Kufa, Arabia, and Rhazes in Tehran. Geber, in *The Sum of Perfection*, spoke of sulfur and mercury as the two principles of matter. When combined in ideal proportions they produced gold. Rhazes believed that all metals derived from mercury and sulfur and in his *Book of Secrets* added a third principle of salt. Natural philosophers held that the Muslims' three principles of sulfur (flammability), salt (fixity), and mercury found mainly as the sulfide HgS (volatility, fluidity), described the composition and properties of matter.

Alchemy continued to flourish in the Middle Ages. Alchemists developed new apparatus, prepared new substances, and distilled alcohol, vitriol ($H_2SO_4$), and aqua fortis ($HNO_3$). Most European alchemists and Muslim alchemists, such as Geber and Rhazes, accepted transmutation, whereas Avicenna, questioned transmutation.

Muslim alchemists believed they could determine the proportion of elements in minerals by numerology and through laboratory procedures bring them into superior balance, even converting them into gold. Geber placed the elements and all things inanimate and animate in a dynamic, zodiacal framework. He wanted to probe the *souls* of materials and save those of human beings.

The alchemist Raymond Lull (1232-1315) was a Christian mystic in Mallorca, a group of Atlantic Ocean islands belonging to Spain. He had an extraordinary zeal for the conversion of Muslims to Christianity. Unfortunately, while in Tunis, North Africa, where Lull spent considerable time on conversions, Saracens (Arab and Turkish Muslims) stoned him to death.

Lull constructed a universe of mercury in which the angels, heavens, and terrestrial objects consisted of different forms of mercury. The quintessence (mercury) permeated the entire universe and represented the universal manifestation of God. Lull's pantheistic interpretation of God was incompatible with Christian teaching. In his alchemical investigations, Lull used alcohol to extract the essence of plant substances for use in medicines. This initiated the application of alchemy to medicine, or iatrochemistry.

European universities did not permit the teaching of alchemy because it was mystical and a mechanical art. Pope John XXII (1249-1334) in a bull issued in 1317 banned all alchemical practice.

**Figure 2**
**Alchemist's Laboratory**
Marzolino/Shutterstock.com (top)
Algol/Shutterstock.com (bottom)

*Gunpowder*

Gunpowder (black powder) was a medieval invention. Wu Ching Tsung Yao in China invented gunpowder in 1044. It consisted of 10 percent sulfur (S), 15 percent charcoal (C), and 75 percent potassium nitrate ($KNO_3$) or saltpeter. Gunpowder producers obtained sulfur from its natural deposits found in mines, nitrates from natural deposits or from animal wastes (manure), and charcoal from the incomplete combustion of wood. The Chinese initially used gunpowder in fireworks and in primitive rockets that they launched to frighten their enemies.

Gunpowder produced an explosion because when sparked or impacted potassium nitrate breaks down, releasing oxygen gas which combines with carbon (charcoal) forming carbon dioxide gas ($CO_2$) and nitrogen gas. Some nitrogen oxides ($NO_2$) also may result. The volume of the gases is 1,000 times, perhaps 3,000 times, as great as the original volume, producing the explosion.

Roger Bacon (1214-94), a Franciscan priest who studied under Robert Grosseteste (1168-1253) at Oxford, and independently Peter Peregrinus (b. 1220) in Paris, introduced gunpowder to Europe in 1250. European armies used gunpowder in firearms. Miners used it in blasting.

***Magnetism and the evolution of the compass***

Thales (640-546 BC) in Miletus discovered the magnetic properties of the naturally-occurring lodestone (magnetite, iron oxide, $Fe_3O_4$). He observed a magnet's *action at a distance* resulting from the attraction or repulsion of magnetic substances.

Shen Kua (1031-95) in Yangchow and Jingkou, China, in 1095 in *Dream Brook Essays* described the use of a needle shaped from lodestone as a mariner's compass that pointed north and indicated direction. A few years later, around 1100, Muslim sailors constructed a mariner's compass consisting of a lodestone needle stuck through a cork, straw, or wood splinter, and set floating in a basin of water. The mariner's compass guided English sailors in the 1100s and Italian sailors in the 1200s.

Peregrinus in *Epistola de Magnete* (ca. 1269) described the mariner's compass. He made it practical by placing the compass on a pivot, putting a graduated circular scale around it, and setting it in a box. Peregrinus gave an explanation of magnetic dip saying that it resulted because a magnet pointed to the celestial poles. He determined the law of attraction and repulsion for magnets and recognized that dividing a magnet into several pieces gave smaller magnets.

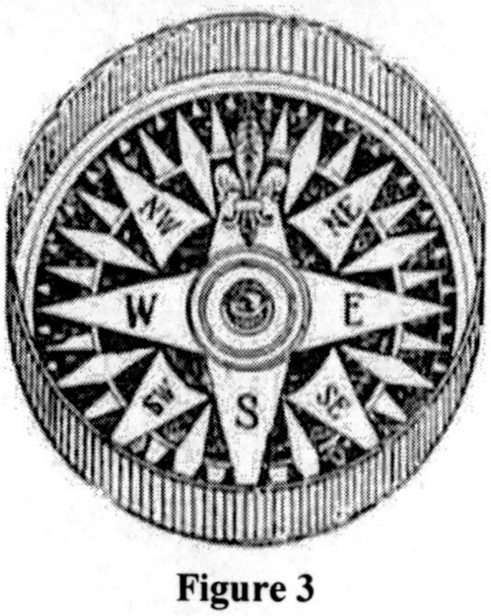


**Figure 3**
**Mariner's Compass**
Source: William H. Snyder. *Everyday Science*. New York: Allyn and Bacon, 1919.

### *Geographic science: maps and globes*

Early maps date from 2300 BC in Iraq. The first mapmakers oriented their maps to the east with Asia at the top because this was the direction in which the Sun rose. Later mapmakers divided maps according to 8 primary winds, 8 half primary winds, and 16 quarter winds.

Eratosthenes in the third century BC was the first to construct a world map using lines of latitude and longitude. Hipparchus in the second century BC first used lines of latitude and longitude to specify the locations of places.

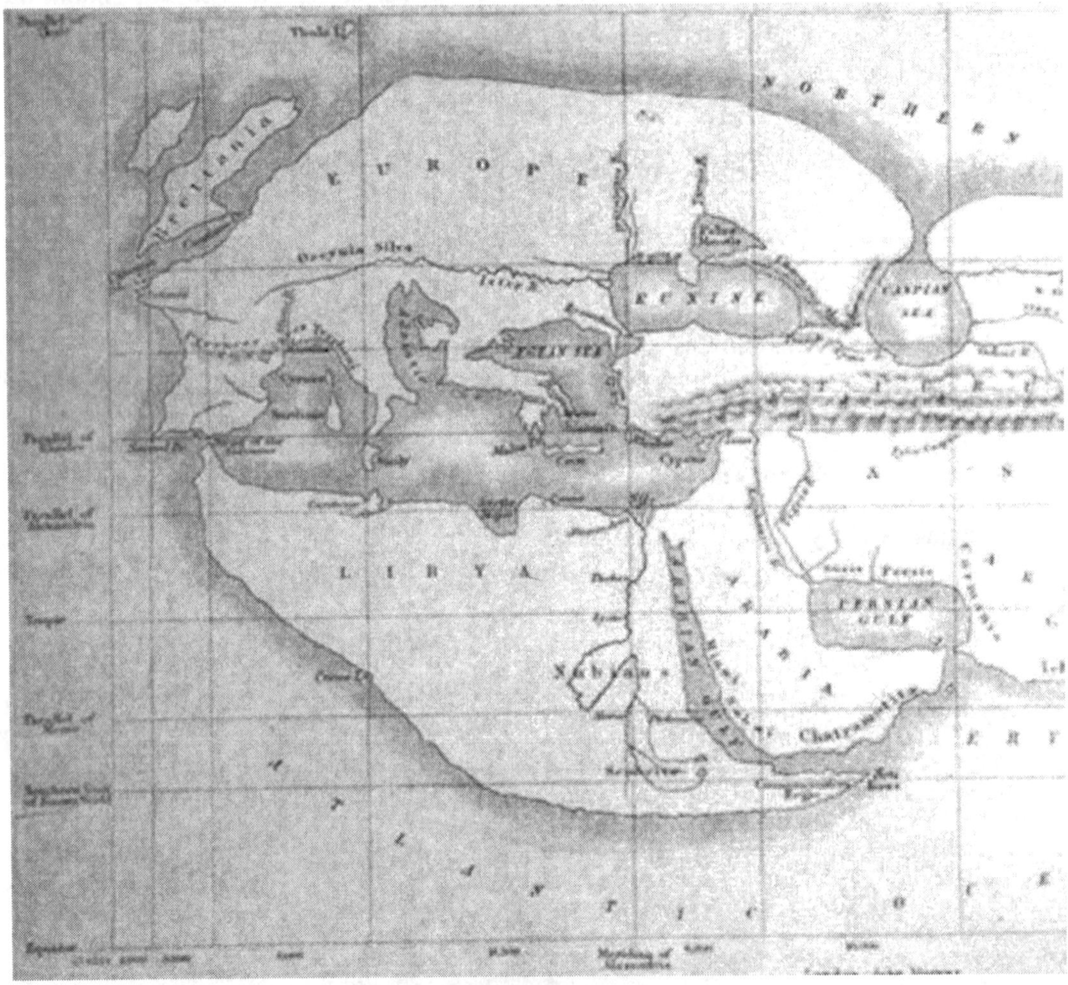


**Figure 4**
**Eratosthenes' Map of the Inhabited World ca. 194 BC**
Source: E. H. Bunbury. *A History of Ancient Geography among the Greeks and Romans from the Earliest Ages till the Fall of the Roman Empire.* London: John Murray, 1883.

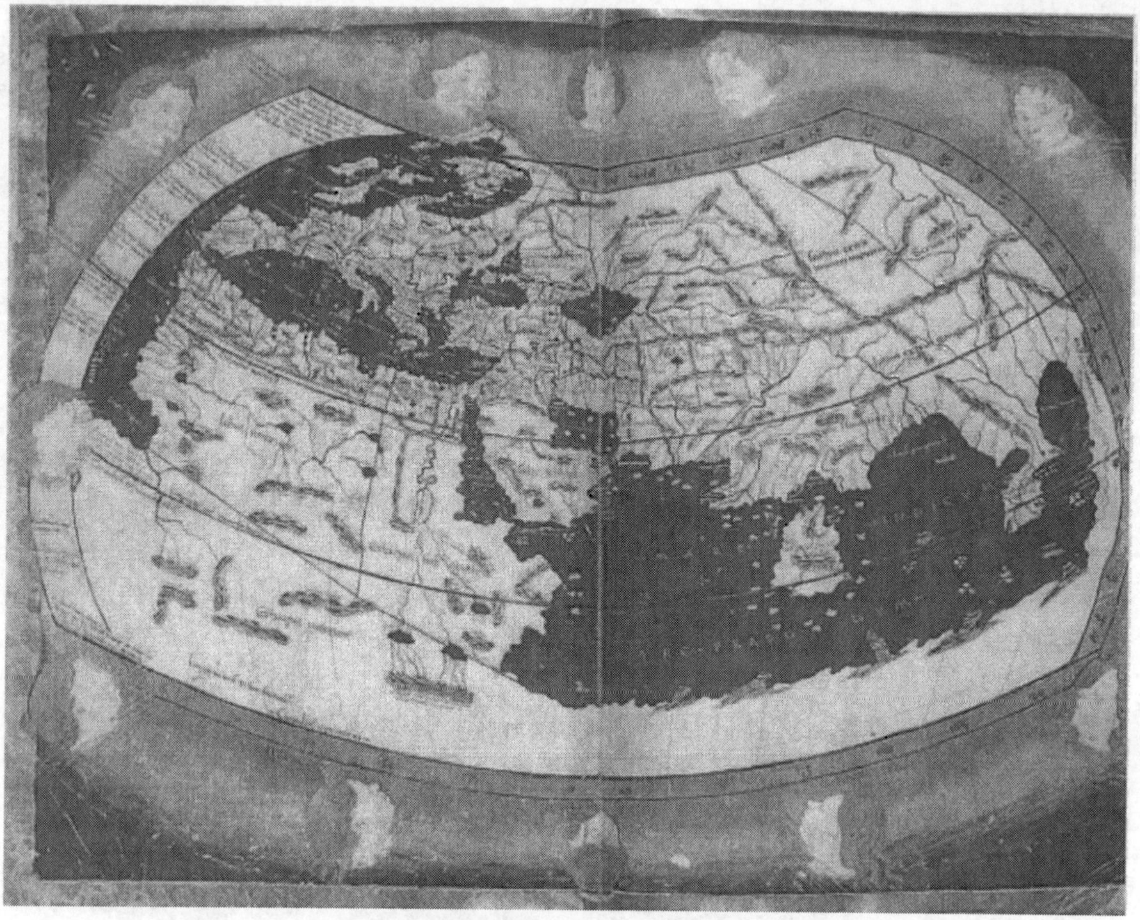

**Figure 5**
**Ptolemy's World Map**
Source: Claudius Ptolemaeus. *Cosmographia Claudii Ptolomaei Alexandrini*. trans. Jacob d'Angelo. National Library,Warsaw, Poland, 1467.

Ptolemy in his *Geographia,* written around 129 AD, adopted the north orientation of a map because the better known regions were in the northern latitudes. His maps had three continents: Europe, Asia, and Africa. The Christian sacking of Alexandria in 391 led to the destruction of Ptolemy's maps.

Ptolemy was the first to project lines of latitude and longitude from a spherical surface or a globe, onto a cone, making a conversion from spherical to planar coordinates. In Ptolemy's conical projection the parallels of latitude, the east-west lines, were concentric circles and the meridians, the north-south lines, were equally spaced radii. Ptolemy also developed two kinds of azimuthal (planar) projections, which resulted from projecting a spherical surface onto a plane. The first, or orthographic projection, gave the appearance of a globe viewed from space. Its projecting lines were perpendicular to the plane of projection. The second kind of azimuthal projection was the stereographic projection in which the meridians were straight lines and the parallels of latitude were arcs of circles. Johann Werner (1468-1522), a Catholic priest in Nüremberg, in his 1514 publication *Geographicae Cl Ptolomaei*, discussed Ptolemy's stereographic projection.

Medieval maps used a T-O design that Isidore of Seville (560-636), a Catholic bishop and scholar, in his very popular *Etymologiae* introduced to Europe. Isidore's *Etymologiae* consisted of 20 books written shortly before his death and printed in 10 editions from 1470 to 1530. It showed the land (terra, T) surrounded by oceans (O) and the three continents divided by a T representing major waterways, with Asia at the top, Europe at the lower left, and Africa at the lower right. Christianity had adopted maps with an eastern orientation, putting Asia at the top because the Middle East was the home of Christ.

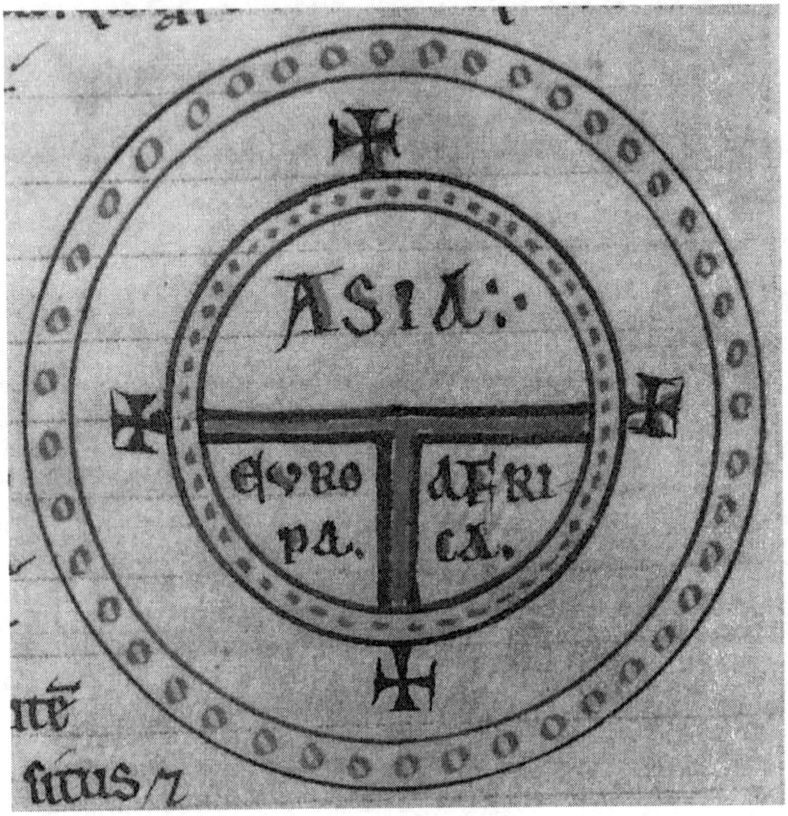

**Figure 6**
**T-O Map with East at Top**
Don River separates Europe from Asia
Nile River separates Africa from Asia
Mediterranean Sea separates Europe from Africa
Source: Isidore, Bishop of Seville. *Etymologiae*. several eds., 1470-1530.

Gerardus Mercator (1512-94), a geographer in Flanders and Germany, in 1569 devised a cylindric method of projecting the globe's curved surface onto a flat map and produced the most famous world maps. In making his projection Mercator placed a paper cylinder around a globe touching at the equator and drew the lines of latitude and longitude on the paper as if looking from the center of the globe toward the paper (map). His cylindrical projection had straight meridians and parallels of latitude that intersected at right angles. The scale is true at the equator, but became increasingly distorted (larger) in the Polar Regions.

Martin Behaim (1459-1507), a geographer in Nüremberg, in 1492 constructed the first known terrestrial globe. His globe of 20 inch (5.1 cm) diameter and the use of projection in map making provided clear evidence of the widespread acceptance of a spherical Earth in Columbus's time.

Behaim lived for a long time in Portugal where he learned about voyages of discovery. From 1491 to 1493 he once again lived in Nüremberg his home city for some time because of an inheritance issue. To show interested townspeople the importance of the new sea routes and countries and to encourage them to finance a new expedition westwards to China, he had local craftsmen construct a globe. His globe is the world's oldest existing terrestrial globe. Its cartography is a mixture of the ancient Ptolemaic maps, medieval European maps, and the so-called *portulan* charts, filled with the results of the most recent Portuguese discoveries. Behaim's globe has never left Nüremberg for any length of time in five centuries. Manufacturers in Nüremberg made globes until the nineteenth century.

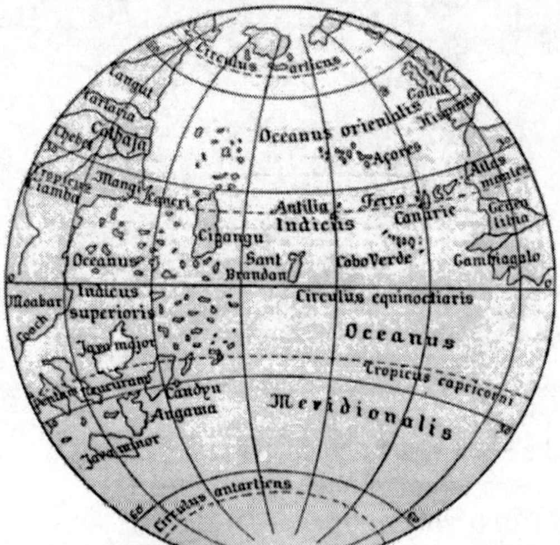

**Figure 7**
**Martin Behaim Globe 1492**
Source: *Encyclopedia Larousse illustree*, 1898 (left). Bocman1973/Shutterstock.com (right)

***Astronomy***

Ptolemy's geocentric model underwent no change in the Middle Ages and remained the accepted model of the universe. John of Hollywood, or Johannes de Sacrobosco (1195-1256) in England and France, published his geocentric model in *Tractatus de Sphaera* in 1220. He discussed the shape and central place of Earth, spheres in the sky, planets, and eclipses. His book was the first astronomy text printed in 1472 and continued in wide use until the 1600s. Sacrobosco used Arabic numbers in his arithmetical studies on addition, subtraction, and multiplication.

The Spanish monarch, Alfonso X (1221-84) in Toledo, sponsored the completion of revised planetary tables known as the Alfonsine Tables. The tables published in 1252, on the day he became king, represented the best of the Middle Ages and not replaced for three centuries.

Regiomontanus (1436-76), a German astronomer in Nüremberg and Rome and, probably the greatest astronomer of the fifteenth century, improved the Alfonsine Tables. He accepted the geocentric universe. In 1476 Regiomontamus completed the Latin translation of Ptolemy's *Almagest*. His teacher, the Austrian astronomer George Peuerbach (1423-61) in Vienna, had begun an abridged translation in 1460.

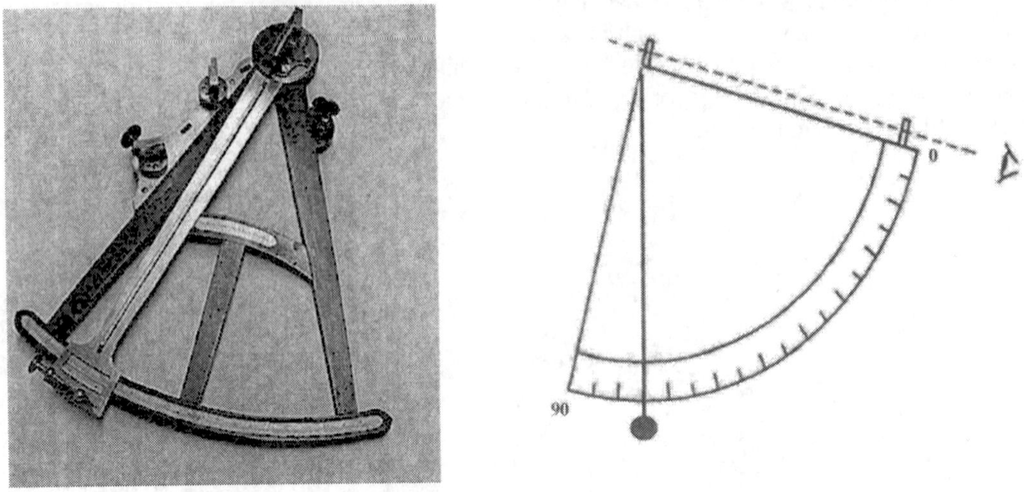


**Figure 8**
**Marine Quadrant**
The observer read the altitude from the point where thread cuts arc.
Source: National Park Service, US Department of Interior (left)
Courtesy of the author (right)

***Physics***

*Light*

Alhazen (965-1039), the foremost authority on light in the Middle Ages, lived in Iraq and in Cairo during the reign of the Egyptian caliph Al-Hakim (996-1020), and later in Spain. In 1008 Alhazen proposed constructing a dam at Aswan to store water from the Nile River for irrigation in the dry season and to control the river's annual flooding in the summer season. He obtained Al-Hakim's financial support for the project, but a visit to Aswan convinced Alhazen that he had underestimated the project's size. Fearing the wrath of the madman Al-Hakim, Alhazen spent the next 12 years, until Al-Hakim's death in 1020, feigning madness.

Alhazen's theory of vision held that an object reflects rays of light into the eye enabling an observer to see the object. Democritus and Aristotle held the same theory. Sight or vision did not occur because of rays that the eye emitted, as Ptolemy, Plato, and Euclid believed.

In his physiological studies on light, Alhazen described and named several parts of the eye including the retina, humor, and cornea. He explained that we saw because the lens of each eye focused light on the retina, a light-sensitive layer of tissue that lined the inner surface of the eye. It functions like the film in a camera.

Alhazen investigated mirrors, lenses, and image formation and constructed the first pinhole camera, an enclosed box with a tiny opening into which light enters and forms an inverted image on the opposite wall. Passing the light through a convex lens reversed the image.

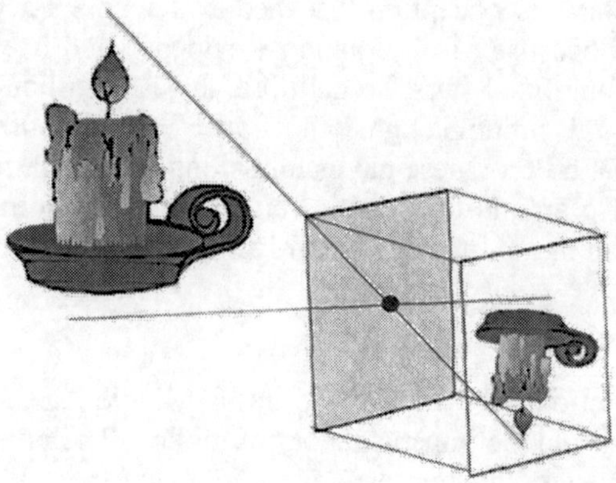

**Figure 9**
**Pinhole Camera**
Courtesy of the author

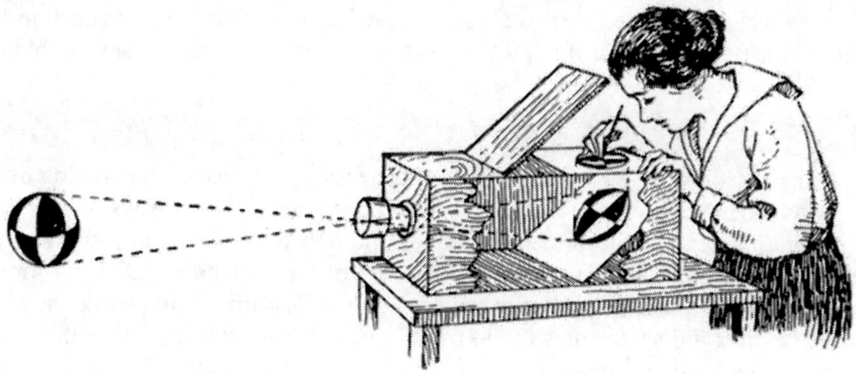

The pinhole reflex *carmera obscura introduced* in the seventeenth century, incorporated a mirror (a), which directed the image forward by the lens (b) onto translucent paper (c) supported on glass.

**Figure 10**
**Camera Obscura**
Source: Ellsworth D. Foster, ed. *The American Educator.* Vol. 2.
Chicago: Ralph Durham Company, 1921.

Alhazen gave the law of reflection, as Plato and others had done, and he showed that the incident ray and the reflected ray lay in the same plane. He explained the formation of a rainbow using refraction. The raindrops refract or bend the rays of sunlight as the rays pass from one medium to another. The raindrops break up or disperse the sunlight into its observed component colors, the colors making up white light.

Alhazen's *Optics* remained unpublished until the publication of a Latin translation 500 years later in 1572. It consisted of seven books of experimental and mathematical discussions on light and was the most complete publication on light to that time.

In Europe Roger Bacon emphasized mathematics and the experimental method in his investigations on light. Bacon studied mirrors, convex and concave lens. He wanted to put scientific knowledge in the service of the Catholic Church in the tradition of Augustine.

Contrary to the popular misconception that medieval scholars accepted a flat Earth, Bacon believed that Earth was round because of observational evidence and because a sphere had the most perfect shape. The sphere of the fixed stars, he believed, lay 130 million miles away. Bacon was the first to suggest that man could circumnavigate the Earth. About 300 years later, in 1519-21, Ferdinand Magellan (1480-1521), a Portuguese navigator sailing for Spain, proved Bacon correct when his crew completed the first trip around the world. Magellan never returned home. Natives in the Philippine Islands killed Magellan in a skirmish in April 1521.

*Motion*

Several theories of motion were widespread in the Middle Ages. The main debate occurred between the Aristotelian theory and the impetus theory of motion. The impetus theory, along with Geber's and Rhazes' three principles of matter, was one of the few new ideas introduced in medieval natural philosophy which remained essentially Aristotelian.

• *Natural motion*

Medievalists generally believed that the heavens had circular motion. For the atomists the random motion of atoms resulted in collisions that led to the formation of all terrestrial bodies and to the formation of the celestial bodies. Aristotelians believed that the four elements had up or down motion and sought their natural place. For falling bodies, a heavy body fell faster than a light body.

• *Projectile motion*

Aristotle claimed that a motive (external) force was necessary to sustain the motion of a projectile and that the air was the motive force. The source, or force, of the motion always had to remain in contact with the moving object. As the projectile moved through the air, air passed from the front of the projectile along the length of the projectile to the rear of the projectile pushing it forward. The air in front of the projectile offered resistance to the projectile's forward motion while the air behind pushed it forward. At no time did a vacuum exist in front or behind the projectile. Therefore projectile motion in a vacuum was impossible, a moot point because Aristotelians rejected the existence of a vacuum.

• *Impetus theory*

According to the impetus theory of motion that John of Philoponus (late 400s-500s AD) in Alexandria proposed in his *Physica* in 517 the motion of a body did not require continuous contact between the mover and the object it moved. Instead, a force applied to a body conferred an impetus upon the body and put the body in motion. The impetus gradually diminished because of air resistance, and the body eventually came to rest.

An impetus accounted for the motion of projectiles, motion in a vacuum, and motion of the heavenly bodies on which God had conferred an impetus. Because the heavens were perfect the impetus did not diminish or wear out, and the heavens therefore had eternal motion. Also no air was present to offer resistance to the motion of the heavenly bodies. The heavenly bodies did not have natural circular motion.

Few medieval scholars, other than its main supporters given below, accepted the impetus theory. William Ockham (1295-1349), a Franciscan priest at Oxford, did not believe that motion required continuous contact between the mover and the object moved. Magnetism demonstrated the reality of action at a distance in which the objects in motion were not in contact with each other. Motion in a vacuum did occur. Ockham was also the source of *Ockham's Razor*, which says that the simplest explanation of an observation, event, or development is likely the correct explanation.

Jean Buridan (1295-1358), a Catholic priest at the University of Paris, gave two arguments against the Aristotelian theory of projectile motion. A spinning top did not move forward as a projectile did, and a javelin with a wider flat end did not move faster than one with a pointed end.

Another Aristotelian critic, Nicholas Oresme (1320-82), a Catholic priest who studied with Buridan, resided in Navarre, France, in the 1350s and later became Bishop of Lisieux. Oresme gave the first graphical representation of velocities, in which he plotted velocity (y vertical axis) against distance (x horizontal axis) and introduced the idea of uniform and difform (non-uniform) motion.

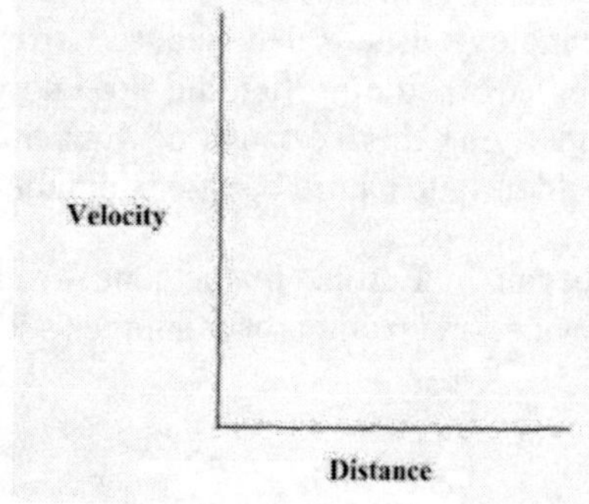


**Figure 11**
**Oresme's Graphical Representation of Motion**
Courtesy of the author

Oresme proposed that Earth rotates daily on its axis from west to east and that its rotation accounts for day and night. Perfection lay in repose (rest), and therefore the sphere of the fixed stars did not rotate daily. He discussed the relative motion of Earth and the fixed stars and said the universe might be infinite in size.

A final critic Nicholas of Cusa (1401-64), a bishop in Austria, in 1446 appointed a cardinal and in 1450 papal legate for Germany, wrote *Docta Ignoratia* in 1440. Cusa discussed the daily rotation of Earth on its axis. He believed that heaven and Earth (the universe) consisted of the same four elements and considered the possibility of life on other planets.

***Medicine***

Medical science passed unchanged from ancient times and was that of Galen. Avicenna's *Canon of Medicine* remained a standard medical text in Europe until the Renaissance. Mortality rates also remained high. A 50 percent mortality rate for children under one year of age prevailed.

Most physicians were men, but a few notable women, who were sometimes wives or daughters of physicians, practiced medicine in the Middle Ages. One of them, Heloise (1101-64), a well-educated nun in Paris, is better known for her earlier affair with the empiricist philosopher and theologian Peter Abelard (1079-1142) than for her later practice of medicine.

Hildegard (1098-1179), a German nun in Bingen, wrote *Causae et Curae* and *Physica.* She assimilated Galen's teachings and German folk remedies, including homeopathic principles and the doctrine of signatures, into her medical teachings. Hildegard also studied cosmology.

Trotula of Salerno (d. 1097) practiced Hippocratic-Galenic medicine. She dealt with childbirth and infertility. Sarah of Wurzburg (1400s) was a physician in Bavaria.

On the issue of abortion the Catholic Church which was the dominant European church, in its official position favored procreation. St. Jerome (347-420) condemned those who used abortifacients, while his contemporary Augustine believed Christians should practice passionless procreation. During the period 1000-1300 the Catholic Church opposed abortion after the fetus had formed, up to 8 weeks. Nevertheless, many writers, such as Peter of Spain (1210 or 1220-77), a physician and later Pope John XXI (or XX) from 1276 to 1277, related practical information on contraceptives and abortifacients such as artemisia.

## IV. MEDIEVAL ENGINEERING

In the medieval period technology usually led science. Artisans and mechanics knew how machines and other mechanical devices worked but often did not know why. They did not know or understand the scientific principles underlying the workings of a machine or mechanical device. They believed that the mechanical arts could imitate nature but not supplant or improve on nature.

An industrial revolution began in Europe in the tenth-twelfth centuries (900-1100) as a result of introducing labor saving devices and using power sources such as water, wind, and horse power.

### *Water and wind power*

The main uses of water and wind power were for grinding grains, pressing seeds, hammering rocks, and pumping water.

#### *Water wheels*

The Greeks as early as 100 BC used water wheels in their pageants and festivals but did not make widespread use of them as labor saving devices. Roman engineers such as Vitruvius Pollio (90 BC-25 BC) in Italy made greater application of water wheels, and for centuries their use continued in the Roman Empire. A typical water wheel had a horizontal axle that rotated like the axle of a cart and produced 3-5 horsepower.

Medieval European society and other contemporary societies used two kinds of water wheels, the overshot and undershot wheel. In the overshot wheel, water flowed down and over the wheel and caused the wheel to rotate. In the undershot wheel, water flowed under the wheel, causing it to rotate. Centuries later, in 1759, John Smeaton (1724-92), an English civil engineer in London, established the overshot wheel's two-to-three times (22 to 63 percent) higher efficiency compared to the undershot wheel.

The *Domesday Book* (1086) was a survey of England's lands that William the conqueror (1027-87) initiated and undertook from 1066 to 1085. It counted 5,624 water mills in England versus 100 water mills one hundred years earlier. The number of water mills in central Europe also increased significantly.

**Undershot Water wheel**

**Overshot Water wheel**

**Undershot Water wheel**

**Figure 12**
**Water wheels**
Source: Edwin J. Houston. *The Elements of Natural Philosophy.*
Philadelphia: Eldridge & Brother, 1900 (top).
Source: William H. Snyder. *Everyday Science.* New York: Allyn and Bacon, 1919 (bottom).

*Windmills*

Windmills date from the seventh century (600s). A typical windmill provided 10 horsepower. Two kinds of windmills, vertical axis and horizontal axis, were in use.

The eastern and Arab lands developed the vertical axis windmill. It had 8-12 sails in a circle rigged between two sets of horizontal spokes emerging from its vertically-turning axle.

The horizontal axis windmill, found in Europe, was similar to the waterwheel. It had a horizontally-turning axle and 8-12 sails.

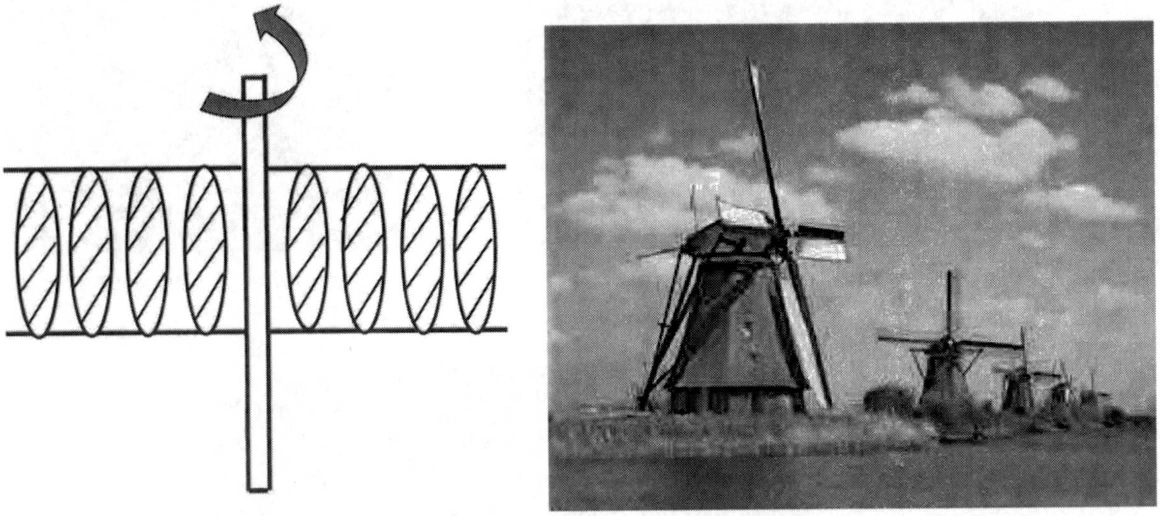

**Figure 13**
**Arabian Windmill Early 1300s (left), Western Windmill Horizontal Axis (right)**
Courtesy of the author (left), Fedor Selivanov/Shutterstock.com (right)

***Tenth-century revolution in agriculture: plows, harnesses, and horseshoes***

To slow down the depletion of crop lands, agriculturalists began the rotation of crops using a two or three field system. With the introduction of heavier plows, fields acquired a rectangular shape to reduce the number of turns a plow made.

The medieval plow consisted of three main components: a coulter or vertical knife, a plowshare or horizontal cutting blade, and a moldboard or curved iron plate to turn over the soil.

The plow increased in size and weight and required as many as eight oxen to pull it. The increased size led to communal sharing of oxen and plows and to the formation of the cooperative agricultural community, or manor, and to the formation of villages. The introduction of the shoulder (horizontal) padded harness for horses eliminated the choking caused by the misapplication of the older oxen's harness to horses. When pulling a load, the straps of the ox's harness pressed against the horse's windpipe, choking the horse. The introduction of wrought iron horseshoes protected the hoofs and provided the horse with better traction.

With these innovations a horse became 3-4 times as efficient as previously. Horses gradually replaced oxen despite its costing more to feed the horse grains, such as oats, barley, and corn, instead of the hay used for oxen.

**Ancient Horse Throat and Girth Harness**

**Medieval Horse Breast Strap Harness**

**Figure 14**
**Horse Harnesses**
Courtesy of the author

## V. MEDIEVAL LIVING CONDITIONS AND PRACTICAL SCIENCE

### *Feudalism and slavery*

Feudalism originated in medieval Europe in the eighth century (700s) and flourished from the tenth to the thirteenth centuries (900-1200), particularly in France, Germany, and England, but not in Italy. It was a social, political, and economic system, a hierarchy with the king at the top. His vassals (subjects, dependents) were the great nobles, whose vassals were the lords, whose vassals were the serfs. Each group pledged obedience and service to its master in return for protection, maintenance, and a tenancy of land called a fief. The growth of trade and the rise of economically powerful towns gradually caused the feudal system to die out by the fifteenth century (1400).

The slave trade flourished around 1000 with Slavs and Africans the major sources of slaves. About 10 percent of England's population were slaves. Many families sold their children into slavery because of economic hardship.

### *Homes in the village, town, or city*

Villages, towns, and cities were often noisy, poorly lighted, and dirty. Most people lived in hovels with dried mud walls called wattle, dirt floors, timber frames, and thatched roofs, although in London in 1212 tiles began to replace thatched and wooden roofs. Most hovels and homes lacked specific rooms and consequently lacked comfort, intimacy, and privacy.

The homes of the wealthier class had several stories: a cellar that contained casks of wine and ale, and possibly a well; a ground floor with a kitchen and workshops; a second floor where the inhabitants lived and slept; and an attic for storing clothes, dried food, and anything else. Furniture, such as the trestle table, was basic and limited, easy to take apart and transport. People stored their valuables in chests. Candles, torches, and oil lamps were the common lighting sources.

Homes had a hearth (fireplace) with a chimney by 1200. Some homes had glass windows, which the townspeople or artisans made by heating silicates (sand) with alkali (sodium or potassium hydroxide, NaOH, KOH), although most homes still relied on wooden shutters and oiled parchment to cover window openings. Medieval homes had no bathrooms. The homes were unsanitary and so were the people, despite the availability of soap by 1259. Medievalists manufactured soap by leaching (washing) plant ashes ($Na_2CO_3$) to give an alkaline product and mixing it with vegetable or animal fat or oil. Soap for bathing was a common household item in Islam, whereas Europeans considered bathing unhealthy.

Because privacy did not-exist in most medieval homes children shared many of the same experiences, both pleasant and unpleasant, as adults. As a result society increasingly regarded them as young adults, especially from the time of Elizabeth I to the early nineteenth century. This helps to explain why the nursery rhymes and stories that date from the late 1600s, such as *Sleeping Beauty* which the Frenchman Charles Perrault (1628-1703) wrote in 1697 and *Little Red Riding Hood* which Perrault wrote in 1697 and the German writers Jacob (1785-1863) and Carl (1786-1859) Grimm later modified, are tales that contain both the good and evil circumstances of life. Children of that time experienced good and evil in real life and writers commonly included both circumstances in children's literature.

**Figure 15**
**Medieval Home**
Panos Karas/Shutterstock.com (left)
3drenderings/Shutterstock.com (right)

***Food, drink, and etiquette***

Food consisted of bread, fish, grains (oats, wheat, and rye), vegetables, fruits, cheese, poultry, and meat which the townspeople preserved by smoking, salting, and pickling.

The spoon, knife, and the fork introduced in Italy in the 1400s, were the main eating utensils. Travelers usually carried their own knives. Louis XIV of France (1638-1715), banned pointed knives at dinners as a safety measure, which forced the introduction of blunt knives.

## EXAMPLES OF THE RULES OF ETIQUETTE IN THE MIDDLE AGES

***Thirteenth Century***

A number of people gnaw a bone and then put it back in the dish. This is a serious offense.

Refrain from falling upon the dish like a swine while eating, snorting disgustingly, and smacking the lips.

Do not spit over or on the table in the manner of hunters.

When you blow your nose or cough, turn round so that nothing falls on the table.

***Fourteenth Century***

A man who clears his throat when he eats, and one who blows his nose in the tablecloth, are both ill-bred, I assure you.

You should not poke your teeth with your knife, as some do, it is a bad habit.

I hear that some eat unwashed. May their fingers be palsied!

***Fifteenth Century***

Do not put back on your plate what has been in your mouth.

Do not chew anything you have to spit out again.

It is bad manners to dip food into the salt.

During these centuries, etiquette books offered much advice on the proper way of blowing one's nose. Tissues did not exist, and handkerchiefs had not yet come into common use. Society frowned upon the practice of blowing into a tablecloth or coat sleeve and accepted the practice of blowing into fingers. Painters and sculptors reproduced these gestures. Among the knights depicted on the tombstone of French King Philip the Bold (1245-85) at Dijon, France, one is blowing his nose into his coat, another into his fingers.

Erasmus (1466-1536) of Rotterdam, Holland, the scholar and humanist who wrote one of history's early etiquette books, provided some of the first recorded rules of behavior for the bathroom and bodily functions. He cautioned that "It is impolite to greet someone who is urinating or defecating," and on breaking wind, he advises the offender to "let a cough hide the explosive sound . . . . Follow the law; Replace farts with coughs."

Source: Charles Panati. *Extraordinary Origins of Everyday Things.*
New York: Harper & Row, Publishers, 1987.

With the gradual breakdown of the Roman aqueduct system a scarcity of good supplies of water arose. City water usually was unsafe to drink. The brewing of beer, a name derived from the Saxon world *baere* for barley and originally made from barley and yeast, underwent a major change with the introduction of hops in the tenth century. Hops kept beer from going bad quickly and gave beer its bitter taste. Hop-flavored beer replaced the existing fruit beers. By the mid-1600s great brewing centers had arisen in Oxford and Munich.

The Romans discovered that wine, produced from fermenting grapes, and kept in a container sealed tightly with a cork, did not go bad and improved with age. This helped winemaking spread west and north from the Mediterranean Middle East region to central Europe in 800-1200. By the early seventeenth century improved glassmaking led to stronger and cheaper bottles and allowed bottles to replace pottery or leather jugs as containers.

King James I (1566-1625) of England in the early 1600s ordered glassmakers to stop using wood to heat their furnaces in order to preserve England's dwindling forests. Their switch to coal produced a hotter fire and enabled Sir Kenelm Digby (1603-65), a natural philosopher in London, to produce for the first time thicker, stronger, darker, and cheaper wine bottles, and thereby invent the modern wine bottle. The monasteries, such as the Benedictines at Monte Cassino in Italy and Cluny in Burgundy (France) and the Cistercians in Burgundy and Germany, owned or created most of Europe's finest vineyards.

***Clocks: water and mechanical***

The earliest water clocks date from 1500-1600 BC and with sun dials and gnomon sticks represent the earliest time keeping devices. The Egyptians, Babylonians, Chinese, Greeks, Romans, and other ancient and medieval societies constructed simple as well as elaborate water clocks. All of the water clocks determined time by regulating the flow and measuring the volume of water passing from one container into another. Markings on the inside of the containers indicated the passing of the hours as the water level fell in a container (outflow water clock) or rose in a container (inflow water clock). The tomb of Amenhotep, dating from 1500 BC, contained a water clock. The Greek astronomer Andronicos of Cyrrhus (ca.100 BC) in the first century BC constructed a tower in Athens called the Tower of the Winds that had sundials and a water clock, which the Greeks called a clepsydra. In Kaifeng, China, Su Sung (1020-1101) in 1088 constructed an elaborate 40-foot water-powered mechanized clock tower. The main problem with a water clock was to control accurately the rate of water flow. This difficulty led to the search for better time keeping devices and to the construction of the first mechanical clocks that did not use water flow to measure time.

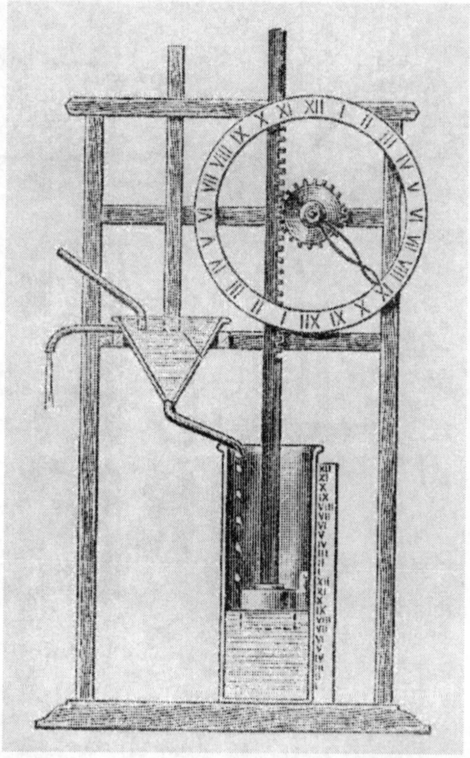

The water clock's reservoir (top left) had a constant supply of water and an overflow outlet to maintain a steady pressure. The reservoir dripped into a main tank (center) which had a baffle wall to stop ripples and a float carrying a pointer. As the water level rose in the tank, the pointer indicated the time passed on the circular scale showing the hours.

**Figure 16**
**Water Clock**
Nicku/Shutterstock.com

In the first mechanical clocks, which artisans constructed in the thirteenth and fourteenth centuries (late 1200s-1300s), a slowly falling weight replaced slowly falling water. The first mechanical clock, constructed in 1285, had an escapement with a verge and a foliot that rang a bell. The escapement controlled the escape of power from the spring or weight. These were the bell towers found in medieval towns and in monasteries where monks needed clocks to indicate their prayer times. The hour hand appeared on mechanical clocks for the first time in Italy: in 1335 in the city hall tower in Milan, in 1344 in Padua, and possibly earlier in Orvieto in 1307 and Ragusa in 1322. Mechanical clocks beginning in 1400 used a balance beam with a crown wheel in addition to the verge-and-foliot escapement to regulate their movement but did not keep accurate time.

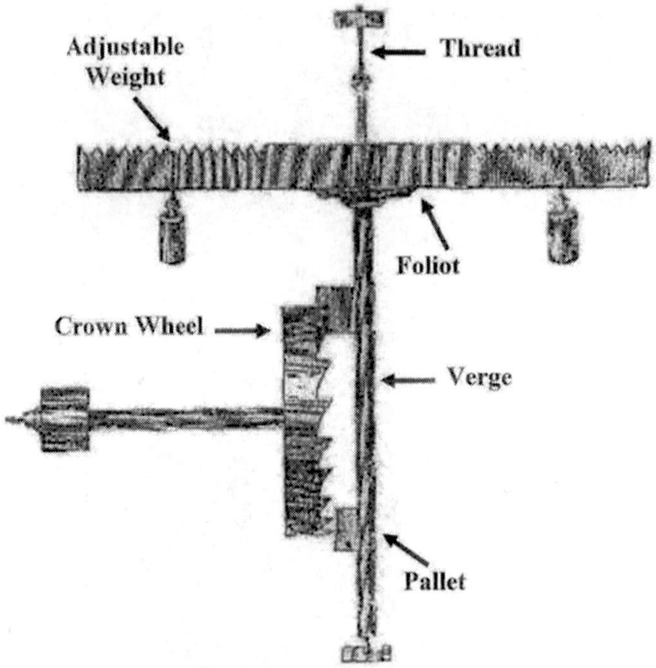


**Figure 17**
**Verge-and-Foliot Escapement**
Source: Pierre Dubois. *Historie de l' Horlogerie*, 1849.

The addition of the minute hand in 1577, which Jest Bürgi (1552-1632), a Swiss clock maker in Lichtensteig likely invented, improved the mechanical clock's practicality. Bürgi constructed his minute hand clock for Tycho Brahe (1546-1601), the Danish astronomer. Improved accuracy occurred only in 1656 when Christiaan Huygens (1629-95) in Holland designed the first weight-driven clock with a pendulum instead of a verge-and-foliot escapement or balance wheel. The pendulum clock made practical the minute hand and the second hand which clocks first added in the 1690s, after Robert Hooke (1635-1703) in London invented the anchor escapement in 1687, although 1657, 1664, 1666 are other suggested years for Hooke's invention. Huygens' 1656 clock had both a minute hand and a second hand.

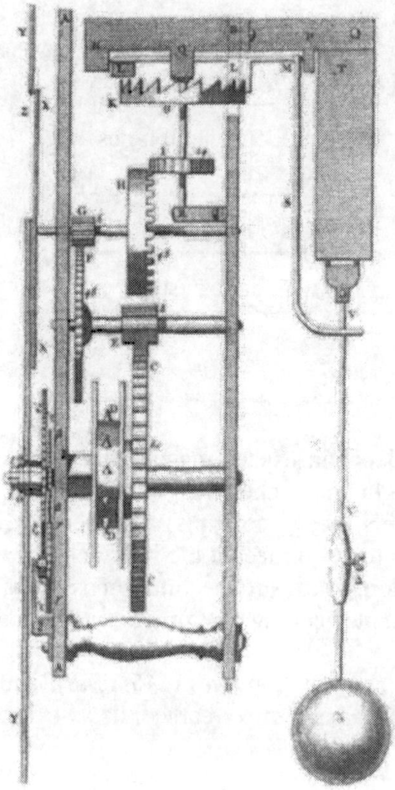

**Figure 18**
**Huygens' Verge Pendulum Clock**
Source: Christian Huygens. *Horologium Oscillatorium.* Paris, 1673.

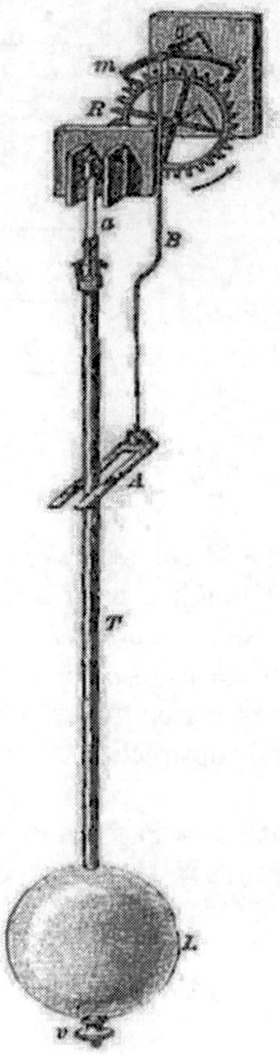

**Figure 19**
**Anchor Escapement**
Source: J. Dorman Steele. *Popular Physics.* New York: American Book Company, 1896.

***Black Death or plague 1347-61***

A complex combination of climate change, such as temperature change, and ecological factors, such as an imbalance in nature that altered a clean water supply or the atmosphere and allowed bacteria to thrive, usually triggered an outbreak of plagues and other epidemics.

The plague known as the Black Death gained its name because its victims' blood clotted in the capillaries and turned black. The European plague spread from an outbreak of the plague in China in 1331 into the Crimea, then into Sicily, and north into Europe. Ships traveling from one port to another provided the chief means of transporting the contagion.

Europeans believed the plague was God's punishment for sin, but bacteria spread by rats and fleas caused the plague. Fleas bit the rats infected with bacteria, and then bit other rats infecting those rats, thereby carrying the disease to other rats. The fleas also bit humans and infected humans who then infected others humans by spreading the bacteria with their breath.

Three strains of the plague devastated global population. It disrupted trade, resulting in a depressed economy and labor shortage. About 1/3 of Europe's 50 million inhabitants died. The huge loss of life led to a higher standard of living for the survivors and to the increased use of labor saving devices.

| BLACK DEATH | |
|---|---|
| ***Strain*** | ***Body Region Affected*** |
| Bubonic | penetrates the mucous membranes and skin |
| Pneumonic | attacks the lungs |
| Septacemic | attacks the lymphatic system |

***Cathedral building***

During the Middle Ages expert craftsmen and builders constructed magnificent stone cathedrals. The cathedral builders used arches to deflect the weight the arches supported down the sides of the cathedral to 25-foot thick masonry foundations. Piers (heavy columns) and pier buttresses (towers), which by means of a stone bridge called a flying buttress, deflected the horizontal forces of the vault (the arched masonry roof or ceiling) to the foundation. These architectural innovations resulted in the successful construction of cathedrals with soaring vaulted ceilings on top of narrow walls.

Notre Dame in Paris (1163-1313), Westminster Abbey in London (1245-1745), and St. Peter's in Cologne (1248-1880) represent Gothic (pointed arches) style cathedrals. St Peter's in Rome (1506-1626) is a Romanesque (round arches) style cathedral.

## V. LAND AND WATER TRANSPORTATION

***Roman and medieval road construction***

The Romans constructed 50,000 miles of roads, more than twice Earth's circumference, mainly in the eastern empire where most of the population lived. They built 29 military highways and 343 other highways and catalogued their road system in 200 AD.

The Roman road system varied according to the situation and the materials available. The Romans constructed different classes of roads depending on the road's importance. Their roads were public, military, local, or private.

In constructing their best roads, Roman road builders dug down trenches to a firm foundation and then added five layers, each 10-15 inches (25-40 cm) thick. The layers from bottom to top were fine dry earth, crushed rocks (gravel), concrete, crushed rocks, and stone slabs. The depth of the trenches often measured 3-4 feet.

The Romans constructed the first two-lane vias in 312 BC from Rome to Capua in southern Italy, north of Naples. The two-lane vias and countryside roads measured 12-20 feet (2.6-6 meters) wide to allow passage of four-wheeled carts whose wheels were about 4 feet 8 ½ inches apart. Bypass roads had an 8-foot width, the widest roads a 40-foot width. Pompeii's main roads measured 13 feet wide.

The extensive Roman road system broke down during the Middle Ages because individual communities could not afford its upkeep and because no central authority existed to maintain the roads. Medieval roads were inferior to the Roman roads. They usually had two layers: a loose foundation of sand with broken stone on top. Some roads had blocks cemented with mortar on a base of sand. This construction enabled two- and four-wheeled carts to travel on them.

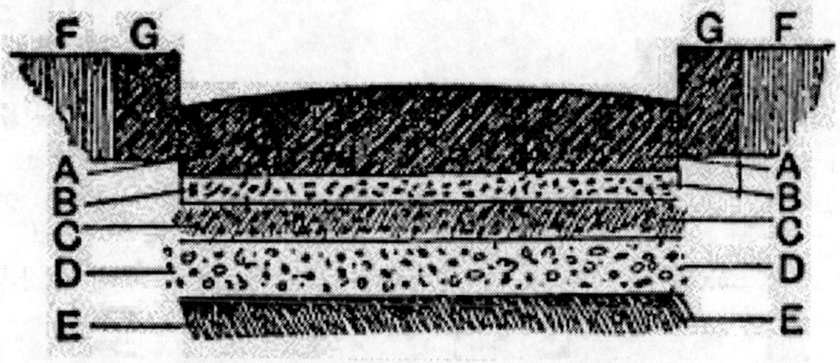


The Romans completed new roads and bridges before work began on the city itself. Surveyors marked out a road with stakes, workers dug a ditch on each side into which they set a row of curbstones. They then dug a deeper ditch between the two rows of curbstones and filled it with layers of stones of varying sizes. The top layer formed the pavement of the road and rose slightly in the center to force the rainwater into side ditches. They constructed the pavement of flat stones carefully fitted together and filled any spaces between the flat stones with smaller stones or pieces of scrap iron.

**Figure 20**
**Roman Road Construction**
Alfredo Cerra/Shutterstock.com (left)
docstockmedia/Shutterstock.com (right)
Source: Harold Whetstone Johnston. *The Private Life of the Romans*. Chicago: Scott, Foresman, and Company, 1909 (bottom).

***Water transportation: ship design, speed, and latitude-longitude determination***

*Ship design advanced in four ways during the Middle Ages*

Ships had the *lateen (triangular) sail* by the eighth century. It provided maneuverability and enabled a ship to sail into the wind. This happens because the slanted sail deflects the wind blowing against it and produces a force acting perpendicular to the sail. Resolution of the wind's force into its two components gives one useful component that points forward parallel to the ship's keel (the curved wooden piece extending along the entire length of the bottom of a ship) and one useless component that is perpendicular to the keel and tends to tip the boat. The ship's keel neutralizes the perpendicular component leaving only the parallel component and allows the ship to sail into the wind. Columbus's three ships, *Nina*, *Pinta*, and *Santa Maria* had lateen sails.

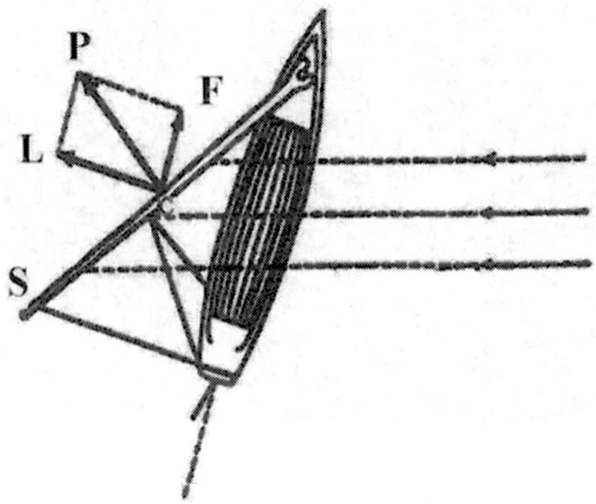


The diagram shows how a wind pushes a boat ahead instead of forcing it backward. The wind blowing against the slanting sail S is deflected and causes a force perpendicular to the surface. The force represented by the arrow CP in the diagram, is the force that we can resolve into two components, one useful, CF, which points forward parallel to the keel of the boat, and the other useless, CL, which tends to tip the boat and to move it leeward. A deep keel or a centerboard largely prevents the leeward or side ward movement. The net effect of the wind is to drive the boat forward.

**Figure 21**
**How a Boat Sails Into the Wind**
Source: Newton Henry Black and Harvey Nathaniel Davis. *New Practical Physics*. New York: Macmillan Company, 1913.

The introduction of the *stern (rear) rudder* for better steering occurred by the end of the thirteenth century. It replaced the starboard (right side) rudder or oar used previously to steer a ship.

Ships carried a *mariner's compass* by the end of the thirteenth century. The compass had 32 points on its circumference. Navigators recognized 12 winds in ancient and medieval times, but reduced the number to 8 in Columbus's time. The compass sat on a pivot in a circular bowl and turned freely in any direction with the ship's motion. Ships carried plenty of compass needles. Magellan had 35 needles on his flagship.

Ship builders introduced the *carvel (caravel)*, a smooth inlaid type body, by the fourteenth century. It provided an alternative to the early clinker (layered like the shingles on a roof) siding. Columbus's *Nina* and *Pinta* were caravels.

Typical ship dimensions for beam width, keel length, and overall length were in a 1:2:3 ratio, or 25 x 50 x 75 feet. The *Nina* had a 21-foot beam width, a 49-foot keel length, and a 67-foot length. The *Pinta*'s dimensions were 22 feet, 52 feet, and 70 feet. The *Santa Maria's* three dimensions measured 26 feet, 55.5 feet, and 78.5 feet.

Folklore about magnetic mountains that attracted ships with iron nails and led to wrecks was common. So were stories of the Lorelei (sirens) in the Rhine River luring sailors to their deaths.

*Measuring a ship's speed and water depth*

Navigators in the late 1500s measured their ship's speed in knots. They used a light rope with equally spaced knots on the rope, every 47 feet, 3 inches (47.25 feet) or sometimes 50.4 feet, to measure distance. At the end of the rope they attached a billet of weighted wood that they flanged so that the wood would float without moving in the water. The number of knots that went over the rail in 28 or 30 seconds equaled the number of nautical miles per hour at which the ship was moving. The emptying of a 28-second or a 30-second sandglass, called an ampolleta, measured the time.

The spherical Earth is 360° of longitude. Each degree consists of 60 minutes, giving a total of 360 x 60 = 21,600 minutes of longitude around Earth. Navigators set 1 minute of longitude at the equator equal to 1 nautical mile so that by definition Earth at its equator measured 21,600 nautical miles. Earth's circumference calculated in Roman miles is 24,840 miles. The nautical mile is therefore about 1.15 times larger than the measured Roman statute mile (5,280 feet) or 1.15 x 5,280 = 6,080 feet.

Navigators established the spacing of the knots by dividing the number of feet in a nautical mile by 60 x 60, or 6,080 feet divided by 3,600 seconds = 1.68, to obtain the number of feet the ship moved in one second. Multiplying by 30 seconds, or 1.68 x 30, gave the spacing between the knots, 50.4 feet or 47.25 feet (1.68 x 28) using a 28-second sandglass. A sailor counted the number of knots that unwound from the rope hanging in the water as the ship moved forward. Because the spacing between knots on a rope represented the distance travelled in a 30-second interval, or 50.4 feet, the distance travelled in one hour (3600 seconds) corresponded to the number of feet in one hour which is the nautical mile, or 6080 feet (50.4 x 3,600 = 18,344 ÷ 30 = 6,080). Columbus estimated his speed at about 4 Roman miles per hour, the equivalent of 3.18 nautical miles per hour.

Often sailors estimated their speed by watching the water bubbles or gulf weed float by.

Sailors measured water depth with a lead line. It had a standard length of 40 fathoms, but the lines sometimes were 100 fathoms long (1 fathom = 6 feet). The depth of the Atlantic Ocean is about 5 miles at its midpoint. When Columbus had sailed far into the Atlantic he tried, in vain, to measure the water's depth and therefore estimate his distance from land. He used a 100-fathom lead line.

*Determining location at sea: latitude and longitude the main problem*

*Latitude* is the north-south distance from a reference point (the equator) measured in degrees from 0 to 90. The equator is 0° latitude, each pole is 90° latitude. Ancient and medieval navigators used an astrolabe or quadrant to measure the angle between the horizon (0°) and a reference point, such as the Sun or Polaris, the North Star whose position nearly coincides with the north celestial pole and therefore appears in the same place in the night sky. The navigators' position would change depending on the ship's north-south location and consequently so would the angles they measured for a particular latitude. The angle gave the latitude at that location. Navigators began to compile tables of latitude at different north-south locations at sea or ports and made them available for use. Their longitudinal position remained highly uncertain.

*Longitude* is the east-west distance from a reference point and measured in degrees from 0° to 180° east and from 0° to 180° west. It remained an unsolved problem in the Middle Ages and Renaissance and not solved until the construction of accurate clocks (chronometers) in the 1700s.

During the Middle Ages prior to the introduction of accurate clocks, dead reckoning, star method, and several lunar methods were the known methods of determining longitude.

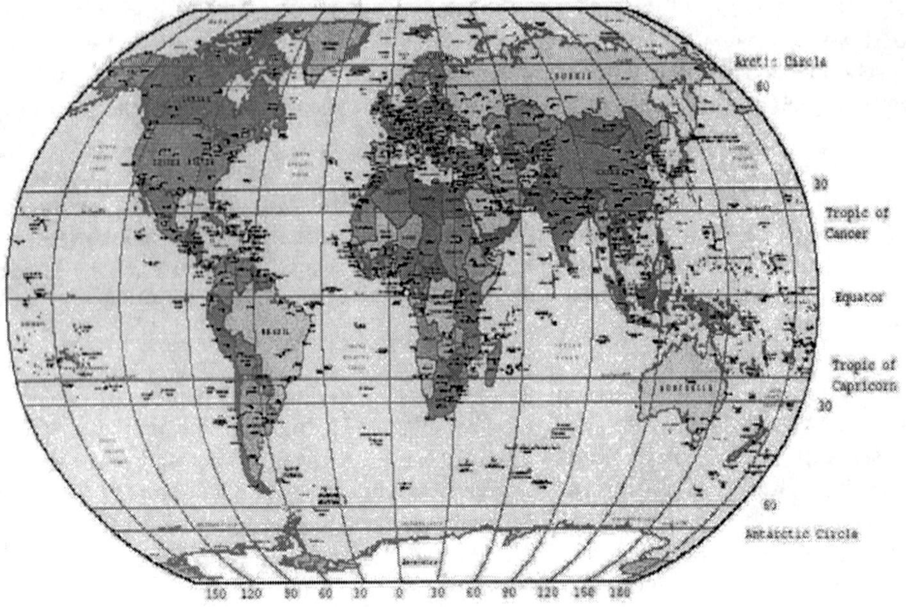


**Figure 22**
**Latitude and Longitude**
Source: *The World Factbook.* Washington, DC: Central intelligence Agency, 2013.

Dead reckoning required recording the direction of the voyage with a compass and calculating the distance (d) sailed by measuring the time (t) and speed (v) traveled in the voyage and inserting them into the equation d = v/t. In dead reckoning the navigator calculated the ship's daily position by measuring the distance and direction the ship traveled since its last accurately determined position. Starting from a known point, such as the ship's home port, the navigator measured his course and distance from that point on a chart, pricking the chart with a pin to indicate the new position and repeating the procedure throughout the voyage. Ocean and tidal currents, winds, and heavy seas affected dead reckoning's accuracy.

The star method compared star observations at two selected locations with local time at the two locations. This worked well on land especially after 1657 with the invention of the telescope (1608) and the pendulum clock (1650). It calculated longitude using the same methodology as the lunar eclipse method discussed below.

A third method, the lunar eclipse method, required knowing the time a lunar eclipse occurred in two locations. Hipparchus in ancient times proposed this method. Centuries later Regiomontanus in his *Ephemerides* (1474) and Abraham Zacuto (1450-1522), an astronomer in Salamanca in 1492 and later in Zaragoza and Cartagena, in his *Almanach Perpetuum* (1496) used the lunar eclipse method. They recorded eclipses predicted for Nüremberg (12° east of Greenwich) and Salamanca (6° west of Greenwich). Knowing the two times and that Earth rotates on its axis from west to east, or the Sun from east to west, 360° in 24 hours, or 15° in one hour, they calculated the longitude, the east or west distance between the two locations. For Salamanca and Nüremberg, which have an 18° difference in longitude, the eclipse would occur about one hour earlier in Salamanca. If the eclipse occurred at 3:00 am in Nüremberg, observers in Salamanca would see the same eclipse at 2:00 am, their time. If the two clock times for an eclipse were 2:00 pm in one location and 5:00 pm in the other, the two locations or longitudes would be 3 x 15° or 45° apart. The 5:00 pm location would be 45° east of the 2:00 pm location.

Werner in 1514 proposed the lunar distance method in his commentary on chapter four of Ptolemy's *Geographia.* Werner pointed out that the Moon moves across the background of the fixed stars very quickly, moving through an angular distance equal to its diameter in one hour (30' or ½° per hour or 12° per 24-hour day). Tables that gave the angular distance (the number of degrees) of the Moon from a fixed star or stars at a designated reference location (0° longitude) for specific times enabled Werner to designate an absolute time for the place of observation and determine other longitudes. An observer measured the angular distance of the Moon from the same fixed star(s) at each of these other longitudes, recorded the time of each measurement of angular distance, and noted the time at the two locations when the angular distances were the same. Knowing the time interval for the same observed angular distance enabled the observer to calculate each longitude. Werner's lunar distance method was similar to what Hipparchus had proposed in his lunar eclipse method. Werner suggested using a cross staff to measure the angular distance between the Moon and selected stars.

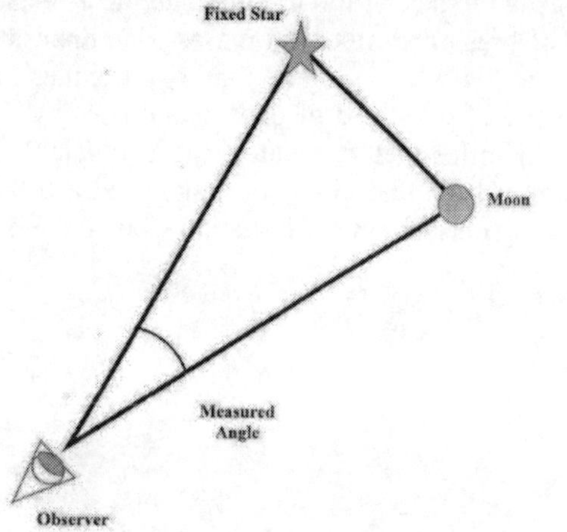


Astronomers knew that the Moon moved across the background sphere of the fixed stars quite rapidly. The Moon moved a distance equal to its diameter in one hour. This distance equaled an angular distance of 30' in one hour, 1° in two hours, or 12° in 24 hours. The angular distance is the angle measured in degrees, minutes, or seconds (60' = 1°, 60" = 1'), using a cross staff, astrolabe, quadrant, sextant, or any other suitable instrument, that corresponds to the linear distance between the Moon and a fixed star chosen as a reference point. The linear distance is an arc, or part of the circumference of the sphere of the fixed stars, but it is long enough for astronomers to consider it a straight line. The arc is the base of a right triangle and measures the distance of the Moon (M) from the star (*S*) chosen as the reference point. The other side of the right angle is the distance from the star (*S*) to the observer (*O*) and the hypotheses is distance from the Moon (*M*) to the observer (*O*). The right triangle has the side *SM* which is the distance or arc of the fixed sphere, *OS* which is the distance from the star to the observer, and *OM* which is the hypotheses of the triangle. The observer measures the angle *SOM* which is the number of degrees that corresponds to the lunar distance *SM*.

To calculate longitude the observer measures, or obtains from previously measured distances in a table, the angular (lunar) distance of the Moon from a fixed star at a specific time and location and then measures the lunar distance at his unknown longitude. Knowing the difference in degrees between the two locations, say 3°, and knowing the rate at which the Moon moves against the background of the fixed stars, 1° in 2 hours, enabled him to calculate the difference in time between the two locations. If the time at the first location was 9:00 am the time difference at the unknown location is 3° x 2 hours per degree or 6 hours. The sphere of the fixed stars like the Sun, rotates daily from east to west, 360° in 24 hours or 15° in one hour. If the first location is Greenwich and designated 0° longitude, the observer's longitude is 6 x 15 degrees or 90 degrees east or west of Greenwich, depending on whether the ship was traveling east or west from Greenwich. If Greenwich time was 12 pm, or noon, the time on the ship traveling east was 6 pm, if traveling west the time was 6 am.

**Figure 23**
**The Lunar Distance Method**
Courtesy of the author

To establish a reference location from which to measure longitude, the British government constructed Greenwich Observatory in 1674 and designated it as 0° longitude. Greenwich remains the standard 0° reference to this day. The first clocks that maintained their accuracy at sea and enabled navigators to know the time at Greenwich (0° longitude) were those that John Harrison (1693-1776), a carpenter and clockmaker in Yorkshire, England, built in the years from 1737 to 1770. His best clocks were those he built after Benjamin Huntsman (1704-76) in Sheffield in 1745-50 learned how to make good steel alloys that weather conditions did not affect.

Harrison constructed five chronometers: H-1 in 1737, H-2 in 1741, H-3 in 1759, H-4 in 1760, and H-5 in 1770. H-1 and H-2 had a brass and steel gridiron pendulum consisting of alternating rods of brass and steel which expanded in a 3:2 ratio to eliminate the effects of temperature changes, H-3 had a bimetallic construction of brass and steel known as a compensated construction to eliminate the effects of temperature changes. H-4 lost no more than two minutes on two, five-month sea voyages to Jamaica. H-5 had an accuracy of one-third of one second per day. Each minute a clock was in error represented a distance of 15 miles at the equator, the equivalent of 68 miles for one degree. The eight-member British Board of Longitude in 1714 announced a £20,000 prize (possibly $2 million in today's dollars) for an accurate chronometer. Harrison received the prize money over a ten-year period 1763-73.

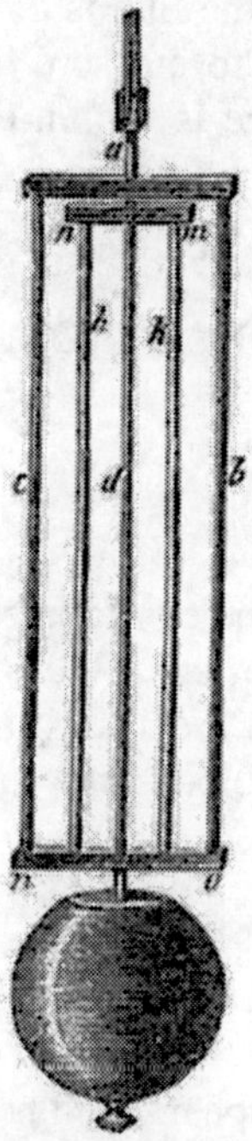

**Gridiron Pendulum**

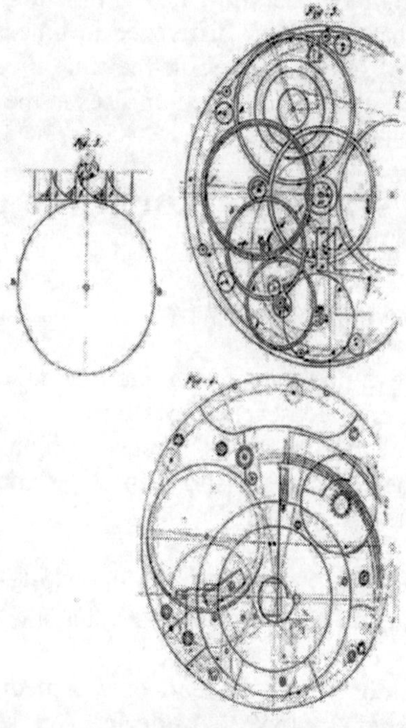

**Inner Mechanism of H-4**

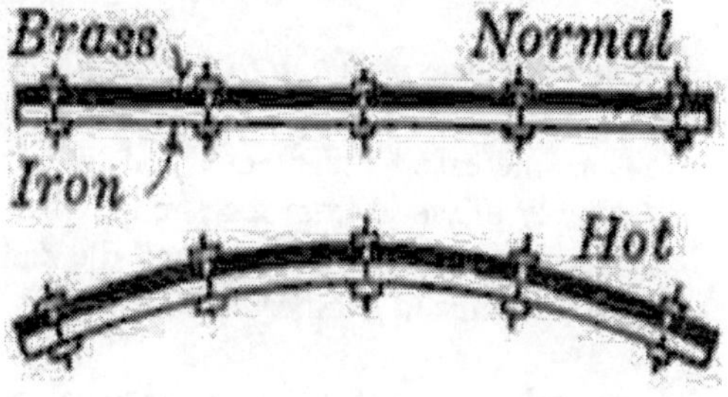


**Bimetallic Strip**

**Figure 24**
**Harrison's Chronometers**

Source: J. Dorman Steele. *Popular Physics.* New York: American Book Company, 1896 (upper left).
Source: Ferdinand Berthoud. *The Principles of Mr. Harrison's Time-keeper.*
London: Order of the Commissioners of Longitude, 1767 (upper right).
Source: Newton Henry Black and Harvey Nathaniel Davis. *New Practical Physics.*
New York: Macmillan Company, 1913 (bottom).

Having an accurate chronometer to give the time at the Greenwich fixed reference point of 0° longitude and knowing the time at your location (12:00 noon when the Sun is directly overhead), and that each hour difference in time equals 15°, allowed a navigator to measure longitude east or west of Greenwich. So if the time in Greenwich is 6 pm and your time is 12 pm or noon, 6 hours earlier, and because the Sun rises in the east and sets in the west, your longitude is 6 x 15° or 90° west of Greenwich.

## VI. INDUSTRIAL DEVELOPMENTS: THE ENVIRONMENT, TEXTILES, IRON, AND PRINTING

### *Environmental awareness*

Medieval legislators were the first to show an awareness of protecting the environment for the future. They enacted several laws during this period. In Verona, Italy, in the late 1200s, a law passed to control the fish supply banned fishing in February. In Ferrara, Italy, in the 1200s, a law limited the number of butcher shops in a designated area to control the waste (blood and entrails) the city's butchers discarded.

In Piran, in east Turkey on the Tigris River (now Egil), in the early 1300s, the law allowed animal grazing on land for four years at a time and then the land had to return to cultivation.

A century after the end of the medieval period a significant atmospheric temperature reduction occurred. In 1550, a Little Ice Age set in and lasted until 1850. During this time temperatures in Europe and in other locations outside of Europe dropped 1-2° C (2-4° F). An ice age may result from a slight change in Earth's tilt (obliquity) or its wobble. Because Earth is not a perfect sphere but flattened at its poles, it wobbles, causing Earth's precession. A change in Earth's mass distribution can produce a change in its wobble and a change in its temperature producing an ice age.

#### *Atmospheric pollution in medieval Europe*

The increased use of coal led to increased smoke pollution in medieval Europe. The diminishing supply of wood from forests, increasing its cost, made coal when available a cheaper substitute. Coal also provided more heat than the same amount of wood. The English smoke problem exemplified what occurred in other medieval cities that began burning increasing amounts of soft bituminous coal.

A well-known story about medieval smoke pollution in England dates from 1257 when Queen Eleanor (d. 1290) wife of the King Edward I (1239-1307), also known as Longshanks (reigned 1272-1307), had to leave her residence at Nottingham Castle because of the objectionable smoke produced from burning sea (soft or bituminous) coal.

England's air quality continued to deteriorate, and the soot and smoke the air contained aroused the emotions of many town dwellers. Londoners were particularly angered in 1288 when the lime burners began substituting soft (bituminous) coal for the wood and charcoal previously used in their burners. Because of public complaints about the smoke nuisance and the widely-held belief that smoke was poisonous, King Edward notified all artisans that they faced severe punishment, even death, if they persisted in burning such a malodorous substance and ordered his officers and city sheriffs to enforce his command and remedy the problem. Edward also appointed the first smoke commission, instituted fines for first-time smoke polluters, and threatened the destruction of furnaces if the owners failed to suppress the smoke.

Little improvement occurred, and in 1306 Edward, at the request of Parliament, issued a proclamation prohibiting the combustion of soft coal in London. A few years later in 1343 Parliament passed the country's first anti-smoke regulation, making the production of smoke a capital offense. During this time period, in 1307, London officials reputably hanged a blacksmith for burning soft coal.

Despite Edward's proclamation and threats, English artisans and the poor who could not afford to buy wood continued to use coal for another 300 years, through the reign of Elizabeth I (1533-1603).

***Textile industry***

Florence, Italy, was the world's textile center. It had 200 textile workshops and employed 30,000 workers. Bologna was a second center. The industry was very secretive about its manufacturing techniques, but five steps gradually developed for the manufacture of textiles. They were: combing and carding (brushing) the woolen or cotton fibers on a distaff, twisting or spinning into thread the fibers from the distaff onto a spindle, weaving the thread to a finished cloth on a loom, fulling (beating) the cloth with clay under water to remove grease and oil, and dyeing the cloth.

Mechanization of the textile industry occurred in the twelfth and thirteenth centuries using waterpower. The mechanization began with fulling, followed by weaving, spinning, and combing. Centuries later, Richard Arkwright (1735-92) in Cormford, after about three years of experimenting, patented in 1771 a practical mechanical water-powered spinning process using rollers to draw out the cotton fibers. Edmund Cartwright (1743-1823) in Manchester in 1789 and 1790 patented a wool-combining machine that did the work of 20 hand-combers. In 1791 he installed 24 steam-powered looms for weaving.

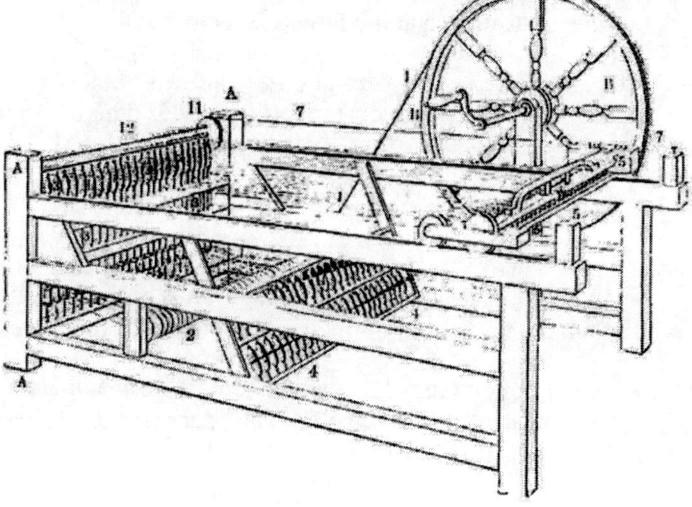

**Figure 25**
**Spinning Jenny**
Source: Richard Marsden. *Cotton Spinning: its Development, Principles, and Practice.* London: George Bell and Sons, 1884.

***Iron industry***

By the Middle Ages iron replaced copper, bronze (a copper and 1 to 10 percent tin alloy), and brass (a copper and 15 to 40 percent zinc alloy) as the major structural material. Little production of steel occurred. The chronology leading up to the introduction of iron appears below.

| HISTORICAL AGES | |
|---|---|
| ***Age*** | ***Time*** |
| New Stone Age (Neolithic) | 8000 BC-2000 BC |
| Copper Age | 6500 BC |
| Bronze Age | 4000 BC-3500 BC |
| Iron Age | 1000 BC |

*Manufacturing methods*

Iron manufacturing in medieval times was an art not a science. The iron master learned by experience and by trial and error what proportions of the reactants to use and how long to heat them. He had little knowledge of the chemical composition of the iron ores and other reactants. Medieval iron masters manufactured iron and some steel by two different methods.

In the first method, naturally-occurring iron ore (iron oxides, $Fe_2O_3$, $Fe_3O_4$,) heated with charcoal (C, beginning in 1709 with coke) and with limestone ($CaCO_3$, calcium carbonate) added to remove impurities, produced pig iron (4 percent carbon), and in separate steps wrought iron (less than 1 percent carbon), or steel (1-2 percent carbon). The final product depended on the proportions of reactants used, the time of heating, and the furnace's temperature.

In the second method iron ore heated with pig iron, and with limestone added to remove impurities, gave wrought iron or steel, depending on proportions used, time of heating, and temperature of the furnace.

*Furnace construction and uses of iron*

Iron workers constructed the furnaces of fired (baked or heated) clay bricks. They added alternating layers of iron ore, charcoal, and limestone to fill the furnace, and then heated the furnace at the top. The molten product ran out the bottom of the furnace into molds of various shapes.

Medievalists used wrought iron (soft, low carbon content) to manufacture horseshoes and nails; cast iron (hard, high carbon content) to manufacture pots, pans, tools, and weapons; steel to manufacture high quality utensils, weapons, and tools.

*Chinese iron manufacture*

The Chinese used iron for the blades of quality (expensive) weapons as early as the eleventh century BC. Iron smelting dates from the sixth century BC in the southeast regions, although the process may have begun as early as the eighth century BC, brought from the West by Scythian nomads in Central Asia, in Russia north and northeast of the Black Sea. The Chinese, as in the West, learned how to produce bronze, and in the period from the eleventh century BC to the sixth century BC (1000 BC-500 BC) they learned how to smelt iron. Archaeologists have found artifacts dating from this period that are of both meteoric iron and smelted iron. Compared to the West, iron came late to China, but the Chinese were casting iron almost immediately after its introduction.

Chinese steel production dates from this same period. The process used was similar to the nineteenth- century European Siemens-Martin open-hearth process that used iron ore, scrap iron, and steel as its reactants. Chinese production of iron and steel continued into the Middle Ages and apparently was the world's largest and most efficient until the 1700s. At that time technical improvements in the West, resulting in lower production costs, put an end to China's leadership.

Source: Georgius Agricola. *De Re Metallic*. Basel: Froben, 1556. English translation 1912.

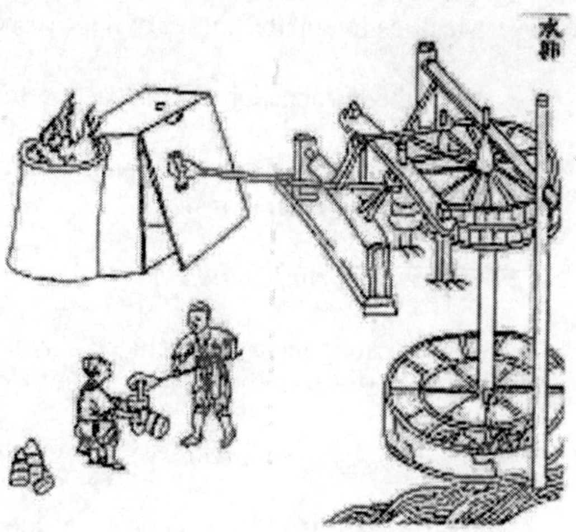

Source: Wang Zhen. *Nong Shu*, 1313, Yuan Dynasty.

**Figure 26**
**Medieval Iron Smelting**

*Important advances that were part of the medieval iron industry*

The introduction of the drawplate in the tenth century enabled iron workers to manufacture wire and nails.

**Figure 27**
**Drawplate**
Source: William H. Snyder. *Everyday Science*. New York, Allyn and Bacon, 1919.

They used waterpower, introduced in the iron industry in the eleventh and twelfth centuries (1000-11000), to crush iron ores and to operate bellows to achieve higher temperatures.

Iron manufacturers introduced cast iron firearms in the fourteenth and fifteenth centuries (1300-1400). Previously, forging (hammering) was the only way to produce iron firearms, although metal workers knew how to cast bronze. Firearms replaced the older catapult and sling artillery. Christians and Muslims fought the battles of the Crusades (1096-1291) without firearms.

With the introduction of iron casting in molds, standardization in the manufacture of firearms, cannon balls, tools, and other items occurred. A firearms industry developed in Milan, Italy; Nüremberg, Germany; Paris, Tours and Lyon, France; and London, England. The study of ballistics began with the introduction of firearms.

***Paper and printing industry***

Early writing materials included clay tablets, papyrus, bark, vellum (lamb or calf skin), and parchment (goat or sheep skin).

*Invention of paper and inks*

Tsai Lun (50-118), a eunuch at the court of the Chinese Emperor Ho Ti Liu Chao (79-106), manufactured paper in 105 from the bark of a mulberry tree mixed with scraps of linen and hemp. Paper making had advanced sufficiently so that in 1000 the Chinese were using paper money in addition to silver and gold coins.

The Mayans, independent of the Chinese, made paper from tree bark in 500. The paper process spread to Islam in 751 when the Muslims captured Samarkand in East Uzbekistan, Russia, and to Europe in 1000-1200 where Europeans used cotton, flax, and cellulose for paper manufacture.

*Manufacturing paper and inks*

The manufacture of paper in ancient and medieval times required immersing cotton or flax in water, beating and pressing it into sheets enclosed in a wooden frame with a screen fitted horizontally across the bottom of the frame, and draining the sheets to remove the water. Centuries later, in 1798, Nicholas Louis Robert (1761-1828) in France built the first paper manufacturing machine. His hand-cranked machine was a large endless wire machine that filtered the pulp, replaced hand-dipping, and produced paper in a continuous roll.

Early printers used inks made from pine resin, later carbon black (soot) and pigments from plants or minerals mixed with linseed oil from the flax plant. Johann Gutenberg (1398-1468) in Mainz in the 1400s used a copper-lead slurry as his ink.

Manufacturers from 1300 to 1700 sized or coated paper to prevent ink from smearing by dipping each sheet in gelatin obtained by boiling skin, tendons, and ligaments from animals and then hydrolyzing them. A worker could size 750 sheets per minute.

*Printing and movable type*

The Babylonians and Sumerians used block printing (incising). The first printed book was the *Diamond Sutra* published in 868. The 16-foot roll discovered in 1907 is in the British Museum. The *Diamond Sutra*, a scriptural text, is one of the most important Buddhist sutras (precepts, teachings). It reveals the subtle nature of reality and is a sutra of ultimate truth.

Pi Sheng (990-1051), a Chinese alchemist, in 1045 used ceramic (clay) type embedded in a layer of resin, wax, and paper ash in an iron form in his movable type printing press. Each piece of movable type had one Chinese character that Pi Sheng carved in relief on the piece. Printing Chinese characters was extremely difficult, and because a large treatise might require 60,000 characters, the Chinese failed to develop a printing industry.

Gutenberg used type metal, a tin-antimony-lead alloy, in his movable type printing press and a copper-lead slurry for his ink. He printed the *Bible* in 1455. It was the first book printed in Europe. The Bible consists of 2 volumes, 42 lines per page, and 1282 pages. Gutenberg had financial troubles and supposedly had his partner Johann Fust (1400-66) do the actual printing. They printed 185 Bibles, 48-49 have survived. Laurens Jansz Coster in Haarlem, Holland, also printed books in 1455.

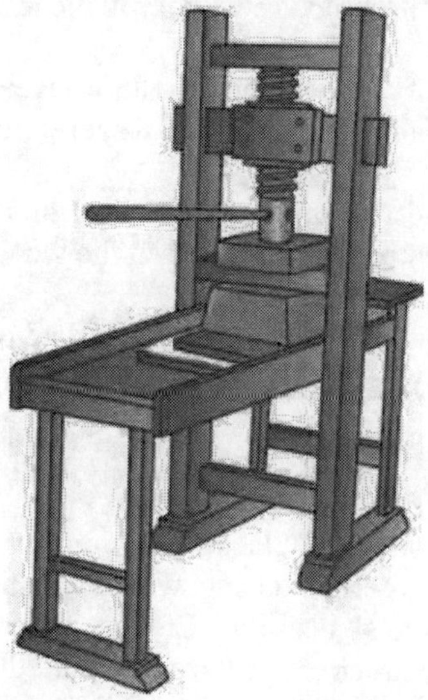

In the traditional press shown above, the printer placed the paper in the rectangular holder (right), folded a protective frame over and it, and then screw-pressed the paper on to the prepared box of inked type.

To make uniform-sized movable type, the printer prepared a steel letter punch, then used it to hammer the impression of the letter into a copper matrix. He poured the lead-tin-antimony alloy into the mold, structured to produce the typeface raised on a shoulder and stalk of uniform height.

**Figure 28**
**Gutenberg's Printing Press**
crixtina/Shutterstock.com

In the 1800s as the paper and printing industry continued to grow, a shortage of cotton, flax, and linen arose and resulted in the use of wood pulp from trees. But problems with the quality of pulp obtained from trees arose because of the presence of lignin (17-30 percent of wood), which releases organic acids on exposure to air, and because of the sizing solution that the German paper manufacturers started using in 1807. The new sizing solution consisted of rosin, a resin obtained from pine trees, and alum added for adhesion which releases sulfuric acid in solution. This acid-releasing sizing process became widely used after 1836. It remains the main cause of the deterioration of many books printed with this acid-containing paper. Better quality books contain acid free paper.

*Impact of printing and later developments and problems in the paper industry*

By 1500, 1,050 presses were operating in Europe, half of them in Italy. Venice was the world's printing capital. The presses printed 10 million books, comprising 40,000 different titles.

| NINETEENTH-CENTURY ADVANCES IN PAPER PROCESSING |
|---|
| Case binding, a technique developed in the early nineteenth century, made possible the mass production of books with uniform bindings. |
| Friedrich Gottlob Keller (1816-95) in Germany in 1850 designed a wood grinder to pulp wood by grinding away its structure with a roughened stone. |
| Hugh Burgess (1825-92) in England in 1852 developed a charcoal pulping process using sodium hydroxide. He built a paper mill at Roger's Ford in Pennsylvania in 1854. |
| C.B. Tilghman, a chemist in the United States, in 1867 developed a sulfite pulping device that dissolved the unwanted resins present in the wood. |
| C.F. Dahl in Sweden in 1879 developed the sulfate process using sodium sulfate. This is the Kraft process used today. |

## VII. CONCLUSION

Science and learning were not entirely lacking during Europe's Dark Age. Nevertheless, with the end of the Dark Age in 1000, a period during which Muslim scholars such as Rhazes and Avicenna preserved much of the ancients' learning, Europe underwent a more rapid revival of science and learning. Little scientific instruction occurred within the newly established universities in Paris, Oxford, Cambridge, and Bologna, although they did offer a professional degree in medicine and law and the PhD in philosophy and theology, and an undergraduate curriculum that included astronomy and mathematics. The environment for learning at least had begun to emerge and would spread rapidly in the next 200 years during the Renaissance period of 1400-1600. Science and medicine, however, remained essentially Aristotelian and Hippocratic-Galen.

Advances occurred in map making, magnetism, in astronomy, with Al Farghani's calculations of planetary distances and publication of the Alfonsine Tables in 1252; gunpowder production; and in optics, mainly Alhazen's studies on reflection, dispersion, the pin hole camera, the eye's physiology, and those of Bacon. Geber, Rhazes, and Lull continued investigations on the composition of matter and the combustion of matter. The impetus theory of Ockham, Buridan, Oresme, and Cusa, although it had few followers, offered an alternative to Aristotle's theories of natural and projectile motion.

Astrolabes and quadrants enabled navigators to determine latitude. The lunar eclipse method and the lunar distance method of Werner and others provided a reliable but laborious way of making the more difficult longitude determination. The solution of the longitude problem occurred only in the 1700s with Harrison's construction of accurate sea-suitable chronometers.

Craftsman assisted the birth of modern science with their contributions to agriculture, particularly the harness and horseshoe; mechanical clocks, water and wind power; and industries such as iron, textiles, and printing, most importantly Gutenberg's printing press in 1455; and navigation. Technology led scientific understanding in most of the inventions that occurred in the medieval years.

Both scholars and craftsmen accepted and practiced the new method of inquiry and experimentation of nature that contributed to the intellectual transformation that arose in Europe after 1000.

## REVIEW QUESTIONS ON MEDIEVAL SCIENCE

1. What is the main theme of this chapter?

2. How should we describe the science of the Middle Ages?

3. After the fall of Rome who preserved the science of the ancients? What was the geographical pathway the science of the ancients followed?

4. What were the main characteristics of medieval universities?

5. Which medical theory dominated medieval medicine?

6. What were the three chemical principles of the Muslims?

7. Discuss Raymond Lull's universe.

8. What is gunpowder and who invented it?

9. Identify: Gerard of Cremona, Shen Kua, Peregrinus, Roger Bacon, Alhazen, William Ockham, Nicholas Oresme, Nicholas of Cusa.

10. Discuss the design of maps in the medieval period.

11. Was medieval astronomy geocentric or heliocentric?

12. Discuss medieval theories of motion, the impetus theory.

13. In medieval engineering, technology led science. Explain the meaning of this statement.

14. Discuss the advances in power engineering and agriculture.

15. When did the Black Death occur and what was its cause?

16. Discuss feudalism, medieval living and culture, the Domesday Book of 1086.

17. Compare Roman road construction with medieval road construction.

18. What advances occurred in ship design and construction, and in navigation (latitude and longitude)?

19. Discuss medieval iron and steel production.

20. What country was the center of the medieval textile industry?

21. What advances occurred in the printing industry, paper, inks, and press?

22. Who was Johannes Gutenberg?

23. What were some of the later problems of the paper and printing industry?

# CHAPTER 4
# SCIENCE IN THE RENAISSANCE 1400-1600

Educatio Superat omnia.
Renaissance motto

## I. INTRODUCTION

Nineteenth-century historians, such as the Frenchman Jules Michelet (1798-1874) and the Swiss Jakob Burkhardt (1818-97), were the first to use the term *renaissance* to refer to the rebirth of learning that occurred in the period from 1350 or 1400 to 1600. The Renaissance began first in Italy because of the trade-based economic prosperity of the Italian city states. Industries, such as textiles, glass, and armaments, provided the wealth and leisure for a cultural revival and enlightenment.

Several developments characterized the Renaissance. A revival of antiquity occurred. The desire to study the Greek and Roman classics in their original language led to recovery of the lost or forgotten works of the ancient scholars. During the Middle Ages the scholarly texts available were usually Latin translations of Arabic texts, which were translations of the Greek and Roman texts. This gave rise to the possibility of errors in translating, or of a translator's misrepresenting what the original author had written, or of the translator's adding commentary that later readers could have mistaken as the original author's work. The original Greek and Roman texts that survived were sometimes found in the monasteries of the Middle East or Western Europe or were in the possession of prominent individuals or patrons of the arts.

Despite their reverence for the past, Renaissance scholars challenged the scientific authority of the ancient scholars, particularly Aristotle, Ptolemy, and Galen. They also denied any progress in the immediate past and demeaned the Middle Ages as the Dark Ages. The rejection of the tradition, authority, doctrines, and rationalism of medieval scholars, led some Renaissance scholars, such as Paracelsus, to subjectivism and irrationality and to a fascination with magic and the occult.

A renewal of learning and cultivation of the arts took place. *Humanism*, or the study of grammar, rhetoric, poetry, history, and moral philosophy, became an important school of learning. The study of these subjects would elevate man to his highest level and allow him to achieve the perfections of life. Humanistic studies differed from the trivium and quadrivium which included some mathematics and science. Prominent Renaissance humanists included Francesco Petrarch (1304-74) in Avignon and northern Italy, who introduced the phrase the *Dark Age* in about 1340, and Desiderius Erasmus (1466-1536) in Rotterdam, Holland.

Music, one of the subjects of the quadrivium and seven liberal arts, experienced significant change during the Renaissance. A close relation existed between the text (words) and music of the Renaissance. The music was to enhance the text's meaning and emotion. Vocal music was more important than instrumental music whose popular instruments were the lute, harp, organ, trumpet, and drums.

Renaissance music was chiefly polyphonic. It was fuller and had richer harmonies than medieval music of single line melodies (monody) of chant. The change was similar to that taking place in art (painting) with the introduction of perspective.

Ottavio Petrucci (1466-1539) in Italy around 1498 printed the first complete collection of part-song (two or more voice parts). Other composers were Nicolas Gombert (1495-1556), a cleric (priest) who served at several cathedrals in France; Orlando Lassus (1532-94) in Munich, Germany, and Claudio

Monteverdi (1567-1643) in Venice, Italy. Jacopo Peri (1561-1633) wrote *Daphne*, the first real opera. It premiered in Florence in 1597, followed by his *Eurydice* in 1600.

Clearly the best known Renaissance figure was Leonardo da Vinci. He made everlasting contributions to art with his paintings of *The Last Supper* and the *Mona Lisa*. His medical, scientific, and engineering researches were voluminous and notable except that Leonardo did not publish any of them. Nor did he attempt to disseminate the results of his studies by teaching or communicating them to others in an academic environment. Consequently, most of his work remained unknown until the nineteenth century and was no longer original. Leonardo's impact on the development of science remains therefore minimal.

The significant advances instead came in the 1543 publications of Copernicus and Vesalius and to some extent in the work of Paracelsus. Copernicus introduced a detailed astronomical model of a Sun-centered universe that presented the first real challenge to the Aristotelian-Ptolemaic geocentric universe. Vesalius's book on anatomy seriously challenged the Galenist school of medicine and for the first time provided accurate anatomical diagrams of the human body. Paracelsus also challenged the Galenist four humor theory and proposed instead that chemistry applied to medicine and the use of plant and mineral extracts in the treatment of diseases (iatrochemistry) would lead to good health.

Alchemy continued to flourish in the Renaissance, but in contrast Renaissance metallurgists such as Georgius Agricola provided straightforward accounts, devoid of mysticism, on how to extract metals from their ores, assaying techniques, and the manufacture of essential apparatus.

The increased amount of mining and smelting brought with it increased atmospheric pollution and the first attempt by communities to control the spread of pollution.

## II. LEONARDO DA VINCI

Leonardo da Vinci (1452-1519) is probably the best known Renaissance figure. He was born in Vinci, Tuscany, the illegitimate son of Piero Fruosino da Vinci (1426-1504), a wealthy Florentine notary, and Caterina di Piero, a peasant woman. Piero, who raised Leonardo, later married another woman Albiera Amadori (1436-64), who was childless, and after her death married Francesca Lanfredini (1441-75) who was also childless. After her death he married Margherita Guglielmo with whom he had four sons and two daughters, and after her death Lucrezia Cortigiano with whom he had five sons and one daughter. They were Leonardo's nine half-brothers and three half-sisters.

Leonardo spent his early years as an apprentice to Andrea del Verrocchio (1436-88) in Florence. He never mastered Greek or Latin, and like other Renaissance artists, sculptors, and architects, he earned his living by working on projects for wealthy benefactors, such as the Medicis or Sforzas, or for the Popes and other members of the religious community. He spent the years 1482-99 in Milan, a city made wealthy by the textile and armaments industries, where he studied mathematics, mechanics, anatomy, and optics, and was in the employ of Ludovico Sforza (1452-1508), known as il Moro, the Duke of Milan. Following Sforza's fall and Milan's occupancy by the French king Louis XII (1462-1515) in October 1499, Leonardo resided in Florence from 1501 to 1503 working for Cesare Borgia (1476-1507) and in Rome from 1513 to 1516. Leonardo lived his last years, from late 1516, in France in the service of Francis I (1494-1547), crowned king of France in 1515. He is buried in the cloister of the Church of St. Florentin in Amboise, on the south bank of the Loire River about 10 miles east of Tours in central west France.

***Artist: the introduction of perspective in art***

Perspective results when there is a vanishing point or single point at which all parallel lines passing from near to far appear to converge or meet, thereby giving the sense of depth.

Leon Alberti (1404-72), an Italian artist in Rome, first discussed and demonstrated perspective in art in his 1435 work *On Painting*. He used an arrangement of grids placed in front of the scene or site he intended to paint in order to represent distances accurately. Medieval art was flat, two dimensional, and did not take relative size and distance into account. Because the purpose of medieval life was to achieve eternal happiness with God in heaven, in medieval art those beings who were closer to God in the hierarchy were larger in the paintings. Angels were larger than humans. With the introduction of perspective in art, beings appeared in their proper relation to each other and not to God.

Filippo Brunelleschi (1377-1446), an Italian architect in Florence, used perspective in preparing his designs of the dome of St. Mary of the Flower Church (1420-36), the famous dome in Florence. Brunnelleschi demonstrated perspective in this way. He painted on a small piece of wood a reverse image of the Baptistry of St. Giovanni which is in front of the church. He reflected the image of the Baptistry from a mirror to obtain the reverse image. He then drilled a hole through the door of the image which was at the center of the painted piece. Brunnelleschi stood in the piazza in front of the church, held the painted piece in front of his face, and looked through the hole from the unpainted back of the wooden piece. He saw the Baptistry. Then with his other hand he held the mirror behind the painted side and looked through the hole and saw the image of the painted picture of the Baptistry in the mirror. When he removed the mirror and looked at the real Baptistry the view was identical to the image in the mirror. It was impossible to tell if the viewer was looking at the real Baptistry or its mirror image.

Leonardo used perspective and geometry extensively in his art work. No antagonism existed among science, mathematics, and art. Perspective linked them, painting was science.

Brunelleschi viewed the Baptistry's painting reflected in a mirror by looking through the hole in the center of the door. He probably covered the sky area over the Baptistry with a mirror to reflect the clouds.

**Figure 1**
**Baptistry of St. Giovanni, Florence 1878**
Source: Giacomo Brogi (1822-81). *Catalogue of Photographs Published by Giacomo Brogi.* Florence, 1912.

Leonardo's *Last Supper* (1495-97) is a 30 feet x 14 feet painting on the refectory (dining hall) wall of the Dominican monastery of Santa Maria delle Grazie in Milan. Leonardo painted it for Ludovico Sforza, as a gift to the Dominicans. It is not a true fresco because Leonardo painted it on dry finished wall rather than on wet stucco. His painting portrays not so much the institution of the Eucharist but Christ's announcement of his betrayal by Judas Iscariot (d. 30). Leonardo showed the humanity of Christ and his disciples, he gave the painting a feeling of this worldliness rather than other worldly. This artistic realism was one of the key features of the High Renaissance.

**Figure 2**
**The Last Supper 1495-97 Milan**
Source: Corrado Ricci. *Art in Northern Italy.* New York; Charles Scribner's Sons, 1911.

Leonardo's *Mona Lisa* or *La Gioconda*, painted in 1504 (1503-06), was likely the wife of Florentine banker Zanobi del Giocondo. After Leonardo's death in 1519, Francis I paid Leonardo's estate 4,000 ducats (gold coins possibly worth $50,000) for the painting. The *Mona Lisa* is now in the Louvre in Paris.

**Figure 3**
**Mona Lisa 1504 Louvre**
Source: Selwyn Brinton. *Leonardo at Milan.* Vol. 7.
London: George Bell & Son, 1911.

### *Architect and engineer*

Leonardo designed churches and church interiors. He proposed plans for an ideal elevated city with a river flowing beneath to carry off waste. His city contained arcades, gardens, squares, and fountains with elaborate waterworks. Leonardo sketched his plans for an ideal city during an outbreak of the 1484-85 plague in Milan that killed 50,000.

He designed locks, pumps, canal systems, and weapons of war for his patrons, such as Ludovico Sforza in Milan and Cesare Borgia in Rome. Leonardo's designs of military vehicles, which date from 1489, were good designs, but they lacked a prime mover.

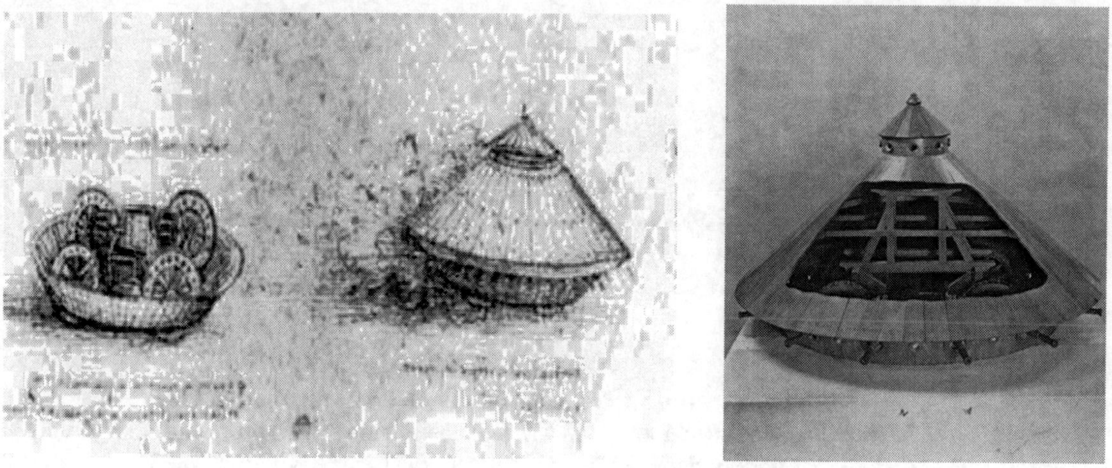

**Figure 4**
**Leonardo's Sketches: Tank**

Source: Jean Paul Richter, trans. *The Literary Works of Leonardo da Vinci.* 2 vols. London: Sampson Low, Martsen, Searle & Rivington, 1883 (left). Courtesy of the author, History of Science Museum, Florence (right).

### *Religious beliefs and philosophy of science*

*Religious beliefs*: Leonardo never clearly revealed his religious views. He accepted an immanent providence, a God who refrained from intervention in the universe like the God of Lucretius or the Stoics. He did not accept the reality of miracles. That which was of God was for the clergy, that which was of nature was for man. Leonardo believed that the body had a soul, but like Aristotle's soul, his soul died with the body.

*Ontology*: In Leonardo's view of nature the variety found in nature and nature's infinite adaptability and dynamic movement fascinated him.

*Epistemology*: Leonardo did not accept the authority of the ancient scholars. He placed emphasis on experiment. We must observe and experience nature directly, not rely on the authority of the ancient philosophers. Ideas that begin and end or remain in the mind cannot contain truth, we must test ideas by experience. Science was the union of mathematics and experimental activity.

***Main fields of scientific study date mainly from his last 30 years, remained largely unpublished until the 1800s***

*Flight*: Some of the information in Leonardo's manuscript, *Codex on Flight of Birds* (1503-6), is speculative. It contains his design of an ornithopter, or flapping wings model of an aircraft, a helicopter and a parachute.

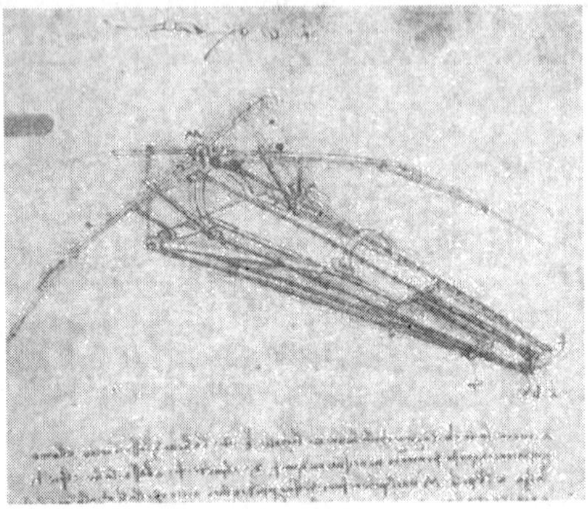

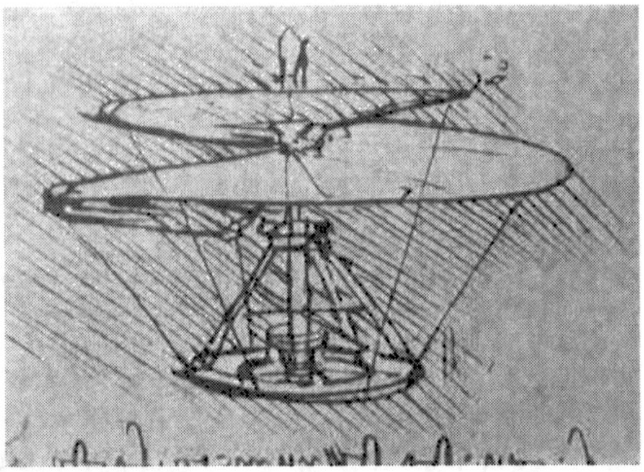

**Figure 5**
**Leonardo's Sketches: Airplane (left), Helicopter (right)**
Source: Jean Paul Richter, trans. *The Literary Works of Leonardo da Vinci*. 2 vols. London: Sampson Low, Martsen, Searle & Rivington, 1883.

*Optics*: Leonardo wrote on lenses, perspective and spreading of light, and made a physiological study of light. Sight was the queen of the senses.

*Geology and astronomy*: He accepted the four elements. Earth was in a state of continuous change. Leonardo described a Sun-centered universe in which Earth rotated on its axis and the Moon shone by reflected light. This was similar to what Aristarchus and the impetus school proposed.

*Motion and mechanics*: Leonardo recognized the difference between potential and kinetic energy as a result of his studies on falling water. He rejected perpetual motion, accepted the impetus theory and believed that motion was possible in a vacuum.

*Botany*: Leonardo gave mainly descriptive accounts of plants and produced excellent illustrations of the plants he studied.

*Anatomy and Physiology*: Leonardo's anatomy was largely Galen's. In carrying out his nine dissections from 1487 to 1493 while in Milan and in 1501-2 while in Florence, Leonardo had no refrigeration or preservatives other than alcohol to use. He studied the skull, heart, appendix, and showed that the muscles act along lines of force. He had an aversion to depicting the female sex organs and left their place blank in his drawings. Most artists of Leonardo's time, such as Michelangelo Buonarroti (1475-1564) with whom Leonardo never got along, carried out dissections so they could draw or sculpt the body accurately.

Leonardo wanted to systematize the study of anatomy. He made careful and accurate investigations and his anatomical studies were the best up to that time. He studied the physiology of the whole body as a mechanism and regarded the body as a soft machine. Leonardo employed the scientific method in his work and at first used anatomy to enhance his art. Later, anatomy became an end in itself.

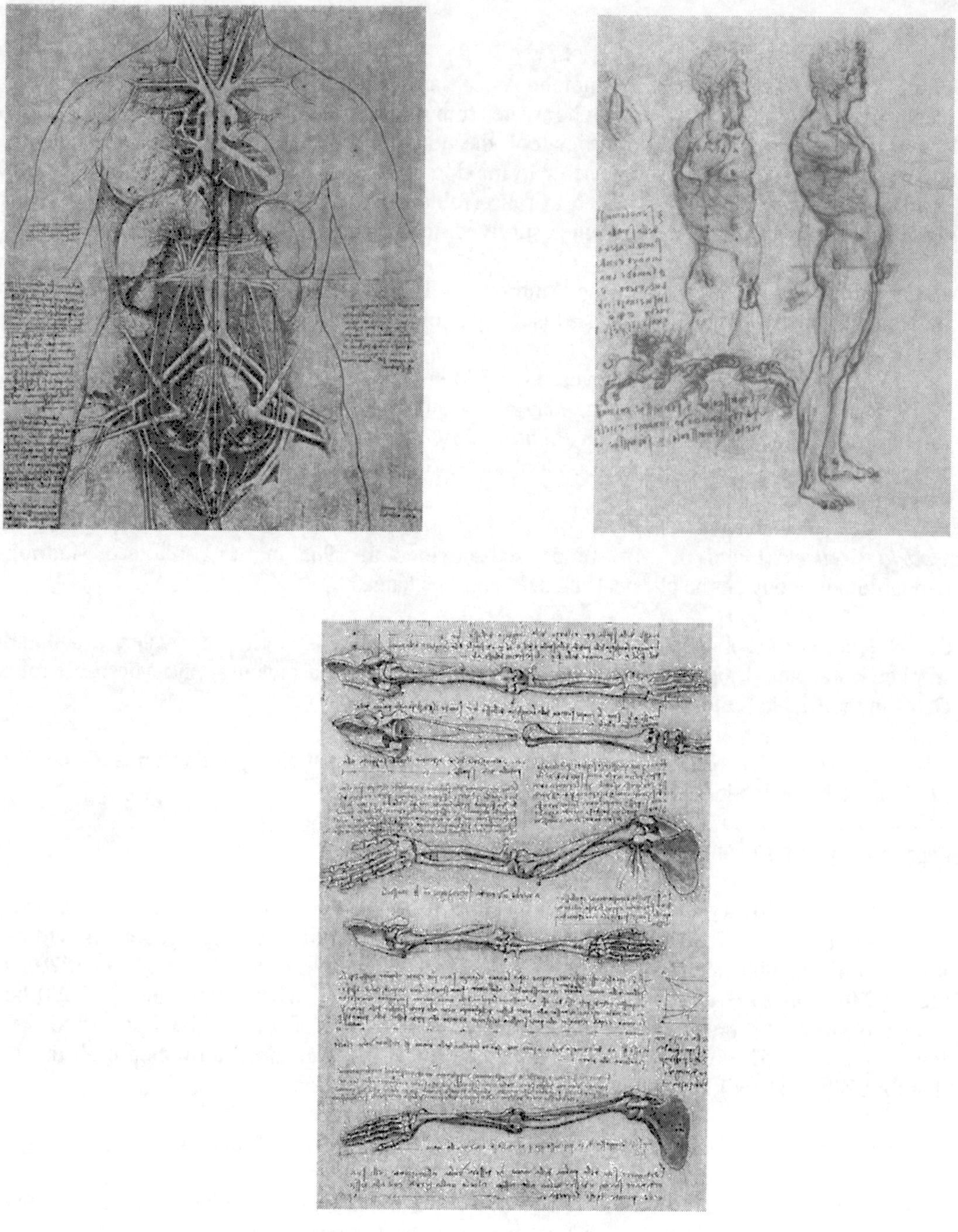

**Figure 6**
**Leonardo's Sketches: Anatomy**
**Internal Organs (top left), Male Form top right), Arm (bottom)**
Source: Jean Paul Richter, trans. *The Literary Works of Leonardo da Vinci*. 2 vols. London: Sampson Low, Martsen, Searle & Rivington, 1883.

***Notebooks***

Leonardo's investigations filled 50 notebooks of various sizes, some as small as a book of matches. They total 5,000 pages of text, which he wrote from right to left on the pages, and 600 drawings, all of it unpublished in his lifetime. He never took the time to organize his writings and had no interest in making them available to anyone else or in making a contribution to science. Leonardo's first work, *Treatise on Painting*, was first printed in folio (folded pages) in Paris in 1651. A brief description of his main works, about one third of which survived, follows.

*Codex Atlanticus* (1478-1518) in the Ambrosiana Library, Milan, contains Leonardo's writings on mathematics, physics, and mechanics. It is a huge collection of 393 folio pages.

*Windsor Codex* (1478-1518) rediscovered in 1778 in Windsor Castle, near London, and first published in 1935, deals with biological and anatomical studies. His investigations on the heart, done in 1508-15, are in the 1651 *Treatise on Painting* now in Windsor Castle. Another three volumes are at the South Kensington Museum in London, and a 566-page manuscript is at the British Museum in London.

*Madrid Codices* (I and II, 1491-1505), rediscovered in 1965 in the Biblioteca Madrid, contain Leonardo's writings on the physical sciences and mechanics.

*Codex Leicester* (1506-10): 72 pages, 300 diagrams, contains studies on water, a submarine, and snorkel; parachute design; geology and fossils, astronomy, and meteorology. Microsoft founder Bill Gates in the United States presently owns it.

*Codex on Flight of Birds* (1503-6) and 12 other manuscripts that Napoleon's troops looted from Milan in 1796 are in the Institute of France in Paris.

***Impact on scientific development***

Leonardo exemplifies the importance of curiosity in science. Leonardo had a curiosity about everything in science, but he was a procrastinator and a perfectionist. When Leonardo died in 1519 he left all of his unpublished notebooks to his trusted companion Francesco Melzi (1490-1570), whom he met in Milan in 1513 and who guarded them for fifty years. When Melzi died in 1570 he left the notebooks to his nephew Orazio Melzi who gave away some of them. Others passed to the sculptor Pompeo Leoni (1533-1608) in Milan. Leonardo's notebooks remained unstudied until the 1870s. By that time others already had published what Leonardo had written.

*Universality of learning versus specialization of learning as seen in Leonardo's career*

Leonardo never would have promoted specialization in learning, which later became characteristic of Western learning or education. For him there was no separation of art from science. Mathematics, geometry, and perspective in art were the pathway to the unification of learning. Yet the vastness of what Leonardo tried to accomplish showed that the universality of learning so common in the Middle Ages and Renaissance was becoming impossible and would give way to the idea of specialization in learning.

## III. NICOLAS COPERNICUS AND THE HELIOCENTRIC MODEL OF THE UNIVERSE

Nicolas Copernicus (1473-1543) was born in Torun, Poland, a well-preserved Gothic town on the banks of the Vistula River and famous today for its Copernicus House Museum. Copernicus's father died when Copernicus was ten, his mother died earlier, and his uncle, a Catholic bishop, raised him and his older brother. Copernicus lived mostly in Frombork, where he was a canon at the Frombork

Cathedral except for his early years in Kracow (from 1491) and when he studied law, medicine, art, and theology in Bologna (1496-1500), Padua (1501), and Ferrara (1503).

Copernicus accepted neoplatonist ideas. Neoplatonism was a school of thought that Plotinus (205-70) in Alexandria and Rome in 245 developed and formulated in his *Enneads* (6 volumes). It remained dominant from the third to the sixth centuries.

As a neoplatonist, Copernicus believed that you acquired true knowledge by ascending upward from matter at the lowest end of the scale and the world of life (the World Soul) to the Divine Mind (Nous or intelligible world) and finally to the One or God. The acquisition of true knowledge was an intuitive process that bypassed the senses.

***First writings and publications on his heliocentric model***

*Commentariolus* (*Little Commentary*) was a manuscript Copernicus wrote between 1502 and 1510 and circulated in 1510 but never published. Copernicus used 34 spheres (deferents and epicycles) and later added 14 more to improve the accuracy of his model. His disciple Georg Rheticus (1514-74) who spent two years with Copernicus in Frombork and later taught in Kracow, published a 40-page account of Copernicus's model, the *Narratio Prima* (*First Report*), in 1540.

***Reasons for proposing a heliocentric model of the universe***

Copernicus accepted neoplatonic beliefs, and because of his neoplatonism he envisioned the Sun as the physical manifestation of God. It represented the light and life of the universe and remained at rest at the center of the universe. Neoplatonists related God to the principle of unity. They made God transcendent and connected to the world through a series of intermediaries who derive from God by a principle of emanation. Reality was a series from the divine to the material, and man who has in him some part of the divine, strives for union with God, the eternal source of all things.

As an astronomer trying to account for the motions of the planets, Copernicus showed that a simpler, astronomical model of the heavens resulted with the Sun at the center of the universe.

Copernicus argued in *Little Commentary* that his placing of the Sun at the center of the universe and Earth in motion around the Sun provided a natural explanation for retrograde motion of the three exterior planets, Mars, Jupiter, and Saturn. His explanation for retrograde motion became the first of Copernicus's main arguments for a Sun-centered universe.

Because Earth's orbit was smaller than the orbits of the exterior planets, Mars, Jupiter, and Saturn, and because Earth moved faster than those planets, Earth at certain times was *behind* the exterior planet. At other times it caught up to the exterior planet, and then passed it, giving the appearance of retrograde motion of the exterior planets when viewing the exterior planets from Earth. If Earth is stationary no physical explanation is possible, although a mathematical explanation is possible using deferents and epicycles of various sizes and revolving at various rates as Apollonius had shown in the third century BC.

Copernicus noted that in an Earth-centered universe the planetary shell carrying Mars intersected the shell carrying the Sun. But the material shells could not penetrate or pass through one another, and if they did the celestial mechanics became unworkable. A Sun-centered universe with Earth in motion in an orbit between Venus and Mars eliminated the problem of intersecting spheres.

Copernicus pointed out in his second main argument in *Little Commentary* that by placing Earth in orbit around the Sun, the calculated planetary distances from the Sun and their periods of revolution around the Sun fell into a regular order, increasing from Mercury to Saturn.

Copernicus's heliocentric model required Earth to have three motions, an annual revolving motion, a daily rotating motion, and a slow motion of precession.

In its first motion, Earth with its Moon and all the other planets revolved around the Sun annually from west to east, counterclockwise.

In its second motion, Earth rotated on its axis from west to east, counterclockwise, once every 24 hours.

Precession was Earth's third motion, which we know today results from the action of the Sun and Moon upon the protuberant matter about Earth's equator. Earth slowly wobbled or twisted through one revolution in 26,700 years (50 seconds of arc per year) like a slowly spinning top. Earth's axis slowly moves in a circle from east to west clockwise through the 12 constellations of the ecliptic, or zodiac, spending about 2,200 years in each constellation. Earth's angle of inclination (23½°) does not change very much, only the location of its axis with respect to the 12 constellations slowly changes.

Because of precession, the tropical (solar) year is shorter than the sidereal year by about 20 minutes. The tropical year measures the time the Sun in its apparent annual west-to-east journey around the equator takes to return to the same position on the equator. The sidereal year is the time the Sun in its apparent annual journey through the constellations making up the ecliptic takes to return to its same position relative to the constellations. The tropical year measures the calendar year. Precession therefore required the removal of days from the civic and liturgical calendar to keep the solar and sidereal years nearly equal and to have the equinoxes (21 March, 22 September) and solstices (21 June, 22 December) which determined the start of the seasons and Easter to occur at the same time and close to their accepted dates. If precession over a long time period had advanced the date of the spring (vernal) equinox from 21 March to 11 March, removing 10 days from the calendar (12 through 20 March) and going directly from 11 March to 21 March, would bring the tropical year and the calendar year into agreement with the sidereal year.

In early Christian times Earth's axis moved from *Aries* to *Pisces* and now is moving towards *Aquarius* and then *Capricorn*. This is precession and the reason why the equinoxes occur earlier each year. Earth's precession also caused Easter, which the Council of Nicaea established in 325 as the first Sunday after the first full Moon on or following the vernal equinox, to occur earlier and earlier. Precession makes necessary the calendar changes to compensate for the advancing of the seasons.

**Figure 7**
Copernicus, Conversations with God, late nineteenth century
Jan Matejko (1838-93)

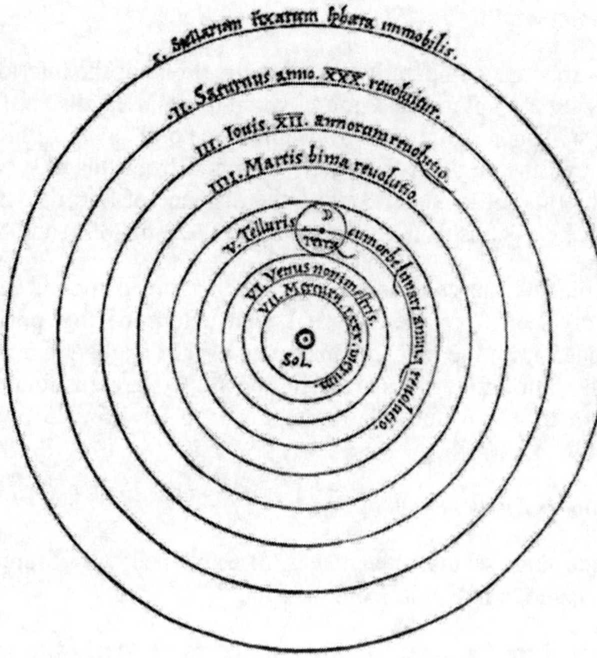


Copernicus's solar system reproduced from his *De Revolutionibus Orbium Coelestium*. The Sun is the center. Mercury, Venus, Earth (Telluris) and Moon, Mars, Jupiter, and Saturn move around it.

**Figure 8**
**The Copernican Universe**
Source: Arthur Berry. *A Short History of Astronomy: From Earliest Times Through the Nineteenth Century.* London: John Murray, 1898.

\

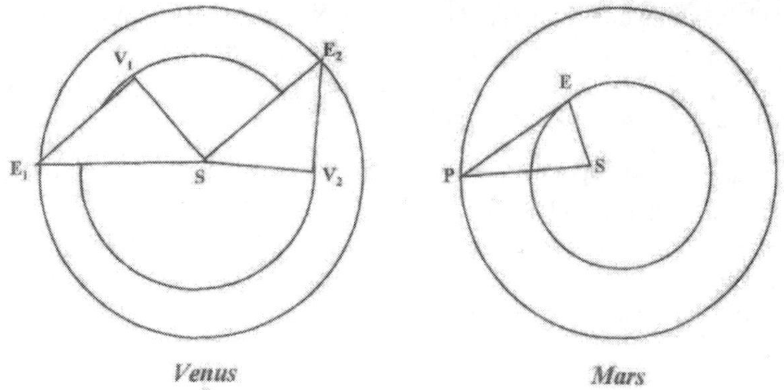


**Figure 9**
**Calculating the Relative Distance of an Interior (Venus) and Exterior Planet (Mars)**
Source: Arthur Berry. *A Short History of Astronomy: From Earliest Times Through the Nineteenth Century.* London: John Murray, 1898.

*Sizes of the planetary orbits*

Copernicus determined the comparative sizes of the orbits of the interior planets, Venus and Mercury, to Earth by observing each planet and noting when it was most distant from the Sun. At this distance a line from the most distant point and from another point on its orbit forms a right angle triangle. Copernicus measured the angle at the other point and from this observation established the shape of the triangle and the ratio of its sides. Setting the distance of Earth from Sun at 100, he calculated that the average distance of Venus from the Sun was 72 (72.3 in 2000) and Mercury 36 (38.7 in 2000).

For the exterior planets the calculation was much more difficult. Copernicus had to compute the planet's position as seen from the Sun and then determine this position when Earth and Sun are farthest apart as seen from the planet. This gives a right angle and a measurable angle. The shape of the triangle and the ratio of the sides are then known. Copernicus found the average distance of Mars, Jupiter, and Saturn from the Sun relative to Earth to be 1.5 (1.5 in 2000), 5 (5.2 in 2000), and 9 (9.5 in 2000).

*Advantages of Copernicus's model*

Copernicus's model had several advantages. It explained why Mercury and Venus, being interior planets, never appeared to move far from the Sun.

His model accounted for retrograde motion, the apparent backward motion of the planets that were farther from the Sun than Earth.

His model accounted for the precession, the earlier yearly occurrence, of the equinoxes. This is a slow westward (clockwise) motion of the equinoctial points (equinoxes) along the ecliptic (50 seconds of arc per year), rather that the slow westward (clockwise) turning of the celestial sphere as Hipparchus, Ptolemy, and other geocentrists believed. If not accounted for precession causes the seasons to begin earlier each year.

Copernicus's model had fewer spheres (epicycles and deferents) compared to Ptolemy's model, which had 80 spheres, versus 48 spheres for Copernicus, because Copernicus did not have to account for retrograde motion. Yet, it provided the same accuracy as Ptolemy's model.

*Delay in publication of De Revolutionibus Orbium Coelestium (1543)*

Copernicus delayed publication because he feared ridicule. Martin Luther (1483-1546) in Wittemberg in 1539 called Copernicus both a fool and a heretic. Religious issues were not the main problem at that time. He dedicated his book to Pope Paul III (1468-1549, elected 1534). Of the 400-500 copies of the 400-page book printed, 251 survived. Copernicus apparently received a printed copy of his book on his death bed on 24 May 1543. A handwritten copy dating from the 1500s is in the rare book collection at Karl-Franzen University in Graz, Austria.

Andreas Osiander (1498-1552), a Lutheran minister in Nüremberg whom Copernicus's disciple Rheticus put in charge of publication in 1543, added anonymously a *false* preface to Copernicus's book. Osiander wrote that Copernicus's model was only a hypothesis; it did not describe the real or true arrangement of the planets, and was only a calculating model. Copernicus believed his model did describe reality and was ontologically correct. He never knew what Osiander had done. Johannes Kepler in 1609 discovered that Osiander had written the preface, not Copernicus.

*Objections to the Copernican model*

Copernicus's critics offered several objections to his model. They said no evidence supported Earth's motion. In Copernicus's time convincing evidence of Earth's motion was lacking. We know today that we don't feel Earth's motion because Earth's orbital motion is nearly constant or uniform. Because the orbit is large, the sense of flying off Earth as it revolves in its orbit is small compared to the gravitational force or our weight. Because uniform motion and rest are equivalent and because we are sharing in Earth's motion we do not sense Earth's motion. Our bodies have become adjusted to Earth's motion because we never have experienced a different situation. Yet Earth is revolving in its orbit at a speed of 18.5 miles per second and rotating on its axis at a speed of 2 miles per second.

Critics asked why are the birds not left behind as Earth moves through the heavens?

What is pushing Earth? The other planets are heavenly and therefore perfect spheres that had naturally circular motion. Earth did not, so why does it move?

Why do stones fall to Earth if Earth is not at rest at the center of the universe?

Why doesn't Earth fall apart as it revolves and rotates through the heavens?

Finally, the critics asked, as Aristotle did, why was there no evidence of parallax, which should occur if Earth revolved in an orbit around the Sun. Friedrich Bessel (1784-1846) in Königsberg, Germany, first detected parallax because of Earth's motion three centuries later in 1838. He used a Fraunhofer heliometer, an instrument named after Josef Fraunhofer (1787-1826), a Bavarian spectroscopist, to measure the angular distance between stars.

The Foucault pendulum that Jean Foucault (1819-68) constructed in 1851, the original of which is in the Pantheon in Paris, provided evidence of Earth's daily rotation. Its lead-coated brass bob weighs 28 kg and hangs by a 67 m wire from thc Pantheon's dome. The pendulum's swing appears to rotate with time, but the plane of the pendulum's swing, as a gyroscope does, remains fixed in space as Earth rotates beneath it. Earth rotates clockwise 11 degrees per hour, rotating 360 degrees, a complete circle, in 32.7 hours. At the north and south poles one complete rotation takes one sidereal day, 24 hours. The time of rotation depends on the mass of Earth beneath the pendulum.

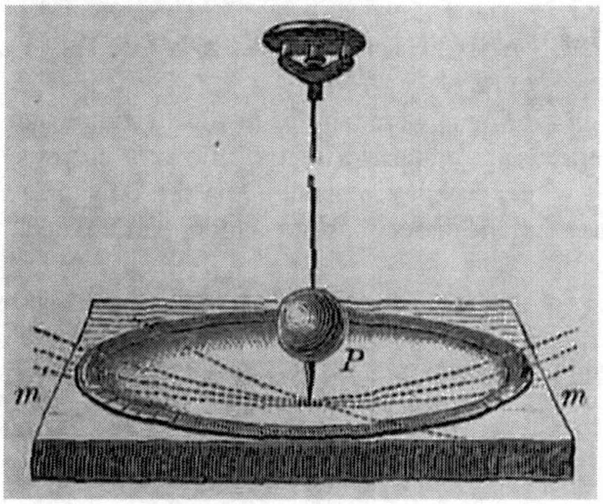


The pendulum's tip (P) traced its vibrations north and south on a table beneath sprinkled with fine sand. The paths did not coincide, but at each return to the outside the pendulum marked a point to the right. The shifting of the plane of vibration increases as the pendulum moves north or south from the equator.

At Earth's poles the pendulum constantly vibrates in the same vertical plane and completes one revolution in 24 hours. At the equator the path does not move east or west because the plane of vibration moves forward with Earth's daily motion.

**Figure 10**
**Foucault's Pendulum**
Source: J. Dorman Steele. *Popular Physics.* New York: American Book Company, 1896.

### *Significance of the Copernican model*

The Copernican model dealt a blow to astrologers and their macro-micro cosmic link because Earth no longer occupied a central position in the universe. It was just another planet revolving around the Sun. Its loss of status upset the entire medieval system of thought. The distinction between celestial and terrestrial or sublunar matter and motion became meaningless. Nicholas of Cusa 100 years earlier had proposed that heaven and Earth consisted of the same four elements.

The relative motion of Earth and Sun became a possibility. Therefore we could no longer be certain what was moving and what was at rest.

Finally, because the sphere containing the fixed stars no longer rotated from east to west daily, was the sphere still necessary? Maybe the fixed stars were not really fixed and equidistant from the center of the universe, but were at different distances. Were we living in a closed world or an infinite universe?

### *Conclusion: Copernicus's shortcoming*

Copernicus's heliocentric model lacked the physical laws of motion (inertia and acceleration) needed to support it and make it acceptable to the scientific community. Copernicus provided the astronomy but not the physics to make a heliocentric model acceptable. This is what Galileo began with his quantitative investigations on acceleration and falling bodies and what Newton completed with his three laws of motion and law of universal gravitation.

Nevertheless Copernicus's *De Revolutionibus* published in 1543 initiated the scientific revolution of the sixteenth and seventeenth centuries. It dramatically upset the scientific beliefs of its day and raises the question of whether we could have such a dramatic upsetting of our scientific beliefs today.

## IV. ANDREAS VESALIUS AND THE FOUNDATION OF ANATOMY

In the 1500s Padua and Bologna were famous medical schools and centers for surgery. A physician at these universities carried out or supervised dissections of corpses on a marble table in a lecture hall, usually in cooler or winter weather. Alcohol was the only preservative available to the physician. Medical students observed the dissections but seldom participated.

The Flemish anatomist Andreas Vesalius (1514-64), the foremost Renaissance anatomist, studied medicine in Brussels, Belgium, and later in Padua, Italy, and Spain, where he served as court physician to Philip II (1527-98) in Madrid, Spain.

**Figure 11**
**Andreas Vesalius**
**US National Institutes of Health, National Library of Medicine**
Source: Stephen van Calcar. Woodcut of Vesalius in *De Humani Corporis Fabrica.* Padua, 1543.

***Anatomy***

Vesalius challenged Galen's anatomical conclusions, arguing that Galen performed his dissections on apes, pigs, and other animals, not humans. He emphasized *hands on* dissection by physicians and their students rather than the physician's reading from a book on surgery and students' watching while a demonstrator did the dissecting. Instead of looking to Aristotle and Galen for answers, the physician must look to nature. Vesalius stressed demonstrations, experiments, and comparative anatomy and carried out public dissections on corpses while in Paris, Louvain, and Padua. His insistence on student experimentation, which required corpses, sent medical students on a wave of body snatching and grave robbing.

In *De Humani Corporis Fabrica* which Vesalius published in 1543 when in Padua, he introduced the modern anatomical terms for muscles, bones, brains, and abdominal parts. Alhazen already had named the parts of the eye. *De Humani* was the first medical book to have accurate anatomical diagrams which the German artist Johan Stephen van Calcar (1499-1546) in Padua may have produced from woodcuts. A World War II bombing raid over Munich in 1943 destroyed all (about 200) the woodcuts which were in the Bavarian State Library or possibly in the library of the University of Munich.

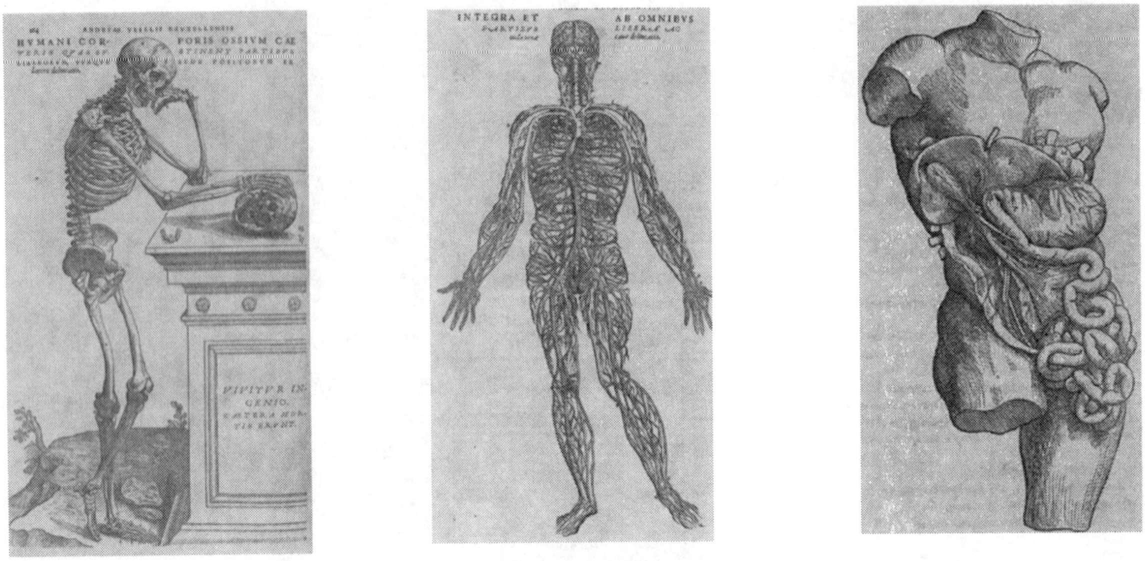

**Figure 12**
**Vesalius's Medical Diagrams: Skeleton, Nervous System, Internal Organs**
Source: Andreas Vesalius. *De Humani Corporis Fabrica*. Padua, 1543.

### *Physiology*

Vesalius at first accepted the non-circulatory ebb and flow motion that Galen in the second century had proposed for the pathway of blood in the body. By 1555 he doubted its correctness. Vesalius believed the brain was the source of life, mind, and emotion, not the heart as Aristotle thought.

Vesalius's life ended tragically. He died in the spring of 1564 in a shipwreck off the coast of Zakynthos, Greece, (or on an island off the coast, of illness) while returning from a pilgrimage he made to Jerusalem and Mount Sinai in the Holy Land. According to a letter written in Paris and dated 1 January 1565, Vesalius made the pilgrimage when in Spain because of a death resulting from his dissection of a distinguished, but not yet dead, gentleman whose heart was still beating. This story, however, is a fabrication because Vesalius never did any dissections while in Spain. He very likely made the pilgrimage for religious reasons.

## V. CHEMISTRY IN MEDICINE: IATROCHEMISTRY

Chemistry did not exist as a separate discipline at this time. It was associated with medicine, pharmacy, metallurgy, and alchemy. Paracelsus (1493-1541), a colorful Renaissance figure, embodied all of these practices in his teachings, researches, and writings.

### *Paracelsus*

Paracelsus was born in Switzerland and studied medicine with his father Wilhelm and in Ferrara, Italy. His real name was Theophrastus Bombastus von Hohenheim, but he chose Paracelsus because he considered himself the equal of the renowned Roman physician Aulus Celsus (25-50 AD). Paracelsus practiced medicine in Basel, in other Swiss cities, and in Austria. He was disdainful of medical authority and tradition and taught his classes in German rather than use the scholar's Latin.

Paracelsus began his teaching in Basel by publicly burning the books of Galen and Avicenna and announcing that they were undergoing the same fate in their afterlife. Paracelsus lost his position in Basel in 1528 because of his contempt for the Galenist medical establishment. Speaking of the physicians he said that you are not worthy that a dog should raise his hind leg against you. He became a wanderer and an alcoholic and while in Salzburg, Austria, died accidentally from a fall.

**Figure 13**
**Paracelsus 1528 Louvre**
Source: Quentin Massys (1466-1530)

*Religious and philosophical views*

Paracelsus was a neoplatonist and gnostic. Neoplatonism, a philosophical belief originating with Plotinus in the third century AD, held that from God emanates the Divine Mind, from which the World Soul emanates and comprehends the visible world. Man's ideal is to rise upwards towards union with God through contemplation, bypassing the senses.

Gnosticism was a theological-philosophical belief dating from Hellenistic times (after 323 BC) and reached its high point in the second century AD. A Gnostic has acquired spiritual magic and can manipulate the non-mechanistic forces inherent in nature, such as the motions of the four elements, the animistic forces in the universe, and vitalism, through intuition, not through the senses. Matter is evil and spirit is good (dualism), salvation comes from secret knowledge. Prophecy was the highest form of magic. Gnosticism is a form of dualism as are the eastern religions of Manichaeism and Zoroastrianism.

*Theory of the elements and medical beliefs*

Paracelsus accepted the four element theory of the Greeks and the three principles of the Arabs.

He rejected most earlier medical beliefs, those of Hippocrates, Galen, and Avicenna, including the four humors theory of phlegmatic (phlegm), choleric (yellow bile), sanguine (blood), and melancholy (black bile).

Other theories of disease that Paracelsus challenged were those of signatures and sympathetic remedies.

According to the medical tradition of signatures God imparted a signature, some sign or mark or shape, to a plant to indicate which part of the body to treat with the plant. Walnuts were good for brain disease, green onions for the heart, liverwort and lungwort aided the liver and lungs.

To apply a sympathetic remedy in treating a patient, such as healing an axe wound or cut, the physician would clean and bandage the wound but apply the salve to the axe that caused the wound. Similarly a patient would drink a soup made of a viper's liver and heart to cure chicken pox. The playing of musical instruments, such as the flute, cured gout, whereas playing a trumpet cured sciatica, a pain in the lower back, buttocks, and hips.

Paracelsus believed life was a chemical process, the body was a chemical factory or system. The *archaeus*, the active principle in digestion, separated nutrients from waste in the body.

During the Renaissance the purpose of alchemy shifted toward the preparation of medicines (*arcana*) from plants and minerals to cure diseases. This led to the experimental testing of many substances to determine their effectiveness in treatment.

Paracelsus contributed to a new medical tradition called iatrochemistry, meaning medical chemistry, which had its roots in Raymond Lull's earlier work in using plant extracts or essences for treating disease. Paracelsus discussed his medical ideas in *Archidoxies* which he wrote in 1525 and which his followers published in 1570, after Paracelsus's death. The term iatrochemistry appeared for the first time in the translation of Paracelsus's *Archidoxies.*

In his study of diseases and their natural causes Paracelsus concluded that disease, including women's diseases and illnesses, resulted when foreign parasites invaded the body. He used opium and compounds of lead, mercury, arsenic, and zinc in the treatment of disease. All of them are dangerous in small quantities except zinc.

**Figure 14**
**Therapeutic Bathing and Cupping**
Everett Historical/Shutterstock.com

***Doctrines of fermentation and fertilization***

In iatrochemistry, fermentation was a natural (harmless) mechanical-chemical process that did not result in disease, whereas putrefaction was a fermentation that caused disease in some unnatural way. Fertilization was a natural fermentation of male and female seeds in a woman's womb and generated a healthy fetus. Syphilis resulted because a woman had multiple sex partners. This gave a preternatural (abnormal) mixture of different men's seeds, and this unnatural mixture resulted in putrefaction, causing syphilis.

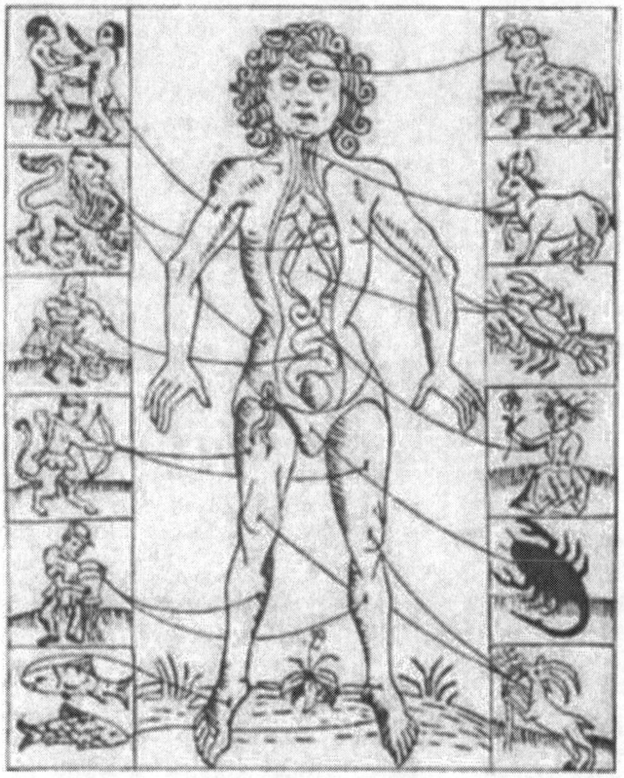

**Figure 15**
**Zodiacal Man**
Public Domain.

***Other medical chemists (iatrochemists) of the Renaissance and 1600s***

Oswald Croll (1560-1609), Daniel Sennert (1572-1637), Andreas Libavius (1560-1616), and Johann Glauber (1604-1670) practiced in Germany. Thomas Willis (1621-75) practiced in London. Franciscus Sylvius (1614-1672) in Holland believed that for a body to remain healthy it had to maintain an acid-base balance in the body. An imbalance produced disease.

## VI. WITCHCRAFT AND VAMPIRES

***Witchcraft***

Germany was the center of persecution of witchcraft. Villagers often suspected red-headed women with body moles of being witches (wise women) and of practicing witchcraft. Some of these suspected women used incantations and herbal remedies made from plants that they grew in their gardens for use in childbirth and treatment of illness. The clergy and physicians viewed them suspiciously.

Renaissance stories claimed that witches could fly by rubbing on themselves an oil made from the fat of babies that they had kidnapped and murdered. Witches used their witchcraft to cause plagues, hail, and thunderstorms that ruined crops.

Heinrich Kramer (1430-1505) and Jakob Sprenger (1436-95), Dominican priests in Germany, in 1484 wrote *Hexenhammer*, the best known account of Renaissance witchcraft. From 1500 to 1650 religious and civil authorities executed by hanging, by the sword, or by burning at the stake, thousands of witches for practicing black magic and fornicating with the devil.

In 1584 Reginald Scot (1538-99) in Smeeth, Kent County, England, published *Discoverie of Witchcraft*, in which he exposed the tricks of witches to show that the tricks were merely human manipulations. He hoped his exposure would stem the popularity of witch hangings in England, but it had little effect.

***Vampire legend***

Count Dracula, the vampire that Bram Stoker (1849-1912) the Irish writer wrote about in 1897, was actually Vlad Tepes (1400-76) or Vlad the Impaler, a Romanian nobleman and a member of the Order of Dragon (Dracul), an order dedicated to fighting the Turks and heretics. Vlad killed or executed 20,000 of his enemies and resorted to mass impalements in his military campaigns. He was a Romanian hero. From these exploits emerged Stoker's tale of Count Dracula, the bloodthirsty vampire.

**Figure 16**
**Prince Dracula and His Castle**
Public domain (left); courtesy of the author (middle); Mikadun/Shutterstock.com (right)

## VII. BEGINNING OF METALLURGY

Metallurgy dealt with the separation and recovery of metals from their ores. Renaissance metallurgists were interested in how much metal they could recover rather than how much was present in an ore. To carry out their investigations they introduced quantitative measurements, constructed balances sensitive to 0.1 milligram, and developed new equipment, such as flasks, stills, and furnaces.

Vannoccio Biringuccio, Georgius Agricola, and Lazarus Ercker (1530-94) in Bohemia, were well-known Renaissance metallurgists. Their texts were empirical rather than theoretical in method. They wrote practical guides to mining and assaying techniques, their texts differed completely from the mystical and mysterious alchemical texts.

Vannoccio Biringuccio (1480-1539), in northern Italy, published *Pirotechnia* (1540) in which he expressed skepticism about alchemy and gave a formula for gunpowder.

Georg Bauer or Georgius Agricola (1490-1555) in Chemnitz, Germany, was the father of mineralogy and the author of *De Re Metallica* (1556). His book contained 292 woodcuts and described assaying methods, separation techniques, and gunpowder preparation. Agricola also dealt with occupational hazards in *De Re Metallica.* He wrote that the Italian city states had passed laws against mining because of its destructive effects on woodlands, fields, vineyards, and olive groves. Critics of the mining industry complained that mining was a perilous occupation to pursue because the pestilential air that miners breathed sometimes rotted their lungs away and sometimes killed them. Agricola, however, disagreed with the critics. He remarked that such things rarely happened and only because the miners were careless. Herbert Hoover (1874-1964), the American president who was also a mining engineer, and his wife in 1912 translated Agricola's book into English.

**Figure 17**
**Smelting Ore**
Source: Georgius Agricola. *De Re Metallic*. Basel: Froben, 1556. English translation 1912.

**Air pollution *in renaissance Europe***

European cities or regions enacted legislation to combat air pollution, but enforcement seemed rather limited except in a few isolated cases. In Renaissance Cologne, a flourishing commercial and artistic center and prominent member of the Hanseatic League, city officials on 11 June 1461 passed Germany's first anti-smoke legislation. Local inhabitants complained of the harmful fumes emanating from a copper and lead smelter and forced its owner to stop the pollution. The account does not make clear whether the smelter actually shut down or relocated, which at that time were the usual ways to stop factories from polluting.

Similarly, in sixteenth-century France, boats ascending the Seine River brought coal to the port city of Rouen, an artistic center and site of Joan of Arc's burning in 1431, and for at least a century its inhabitants burned coal in their homes uninhibited by any restrictions. But in the period 1500-10 when

Rouen's textile industrialists began to burn coal in their factories the protests began. A city commission looking into the smoke problem in 1510 summoned 40 witnesses including medical doctors who spoke against the evil of burning coal because of the dangerous smells it produced. The report the commission issued made two recommendations: (1) coal consuming industries should move from Rouen and (2) that they should have chimneys sufficient to remove the pollution.

Like several of her royal predecessors, Elizabeth I (1533-1603) in England found coal fires repugnant. So did many of the female nobility, and they refused to enter homes that burned coal or cooked meals over sea coals. The most significant improvement to occur at that time was the construction of chimneys to carry off and dissipate the smoke. Those homes that had chimneys, had expensive iron chimneys, but the late fifteenth-century introduction of easily and cheaply made bricks in coal furnaces allowed everyone to have chimneys and to burn coal in their hearths.

## VIII. CONCLUSION

Renaissance scholars revered as well as challenged the dogma of the ancient scholars. Medieval scholars began an examination of the ancients' writings, but for the most part they had available only Arab translations of the original works. Renaissance scholars instead recovered these works in the original Greek and Latin and undertook their analyses on the original works and not on their translations.

A fusion of the craft tradition with the new world view of the natural philosophers occurred together with a resurgence of mathematics in measurement.

Industry and commerce, as in the Florentine and Bolognese textile industry, now became sources of wealth. Wealth no longer depended solely on land ownership.

Paracelsus challenged the four humor theory of Hippocrates and Galen and other popular Renaissance medical beliefs such as signatures and sympathetic remedies. He contributed significantly to the establishment of a new medical tradition called iatrochemistry, meaning medical chemistry, in which the emphasis in alchemy shifted from the transmutation of metals to the preparation of medicines from metals and plants. He believed life was a chemical process.

The 1543 publications of Vesalius and Copernicus represented the most serious challenge to the long dominant Aristotelian-Ptolemaic-Galenist science and marked the beginning of a new world view known as the mechanical philosophy of nature. Nature consisted of matter in motion, and the natural philosopher's objective aimed to discover the mathematical laws of motion that accounted for the operation and composition, that is the macroscopic and microscopic behavior, of nature. The publication of Isaac Newton's *Principia* in 1687 magnificently summarized the new mechanical philosophy.

## REVIEW QUESTIONS ON THE RENAISSANCE

1. What is the main theme of this chapter?
2. What are the characteristics often used to describe the Renaissance?
3. What is neoplatonism, gnosticism?
4. What is the most significant difference between medieval art and Renaissance art?
5. Discuss Leonardo's religious beliefs, his epistemology, ontology.
6. Do you see any Aristotelian influence on Leonardo, any of Galen's influence?
7. Is Leonardo a good example of what science was capable of accomplishing?
8. Was Leonardo's lack of education an advantage or a disadvantage?
9. How did Leonardo regard the relation of science and the arts?
10. Discuss Leonardo's contributions to science.
11. Discuss Leonardo's contributions to art.
12. What were Leonardo's contributions to anatomy and physiology?
13. How did Leonardo earn a living?
14. Why didn't Leonardo publish his voluminous collection of manuscripts?
15. Whatever happened to Leonardo's manuscripts after Leonardo's death?
16. What was Copernicus's astronomical model of the planets?
17. Why did he propose his model, and when did he publish it?
18. Discuss Vesalius's work in anatomy. Compare his methodology with that of Copernicus.
19. What new approach or outlook did Paracelsus bring to alchemy?
20. What did Paracelsus believe was the cause of disease and how did he propose to treat disease and illness?
21. What differences existed between the Renaissance alchemists and metallurgists? How did their books and other publications differ?
22. Identify: Mona Lisa, Francesco Melzi, Michelangelo Buonarroti, Cesare Borgia, Francis I, the philosopher's stone, alchemy, witchcraft, Count Dracula, Agricola.

# CHAPTER 5
# THE SCIENTIFIC REVOLUTION OF THE SIXTEENTH AND SEVENTEENTH CENTURIES
# 1600-1700

Theories are very thin and unsubstantial;
experience only is tangible
Hosea Bailey (1771-1852) American clergyman

## I. INTRODUCTION

In the seventeenth century the center of science shifted from Italy to northwest Europe, to England, France, and Germany.

The mechanical philosophy, or the study of matter in motion, dominated scientific thinking and characterized the scientific revolution of the sixteenth and seventeenth centuries. The major objective of the mechanical philosophers was to discover and to quantify with mathematical equations the laws that accounted for the motion of matter, from the invisible atoms to the visible planets. They succeeded in accounting for the motions of the planets and falling objects but not for the motion of atoms.

Other developments and advances of the scientific revolution included a continuing increase in experimentation; invention of new instruments, such as the telescope, microscope, barometer, vacuum pump, and thermometer; mathematical advances that led to the introduction of logarithms (and the slide rule), analytical geometry, and calculus. The number of scientists increased, scientific societies such as England's Royal Society and France's Academy of Sciences arose. They began publication of new journals, particularly the *Philosophical Transactions* of England's Royal Society in 1665 and the *Comptes Rendu* of the French Academy of Sciences in 1666.

Astronomy and physics advanced significantly. Galileo Galilei, Johannes Kepler, and Isaac Newton discovered laws of motion that accounted for the acceleration of falling bodies and the motion of the planets. These laws culminated in Newton's law of universal gravitation.

Investigations on light led to the construction of the first telescope and microscope and to an understanding of refraction. Newton's support of the particle theory of light made it the dominant theory compared to the wave theory which had only minimal support. Newton showed that white light was really a composite of seven colors, and for the first time astronomers calculated a value for the speed of light.

Physicists made the first measurement of the speed of sound in the mid-1600s. Robert Boyle demonstrated that sound waves, unlike light, required a medium for their transfer.

Evangelista Torricelli's studies on the atmosphere and on the existence of a vacuum, or whether nature abhors a vacuum as Aristotle claimed, led to his invention of the barometer. Investigations on heat, which seventeenth-century physicists believed was mechanical, or a consequence of matter in motion, led later to the invention of useful thermometers. The first useful thermometers appeared only in the 1700s.

Jan Baptiste van Helmont and Robert Boyle began the study of gases, but they had no suitable collecting device, which hampered their investigations. Nevertheless their investigations were the first stage in solving the problems of chemical identity which advanced significantly in the 1700s. Boyle in 1662 also established an inverse relation between pressure and volume, known today as Boyle's Law.

In the life sciences William Harvey's studies in physiology led to his discovery in 1628 of the circulation of the blood in the body and complemented Vesalius's publication in 1543 on the anatomy of the human body.

Instruments such as the telescope, microscope, barometer, and vacuum pump enabled the mechanical philosophers and atomists to promote empirical science and the scientific method and to measure the mechanical forces in nature. Advances in mathematics, such as the invention of logarithms (and the slide rule), decimals, analytical geometry, and calculus, allowed scientists to simplify calculations, to make calculations that they could not have made prior to these advances, and to put their experimentally obtained results in new quantified relations and equations.

The influence of those in society who earned their living from industry and commerce rather than from land ownership, as the old aristocrats did, began to grow. This was the case in England, France, Germany, Holland, Italy, and elsewhere in Europe.

In England during the reign of Elizabeth I (1533-1603) and the emergence of England as a world power, industry, agriculture, and the arts flourished. So did air pollution. It achieved new heights, reaching a point where Elizabeth, around 1590, complained of the great annoyance coal smoke caused in Westminster Palace. Her successor James I (1566-1625) tried to control the smoke problem by importing clean-burning anthracite coal from Scotland to replace the smoky bituminous coal from Lancashire, Durham, and Cornwall.

A telling account of England's serious air pollution problem appeared in 1661. John Evelyn (1620-1706), an English writer and philosopher, published *Fumifugium or the Inconvenience of the Aer and Smoak of London Dissipated* in which he proposed remedies, such as the creation of large public parks, to reduce London's air pollution problem. The air was so bad, he said, that you could hardly see across the street and scarcely breathe. In fact by 1700, 600 ships regularly transported sea coal (bituminous coal) from Newcastle to London compared to two ships in 1650. The problem in England, and soon to occur in other countries that underwent rapid industrialization, resulted because of the demand for coal by the iron industry and the navy. Both required huge amounts of coal for fuel which the dwindling forests of Europe could no longer provide.

## II. EXPERIMENTAL METHOD

### *William Gilbert: studies on magnetism*

William Gilbert (1544-1603) in London served as physician to Elizabeth I and James I. His research on magnetism exemplified the experimental approach to studying nature. It built on centuries-old knowledge of magnetic behavior.

From the time of Thales (640-546 BC) natural philosophers had recognized the magnetic behavior of lodestone (magnetite), a naturally-occurring iron ore, $Fe_3O_4$. Peter Peregrinus (1269) had investigated the attraction of unlike poles and the repulsion of like poles and constructed compasses. He believed incorrectly that a compass needle dipped because of its north pole's attraction to Polaris, the North Star.

In *De Magnete*, published in 1600, Gilbert offered a different explanation of magnetic dip. He showed that Earth acted as a magnet by noting the similar behavior of a pivoted compass needle toward Earth and toward a lodestone magnet that he shaped into a sphere and located its two poles. The compass needle increasingly dipped as he moved it from the equator of his spherical magnet to its top and bottom poles just as it did when it moved from Earth's equator to Earth's poles. Gilbert concluded that the magnetic dip of a pivoted compass needle increasingly dipped because of Earth's magnetism and that the attraction was strongest at its poles. A compass needle did not dip because its pole pointed toward Polaris, the North Star, as Peregrinus believed.

**Figure 1**
**Magnetic Dip**
Source: Newton Henry Black and Harvey Nathaniel Davis. *New Practical Physics.* New York: Macmillan Company, 1913.

*Cause of Earth's magnetism*

Gilbert concluded that Earth contained a fixed magnetic region or core within it. His explanation failed to account for a varying magnetic declination, and the discovery of the Curie point by Pierre Curie (1859-1906) in 1895 proved Gilbert incorrect. The Curie point is the temperature at which a substance loses its magnetism. For iron the Curie point is 760° C, whereas Earth's interior is 1000° C, making it too hot to have a fixed iron core that produces Earth's permanent magnetism. Iron has a melting point of 1539° C.

Gilbert extended his beliefs on magnetism to the heavens. He accepted the Copernican theory and believed that a magnetic force held the planets in their orbits. Gilbert also investigated static electricity. He rubbed amber with wool, glass with silk, and noted the similarities and differences between static electricity and magnetism.

Beginning with the nineteenth-century and early twentieth-century studies of Carl Gauss (1777-1855) and Wilhelm Weber (1804-91) at Göttingen in the 1830s, Bernard Brunhes (1867-1910) at the Puy-de-Dôme Observatory in Auvergne (France) in 1906, Joseph Larmor (1857-1942) at Cambridge in 1919, Walter Elessar (1904-91) then with the Army Signal Corps and Radio Corporation of America in 1946-47 and Edward Bullard (1907-80) at the University of Toronto and the National Physical Laboratory in London in 1950 and 1954, scientists have shown that Earth is a weak magnet with a magnetic field strength of 0.25-0.75 oersteds (1 Oersted = $250/\pi$ amperes per meter), whereas a good horseshoe magnet has a magnetic field strength of 1,000 oersteds.

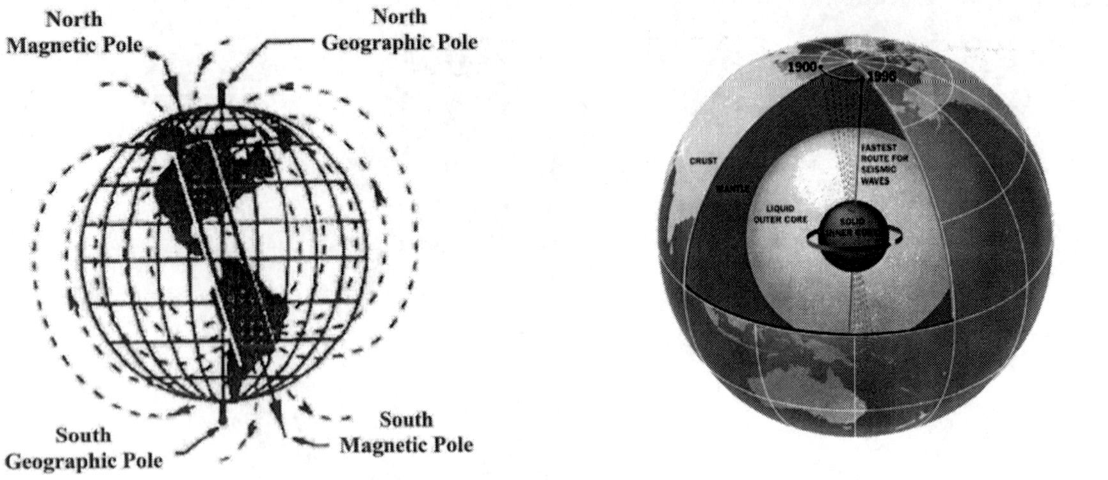


**Figure 2**
**Earth's Magnetic Field Showing Declination (left), Earth's Core (right)**
Source: US Naval Administration.tpub.com (left); US Geological Survey (right)

The twentieth-century dynamo theory that originated in the work of Elasser and Bullard holds that Earth's iron-nickel magnetic region, or core, is fluid-like and movable, consisting of whirlpools or eddies circulating in the core. According to the theory, Earth's rotation causes a convective motion of the iron-nickel fluid core, and the core's motion induces a weak electric current that generates a self-sustaining magnetic field. Brunhes in 1906 first observed that the magnetic poles reversed, or switched, their positions. The dynamo theory explains why the magnetic poles change their locations and reverse their positions every 100,000 years.

*Observations on magnetic declination*

Magnetic declination is the difference in degrees between geographic north (true north or the northernmost point on Earth's surface located at 90° north latitude) and magnetic north. Columbus noted the changing declination of his compass during his voyage of 1492. It went from east of geographic north in Spain, shifted westward as he crossed the Atlantic Ocean, pointed due north at mid-ocean, and then west of geographic north thereafter. In Gilbert's day magnetic north in London was 11° east of geographic north in 1580 and 6° east in 1600. In 1657 the two poles lined up, in 1820 the declination was 24° west of geographic north, in 1958 magnetic north was 7° west of geographic north. In 2004 magnetic north was 6° west of geographic north.

### *Francis Bacon*

Francis Bacon (1561-1626) was a philosopher of science in London. He rose to the highest ranks at the English court under Elizabeth I and James I before his downfall in 1621 because of accepting bribes. Bacon pleaded guilty but said the bribes never influenced his judgment on issues he decided while holding crown positions as Lord Keeper, Lord Chancellor, and Viscount St. Albans. The court sentenced Bacon to the London Tower and fined him £40,000, an enormous sum, but James I commuted Bacon's punishment. Bacon died a few years later of pneumonia, allegedly brought on by a chill he caught when testing ice as a preservative for food.

*Philosophy of science: the inductive method*

In *The Advancement of Learning* published in 1605 Bacon attacked mysticism and its belief that through intuition the mystic can gain control over the non-mechanical forces in nature, such as the neoplatonists and gnostics believed, or the belief that the universe was a living organism (animism), vitalism, and the natural motions of the four elements.

Bacon in 1620 wrote *Novum Organum* in opposition to Aristotle's *Organum* and the use of deductive reasoning in logic. He argued correctly that if the initial premise of a syllogism is incorrect it will lead to an erroneous conclusion. Therefore natural philosophers must establish beyond doubt the correctness of their starting position. Experiment and the inductive method in which we obtain knowledge from the senses, he said, offered the best way to accomplish this. Knowledge is empirical and results in control over nature. Knowledge is power.

*Program for doing science*

Bacon believed scientists had adopted the wrong scientific methodology. They separated theory, observation and experiment, and practical application. They used deduction rather than induction to learn about nature, and they appealed to the infallible authority of Aristotle instead of investigating nature themselves. Books must follow sciences, he said, not sciences following books.

Bacon warned that we must beware of idols, by which he meant blindly accepting the authority of others or accepting what everyone else believed.

His never-completed *New Atlantis* of 1618 (published posthumously in 1627) was a utopian work in which he laid out his program for the complete investigation of nature. The House of Salomon was a research center in which nine groups of workers, planners, practitioners, recorders, and theorists, investigated nature. There were 36 researchers altogether, 12 collectors and 3 each of the other 8 groups. Bacon's program for doing science contributed to the guiding philosophy of the Royal Society at the time of its formation in 1660, *nullius in verba*, or *take nobody's word for it.*

*Significance*

Bacon gave respectability to experimental science. Not only artisans and craftsmen but also scientists should work with their hands. He failed to see the importance of mathematics in science because much of it, such as geometry and trigonometry, was deductive, and he allowed no role for creativity and imagination. Bacon's scientific program placed too much emphasis on methodology as the key to scientific understanding or discovery.

He rejected the Copernican model of the universe because he believed no knowledge obtained from the senses or from experimentation supported Copernicus's heliocentric universe.

## III. GALILEO GALILEI

Galileo (1564-1642) was born in Pisa, the same year as William Shakespeare (1564-1616) in Stratford and London, and the year Michelangelo Buonarotti (1475-1564) in Rome died. Galileo studied medicine and then mathematics at the University of Pisa from 1581-85. He taught at Pisa (1590-92), Padua (1592-1610), and served as ducal philosopher to the Medicis in Florence from 1610 to 1642.

Galileo never married but lived from about 1599 to 1610 with his mistress Marina Gamba (1570-1612) with whom he had three children, two daughters both of whom became nuns, and a son Vincenzo, a talented musician like Galileo's father.

Galileo believed in quantifying science. He held neoplatonic views and asserted that mathematics provided the key to understanding the universe. In *Il Saggiatore (The Assayer)*, which he published in 1623 and dedicated to his friend Pope Urban VIII (1568-1644), Galileo divorced natural philosophy from science (mathematics). Natural philosophy, he said, may ask questions about the ultimate causes of reality, but it can never answer them with certainty. Only mathematics provides the human mind with the certainty it seeks. Galileo stressed the importance of quantitative experiments and hypothetical deductive reasoning as a philosophy of science. He adopted an idealized description of nature to study and understand the workings of nature, particularly the laws of falling bodies. Galileo developed the laws of motion to support the heliocentric model of the universe, which he accepted some years prior to 1597. He mentioned his acceptance in a letter he wrote on 4 August 1597 to Johannes Kepler. His early publications on mechanics (dynamics) include *De Motu* (1590) and *Le Meccaniche* (editions of 1593, 1594, 1600).

**Figure 3**
**Bust of Galileo**
Courtesy of the author
History of Science Museum, Florence

***Experiments on mechanics: falling bodies***

*Leaning Tower of Pisa*

The architect Bonannus of Pisa (Bonanno Pisano) began construction of the bell tower of the Cathedral of St. Mary of the Assumption in 1174, its completion occurred in 1350. The 14,000-ton tower stands 180 feet tall and leans about 13 feet off the perpendicular. It started to lean after workers completed only 35 feet because it sits on a water table 50 yards underground. The tower has 294 spiral steps leading to the top. Contrary to popular belief, when Galileo lived in Pisa, he never dropped weights from the tower to demonstrate the laws of falling bodies. He did drop weights from other heights.

**Figure 4**
**Leaning Tower of Pisa**
Source: Library of Congress (left); 135pixels/Shutterstock.com (right)

*Isochronism of a pendulum*

Galileo first demonstrated the isochronicity of a swinging object in 1583 when he watched a large lamp swinging in the cathedral in Pisa. He used his pulse to measure time. In subsequent laboratory experiments Galileo showed that for pendulums of equal length the time for each complete swing (period) was the same for large and small amplitudes (distances). Galileo recognized that the pendulum is a falling body. Its motion showed that a falling body accelerated uniformly as it fell.

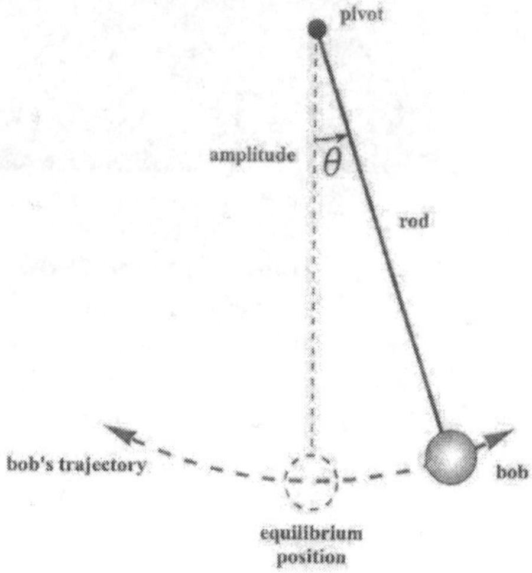


**Figure 5**
**Isochronicity of a Pendulum**
Courtesy of the author

*Inclined plane experiments*

Jordanus de Nemore (1225-60), possibly an Italian mathematician or a German Dominican priest in Paris, first used an inclined plane, down which he rolled bodies to study their motion. Much of his work remained unpublished until the sixteenth century.

Simon Stevinus (1548-1620), a Belgian mathematician-physicist-engineer in Holland, in 1586 experimented with an inclined plane and also dropped different weights simultaneously from a height in trying to determine the laws governing the motion of falling bodies. His book on astronomy *De Hemelloop* (1608) was one of the first to support the Copernican model of the universe.

Galileo's inclined plane experiments of 1602-5 showed that a body, such as a ball, accelerated uniformly as it rolled down the plane. He proved in 1605 that the rate of fall did not depend on the body's weight, and he believed that in the absence of external resistance, such as the air, all freely falling bodies, falling for equal distances, should reach the ground at the same time. Galileo measured the distance and used his pulse, a sandglass, or a water glass to measure time. With this experimental data he developed equations or laws for calculating the final velocity of the falling body and the total distance the body traveled.

**Figure 6**
**Galileo's Inclined Plane**
Courtesy of the author
History of Science Museum, Florence

Galileo was extending Archimedes's reasoning that a body sank (fell through a liquid such as water) or floated depending on its relative density compared to the liquid's density, to argue that for a falling body its weight did not affect its rate of fall. The density of the body and the density of the medium, which offered resistance to the fall, determined the rate of fall so that if no medium existed to affect the rate of fall, as in a vacuum, all bodies would fall at the same speed.

The equations for a falling body that Galileo derived for calculating final velocity ($v_f$), acceleration (a), and distance (d), where $v_a$ = average velocity and t = time, follow.

*Uniformly accelerated motion*

Galileo clearly explained acceleration and its effect on the motion of bodies. He performed many experiments to determine the rate at which a body fell by measuring the rate at which a ball rolled down an inclined plane.

In one of Galileo's experiments he raised one end of his inclined plane steeply enough to allow a ball to roll down at the rate of one foot in the first second. As the ball rolled down the plane, he determined its position at one-second intervals. Galileo's data appears in the accompanying table.

| GALILEO'S ACCELERATION EXPERIMENTS | | | |
|---|---|---|---|
| ***Elapsed time (sec) Measured*** | ***Total Distance (ft) Measured*** | ***Distance Each Second (ft) Calculated*** | ***Final Velocity (ft/sec) Calculated*** |
| 1 | 1 | 1 | 2 |
| 2 | 4 | 3 | 4 |
| 3 | 9 | 5 | 6 |
| 4 | 16 | 7 | 8 |
| 5 | 25 | 9 | 10 |

The distance the ball rolled each second is 2 feet greater than it rolled in the previous second so that its velocity after 1 second is 2 ft/sec and its acceleration, or gain in velocity for each second, is therefore 2 ft/sec or 2 ft/sec². The initial velocity of the ball is zero. Its acceleration, or gain in velocity, is 2 ft/sec² and its velocity at the end of *the first second* is therefore 2 ft/sec. If at that time the ball stopped accelerating, it would continue to move with a constant velocity of 2 ft/sec. At the end of the next second, its velocity has increased to 2 x 2 ft/sec, or 4 ft/sec. In five seconds its original velocity has increased to 5 x 2 ft/sec, or 10 ft/sec, and its velocity at the end of the fifth second is 10 ft/sec. Therefore, *the final velocity of an object starting from rest equals the product of the acceleration and the elapsed time. If an object does not start from rest, its final velocity will equal the sum of its initial velocity and the increase in velocity produced by the acceleration.*

To understand the difference between average velocity and final velocity think of yourself sitting in a uniformly accelerating automobile where you can read the velocity on the speedometer at each selected time of travel. You can record the final velocity at each selected time and you can calculate the average velocity from the known distance traveled and the total time required to travel the known distance.

The final velocity of a uniformly accelerating body is $v_f = at$.

The average velocity of a uniformly accelerating body is

$$v_a = \frac{(initial\ velocity + final\ velocity)}{2} \quad \text{or} \quad v_a = \frac{(o + a\,t)}{2}$$

The distance traveled by a uniformly accelerating body,

$$d = v_a t \ or\ d = \frac{at^2}{2} \ or\ d = \frac{1}{2}at^2$$

For an inclined plane, the acceleration depends on the ratio of $H$, the height of the plane (increasing it), to $L$, the length of the plane (decreasing it), so that

$$a \approx \frac{(H)\ height\ of\ plane}{(L) length\ of\ plane}$$

where $g$ (for gravity) is a proportionality constant, $a = g\frac{H}{L}$.

In the diagram below Θ is the angle of elevation and

$$\sin \Theta \approx \frac{opposite\ side}{adjacent\ side} \text{ or } \frac{H\ height}{L\ length}.$$

Rearranging the equation gives

$$g = a\frac{L}{H}.$$

Measuring $a$, $L$, and $H$ experimentally enabled Galileo to calculate $g$.

For a vertically inclined plane, which is equivalent to a freely falling body, $H = L$ and $g = a$, or 32 ft/sec², or 9.8 meters/sec². Galileo's experiments demonstrated that a freely falling body experiences uniform acceleration.

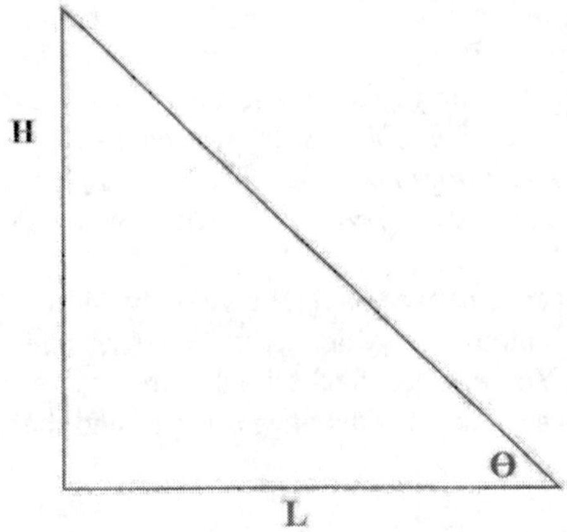


**Figure 7**
**Galileo's Calculation of Acceleration**
Courtesy of the author

*Relative motion*

Galileo believed that a falling body obeyed the same laws of motion whether Earth was at rest or in uniform (non-accelerated) motion. His early statement of the *law of inertia* correctly proposed that a falling body behaved the same way whether Earth was at rest or in uniform (non-accelerated) motion. Galileo argued correctly that a weight dropped from a height such as a ship's mast would fall in a straight line path regardless of whether the ship was moving uniformly or at rest. This experimental result is correct for the motion of an object on any uniformly moving body, such as a train or airplane. Galileo used his law of inertia to argue that the laws of motion we observed were the same whether Earth was at rest or in uniform motion.

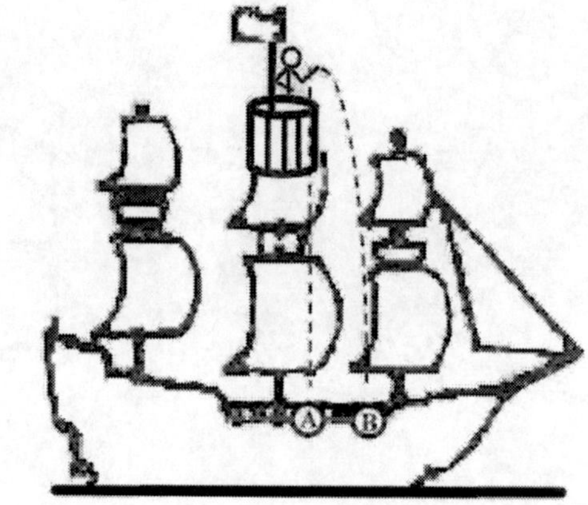


The sailor in the crow's nest drops a stone. Does the stone fall straight down A or follow a curving path B as a result of the ship's forward motion? It falls straight down.

**Figure 8**
**Galileo's Law of Inertia**
Courtesy of the author

*Composition of forces*

Galileo recognized in 1608-9 that the actually observed motion of a projectile is the result of a composition or sum of forces acting on the projectile. Horizontal and vertical (gravity) forces act simultaneously on a projected body such as a cannon ball. The resulting path is a parabola (b) below.

The combination of a horizontal and vertical force acting on a body explained why a moving Earth did not leave behind objects not firmly attached to it. Galileo had answered the anti-Copernican argument that an Earth rotating and revolving through space would leave the birds behind.

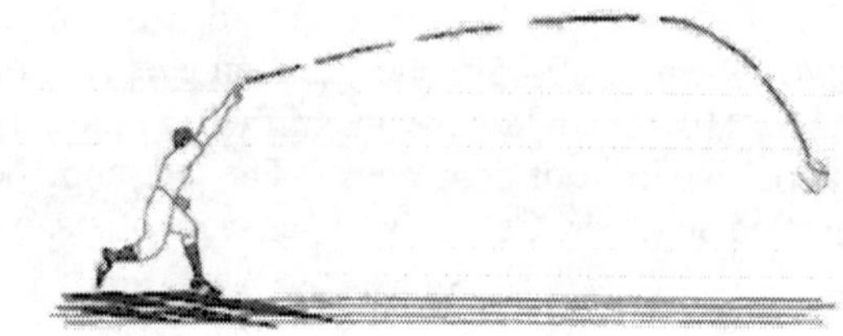

a

b

**Figure 9**
**Composition of Forces: Select the Projectile's Pathway**
Courtesy of the author

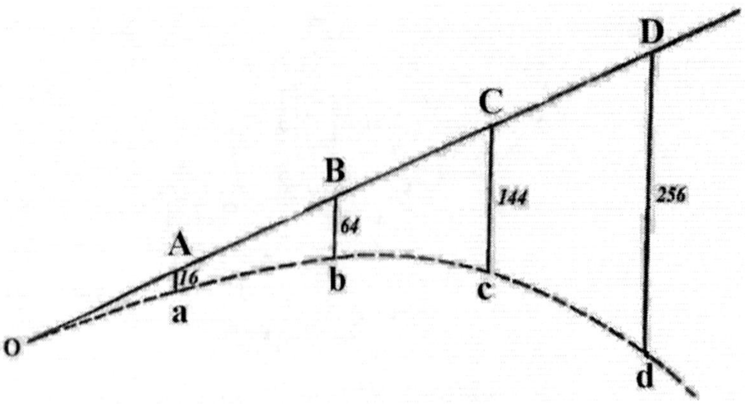


**Figure 10**
**Parabolic Orbit of a Projectile**
Courtesy of the author

***Defense of the Copernican model: casting doubt on Aristotle***

*Invention of the telescope*

Hans Lippershey (1570-1619), an optician and spectacle maker in Middleburg, Holland, constructed the first refracting telescope and petitioned for a 30-year patent on 2 October 1608. The General States of the Hague declined Lippershey's petition because other spectacle makers had made the same claim. Lippershey arranged a concave eye lens and a convex objective lens (the lens toward the object) in an opaque (not transparent to light) tube 50 cm long and 3-4 cm in diameter. It magnified about 3-4 times and was similar to an opera glass. He thought the telescope would find primary use by the military.

Galileo and many other astronomers soon heard of Lippershey's invention, and in May 1609 Galileo constructed a refracting telescope in which he enclosed two convex lenses, a plano convex objective of long focal length (76-89 cm or 30-35 inches) and an eyepiece of short focal length (about 5 cm or 2 inches) in a lead tube. His first telescope magnified 3 times. He built two others that magnified 32 times which was about the limit of their power. The telescope he used to view the Moon and other heavenly bodies magnified 30 times. Galileo's lenses were of exceptional quality.

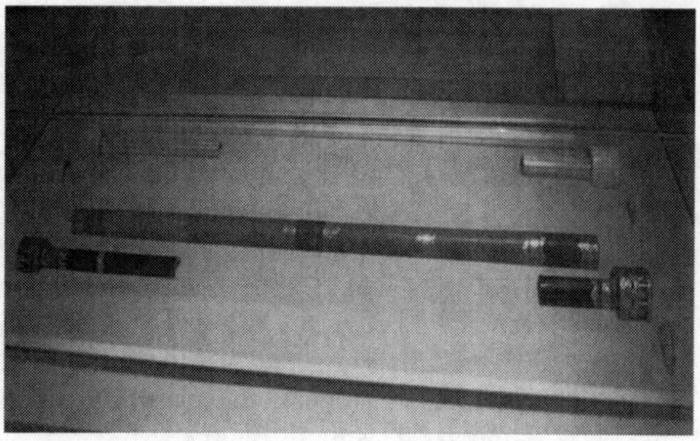

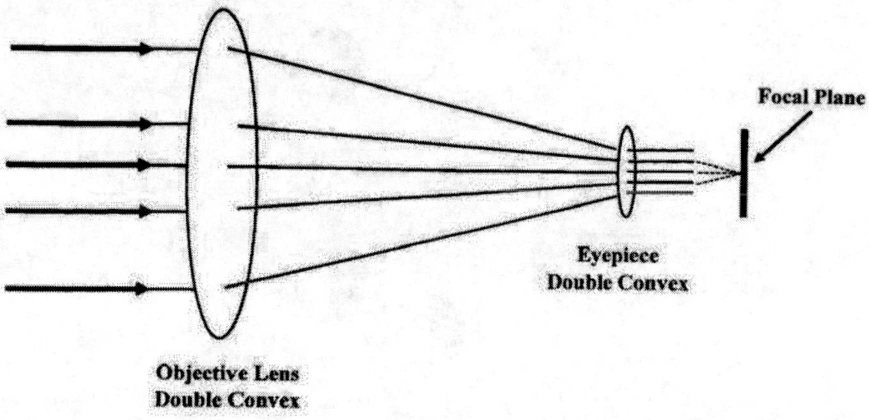


The observer places his eye at the focal plane of a double-convex lens to see the illuminated objective. An eyepiece magnifies the real image produced by the objective.

**Figure 11**
**Galileo's Telescopes**
Courtesy of the author
History of Science Museum, Florence

*Galileo's discoveries with the telescope*

Galileo in 1609 showed that the Moon had a rough surface and was not a perfect sphere. This suggested that perhaps the other heavenly bodies did not have perfect surfaces and were similar to Earth. This belief is implicit in Democritus's atomic theory and in Nicholas of Cusa's writings. Galileo showed next that the Milky Way consisted of many stars and that Jupiter had four moons or satellites. He named the four moons the Medicean Stars in honor of his soon-to-be patron, the Medici family in Florence. This last discovery had great significance because it demonstrated that not all heavenly bodies circled Earth. Nor did Jupiter leave its moons behind as it moved through the heavens. Galileo therefore argued that if Earth moved through space it would not leave its Moon behind.

Galileo published these discoveries in *Siderius Nuncius* (*Starry Messenger*) in 1610. He dedicated the book to Cosmo II de Medici (1590-1621), the Grand Duke of Tuscany, who then appointed Galileo ducal philosopher and mathematician with an annual salary of 1,000 gold florins (one gold florin weighs 3.5 grams or 0.0123 ounces). Galileo now left Padua and in 1610 moved to a villa at Arcetri near Florence where he remained for the rest of his life.

The publication of the *Starry Messenger* brought Galileo immediate fame, and in 1611 he went to Rome where Jesuit astronomers, such as Christopher Clavius (1537-1612), confirmed his telescopic discoveries.

Galileo announced two additional telescopic discoveries in *Letters on Sunspots* published in 1613. In this book Galileo endorsed the Copernican model and pointed out that the two discoveries, the phases of Venus and the seasonal variation of the sunspots' paths, supported a Sun-centered model. Venus (and the other interior planet Mercury) would show phases, he said, only if it revolved around the Sun. It always would appear as a crescent in the Ptolemaic model of the universe because it always remained between Earth and Sun. This discovery also established that the planets shone by reflection of sunlight. He noted that the sunspots rotated every 27 days. Galileo intended in writing *Letters on Sunspots* to counter the planetary model that Tycho Brahe proposed in 1599 and had gained the support of Jesuit astronomers.

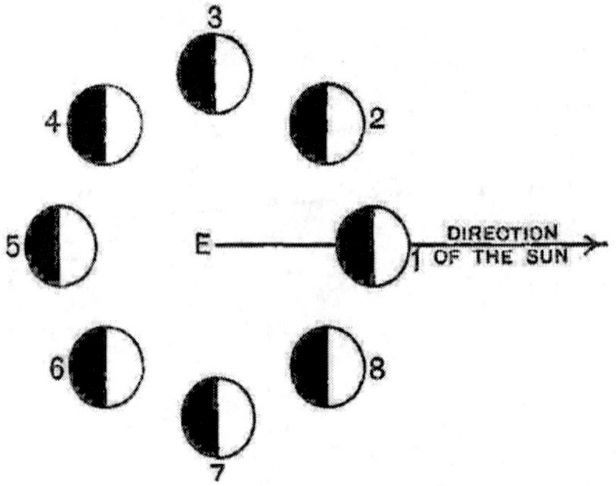


**Figure 12**
**Phases of the Moon or a Planet**
Source: Arthur Berry. *A Short History of Astronomy: From Earliest Times Through the Nineteenth Century.* London: John Murray, 1898.

*Biblical issues*

At this time the focus of the Earth-centered Sun-centered debate shifted from astronomy to theology mainly because Galileo's opponents failed to discredit his scientific discoveries and because of the ongoing religious controversy between Catholicism and Protestantism, each of which considered itself the true Christian Church. The Protestant Reformation had begun in 1517 with Martin Luther and his 95 theses attacking the Catholic Church's granting of indulgences and its other doctrines. The Catholics responded with the Counter Reformation and the Council of Trent in 1545-63. The outbreak of the Thirty Years War of 1618-48 aggravated further the tense situation.

The debate shifted to biblical statements, such as a stationary Earth in *Psalms 18:6-7, 92:1, 103:1 and 5*, where God fixed Earth upon its foundation not to be moved forever, and the Sun's rising and setting in *Ecclesiastes 1:5-6*, where the Sun rises and the Sun goes down, then it presses on to the place where it rises; and in *Josue (Joshua) 10:12-13*, in which Josue commands the Sun to stand still.

As a consequence of the Bible's statements about a stationary Earth and a moving Sun, Pope Paul V (1550-1621) and the Congregation of the Index, established in 1571 by Pope Pius V (1504-72), suspended Copernicus's book from 1616 to 1620 during which time the Congregation made corrections. It revised 10 lines in the preface to make the Copernican model more hypothetical and to indicate that the Copernican model was not ontologically true but was only a calculating model. At the same time the Congregation also prohibited and suspended two other writings. Diego de Zúñiga (1536-97), an Augustinian friar from Salamanca, Spain, in 1584 had written *On Job*, a commentary on the *Book of Job*, in which he supported Earth's motion. Thirty years later in early 1615 Paulo Antonio Foscarini (1580-1616), a Carmelite friar at the University of Messina, published *Letter of opinion over the Pythagorean and Copernican opinion concerning the mobility of the earth and the stability of the sun* (English translation 1661), in which he tried to reconcile Copernicanism and the Holy Scriptures. Foscarini's book appeared one year before the Congregation banned Copernicus's *De Revolutionibus*.

Because of the impending conflict with church authorities, Galileo decided to go to Rome to discuss the Copernican issue. The essence of his argument was that the Bible shows us the way to go to Heaven, not the way the Heavens go. At his first hearing in Rome on 26 February 1616 Cardinal Roberto Bellarmine (1542-1621), the leading theologian in Rome and the Pope's spokesman, told Galileo not to hold or defend the Copernican theory because it contradicted scriptures. Galileo agreed to this and received a written statement from Bellarmine indicating that Galileo had not recanted his position nor had he received a penance.

But at this same hearing Michelangelo Segizi (1585-1625), a Dominican and the Commissary-General of the Congregation of the Holy Office (the Congregation of the Inquisition or Roman Congregation), which the Catholic Church established in 1542 to combat heresy, intervened and ordered Galileo that he must not teach, hold, or defend Copernicanism verbally or in writing. Bellarmine may have told Galileo to ignore Segizi's order because he was in charge of the hearing and that Segizi was to intervene only if Galileo did not agree to Bellarmine's pronouncement. Segizi or someone else included the order and a report of the 1616 hearing in Galileo's file in the Holy Office. It became the major source of controversy 17 years later at Galileo's trial in 1633 and has remained controversial.

In his hearing with Galileo, Bellarmine told Galileo that for the Catholic Church to accept the Copernican model, he would have to give a demonstration of Earth's motion, something neither Augustine nor Thomas Aquinas, the Catholic Church's foremost authorities, had asked for in situations where faith and science appeared to conflict. Augustine and Aquinas said the Bible was not intended to teach science, and the Church's scholars should not invoke the Bible's authority in scientific disputes. Though most of the Church fathers accepted Earth's immobility, not one of them held that Catholics had to accept Earth's immobility as revealed truth. Room for an alternative exegesis (interpretation) still existed. Bellarmine erred in his demand. If he and Galileo had engaged in scientific debate, Bellarmine had made a reasonable, in fact, necessary demand that Galileo provide evidence supporting Copernicanism. But Bellarmine and Galileo were debating whether the Bible was an authoritative book of science and whether scholars had to accept everything in it as literal truth and divinely inspired.

Galileo did concede to scripture, being divinely inspired, superior authority even in science unless contrary physical arguments were demonstrative. Nevertheless the 1616 ruling represented only the fallible decision of a Roman congregation.

*Trial and condemnation: second hearing in Rome*

In the years following his 1616 hearing with Bellarmine, Galileo wrote *Il Saggiatore* (*The Assayer*), his 1623 publication on atoms. He believed that it was necessary to assign two kinds of qualities to atoms in order to understand their behavior. *Primary* (inherent) qualities were properties, such as size, shape, and motion, present in the atom. *Secondary* (subjective) qualities, such as taste, color, hotness or coldness, were in the observer or what the observer experienced. *Il Saggiatore* served another purpose. Galileo used it to attack the Jesuit astronomer, mathematician, and architect, Orazio Grassi (1590-1654), which cost him the support of the Jesuits. Grassi in 1619 had written on the three comets that appeared in 1618. Galileo believed Grassi's paper supported Brahe's geocentrism, not Copernicus's heliocentrism.

The Copernican issue, nevertheless, remained quiet for seven years until Galileo met six times in 1623 with his friend Maffeo Barberini (1568-1644), the new Pope Urban VIII (1623-44), to discuss his writings on Copernicanism. Urban told Galileo that as long as Galileo treated Copernicanism hypothetically and did not try to prove the correctness of the Copernican model he could write about it. Galileo could treat the Copernican model as a calculating device, but not as ontologically true. Galileo then began work on a book about Copernicanism, and upon its completion he sent it to the Church's censors who approved it after revising the preface and conclusion. It received an imprimatur (the Catholic Church's seal of approval) and went to press in Florence in 1632. Entitled *Dialogue on the Two Chief World Systems*, the book produced an uproar from the Pope and Catholic Church authorities. The Inquisitor in Florence tried to stop publication, but the book sold out. Galileo received a summons to appear in Rome in 30 days to answer the charges made against him by the Congregation of the Index and the Congregation of the Holy Office.

In the *Dialogue on Two Chief World Systems* three participants, an Aristotelian (Simplicio), a Copernican (Salviati), and a neutral speaker (Sagredo), engaged in a debate in which Galileo tried to prove the Copernican model. The two proofs he offered for Earth's motion were the seasonal variation of sunspot paths and the tides which he thought, incorrectly, resulted from Earth's daily and annual motion. Galileo disregarded Kepler's argument that the tides resulted from the Moon's motion.

*The trial, two charges made against Galileo in 1633*

The 1633 hearings before the seven cardinals of the Congregation of the Holy Office lasted from 21 April to 21 June. The first charge was that Galileo had disobeyed the order not to teach, hold, or defend Copernicanism verbally or in writing. He said nothing of this order, which the Commissary-General of the Holy Office, Michelangelo Segizi, had given him in 1616, and a report of the 1616 hearing found in Galileo's Inquisition file. The original copy of the order is missing. Galileo's Inquisition file contains only a transcribed report. Bellarmine died in 1621 and Segizi in 1625, leaving no one to explain what happened at the 1616 hearing. The reason for Galileo's silence is unclear.

**Figure 13**
**Roberto Bellarmine, Embalmed Body in Church of St Ignatius in Rome, not far from the Vatican**
Courtesy of the author

The second charge was that Galileo had presented Copernicanism as a fact rather than a hypothesis.

At the conclusion of the hearings the Congregation of the Holy Office found Galileo guilty of Copernicanism and suspected of holding a heretical doctrine. After some persuasion, Galileo recanted and received a salutary penance of having to recite the seven penitential psalms once a week for three years. His daughter Virginia Galilei (1600-34), who became a nun, Sister Maria Celeste in Arcetri, recited the Psalms for him. Galileo also received a prison sentence of house arrest in Florence for the rest of his life. Church authorities commuted the sentence after the trial.

Had the religious wars in Europe, specifically the Thirty Years War of 1618-48, not occurred and the political struggles within the Catholic Church among Jesuits, Dominicans, and the Pope not arisen, the debate over Copernicanism and Galileo would have remained a far less serious issue.

*Aftermath of the trial*

Pope Urban VIII forbade the construction of a monument on Galileo's casket in Santa Croce Church in Florence. The first monument appeared only in 1737.

**Figure 14**
**Galileo's Index Finger and Thumb of Right Hand ca. 1737**
Courtesy of the author
History of Science Museum, Florence

The Catholic Church acknowledged heliocentrism and abrogated the general ban against Copernicanism in 1758, but the *Dialogue* remained banned (remained on the Index) until 1822. The Catholic Church abolished the Index in 1966.

Only recently, in 1992, has Pope John Paul II (1920-2005) and the Catholic Church cleared Galileo of heresy charges, although it accepted Copernicanism in1758 and had actively discussed the Galileo issue since 1964. At most the Holy Office in 1633 should have charged Galileo with disobedience not heresy.

***Final years***

Galileo lost his sight in 1637. The next year Galileo published *Two New Sciences* (1638) in Leiden, Holland, in which he summarized his years of study on mechanics. *Two New Sciences* contained proof of the law of fall in a vacuum, the principle of the independence of forces, and the theory of parabolic ballistics.

An alternative interpretation of Galileo's trial by Pietro Redondi in *Galileo Heretic* published in 1987 suggested that the real issue was Galileo's belief in atoms as unchangeable. This implied that no real change of bread and wine (transubstantiation) took place in the Eucharist and therefore challenged the central belief of Catholicism. Galileo's rival, the Jesuit Grassi, may have been the source who raised this charge against Galileo.

## IV. PROBLEM OF CHEMICAL IDENTITY IN SEVENTEENTH-CENTURY CHEMISTRY

Chemistry began to emerge as a separate discipline and with its emergence came the principal problem of chemical identity. Natural philosophers knew 14 of today's elements, among them gold, silver, iron, copper, mercury, tin, lead, zinc, and sulfur, but they had no clear understanding of the difference between an atom and a molecule, or between a simple substance and a compound substance. They could prepare the same substance, such as carbon dioxide gas or sulfuric acid, in different ways but had no sure way of establishing its chemical composition and the identity of substances.

Confusion existed over transmutation versus displacement. Did the reaction of iron filings with a copper sulfate solution convert iron into copper (transmutation) or did the iron merely displace the copper (ions) from the solution (displacement)?

***Jan Baptiste van Helmont***

Jan Baptiste van Helmont (1579-1644), the physician, chemist, and mystic in Brussels who shunned academic titles, represents the beginning of the transition from alchemy to chemistry and of investigations that contributed to solving the problem of chemical identity. His son Franciscus van Helmont (1614-98-99) edited and published his father's writings entitled *Ortus medicinae* in 1648. *Ortus Medicinae* appeared in 1662 in English translation as *Physick Refined.*

Helmont accepted alchemical teachings and gave an account in *Physick Refined* in which the philosopher's stone converted 2,000 times its weight of mercury into gold. He also believed in spontaneous generation, which was the generation of the living from the non-living, such as mice arising from dirty wheat, bugs arising from flour or meal, or larvae from putrefaction or decay of foodstuffs.

*Quantification in chemistry*

Despite his belief in alchemy and spontaneous generation, Helmont practiced quantification in chemistry, and in his experiments he used an analytical balance to weigh the amounts of the reactants and products.

In the willow tree experiment published posthumously in 1648 Helmont grew a willow tree for five years adding only water. The tree gained 164 pounds, the soil lost only 2 ounces. He maintained that water and not earth accounted for the tree's growth and did not recognize the importance of carbon dioxide and the photosynthesis reaction in plant growth. The willow tree experiment convinced Helmont that water was the only element.

$$CO_2 + H_2O \rightarrow CH_2O + O_2$$

In a second quantification experiment Helmont heated earth (sand, silicates) and alkali to give a glass which when reacted with or exposed to air gave a watery solution. Adding nitric acid restored the original earth. He compared the weights of the reactants and products and found them to be nearly the same. This experiment also provided evidence that earth did not react as a fundamental substance or an element.

*Other chemical experiments: study of gases*

Helmont recognized the difference between displacement and transmutation. In a displacement reaction one metal displaced another from a solution of a salt, such as iron displacing copper when added to a copper sulfate solution. He showed that metals dissolved in acids.

$$Fe + CuSO_4 \rightarrow Cu + FeSO_4$$

$$\text{solid} + \text{solution} \qquad \text{solid} + \text{solution}$$

Helmont was one of the first to study gases (chaos), including those that the human body emitted. He believed that gases consisted of invisible atoms. He prepared carbon dioxide, sulfur dioxide, hydrogen, methane, and the gases produced upon igniting gunpowder (nitrogen, nitrogen oxides, carbon dioxide, sulfur dioxide). He had no suitable apparatus for collecting the gases other than animal bladders and could not study the properties of the gases adequately. For this reason Helmont has not received credit for the discovery of the gases he investigated.

Helmont recognized, sometimes from smell or color, that he could prepare the same gas from different sources. Sulfur dioxide resulted from burning sulfur or distilling metallic sulfates of copper, zinc, or iron. He dissolved the gas in water. Carbon dioxide resulted from burning wood, fermentation, or decomposition of carbonates.

He observed the fixation of a gas, such as carbon dioxide, in another substance. Lime ($CaO$) fixed (reacted with) carbon dioxide to give limestone ($CaCO_3$), heating the limestone product released the fixed gas. Heat, fermentation, and chemical reaction released fixed gases.

Contrary to what the Aristotelians claimed, Helmont, like Galileo and Evangelista Torricelli, believed in the possibility of a vacuum. A lighted candle set on a dish that floated on water and then enclosed in an airtight container, produced a vacuum, because, as Helmont recognized, the burning candle consumed the air in the container. The water, in which the carbon dioxide gaseous product resulting from the burning dissolved, rose to fill the empty space.

*Teachings on medicine*

Helmont accepted Paracelsus's theory of the specificity of disease caused by foreign particles getting into the body.

***Robert Boyle***

Robert Boyle (1627-91), an independently wealthy London chemist who experimented in his private laboratory, helped to separate chemistry from medicine and to establish it as a distinct science with his publication of the *Sceptical Chymist* in 1661.

Boyle adopted a Baconian (inductive) approach in science. He believed in atomism, or the corpuscular theory, which he derived as a result of his studies on gases, and gave a clear statement of his atomic views in 1666. Atoms were the fundamental particles that composed everything. Boyle, like Galileo, assigned two kinds of qualities to atoms, primary and secondary, in order to classify them and to understand their behavior.

*Primary* (inherent) qualities of atoms were properties, such as size, shape, and motion that existed in the atom.

*Secondary* (subjective) qualities, such as taste, color, and hotness or coldness, were properties that resided in the observer or that the observer experienced.

Boyle carried out significant qualitative analyses of many chemical substances. He introduced acid-base indicators, such as syrup of violets and litmus paper, and used them to classify substances as acid (from the Latin *acere* meaning sour), alkali (from the Arabic *al-qily* meaning to roast in a pan), and neutral. He used flame tests and precipitation as other chemical identification techniques.

Boyle defined an *element* as any substance not consisting of or broken down into simpler substances and believed that natural philosophers must perform experiments to establish if a substance was an element.

Boyle accepted the transmutation of base metals into gold. He favored repeal of the British law against transmutation of base metals into silver and gold, which he tried to do in his laboratory because he saw transmutation as a possible way of increasing England's wealth. Parliament passed the Act against Multipliers (transmutation of base metals into silver and gold) in 1404 during the reign of Henry IV (1366-1413) who reigned from 1399 to 1413. The repeal, called the Mines Royal Act 1689 (An Act to prevent Disputes and Controversies concerning Royal Mines), passed in 1689 under William III (1650-1702).

## V. STUDIES ON AIR: BAROMETER AND AIR PRESSURE EXPERIMENTS

### *Contributions of Torricelli, Pascal, and Guericke*

Aristotelians argued that water rose in a lift pump because nature abhors a vacuum. Therefore they accepted the possibility of pumping water to any height in order to prevent a vacuum from forming. Yet anyone who actually pumped water knew that the pumping elevated water only to a maximum height of about 33 feet. Many experimenters accepted a maximum height of 33 feet because they believed the water column became so heavy that it collapsed from its own weight.

Evangelista Torricelli (1608-47) in Florence, a Copernican who served for three months in 1641 as Galileo's assistant and after Galileo's death as mathematician to the court in Tuscany and professor at Pisa, discussed the vacuum problem with Galileo. They agreed that if air had weight, the weight of the air pushing against the water in a force or lift pump raised the water in a well or a tube but only to a height of 33 feet. Additional pumping would not raise the water any higher. The weight of the air pushing against the water, they said, established the height of the water column not nature's abhorrence of a vacuum.

Torricelli decided to test this hypothesis and used mercury in his experiments because of its convenience compared to pumping water. Mercury is 13.6 times as dense as water, which enabled him to use a glass tube 13.6 times shorter. He performed his experiment in 1643 and in so doing constructed the first mercury barometer.

Torricelli poured mercury into a glass tube about 40 inches long and then inverted the tube over a dish containing mercury. The mercury inside the tube fell a short distance. Torricelli observed that the weight of the air supported a mercury column in the tube that measured about 30 inches high and that the height varied daily. He concluded that a vacuum existed in the upper empty space of the tube.

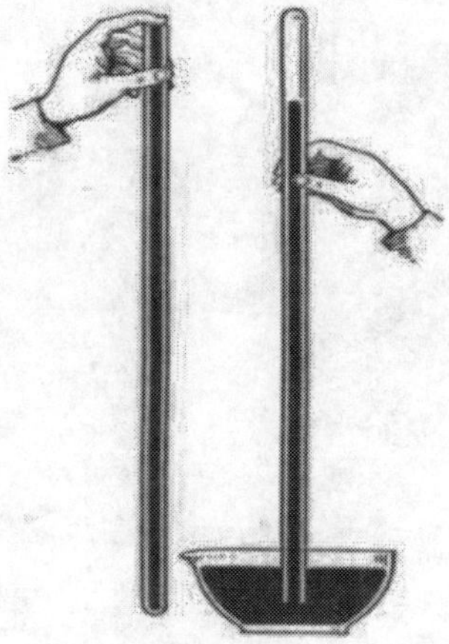

Mercury column in a glass tube 40 inches long

**Figure 15**
**Torricelli's Barometer**
Edwin J. Houston. *The Elements of Natural Philosophy.* Philadelphia: Eldridge & Brother, 1900.

Torricelli's experiment showed that air has weight (exerts a pressure) and that its weight (pressure) pushed water only to a height of 33 feet or mercury to a height of about 30 inches. His experiment suggested that air, or the atmosphere, surrounding Earth, rises or extends to a definite height (known today to be about 33 miles) and not indefinitely.

Torricelli advanced science in three ways. First, he disproved the ancient claim that nature abhors a vacuum. Second, he demonstrated that the weight of the atmosphere raised water to a maximum height of 33 feet, or mercury to a height of about 30 inches. Third, he introduced to science the barometer, a valuable instrument, in which the slight rise and fall of the mercury column indicated changes in atmospheric pressure. Torricelli's investigations on air pressure and vacuums led to the 1662 pressure-volume law of Robert Boyle (1627-91) in Oxford and Edmé Marriotte (1620-84) in Paris, and eventually in the mid-1700s to the steam engine of the Scottish inventor James Watt (1736-1819) in Birmingham.

One year after Torricelli's death from typhoid fever, Blaise Pascal (1623-62) then in Rouen, France, confirmed Torricelli's barometer experiment. In 1648, Pascal had his brother-in-law Florin Périer carry a barometer to the top of Puy-de-Dôme, a mountain 4,705 feet high in south central France's Auvergne Mountains. The mercury column fell two inches. The atmosphere extended to a definite height and did not extend upward indefinitely. The weight of the air decreased as the altitude increased. A vacuum may exist at higher altitudes. We know today that most of space is a vacuum.

In a related experiment Otto von Guericke (1602-86), a military engineer in Magdeburg, Germany, in 1650 used a suction (water) lift pump to remove air from an empty metal barrel. Guericke connected the barrel to a pump with a pipe containing a stopcock. His hemisphere (vacuum) experiment before the Reichstag in Regensburg in May 1654, demonstrated the force or weight (pressure) of air acting on the hemispheres. Sixteen horses (two teams of 8 horses) failed to pull apart two 22-inch diameter copper hemispheres fitted together and the air inside them removed. The atmosphere exerted a pressure equal to a force or weight of 2,700 pounds. Guericke published his studies in *De Vacuo Spatio* (1672).

**Figure 16**
**Guericke's Hemispheres**
Source: Otto von Guericke. *Experimenta nova (ut vocantur) Magdeburgica de vacuo spatio.* Amsterdam, Netherlands: Johann Jansson, 1672.
Gaspar Schott published the same engraving in 1657.

***Boyle's law of 1662, demonstration of air pressure, and related experiments on air***

Upon learning of Torricelli's and Guericke's experiments, Robert Hooke (1635-1703), Boyle's assistant, in 1657 constructed the first practical vacuum pump. Boyle used this and other pumps in his studies on air and showed that the air pressure (the force or weight of air per unit area) and not simply the weight of the air caused mercury to rise in a Torricellian tube. He demonstrated this by putting the mercury tube inside a glass container and plugging the opening of the container. This blocked air from entering the container and left unchanged the weight of the air inside the container acting on the mercury's surface area. Nothing happened to the height of the mercury in the tube, it fell only when Boyle removed the air from the container.

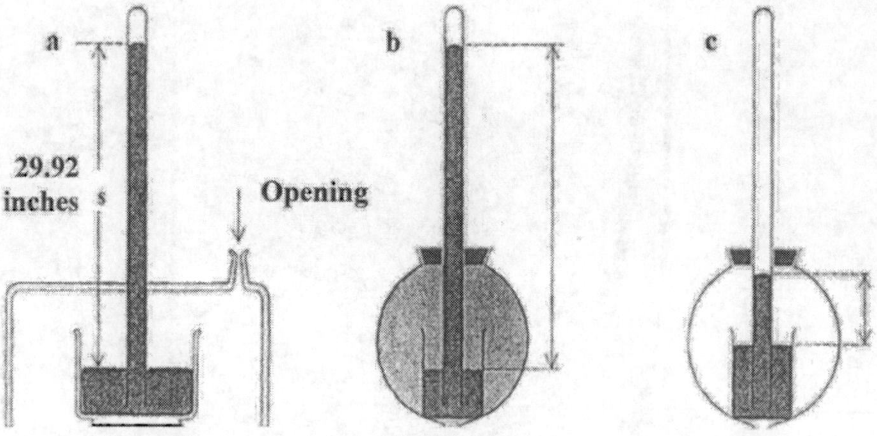


Torricelli's experiment showed that the *pressure* of the atmosphere supported 29.92 inches of mercury, **a**. Pressure, not the air's weight, affected the mercury's height in the tube because no difference in height resulted when Boyle plugged the opening and thereby changed the total weight of air on the mercury's surface. Boyle constructed an air pump to reduce the atmospheric pressure on the exposed mercury surface. He placed a Torricellian tube inside the sealed receiver of his apparatus, **b**, removed air from the receiver, and observed that the level of the mercury column fell, **c**.

**Figure 17**
**Boyle's Demonstration of Air Pressure**
Courtesy of the author

Using an evacuated glass globe of 30 quarts volume Boyle experimented from 1657 to 1659 on combustion, respiration, and transmission of the sound of a watch. He showed that all of these processes required the presence of air. A burning candle extinguished, mice and small birds died, and he could not hear the watch's ticking after he removed air from the globe. Boyle suffered from poor health, and he did the experiment on respiration for possible health benefits.

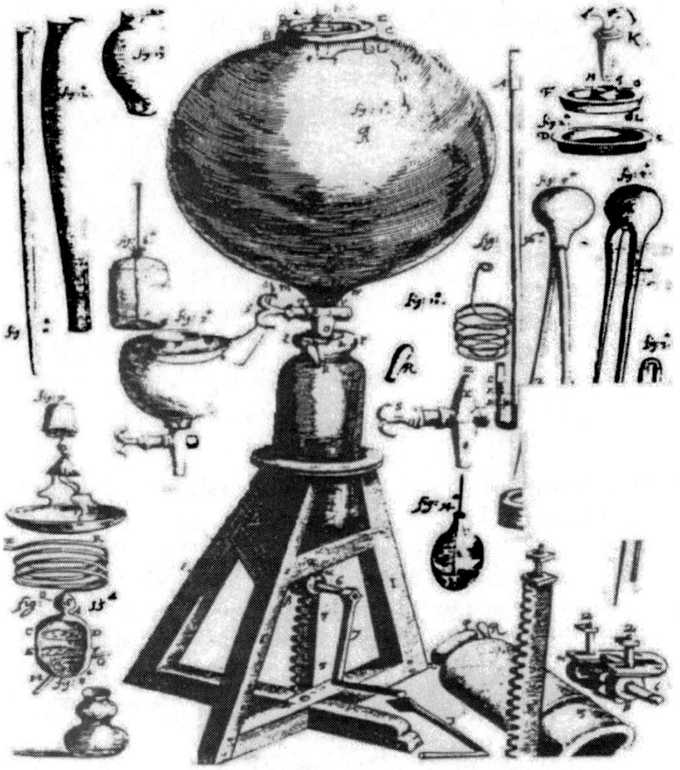

Boyle's air pump consisted of a 30-quart glass receiver A connected by a stopcock N to a 14-inch long brass pumping cylinder through which a toothed shaft 5 drew a padded piston 4. To remove air from the receiver Boyle closed the stopcock, cranked down the piston, and opened the stopcock. "Part of the Air formerly contained in the Receiver," Boyle wrote, "will nimbly descend into the Cylinder." Next, he closed the stopcock, removed the brass plug, and raised the piston, expelling air from the cylinder. Boyle repeated the procedure to remove more air from the receiver. The diagram shows other accessories that Boyle used in his experiments.

**Figure 18**
**Boyle's Air Pump**
Source: Robert Boyle. *New Experiments Physico Mechanical: Touching the Spring of the Air and their Effects.* Oxford: H. Hall for Tho. Robinson. 3d ed., 1682.

In another experiment, Boyle verified Galileo's claim about the rate of fall of bodies of different weight in a vacuum. A feather and a piece of lead inside an evacuated glass tube fell at the same rate when dropped from the same height. They reached the ground at the same time.

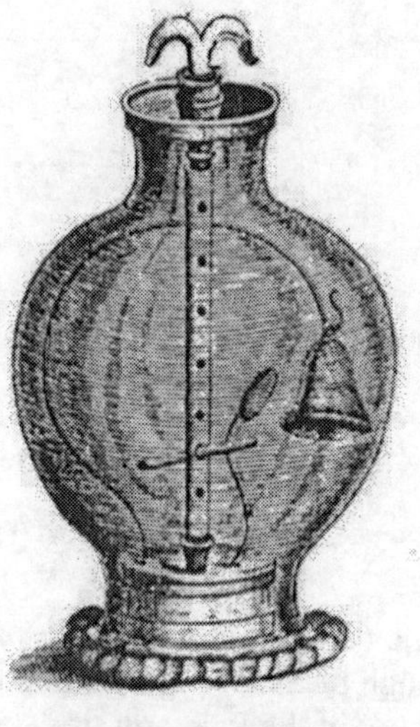

**Wood engraving from *New Experiments* shows how Boyle connected a cylinder or axle tree to a turning key. The connection enabled Boyle to strike a bell enclosed in a vacuum.**

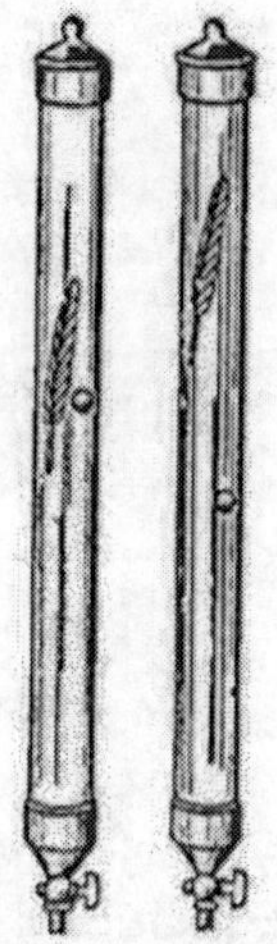

**Experiment showing a feather cross falling in the evacuated tube on left.**

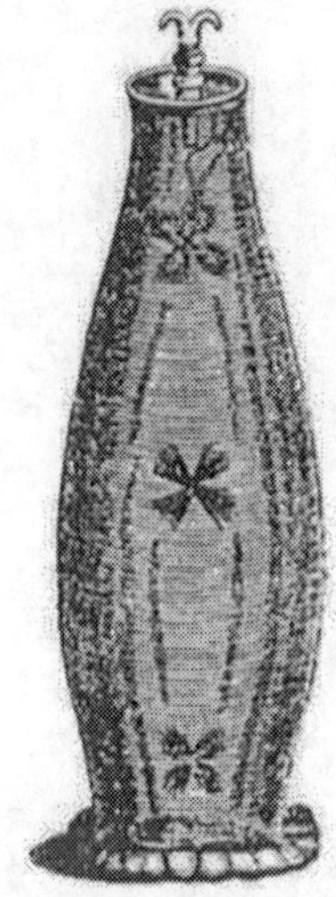

**Boyle's arrangement for allowing a feather cross to fall in an evacuated receiver.**

**Figure 19**
**Boyle's Vacuum Experiments**
Source: Robert Boyle. *New Experiments Physico Mechanical: Touching the Spring of the Air and their Effects.* Oxford: H. Hall for Tho. Robinson. 3d ed., 1682 (left and right).
Source: Newton Henry Black and Harvey Nathaniel Davis. *New Practical Physics.* New York: Macmillan Company, 1913 (middle).

He established Boyle's Law, the pressure-volume inverse relation, showing in 1662 that the product of the pressure P and the volume V is a constant, or that PV = k at constant temperature. An increase in P decreases V, doubling P reduces V to one-half its original volume, tripling P reduces V to one-third its original volume, and similarly for other pressures.

| BOYLE'S LAW | | |
|---|---|---|
| ***Pressure*** | ***Volume*** | ***Pressure x Volume*** |
| **P** | **V** | **PV** |
| 1 | 48 | 48 |
| 2 | 24 | 48 |
| 3 | 16 | 48 |
| 4 | 12 | 48 |
| 5 | 9.6 | 48 |

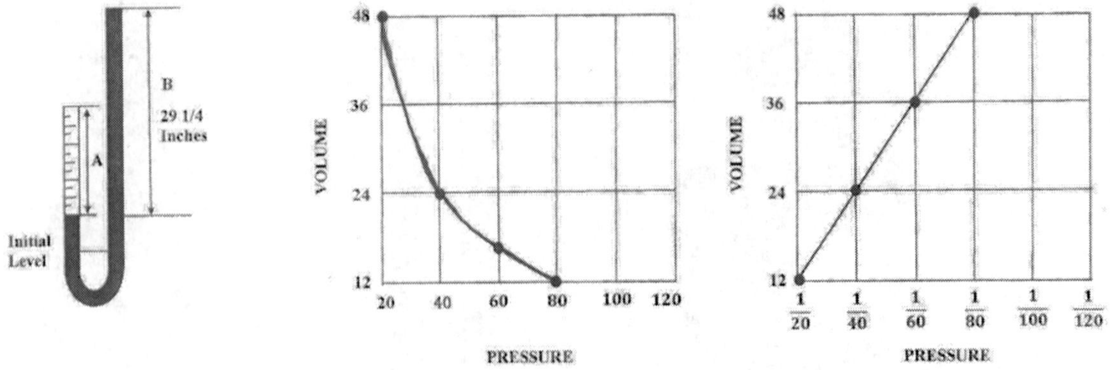


The air above the mercury on the closed (left) side of the glass tube had an initial volume of 48 units. When Boyle added sufficient mercury through the open (right) end to cut the volume of the trapped air in half, 24 units A, the mercury in the long arm of the tube B measured 29 1/4 inches higher than in the closed side A. Boyle made other pressure-volume measurements and obtained the curve shown above middle. He concluded that at constant temperature *the pressures and volumes are in reciprocal proportions*, or, the volume of a gas is inversely proportional to the pressure on it, $P_1V_1 = P_2V_2$.

**Figure 20**
**J-Tube Experiment Demonstrating Boyle's Law**
Courtesy of the author

The compressibility of air supported Boyle's belief in an atomic theory of matter in which atoms move randomly in largely empty space.

Later Jacques Charles (1746-1823), a French physicist in Paris, carried out temperature-volume studies on gases. His law of 1787 showed that a gas expanded 1/273 of its volume at 0°C for every one degree temperature increase and decreased by 1/273 of its volume at 0°C for every one degree decrease in temperature. Volume varied directly with temperature, or $V = kT$ at constant pressure.

## VI. ADVANCES IN MATHEMATICS

### *Decimal fractions*

In 499-500, Aryabhata (476-550 AD) in Kusumapura, modern Patna, described Hindu methods of multiplication, long division, and algebra that Western mathematicians learned of and adopted. The Hindus, like the Babylonians, created exercises to help them with calculations of taxation, debt, and interest.

Nearly 1,000 years later Simon Stevinus in Holland, published *Le Practique d'Arithmetique* in 1582 and *De Thiende* in 1585 in which he introduced decimal fractions and proposed the use of a decimal system. The Arabs and Chinese had long used decimals, but Stevinus brought decimal fractions to European mathematics, and others such as John Napier adopted his notation. Using Stevinus's decimal notation, 0.5 = 5/10, 0.05 = 5/100.

About the same time Stevinus introduced decimal fractions, Francois Viéte (1540-1603) in Paris in 1591 invented the algebraic symbolism used today. Viéte designated vowels for variables and consonants for constants. He introduced the algebraic terms, coefficient and negative.

| **AN EARLY HINDU ALGEBRAIC PROBLEM** | |
|---|---|
| A merchant pays duty on certain goods at three different places. | At the first he gives 1/3 of his coins, at the second ¼ of what he has left, and at the third 1/3 of the remainder. The total equals 24 coins. |
| How many coins he have at first? | $x$ = what he had at first<br>1 gives up 1/3 $x$<br>2 gives up ¼ (2/3 $x$)<br>3 gives up 1/3[ (3/4 (2/3 $x$)]<br>24 1/3 + 1/6 $x$ + 1/6 $x$<br>$x$ 36 coins |

***Logarithms***

John Napier (1550-1617), a mathematician in Merchiston, near Edinburgh, Scotland, invented logarithms. In 1614 Napier published *A Description of the Wonderful Law of Logarithms*, in which he introduced a logarithmic scale similar to natural log base e (2.118). The thought process that led Napier to his invention depended on the movement of numerical positions in a straight line and on a correspondence of a geometrical series (2, 4, 8, 16, 32, 64), which is equivalent to $2^1$, $2^2$, $2^3$, $2^4$, $2^5$, $2^6$, and an arithmetical series (2, 4, 6, 8, 10, 12, or 2 x 1, 2 x 2, 2 x 3, 2 x 4, 2 x 5, 2 x 6). His process, nevertheless, remains quite obscure. The use of logarithms enabled Napier to multiply numbers by adding exponents and to divide numbers by subtracting exponents.

$$8 \times 16 = 2^3 + 2^4 = 2^7 \text{ or } 128 \text{ and } 2^6 - 2^4 = 2^2 \text{ or } 64 \div 16 = 4$$

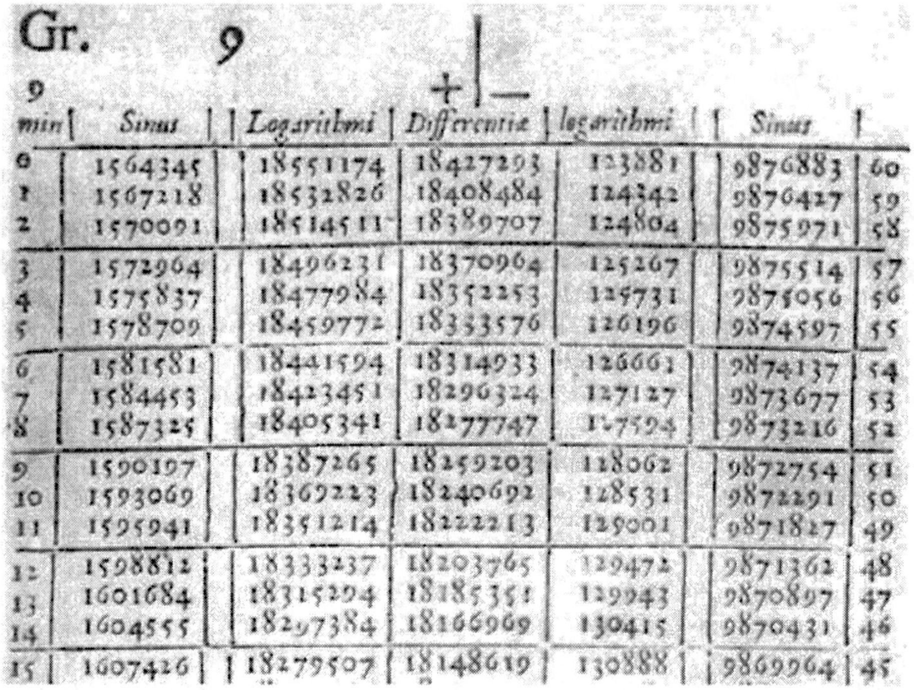
Gr. 9

9 + | −

| min | Sinus | Logarithmi | Differentiae | logarithmi | Sinus | |
|---|---|---|---|---|---|---|
| 0 | 1564345 | 18551174 | 18427293 | 123881 | 9876883 | 60 |
| 1 | 1567218 | 18532826 | 18408484 | 124342 | 9876427 | 59 |
| 2 | 1570091 | 18514511 | 18389707 | 124804 | 9875971 | 58 |
| 3 | 1572964 | 18496231 | 18370964 | 125267 | 9875514 | 57 |
| 4 | 1575837 | 18477984 | 18352253 | 125731 | 9875056 | 56 |
| 5 | 1578709 | 18459772 | 18333576 | 126196 | 9874597 | 55 |
| 6 | 1581581 | 18441594 | 18314933 | 126661 | 9874137 | 54 |
| 7 | 1584453 | 18423451 | 18296324 | 127127 | 9873677 | 53 |
| 8 | 1587325 | 18405341 | 18277747 | 127594 | 9873216 | 52 |
| 9 | 1590197 | 18387265 | 18259203 | 128062 | 9872754 | 51 |
| 10 | 1593069 | 18369223 | 18240692 | 128531 | 9872291 | 50 |
| 11 | 1595941 | 18351214 | 18222213 | 129001 | 9871827 | 49 |
| 12 | 1598812 | 18333237 | 18203765 | 129472 | 9871362 | 48 |
| 13 | 1601684 | 18315294 | 18185351 | 129943 | 9870897 | 47 |
| 14 | 1604555 | 18297384 | 18166969 | 130415 | 9870431 | 46 |
| 15 | 1607426 | 18279507 | 18148619 | 130888 | 9869964 | 45 |

**Figure 21**
**Page from Napier's 1614 book**
John Napier. *Mirifici logarithmorum canonis description* (*A Description of the Wonderful Table of Logarithms*). Edinburgh: Hart, 1614.

Napier adopted the decimal notation (decimal point) for fractions after Stevinus in 1585 introduced the decimal fraction: 0.1 = 1/10, and 0.01 = 1/100, and similarly for other decimal fractions. Napier wrote on prophecy, and in a book published in 1594 he predicted that the world would end in 1688, the year of the Glorious Revolution in England.

After discussing logarithms with Napier, the London mathematician Henry Briggs (1556-1631) in *Arthimetica Logarithmica* (1624) introduced log base 10 where $10^1 = 10$, $10^2 = 100$, $10^3 = 1{,}000$, and so on. Briggs worked out the logarithms for numbers up to 20,000 and from 90,000 to 100,000. Other mathematicians filled in the remaining logarithms. Briggs invented long division in 1617, and in 1621 he invented the slide rule which consisted of two movable rulers marked off with logarithmic scales.

| INTRODUCTION OF MATHEMATICAL SIGNS | |
|---|---|
| Johannes Widmann (1462-98?) in Bohemia | First used + and – signs in 1489 to indicate more and deficiency; + is a contraction for *et*, - is a contraction for *m* meaning minus. |
| Giel van der Hoeke, a Dutch mathematician | Introduced + and – signs in 1514 |
| Robert Recorde (1510-58) in England | First used the equal sign in 1557 |
| William Oughtred (1575-1660), English minister and mathematician | Introduced mathematical signs for multiplication, proportion, and trigonometric functions in *Clavis Mathematicae* (1631) |
| Johann H. Rahn (1622-76) in Switzerland | Introduced the division sign in 1659 |
| Leonhard Euler (1707-83) in Berlin | First used e the symbol for natural logarithms in 1727, the symbol π in 1737, signs for trigonometric functions in 1748, parentheses ( ) in 1748, Σ for summation in 1755 |

## VII. PHYSICS OF LIGHT

### *Law of refraction*

Observers dating from ancient times and Alhazen in medieval times had observed refraction, which is the bending of light when it passes from one medium (substance) to another of different density. When light passes obliquely (at an angle) from a medium of lesser density to a medium of greater density, such as from air to water, the light ray bends toward the normal, or perpendicular. When the light ray passes from a medium of greater density to one of lesser density, from water to air, the ray bends away from the normal.

Ptolemy believed incorrectly that for the two media the ratio of the angle of incidence $i$ (the angle between the incident or incoming ray and the normal) to the angle of refraction $r$ (the angle between the refracted or bent ray and the normal) did not change, it remained constant. This is approximately correct only for small angles.

$$\frac{\angle i}{\angle r} = k$$

The discovery of the correct law did not come until 1621 when Willibrord Snell (1591-1626) at Leiden, gave a statement of the law of refraction, now called Snell's Law. He derived the law as the ratio of lengths in the two media and interpreted the lengths as the ratio of cosecants of the angles of incidence and refraction. Snell did not publish his discovery. Christian Huygens (1629-95), a Dutch physicist in Paris, called attention to Snell's work in 1703.

Snell showed that for any angle of incidence the ratio of the cosecants (hypotenuse divided by opposite side) of the angles of incidence and refraction remains always the same, or a constant. Because the cosecants vary inversely as the sines of angles, defined as the opposite side (of the acute angle in a right triangle) over the hypotenuse, Snell's statement is equivalent to the way physicists express the law. Snell apparently verified the law experimentally.

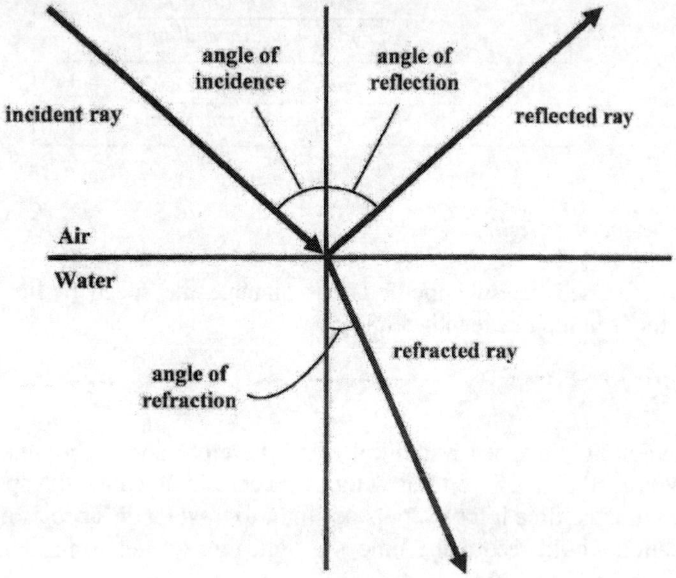


**Figure 22**
**Reflection and Refraction of Light**
Courtesy of the author

René Descartes (1596-1650) in France independently discovered the law of refraction and published it in *La Dioptrique* (*Dioptric*) in 1637. He gave the first published statement that for any angle of incidence at which light passed from one medium to another the ratio of the sine of the angle of incidence $i$ to the sine of the angle of refraction $r$ had a constant value.

$$\frac{\sin \angle i}{\sin \angle r} = k$$

Descartes performed no experiments but deduced the law theoretically from three assumptions: the speed of light is greater in a denser medium (incorrect), for the same two media these speeds have the same ratio for all angles of incidence, and the speed component parallel to the refracting surface remains unchanged during refraction (incorrect). The improbability of the correctness of these assumptions led to attacks on Descartes' derivation from the mathematician Pierre Fermat (1601-85) in Toulouse. Fermat was one of the founders of analytical geometry, calculus, and modern arithmetic. He deduced the law in 1662 from the assumption that light travels from a point in one medium to a point in another medium in the least time and that speed of light decreased in the denser medium, a correct assumption.

The law for refraction for any two media has the form below, where $n_1$, $n_2$ is the index of refraction for each medium. The practical significance of refraction and the law of refraction in constructing refracting telescopes soon became apparent.

| LAW OF REFRACTION |
| --- |
| $\frac{sini}{sinr} = \frac{n_1}{n_2} = n$ |
| $n_1 = \frac{speed\ of\ light\ in\ air}{speed\ of\ light\ in\ medium_1}$ |
| $n_2 = \frac{speed\ of\ light\ in\ air}{speed\ of\ light\ in\ medium_2}$ |

***Calculating the speed of light***

Astronomers first used celestial methods to calculate the speed of light. They developed the first terrestrial methods in the nineteenth century.

*Galileo's method*

Descartes thought light's speed was infinite and therefore not measurable. Galileo thought otherwise, and in his *Discourse* of 1638 he tried without success to calculate the speed of light by using a slide lantern to measure the time it took a pulse of light to travel a distance of one mile. His assistant at this one-mile location would record the time the light took to get from Galileo to his location. Galileo failed to make any measurements.

*Olaus Roemer's calculation*

Giovanni Cassini (1625-1712) an Italian astronomer in Bologna and from 1669 director of the Paris Observatory that Louis XIV (1628-1715) founded in 1671, completed tables of the motions of Jupiter's moons or satellites in 1668. He established the period of revolution for each moon.

Using Cassini's tables Olaus Roemer (1644-1710), a Danish astronomer in Paris and one-time assistant to Cassini, in 1676 calculated 227,000 km/sec (132,682 mi/sec) for the speed of light. Over an eight-year period, Roemer timed successive eclipses of the first of Jupiter's moons. He found that the eclipses occurred 42 ½ hours apart when Earth had moved closest to Jupiter but occurred 22 minutes later after a six-month interval when Earth had moved farthest from Jupiter.

Earth's changing position in its annual orbit around the Sun accounted for the different times it took the Sun's reflected light from Jupiter to reach Earth. Roemer knew the time difference (22 minutes or 1,320 seconds) and the distance the light traveled (186,000,000 miles) in those 22 minutes, which was the diameter of Earth's orbit. Substituting these figures into the equation = $\frac{distance}{time}$, $v = \frac{d}{t}$, he calculated the speed of light. Roemer used leagues in his calculations to measure the radius of Earth's orbit around the Sun, where 1 league = 2.4 - 4.6 miles.

Kepler's third law, $\frac{R^3}{T^2} = K$ where R was the radius of the planet's orbit, T the time it took to complete an orbit, and K a constant which had the same value for all the planets, enabled astronomers to calculate the radius for Earth's orbit and for the orbits of all planets.

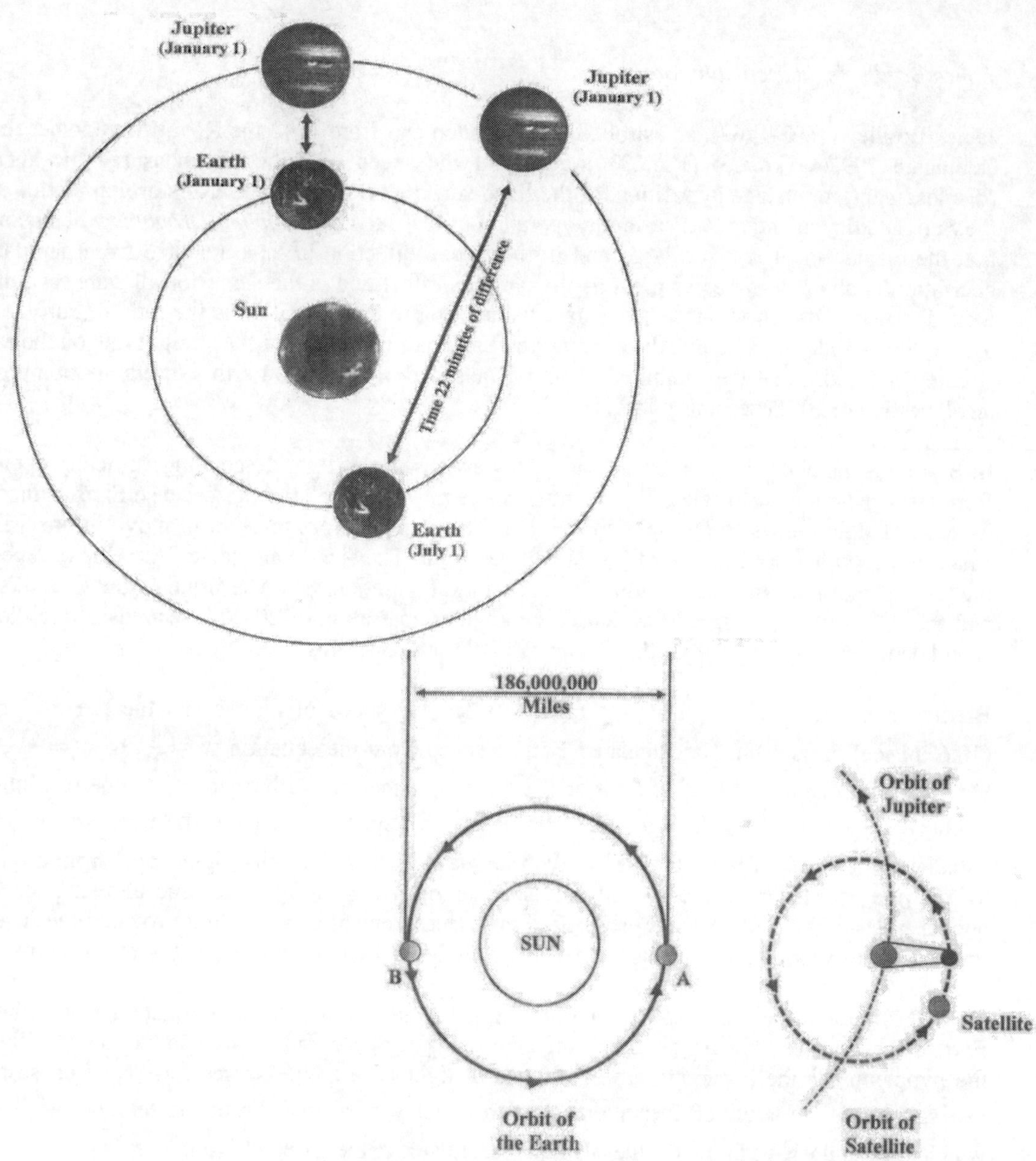


Earth is at position A during an eclipse of Jupiter's satellite. If Earth remained at position A, the next eclipse occurs 42½ hours later, the next one 42½ hours after the previous eclipse, etc. But because Earth revolves in its orbit about the Sun, Earth leaves position A and moves away from Jupiter. The light from Jupiter therefore requires a longer time to reach Earth, and the eclipses of Jupiter's satellite recur at intervals slightly greater than 42½ hours. Roemer noticed that six months later, when Earth had moved to position B, the satellite's eclipses occurred 1,320 seconds (22 minutes) late. He reasoned correctly that the 1,320 second delay was the time light took to travel the additional distance from A to B. This distance is the diameter of Earth's orbit and equal to 186,000,000 miles. Because light travels 186,000,000 miles in 1,320 seconds, Roemer calculated the speed of light from the equation $v = d/t$, or 186,000,000 divided by 1,320, giving $v = 132{,}680$ miles per second. Later measurements showed that Roemer obtained a low value.

**Figure 23**
**Roemer's Method of Measuring the Speed of Light 1676**
Courtesy of the author

*James Bradley's later calculation*

James Bradley (1692-1762), an astronomer in London and from 1742 the Royal Astronomer, in 1729 calculated 299,714 km/sec (186,233 mi/sec) for the speed of light. He measured the aberration (displacement), or change in position, of the fixed stars that results from Earth's orbital motion around the Sun. Bradley intended to measure the parallax of the fixed star *Gamma Draconis* when he noticed that the displacement was too large and in the wrong direction to have resulted from parallax. The aberration is about 30 times as great as the parallax effect and is the same for all stars regardless of their distance or direction. It has a constant value because the aberration is the ratio of half the major axis of the ecliptic (the circle at the ecliptic pole) and half the length of the straight line on the ecliptic. Bradley concluded that the combined effect of the speed of light and Earth's orbital motion produced an aberration of 20-25 seconds of arc.

In Bradley's method, an observer on Earth, to view a beam of light descending vertically upon Earth from a star near the north celestial pole, must angle his telescope slightly, because Earth is moving in its orbit at right angles to the light's direction. The star therefore appears to move in one year in a small ellipse in the sky because of Earth's curved orbit. The size (diameter) of the ellipse depends on the ratio of the speed of Earth's motion to that of light's motion. Because Bradley knew Earth's speed and the ratio of the two speeds, which is the angular aberration of 20-20.5 seconds of arc, he could calculate the speed of light from the known ratio of the two speeds.

Bradley measured the aberration, the ratio between the speed of light $c$ and the speed of Earth $v$ (18.6 mi/sec) in its orbit. He calculated Earth's speed from the equation $v = \frac{d}{t}$, or $v = \frac{2\pi r}{t}$, where $d$ = the circumference of Earth's orbit or $2 \pi r$ and $t$ = time for Earth to complete one revolution. He obtained a value $\frac{v}{c} = 1 \times 10^{-4}$, for the ratio of the two speeds, and from the value of this ratio he calculated the speed of light. Bradley asked the question, how fast must light travel from a fixed star to give the detected aberration of 20-20.5 seconds of arc, while at same time allowing for Earth's annual west-to-east motion in its orbit. Bradley's measurement of aberration provided the first direct proof of Earth's motion. If Earth is not moving or light's speed is infinite, no aberration occurs.

To make the calculation, Bradley set up a right triangle in which the horizontal side represented Earth's speed $v$, the vertical side represented the speed of light $c$ and the true path of a ray of light, and the hypotenuse $b$ the apparent path of the ray of light. The angle between the hypotenuse and the vertical side is the angle of aberration $a$ which Bradley measured. From the relation $\tan a = \frac{v}{c}$, in which $\tan a$ and $v$ have known values, Bradley calculated $c$, the speed of light.

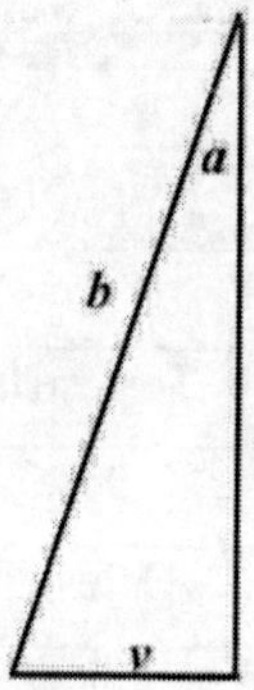


*speed of Earth (v) = 18.6 mi/sec*
*angle of aberration (a)* = 20-20.5 seconds of arc

$$\tan a = \frac{speed\ of\ Earth\ (v)}{speed\ of\ light\ (c)}$$

$$1 x\ 10^{-4} = \frac{18.6\ mi/sec}{c\ mi/sec}$$

$$c = 18.6 \times 10^4$$

$$c = 186{,}000 \text{ mi/sec}$$

Bradley's method depends on the effect called aberration. *c* is the direction of a telescope. While Earth's motion carries the observer along *v*, the ray of light travels the length of the telescope *c*. When the ray reaches the eye, the telescope is in the direction *b*.

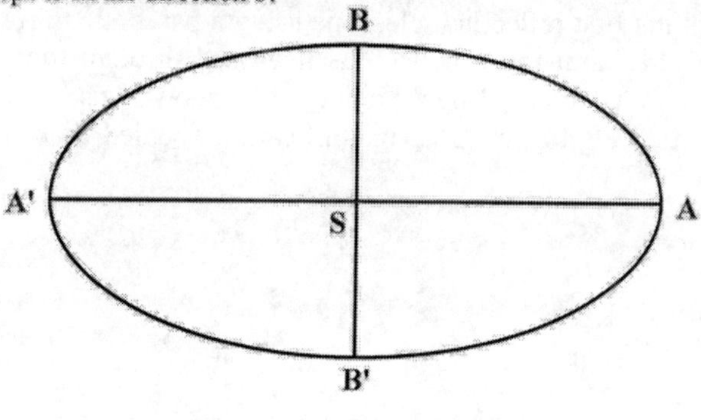


**The Aberrational Ellipse**

| | |
|---|---|
| S: | Position star occupies if no aberration exists. |
| AA' : | Twice a year, at six month intervals, Earth's motion is at right angles to the direction of the star. At these times the star receives the greatest displacement from its mean position and is at the ends of the major |
| BB: | axis of the ellipse. |
| SA: | The star undergoes its least displacement. |
| | Is one-half of AA' and has the same value for all stars (the constant of aberration). |

**Figure 24**
**Bradley's Calculation of the Speed of Light 1729**
Source: Courtesy of the author (top); Arthur Berry. *A Short History of Astronomy: From Earliest Times Through the Nineteenth Century.* London: John Murray, 1898 (bottom).

| MEASURING THE SPEED OF LIGHT | | | |
|---|---|---|---|
| ***Scientist*** | ***Year*** | ***Method*** | ***Value*** |
| Descartes | 1630s | | Infinite |
| Galileo | 1638 | Terrestrial | No value obtained |
| Roemer | 1676 | Celestial | 227,000 km/sec |
| Bradley | 1724 | Celestial | 229,714 km/sec |

***Particle and wave models of light***

*Light as a particle*

The particle theory of light dominated in the 1600s and 1700s. It fit nicely into the mechanical philosophy, or matter in motion, model of the universe. Most scientists including Boyle, Galileo, and Isaac Newton (1642-1727) at Cambridge, England, accepted the particle, or corpuscular, theory of light.

Newton was the foremost advocate of the particle theory, which he discussed in *Opticks*, his 1704 summary of his studies on light. He accepted the particle theory because light traveled in straight lines and cast sharp shadows. Waves did not behave this way according to Newton. He believed incorrectly that light traveled faster the more dense the medium, because the particles of light experienced greater attraction from a denser medium consisting of particles closely packed together.

Newton is 1668 built the first reflecting telescope using a parabolic mirror. He wanted to eliminate the chromatic aberration the light ray produced as it passed through the convex lenses of a refracting telescope. His telescope, which the Royal Society of London owns, measured 6 inches long, 1 inch in diameter, and magnified 30-40 times. Its magnification equaled that of a refracting telescope about 8 times as long.

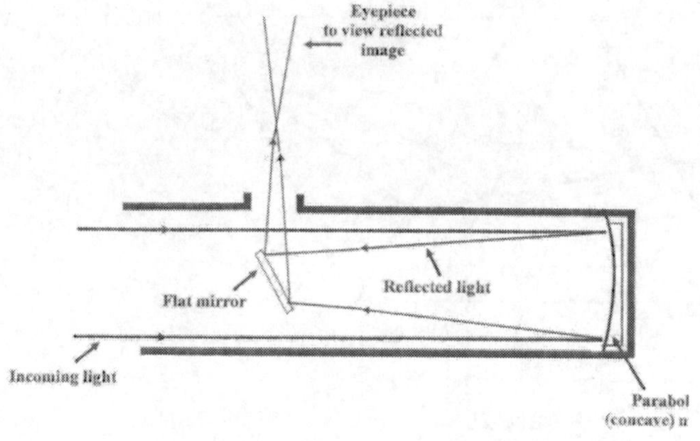


**Newtonian Principle Reflecting Telescope**

**Newton's Telescope**

**Figure 25**
**William Lassell's 48-inch Reflecting Telescope at Malta**
Source: Courtesy of the author (left); "The Progress of Science."
*Popular Science Monthly* 69 (November 1906) (right).
*Nature* 9 (1873) (bottom).

In Newton's prism experiment of 1672, Newton separated and recombined white light by refraction, passing a ray of light through two triangular prisms arranged one behind the other and inverted with respect to the other. He showed that white light consisted of seven colors, red, orange, yellow, green, blue, indigo, and violet, and that the prism separated but did not produce the colors, as Aristotle, Descartes, Hooke, and most other investigators believed.

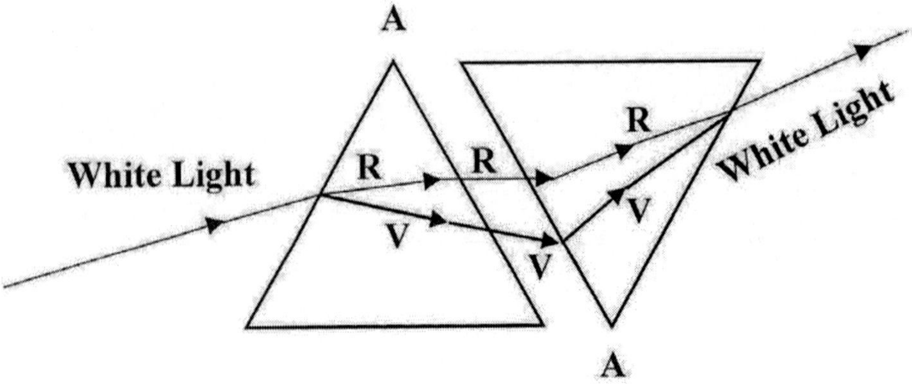


**Figure 26**
**Newton's Prism Experiment**
Courtesy of the author

*Light as a wave*

The wave theory of light had only a small number of supporters. One of them was Francesco Grimaldi (1618-63), a Jesuit priest in Bologna, Italy. The posthumous publication of his diffraction experiments in 1666 demonstrated two new properties of light. First, light passing through a narrow opening or slit made in a screen produced colored fringes, or shading, at the opening or slit and second, the image of the illuminated opening or slit that appeared on a second screen, placed behind the first, was larger than expected. Grimaldi called the bending of light around an opening or edge, diffraction. He believed correctly that only a wave motion could have produced these results.

Hooke also adopted a wave theory of light which he discussed in his *Micrographia* published in 1665. Huygens, a third wave theory proponent, published his wave model in *Traité de la lumiére* in 1678. He proposed that light, just like sound, consisted of longitudinal waves or pulses vibrating parallel to the direction in which the light traveled. Huygens adopted a wave theory because he believed only waves could travel so rapidly.

## VIII. ESTABLISHMENT OF SCIENTIFIC SOCIETIES AND ACADEMIES

In the period from 1323 to 1599, 29 scientific societies existed in Europe, the first in Toulouse, France. From 1600 to 1699 another 49 societies came into existence. From 1700 to 1739 an additional 33 societies arose, and from 1740 to 1799, 157 new societies emerged.

Duke Federico Cesi (1585-1630) in Rome in 1603 established the Academia dei Lincei (Academy of the Lynxes), one of the early European scientific academies. Galileo became a member in 1609. Pope Pius IX (1792-1878) revived the academy in 1847, and its name since 1903 is the Pontifical Academy of the New Lynxes in Rome.

In England the death of Oliver Cromwell (1599-1650) led in 1660 to the restoration of the monarchy under Charles II (1630-85). That same year Charles II issued a royal charter establishing the Royal Society of London, one of Europe's oldest scientific societies; and in 1675 through an act of Parliament Charles II established the Greenwich Observatory.

The Royal Society received no support from the English government. It had less restrictive membership requirements and lacked the prestige of the French Academy whose senior members were civil servants who had to live in Paris and received a salary of 20,000 livres ($80,000 in 2013) from the king. Members of the French Academy had to publish their research regularly. The Royal Society did not impose such requirements. Members of the Royal Society included Hooke, Boyle, and Newton.

More than a century passed before Count Rumford's initiative led to the founding of the Royal Institution in London in 1800 and to Britain's scientists organizing the British Association for the Advancement of Science in York in 1831.

**Figure 27**
**Royal Society Meeting Room**
Source: "The Progress of Science." *Popular Science Monthly* 69 (November 1906).

Louis XIV (1638-1715) established the Académie Royale des Sciences in Paris in 1666 and the Paris Observatory in 1667. The Academy's membership included Cassini, Huygens, Mariotte, and Charles.

The École Nationale des Ponts et Chaussées (Bridges and Roads) followed in 1747 and the École des Mines in 1783, both in Paris. Five years after the overthrow of the French monarchy in 1789, the revolutionary government in 1794 established the École Polytechnique in Paris to recruit and educate military professionals and engineers for the army, the Écoles Normale Supérieure for the preparation of secondary school teachers, the Conservatoire Nationale des Artes et Métiers (Arts and Trades) in Paris, and several military and medical schools.

When Napoleon I Bonaparte (1769-1821) assumed control of France in November 1799, although only officially in December 1804 when crowned emperor, he established additional écoles polytechniques, or technical schools, for engineering, mining, and the practical arts. French scientists established the French Association for the Advancement of Science in Paris much later, in 1870, the year of France's defeat in the Franco-Prussian War and the end of the Second Republic of 1848-70.

By 1750, scientific communities in Bologna (1711), St. Petersburg (1725), and Stockholm (1735) established major scientific academies. The Germanic states established several technical institutes in the 1700s-1800s. These included the Royal Prussian Academy of Sciences in Berlin in 1700, the Mining Academy in Berlin in 1770, the German Society for Scientists and Physicians in Leipzig in 1822, and the Saxon Technical School in Dresden in 1828. Similar educational, scientific, and technological institutions existed or arose in other European states or countries including the Netherlands. The United States established its National Academy of Sciences in 1863.

## IX. MEASURING HEAT

### *What is heat?*

The particle or mechanical theory of heat dominated the study of heat and included Bacon, Boyle, and Newton among its supporters. According to mechanical theory, matter in motion produced heat.

*Thermometers*

The development of thermometers to measure heat lagged at this time. Their low level of development is striking in comparison with the development of telescopes and microscopes. The construction of accurate thermometers had to meet several requirements. The liquid used had to have a uniform expansion, a high boiling point and a low freezing point, and not stick or adhere to the glass walls of the tube that held the liquid. This was the major reason for using mercury, a liquid metal that has a boiling point of 356.7°C and a freezing point of -38.67°C. The glass tube required a uniform bore in order to measure the expansion or contraction accurately.

Galileo's air thermometer, or thermoscope (1593, 1603), an early version of a thermometer, consisted of a glass tube with a bulb at one end, and the other end immersed in a tube filled with water. Galileo used a water scale. His thermometer had no fixed reference points. Atmospheric pressure affected its temperature values.

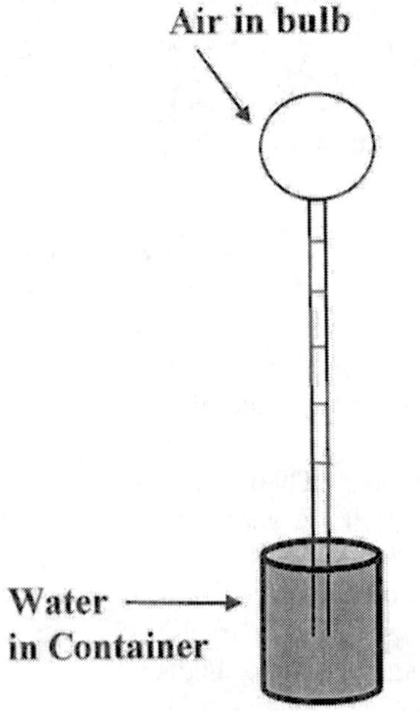


**Figure 28**
**Galileo's Thermoscope**
Courtesy of the author

Galileo's air thermometer or thermoscope (1593, 1603), an early version of a thermometer, consisted of a glass tube with a bulb at one end, and the other end immersed in a tube filled with water. Galileo used a water scale. His thermometer had no fixed reference points. Atmospheric pressure affected its temperature values.

Newton in 1701 used a mercury scale thermometer in which he set the freezing point of water at 0°. He selected the body temperature, which according to Newton was the armpit temperature of a healthy Englishman and set at 12°, as his second fixed point. On Newton's scale water boiled at 34°.

Guillaume Amontons (1663-1705), a physicist in Paris who lost his hearing as a teenager, in 1702 constructed a J-shaped air thermometer containing mercury. He set the boiling point of water as his fixed point. Amontons noticed in 1699 that for a specific change in temperature the volume of a gas changed by a definite amount. His observation represents an early statement of Charles's law of 1787.

*Later improved thermometer scales of the eighteenth century*

René Réaumur (1683-1757), a scientist and technologist and director of the Academy of Sciences in Paris, in 1730 used an ethyl alcohol-water solution in a 4:1 proportion. He introduced a scale of 0-80 degrees for the freezing and boiling points of water. Réaumur used alcohol because it had a greater coefficient of expansion and a lower freezing point (-117.3°C) than mercury, although it boiled (78.5°C) at a lower temperature than mercury. Réaumur could not prepare a perfectly anhydrous (water-free) alcohol, and so he combined alcohol and water in a mixture that gave an uniformly divisible scale.

Daniel Fahrenheit (1686-1736), an instrument maker in Amsterdam, developed a mercury scale thermometer in 1724. The freezing and boiling points of water were the fixed points. Fahrenheit refined Newton's earlier scale. He used an ice-water-salt mixture to get the lowest temperature that he could and marked it as 0°. He then set the freezing and boiling points of water as his fixed temperatures of 32°and 212°, after adjusting the numbers to get 180° difference between the two fixed points. Fahrenheit had to adjust his original body temperature set at 96° or 8 x 12°, from Newton, once he set his fixed temperatures at 32°and 212°.

Anders Celsius (1701-44), an astronomer at Uppsala, Sweden, used a mercury scale in 1742. He adopted a decimal scale with the freezing and boiling points of water, 0° and 100°, his fixed points. His original thermometer had the reference points reversed, 0° the boiling point, 100° the freezing point. Martin Strömer (1707-70), Celsius's successor at the observatory in Uppsala, inverted the two reference points in 1750.

*Significance of fixed temperature points, the freezing and boiling points*

The use of fixed reference points on a thermometer scale, specifically the freezing and boiling points of a substance, enabled scientists to clarify the difference between quantity of heat and temperature (intensity of heat). Their introduction showed that an increase in the quantity of heat added to a body did not always produce an increase in the intensity of heat, or temperature.

Amontons in 1699 first noticed the uniform expansion of a gas when heated, but he did not measure the amount of expansion. Later, in 1787, Charles carried out temperature-volume studies that led to the law of gaseous expansion. He showed that different gases expanded by the same amount, 1/273 of its volume at 0°C for each degree rise in temperature, and that a gas decreased by 1/273 of its volume at 0°C for each degree decrease in temperature. Volume varied directly with temperature or $V = kT$ at constant pressure (Charles Law).

## X. SOUND AND ITS MEASUREMENT

Pythagoras, Aristotle, and other natural philosophers recognized that sound consisted of pulses or vibrations in air or another medium and that the transmission of sound appeared to require a medium. The vibrations of a stringed musical instrument, a bow string or a drum head, provided easily observable experimental evidence of the vibrations producing sound.

Based on these earlier observations and their own observations of vibrating strings of musical instruments such as a harp, lyre, or violin, seventeenth-century natural philosophers believed that vibration of the particles of air or other medium produced sound. A longitudinal wave, a wave pattern that had the wave vibrations moving parallel to the direction in which the wave traveled, as indicated by observing the vibrating body, produced sound. The sound wave required a medium for transmission, as Boyle's vacuum pump-clock experiment demonstrated.

### *Calculating the speed of sound*

By the mid-1600s experimenters had not yet calculated the speed of light, but they knew light traveled much faster than sound, and they had noted the time difference between seeing the flash of a cannon when fired, which occurs almost instantly, and hearing the sound produced by the cannon's firing. Cannon firing became the most common method of measuring the speed of sound in air. Marin Mersenne (1588-1648), a Minim priest in Paris, used a cannon's firing in his experiments of 1636. He measured the difference in time between the observer's seeing the cannon's flash, which indicated the time that the firing occurred, and hearing the sound produced by the firing. The observer carefully measured the distance between him and the cannon, recorded the time at which he saw the flash and how much more time elapsed before he heard the sound. Using the formula, velocity = distance divided by time, or v = d/t, where d is the measured distance between cannon and observer and t the time the sound waves took to travel from cannon to observer, Mersenne calculated v = 1,380 ft/sec for the speed of sound in air.

Pierre Gassendi (1592-1655), a priest in Paris who adopted an atomic theory of matter in 1649, also used the cannon method to determine the speed of sound. His calculation in 1660 gave v = 1,473 ft/sec.

Newton in the *Principia* (1687) theoretically determined the speed of sound, v = 979 ft/sec, based on measurements of air pressure (stiffness) and the density of the medium through which the sound traveled. He used the equation $v = \sqrt{\frac{pressure}{density}}$, which showed that as the medium's density increases, sound velocity decreases, and as the medium's pressure increases the sound velocity increases. His value was too low because the compression of a sound wave does not occur isothermally (at constant temperature) as Newton assumed. In his studies on the wave motion of sound published in the *Principia*, Newton derived the relation showing that the speed of a wave equaled its frequency multiplied by the wavelength.

| WAVE EQUATION FOR SPEED |
|---|
| $v = f\lambda$ |
| v = speed |
| f – frequency |
| $\lambda = wavelength$ |

The accepted speed of sound in air is 1089 ft/sec at 0° C, changing by 2 ft/sec for each degree of temperature change. Knowing the speed of sound now made it possible for investigators to calculate the approximate distance between an observer on the ground and a lightning discharge in the atmosphere by timing the difference between seeing the discharge which occurs almost instantly and hearing the sound, crack, or thunder it produced. Using the equation d = v/t, a five second time interval corresponds to about one mile, 5 x 1,089 = 5,445 feet. Lord Rayleigh (1842-1919), a physicist at Cambridge in England, summed up classical acoustics in the *Theory of Sound* published in 1877-78.

## XI. REVOLUTION IN PHYSIOLOGY: WILLIAM HARVEY AND THE CIRCULATION OF THE BLOOD

### *Earlier beliefs*

Aristotle did not believe the blood circulated in a living body because only the perfect bodies, the heavenly spheres, had circular motion. Galen thought the human body contained two kinds of blood, which made unnecessary a circulation of the blood.

### *Minor circulation of the blood*

Michael Servetus (1511-53), a Spanish biologist, philosopher, lawyer, and astronomer who studied medicine in Paris, discovered the minor circulation of the blood. Servetus speculated on the minor circulation of the blood through the lungs, proposing that blood flowed from one side of the heart to the arteries, to the lungs, to the veins, and back to the other side of the heart. He discussed the minor circulation in six pages in *The Restoration of Christianity* which he published secretly in early 1553. Servetus, who abandoned his Catholicism and by 1531 had become a Unitarian, believed that respiration purified the blood. The soul was present in the blood and perished with the body. His heretical position offended both Catholics and Protestants.

The Protestant reformer John Calvin (1509-64) had corresponded with Servetus beginning in 1546. He knew that Servetus used the alias Michel de Villeneuve or Villanovanus (from the family home in Villanueva, Spain). When Servetus escaped by friendly connivance from imprisonment by the Catholic Inquisition in Vienne, France, and foolishly went to Geneva on his way to northern Italy, Calvin recaptured him. Calvin had Servetus burned at the stake for his heretical views, particularly his Unitarianism. Calvin had all but three copies of Servetus's book burned. Only in 1694 did William Wotton (1666-1727), an English scholar and author in Buckinghamshire, notice Servetus's discussion on the minor circulation of the blood.

### *Major circulation of the blood*

In his physiological studies on the human body Hieronymus Fabricius (1533-1619), William Harvey's teacher at the medical school in Padua, in 1574 discovered the one-way valves in the veins. He believed they slowed down the blood's motion in the body.

Harvey (1578-1657) in London, who served as the royal physician from 1618 to 1647, continued the physiological studies and demonstrated the major and minor circulation of the blood. He published his results in 1628 in *Exercitatio Anatomica de Motu Cordis et Sanguinis in Animalibus*. Harvey noted in his dissections that the valves separating the two upper chambers of the heart (auricles) from the two lower chambers (ventricles) were one way. Blood went only from auricle to ventricle. Harvey experimented with cold-blooded animals, cadavers, sheep, and deer. He did not use a microscope probably because he did not think it would reveal any new discovery or information.

Harvey showed that blood flows from one side of the heart to the arteries and then to the veins and back to the other side of the heart (H→A→V→H) in two separate circulations, one through the lungs (minor), the other to and through the body's extremities (major). This demonstrated the circulation of the blood in the body. In his classic experiment Harvey tied a tourniquet on an arm below the elbow tightly enough to stop the blood flowing in a major vein (veins are closer to the surface than the arteries) so that a bulge formed in the vein on the side of the tourniquet away from the heart. When he tied a tourniquet tightly enough to stop the flow from the heart into an artery, the bulge formed on the side of the tourniquet closer to the heart. In normal activities the heart, which weighs 8-12 ounces, pumped 5-6 quarts of blood per minute.

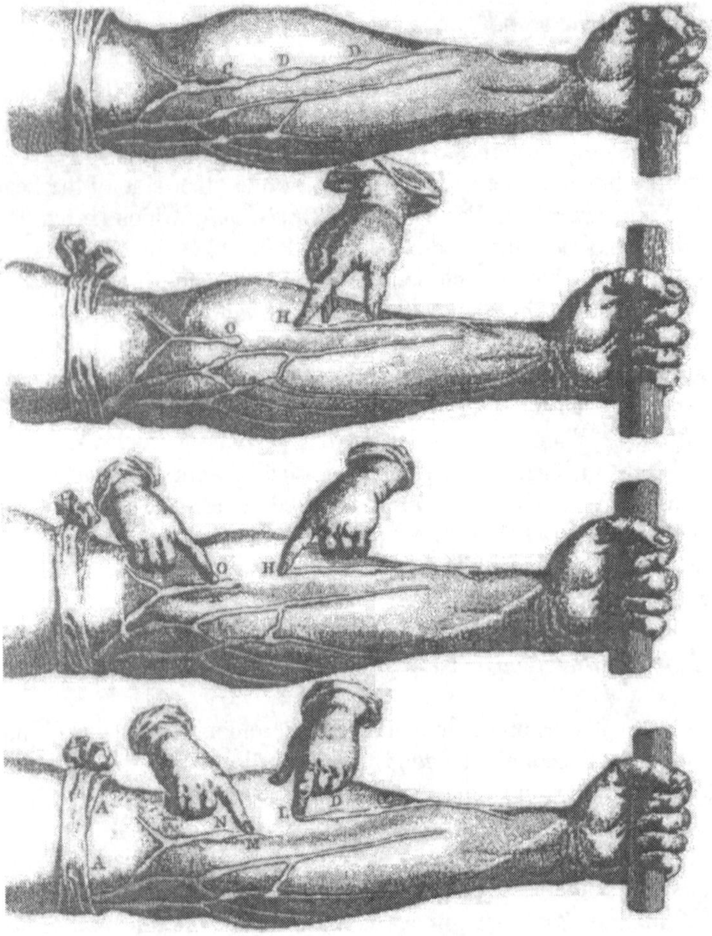

Top drawing shows that tying off the arm above the elbow (A) causes nodes to appear at the location of valves in the veins (BCDEF). Second drawing: pressing the blood out of the vein between O and H and placing a finger at H, the vein remains empty. Third drawing: pushing a second finger down the vein toward O, the vein still remains empty. Fourth drawing: placing a finger at L and pushing a second finger along the vein from L to N, the vein likewise remains empty. Harvey deduced from these experiments that the blood flows along the veins towards the heart. Engravings are from Harvey's 1766 edition of *Exercitatio Anatomica de Motu Cordis et Sanguinis in Animalibus.*

**Figure 29**
**Harvey's Experiment on Circulation of the Blood**
Source: William Harvey. *Exercitatio Anatomica de Motu Cordis et Sanguinis in Animalibus.* Frankfurt, 1628.

***Malpighi and the capillaries***

Harvey suspected the existence of connections between the arteries and veins, but without a good microscope he failed to see them. Marcello Malpighi (1628-94) provided the experimental confirmation. Malpighi, an anatomist at Bologna and after 1691 physician to the newly elected Pope Innocent XII (1615-1700) in Rome, in 1660 discovered the capillaries that complete the circulation by connecting the veins and arteries. He discovered the capillaries while studying the lungs of a frog with a microscope.

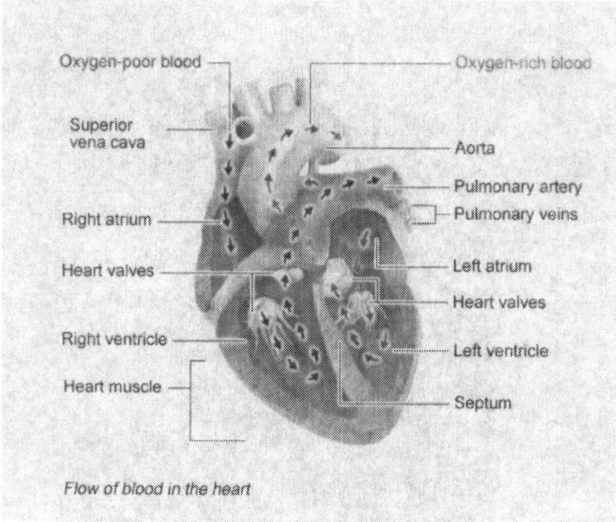


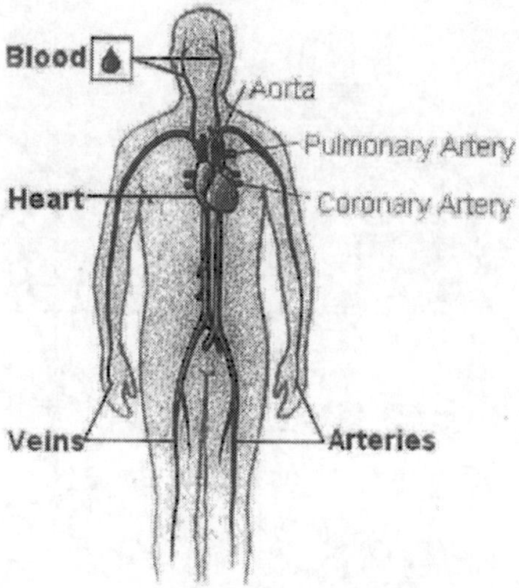


Flow of the blood coming from the extremities and entering from superior vena cava, and from inferior vena cava; to right atrium; to right ventricle; to pulmonary artery; to lungs (not shown); to pulmonary veins; to left atrium; to left ventricle; leaving by the extremities (not shown).

**Figure 30**
**Circulation of the Blood**
Courtesy of US National Library of Medicine, National Institutes of Medicine

***Related medical developments***

Richard Lower (1631-91), a London physician, carried out a direct blood transfusion with dogs in 1666. He transfused blood from the artery of one dog to the vein of another. He successfully transferred blood from a sheep to a man in 1667. Similar experiments in France in 1667 resulted in a patient's death, and the subsequent legal action held up transfusion studies for 150 years. Lower in 1669 demonstrated the difference in color between venous (dark) and arterial (bright) blood.

***Construction and applications of the microscope***

Hans Jansen, a spectacle maker, and his son Zacharias Jansen (1580-1638) in Middleburg, Holland, also Lippershey's home, built the first compound microscope in about 1595. Their early microscopes consisted of 2 convex lenses, 3 metal sliding tubes, and measured about 18 inches in length and 2 inches in diameter. It magnified about 3 times.

Hooke in London designed and had Christopher Cock, an English instrument maker, construct the first practical compound microscope which Hooke described in *Micrographia* (1665). It had 3 convex lenses and magnified enough to reveal the fine structure of cork.

Jan Swammerdam (1637-80), an anatomist in Leiden, used the microscope to discover the red blood corpuscle in 1663. Publication of his results occurred posthumously in 1738.

Anton van Leeuenhoek (1632-1723), a microscopist in Delft, Holland, studied red blood cells, and discovered protozoa in 1677 and bacteria in 1683 with his microscope that magnified 270 times.

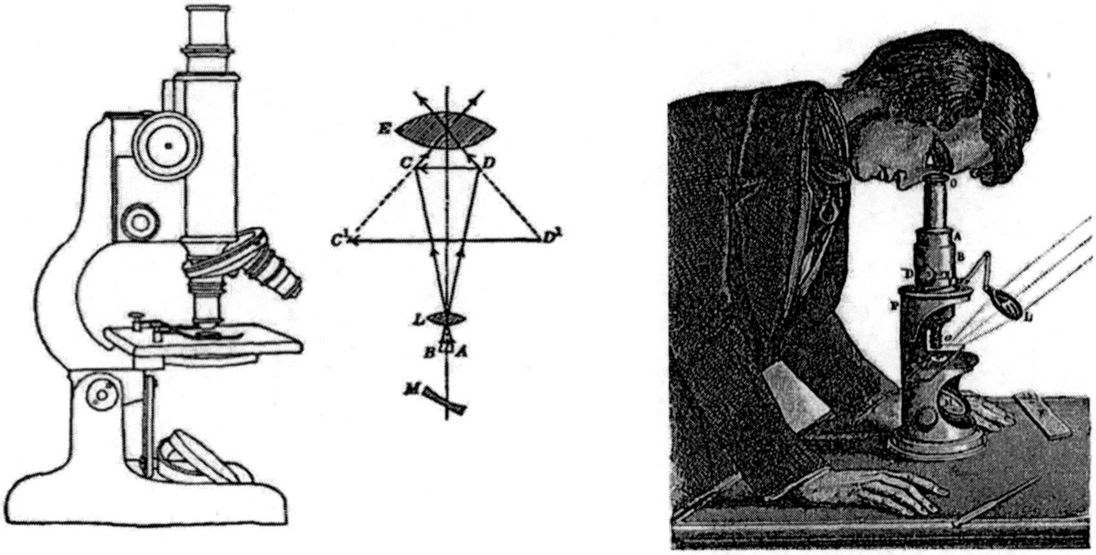

**Figure 31**
**Early Microscope**
Source: Newton Henry Black and Harvey Nathaniel Davis. *New Practical Physics.* New York: Macmillan Company, 1913 (left).
Source: J. Dorman Steele. *Popular Physics.* New York: American Book Company, 1896 (right).

## XII. RENÉ DESCARTES AND THE MECHANICAL PHILOSOPHY

René Descartes (1596-1650), a seventeenth-century French philosopher and mathematician who resided in Holland, Paris, and Stockholm where he died, believed the time had come for a grand synthesis of natural philosophy and immodestly claimed that he had the capability to do it. His starting point in 1620 required him to doubt knowledge of everything except knowledge that he acquired as a result of clear and fundamental ideas.

### *Beginning of Descartes' program: mind and matter*

Descartes began his grand synthesis by concluding that he could not doubt that he was doing the doubting, therefore mind exists. This is the origin of Descartes' famous expression, *I think therefore I am, Cogito ergo sum.* Descartes then argued that God does not deceive us, and because Descartes had a clear and distinct idea of a quality-less matter whose only property was extension, matter must also exist. He had a clear and fundamental idea of mind and of quality-less extended matter, or of thought and extension. Mind and matter were fundamental. His matter had extension yet it had no mass.

What connected mind and matter? According to Descartes, the pineal gland, a small structure attached to the brain and which he erroneously believed animals did not have, made the connection.

### *Descartes' science*

Cartesian science was deductive. Descartes wanted to reduce nature to mathematical law through deduction and to use experiment to distinguish among the consequences deduced.

*Astronomy and cosmology:* Extension or extended matter in motion filled all space. Descartes proposed a vortex theory of the planets in which the ethereal particles formed whirlpools or vortexes around the Sun and carried the planets within them. Each planet remained at rest at the center of a vortex that revolved around the Sun. This enabled Descartes to argue that Earth remained at rest, only the vortex actually moved and carried Earth around the Sun. Descartes published these ideas in *Principles of Philosophy* in 1644. He planned to publish his work earlier but delayed because of the controversy involving Galileo in 1633.

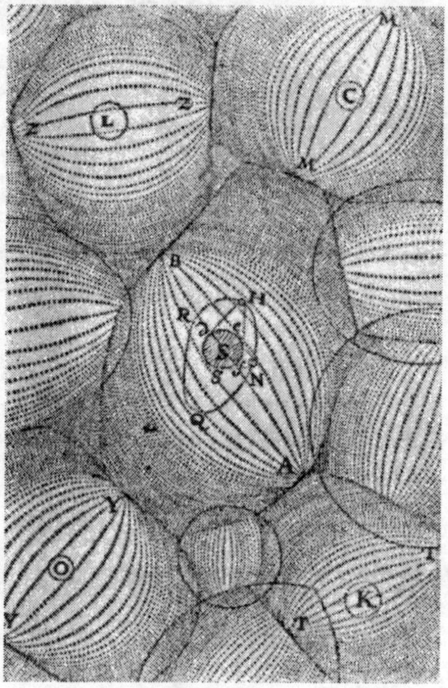

**Figure 32**
**Descartes' Heliocentric Vortex Universe**
Source: Library of Congress.
René Descartes. *Principia Philosophiae*. Amstelodami: Apud Lodovicum Elzevirium, 1644.

*Mechanics:* Descartes recognized conservation of motion and the principle of inertia.

*Structure of matter:* Descartes adopted an atomic theory, but he did not accept action at a distance. Contact among atoms produced all activity. He completely filled space with finer and finer particles, or ether, leaving nothing unoccupied.

*Physics:* Light was mechanical, an instantaneous pressure transmitted through ether. Descartes published his interpretation of light in *Meteors*, which he included with his *Discourse on Method* published in 1637. He published the law of refraction in 1637 in *Dioptric*, including it in *Discourse on Method.*

*Biology*: According to Descartes, animals were machines, matter in motion. They had no feelings, no mind, and acted according to their instincts, not rationally.

***Cartesian mathematics***

Descartes in 1619 invented analytical geometry, which he said came to him in a dream, and published it in *Geometry* in 1637 in *Discourse on Method*. Descartes reasoned that he could represent every point in a plane with two numbers and every point in space with three numbers. He showed that an algebraic equation represented a geometric shape. This breakthrough linked algebra, also called analysis, and geometry. Some of the equations and the shapes they represent appear below.

Fermat independently invented analytical geometry in 1629. He did not follow up his work which remained unpublished until 1679.

| DESCARTES' ANALYTICAL GEOMETRY | |
|---|---|
| Straight line ($y = ax + b$) | Measurements of density and of velocity when plotted give a straight line graph |
| Circle ($x^2 + y^2 = c^2$) | Size of deferents and epicycles |
| Parabola ($y = ax^2 - b$) | Path of a projectile |
| Ellipse ($\frac{x^2}{a^2} + \frac{y^2}{b^2} = 1$) | Kepler's planetary paths |

***Descartes' failure***

Descartes' program failed because of its vast breadth or extension. Descartes saw himself as an artist, an architect, creating a model of the universe that he had to complete in its totality. He never tested many of his deductions, and when others tested his deductions, they proved Descartes' scientific claims largely incorrect. Newton investigated Descartes' vortexes, treating them as a problem of fluid mechanics, and showed that his vortexes could not exist.

***An alternative to Descartes' mechanical philosophy***

Gottfried Leibniz (1646-1716), a German mathematician in Hanover, rejected the mechanical philosophy. He believed nature had inbuilt purposes and that matter contained units of energy called monads. In conscious beings, monads were the souls, and God the supreme monad who created the world of monads. Everything had its place, and we lived in the best of all possible worlds for God's purpose or for the fulfillment of his ends.

## XIII. THEORY OF UNIVERSAL GRAVITATION

***Tycho Brahe***

Tycho Brahe (1546-1601), an astronomer in Denmark, later Prague, had the reputation of a somewhat difficult person with whom to have a relationship and sometimes had difficulties in his relations with others. In one of his youthful encounters at age 20 he lost the tip of his nose in a fencing duel with a cousin. A glued on gold-silver alloy tip (actually brass) replaced the lost piece of nose. He ranks as the greatest and last of the naked-eye astronomers, whose observations of the planetary motions remained unsurpassed in their time. A sister, Sophia Brahe (1556-1643) assisted him in making some of his observations. Brahe died of uremia (kidney failure), a story that includes much hearsay about the circumstances of his death. A study in 2001 claimed he died of mercury poisoning. Czech and Danish scientists from Aarhus University in Denmark and at the University of Rostock in a 2012 study convincingly disproved the mercury poisoning claim.

In 1576 Tycho with the financial assistance of Frederick II (1534-88), King of Denmark, constructed his observatory called Uraniborg on the island of Hven off the Danish coast. He equipped it with excellent instruments, such as the large and precision measuring quadrant made of wood, brass, and iron. Tycho's observations of the planets were correct to 2 minutes of arc, 120 seconds, some to 30 seconds, whereas Ptolemy's observations were correct to 10 minutes of arc. A circle contains 360°, 60 minutes = 1 degree, 60 seconds = 1 minute. The invention of the telescope in 1608 soon made Tycho's instruments obsolete.

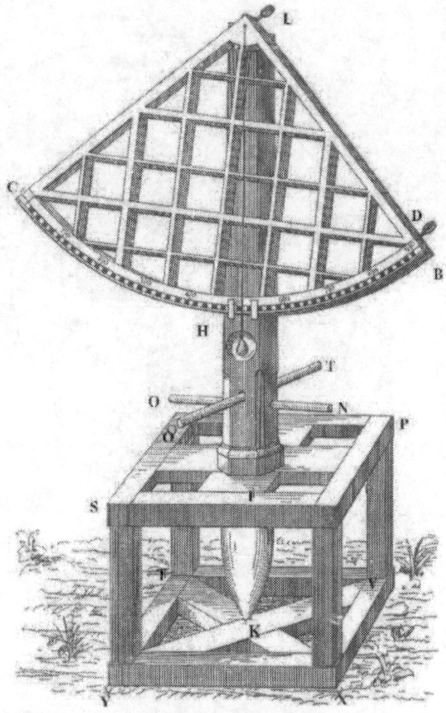


Tycho's early outdoor quadrant for observing altitudes using the sight D, L, and the plumb line.

**Figure 33**
**Tycho Brahe's First Observatory at Augsburg 1569-70**
Source: Tycho Brahe. *Astronomiæ instauratæ mecanica*. Wandsbek, 1598.

***Tycho's contributions***

Tycho introduced the term *nova* to designate a new star that appeared in 1572. This showed the imperfection of the heavens, contrary to what the Aristotelians claimed. In his system of the world, published in *De Mundi Aetherei* (1588), the Moon and Sun revolved around a stationary Earth, all the remaining planets revolved around the Sun. Christoph Rothmann (1550-1605), an astronomer in Hesse, prior to 1588 proposed the same model. Rothmann corresponded with Tycho in 1596. Nicolaus Reimers or Raimarus (1551-1600), court mathematician to the Holy Roman Emperor Rudolf II (1552-1612) in Prague, proposed the same model in his *De Astronomicis Hypothesibus* of 1597. A question of priority arose as a result. Most astronomers favored Tycho's model until about 1650 because it avoided the problems associated with a moving Earth. It accounted for the phases of Venus, but not for the later discovery of parallax.

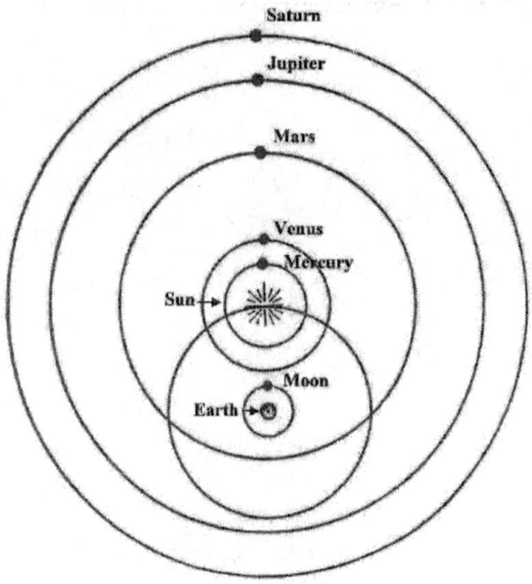


**Figure 34**
**Tycho Brahe's Geocentric System**
Courtesy of the author

Tycho moved to Prague in 1599 after the Danish King Christian IV (1577-1648) took away his pension and his rights to Hven. He had his instruments shipped to Prague, but the Thirty Years War of 1618-48 led to their burning and destruction. Tycho compiled a catalogue of 777 fixed stars that Johannes Kepler, then his assistant, published posthumously in 1602.

***Calendar reform of 1582***

Julius Caesar in 46 BC introduced the Julian calendar to replace the complicated existing Roman calendar that required the insertion of an intercalary (thirteenth) month of variable length and in which the calendar (civil) equinoxes preceded the astronomical equinoxes by three months. The Julian calendar, first used in January 45 BC, in its final form had 12 months totaling 365 days and added an extra day to February every fourth year to give a leap year. The mean length of the year was 365¼ days. Caesar ended use of the lunar year, the intercalary month, and regulated the civil year solely by the Sun. Despite these improvements that improved the calendar's accuracy the Julian calendar had an error of one day every 128 years and required additional changes by the late 1500s. Because of the precession of the equinoxes, Easter continued to occur earlier and earlier in the year, making additional correction necessary to prevent this from happening.

This correction came when Pope Gregory XIII (1502-85) introduced the Gregorian calendar in 1582. The Gregorian calendar removed 10 days from the year and went from Wednesday 4 October to Thursday 15 October to correct for the 10 days the vernal equinox and Easter had advanced since the Council of Nicaea in 325. It moved the vernal equinox from 10-11 March to 21 March To keep the calendar in agreement with the seasons, 1700, 1800, and 1900 were no longer leap years, and only centennials divisible by 400 became leap years. The Gregorian calendar made 2000 a leap year, but not 2100.

The Gregorian calendar quickly replaced the Julian calendar in Catholic countries, such as Italy, Spain, Portugal, Poland, France, and Belgium. Protestant countries gradually adopted the new calendar in the 1700s, Holland and parts of Germany in 1700-01, England and its colonies in 1752, and Sweden in 1753. The Eastern Orthodox and Greek Catholic churches in Russia, Ukraine, and Eastern Europe still use the Julian calendar for liturgical purposes such as celebrating Easter and Christmas. Russia adopted the Gregorian calendar in 1918, Romania in 1918, Greece in 1924.

***Johannes Kepler and the planetary paths***

Johannes Kepler (1571-1630), a German astronomer and ordained Lutheran minister carried out astronomical investigations while in Graz and later in Prague. His teacher, Michael Maestlin (1550-1631) at Tübingen, was one of the first astronomers to teach publicly the Ptolemaic and Copernican models, and in the last edition of his *Epitome Astronomiae* published in 1588 he included appendixes that discussed the Copernican model.

Kepler became Tycho's assistant in Prague from 1600 to 1601. They were a good team that paired Tycho's excellent observational data with Kepler's considerable mathematical skills. During the last eighteen months of Tycho's life Kepler used Tycho's tables of planetary motions, especially those of Mars, to establish his (Kepler's) three laws of motion that accounted for the planetary orbits and supported his Copernican belief.

*Mysterium Cosmographicum* and *Harmonice Mundie*

In the *Mysterium Cosmographicum*, published in 1596, Kepler tried to show that a law of planetary orbits related the five perfect or regular geometric solids and the six planets, and therefore explained why only six planets existed. He had found the grand scheme of the universe. Upon circumscribing the known size of each planet's orbit around one of the geometric solids, Kepler obtained a reasonable fit for each planet's orbit and for the geometric shape. But the relation was purely *ad hoc*. His scheme from outermost to innermost planet follows.

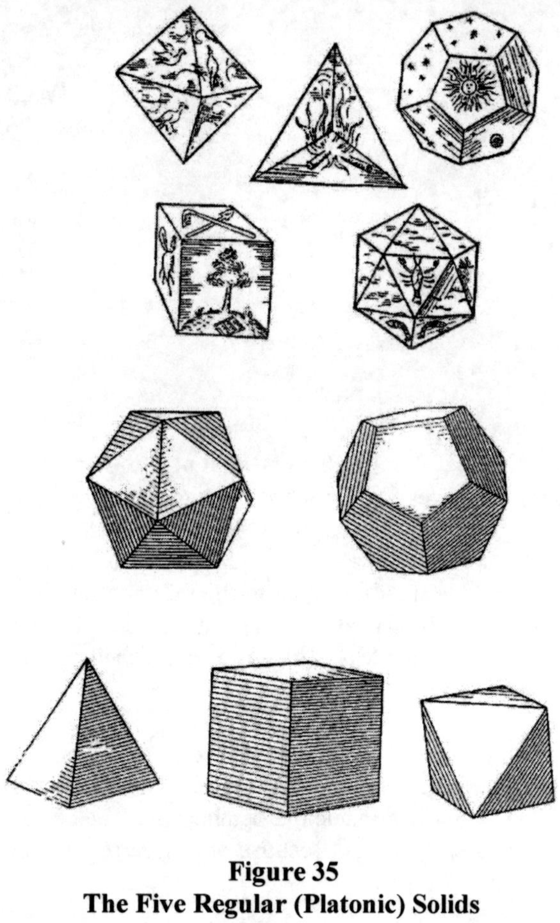

**Figure 35**
**The Five Regular (Platonic) Solids**
Source: Johannes Kepler. *Mysterium Cosmographicum*. Tübingen, 1596.

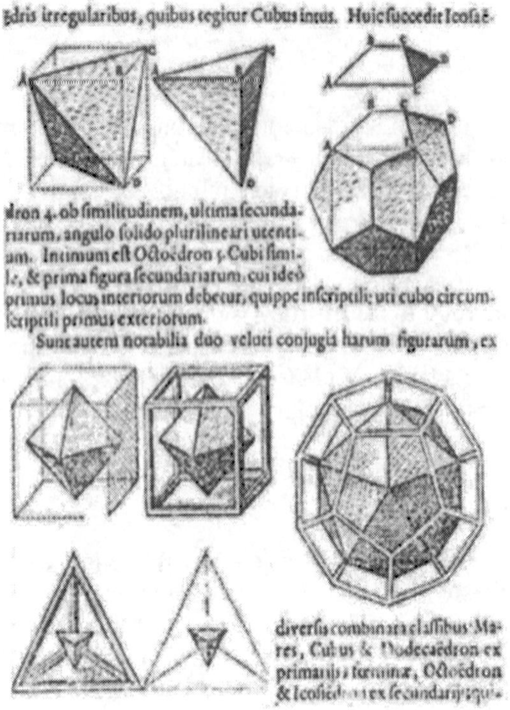

ëdris irregularibus, quibus tegitur Cubus intus. Huic succedit Icosaë-
dron 4. ob similitudinem, ultima secunda-
riarum, angulo solido plurilineari utenti-
um. Intimum est Octoëdron 5. Cubi simi-
le, & prima figura secundariarum, cui ideò
primus locus interiorum debetur, quippe inscriptili: uti cubo circum-
scriptili primus exteriorum.
Sunt autem notabilia duo veluti conjugia harum figurarum, ex
diversis combinata classibus: Ma-
res, Cubus & Dodecaëdron ex
primariis, feminæ, Octoëdron
& Icosiëdron ex secundariis, qui-

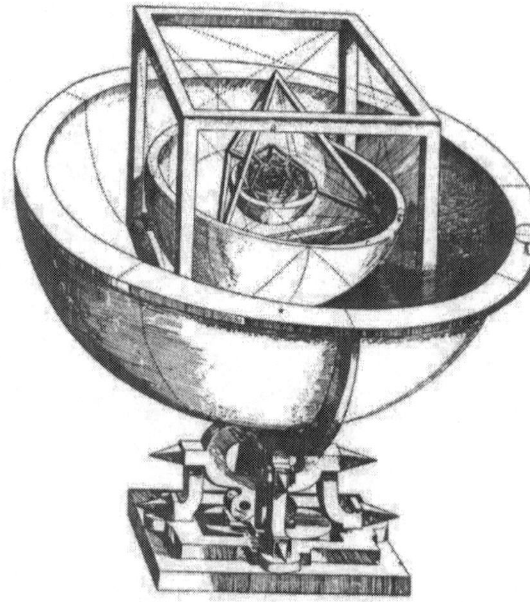

**Figure 36**
**Kepler's Model of the Universe**
Source: Johannes Kepler. *Mysterium Cosmographicum.* Tübingen, 1596.

Kepler circumscribed a sphere with a radius equal to that of Saturn's orbit about a cube (a). A sphere inscribed within this cube has a radius equal to the radius of Jupiter's orbit. The sphere of the orbit of Jupiter (b) has a tetrahedron inscribed. A sphere inscribed within the tetrahedron gives the radius of the orbit of Mars. The sphere for Mars (c) has a dodecahedron inscribed. A sphere inscribed within Mars' orbit gives Earth's orbit (d).

Kepler continued this process of alternately inscribing of spheres and regular solids, using the icosahedron (20-sided) and the octahedron (8-sided). These gave the orbits of Venus (e) and Mercury (f), which is on a sphere inscribed inside the octahedron. Kepler considered the five solids as bridging the intervals between planetary orbits. Because only five regular solids existed, Kepler believed only six planets existed in the universe.

| KEPLER'S PLANETARY SCHEME | |
|---|---|
| ***Planet*** | ***Solid Inscribed*** |
| Orbit of Saturn | cube (6 sides) inscribed |
| Orbit of Jupiter | tetrahedron (4 sides) inscribed |
| Orbit of Mars | dodecahedron (12 sides) inscribed |
| Orbit of Venus | icosahedrons (20 sides) inscribed |
| Orbit of Mercury | octahedron (8 sides) inscribed |

The discovery of three new planets rendered Kepler's scheme irrelevant. William Herschel (1738-1822), a German-born astronomer in London, discovered Uranus in 1781. Johann Galle (1812-1910), an astronomer in Berlin, discovered Neptune in 1846. Clyde Tombaugh (1906-97), an astronomer at the Lowell Observatory in Arizona, discovered Pluto in 1930, although since 2006 astronomers no longer define Pluto as a planet.

In another unfulfilled adventure published in *Harmonice Mundie* in 1619, Kepler tried to calculate the frequency of musical notes each planet supposedly emitted or sounded as it revolved in its orbit and thus discover the harmony of the spheres. According to Kepler, Earth's frequencies, or notes, corresponded to mi, fa, mi, which he thought indicated famine and misery, two misfortunes clearly characteristic of his time.

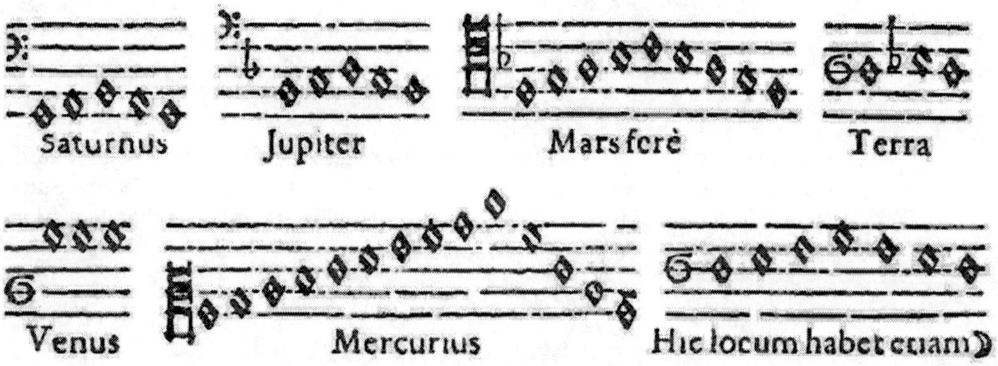


**Figure 37**
**Kepler's Harmony of the Spheres**
Source: Arthur Berry. *A Short History of Astronomy: From Earliest Times Through the Nineteenth Century.* London: John Murray, 1898.

*Kepler's three laws of planetary motion*

Kepler's planetary studies led to his discovery of three laws of planetary motion. His first two laws are in *Astronomia Nova* (1609). He derived them using observational data Tycho had compiled for the motion of Mars. Kepler found that Tycho's data would not give a circular path, but when he substituted an elliptical path for each planet's movement he obtained excellent agreement with the data.

First Law: Planets move in elliptical paths with the Sun at one focus (the other focus is empty). Kepler had proposed for the first time that the planets did not move in circles. By doing so he eliminated the complicated array of deferents and epicycles that from the time of Apollonius and Hipparchus astronomers had incorporated in their models. Kepler believed that a magnetic force emanating from the Sun kept the planets in their orbits. The world was a closed system, the stars occupied a thin shell outside the solar system.

Second Law: A line joining the Sun and a planet sweeps out equal areas in equal time periods. The second law showed that the planets did not move uniformly in their orbits but sometimes sped up or slowed down depending on the distance from the Sun. Kepler's second law gave astronomers a way of determining the planet's complete orbit.

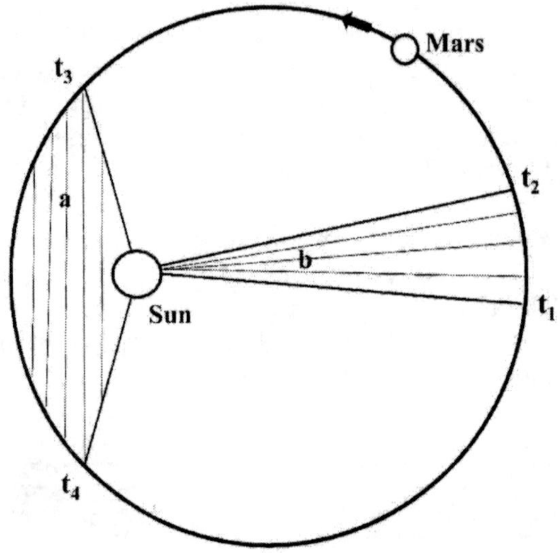


Mars travels along its orbit with varying speed, moving faster when closer to the Sun. Kepler found that for equal time intervals, $t_2 - t_1 = t_4 - t_3$, a line between the Sun and the planet swept out equal (area b = area a). The drawing exaggerates the elongation of the ellipse to illustrate the equal area law more clearly.

**Figure 38**
**Kepler's Second Law (1609)**
Courtesy of the author

Kepler's *Harmonice Mundie* of 1619 contains the third law: $\frac{R^3}{T^2} = k$, where

$R$ = radius of the planet's orbit, the average of its major and minor axis
$T$ = time for the planet to complete its orbit
$K$ = a constant that is the same for each planet.

This law also made possible the calculation of the planetary orbits. A graph showing $T^2$ as the x value and $R^3$ as the y value gives a straight line. In Kepler's day astronomers calculated radii only in multiples of the radius of Earth's orbit. The radius of Earth's orbit is an astronomical unit (AU) of length. Kepler's nearly constant values of $R^3/T^2$ illustrate his third law. The last column gives later measurements of the orbits and periods.

| KEPLER'S $R^3/T^2$ RATIOS | | | | |
|---|---|---|---|---|
| ***Planet*** | ***Radius R of Orbit of Planet in AU*** | ***Period T in Days*** | ***$R^3/T^2$ in $(AU)^3/Day^2$*** | ***$R^3/T^2$ Modern Values in $m^3/sec^2$*** |
| Mercury | 0.389 | 87.77 | $7.64 \times 10^6$ | $3.354 \times 10^{18}$ |
| Venus | 0.724 | 224.70 | 7.52 | 3.352 |
| Earth | 1.000 | 365.25 | 7.50 | 3.354 |
| Mars | 1.524 | 686.98 | 7.50 | 3.354 |
| Jupiter | 5.200 | 4,332.62 | 7.490 | 3.355 |
| Saturn | 9.510 | 10,759.20 | 7.430 | 3.353 |

Source: *Physics* (Physical Science Study Committee).
Toronto: Copp Clark Publishing Company, 1965.

Kepler was probably the first person to use logarithms in mathematical calculations. He published the Rudolphine Tables of planetary motions in 1627 in honor of his patron Rudolph II (1552-1612), king of Bohemia and Hungary and Holy Roman Emperor.

Kepler's laws give a kinematic description of the planetary motions, not a dynamic interpretation. They describe the motions but do not account for the forces producing the motion. The only dynamics that accounts for the motions of the planets requires a Sun-centered universe. No one has developed a dynamics and a law of force for an Earth-centered universe as Newton with his law of universal gravitation did for a Sun-centered universe.

***Isaac Newton, the laws of motion and universal gravitation***

Newton, England's foremost physicist, who taught at the Church of England's Trinity College, Cambridge, disavowed the Church's Trinitarian interpretation of God and secretly held a Unitarian understanding of God. Accepting the Trinity, according to Newton, violated the first commandment, not to have other Gods before me.

Newton received his master's degree from Cambridge in 1668 and in 1669 accepted appointment as Lucasian professor of mathematics at Cambridge. He published two important papers on light in 1672 and 1674 and the *Principia*, his classic work on astronomy, in 1687. Newton won election to Parliament, representing Cambridge, in 1689 and again in 1701. Praised for his astronomy and physics, Newton practiced alchemy, and his experimentation with mercury most likely caused his breakdown and illness of 1692-93. During his lifetime he engaged in well-known controversies with Hooke, Leibniz, and John Flamsteed (1646-1719), the first Astronomer Royal. He received appointment as warden of the mint in 1696 and master of the mint in 1699 in London. During his years at the mint Newton introduced a procedure to prevent clipping of gold and silver coins by minting them with a slim ridge around a coin's circumference, making obvious any clipping that had occurred. Newton won election to the presidency of the Royal Society in London in 1703 and served continuously from 1703 to 1727. He published *Opticks* in 1704, Queen Anne (1665-1714) knighted him in 1705. Newton never married, he gave away most of his wealth to his relatives and the Royal Society. He died from kidney stone complications, a magnificent monument holding his remains is in Westminster Abbey.

**Figure 39**
**Newton's Death Mask**
Source: "The Progress of Science." *Popular Science Monthly* 69 (November 1906).

*Invention of calculus: controversy with Leibniz over the invention of calculus*

Early in his career in 1664-66, Newton invented calculus because he had to show that Earth's attractive force (gravity) acting on the Moon acted as if Earth had its entire mass concentrated at its center. He did not publish his discovery until 1704 in *Opticks*. Newton claimed Leibniz saw his (Newton's) letter of 1671 in which Newton discussed how he developed calculus.

In Newton's notation for differential calculus, or differentiation, he indicated the first derivative of a function with a dot placed over the function, which he called a fluxion: $\frac{dx}{dt} = \dot{x}$ , and the second derivative: $\frac{d^2}{dt^2} = \ddot{x}$. His notation proved useful in mechanics and other engineering studies that use higher derivatives sparingly.

Leibniz invented calculus independently in 1675 and published it in 1684, twenty years before Newton's publication. Leibniz represented the derivative of a function, how $y$ changes with respect to $x$, as $\frac{dy}{dx}$. His notation, not Newton's, acquired wide use. Leibniz studied the problem of deriving the equation of a curve from the properties of its tangents and recognized that the method was integration. This enabled him to make the connection between integration and differentiation. Leibniz introduced the notation for integration, the elongated S for sums.

In the Leibniz controversy on the priority of who invented calculus Newton went to extreme and devious means to establish his priority. In 1712, while president of the Royal Society, he packed the Royal Society's eleven-member investigative committee with his supporters, anonymously wrote the preface to the report the investigative committee compiled, directed the committee's every activity, and produced a report that clearly established Newton's priority. Newton took these extreme measures because of the newly institutionalized value attached to priority in science.

*Astronomy*

Astronomers in the seventeenth century focused on a central question: what kind of attractive force the Sun exerted on the planets and caused them to move according to Kepler's laws. The emphasis had shifted from a kinematic description to a dynamic explanation that gave an equation for the attractive force.

In trying to answer the question, Newton started with the idea that the Moon was a falling body and that the same force attracted the Moon and a body (an apple) falling to Earth except that the Moon was much farther (240,000 miles) and fell much more slowly. Newton suspected that the attracting force decreased according to an inverse square law. This followed from Hooke's equation for centripetal acceleration and Kepler's third law.

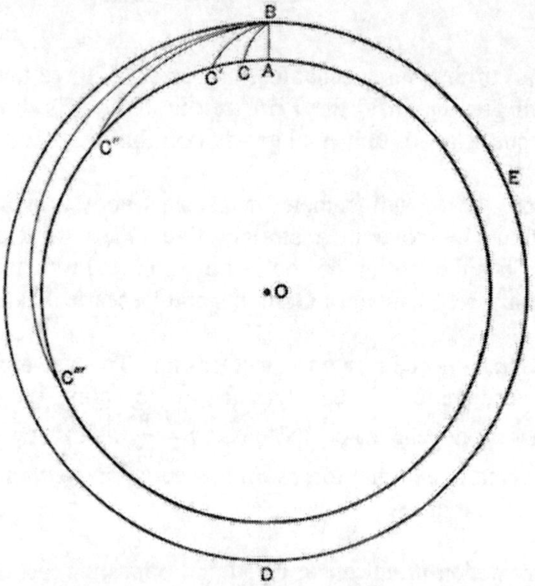


**Figure 40**
**The Moon as a Projectile**
Source: Arthur Berry. *A Short History of Astronomy*: *From Earliest Times Through the Nineteenth Century.* London: John Murray, 1898.

To calculate the force Newton had to know how many times more distant the Moon was from the center of Earth than the falling body was from the center of Earth, or how many times Earth's radius the Moon's center was from Earth's center. Newton found that the distance from Moon to Earth was 60.3 times Earth's radius.

Newton's astronomical synthesis required that he start with the idea that straight line or linear motion, and not circular motion, was the natural motion of the planets or of any moving body. He also needed the results Galileo, Hooke, and Kepler previously had obtained. Galileo calculated acceleration, a, or g, and Hooke in 1679 calculated the centripetal acceleration for a body moving uniformly around a circle, such as a planet orbiting the Sun.

Hooke recognized that the motion of such an orbiting body had two components: its inertial straight line motion and its centripetal or central-seeking motion. He derived the equation below to account for the acceleration a, where R is the radius of the body's orbit and T the time it took to complete its orbit, or its period. Hooke included this equation in a letter he sent to his rival Newton, and he suggested that the attractive force between the orbiting body and the central body varied inversely as the square of the distance between them. But unlike Newton he lacked the mathematical skill to show this quantitatively.

$$a = \frac{4\pi^2 R}{T^2}$$

Hooke's equation enabled Newton to calculate the Moon's acceleration from the radius of its orbit and its period of rotation, obtaining the value a = $2.7 \times 10^{-3}$ m/sec$^2$. Newton then calculated the force acting on the Moon from his second law of motion F = ma, where F = force, m = mass of Earth, and a = acceleration of the Moon.

Using Kepler's third law, $\frac{R^3}{T^2} = K$, Newton substituted $\frac{R^3}{K} = T^2$ to eliminate T in Hooke's equation above for acceleration and combined it with his second law to show that F is proportional to

$$\frac{1}{R^2} \quad or \quad F = \frac{4\pi^2 Km}{R^2}.$$

Newton published all of this, his mechanical universe, his three laws of motion, and his derivation of the law of universal gravitation in the *Principia* in 1687. His three laws of motion that he used to derive the inverse square law of universal gravitation follow.

First Law: A body at rest will remain at rest or a body moving uniformly in a straight line will continue to move in a straight line unless an external force acts on it. This law describes the motion of bodies not subjected to external forces. This is the principle of inertia, a restatement of Galileo's and Descartes' explanation of inertia.

Second Law: $F = ma$, or force = mass x acceleration. The rate at which the speed of a body changes (its acceleration) is equal to the force acting on the body divided by its mass, or inversely dependent on its mass, $a = \frac{F}{m}$. This law describes how bodies move when subjected to external forces. In this equation Newton made the first distinction between weight and mass.

Third Law: Every action produces an equal and opposite reaction. Earth attracts the Moon and the Moon attracts Earth. This reciprocal attraction was true for all the planets and for all bodies on Earth.

The final form of the inverse square law of universal gravitation that Newton derived for any two bodies attracting each other appears below.

$$F = \frac{G\, m_1 m_2}{R^2}$$

F = force of attraction
G = gravitational constant which incorporates Kepler's constant K times $4\pi^2$
$m_1$ $m_2$ = mass of the bodies that attract each other
R = distance between the bodies.

In order to calculate F, the attractive force between Earth and an object of mass m at Earth's surface, Newton had to have a value for G. To calculate G, Newton estimated Earth's mass by assuming that Earth had a density five times the density of water and multiplying Earth's (spherical) volume by 5. He used the well-known formulas below for the volume of a sphere (Earth) and for density.

$$V_{Earth} = \frac{4\pi\, r^3}{3}$$

$$Density = \frac{mass}{volume}$$

He then substituted the values of $m_e$ (mass of Earth), R (radius of Earth), and F = mg (for the acceleration of anybody near or at Earth's surface) to eliminate F and m, into the inverse square law equation and calculated G, obtaining the value $G = 1.0 \times 10^{-11}$ $m^3/kg\text{-}sec^2$.

$$F = G\frac{m_e m}{R_e^2}$$

$m_e$ = $6 \times 10^{24}$ kilograms

$R_e$ = $6.38 \times 10^6$ meters

F = mg

G = F/m = 9.8 meters/second$^2$

dividing both sides by m to eliminate m

$$9.8\frac{meters}{second^2} = \frac{G \times 6 \times 10^{24}\ kilograms}{(6.38 \times 10^6\ meters)^2}$$

rearranging the equation to solve for G

G = 1.0 x 10-11 meters3/kilograms-seconds$^2$.

European physicists, such as Leibniz, rejected Newton's conception of force and gravity during the first decades of the eighteenth century. They regarded it as occult because it implied action at a distance and not a mechanical interpretation, as were Descartes' vortices. Newtonian mechanics did not dominate physics and astronomy until the 1750s.

Emile du Chátelet (1706-49) in Paris studied Newtonian physics with Pierre Maupertuis (1698-1759) and Alexis-Claude Clairaut (1713-65). She became the mistress of Voltaire (1694-1778), the French essayist and critic, who held Newtonian views. Chátelet began a translation of the *Principia* into French in 1744. Publication of her completed translation occurred in 1759, ten years after her death from puerperal fever which she contracted after giving birth. Her translation made Newtonian ideas available in France. It remains the standard translation today.

Newton's *Principia* was very likely the last major scientific work written in Latin. In the 1700s European scientists wrote in their native languages. Abandoning Latin as the language of the scholar they nationalized science and commonly referred to an English, an Italian, a French, or German school of thought.

*Henry Cavendish's determination of Earth's mass in 1798, calculation of G*

Henry Cavendish (1731-1810), an English physicist who lived near London, provided the first accurate measurement of Earth's mass. He used Hooke's law of 1678 which showed that the strain an object experienced is proportional to the stress applied. A sensitive torsion balance measured the twist (strain) produced when he subjected the two small lead spheres of his balance, each weighing 1.61 pounds and 2 inches in diameter, to small forces (stress) upon bringing two smaller metal spheres of known weight close to each of the lead spheres. Cavendish placed one of the smaller spheres in front of the first lead sphere, the other behind the second lead sphere. A small mirror attached to the suspended fine quartz fiber of the balance reflected a beam of light onto a scale that measured the angle through which the fiber and mirror twisted. The angle of twist measured the torque resulting from the attraction between each pair of spheres. The two torques act in the same direction to produce the twist and measure the gravitational force exerted by the two pairs of spheres. Changing the positions of the smaller metal spheres so the one in front was now behind and vice versa made the fiber twist in the opposite direction. The angle of twist, Cavendish demonstrated, depended only on the masses of the spheres and on the gravitational force acting between each pair of spheres. The twisting motion stopped when the gravitational force attracting the two spheres balanced the twisting motion of the spheres.

Repeating the experiment by bringing two large lead spheres, each weighing 348 pounds and 12 inches in diameter, close to the small lead spheres of the balance, Cavendish produced a greater twist that measured, as he had shown, the torque, and was proportional to the gravitational force (F), between each pair of spheres. He determined the values of the torque required to produce a given angle of twist in a separate experiment using Hooke's law, which gave him a value for F the gravitational force. He determined the gravitational force of Earth acting on the small lead sphere by weighing it and calculating its mass from Newton's second law F= mg. He had a value for the distance between the small lead sphere and Earth (R=radius of Earth), and from the ratio of the two force measurements Cavendish in 1798 determined Earth's mass (mean density). Having an accurate value for Earth's mass, and a calibrated torsion balance scale that provided values for the gravitational force between a pair of objects, nineteenth-century physicists calculated the first precise value for G, obtaining values of $G = 0.667 \times 10^{-10}$ $m^3/kg\text{-}sec^2$.

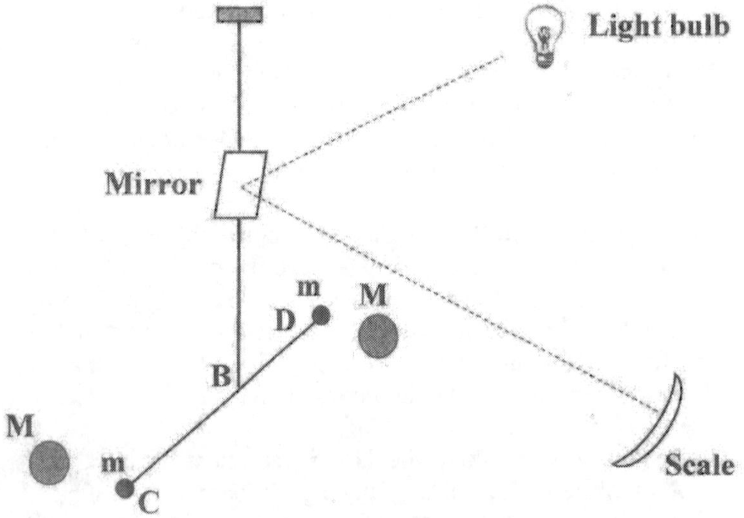


A simplified diagram of the apparatus Cavendish used to verify the law of universal gravitation for small objects and for later measurements of the gravitational constant G.

**Figure 41**
**Calculation of G**
Courtesy of the author

| THE SOLAR SYSTEM | | | | | |
|---|---|---|---|---|---|
| *Objects* | *Mass (Kilograms)* | *Radius (Meters)* | *Period of Rotation (Seconds)* | *Mean Radius of Orbit (Meters)* | *Period of Revolution (Seconds)* |
| Sun | $1.98 \times 10^{30}$ | $6.95 \times 10^{8}$ | $2.14 \times 10^{6}$ | | |
| Mercury | $3.28 \times 10^{23}$ | $2.57 \times 10^{6}$ | $7.60 \times 10^{6}$ | $5.79 \times 10^{10}$ | $7.60 \times 10^{6}$ |
| Venus | $4.83 \times 10^{24}$ | $6.31 \times 10^{6}$ | $2.6 \times 10^{6}$ (?) | $1.08 \times 10^{11}$ | $1.94 \times 10^{7}$ |
| Earth | $5.98 \times 10^{24}$ | $6.38 \times 10^{6}$ | $8.61 \times 10^{4}$ | $1.49 \times 10^{11}$ | $3.16 \times 10^{7}$ |
| Mars | $6.37 \times 10^{23}$ | $3.43 \times 10^{6}$ | $8.85 \times 10^{4}$ | $2.28 \times 10^{11}$ | $5.94 \times 10^{7}$ |
| Jupiter | $1.90 \times 10^{27}$ | $7.18 \times 10^{7}$ | $3.54 \times 10^{4}$ | $7.78 \times 10^{11}$ | $3.74 \times 10^{8}$ |
| Saturn | $5.67 \times 10^{26}$ | $6.03 \times 10^{7}$ | $3.60 \times 10^{4}$ | $1.43 \times 10^{12}$ | $9.30 \times 10^{8}$ |
| Uranus | $8.80 \times 10^{25}$ | $2.67 \times 10^{7}$ | $3.88 \times 10^{4}$ | $2.87 \times 10^{12}$ | $2.66 \times 10^{9}$ |
| Neptune | $1.03 \times 10^{26}$ | $2.48 \times 10^{7}$ | $5.69 \times 10^{4}$ | $4.50 \times 10^{12}$ | $5.20 \times 10^{9}$ |
| Pluto | ? | ? | ? | $5.9 \times 10^{12}$ | $7.82 \times 10^{9}$ |
| Moon | $7.34 \times 10^{22}$ | $1.74 \times 10^{6}$ | $2.36 \times 10^{6}$ | $3.8 \times 10^{8}$ | $2.36 \times 10^{6}$ |

Source: *Physics* (Physical Science Study Committee). Toronto: Copp Clark Publishing Company, 1965.

## XIV. CONCLUSION

Newton's mechanical universe completed the scientific revolution begun by Copernicus. What defined the scientific revolution? The invention of new instruments or apparatus? New developments in mathematics? A new way of examining nature, that is, viewing nature as matter in motion, or as a machine, an interpretation called the mechanical philosophy?

The new instruments that scientists invented, such as the vacuum pump, barometer, telescope, and microscope, enabled them to experiment and investigate the behavior of nature using procedures that had eluded earlier scientists. Their experimental investigations led to Harvey's discovery that blood circulates in the body; that white light is a composite of colors and has a finite velocity; to the establishment of scientific laws, those of Boyle and Snell, Galileo's and Kepler's laws of motion, and most significantly Newton's three laws and law of universal gravitation. The new mathematical calculating methods, such as Napier's and Briggs's logarithms and slide rule, Descartes's analytical geometry, and Newton's and Leibniz's calculus, enabled scientists to make new kinds of measurements, to quantify their experimental results, and to convert their experimental observations into mathematical equations. Van Helmont's and Boyle's quantitative studies on the composition of matter, especially the study of gases, combustion, and chemical classification and identification of substances established a foundation for a revolution in chemistry that occurred in the next century.

Instruments and quantitative measurements were both crucial to advancing the mechanical philosophy. They represent two essential components that made possible the mechanical interpretation of nature. Neither new instruments nor new mathematics sufficiently account for the significant sixteenth- and seventeenth-century scientific advances that constituted the scientific revolution, but together they enabled the mechanical philosophy to become the dominant view of nature that defined the scientific revolution.

## REVIEW QUESTIONS ON THE SCIENTIFIC REVOLUTION

1. What is the main theme of this chapter?

2. Historians of science have defined the scientific revolution in different ways. What are they and what is the currently accepted definition?

3. What did William Gilbert conclude about Earth and magnetism?

4. Why is Francis Bacon significant in the history of science? What was his philosophy of science?

5. What were Galileo's contributions to the study of motion and to astronomy? What was his philosophy of science?

6. Discuss Galileo's conflict with the Catholic Church regarding the system of the world? What is your view regarding the controversy?

7. What did Johann Baptiste van Helmont contribute to the study of the composition of matter?

8. What was Robert Boyle's contribution to the study of gases and the composition of matter?

9. Who invented the barometer? What does the expression *nature abhors a vacuum* mean?

10. What major advances occurred in the study of mathematics in the sixteenth and seventeenth centuries?

11. What was the dominant model of light in the sixteenth and seventeenth centuries? Who supported the rival theories?

12. What three major advances occurred in the study of light?

13. Which important scientific societies arose in the 1600s?

14. What was the seventeenth-century model of heat?

15. Discuss the construction of the early thermometers.

16. What was sound and how did seventeenth-century experimenters measure it?

17. What did Michael Servetus, William Harvey and Marcello Malpighi contribute to physiology?

18. Identify: René Déscartes, Tycho Brahe, Robert Hooke, Gottfried Leibniz.

19. What are Johannes Kepler's three laws of motion?

20. What was Isaac Newton's stroke of genius?

21. What are Newton's three laws of motion and his law of universal gravitation?

22. Why did Newton pursue the study of alchemy?

# CHAPTER 6
# THE PHYSICAL AND NATURAL SCIENCES IN THE EIGHTEENTH CENTURY: THE ENLIGHTENMENT

Speech is the representation of the mind,
and writing is the representation of speech.
Aristotle (384-322 BC)

## I. INTRODUCTION

The eighteenth century was a troubled period with social revolutions occurring in Britain's American colonies and in France. The American Revolution lasted from 1776 to 1783, the French Revolution from 1789 to 1799, followed by the rise of Napoleon Bonaparte (1769-1821) in 1799. Napoleon promoted science-based industry and the fundamental research required for industrial development.

Identification and classification in science was a major activity in both the physical and life sciences. The study of gases as a new form of matter in addition to solids and liquids became increasingly important. Investigators studied the properties of gases intending to identify and classify them. Helmont and Boyle in their studies on gases in the previous century had begun the identification and classification. Boyle's work on acids and bases contributed significantly to the classification of substances.

In the physical sciences the mechanical philosophy still dominated. Studies on matter emphasized an understanding of composition, combustion, and imponderables (fluids) present in matter. Imponderables consisted of self-repulsive particles that were strongly attracted to matter. They sometimes had weight, although never much, and sometimes were weightless. Scientists used imponderables to account for combustion, heat, electricity, and magnetism. Quantification was not yet essential to the study of composition but became so in the investigations of the experimentalists Stephen Hales, Joseph Black, Henry Cavendish, Joseph Priestley, and Antoine Lavoisier. Lavoisier's emphasis on quantification enabled him to arrive at the oxygen theory of combustion and the law of conservation of mass. He also gave a useful definition of a chemical element and contributed to the binomial system of nomenclature that chemists have continued to use.

Lavoisier's studies on heat led him to introduce caloric as the element, or fluid, of heat. The caloric theory of heat dominated the eighteenth century and much of the nineteenth century until the 1860s. Black and most other scientists adopted the caloric theory. Despite his accepting the incorrect caloric theory, Black in the 1760s carried out quantitative experiments on heat. He introduced and calculated for the first time values for the quantity of heat, specific heat, the heat of fusion, and the heat of vaporization. Count Rumford (Benjamin Thompson) rejected the caloric theory. He proposed instead a mechanical theory of heat, arguing that particles, or matter, in motion produced heat. Rumford's mechanical theory gained few followers despite his experiments that strongly supported it.

The studies on heat resulted in the development and use of steam power in steam engines. The application of steam engines for locomotion and in factories, replacing water power, initiated the industrial revolution, first in England and then on the European continent and the United States.

Electrical studies advanced greatly with Benjamin Franklin's one-fluid theory of static electricity in 1747 and in 1800 with Alessandro Volta's invention of the first voltaic cell, or battery, and the production of current electricity. The battery resulted in the introduction of a new and powerful investigative tool in science.

John Ray and Carl Linnaeus developed natural and artificial systems of classifying plants and animals. Linnaeus perfected the binomial system of nomenclature used in the two classification systems. The binomial system remains in use.

Most scientists adopted the empirical philosophy (empiricism) or sensationalism of John Locke (1632-1704) in London. Locke argued that experience and the phenomena (sense knowledge) were the source of all knowledge including our ideas.

The alternative was the idealism of Immanuel Kant (1724-1804) at Königsberg, Germany, who believed there exists in the mind *a priori* fundamental categories such as space, time, force, and motion that we use to organize knowledge obtained from the senses.

A revolution also was occurring in industry and economics. James Watt's steam engine in 1776 ushered in Europe's industrial revolution. Adam Smith (1723-90), the Scottish economist, published *The Wealth of Nations* in 1776. It represented the beginning of a revolution in economic thought, the new laissez-faire economic philosophy, replacing mercantilism.

## II. THEORIES OF COMBUSTION AND THE DISCOVERY OF GASES

In the investigation of matter the major change from the previous century was the use of imponderable fluids, such as phlogiston and caloric, to account for combustion and heat and the properties of matter.

Eighteenth-century discoverers of gases named each gas according to its most obvious property. Carbon dioxide was fixed air, hydrogen was inflammable air, and oxygen was dephlogisticated air. Because scientists named each newly-discovered gas an air, their nomenclature indicated that they considered each gas a modification of ordinary air. They remained uncertain whether the gases were fundamental substances, or elements, as defined by Boyle and later Lavoisier.

### *Phlogiston theory of combustion*

The roots of the phlogiston theory go back to the element fire and to J.J. Becher's *terra pinguis* in the seventeenth century. Becher (1635-82) then at Mainz and later in Munich, Germany, published *Physicae Subterranea* in 1667. He classified substances into three earths, one of which, *terra pinguis*, became the principle of combustion.

| BECHER'S EARTHS | |
|---|---|
| *terra lapidae* | vitreous glassy earth |
| *terra mercurialis* | mercurial earth |
| *terra pinguis* | fatty earth and the principle of combustion |

Becher lived in Holland beginning in 1678. As an alchemist he claimed he could convert iron, lead, copper, sea sand, and other substances into gold. Unable to fulfill his claim to the Dutch Estates General he left Holland for London in 1680. Becher also had a plan to connect the Rhine and Danube Rivers with a canal to improve transportation and trade between Holland and Austria. He never carried out his plan.

George Stahl (1660-1734) in Halle and later Berlin introduced the phlogiston theory of combustion in *Fundamenta Chymiae* published in 1723. Stahl replaced Becher's *terra pinguis* with phlogiston meaning inflammable, or flammable, which, he said, was a fluid present in a combustible substance.

Stahl and others recognized that air was essential for combustion, but they believed air served as the carrier for phlogiston.

Stahl realized that rusting and combustion (burning) were the same process, both are oxidations, just as Boyle had concluded that combustion and respiration were the same process. The difference is that rusting occurs slowly, combustion is rapid. Because metals contained phlogiston, they were not simple substances. When burned (oxidized) they gained weight.

*Weakness of the phlogiston theory*

Phlogistonists made no attempt to quantify their experimental results and to account for weight changes in combustion. Metals such as iron, copper, lead, and tin, gained weight when burned, whereas many non-metals and other substances, such as wood and coal, lost weight when burned because some of the products of combustion escaped as gases while some remained as ash. Yet each of these substances lost phlogiston when burned.

Later, phlogistonists introduced the principle or property of levity (buoyancy) to account for the gain in weight in some combustion experiments. Phlogiston's property of levity made a metal that contained phlogiston appear lighter and weigh less than it did before it lost phlogiston as in an oxidation reaction. The argument for levity is similar to comparing the weight of a balloon before and after filling it with air. Which *appears* heavier?

Stahl at first accepted the alchemical belief in conversion of base metals into gold although later he questioned the possibility of such a change.

His physiological studies convinced him that the body was a living system, not chemical or mechanical, but possessing a vital force (vitalism) different from the forces present in nonliving systems.

***Discovery of gases (airs): carbon dioxide, hydrogen, oxygen***

Stephen Hales (1677-1761) in Teddington (London) adopted a physical quantitative approach, that is a Newtonian approach, in his investigations on gases. In the 1720s he introduced the pneumatic trough to collect and measure the volume of gases produced in a chemical reaction, solving the problem that hindered Helmont. His trough was a container filled with water or mercury and in later designs had a horizontal shelf on the bottom. The shelf had a groove down the center over which an experimenter placed an inverted glass jar filled with water or mercury. A metal tube, later a rubber tube, from the gas generator that fit in the groove and passed into the open mouth of the glass jar allowed the gas to pass into the jar where it displaced the liquid in the jar.

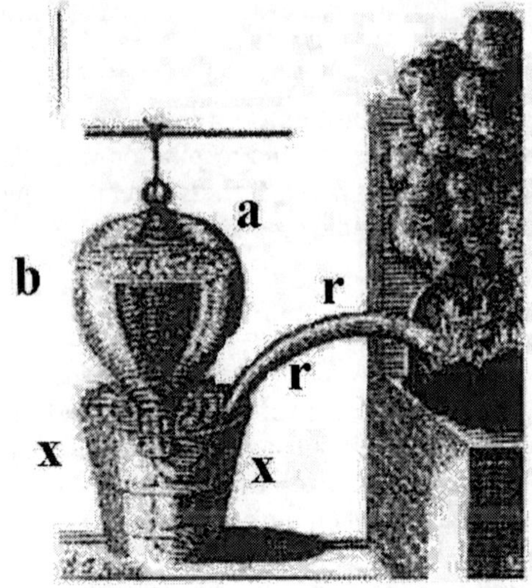


**Figure 1**
**Hales' Pneumatic Trough**
Source: Stephen Hales. *Vegetable Staticks.* London: W. and J. Innys, 1727.

Hales experimented with air and respiration and recognized that a part of the air (carbon dioxide, $CO_2$), and not water only, as Helmont believed, contributed to a plant's growth. He published his results of 124 experiments in *Vegetable Staticks* (1727). Hales showed that metals could fix (combine with) air when heated in air. He did not recognize that the gases he worked with, such as carbon dioxide and oxygen ($O_2$), were distinct substances and thought they were modifications of air.

Joseph Black (1728-99) at Edinburgh discovered fixed air (carbon dioxide) in 1754 as a result of heating calcium carbonate. He carried out quantitative calculations on the production of carbon dioxide by calcium carbonate decomposition and investigated neutralization reactions with acids.

$$CaCO_3 \rightarrow CaO + CO_2$$

Black studied the properties of carbon dioxide and showed that carbon dioxide did not support combustion by pouring carbon dioxide over a lighted candle, causing its flame to extinguish. He demonstrated that carbon dioxide was a component of the atmosphere by converting calcium oxide to calcium carbonate on exposure to air. Calcium oxide reacted with the carbon dioxide that is present in the atmosphere.

$$CaO + CO_2 \rightarrow CaCO_3.$$

Hydrogen was the next important gas discovered. Henry Cavendish (1731-1810) in London discovered hydrogen gas in 1766. Cavendish was independently wealthy as a result of his inheritance at age 40 and had the time to carry out investigations in his private laboratory. At his death Cavendish left a fortune of one million pounds ($50.5 million in 2005). The Cavendish family endowed the Cavendish Laboratory at Cambridge University in 1871.

Cavendish supported the phlogiston theory, but he adopted a quantitative methodology in his experiments on gases. He discovered a gas he called inflammable air (hydrogen) upon reacting an acid, such as sulfuric or hydrochloric acid, with a metal, such as zinc or tin. He thought hydrogen was phlogiston because it burned so easily and was so light (its density had a low value).

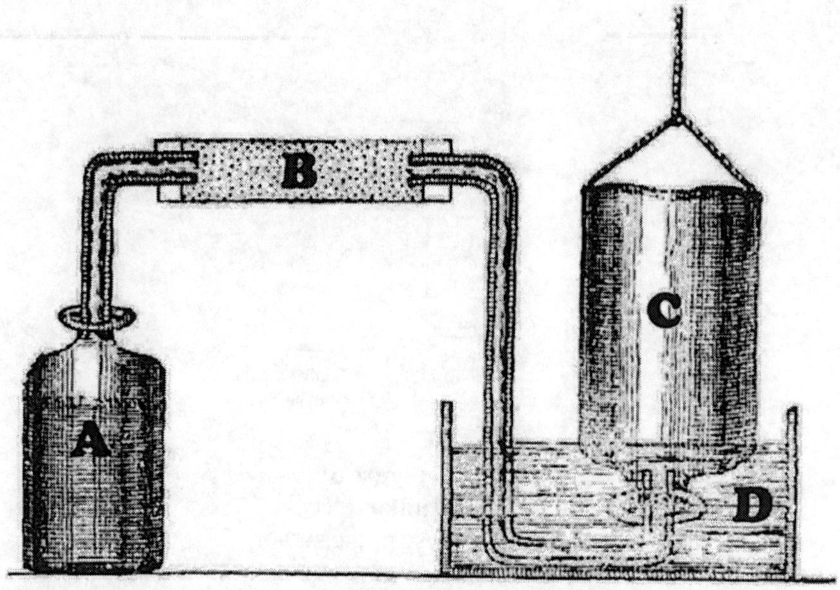


A gas prepared in A (left) passes via the tube C (center) into the water in D (right) where it rises, displacing the water, which passes into the trough E.

**Figure 2**
**Cavendish's Apparatus for the Collection of Gases (1766)**
Source: Henry Cavendish. "Three Papers containing Experiments on factitious Air." *Philosophical Transactions of the Royal Society* (London) 56(1766): 77-101.

Cavendish carried out density measurements of hydrogen and other gases. He showed that the density of hydrogen was 1/14 that of air. His experiment in 1783, in which he produced water by burning hydrogen in air, demonstrated that water was not an element.

In an experiment in 1785 Cavendish, upon sparking air in a eudiometer, discovered argon gas, one of the rare gases found in the atmosphere. The eudiometer is a thick-walled calibrated glass tube sealed at one end and tightly stoppered at the other end. A pair of metallic electrodes extending from inside the eudiometer, each connected to an insulated electrical wire, led to an external high voltage source. The sparking produced inside the tube by the high voltage source combined the nitrogen and oxygen gases present in the air inside the eudiometer. Cavendish added sufficient oxygen to consume all the nitrogen and dissolved the product, $NO_2$, in water. Argon was the insoluble unreactive gas left behind. No one at that time recognized the significance of this experiment, that Cavendish had discovered a new element.

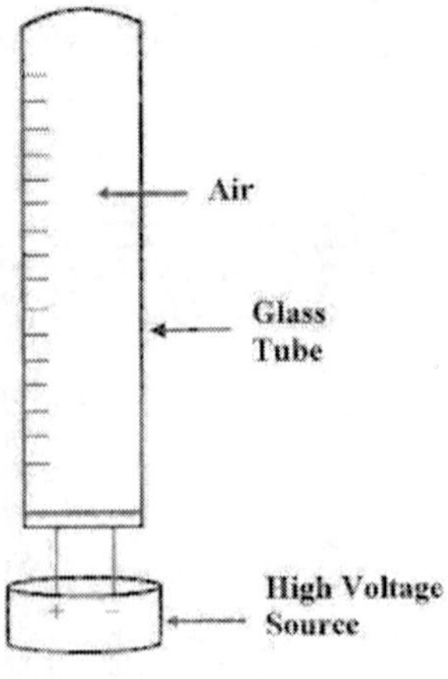


**Figure 3**
**Eudiometer**
Courtesy of the author

James Clerk Maxwell at the Cavendish Laboratory noted this anomalous result in 1879. Five years later, in late 1894 and again in early 1895, William Ramsay (1852-1916) at University College in London and Lord Rayleigh, John William Strutt (1842-1919), working in his private laboratory in Terling Place, Essex County near London, announced their discovery of argon. Rayleigh, who in 1892 followed up Cavendish's 1785 experiment, received the Nobel Prize in physics in 1904. Ramsay received the Nobel Prize in chemistry that same year.

In measuring the gravitational force using a sensitive torsion balance, Cavendish in 1798 accurately calculated Earth's density (mass). His calculation led to an accurate determination of the gravitational constant *G* in Newton's inverse square law of universal gravitation.

Joseph Priestley (1733-1804) discovered oxygen, the third significant gas in 1774 but remained committed to the phlogiston theory of combustion. A Unitarian minister and religious dissenter and a supporter of the American and French revolutions, Priestley at the time of his discovery earned £250 per year as librarian and family tutor for William Perry (1737-1805), the wealthy second Earl of Shelbourne and future British prime minister (1782-83) in Caine, England. Priestley, who lived in Birmingham until 1781 and then in London, had to leave England in 1794 because of his political views and spent his last ten years in Northumberland, Pennsylvania.

In the United States chemists, such as James Woodhouse (1770-1809), who studied under Benjamin Rush (1746-1813) at Pennsylvania and taught at Pennsylvania, and the Quaker Samuel Mitchill (1764-1831) at Columbia, overwhelmingly adopted Lavoisier's new chemistry and oxygen theory of combustion, not phlogiston. Woodhouse, who in 1792 founded the Chemical Society of Philadelphia, the first chemical society in the United States, debated combustion theories with Priestley for 10 years. Their debates, published mainly in *Medical Repository*, a journal Mitchill founded in 1797 and published until 1824, began after Priestley and family left London for Northumberland to escape political persecution because he sympathized with the French Revolution. Priestley failed to convert a single American chemist to the phlogiston theory.

Priestley believed gases were different kinds of airs not different elements and never abandoned the phlogiston theory of combustion. In his experiments on gases in 1774 he discovered a gas he called dephlogisticated air (oxygen) upon heating red mercury calx, which is HgO, mercury (II) oxide, and collecting the liberated gas. Priestley also found that heating mercury in air produced the red calx. He investigated the properties of the gas and showed that substances burned more brightly in it. This occurred, he said, because dephlogisticated air (oxygen) was so lacking in phlogiston that the phlogiston present in other substances rapidly rushed into it (dephlogisticated air), causing the brighter burning.

Priestley discovered several other gases, specifically ammonia ($NH_3$) in 1774, nitrous oxide ($N_20$) in 1774, nitrogen dioxide ($NO_2$) in 1774, and sulfur dioxide ($SO_2$) in 1775, using a pneumatic trough to collect water-soluble gases over mercury. He prepared carbonated, or soda, water in 1767 (published in 1772) by passing carbon dioxide into water and noted its pleasant taste. He thought carbonated water might cure scurvy. Priestley published his studies on gases in six volumes, *Experiments and Observations on Different Kinds of Air* (vol. 1, 1774; vol. 2, 1775; vol.3, 1777; vol. 4, 1779; vol.5, 1781; vol. 6, 1786; and again in 1790 in three volumes with the same title.

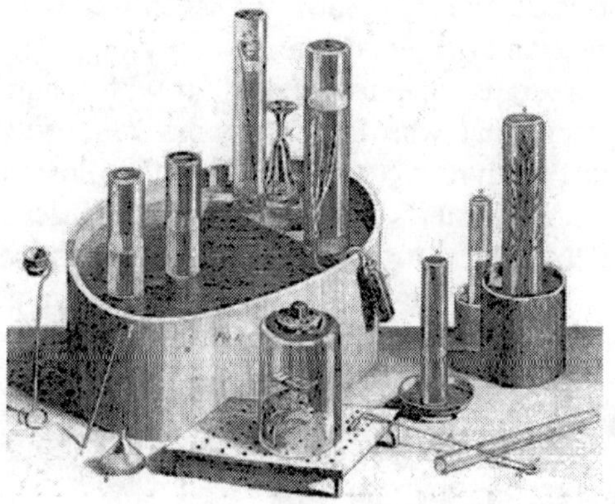

**Figure 4**
**Pneumatic Trough (left), Priestley's Laboratory (right)**
Source: Joseph Priestley. *Experiments and Observations on Different Kinds of Air*. London, 1775 (left).
Source: F.J. Moore. *History of Chemistry*. New York: McGraw-Hill Book Company, 1918 (right).

## III. LAVOISIER AND THE EIGHTEENTH-CENTURY REVOLUTION IN CHEMISTRY: THE STUDY OF COMPOSITION

Antoine Lavoisier (1743-94), the French empirical and quantitative chemist in Paris, used science to promote public works, such as town lighting, and the reform of education, hospitals, prisons, and agriculture.

Lavoisier improved the French gunpowder industry which prior to the 1770s produced a poor product. He served as a commissioner of the Royal Gunpowder Administration in 1775 and taught Iréné Du Pont (1771-1834) how to manufacture gunpowder. Du Pont and his family left France because of the French Revolution and in 1799 came to the United States, where in 1802 he established the DuPont Company in Wilmington, Delaware.

In 1768 Lavoisier became a member of the Ferme Générale (1726-90), a 40-partner (as many as 60 earlier) private tax collecting agency that worked for the government and which the French citizenry loathed. Lavoisier used the income from business investments, such as the Ferme, to operate his experimental farm. His membership in the Ferme unfortunately led to his arrest and execution by the guillotine in May 1794, during the French Revolution's reign of terror in 1793-94. The revolutionaries charged Lavoisier with selling watered-down tobacco and of adding ingredients injurious to the health of citizens. They condemned Lavoisier and his colleagues for counter-revolutionary conspiracy and for supplying the Republic's enemies with large sums of money withheld from the National Treasury. All charge were untrue.

Lavoisier and his father-in-law, Jacques Paulze (1723-94) a lawyer and senior member of the Ferme, surrendered to the revolutionaries on 28 November 1793 at Port Libre Prison (Port Royal) in Paris, a monastery converted into a prison. The revolutionaries transferred them to the Ferme's former headquarters on 25 December and then on 6 May to the Conciergerie, a medieval fortress on the Isle of the City (Paris). Their trial began at 10 am on 8 May, their execution occurred the same day at 5 pm at the Place de la Révolution (Place de la Concorde). During the reign of terror revolutionaries also guillotined the astronomer Jean-Sylvain Bailly (1736-93), and André Ampere's father. The mathematician Marie Jean Condorcet (1743-94), after his arrest in 1794, was found dead in his prison cell possibly a suicide by poisoning, although no evidence ever confirmed the cause of his death. The revolutionaries executed seven scientists, 18 partners of the Ferme, and 18,000 other enemies during the reign of terror.

They also imprisoned Lavoisier's wife, Marie Anne Paulze (1758-1836), for seven weeks from June to August 1794. Marie Anne, who was only fourteen when she married Lavoisier in 1771, assisted Lavoisier in his experiments and kept the laboratory records. She had studied art with Jacques-Louis David (1748-1825), the famous Parisian neo-classical artist who in 1788 painted the only known portrait of Lavoisier and his wife. It cost 7,000 livres (1 livre = $40). She drew the illustrations for Lavoisier's experiments, particularly for the plates in Lavoisier's 1789 textbook. Because Lavoisier knew no English, Marie Anne learned English and translated the English language publications of Priestley, Cavendish, and other scientists. After Lavoisier's execution she edited and published Lavoisier's memoirs and sent copies to the European scientific societies, academies, and to a number of scientists. Beginning in 1781 she carried on a ten-year affair with Pierre-Samuel Du Pont (1739-1817), Iréné's father and Lavoisier's colleague.

**Figure 5**
**Jacques-Louis David Portrait of Lavoisier and Marie Anne 1788**
Public Domain. Original in Metropolitan Museum of Art, New York City.

***Lavoisier's quantitative and combustion experiments***

Lavoisier was an empiricist who emphasized *quantitative measurements*. His 101-day pelican experiment in 1768 in which he disproved the widely held erroneous belief in the conversion of water to earth by extensive heating was a classic example of his quantitative approach. Lavoisier weighed separately the glass apparatus, called a pelican, and its water content before and after the experiment to show that the weight of the water did not change. Instead, the weight of the apparatus decreased by an amount equal to the weight of the sediment (earth) deposited inside the apparatus. No conversion of water to earth occurred. The hot water reacted with the glass apparatus and caused small pieces of glass to flake off and appear as sediment.

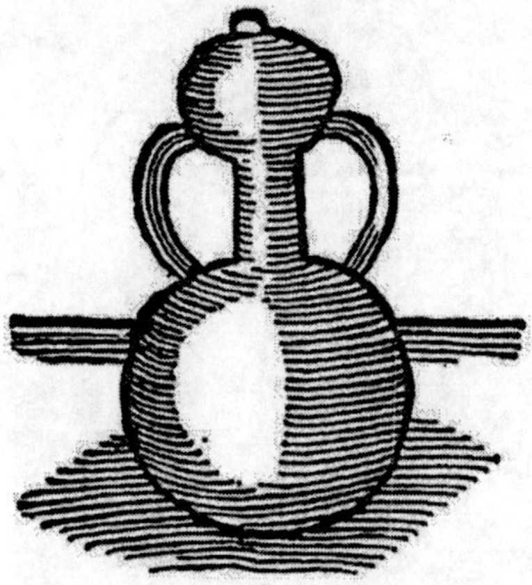

**Figure 6**
**The Pelican**
Source: M. M. Pattison Muir. *The Story of Alchemy and the Beginnings of Chemistry.* London: Hodder and Stoughton, 1913.

In combustion studies of 1772-74 Lavoisier in separate experiments weighed and then burned each of the nonmetals carbon, sulfur, and phosphorus in a sufficient quantity of air contained in a weighed closed container to form their oxides and then added a known quantity of water to form their acid solutions. Phosphorus gained 2.5 times its weight. He also burned weighed amounts of the metals lead, tin, and mercury in a sufficient amount of air contained in a weighed closed container to form their solid oxides. He showed that in every case the product weighed more than the reactant. Each substance that he burned gained something from the air (oxygen) and left a partial vacuum in the container. Previously Lavoisier believed that in a combustion reaction each substance combined with common air in a pure state.

Priestley's coming to Paris in 1774 and relating to Lavoisier the results of his experiments on heating mercury in air provided the clue Lavoisier needed to realize that something in the air (oxygen), and not common air, had caused the combustion reactions. This unknown component in the air had combined with mercury to form its red calx (oxide of mercury), and when heated the calx gave off a gas (Priestley's dephlogisticated air) that caused substances to burn more brightly in it than in air. The gas was oxygen.

Daniel Rutherford (1749-1819) in Edinburgh already had discovered nitrogen in 1772 in combustion experiments that removed oxygen from air. He then absorbed the carbon dioxide present, leaving nitrogen. But neither Rutherford nor Priestley understood that oxygen present in the air was responsible for the combustion.

In Lavoisier's interpretation of Priestley's and his own combustion reactions he proposed that dephlogisticated air (oxygen) was a component of air and the substance responsible for combustion. To test his theory Lavoisier in 1777 carried out a twelve-day mercury experiment in which he heated four ounces of mercury in a closed container (A) to show that it combined with oxygen from the air in the container (E) to form red mercury calx. Lavoisier then calculated that the remaining air had lost 20 percent of its volume and did not support combustion. It was essentially nitrogen. Decomposing the red calx and returning the gas (oxygen) it gave off to the container, restored the air inside the container to its former condition. The experiment demonstrated the validity of Lavoisier's combustion theory and also established the composition of air: $O_2$ = 20 percent and $N_2$ = 79 percent plus small amounts of carbon dioxide and other gases.

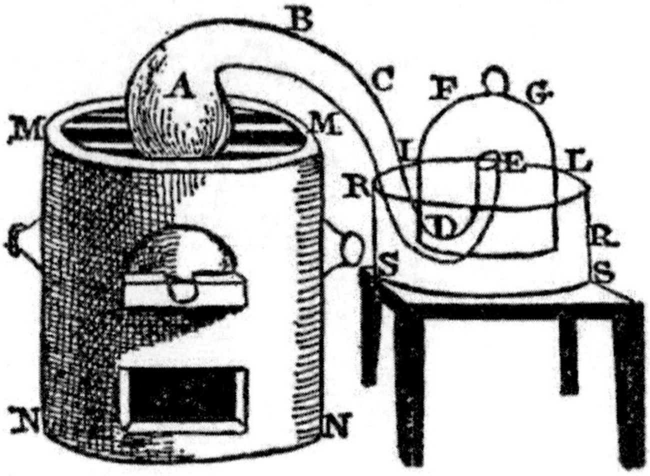


**Figure 7**
**Lavoisier's Analysis of Air**
Source: M. M. Pattison Muir. *The Story of Alchemy and the Beginnings of Chemistry.* London: Hodder and Stoughton, 1913.

In 1778 Lavoisier proposed incorrectly as a result of his combustion experiments that all acids contained oxygen, and for this reason he renamed Priestley's dephlogisticated air, *oxygen*, which means *acid former*. Humphry Davy (1778-1829) at the Royal Institution in 1810 proved in electrolytic experiments that all acids, such as hydrochloric acid, contained hydrogen but not oxygen. Davy carried out his electrolytic experiments because of the availability of large batteries and the production of current electricity beginning in 1800.

Lavoisier determined the composition of water in 1783. He burned hydrogen in oxygen and showed that water consisted of 2 volumes of hydrogen and 1 volume of oxygen.

In his physiological experiments of 1772 and 1773, Lavoisier studied respiration and measured the amount of heat produced. He recognized that respiration was a slow combustion. It required oxygen and provided heat for the body according to the reaction below.

$$CH_2O + O_2 \rightarrow CO_2 + H_2O.$$

Lavoisier published *Traité Élémentaire de Chimie* in 1789 (*Elements of Chemistry*, 1790) summarizing his years of experimental research. It contained a table of 33 elements that included oxygen, hydrogen, nitrogen, sulfur, iron, tin, lead, and other metals. Lavoisier apparently believed heat (caloric) and light were elements and listed them in his table, although he may have included them for convenience rather than a result he established from quantitative experimental evidence. Newton's 1672 prism experiment had shown that white light consisted of the colors of the rainbow and therefore not a simple substance or the last point of analysis. Lavoisier likely knew this about white light.

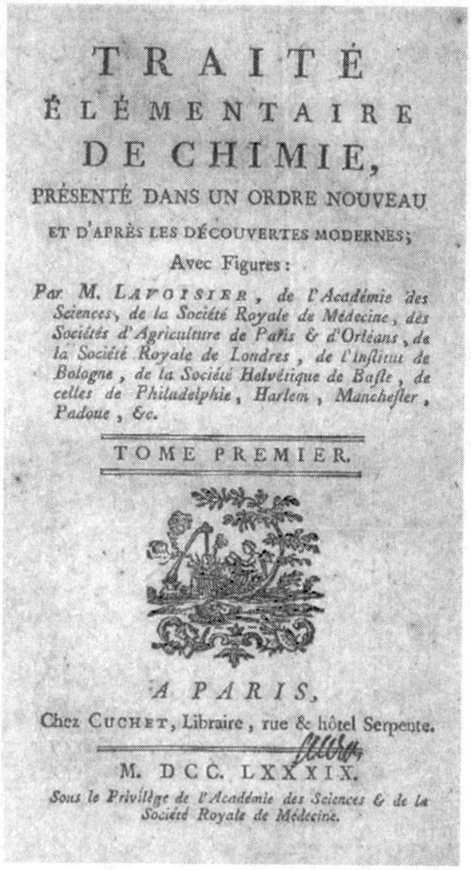

TRAITÉ
ÉLÉMENTAIRE
DE CHIMIE,
PRÉSENTÉ DANS UN ORDRE NOUVEAU
ET D'APRÈS LES DÉCOUVERTES MODERNES;
Avec Figures:
*Par M. LAVOISIER, de l'Académie des Sciences, de la Société Royale de Médecine, des Sociétés d'Agriculture de Paris & d'Orléans, de la Société Royale de Londres, de l'Inſtitut de Bologne, de la Société Helvétique de Baſle, de celles de Philadelphie, Harlem, Mancheſter, Padoue, &c.*

TOME PREMIER.

*A PARIS,*
Chez CUCHET, Libraire, rue & hôtel Serpente.

M. DCC. LXXXIX.
*Sous le Privilége de l'Académie des Sciences & de la Société Royale de Médecine.*

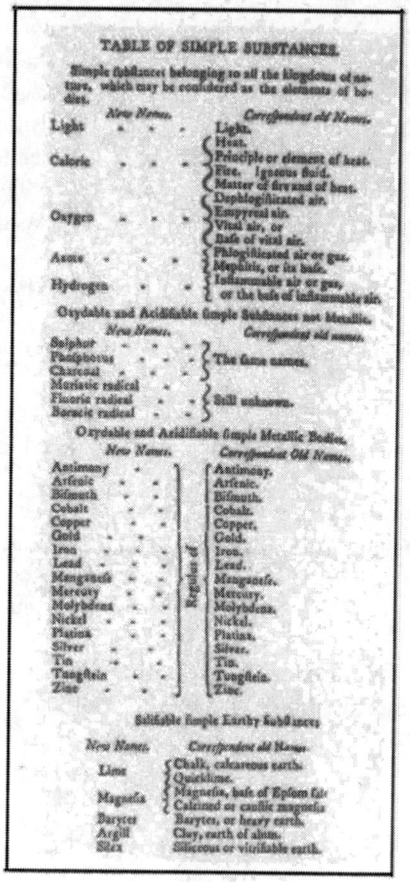

TABLE OF SIMPLE SUBSTANCES.

Simple fubftances belonging to all the kingdoms of nature, which may be confidered as the elements of bodies.

| *New Names.* | *Correfpondent old Names.* |
|---|---|
| Light | Light. |
| Caloric | Heat. Principle or element of heat. Fire. Igneous fluid. Matter of fire and of heat. |
| Oxygen | Dephlogifticated air. Empyreal air. Vital air, or Bafe of vital air. |
| Azote | Phlogifticated air or gas. Mephitis, or its bafe. |
| Hydrogen | Inflammable air or gas, or the bafe of inflammable air. |

Oxydable and Acidifiable fimple Subftances not Metallic.

| *New Names.* | *Correfpondent old names.* |
|---|---|
| Sulphur, Phofphorus, Charcoal | The fame names. |
| Muriatic radical, Fluoric radical, Boracic radical | Still unknown. |

Oxydable and Acidifiable fimple Metallic Bodies.

| *New Names.* | *Correfpondent Old Names.* |
|---|---|
| Antimony | Regulus of Antimony. |
| Arfenic | Arfenic. |
| Bifmuth | Bifmuth. |
| Cobalt | Cobalt. |
| Copper | Copper. |
| Gold | Gold. |
| Iron | Iron. |
| Lead | Lead. |
| Manganefe | Manganefe. |
| Mercury | Mercury. |
| Molybdena | Molybdena. |
| Nickel | Nickel. |
| Platina | Platina. |
| Silver | Silver. |
| Tin | Tin. |
| Tungftein | Tungftein. |
| Zinc | Zinc. |

Salifiable fimple Earthy Subftances.

| *New Names.* | *Correfpondent old Names.* |
|---|---|
| Lime | Chalk, calcareous earth. Quicklime. |
| Magnefia | Magnefia, bafe of Epfom falt. Calcined or cauftic magnefia. |
| Barytes | Barytes, or heavy earth. |
| Argill | Clay, earth of alum. |
| Silex | Siliceous or vitrifiable earth. |

**Figure 8**
**Title Page and Table of Elements from Lavoisier's *Traité Élémentaire***
Source: Antoine Lavoisier. *Traité Élémentaire de Chimie*. Paris: Cuchet, 1789.

***Law of conservation of mass and the definition of an element***

Lavoisier showed clearly in his experiments of 1768-72 that in a chemical reaction the mass of the reactants was equal to the mass of the products. This was the first clear statement of the law of conservation of mass, the third conservation law in science, after Newton's conservation of momentum in 1687 and Benjamin Franklin's conservation of electrical charge in 1747.

According to Lavoiser, if a substance resists further analysis or when you have reached the last or end point of its analysis, the substance is an element. Lavoiser enunciated this practical or working definition of an element in *Méthode de Nomenclature Chimique* published in 1787 with Louis Guyton de Morveau (1737-1816), Claude Louis Berthollet (1748-1822), and Antoine Fourcroy (1755-1809).

***Reform in nomenclature***

Lavoisier introduced the binomial system of nomenclature in *Méthode de Nomenclature Chimique*. In the binomial system the name identified the components of a compound, such as the oxide of mercury, or the oxides of iron, lead, tin, phosphorus, carbon, and other compounds.

He renamed dephlogisticated air, oxygen, meaning acid former. Inflammable air became hydrogen meaning water former. Other common names, such as quicksilver, lime, rock salt, Glauber's salt, dolomite, potash, saltpeter, and magnetite also changed, so that the compound's name indicated its composition. Lime became calcium oxide, rock salt became sodium chloride, Glauber's salt became sodium sulfate, potash became potassium carbonate, saltpeter became potassium nitrate (sometimes sodium nitrate). Phlogistonists such as Antoine Baumé (1728-1804) in France, Black, Priestley, and Cavendish in England; Franklin and Thomas Jefferson (1743-1826) in the United States remained unimpressed or disapproved of the new nomenclature.

## IV. INTRODUCTION OF THE METRIC SYSTEM

***French scientists introduced the metric system of measurement in the period from 1791 to 1798***

Because of the many confusing units or systems of measurements that eighteenth-century European scientists used, such as the English and the French units for weights and lengths, the Académie Royale des Sciences in 1790 set up a commission to devise a standard set of units that the international scientific community would adopt. The commission consisted of Marie Jean Condorcet, Pierre Simon Laplace (1749-1827), Joseph Louis Lagrange (1736-1813), Jean C. Borda (1733-99), and Gaspard Monge (1744-1818). Lavoisier replaced Laplace on the commission. The system they proposed was the metric system, which had units that were multiples or fractions of ten. France made the metric system legal in 1799, and other countries adopted it in the nineteenth and twentieth centuries. Belgium, Luxembourg, and the Netherlands adopted the metric system in 1820, Japan in 1868, Russia in 1917, Britain from 1965 to 1980. The United States never adopted the metric system.

**Figure 9**
**Standard Kilogram Weight of Platinum-Iridium Cylinder**
Source: Newton Henry Black and Harvey Nathaniel Davis. *New Practical Physics.* New York: Macmillan Company, 1913.

| THE METRIC SYSTEM | |
|---|---|
| ***Unit*** | ***Description*** |
| Meter the unit for length | The commission in 1795 defined the meter as one ten millionth of Earth's quadrant (the distance) between the north pole and equator. Borda suggested the name meter. |
| Gram the unit for mass | A gram (small weight) was the weight of a cubic centimeter (a cube that was 1/100 of a meter on each side) of water at maximum density. A kilogram was 1,000 times that weight. |
| Liter the unit for volume | A liter was 1/1000 the volume of a cubic meter, the volume of a cube 1/10 meter on each side. |

## V. THEORIES OF HEAT

### *Caloric theory*

Lavoisier introduced caloric as the element of heat in the 1787 publication *Méthode de Nomenclature Chimique*. Most scientists including Black, Priestley, Benjamin Franklin, and Amedeo Avogadro (1776-1856) in Turin, Italy, accepted the caloric theory. Caloric was the dominant theory of heat in the late eighteenth century and for much of the nineteenth century.

According to the caloric theory heat was a fluid that consisted of particles that were self-repulsive but strongly attracted to matter. Adding or removing caloric from a substance accounted for the expansion and contraction of substances and for the shapes of solids, liquids, and gases. A solid contained the least amount of caloric, and the small repulsive force present allowed a solid to have a definite shape. A liquid contained more caloric, and the increased repulsion caused the liquid to take the shape of its container. A gas contained the most caloric, and the large repulsive force resulted in a gas having no shape.

*Joseph Black's studies on heat 1760-64*

Black first distinguished between the quantity of heat (H) a substance gained or lost, and the intensity of heat, which was the temperature (*t*) of a substance. He also defined and calculated for ice, water, and other substances their heat of fusion, heat of vaporization, specific heat, and quantity of heat gained or lost when heated or cooled.

Black found that when he heated ice at a temperature below the melting point of ice the temperature rose steadily until the ice began to melt and remained at that temperature (0°C) until all the ice melted to water. It then rose steadily until the water began to vaporize or boil and remained at that temperature (100°C) until all the water vaporized, after which the temperature began to rise again. This experiment made clear that the heat added at the melting point, the heat of fusion, and at the boiling point, the heat of vaporization, did not increase the temperature but instead brought about the change from a solid to a liquid and from a liquid to a vapor. The heat, which Black believed was caloric, was hidden or latent in the liquid and in the vapor. For water, Black's calculation of the heat of fusion was 139 Btu per pound. The accepted value is 144 Btu per pound or 80 calories per gram. For the heat of vaporization of water he calculated 810 Btu per pound. The accepted value is 970 Btu per pound or 540 calories per gram.

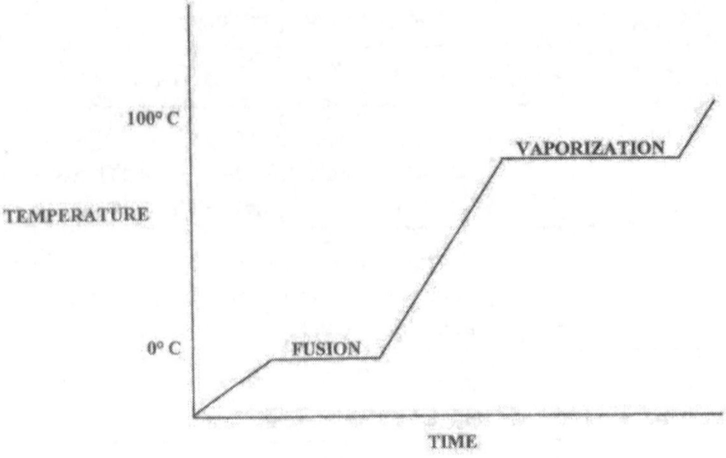


**Figure 10**
**Changes of State**
Courtesy of the author

Black in 1761 introduced the relation for calculating the quantity of heat gained or lost by a substance and carried out his first measurements of the heat of fusion.

| **BLACK'S CALCULATION OF QUANTITY OF HEAT** | |
|---|---|
| Quantity of heat (H) = specific heat x weight of the substance x temperature change | |
| H = | $s\, w\, \Delta t$ |
| s = | specific heat, the amount of heat required to raise a unit mass (one gram, one pound) of a substance one degree (C°, F°). Water was the substance selected as the standard with a value of one, and the specific heats of all other substances were relative to its value. |
| w = | weight of the substance |
| $\Delta t$ = | change in temperature |

Black defined the *heat of fusion* in 1761 as the heat required to change a unit mass (one gram, one pound) of a solid to a liquid at the freezing point of the liquid. He calculated the heat of fusion by assuming that the heat gained by the solid substance when melting must equal the heat lost by the substance in its liquid state and by the calorimeter which contained the solid and liquid. A calorimeter was an insulated container that did not lose its heat to the surroundings or gain heat from the surroundings.

Black defined the *heat of vaporization* in 1762 as the heat required to change a unit mass (one gram, one pound) of a liquid to a vapor at the boiling point of the liquid. He calculated the heat of vaporization by measuring the heat lost by a vapor upon condensing to a liquid and recognizing that it was equal to the heat gained by the substance in its liquid state and by the calorimeter which contained the vapor and liquid.

Black defined *specific heat* in 1764 as the heat required to raise a unit mass (one pound) of a substance, one fahrenheit degree (or one gram of a substance one celsius degree). He measured the specific heat of a substance by mixing weighed amounts of two different substances, such as water whose specific heat is one, and lead shot, each at a different temperature, in a calorimeter, recognizing that upon mixing them the quantity of heat gained by one is equal to the quantity of heat lost by the other, and solving for the unknown *s* (specific heat) in his equation for calculating H, the quantity of heat.

$$\Delta H_{lead} = \Delta H_{water}$$

Black constructed an ice calorimeter consisting of a block of ice hollowed out in the center and covered with a slab of ice to determine the specific heat of a substance. He weighed the substance $M_s$, heated it to a temperature $t_1$, dropped it into the ice calorimeter, and covered it.

Black allowed enough time for the substance to cool to the ice temperature $t_0$, removed the water formed by the heat given to the ice, and weighed the water. He equated the heat the substance gave up to the heat the ice absorbed to get the relation

$$M_s (h_i - h_0) = M_w L.$$

where M is the molecular weight of a substance. L is the amount of heat required to melt a unit mass of ice called the latent heat of fusion of ice, and $h_i$ - $h_0$ is the amount of heat a unit mass of the substance gave up in cooling from $t_1$ to $t_0$. The mean specific heat $C_m$ of the substance over the temperature interval is

$$C_m = (h_i - h_0 / (t_1 - t_0) = M_w L / M_s (t_1 - t_0).$$

The symbol h represents the enthalpy or heat content per unit mass. The equation $H = M_s h$. relates h to the enthalpy H.

Johann Carl Wilcke (1732-99), the son of a German minister in Stockholm and a co-worker of Franz Aepinus (1724-1802) then director of the Astronomisches Rechen-Institute in Berlin, independently calculated a substance's heat of fusion. He recognized in 1772, a few years after Black, that to convert a solid to a liquid at the melting point of the solid required adding a specific amount of heat. In 1781 he carried out investigations on the specific heats of substances. Wilcke published his results, whereas Black published very little of his earlier thermal studies in books or journals. Black discussed and demonstrated his thermal studies in his lectures. Students' copies of his lecture notes have survived.

***Mechanical theory of heat***

The alternative theory that mechanical motion, or particles of matter in motion, produced heat had few supporters in the eighteenth and nineteenth centuries. They included Boyle, Newton, and Davy in London, and Benjamin Thompson (1753-1814), better known as Count Rumford.

Rumford, though born in Massachusetts, spent most of his life as a scientist and military engineer in London, Bavaria, and France. Rumford remained loyal to the British during the Revolutionary War of 1775-83, becoming a British spy. He included secret messages written in invisible ink between the lines of innocuous everyday reports he sent to the British military. With the discovery of his loyalist sympathies, Rumford fled the colonies for England in the summer of 1776. He left behind his wife Sarah Walker Rolfe (1735-99), a wealthy Concord, New Hampshire, widow he married in 1772, and their daughter Sarah (Sally) whom he last saw two months after her birth in 1774. Rumford returned to the United States once more in 1781-83 as a lieutenant-colonel in the British army.

Always the opportunist, Rumford, the wartime Undersecretary of the Northern Department and Inspector General of the British Forces, enriched himself during the war by selling, at marked-up prices, horses, swords, and silk for the British army's uniforms. The silk sales case was particularly ingenious. Silk is hygroscopic, which means it absorbs moisture. Rumford bought silk in England, and shipped it across the Atlantic Ocean to the colonies for manufacture into uniforms. By the time the silk cargo arrived in the colonies, each pound of silk had absorbed eight times it weight in moisture, so that when Rumford received payment based on the weight of silk he had a profit of 800 percent.

In late 1784 with the war over, Rumford left England for Bavaria and served as military engineer and minister of war to Karl Theodor (1742-99), the elector of Bavaria in Munich. There he carried out cannon boring experiments that led him to question the caloric theory and propose a mechanical theory of heat. Rumford received the title of count in 1791 from the elector for his military contributions to Bavaria and chose the name Rumford, the original name for Concord, New Hampshire, his boyhood home.

Rumford returned to England in 1798, won election to the Royal Society in 1799, established the Royal Institution in London in January 1800, and hired Davy as the Institution's chemical and electrical demonstrator. Rumford wanted the Institution to focus on applied science, on the application of science for the benefit of society, but the other Institution directors did not support Rumford's program, and in 1801 he left London to travel in Europe. He eventually settled in Paris in 1804 where he continued his scientific research.

Rumford had long wished to return to the United States, and in 1799 prior to his organizing the Royal Institution, he proposed as part of his repatriation plan, establishing a military institution at the West Point, New York, military fort. The plan, involving the United States minister to England Rufus King (1755-1827) and the secretary of war James McHenry (1753-1816), called for him to head the institution. It never materialized. The British government informed King and the United States Department of War that Rumford had spied for the British during the Revolutionary War, and in a pre-arranged agreement Rumford declined West Point's headship. Nevertheless, Congress in 1802 established the first military institution in the United States at West Point.

Rumford's abandoned first wife Sarah had died in 1799. In Paris he met Lavoisier's widow, Marie Anne Paulze. They married in 1805, and as a result of the marriage Rumford became a very wealthy man, having an annual income of £8,000 ($34,800), equivalent to £320,000 ($480,000) in 2000. Marital discord soon arose and the marriage ended in divorce after four mainly unhappy years. Rumford's daughter Sarah, with whom he later established a warm friendship, lived for a time with him in London and Europe. A relationship with Victoire Lafevre while in Auteuil, a suburb of Paris where Rumford moved after his divorce, produced a son, Charles. She was his servant and the alleged daughter of Marie's gardener. An earlier liaison with the elector Theodor's mistress, Countess Baumgarten, in Munich produced a daughter Sophie in 1798. Rumford became ill with a high fever in 1814 and died suddenly from the fever. His grave is in Auteuil, today a wealthy and prestigious borough of Paris, incorporated into Paris in 1859.

Rumford carried out his cannon boring experiments on measuring heat and the mechanical equivalent of heat in 1798 while in Munich. He believed his experiments established that mechanical motion produced heat because the boring action produced the heating effect. The boring generated so much heat that the workers at the government's arsenal had to bore the iron cannon barrels under water. Once the boring stopped, so did the release of heat. Rumford concluded that the caloric present in the metal clearly failed to account for the large quantity of heat released.

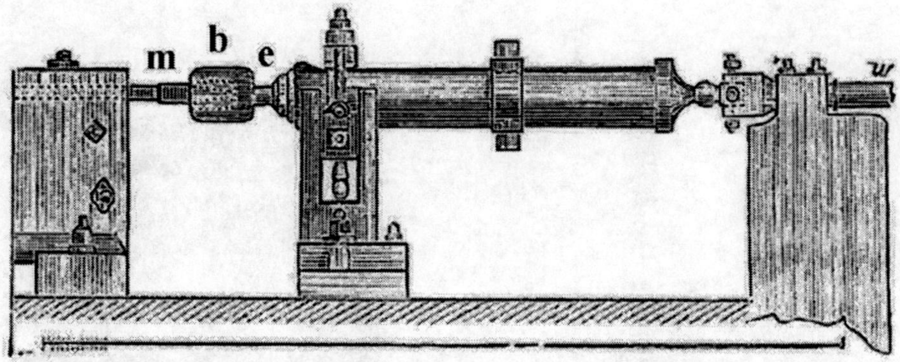


In 1798 at a Munich munitions factory, Rumford set up the apparatus shown above. He fit a steel boring rod (m) snugly into a fat cylinder and joined it to a rotating cannon barrel. A water-filled jacket (not shown) enclosed the turning cylinder. Heat from the friction between the cylinder and borer made the water boil. Because the water boiled as long as the cylinder turned Rumford argued that the heat supply was inexhaustible and therefore heat was not a material substance (caloric) stored in the cylinder.

**Figure 11**
**Rumford's Cannon-Boring Experiment**
Source: Porter Pournier. "History of the Dynamical Theory of Heat 1." *Popular Science Monthly* 12 (December 1877).

Supporters of the caloric theory rejected Rumford's conclusion about the source of heat released because they believed the boring motion released caloric, which produced the heating effect. Once the boring stopped so did the release of caloric, resulting in a gradual cooling. Rumford questioned how a metal could contain so much caloric and yet not melt and remain cool to the touch. Apparently the heating effect manifested itself only when caloric left the body containing it. No matter how much caloric resided inside a body, the body remained cool on the outside. Upon release of its caloric the body was hot. In the boring process, despite the presence of significant caloric inside the iron cannon barrel, the barrel was cool before boring and became hot only upon release of caloric during boring.

Rumford's belief that mechanical motion produced heat led him to investigate the relation between heat and motion, and to ask how much heat a known amount of mechanical motion or work produces. The relation is the *mechanical equivalent of heat.* In his experiments of 1798 Rumford provided the measurements required to calculate a mathematical value for the relation, but he never made the calculation. James Joule (1818-89), an English physicist in Manchester, used Rumford's measurements and in 1843 calculated the relation between heat and mechanical motion.

In Rumford's experiment, a horse (actually two horses) walked on a treadmill and turned a paddle wheel set inside a barrel filled with a known amount of water. Rumford measured the change in the water's temperature in a known time interval, and from Rumford's measurements Joule showed that a fixed amount of mechanical motion, or work (a force acting over a distance, or $W = F\ d$), always generated the same quantity of heat energy. This is the mechanical equivalent of heat. Joule's calculation of the mechanical equivalent of heat follows.

**Figure 12**
**Rumford's Mechanical Equivalent of Heat**
Source: Jacob Abbott. "The New Theory of Heat."
*Harper's New Monthly Magazine* 231 (August 1869).

Rumford estimated that the heat capacity of the water and metal (paddle wheel), ignoring the heat capacity of the wooden box container which Rumford considered negligible, was equivalent to the heat capacity of 26.58 pounds (Lb) of water.

One horse on a treadmill could push the paddle wheel and raise the water's temperature from 33° F to 212° F or 180 degrees in 2.5 hours (150 minutes). Therefore the rate of increase in temperature, $\Delta t = 1.2°$ F per minute.

Using Black's equation for $\Delta H$, the quantity of heat gained or lost by a substance, Joule calculated the number of British thermal units (Btu) of heat generated by one horse (1 horsepower, Hp) in one minute.

$$\Delta H = \text{mass} \times \Delta t \times \text{specific heat of water}$$
$$= 26.58 \times 1.2 \times 1$$
$$= 31.92 \text{ Btu}$$

James Watt (1736-1819) in Birmingham, England, had observed that a horse could raise a 150-pound weight four feet in one second, and in 1806 he defined one horsepower as 550 ft-Lb per second or 33,000 ft-Lb per minute.

| | | |
|---|---|---|
| If 1 Hp | = | 31.92 Btu |
| and 1 Hp | = | 33,000 ft-Lb |
| then 31.92 Btu | = | 33,000 ft-Lb |
| and 1 Btu of heat | = | 1,034 ft-Lb (774 ft-Lb = correct value) of mechanical energy |

**1 HP = 550 foot pounds per second (FPS)**

**Figure 13**
**Watt's Determination of Horsepower**
Courtesy of the author

Rumford's research on heat led to one other important discovery of great significance. He discovered that water had a greater density at 4° C than at 0° C. The volume of water increases as water cools from 4° C to 0° C. Consequently, in a body of water the warmer water at 4° C sinks to the bottom and the colder water at 0° C rises to the top, where it freezes to ice. This explains why ice forms at the top of a river, lake, or pond and not at the bottom.

***End of the caloric theory***

James Clerk Maxwell (1831-79) at Aberdeen and in 1871 at the Cavendish Laboratory at Cambridge, England, and Ludwig Boltzmann (1844-1906) at Graz and Vienna retired the caloric theory when they provided a mathematical foundation for the kinetic-molecular theory of heat. In experiments with gases in the 1860s-70s Maxwell and Boltzmann showed that the motions of molecules and nothing else accounted for the laws and properties of heat. Caloric was unnecessary, and it disappeared from thermodynamics and science. The kinetic-molecular theory now became the accepted theory of heat.

## VI. APPLICATIONS OF HEAT ENERGY, DEVELOPMENT OF STEAM POWER

***Steam engines: Savery, Newcomen, and Watt***

The single-action atmospheric-steam and steam engines of Thomas Savery, Thomas Newcomen, and James Watt; and the double-action reciprocating steam engine of Watt in 1788 ushered in the industrial revolution. Steam power as used in the steam engine replaced water power as the main source of energy.

The first use of steam engines as labor-saving devices occurred in mines and then later in ships. Thomas Savery (1650-1715), a sea captain and military engineer in London, described in *The Miner's Friend* (1702), an early single-action atmospheric-low-pressure-steam engine with no moving parts that he said did the work of two horses. Savery gave the first implicit statement of the expression *horsepower*.

Savery received a 14-year patent on his steam engine (British Patent 365) in 1698. The Fire Engine Act of 1699 extended his patent to 1733. Savery's engine consisted of an upright brass cylinder connected to a round brick boiler (similar to what brewers used) usually 2½ feet or less and a rising main (pipe). The main descended in one direction to the well or water source and ascended in the other direction to the delivery cistern or outlet pipe through which passed the water removed from the well. His engine did not have a separate condenser (steam condensing chamber).

To operate the engine, Savery directed low-pressure steam at 1-3 atmospheres (14.7-44.1 pounds per square inch [psi] where 1 atmosphere [atm] = 14.7 psi) into the cylinder, forcing out the air. Cold water from an external source sprayed into the cylinder condensed the steam, leaving a partial vacuum. By opening the main's lower valve, atmospheric pressure forced water from the well into the cylinder where, after closing the lower valve and opening the upper valve on the main, steam from the boiler forced the water into the main to repeat the cycle. Savery's engine pumped water to a height of 25 feet (32 feet in a perfect vacuum) at 1 atm and 70 feet at 3 atm, but pumping water to country homes and gardens was its chief use.

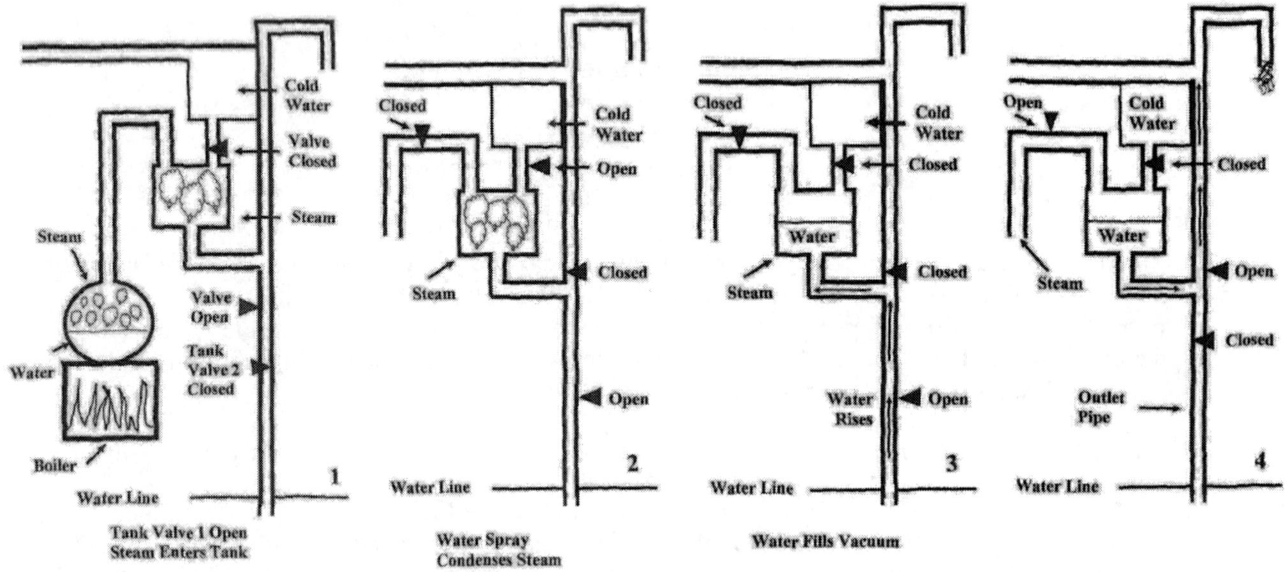


Steam filled the rectangular tank and forced out the air. Upon cooling the steam with a water spray, the steam condensed to a liquid, leaving a partial vacuum. The operator now opened the tank valve, and the pressure of the atmosphere on the water in the mine or well forced the water into the tank. Water rises 32 feet with a perfect vacuum, but because the vacuum was not perfect, the engine operating at 3 atm pumped about 70 feet at a time. In the final step steam empties the tank, forcing the water up the outlet pipe. Mines often were 200 feet deep, and pumps raised the water in the mine in stages, using several pumps and boilers placed underground at 60-80 foot intervals. Pumps sometimes raised water only 20 feet.

**Figure 14**
**Savery's Miner's Friend 1698**
Courtesy of the author

Thomas Newcomen (1663-1729), a Baptist lay preacher, metal worker, and toolsmith in Dartmouth, England, in 1712 after about 14 years of experimenting, constructed, a Savery-like 5-horsepower single-action atmospheric-low pressure steam engine at the Conygree Coalworks, Tipton near Dudley Castle and earlier likely constructed another engine near Wolverhampton. In his engines a tightly-fitting brass piston moved up and down in a brass cylinder 12 times (cycles) per minute and pumped 10 gallons of water per minute to a height of 150 feet. A 1760 Newcomen engine produced 20 horsepower at 14 cycles per minute. His engine, like Savery's, did not have a separate condenser, only a single upright brass cylinder with a 21-inch internal diameter and a height of 7 feet 10 inches, into which steam from the brick boiler passed, raising the piston. Adding cold water to the cylinder condensed the steam and caused the piston to fall down.

One of Newcomen's steam engines had run through its cycles much faster than previously. In searching for an explanation Newcomen discovered that its piston had a hole that allowed the cylinder's cooling water to pass through the hole directly into the cylinder and condense the steam inside the cylinder. Newcomen previously had cooled the cylinder by surrounding it with a cold water jacket. Newcomen's discovery led to his adding the cold water directly to the hot cylinder thereby eliminating the cold water jacket. The addition of more steam to the cylinder resumed the cycle.

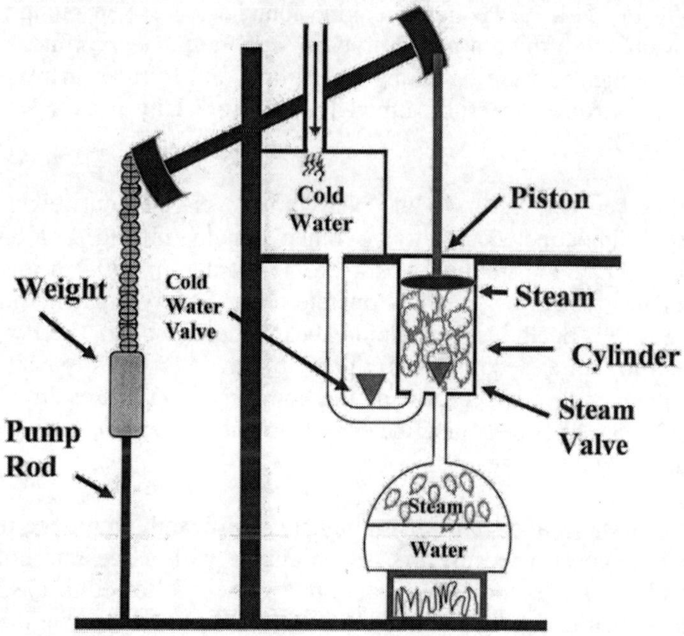


When the piston is at the bottom of the cylinder, the steam valve is open, thereby allowing low-pressure steam to flow into the cylinder and raise the piston. The steam can lift the piston because the counterweight on the side of the beam almost exactly counterbalances the piston's weight. The piston rises and when it reaches the top, the cylinder is full of steam. The pump operator now opens the cold water valve. The cold water entering the cylinder condenses the steam in the cylinder, creating a vacuum beneath the piston. This enables the pressure of the atmosphere on the top of the piston to force down the piston. As the piston falls, it gives a strong stroke to the water pump at the other end of the beam. When the piston reaches the bottom of the cylinder, the process begins again. The cycle repeats itself 12 times per minute. The pump could raise water 150 feet. The operator could extend the pump rod to any reasonable depth.

**Figure 15**
**Newcomen's Pump 1712**
Courtesy of the author

In the single-cylinder Savery and Newcomen engines, alternately heating and cooling the same cylinder in the operating cycle contributed to their low efficiency. In the heating stage of a cycle, some of the heat energy from the boiler heated the brass cylinder instead of converting the water in the cylinder to steam. In the cooling stage of the cycle, some of the cold water squirted into the now-hot cylinder cooled the cylinder instead of condensing the steam. Subsequent heating stages once again required more heat to convert the water to steam because the now-cold cylinder acted to condense the steam and counteract the converting of water to steam.

Because the efficiency of a steam engine, or any engine, depends on the temperature difference between the boiler and the condenser, or the engine's high and low temperatures $T_1$-$T_2$, an engine that used the same cylinder for the heating and cooling stages would have greater difficulty in achieving and maintaining a significant difference between its high and low temperature. It also requires more fuel to achieve and maintain a temperature difference. The lack of separate hot and cold cylinders resulted in the Savery and Newcomen engine's low efficiency.

Newcomen's engine had an efficiency of only 0.5-1.5 percent. It lifted 2,919 pounds of water one foot for each bushel (84 pounds) of coal consumed, consuming 12 tons of coal per day. Despite its high cost of £1,000 in 1714, Newcomen's engine found wide use in pumping water from English coal mines, Cornwall's tin mines, and European coal and metal mines. Between 1712 and 1733, 110 Newcomen engines were operating in England and Europe, increasing to 2,000 in 1800. They found little use in North America, although New York City used a Newcomen engine in its early waterworks in 1705.

The similarities between Savery's and Newcomen's engines outweighed their differences, and an English patent ruling concluded that Newcomen's engine infringed on Savery's 1698 patent that did not expire until 1733. The ruling had forced Newcomen in 1705 to join Savery in the commercial application of their steam engines and operate under Savery's patent, and to pay an annual sum of £182. Upon Savery's death in 1715 a joint stock company called The Proprietors of the Invention for Raising Water by Fire acquired the rights to Savery's 1698 patent. The Proprietors awarded Newcomen, their chief engineer, 20 of the company's 80 shares in 1716. When Savery's patent expired in 1733, Savery's steam engine came into public domain, ending royalties that required users to pay as much as £420 per year.

James Watt in Glasgow and later in Birmingham significantly improved the steam engine's efficiency in 1765 when he constructed the first steam engine with a separate hot (steam) cylinder and cold chamber (condenser). A second significant improvement followed in 1782 when Watt constructed the first true steam engine, in which unlike the Savery and Newcomen engines, steam drove the piston in the cylinder in both directions, backward and forward, or up and down. His double-action low-pressure engine produced up-and-down motion for use in mines and rotary motion for use in factories and on steamboats. It had an efficiency of 2.7-4.5 percent and developed 15-16 horsepower. A 1786 engine developed 45 horsepower at 12 cycles per minute. Watt in 1789 patented his third improvement, the fly-ball governor to control the steam engine's speed.

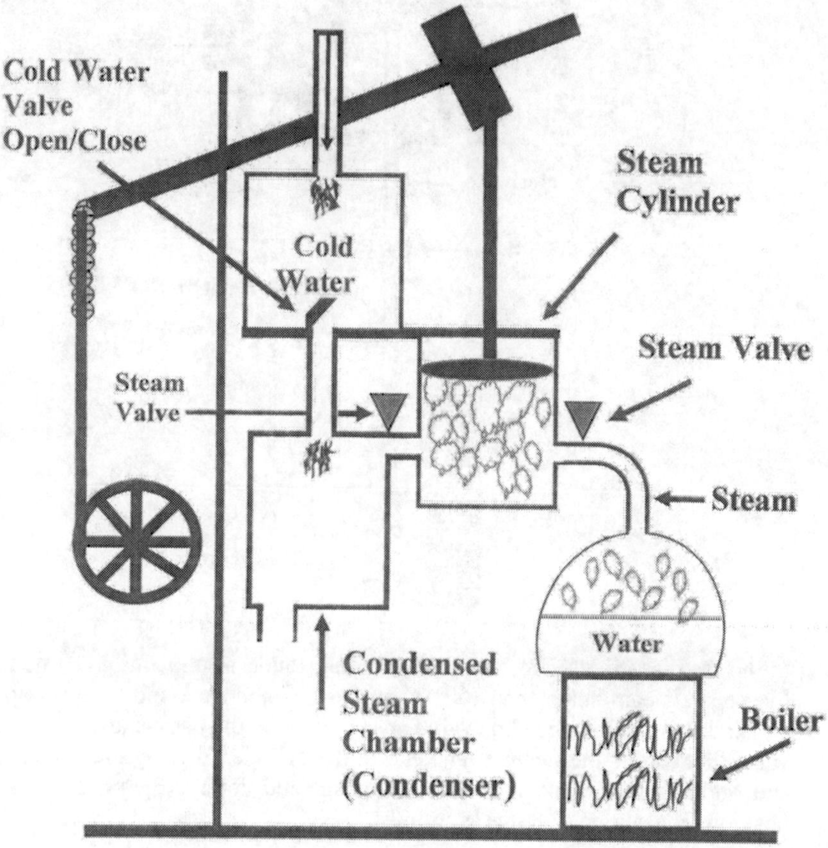


WATT'S STEAM ENGINE 1765

Watt's steam engine had a separate chamber in which it condensed the steam. He realized that Newcomen's engine lost much energy and was extremely inefficient because it alternately had to cool and heat the same cylinder. It wasted most of the engine's heat in warming the cylinder walls. Watt solved the problem by using a separate steam cylinder and condensing chamber. He kept the steam cylinder always hot and the condenser always cool.

For each power stroke of the Watt engine, the steam valve opened and forced the piston up. When the piston neared the top of the cylinder, the steam valve closed and the cold water (condenser) valve opened. Steam then left the cylinder and entered the condenser. The condenser remained cool because water flowing over it condensed the steam. As steam left the cylinder the pressure decreased. Atmospheric pressure, helped by the movement of the flywheel, pushed the piston down. When the piston reached the bottom of the cylinder, the condenser valve closed, the steam valve opened, and the process repeated itself.

**Figure 16**
**Watt's Steam Engine 1765**
Courtesy of the author

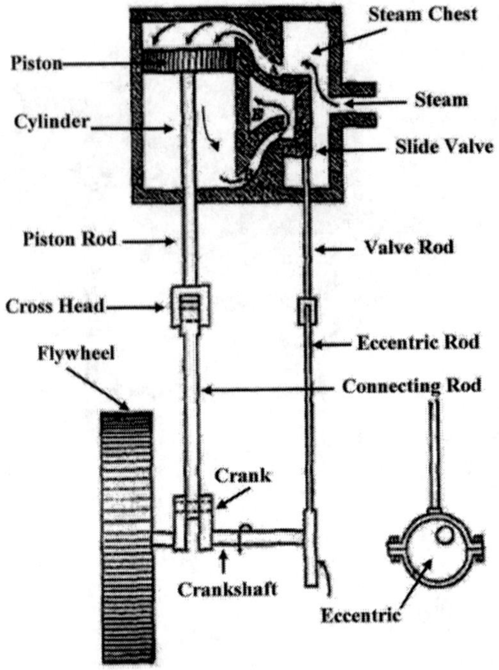


In the reciprocating steam engine, commonly found in a locomotive, the pressure of the live steam (steam hotter than 100° C) pushes a piston back and forth (or up and down) in a cylinder. An automatic slide valve makes possible the back-and-forth motion. The slide valve first admits the steam at one end of the cylinder, driving the piston to the opposite end, and then admits the steam at the opposite end, forcing the piston back to its original position. The engine operates as follows.

Coal, or some other fuel, burned in a high-pressure boiler provides a steady supply of live steam which travels to the steam chest, from which it passes through the top Port A into the cylinder. The steam expands and drives the piston towards bottom of the cylinder. The piston passes the movement to the connecting rod and the crank shaft, causing the flywheel to turn.

As soon as the piston nears the bottom of the cylinder, the eccentric, which Watt mounted on the axle and turns with the axle, pushes the valve rod to the top. This, in turn, pushes the slide valve towards the top of the steam chest. Steam now enters the cylinder through the bottom Port B, and drives the piston back to the top of the cylinder. At the same time, the relatively cold used steam left in front of the piston from the previous stroke, leaves through Port A and passes to the outside through the exhaust pipe.

When the piston nears the top of the cylinder, the eccentric arrangement again pulls the slide valve back to its first position. As a result, steam passes into the cylinder through the top Port A, and drives the piston back to the left. This process continues in such a way that the piston's regular up-and-down motion keeps the flywheel turning. The heavy flywheel's inertia enables the engine to run smoothly.

**Figure 17**
**Watt Slide Valve (Reciprocating) Steam Engine 1782**
Source: Newton Henry Black and Harvey Nathaniel Davis, *New Practical Physics*. New York: Macmillan Company, 1913.

*Impact of Watt's double-action reciprocating steam engine*

The introduction of Watt's double-action steam engine launched Europe's eighteenth-century industrial revolution and the use of steam-powered manufacturing in the iron and textile industries, in mining, and in propelling ships. Watt, however, opposed the use of high-pressure steam engines, meaning any engine operating above 1-3 atm, because of possible boiler explosions. In 1765-66 he went into partnership to manufacture steam engines with James Roebuck (1718-94) at Kinneil, Scotland, who went bankrupt in 1773, and in 1775 with Matthew Boulton (1728-1824) in Birmingham. By 1800 England had 500 Watt engines operating. A strict embargo that began in 1785, with few exceptions, banned the export of any English-invented machinery including steam engines.

The first high-pressure steam engines appeared in 1804, after Watt's main patent expired in 1800. Richard Trevithick (1771-1833), an English inventor in Cornwall, using improved metallurgical and alloying techniques constructed structurally stronger steam engines that operated at temperatures between 177° C and 100° C and at high pressures of 50-100 psi (ca. 3-7 atm). His engine expelled steam directly from the cylinder(s) to the atmosphere. Trevithick eliminated the condenser thereby circumventing Watt's patent. Higher pressures reduced engine size 5-10 times and increased efficiency to 17 percent. Trevithick constructed a number of high-pressure steam engines for use in sugar and corn mills, iron works, and to propel carriages.

Trevithick spent the years 1816-27 in Peru where he solved problems associated with assembling the steam engines he sold for £600 to the Pasco Mining Company. Pasco used Trevithick's steam engines in its silver mines at Cerro de Pasco, high in the Peruvian Andes. While in Peru, Trevithick advised other mining interests and the Peruvian government. He died of pneumonia six years after returning to England. His friends buried the now penniless Trevithick in an unmarked grave in Dartford, England.

In the United States, Oliver Evans (1755-1819), a tenacious, combative, self-educated Philadelphia inventor, in 1805 built a light, high-pressure double-action steam engine that operated at 30 psi and developed 5 horsepower. Evans attached his steam engine to his *amphibious digger*, the *Orukter Amphibolos*, a $4,000, 15-ton 30-foot long four-wheeled wagon with a rear paddle wheel. His steam-propelled digger represented several firsts: the United States' first self-propelled wheeled vehicle, the world's first amphibious steam-propelled vehicle, and one of the world's first high-pressure steam vehicles. Evans constructed the digger for the Philadelphia Board of Health which wanted a steam-powered vehicle to dredge and clean the city's docks. His digger worked, but its size overwhelmed its efficiency, and it failed to interest the public.

*Later development of the steam turbine and diesel engine*

By the last decades of the nineteenth century the Watt steam engine used to power ships gave way to two new kinds of pcwer sources, or engines, with the invention of the steam turbine and diesel engine. Charles Parsons (1854-1931), an English engineer who founded the Parsons Works in Newcastle-upon-Tyne, invented the steam turbine in 1884. His 1884 turbine, in which steam at 60 psi blown through brass fan blades turned or rotated a shaft uniformly, produced 10 horsepower at 18,000 revolutions per minute. Parsons perfected the turbine in 1894, using gears to reduce the turbine wheel's frequency to 180-1,800 revolutions per minute, and thereby enabled it to propel both slow heavy and fast light ships. He invented the much more efficient turbine because he believed fossil fuel reserves were running out.

Parson's 3-turbine 2,300 horsepower *Turbinia*, a steel ship 100 feet long, 9 feet wide that displaced 44.5 tons, had a speed of 32.75-34.5 knots, surpassing the 31-32 knots of the fastest steam-engine-powered ships. The *Turbinia,* built in 1897, demonstrated its high speed during its first-time appearance at Britain's June 1897 Spithead Navy Review in the Solent, a channel between England

and the Isle of Wright. The Royal Navy's 165 steam-engine powered ships were steaming in review to honor Queen Victoria (1819-1901) on her 60-year jubilee when the *Turbinia* with a two-man crew and Parsons driving streaked in and past the Royal Navy's ships.

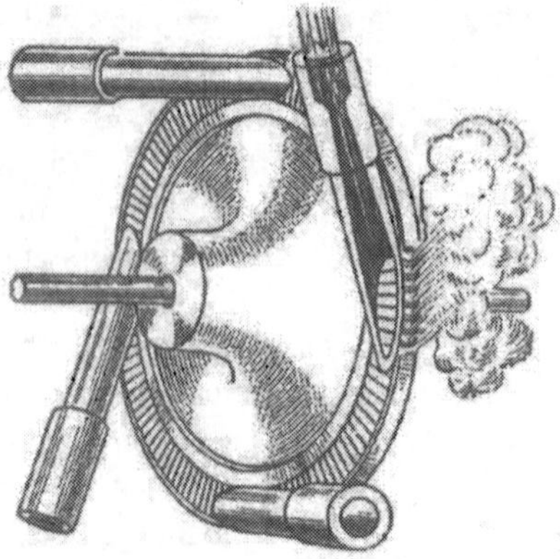

**Figure 18**
**Steam Turbine**
Source: Newton Henry Black and Harvey Nathaniel Davis. *New Practical Physics*. New York: Macmillan Company, 1913.

Steam turbines have an efficiency of 50 percent and higher, triple the 17 percent efficiency of a steam engine. A turbine operating between 1000° F (535° C) and 70° F (20° C) has an efficiency of 63.7 percent. The higher efficiency resulted for two reasons: because of the large operational temperature difference and because the steam striking the turbine wheel produced rotary motion directly. The turbine did not convert the steam engine's up and down movement into a rotary motion.

The British navy and merchant-owned ships installed turbines instead of the slower steam engines in 1906, in the ocean liner *Mauretania* launched in January 1906, in the battleship *HMS Dreadnought* launched in February 1906, and in the *Mauretania's* sister ocean liner *Lusitania* launched in June 1906. They were the first turbine-powered ships. Each 785-foot, 40,000-ton ocean liner had four Parsons turbines that generated 74,000 horsepower at 180 revolutions per minute, and a speed of 26 knots. The Navy's 490-foot, 17,900-ton *Dreadnought* had four Parsons turbines that produced 24,712 horsepower and a speed of 21.25 knots. Parsons said he spent about £24,000 ($96,000) before he received the first order for a turbine.

The diesel engine appeared about the same time as the turbine. Rudolf Diesel (1858-1913), a Paris-born (of German parents) engineer in Berlin, patented the diesel engine in 1892 and in 1897 perfected its design. It operated by compressing air in a cylinder to a temperature high enough to ignite a petroleum-derived hydrocarbon fuel (diesel oil) injected into the cylinder. The fuel's combustion initiated a cyclical up-and-down movement of a piston in the cylinder. The diesel engine's efficiency operating between 1527° C and 527° C is about 56 percent whereas the efficiency of an internal combustion engine operating between 1000° C and 500° C is only 39 percent. The diesel engine's large size restricted its first uses to large ships.

Diesel, an internationalist and pacifist, suffered from poor health, neurotic behavior, and breakdowns. Despite having acquired considerable wealth, Diesel's lavish lifestyle and some bad investments left him near financial ruin. In September 1913 while traveling by ship for consultations with the British Admiralty and the Consolidated Diesel Manufacturing Company in London, Diesel fell or jumped overboard from the *Dresden*, the Antwerp to Harwich mail steamer, and drowned in the English Channel. Ten days later crew members on a passing Dutch ship found his unrecognizable body floating in the channel, removed a wallet, identification card, an eye glass case and coin purse identified as Diesel's and gave him a sea burial in the channel.

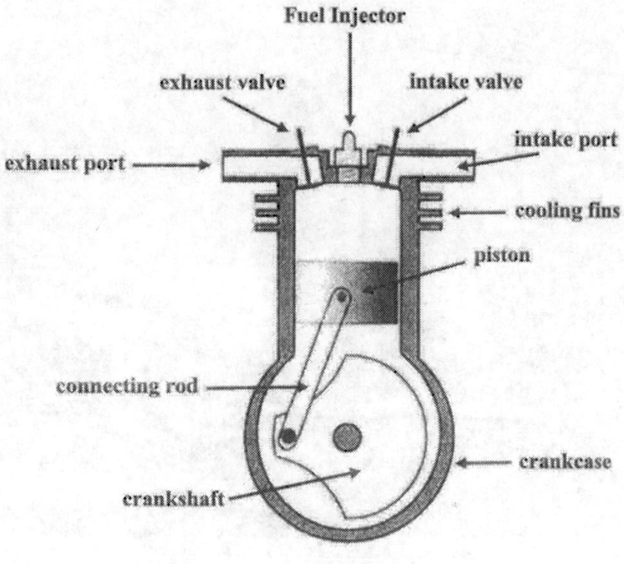


**Figure 19**
**Diesel Engine**
Courtesy of author.

***Environmental effects of steam power***

The steam engine not only ushered in the industrial revolution it also led to industrial-scale consumption of coal. Coal already had begun to replace wood as the domestic fuel and for cooking and as a source of coke in the iron industry, and it now became the main source of fuel to provide the steam to operate the steam engines in factories. The result was soot-covered buildings in industrial cities such as Birmingham, Manchester, and London; air pollution resulting from the discharge of carbon dioxide, sulfur dioxides, and nitrogen oxides, and from dangerous metals such as mercury and lead; and land and water pollution from the dumping of industrial wastes such as ash and slag from coal burning furnaces. Rumford, when he resided in London, criticized London's and England's widespread coal-produced smoke pollution.

## VII. ADVANCES IN ELECTRICAL THEORY AND APPLICATIONS

***Background***

The discovery and early studies on static (friction) electricity date from Thales (640-546 BC) in Miletus and included Theophrastus (372-287 BC) Aristotle's successor at the Lyceum in Athens, Peter Peregrinus (ca. 1269) in Europe, and William Gilbert (1544-1603) in England. Otto von Guericke (1602-86) in Magdeburg, Germany, in 1660 constructed an early static electric machine that consisted of a rotating sulfur globe mounted on an axle and noted the attraction and repulsion of electrically-charged objects. No further significant advances in electrical investigations occurred in the 1700s until the construction of condensers or capacitors for storing electricity.

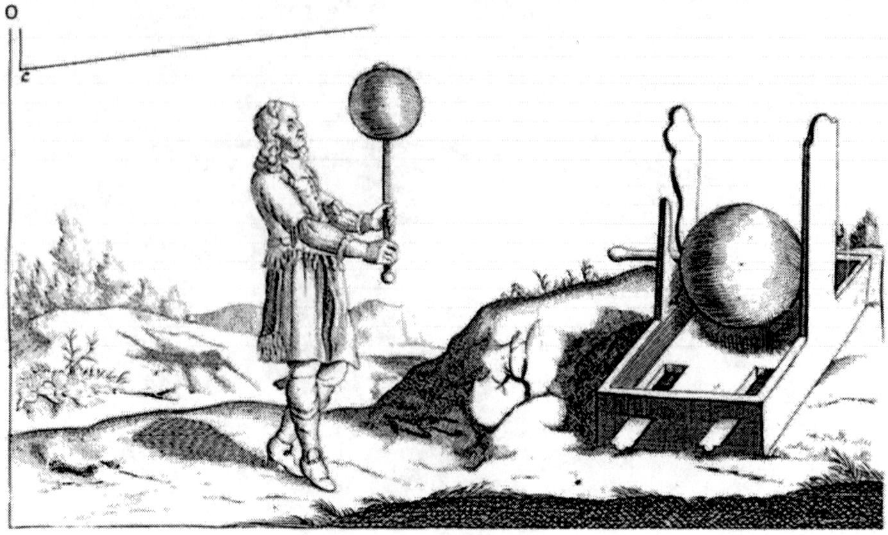


**Figure 20**
**Production of Static Electricity**
Source: Otto von Guericke. *Experimenta Nova (ut vocantur) Magdeburgica De Vacuo Spatio.* Amstelodami: Jansonnius, 1672.

Ewald von Kleist (1700-48) in Pomerania, Prussia, in November 1745 and independently Pieter van Musschenbroek (1692-1761) at Leiden, in January 1746 constructed the first condensers, since that time known as the Leiden jar. Their condensers, having voltages that ranged from 6 to 500 volts but low amperages (small electric currents), led to sensational and sometimes dangerous demonstrations. Jean Nollet (1700-70) in Paris in 1746 performed some of the best known public demonstrations. He showed that the electrical discharge in a Leiden jar went both ways (positively-charged source ↔ negatively-charged source) and generally followed the path of least resistance. In other static discharge experiments, Nollet assembled rings (circles) of humans, one of which had 180 palace guards (gendarmes) and another 200 monks. He shocked them upon sending an electrical discharge through the human ring.

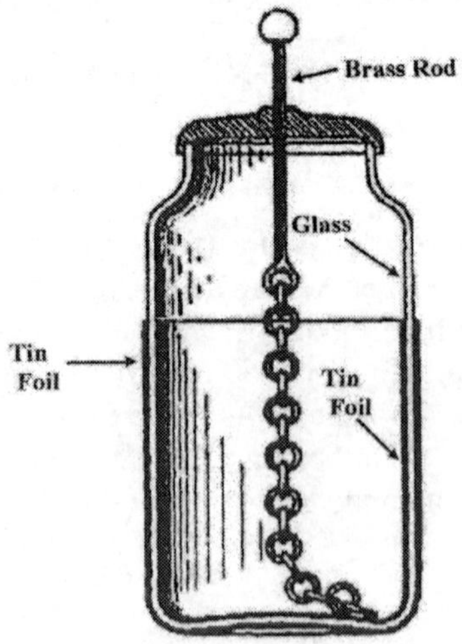


**Figure 21**
**Leiden Jar Condenser**
Source: William Henry Marchant. *Wireless Telegraphy: A handbook for the use of operators and students*. New York: Whittaker and Co., 1914.

**Figure 22**
**Static Electricity Machine**
Source: Augustin Privat Deschanel. *Elementary Treatise on Natural Philosophy*, trans. Joseph David Everett. New York: D. Appleton and Company, 1889. Original diagram in Abbé Nollet. *Leçons de Physique Experimentale.* Paris: Chez les Freres Guerin, 1767.

***Fluid theories of electricity: Du Fay and Franklin***

In the eighteenth century, natural philosophers believed that electricity, magnetism, combustibility, and heat resulted from the presence of different kinds of fluids in matter. These fluids, the electric fluid, the magnetic fluid, the phlogiston fluid (combustion), and the caloric fluid (heat), consisted of particles (atoms). The particles of each kind of fluid repelled its own kind of particle but strongly attracted particles of matter.

To account for the electrical behavior of matter, Charles Du Fay (1698-1739) in Paris in 1734 proposed a two-fluid theory of static electricity. Either a vitreous fluid identical to the kind of fluid generated by rubbing a glass rod with silk, or a resinous fluid identical to the kind of fluid generated by rubbing an amber rod with wool or fur, was present in matter and attracted to matter. Substances that contained the same kind of elastic fluid particles, vitreous-vitreous or resinous-resinous, repelled each other, whereas substances containing unlike electric fluid particles, vitreous-resinous, attracted each other.

In the years before he began his electrical researches, Du Fay served in the French army, enlisting at age fourteen. After retiring with the rank of captain, he held the position of chemist at the Academy of Sciences and in 1732 became director of the Royal Garden of Plants in Paris. He began his electrical studies at that time, establishing the idea of conductors and insulators and effectively promoting the two-fluid theory. Nollet and Charles Coulomb (1736-1806) in Paris adopted Du Fay's two-fluid electrical theory.

Franklin, who considered himself both a theoretician and a practical scientist, proposed a one-fluid theory of electricity because he believed a single fluid could account for electrical behavior. Franklin had no scientific reputation in Europe in the years before beginning his electrical studies, but he already had achieved fame and significant wealth in the American colonies. In 1731 Franklin established the colonies' first circulating library called the Library Company of Philadelphia, and in

December 1732 he first published *Poor Richard's Almanack.* It sold 10,000 copies annually until its publication ended in 1757. He helped found the American Philosophical Society in Philadelphia in 1743, whose early membership included George Washington (1732-99), John Adams (1765-1823), Thomas Jefferson, Alexander Hamilton (1755-1804), and Benjamin Rush. Franklin served as its first secretary and then president in 1769.Volume one of the Society's *Transactions*, covering the years 1769-71, coincided with Franklin's presidency. Following the publication of his electrical researches, the British Royal Society in 1756 and the French Academy of Sciences in 1772 elected him a member.

With Rush, Franklin in April 1775 organized the first American society for the abolition of slavery, and with John Jay (1745-1829) and John Adams, Franklin in 1783 negotiated the Treaty of Paris that ended the American Revolutionary War. He was musical and played the violin, harp, and guitar. He invented the glass harmonica, a simple instrument played by touching with dampened (moist) fingers, the edge of spinning glasses mounted on an axle. Franklin suffered from bladder stones in his later years and died from pleurisy accompanied by suppurated (pus-filled) lungs.

Franklin proposed his one-fluid theory of electricity in 1747 and published it in 1750. Like the other fluids, the self-repulsive particles of the electric fluid, or element of electricity, resided in all matter and strongly attracted particles of matter. Friction collected the electric fluid and transferred it from one substance to another. The law of conservation of charge, which he proposed in 1747, followed directly from the friction process: what one substance gained in electric charge, the other lost.

Franklin's one-fluid theory required a new electrical nomenclature, according to which positive charge (plus) indicated the gain of electric fluid and negative charge (minus) indicated the loss of electric fluid. Positive and negative charge replaced Du Fay's vitreous and resinous electricity, but they were the opposite of later electrical nomenclature. Franklin also redefined Du Fay's laws of attraction and repulsion for static charges, using the terminology: like charges repel one another, unlike charges attract one another. He introduced the terms battery, condenser, conductor, charge, discharge, armature, electric shock, and electrician.

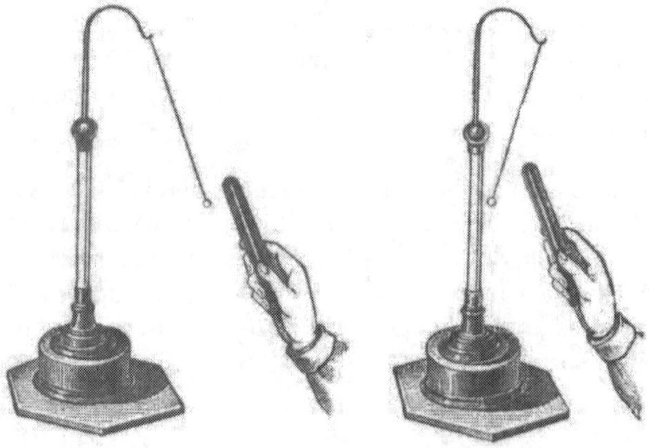

**Figure 23**

**Like Charges Repel Each Other - Unlike Charges Attract Each Other**

Source: J. Dorman Steele. *Popular Physics*. New York: American Book Company, 1896.

Franklin's theory had one serious weakness. It explained the repulsion of two positively-charged substances, each of which had gained particles of the self-repulsive electric fluid, and the attraction between a positively-charged substance which had an excess of the self-repulsive electrical particles and a negatively-charged substance which had an excess of matter or a deficiency of the self-repulsive electrical particles. It failed to explain the repulsion of two negatively-charged substances, each of which had lost the self-repulsive electric fluid and therefore the cause of repulsion.

| ATTRACTION AND REPULSION | | |
|---|---|---|
| ***Particle 1*** | ***Particle 2*** | ***Effect*** |
| Electric | Electric | Repulsion |
| Electric | Matter | Attraction |
| Matter | Matter | Attraction (Newtonian gravity)<br>Repulsion (Franklin, Aepinus) |

To account for the repulsion of two negatively-charged substances, Franz Aepinus (1724-1802) in St. Petersburg (Leningrad), Russia, in 1759 proposed a complete duality of matter in which like repelled like. Particles of the same kind of fluid were self-repulsive and particles of matter were self-repulsive, so that positive fluid repelled positive fluid, negative fluid repelled negative fluid, and matter repelled matter. Unlike particles of matter and fluid, and unlike positive and negative electric fluids attracted each other.

Aepinus's duality accounted for the repulsion of the negatively-charged substances, but it violated the law of gravity which held that particles of matter (like particles) attracted each other. To deal with this dilemma, Aepinus concluded that in ordinary matter, particles of matter and particles of the electric, magnetic, and caloric fluids, neutralized all the attractions and repulsions and made their effects non-evident.

The problem of fluid and matter coexisting in a substance remained unresolved until 1897 with the discovery of the electron and the electrical atom. The properties of electricity and magnetism, previously belonging to fluids *separate* from matter, now belonged to matter. Electricity and magnetism were properties of the particles of matter that constituted the new electrical structural atom.

Two years after proposing his one-fluid theory of electricity, Franklin devised an experiment to establish the long-suspected identity of lightning and electricity. He already pointed out in 1743 that storms in the colonies generally moved from west to east or from southwest to northeast and concluded that when a thunderstorm occurred, three kinds of particles - air, water, and electricity - were present. When a cloud loaded with electrically-charged particles struck a mountain or another cloud system, a snap of lightning released the electricity and restored equilibrium.

Many European scientists, such as Nollet in France, also accepted the identity of electricity and lightning because they behaved similarly. Like Franklin, they recognized that both emitted light of the same color, gave off a sulfurous smell, traveled in a crooked direction, passed through water, destroyed animals, and melted metals. Franklin, however, in 1749 first proposed an experiment to test the lightning-electricity identity. He arranged for Peter Collinson (1694-1768), his correspondent in London, to publish a description of his proposed experiment in 1751. Later translations appeared in French, German, Italian, and Latin.

Franklin's experiment required a sentry box with a pointed iron rod extending 20 or 30 feet upward from an insulating wooden or cork pad to attract electrical discharges occurring in the atmosphere. The box sat on a high tower or steeple. An experimenter stood inside the box on the insulated wooden or cork pad. He held an insulated metal key to the rod intending to draw off sparks of electricity, which Franklin thought posed no danger and would not affect him because he was standing on an insulating pad. The sparks indicated that an electrical discharge had struck the rod and passed down its length where the experimenter drew it off. The experimental arrangement introduced the now-familiar lightning rod.

**Figure 24**
**Franklin's Sentry Box 1750**
Source: http://www.revolutionary-war-and-beyond.com/benjamin-franklin-and-electricity.html

The French Academy learned of Franklin's proposed experiment the next year and decided to do the experiment. In the first test, at Marly near Paris on 10 May, 1752, the naturalist Thomas Francois Dalibard (1703-79) and other French investigators placed a covered sentry box on the ground and extended a 40-foot iron rod from the box into the sky. An insulated brass wire drew off electricity. Their experiment showed, moreover, that the lightning discharge had a negative charge rather than a positive charge as Franklin believed. A week later on 18 May, Dalibard's collaborator in Paris, identified only as Monsieur Delor, erected a lightning rod on his house and successfully repeated the experiment.

**Figure 25**
**First Test of Lightning-Electricity Identity**
Source: Illustration of experiment from Louis Figuier. *Les Merveilles de la Science.*
Paris: Libraire Furne, Jouvet et Cie, Editeurs, 1867-1891.

The French King Louis XV informed the Royal Society of the Franklin experiment. John Canton (1718-72), an English physicist in London and inventor of the pith ball electroscope (1753), successfully repeated the sentry box experiment on 20 July 1752. Franklin learned of the French and English successes in October 1752. But already in Philadelphia in June 1752, Franklin and his illegitimate son William (1731-1813) carried out a similar experiment. William later became New Jersey's last royal governor. He remained a British loyalist, and spent 38 months as a revolutionary war prisoner.

In Franklin's demonstration, he constructed a kite from two light strips of cedar in the shape of a cross that had the four corners of a large thin silk handkerchief tied to the four ends of the cross. He attached a foot-long pointed metal spike to the top of the kite and a long string of silk twine to the kite's bottom. Franklin attached the other end of the silk twine to an iron key. He then tied a thin metal wire to the key and inserted the wire into a Leiden jar, enabling him to determine whether the lightning discharge had a positive or negative charge. He attached a silk ribbon to the key and held on to the kite with the silk ribbon. He used a kite because it ascended much higher in the sky than the sentry box and more readily made contact with lightning. Franklin sent the kite aloft during a thunderstorm in an open field on the banks of the Schuylkill River in Philadelphia, and when a storm cloud approached he and William took shelter in a nearby barn to keep dry.

When lightning struck the kite the discharge passed down the wet silk twine to the Leiden jar, charging it. The discharge did not pass through the dry silk ribbon that Franklin believed insulated him from the charge on the iron key. He saw that the fibers on the wet twine bristled during the thunderstorm because, as Franklin noted, they were repelling one another, and when he brought his knuckle near the key he felt an electrical shock. These two observations, the shock that Franklin received and the bristles that stood out on the wet silk twine, and the evidence that the Leiden jar had a negative charge, clearly indicated the presence of electricity and its identity with lightning. Franklin described his observations in his newspaper, the *Pennsylvania Gazette*, on 19 October 1752, later in Joseph Priestley's *History and Present State of Electricity* published in 1767, and within a collection of other letters he had written to Collinson who published them in *Experiments and Observations on Electricity made at Philadelphia in America* (London, 1769).

Despite its success, Franklin had performed an extremely dangerous experiment because a lightning discharge can reach 100 million volts and 200,000 amperes. For comparison, the electric chair first used more than 100 years later in criminal executions delivered 2,200 volts and 4-8 amperes in 30-second intervals for about two minutes. The human body has a resistance of 400-500 ohms. An alternating current (*ac)* of 0.06-0.10 amperes can cause serious injury or possibly death because of heart fibrillation, 0.10 amperes *ac* passing through the body for two seconds usually kills.

Neither Franklin nor those he corresponded with or wrote about the kite experiment, such as Collinson and Priestley, expressed serious concerns about the dangers of the kite experiment. As long as the experimenter stood on an insulated pad and grounded the lightning rod, Franklin believed in the experiment's safety. The thrill of discovery also may have contributed to Franklin's unwarranted sense of safety. The criticisms about the dangers came much later in the twentieth century from scientists who recreated the experiment in a controlled setting and concluded that a direct lightning strike would have killed Franklin or anyone else who did the experiment. Tom Tucker, an American author in North Carolina who writes on the history of invention, claimed in his 2005 book *Bolt of Fate*, without providing unquestioned evidence to support his claim, that Franklin never did the kite experiment.

Other sentry box experiments allegedly succeeded in Germany, Russia, and Belgium in 1752 or 1753, but lightning killed Georg Wilhelm Richmann (1711-53), a German physicist at the St. Petersburg's Academy of Science, Russia, in August 1753 while experimenting with a variation of Franklin's sentry box. Richmann mounted an iron wire vertically on the top of his home and ran the wire inside his home. He attached the end of the wire to the end of a 1-inch thick, 1-foot long iron bar inside his home and vertically suspended the iron bar above a bowl of water partly filled with brass filings. An electrically-charged (compass) needle placed near the bowl indicated the proximity of a lightning discharge. The likely possibility of lightning on that August day led Richmann to stand about a foot away from the needle and bowl to watch for evidence of an electrical discharge. While observing the needle for any deflection, a four-inch blue-white globe of fire or electric spark that escaped from the bar, struck Richmann in the forehead. It knocked him to the ground, killing him. The electrical discharge knocked down Richmann's colleague, identified as M. Sokolaw the engraver at the Academy, standing nearby, but did not seriously injure him. Franklin published a notice of Richmann's death, apparently the first from repeating the sentry box experiment, in the *Pennsylvania Gazette* in March 1754. He noted that in performing the experiment Richmann stood on the floor and not on an insulated stand.

Franklin's invention of the lightning rod in 1753 and its rapidly growing application had an unintended consequence: a probable cause of earthquakes. Earthquakes struck New England in 1727, 1755 and again in 1761. At this time a favorite geologic theory of earthquake formation, which John Winthrop IV (1714-79) and other colonial scientists held, described Earth as a spherical body containing subterranean holes or caverns of flammable materials that exploded on heating, and not as a solid mass.

The 1755 earthquake struck on 18 November, only seventeen days after the Lisbon quake that killed an estimated 60,000 people. For the colonists, the two catastrophes seemed like simultaneous events. The New England quake caused considerable damage, as accounts of its occurrence recorded four minutes of shaking and the destruction of many chimneys. The Reverend Thomas Prince (1687-1758), a Harvard graduate (1707) and pastor of Boston's Old South Church (1718-58) who experienced the 1727 earthquake and published a sermon on it, republished his sermon on earthquakes in 1755.

In the re-issue of *Earthquakes the Works of God*, Prince added an appendix in which he linked an earthquake's cause to Franklin's lightning rod and the proliferation of lightning rods in Boston as the reason for the severity of the Boston quake. The Boston earthquake, Prince wrote, shows God operating *by means of the electrical substance:* The more Points of Iron are erected round the Earth to draw the Electrical Substance out of the Air; the more Earth must need be charged with it. And therefore it seems worthy of Consideration, Whether any Part of the Earth's being fuller of this terrible Substance, may not be more exposed to more shocking Earthquakes. In Boston there are more erected than anywhere else in New England; and Boston seems to be more dreadfully shaken. Oh! There is no getting out of the mighty Hand of God! If we think to avoid in the Air, we cannot in the Earth.

Other New Englanders agreed with Prince. They opposed lightning rods, and believed that thunder and lightning, and earthquakes signified divine judgments and warnings, and that to divert lightning robbed the Almighty of his tools of vengeance. Winthrop IV feared that these beliefs would reduce the use of lightning rods and thereby increase the danger of lightning strikes.

During this period of electrifying excitement Laura Bassi (1711-78) in her laboratory at the University of Bologna carried out static electric experiments on neurophysiology. She and her husband installed the first lightning rod at the university. Bassi, at Bologna since 1732, was the first woman to teach in a European university and the second to obtain a doctorate in philosophy. By 1760 she received a salary of 1,600 lire, more than that of any other science professor at Bologna. The mathematician Maria Agnesi (1718-99) became the second woman scientist at Bologna.

***Law of magnetic attraction and Coulomb's law***

Johann Tobias Mayer (1723-62) at the George-August Academy in Göttingen in 1760 demonstrated that magnetism obeyed an inverse square force law of attraction and repulsion, having the same form as Newton's law of gravitation and the law of electrostatics that Coulomb discovered in 1785. As occurred with static electricity, investigators disagreed on the number of magnetic fluids. Leonhard

Euler in Berlin accepted a one magnetic fluid theory, whereas Coulomb believed two different kinds of magnetic fluids resided in substances.

Studies on the electric fluid became a quantitative science in 1785 when Coulomb measured the strength of friction (static) charges. To make the measurements Coulomb used a torsion balance, an instrument he had perfected since 1777. Coulomb's torsion balance measured the electrical force by recording the amount of twist produced in a thin vertically-hanging wire. A thin horizontally- arranged rod attached at its midpoint to the wire had a metal sphere fixed at each of its ends. Coulomb charged one of the rod's spheres negative, brought a negatively-charged metal sphere close to it, and calculated the repulsive force between the two spheres by measuring the amount of twist in the torsion balance wire. The force was proportional to the product of the charge on each sphere and inversely proportional to the square of the distance between the spheres. In a second paper in 1785 Coulomb proved the validity of the inverse square force law for the more difficult-to-measure electric attraction of unlike charges. Coulomb's Law had a form similar to Newton's law of universal gravitation.

The British geologist and reverend, John Michell (1724-93) in Thornhill near Leeds, may have developed the idea of a torsion balance independently of Coulomb. He intended to measure the gravitational constant in Newton's equation for gravitational force. Michell died before completing the experiment, and his countryman Henry Cavendish (1731-1810), using Michell's torsion balance design, in 1798 determined Earth's mass, which led to the later calculation of the gravitational constant.

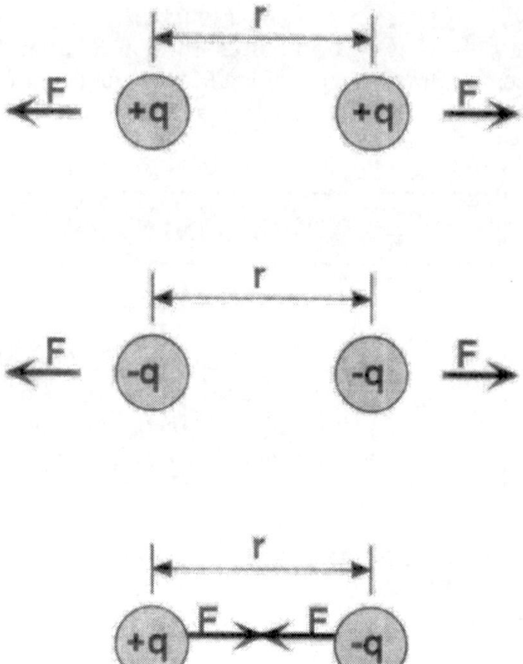


**Figure 26**
**Coulomb's Law**
Courtesy of the author

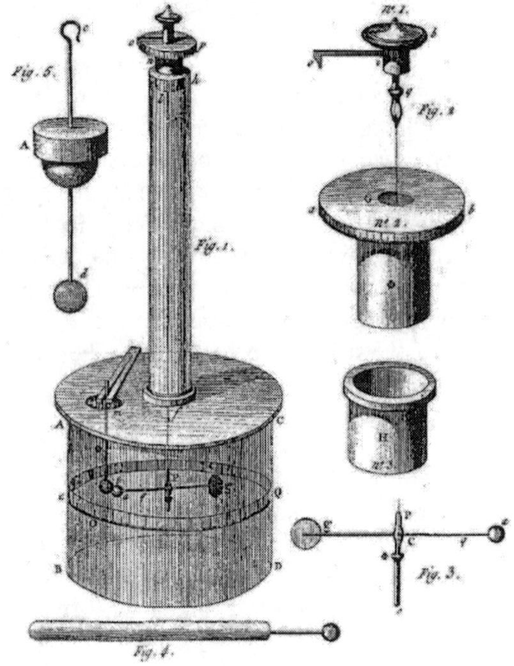

**Figure 27**
**Coulomb's Torsion Balance**
Source: Charles Coulomb. "First Memoir on Electricity and Magnetism." *History of the Royal Academy of Sciences*. Paris, 1785.

| **COULOMB'S LAW** |
|---|
| $F = K\frac{q_1\, q_2}{r^2}$ |
| *q* = the static electric charge on an object<br>*r* = distance between the charged objects<br>*F* = force of attraction or repulsion<br>*K* = a constant of proportionality |

***Discovery and generation of current electricity in 1800***

*Luigi Galvani and animal electricity*

Galvani (1737-98), an anatomist at Bologna, began an experimental study of static electricity in 1780. In his experiments Galvani observed the twitching of a frog's legs when they accidentally became part of an electric circuit containing a static electricity machine. Galvani also expected the frog's legs to twitch when he hung them outside on a brass hook against an iron lattice during a lightning (electrical) discharge and thunderstorm. He expected them to twitch during a thunderstorm because the arrangement was the same as the laboratory setup in which he observed the initial twitching. Galvani continued his experiments and found that the frog's legs even twitched when hung outside on the fence when no source of electricity (lightning) appeared in the sky. Although Galvani did not realize it, he had constructed an electrolytic cell and an electric circuit, which accounted for the frog's legs twitching as the electric current flowed through them.

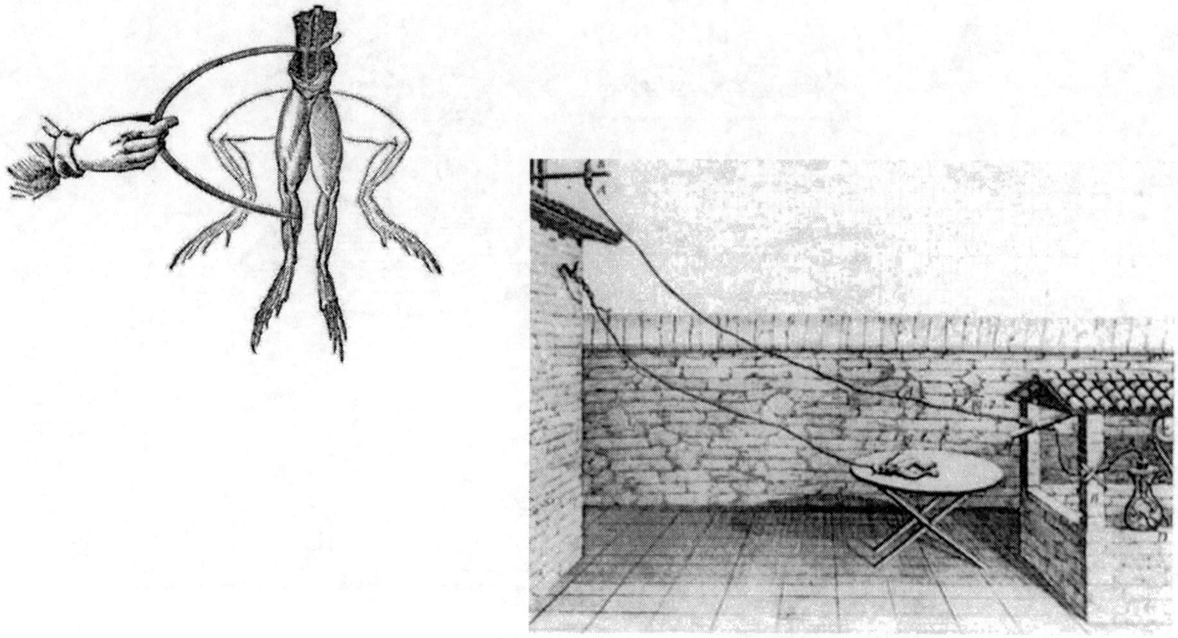

**Figure 28**
**Galvani's Frog's Legs Experiment**
Source: Luigi Galvani. *De Viribus Electricitatis in motu musculari commentaries.* Bologna: Academy of Sciences of the Institute of Bologna, 1791.

Galvani thought the electricity originated in the frog's legs (muscles), or animal electricity. The outside influence (lightning) enabled the electricity to manifest itself, but as he had shown already, the production of electricity did not require a lightning discharge. As an anatomist he believed the animal and not the two metals caused the electricity to flow in the circuit he had constructed. His discovery of animal electricity had resulted in his conviction that he had discovered the secret of life. Galvani published his results in 1791 in a 91-page book, *De Viribus Electricitatis in motu musculari commentarius (Commentary on the Effects of Electricity on Molecular Motion).*

*Alessandro Volta and contact electricity*

Volta (1745-1827), a physicist at Pavia, Italy, did not agree with Galvani's conclusion that the animal caused the electricity and proposed instead a theory of metallic electricity in which contact between two different metals produced electricity. This initiated a long dispute between Volta and Galvani.

Volta, who adopted the one-fluid theory of electricity, already in 1775 had constructed an electrophorus, a sensitive device for detecting an electric charge. It consisted of a circular metal plate with an insulated handle set on top of a resin slab and a second circular metal plate on which sat the resin. To charge the electrophorus Volta removed the top metal plate, charged the resin by friction, replaced the top metal plate, and then grounded the top metal plate by touching it. The charges (electrons) in the negatively-charged resin repelled the negative charges (electrons) in the top metal plate. The negative charges passed out of the top metal plates (upon touching it, grounding), traveling to the ground, and left the top metal plate with a positive charge by induction.

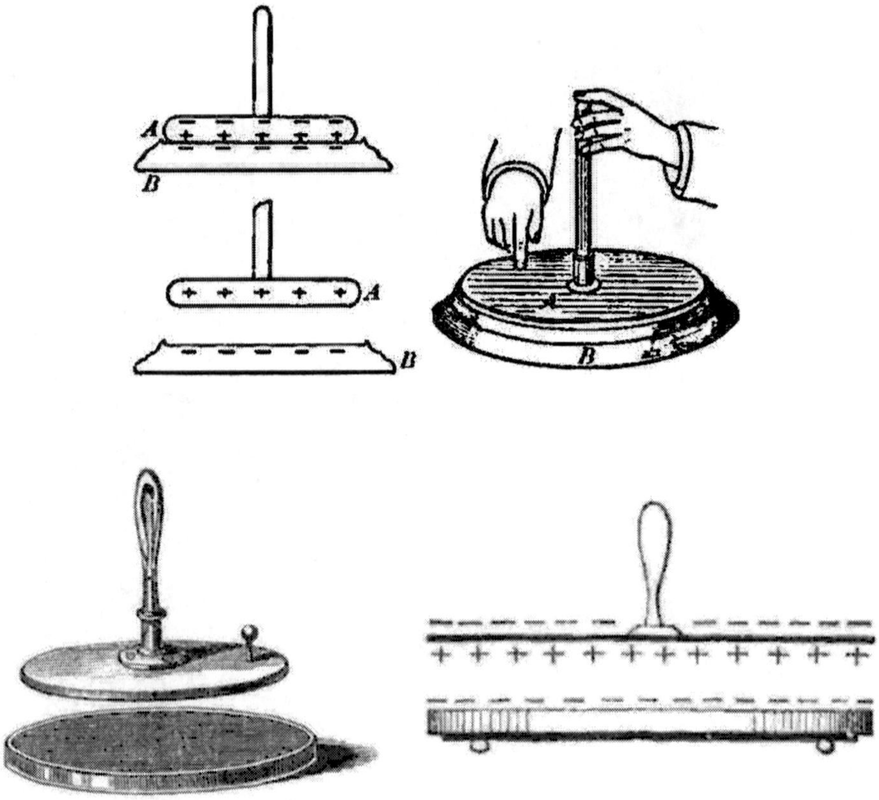


Volta announced his first invention, the electrophorus, in a 1775 letter to Joseph Priestley. The electrophorus consisted of an upper metal plate A and a disk B of non-conducting substance such as resin. Volta gave the resin B a negative charge by friction and induced a positive charge in the upper plate A when he placed it on the resin, grounded it, and then lifted it. He then could charge the upper plate A repeatedly by placing it on the resin B and then lifting it. The electrophorus provided a reproducible source of static electricity which earlier investigators had obtained only by direct friction.

**Figure 29**
**Volta's Electrophorus**
Source: Newton Henry Black and Harvey Nathaniel Davis. *New Practical Physics.* New York: Macmillan Company, 1913 (top).
Source: Robert Hare. *A Compendium of the Course of Instruction in the Medical Department of the University of Pennsylvania.* Philadelphia: J.G. Auner, 1840 (bottom).

Volta constructed a condensing microscope in 1782. It also worked by induction and could detect smaller charges than the electrophorus detected. He used it in 1792 to detect the presence of an electric charge produced by a copper plate when it made contact with a zinc plate. The gold leaf electroscope that Abraham Bennett (1749-99), a clergyman, in Wirksworth, England, constructed in 1787, provided a third device for the detection of an electric charge.

When charged either by contact or by induction, the two leaves of the electroscope acquired the same charge, either positive or negative, and repelled each other. An unknown charge brought close to the electroscope would cause the leaves to repel even more if it had a like charge, or collapse if it had an unlike charge.

**Figure 30**
**Gold Leaf Electroscope**
Source: Edwin J. Houston. *The Elements of Natural Philosophy*.
Philadelphia: Eldridge & Brother, 1900.

*Volta's experiments to disprove animal electricity*

To challenge Galvani's animal electricity, Volta tested an electrical circuit consisting of one metal and a frog's legs, and a circuit with no metals and only a frog's legs. No twitching (electricity) resulted each time. In these and other experiments Volta intended to show that the twitching resulted from stimulation of the frog's nerves, not the muscles.

In related experiments with metals, Volta placed two different metals, gold and silver connected with a copper wire, on his tongue and experienced a bitter taste. He associated the taste with the production of electricity in the circuit he had constructed.

In a second experiment Volta fused together one end of each of two different metals strips such as copper and zinc. He put the copper end on his tongue and the zinc end in contact with his eye. He observed a light flash which he interpreted as evidence of having produced an electric current.

*Voltaic pile and Volta's contact theory of current electricity production*

While performing his electrical experiments Volta found that when he placed two different metals in contact with one another they produced an electric current. Starting with this observation, in 1800 he constructed the first electrical pile or battery. Volta alternated plates of copper or silver and zinc to give a copper-zinc or silver-zinc pair, or couple, each pair separated by a cardboard disk soaked in acid solution. In this way he multiplied the voltage produced by each metallic couple. This arrangement resulted in the first practical cell (battery) to produce a steady flow, or current electricity, rather than a one-time static discharge. Volta placed the two metals in contact with each other because he believed the contact of the two metals produced electricity and that the pile operated similarly to the action of a Leiden jar.

The piling of the metallic couples multiplied the electric charges produced by a single pair or couple of metals in contact. The solution allowed the charge to pass from one pair of plates to the next pair. Volta drew off the charges at the two ends of the pile with metal wire conductors. He distinguished tension (voltage) from the quantity of electrical charge (current) produced in his pile. He announced his discovery in a March 1800 letter to Sir Joseph Banks, president of the Royal Society of London, and also published it in the *Philosophical Transactions of the Royal Society* in 1800.

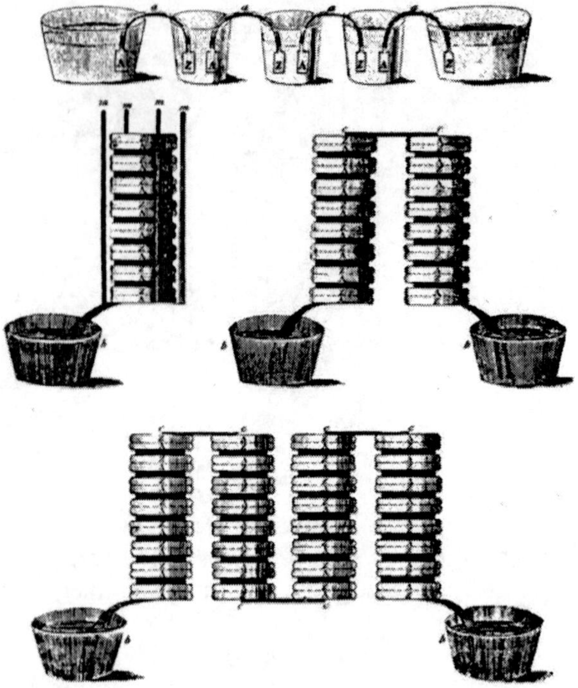

The first battery, which Volta called a pile, appeared in the illustration that accompanied his announcement of the invention. At the top, the battery appears in a form that Volta called a chain of cups. It consisted of different metals, such as silver A for argentum and Z for zinc, immersed in salt water. Volta connected them in a chain, by means of several metallic arcs. The lower figures show the column form of the battery, in which a disk of moistened pasteboard separated the silver-zinc pair.

**Figure 31**
**Volta's Pile, the First Electric Battery**
Source: Alexander Volta. "On the Electricity Exited by the Mere Contact of Conducting Substances of Different Kinds." *Philosophical Transactions of the Royal Society* (London) 90(1800): 403-31.
Source: F.J. Moore. *A History of Chemistry.* New York: McGraw-Hill Book Company, 1918.

Only a few weeks after Volta's announcement, Anthony Carlisle (1768-1840) and William Nicholson (1750-1815) in London, constructed a voltaic pile and electrolyzed water to produce oxygen and hydrogen.

A short time later, Davy, whom Rumford had invited to accept an appointment at the Royal Institution in London in 1801, constructed powerful voltaic piles consisting of 100 metal plates. Davy showed that a chemical reaction between the metal plates and the acid solution produced electricity, not the metallic contact of the two plates, as Volta believed. Davy used a 100-plate pile of zinc and copper to produce an electric current, and with it in 1807-8 he discovered six new elements: sodium, potassium, calcium, magnesium, strontium, and barium.

Later, in 1881, the International Electrical Congress meeting in Paris honored Volta when it named the volt the unit of electromotive force. Other units named at this time for those who contributed to the science of electricity included Ampere, Coulomb, Faraday, Henry, and Ohm.

*Electricity in the popular literature*

The electrical experiments of Volta, Davy, and other experimenters caught the imagination of the public. Galvani's nephew, Giovanni Aldini (1762-1834) at Bologna, in a demonstration done in London in January 1803 at the Royal College of Surgeons, attached electrodes to the corpse of George Forster, an executed murderer. When attached successively to the wrist, legs, and face, the corpse's hand jerked, its feet moved erratically, its eyes opened, its mouth twisted, and its face took on different expressions. Aldini also carried out galvanic experiments on warm-blooded animals, described in his 1794 publication *De animali electricitate dissertationes duae.*

The following May 1804, Mr. Hardie, identified in an advertisement as a lecturer at the Pantheon in London, in addition to routine electrical experiments, such as the production of gases and the deflagration of metals, performed more sensational experiments on dissected animals. Hardie described his experiments as inducing the functions of smelling, biting, chewing, swallowing, and wrinkling in the head of a sheep, ox, or other larger animals, long after its separation from the body.

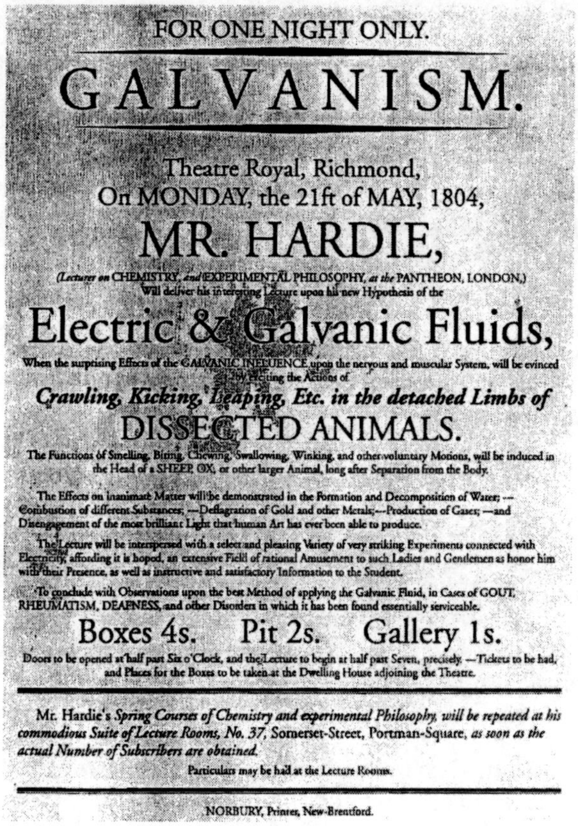


**Figure 32**
**Mr. Hardie Advertisement on Galvanism**
Source: Norbury Printer, New Brentford, May, 1804.

As a consequence of the many sensational public electrical demonstrations, the English writer Mary Shelley (1797-1851), in London, who had some familiarity with Davy's work on electricity, in 1818 published the well-known Gothic novel *Frankenstein, or the Modern Prometheus*. In the story, Victor Frankenstein, a young medical student, used the electric current to bring to life a soulless monster made out of parts of corpses obtained from churchyards and dissecting rooms.

## VIII. CLASSIFICATION IN NATURE

### *Nomenclature*

Already by the 1600s the methods of naming plants and animals had become complex. Plants had names that described their appearance or uses. Naturalists continued to use the two systems of classification, the artificial and natural, that Aristotle had introduced nearly 2,000 years earlier.

Most eighteenth-century naturalists believed in the separate creation and fixity of plant and animal species and a balance in nature. Such beliefs provided evidence of a rational creator.

### *Artificial or discontinuous classification system*

This system used one, or a few, characteristics to classify plants and animals, such as pistils and stamens for plants and toes and teeth for animals. It found wider use than the natural system because it was simpler. Andrea Cesalpino, Marcello Malpighi, Nehemiah Grew, and Rudolph Camerarius adopted an artificial classification.

Andrea Cesalpino (1524-1603) at Pisa, later Rome, in *De plantis libri XVI* published in 1583 adopted an artificial classification system using roots and fruit organs. He made a plant's vegetative soul the basis of classification and responsible for its nutrition and reproduction.

Marcello Malpighi (1628-94) at Bologna adopted an artificial system by 1675. He introduced a vertical scale with plants at the bottom and humans at the top. The size of an organism's air passages (air holes, gills, lungs) ranked an organism.

Nehemiah Grew (1641-1712) in London and Rudolph Camerarius (1665-1721) at Tübingen, Germany, adopted an artificial classification in their plant studies. Grew discovered in 1692 that plants reproduce sexually. Camerarius showed in his *Epistola* of 1694 that pistils were the female organ and that stamens were the male organ.

### *Natural or continuous classification system*

This system included as many characteristics as possible, such as roots, stems, leaves, flowers, and reproductive organs, in classifying plants; and toes, hoofs, teeth, and horns in classifying animals. Gaspard Bauhin and John Ray adopted a natural classification system.

Gaspard Bauhin (1560-1624) at Basel, Switzerland, adopted a natural classification in which he introduced a binomial system of naming plants. Bauhin named 6,000 plants. He published his natural system in *Pinax* in 1623.

John Ray (1627-1705) in Cambridge, England, published *Historia Plantarum*, 3 volumes, in 1686, 1688, and 1704. In this work he classified 18,600 plant species using a natural classification of plants he developed in 1693 and which he based on roots, leaves, stems, flowers, and organs.

Ray used hoofs, toes, and teeth to classify animals. He had 5 classes of animals: quadrupeds, birds, fishes, snakes, and insects. Infertility of offspring indicated that the parents were separate species.

Ray, a Neoplatonist, regarded nature as a manifestation of the divinity. He discussed these ideas in *The Wisdom of God*, published in 1691.

### *Carl Linnaeus*

Linnaeus (1707-78) at Uppsala, Sweden, ranks as the greatest eighteenth-century classifier, classifying plants, animals, and human races. In *Systema Naturae* (1735) he used an artificial (sexual) classification. He published twelve editions of his *Systema* up to 1768. During that time it increased in length from 14 to 2,500 pages and described 18,000 species. Collectors from Europe and from across the Atlantic Ocean sent him descriptions of species to include in his classification system. In Long Island, New York, Cadwallader Colden (1688-1776), a graduate of the University of Edinburgh's medical school and later New York State's lieutenant-governor, who owned 3,000 acres of wilderness west of Newburgh, catalogued and described a collection of American plants which he sent to Linnaeus. Linnaeus published the collection in 1743. Cadwallader's daughter, Jane Colden (1724-66) in New York City, probably the British colonies' first woman botanist, also studied New York's natural history. She knew Latin and in her 1753-58 manuscript *Flora of New York* adopted Linnaeus's sexual classification system to describe 400 plant species indigenous to the New York colony. Her manuscript included morphological (structural) descriptions and information on the domestic and medicinal uses of the plants, all of which she obtained from local inhabitants including Indians.

Cadwallader Colden's friend, John Bartram (1699-1777), remains the best known British colonial naturalist. Bartram, an amiable self-taught farmer and experimental botanist who lived near Philadelphia, and William (1739-1832), his son and sometimes partner, traveled extensively throughout the colonies after the fall harvest collecting plant specimens. Bartram carried on an extensive correspondence with the London Quaker, woolen-maker and naturalist Peter Collinson (1693/94-1768), who paid him (Bartram) to gather plants for his collection, and corresponded with many other European and colonial naturalists. Linnaeus called John Bartram the world's greatest natural botanist.

Bartram's 600 surviving letters tell us much about colonial natural history, such as plant hybridization and the use of herbal remedies that included ginseng and rattlesnake root in the treatment of diseases and ailments. He discovered a small wild tea tree, specimens of which he named *Franklinia* and planted in his garden. The tea tree no longer grows in the wild, all specimens today descended from those in Bartram's Garden. The historic garden the Bartrams established in Kingessing, Pennsylvania, in 1728 remains North America's oldest continuous botanical garden.

Bartram's letters discuss his neptunist belief and apparent acceptance of the biblical universal flood that altered existing mountains, created new river channels, deposited soils, changed Earth's strata, and extinguished species. In adopting neptunism, Bartram agreed with mainstream colonial and European thinking that Georges Cuvier (1760-1832) in Paris later dominated.

Bartram from 1765 held the title of King's Botanist which Franklin and Collinson helped him obtain. He received an annual salary of £50 ($227 in 1765, £4,127 or $6,200 in 2000). The Bartrams' accounts of their many travels, some as far as Georgia and Florida, most notably William's *Travels through North and South Carolina, Georgia, East and West Florida, the Cherokee County, etc.* published in 1791, influenced early nineteenth-century romantic writers. The English poet Samuel Taylor Coleridge (1772-1834) borrowed scenery descriptions for his unfinished 1816 dream poem *Kubla Khan.*

Linnaeus accepted the separate creation and fixity of species and a balance in nature. This provided evidence of a rational creator. Later, because of the evidence from hybridization, he accepted that species could change within genera. Linnaeus grouped plants in decreasing generality according to classes, orders, genera, and species. He established 24 classes of plants according to the number of stamens. Pistils determined a plant's order. He established 6 classes of animals: quadrupeds, birds, amphibia, fishes, insects, and worms.

Linnaeus proposed four races, possibly based on the four humors, in his 1740 edition of *Systema Naturae*. He gave descriptive labels of the races in his 1759 edition. Linnaeus in 1758 introduced the well-known classification terms *primate* and *homo sapiens* to the naturalist's nomenclature.

| LINNAEUS'S RACES | |
|---|---|
| ***Race*** | ***Characteristics*** |
| Red | Choleric |
| White | Sanguine |
| Black | Phlegmatic |
| Yellow | Melancholic |

In *Species Plantarum* (1753), Linnaeus used a binomial system of nomenclature to indicate genus and species. *Mentha spicata (spearmint)*, *rosa gallica*, *canis lupus*, and *homo sapiens* represent other examples of his system. He eventually named 7,700 plant species and 4,400 animal species.

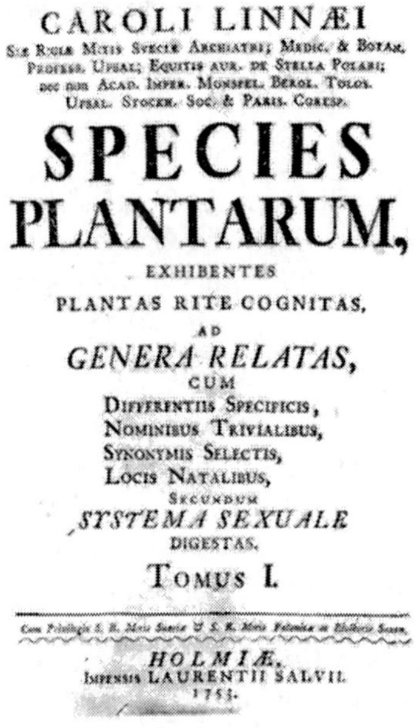
CAROLI LINNÆI
S:Æ R:GIÆ M:TIS SVECIÆ ARCHIATRI; MEDIC. & BOTAN. PROFESS. UPSAL; EQUITIS AUR. DE STELLA POLARI; nec non ACAD. IMPER. MONSPEL. BEROL. TOLOS. UPSAL. STOCKH. SOC. & PARIS. CORESP.

SPECIES PLANTARUM,
EXHIBENTES PLANTAS RITE COGNITAS, AD *GENERA RELATAS,* CUM DIFFERENTIIS SPECIFICIS, NOMINIBUS TRIVIALIBUS, SYNONYMIS SELECTIS, LOCIS NATALIBUS, SECUNDUM *SYSTEMA SEXUALE* DIGESTAS.

TOMUS I.

Cum Privilegio S:æ R:æ M:tis Sueciæ & S:æ R:æ M:tis Polonicæ ac Electoris Saxon.

*HOLMIÆ,*
IMPENSIS LAURENTII SALVII.
1753.

Plantarum , Uppsala 1753

**Figure 33**
**Cover and Page from Linnaeus *Species Plantarum***
Source: Carl Linnaeus. *Species Plantarum*. Uppsala: Lars Salvius, 1753.

## IX. CONCLUSION

The mechanical philosophy continued to dominate in the eighteenth century, and classification was an organizing force in both the physical and life sciences. The introduction of imponderables, different kinds of fluids, to account for heat, combustion, electricity, and magnetism allowed investigators to classify substances according to their properties and provided some understanding of a substance's composition. In chemistry, Black, Cavendish, and Priestley's investigations on the composition of matter, the main field of chemical study, led to the discovery of the gases carbon dioxide, hydrogen, oxygen, and others and to their classification as a new third state of matter. Lavoisier contributed greatly to the development of chemistry with the oxygen theory of combustion and the distinction he made between simple and complex substances, that is, between elements and compounds. He gave an *experimental* definition of a chemical element, a definition that experiment could test. His role in developing a system of nomenclature greatly assisted in clarifying the composition of matter. His statement of the law of conservation of matter remains the starting point for the study of chemical reactions and analysis and later for the link between the conservation of matter and the conservation of energy.

In electrical studies Franklin's one-fluid theory of electricity replaced Du Fay's two-fluid theory. Franklin also proposed the sentry box experiment that in 1752 demonstrated the identity of lightning and electricity. With Volta's construction of the first batteries in 1800, current electricity became a new analytical tool that helped investigators determine whether a substance was an element or a compound and thereby classify it.

Lavoisier's studies on heat led him to introduce caloric as the element, or fluid, of heat. The caloric theory of heat dominated the eighteenth century and much of the nineteenth century until the 1860s. Black and most other scientists adopted the caloric theory. Despite his accepting the incorrect caloric theory, Black in the 1760s carried out quantitative experiments on heat. He introduced and calculated for the first time values for the quantity of heat, specific heat, the heat of fusion, and the heat of vaporization. Count Rumford (Benjamin Thompson) rejected the caloric theory. He proposed instead a mechanical theory of heat, arguing that particles, or matter, in motion produced heat. Rumford's mechanical theory gained few followers despite his experiments that strongly supported it. His experimental data nevertheless exposed the weakness of the caloric theory and enabled Joule to calculate the mechanical equivalent of heat. Rumford's experiments qualified as an original contribution to the study of thermodynamics.

The studies on heat resulted in the development and use of steam power in steam engines. The application of steam engines for locomotion and in factories, replacing water power, initiated the industrial revolution, first in England and then on the European continent and the United States.

In the life sciences Ray and especially Linnaeus developed classification methods. A binominal system of nomenclature enabled naturalists to classify many species of the natural world.

The mechanical philosophy and the search for the mathematical quantitative laws of nature continued to dominate the physical sciences in the eighteenth and into the nineteenth century. In the life sciences naturalists focused their activity on classifying plant and animal species.

## REVIEW QUESTIONS ON EIGHTEENTH-CENTURY SCIENCE
## THE ENLIGHTENMENT

1. What is the main theme of this chapter?

2. Why do historians of science often characterize the eighteenth century as a materialistic age in science?

3. What branch of investigation was the essence of the eighteenth-century chemical revolution?

4. Explain the phlogiston theory of combustion. Who proposed this theory?

5. Discuss Joseph Black's work on carbon dioxide.

6. Who is Henry Cavendish?

7. Who discovered oxygen gas and who proposed the oxygen theory of combustion?

8. What were Antoine Lavoisier's contributions to eighteenth-century chemistry?

9. What is the origin of the metric system?

10. What was the caloric theory of heat?

11. Which eighteenth-century scientist quantified and clarified the meaning of heat of combustion, heat of vaporization, specific heat, and other heat-related quantities?

12. Who was Count Rumford, what was his theory of heat, and what was the significance of his 1798 cannon-boring experiments?

13. Did the steam engine play an important part in the industrial revolution? How?

14. Why was James Watt's steam engine an improvement over Thomas Newcomen's steam engine?

15. What is the difference between Charles Dufay's and Benjamin Franklin's theories of electricity? What was the most serious weakness of Franklin's theory?

16. What is Coulomb's law of static electricity?

17. What was the main issue in electricity that Luigi Galvani and Alessandro Volta argued about and how did they settle their issue?

18. What was the significance of Volta's electric pile?

19. How did eighteenth-century naturalists name species?

20. How did John Ray and Linnaeus classify species?

# CHAPTER 7
# DARWIN'S CENTURY

Darwinian man, though well-behaved, at best is only a monkey shaved!
William Gilbert (1836-1911) and Arthur Sullivan (1842-1900)
*Princess Ida*, 1884

## I. INTRODUCTION

Eighteenth-century naturalists had few theories or laws that governed the behavior of their living world, comparable to the laws of the physical sciences that governed the behavior of light, electricity, magnetism, and matter in motion. Classification of plant and animal species dominated the activity of eighteenth-century European and American naturalists such as John Ray, Carl Linnaeus, and John Bartram. Most of them, with the exception of the evolutionary naturalists Georges Buffon, Erasmus Darwin, Jean Baptiste Lamarck, and a few others, accepted the immutability of species.

In the late eighteenth and nineteenth century the interests of naturalists expanded beyond classification. They began to develop theories about the origin of species and whether species had evolved over long time periods, or had remained unchanged since God created them and the universe 6,000 years ago. The two main schools of geologists, the catastrophists and the uniformitarians, developed theories to account for the changes in Earth's strata. Catastrophists, such as Georges Cuvier, accepted a young Earth whereas Buffon accepted an older Earth. The uniformitarians James Hutton and Charles Lyell studied Earth's strata and concluded that Earth was old, about 240-300 million years old, and that Earth's strata and the plant and animal species living on the strata had changed over time. Until the introduction of radioactive dating in 1905, scientists had no accurate way of assigning an age to Earth.

Naturalists proposed two theories to account for the change or evolution of species: the Buffon-Erasmus Darwin-Lamarck influence of environment-use and disuse theory and Charles Darwin's theory of natural selection. Of the two theories only Charles Darwin's theory offered naturalists a testable mechanism. Buffon, Erasmus Darwin, and Lamarck independently proposed that a changing environment drove an organism to change (evolve) into a new organism better adapted to survive in the new environment. An organism's internal driving force enabled it to respond to the changing environment and adapt to the change in order to survive. Charles Darwin rejected this theory and proposed that some organisms were born with certain characteristics that made them better adapted to their environment. They had an advantage that enabled them to survive and gradually dominate, whereas those not born with the advantage failed to adapt, perished, and eventually became extinct.

Charles Darwin's theory that he published first in 1858 and in 1859 in *The Origin of Species*, evoked controversy because it denied society's long-standing belief in progress and advocated a random, yet directed, materialistic process that had no place for God. His theory was incomplete and required Gregor Mendel's discoveries in genetics in 1866, which showed how a characteristic passed unchanged from one generation to the next, to make it an acceptable mechanism of evolutionary change.

Thomas Huxley in England vigorously defended Darwin's theory against critics such as Richard Owen. Several German scientists Mathias Schleiden, Ernst Haeckel, and August Weismann also defended Darwinism. In the United States, Asa Gray at Harvard defended Darwin in his discussions and debates with Louis Agassiz his Harvard colleague and dedicated anti-Darwinist. Gray, unlike Darwin, believed that God directed the evolutionary process and proposed a theistic theory of natural selection.

Since the time Darwin published his theory of natural selection in *The Origin of Species*, evolutionists, evolutionary skeptics and deniers, theologians, pseudo-theologians, and biblical creationists have offered different interpretations of when, why, and how the universe originated and developed. They debated whether the universe and its living organisms were the product of a Godless evolution, the product of a God-directed evolution, or the unchanged product of God's one-time creation 6,000 years ago.

Biblical creationism in its most recent revival, called intelligent design, has continued to argue that evolution is *just* a theory. Intelligent design, the creationists say, is an equally acceptable alternative and belongs in science textbooks and science curricula. Most scientists disagree with that viewpoint.

## II. HOW OLD IS EARTH

***What did earlier cultures or societies believe about the origin of the universe?***

The cosmologies of earlier European, American, and Asian cultures described a simultaneous creation of Earth and universe. They made no distinction between the two creation times and accepted cyclical theories of cosmology in which the universe underwent cycles of creation and destruction.

The Mayans calculated that the most recent cyclical creation of the universe occurred in 3113 BC, a date that is the modern calendar equivalent of their year one. In the Chinese cosmological cycle, creation and destruction of the universe occurrs every 23.6 million years. The Brahmins of India in their Vedas (1500 BC - 300 BC), their four sacred books, described a complicated cosmology in which an eternal universe went through millions of cycles of creations, deaths (destruction) by fire or water, and re-creations (rebirths), or millions of k*alpas*. Each *kalpa* was 4.32 billion years and represented one day in the life of Brahma, the god of creation.

Greek society believed that natural history repeated itself perpetually. In Plato's *Republic*, Earth had a 72,000 year cycle, a 36,000 year Golden Age followed by 36,000 years of chaos and disorder. Plato accepted a time of creation for the universe, whereas Aristotle believed the universe always existed. Aristotle's belief in an eternal universe enabled him to avoid having to answer the question of what caused the universe or what existed before the universe. When asked essentially the same question, '*what was God doing before he created the universe?*' Augustine (354-430), the great Catholic theologian and neo-Platonist, allegedly replied that God was busy creating hell for those who asked such dumb questions.

***Biblical genealogy and the age of Earth***

Prior to the discovery of radioactivity in 1896 scientists had no way of accurately calculating Earth's age. Archeological evidence, a salinity test, and thermodynamic measurements of Earth's rate of absorbing and emitting radiation provided estimates that varied widely and lacked certainty. Biblical scholars looked instead to the genealogy found in the Old Testament, in *Genesis*, chapters 5 and 11, where by connecting the life spans of the biblical patriarchs they could move back through the generations to the time of Adam and Eve. Using biblical genealogy, Adam and Eve's creation became the year of Earth's creation, and Earth's creation became the year of the universe's creation. For the biblical scholars no distinction existed among the three creations. Time was now linear, not cyclical.

What is the historical accuracy of the Old Testament on which these scholars based their genealogy? How accurate is the New Testament? Different authors wrote the Old and New Testament of the Bible, a word derived from the Greek *Biblia* (*Books*), over a 1,000-1,300 year period. They include Moses (1400-1200 BC) the Hebrew prophet-lawgiver to whom God first revealed the religion of Israel and who led the Israelites out of Egypt to the promised land, the Apostles John (6-101 AD) and Matthew (1 BC-60 AD), and the Christian authors Mark (ca. 62 AD) and Luke (died ca. 84 AD). It describes events prior to (1900-1800 BC), which most biblical scholars consider pre-historical, and events from Abraham to 100 AD which they consider historical.

The compilation of the Old Testament, the 24 books of the Hebrew Scriptures, or Tanakh, took place from 900 BC to 1 BC. The first five books of the Old Testament, the five books of the Law of Moses, *Genesis, Exodus, Leviticus, Numbers, and Deuteronomy*, are the Pentateuch (the Hebrew Torah). Biblical scholars traditionally attributed the Pentateuch's authorship to Moses. Later scholars consider them a compilation of at least four documents identified as J, E, P and D, dating from 900 to 400 BC and differing stylistically and theologically.

*Genesis* is the first of the five books. It has 50 chapters. Chapters 1 through 11 of *Genesis* form the introduction to the historical account (the pre-historical account). They summarize events from humanity's origin in 4004 BC to Abraham, which is the beginning of the historical account and of which no original Hebrew manuscripts have survived. The earliest extant manuscripts date from 250 BC in Greek (Septuagint or pre-Christian Greek version, referring to its 70 translators) and from 900-1000 AD in Hebrew (Massoretic text named for the Jewish compilers, the Massorenes).

The Protestant version of the Old Testament is the same as the Hebrew but divides the scriptures into 39 books. The Catholic version has 46 books. For over 1,000 years the Catholic Church included seven books (Tobias, Judith, Wisdom, Sirach or Ecclesiasticus, Baruch, and 1 and 2 Machabees) and small portions of Daniel and Esther in the Greek translation of the Old Testament. It considered these books inspired. Martin Luther (1483-1546) disagreed and removed the seven books around 1521 during the Protestant Reformation.

The New Testament written in Greek chronicles Christianity's first century (1-100 AD). The earliest extant manuscripts date from the 200s-300s AD. Both Protestant and Catholic versions of the New Testament have 27 books.

The genealogy found in the pre-historical chapters of *Genesis*, in Chapters 5 and 11, indicates the number of generations since the time of Adam and Eve. The earliest date calculated became the year that biblical scholars, such as the German astronomer Johannes Kepler (1571-1630) and the theologians Bishop John Lightfoot (1601-75) at Cambridge and Archbishop James Ussher (1581-1656) in Dublin, Ireland, assigned to the age of Earth.

| BIBLICAL GENEALOGY | |
|---|---|
| ***Name*** | ***Date (Estimated)*** |
| Adam | 4004-3074 BC |
| Eve | 4004 BC |
| Methuselah | 3317-2348 BC |
| Noah | 2948-1998 BC (flood 2348 BC) |
| Abraham | 1900-1800 BC, 2126-1881 BC or 2056-1983 BC |
| Isaac | 1956-1776 BC |
| Jacob | 1896-1749 BC |
| Joseph | 1800 BC |
| Moses | 1400-1200 BC |
| Saul | 1100-1000 BC |
| David | 1085-1015 BC |
| Solomon | reigned 971-931 BC |
| Jesus Christ | 4 BC-29 AD (from Matthew 1:17 and Luke 3:22-38) |

The long lifetimes the chronology in *Genesis* assigns to individuals, such as Adam's 930 years, Adam's son Seth's (4134-3942 BC) 912 years, and Seth's descendant Methuselah's 969 years, are artificial. The biblical authors used them to indicate the passage of time, which they knew to be longer than a few generations. They magnified the ages of individuals which was an accepted literary device of the time. The biblical authors also assigned an even number of generations between certain individuals, often 10 generations or 2,000 years. Methuselah bridged the gap between the time of creation and Noah, 2,000 years passed from the time of creation to Abraham. The years associated the great men of the past with a specific time period or age.

Kepler in 1614 used a biblical chronology and astronomy in calculating Earth's age. Three astronomical events, the single (one-time) planetary conjunction of Jupiter and Saturn in December 1603 when they were in the same celestial longitude or degree of the Zodiac and had least separation; the massing of Mars, Jupiter, and Saturn in September-October 1604 when they appeared close together in the same region of the sky, or close longitudinally; and the appearance on 10 October 1604 of a bright new star (nova) between Jupiter and Saturn led him to conclude that Christ's birth occurred in 7 or 6 BC and Earth's creation in 3993 BC.

Jupiter-Saturn conjunctions occur every 20 years. The conjunctions shift westward (clockwise) through the Zodiac, returning in 800 years to their original position with respect to the vernal equinox. Kepler associated the 800-year cycle with great historical events, such as the death of Charlemagne (742-814) and the Protestant Reformation beginning in 1521. Kepler calculated that a triple conjunction (3 conjunctions) of Jupiter and Saturn had occurred in 7 BC in May, October and December; a massing of Mars, Jupiter, and Saturn had occurred in 6 BC; and the nova that appeared in 6 BC were the three heavenly signs that sent the Magi to Bethlehem to find the newborn king Jesus. A new star appeared at the time of the single conjunction in 1603 and massing in 1604 just as the new star of Bethlehem appeared at the time of the triple conjunction in 7 BC and massing in 6 BC.

Lightfoot in *A Few and New Observations upon the Book of Genesis* published in 1642 and in *The Harmony of the Four Evangelists* published in 1644, used the Julian calendar of 12 months, each having 30 days, genealogies, and history, and calculated that Earth's creation occurred on the autumnal equinox, Sunday, 12 September 3928 BC. Adam and Eve's creation occurred about the third hour of the day or nine o'clock in the morning, on Friday, 17 September 3928 BC. Both Lightfoot and Ussher chose the autumnal equinox rather than the vernal equinox because the Jewish year began in autumn. They chose Sunday as the first day of the six days of creation because Saturday, the Sabbath, was the Jewish traditional day of rest.

Ussher in his 1650 publication, *Annales Veteris Testamenti a Prima Mundi Origine Deducti* (*Annals of the Old Testament, Deduced from the First Origins of the World*, 1658), using the Julian calendar, calculated the creation of Earth in 4004 BC. He worked backward through the Hebrew *Book of Judges*, which spanned a 330-year period, dated Abraham's arrival in Canaan in 2126 BC, estimated that Joshua conquered Canaan in 1451 BC, and used the ages of pre-flood biblical characters to calculate the time from creation to the flood. The beginning of time, by which Ussher meant the creation of Heaven and Earth, started at sundown on Saturday 22 October 4004 BC. The next day, Sunday 23 October, was the first Sunday after the autumnal equinox. On the sixth day, Friday 28 October, God created Adam and then Eve.

Ussher relied on history, chronologies, and genealogies beginning in *Genesis* 11:10, analysis of astronomical and calendrical cycles, and the traditional belief that 6,000 years had to pass between Earth's creation and the Second Coming. The 6,000 year interval refers to the Bible's *2 Peter 3:8* where "one day is with the Lord as a thousand years, and a thousand years as one day." The time between Earth's creation and the universal flood was 1,000 years. He believed that Solomon completed his temple 3,000 years after Earth's creation and that Jesus appeared exactly 1,000 years later, or 4,000 years after Earth's creation. Ussher then added four years to 4000 BC because Herod (73-4 BC), king of the Jews, was alive at the time of Jesus's birth. Ussher acknowledged that his dates were provisional and depended on the textual readings.

Bishop William Lloyd (1627-1717) in Winchester, England, in 1701 added the creation date of 4004 BC to the King James Version of the Bible at the beginning of *Genesis*, where it remained until the 1890s-early 1900s. Cambridge University Press removed the date in 1900, Oxford University Press in 1910. The creation date no longer appears in any Bibles.

***Archeological evidence***

The nineteenth-century deciphering of Egyptian hieroglyphics found on the Rosetta Stone and of Babylonian cuneiform, the discovery of a Ramses II statue and the foundation on which it once rested, provided new information for archeologists to use in establishing Earth's age. Captain Pierre Bouchard (1776-1832), a French military engineer with Napoleon's troops in Egypt, in August 1799 discovered the Rosetta Stone, an irregularly shaped somewhat damaged black basalt stone 118 cm high, 77 cm wide, 30 cm thick weighing 773 kg (1,700 pounds). His engineers were making repairs to Fort Julien near Rosetta (Rashid) in the Nile delta, about 30 miles from Alexandria, and found the stone lying on the ground, although it may have been part of a wall undergoing demolition.

The triple inscription that priests in Memphis wrote was from bottom to top in ancient Greek; Egyptian demotic, a cursive script that evolved from hieroglyphics and dates from 643 BC; and Egyptian hieroglyphics, developed in 3000 BC during the First Dynasty. The inscriptions praised Ptolemy V Epiphanes (reigned ca. 204-181 BC) because he remitted their dues. The Rosetta Stone, which the Memphis priests erected in 195 BC, passed into British hands, and to the British Museum in London in 1802 following Napoleon's defeat and the French surrender of Egypt in 1801.

Thomas Young (1773-1829), a London physician, in 1818 and Jean-Francois Champollion (1790-1832), a French linguist at Grenoble and later at the College of France in Paris, in 1821-22 deciphered the Egyptian hieroglyphics by comparing the script of the three writings. They knew the Greek and demotic scripts, and assuming each text contained the same passages, they deciphered the hieroglyphics. Young discovered in 1814 that the direction in which the birds and animals in the pictorial script faced revealed the direction in which to read the hieroglyphics. He found that the Egyptians wrote any significant word, such as the Pharaoh Ptolemy I Soter (367-283 BC) and his wife the Ptolemaic queen Berenika (Berenice), whose names appeared in the Greek script, within an oval. The French called this enclosed hieroglyphic a *cartouche* (French for cartridge). Sneferu (reigned 2613-2589 BC), the first pharaoh of the Fourth Dynasty and Egypt's greatest pyramid builder, introduced the oval to replace the older rectangular serekh that enclosed one or more hieroglyphics.

Believing that Ptolemy's name had roughly the same pronunciation in any language, Young matched the letters of Ptolemy's name with the hieroglyphics in that cartouche and then correlated each hieroglyphic with the phonetic sound that he assigned to each letter, P,T,O,L,E,M,Y. He repeated his matching strategy on another cartouche containing Berenika's name and identified each sound of its hieroglyphics. Young applied a method of spelling out names phonetically because the Ptolemaic Dynasty was of Macedonian and not of Egyptian descent. Its names were foreign, and the Egyptian scribes therefore had to spell out each name because the name had no symbol in the collection of hieroglyphics and then assign the sound a symbol. He was not sure his method applied to traditional Egyptian names.

Champollion in 1822 examined cartouches that contained traditional Egyptian names and demonstrated that Young's phonetic spelling out also deciphered the traditional names. This established beyond doubt the validity of Young's idea that the hieroglyphics were phonetic and corresponded to sounds, not symbols or allegories and were the root of current Egyptian Coptic script. Champollion continued Young's work and found that some of the hieroglyphics were alphabetic, some syllabic, and some determinative, representing the whole idea or object previously expressed.

He studied a very old cartouche with four symbols. The first symbol was a circle with a dot in the center and appeared to represent the Sun which the Egyptians called Ra. The second symbol resembled the letter m. The third and fourth symbols were the same and Champollion assumed each was an s. The cartouche spelled the name Ramses which proved correct. Champollion compared the cartouche for Ptolemy with an unknown nine symbol cartouche that contained some of the same symbols found in the Ptolemy cartouche, such as the symbol for p (a square), σ (as in a noose), and l (lion), and identified the nine-symbol cartouche as spelling Cleopatra. Champollion concluded from his study of the three inscriptions that Egyptian scribes had translated from the Greek text to the hieroglyphics text and not the reverse as previously thought.

Champollion died unexpectedly of a stroke only ten years after his major breakthrough. His deciphering of the hieroglyphics enabled later researchers to establish the chronology of ancient Egypt, and provided independent confirmation of biblical events and time periods, such as the *Exodus* account of Moses and the Israelis' escape from slavery in Egypt.

| AN EXAMPLE OF PHONETICS APPLIED TO HIEROGLYPHICS | | |
|---|---|---|
| ***Spelling*** | ***Sound*** | ***Hieroglyphic*** |
| Belief | Bee<br>Leaf | Symbol for a Bee<br>Symbol for a Leaf |

An inscription glorifying the reign of Darius I (reigned 522-486 BC), king of the Persians, provided the clue to deciphering cuneiform (Latin *cuneus*, wedge), a script that the Sumerians invented and used about 3000 BC. Darius had the inscriptions, 15 meters high 25 meters wide, carved in 519 BC in the Old Persian Empire's three official languages, Old Persian, Elamite, and Babylonian (Akkadian), on a limestone cliff face 100 meters above a main road at Behistun, Persia (Iran). Henry Rawlinson (1810-95), a British army officer and government official who spent the years 1833-39 in Persia organizing the Shah's army, began serious study of the inscriptions in 1835. Rawlinson, very likely a skilled mountain climber, had to scale the cliff to copy the Old Persian inscription. When he returned to Persia in 1843 he copied and studied the less accessible Elamite and Babylonian inscriptions.

Rawlinson suspected the three inscriptions contained the same message. He recognized the word for Darius and had available about one-third of the alphabet-based Old Persian syllabary that the German cuneiform expert Georg Friedrich Grotefend (1775-1853) in Hanover deciphered beginning in 1802. Rawlinson also found that the first section of the text that Darius had written, contained a list of Persian kings (Darius's genealogy) identical to a list found in Herodutus's *Histories*. By matching the names and the characters, Rawlinson in 1837 established the meaning of the 42 signs used in the Old Persian alphabet. He published his results in 1846.

Neils Ludvig Westergaard (1815-78), a Danish linguist, in 1844 and Edwin Norris (1795-1872), an English philologist, in 1852 began the successful deciphering of the Elamitic's 131 characters. Elamite is a dead language whose cuneiform contains 96 syllabic signs, 16 logograms, and 5 determinants with no relation to a spoken language. The agreement of linguists since 1777 that the three cliff face inscriptions repeated the same text word for word made the deciphering much easier. Rawlinson built on the work of earlier French and Irish scholars and in 1852 succeeded in deciphering the 500 characters of the Babylonian inscription. Their combined efforts enabled scholars for the first time to read the cuneiform on clay tablets, such as those uncovered in the excavations at Nineveh, the magnificent Assyrian capital, located in modern Iraq.

By 1847 archeologists had discovered many of the statues that the Egyptian pharaoh Ramses II (1304-1237 BC) dedicated to himself. One of them was a 90-ton red granite statue, originally 13.5 meters (45 feet) high and of which 10.5 meters remained. The Italian sea captain Giovanni Caviglia (1770-1845) discovered the statue near Memphis in 1820. Its foundation, excavated in 1859, lay 9 feet or 108 inches below the surface. Knowing that Ramses II lived about 3,200 years ago, indicated that the sediment covering the foundation had accumulated at the rate of 3.4 inches per century (108/3200 inches per year x 100 = inches per century). This enabled archeologists to estimate the age of other nearby buried objects, such as fired bricks found 40 feet (480 inches) below the surface that corresponded to an age of 13,700 years.

This 1847 illustration by Joseph Bonomi (1796-1878) of the British Museum in London shows the Ramses statue as it lay face down, having fallen from its buried pedestal. About five years later, excavations established the thickness of river-laid sediment about its base. The Pyramids of Gizeh appear in the distance at the left and the Nile River is near the horizon at the right. Two observers, one standing and one sitting, indicate the size (45-42 feet) of the statue.

**Figure 1**
**The Colossal Statue of Ramses II at Memphis**
Source: Carl O. Dunbar. *Historical Geology*, 2d ed. New York: John Wiley &Sons, 1949, 1960.

***Salinity and thermodynamic methods of dating Earth's age***

John Joly (1857-1933), a physicist at Trinity College, Dublin, following a suggestion Edmund Halley (1656-1742) made in 1715, used the salinity of oceans, which is nearly the same for all oceans, to measure Earth's age. In his studies, Joly determined the present salt concentration in the oceans and used salt traps to measure how much salt the oceans collected annually from salt runoff flowing into the oceans. If the oceans had no salt in the beginning, and if all the salt in the oceans came from salt runoff, assumptions that Joly acknowledged he could not verify but were as possible as any other assumptions, he estimated how many years the oceans needed to reach their present salt concentration. Using a uniformitarian argument of regular and continuous salt deposits over a long time, Joly in 1899 calculated 80 to 90 million years for Earth's age and shortly afterwards increased Earth's age to 100 million years. Beginning in 1908 Joly used radioactive dating to calculate Earth's age, concluding with Ernest Rutherford in 1913 that the Devonian period occurred 400 million years ago.

Another physicist, William Thomson, Lord Kelvin (1824-1907), at Glasgow, improved on the earlier thermodynamic calculations that Newton and Buffon used to calculate Earth's age. In studies from 1862 to 1897 Kelvin measured Earth's rate of heating and cooling, arguing that gravitational contraction that converted kinetic energy (energy of motion) to thermal energy had generated almost all of Earth's heat, with a small contribution coming from the Sun.

Two laws, a law that the mathematician Joseph Fourier (1769-1830) at Grenoble in 1807-10 developed to account for the conduction of heat in solids, and the second law of thermodynamics, first proposed in 1850, enabled physicists to calculate how Earth radiated heat into the surrounding colder space. The two laws showed that the amount of heat flowing out of Earth as it cooled decreased with time, and that Earth lost considerable heat and cooled rapidly early in its history because of the large temperature difference between a hot Earth and cold space. Starting with these premises, measurements of the heat capacity of rocks and of Earth's temperature variations with depth from Earth's surface, Kelvin in 1862 calculated that the Sun was 500 million years old and that Earth solidified from a molten state 98 million years ago. In his final calculation in 1897, Kelvin lowered his estimates of Earth's age to 20-40 million years old, and because of Earth's young age rejected Darwin's theory of natural selection.

Darwin and others had speculated that Earth was possibly 300 million years old, or many times older than Kelvin's 20-40 million years, in order for Earth and its species to have evolved to their present state. Kelvin's quantitative calculations indicated a young Earth with insufficient time for different species to have evolved.

Thomas Huxley (1825-95), a naturalist at the Royal College of Medicine in London and an early critic of Kelvin's calculations, argued in his 1869 presidential address to the Geological Society of London that Kelvin made false assumptions in calculating his cooling rates. Kelvin's error and the rejection of his cooling rates came much later however, after the discovery of radioactivity in 1896 and Rutherford's experimental proof in 1904 that the radioactive decay of naturally-occurring elements, such as uranium present in Earth, released heat. Eugene Péligot (1811-90), a chemist at the Conservatory of Arts and Measures in Paris, first isolated uranium in 1841 by reducing anhydrous uranium (IV) chloride ($UCl_4$) with potassium. Uranium's radioactive decay, however, remained unknown to scientists until the discovery of radioactive elements and radioactivity in 1896. Rutherford's radioactive decay experiments a few years later year in 1904 confirmed that Kelvin erred by not accounting for the heat released by the radioactive decay of uranium ($^{238}_{92}U$) present in Earth. The decay process heated Earth and made it appear warmer and therefore younger than its true age.

Recent experiments that the Japanese physicist Itaru Shimizu at Tohoku University conducted from 2002 to 2009 have established how much of Earth's heat comes from radioactive decay. Shimizu used neutrino sensors buried deep inside Mount Ikenoyamato near the town of Kamioka to detect the number of geoneutrinos released from the radioactive decay of uranium-238 and thorium-232. His sensors counted 106 neutrinos resulting from radioactive decay. They generated about 20 terawatts, and account for 54 percent of the heat flow from Earth's interior. Radioactive decay also accounts for Earth's underground deposits of lead and helium. Earth's cooling (primordial heat loss) accounts for the remaining 46 percent of heat flow from its interior. These latest experimental results agree with previous estimates of radiogenic heat loss. The combined radiogenic and primordial heat loss show that Earth is presently cooling at the rate of 100° C every 1 billion years.

Other nineteenth-century physicists, such as Hermann von Helmholtz (1821-94) at Berlin in 1854, and in 1893 Simon Newcomb (1835-1909), superintendent of the Nautical Almanac Office in Washington, DC, and mathematics and astronomy professor at Johns Hopkins, estimated Earth's age at 100 million years. They calculated the time it took for a nebular cloud of gas and dust to condense gravitationally to form a new star the present size of the Sun, and assumed that Earth had not existed before the Sun. Their 100 million year age became Earth's accepted upper age limit prior to the introduction of radioactive dating methods in the early 1900s.

### ***Radioactive dating methods***

#### *Uranium dating*

Recognition of radioactive dating began in 1905 when Rutherford suggested in the Silliman lectures at Yale that uranium's radioactive decay gave measurable helium-uranium and lead-uranium ratios and provided a new way of dating the age of rocks and the age of Earth. Already in a 1900 paper, Rutherford and the chemist Robert Owens (1870-1940) at McGill University in Montreal, had studied thorium oxide's (thorium dioxide, $ThO_2$) decay rate. He enclosed a 1 mm thick layer of thorium oxide inside a narrow rectangular paper container constructed of two thin sheets of writing paper (A) and placed the thorium oxide inside a horizontally-arranged long metal tube (B). Rutherford attached one end of the tube to a large horizontally-arranged insulated cylinder (C) that had a small number of holes at its ends. The holes allowed air to pass by an insulated electrode (D) fixed inside the cylinder. He connected the insulated electrode to an electrometer and grounded (connected to Earth) the electrometer, connected the cylinder to the positive terminal of a 100-volt battery and grounded (connected to Earth) the negative battery terminal. A slow current of air passed through the metal tube

(B) and by the thorium oxide. The air current carried the gaseous radioactive particles resulting from thorium oxide's decay to the cylinder (C) where they ionized the air and set up an electric current between the insulated electrode (D) inside the cylinder and the cylinder that served as the positive electrode.

Rutherford observed that the current, indicated by the deflection of the electrometer's needle, decreased geometrically with time and was proportional to the intensity of the radiation the gaseous radioactive particles, which Rutherford called thoron (radon, $^{220}_{86}Rn$ ), emitted. He found that it took one minute for the intensity of the radiation from thorium oxide to fall to one-half its value. Its present experimentally-measured half-life is 52 seconds.

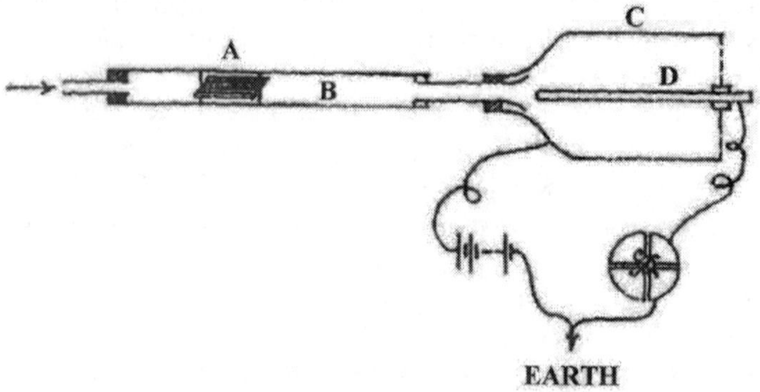


**Figure 2**
**Thorium Oxide Decay**
Source: Ernest Rutherford. "A Radioactive Substance Emitted from Thorium Compounds." *Philosophical Magazine* 49 (January 1900).

Following the thorium oxide decay rate experiments, Rutherford and his graduate student Harriett Brooks (1876-1933) in studies from 1902 to 1905 at McGill showed that the uranium decay products, RaB and RaD, were likely lead ($^{214}_{82}Pb$, half-life 26.8 minutes and $^{210}_{82}Pb$, half-life 20 years). To indicate the time it took for RaB and RaD or for any radioactive element to lose one-half of its radioactivity, Rutherford in 1902 introduced the term half-life. The uranium dating method, he concluded, depended on two factors: the calculated decay rate and the decay reaction reaching equilibrium so that the quantitative weight ratio of the elements present equaled the ratio of their decay rates.

Two years after Rutherford's Yale address, Bertram Boltwood (1870-1927), a physicist at Yale, began uranium dating experiments. Considerable confusion existed in the new field of radioactivity at that time, particularly in identifying whether the decay products were new elements or different forms (isotopes) of existing elements, and in discovering the paths of decay. In less than 20 years after the discovery of radioactivity, scientists largely had resolved the identity problem and the paths of radioactive decay.

Pierre (1859-1906) and Marie Curie (1867-1934) in Paris in 1898 found that pitchblende, a uranium compound about 80 percent uranium dioxide, $UO_2$, contained two new elements that they identified as polonium ($^{210}_{84}Po$) and radium ($^{226}_{88}Ra$). Rutherford and Owens at McGill in 1899 found that thorium compounds emitted a radioactive gas that Rutherford in 1900 named thoron. William Crookes (1832-1919) in London showed in 1900 that uranyl nitrate, $UO_2(NO_3)_2.6H_2O$, contained a radioactive component he called uranium $X_1$ (thorium, $^{234}_{90}Th$). Friedrich Dorn (1848-1916) at Halle, repeating the Curies' 1899 experiments, in 1900 observed another radioactive gas that radium emitted which he called niton ($^{222}_{86}Rn$). Rutherford and Frederick Soddy (1877-1956) in 1902 discovered still another radioactive emanation from thorium that they named thorium X ($^{224}_{88}Ra$). In 1904 Rutherford and Owens observed a radioactive emission from radium ($^{219}_{86}Rn$) that they named actinon, An.

| RADIOACTIVE DECAY | | | | |
|---|---|---|---|---|
| ***Reactant*** | ***Product*** | ***Half-Life*** | ***Discoverer*** | ***Year*** |
| $UO_2$, Uranium dioxide | $^{210}_{84}Po$, Polonium<br>$^{226}_{88}Ra$, Radium | 138.4 days<br>1,626 years | Marie and Pierre Curie | 1898 |
| $ThO_2$, Thorium dioxide | $^{226}_{86}Rn$, Radon | 52 seconds | Rutherford and Owens | 1899-1900 |
| $UO_2(NO_3)_2 \cdot 6H_2O$ Uranyl nitrate | $^{234}_{90}Th$, Thorium | 24.1 days | Crookes | 1900 |
| Radium | $^{222}_{86}Rn$, Radon | 3.8 days | Dorn | 1900 |
| Thorium | $^{224}_{88}Ra$, Radium | 3.64 days | Rutherford and Soddy | 1902 |
| Thorium | $^{219}_{86}Rn$, Radon | 3.9 seconds | Rutherford and Owens | 1904 |

In the United States, Boltwood's 1906 studies on thorium compounds, those of Herbert McCoy (1870-1945) and William H. Ross (1875-1947) at Chicago, and of Armenian-born Haroutune Dadourian (1878-1974) at Yale that same year provided additional information on the paths of radioactive decay. Boltwood, and McCoy and Ross measured the total radioactivity of several thorium minerals, such as organite and monazite. They found that after deducting the amount of radioactivity produced by the small amount of uranium present, the remaining radioactivity and the radioactive components in the minerals resulted from thorium's radioactive decay. Dadourian measured the amount of radiation emitted by thorium minerals and salts of known composition and found that the radiation depended on their thorium content.

With this information, Boltwood calculated the lead-uranium ratio in 10 uranium compounds collected from around the world and concluded from the value of the ratio that it had taken 410 million to 2.2 billion ($4.10 \times 10^8$ to $2.2 \times 10^9$) years for the amount of lead to reach the amount found in the uranium compounds. Earth was possibly 2.2 billion years old. Boltwood in 1907 discovered ionium, which was not a new element as he first suspected, but an isotope of thorium, $^{230}_{90}Th$ or Th-230, resulting from the decay of uranium. Th-230 is also unstable with a half-life of 80,000 years and slowly decays into radium. Boltwood spent the academic year 1909-10 with Rutherford in Manchester. He suffered a breakdown in 1924 caused by overstrain and in 1927 committed suicide.

## SIMPLIFIED CALCULATION OF URANIUM-LEAD DATING METHOD

In one year, 1 gram of a uranium compound (U) decayed to give $1.32 \times 10^{-10}$ grams of lead (Pb).

In one year, $U$ grams of uranium will decay to give $U \times 1.32 \times 10^{-10}$ grams of lead.

In $t$ $years$, $U$ grams of uranium will decay to give $t \times U \times 1.32 \times 10^{-10}$ grams of lead.

The amount of lead formed is then $Pb = t \times U \times 1.32 \times 10^{-10}$ grams.

Rearranging the equation to solve for $t$ gives $t = \frac{Pb}{U} \times 7.6 \times 10^9$ years.

The experimentally determined ratio of grams of lead to grams of uranium Pb/U is $5 \times 10^{-2}$, and the age of the sample is

$$t = 5x10^{-2} \times 7.6 \times 10^9 \text{ years},$$

$$t = 38x10^7 \text{ years},$$

$$t = 3.8x10^8 \text{ years}.$$

The English geologist Arthur Holmes (1890-1965), then at Imperial College in London, using mainly Boltwood's measurements for the lead-uranium ratio and an improved value for uranium's rate of decay, in 1911 calculated an age of Earth. His ages ranged from 340 million years based on a sample from the Carbonaceous age to 1.640 billion years for a sample from the Precambrian age.

Both Boltwood and Holmes obtained inaccurate values for several reasons, although some of Holmes's values were within 20 percent of currently accepted ages. First of all, in 1911 scientists had not yet established the idea of isotopes, or atoms of the same element that had different atomic weights. They introduced and clarified the meaning of isotopes only in 1913, after about a decade of confusion. Secondly, scientists had not established the sequence or the identity of the products formed in uranium's decay to lead (U → Th →Ra → Rn→ Po →Pb), nor did they know that thorium decayed to give lead. Finally, they did not know whether all the lead present in the rock samples had come from the radioactive decay of uranium or thorium, or existed naturally or independently of any radioactive decay, and that the rock samples, the Precambrian being the oldest, had formed at the same time as Earth. Despite having to work with several disadvantages, in 1927 the astronomer Henry Russell (1877-1957) at Princeton and Holmes, using uranium-lead ratios, independently calculated ages of 4 billion years and 1.6-3.0 billion years, respectively. Ten years later Rutherford calculated 3.4 billion years as Earth's age.

The most significant breakthrough on the study of isotopes occurred in 1919 when Francis Aston (1877-1945) at Cambridge invented the mass spectrograph. The mass spectrograph separated positively-charged ions (charged atoms) of different elements and isotopic ions of the same element according to their atomic masses. An electric field accelerated the ions, and a magnetic field deflected each ion in a parabolic path. The magnetic field bent the lighter ions more than the heavier ions so that each ion struck a photographic plate placed in its path at a different location, thereby indicating its presence in the sample. Aston studied lead's isotopes in 1927. He confirmed that the lead isotope found with uranium and produced as a decay product of uranium, U-238, had a lower atomic weight, Pb-206, than the atomic weight of naturally-occurring or ordinary lead, 207.2.

Further clarification of the lead isotope problem came in 1936 when Alfred Nier (1911-94), then a National Research Council fellow at Harvard and from 1936 at Minnesota, greatly improved the resolution of the mass spectrometers then available. The mass spectrometer, which dates from the 1918 experiments of Arthur Dempster (1886-1950) at Chicago, differed from the mass spectrograph in that it collected the separated positive ions in a metal box (often copper) as they emerged from the magnetic field. The flow of electrons present in the metal of the box produced an electron flow (electric current) in a metal wire inside the box. The electric current also neutralized the positive ions in the box. Measuring and recording the electric current with an electrometer tube amplifier indicated how many positive ions entered the metal box. The more positive ions entering the metal box, the greater the current required to neutralize the positive ions. A vacuum pump removed the neutralized ions from the metal box.

Nier's 180° mass spectrometer of 1936 enabled him to measure accurately the isotopic composition of uranium and lead (and other elements) and thereby establish their atomic weights. Over the next decade, specifically in 1940 and 1947, Nier developed a far more compact mass spectrometer called the 60° magnetic sector field mass spectrometer. Its simpler design, lower energy consumption, and high resolution revolutionized mass spectrometry and made the mass spectrometer one of science's most widely used analytical instruments.

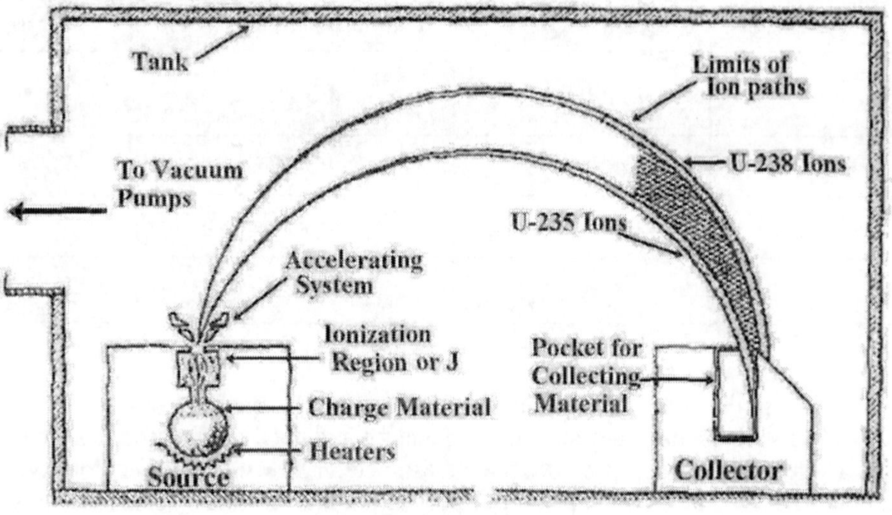


**Figure 3**
**Alfred Nier's Mass Spectrometer**
Public domain

With his improved mass spectrometer, Nier in 1938 experimentally determined the percentage of the lead isotope Pb-204, lead that experimental evidence suggested was not a product of radioactive decay and already present in the samples tested, and the percentages of the lead isotopes, Pb-206, Pb-207, and Pb-208. The three heaviest lead isotopes were the end products of the decay of the natural radioactive isotopes U-238, U-235 (Ac-227), and non-naturally occurring Th-232, respectively. The percentages of the four lead isotopes that he calculated in lead compounds were Pb-204: 1.26 and 1.48 percent; Pb-206: 23.59 and 27.31 percent; Pb-207: 22.64 and 20.00 percent; and Pb-208: 52.29 and 51.43 percent. Other later investigators obtained essentially the same results.

Nier then calculated the ratios of Pb-204 to the other three lead isotopes, Pb-206/Pb-204, Pb-207/Pb-204, and Pb-208/Pb-204, and found that they varied in different samples. The ratios depended on the relative amounts of uranium and thorium they contained and on the age of the samples. A sample of lead sulfide, PbS from Ivigtut, Greenland, had the highest amount of non-radioactive occurring (primeval) lead, Pb-204. The ratio he determined for the four isotopes of his Greenland sample was 1:14.54: 14.60: 34.45.

Nier next confirmed that although the isotopic ratios in a sample of ordinary lead varied, the atomic weight of ordinary lead, which is the average weight of all its isotopes, remained constant. This occurred because the most abundant lead isotopes, Pb-208 from Th-232, Pb-207 from U-235, and Pb-206 from U-238, had the same average atomic weight as the atomic weight of ordinary lead, 207.2. As long as each of the three radioactive isotopes formed at the same rate in all lead compounds, its isotopic composition relative to any non-radioactive-occurring lead already present (Pb-204), always increased, but the atomic weight did not. Nier reasoned that if he knew the amount of non-radioactive-occurring lead Pb-204 present in the sample, he could use the isotope abundance and known decay rate (half-life) of the three radioactive lead isotopes to calculate the amount of lead resulting from the radioactive decay of Th-232 and from naturally-occurring uranium, which consisted of 99 percent U-238 and 0.7 percent U-235. From these two measurements he could make three independent calculations of the age of the sample.

| URANIUM, THORIUM, AND LEAD ISOTOPES | | | |
|---|---|---|---|
| ***Isotope*** | ***Half-Life*** | ***Stable Decay Product*** | ***Primordial Lead*** |
| U-235 | 704 million years | Pb-207 | |
| U-238 | 4.5 billion years | Pb-206 | |
| Th-232 | 14.0 billion years | Pb-208 | |
| | | | Pb-204: not formed from radioactive decay |

In 1941 Nier obtained measurements of the lowest possible ratios for Pb-206/Pb-204 and Pb-207/Pb-204, indicating the smallest amount of radioactive lead formed from U-235 and U-238 decay. These measurements provided a good estimate of how much non-radioactive naturally-occurring lead Pb-204 existed at the time of Earth's formation. Nuclear reactions in the stars probably produced Pb-204 before Earth's formation. When used in conjunction with other lead samples obtained from radioactive decay, the ratios indicated an Earth 2.57 billion years old, revised to 3.2 billion years old in 1947 or about 70 percent of Earth's most recently calculated age. Studies in 1946 by Holmes then at Edinburgh and Fritz Houtermans (1903-66) then at Göttingen gave an age of 2.9 billion years, essentially the same as Nier's age. Holmes in 1947 calculated 3.3 billion years for Earth's age.

The culmination of these studies began shortly after World War II. Harrison Brown (1917-86), a geochemist at Chicago from 1946 to 1951 and from 1952 at Caltech, proposed that the lead present in iron meteorites, Pb-204, was the lead present in the solar system at the time of its formation, or primordial lead. The quantity of lead found in iron meteorites remained unaffected by uranium decay because only a trace amount of uranium existed in iron meteorites. Calculating this amount of lead therefore indicated the amount of lead present in Earth at the time of its formation and before the production of lead from the radioactive decay of uranium.

Brown's student at Chicago, the geochemist Clair C. Patterson (1922-95) who went with Brown to Caltech, confirmed Brown's belief. Using the isotope dilution method, Patterson in 1953 measured the isotopic concentration of primordial lead, Pb-204, present in the troilite sulfide (FeS) phase of the Canyon Diablo meteorite that caused the famous Meteor Crater near Flagstaff, Arizona. It contained less than 10 parts per billion of uranium, indicating that the lead present, Pb-204, had been there from the time of the meteorite's formation and did not result from uranium decay. Because the meteorite and Earth samples, one of which, galena (PbS), contained practically no uranium or thorium and consequently no radioactive lead isotopes, originated at the same time, they had the same age. Patterson knew the amount of primordial lead, the total amount of lead, and therefore the amount of lead formed from radioactive decay, his samples contained. From the lead isotopic ratios Pb-206/Pb 204 and Pb-207/Pb 204, he calculated the age of Earth (and the meteorite), obtaining an age of 4.5 ± 0.3 billion years. He published his study in 1956.

Nier's discovery of potassium, ${}^{40}_{19}K$, the rare potassium isotope, in 1935 and his separation of isotopes with the mass spectrometer led to the postwar introduction of two other radioactive methods for dating the age of rocks. The first method, potassium-argon dating, originated in the 1937 studies of Carl von Weizsäcker (1912-2007) at Berlin and other physicists. Their studies indicated a causal relation between the under abundance of K-40 and the overabundance of argon, Ar-40 ( ${}^{40}_{18}Ar$) in the atmosphere. Nier and Lyman T. Aldrich (1917-2013) in 1948 measured the isotopic composition of argon obtained from potassium-containing minerals and found an excess of Ar-40. This proved K-40 was slightly radioactive and decayed to give Ar-40. It led to the development of the potassium-argon dating method which gives effective results in the range 4.6 x $10^9$ years (Earth's age) to $10^4$ years and applicable to minerals such as muscovite (white mica, a silicate), biotite (black mica, a silicate), and glauconite (a silicate).

The introduction of the second method, rubidium-strontium (Rb-Sr) dating, occurred after Aldrich moved from Minnesota to the Carnegie Institution in Washington around 1950. The Rb-Sr dating method (${}^{87}_{37}Rb - {}^{87}_{38}Sr$), to which Aldrich contributed significantly, remains the most reliable and useful method for dating older rocks and minerals, such as muscovite, biotite, and potassium-feldspar (potassium aluminosilicate) of the Cenzoic era. It has an effective range of 4.6 x $10^9$ years (Earth's age) to $10^8$ years.

*Carbon-14 dating*

Willard Libby (1908-80), a physicist at Chicago, invented carbon C-14 ( ${}^{14}_{6}C$) dating in 1946. He received the 1960 Nobel Prize in chemistry for his invention. The method depends on the rate of formation and decay of C-14 in the atmosphere. C-14 results from the cosmic ray (proton) bombardment of naturally-occurring atmospheric nitrogen N-14 ( ${}^{14}_{7}N$) to form radioactive C-14 and the subsequent decay of C-14 to N-14.

$$ {}^{14}_{7}N \quad + \quad \underset{formed\ from\ cosmic\ rays}{{}^{1}_{0}n \longrightarrow {}^{14}_{6}C + {}^{1}_{1}H} $$

$$ {}^{14}_{6}C \longrightarrow {}^{14}_{7}N + {}^{0}_{-1}e \quad + \quad \underset{antineutrino}{\bar{\nu}_e} $$

The ratio of radioactive C-14 to non-radioactive C-12, both of which are present in the atmosphere as carbon dioxide, $CO_2$, and subsequently in a plant because of photosynthesis or in an animal that consumed the plant, remains steady throughout the plant's or animal's lifetime. Once the plant or animal dies, the C-14 / C-12 ratio decreases because the plant or animal no longer takes in C-14. The method works for any carbon-containing substance that died less than 30,000 years ago.

The traditional C-14 dating method determined the C-14 concentration resulting from the radioactive decay of C-14 to N-14 by recording with a geiger counter the number of beta particles (radiations) or electrons emitted per minute per gram of sample. It required sample sizes of at least one gram and long counting times.

A newly-formed sample of C-14 emits about 12-13.5 beta radiations per minute per gram of sample. It has a half-life of 5,568 years, which is the time it takes for one-half of the radioactive C-14 to decay to N-14. Therefore, if a plant or animal fossil emits 7.6 radiations per minute per gram of fossil, only one-half of the expected amount of C-14 is present in the fossil, and the fossil is 5,568 years old. If the plant or animal fossil emits 3.8 radiations per minute per gram of fossil, one-fourth the expected amount of C-14 is present and the fossil is twice as old, 2 x 5,568 years, and so on.

C-14 dating is effective to an age of 30,000 years. After that the amounts of C-14 are too small to permit accurate determinations (0.47-0.23 radiations per minute). An experimenter can check the accuracy of C-14 dating, because it deals with young ages, by comparing the ages calculated with other independent methods, such as tree ring dating or the written historical record. In most cases the correlation is very good.

C-14 dating has dated samples of charcoal, wood, bone, soil, hair, pottery, and avian eggshells and in 1951-54 resolved the Mayan Long Count cycle chronology used to establish an accurate history of Mayan society. Stonehenge samples gave 9.7 disintegrations per minute, indicating an age of about 4,900 years. A cypress beam found in the tomb of the Egyptian pharaoh Sneferu, in the reddish sandstone-limestone or North Red Pyramid in Dashur, gave 8 disintegrations per minute. This indicates that the tomb has an age of about 5,500 years, whereas the accepted date of the tomb is 2600 BC. Artifacts from Folsom man in New Mexico gave 4.5 disintegrations per minute corresponding to an age of about 9,800 years. Giant sloth dung recovered from a cave near Las Vegas Nevada, gave 4.2 disintegrations and an age of 10,500 years.

A newer and faster C-14 dating method called the Accelerator Mass Spectrometry Method (AMS) originated in the 1939 helium isotope experiments of the physicists Luis Alvarez (1911-88) and Robert Cornog (1912-98) at Berkeley, but became widely used beginning only in 1977. AMS directly detects C-14 atoms in a sample rather than waiting for them to decay, and can detect C-14 atoms in samples having C-14 / C-12 ratios as low as $1.2 \times 10^{-12}$ and containing less than 0.1 mg carbon. It has an age uncertainty as low as 0.2 percent or 20-30 years.

AMS investigations at the University of Arizona have confirmed the date of the Dead Sea Scrolls. The Dead Sea Scrolls are the approximately 400 fragments of parchment and papyrus manuscripts of Old Testament books. Bedouins discovered the scrolls in 1947-52 in caves around the Qumram Valley and Judean Desert on the northwest coast of the Dead Sea, a 550-square mile salt lake on the Israel-Jordan border. The Essenes, an ascetic Jewish sect, wrote the scrolls and hid them in caves to keep them from their Roman conquerors. Paleography, or analysis of writing style, the method used originally to date the scrolls, placed their writing at 150-30 BC. AMS calibrations assigned a date of 150 BC to 80 AD. The 230-year date range resulted because of the shape of tree-ring calibration curves for the period 200 BC-1 BC.

## III. HOW OLD IS HUMANITY

As the universe evolved over billions of years, so did humans, although their evolution occurred over a comparatively shorter time period of millions of years. A summary of human development appears in the accompanying table.

| CHRONOLOGY OF HUMANKIND | | | |
|---|---|---|---|
| ***Name and Age*** | ***Discoverer When and Where*** | ***Physical Dimensions*** | ***Comments*** |
| Ardipithecus ramidus<br>4.4 million years ago | Gen Suwa and Tim White in 1992 in Aramis, Ethiopia. | Height: 1.2 m (4 ft)<br>Weight: 45.5 kg (100 lb)<br>Brain Size: 300-350 $cm^3$ (18.3-21.4 $in^3$) | This is the earliest hominid-ape form, chimpanzee-like, bipedal. |
| ***Australopithecus genus*** | | | |
| Australopithecus anamensis<br>4-3.2 million years ago | Bryan Patterson (1909-79) in 1965 in Kanapoi, Kenya. | Height: 1.52 m (5 ft) male<br>1.07 m (42 in) female<br>Weight: 54-68 kg (120-150 lb) male<br>27 kg (60 lb) female<br>Brain Size: unknown but small | A pre-human form that later discoveries have shown may have been bipedal. |
| Australopithecus afarensis<br>3.4 million years ago<br><br>Lucy<br>3.2 million years ago | Donald Johansson (b. 1947) in 1973 in Hadar, Ethiopia<br><br>Donald Johansson and Tom Gray in 1974 in Hadar, Ethiopia. | Height: 0.92-1.35 m (36-53 in)<br>Weight: 30.5-80.5 kg (67-177 lb)<br>Brain Size: 310-485 $cm^3$ (19-29.7 $in^3$)<br><br>Height: 1.07 m (42 in)<br>Weight: 28 kg (62 lb) which was small for her species<br>Brain Size: 400 $cm^3$ (24.5 $in^3$) | Johansson and Gray found about 40 percent of Lucy's skeleton, including her pelvis, femur, and tibia, showing that she was bipedal and female, and had lived about 25 years. |
| Austrolopithecus africanus<br>2-3 million years ago | Raymond Dart (1893-1988) in 1924 in Taung, South Africa. | Height: 1.37 m (54 in) male<br>1.14 m (45 in) female<br>Weight: 41 kg (90 lb) male<br>27 kg (60 lb) female<br>Brain Size: 440 $cm^3$ (26.8 $in^3$) adult | Dart was an Australian anatomist and paleoanthropologist. The fossil he discovered was of a 3-year old child. Africanus was bipedal. |
| Australopithecus aethiopicus<br>2.5 (2.7-2.3) million years ago | Alan Walker at Penn State, a co-worker of Richard Leakey (b. 1944), in 1988 near West Turkana, Kenya. | Height: 1.37 m (54 in) male<br>1.09 m (43 in) female<br>Weight: 40-49 kg (88-108 lb) male<br>32-34 kg (70-75 lb) female<br>Brain Size: 410 $cm^3$ (25.1 $in^3$) | |
| Australopithecus robustus<br>1.5-2.0 million years ago | A schoolboy, Gert Terblanche, at Swartkrans, South Africa, in 1938. | Height: 1.5 m (59 in)<br>Weight: 45 kg (50-70 lb)<br>Brain Size: 400 $cm^3$ (500-600 $cm^3$), 24.5 $in^3$ (21-27 $in^3$) | |
| Australopithecus bosei<br>1.8 (2.3-1.8) million years ago | Mary Leakey (1913-96) in 1959 at Olduvai Gorge, Tanzania. | Height: 1.4 m (55 in) male<br>1.1 m (43 in) female<br>Weight: 49 kg (108 lb) male<br>34 kg (75 lb) female<br>Brain Size: 530 $cm^3$ (500-600 $cm^3$), 32.5 $in^3$ (31-37 $in^3$) | Mary Leakey was a British paleoanthropologist. *Bosei* was the first hominid accurately dated using K-Ar (potassium-argon) dating. Leakey's son, Richard Leakey, discovered a complete intact *bosei* cranium that lacked only teeth. |

| CHRONOLOGY OF HUMANKIND | | | |
|---|---|---|---|
| ***Name and Age*** | ***Discoverer When and Where*** | ***Physical Dimensions*** | ***Comments*** |
| | ***Homo genus*** | | |
| Homo habilis (handyman, skillful) 2.2-1.6 million years ago | Louis Leakey (1903-72) in 1959 in Olduvai Gorge, Tanzania. | Height: 1-1.5 m (39-59 in)<br>Brain Size: 750 $cm^3$ (46.5 $in^3$) | Louis Leakey was the son of a British missionary in Kenya, husband of Mary Leakey, and patriarch of the Leakey family of paleoanthropologists. |
| Homo erectus (Java man) 1.16-0.4 million years ago | Eugene Dubois (1858-1940) in 1891 near Trinil on the Indonesian island of Java. | Height: 2 m (79 in)<br>Brain Size: 800-1,200 $cm^3$ (49-73.5 $in^3$) | *Erectus* (Java man) controlled fire. Other anthropologists found remains in Kenya, Tanzania, China, Pakistan, and Germany (Heidelberg man). |
| Homo erectus (Peking man) | Davidson Black (1894-1934) at Choukoutien near Beijing (Peking). | Brain Size: 915-1,255 $cm^3$ (56-73.5 $in^3$) | Black discovered a molar tooth in 1927, a complete skull cap in 1929, and another skull in 1930. |
| Homo sapiens archaic (wise, early) 400,000-30,000 years ago | A laborer in 1821 discovered a fossil (a complete cranium) at Broken Hill in Northern Rhodesia, now Kabwe, Zambia. | Height: as tall as modern humans<br>Brain Size: 1,200 $cm^3$ (73 $in^3$) | Carl Linnaeus introduced the name *homo sapiens* for the human species in 1758 before any discovery and classification of the *homo* species had occurred. |
| Homo sapiens neanderthalensis (Neanderthal man) 130,000-35,000 years ago | Johann Fuhlrott (1804-77), a German high school science teacher, in 1856 found an almost complete skeleton in a cave near Düsseldorf, near the Neander River in Rhineland. | Height: 1.68 m (66 in)<br>Brain Size: 1,450 $cm^3$ (91.5 $in^3$) slightly larger than that of modern humans | Two earlier finds, one in Belgium in 1829 and one in Gibraltar in 1848, were not identified as *Neanderthal* until 1936 and 1864 respectively. |
| Homo sapiens sapiens (modern man) 30,000 years ago and continues to the present | | Height: modern humans<br>Brain Size: 1,300 $cm^3$ (82 $in^3$) | |
| Homo sapiens sapiens (Cro-magnon) | Louis Lartet (1840-99) studied the skeleton found in 1868 in the Cro-magnon cave in Vézère, Bordeaux, southern France. | Height: modern humans<br>Brain Size: 1,300 $cm^3$ (82 $in^3$) | The site contained stone tools, shells, ivory pendants, and carved reindeer antlers. |

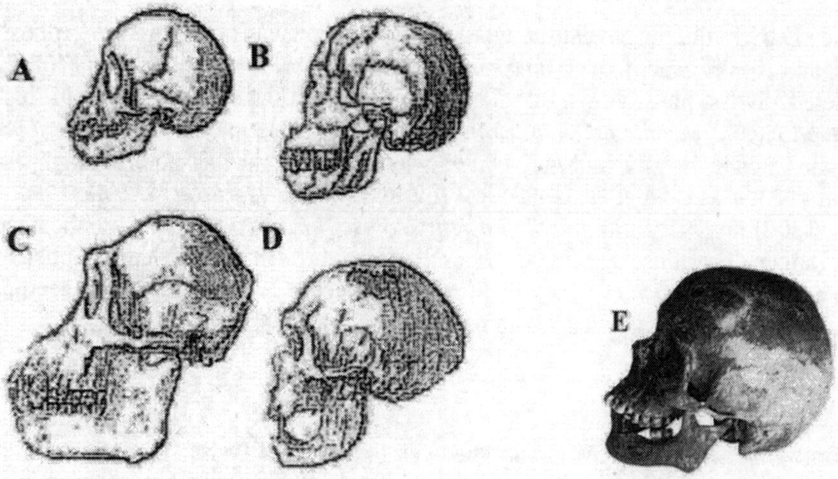


**A. Young chimpanzee skull**
**B. Piltdown Man skull**
**C. Adult male chimpanzee**
**D. Neanderthal man**

**E. Homo sapiens**

**Figure 4**
**Evolution of Humanity**
Source: William King Gregory. "The Dawn Man of Piltdown, England." *American Museum Journal* 14(1914) (left).
© David Davis, 2014. Used under license from Shutterstock, Inc. (right).

## IV. ARGUMENT FOR DESIGN IN THE UNIVERSE

***William Paley's argument***

In the half century before Darwin published his theory of evolution the argument for a God-designed universe dominated Western and Eastern social thought. Evolutionary ideas appeared in the published literature but had not achieved widespread acceptability.

William Paley (1743-1805), an Anglican archdeacon in Cambridge, gave the definitive statement of the argument for design in the universe in his 1801 classic *Natural Theology*. Paley did not originate the argument for design in the universe. The same argument appeared in the writings of Augustine, briefly in *City of God*, and Thomas Aquinas (1225-74), in *Summa Theologica*, Article 3, Question 2. It was one of Aquinas's five proofs for the existence of God. Paley's example of finding a watch on the ground led to the conclusion that a watchmaker had constructed the watch and that the parts of the watch had not come together in a working design by chance. Because the watch had a design, it must have had a designer. Paley compared the eye's intricate design with a telescope and suggested that both must have a designer. Living organisms are more complicated than watches, and only an intelligent designer could have created them. The designer is God.

Paley's argument received universal praise as convincingly demonstrating the existence of an intelligent creator. Many American colleges and seminaries in the period from 1825 to 1850 used Paley's book in their theology courses. His argument for design also appeared in science textbooks, such as Alonzo Gray's *Elements of Chemistry* (1840) used in the Teacher's Seminary in Andover, Massachusetts.

## V. EARLY THEORIES OF EVOLUTION

Charles Darwin did not invent the word *evolution*, nor was he the first to propose a theory of evolution or organic development over long time periods. Albrecht von Haller (1708-77), a well-known and respected Swiss physiologist at Göttingen, used the word *evolution* in his commentary on the *Praelectiones academicae* (*Academic Lectures*) of Hermann Boerhaave (1668-1738), volume 3, published posthumously in 1744. Haller was a pioneer in the experimental study of physiology and anatomy. Charles Bonnet (1720-93) in 1762 in *Considerations on Organic Bodies* and Erasmus Darwin (1731-1802) in 1791 in *Botanical Garden* also used *evolution* to describe the change or development of an organism. Erasmus Darwin, Georges Buffon (1707-88), and Jean Baptiste Lamarck (1749-1829) were among the first naturalists to propose a theory, or mechanism, to account for their belief that evolutionary change had occurred in nature.

### *Georges Louis Buffon*

Buffon was a naturalist in Montbard and Paris and one of the best known early evolutionists. Educated as a lawyer and physician who in 1732 inherited considerable wealth from his mother, he served as keeper of the King's Garden in Paris beginning in 1739. Louis XV (1710-74) made him a count in 1771. Buffon died of kidney stones one year before the French Revolution. His only son Georges (1764-94), who served in the army, lost his life during the revolution, possibly by the guillotine.

Buffon's widely-read major publication *Natural History*, 44 volumes, 1749-1804, included *The History and Theory of Earth*, Volume I, 1749; *The Degeneration of Animals*, Volume 14, 1766; and *Epochs of Nature*, Supplementary Volume 5, 1778.

#### *Age of Earth*

Buffon was one of the first scientists to distinguish between the age of the universe and the age of Earth, publishing his conclusion in 1749. He believed that a collision between a comet and the Sun produced Earth and the other planets. The Russian chemist Mikhail Lomonosov (1711-65) in St. Petersburg reached the same conclusion in the years 1757-63, setting the time interval at several hundred thousand years. Buffon accepted a gradual evolution of Earth's strata and an Earth that was about 75,000 years old. Like Newton, he derived Earth's age by heating and cooling spheres of different composition.

Buffon described his experiments on calculating Earth's age in *Epochs of Nature*. He heated to white heat 10 iron spheres 0.5-5 inches in diameter, other spheres that had a composition similar to Earth's composition, cooled them, and estimated how long a sphere with a radius equal to Earth's radius would take to cool. His calculations showed that Earth took 40,000 years to reach a temperature where life could exist, which he established by touching the sphere to determine when it had cooled enough; 75,000 years to reach its present temperature; and 93,000 years for it to become too cold for life to exist. In his calculations on cooling, Buffon assumed that most of Earth's heat was residual heat remaining from its original molten state and that only a small amount of its heat came from solar radiation. This is not a correct or complete explanation.

Newton described his cooling sphere method to calculate Earth's age in the *Principia* (1687). He heated a one-inch diameter iron sphere to red heat and then cooled it. Unlike Buffon, Newton rejected his calculation of an Earth 50,000 years old as too old. Instead, he accepted *Genesis's* chronology and estimated that Earth's creation occurred in 3500 BC.

Buffon, a Deist, believed the Bible was not a book of science and that the days in *Genesis* were epochs ranging from 3,000 to 35,000 years. Because God created the Sun on the fourth day, and the succession of a 24-hour day and night must occur after the Sun's creation, Buffon concluded that the biblical days were not 24 hours long. This was similar to the day-age creation that many nineteenth-century American geologists and naturalists adopted.

*Genesis* actually contains two creation accounts of Earth and heavenly bodies. The first is a very primitive description in which a primordial mass of water covers a flat-surfaced Earth, over which is a moving firmament, the sky. The Sun, Moon, and stars are fixed on the moving firmament (*Genesis* 1:7, 1:9, 1:10, 1:14-19). This is the six-day creation in which God separated light from darkness on the first day and created the Sun and celestial bodies on the fourth day, separated the waters on the second day and created the denizens (occupants) of the waters and the air that separates them on the fifth day, and created plant life on the third day and the occupants of Earth and man and woman on the sixth day (*Genesis* 1:27).

The second creation account (*Genesis* 2:4), describes only one day and contains no separation of the primordial chaos of water. God created man first (2:7), then plants (2:9), animals (2:19), and finally woman (2:22). The second account includes the fall of man which, according to biblical scholars, is the reason for its inclusion in *Genesis*.

*Degeneration of animals*

Buffon in 1766 proposed that Earth's changing environment caused species to change. He adopted a neptunist position, believing that water or floods accounted for Earth's strata changing over time. Therefore the species occupying Earth had to change in order to survive. An organism would develop those body parts that were useful for its survival, those body parts not useful or no longer used, withered and disappeared. The lateral toes of a pig, three bone splints in a horse's leg but only one hoof, remnants of hind legs on a whale, and human tail bones provided evidence of non-useful body parts that resulted in the change or evolution of species.

Succeeding generations inherited and passed on the acquired useful characteristics of preceding generations to their offspring based on the animals' internal needs acting through a process of hereditary memory. Buffon and other contemporary naturalists had no testable mechanism for the transmission of acquired characteristics to succeeding generations and relied on inheritance as the only way.

Buffon accepted the epigenesis theory, the evolution of an organism from simple to complex form, rather than preformation theory in which a tiny but already completely formed organism grew to full size. He adopted a natural classification system, one based on many characteristics, of an organism such as a plant's roots, stems, flowers, and leaves. The fertility of offspring provided the test for different species. Because a mare (female horse) and a jackass (donkey jack or male donkey) produced a sterile male or female mule, they were different species.

***Erasmus Darwin***

Darwin, a wealthy English physician in Lichfield, was the grandfather of Charles Darwin and Francis Galton (1822-1911) the eugenicist. Galton laid out the principles of eugenics, or the selective breeding of humans, in his book *Hereditary Genius* (1869) and coined the word *eugenics* in 1883 in *Inquiries into the Human Faculty*. Galton in 1888 invented the correlation coefficient used in statistical measurements, and in 1892 introduced finger printing to identify individuals in criminal investigations.

In *Zoonomia*, or *The Laws of Organic Life,* published in 2 volumes in 1794 and 1796, Darwin, like Buffon, argued that the influence of the environment and an organism's use or non-use of its body parts to adapt to a changing environment brought changes to an organism. These two factors caused species to evolve over time. Future generations inherited any characteristics acquired by their ancestors. *Zoonomia* was very popular in the United States, going through four editions before 1820. Darwin accepted Linnaeus's artificial system of classifying species, which used only a few characteristics, such as pistils and stamens, to classify plant species.

Darwin was a member of the Lunar Society, an intellectual group whose membership included Priestley, Watt, and Boulton. The rather corpulent Darwin, who enjoyed a long active intellectual and social life, died of heart disease.

***Jean Baptiste Lamarck***

A former soldier and revolutionary, Lamarck was a materialist and an opponent of Lavoisier's new quantitative chemistry. He was one of a diminishing group of universalist natural philosophers who gradually disappeared because of the increasing specialization occurring in science. Early in his career Lamarck tutored Buffon's son and served in Paris as botanist of the King's Garden from 1782 until its closing in 1790 during the French Revolution.

In *Zoological Philosophy* published in 1809, Lamarck, like Buffon and Erasmus Darwin, concluded that the influence of the environment forced an organism to use or no longer use some of its body's parts and produced change in a species. All species gained their energy from heat and electricity. An inner life force or energy drove a species to increasing size, complexity, perfection, and progress and to acquire the characteristics it needed in order to survive in a changing environment. The environment deflected the organism from its progress, which resulted in the emergence of new species. Species did not become extinct, they evolved by branching into other species.

This is how Lamarck accounted for the evolution of the long-legged and long-necked giraffe from the okapi and how the giraffe acquired its spots for protection from predators. Similarly an inner life force led to an animal's acquiring a fur coat in a cold climate. Succeeding generations then inherited the acquired characteristics. The main weakness of Lamarck's theory was that scientists had no way to test it. No evidence suggested that evolutionary changes arose in response to an organism's needs.

Linnaeus, the great eighteenth-century classifier, did not classify the invertebrates, identifying only insects and worms as classes. Lamarck identified 10 classes which he described in his seven-volume *History of Invertebrates* (1815-22) and was the first naturalist to use the terms *vertebrate* and *invertebrate*. He popularized the word *biology* in 1802 in *Hydrologie* after the German physiologist and nature philosopher Karl Burdach (1776-1847) at Königsberg introduced the new word *biology* to natural history in 1800.

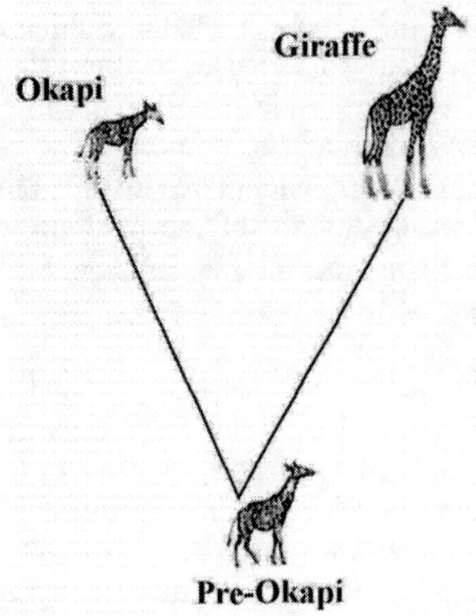


The morphological divergence of the okapi and the giraffe began long before they developed into separate species and continued long after. It began when a geographic barrier divided the pre-okapi species into two populations. Eventually, enough genetic differences accumulated to render the two groups incapable of interbreeding, thereby qualifying them as separate species. Evolution has followed both patterns, and both probably figured in the development of the okapi and the giraffe.

**Figure 5**
**Morphological Divergence of the Okapi and Giraffe**
Courtesy of the author

***Recent Lamarckians***

Paul Kammerer (1880-1926), an Austrian geneticist at the Institute for Experimental Biology in Vienna, claimed in his amphibian experiments of 1904-26 to have demonstrated that black and spotted salamanders (a small four-inch long lizard-like animal related to frogs and toads) and the midwife toad had inherited acquired characteristics that became more prevalent in each generation. The salamanders inherited a color adaptation.

In the toad experiments, Kammerer said that when he changed the breeding environment from land to water and warmed the water, the male midwife toad developed black thumb pads, or calluses, called nuptial pads. The pad's tiny spikes gave the male a better grasp of the female when mating in water, whereas when mating on dry land it did not need the pads to hold on to the female. According to Kammerer, male offspring inherited the acquired thumb pads even when he returned them to their original environment on land where they did not need the pads. Kammerer knew that prehistoric ancestors of the midwife toad had black nuptial pads but claimed his experiments demonstrated the inheritance of acquired characteristics. He allegedly bred the toads over six generations before the lineage died out.

Outside of the Institute, Kammerer, a socialist and an atheist, saw his life unravel precipitously by 1926. He survived several extramarital affairs, but World War I and its aftermath left him nearly destitute. He lost most of his specimens during the war, and in December 1922 he lost his position at the Institute for lack of funding. To earn money, he gave a series of lectures on his breeding experiments at Cambridge, England, in 1923, and in 1923 and 1924 he came to the United States to speak at Yale, Johns Hopkins, and other American universities.

Kammerer's lectures on Lamarckian inheritance of acquired characteristics evoked considerable skepticism and criticism. William Bateson (1861-1926), the Cambridge Darwinian evolutionist, geneticist, and Kammerer's longtime opponent, challenged the results that Kammerer had obtained ten years earlier. Another skeptic, Gladwyn Kingsley Noble (1894-1940), an American biologist at the American Museum of Natural History in New York City, in early 1926 obtained permission to visit Kammerer's laboratory at the Institute and examine the one remaining male midwife toad specimen that Kammerer had preserved in alcohol. Kammerer lost the others during the war. The biologist Hans Przibram (1874-1944), founder and director of the Institute who later died in a Nazi concentration camp at Terezin (Theresienstadt) in Bohemia, accompanied Noble. They were the first scientists to examine Kammerer's remaining toad.

During his examination Noble discovered that someone had injected the midwife toad's thumb pad with India ink to enhance the inherited acquired color change. Noble failed to locate the tiny spikes, and on 7 August 1926 he published an article in *Nature*, accusing Kammerer of scientific fraud. Kammerer, who left the Institute in 1922, blamed an assistant at the Institute and prepared to leave Austria to take a position in Ivan Pavlov's institute at the University of Moscow. Kammerer never left Austria. Arthur Koestler (1905-83), a Kammerer defender and author of *The Case of the Midwife Toad* (1971), wrote that on 24 September, Kammerer went for a walk on the narrow Schneeberg (Austrian) mountain path that led from Puchberg to Homberg, sat down at Theresa Rock, and shot himself in the head with a revolver.

A second Lamarckian, Trofim D. Lysenko (1898-1976), a Ukrainian agronomist at the Gandzha Experiment Station in Azerbaijan and in 1940 director of the Institute of Genetics in Moscow, began experiments on wheat production in 1929. Lysenko claimed that by moistening and refrigerating the seeds of spring wheat that wheat growers sowed (planted) in the spring and harvested in autumn, he had developed a new cold-weather-resistant winter wheat that growers could sow in autumn, thereby significantly increasing wheat production for the next spring harvest and for all future harvests. Lysenko called this process vernalization and claimed it would solve the Soviets' chronic wheat crop failure. He did not believe in genes and said that he produced the new species with its acquired inheritable characteristics by modifying the environment in which the wheat grew.

Lysenko was particularly influential under Joseph Stalin (1879-1953) and the Soviets who embraced Lamarckianism because it held that the environment, in this case communism, and not genetics determined an individual's and hence society's behavior. With Stalin's death in 1953, Lysenko's influence waned, and in 1965, the year after the ouster of the Soviet leader Nikita Khrushchev (1894-1971), Lysenko lost his position as director of the Institute of Genetics. He was primarily responsible for the decline of genetics research in the Soviet Union.

## VI. GEOLOGICAL DEVELOPMENTS: CATASTROPHISTS AND UNIFORMITARIANS

The nineteenth-century's central geological issue was to discover the forces that changed Earth's strata. Two schools of thought, catastrophists and uniformitarians, emerged to answer this question. Catastrophists believed that violent natural forces produced the changes. They did not agree among themselves on the natural force and fell into two groups: vulcanists, who claimed that heat and volcanic action produced changes in Earth's strata, and the far more numerous neptunists, who argued that water or floods had deposited the strata. Uniformitarians, as the name indicates, believed gradual natural forces produced the changes in Earth's strata.

### *Catastrophists: Vulcanists and Neptunists*

The much less influential vulcanist school that Jean-Étienne Guettard (1715-86) in Paris founded in 1752 included Sir William Hamilton (1805-65) at Dublin, John Playfair (1748-1819) at Edinburgh, and Roderick Murchison (1792-1871) in Durham and London; the Italian Abbé Antonio Moro (1687-1764) in Venice; and Henry Rogers (1808-66) at Pennsylvania later Glasgow and his brother William Rogers (1804-82) at MIT.

Abraham Werner (1750-1817) at the Mining Academy in Freiburg, founded the neptunist school in 1787. Charles Bonnet in Geneva, Buffon and Georges Cuvier in Paris promoted the neptunist viewpoint. Most Americans, William MacClure in Philadelphia, Benjamin Silliman at Yale, Amos Eaton at Rensselaer Institute of Technology, James Hall New York State geologist in Albany, and James Dwight Dana at Yale, were neptunists, who later adopted uniformitarianism.

Werner believed that a primeval ocean covered all Earth and that sedimentation laid down by the receding waters formed the continents. Strata arose in a definite succession and changed over time. Werner paid little attention to the evidence of lava flows and therefore of vulcanism in other regions of Europe. He held deistic beliefs and did not link his neptunist theory to the biblical flood.

Werner was the first to develop a complete system of geology, which he published in *Kurze Klassifikation und Beschreibung der verschiedenen Gebirgsarten* (*Brief Classification and Description of different Mountain Forms*), a 1786 article republished as a book in 1787. His most famous student, Alexander von Humboldt (1769-1859), explored the Americas, including Mexico, Ecuador, Venezuela, and Cuba from 1799 to 1804 and discovered 6,300 new plant species. Humboldt published his results in a 23-volume series *Voyages aux regions Equinoxiales du Nouveau Continent* (*Voyage to the Equinoctial Areas of the New Continent)* from 1805 to 1834.

Charles Bonnet (1720-93), a Swiss naturalist in Geneva, accepted neptunism and with neptunism acting as the natural force of change, proposed a preformation theory of organisms to account for the fossils present in Earth's strata. In *Considerations on Organic Bodies* published in 1762, Bonnet, who lost his hearing at age seven and was now nearly blind, wrote that the already-formed, or preformed, organism existed in miniature in either the egg or sperm. God put the preformed human beings there at the beginning of creation and established the precise moment for each one to come to life.

Bonnet's preformation theory implied that species remained fixed and required that he explain the obvious changes in some species and the extinction of others, as the fossils present in strata clearly indicated. According to Bonnet, a series of floods wiped out existing species, the remains of which were the fossils present in strata. The fossils were more primitive the deeper the strata because with each successive divine creation after a flood that wiped out all existing species, each species moved up the scale of perfection and the scale of time. Bonnet's belief in God's successive creations enabled him to reconcile his belief in the fixity of species with the evidence of fossil change. He published these ideas in 1770 in *Philosophical Palingesis* or *Ideas on the Past and Future States of Living Beings*.

Early in his career, in 1740, Bonnet discovered parthenogenesis, virgin birth or asexual reproduction, in aphids. In parthenogenesis, the male sperm cell does not fertilize the female egg cell. The nucleus of the unfertilized egg cell already contains the full number of chromosomes. The cell divides, just as it does in sexual fertilization, when one of the nuclei comes from the sperm. The cells continue to divide and form a look-alike early stage embryo or blastocyst.

Parthenogenesis, which gives both male and female offspring, occurs not only in aphids, but is common among invertebrates, including insects such as rotifiers, honeybees, and wasps. It also occurs in other lower forms of life, such as roundworms, flatworms, the whiptail lizard, blind snake, garter snake, farm turkey, molly (fish), and apparently sharks. Some of the whiptail lizard species, except for one snake species, are the only known vertebrates to have an entirely female species. Sharks also produce only females. Parthenogenesis occurs in many plants, such as dandelion and hawkweed. The stimuli, such as an abundance of food and a favorable environment, a lack of males, and mechanisms for parthenogenesis vary among the different kinds of parthenogenetic organisms.

Normal male-female sexual conception received an explanation a century later in the microscopic experiments of the zoologist Oscar Hertwig (1849-1922) at Jena, Germany, and the French biologist Hermann Fol (1845-92) at Geneva. They established definitively the presence of both sperm and egg in the fertilized egg. Hertwig, a neo-Lamarckian and student of Ernst Haeckel (1834-1919) at Jena, in 1875 observed sea urchin sperm entering an egg and the sperm and egg nuclei fusing. Fol in 1879 observed fertilization in the starfish. The French government in 1891 sent Fol on an expedition to the coast of Tunisia to study the distribution of sponges. Fol and a crew of two set sail on his new yacht, the *Aster*. They disappeared mysteriously. An investigation into Fol's disappearance provided no explanation.

Georges Cuvier (1769-1832), who spent the reign of terror years as a private tutor in Normandy and in 1795 held the chair of comparative anatomy at the Garden of Plants in Paris, was the foremost of the nineteenth-century neptunists. He made extensive geological investigations of the Paris Basin, founded comparative anatomy and paleontology (study of fossils), and began the practice of reconstructing extinct animals from fossil remains. His major publications included *Researches on Fossil Bones* in 1812 and *Theory of Earth* in 1815.

Cuvier believed that God created each species, they did not evolve, as Buffon, Erasmus Darwin, and Lamarck claimed, because they already were perfectly adapted. Species, such as the mastodon, became extinct because of catastrophes, and the fossils provided evidence of their extinction. Evolution, he said, contradicted moral law, the Bible, and natural science. Cuvier adopted functionalism, or a functional purpose, in the design of an organism. God gave an organism's parts a specific design because each part had to perform a specific function.

Cuvier held that the biblical flood in 3000 BC was the last of four regional floods, not universal floods, and that the Bible did not include any mention of earlier occurring floods that wiped out all life and led God to recreate life. In this way Cuvier, like Bonnet, reconciled the old age of the deepest fossils, which were clearly different in appearance from the younger fossils, with the Bible's young age of Earth. Cuvier's followers proposed 27 floods to account for the deeply buried old fossils.

***Uniformitarians: Hutton and Lyell***

The Edinburgh geologist, James Hutton (1726-97), friend of Black and Watt, founded the uniformitarian school of geologists. Hutton's observations in the 1770s-80s of Earth's strata in Scotland, England, and Wales revealed a regularity or uniformity of the various layers. They led him to conclude that the strata resulted from uniform or steady forces acting on Earth's surface and not from catastrophic disturbances. He observed that the 74-mile long Hadrian's Wall, which the Romans constructed from 120 to 123 AD across northern England to keep out the Picts and other dangerous northern tribes, remained unchanged after 1,600 years. He also noted that the angular unconformities found in the slate, shale, and sandstone formations at Siccar Point in Scotland required a very long time period. These and other observations, according to Hutton, suggested an Earth of considerable age.

Hutton published these conclusions in 1795 in the two-volume *Theory of Earth* in which he gave the fundamental tenets of uniformitarianism. Earth's surface (strata) underwent slow gradual change resulting from the action of water, heat, and wind. The same forces were acting now as they did in the past and would continue to act in the future. No evidence of their beginning or of their end existed. The processes of erosion and sedimentation accounted for the present appearance of Earth's surface.

A second Edinburgh geologist, James Hall (1761-1832), in papers published from 1798 to 1812 provided additional experimental evidence to support uniformitarianism. His experiments showed that slowly heating and cooling rocks or glass gave crystalline structures instead of the amorphous (structureless) masses produced on rapid heating, and challenged the neptunists' argument that only precipitation from water produced crystalline structures. In other experiments Hall slowly heated carbonates ($CaCO_3$, $MgCO_3$) such as dolomite and limestone, causing them to melt and not decompose. Because their melting occurred quite easily, the uniformitarian theory could account for the presence of the many carbonates in Earth's strata.

Geology continued its growth as a historical science into the nineteenth century thanks to the English geologist William Smith (1769-1839) in Somerset and later in London. In his role as a surveyor and superintendent working on the construction of canals beginning around 1791, Smith studied the fossils discovered throughout England. He found the same relative order of strata and the fossils they contained in England, Europe, the American continent, and probably in other continents, and that fossils were more primitive as the strata's depth from the surface increased.

Smith's work, including the construction of geological maps, made geology a historical study and contributed to the uniformitarian tenet of continuity and slow gradual change. He published his results in *A Delineation of the Strata of England and Wales, with Part of Scotland* in 1815, in *Strata Identified by Organized Fossils* in 1816, and in *Stratigraphical System of Organized Fossils* in 1819. The American geologist James Hall (1811-98) at the Rensselaer in Troy, New York, in 1843 confirmed Smith's conclusions.

Charles Lyell (1797-1875), a London lawyer turned geologist at King's College, provided the final evidence that enabled uniformitarianism to become the accepted theory of geological change when he published his three-volume *Principles of Geology* in 1830, 1832, and 1833. His studies of strata in France and Italy provided a thorough presentation of the uniformitarians' case and a rejection of any catastrophes as the cause of stratigraphic change. His later visits to Canada and the United States in 1841-42, 1845-46, and in 1853 supported his uniformitarian belief. Lyell gave four principles of uniformity: (1) uniformity of law, natural laws have not changed; (2) uniformity of processes, the same causes are acting now as were acting in the past; (3) uniformity of rate, changes are occurring in the present at the same rate they occurred in the past; (4) uniformity of state, Earth is essentially the same now as it existed in the past.

While in Canada and the United States Lyell examined geological sites with Benjamin Silliman (1779-1864) at Yale and the American geologist James Hall. His visits took him to Niagara Falls in October 1841 where with Hall he estimated the Falls' rate of recession at about 1 foot per year (7 miles total) and the Falls and Niagara Gorge's age at about 35,000 years old. Lyell underestimated the rate of recession which for the last 560 years has averaged about 3-4.5 feet per year (1.0-1.5 meters per year) but has since slowed to 1 foot per year, in agreement with Lyell's earlier calculation. The 7-mile erosion of Niagara Falls gives an estimated age of 12,300 years and that its formation began at the time of the last ice age about 18,000 years ago. Lyell also visited the Mississippi River delta, where he estimated the annual sediment accumulation, and to the coal formation in Nova Scotia. Largely as a result of Lyell's convincing investigations most American geologists among them Silliman, William MacClure (1763-1840), president of the Academy of Natural Sciences in Philadelphia, Amos Eaton (1776-1824) at Rensselaer, Hall, James Dwight Dana (1813-95) at Yale, and Asa Gray (1810-88) the Harvard botanist, slowly shifted their support from neptunism to uniformitarianism.

Lyell's published account of his first visit entitled *Travels in North America in the Years 1841-1842* contains his comments on southern slavery. A polygenesist who accepted multiple centers of creation occurring at different times and locations, he described slavery favorably and disagreed with the North's abolitionist propaganda. Slavery was a mild, kindly, inefficient, and uneconomical institution in which the slaves ate better than a large part of Europe's laboring class.

Lyell accepted an old Earth, about 240 million years based on fossils of marine mollusks. He went too far in rejecting catastrophes as causes of change in Earth's strata despite convincing evidence that catastrophic events, such as glaciers, ice ages, and earthquakes, had occurred throughout Earth's history.

In *Geological Evidences of the Antiquity of Man*, published in 1862, Lyell suggested that man had existed long before biblical times and had evolved over time. The discoveries of Jacques Boucher de Perthes (1788-1868), an amateur paleontologist and customs director in Abbeville, France, published in 10 volumes from 1846 to 1857, convinced Lyell of man's antiquity. Beginning in 1825 Boucher de

Perthes' excavations from gravel pits at Saint-Acheul on the banks of the Somme River uncovered deeply buried crude manmade stone axes and tools that were very likely 300,000 to 1,000,000 years old. A few years later, in 1864, Lyell adopted Darwin's theory of evolution.

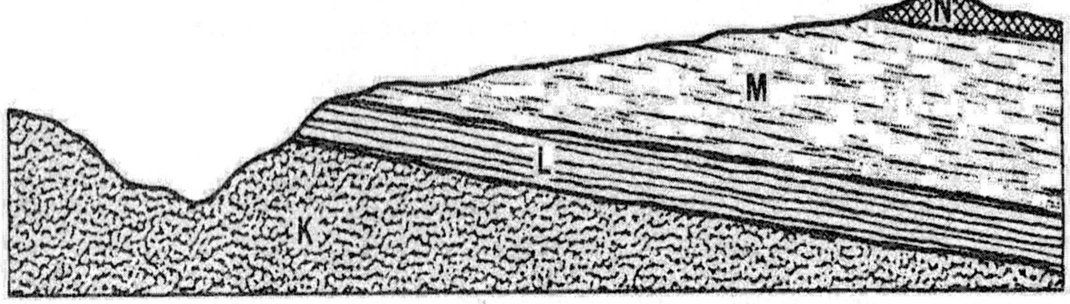


Each rock bed is younger than the strata under it.

**Figure 6**
**The Law of Superposition**
Source: US Atomic Energy Commission/Division of Technical Information

## VII. EVOLUTION BY NATURAL SELECTION

### *Charles Darwin*

The son of a wealthy physician, Robert Darwin (1766-1848), and grandson of Erasmus Darwin, Charles Darwin (1809-82) briefly studied medicine at Edinburgh from 1825 to 1827 but then left for Cambridge from which he graduated in 1831. Darwin described his Cambridge years, January 1828 - June 1831, during which he earned a three-year (ten terms or 3 1/3 years) Bachelor of Arts degree, among the happiest of his life.

At Christ's College, Darwin lived in the most expensive college rooms available to an undergraduate. He had a shoe blacker (boot polisher), bed maker, laundress, dishwasher (scullion), chimney sweep, coalman, tailor, hatter, and a barber who kept him well groomed. Darwin supplemented his college's standard cuisine of meat and beer with vegetables, for which he paid 5½ pence (£1.60 in 2010) daily. Cambridge also provided a valet or servant, called a *gyp*, for its students. All of these expenses cost Darwin just over £636 (£45,000 or $69,750 in 2010) for the three years, plus £14 (£978 in 2010) for his Bachelor of Arts Commencement in 1831 and £12 (£838 in 2010) for his Master of Arts Commencement in 1837. Two hours of daily lectures on the classics and mathematics occupied his mornings, allowing Darwin to spend the rest of the day on his studies, his interests that included shooting and collecting beetles, and enjoying a college life of socializing and drinking wine with his friends. (£1 in 1830 = £70 in 2010 and £1 in 2010 = $1.55 in 2010)

The same year that he graduated from Cambridge, Darwin accepted the opportunity to sail on the *Beagle*, on a five-year surveying trip around the world. The *Beagle*, a frigate only 90 feet long, 34 feet wide, and weighing 235 tons, had a crew of 74 men. The trip, lasting from 1831 to 1836, marked the beginning of his career as a naturalist pondering the origin of species and seriously impaired his health. Darwin appeared to suffer from trypanosomiasis, a disease caused by a parasite living in the victim's blood, similar to sleeping sickness. It forced him to spend the remainder of his life on his Down House Estate near the village of Downe, 16 miles southeast of London. His illness also may have had a psychological cause resulting from the many years of suppressing what he knew was a controversial theory of the origin of species.

Darwin's mother Susannah Wedgwood (1765-1817) was the daughter of Josiah Wedgwood (1730-95), the famous English pottery maker and abolitionist and good friend of fellow abolitionist Erasmus Darwin. Darwin's wife, Emma Wedgwood (1808-96) whom he married in January 1839, was Josiah's granddaughter. Although raised in a less than devout religious family of two sons and four daughters, Darwin as a young man accepted the *Genesis* creation account and fixity of species. By 1859 he still held some kind of theistic belief (a Deist) about the origin of species, although the death of his ten-year old daughter Anne in 1851 from tuberculosis led him to challenge seriously the existence of a benevolent Christian God. By 1870 he was an agnostic debating God's existence and his role in creating the universe. Darwin suffered several minor heart attacks in early 1882 and died of heart disease in April. His remains lie beneath a simple floor plaque in Westminster Abbey.

In 1839 Darwin published *A Naturalist's Voyage on the Beagle*, an account of his five-year trip around the world. In collaboration with five other leading naturalists, he edited *The Zoology of the Voyage of HMS Beagle* (5 parts, 1838-48) which described the living and extinct creatures of South America. The *Beagle's* captain, Vice-Admiral Robert Fitzroy (1805-65), a fourth great grandson of Charles II and a strong believer in special creation, later said he regretted taking Darwin as his companion on the *Beagle*. Fitzroy also quarreled with Darwin over slave trade, which Fitzroy justified and Darwin rejected. Darwin believed in the common origin of humanity and that treating humans as property was unethical. Fitzroy suffered from melancholy (depression) and in 1865 committed suicide by slitting his throat in his home outside London.

Darwin's voyage took him to the Cape Verde Islands, South America, Tahiti, New Zealand, and Australia before returning to England. He collected a large number of plant and animal species in these locations and shipped them to England preserved in alcohol in wooden barrels his ship mates constructed for him. During his visit in 1835 to the Galapagos Islands in the Pacific Ocean about 600 miles off the coast of Ecuador, Darwin studied the wildlife. He observed the presence of 13 kinds of visibly-related birds (blackbirds, grossbills, and finches) classifying them according to the different appearances of their beaks. Darwin questioned whether the different kinds of beaks resulted from a natural cause or a supernatural cause, God having created them that way, but at this time he had no explanation of what caused their beaks to differ.

Upon his return to London, Darwin in January 1837 donated his Galapagos birds to the Geological Society of London. The Society asked the ornithologist John Gould (1804-81), the London Zoological Society's first curator and preserver, to examine the birds, and in March 1837 Gould concluded that Darwin's birds were really 13 species of finches. Gould published his study of the Galapagos finches entitled *Birds Part 3* of *The Zoology of the Voyage of HMS Beagle,* 1838, 1839, and 1841.

Gould's classification led Darwin to recognize that the 13 species of finches existed only on the Galapagos Islands and not on the mainland. Yet in early 1838, two years after returning from his voyage, Darwin still could not account for the differences in the finches' beaks. His reading in autumn 1838 of Thomas Malthus (1766-1834) *Essay on the Principle of Population* (1798), and Adam Smith (1723-90) *Wealth of Nations* (1776), provided the clues. They contributed to his accepting that competition for food among species was the reason why the Galapagos finches had different beaks. These differences, Darwin believed, enabled some finches to eat a greater variety of foods that included soft seeds, hard seeds, and insects, and even to prick seabirds to suck their blood. Some finches adapted better because of their more varied diet and survived, whereas those finches that could eat only a few foods, such as soft seeds or hard seeds, were more likely to experience starvation and death and possibly eventual extinction over a long period of time. The 13 different kinds of finches represented the 13 species that had adapted best to the food supply in the environment in which each existed.

Mainland finches usually ate seeds, but the Galapagos finches ate other foods. Those with a short stubby beak cracked hard seeds, those with a longer thinner beak ate insects, and those with a long sharp beak drank blood. But how had the finches acquired the different beaks that enabled them to adapt to their environment and thereby survive? Darwin concluded they were born that way. Those favorable variations, traits, or characteristics that some members of a species acquired randomly (by accident) at birth enabled those members to adapt to their environment and survive, whereas those members lacking these advantages failed to adapt and eventually died off. Given enough time (evidence from molecular biology has confirmed that the mainland migration occurred about 23 million years ago), those members with the favorable variations became the dominant variety, giving rise to a new species.

Each of the 13 species that Darwin observed on a specific Galapagos Island was the species that had adapted best to its geographic location and had survived. Any other species not so well adapted gradually died off and became extinct. Giraffes did not stretch their necks, as Lamarck said, some were born with the advantage of having longer necks and survived better in their environment, especially when the food supply, consisting of tree leaves on lower branches, dwindled. Their long-necked offspring who could reach the higher branches with leaves became the dominant type of giraffe.

1. Geospiza magnirostris 2. Geospiza fortis
3. Geospiza parvula 4. Certhidea olivacea

Finches from Galapagos Archipelago

Without competition from other species, the Galapagos finches evolved to take advantage of every available habitat and foodstuff. The 13 species are all 4-8 inches (10-20 cm) long, and brown or black in color. Birds differ in the shape of their beaks. Some have a slender curved beak for probing flowers, others a parrot-like beak for crushing nuts and seeds. Two species have perfected the ability to hold cactus spines to search for grubs.

**Figure 7**
**Darwin's Finches**

Source: Charles Darwin. *Journal of Researches into the Natural History and Geology of the Countries Visited During the Voyage of H.M.S. Beagle Round the World.* London: John Murray, 1845.

Darwin's theory of natural selection that he outlined in an 1842 unpublished 42-page manuscript grew to 221 pages in 1844, and he published it in 1858 in the *Proceedings of the Linnaean Society*. His theory consisted of five parts: common ancestry; descent with modification; gradualism or change by small, incremental steps; multiplication of species by splitting of lineages; and natural selection acting on random variations, which worked best in small, isolated populations and known today as the allopatric, or in another place, theory.

The next year, in 1859, Darwin published his long-awaited and for some still controversial *The Origin of Species*. In *The Origin* Darwin dealt only with plants and animals. He gave no real examples of evolutionary change, and for every claim he made, other theories offered alternative explanations. Two other books on evolution followed: *The Descent of Man* in 1871 and *The Expression of Emotions in*

*Man and Animals* in 1872. His evolutionary trilogy showed a gradual expansion of his evolutionary beliefs from plants and animals to man, and to man's emotions.

Darwin mentioned a creator in the last chapter of the 1860 second edition of *The Origin*, but the creator disappeared in later editions. In the fifth edition of *The Origin* published in 1869, he replaced *descent with modification* with *survival of the fittest* to describe his theory. The English philosopher, sociologist, and know-it-all Herbert Spencer already coined and popularized *survival of the fittest* in his 1852 publication "A Theory of Population."

Spencer claimed his universal law of evolution held for everything in the universe, including psychology, religion, and society. His theory of evolution was progressive and directed, unlike Darwin's natural selection in which no directive, progressive force existed. Only the best adapted survived. Spencer's survival of the fittest meant survival of the best, and his theory of evolution also provided the rationale for the misguided and misnamed Social Darwinism. Darwin understood that his theory of evolution provided an explanation or justification for the ills of society, such as racism and colonialism, but he did not condone racism and colonialism. He made a mistake in adopting Spencer's survival of the fittest. Unfortunately, what society calls Social Darwinism is really Spencerism.

In *Descent of Man,* Darwin applied evolution to humanity's physiological development, and wrote that man had descended (evolved) from ape-like creatures. In *The Expression of Emotions,* he applied evolution to humanity's psychological behavior, to all human emotions and feelings, morality, and religion. No unchanging qualities remained in humanity. He first used the word evolution in *Descent of Man* and the next year in the sixth and last edition of the *Origin* published in 1872.

Darwin's theory of natural selection attracted naturalists because it provided a mechanism for them to test. The main weakness of Darwin's theory was its inability to explain how an organism or body preserved and passed its characteristics from one generation to the next. His pangenesis theory, developed in 1865-66 and published in 1868 in *The Variation of Animals and Plants under Domestication,* unsuccessfully dealt with this problem. According to Darwin, particles, or gemmules, distributed throughout each parent's body, recorded each parent's characteristics so that all characteristics acquired and recorded during the parents' lifetime passed on to their descendants. His pangenesis theory was similar to Lamarck's inheritance of acquired characteristics.

In *Variation*, Darwin discussed his belief, and the widely-held belief, in the blending of characteristics, which over time would dilute the favored characteristics in successive generations. Fleeming Jenkins (1833-85), an English engineer at University College, London, and at Edinburgh, pointed out this flaw in the blending theory of inheritance in 1867 in a long review of the fourth edition (1866) of *The Origin.* In *Descent of Man,* Darwin proposed that sexual selection contributed to the evolution of species. The female, by selecting the male with the most desirable characteristics, such as flamboyancy or color, helped preserve a particular characteristic. Darwin needed Mendel's genetics to show how the characteristics an organism randomly acquired at birth did not blend and passed unchanged from generation to generation, but he very likely never knew of Mendel's 1866 publication.

Objections to Darwin's theory emerged because natural selection denied nineteenth-century's established and cherished idea of progress. His theory was materialistic, a random, yet deterministic, process by which natural selection acted to preserve the favored or best adapted characteristics of a species. It guaranteed survival through adaptation, but did not guarantee progress or the inevitability of progress. Naturalists accepted Darwin's evolutionary theory because they could test its validity. No test existed for Lamarck's belief in an organism's internal force driving it towards perfection. Nevertheless Lamarckians, such as Kammerer and Lysenko, claimed they had demonstrated the environment's influence on the members of a species, resulting in acquired characteristics that the species passed on to later generations.

***Proponents of natural selection: Wallace, Huxley, Haeckel, Weismann***

The English naturalist Alfred Russel Wallace (1823-1913) was a co-discoverer of natural selection. Unlike Darwin, Wallace had little formal education and wealth. He was an extremely modest person, a socialist who supported land and labor reform, home rule for Ireland from England, and opposed Social Darwinism and eugenics. Wallace spent the years from 1848 to 1852 in the Amazon rain forest in South America studying and collecting plant and animal species, but lost all his collections at sea when the ship on which he was returning to England caught fire and sank. Wallace escaped with a single box containing drawings of fish and palm trees. Two years after returning from South America, Wallace departed for the Malay Archipelago (Malaysia, Indonesia, Borneo, New Guinea) where from 1854 to 1862 he continued his studies even while suffering from malaria. During those eight years he traveled 14,000 miles and collected about 125,000 specimens.

Wallace wrote a brief 4,000 word paper outlining his views on the evolution of species in February 1858 while in the Spice Islands (Moluccas in East Indonesia), and sent it for comments to Darwin, with whom he had corresponded since 1856. Wallace had read Malthus, and Wallace's paper "On the Tendency of Varieties to Depart Indefinitely from the Original Type," closely resembled in its conclusions Darwin's 1842 unpublished manuscript. He probably would have beaten Darwin to publication, but Darwin's friends, Lyell and the botanist-explorer Joseph Hooker (1817-1911) at the Royal Botanical Gardens in Kew near London, intervened. They knew for years of Darwin's ideas on evolution and arranged for a joint presentation of both papers on 1 July 1858 before the Linnaean Society in London. Their intervention established Darwin's priority in arriving at the theory of natural selection. Neither Darwin nor Wallace attended the July meeting. The society's secretary read Wallace's 1858 paper and parts of Darwin's 1844 manuscript. The Linnaean Society then published both papers in Volume 3 (1858) of its *Proceedings.*

The two theories had differences. Wallace's theory was not as harsh as Darwin's. Wallace believed that lesser adapted species could coexist with better adapted species and were not necessarily doomed to extinction. Nevertheless, Wallace acknowledged a "nature red in tooth and claw" that Alfred Lord Tennyson (1809-92) mentioned in his long 1850 poem *In Memoriam*, a philosophic elegy and statement lamenting the death of his best friend.

A few years later, in 1867, Wallace accepted spiritualism. He published *On Miracles and Modern Spiritualism* in 1875, and concluded that natural selection or evolution did not apply to man's spiritual and intellectual qualities. They had not evolved. God created those qualities. Despite these differences from Darwin, Wallace summarized his ideas on evolution in his 1889 book, *Darwinism.*

Wallace, who lived in London, Essex, Broadstone, and other English cities, financed his research by selling the rare specimens he collected and supported himself on meager earnings of £60 a year and a small £200 annual pension that Darwin in 1881 helped secure in recognition of Wallace's scientific contributions. From autumn 1886 to autumn 1887, Wallace made his only lecture tour, a ten-month tour of the United States and Canada, stopping in Boston to deliver the Lowell Lectures; the New York City area; Washington, DC; San Francisco area; Toronto, and Kingston, Ontario, the homes of the University of Toronto and Queen's University.

Thomas Huxley (1825-95), a materialist, skeptic, religious agnostic, a term he coined in 1860, and a Darwinian published *Man's Place in Nature* in 1863. Huxley, at the School of Mines in London, appointed himself Darwin's bulldog and engaged in numerous debates on evolution and on other topics, among them education. His most memorable debate was with Bishop Samuel Wilberforce (1805-73), Bishop of Oxford. Wilberforce's father William Wilberforce (1759-1833), an evangelical Christian, philanthropist, and Member of Parliament for Yorkshire, in 1807 secured the abolition of the slave trade in Britain and the British Empire. Abolition followed in the United States in 1808. In 1833 he secured the emancipation of slaves in Britain and most of the British Empire. Emancipation followed in the United States on 1 January 1863.

Slavery in Europe had declined drastically centuries earlier with the fall of the Western Roman Empire and the rise of feudalism (serfdom) from the 900s to the 1200s. In some countries, such as Italy, slavery was never a legal or formal institution, and condemned by the Catholic Church. Poland abolished slavery in the 1400s, Lithuania in 1588, Russia in 1669 and 1723, Scotland in 1778, and France in 1794 (revived in 1802, banned permanently in 1848). Slave trade continued into the 1800s, although by mid-century most of the European nations abolished slavery in their colonies in the Americas and elsewhere. Sweden acted in 1847, France in 1848, Denmark in 1848, Holland in 1863, and Spain in 1873-76. Darwin believed the treatment of humans as property was unethical. He witnessed firsthand the cruelty and indignity of slavery in Brazil when the *Beagle* arrived there in 1832. He did not live to see the end of slavery in Brazil in 1888.

The Huxley-Wilberforce debate on Darwin's theory of evolution occurred at the 1860 meeting of the British Association for the Advancement of Science in Oxford and attracted 700 people. During the debate, Bishop Wilberforce, also known as Slippery Sam or Unctious Sam, innocently asked Huxley whether he had descended from apes on his grandmother's side or on his grandfather's side. Huxley responded that if he had to choose between Wilberforce and an ape for his ancestor, he'd choose the ape. Fitzroy attended, and he rose, enraged and waving his Bible, to shout his regrets at having taken Darwin aboard the *Beagle*.

In the years after the Oxford debate, Huxley proposed his theory that birds had descended directly from dinosaurs. He published his theory in 1868. The discovery of the 155-million year old Jurassic fossil of *archeopteryx lithographica*, the winged dinosaur, had occurred in the Solnhofen limestone of Bavaria in southern Germany in 1861, and led to discussion about its origin. Beginning in 1926 and continuing until about 1964, ornithologists and other scientists believed, however, that *archeopteryx* was a full-fledged bird that developed independently of dinosaurs and capable of limited flight. With the discovery of a new theropod dinosaur in Montana in 1964, and a renewed examination of dinosaur and bird fossils, Huxley's theory, beginning in 1972, has received almost unanimous support from paleontologists, ushering in the Dinosaur Renaissance.

Mathias Schleiden (1804-81), a botanist at Jena, was among the first of the German scientists to accept Darwinism. In *Grundzüge der Wissenschaftlichen Botanik* (*Principles of Scientific Botany*), his 1842 textbook on botany, Schleiden challenged nature philosophers (naturphilosophie) such as Karl Burdach at Königsberg and Lorenz Oken (1779-1851) at Jena who believed in an organism's progressive evolution to a higher form.

Another German Darwinian Ernst Haeckel, the loud, boisterous, and excitable anatomist-zoologist at Jena, studied under Rudolf Virchow (1821-1902), an anti-Darwinist physician at Berlin. Haeckel in 1866 proposed his theory that *ontogeny* (the biological development of a species) recapitulates *phylogeny* (the evolutionary development of a species). This does not occur, and only superficial similarities exist. Haeckel was a very good artist, and some of his drawings supporting his recapitulation theory are more works of art than they are scientifically accurate representations. He later acknowledged that he relied on memory and used artistic license in preparing his drawings. Haeckel published *The Evolution of Man* (2 volumes) in 1897. Although Haeckel supported Darwin's natural selection, he accepted Lamarck's inheritance of acquired characteristics.

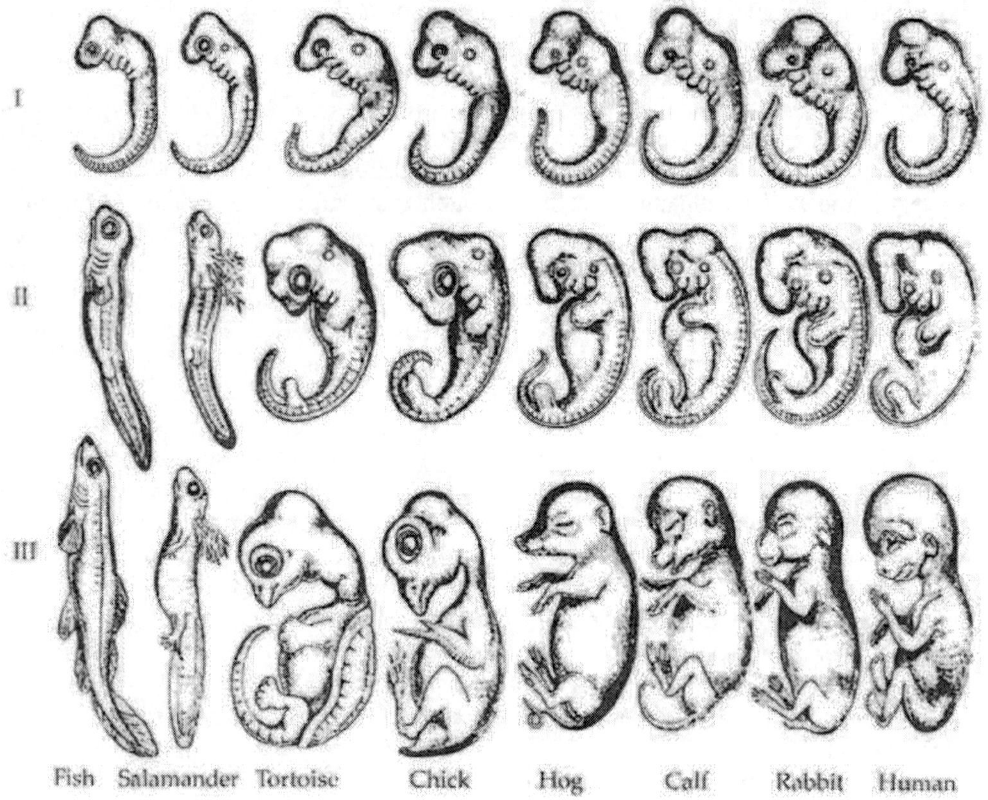


**Figure 8**
**Haeckel's Diagrams of Embryos from *The Evolution of Man***
Source: Ernst Haeckel. *The Evolution of Man*. Vol. 1. New York: D. Appleton and Company, 1897.

The Darwinian August Weismann (1834-1914), a zoologist at Freiburg, carried out mice breeding experiments in 1883, allegedly to disprove Lamarckian evolution and the inheritance of acquired characteristics. In experiments with 1,582 mice, he cut off their tails and then bred them over 22 generations. (The gestation period for a mouse is 19-21 days). The offspring of the mice produced in his experiments did not acquire cut-off tails, but his experiments proved nothing because Lamarck believed the animal's internal striving to adapt to its environment produced changes that it inherited, not its mutilation. Weismann's purpose remains unclear. He and other scientists had plenty of evidence and knew that a mutilation made on a plant, animal, or human such as the loss of an eye, ear, or limb never appeared in the offspring of the mutilated parent.

Weismann's studies in 1883 on *diptera* (insects such as the fly and mosquito having one pair of wings), small crustaceans, and *hydromedusae* (jellyfish) led him and other scientists to conclude in 1884 that the cell's nucleus (chromosomes) played the main role in heredity. In 1885 he located the germ plasm (genes) that is present in the nucleus of germ cells (the body's reproductive cells, its egg and sperm cells) and made a distinction between germ cells and somatic (body) cells. The organism's germ cells passed on unchanged to succeeding generations unaffected by the environment (the hereditary characteristics), its somatic cells, which are present everywhere in the body, differentiated themselves into organs, tissues, and other bodily parts. The environment could modify the somatic cells (acquired characteristics), but the organism could not pass the modifications on to succeeding generations.

**Figure 9**
**Whale Family Tree**
Courtesy of NOAA

***Opponents of natural selection: Owen***

Richard Owen (1804-92), a comparative anatomist at the British Museum in South Kensington (London) who studied briefly with Cuvier in Paris, published his rejection of natural selection in a lengthy anonymous criticism in the *Edinburgh Review* of April 1860 and in later publications. He accepted the progressive evolution of an organism to a higher form and design and purpose in nature. His ideas closely resembled the beliefs of the continental nature philosophers Johann Goethe (1749-1832) in Weimar, Burdach, Oken, and Augustin Saint-Hilaire (1799-1853), a French botanist who traveled extensively in South America, especially Brazil. Owen, sometimes described as vain and arrogant, advised Wilberforce in his debate with Huxley and engaged Huxley in debates on evolution.

Owen disagreed with Huxley's argument that a human was a primate, accepting instead, a human's distinctiveness because of the human brain's structure, particularly *hippocampus minor*, the small internal ridges at the back of the brain. Owen in 1857 incorrectly claimed that the gorilla, a new species of ape discovered in 1847, lacked the ridges, whereas Huxley in an 1861 paper reprinted in 1863 in *Man's Place in Nature* proved their presence in the gorilla. Owen coined the word *dinosaur,* meaning giant lizard, in 1841 at the meeting of the British Association for the Advancement of Science in Plymouth.

In Germany, anti-Darwinists included Rudolf Kölliker (1817-1905), an embryologist at Würzburg who believed in evolution by jumps and rejected Darwin's natural selection; Virchow at Berlin; and the embryologist Karl von Baer (1792-1876) at Königsberg and later in St. Petersburg, Russia. Hertwig in his 1916 publication *Das Werden der Organismen* (*The Origin of Organisms*) criticized the idea of chance variations and natural selection. He believed in the heredity of acquired characteristics. In France, the physiologist Claude Bernard (1813-78) at the Collège de France in Paris and Pasteur at the École Normal Supérieure and the Sorbonne in Paris ignored or doubted, and rejected Darwinism. The conservative Pasteur did not accept evolutionary changes in species.

Compared to European society, in late nineteenth-century Japan, Darwinism was not a religious issue. It gained acceptance because Japanese military and political leaders adopted Darwinism, or more correctly, Spencer's survival of the fittest, to justify Japan's military and industrial buildup. Darwinism represented to Japanese leaders an eternal and unchangeable natural law and therefore an acceptable natural law.

## VIII. AMERICAN DEBATE OVER DARWINISM: LOUIS AGASSIZ AND ASA GRAY

***Racial issue and debate on the origin of the races: monogenesis vs. polygenesis***

Monogenesis, or a single creation of the human races as described in *Genesis*, versus the polygenesis belief that God created each race as an immutable separate species, preceded debate on evolution in the United States. Because of the Adam and Eve story in *Genesis*, polygenesis had few supporters in Christian Europe until the seventeenth century, when northern Europeans began receiving more information on Africans and Native Americans. Buffon, who in 1749 allegedly introduced the word *race* to the natural sciences and invented the human categories of European, Mongolian, and southern Asiatic, accepted monogenesis and believed climate caused the different colors of humans. Similarly, Johann Blumenbach (1752-1840), an anatomist at Göttingen and the founder of anthropology, in 1775 proposed five varieties (races) of humans that he called Caucasian (white), Mongolian (yellow), Ethiopian (black), American (red), and Malay (brown). Degeneration produced the differences, but Blumenbach remained committed to monogenesis.

By the early 1800s opinions had changed. Significant differences among human varieties and a biblical chronology that allowed too few generations from the time of Adam and Eve in 4004 BC to account for the varieties led some scholars to question monogenesis and propose that polygenesis accounted for the human varieties or races. These scholars also doubted that climate could produce different varieties.

Voltaire (1694-1778) in Paris discussed his conversion to polygenesis in 1756 in *Essay on General History and on the Customs and Spirit of Nations since Charlemagne until our Times* and in 1765 in *The Philosophy of History*. He combined the two works in 1769 entitling it *Essay on the Customs and Spirit of Nations*. The Scottish jurist Henry Home or Lord Kames (1692-1782) in 1774; Charles White (1728-1813) a Manchester surgeon in 1799; and the French physician Julien Joseph Virey (1775-1846) at the Val-de-Grace Military Hospital in Paris in 1801 also abandoned monogenesis and embraced polygenesis. White's 1799 book *An Account of the Regular Gradation in Man and in Different Animals and Vegetables* depicted the black African as the lowest form of human life and the Greek ideal at the top. The lower humans appear in close proximity to the lower animals and the black African nearer the ape. Virey believed all other races intellectually, scientifically, and artistically incapable compared to Europeans, and barbaric. The races were different species along a spectrum. Virey believed two black races existed. He placed them closest to apes and incapable of civilization. In the United States, Samuel Morton (1799-1851), Josiah C. Nott (1804-73), George Gliddon (1809-57), Charles Coldwell (1772-1853) a Kentucky physician, and later Agassiz took up the polygenesis cause.

Morton, a Quaker-raised physician at Pennsylvania, provided cranial evidence that he said supported the polygenesis interpretation of racial origins. Morton possessed the world's largest collection of skulls, having collected 1,000 skulls from 1828 to 1851. In the 1830s he carried out craniology studies in which he made lead shot measurements of white, Native American, black, Asian, and Egyptian skull capacities, showing that the white race had the largest capacity and therefore larger brains than the black, red, and yellow races. Morton published his highly-selective doctored results, in which he ignored body size, in *Crania Americana* in 1839, claiming that they provided scientific justification for the white race's superiority.

Stephen Gould in his 1981 book *The Mismeasure of Man* criticized Morton's results, arguing that the skulls he measured did not represent the groups he studied and that Morton selected them to highlight cranial capacity differences among the races. John S. Michael, an undergraduate student researcher at Macalaster College in St. Paul, Minnesota, in 1988 and Jason Lewis and his team of physical anthropologists at Pennsylvania in 2011 challenged Gould's conclusion. Lewis now at Stanford argued that Morton did not doctor his results. The 2011 investigation, he wrote, showed that Morton's results differed significantly from their measurements of the same skulls in only 2 percent of the cases. Lewis did not counter Gould's argument that the skulls were selective and not representative of the races Morton studied.

| MORTON'S RANKING AND RACIAL DESCRIPTIONS FROM *CRANIA AMERICANA* | |
|---|---|
| ***Race*** | ***Characteristics*** |
| Europeans | Distinguished for the facility with which it attains the highest intellectual endowments, have peopled the finest portions of Earth, and given birth to its fairest inhabitants. |
| Asians | Ingenious, imitative, and highly susceptible of cultivation (learning), they have been compared to the monkey race, whose attention is perpetually changing from one object to another. |
| Native Americans | In the mental character, they are averse to cultivation and slow in acquiring knowledge; restless, revengeful, and fond of war; crafty, sensual, ungrateful, obstinate. |
| Africans | The many nations which compose this race present a singular diversity of intellectual character, of which the far extreme is the lowest grade of humanity; fond of amusements, not infrequently characterized by superstition and cruelty; have little invention, strong powers of imitation so they readily acquire mechanic arts; have a great talent for music. (He forgot to include their skill in sports.) |

Ten years later, in 1849, Nott, a pro-slavery physician in Mobile, Alabama, and lecturer on *niggerology*, published *Biblical and Physical History of Man* in which he concurred that Morton's craniology studies provided a scientific basis for polygenesis. Nott, who studied under Morton and received his medical degree at Pennsylvania, maintained that the races were separate species and not equal. The enslavement of blacks, a clearly lower and inferior species, was neither immoral nor contrary to God's will. The fact that white and black, white and red, and white and any skin color had interbred and produced fertile offspring, which scientists accepted as the standard for differentiating among species, failed to deter Nott.

Nott published two books with Gliddon, an Egyptologist and in 1832 American vice-counsel in Alexandria, entitled *Types of Mankind* in 1854 and *Indigenous Races of Earth* in 1857. In these and in his other publications, Nott argued that the Bible gave an incomplete and inaccurate account of human creation. Only one of the several human races had descended from Adam and Eve. His questioning of the Bible's accuracy generated considerable controversy among the same biblical literalists who championed Nott's support of polygenesis to justify black inferiority and slavery. Gliddon believed in the divine ordination of the white race's superiority. He died, possibly a suicide of an apparent overdose of opium, in a Panama hotel where he had gone to work for the British Honduras Railroad.

Later studies on brain size, particularly those of the American anatomist Edward A. Spitzka (1876-1922) at the Jefferson Medical College in Philadelphia, supported Morton's racist beliefs and provided other revealing brain size (weight) comparisons. Spitzka published his studies in 1903 and 1907 in the *American Anthropologist* and the *Transactions of the American Philosophical Society*. In one study he compared the brain of an eminent white male, an African bushwoman, and a gorilla and found that the same difference existed between the white male and the smaller black female and between the black female and the gorilla.

Spitzka also found that the average brain weight of 1,300-1,400 grams (454 grams = 1 pound) for Europeans or Americans and the brain weight of 134 eminent scientists and intellectuals he measured, hardly differed. The German physician Franz Gall (1758-1828) in Paris, the founder of phrenology, had a brain that weighed 1,198 grams. Paul Broca (1824-80), an influential Parisian surgeon, Darwinian, anthropologist, craniometrician, and the leading advocate of the brain size-intelligence theory, had a brain weight of 1,424 grams. Karl Gauss (1777-1855), the physicist, astronomer, and mathematician at Göttingen, whose intellectual genius no one doubted, had a brain that weighed 1,492 grams. Liebig's brain weighed 1,352 grams. The heaviest of the scientists' brains, that of Cuvier, weighed 1,830 grams.

Studies on writers and statesmen showed the same range in brain weight. In the United States, Walt Whitman (1819-92) in New York City had a brain that weighed 1,282 grams, that of the Boston lawyer Daniel Webster (1782-1852) weighed 1,735 grams. The 1921 Nobel Prize winner in literature Anatole France (1844-1924) in Paris had a brain weight of only 1,017 grams. The biggest brain of all those studied belonged to the Russian writer Ivan Turgenev (1818-83) who spent most of his years outside of Russia in Baden-Baden or Paris. It weighed 2,012 grams.

An 1880 study of Theodor Bischoff (1807-82), a physiologist and anatomist and anti-Darwinist at Munich, provided another startling set of brain weights. Bischoff found that the 119 criminal brains he weighed exceeded those of non-criminals. On average, the bad brains exceeded the good brains by 11 grams, 14 bad brains weighed more than 1,500 grams, 5 bad brains weighed more than 1,600 grams, one weighed 1,809 grams. The female brain of a husband killer, the largest female brain ever examined, weighed 1,565 grams. An autopsy that Spitzka performed on the brain of Leon Czolgosz (1873-1901), the assassin of the president William McKinley (1843-1901), found that it slightly exceeded average brain weight, 1,460 grams (51.5 ounces).

| COMPARISON OF WHITE, BLACK, AND GORILLA BRAIN WEIGHT | | |
|---|---|---|
| ***Subject*** | ***Brain Weight (Grams)*** | ***Approximate Ratio*** |
| Ivan Turgenev | 2,012 | 1.0 |
| George Cuvier | 1,830 | 1.0 |
| Benjamin Butler | 1,758 | 1.0 |
| William Thackeray | 1,658 | 1.0 |
| | | |
| Zulu | 1,050 | 0.50 |
| Australian | 907 | 0.50 |
| Bushwoman | 794 | 0.50 |
| | | |
| Gorilla | 425 | 0.25 |
| Orangutan | 400 | 0.25 |
| Chimpanzee | 390 | 0.25 |

| BRAIN WEIGHTS OF EMINENT MALES | | | |
|---|---|---|---|
| ***Name*** | ***Occupation*** | ***Age*** | ***Brain Weight (Grams)*** |
| Ivan Turgenev | Poet and Novelist | 65 | 2,012 |
| George Cuvier | Naturalist | 63 | 1,830 |
| Benjamin F. Butler | General and lawyer | 74 | 1,758 |
| Walter V. Siemens | Physicist | 68 | 1,600 |
| Edward D. Cope | Paleontologist | 57 | 1,574 |
| James Young Simpson | Physician | 56 | 1,531 |
| Daniel Webster | Statesman | 70 | 1,518 |
| Chauncey Wright | Philosopher | 45 | 1,516 |
| Louis Agassiz | Naturalist | 66 | 1,495 |
| Karl F. Gauss | Mathematician | 78 | 1,492 |
| John Wesley Powell | Geologist and ethnologist | 68 | 1,488 |

| BRAIN WEIGHTS OF EMINENT MALES | | | |
|---|---|---|---|
| ***Name*** | ***Occupation*** | ***Age*** | ***Brain Weight (Grams)*** |
| Paul Broca | Anthropologist | 55 | 1,484 |
| Hermann von Helmholtz | Physiologist | 73 | 1,440 |
| Joseph Leidy | Morphologist | 67 | 1,415 |
| Charles Babbage | Mathematician and inventor | 79 | 1,403 |
| Justus von Liebig | Chemist | 70 | 1,352 |
| Walt Whitman | Poet | 72 | 1,282 |
| Franz J. Gall | Anatomist and phrenologist | 70 | 1,198 |
| Anatole France | Writer | 80 | 1,017 |

| BRAIN WEIGHTS BY DISCIPLINE | | | |
|---|---|---|---|
| ***Categories*** | ***Number of Cases*** | ***Average Age*** | ***Average Brain Weight (Grams)*** |
| Exact Sciences | 12 | 67.6 | 1,532.0 |
| Natural Sciences | 48 | 63.8 | 1,440.0 |
| All Sciences | 60 | 64.5 | 1,485.0 |
| Fine Arts, Philosophy, other humanities | 26 | 59 | 1,485.0 |
| Men of Action (government, politics, military, and others) | 14 | 65 | 1,490.0 |
| Total Averages | 100 | 62.4 (97 cases) | 1,469.65 |

The monogenesis-polygenesis debate attracted Agassiz the Swiss naturalist who had achieved a reputation in the 1830s and 1840s as an expert on the study of fish fossils and glaciers. His *Études sur Glaciers (Studies on Glaciers)* published in 1840, clearly established the existence of previous ice ages. Agassiz came to the United States in autumn 1846 on a speaking tour to give the Lowell Lectures in Boston and stayed to accept a professorship at Harvard's Lawrence Scientific School, following its establishment in 1847. He left behind a wife who died in 1848 and three children. Agassiz in 1850 married Elizabeth Cabot Cary (1822-1907), a prominent Bostonian and one of the founders of Radcliffe College (est. 1879). He remained at Harvard from 1848 until his death from a stroke.

Agassiz was a catastrophist, and like his mentor Cuvier, believed that after global catastrophes had destroyed all life on Earth, God recreated all the species. He did not accept the *Genesis* account of creation, ridiculing the Adam and Eve story as absurd and rejecting a literal six-day creation that occurred 6,000 years ago.

Gould, in his *Mismeasure of Man,* vividly relates how Agassiz's understanding of racial origins underwent a conversion from monogenesis to polygenesis in 1846, shortly after his arrival and initial contact with black waiters in a Philadelphia restaurant. It resulted from his personal feeling of repugnance of blacks and of their racial inferiority, and not from any scientific studies he had done. His visit to Morton and inspection of Morton's skull collection in December 1846 likely contributed to his conversion.

Describing his first meeting with blacks in a long letter to his mother in 1846, Agassiz wrote that the sentiment that they inspired in him was contrary to all our ideas about the confraternity of the human type and the unique origin of our species. But truth before all, Agassiz said he experienced pity at the sight of this degraded and degenerate race, their lot inspired compassion in him in thinking that they were really men, and he could not repress the feeling that they are not of the same blood as us. Agassiz gave a particularly unflattering description of the waiters' faces, teeth, hair, and hands and wished he could have dined elsewhere. What unhappiness for the white race to have tied its existence so closely with that of Negroes in certain countries. God preserve us from such a contact.

Four years later, in 1850, Agassiz published another defense of polygeny and the self-evident hierarchy of races in the *Christian Examiner*. Agassiz wrote that "There are upon Earth different races of men, inhabiting different parts of its surface . . . and this fact presses upon us the obligation to settle the relative rank among these races. The indomitable, courageous, proud Indian– in how different a light he stands by the side of the submissive, obsequious, imitative Negro, or by the side of the tricky, cunning, and cowardly Mongolian! Are not these facts indications that the different races do not rank upon one level in nature?"

Agassiz proposed keeping the races separate because breeding with blacks would lead to blending and dilute white superiority. In four letters written in August 1863 during the Civil War, seven months after Lincoln on 1 January 1863 signed the Emancipation Proclamation abolishing slavery in the rebel states, Agassiz wrote that "the separation should occur naturally because mulattoes, as a weak strain, would eventually die out. Blacks will leave the northern climates so unsuited to them because God created them as a separate species for Africa. They will move south in droves and will eventually prevail in a few lowland states, although whites will maintain dominion over the seashore and elevated ground. We shall have to recognize these states, even admit them to the Union, as the best solution to a bad situation. After all we do recognize Haity and Liberia."

"In everything unlike other races, they may but be compared to children, grown in the stature of adults, while retaining a childlike mind. How shall we eradicate the stigma of a lower race when its blood has once been allowed to flow freely into that of our children? The production of half-breeds, is as much a sin against nature, as incest in a civilized community is a sin against purity of character. . . . I hold it to be a perversion of every natural sentiment."

"As soon as the sexual desires are awakening in the young men of the South, they find it easy to gratify themselves by the readiness with which they are met by coloured [mulatto] house servants. [This contact] blunts his better instincts in that direction and leads him gradually to seek more spicy partners, as I have heard the full blacks called by fast young men. One thing is certain, that there is no elevating element whatsoever conceivable in the connection of individuals of different races; there is neither love, nor desire for improvement of any kind, it is altogether a physical connection."

Agassiz did not support slavery, but his acceptance of polygenesis provided *scientific* justification for slavery's supporters who already believed blacks were a lower and inferior species or race. At this time, in 1850, the Southern Baptists in the United States split with the Northern Baptists because the southern branch wanted to continue the support of slavery. The south and the Southern Baptists lost their fight for slavery in 1865, but racial attitudes have never disappeared. One of the less-known examples of government-sanctioned racism appeared in a 1925 Army War College study. *The Use of Negro Manpower in War* concluded that black men were cowards and poor technicians and fighters, lacking initiatives and resourcefulness. The report said they were a subspecies of the human population. Studies as recent as 2010 show that about 50 percent of the US population has racist beliefs.

***Evolutionary issue and debate: Agassiz and Gray***

Agassiz never abandoned his anti-evolutionist and anti-Darwinist position. A Unitarian and a creationist, Agassiz believed that the creation of each plant and animal represented a thought of God. Similarities existed, but they were associations of ideas in God's mind, not evidence of evolution or descent with modification. Members within a particular species (varieties) varied, but each species remained fixed and static and did not change. God eliminated old species and had done so during the glacial periods, and then created new species (successive creations). This resulted in progress. Agassiz remained convinced to the end that no change of species ever had occurred.

Asa Gray, the Harvard botanist, was Agassiz's leading critic and Darwin's main supporter in the United States. He met Darwin briefly in England in 1839, and corresponded with him from 1855. In their correspondence of September 1857, two years before the *Origin's* publication, Darwin described his theory of natural selection, and in 1860 after its publication, Gray and Agassiz, in print and orally, vigorously expressed their opposing positions on natural selection. A practicing Congregationalist, Gray could not accept Darwin's naturalistic evolution of species and believed that God directed the evolutionary process. God used evolution to create the universe and the species that existed on Earth. In his review of *The Origin,* published in March 1860 in the *American Journal of Science,* Gray urged Darwin to adopt a theistic evolution.

Darwin rejected Gray's advice, arguing that no evidence supported a God-directed evolutionary process. Darwin also thought too much suffering existed in the world to justify evolution as God's creative plan. Gray countered that evolution gave us a better understanding of one of theology's most difficult problems, namely, to reconcile the existence of pain and suffering in a world a benevolent God created. Because competition and struggle were the conditions that led to evolutionary changes, pain and suffering were the cost required for the emergence of intelligent creatures capable of reflecting on their origins. Gray published *Darwiniana* in 1876, arguing for the compatibility of natural selection and Christianity, or theistic evolution. He died twelve years later from a paralytic stroke.

***More American anti-Darwinists: Silliman, Dana, Hitchcock***

Despite their opposition to natural selection but not to the likelihood of an old Earth, several American anti-Darwinists sought some way to reconcile their biblical belief with the abundant geologic evidence supporting an old Earth. Benjamin Silliman, Arnold Guyot, John William Dawson, James Dwight Dana, and Josiah Nott accepted a day-age creationism, or a progressive old Earth creationism, in which the days of creation were long time periods during which God created separately plants, animals, and humans.

Silliman, a Yale chemist and geologist and a Congregationalist, was the United States' leading spokesman for science in the first half of the nineteenth century. He believed that the study of science revealed God's majesty and glory and tried to reconcile science with the Bible. In his day-age creationism, each day mentioned in *Genesis* corresponded to an age in time. Silliman's son-in-law, the Yale geologist James Dwight Dana, in *Science and the Bible* (1856) endorsed day-age creationism as a way of reconciling geological evidence for an old Earth with the Bible. Dana, a Congregationalist, adopted a theistic evolution in the 1870s and believed in God's special creation of humans. His ideas were Lamarckian.

Arnold Guyot (1807-84), a Swiss-born geologist at Princeton and devout Presbyterian, believed that God created the universe, and what unfolded in the universe, including the hierarchy of races, was God's plan. In *Creation or the Biblical Cosmology in the Light of Modern Science* (1884), Guyot wrote that in *Genesis* creation occurred three times: in matter, animal life, and humans. The fundamentalist John William Dawson (1820-99), a geologist at McGill University in Montreal, strongly opposed evolution and Darwinism early in his career. But in *Facts and Fancies in Modern Science* published in 1882 he allowed for the possibility of theistic evolution. Dawson rejected a worldwide flood, but saw many parallels between *Genesis* and geology such as oceans preceding dry land and man the culmination of all creation.

Another geologist Edward Hitchcock (1793-1864), one of Silliman's students, a committed anti-slavery Congregational minister, Amherst professor and president from 1845 to 1885, tried to prove the compatibility of the Bible and geology. He accepted an ancient Earth and gap theory creationism. According to the gap theory that the Scottish Presbyterian theologian Thomas Chalmers (1780-1847) proposed in 1814, God created a perfect Earth in six days, but it became corrupt, and God destroyed it. A tremendous gap in time of billions of years followed in which a number of catastrophes occurred, as suggested from a reading of *Genesis* 1:1, 1:2, and 1:3, and God recreated Earth in six literal days. In his very popular book, *The Religion of Geology and its Connected Sciences* (1851), Hitchcock argued for the truth of both science and theology, or the unity of truth. Scientific truth is religious truth and to use science against religion was a perversion of science. He believed that geology more than any other science proved that God intervened directly to guide and alter Earth's history.

Timothy Conrad (1803-77) and Matthew Maury (1806-73) rejected Darwinism. Conrad, a Quaker minister's son, botanist at Pennsylvania, and in 1837 New York State geologist, was Agassiz's devoted follower. Like Agassiz, he believed in successive creations in which God created all new life. Maury, an oceanographer at the National Observatory, later a Confederate naval commander, and from 1868 at the Virginia Military Institute, was a biblical literalist and trusted only the Bible. He supported slavery and secession. In 1850-51 Maury proposed that southern slave owners and their slaves relocate in the Amazon Valley in Brazil, a slave country (until 1888), and thereby provide a safety valve for what he believed would lead to an inevitable conflict in the United States. After the Confederate defeat in the Civil War, Maury proposed a similar exodus to Mexico, which abolished slavery in 1824, where the former slaves would enter the country as peons.

***American evolutionists: Marsh and Cope***

Most of those who accepted the evolution of species, including lay people, clergymen such as the Congregationalist Henry Ward Beecher (1813-87) in Brooklyn and Presbyterian James McCosh (1811-94) at Princeton, and scientists such as Gray, Henry, and Agassiz's former student Alpheus Hyatt (1838-1902) at MIT, were not Darwinists. They were theistic, Spencerian, or neo-Lamarckian evolutionists and saw evolution as an orderly progressive process leading to more highly developed forms. Darwin's theory of natural selection gained acceptance within the scientific community much later, only in the 1940s-50s. At that time, scientists, such as Theodosius Dobzhansky (1900-75) at Caltech and Columbia in 1937, and in 1942 Julian Huxley (1887-1975) in London, Thomas Huxley's grandson, combined natural selection, Mendel's genetics, and other developments in the life sciences to give a more complete theory of evolution. They called the new theory neo-Darwinism or to use Huxley's words, the new synthesis or evolutionary synthesis.

Joseph Henry at the Smithsonian Institution believed evolution was a good working hypothesis. His understanding of evolution, like that of Jeffries Wyman (1814-74), a Harvard anatomist and practicing

Unitarian who suffered from a chronic lung problem (probably pneumonia) that killed him, was similar to Gray's theistic evolution. Lamarckian and neo-Lamarckian evolutionists included Joseph Le Conte, John Wesley Powell, Othniel C. Marsh, Edward Drinker Cope, and Henry Fairfield Osborne.

Le Conte (1823-1901), a physiologist and geologist and one of the organizers of the University of California (est. 1868) and the Sierra Club (est. 1889), accepted the evolution of species in 1873 and became a neo-Lamarckian. He equated evolution with progress and believed that the theory when applied to social policies would lead to a better society. Powell (1834-1902) of the United States Geological Survey, although raised a Methodist, later became an atheistic Lamarckian.

Marsh (1831-99) at Yale, the United States' first professor of paleontology, member of the United States Geological Survey from 1879, and president of the National Academy of Sciences from 1883 to 1895, said he was a Darwinist, but like Lamarck he believed that an inherent progressive force drove evolution. Marsh studied the evolution of the horse, reconstructing its growth over 60 million years; the evolution of dinosaurs, discovering 86 species to add to the 16 known species; and was likely the first American paleontologist to argue, like Huxley earlier, that birds had descended from dinosaurs. Marsh accepted the dinosaur-bird descent in 1877.

From 1868 Marsh was Cope's longtime bitter rival in the search for fossils in the midwest and far west. The methodical Marsh was primarily responsible for acquiring the vast collection of fossils at Yale's Peabody Museum of Natural History. His uncle, the Massachusetts born London financier and banker George Peabody (1795-1869), in 1866 provided $150,000 to establish the museum. Peabody also donated $150,000 in 1866 to establish Harvard's Peabody Museum of Archeology and Ethnology.

Cope (1840-97), a wealthy Quaker-born Unitarian and paleontologist at Pennsylvania, in 1871 proposed the still controversial rule that plant and animal lineages increase in size over time. He was a member of geological surveys from 1871 to 1879. Cope summarized his Lamarckian beliefs along with his racist and sexist views in 1887 in *The Origin of the Fittest*, in which he wrote "we all admit the existence of high and lower races, the latter being those which we now find to present greater or less approximations to the apes." The "peculiar structural characters that belong to the Negro in his most typical form are of that kind." Their characteristics resemble "those of immature stages of the Indo-European race."

Cope had no earned academic degrees but wrote prolifically, publishing over 1,200 scientific papers. Like Marsh, Cope studied the evolution of dinosaurs, discovering 56 species, and the evolution of the horse using the considerable fossil evidence he collected in the western states, such as Wyoming and Oregon. Cope claimed that on his many fossil hunts, he often pacified hostile Indians and escaped danger by repeatedly removing his false teeth, snapping them a few times, and then replacing them.

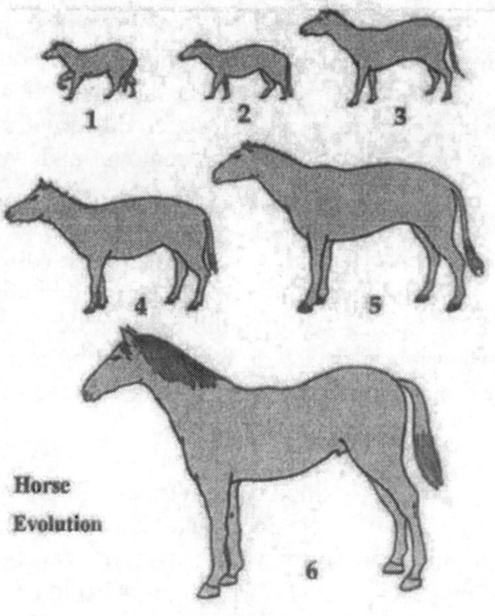


1. Four-toed horse, Echippus, about one foot high. Lower Eocene. N. America.
2. Another four-toed horse, Orohippus, a little over a foot high. Middle Eocene, N. America.
3. Three-toed horse, Mesohippus, about the size of a sheep. Middle Oligocene, N. America.
4 Three-toed horse, Merychippus, Miocene, N. America. Only one toe reaches the ground on each foot, but the remains of two others are prominent.
5. The first one-toed horse, Pliohippus, about forty inches high at the shoulder. Pliocene, N. America.
6. The modern horse, running on the third digit of each foot.

**Figure 10**
**Evolution of the Horse**
Public domain

Cope's longtime, bitter, and financially expensive rivalry with Marsh, known as the *bones wars*, began in 1868 and quickly escalated. The two were originally friends, but had a falling out in 1869 when Cope, in one of his *American Naturalist* journal articles, published an incorrect diagram of elasmosaurus, the longest (46 feet) of the plesiosaurs, a long-necked Cretaceous period (65 million years ago) fish-eating reptile. Cope, who identified and named elasmosaurus in 1868 from its fossil remains found in Wyoming, put the skull on the tail instead of the neck of a reconstructed elasmosaurus.

Marsh exploited Cope's error, and allegedly tried to bribe Cope's student field workers to send him some of Cope's latest fossils and to eliminate Cope's federal funding. Cope struck back and accused Marsh of stealing fossils. When Marsh joined the newly organized United States Geological Survey in 1879, Cope left the survey. Their war ended with Cope's death in 1897, and after each had spent personal fortunes of $80,000 to $100,000, including Marsh's $100,000 inheritance from his uncle in 1869.

The younger cocky and combative Cope, who spent his last years in poverty, considered himself one of the fittest and willed his brain and skeleton for scientific study upon his death in 1897. The Anthropometric Society established in 1889 in Philadelphia received Cope's remains. It funded the

Wistar Institute at the University of Pennsylvania the final resting place of many brains and skeletons, including Cope's. Edward A. Spitzka's father, Edward Charles Spitzka (1852-1914), a New York City neurologist and one of the society's eight founding members, did the study. He found that Cope's brain weighed 1,545 grams, more than average, but his skeleton fell far short of fitness and revealed clear evidence of syphilis, which may have killed Cope. Edward Charles Spitzka wanted his own brain weighed after his death. His son performed the removal and weighing, obtaining a weight of 1,400 grams.

Osborne (1857-1935), a paleontologist at Columbia and the American Museum of Natural History in New York City and an Episcopalian, embraced neo-Lamarckianism. Osborne, who aided Cope in his attacks on Marsh in 1890, studied the evolution of reptiles and mammals and originated the term *adaptive radiation* in discussing species. He believed the human races were different species and that the standard of intelligence of the average adult black equaled that of an eleven-year old youth of the species *Homo sapiens*.

The few Darwinian evolutionists included William Rogers at MIT and David Starr Jordan (1851-1931), a biologist and in 1913 president and chancellor of Stanford Junior College. Agassiz, the anti-evolutionist, had inspired Jordan to study fossils of fish, but Jordan later accepted a Darwinian evolutionary theory and eugenicist beliefs. Although raised in a Universalist-Unitarian environment, he remained removed from organized religion. Rogers, whose religion leaned toward Unitarianism, held evolutionary ideas that were more Darwinian than those of any other evolutionary theory.

A few well-known scientists never expressed publicly an acceptance or rejection of evolution. They included the Presbyterian John Torrey (1796-1873), a chemist and botanist at Princeton and later at the United States Assay Office in New York City; James Hall, a Catholic geologist and paleontologist in Albany, New York; and Spencer Baird (1823-87), an agnostic zoologist at the Smithsonian Institution. Torrey, whose anatomical studies indicated significant differences between humans and gorillas and only slight differences between blacks and whites, strongly influenced Gray, encouraging him to study botany, and collaborated with Gray on several publications. Hall, who taught at Rensselaer, was one of the founders of the American Association of Geologists and Naturalists established in 1840, and in 1856 president of the American Association for the Advancement of Science. Baird, under whom Cope studied, succeeded Henry at the Smithsonian in 1878.

## IX. THE BEGINNING OF GENETICS: SOLVING THE INHERITANCE PROBLEM

### *Gregor Mendel and the laws of inheritance*

Gregor Mendel (1822-84), an Augustinian priest in Brno (Brünn), Austria, today Czech Republic, and the founder of genetics, had a longtime interest and familiarity with the work of local plant and animal breeders. While attending the University of Vienna from 1851 to 1853, Mendel studied physics with Christian Doppler (1803-53), famous for discovering the Doppler Effect, and statistics, then a relatively new field and not taught at many universities. Mendel's use of statistics and quantification in plant studies was rare and enabled him to discover and demonstrate the mathematical relations present in his plant breeding experiments. Fortunately for Mendel, the Abbot Cyrill Franz Knapp (1792-1867) already had initiated plant breeding experiments when Mendel arrived at the Brno St. Thomas monastery and began teaching science at the Modern School.

Mendel wanted to learn how the characteristics of plants, such as plant height and color of seed and pod, passed from generation to generation. For this reason, he continued the breeding experiments with the garden pea plant (*Pisum sativum*), which he could grow in large numbers and easily manipulate, and either cross-pollinate or let self-pollinate. From 1856 to 1863 he studied the inheritance of characteristics, which Mendel called *elemente* (genes), in 28,000 pea plants in the monastery's 4.9 acre garden.

Mendel published his results in an 1866 paper "Experiments with Plant Hybrids" in *The Transactions of the Brünn Natural History Society*, establishing what scientists as early as 1903 called Mendel's laws. The first law, the law of segregation, showed that in the breeding process a specific *element* (gene) determined each of the different characteristics (traits) of the parents and their offspring. One *element* determined plant height, another *element* pod color, and similarly for the other characteristics.

The second law, the law of independent assortment of characteristics, showed that an offspring inherited an *element* separately, and that one inherited *element* did not affect any other inherited *element*. The *element* for plant height would not affect the *element* for pea color, or any other *element*, although the work of Thomas Morgan at Columbia on gene linkage in 1910-11 showed that Mendel's second law does not always hold. Genes located closely to each other on the same chromosome (that part of the cell's nucleus containing the genes) may be inherited together.

Mendel's third law showed that each plant (just as each animal and human) has two sets of *elemente*, one set inherited from each parent, and that the interaction of the *element* for a particular characteristic, one from each parent, determined the inheritance of the characteristic, such as plant height or pod color. The pea plant's characteristics did not blend in breeding but passed on intact from generation to generation, and one *element* (an *allele*) of the pair always predominated over the other *element* (*allele*). This was Mendel's principle of dominance. Dominance occurs because one of the alleles makes a broken protein, a protein that doesn't do anything. When this happens, the other allele, the working allele, usually is the dominant allele. The non-functioning allele is the recessive allele.

In one series of experiments, Mendel selectively bred his pea plants until he obtained tall plants (T) that consistently bred tall generation after generation, and short plants (t) that always bred short generation after generation. Then he bred a tall breeder (TT) with a short breeder (tt), and in the first generation obtained all tall plants, all of which were hybrids (Tt). When he self-pollinated one of these hybrids the next generation consisted of 3 tall plants for every 1 short plant. One of the tall plants was a pure tall (TT), two were hybrids (Tt, Tt), and the short plant was pure short (tt).

In identifying the inherited characteristics of a species, Mendel defined as *dominant* a characteristic that when present in only one parent appeared in a later generation. Tallness in the pea plants was dominant because in breeding pure tall with pure short plants, all the plants in the first generation were tall, even though each contained the short characteristic and was a hybrid (Tt). *Recessive* meant that both parents had to possess the characteristic in order for it to appear in later generations. Shortness in the pea plants was recessive. Shortness was present in each first generation hybrid (Tt), all of which were tall, but did not reappear until the second generation, and only as one-fourth of all the plants and only when not combined with the dominant tall characteristic (TT, 2 Tt, tt). Mendel had identified two important patterns: the pattern in which characteristics, later identified as genes, passed on to succeeding generations and the pattern of hybrid formation according to the Mendelian 3:1 ratio.

Mendel's plant breeding experiments succeeded for three reasons. First, the garden pea plant that he studied contained only seven different chromosomes (seven chromosome pairs). Each of five chromosomes carried a different characteristic or five total, the other two characteristics were on the sixth chromosome and widely separated. (Later studies have shown that *Pisum sativum* contains at least 40 genes and 4.3 billion base pairs). The seven characteristics were height of plant (tall or short 6-7 feet or 9-18 inches), color of unripe pods (green or vivid yellow), texture of seed (smooth or wrinkled), color of pea (yellow or bright green), color of seed coat (gray or white) and color of flower (purple or white), form of the ripe pods (inflated or pinched between the seeds), position of flowers (along the stem or on top). He established in each pair that the dominant characteristics were tall, green pod, smooth seed, yellow pea, gray seed coat, purple flower, inflated pods, and flowers on stem.

Second, Mendel carefully controlled the plants' breeding, and third he applied quantitative measure to establish the rules of inheritance. He was fortunate in choosing the pea plant because it breeds true, meaning it had a closed flower and fertilized itself, and only a single gene, the dominant gene of each inherited pair of genes determined each of the seven inheritance patterns, one for height, one for pod color, one for seed texture, one for pea color, one for seed coat color and flower color, one for pod form, one for flower position.

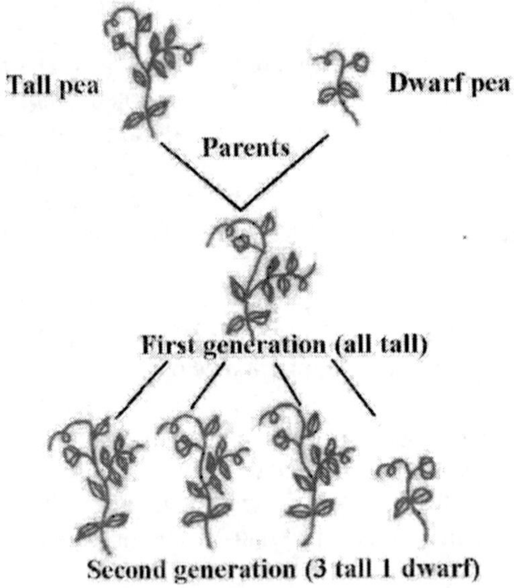


When Mendel cross-bred pure tall pea with pure dwarf pea plants, he found that the first generation of offspring were all tall. When he cross-bred the offspring he obtained one dwarf plant for every three tall plants.

**Figure 11**
**Mendel's Pea Plant Experiments**
Courtesy of the author

Mendel sent his results to Karl Nägeli (1817-91), the famous botanist and authority on evolution, at Munich. His use of mathematics (statistics) failed to impress Nägeli because in the 1860s botanists seldom, if ever, provided a mathematical presentation and interpretation of their experimental results. Nägeli's comments discouraged Mendel, an unknown in the scientific community, and he made no attempt to publish the results of other plant breeding experiments he described in letters to Nägeli. Their publication would have given his discoveries greater exposure within the scientific community.

As it turned out Mendel's use of mathematics enabled him to demonstrate the pattern of inheritance and to disprove the idea of blending of characteristics, but mathematics was the major reason why naturalists ignored his 1866 paper for 34 years. Mendel purchased copies of Darwin's books, including *The Origin*. The German translation of *The Origin* appeared in 1862, the monastery had a second edition dated 1863 with Mendel's handwritten notes on two pages and vertical lines drawn on eighteen pages, but whether he made any connection between the inheritance patterns he discovered and Darwin's theory of evolution remains unclear. Mendel briefly mentioned Darwin only four times after publishing his 1866 paper, all in 1870, and indicated neither strong support nor disagreement with Darwin's theory. He rejected the inheritance of acquired characteristics and that a changing environment gave rise to new species, both of which Darwin accepted. Mendel in contrast wrote of a constancy in nature, constant characteristics, constant species, and species having limits beyond which they could not change.

Different authors have reached different conclusions, suggesting that Mendel considered his experiments unrelated to Darwin's theory, and alternatively that Mendel accepted Darwinian natural selection, recognized the connection between his work and Darwin's, and attempted to provide a more complete understanding of the evolutionary process. They have proposed that Mendel's colleagues in Brno suspected he held Darwinian beliefs and was a free-thinker, which Mendel as a Catholic priest would have refrained from acknowledging. He may, like Gray, have adopted a theistic natural selection, but in his day the Catholic Church had not yet accepted any kind of evolutionary belief, even theistic evolution.

Conversely, Darwin very likely never knew of Mendel's conclusions. Mendel sent 40 copies of his 1866 paper to European scientists including Schleiden, Boveri, and Nägeli. The Royal Society, Linneaen Society, and Greenwich Observatory in England had copies, leaving about 30 copies unaccounted for, among them a possible Darwin copy. Darwin and Mendel, who visited London in July-August 1862 and attended Britain's Industrial Fair, never met to discuss their ideas. Mendel died of kidney problems and an enlarged heart two years after Darwin's death. His obituary did not mention his work on heredity. Mendel's successor as abbott at the Brno monastery burned many of Mendel's papers because he considered them worthless.

| DOMINANT AND RECESSIVE CHARACTERISTICS IN HUMANS | |
|---|---|
| ***Dominant trait*** | ***Recessive trait*** |
| Brown eyes | Blue eyes |
| Curly hair | Straight hair |
| Extra digits | Normal digits |
| Fused fingers | Normal fingers |
| Dwarfism | Normal growth |
| Normal hearing | Deafness from birth |
| Normal pigmentation | Albinism |
| Normal sight | Blindness from birth |
| Normal speech | Muteness from birth |

***Rediscovery of Mendel's experiments and the introduction of the Punnet Square***

Three scientists independently carrying out plant breeding experiments in 1900 obtained the same results that Mendel published in his 1866 paper. Their experiments, although clearly significant, resulted in a rediscovery of Mendel's publication that for 34 years lay largely unnoticed mainly because of its mathematical statistical presentation.

Hugo de Vries (1848-1935), a plant physiologist at Amsterdam, on 24 March, 1900 became the first scientist to rediscover Mendel's paper on pea plant experiments. In de Vries's 1880s breeding experiments with a type of thistle plant called the *American Evening Primrose* and with other plants, he obtained the same 3 to 1 Mendelian ratio but failed to recognize that the ratio had general application in the plant kingdom.

De Vries' experiments with the *American Evening Primrose* also showed that the breeding process sometimes gave rise to a significantly different and new variety that when selectively bred led to a new species. De Vries introduced the term *mutation* meaning change, from the Latin verb *mutare* to change, to account for his discovery, and popularized it in his two-volume book *Die Mutationstheorie*, published in 1901 and 1903. The change he observed in the *American Evening Primrose* was not a true mutation, but a reversion of a hybrid to its ancestral-type chromosomal arrangement.

Animal and pigeon breeders frequently had observed the same sudden change in breeding experiments, and referred to these offspring as sports or freaks. One of these changes that a Dover, Massachusetts, farmer Seth Wright first observed in 1791 in a single ram in his flock resulted in a short crooked-legged offspring called *ancon* (Greek, hook or bend) sheep. Wright and other sheep breeders bred the *ancon*, all of them short-legged, which prevented them from jumping over stone walls or fences, thereby reducing the number of lost sheep. Their short legs also limited the *ancon's* ability to run, making them less active and more likely to gain weight quickly. Darwin referred to *ancon* sheep as evidence of evolution in *The Origin*, Chapter 1, in the fifth section, Principles of Selection Anciently Followed and Their Effects, paragraph one. Wright's breeding experiment indicated that *ancon* was the result of a recessive Mendelian trait and of inbreeding.

Karl Correns (1864-1933), a botanist at Berlin, while experimenting with corn and pea plants, rediscovered Mendel's paper one month after de Vries on 24 April 1900. Correns, who married Nägeli's niece, studied under Nägeli at Munich. Erich Tschermak von Seysenegg (1871-1962), an Austrian botanist at the Hochschule für Bodenkultur in Vienna, made the third rediscovery in June 1900 while experimenting with pea plants.

Mendel's paper guided the three rediscoverers' interpretation and understanding of their plant breeding experiments. None of them recognized, as Mendel did, that *elemente* (genes) governed the transmission of separate characteristics to offspring in the 3 to 1 mathematical ratio.

The Punnett Square, which remains today the simplest and clearest representation of the Mendelian 3:1 ratio and the dominant and recessive pattern of genetic behavior, first appeared in 1905. Its inventor Reginald Punnett (1876-1967) at Cambridge, England, author of the first genetics textbook *Mendelism* in 1905 and named the world's first professor of genetics in 1910, showed in simple graphic form that a pure tall TT when bred with a pure short tt gave all tall in the first generation even though all four offspring were hybrids: Tt, Tt, Tt, Tt. In the second generation Tt bred with Tt gave TT, Tt, Tt, tt, which is the 3:1 ratio for tall, although only one was pure tall TT, and the other two were hybrids, Tt, Tt.

| **Punnett Square Showing the Crossing of Pure Tall Peas with Pure Dwarf Peas** | | |
|---|---|---|
| ***Eggs*** | ***Sperms*** | |
| | ***t*** | ***t*** |
| | | |
| ***T*** | **Tt (Tall)** | **Tt (Tall)** |
| ***T*** | **Tt (Tall)** | **Tt (Tall)** |

| **Punnett Square Showing the Crossing of Two Hybrid Peas** | | |
|---|---|---|
| ***Eggs*** | ***Sperms*** | |
| | ***t*** | ***t*** |
| | | |
| ***T*** | **Tt (Tall)** | **Tt (Tall)** |
| ***T*** | **Tt (Tall)** | **Tt (Dwarf)** |

The 1900 rediscovery of Mendel's 1866 paper brought Mendel to the attention of the scientific community. A decade later, new controversy emerged. In a 1911 article and in a more detailed article published in 1936, the English geneticist Ronald Fisher (1890-1962), then at University College

London, challenged the authenticity of Mendel's results. He concluded that Mendel's chance of obtaining his published results was one in 30,000.

Fisher had his supporters. One of them, the American physiologist Robert S. Root-Bernstein (b. 1953) at Michigan State, pointed out in a 1983 paper that the numbers from two of Mendel's experiments, 5,474 round seeds to 1,850 wrinkled seeds giving a 2.96 to 1 ratio, and 6,002 yellow seeds to 2,001 green seeds giving a 3 to 1 ratio, were too good. Mendel must have manipulated his results.

Most geneticists accepted Mendel's results and sided with Franz Weiling, a German geneticist at Bonn, who in his 1966 study of breeding experiments questioned Fisher's methodology. Sewall Wright (1889-1988), a geneticist at Chicago and Wisconsin, spoke for the scientific community, and concluded in 1966 after a thorough study of Mendel's work that Mendel had not deliberately falsified his results. Wright, who with Dobzhansky, Huxley, and others beginning in the late 1930s, was one of the foremost synthesizers of Darwinian evolution and Mendelian genetics and a founder of population genetics.

## X. A NEW THEORY OF EVOLUTION: PUNCTUATED EQUILIBRIUM

To deal with the question of the so-called incomplete fossil record, that is the supposed lack of fossils' showing a gradual evolutionary change of one species into its daughter species (phyletic gradualisms), the paleontologists Gould at Harvard and Niles Eldridge (b. 1943) at the American Museum of Natural History in New York City, in 1972 introduced a theory of *punctuated equilibrium*. Their theory, which they developed after studying the fossils of shellfish, proposed the existence of long periods, millions of years, during which species remain practically unchanged. This is the equilibrium phase. Relatively short periods of 10,000-20,000 years interrupt this phase, during which rapid change, leading to the formation of new species, occurs. This is the punctuated period.

Gould and Eldridge concluded that evolutionary changes occurred rapidly at specific times, and that during these times environmental catastrophes or disasters or significant environmental changes occurred and produced evolutionary changes. Instead of changing gradually from one generation to the next, Gould and Eldridge see evolution proceeding in discrete steps. Punctuated equilibrium says no transitional forms exist between species and therefore no missing links. The fossils of invertebrates found in the Cambrian strata (570-500 million years ago) appear to have no known ancestors and support Gould and Eldridge's rapid change or punctuation. The English zoologist and Darwinist, Richard Dawkins (b. 1941) at Oxford, has criticized punctuated equilibrium, preferring to support the gradualist or Darwinist evolutionary change.

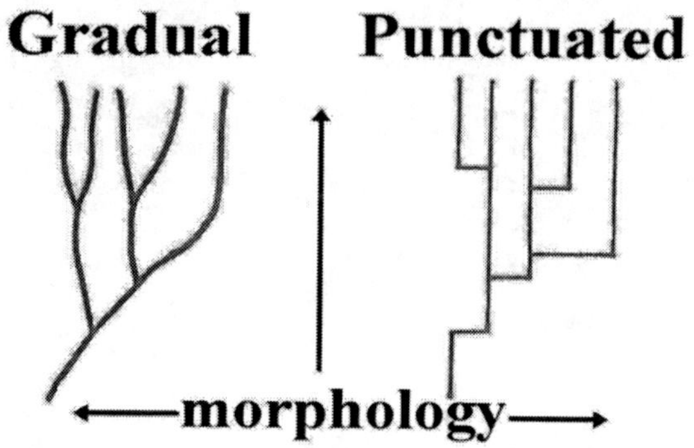


**Figure 12a**
**Comparison of Continuous Change Theory and Gould's Punctuated Equilibrium**
Courtesy of the author

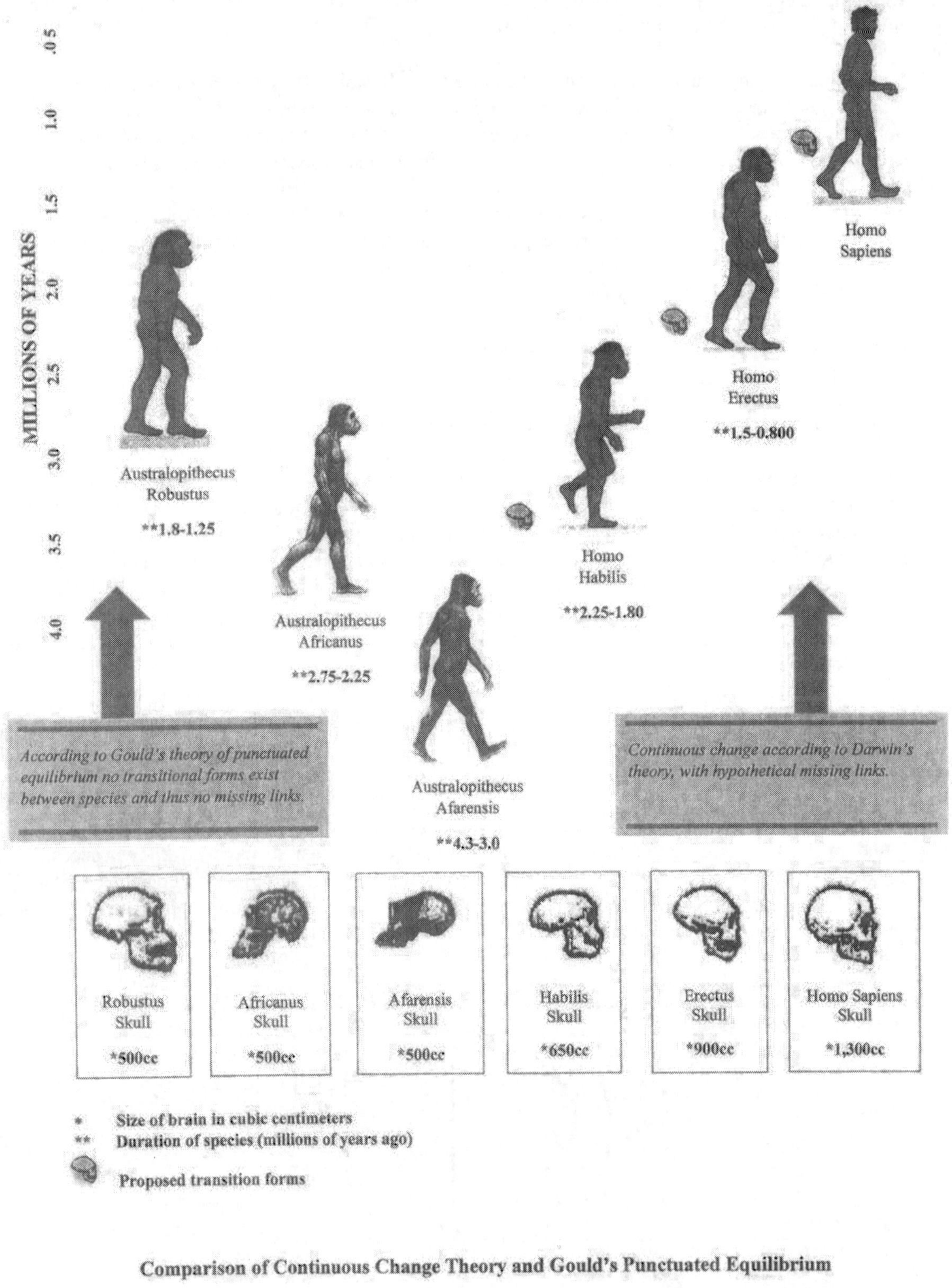


**Figure 12b**
**Comparison of Continuous Change Theory and Gould's Punctuated Equilibrium**
Courtesy of the author

## XI. EVIDENCE SUPPORTING EVOLUTION

Because of increasing biblical literalist-fundamentalist-creationist challenges to biological evolution, the National Academy of Sciences in 1999 published *Science and Creationism: A View from the National Academy* (revised and retitled in 2008 as *Science, Evolution,* and *Creationism*), a 48-page report that summarized the scientific evidence supporting evolution. The 15-member committee, whose members included Francisco Ayala at the University of California, Irvine, a former student of Dobzhansky, and Bruce Alberts at the University of California, San Francisco, examined evidence from paleontology, comparative anatomy, biogeography, embryology, and molecular biology. The following paragraphs taken from the report present its four main arguments.

***Paleontology: the fossil record***

Hundreds of thousands of fossil organisms found in well-dated rock sequences represent successions of life forms through time and reveal many evolutionary transitions. The organisms indicate that the simplest type of microbial life existed 3.5 billion years ago. Eukaryotic cells, the oldest evidence of organisms more complex than bacteria that scientists have discovered in fossils sealed in rocks, date from 2 billion years ago. Multicellular organisms, including fungi, plants, and animals, appear only in younger geological strata.

| ORDER OF INCREASINGLY COMPLEX LIFE FORMS | |
|---|---|
| ***Life Form*** | ***Millions of Years Since First Known Appearance (Approximate)*** |
| Microbial (prokaryotic cells) | 3,500 |
| Complex (eukaryotic cells) | 2,000 |
| First multicellular animals | 670 |
| Shell-bearing animals | 540 |
| Vertebrates (simple fishes) | 490 |
| Amphibians | 350 |
| Reptiles | 310 |
| Mammals | 200 |
| Nonhuman primates | 60 |
| Earliest apes | 25 |
| Australopithecine ancestors of humans | 4 |
| Modern humans | 0.15 (150,000 years) |

Scientists have discovered many intermediate forms between fish and amphibians, between amphibians and reptiles, and between reptiles and mammals so that no clear line of transition occurs from one species to another. Most fossils represent life forms that came between the form that preceded them and the form that followed them. The fossil record has enabled scientists to predict that future paleontological studies will not uncover any reversals in fossil transitions. Amphibians will not appear before fishes, mammals will not appear before reptiles, and no complex life will appear before the oldest eukaryotic cells.

***Comparative anatomy: common structures***

Comparative anatomy has reinforced the evidence from paleontology. Humans, mice, and bats have strikingly similar skeletons despite their different ways of life and diverse environments. They have an obvious bone-by-bone correspondence, yet a person writes, a mouse runs, and a bat flies with structures that differ in detail but show general similarities.

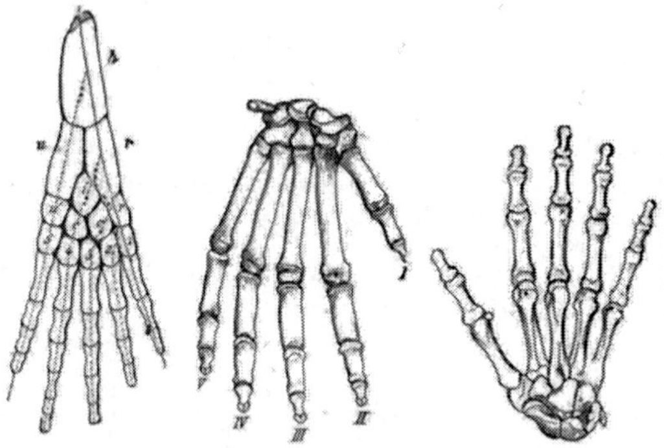

The bones in the forelimbs of terrestrial and some aquatic vertebrates are remarkably similar because they all evolved from the forelimbs of a common ancestor. From left to right, fore limb of an amphibian, hand bones of gorilla, human hand bones.

**Figure 13**
**Common Ancestry**
Source: Ernst Haeckel. *The Evolution of Man.* Vol. 2. New York: D. Appleton and Company, 1897.

Scientists call these similar structures homologies and have concluded that a common descent explains their similarities. Paleontology and comparative anatomy show the common ancestry of the mammalian ear and jaw through transitional stages. The lower jaws of mammals contain only one bone, whereas those of reptiles have several bones. The other bones in the reptile jaw are homologous with bones now found in the mammalian ear.

***Biogeography: distribution of species***

Biogeography has contributed evidence for descent from common ancestors, giving a tremendous diversity of life. Naturalists have described and named 250,000 species of living plants, 100,000 species of fungi, and one million species of animals, each occupying its own ecological setting, or place in nature. The plant and animal species found on the Galapagos Islands are similar yet belong to different species. Evolutionary theory claims the biological diversity resulted from the descendants of local and migrant predecessors adapting to their diverse environments.

Paleontology and molecular biology have confirmed that a geographic region's existing species and local fossils have similar structures. These studies have provided evidence that species without an established local ancestry migrated into the region. Such migration occurred in North and South America. Very different native organisms evolved on North and South America until the isthmus of Panama emerged 3 million years ago. After that time, South American mammals, such as armadillos, porcupines, and opossums, migrated north whereas the North American mountain lion moved south.

The Hawaiian Islands reveal how geographic remoteness produces biological diversity. Because of their isolation, located farther from a mainland than any group of islands, and the geologic evidence that the islands were never part of any mainland, Hawaii lacks many types of organisms and has greater numbers of fewer types of organisms. This occurred because many types of organisms never reached the islands, and those that did diversified over time because related organisms competing for the same resources were absent. Examples include the fly species *Drosophila*, of which 25 percent of the 2,000 species that exist throughout the world exist only in Hawaii, and the 1,000 species of snails and other land mollusks that reside only in Hawaii. The biological explanation for the multiplicity of related species, their great diversity, in remote localities, is that these species evolved from a few common ancestors that colonized an isolated environment.

***Embryology: similarities during development***

Embryology, the study of biological development from the time of conception, has provided another source of independent evidence for common descent. Barnacles represent sedentary crustaceans with little apparent similarity to such other crustaceans as lobsters, shrimps, or copepods. Yet barnacles pass through a larval stage in which they look like other crustacean larvae. The similarity of larval stages supports the conclusion that all crustaceans have homologous parts and a common ancestry.

Widely varying organisms, from fruit flies to worms, to mice, to humans, have very similar sequences of genes that are active early in their development. Those genes influence body segmentation or orientation in all their diverse groups. A very early common ancestry in all these groups best explains the presence of such similar genes doing similar things across a wide range of organisms.

***Evidence from molecular biology***

The discoveries of modern biochemistry and molecular biology have reinforced the unifying principles of common descent.

The code that translates nucleotide sequences into amino acid sequences, the genetic code, is essentially the same in all organisms. Proteins in all organisms invariably contain the same set of 20 amino acids. The unity of composition and function strongly favors a common descent of the most diverse organisms. Comparison of the human and chimp DNA sequences for the gene that encodes the hormone leptin, which is involved in the metabolism of fats, reveals only five differences in 250 nucleotides. Where the human and chimpanzee sequences differ, the corresponding nucleotide in the gorilla (shaded bars) enables scientists to derive the nucleotide that likely existed in the common ancestor of humans, chimpanzees, and gorillas. In two cases, the gorilla and human nucleotides match, while in the other three cases, the gorilla and chimpanzee sequences are the same. The common ancestor of the gorilla, chimpanzee, and human is most likely to have had the nucleotide that is the same in two of the three modern-day organisms because this would require just one DNA change rather than two.

```
human GTGCCCATCCAAAAAGTCCAAGATGACACCAAAACCCTCATCAAGACAATTGTCACCAGG
chimp GTGCCCATCCAAAAAGTCCAGGATGACACCAAAACCCTCATCAAGACAATTGTCACCAGG
                         A gorilla

human ATCAATGACATTTCACACACGCAGTCAGTCTCCTCCAAACAGAAAGTCACCGGTTTGGAC
chimp ATCAATGACATTTCACACACGCAGTCAGTCTCCTCCAAACAGAAGGTCACCGGTTTGGAC
                                                   G gorilla
human TTCATTCCTGGGCTCCACCCCATCCTGACCTTATCCAAGATGGACCAGACACTGGCAGTC
chimp TTCATTCCTGGGCTCCACCCTATCCTGACCTTATCCAAGATGGACCAGACACTGGCACTC
                         C gorilla

human TACCAACAGATCCTCACCAGTATGCCTTCCAGAAACGTGATCCAAATATCCAACGACCTG
chimp TACCAACAGATCCTCACCAGTATGCCTTCCAGAAACATGATCCAAATATCCAACGACCTG
                                          A gorilla

human GAGAACCTCCGGGATCTTCTTCAGGTGCTGGCCTTCTCTAAGAGCTGCCACTTGCCCTGG
chimp GAGAACCTCCGGGACCTTCTTCAGGTGCTGGCCTTCTCTAAGAGCTGCCACTTGCCCTGG
                    C gorilla
```

**Figure 14**
**Human and Chimpanzee DNA**
Courtesy of the author

As the ability to sequence the nucleotides composing DNA improved, scientists have begun to use genes to reconstruct an organism's evolutionary history. Because of mutations, which are random events, the nucleotide sequence in a gene gradually changes over time. The more closely the relation between two organisms, the less their DNA differs. Because humans and other organisms have tens of thousands of genes, DNA contains a tremendous amount of information about each organism's evolutionary history.

Genes evolve at different rates because some proteins can tolerate change in their amino acid sequence better than other proteins. The genes that encode these more tolerant, less constrained proteins, evolve faster. The average rate at which each kind of gene or protein evolves gives rise to a *molecular clock.* The molecular clock determines evolutionary relationships among organisms and indicates the time in the past when species started to diverge from one another. Once scientists calibrate the clock for a particular gene or protein by reference to some events whose time they know, they can determine the actual chronological time when all other events occurred by examining the protein or gene tree. A 1 percent change in amino acid sequence requires 20 million years.

Additional evidence supporting evolution involves sequences of DNA known as *pseudogenes.* Pseudogenes are remnants of genes that no longer function but which DNA carries along as excess baggage. Pseudogenes change through time, as they pass from ancestors to descendants, and they offer an especially useful way of reconstructing evolutionary relations.

With functioning genes, one possible explanation for the relative similarity between genes from different organisms is their similar ways of life. The greater similarity of horse and zebra genes compared to horse and tiger genes may account for the horse's and zebra's similar habitats and behaviors. This possible explanation does not work for pseudogenes, because pseudogenes perform no function. The degree of similarity between pseudogenes reflects their evolutionary relatedness. The more remote the last common ancestor of two organisms, the more dissimilar are their pseudogenes.

The evidence for evolution from molecular biology is overwhelming and is growing quickly. In the study of whales, the molecular evidence has enabled scientists to go beyond the paleontological evidence. Scientists long have postulated that whales descended from land mammals that returned to the sea. From anatomical and paleontological evidence, the whale's closest living land relatives are the even-toed hoofed mammals, such as modern cattle, sheep, camels, and goats.

***Creationism and the evidence for evolution***

Creationists cite what they say is an incomplete fossil record as evidence for the failure of evolutionary theory. The fossil record was incomplete in Darwin's time, but later paleontological research filled many of the existing gaps. The most persuasive fossil evidence for evolution is the consistency of the fossil sequence from early to recent. Nowhere on Earth do we find mammals in Devonian (the age of fishes) strata, or human fossils coexisting with dinosaur remains. Undisturbed strata with simple unicellular organisms predate those with multicellular organisms, and invertebrates precede vertebrates. Scientists have never found this sequence inverted. Fossils from adjacent strata are more similar than fossils from temporally distant strata. The most reasonable scientific conclusion drawn from the fossil record is that descent with modification has occurred.

Creationists argue that no one has seen evolution occur. This misses the point about how science tests hypotheses. We do not see Earth going around the Sun, the atoms that compose matter, the electrons flowing in a conductor, or the genes on a chromosome, we see their consequences. Scientists infer that atoms exist and Earth revolves because their extensive observations and experimentation have tested predictions derived from these concepts. By examining the fossil record, we see the obvious changes in fossils present in the strata, we do not have to be physically present to witness the changes.

On a different scale, we experience evolution occurring every day. Evolutionary forces have produced the annual changes in influenza viruses and the emergence of antibiotic-resistant bacteria. The rapidity which organisms with short generation times, such as bacteria and viruses, evolve under the influence of their environments is of great medical significance. Many laboratory experiments have shown that because of mutation and natural selection, such microorganisms change in specific ways from those of immediately preceding generations.

Molecular evolutionary data counter the recent arguments from intelligent design, whose proponents argue that structural complexity proves God's direct hand in specially creating organisms as they exist today. Their arguments echo those of Paley who held that an omnipotent creator specially designed the intricately-organized vertebrate eye in its present form. Modern-day intelligent design proponents argue that DNA and other molecular structures and molecular processes, such as the multi-stepped blood clotting process, are so irreducibly complex that they can function only if all the components operate at once. Proponents of intelligent design say that these structures and processes could not have evolved in the stepwise mode characteristic of natural selection.

Alleged *irreducibly* complex structures and processes typically are not, on closer inspection. A complex structure or biochemical process can function even if all its components are not present and functioning. Natural selection can build biochemical systems from simpler systems. Scientists can trace the history of a protein through simpler organisms. Jawless fish have a simpler hemoglobin protein than jawed fish, which in turn have a simpler hemoglobin than mammals.

Similarly, evolutionary mechanisms can explain the origin of supposedly irreducible structures or highly complex anatomical structures. The supposedly irreducible flagellum (a hair-like structure that protrudes from a bacterium and gives it motion) has components, each having its own individual function, and intermediate forms of flagella exist. Eyes may have evolved independently many times during the history of life on Earth. The steps proceed from a simple eye spot consisting of light-sensitive retinula cells now found in the flatworm, to formation of individual photosensitive units (ommatidia) in insects with light focusing lenses, to the eventual formation of an eye with a single lens focusing images onto a retina. In humans and other vertebrates, the retina consists not only of photoreceptor cells but also of several types of neurons that begin to analyze the visual image. Through such gradual steps, very different kinds of eyes have evolved, from simple light-sensing organs, to highly complex systems for vision.

***One more proof***

A recent experiment that Yousif Shamoo and his two-man team at Rice University in Houston published in *Molecular Cell* in May 2006 has provided additional support that natural selection is the mechanism that best explains evolutionary change. Using a glass-and-stainless steel fermentor the size of a coffee pot, Shamoo demonstrated in a *quick time* microcosm how the slow natural selection process produces evolutionary change. Rice University's news release called Shamoo's experiment, *capturing natural selection in a flask.* His experiment established when and how a specific gene mutated and how that single genetic change resulted in the domination of one population over others. To paraphrase Shamoo, if you can make genetic copies faster than the competition, you win. He provoked a war for dominance among competing bacteria and proved natural selection correct.

Shamoo used a mutant strain of the harmless heat-loving bacterium *G stearothermophilus*, which thrives at temperatures up to 73°C, but cannot make a key protein necessary for regulating its metabolism at high temperatures. Shamoo grew the bacteria for one month in the fermentor, raising the temperature 0.5°C each day. He sampled the fermentor for new strains every other day. The experiment yielded 1,500 generations of the rapidly produced bacteria, all of which either perished or survived. Eventually one mutant strain eliminated most of its competition and survived because it efficiently metabolized food (brewer's yeast) at high temperatures. Shamoo isolated the protein necessary for the bacteria to thrive as the temperature increased, locating it at a single spot (gene) in the bacteria's DNA and then monitored that spot.

The average number of individual bacteria in the fermentor totaled about 100 billion, leading Shamoo to estimate that about 20 million mutations occurred, of which only 700 had the capacity to produce new variants of the targeted gene. About 343 different strains actually developed, each contained one of six variants of the essential gene. The first variant, Q199R arose almost immediately and remained dominant through the 500th generation. At 62°C it could compete no longer and five new variants, each a mutation of Q199R that successfully adapted to the rising temperature, arose. Three of the five variants died off within several days, the remaining two competed for the next three weeks.

Shamoo found that the mutations did not arise where expected within the gene. Four of the six mutations occurred in gene regions that were identical in both heat-resistant and non-heat resistant forms of *G stearothermophilus*, showing that evolution acts at the atomic and molecular level. Shamoo's team repeated the experiment and obtained the identical mutants. The repeat test produced the same mutant genes, suggesting to Shamoo that the pathways of molecular adaptation, which represents the essence of evolution, are not highly variable under ideal conditions.

## XII. HOW MANY SPECIES ARE THERE

The United Nations' *Global Biodiversity Assessment*, a 1,100 page report published in 1995, suggested that 13-14 million different kinds of living things probably inhabit Earth, and that scientists have classified 1.75 million of them. The report says that since 1700, the world's cropland has increased fivefold as a result of adding more farm fields, and this has led to a loss of 54 percent of the United States' wetlands, 90 percent of Brazil's Atlantic forests, and 98 percent of the dry forests on Central America's Pacific Coast. Accompanying the wetlands losses since the start of the nineteenth century has been the loss of the best known species, the flowering plants and higher animals that have been disappearing 50-100 times faster than they did in prehistoric times. At least 112 kinds of birds and mammals have become extinct in the last 200 years. Today about 90 percent of the world's food supply, according to the report, comes from only 100 plant species.

In another international summit on bioactivity held at the National Academy of Sciences in October 1997, Robert M. May (b. 1936), a mathematical biologist at Oxford and the British government's science adviser, estimated that the total number of species on Earth was anywhere from 5 to 15 million. He believes that scientists have described about 1.5 million species, or about 200 times the 9,000 species Linnaeus classified (plants in 1753 and animals in 1758) and have introduced about 10,000 new species annually.

What is the rate of species' extinction? Stuart L. Pimm (b. 1949), an ecologist and evolutionary biologist at the University of Tennessee in Knoxville and now at Duke University, from paleological evidence concluded in a 1993 publication and in other publications in the 1990s-2000s that a species has an expected survival time of about one million years. But because of habitat destruction, global warming, acid rain, and other environmental changes, the expected survival time has fallen to about 1,000 years. Human beings, using only Stone Age technology, over a few thousand years have wiped out 10 percent of the world's bird species, and in Hawaii humans made extinct more than 90 percent of the bird species. Pimm says that his calculations and those of others on the effects of the continued splintering and shrinking of forest areas suggest that 50 percent of the world's bird species in the next half-century, and 75 percent in the next century will disappear, based on what we humans already have done. The increase in extinction rate, with the increase in human population and the emerging industrialization of the non-industrial countries, will continue, so that in the next century the life span of a typical species will decrease from 1,000 years to only 100 years.

Other studies indicate that the significant loss of population (numbers) in plant or animal species, may threaten an ecosystem more than the total loss of a species, mainly because the health and survival of many other life forms within an ecosystem depend on having a plentiful population of a particular species in the ecosystem. A study published in the 24 October 1997 issue of *Science* showed that the extinction of the separate populations that make up species has occurred at a rate 3 to 8 times faster than the extinction rate of species. Using known rates of tropical deforestation, the study suggested that populations have disappeared at an annual average rate of 0.8 percent, or 1,800 populations per hour in tropical regions alone. Species have disappeared in the tropics at an annual rate of 0.1-0.3 percent, or 2-5 species hourly. The five known mass extinctions that have occurred on planet Earth in the last billion years resulted from asteroid impacts or global climatic shifts. The sixth extinction that some scientists see as imminent will have a human cause. This both worries and offers hope to scientists.

## XIII. EVOLUTION AND RELIGION

The emergence of evolutionary ideas in the second half of the nineteenth century generated strong and different opinions on the role or place of God in the origin of species, including humans, and on the origin of the universe. In England Charles Kingsley (1819-75), an Anglican clergyman at Cambridge, writer, advocate of social reform, and Richard Owen lampooner, believed that a God who made all things that make themselves was much wiser than a God who simply made all things. Similarly, Frederick Temple (1821-1902), archbishop of Canterbury from 1896 to 1902, thought that the evolutionary process reflected more favorably on the unity of a creator, than did belief in separate creations. Their beliefs, and those of other theologians, commentators on religion, philosophers, and scientists, fell into schools of thought often described as traditionalists, modernists, and naturalists.

### *Traditionalists*

Traditionalists of every religious denomination, Protestant, Catholic, Jew, Muslim, and some Hindus, although sometimes holding different interpretations of their faith's main religious texts, whether the Bible, Koran, Vedas, or other religious text, stressed God's transcendence, separation, removal, or externality from the material world.

Charles Hodge (1797-1878), a conservative Presbyterian clergyman at Princeton, in *What is Darwinism*? published in 1874, argued that either Darwin was wrong or God did not exist. Man has not evolved, although lower forms of life may have evolved. Hodge said theologians must distinguish between what the Bible intended to teach, that God created the universe, and what the biblical authors believed and wrote in the Bible. The first is inerrant. The second, such as a stationary Earth around which the Sun moves and which implies a geocentric arrangement of the planets mentioned in *Genesis*, is not. Similarly, Hodge believed the Bible permitted slavery (various passages, Leviticus 25:44-46; Exodus 21:2-6, 7-14, 20-21; Paul in Ephesians 6:5-8, Philemon 1:1-25), but his opposition to slavery made him no less a Christian believer. He deplored the mistreatment of slaves but did not attack the institution of slavery itself.

James McCosh, Presbyterian clergyman and president of Princeton in 1868, had orthodox beliefs, but he accepted the human body's evolution and God's special creation of the soul or spirit within the body. He was one of the few ministers in the United States at the time who said that to accept evolution did not require a denial of God.

Protestant fundamentalist Christians differed from the traditional Christianity of Hodge and McCosh and believed in the Bible's literal interpretation. Prominent fundamentalists included the Chicago minister Clarence Dixon Ames (1854-1925), who assisted from 1910 to 1915 as an editor in publishing all twelve volumes of *The Fundamentals*, a series of brief 125-30 page religious pamphlets. The $300,000 financing for the publication came from the Union Oil Company president and millionaire Lyman Stewart (1840-1923) and his brother Milton Stewart (1838-1923), two Presbyterian laymen in Santa Paula, a city 100 miles northwest of Los Angeles, California. The series totaled 90 essays written by 64 authors.

William Bell Riley (1867-1941), a Baptist minister in Minneapolis and from 1919 to 1929 head of the World's Christian Fundamentals Association established in 1919, and Billy Sunday (1862-1935) in Chicago and Winona Lake, Indiana, a former baseball player and ordained a Presbyterian minister in 1903, were other well-known fundamentalists. They held that God created all the species in six 24 hour days and that no evolution ever occurred. William Jennings Bryan (1860-1925), former House member from Nebraska and three-time Democratic presidential candidate, accepted some evolution of lower life forms and an older age of Earth (day-age creationism).

Nineteenth-century Catholic theologians and scholars under the leadership of Pope Pius IX (1792-1878), pope from 1846 to 1878, generally were hostile to evolutionary ideas, especially their application to humans. Critical articles appeared in Pius IX's December 1864 encyclical *Quanta Cura* and its accompanying *Syllabus errorum*, and in publications such as the *American Catholic Quarterly Review* and the *Catholic Word*. Orestes Brownson (1803-76), a philosopher in Boston and later in New York City who converted from a Presbyterian (1823), to a Universalist (1824), to a Unitarian (1832), and finally to Catholicism in 1844, argued that if evolution were true, Christianity was false. On the issue of slavery, Brownson disagreed with abolitionists and believed preservation of the Union ranked higher than the abolition of slavery. The Negro should receive the natural rights of man even though he was not the equal of the white man.

Pasteur accepted a creationist origin of species. He successfully rebutted claims for spontaneous generation in his1864 debate before the French Academy of Sciences in Paris with Félix-Archiméde Pouchet (1800-72), a biologist at the School of Medicine at Rouen, and publicly attacked materialism and atheism. Pasteur condemned the positivists for banishing the infinite and the spiritual world from their philosophizing.

Other Catholic scholars tried to reconcile evolution and religion. St. George Jackson Mivart (1827-1900), a biologist-natural historian who converted from evangelism to Catholicism in 1844, held different academic positions in London. He wrote in *On the Genesis of Species* (1871) that species were not fixed but had an innate force capable of suddenly generating a new species. In *Contemporary Evolution* (1876), Mivart argued that the evolutionary process had pre-ordained ends and purposes. Mivart vehemently attacked Darwinians. Thomas Huxley called his beliefs Lamarckian. The Catholic Church did not agree with Mivart's evolutionary beliefs and in 1900 excommunicated Mivart for his evolutionary beliefs. Four years later, in 1904, the Catholic Church reinstated him.

Edouard LeRoy (1870-1954), a mathematician-philosopher and priest at the Collége de France in Paris, held that a psychic factor was the basis of life. His beliefs resembled those of Mivart and his friend Pierre Teilhard de Chardin (1881-1955), a Jesuit priest and palaeontologist in Paris, who later spent time (1926-46) in China and the United States (1951-55). Teilhard believed in teleological or purposeful directed evolution leading to the increasing complexity of organisms, and in a consciousness of the cosmos and life. Scholars have interpreted his ideas as both neo-Lamarckian and neo-Darwinian.

The Catholic Church did not accept the evolution of species until the twentieth century. It teaches that God directs evolution and that the universe is in a state of journeying toward a yet unattained ultimate perfection. God also immediately creates the human soul. Pope John Paul II (1920-2005), repeating the words of Pope Pius XII (1876-1958) in his 1950 encyclical *Humani Generis,* said in a message to the Pontifical Academy of Sciences in Rome in October 1996, that no conflict exists between Catholicism and evolution as long as evolution deals with the development of organisms including humans, and not the spiritual and moral development of humans. Pope Francis (b.1936) reaffirmed this message in October 2014. Catholicism teaches that the Bible is not a book of science, the lesson of *Genesis* is God's creation of the universe, not the author's description of the universe, which may be in error. Catholics, like most Protestants, must therefore distinguish between what the Bible teaches and what the biblical authors believed.

Initially, nineteenth-century Judaism regarded evolution as incompatible with the *Torah,* the Pentateuch or first five books of the Bible, which orthodox Jewish theologians maintained was complete and unchangeable. Rabbi Kaufman Kohler (1843-1926) in New York City, in the 1860s-70s led the break with Orthodox Judaism. By 1875 other Jewish theologians, such as Rabbi Isaac Mayer Wise (1819-1900) in Cincinnati, while rejecting Darwinism, declared in *Cosmic God* (1875) that evolution was a developmental process directed from within and dependent on a mystical vitalism.

Wise believed that design, purpose, and final cause existed in nature. In the three-day November 1885 Pittsburgh Conference (Declaration of Principles, the Pittsburgh Platform) that Kohler called, Wise and other reform-minded rabbis announced that they saw no conflict between the Bible and the new developments in science, such as evolution. Natural science was not antagonistic to the beliefs of Judaism. Most Jewish theologians today hold that God directs evolution and that the Bible is not a book of science.

Hindu sages in India believed in an eternal universe that had no one beginning and will have no end. As the fifth-century Hindu philosopher-astronomer Aryabhata I (b. 476) pointed out, the universe passed through millions of cycles *(kalpas)* of creation, destruction by fire or water, and re-creation. They believe that the Supreme Being revealed to the Vedic (Hindu) sages the laws of nature, including the cycles of creation, destruction, and re-creation, during their meditations. Because the knowledge was beyond the comprehension of ordinary humans, the sages coded the knowledge into a simple form that everyone could understand, and preserved it in the Vedas, the four sacred books that are the foundation of Hinduism. Composed and passed down orally from about 1500 BC and first written in Vedic Sanskrit, an early form of Sanskrit, the Vedas acquired their present form about 300 BC. The sages also preserved the laws of nature in the 17 or 18 post-Vedic Purana texts which date from 700-800 AD to 1500 AD.

In the Veda and Purana texts, a universe lasts for one *kalpa,* which is 4.32 billion years, or one day in the life of Brahma, the god who created the universe. The destruction of the universe by fire or water follows, during which Brahma rests for one *kalpa*, 4.32 billion years, or one night of Brahma. Upon awakening, Brahma recreates the universe. A Brahma day and night was therefore 8.64 billion years.

The sacred book Manu Smriti (Laws of Manu, 150-120 BC) described a complicated cosmology in which each *kalpa* required the passage of 14 *manus*, where each *manu* lasted 308,448,000 years, plus a final twilight period *(krita yuga)* of 1,728,000 years. The 14 *manus* plus one *krita yuga* represented one *kalpa,* or 4.32 billion years.

The cycle of creation-destruction-recreation repeats itself for 100 years of Brahma which is the lifespan of one Brahma. Each year of Brahma is 360 days (12 months) x 8.64 billion years, or 3.11 trillion years, and the lifespan of one Brahma is 100 x 3.11, or 311 trillion years. At this time destruction occurs and a new Brahma emerges to begin another creative cycle.

Brahma did not create the universe out of nothing, because if he did, the inequities in creation become his doing thereby making Brahma an unjust creator. This is in disagreement with the Hindu belief that human actions determine human conditions. Brahma is the creator of the universe and of all living things, and one of the three Gods of the Hindu trinity or *trimurti* that includes Vishnu, the preserver and Shiva, the Destroyer.

Muslim theologians accept a strict creationism. A day does not mean a 24-hour period and has different meanings, such as 1,000 years, in different contexts. They do not permit an evolutionary interpretation of God's creation act. God created the universe, the celestial bodies, including Earth, out of a vapor and all living things from water. The Koran does not give an order of creation as in *Genesis*.

Muslims reject the order in *Genesis* because of its contradictions, such as the separation of day and night on the first day of creation, creation of vegetation on the third day, and the Sun's creation on the fourth day. The Koran says that the heavens and Earth were one mass that exploded into existence, but it does not define the age of the universe.

***Modernists***

Modernism in Christianity emerged in the last half of the nineteenth century as a Lamarckian-Pantheist alternative to Darwinian evolution. Modernists stress God's immanence or presence in nature and that evolution is progressive. They believe in the unity of God, man, and nature. Henry Ward Beecher (1813-87), a Congregational minister at Plymouth Church in Brooklyn, New York, and Lyman Abbott (1835-1922), his successor at Plymouth Church, were well-known modernists. The modernist John Fiske (1842-1901), a Harvard philosopher and historian, was the chief American exponent of progressive evolution.

***Naturalists***

Naturalists include materialists, atheists, and agnostics. Critics of naturalism, such as the Catholic homeopathic physician Pierre Jousset (1818-1910) at St. Jacques Hospital in Paris, regarded naturalists as very dangerous anti-Christian scientists because they had imposed their theories as dogmas and their hypotheses as incontestable truths. The intolerance that naturalists blamed on the Catholic Church, Jousset said, had become their supreme law.

Naturalists such as Darwin, Huxley, and Spencer argued that neither reason, nor science could prove the existence of God. Traditionalists and modernists argued that neither science, nor reason could disprove God's existence. Haeckel, who published *The Riddle of the Universe* in 1900, accepted a materialist philosophy and believed that only matter in motion existed. Friedrich Nietzsche (1846-1900), who taught at the University of Basel and died after 11 years of insanity, wrote in *Beyond Good and Evil* (1886) that God does not exist, only matter in motion existed. Natural causes, not supernatural causes, acted in nature. Nietzsche's also discussed his belief in eugenics in *Beyond Good and Evil*.

Marcellin Berthelot (1827-1907), the French chemist, public servant, materialist, and militant atheist in Paris, like Haeckel wanted to banish spirituality from the universe and reduce the study of nature to matter and motion. Beginning in the 1850s Berthelot synthesized organic compounds, such as benzene, acetylene, naphthalene, formic acid, and methyl alcohol from their elements or simpler compounds. He claimed that their syntheses, rather than their isolation or separation from natural products, eliminated any mystery or vital force associated with living organisms, which were organic rather than inorganic in composition. Berthelot did not want chemistry to degenerate into a religion, and for chemists to believe in atoms as the Catholic believes in the existence of Christ in the communion host (wafer).

***The Scopes Trial: Dayton, Tennessee, 10-21 July 1925***

In the years immediately following World War I, Christian fundamentalists and social conservatives such as Riley and Sunday, complained the United States was losing its moral standards. The war, they said, already revealed Europe's decayed state, especially in Germany where belief in a Darwinian survival of the fittest had contributed to Germany's aggression and the outbreak of war. In the United States, they said the moral decline had led to considerable labor unrest, which they associated with workers' unions and socialism and the communist Red Scare of 1920, to widespread drinking despite Prohibition's beginning in 1920, to sexual immorality as typified in women's dress such as the *flapper girl*, and to women's smoking in public.

In Texas, Miriam A. Ferguson (1875-1961), governor from 1924 to 1926 and from 1932 to 1934, condemned the declining moral values. Called *Ma* by her supporters partly because of her initials, Ferguson allegedly said that she wanted no more of those sinful languages in the public schools. Stop learning our kids that dirty rotten French and Spanish. If the English language was good enough for Jesus Christ (*Ma* allegedly said) it's good enough for Texas. *Ma* warned against the influx of intellectual crap from the north and the loss of morality. Known for her devotion to two daughters and her husband, James Ferguson (1871-1944), a former governor (1915-17) whom the Texas legislature impeached in 1917, she cautioned that we can't recline on no flowery beds of roses. We gotta save our daughters, gotta save their bodies, gotta stop fornication. To the fundamentalists, evolutionary ideas taught in public schools were just another sign of the country's growing decadence.

The Scopes trial of 1925 was clearly not an isolated event, but a part of the reaction by Christian conservative moralists and fundamentalists to the period known as the *Jazz Age* or the *Roaring Twenties.* Ironically, the most severe critics of biological evolution and especially Darwinian evolution, such as Sunday and Riley, were often the most ardent supporters of Social Darwinism. These critics with few exceptions, among them William Jennings Bryan and social and religious workers, ignored the injustices and degradation of humanity that Social Darwinism forced on American society.

The Scopes trial was a test case on the constitutionality of the Butler Act, a bill that Tennessee Democratic state representative John Washington Butler (1876-1952) introduced in 1925. Butler, a member of the small Primitive Baptist sect, prosperous corn, tobacco, and wheat farmer, and admirer of Bryan, served one term in the Tennessee House representing Macon, Trousdale, and Sumner Counties northeast of Nashville from 1923 to 1927. He introduced his bill after he heard the story of a young woman in his district who had her faith shaken upon attending a nearby university and learning about evolution in a biology class. That same godless evolutionary theory, he thought, might turn his neighbors' children, perhaps one of his three sons, into atheists. His reading of Darwin's *The Origin* and *Descent of Man* led to the same conclusion. The Tennessee House (75 to 5) and Senate (24 to 6) passed his bill that banned teaching evolution in Tennessee's public schools, including universities and normal schools, and imposed a maximum fine of $500 for any teacher who violated the law. The governor Austin Peay (1876-1927) signed the Butler Bill into law on 21 March 1925.

Bryan, very much a fundamentalist but a day-age creationist and not a six-day literalist whom Riley asked to assist in the trial, and two other lawyers prosecuted Scopes. Clarence Darrow (1857-1938), Dudley Field Malone (1882-1950), and Arthur Garfield Hays (1881-1954) were the defense attorneys the American Civil Liberties Union (ACLU) provided to defend Scopes and to argue that teachers should include evolution in science instruction. Butler attended the trial.

John Thomas Scopes (1901-73), a graduate of the University of Kentucky and a first year science teacher at Rhea County Consolidated High School, admitted teaching evolution from page 192 of George Hunter's text *Civic Biology* (1914) on 21 April 1925 and therefore breaking the law. Scopes' lawyers wanted to transform the trial into a national referendum on the teaching of evolution in American schools, whereas Judge John Raulston (1868-1956), a religious man who quoted scriptures, successfully restricted the trial to a single issue: Did Scopes break the law? Despite the arguments of Darrow and Malone, the jury found Scopes guilty after deliberating for eight minutes. Raulston imposed a fine of $100 on Scopes, but according to state law only a jury could impose fines greater than $50. The Tennessee Supreme Court, therefore, reversed the judgment against Scopes in 1927 on a technicality, but it upheld the constitutionality of Tennessee's anti-evolution law.

Although Scopes agreed to participate in contesting the Butler bill in court, once the trial ended he left teaching after his first year. He attended the University of Chicago, studying geology for two years on a scholarship from newsmen and scientists, worked for Gulf Oil in Venezuela, and later for United Gas Company in Houston and Shreveport, Louisiana, until retiring in 1963. Bryan died of apoplexy (stroke) five days after the trial. Raulston lost his next judicial election.

Several states in the 1920s passed anti-evolution legislation, including Oklahoma in 1923 but repealed it in 1925, Florida in 1923, Mississippi in 1926, and Arkansas in 1928 although in 1968 the Supreme Court ruled the legislation unconstitutional. Tennessee repealed the Butler Bill in 1967 but in 1973 passed an equal time resolution for teaching evolution and creationism that a federal district court in 1975 ruled unconstitutional.

***Creationism today***

By 1930 an estimated 70 percent, and as late as 1942, about 50 percent of all American biology teachers avoided teaching evolution. The 1927 edition of Hunter's *Civic Biology* called *A New Civic Biology* removed its 6 pages and charts on evolution and replaced the word evolution with development. Few biology textbooks published between 1928 and 1960 emphasized evolution.

Harry Rimmer (1890-1952) and George McCready Price (1870-1963) led the anti-evolutionist crusaders of that time. Rimmer, a Presbyterian evangelist and author of *The Theory of Evolution and The Facts of Science* (1935), headed the one-man Research Bureau in Los Angeles and later served as field secretary for Riley's World's Christian Fundamentals Association. Price, a Canadian-born Seventh Day Adventist, self-educated geologist and author of *The New Geology* (1923), taught at Pacific Union College, an Adventist college in Angwin, California; Union College in Lincoln, Nebraska; and other Adventist colleges. He was the only living creation scientist Bryan could name at the Scopes trial but did not testify because he was teaching in England. Darrow called Price a charlatan. Another anti-evolutionist Harold W. Clark (1891-1986), a young Earth creationist, studied under Price at Pacific Union College and taught biology there for 35 years. Clark published *Back to Creationism* in 1929 and other creationist books.

In 1958, following the Sputnik scare of October 1957 and the fear that the United States was losing its scientific leadership to its Cold War opponent, the Soviet Union, Congress passed the National Defense Education Act, that from 1958 to 1962 provided $887 million for education that supported US national security, of which $300 million was for science and mathematics education, teaching, and literacy programs. The funds enabled the National Science Foundation to develop a series of up-to-date textbooks in physics, chemistry, biology, and earth sciences. The first of the high school science textbook series, the biology texts, appeared in 1963. A team of scientists and educators directed by the Biological Sciences Curriculum Study (BSCS) with its headquarters at the University of Colorado in Boulder prepared and tested the biology texts at a cost of $738,000.

The biology texts emphasized evolution, and sparked a renewed effort by anti-evolutionists, some of whom now called themselves scientific creationists, to eliminate evolutionary ideas from the textbooks or at least give creation theory equal time in the biology curriculum. Henry Morris (1918-2006), a professor of hydraulic engineering at Virginia Polytechnic Institute in Blacksburg, Virginia, from 1963 to 1970, was the leading supporter of the most recent anti-evolutionary campaign. Morris, co-author of *The Genesis Flood* (1961) with John Whitcomb (b. 1924), was one of the ten founders of the Creation Research Society, and founder and president of the four-member Institute for Creation Research in 1970 in Santee, California. Whitcomb, who taught until 1990 at Grace Theological Seminary in Winona Lake, Indiana, is the president of Whitcomb Ministries in Indianapolis.

Well-known creationists who were co-founders of the Creation Research Society when it first met in Midland, Michigan, (another source says Ann Arbor) in 1963 include Frank Lewis Marsh (1899-1992), a biology professor at Union College and author of *Variation and Fixity in Nature* (1976); John Grebe (1900-84), a chemist who held various US government positions; John N. Moore, who taught at Michigan State and was an editor of the *Creation Research Society Quarterly*; Walter Lammerts (1904-96), a plant geneticist with a PhD in genetics from Berkeley who formerly taught at UCLA; and Duane Gish (1927-2013), a biochemist who holds a PhD in biochemistry from Berkeley and served as vice-president at the Institute for Creation Research now located in San Diego, California. Gish, who debated Michael Shermer (b. 1954), a historian of science in Altadena, California, and editor of *Skeptic*, and Massimo Pigliucci (b. 1964), a philosopher of science at CUNY-Lehman College in New York City, claimed he never lost a debate to an evolutionist.

Morris and his associates supported their arguments against evolution by calling on the teachings of Francis Bacon, who stressed that true science is factual not theoretical, and that because evolution is only a theory, it is not true science. They have ignored what every scientist knows, that a scientific theory is not a guess, but a well-developed explanation supported by observations and tests, as in the case of the atomic theory or quantum theory. They also have adopted Thomas Kuhn's argument that scientific beliefs change as a result of competing models or paradigms, in this case the creationist and evolutionary models, to demand equal textbook treatment and classroom time for both models.

Creationists recently have employed Karl Popper's falsification method that holds that all scientific theories are ultimately falsifiable. They rallied behind Popper's argument that because Darwinian natural selection is so broad in its interpretation, it is ultimately unfalsifiable and therefore metaphysical and not scientific or subject to scientific testing. Creation scientists apparently have not thought of the consequences of using Popper's falsification test. If scientific ideas are ultimately falsifiable, so is creation science and the author of creation science, who is God, falsifiable. This is not a consequence that creation scientists intended or can accept.

The 1960s creationist movement that Morris, Gish, and others sparked, experienced another revival in the late 1980s - early 1990s, highlighted by a name change from creationism or creation science to intelligent design movement. Its proponents consider themselves a scientific group that uses scientific evidence to support its belief that an intelligent designer directed the formation of the universe and the development of life. Intelligent design, they say, is a scientific theory that does not identify the designer other than to claim that one exists, and may or may not be God. It avoids discussions on Earth's age and whether humans descended from primates. The selection of the name intelligent is to avoid any religious connotation that creationism and creation science implied.

The first significant intelligent design publication since its revival appeared in 1989. Percival Davis, who teaches the life sciences at Hillsborough Community College in Tampa, Florida, and Dean Kenyon, a biologist at San Francisco State University wrote *Of Pandas and People*: *The Central Question of Biological Origins*, a 170-page creationist text published by the Foundation of Thought and Ethics, a publisher of creationist books in Richardson, Texas. Phillip Johnson (b. 1940), a Berkeley law professor, embraced the theory and soon became intelligent design's leading spokesman. Johnson had undergone a crushing divorce and was searching for meaning in life. He found it in intelligent design, and in 1991 he published *Darwin On Trial* in which he labeled evolution an atheistic religion founded on philosophical materialism.

Johnson's book inspired other intelligent designers who met at Southern Methodist University in Dallas in 1992. Its participants included Johnson, William Dembski (b. 1960), a mathematician and philosopher now at Southwestern Baptist Theological Seminary in Fort Worth, Texas, and author of *Intelligent Design* (1999); Stephen Meyer (b. 1958), a program director and senior fellow at the Discovery Institute's Center for Science and Culture in Seattle, Washington; and Michael Behe (b. 1952), a Catholic, probably the best known intelligent design scientist.

Behe, a biochemist at Lehigh University, argued in his 1996 book *Darwin's Black Box* that the irreducible complexity of all organisms, such as the eye or bacterial flagellum (a hair-like part that gives a bacterium motion), is evidence of an intelligent designer. Scientists, he says, have yet to show how genes originated. The Discovery Institute, founded in Seattle in 1990, and its Center for Science and Culture established in 1996, is the home of intelligent design.

A 2011 survey of 1,000 biology teachers in the United States showed that the intelligent design movement, just as creation science earlier, continues to have its believers in the educational community. A significant number, 13 percent of high school biology teachers promote and teach creationism in the classroom, 30 percent unequivocally teach evolution, and a majority 57 percent, do not take an active stance on teaching evolution in the classroom. Within the general population 40 percent hold some kind of creationist beliefs.

***Scientists and atheism***

Are scientists more atheistic in the 2000s than they were in the past? Creationists claim they are, and for decades have applied the label of atheism, or godless, to discredit scientific theories such as natural selection. Yet in a 1997 survey that Edward J. Larson (b.1953), a history professor at Georgia and later at Pepperdine University in Malibu, California, and Larry Witham (b.1952), a Washington, DC, author and journalist, carried out, no evidence supported the creationists' position. Larson and Witham repeated a survey that James Leuba (1867-1946), an atheist and sociology professor at Bryn Mawr College in Pennsylvania, conducted in 1916. Leuba asked 1,000 scientists randomly chosen from *American Men of Science* if they believed in a God to whom one prayed in expectation of receiving an answer, that is, in a personal God. His results, which scandalized William Jennings Bryan and religious leaders of that day, revealed that only 41.8 percent of the scientists answered with yes.

Leuba published his survey in 1916 in *The Belief in God and Immortality: A Psychological Anthropological, and Statistical Study*. He predicted that disbelief in God would grow as education spread. Larson and Witham's repeat of the survey, which was identical in every way, failed to confirm Leuba's prediction. They found that 39.3 percent of the 1,000 scientists surveyed believed in a personal God, a result statistically the same as Leuba's. Other surveys in the 1990s-2000s have shown that about 93 percent of Americans believe in some kind of God, a figure that has varied little in 50 years.

## XIV. CONCLUSION

In the nineteenth century, geologists and naturalists progressed from the classification of rocks, strata, plants, and animals, to the development of theories and laws to account for the formation of strata, Earth's age, and the origin of plant and animal species. Experimental tests supported their theories, which were mainly qualitative rather than quantitative, as in the physical sciences. As geology and natural history matured and advanced, they, like the physical sciences, became increasingly quantitative in the twentieth century.

Contributions from uniformitarian geologists and naturalists, such as Hutton and Lyell, whose geological school replaced the vulcan-neptune catastrophists, enabled Darwin to develop his theory of natural selection and provide a mechanism for evolutionary change that scientists could test. The testability of Darwin's natural selection is why many naturalists adopted natural selection rather than the untestable inner force of Buffon's, Lamarck's, and Erasmus Darwin's evolutionary theory. Darwin published his theory of natural selection in *The Origin of Species* in 1859, followed by two other major works, *The Descent of Man* in 1871 and *The Expression of Emotions in Man and Animals* in1872. Darwinian spokesmen such as Thomas Huxley promoted natural selection. Mendel's work in genetics eliminated the major weakness of natural selection, the inheritance of favored characteristics. He showed in 1866 how the favorable characteristics that allowed an organism to adapt best to an environment and survive passed intact from generation to generation. No blending of the favorable characteristics occurred. The synthesis of Darwinian natural selection and Mendel's genetics that began in 1937 with Dobzhansky and in 1942 with Julian Huxley led to a more complete theory of evolution know as neo-Darwinism or the evolutionary synthesis. Radioactivity's discovery in 1896 and the development of radioactive dating methods in the 1900-50s enabled scientists to establish with considerable certainty Earth had existed for 4.6 billion years.

Darwin's theory denied that evolution implied progress and that God directed the evolutionary process. For this reason, many nineteenth-century scientists, such as Gray, as well as theologians, philosophers, educators, and much of nineteenth-century society never accepted a Darwinian natural selection theory of evolution. They preferred theistic evolution or some kind of neo-Lamarckian force that guided the evolutionary process, or Spencerism.

In the United States, in the decade before Darwin published his theory of natural selection and the outbreak of the Civil War, the monogenesis-polygenesis debate on the origin of the human races arose and augmented the racial anxiety and slave issue of that time. Polygenesis provided a rationale for racists and slave proponents, and some scientists in the United States and Europe converted from monogenesis to polygenesis, more because of their racial prejudice than because of scientific evidence. The most notable convert in the United States was the Swiss-born Agassiz who changed his belief after his first meeting with blacks in the United States. Agassiz, a strongly committed anti-evolutionist, achieved a high standing within the creationist community for his spirited debates and attacks on evolutionists such as Gray.

The debate continued into the twentieth century, most notably with the Scopes trial in 1925, and the revival of the creationist movement in the 1960s following the Sputnik scare and the introduction of new secondary school science curricula, particularly BSCS, the biological sciences curriculum with its emphasis on Darwinian evolution.

Essentially, the same debate has continued into the 2000s with most of the established western religions, Catholicism, Protestantism, and Judaism, accepting the organic evolution of life, including human life, but maintaining a role for God in the evolutionary process. Only fundamentalist religious believers, with some exceptions, continue to deny any evolution of species and an Earth much older than 6,000 years despite the overwhelming evidence supporting the evolution of species and an old Earth. Ironically, the most fervent deniers of biological Darwinism are the fiercest adherents of the misguided Social Darwinism. For the deniers, the maxim "you're entitled to your own opinion but not your own facts" now means "you're entitled to your own opinion and your own facts."

## PROPOSED DIVISIONS OF THE PRECAMBRIAN TIME PERIOD

Geologists have not accepted universally the name *Archean* for an eon and currently use several nomenclature systems. Precambrian, which they divide into late, middle, and early, is the oldest designation for this time period. All dates are billion years ago.

| GEOLOGICAL DIVISION OF PRECAMBRIAN TIME PERIOD | |
|---|---|
| Proterozoic Eon | 2.5-0.570 |
| Archean or Archeozoic Eon | 3.8-2.5 |
| Hadean Eon | 4.5-3.8 |

| THE UNITED STATES GEOLOGICAL SURVEY DIVISION OF PRECAMBRIAN ERA | |
|---|---|
| Proterozoic Period | 0.570 |
| Archean Period | 3.6 |
| Hadean Period | 4.5 |

| CANADIAN GEOLOGICAL SURVEY DIVISION OF PRECAMBRIAN TIME PERIOD | |
|---|---|
| Proterozoic upper eon | 0.570 |
| Archean lower eon | 4.5 |

| L.J. SALOP PROPOSED PRECAMBRIAN DIVISION (1973) | |
|---|---|
| Eocambrian | 0.68 |
| Epiproterozoic | 1.0 |
| | 1.3 |
| Neoproterozoic upper | 1.6 |
| Neoproterozoic middle | 1.9 |
| Neoproterozoic lower | |
| | 2.6 |
| Mesoproterozoic | 3.5 |
| Paleoproterozoic | |

| GEOLOGICAL TIME CHART | | | | | |
|---|---|---|---|---|---|
| ***Eon*** | ***Eras*** | ***Periods (of time) or Series (of rock)*** | ***Epochs (of time) or Series (of rock)*** | ***Approximate time in millions of years*** | ***Physical and biological features*** |
| **PHANEROZOIC** John Phillips (1800-74) in 1841 | **CENOZOIC** John Phillips in 1840 | **QUATERNARY** Jules Desnoyers (1800-87) in 1829 | HOLOCENE | 0.0117 to Present | Development of modern humans. |
| | | | PLEISTOCENE Charles Lyell (1797-1875) in 1839 | 2.58 to 0.0117 | Ice sheets cover Europe and North America; appearance of early humans. |
| | | **NEOGENE** | PLIOCENE Charles Lyell in 1833 | 5.33 to 2.58 | Development of modern plants and animals; formation of mountains in western America. |
| | | | MIOCENE Charles Lyell in 1833 | 23.03 to 5.33 | Highest development of large mammals; formation of mountains, including the Alps, Andes, and Himalayas. |
| | | **PALEOGENE** | OLIGOCENE Heinrich E. von Beyrich (1815-96) in 1854 | 33.9 to 23.03 | Development of higher mammals |
| | | | EOCENE Charles Lyell in 1833 | 55.8 to 33.9 | Rise to dominance of mammals, appearance of ancentral horse and primates |
| | | | PALEOCENE Charles Lapworth (1842-1920) in 1879 or William P. Schimper (1808-80) in 1874 | 65.5 to 55.8 | Marine invertebrate animals (mollusks, plankton), large mammals |

| GEOLOGICAL TIME CHART | | | | | |
|---|---|---|---|---|---|
| ***Eon*** | ***Eras*** | ***Periods (of time) or Series (of rock)*** | ***Epochs (of time) or Series (of rock)*** | ***Approximate time in millions of years*** | ***Physical and biological features*** |
| **PHANEROZOIC** John Phillips (1800-74) in 1841 | **MESOZOIC** John Phillips in 1840 | **CRETACEOUS** Jean Baptiste Julien D'Omalius d'Halloy (1783-1875) in 1822 | | 145.5 to 65.5 | Extinction of the dinosaurs; development of early mammals and flowering plants; deposit of chalk beds |
| | | **JURASSIC** Alexander von Humboldt (1769-1889) in 1799 | | 199.6 to 145.5 | Appearance of flying reptiles and birds; dominance of dinosaurs; appearance of primitive mammals; abundance of coniferous trees |
| | | **TRIASSIC** Friedrich von Alberti (1795-1878) in 1834 | | 251.0 to 199.6 | Appearance of dinosaurs; dominance of reptiles; appearance of cycadaceous trees |

| GEOLOGICAL TIME CHART | | | | | |
|---|---|---|---|---|---|
| ***Eon*** | ***Eras*** | ***Periods (of time) or Series (of rock)*** | ***Epochs (of time) or Series (of rock)*** | ***Approximate time in millions of years*** | ***Physical and biological features*** |
| **PHANEROZOIC** John Phillips (1800-74) in 1841 | **PALEOZOIC** John Pillips and Adam Sedgwick (1788-1873) In 1838-40 | **PERMIAN** Roderick Murchison (1792-1871) in 1841 | | 299 to 251 | Development of reptiles; decline of huge plants of the Carboniferous |
| | | **CARBONIFEROUS** William Conybeare (1787-1857) and William Phillips (1775-1828) in 1822 | PENNSYLVANIAN Henry Shaler Williams (1847-1918) In 1891 | 359.2 to 299 | Age of coal formation of coal beds from luxuriant plant life in warm, swampy forests, great fernlike trees; appearance of primitive conifers; abundance of insect life; first appearance of reptiles; development of amphibians |
| | | | | 320 | |
| | | | MISSISSIPPIAN Alexander Winchell (1824-91) In 1870 | 360 | |
| | | **DEVONIAN** Adam Sedgwick and Roderick Murchison in 1839 | | 416 to 359.2 | Age of the fish; appearance of primitive amphibians; development of primitive plant life on dry continents |
| | | **SILURIAN** Adam Sedgwick and Roderick Murchison in 1805 | | 443.7 to 416 | Appearance of scorpions, the first animals to live on land; extensive coral reefs. |
| | | **ORDOVICIAN** Charles Lapworth in 1879 | | 488.3 to 443.7 | Floods and recession of shallow seas; deposits of limestone, lead and zinc ores; abundance of marine invertebrate life; appearance of a few primitive fishlike vertebrates. |
| | | **CAMBRIAN** Adam Sedgwick and Roderick Murchison in 1835 | | 542 to 488.3 | Shallow seas over much of the land; formation of sedimentary rocks; development of marine invertebrate life, including brachiopods, snails, sponges, and trilobites |

| GEOLOGICAL TIME CHART | | | | | |
|---|---|---|---|---|---|
| *PRECAMBRIAN* | | | | | |
| ***Eon*** | ***Eras*** | ***Periods (of time) or Series (of rock)*** | ***Epochs (of time) or Series (of rock)*** | ***Approximate time in millions of years*** | ***Physical and biological features*** |
| **PROTEROZOIC** Samuel F. Emmons (1841-1911) in 1888 | | | | Neo: Late 542-1,000 | Oldest animal fossils, formation of mountains |
| | | | | Meso: Middle 1,000-1,600 | deposits of iron ore abundance of lime secreting algae |
| | | | | Paleo: Early 1,600-2,500 | appearance of sponges, transition to oxygen atmosphere |
| **ARCHEOZOIC or ARCHEAN** James D. Dana (1813-95) in 1872 | | | | Neo: 2,500-2,800 | Great volcanic activity |
| | | | | Meso: 2,800-3,200 | formation of igneous rocks |
| | | | | Paleo: 3,200-3,600 | some microscopic algae, probably some protozoa |
| | | | | Eo: 3,600-3,800 | Oldest Earth rocks |
| **HADEAN** Preston Clous | | | | 3,800-4,500 | Formation of Earth and solar system. |

## REVIEW QUESTIONS ON DARWIN'S CENTURY

1. What is the main theme of this chapter?
2. According to biblical genealogy, how old is Earth?
3. What are other methods of dating Earth?
4. Who was William Paley and what was his argument for design in the universe?
5. How did Count Buffon calculate Earth's age?
6. Discuss the evolutionary theories of Buffon, Erasmus Darwin, and Jean Lamarck.
7. Who was Paul Kammerer? Trofim Lysenko?
8. How did the neptunist and vulcanist theories account for changes in Earth's strata? Identify Abraham Werner; Charles Bonnet; preformation theory; parthenogenesis; George Cuvier.
9. What did the uniformitarian school believe caused the changes in Earth's strata? Identify James Hutton, Charles Lyell.
10. What was the most striking example of gradual change in species that Charles Darwin observed during his five-year voyage on the *Beagle*?
11. What theory of evolution did Charles Darwin propose and how did it differ from the earlier theories of Buffon and others?
12. Why did Darwin delay in publishing his theory of evolution?
13. What was the focus of Darwin's discussion on evolution in the *Origins of Species*?
14. What was the major weakness of Darwin's evolutionary theory?
15. Who was Alfred Russel Wallace?
16. Identify: Thomas Huxley, Ernst Haeckel, August Weismann, Bishop Wilberforce, Asa Gray, Louis Agassiz.
17. How did Asa Gray modify Darwin's theory of natural selection?
18. What was the central issue in the monogenesis-polygenesis debate in antebellum United States?
19. What did Gregor Mendel's pea plant experiments demonstrate? Why did he succeed?
20. Why didn't Darwin use the results of Mendel's experiments and why didn't Mendel apply his results to correct the major weakness of Darwin's theory?
21. Why do most scientists today accept natural selection as the origin of species?
22. What was the specific issue contested at the 1925 Scopes trial?
23. Discuss today's evolutionary-creationist debate.
24. Does scientific evidence support Darwin's theory of evolution? If so, what is the evidence, or is there no evidence of Darwin's theory or of any evolutionary theory?

# CHAPTER 8
# MOLECULAR BIOLOGY AND THE DOUBLE HELIX

> But now, to my delight and amazement, the answer was turning out to be profoundly interesting. For over two hours I happily lay awake with pairs of adenine residues whirling in front of my closed eyes. Only for brief moments did the fear shoot through me that an idea this good could be wrong.
>
> James D. Watson, *The Double Helix*

## I. INTRODUCTION

Molecular biology emerged as a new scientific discipline in the 1950s, building on nineteenth- and twentieth-century cell, protein, chromosome, and gene investigations. These investigations included the Swiss biologist Friedrich Miescher's 1869 discovery of the two nucleic acids deoxyribonucleic acid (DNA) and ribonucleic acid (RNA); the chromosomal studies of Edmund Wilson, Thomas Morgan, and their school at Columbia; the amino acid and protein studies of Emil Fischer at Berlin; the DNA and RNA analyses of Albrecht Kossel at Berlin and Phoebus Levene at the Rockefeller Institute; the contributions of Frederick Twort at London and Felix d'Herelle in Paris that marked the beginning of bacterial virus (bacteriophage) studies; and the work of John Northrop at the Rockefeller Institute, Max Schlesinger in Frankfurt-am-Main, and John Bernal at the Cavendish Laboratory that established the chemical composition of bacteriophages, showing that they consisted of protein and nucleic acid.

The infusion of physical scientists, such as Maurice Wilkins and Francis Crick, and the introduction of the mathematical-quantitative methods and the instruments of the physical sciences, such as the electron microscope, chromatography, radioactive tracers, and x-ray diffraction, particularly after World War II, greatly aided molecular biology's development. The electron microscope provided the first images of bacteriophages and later, images of DNA. Erwin Chargaff at Columbia in 1950 used paper chromatography and revealed the one-to-one ratio of the organic base pairs adenine-thymine and cytosine-guanine, A:T and C:G, in DNA. The radioactive phosphorus and sulfur tracers that Alfred Hershey and Martha Chase at Cold Spring Harbor introduced in their crucial experiment in 1952 proved that DNA, and not protein, was the gene and the carrier of hereditary material. X-ray diffraction, as Rosalind Franklin and Wilkins at King's College demonstrated, provided convincing evidence that DNA had a double helix structure with its four organic bases, A-T and C-G, paired inside the helixes. The cumulative evidence enabled James Watson and Crick at the Cavendish Laboratory in 1953 to solve DNA's molecular structure, a discovery that may rank as the greatest scientific achievement of the twentieth century.

During this century of fruitful investigation, from about 1838 to 1953, scientific understanding of the unit of life and the carrier of hereditary characteristics advanced from the cell, to the chromosomes in the cell's nucleus, to the genes located on the chromosomes, and to the DNA molecules whose sequence of bases determine the gene makeup of all living organisms.

In the sixty years since that time scientists have continued to enrich our understanding of chromosomes and genes. Marshall Nirenberg at the National Institutes of Health in Bethesda, Maryland, and Har Gobind Khorana at Wisconsin in 1961 broke the genetic code. Genetic engineering, cloning, and stem cell research advanced significantly, and the Human Genome Project in 2001 mapped and sequenced the human genome. Not surprisingly the advances in genetics brought controversy and generated much debate inside and outside the scientific community. Human cloning and stem cell research have become the two most controversial scientific, social, and moral issues. Animal cloning experienced little opposition and gained wide acceptance, but what scientists and society recognize as the next step, the possibility of human cloning, or the creation of human life, has triggered considerable opposition.

Similarly, stem cell research, particularly the use of embryonic stem cells that have not yet begun to differentiate into specific kinds of body cells such as skin cells or hair cells, has met with resistance particularly outside the scientific community. Both of these issues relate to the fundamental question that the Austrian physicist Erwin Schrödinger (1887-1961) then at Dublin asked in his 1944 book, *What is Life?* Until someone answers Schrödinger's question, and its logical derivative, when does intelligent life begin, the debate will continue.

## II. BEGINNING OF GENETICS

### *Developments in cell theory*

Europeans, such as Robert Hooke (1635-1703) and Robert Brown (1773-1858), provided the earliest advances in cell theory. Hooke, a London physicist, curator and a secretary to the Royal Society, published in *Micrographia* (1665) the first clear drawings of plant cells that he made from microscope observations of thin sections of cork, the elastic outer bark of a Mediterranean area oak tree. Later studies showed that the diameter of the average plant cell is 30 micrometers ($3x10^{-6}$ meters).

Brown, a Scottish botanist and the Linnaean Society's librarian in London, in 1831 gave the name *nucleus*, meaning little nut, to the small body that he and others had observed in plant cells. In his microscope experiments of 1827 Brown noted the random movement of pollen grains in water, which he thought provided evidence of life present in pollen grains. His discovery, since called the Brownian movement, occurred because of the atomic or particle makeup of water and in general of matter's atomic structure.

Two German scientists, Mathias Schleiden (1804-81), a botanist at Jena, and his friend Theodor Schwann (1810-82), a zoologist at Berlin, proposed that the cell was the unit of life, the basic structural unit of living things. Quick-tempered and combative, Schleiden in 1838 held that all *plants* consisted of cells. Schwann, unpretentious and genial, who in his later years accepted mysticism, expanded Schleiden's belief and in 1839 maintained that the cell was the basic structural unit of *plants* and *animals*, and therefore the origin of all living things. The definitive statement on cell theory came from Rudolph Virchow (1821-1902), a politically-active pathologist at Berlin in his 1858 book *Cellular Pathology*. Virchow wrote that every cell had in itself the complete characteristics of life and was the basic metabolic unit as well as the structural unit of all living things. All cells came from preexisting cells.

As a result of the studies of Schleiden, Schwann, Virchow, and others, scientists accepted that all organisms have the cell as their basic structural unit. They range from one-celled amoebae or one-celled blue-green algae and bacteria that lack a nucleus to multi-celled organisms, among them the human body. Cells that have no nucleus when fully developed, such as most mammalian red blood cells, cannot reproduce and have short lifetimes. In humans a red blood cell has a diameter of about 8 micrometers.

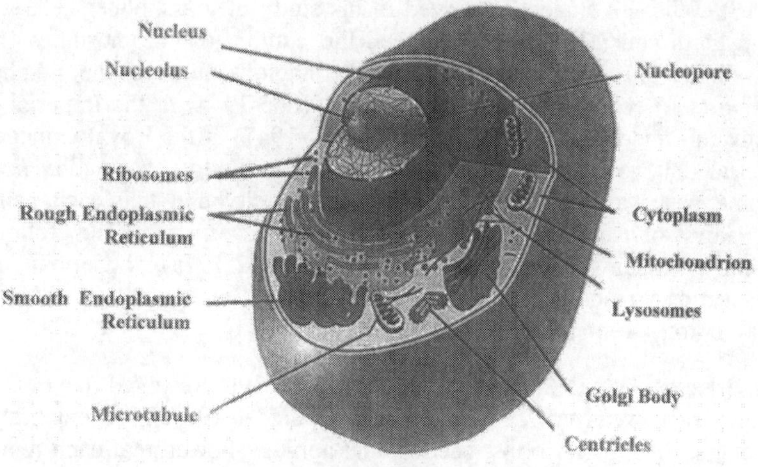


**Figure 1**
**Composite Cell**
Courtesy of the Human Genome Project, National Institutes of Health

***Chromosome studies***

Significant advancement in chromosome and gene studies began in the 1870s-80s with the investigations of the German cytologists Walther Flemming (1843-1905) at Kiel and Eduard Strasburger (1844-1912) at Bonn. Flemming in 1879 found that a structure within the cell's nucleus that he called *chromatin* (a rod-shaped chromosome) absorbed aniline dyes and dyes such as picric acid, chromic acid, and iodine. The absorption enabled experimenters to observe the actual division of a cell into two new cells, each containing a complete set of chromosomes. Flemming in 1882 named the cell division *mitosis*. He was unaware of Mendel's hybridization experiments with pea plants in 1866 and did not realize the connection between his investigations and genetic inheritance. The rediscovery of Mendel's paper in 1900 revealed the importance of Flemming's work.

Strasburger in 1882 distinguished between protoplasm within the nucleus, called *nucleoplasm* and essentially proteins and DNA, and the protoplasm outside the nucleus, *cytoplasm* which later research showed consists of carbohydrates, fats, and proteins. He discovered in 1888 that the germ (gamete, reproductive, or sex) cells, sperm and egg, have half the number of chromosomes that body (somatic) cells have, and introduced the names *haploid* to indicate a cell having half the complete number of chromosomes and *diploid* to indicate a cell having the complete number of chromosomes. Strasburger's work later helped explain sexual reproduction.

The anatomist Heinrich Waldeyer-Hartz (1836-1921) at Berlin in 1888 introduced the word *chromosome* to describe the chromatin structures Flemming worked with in his cell experiments. Two decades later William Bateson introduced the word *genetics* to biology in a letter of 18 April 1905 to Adam Sedgwick and publicly in 1906 at the Third International Congress on Genetics held in London. Three years later Wilhelm L. Johannsen (1857-1927), a botanist at Carlsberg (Denmark), in *The Elements of Heredity* (1909) introduced the word *gene* (Greek meaning *to give birth to*) to indicate the location of the heredity unit on a chromosome.

Rediscovery of Mendel's pea plant experiments in 1900 effectively launched the study of genetics. In the following thirty years, Columbia University in New York City, led by the geneticists Edmund B. Wilson (1856-1939) who left Bryn Mawr for Columbia in 1891, Thomas A. Morgan (1866-1945) who taught at Bryn Mawr from 1891 to 1904 before accepting a position at Columbia in 1904, and Herman Joseph Muller (1890-1967) who obtained his PhD under Morgan in 1916, became the world center for genetics research. The first significant discovery at Columbia occurred in 1902. Walter Sutton

(1877-1916), Wilson's student, observed in his study of grasshopper cells a chromosomal (nuclear) division in germ cells. The reaction halved the number of chromosomes from the diploid number found in somatic or body cells (two) to the haploid number, the number found in gamete or reproductive (sex) cells (one). John B. Farmer (1865-1944) at the Imperial College of Science and Technology in London and John E. Moore (1870-1947) at the Royal College of Science in London, now Imperial College, in 1905 named the reaction that Sutton observed *meiosis*. Sutton noted that the chromosome pair was the probable carrier of heredity and that each pair behaved according to Mendel's laws of segregation and independent assortment. The following year Sutton and independently Wilson's friend Theodor Boveri (1862-1915), a German cytologist at Würzburg, proposed, just as Strasberger had, that each sperm cell or egg cell contained only one chromosome instead of a chromosome pair.

Wilson and Nettie M. Stevens (1861-1912), who in 1903 received her PhD under Morgan and was later a biologist at Bryn Mawr, continued these studies in 1905. Their experiments done independently with insects, such as mealworms, beetles, and aphids, showed that each female egg cell contained a single chromosome that Wilson by 1910 called the X chromosome and that each male sperm cell contained either an X chromosome or a second shorter (smaller) chromosome that he called the Y chromosome.

Stevens's 1905 study of the mealworm *Tenebrio molitor* revealed its two kinds of sperm cells: one in which the nucleus had ten large chromosomes and a second in which the nucleus had nine large and one small chromosome. All the egg nuclei were the same and had ten large chromosomes. When she examined the mealworm's somatic cells she found that each somatic cell of the female offspring had 20 large chromosomes, whereas each of the male offspring's somatic cells had 19 large and one small (shorter) chromosome. These results led Stevens to conclude that an XX fusion of nuclei resulted in a female, and an XY fusion produced a male, thereby establishing the genetic basis of sex determination.

The X and Y sperm chromosomes that Stevens originally called unequal partners because X is larger than Y, established gender, the one-to-one male-female ratio, and obeyed Mendel's laws of segregation and reunion of the X and Y chromosomes. When an X-bearing sperm chromosome fertilized an egg the offspring was a female (XX), whereas when a Y-bearing sperm chromosome fertilized an egg the offspring was a male (XY). Females normally have an XX chromosome pair and males an XY chromosome pair.

Stevens, whose research career began in her mid-thirties, died of breast cancer in 1912. Wilson continued his studies on chromosomes and sexual inheritance in eight papers published from 1905 to 1912, locating precisely for the first time in 1911 an inherited trait on a specific chromosome. The X chromosome contained the gene for red-green colorblindness, the most common type of colorblindness.

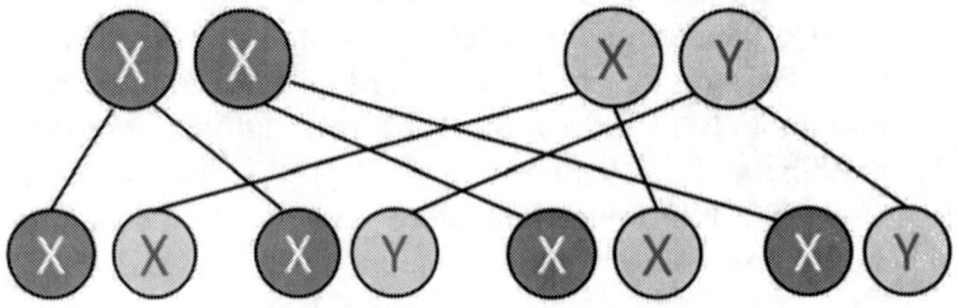


**Female - XX** **Male - XY**

**Figure 2**
**Sex Determination X and Y Chromosomes**
Courtesy of the author

Morgan, Wilson's colleague at Columbia and initially skeptical of Mendel's conclusions, continued the chromosomal research, showing that the genes arranged themselves linearly on chromosomes. Beginning in 1908 he studied generations of fruit flies (*Drosophila melanogaster*), which had only four pairs of chromosomes in each cell and a 12-day life cycle, and in 1911 began drawing the first chromosome maps for fruit flies. Morgan bred the fruit flies in his laboratory, starting with a white-eyed naturally-mutated male and a red-eyed female. He kept them in cotton-plugged milk bottles and other vials, anaesthetized them with ether, and examined them under a lens. His investigations on the 15 kinds of mutations that naturally occurred during their breeding cycle, such as body color, eye color, and wing shape, established that chromosomes carried the hereditary traits (genes) and that the genes responsible for eye color, body color, wing shape, and sex, were on the same chromosome and therefore linked and inherited together.

Morgan summarized his conclusions on the lineal arrangement of genes in chromosomes in *The Mechanism of Mendelian Heredity* in 1915 and his conclusions on mutations in *A Critique of the Theory of Evolution* in 1916. He received the Nobel Prize in medicine in 1933 for his work in genetics. In the years since Morgan first established genetic linking on chromosomes, geneticists have discovered many other cases of gene-linked characteristics, such as the link between red-green colorblindness, hemophilia A and B, Duchenne and Becker's muscular dystrophy, and ichthyosis (dryness and scaling of skin), on the X-chromosome in humans.

Muller, Morgan's student at Columbia, continued the work on mutations, and in his 1927 paper "Artificial Transmutation of the Gene," published in *Science,* he reported that x-rays caused mutations in fruit flies and increased mutation rates by 1,500 times. He defined mutation in the modern sense as the alteration of a gene and constructed gene maps of the chromosomes. By increasing mutation rates, Muller provided a method of rapidly producing all possible mutations and investigating whether they were beneficial, harmful, or neutral. Muller accepted eugenics, the selective breeding of humans, and believed in establishing sperm banks for selective breeding. He spent most of the 1930s in Germany, the Soviet Union, and Edinburgh, Scotland, before returning to the United States in 1941. Muller received the Nobel Prize in medicine in 1946.

Wilson's, Morgan's, and Muller's chromosomal investigations never established the number of chromosomes in a human cell, and until the investigations of Jo Hin Tjio, pronounced Chee-oh (1919-2001), decades later, the number remained an estimate. Tjio, a geneticist born in Java, Indonesia, of Chinese parents and a World War II Japanese prisoner, did plant research for the Spanish government in Zaragoza, Spain, and at the University of Lund in Sweden from 1948 to 1959 before joining the National Institutes of Health in Bethesda in 1959. On 22 December 1955 while working with cells of human embryonic lung tissue magnified 2,300 times Tjio saw that each cell had 46 chromosomes and not the commonly accepted 48.

Herbert Evans (1882-1971), an American anatomist at Berkeley, first proposed 48 as the total number of chromosomes in 1918, and since that time it had remained the best estimate. Uncertainty in counting the number of chromosomes existed because the tissues examined often came from corpses. Mammalian chromosomes tended to cluster after death, making difficult an exact count of the number of chromosomes present in the living cell tissue. Tjio said he did not specifically begin the experiments to count the number of chromosomes. He saw clearly the number present while testing a technique to separate chromosomes, which he identified by their characteristic shapes on glass slides. Tjio published his discovery on 26 January 1956, reluctantly having to share credit with Albert Levan (1905-98) at Lund, only because Levan directed the laboratory in which Tjio spent his summers and holidays, away from his laboratory in Zaragoza.

## III. DISCOVERY OF PROTEINS AND NUCLEIC ACIDS

### *Protein analyses*

Protein analysis began in the 1830s with the investigations of Gerardus Mulder (1802-80), a chemist and physician who practiced and taught in Amsterdam, Rotterdam, and from 1840 at Utrecht, Holland. Prior to his investigations chemists had identified the amino acids leucine as a fermentation product in cheese in 1819, isolated it from muscle fiber and wool in 1820, and separated glycine from gelatin in 1820. Mulder studied the albumins (globular proteins) found in living cells, so named because of their isolation from albumin in egg whites, blood, milk, and plants. In 1838 he isolated the amino acids lysine and glycine from the protein in gelatin and meat using alkaline hydrolysis and showed that the amino acids consisted of carbon, hydrogen, oxygen, and nitrogen atoms. Mulder's work attracted the attention of the Swedish chemist Berzelius, who suggested in a letter of 10 July 1838 that Mulder name the albumins *protein* (Greek for *primary importance*) because they were likely the prime substance of animal nutrition. Mulder introduced the word *protein* in an 1839 article.

Justus von Liebig (1803-73), at Giessen, one of the nineteenth-century's greatest chemists although sometimes described as opinionated and arrogant, continued the analysis of protein molecules in the 1840s. He tried to establish their composition and formulas by hydrolysis with either hydrochloric acid, sulfuric acid, or sodium hydroxide, but only the most stable proteins survived his harsh hydrolytic methods. Liebig isolated the amino acid tyrosine in 1846. That same year the American chemist Eben Horsford (1818-93), while studying in Liebig's laboratory, determined glycine's composition. Horsford returned to the United States to direct the Lawrence Scientific School at Harvard when it opened in 1847. Auguste Cahours (1813-91) in Paris continued the research on amino acids, and in 1861 he determined the structure of glycine, showing that it had the simplest amino acid structure, $H_2N\text{-}CH_2\text{-}COOH$. A few years later in 1866 Liebig's student Heinrich Ritthausen (1826-1912) at Munich isolated another amino acid, glutamic acid.

Better methods of analysis using milder acid and alkali solutions became available beginning in the late 1870s-80s. They enabled chemists to establish the composition and molecular weights and the structure of amino acids and proteins. Emil Fischer (1852-1919) at Berlin synthesized a protein molecule by linking together amino acids and thereby established experimentally the composition of proteins. Fischer had discovered the amino acid-amino acid linkage in proteins (called a peptide bond or polypeptide bonds), and in 1907 he synthesized a protein molecule that linked eighteen amino acids.

| **A POLYPEPTIDE CHAIN** |
|---|
| $H_2N\text{- } CH \text{ - } COOH + H_2N \text{ - } CH_2 \text{ - } COOH \rightarrow H_2 N \text{ - } CH_2 \text{ - } CO.\ HN \text{ - } CH_2 \text{ - } CO.\ HN \text{ - } CH_2 \text{ - } CO.\ HN \text{ - } CH_2 \text{ - } CO. + HOH$ |

Fischer, one of the greatest organic chemists and the father of biochemistry, did much to lay the foundation of structural organic chemistry, establishing the structures of amino acids, sugars, carbohydrates, purines, and other organic compounds. For this work, in 1902 he received the second Nobel Prize awarded in chemistry. Having already lost his wife in 1895, suffering himself from mercury poisoning and exposure to highly toxic phenylhydrazine the likely cause of his intestinal cancer, Fischer grew more despondent following the loss of two of his three sons in World War I. Walter committed suicide in November 1916 at age 25 because of depression resulting from his compulsory military service, and Alfred died of typhus while serving as a medical assistant in Romania in March 1917. Fischer committed suicide by taking cyanide (potassium ferrocyanide) on 15 July 1919. His oldest son Hermann Fischer (1888-1960) who served in a military warfare unit, in 1937 became a chemistry professor at the Banting Institute at the University of Toronto and in 1948 joined the newly-established biochemistry department at Berkeley.

Nutrition studies on proteins and amino acids in the first three decades of the twentieth century established their importance in the production and health of all living cells. They made clear that the proteins present in a cell consisted of twenty-two standard, or natural, amino acids and that protein synthesis therefore required those twenty-two amino acids. The genetic code, determined much later in 1961, required twenty of the amino acids, different mechanisms incorporate the remaining two into protein formation.

In humans, eight amino acids (some scientists say nine) were not naturally present in cells, indicating the cell's inability to produce those eight amino acids. They are the essential, indispensable amino acids that the human body cannot synthesize and which diet had to provide. (Infants and children require two additional amino acids.) Fourteen amino acids are non-essential, the liver produces them. Tryptophan was the first diet-provided amino acid found essential for protein formation and cell growth in humans, multi-cellular animals, and most protozoans. Frederick Hopkins (1861-1947), a biochemist at Cambridge, in diet experiments with mice in 1900 demonstrated that tryptophan was an essential amino acid. Hopkins won the Nobel Prize in medicine in 1929.

At Yale, Thomas Osborne (1859-1929) and Lafayette Mendel (1872-1935) in a series of experiments with rats carried out from 1909 to 1911 demonstrated the necessity in rats' diets of the protein casein and its constituent amino acids, three of which are lysine, leucine, and isoleucine. Following Osborne and Mendel's studies, Harold H. Mitchell (1886-1996), a chemist at Illinois, showed in his nutritional research that the amino acids lysine, histidine, tryptophan, cystine, and tyrosine were essential in rats' diets. All of them except tryptophan were present in casein.

These investigations received confirmation in 1935 when William C. Rose (1887-1985), Mendel's student at Yale and then a biochemist at Illinois, isolated threonine from casein. By 1937 he established in his rat studies that threonine, lysine, tryptophan, histidine, leucine, isoleucine, phenylalanine, methionine, valine, and arginine, were essential amino acids. Rose's experiments on his graduate students in the 1940s established that histidine and arginine were not essential amino acids for humans. Recent research has established that histidine is only semi-essential and arginine is essential only in infants.

***Discovery, analysis, and chemical composition of the two nucleic acids DNA and RNA***

Johann Friedrich Miescher (1844-95), a partially-deaf Swiss biologist who came from a family of distinguished scientists, in 1869 discovered the nucleic acids DNA and RNA, which he called nuclein, in the nuclei of white blood cells. At the time Miescher was a student of the German biochemist Felix Hoppe-Seyler (1825-95) who had established a laboratory in Tübingen, in the Hohentübingen Castle. Hoppe-Seyler had suggested for Miescher's research on how simple and independently-acting animal cells developed into complex living cells, that Miescher study white blood cells (leucocytes). Leucocytes were the principal cellular constituent of laudable pus. The research required a good source of white blood cells, which contained mainly nuclear material and little cytoplasm, and not red blood cells, which have no nucleus. It challenged Miescher to discover how or from where he could obtain a plentiful and constant supply of white blood cells.

In the years before the introduction of antibiotics to treat infections, laudable pus supplied researchers with white blood cells. Laudable pus, pus that oozed from infected wounds onto the cotton dressings used to cover a wound, indicated a patient's improving medical condition because its white blood cells destroyed disease-causing organisms. To obtain laudable pus, Miescher had dressings from a nearby Tübingen surgical clinic brought daily to the laboratory, where he extracted nuclein from the white blood cells by washing them with a sodium sulfate solution. His analyses showed that nuclein dissolved in weak alkalies but not in water and acids. It contained 14 percent nitrogen, 2 percent sulfur (present in protein not DNA), considerable phosphorus (3 percent), an acid (DNA and RNA), and a basic protein component (histones). It had the molecular formula $C_{29}\ H_{49}\ O_{22}\ N_9\ P_3$.

Miescher also found nuclein in yeast, kidney, liver and salmon sperm cells. Later studies showed that all living things have DNA (and RNA) and that any kind of cell that has a nucleus, such as hair and skin cells and plant cells, contains DNA. The method of extraction is essentially the same in every case and requires only a few simple procedures and careful separation of the DNA to prevent its breaking up or shredding.

| PREPARATION OF DNA |
|---|
| In the first step, the preparer blends 50 grams of the DNA source (onion, cauliflower, cheek cells from inside the mouth) in 200 milliliters (mL) of water for 30 seconds. This gives single cells. Strain the cells with muslin cloth, pour 15 mL of the liquid into a screw cap tube with 3 drops of a detergent, and gently shake for 5 seconds. The detergent breaks down the cell membranes by emulsifying the proteins, fats, and carbohydrates. It forms complexes with them and causes them to precipitate from the solution. Carefully add cold ethanol to the top of the tube where it remains in a clear layer. DNA is not soluble in cold ethanol and remains as a precipitate whereas all the other components dissolve in the solution. The DNA molecules appear as white strings or strands in the ethanol layer. A hooked glass rod allows for the DNA's removal from the ethanol layer. |

The 1870-71 Franco-Prussian war and Hoppe-Seyler's reservations that naturally-occurring nuclein, unlike all naturally-occurring substances then known, contained both nitrogen and phosphorus, delayed publication of Miescher's results until 1871. That same year Miescher became head of the physiology department at Basel. Poor health led in 1895 to his early retirement and death from tuberculosis. Richard Altmann (1852-1900), Miescher's student and later a pathologist in Leipzig, renamed nuclein, nucleic acid in 1889.

Albrecht Kossel (1858-1927), a German chemist at Strasbourg, where he was Hoppe-Seyler's assistant before leaving for Berlin in 1883 and later for Marburg and Heidelberg, continued the analyses of nucleic acids. He showed in his 1883 alkali hydrolysis experiments that nucleic acids in yeast consisted of a phosphate group $PO_4^{3-}$, an unidentified sugar, and two classes of colorless crystalline nitrogen-containing organic bases. Fischer in 1884 named the first class of bases purines and established that a purine molecule consisted of two fused rings (a hexagon and a pentagon) containing five carbon atoms and four nitrogen atoms. One of the purines (adenine) had an amine group ($NH_2$) and three hydrogen atoms attached to the two rings. The other purine (guanine) had an amine group, three hydrogen atoms, and an oxygen atom attached to the two rings.

Adolf Pinner (1842-1909) at Berlin coined the name pyrimidines for the second class of bases in 1885. Decomposition and synthesis experiments in the mid-1880s, such as those of Moritz Traube (1826-94) at Berlin, showed that a pyrimidine molecule had a single hexagonal ring. The ring contained four carbon atoms and two nitrogen atoms and had either an amine group, a methyl group, oxygen groups, or hydrogen atoms attached to the carbon atoms in the ring.

Kossel named the two purine bases in the nucleic acids adenine A, $C_5N_4NH_2H_3$ (or $C_5H_5N_5$), molecular weight MW=135.14 amu or Dalton, Da, which he first isolated; and guanine G, $C_5N_4ONH_2H_3$ (or $C_5H_5N_5O$), MW=151.14 amu, which Liebig's student Julius B. Unger (1819-85) at Giessen isolated in 1844. He named the three pyrimidine bases in the nucleic acids cytosine C, $C_4N_2ONH_2H_3$, MW=111.11 amu; thymine T, $C_4N_2O_2CH_3H_3$, MW=126.12 amu; and uracil U, $C_4N_2O_2H_4$, MW=112.1 amu. Kossel's student Alberto Ascoli (1877-1957) isolated uracil by hydrolyzing the RNA of yeast in 1900, and in 1901 Hans Steudel at Marburg determined its composition. Kossel received the Nobel Prize in medicine in 1910.

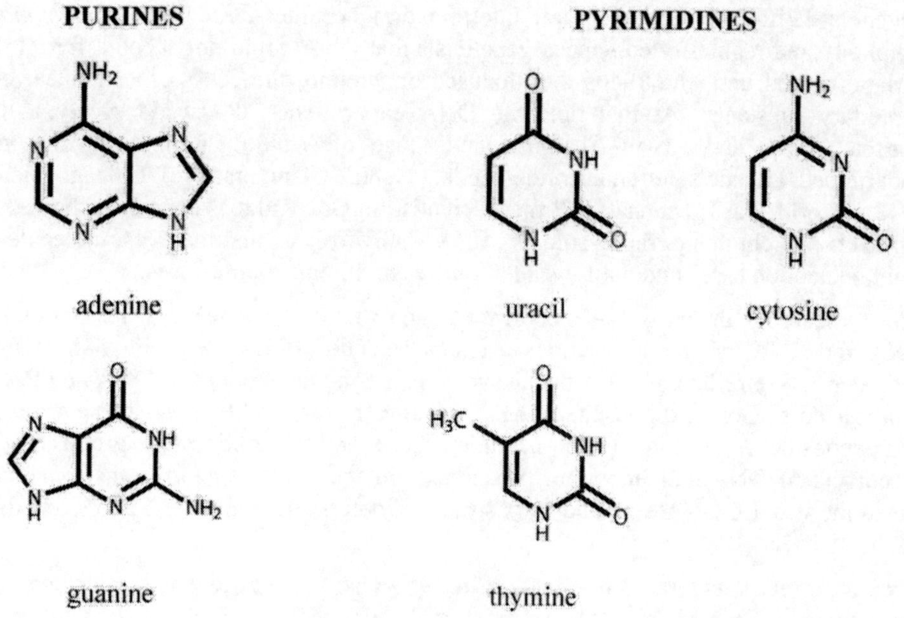


**Figure 3**
**DNA and RNA Bases**
chromatos/Shutterstock.com

***Levene's tetranucleotide structure of DNA***

Research continued on the nucleic acids, and by the end of the 1920s Phoebus Levene (1869-1940), a Russian-born chemist who studied under Fischer and Kossel in Germany, confirmed in hydrolysis experiments that nuclein consisted of two kinds of nucleic acid, DNA and RNA, and determined the composition of each. Levene came from St. Petersburg, Russia, to New York City in 1891. In 1905 he accepted a position at the Rockefeller Institute in New York City, a research institute that John D. Rockefeller (1839-1937) established in 1901. Rockefeller Institute changed its name to Rockefeller University in 1965.

Levene's breakthroughs on RNA and DNA came twenty years apart. In 1909 he identified the five-carbon sugar ribose $C_5H_{10}O_5$ [(CHOH)(CHOH)(CHOH)(CH)($CH_2OH$)O] in RNA obtained from yeast, and in 1929 he identified a second five-carbon sugar deoxyribose $C_5H_{10}O_4$ [(CHOH)($CH_2$)(CHOH)(CH)($CH_2OH$)O] in the DNA of a dog's thymus. It took Levene 20 years and about $300,000 to isolate the deoxyribose sugar because the acid he used to react with the DNA molecule decomposed the sugar, and he had no suitable nondestructive substitute. He succeeded by passing a solution of DNA through the stomach and intestine (the gastrointestinal system) of a dog and removing the solution through a fistula.

Using mild alkaline hydrolysis, Levene isolated DNA chains that were four nucleotides long. Each nucleotide, a word that Levene coined in a 1908 article, consisted of the sugar deoxyribose, a phosphate group, and one of the four different organic bases, adenine, cytosine, guanine, and thymine, or (phosphate-sugar-guanine • phosphate-sugar-cytosine • phosphate-sugar-thymine • phosphate-sugar-adenine). He concluded incorrectly that the structure of the DNA molecule was an unchanging repeating series of the four nucleotide bases that required equal amounts of the four bases. They were present in a 1: 1 : 1 : 1 ratio, $(ATCG)_n$ where n = 1,2,3 ... and represented the number of tetranucleotides in the polymer. The DNA tetranucleotide had the molecular formula $C_{39}H_{51}N_{15}P_4O_{25}$ and a molecular weight of about 1,231 amu. Levene believed RNA had a similar tetranucleotide structure, except that the sugar was ribose and the organic bases were adenine, cytosine, guanine, and uracil instead of thymine.

Levene's 1912 incorrect structural interpretation became the *tetranucleotide model* of DNA. Its simplicity and regularity convinced geneticists that DNA could not account for all the known genetic variations, and until the 1940s they focused on protein, the cell's other nuclear component, as the hereditary substance. At that time the DNA experiments of Oswald Avery, Alfred Hershey and Martha Chase, and Erwin Chargaff, and those of Waldo Cohn (1900-99) on RNA seriously undermined Levene's tetranucleotide model. Cohn, a University of Chicago biochemist and from 1943 a World War II Manhattan Project scientist, in Oak Ridge, Tennessee, showed in 1950 and 1951 ion-exchange chromatography studies of RNA's hydrolysis that the RNA molecule contained at least eight nucleotide bases and not Levene's four repeating nucleotide bases.

Other research in the early 1940s confirmed and extended Levene's conclusions on the *composition* of DNA and RNA, not their structures. Alexander Todd (1907-97), an English chemist at Manchester and later at Cambridge, in 1944 began synthesizing the nucleotides of DNA and RNA, and in 1947 he synthesized adenosine diphosphate (ADP, adenine, ribose, and two phosphate groups, $C_{10}H_{15}N_5O_{10}P_2$) and adenosine triphosphate (ATP, adenine, ribose, and three phosphate groups, $C_{10}H_{16}N_5O_{13}P_3$), two precursors of DNA and important participants in the body's energy transformations. His syntheses paved the way for DNA's double helix structural determination in 1953 and led to Todd's Nobel Prize in chemistry in 1959.

DNA's correct structural determination in 1953 also proved Levene's molecular weight calculation incorrect and established DNA's correct molecular weight. Each of its four organic bases had a different MW, adenine 135.14 amu, thymine 126.12 amu, cytosine 111.11 amu, and guanine 151.14 amu. In the double helix (double stranded) DNA molecule, each nucleotide consists of the deoxyribose sugar MW=134.1 amu, a phosphate group MW=95.97 amu, and one of the four organic bases A, T, C, G, whose average MW=131 amu. This gives each nucleotide an average MW=361 amu. Because DNA contains two strands, the average MW of a nucleotide pair is 2 x 361=722 amu. If a DNA segment whose MW you are calculating contains 1,000 base pairs, its MW=722 x 1,000=722,000 amu, for a segment containing 10,000 base pairs its MW=722 x 10,000 = 7,220,000 amu, and similarly for any other number of base pairs in a DNA segment.

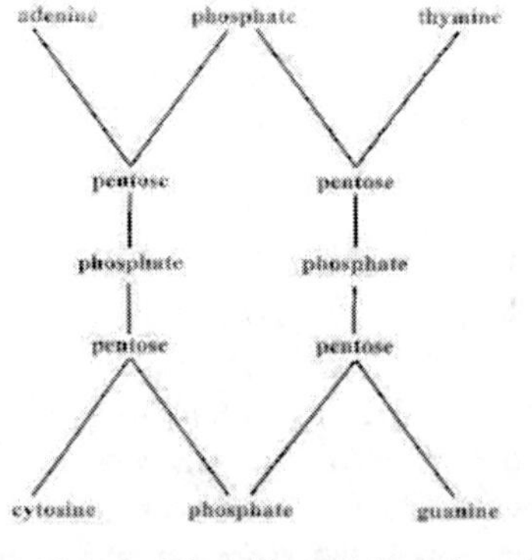


Levene's structure of DNA, widely held until well into the 1940s, pictured the molecule as a tetranucleotide composed of one unit each of adenylic, guanylic, thymidylic, and cytidylic acids. The structure required a molecular weight of 1,231 amu for DNA and that the four nucleotide bases (adenine, guanine, thymine, and cytosine) occur in exactly equal proportions. Even when scientists finally realized in the 1940s that DNA's molecular weight was much higher (in the millions or billions), many still believed that the tetranucleotide was the basic repeating unit of the large DNA polymer. The mistaken belief in the uniform macromolecular structure proved an obstacle to the eventual acceptance that DNA is the genetic material.

**Figure 4**
**Levene's Tetranucleotide Structure of DNA 1919-49**
Courtesy of the author

## IV. USE OF X-RAYS IN ESTABLISHING MOLECULAR STRUCTURE

The history of x-rays began in November 1895 when Wilhelm Röntgen (1845-1923), a German physicist at Würzburg and winner of the first Nobel Prize in physics in 1901, accidentally discovered a new kind of emanation. Röntgen was studying the luminescence cathode rays produced when they struck the glass wall of his high voltage gas discharge tube. He noticed that something emanating from the tube, which he called x-rays, caused a nearby card coated with green barium platinum cyanide crystals, barium cyanoplatinite $BaPt(CN)_4 4 H_2O$, to glow.

The ease with which x-rays penetrated matter, including the human body as Röntgen demonstrated, led to their immediate use in medicine. Within a few years the introduction of x-ray diffraction, the deflecting or bending of a beam of x-rays around an edge of a crystal, provided physicists with a powerful new method of determining the structure of any crystal or near-crystalline substance.

Max von Laue (1879-1960), a German physicist at Munich, discovered x-ray diffraction in 1912. His discovery earned him the 1914 Nobel Prize in physics. Laue concluded that a crystal, which had its atoms or ions arranged in a regular repeating pattern of planes separated by small distances giving a periodic (repeating) arrangement, should diffract a beam of x-rays. The diffraction occurred because the x-ray's wavelength of $10^{-8}$ cm (1 angstrom, A = $10^{-8}$ cm) was comparable to the distance between the planes in the crystal. Walter Friedrich (1883-1968) and Paul Knipping (1883-1935), Laue's assistants at Munich, in 1912 carried out the first successful diffraction experiment, using a crystal of copper sulfate ($CuSO_4.5H_20$) and then repeated the diffraction with a zinc sulfide (ZnS) crystal. They obtained a uniformly distributed pattern of spots on a photographic film placed behind the crystal.

In 1912, shortly after Laue's discovery of x-ray diffraction, the physicists William H. Bragg (1862-1942) and his son William L. Bragg (1890-1971) at Leeds in England, experimentally established the relation known as Bragg's law, $n\lambda = 2d \sin \theta$, the fundamental law of crystallography. With it they calculated the distance between the atomic or ionic planes in crystals of diamond, copper, potassium chloride (KCl), and many other substances. For their x-ray research, the Braggs in 1915 received the Nobel Prize in physics, the only father-son pair to share the prize. William Lawrence Bragg moved to Manchester in 1919 and became director of the Cavendish Laboratory at Cambridge in 1938.

| BRAGG'S LAW |
| --- |
| $n\lambda = 2\ d \sin\theta$<br><br>$n = 0, 1, 2, 3, 4, \ldots\ldots$<br><br>$\lambda$ = wavelength of the incident x-ray<br><br>d = distance between the atomic (ionic) planes in a crystal<br><br>$\theta$ = angle at which the x-ray strikes the crystal plane<br>(angle of incidence) |

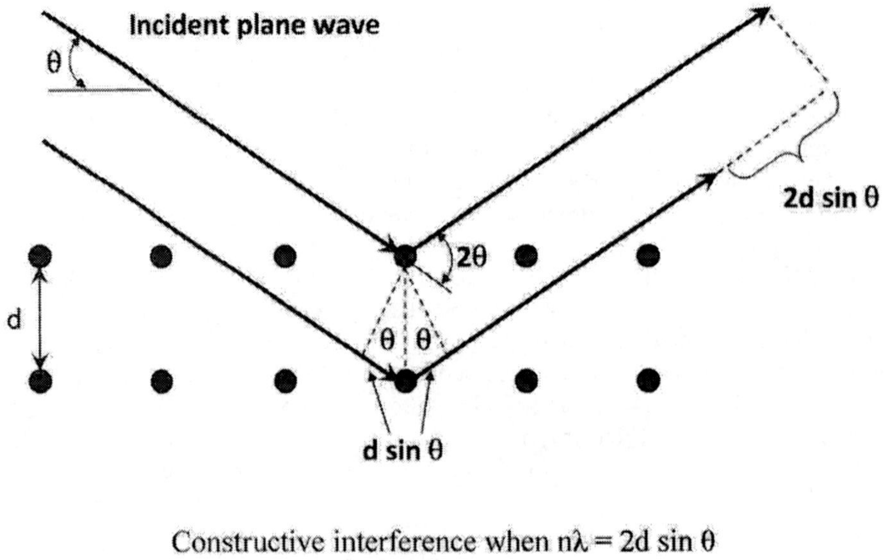


**Figure 5**
**Bragg's Law**
Courtesy of the author

William H. Bragg's one-time associate, William Astbury (1898-1961), while at Leeds in 1938 published the first x-ray diffraction photographs of DNA. His photographs of DNA fibers suggested that the four organic bases were perpendicular to the molecule's long chain. In a 1944 publication he found that the x-ray diffraction pattern repeated along the axis of the DNA molecule at a length corresponding to a fiber thickness (width) of eight nucleotides, or some multiple of eight nucleotides, probably sixteen. The least possible value of the repeating fiber period, the width of eight nucleotides, was 27 A and the distance between successive nucleotides was therefore 3.4 A. Astbury concluded that the insufficient clarity of his x-ray photographs prevented him from determining with certainty the structure of DNA's unit cell. 2 used the term *molecular biology* in 1953, in referring to investigations on the structures of DNA, RNA, proteins, and other molecules involved in life processes. He did not know that fifteen years earlier, in 1938, Warren Weaver (1894-1970) at the Rockefeller Foundation in New York City in his annual report to the foundation's president already called these investigations *molecular biology (The Rockefeller Foundation, Annual Report 1938).*

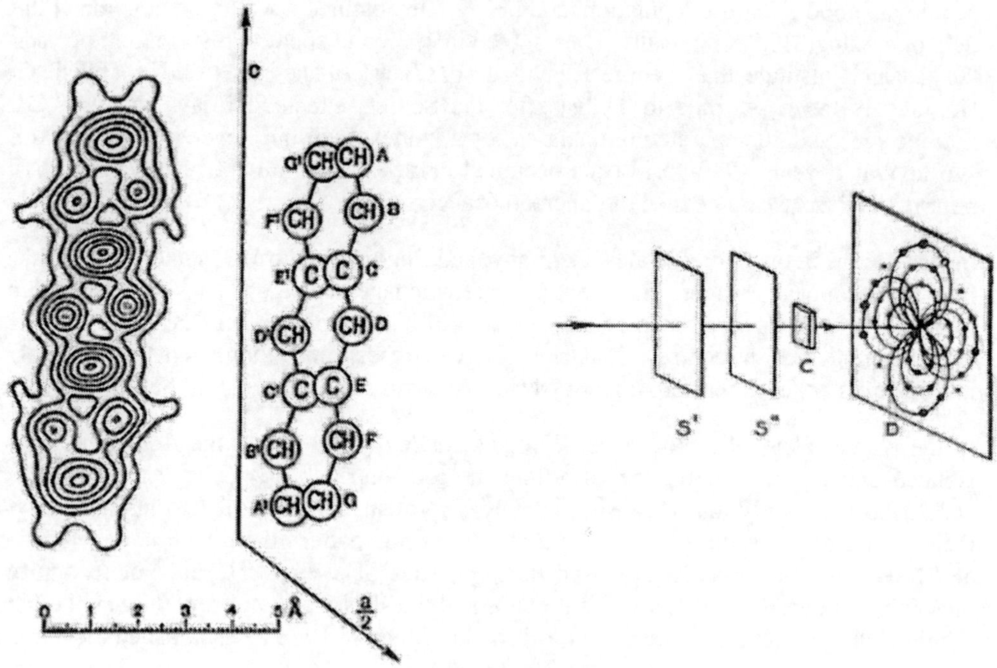


**Figure 6**
**X-Ray Diffraction, left Anthracene $C_{14}H_{10}$**
Source: W. H. Bragg."The Structure of Organic Crystals." *Proceedings of the Physical Society* (London); 34 (1921): 33-50; and "The Crystalline Structure of Anthracene." 35 (1922): 167-69 (left).
Source: William Foster (1869-1937). *The Elements of Chemistry.*
New York: D. Van Nostrand Company, 1925 (right).

## V. CHEMICAL CONTRIBUTIONS TO THE STRUCTURE OF DNA AND RNA: EVIDENCE THAT NUCLEIC ACID WAS THE GENE AND THE INFORMATION CARRIER

### *Bacteriophage composition*

Two bacteriologists, the reclusive and eccentric Frederick Twort (1877-1950) at the University of London and Canadian-born Felix d'Herelle (1873-1949) at the Pasteur Institute in Paris independently discovered bacteriophages. Twort in 1915 was growing viruses in different media when he observed that some of the cultures of *staphylococcus aureus* bacteria he was growing on agar, a polysaccharide gel obtained from seaweed, became transparent. He added a small amount of the transparent cultured bacteria to the unaffected region of his culture, and it also became transparent. In 1915 transparency produced in a bacterial culture proved bacterial destruction. Twort suspected that an invisible microbe, probably a virus, had caused the bacteria's destruction and the transparency. Service in World War I interrupted Twort's virus research, and he never pursued it after the war.

D'Herelle, who in 1917 was studying the feces of patients who had recovered from bacterial dysentery, found that certain regions in his bacterial culture contained no bacteria. Some kind of microbe had destroyed the bacteria, and d'Herelle, like Twort, concluded that a virus had infected the smaller bacterial cells and killed them. He called the virus a *bacteriophage* (Greek, *phago*, to eat), meaning bacterial eater. It infected bacterial cells rather than ordinary cells and was not pathogenic to humans.

D'Herelle had an unusual scientific career. He never attended university, and his lack of a formal education hindered his ability to obtain a permanent position at a university or research institution. At the Pasteur Institute he had no rank other than that of laboratory worker. Nevertheless, d'Herelle acquired a good scientific reputation, and in 1928 he obtained a well-paid position at the Yale Medical School earning $10,000 annually. He resigned in 1933 and spent two six-month periods in 1934-35 at the research institute that George Eliava (d. 1937) established in 1923 in Tbilisi, Georgia (Soviet Union). He never returned to Tbilisi after the Soviets executed Eliava in 1937. D'Herelle in his lifetime received 30 separate nominations for a Nobel Prize but never won the award. He spent the World War II years 1940-45 in Nazi-occupied France under house arrest in Vichy. After the war he returned to Paris where he died of pancreatic cancer.

Investigations in the United States, Germany, and England in the 1930s and 1940s, which included the first bacteriophage images, showed that bacteriophages (phages) consisted of a thin protein coat surrounding a DNA core, or in about 5 percent of bacteriophages an RNA core. They had larger and more complicated structures than non-phage viruses, and their genetic material, which most investigators believed was protein not DNA, played an important role in the cell's chemical processes.

In the United States the biochemist Wendell Stanley (1904-71) at the Rockefeller Institute in 1935 isolated and crystallized for the first time the rod-shaped bacteriophage tobacco mosaic virus and concluded incorrectly that it consisted solely of protein. Stanley infected his tobacco plants, crushed their leaves, grew needle-like crystals from the mosaic pattern the infection developed on the leaves, and found that his crystals possessed the properties of a virus. He did not recognize that tobacco mosaic virus consisted of RNA and protein, and that RNA and not protein was the genetic material. Stanley left the Rockefeller Institute for Berkeley where in 1948 he established the Virus Laboratory.

Stanley's Rockefeller colleague, the biochemist John Northrop (1891-1987) showed in 1938 in his bacteriophage studies with staphylococcus bacteriophage that a bacteriophage consisted of protein and nucleic acid. For their work on bacteriophages, Northrop and Stanley, and the Cornell biochemist James Sumner (1877-1955) for his work on enzymes, shared the 1946 Nobel Prize in chemistry.

Sumner pursued chemical studies despite amputation of his left arm just below the elbow after a friend accidentally shot him while grouse hunting in 1904. He showed in 1926 for the first time that an enzyme was a protein. His crystallization of urease, an enzyme obtained from beans and the first enzyme ever crystallized, gave the pure enzyme and enabled him to identify its properties. Most enzymes have a globular shape. The scientific community mostly doubted Sumner's accomplishment because it challenged the conclusion of the German chemist Richard Willstätter (1872-1942), then in Munich, that enzymes were not proteins. Northrop's investigations of 1930-35 in which he crystallized the enzymes pepsin, ribonuclease, and deoxyribonuclease and showed that they were proteins, confirmed Sumner's experimental results.

In Germany a combination of chemical analyses and instrumentation contributed to the increasing understanding of bacteriophage composition and structure. Martin (Max) Schlesinger (1907-37), a Hungarian chemist at the Institute for Colloid Research in Frankfurt-am-Main, in experiments in 1932, 1934, and 1936 used a high-speed centrifuge to concentrate and separate bacteriophages. He showed that they contained equal amounts of protein and nucleic acid. Schlesinger in the late 1930s fled Nazi Germany for London's National Institute for Medical Research.

In England, John Bernal (1901-71), a crystallographer and molecular biologist who worked with William H. Bragg at the Royal Institution in London from 1923 to 1927 and at the Cavendish Laboratory in Cambridge until 1937, applied x-ray diffraction analysis to investigate the structure of proteins and bacteriophages. With the American crystallographer Isidor Frankuchen (1904-64), Bernal in 1935-36 showed that tobacco mosaic virus consisted of RNA and protein and very likely had a helical structure. Named *the sage* by his contemporaries, the Catholic-turned-atheist Bernal was an early convert to communism, joining the party in 1923, and remained a life-long believer. Frankuchen returned to the United States only months after World War II began in Europe in September 1939 and later went to the Polytechnic Institute of Brooklyn.

Also, at Cambridge, Norman Pirie (1907-97) and Frederick Bawden (1908-72) in 1936 demonstrated that tobacco mosaic virus consisted of RNA and protein.

The invention of the electron microscope made possible the first bacteriophage images. Ernst Ruska (1906-88), then at the Technical College in Berlin, in 1931 constructed the first electron microscope. It magnified 400 times. In 1933 his newly improved electron microscope provided for the first time greater magnification than a light microscope. His brother Helmut Ruska (1908-73), a biochemist at Berlin, in 1939 used the improved electron microscope to obtain the first image of a bacteriophage.

The electron microscope uses an electron lens, which is an electromagnet, to focus a beam of electrons whose wavelength is about 100,000 times shorter than visible light's wavelengths (4,000-7,000 A). Ruska placed several electron lenses in series and passed the electrons through a very thin slice of the sample he wanted to magnify. The sample deflected the electrons onto photographic film or a fluorescent screen where they produced a magnified image of the sample. Ruska's early instruments magnified about 12,000 times compared to 2,000 times for an optical microscope. Today's electron microscopes magnify about one million times and have resolutions in the 0.5-1 A range. Ernst Ruska shared the 1986 Nobel Prize in physics for his invention of the electron microscope.

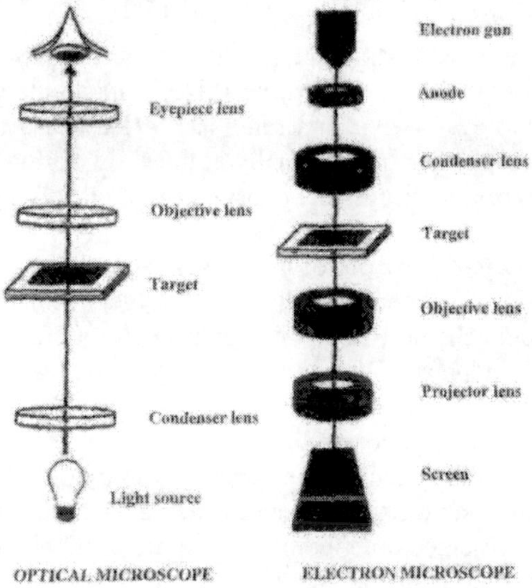


**Figure 7**
**Electron and Optical Microscope**
Courtesy of the author

World War II delayed and sometimes ended non-military research. Nevertheless, in 1944 Gerhard Schramm (1910-69) at the Max Planck Institute in Tübingen confirmed what Schlesinger, Norton, and Bernal, had shown earlier, that bacteriophages such as tobacco mosaic virus, consisted of protein and nucleic acid. Schramm reacted tobacco mosaic virus with mild alkali and broke the virus into free RNA and a considerable amount of nearly-identical or identical protein molecules.

***Determining the genetic material: Avery, Hershey, and Chase***

The direction of bacteriophage research expanded at this time from establishing its composition and structure to examining its role in transmitting biological information and hereditary characteristics, and to identifying a gene and how the gene contributed to life's processes. Max Delbrück (1906-81), a physicist at Berlin who studied with Max Born (1882-1970) at Göttingen and Niels Bohr at Copenhagen, as early as 1935 pondered how the genetic material, or genes, of bacteriophages accounted for the transmission of biological information. The explanation of life, Delbrück said, lay in reducing life to physics and to the discovery of new physical laws that explained life. He believed that Bohr's complementarity principle (duality) of quantum mechanics had application to other scientific fields and would lead to connections between physics and biology. Delbrück's reductionism played a major role in Schrödinger's discussion on the gene as an information carrier of hereditary characteristics in his 1944 classic *What is Life*?.

The emergence of Adolf Hitler (1889-1945) as Germany's chancellor in January 1933 interrupted the research of Delbrück and many Jewish and non-Jewish scientists whom Hitler considered disloyal. The biologist Heinz Fraenkel-Conrat (1910-99), the chemists Schlesinger and Haber, and the physicists Leo Szilard (1898-1964) and Max Born fled Germany. Delbrück, Liebig's great grandson, left Berlin in 1937 and obtained a position at Caltech where Morgan had gone in 1928. His family, like many others, suffered significant losses as a consequence of the war. At Caltech, Delbrück continued his work on viruses and on the even-numbered T-series bacteriophages T2, T4, and T6, trying to understand how bacteriophages reproduced and transmitted biological information. This work led to Delbrück's 1946 discovery that the genes of different viruses exchanged genetic material (combined) to form a new virus different from the original viruses.

For his research on the genetic structure of viruses, bacteriophage genetics, and DNA's replication mechanism, Delbrück shared the 1969 Nobel Prize in medicine with his colleagues, Alfred Hershey and the Italian microbiologist Salvador Luria (1912-91). The trio organized the Phage Group in the 1940s and with other researchers met annually at the Cold Spring Harbor Laboratory founded in 1890 on Long Island, New York, to discuss the latest research on solving the mystery of the gene.

Schrödinger's *What is Life*? grew out of a series of three lectures with the same title that he gave in February 1943 at Trinity College in Dublin. Schrödinger, who won the 1933 Nobel Prize in physics for his theory of wave mechanics, fled Nazi-controlled Graz for Dublin in 1938. His book influenced Crick and Wilkins in London and Watson when he was a student at Chicago in 1945.

Schrödinger proposed that a gene, or the material carrier of life, was an aperiodic or irregularly repeating solid that encoded its genetic message by employing a few different repeating symbols, such as the Morse code's dots and dashes, to convey messages of any size. To paraphrase Schrödinger, the code's transmission represented order coming from order, in which a single group of atoms existing in only one arrangement or copy and acting according to nature's laws, produced orderly events, in tune with each other and their environment.

Schrödinger examined the idea of order arising from disorder or negative entropy, which an organism accomplished by *extracting order* from the environment, and the idea of self-organization, which was a difficult-to-predict process in which a complex system spontaneously appeared. *What is Life?* stimulated new thinking about genetics and the carrier responsible for transmitting an organism's genetic information or hereditary characteristics. It opened the door to the new science of molecular biology.

Most scientists at that time, among them Fraenkel-Conrat who was at the Rockefeller Institute from 1936 to 1938 and since 1951 at Berkeley, Stanley, and Alfred Mirsky (1900-74) a biochemist at the Rockefeller Institute, believed that DNA was too simple a molecule to contain an organism's complete genetic information. Still under the influence of Levene's tetranucleotide model, they suspected that only proteins were sufficiently complex to express all genetic combinations. In the absence of convincing experimental evidence, the issue remained unresolved until 1944, and even then most of the scientific community, such as Mirsky in a 1946 article, continued to doubt evidence that challenged protein's role as the information carrier.

Oswald T. Avery (1877-1955) provided the first experimental evidence that DNA and not protein contained an organism's genetic information. A Baptist minister's son who emigrated from Halifax, Nova Scotia, in eastern Canada to the United States in 1887, Avery obtained his medical degree at Columbia in 1904 and joined the Rockefeller Institute in 1913. With his associates Colin M. Macleod (1909-72) and Maclyn McCarty (1911-2005), Avery in the early 1940s began studies on infecting mice tissue with a pneumococcus bacterium, a spherical-shaped pneumonia-causing bacterium that Pasteur first isolated in 1881.

In their crucial experiment to determine whether the bacterium's DNA or protein component infected the mice tissue, Avery, MacCleod, and McCarty mixed heat-killed cells of *Streptococcus pneumoniae*, a virulent (disease causing) form of pneumococcus which had a smooth polysaccharide protective coat (S form), with a living mutant avirulent (inactive) form, which lacked the smooth coat and had a rough exterior surface (R form). They added both forms to the mice tissue and found that the reaction transformed the R form into the smooth-coated virulent S form. Then in the first of two trials, they removed the protein from the smooth-coated S form infecting bacteria by adding protease and found that the R to S transformation still occurred. In the second trial, when they removed the DNA present in the smooth-coated S form infecting bacteria by adding deoxyribonuclease, no R to S transformation occurred in the R form bacteria. These studies enabled Avery, Macleod, and McCarty to conclude in February 1944 that the pneumococcus bacterium's DNA component was the active factor in the transformation of the R form to the S form and not its protein.

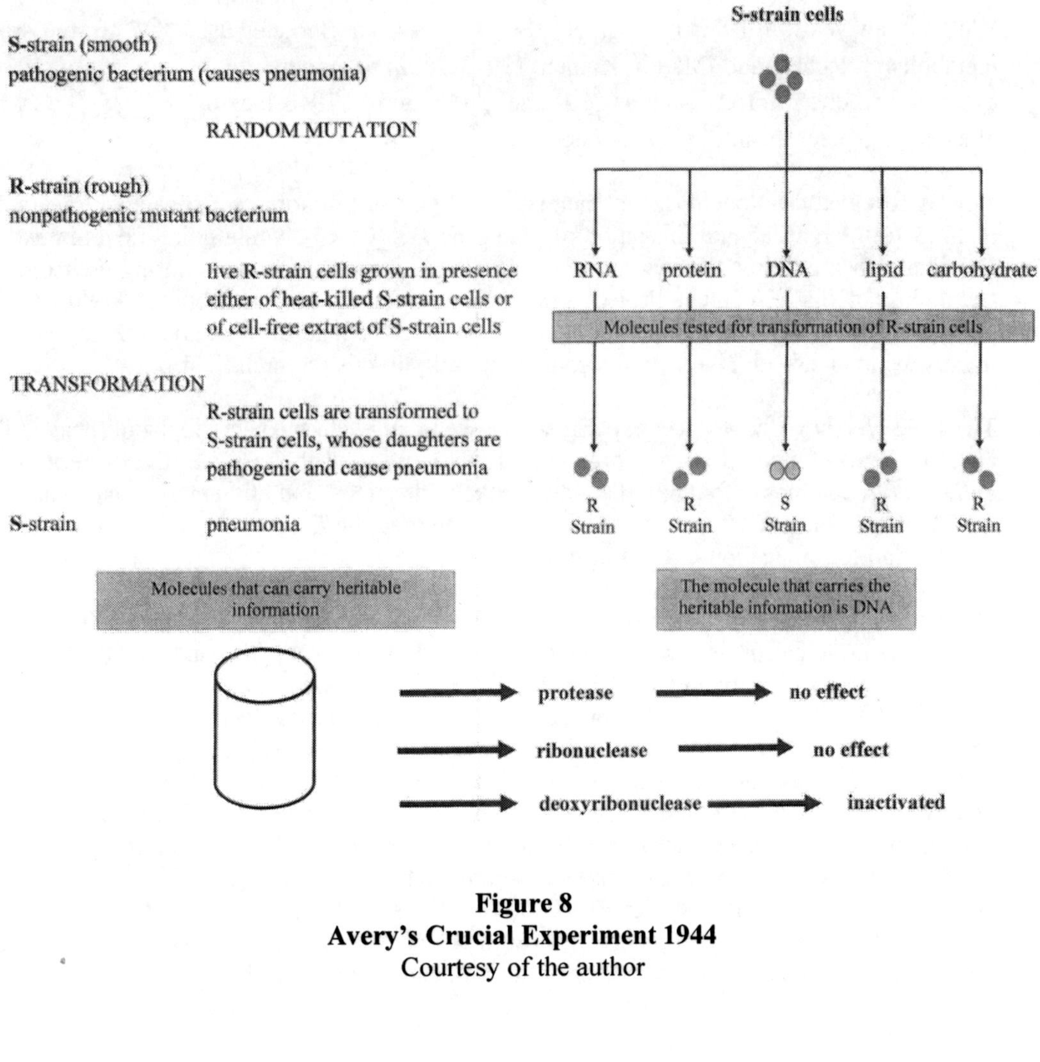


**Figure 8**
**Avery's Crucial Experiment 1944**
Courtesy of the author

Considerable skepticism prevailed in the years following Avery, Macleod, and McCarty's 1944 paper. Watson and Crick never cited their paper as a reason for pursuing the structure of DNA. But eight years later, in 1952, the geneticist Alfred Hershey (1908-97) and his assistant Martha Chase (1927-2003) at Cold Spring Harbor, confirmed Avery's work. Protein contains sulfur and a small amount of phosphorus, DNA contains no sulfur and significant phosphorus present as phosphate. This allowed Hershey and Chase in their famous blender experiment to use cyclotron-produced radioactive sulfur $^{35}_{16}S$ and phosphorus $^{32}_{15}P$ as tracers to follow the reaction pathway. The method required adding a small amount of a radioactive isotope of one or more of the elements involved in the reaction (S and P) and observing the radiation the isotope emitted as it underwent chemical change during the reaction.

Radioactive isotope tracing had its origin in George von Hevesy's inability to separate RaD (radioactive lead, Pb-210) from the large quantities of lead found in pitchblende. Rutherford assigned Hevesy (1886-1966), a Hungarian chemist working in his laboratory, the problem of chemically separating the two substances. Repeated attempts in 1912 failed, but beginning in 1913 the introduction of isotopes, establishing the radioactive decay series, and most importantly identifying RaD as a lead isotope which made impossible a chemical separation, provided the reason for Hevesy's failure and the first recognition of radioactive tracing.

Hevesy's continued to use radioactive isotopes as tracers in chemical reactions, and in 1935 he applied the tracer technique to biological processes, using $^{32}_{15}P$ to study phosphorus metabolism in rats. Within a few years, in 1937-38, Samuel Ruben (1913-43) at Berkeley used $^{32}_{15}P$ to study phospholipid metabolism. Ruben and Martin Kamen (1913-2002) continued these studies in 1939-40 using cyclotron-produced tracers such as $^{32}_{15}P$ and $^{11}_{6}C$ , and in 1940 they discovered $^{14}_{6}C$ which helped elucidate the mechanism of photosynthesis.

Hevesy received the Nobel Prize in chemistry in 1943 for his work on radioactive tracers. Ruben died in 1943 from an accidental inhalation of phosgene gas ($COCl_2$) while doing government research on chemical warfare. Kamen fell victim to the 1950s McCarthy witch hunts during the Cold War period. Beginning in his Manhattan Project years and continuing over a lifetime, Waldo Cohn, another political activist, developed improved methods of preparing, separating, and labeling with radioactive tracers, which made him the world's leading authority in this research field.

The 1952 Hershey-Chase tracer experiment showed that when the bacteriophage (phage) T2, a simple bacterial virus composed only of protein and DNA, infected its host cell, the bacterium *Escherichia coli*, *E. coli*, commonly found in the gut (stomach, intestines, and alimentary canal), the DNA core of the T2 phage, and not its protein capsule or coat, entered the *E. coli* host cell and provided the genetic information for replication of the T2 phage.

Hershey and Chase carried out side by side experiments with separate T2 phage cultures in which they labeled the protein capsule of one T2 culture with radioactive sulfur $^{35}S$ and the DNA core of the other T2 culture with radioactive phosphorus $^{32}P$. They prepared two cultures of *E. coli* bacterial host cells and added the radioactive sulfur phage to one of the bacterial cell cultures and the radioactive phosphorus phage to the other. To increase the reactivity of the phage particles with the bacterial cells, Hershey and Chase agitated each liquid mixture in a blender and centrifuged each mixture. The centrifuging separated any of the lighter T2 phage particles that remained in the liquid (or supernatant) and had not entered the heavier *E. coli* bacterial host cells that fell to the bottom of the centrifuge tube and formed a pellet. In their first mixture the radioactive sulfur of the protein remained mainly in the supernatant and not in the heavier *E. coli* bacterial host cells. In their second mixture the radioactive phosphorus of the DNA was predominantly in the heavier *E. coli* bacterial host cells, resulting in a new generation of T2 phage infected bacterial host cells, and not in the supernatant.

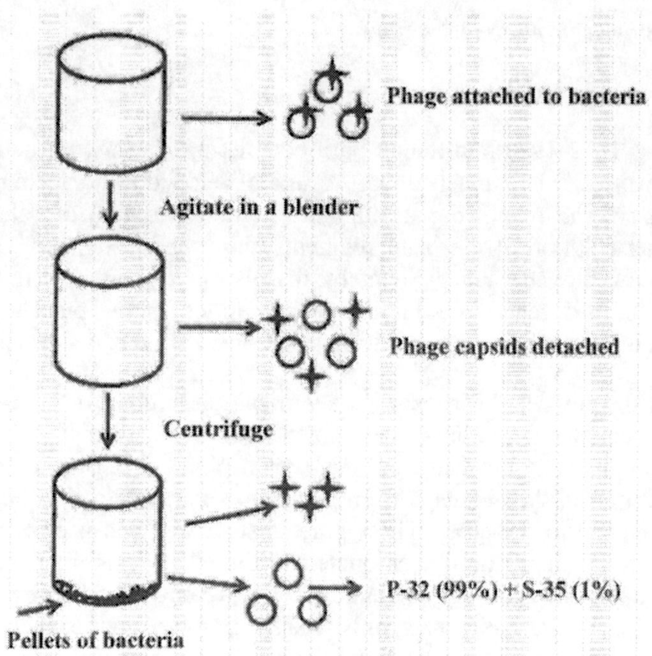


**Figure 9**
**Hershey-Chase Experiment 1952**
Courtesy of the author

Their experiment clearly linked DNA and the transfer of hereditary traits. It proved that genes consisted of DNA and earned Hershey a share of the 1969 Nobel Prize in medicine. Chase received her PhD in 1964, but her career ended in the late 1960s when she experienced a series of financial and personal setbacks. She suffered from dementia in her last decades before dying from pneumonia.

Three years after Hershey and Chase's DNA experiments, Fraenkel-Conrat at Berkeley provided additional evidence about the transmission of genetic information. Fraenkel-Conrat and the physicist Robley Williams (1908-98), an electron microscope expert whom Stanley brought to Berkeley in 1951, broke tobacco mosaic virus into its protein and RNA molecules, and reassembled it without any loss of virulence. Their 1955 experiment showed that the separated RNA molecule and not the separated protein molecule carried the genetic information, and thereby transmitted its virulence. A rival group whose members included Schramm and Alfred Gierer at the Max Planck Institute in Tübingen reached the same conclusion independently in 1956.

Fraenkel-Conrat continued his work on tobacco mosaic virus throughout the 1950s, hoping to determine how DNA produced the different amino acids and how the amino acids combined to form proteins, and thereby solve the genetic code. He determined the complete sequence of the 158 amino acids in the virus's protein coat by 1960 as part of his code-solving method but lost the race to Marshall Nirenberg who learned the protein-coding technique in Fraenkel-Conrat's laboratory.

Concurrent bacterial studies by Joshua Lederberg (b. 1925), a geneticist at Wisconsin, further clarified DNA's role in transferring genetic material. Lederberg showed in his mid-1940s studies on *E. coli* that bacteria reproduce sexually, something that many scientists doubted at that time. In 1952 he demonstrated that bacteriophage virus particles transferred genetic material from one bacterium to another. For his work on bacterial sexual reproduction Lederberg shared the Nobel Prize in medicine in 1958 with the Stanford scientists, microbiologist Edward L. Tatum (1909-75) and geneticist George W. Beadle (1903-89).

## ***Chromatography separation techniques***

*Invention of column chromatography*

Mikhail Tsvett (1872-1920), an Italian-born botanist in Warsaw, introduced column chromatography in a little-known paper he published in 1903 and developed it in papers published from 1906 to 1914. Tsvett used a column of powdered alumina (aluminum oxide) or calcium carbonate to separate a petroleum ether solution of the plant pigments chlorophyll, xanthophyll, and carotene into colored bands and then removed each pigment by dissolving it in ethanol (ethyl alcohol). The separation depended on the difference in solubility or adsorption of the plant's pigments in two immiscible phases, which in Tsvett's experiments were a stationary solid and a moving liquid.

Tsvett packed the powdered carbonate in a vertical glass tube. At the top of the carbonate column he added a petroleum ether solution of the pigments containing a small amount of ethanol to increase the pigments' solubility. He found that although plant pigments easily dissolved in petroleum ether, the ether poorly extracted the pigments from plants and extracted a significant amount of pigments only after adding ethanol to the ether. The pigments separated because of the difference in solubility or adsorption of the pigments in the carbonate and the ether-alcohol solvent. Those pigments that were very soluble in the ether-alcohol solvent and only weakly adsorbed by the carbonate passed quickly through the column. Those pigments, only weakly dissolved in the ether-alcohol solvent but which the carbonate strongly adsorbed, passed slowly down the column. Collecting and analyzing small batches of the ether-alcohol solvent that passed out the bottom of the carbonate column indicated a separation of the pigments or of any compounds inseparable by other techniques. Filter paper strongly adsorbed the extracted plant pigments from petroleum ether.

Chromatography, a sensitive, simple, and fast process, became an extremely important separation technique in chemistry and in solving DNA's structure. Willstätter at the Zurich Technische Hochschule reintroduced and popularized Tsvett's column chromatography in the years from 1906 to 1911, adding to his record of analytical successes. Willstätter already had determined the structure of cocaine in 1898, fourteen years after Sigmund Freud (1856-1939) in Vienna published his famous paper on cocaine's medical use.

In 1906 Willstätter showed that chlorophyll consisted of two closely-related compounds chlorophyll a and b and contained a single atom of magnesium. He also demonstrated the presence of iron in heme, the colored part of hemoglobin. While at the Kaiser Wilhelm Institute in Berlin, Willstätter received the 1915 Nobel Prize in chemistry. His student Richard Kuhn (1906-67) at the Kaiser Wilhelm Institute greatly advanced column chromatography and in 1938 received the Nobel Prize in chemistry for his research. Hitler forced Kuhn to reject the prize.

*Invention of paper chromatography*

A related smaller-scale chromatographic separation method appeared in 1941 and played an important role in amino acid and protein analysis and in determining DNA's structure. The biochemists Archer John Porter Martin (1910-2002) and Richard Synge (1914-94) at the Wool Industries Research Association in Leeds, England, invented paper chromatography and used it to separate and estimate the number of amino acids and the amount of each amino acid in a protein molecule.

In paper chromatography, each of the amino acids present in a drop of an amino acid liquid mixture, placed on the bottom edge of a strip of porous filter paper and dipped into an organic solvent, such as chloroform or phenol, moves up the paper at a different rate. Separation occurs because of the different solubility of the amino acids and the capillary action of the organic solvent front moving up the paper. Martin and Synge shared the 1952 Nobel Prize in chemistry for their invention of this new analytical method. In 1953 Martin perfected gas chromatography for the separation of gaseous mixtures.

The biochemist Frederick Sanger (1918-2013), a member of Cambridge's Biochemistry Department, originally used silica-gel chromatography to separate amino acids. Following Martin and Sanger's invention of paper chromatography, he adopted their simpler method to determine the order of amino acids in the insulin protein molecule and thereby insulin's molecular weight. Scientists at that time believed insulin had a molecular weight of 12,000 amu, or twice the now-accepted 6,000 amu.

In experiments completed in 1945, Sanger found that 1, 2, 4-fluorodinitrobenzene (FDNB), later named 2, 4-dinitrofluorobenezene (Sanger's reagent), under mildly alkaline conditions attached itself to the free amino group at the end of a protein molecule to form a 2, 4-dinitrobenzene protein derivative. He succeeded in attaching FDNB to the free amino group at the end of insulin's amino acid chain and then used hydrochloric acid hydrolysis to break the peptide bonds. The hydrolysis separated the chain into its individual amino acids, while leaving the 2, 4-dinitrobenzene-amino acid bond unbroken.

Sanger separated the amino acids with paper chromatography, using an electric current to move the amino acids according to their electric charges, and noted which amino acid had the yellow color characteristic of the resulting 2, 4- dinitrobenzene-amino acid product. This indicated the amino acid to which FDNB had attached itself and was at the end of the insulin molecule. In 1953 after eight years of investigations, Sanger established the exact order of the 51 amino acids in the long and short chains of insulin molecules from pigs, sheep, horses, and whales. His research on insulin marked the beginning of gene sequencing. Sanger received the 1958 Nobel Prize in chemistry for this work.

In 1980 Sanger shared with Walter Gilbert (b. 1932) at Harvard and Paul Berg (b. 1926) at Stanford a second Nobel Prize in chemistry. He won for determining in 1977 the entire base pair sequence of the DNA molecule in the small virus *phi X174*. It was the first complete DNA sequence, contained 5,386 nucleotides (base pairs), and coded for 11 genes. Sanger, with Marie Curie, Linus Pauling, and John Bardeen (1908-91) then at the Bell Telephone Laboratories in Murray Hill, New Jersey, was one of four scientists to win two Nobel Prizes.

Other applications of chromatography include Cohn's ion-exchange RNA hydrolysis and the photosynthesis studies of Melvin Calvin (1923-97). Calvin, a Berkeley chemist and 1961 Nobel Prize winner in chemistry, used paper chromatography and isotopic tracers in his investigations of 1949-57 to establish the mechanism for photosynthesis.

## VI. SOLVING THE STRUCTURE OF DNA: X-RAY DIFFRACTION, MODELS, AND THE GENETIC CODE

### *Chargaff's paper chromatography breakthrough*

Martin and Synge's invention of paper chromatography enabled Erwin Chargaff (1905-2002), an Austrian biochemist and a later critic of molecular biology and biotechnology, to make a crucial contribution to establishing the structure of DNA and how it transmitted genetic information. Chargaff, who left Europe in 1933 because of Hitler and joined the faculty at Columbia University in 1935, began studying DNA in 1944, after learning of Avery's work.

In 1950 Chargaff used a mild acid hydrolysis to break the phosphodiester bonds of DNA molecules obtained from beef tissue, salmon, humans, and yeast into individual purine and pyrimidine organic bases. Paper chromatography separated the organic bases, and an ultraviolet spectrophotometer measured the amount of each base. He observed that the ratio of bases A-T, adenine-thymine, and C-G, cytosine-guanine, in DNA was constant, indicating that regardless of the sample's source the amounts of A and T were equal, as were the amounts of C and G. Chargaff did not grasp the full significance of his work, but his discovery, since named Chargaff's rule, led to Watson and Crick's recognition of A-T and C-G base pairing and showed definitively that the four bases were not present in equal amounts as required in Levene's tetranucleotide model. Chargaff's discovery convincingly disproved the tetranucleotide model and contributed to its disappearance from molecular biology.

| CHARGAFF'S RULE |
|---|
| $\frac{A}{T} = 1 \quad \frac{C}{G} = 1$ |

Chargaff knew the chemist John Kendrew (1917-97) at the Cavendish Laboratory's Medical Research Council Unit, and when Chargaff visited Cambridge in May 1952 Kendrew invited him along with Watson and Crick to dinner. At that time Chargaff called their attention to his 1950 publication on DNA's base ratios which provided the final clue to DNA's structure. Chargaff later criticized Watson and Crick's model building as a scientifically unsound approach and said of them that he never met two men who knew so little and aspired to so much.

Chargaff believed that Watson and Crick had successfully exploited his rule on the A-T and C-G base ratios without proper attribution or public thanks. He complained that molecular biologists and geneticists in the 1960s-early 1970s no longer required the skills of analytical chemists such as he. The worst was still to come. Administrators of Columbia's College of Physicians and Surgeons in 1974 unceremoniously removed Chargaff from his office and laboratory. They locked him out and exiled him to a distant building because, according to Chargaff, they considered him an old fool and an enemy of scientific progress. Their action, Chargaff said, only reinforced his opinion that some of the scientific community regarded him an outsider.

***Linus Pauling's DNA structure***

Linus Pauling (1901-94), a physical chemist at Caltech and likely the most influential chemist of the twentieth century, received his first Nobel Prize in chemistry in 1954 for his work on chemical bonding and molecular structure. He received his second, the 1962 Nobel Peace Prize, in 1963 for his concerted effort to control nuclear weapons proliferation. Pauling shared neither prize, making him the only two-time recipient of an unshared Nobel Prize. His outspoken criticism of nuclear proliferation led the United States State Department in 1953 to deny him a passport and to his forced resignation as head of Caltech's chemistry division in 1958. In the 1970s-80s Pauling became involved in promoting megadose consumption of vitamin C for cancer prevention, the disease that killed him.

Pauling provided the first structural success in molecular biology. Using x-ray diffraction analysis and molecular models, a technique he pioneered, Pauling in 1951 showed that the structure of keratin, a protein molecule found in hair, fingernails, and animal horn, was an α-helix, the structure later found for most protein molecules.

Applying the same techniques of x-ray analysis and model building, Pauling with Robert Corey (1897-1971) published in *Nature* in February 1953 a one-page announcement and in the *Proceedings of the National Academy of Sciences* a 15-page paper describing a three-stranded intertwined helix structure for DNA. A sugar molecule and an inward-facing phosphate group made up each strand of the helical core or backbone and the nitrogen bases faced out. Pauling calculated the distance between each base pair, or nucleotide (sugar-phosphate-organic base), on the DNA backbone as 3.4 A, and 27.2 A as the distance at which each helix made a complete turn. Astbury had calculated the same distances in 1944, and they were close to the distances Franklin published in April 1953. Pauling's structure did not match DNA's chemical properties, and he quickly abandoned it.

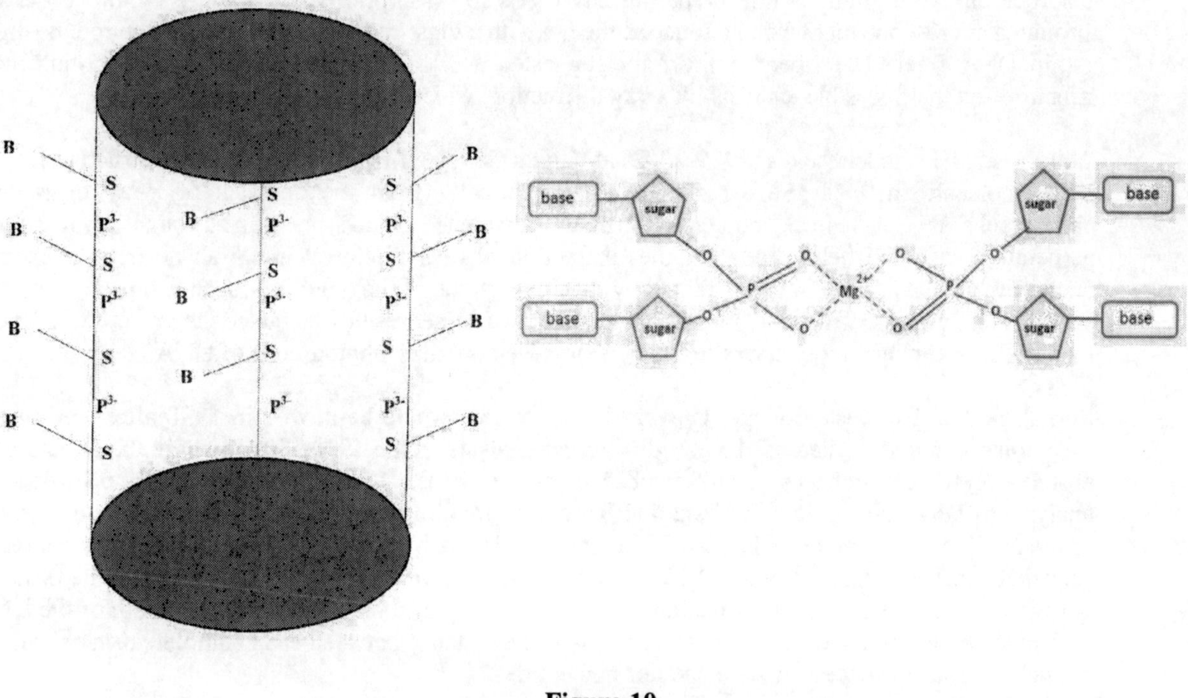


**Figure 10**
**Pauling's Triple Helix Model of DNA; right diagram shows two of the three helixes**
Courtesy of the author

Pauling's structural model failed for several reasons. Pauling packed his model too tightly with the bases pointing out, making it difficult for its strands to unravel or unwind. The oxygen atoms of the phosphate groups facing the middle and stacked on top of one another had negative charges, thereby repelling one another and resulting in an unstable structure. Its hydrogen atoms were too strongly bonded to the phosphate groups and not easily ionized, thereby making his model unlikely to react as an acid. Finally, Pauling's structural model had no predictive value. It offered no explanation for Chargaff's rule that DNA's base pairs A-T and C-G were always present in equal amounts.

Pauling never really said why he proposed a structure that ignored the laws of chemistry and the rationale for model building, sometimes claiming that he used incorrect density data, which accounted for the tight packing, and other times claiming that his x-ray photographs lacked high resolution. Pauling did not know that his x-ray photographs were composites of DNA's A and B forms. Overconfidence may have contributed to Pauling's uncharacteristic faulty reasoning and hastily proposed model.

***X-ray studies on DNA at King's College, University of London***

Pauling was not alone in using x-ray diffraction to determine molecular structures. Maurice Wilkins (1916-2004), a New Zealand-born physicist, directed x-ray diffraction research on DNA at King's College, University of London. During World War II, Wilkins worked on uranium isotope separation first at Birmingham and from 1944 with the Manhattan Project at Berkeley, but turned away from physics because of the postwar emphasis on nuclear weapons research. His interest in DNA began in 1946, while at St. Andrews University, and in 1949 he joined the Medical Research Council Biophysics Unit at King's College. By that time Kossel's and Levene's chemical analyses had established that the DNA and RNA molecules contained a sugar molecule (deoxyribose in DNA and ribose in RNA) a phosphate group, and the same four bases A, T, C, G, except that uracil U in RNA replaced thymine T in DNA. DNA research also showed that the DNA macromolecule selectively absorbed ultraviolet light. While preparing DNA gels for absorption experiments, Wilkins observed through a microscope that when he touched the gel with a glass rod and then removed the rod, he drew a thin DNA fiber. The fibers' uniformity suggested a crystalline arrangement of their molecular structure and their possible analysis by x-ray diffraction, which Wilkins confirmed in 1950.

Wilkins and Alexander Stokes (1919-2003) at King's College, first proposed a helical structure for the DNA molecule in 1951. Stokes showed mathematically from the analysis of x-ray diffraction photographs that a helical molecular structure diffracted x-rays passing through it at angles perpendicular to the helix, and that the diffraction always produced an X shape or image on a photographic plate. The x-rays diffracted by the *zig* portions of the helix produced *zag* patterns on the x-ray plate, the *zag* diffractions produced *zig* patterns. The net result was the X shape that they, Crick, Pauling, Franklin, and other crystallographers observed on x-ray photographs of DNA.

Bernal, now at Birkbeck College, University of London, where he moved in 1938 after leaving the Cavendish Laboratory, had used x-ray diffraction analysis in the 1920s-30s to study the structure of proteins and bacteriophages, such as tobacco mosaic virus. Wilkins applied the same diffraction analysis to DNA, and in late 1951 he and Rosalind Franklin (1920-58) who led a separate research group at King's College established that the DNA molecule had three major periodicities or repeating sequences that occurred every 3.4 A, 20 A, and 34 A. The first periodicity, 3.4 A, gave the distance between successive base pairs or nucleotides on the DNA backbone, 20 A was the width of the DNA molecule about 10-12 atoms across, and 34 A was the distance between each complete turn or twist of the helix, about 10 successive base pairs or nucleotides.

Franklin, a physical chemist who graduated from Cambridge in 1941 and did x-ray diffraction analyses on coal and carbon at Cambridge and at Paris from 1947 to 1950, came to King's College in 1951 to work on DNA's structure. Her investigations with DNA fibers in 1951-53 showed that DNA had a double helix structure with the organic bases on the inside of the helix and the phosphate and sugar groups making up each strand of the helix.

Franklin discovered by November 1951 that DNA had two forms, an A (crystalline) and a B (hydrated) form for which she obtained an excellent photograph on 1-2 May 1952. This proved crucial because it revealed for the first time that earlier x-ray diffraction images, such as Astbury's, were probably composite or mixed images of the A and B forms. The A form provided greater detail than the B form, but the B form, commonly found in human cells, gave the X diffraction pattern characteristic of a helix, and enabled Franklin to confirm existing knowledge of DNA. Most importantly, the B form provided new evidence that DNA had a double helix structure. She labeled the B form photo 51.

The B form showed the presence of four white diamond shapes, above and below and on either side of the arms and legs of the X. This indicated that the x-ray diffraction pattern repeated itself above and below the X shape and that the helix therefore repeated itself and was a macromolecule. The diamond shapes also supported DNA's structural regularity, the regularity of the sugar and phosphate groups that formed the backbone of DNA. Photo 51 showed that the two diamond shapes above and below the X were whiter than those on either side of the X. The whiteness contrast supported a structure that had the sugar-phosphate backbone on the outside of the molecule and the base pairs on the inside and not the opposite arrangement.

The symmetrical pattern of blurry smears that appeared in the diffraction pattern of a helical structure and produced the X shape seen in photo 51, enabled Franklin to measure the three periodicities or repeating sequences of 3.4 A, 20 A, and 34 A present in the DNA molecule. The symmetrical pattern of the smears resulted from the diffraction of x-rays along horizontal layer lines. In photo 51 both of the fourth-layer lines, one line at the top of the X and the other at the bottom, were missing. The missing smears provided the strongest evidence that DNA's structure was a double helix. The missing smear at each of the fourth-layer lines resulted because the x-rays diffracted by the two strands of the double helix where they crossed, canceled each other and eliminated the smear. This occurred at every tenth base pair or at a distance of 34 A.

The B form is the famous photo 51 that Wilkins passed to Watson and Crick in January 1953 without Franklin's knowledge or consent. She strongly suspected the same helical structure for the A, or crystalline, form but did not yet have conclusive evidence. Franklin published her research showing the inside location of the organic bases in the King's College *Medical Research Council Report* for 1952 (December). The Medical Research Council provided the source of her funding, and Crick had access to the *Report*. She also published the B form in *Nature* in April 1953 with the Wilkins and Watson and Crick papers on DNA's structure.

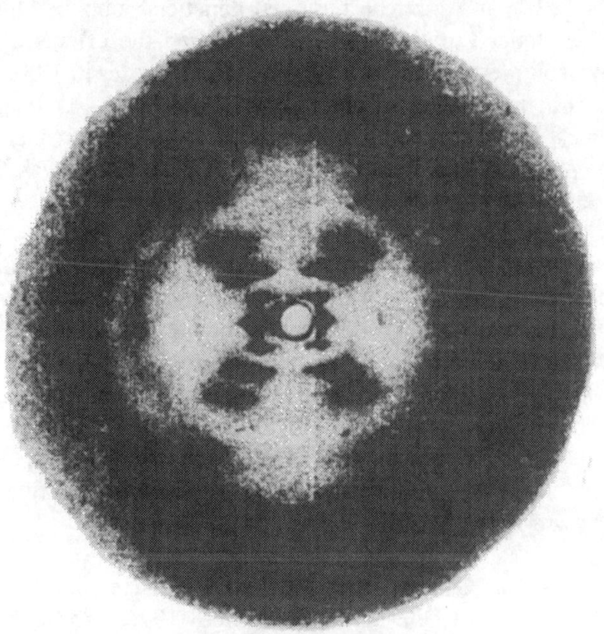

**Figure 11**
**Franklin's B Form of DNA**
Courtesy of the National Human Genome Research Institute and the Smithsonian National Museum of Natural History

Franklin had a troubled relationship with Wilkins, whom his associates described as shy, speculative, and passive compared to Franklin's direct, quick, and decisive personality, and with Watson and Crick. Franklin also believed that her appointment at King's College made her Wilkins's coworker and not his assistant. Her supporters have long claimed that Wilkins, Watson, and Crick never fully acknowledged the significance of her x-ray studies.

Because Wilkins, who directed the x-ray laboratory, had no plans to leave, Franklin left King's College in spring 1953 to work on tobacco mosaic virus and polio virus in Bernal's laboratory at Birkbeck College. There, in papers published in 1955-59, she confirmed Watson's 1954 hypothesis that in the rod-shaped tobacco mosaic virus, consisting of a single kind of protein molecule and RNA, the protein had a helical structure.

Franklin's unpublished x-ray work indicated that the RNA of tobacco mosaic virus also had a long single stranded helix structure embodied inside the protein coat. The 1950s-60s x-ray diffraction studies of Wilkins, Watson, Crick, Aaron Klug (b. 1926), and Donald Caspar (b. 1927) who worked with Watson at Caltech and Harvard, supported a helical structure. It received confirmation in 1974. Klug, Franklin's former postdoctoral student then at the Cavendish Medical Research Council Laboratory of Molecular Biology, and independently Alexander Rich (1924-2015), Pauling's former postdoctoral student then at MIT, determined the helical crystalline structure of transfer RNA (tRNA), a relatively small RNA molecule having about 75 nucleotides. Franklin died in 1958 after a two-year struggle with ovarian cancer. Wilkins died of cancer in 2004.

***Watson, Crick, and the double helix structure of DNA***

James Watson (b. 1928), an American biologist-geneticist, obtained his PhD from Indiana University in 1950 at age 22. Under Luria's direction he studied the effects of hard x-rays on bacteriophage multiplication. Watson met Wilkins at a symposium in Naples in 1951, where he saw for the first time Wilkins' x-ray diffraction images of crystalline DNA. The x-ray images caught Watson's attention, leading to an interest in the structures of nucleic acids and proteins, and he had Luria arrange for him to work under the physicist Max Perutz (1914-2002) at the Cavendish Medical Research Council Unit. While there in 1952 he met Francis Crick (1916-2004).

Crick, an English physicist who worked on magnetic and acoustic mine development during World War II, abandoned physics in 1947 for the new field of molecular biology. Like Wilkins, Crick had no interest in pursuing nuclear weapons research, which at that time was the direction in which research in the physical sciences had moved. Crick, then at University College, University of London, left for Cambridge, and in 1949 he joined Perutz's Cavendish Medical Research Council Unit. Crick obtained his PhD from Cambridge in 1954 for research on the x-ray diffraction of polypeptides and proteins. Working as a team in the Cavendish Medical Research Council Unit, Watson and Crick solved DNA's structure in 1953 and quickly published their results in a brief article "A Structure for Deoxyribose Nucleic Acid." *Nature* 171 (25 April 1953): 737-38.

Watson joined Pauling at Caltech from 1953 to 1955. He returned to Cambridge in 1955 before moving to Harvard in 1956 where he remained until 1976. While at Harvard, Watson, beginning in 1968, did research at Cold Spring Harbor Laboratory on Long Island. He became its director in 1968. Watson directed the Human Genome Project from its start in 1989 to April 1992 before returning to Cold Spring Harbor and serving as its president from 1993 and chancellor from 2004 until his forced retirement in 2007.

Watson had gone to London in October 2007 to give a speech at the Science Museum and to promote his new book, *Avoid Boring People: Lessons from a Life in Science*. In an interview published in *The Times* of London, Watson said that although many people of color have considerable talent "he was inherently gloomy about the prospect for Africa. All our social policies are based on the fact that their intelligence is the same as ours – whereas all the testing says not really." Watson quickly apologized *unreservedly* and said "I cannot understand how I could have said what I am quoted as having said. There is no scientific basis for such a belief."

Crick left Cambridge in 1976 to become a nonresident fellow at the Salk Institute for Biological Studies in La Jolla, California. He became a permanent member in 1977. Crick was also a biochemistry professor at the nearby University of California, San Diego. He died of colon cancer in 2004.

Watson and Crick were theoretical biologists, not experimentalists, and adopted a deductive approach in solving DNA's structure. Their use of geometry to solve DNA's structure followed the tradition of Galileo, Kepler, Descartes, and Einstein, who often sought a geometrical solution to scientific problems. They constructed DNA models according to Pauling's model building tradition, one of them a three-stranded helix that resembled Pauling's incorrect model in having the sugar-phosphate backbones running down the center and the bases on the outside. Franklin pointed out their model's flaws, and they continuously modified their models to agree with the available experimental data. Watson and Crick benefited significantly from Franklin's x-ray data, particularly photo 51 that Wilkins in January 1953 provided them. Her data showing the X pattern indicated the double helix structure of DNA's B or hydrated form.

Jerry Donohue (1920-85), an American organic chemist and crystallographer who had studied with Pauling at Caltech since the 1940s and had come to the Cavendish Laboratory to do research, provided a second important piece of evidence. Donohue told Watson that the *keto* structure (ketone, HC=O) of the organic bases guanine and thymine was the correct structure found in DNA, and not the less common *enol* structure (alcohol, C-OH) that always appeared in organic chemistry textbooks. In the *keto* structure of guanine and thymine, the carbon-oxygen bond was a double bond, C=O, and a hydrogen atom formed a single bond with a nitrogen atom nearby or adjacent to the double bond. The *enol* structure of guanine and thymine had a single bond between the oxygen atom and the carbon atom of the ring, and the hydrogen atom bonded to the oxygen atom and not to the nearby or adjacent nitrogen atom.

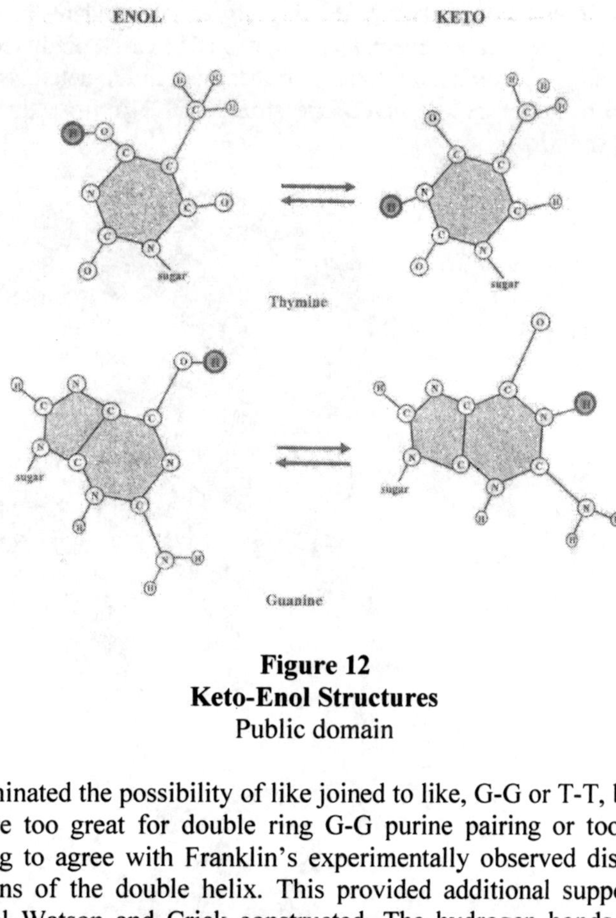


**Figure 12**
**Keto-Enol Structures**
Public domain

The *keto* structure eliminated the possibility of like joined to like, G-G or T-T, because their combined molecular widths were too great for double ring G-G purine pairing or too small for single ring pyrimidine T-T pairing to agree with Franklin's experimentally observed distance (width) of 20 A between the two chains of the double helix. This provided additional support for the interatomic distances in the model Watson and Crick constructed. The hydrogen bonds now formed naturally between A and T and between C and G. Two hydrogen bonds connected A-T, three hydrogen bonds connected C-G, this last structure determination provided by Pauling.

The final piece of structural evidence came when Watson realized that the experimentally-determined quantities of the bases A (a double ring purine structure) and T (a single ring pyrimidine structure) were always equal, as were the quantities of the bases C (a single ring pyrimidine structure) and G (a double ring purine structure). This occurred because A always bonded with T, and C always bonded with G. The A-T and C-G base pairing gave identical three-ring structures that held together the two strands of the helix. The dimensions of their model now matched the atomic dimensions of the DNA molecule obtained from Franklin's x-ray diffraction data. This key insight that explained Chargaff's rule had eluded everyone including Chargaff.

Watson and Crick's model also had to account for the periodicities that Franklin and Wilkins in 1951 observed for their helixes and required the two helical chains of DNA to wrap around the same axis to form a right-handed double helix. This meant that the two chains of DNA ran up and down in opposite directions and were complementary, with the hydrogen bonds of A-T and C-G holding the base pairs together. The periodicities were 3.4 A or 0.34 nanometers (nm) for the distance between successive nucleotides, 20 A (2.0 nm) for the distance between the two chains of the double helix or about 10-12 atoms across, and 34 A (3.4 nm) for the distance between each complete turn of the helix or about 10 successive nucleotides. The complementary pairing of bases, as Watson and Crick predicted in their 1953 paper, ensured that when the two helixes unwound and replicated, the original DNA double helix made one exact duplicate of the genetic information. Each of the old strands served as a template (mold or guide) for the polymerization of a new complimentary strand.

In 1957 Matthew Meselson (b. 1931), Pauling's graduate student, and Franklin Stahl (b. 1929) at Caltech confirmed the replication mechanism Watson and Crick predicted in 1953. They demonstrated experimentally using *E. coli* and $^{15}N$ as a tracer that DNA acted as a template when copied in replication. Each daughter cell received one strand of DNA from the original parent cell and one newly replicated strand.

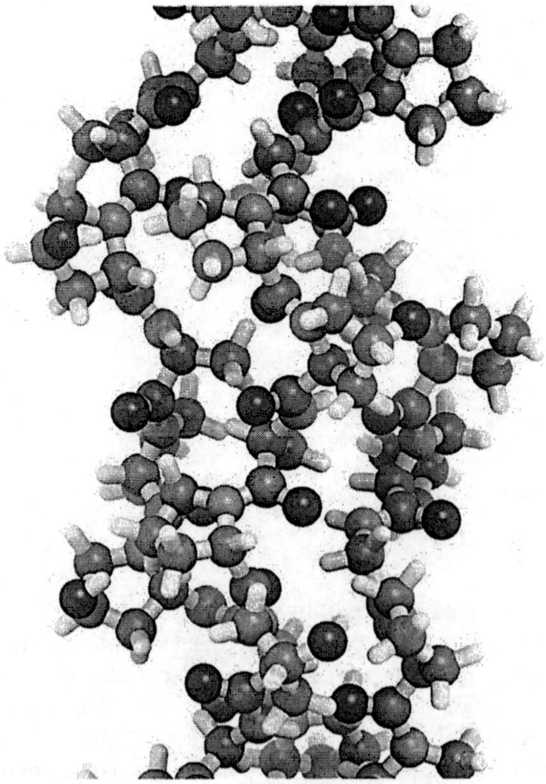

**Figure 13**
**Watson-Crick Model of DNA**
molekuul.be/Shutterstock.com

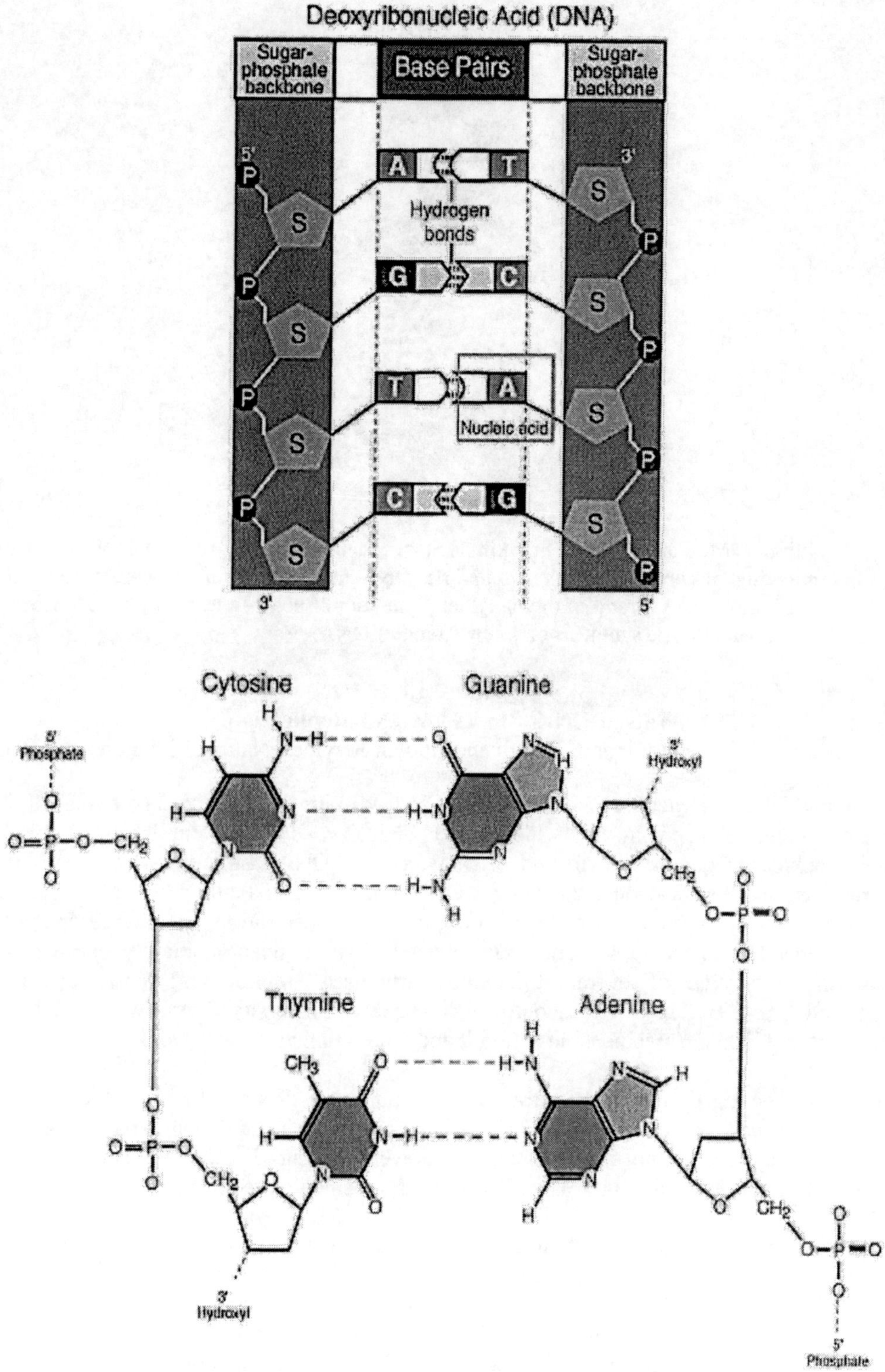


**Figure 14**
**Base Pairing in DNA**
Courtesy of National Human Genome Research Institute and the Smithsonian National Museum of Natural History

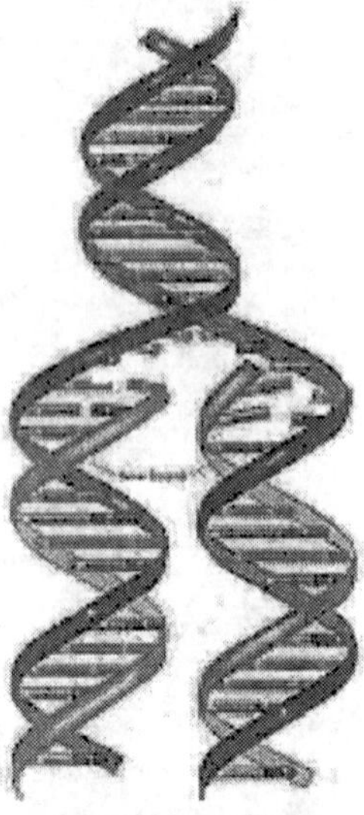

Matthew Meselson and Franklin Stahl demonstrated that DNA replicates semiconservatively, with each strand in a DNA molecule from the parent generation pairing with a new strand in the daughter generation. Each parent strand of DNA served as a template for the synthesis of a new strand of DNA.

**Figure 15**
**Watson-Crick Model for DNA Replication**
Source: US Department of Health and Human Services, National Institutes of Health

For solving DNA's structure, Watson, Crick, and Wilkins in 1962 received the Nobel Prize in medicine. Later, Watson in his widely read *The Double Helix* published in 1968, gave an autobiographical account of how he and Crick determined DNA's structure. Watson described a three way race among Watson-Crick, the King's College group, and Pauling. He portrayed science and scientific discovery as the product of deduction, induction, inspiration, and chance or serendipity. His account generated considerable controversy even before publication, mainly because of Watson's unflattering descriptions of several participants, particularly Franklin, and to a lesser degree Pauling, Bragg, and Crick. Because of the controversy Harvard University Press, Watson's first publisher, canceled its publishing agreement, an action it and other publishers rarely took.

Controversy continued after Watson found a new publisher. Watson believed the book's subtitle, *a personal account*, gave him permission to interpret events as he saw them. His critics considered his personal account inaccurate and unflattering of everyone but Watson. Watson initially wanted to entitle his book *Honest Jim*. In retrospect, *Modest Jim* seemed more appropriate.

To counter Watson's personal and questionable account, in 1975 Anne Sayre (1923-98), an American short story writer, editor, and lawyer, published *Rosalind Franklin and DNA*, a defense of Franklin. Sayre and her husband, David Sayre (1924-2012) a crystallographer who studied crystallography at Oxford and received his PhD from Oxford in 1951, knew Franklin quite well as a result of the two years (1949-51) they spent at Oxford while Franklin was in London. David Sayre later was at Pennsylvania, then at IBM in Yorktown Heights, in New York, and since 1973 an adjunct professor at the State University of New York at Stony Brook. Franklin also visited the Sayres at their home on Long Island when she traveled to the United States. Anne Sayre said that most crystallographers believed that Watson misrepresented Franklin, and in 1969 at the meeting of the International Union of Crystallographers in Stony Brook several crystallographers urged her to write a book to answer Watson. Crick provided much help in completing the Franklin biography. More recently, in 2002,

Brenda Maddox (b. 1932), an American writer living in London, wrote a second biography defending Franklin entitled *Rosalind Franklin: The Dark Lady of DNA*.

Crick published his own account of the discovery of DNA's structure and its aftermath in 1988 in *What Mad Pursuit*. He included a third DNA structure called Z-DNA that Rich at MIT discovered in 1979. Rich and his MIT group and Richard Dickerson, then at Caltech, using x-ray diffraction showed in 1980 that the Z form had a left-handed structure with a zigzag appearance and a special base sequence of alternating purines and pyrimidines. Wilkins, the other major contributor to the controversy, published his autobiography, *Maurice Wilkins, The Third Man of the Double Helix,* in 2003.

***New developments in molecular biology: DNA synthesis, protein structure determinations, Medical Research Council***

Research in molecular biology proceeded at a quickening pace in the 1950s. In 1956 Arthur Kornberg (1918-2007), a biochemist at Washington University and since 1959 at Stanford, developed a three-stage synthesis of DNA. Kornberg isolated for the first time DNA polymerase, later called DNA polymerase l, an enzyme that catalyzed the synthesis of the DNA present in *E. coli*. He combined DNA polymerase 1 with a mixture of nucleotides, each of which had three phosphate groups activated with energy from ATP, and labeled the polymerase 1 nucleotide mix with radioactive $^{14}C$ tracer. Kornberg then added some DNA that acted as primer and template, incubated his prepared sample, and obtained new DNA that contained $^{14}C$ tracer.

$$\textit{DNA triphosphate nucleotide} \xrightarrow{\textit{polymerase catalyst}} \textit{DNA}$$

Kornberg showed first that the prepared sample had to contain DNA in order to produce DNA. DNA polymerase 1 did not initiate DNA synthesis, it only extended a preexisting chain. Second, he showed that the synthesized DNA and the primer *E. coli* DNA had the same ratio of the A-T base pairs and C-G base pairs, or A = T and C = G. He proved the identity of the synthesized DNA and the primer DNA. For these discoveries in which he demonstrated experimentally the transfer of genetic information from one DNA molecule to another, Kornberg shared the 1959 Nobel Prize in medicine with Severo Ochoa, with whom he had previously worked at New York University in 1946. Ochoa received the prize for his synthesis of RNA in 1955.

Kornberg's son Roger (b. 1947), a Stanford chemist, received the 2006 Nobel Prize in chemistry for his detailed crystallographic pictures showing the copying of genetic information from DNA into RNA. Messenger RNA (mRNA) carried the information out of the cell nucleus where it began the construction of proteins.

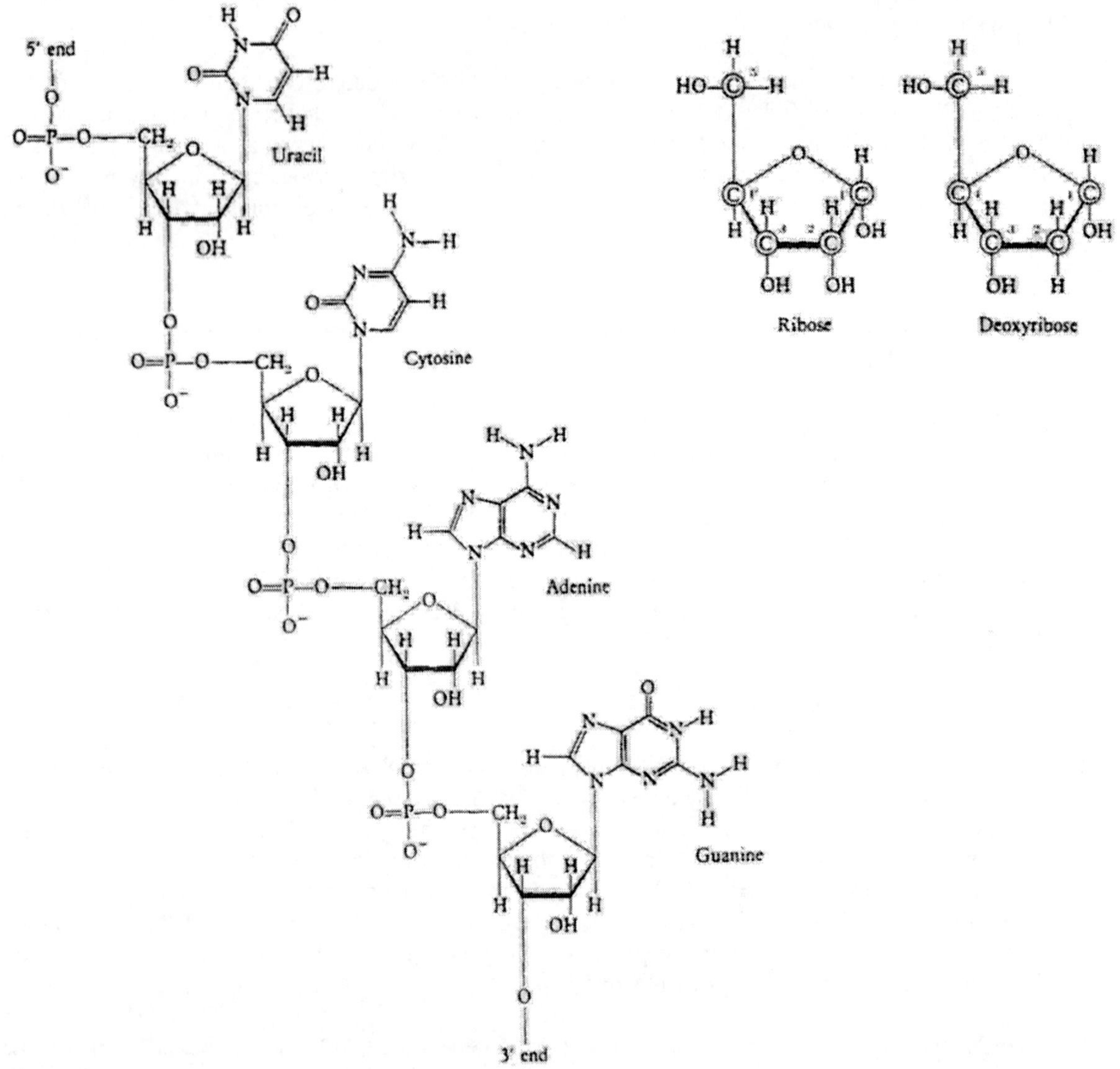


Inset gives detailed structural formulas of the two pentoses, ribose and deoxyribose.

**Figure 16**
**Chemical Structure of RNA**
Public domain

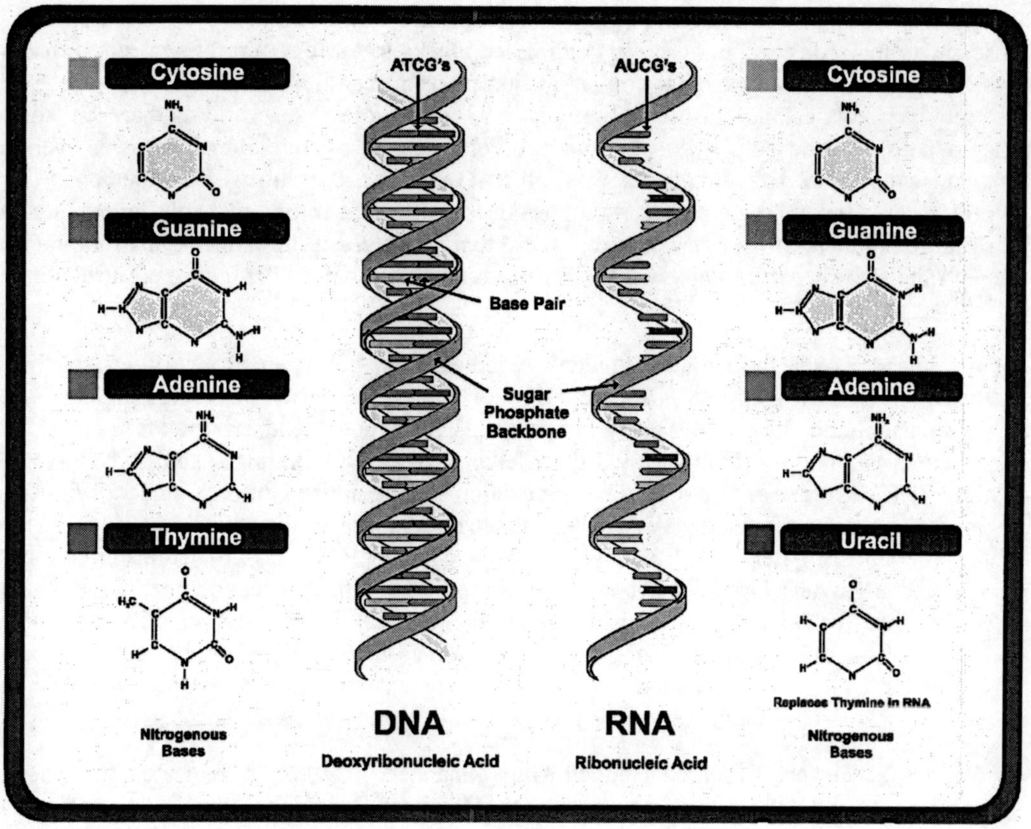


The two strands of DNA make it more stable than the single strand of RNA.

**Figure 17**
**RNA and DNA**
udaix/Shutterstock.com

Protein x-ray studies at the Cavendish Laboratory established the structures of myoglobin and hemoglobin. Kendrew, who worked on radar's development during World War II and became Perutz's research student in 1946, showed in 1957 that myoglobin protein, a polypeptide chain obtained from sperm whale muscle, had a helix structure. Perutz, another of Bernal's students when Bernal directed the Cavendish x-ray group, in 1959 determined the helix structure of the similar but more complicated and easily crystallized hemoglobin molecule. Hemoglobin, the protein of red blood cells containing about 12,000 atoms, consisted of four polypeptide chains folded into eight alpha helixes. His x-ray studies of hemoglobin which began in 1937 were the start of x-ray protein structure-determinations and what later became structural genomics. Perutz and Kendrew shared the 1962 Nobel Prize in chemistry for their x-ray structural determinations.

Perutz, a Jew turned Catholic, left Austria for Cambridge in 1936. Classified an enemy alien when World War II broke out, he spent 1940-41 interned in detention camps in Huyton near Liverpool and on the Isle of Man until July 1940, and from then until January 1941 in Sherbrooke, Quebec, Canada. The cosmologist Hermann Bondi (1919-2005), astrophysicist Thomas Gold (1920-2004), atomic spy Klaus Fuchs (1911-88), and the 1998 Nobel Prize winning chemist Walter Kohn (b. 1923) were other scientists interned in Quebec. Released from internment, Perutz returned to Cambridge where he joined William Bragg. Beginning in December 1942, because of his expertise on glaciers, Perutz worked for the Combined Operations Headquarters first in London and then in Jasper-Lake Louise, Alberta, Canada, on the British government's ultra-secret Habakkuk Project. Bernal supported the project and recommended Perutz.

Proposed to Winston Churchill (1874-1965) by Geoffrey Pyke (1893-1948), a British government scientist, the project required large floating ice blocks to serve as aircraft carriers or bases for aircraft crossing the Atlantic Ocean for combat in Europe or in search of German submarines. Scientists at the Brooklyn Polytechnic Institute brought into the project discovered that ice reinforced with 4-14 percent wood pulp sawdust, a mixture Perutz called pykrete after Pyke, was even more durable than ice alone. Two developments, the high cost (about $100 million) of building an ice aircraft carrier 600 meters long, 90 meters wide, 45 meters deep, with a 2 million ton displacement, and the capability of large bombers to fly nonstop across the Atlantic, rendered the fantastic plan impractical. By early 1944 the British government canceled Project Habakkuk before beginning construction on a full-size ice carrier.

With the war over, the United Kingdom's Medical Research Council in 1947 established its Medical Research Council Unit for Research on the Molecular Structure of Biological Systems at the Cavendish Laboratory and named Perutz as its head and Kendrew a staff member. The Medical Research Council, originally called the Medical Research Committee and Advisory Council, dates from 1913. It was a single research organization for the entire United Kingdom with funds provided under the National Insurance Act of 1911 for medical research. The initial stimulus was to find a cure for tuberculosis. After World War I, in 1919, the Medical Research Council acquired its present name, received a Royal Charter, and became a corporation with a three-member government committee responsible for its funding. Martin joined the Medical Research Council's National Institute of Medical Research at Mill Hill in 1948. Sanger joined the Cavendish Laboratory's Medical Research Council's Unit in 1951. Watson and Crick were early members of the Cavendish Laboratory's Medical Research Council Unit, Crick in 1949, and Watson in 1952.

When the Medical Research Council Unit expanded in 1962, it relocated to a new location in Cambridge, and took on its new name, the Medical Research Council Laboratory of Molecular Biology. Perutz became director of the Laboratory with its three divisions, Cell Biology led by Crick, Structural Studies led by Kendrew, and Protein (Nucleic Acid) Chemistry led by Sanger.

***Proposing a genetic code***

In 1953 Watson and Crick noted that protein synthesis in all cells required 20 amino acids. They were the 20 most abundant amino acids that composed the naturally-occurring proteins, excluding those found in a few odd proteins, or like ornithine and citruline, never found in proteins. Watson, Crick, Sidney Brenner (b.1927), and other geneticists and scientists such as George Gamow, quickly recognized that each of the four bases on DNA's double helix backbone could not code for a different amino acid. Instead, Crick and Brenner thought the unknown code (the genetic code) that specified the number of bases and the base sequence or order required to account for each of the twenty amino acids was most likely a sequence of three bases.

A sequence of one base, $4^1$ or 4, representing a 1 to 1 correspondence between a base and an amino acid on the DNA backbone, accounted for only 4 amino acids: A = amino acid, T = amino acid, C = amino acid, G = amino acid. Similarly, a sequence of two bases to include all combinations, $4^2$, (AA, AT, AC, AG), (TA, TT, TC, TG), (CA, CT, CC, CG), and (GA, GT, GC, GG), accounted for 16 amino acids. A sequence of three bases to include all combinations, $4^3$, (AAA, AAT, AAC, AAG, etc.), accounted for 64 amino acids. The code, therefore, required a sequence of three bases, or a nucleotide triplet. Each triplet specified an amino acid and hence the sequence of chemically linked or bonded amino acids that formed a protein molecule.

Marshall Nirenberg (1927-2010), a biologist and geneticist at the National Institutes of Health (NIH) in Bethesda, Maryland, and Har Gobind Khorana (1922-2011) at Wisconsin provided the experimental evidence of a triplet code that by late 1961 convinced geneticists. Nirenberg in May 1961 showed using cells from *E. coli* bacteria and $^{14}C$ that the nucleotide triplet UUU synthesized the amino acid phenylalanine. Khorana's preparation of synthetic RNA polymers with known base sequences beginning in 1960, also indicated a triplet code. Nirenberg's and Khorana's experimental evidence, and the increasing experimental evidence obtained from *E. coli* mutation studies including Crick and Brenner's own investigations on the bacteriophage T4 that attacked strains of *E. coli*, enabled Crick and Brenner in a December 1961 publication in *Nature* to write with near certainty that the genetic code required a sequence of three bases, such as ATC, AAG, or TGC, on the DNA backbone. A triplet code accounted for each of the twenty amino acids and thereby accounted for the amino acid sequence along the polypeptide chain of a protein. Brenner simplified the language of molecular biology in 1957 when he introduced the word *codon* to replace nucleotide triplet. Brenner followed a new system of nomenclature that Seymour Benzer (1921-2007), a physicist turned molecular biologist at Purdue and later Berkeley, introduced in 1957. Benzer first used the *on* suffix in molecular biology, patterning it on the prevalence of the *on* suffix of the electron, proton, neutron, and many other names adopted by physical scientists.

The naturally-occurring proteins required only 20 amino acids in forming the body's 220 kinds of cells. Crick and Brenner proposed therefore that each of the remaining 44 triplets also coded for one of the 20 amino acids and for stopping or breaking points (stop) in the sequence of the base triplet code. The code, they said, was degenerate, meaning that several different base triplets could code for the same amino acid. Crick and Brenner also proposed that geneticists read the base sequence from a fixed starting point and that the triplets could not overlap for the code to make sense. Brenner in 1957 and Akira Tsugita and Fraenkel-Conrat at Berkeley in their 1960 study on the mutants of tobacco mosaic virus already had provided evidence that the code could not overlap. Tsugita and Fraenkel-Conrat pointed out that if the code were overlapping, a mutation or change in one nucleotide would cause a change in more than one amino acid in the resulting protein. Their experiments showed that this did not happen. When they compared the protein sequences for the mutant and the non-mutant DNA, the change in the mutant DNA occurred in only one amino acid.

Because the DNA of the chromosomes always remained in the nucleus and protein formation occurred in the cytoplasm outside the nucleus, Crick, Brenner, and other geneticists suspected the necessity of an intermediary for the formation of protein molecules from the two nucleic acids, DNA and RNA. The intermediary, a new type of RNA molecule they called transfer RNA (tRNA), decoded DNA and directed the production of an amino acid so that the genetic information flowed in only one direction, from DNA to RNA to amino acid to protein. Experimental evidence beginning in 1956 demonstrated the validity of Crick and Brenner's tRNA-code and protein formation mechanism.

Brenner, a South African molecular biologist who joined the Cavendish's Medical Research Council Unit in 1957 and occupied an office with Crick for 20 years, left Cambridge for Berkeley in 1998. He shared the 2002 Nobel Prize in medicine with Robert Horvitz (b. 1947) at MIT, and John Sulston (b. 1942), the founding director of the Wellcome Trust Sanger Institute near Cambridge, England, and since 2007 at Manchester.

***Role of RNA in understanding the genetic code: transcription and translation***

The first in a series of investigations that led to solving the genetic code and to understanding RNA's intermediary role in the transcription (transfer) of genetic information from DNA to RNA as the initial step in protein synthesis, occurred in 1955. In that year Severo Ochoa (1905-93) synthesized RNA. Ochoa, a Spanish medical doctor who came to the United States in 1940 and to New York University in 1942, recognized that enzymes (proteins) catalyzed the synthesis of RNA and DNA from simple naturally-occurring precursors such as nucleotide di- and triphosphates ADP and ATP. He isolated the enzyme polynucleotide phosphorylase from a strain of *E. coli* bacteria and reacted it with an RNA diphosphate nucleotide. The nucleotides joined together to form a thick jellylike viscous solution that consisted of long, thin, single-stranded RNA molecules.

| | phosphorylase catalyst | |
|---|---|---|
| *RNA diphosphate nucleotide (one kind of nucleotide)* | ⟶ | *RNA (with one kind of nucleotide)* |

Ochoa's method of synthesis allowed him to synthesize an RNA molecule that unlike natural RNA had only one kind of nucleotide base such as uracil or adenine that repeated itself endlessly. His synthetic RNA provided the foundation for Nirenberg's and Khorana's solving the genetic code, and in 1961-62 his laboratory competed with their laboratories in the race to discover the genetic code. Ochoa's method of synthesizing RNA resulted in his sharing the 1959 Nobel Prize in medicine with Kornberg. Four independent investigations in 1960 showed, however, that a different enzyme catalyzed RNA's synthesis and that polynucleotide phosphorylase usually catalyzed RNA's breakdown, not its synthesis. Ochoa returned to Madrid in 1985 to advise the Spanish government on science policy.

Detection of Crick and Brenner's tRNA intermediary required magnification greater than the best optical microscopes, about 2,000 times, but well within the magnification of the electron microscope that the Russian physicist Vladimir Zworkin (1889-1982) constructed in 1939. Zworkin, who came to the United States in 1919, two years after the Bolshevik revolution, worked for the Westinghouse Electric Corporation in Pittsburgh for about seven years. He joined RCA in Camden, New Jersey, in 1929, where he continued development of the electron microscope, achieving 100,000 times magnification in 1939, or 50 times the magnification of the best optical microscope.

Using an electron microscope that like Zworkin's had 100,000 times magnification, George Palade (1912-2008), a refugee physiologist from Soviet-occupied Romania and since 1946 at the Rockefeller Institute, provided the initial detection of the tRNA intermediary. He showed in 1956 that the cytoplasm of a cell contained a large number of tiny spherical structures (10-20 nanometers = 10-20 x $10^{-9}$ meters) that he called microsomes, and that a microsome consisted of RNA and protein (45-60 percent ribosomal RNA and 40-55 percent protein). The microsomes appeared as sheets of black dots. Each sheet measured one microsome thick and contained hundreds of microsomes in repeating groups of four. Palade believed the microsome, which he renamed a ribosome, was the cell's protein-producing location. Ribosomes are present in the cytoplasm of all cells.

Further evidence that RNA was the suspected intermediary came in the early 1960s from researchers in the United States and in France. Mahlon Hoagland (1921-2009), an American biochemist working with Paul Zamecnik (1912-2009) at the Massachusetts General Hospital in Boston, in early 1957 discovered a short-stranded RNA molecule consisting of about 80 nucleotides in the cytoplasm of the cell. What they originally called soluble RNA was Crick and Brenner's tRNA. Hoagland showed using $^{14}C$ radioactive tracers that each of tRNA's base triplets formed a bond with a complementary triplet on a second kind of longer-stranded RNA molecule and combined at its sugar-phosphate backbone with one specific amino acid. Hoagland and Zamecnik had discovered the guidance mechanism that according to Crick enabled an amino acid to attach itself to the ribosome of a cell.

The biochemist Jacques Monod (1910-76) at the Pasteur Institute in Paris in May 1961 named the longer-stranded RNA, messenger RNA (mRNA). In June 1961 Monod, Brenner, and Meselson showed that mRNA was the RNA that carried genetic messages (the genetic triplet code) from DNA to the ribosomes. A ribosome is the specialized site in the cytoplasm where amino acid and protein formation occur. Working in Delbrück's Caltech laboratory with bacteria infected with T2 phage, they used radioactive tracers ($^{13}C$, $^{15}N$, $^{32}P$, $^{35}S$) to follow the reaction pathway and prove that mRNA transcribed the genetic code (the triplets of DNA) from the DNA of the chromosomes (the *operon*) in the nucleus to the ribosomes. This first step in amino acid and protein synthesis is the *transcription* step.

In the second step of amino acid and protein synthesis called the *translation* step, each triplet on mRNA paired with its complimentary tRNA triplet that had a specific amino acid attached to its sugar-phosphate backbone, and in this way synthesized at a ribosome a protein chain consisting of the same amino acid sequence that existed in the original chromosome's DNA molecules. Transfer RNA read off, or translated, the genetic code preserved in DNA and transcribed by mRNA, to produce protein molecules. Twenty kinds of tRNA likely exist, corresponding to the 20 kinds of amino acids that naturally occur in proteins. In this mechanism, the chromosomes of a particular cell produced the protein molecules (enzymes) that directed the chemistry of the cell and therefore all its physical characteristics.

Monod and his associate, the biologist Francois Jacob (1920-2013), in research on the roles of DNA and RNA in amino acid and protein synthesis, in 1960 developed a genetic regulatory mechanism that accounted for the activation and suppression, or the *on* and *off*, of genes, especially those genes that controlled enzyme syntheses in mutant bacteria. Monod introduced the expression *operon* to define a group of functionally-integrated genes on a chromosome (the DNA of the chromosome), controlled by the genes' small end region called the operator or operator site. A different protein molecule, called a repressor, could prevent the expression of the genes by combining with the operator site and thereby turning off the operator.

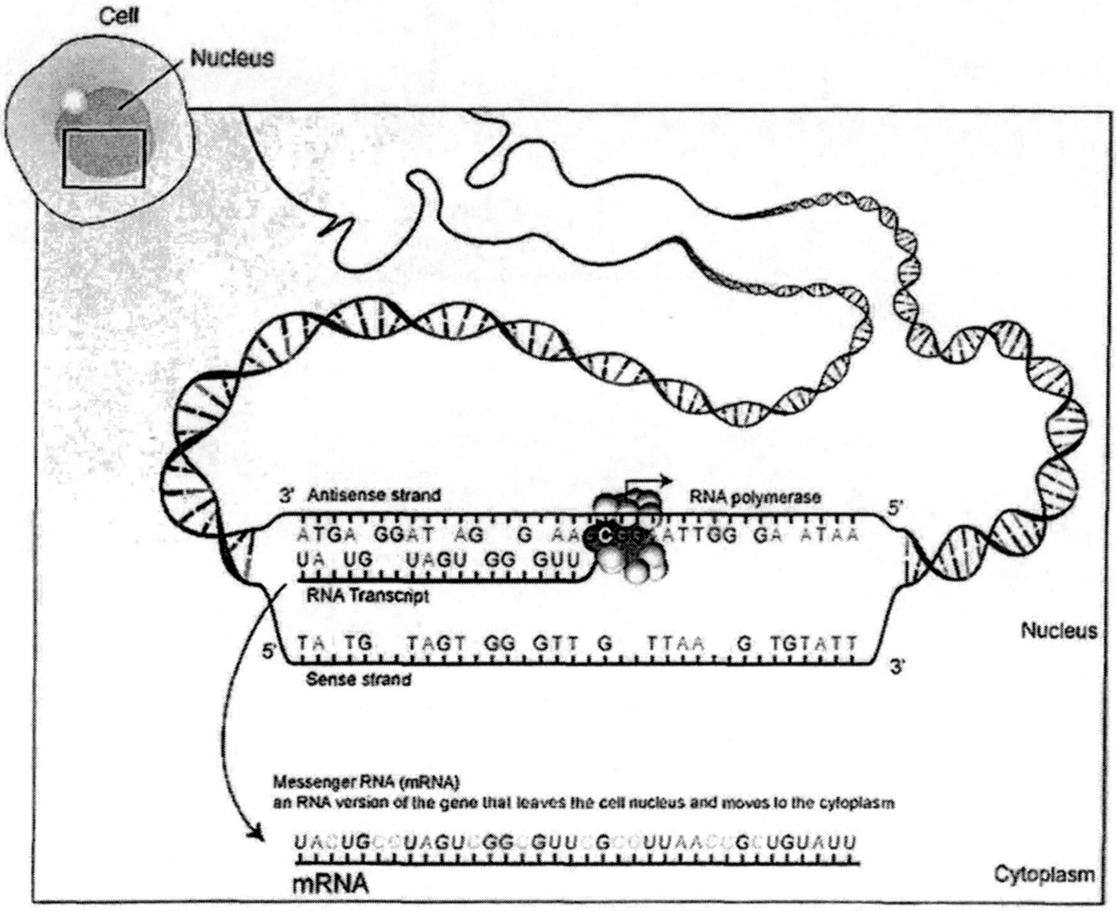


**Figure 18**
**Messenger RNA (mRNA)**
Courtesy of National Human Genome Research Institute, www.genome.gov.

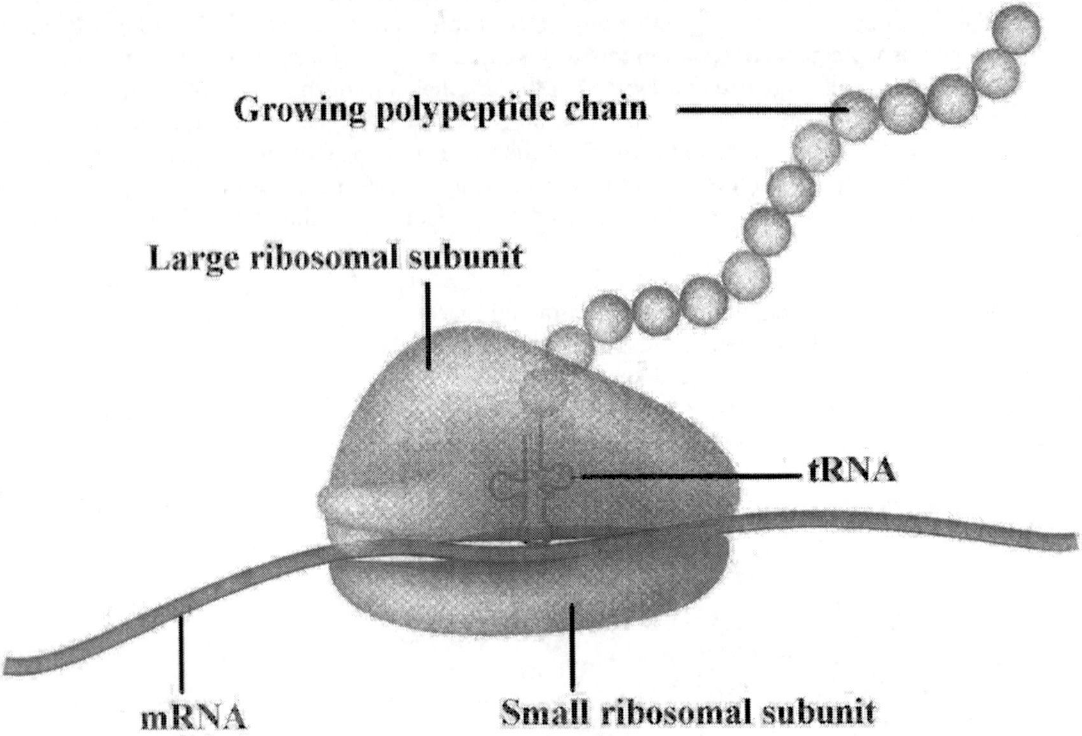


**Figure 19**
**A Ribosome**
Alila Medical Media/Shutterstock.com

| TRANSCRIPTION AND TRANSLATION |
|---|
| DNA ⟶ mRNA (transcription)⟶ tRNA (translation) ⟶ amino acid ⟶ protein ⟶ cell⟶ life |

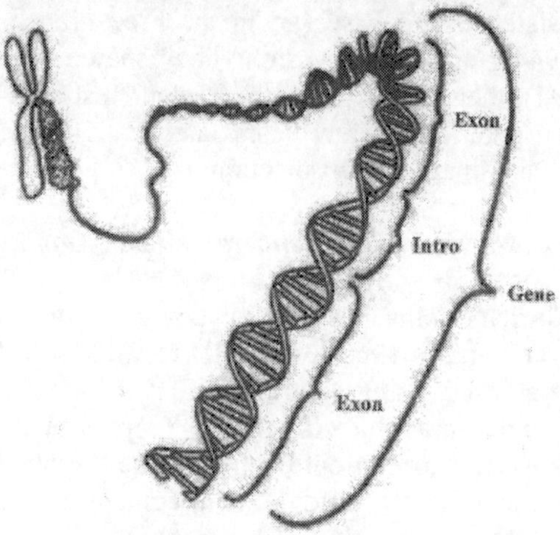


| | | Transcription | Translation | | Amino | Polypeptide |
|---|---|---|---|---|---|---|
| | | DNA | mRNA | tRNA | Acid | Chain |
| | | G | C | G | | |
| | | C | G | C | | |
| | | | U | | | |
| | Exon | A | ribosome | A | | |
| | | C | G | C | | |
| | | G | C | G | | |
| G | | A | U | A | | |
| | | T | A | U | | |
| E | | A | U | A | | |
| | | G | C | G | | |
| N | | T | A | U | | |
| | | T | A | U | | |
| E | Intron | G | C | G | | |
| | | A | U | A | | |
| | | C | G | C | | |
| | | G | C | G | | |
| | | A | U | A | | |
| | | G | C | G | | |
| | | T | A | U | | |
| | | A | U | A | | |
| | Exon | C | G | C | | |
| | | T | A | U | | |
| | | G | C | G | | |
| | | A | U | A | | |
| | | T | A | U | | |

**Figure 20**
**A Gene, Showing Transcription and Translation**
Courtesy of the National Human Genome Research Institute, www.genome.gov. (top)
Courtesy of the author (bottom)

Palade, who went to Yale in 1972, shared the 1974 Nobel Prize in medicine with Albert Claude (1899-1993) at the Catholic University of Louvain in Belgium and Christian de Duve (1917-2013) at the Rockefeller University, for their work on the structural and functional organization of the cell. Monod, whose mother was an American, and Jacob fought against the Nazis in World War II. Monod joined the French resistance, Jacob served in the Free French Army and suffered serious combat injuries during the 1944 Normandy invasion. They shared the 1965 Nobel Prize in medicine with André Lwoff (1902-94), their colleague at the Pasteur Institute. Monod's book on the origin of life, *Chance and Necessity* published a few years later in 1970, argued that life arose by chance and evolved according to Darwinian natural selection.

***Breaking the genetic code: Marshall Nirenberg and Har Gobind Khorana***

Nirenberg, with J. Heinrich Matthaei (b. 1929) a post-doctoral researcher from the University of Bonn who had come to Cornell and in 1960 to the NIH to work with Nirenberg, broke the genetic code in May 1961 when they showed, using cells from *E. coli* bacteria and $^{14}C$, that the mRNA nucleotide triplet, or codon, uracil-uracil-uracil, UUU, represented the amino acid, phenylalanine $C_6H_5CH_2CH.HH_2COOH$. In carrying out phenylalanine's synthesis, Nirenberg used a synthetic RNA polymer, uridylic acid, as mRNA. Uridylic acid contained a single repeating nucleotide (the sugar ribose, a phosphate group, and the organic base uracil), giving it the repetitive structure UUUUUU. . . . . and UUU as its only possible nucleotide triplet. When uridylic acid formed a protein containing a single amino acid, phenylalanine, Nirenberg concluded that UUU corresponded to phenylalanine. This was the first nucleotide triplet discovered and the first demonstration of mRNA. The triplet sequence, or sequence of three base pairs, on the DNA chain corresponded to a specific amino acid and represented the genetic code that Crick had suspected as early as September 1957 and convincingly demonstrated in the 1960 T4 bacteriophage and *E coli.* experiments that he and Brenner published in December 1961.

Nirenberg and his team of twenty researchers continued studies on the mRNA nucleotide triplets, preparing other synthetic RNA polymers, and showed that AAA coded for lysine, CCC for proline, GGG for glycine, GCU, GCC, GCA, and GCG for alanine, and that AAG also coded for lysine. They noted that triplets that coded for the same amino acid usually differed only in the base occupying the third position. By 1965 they had identified the 61 RNA triplets that coded for the amino acids and knew the role of the other three. Nirenberg's team became the first scientists to complete the sequencing of the genetic code and to provide an understanding of DNA's language.

While Nirenberg was identifying the 61 codons that coded for the 20 amino acids, Brenner discovered a different role for the three other codons. In 1965 he found that tRNA did not recognize TAA, TGA, and TAG on DNA (corresponding to UAA, UGA, UAG on mRNA). These triplets did not code for any of the 20 amino acids and instead terminated or stopped the protein chain's growth. Other studies revealed that AUG (methionine) was the only start codon, although GUG and UUG occasionally acted as start codons, and that UAG, AUG, and UGA were the only known exceptions to the rule that no codon specified more than one amino acid. AUG specified methionine and N-formyl methionine. UGA specified the twenty-first amino acid selenocysteine, and UAG specified the twenty-second amino acid pyrrolysine. UGA was also the codon for tryptophan in mitochondria.

Khorana, a molecular biologist born of Hindu parents in Punjab, now West Pakistan, beginning in 1960 prepared synthetic RNA polymers and independently worked out the genetic code for the 64 RNA codons. While at Wisconsin from 1960 to 1970, and since then at MIT, Khorana confirmed, as Nirenberg did, that the nucleotide triplets were the letters of the genetic code, they did not overlap, and were without commas.

Khorana in 1970 achieved the first complete synthesis of an artificial gene. The 77-base-pair gene was a copy of a gene found in the alanine-tRNA molecule of naturally-occurring yeast. Alanine tRNA is the intermediate transfer molecule that directs the amino acid alanine into protein formation. Khorana and his team could not include all the genetic information (control signals) necessary for the synthetic gene to function in a cell and detected no functioning of the synthetic gene. He synthesized a second gene in 1973. The 126-base-pair gene was a copy of a gene found in *E. coli* and contained all the genetic information necessary for the construction of its intermediate transfer molecule, tyrosine tRNA, with 126 base pairs.

Nirenberg and Khorana shared the 1968 Nobel Prize in medicine with the Cornell chemist Robert Holley (1922-93). Holley in 1962 used a protein separation technique, similar to Sanger's, to separate the three small amino acids, alanine, tyrosine, and valine from yeast tRNA. In late 1964 using two ribonucleases (enzymes) and chromatographic analysis, he established the complete 77-nucleotide structure of naturally-occurring alanine tRNA. Holley had determined for the first time the complete chemical structure of a biologically active nucleic acid.

**Second Letter**

| First Letter | U | | C | | A | | G | | Third Letter |
|---|---|---|---|---|---|---|---|---|---|
| U | UUU | Phe | UCU | | UAU | Tyr | UGU | Cys | U |
| | UUC | | UCC | Ser | UAC | | UGC | | C |
| | UUA | Leu | UCA | | UAA | Stop | UGA | Stop | A |
| | UUG | | UCG | | UAG | Stop | UGG | Trp | G |
| C | CUU | | CCU | | CAU | His | CGU | | U |
| | CUC | Leu | CCC | Pro | CAC | | CGC | Arg | C |
| | CUA | | CCA | | CAA | Gln | CGA | | A |
| | CUG | | CCG | | CAG | | CGG | | G |
| A | AUU | | ACU | | AAU | Asn | AGU | Ser | U |
| | AUC | Ile | ACC | Thr | AAC | | AGC | | C |
| | AUA | | ACA | | AAA | Lys | AGA | Arg | A |
| | AUG | Met | ACG | | AAG | | AGG | | G |
| G | GUU | | GCU | | GAU | Asp | GGU | | U |
| | GUC | Val | GCC | Ala | GAC | | GGC | Gly | C |
| | GUA | | GCA | | GAA | Glu | GGA | | A |
| | GUG | | GCG | | GAG | | GGG | | G |

Draw imaginary horizontal and vertical lines to connect the letters. They intersect at the amino acid for which they code.

**Figure 21**
**The Genetic Code**
Courtesy of the National Institutes of Health

Knowing that the triplet sequences of bases determined the formation of amino acids led geneticists to understand that a mutation was an error in the sequence, and resulted from an error in substitution, a deletion or an insertion of a base, or a mix up of a series of bases that Brenner in 1965 first called *nonsense.* (Crick and Brenner used *nonsense* in a different context in their 1961 paper). An example of this kind of genetic error appeared in an article published in *Cell*, March 1993. A team of 58 American and British scientists from Massachusetts General Hospital, MIT, Michigan, California (Irvine), University of Wales in Cardiff, United Kingdom, and Imperial Cancer Research Fund in London, after ten years of work, discovered the defective gene on chromosome 4 that causes Huntington's disease (chorea).

The disease, a nervous disorder which the American physician George Huntington (1850-1916) in Middleport, Ohio, first described in 1872, results in jerky involuntary movements, loss of control of body functions, and deterioration of memory and mental processes. By 1995 researchers had discovered how the gene causes the disease. The healthy gene contains a segment in which the nucleotide triplet CAG, glutamine, repeats itself 38 times and thereby incorporates the amino acid glutamine into a protein 38 times. In the defective gene the CAG segment repeats itself 40 to 150 times. Their studies showed that the more often the segment repeats itself, the earlier the disease strikes and the more serious it becomes. Researchers have yet to find a cure for the disease.

***The beginning of genetic engineering or DNA sequencing***

Hamilton Smith (b. 1931), a molecular biologist at Johns Hopkins, in 1970 isolated the first restrictive enzyme (a protein molecule) that split or cut DNA at a specified site or sites. Smith obtained a site-specific *endonuclease* enzyme from the bacterium *Haemophilus influenzae Rd*, later known as *Hind II.* Smith and Daniel Nathans (1928-99), his collaborator at Johns Hopkins, used a viscometer to measure endonucleolytic cleavage of DNA by cell extracts, phosphorus-32 and phosphorus-33 as tracers, ion-exchange chromatography to cleave (fractionate) the nucleotides, and electrophoresis to analyze the cleaved nucleotides. Isolation of restrictive enzymes permitted the controlled splitting of a gene, giving genetically active fragments which when recombined gave a new gene, or recombinant DNA. This was the beginning of genetic engineering or gene sequencing. Smith and Nathans shared the 1978 Noble Prize in medicine for their discovery.

Three years after Smith's breakthrough, the geneticist Stanley Cohen (b. 1935) at Stanford and the biochemist Herbert Boyer (b. 1936) at the University of California, San Francisco, demonstrated the DNA sequencing technique using restrictive enzymes obtained from *E. coli*. They joined segments of two different plasmids. Plasmids are small ringlets of DNA that float freely in the cytoplasm of certain bacterial cells such as *E. coli* and replicate independently from the coding strand of DNA to give a new recombinant DNA gene. When inserted into *E. coli*, the recombinant DNA possessed the traits present in the two original plasmids and replicated itself. Boyer and Cohen left their university positions and in 1976 founded Genentech, a San Francisco biotechnology company, to commercialize their sequencing technique.

In another successful sequencing experiment carried out in 1981, geneticists transferred rabbit DNA obtained from the hemoglobin gene to mouse DNA to give a recombinant DNA. Thomas Wagner, a geneticist at Ohio University in Athens, directed the research done jointly at Ohio University and the Jackson Laboratory in Bar Harbor, Maine, a laboratory established in 1929 that specialized in genetic research on mice. Ohio University established the Edison Biotechnology Institute in 1984 to continue research on the production of virus-resistant transgenic mice.

As the research pace on DNA and RNA accelerated in the 1980s-90s, geneticists continued to ask how many genes humans had available to split and recombine and to sequence. Until the late 1990s-2000 they believed that the human genome, the 24 *different* chromosomes (22 chromosomes plus either an X or a Y chromosome) present in the nucleus of each of a human body's 10 trillion cells, contained about 100,000 human protein-coding genes or 3,500-5,000 genes on each human chromosome. Walter Gilbert, the Harvard physicist turned molecular biologist and co-recipient of the 1980 Nobel Prize in chemistry, first suggested the 100,000-gene human genome in 1986. Gilbert based his number on his DNA base-pair sequencing research and the research of other geneticists, all of which led to estimates of chromosome size and composition and to the recognition that human gene size varied widely. These studies showed that the protein-coding genes (exon) comprised about 30 percent of each chromosome and that the remaining 70 percent consisted of junk DNA (intron), RNA, and protein. Each gene contained about 30,000 base pairs, giving 3.167 billion base pairs (30,000 x 100,000) in the human genome (the 22 pairs of chromosomes plus X and Y chromosomes). A human chromosome containing 3,500-5,000 genes and 105-150 million base pairs measured $10^{-9}$ meters in length. Gilbert introduced the words exon and intron in 1978.

The dystrophin gene located on the X chromosome, the largest human gene and the cause of muscular dystrophy, has 24 million base pairs. Virus genomes range in size from 3,200 base pairs in the *hepadnavirus MS2*, to 5,386 base pairs in *phi X174* sequenced in 1976, to 7,200 base pairs in the bacteriophage *M13*, to 1,181,404 base pairs in the *Mimivirus*, discovered in 1992. *Mimivirus* identified in 2003 in amoebas, remains at this time the largest of all known viruses. Yeast (fungus) sequenced in 1996 has 12.5 million base pairs, *E. coli* bacterium sequenced in 1997 contains 4.7 million base pairs. The nematode worm, which John Sulston then at the Wellcome Trust Sanger Institute and Robert Waterson (b. 1943) then at Washington University School of Medicine in St. Louis, sequenced in 1998, has 100 million base pairs. The fruit fly sequenced in 2000 has 160 million base pairs, and the mouse genome sequenced in 2002 has 3 billion base pairs.

| **NUMBER OF BASE PAIRS (BASES) IN GENOMES** | | |
|---|---|---|
| ***Genome*** | ***Number of Base Pairs*** | ***Year Sequenced*** |
| *Hepadnavirus MS2* | 3,200 | 1976 |
| *Phi X174* virus | 5,386 | 1976 |
| *M13* bacteriophage | 7,200 | 1980 |
| *Mimivirus* | 1,184,404 | 2003 |
| *E. coli* bacterium | 4.7 million | 1997 |
| Yeast fungus | 12.5 million | 1996 |
| *Caenorhabditis* elegans, Nematode (round) worm | 100 million | 1998 |
| Fruit fly | 160 million | 2000 |
| Mouse genome | 3 billion | 2002 |
| Human genome | 3 billion | 2003 |

Despite the considerable success in determining DNA's molecular structure and chemical properties scientists had not yet obtained photographs of the DNA molecule, and therefore of a gene or genes. This changed in 1987 when the physicist Miquel B. Salmeron (b. 1946), senior staff scientist at the Center for Advanced Materials, Lawrence Berkeley National Laboratory, used a scanning tunneling microscope with a one million times (1 x $10^6$) magnification, a horizontal resolution of 2 A and a vertical resolution of 1 A, to produce the first images of the DNA molecule. Gerd Binnig (b. 1947) and Heinrich Rohrer (1933-2013) at the IBM Zurich Research Laboratory in Rüschlikon, Switzerland, invented the scanning tunneling microscope in 1981. They shared the 1986 Nobel Prize in physics with Ernst Ruska.

In scanning tunneling microscopy, which operates on the quantum mechanical principle of electron tunneling, a probe that narrows to a tip only a single atom wide skims across a conductive surface at a distance of 10 A (1-2 atomic diameters), always maintaining the same electrical current between the probe and the surface atoms. When the probe moves over an atom with a surrounding electron cloud that facilitates electronic conduction, the probe moves up. When it passes over an atom with a surrounding electron cloud that does not facilitate conduction, the probe falls down. A computer receives the up and down probe-produced pattern and converts it to an atom-by-atom contour map of the conductive surface.

## VII. HUMAN GENOME PROJECT

The Human Genome Project had its roots in three meetings held in the early 1980s. David Smith at the Office of Health and Environment (OHER), Department of Energy (DOE),.0 and Mortimer Mendelsohn at the DOE's Lawrence Livermore Laboratory organized the first meeting, a conference in Alta, Utah, 9-13 December 1984. A second meeting that Robert Sinsheimer (b. 1920), then chancellor at the University of California, Santa Cruz, organized, followed in May 1985. Charles DeLisi (b.1941), then OHER's director, and Smith at OHER initiated the third meeting, the First Santa Fe (New Mexico) Conference, held on 3-4 March 1986.

DOE in Washington, DC, sponsored the 1984 meeting with the International Commission for Protection against Environmental Mutagens and Carcinogens (ICPEMC) established in 1977 at Research Triangle Park in North Carolina. DOE was the sole sponsor of the 1986 meeting, and at that time provided $5.3 million for pilot plant projects to begin at its national laboratories. The NIH in Bethesda began funding genome research in 1987. In 1988, after the Office of Technology Assessment (OTA) and the National Academy of Sciences (NAS) recommended a joint research program, DOE and NIH signed an agreement outlining their cooperative plan for genome research. Their plan, the Human Genome Project (HGP), had two objectives: to map, or determine, the location of all protein-coding genes on each human chromosome; and to establish the sequence of the base pairs on each gene and thereby read the entire set of instructions encoded in human DNA.

The National Center for Human Genome Research (NCHGR) established in 1989, and in 1997 renamed the National Human Genome Research Institute (NHGRI), one of 27 institutes and centers that make up the NIH, oversaw the NIH's half of the joint research. With DOE it coordinated the HGP's 15-year $3 billion international research project established on 1 October 1990 to map and sequence all the genes in the human body. The international group, called the International Human Genome Sequencing Consortium (IHGSC), which dates from 1993, included scientists from 20 research centers in the United States and 18 other countries, mainly the United Kingdom, France, Germany, China, and Japan.

Watson in 1988 became the NCHGR's first director, but he resigned in April 1992 after a clash with Bernadine Healy (1944-2011), head of NIH. Watson objected to Healy's decision to award exclusive patent rights on 347 genes that NIH researchers discovered and wanted the gene sequences to belong in the public domain for use by researchers internationally. Watson's critics responded by claiming his investment in biotechnology companies represented a conflict of interest.

Among the early HGP successes were the 1992 mapping of the Y chromosome and chromosome 21, although both maps were incomplete and imperfect. David Page (b. 1957) and his group at the Whitehead Institute of Biomedical Research at MIT, mapped the Y chromosome's 78 genes and their 22.761 million base pairs. The mapping proved a significant challenge because most of the Y chromosome does not undergo recombination and was not amenable to genetic linkage mapping. Genetic linkage mapping analyzes chromosomes when they divide and recombine during formation of egg and sperm and is one of the most efficient tools for localizing genes. Instead, Page's group used a cloning technique known as STS (sequence tagged sites) content mapping.

In this method each STS, or landmark, is a short stretch of DNA that defines a distinct location or reference point in the human genome. A specific chemical test called a polymerase chain reaction (PCR) assay identifies each landmark. Collectively the landmarks, whose locations geneticists have established by comparison with DNA sections from yeast, provide a physical map with which to read the alphabetic sequence of the four DNA bases that identifies each of the genes in the human genome.

Geneticists until recently believed that the little-understood Y chromosome, the male sex chromosome which only the father (the Adam chromosome) can provide, consisted mainly of junk DNA, so called because they do not yet know junk DNA's role. But the studies of Page and Michael Hammer (ca. 1955) at the University of Arizona, in Tucson, have begun to unravel the mystery of the Y chromosome, showing that it remains nearly unchanged from generation to generation and therefore has applications in forensics, evolutionary studies, medical genetics, and genealogy.

Daniel Cohen (b. 1952) and his group at the Centre d'Etude du Polymorphisme Humain (CEPH) and its companion laboratory Généthon in Paris, which Cohen co-founded in 1990, mapped the 355 genes on chromosome 21. Its 33.546 million base pairs contain the genes responsible for Down syndrome, Alzheimer's disease, Lou Gehrig's disease, and several other neurological disorders. People born with Down syndrome, a congenital form of mental retardation named after the London physician John Down (1828-96) who first studied the disease in 1866, have an extra copy of chromosome 21, having three copies instead of the normal chromosome pair.

Serious research on the human genome sequence accelerated internationally in 1996. In the United States DOE-NIH centers, such as those located at Baylor College of Medicine in Houston, Washington University School of Medicine in St. Louis, the Whitehead Institute, Stanford University, and DOE's four national laboratories at Lawrence Berkeley, Lawrence Livermore, Los Alamos, and Oak Ridge, sequenced about 60-70 percent of the human genome. The Sanger Centre, now the Wellcome Trust Sanger Institute, established in 1992 with $95 million and which John Sulston directed; the Genome Sequencing Center at the University of Washington in Seattle which Robert Waterston after leaving Washington University School of Medicine has directed since 2003; and other university research centers including the University of California, San Francisco, and national laboratories around the world completed the remaining 30-40 percent of the sequencing. The laboratories used an automated-Sanger sequencing method and gel electrophoresis to separate the base segments. The gel enables researchers to distinguish easily single base differences in size. Small segments pass more rapidly through the gel's pores than larger segments. Scanning each segment with a laser as the segments move through the gel gives the color and intensity of each DNA band, the color indicates the location of the final 3′ base in each segment.

An international research group of English, American, and Japanese scientists from the Wellcome Trust Sanger Institute, the University of Oklahoma in Norman, Washington University, and Keio University in Tokyo, in December 1999 sequenced the first human chromosome, chromosome 22, with 679 genes and 34.352 million base pairs. The Keio University group, the RIKEN (Rigagaka Kenkyusho) Center Group in Yokohama, and German groups at the Max Planck Institute for Molecular Genetics in Berlin, the German Research Center for Biotechnology in Braunschweig, and the Institute of Molecular Biotechnology in Jena, in May 2000 sequenced the second human chromosome, chromosome 21, with 355 genes and 33.546 million base pairs.

The HGP had made sufficient progress by 1999-2000. Yet as Francis Collins (b. 1950), a physician-geneticist at Michigan who in 1993 succeeded Watson as director of the NCHGR, reported in 1999, the project was only about 10 percent complete. A vigorous sequencing program and competition from Celera Genomics pushed HGP, and in June 2000 it produced a rough 90 percent draft of the human genome. Celera Genomics, a privately-organized company founded in May 1998 by Applera Corporation and J. Craig Venter (b. 1946), a former NIH biologist, and located in Rockville, Maryland, and San Francisco, completed its sequencing of the human genome in June 2000 at a cost of $300 million. HGP and Celera announced their maps on 12 February 2001 at a joint HGP - Celera news conference in Washington, DC. HGP's initial draft appeared in *Nature* on 15 February 2001. Celera's draft appeared in *Science* on 16 February 2001.

Whose genomes did HGP and Celera sequence? HGP scientists used white blood cells from female donors' blood for female samples and sperm samples from males. Sperm cells (X and Y chromosomes) contain all the chromosomes and a higher ratio of DNA to protein than other cell types. Neither the donors nor the scientists knew the identity of the DNA samples sequenced. The exact number of donors remains unknown, although a large number of donors contributed, and scientists selected only a few samples from them. All the samples sequenced were from American donors and the rumor has persisted that a single donor provided most of the DNA sequenced. Celera obtained DNA samples from anonymous European, African, North-Central-South American, and Asian donors that they mixed and then processed for sequencing. Celera's samples included Venter's DNA. Celera published his complete DNA sequence in September 2007.

Marvin Stodolsky at the DOE Office of Biological and Environmental Research, Office of Science, has pointed out that the human genome reference sequences do not represent any one person's genome. They are a starting point for broad comparisons among humans. Knowledge obtained from the sequences applies to all humans because all share the same set of genes and genomic regulatory regions that control the development and maintenance of their biological structures and processes.

Since 2001 scientists at the Wellcome Trust Sanger Institute in December 2001 have produced the third high quality gene sequence, that of chromosome 20 with 727 genes and 60 million base pairs. A complete sequence of chromosome 14, which has 1,050 genes and 87 million base pairs, followed in January 2003 at Genoscope, France's National Sequencing Center in Paris. In April 2003, two years earlier than originally planned, HGP announced it had completed a high-quality sequence of the human genome, 99.99 percent accurate.

Additional chromosomes completely sequenced after the HGP-Celera announcement include Y in June 2003, 7 in July 2003, 6 in October 2003, 13 and 19 in April 2004, 10 and 9 in May 2004, 5 in September 2004, and X in March 2005. The X chromosome is much larger than Y, containing 1,098 genes and 160 million base pairs.

In May 2006 HGP announced completion of the DNA sequence for chromosome 1, the last of the 24 different human chromosomes. More than 250 researchers at Baylor College of Medicine, Wellcome Trust Sanger Institute, Washington University, Max Planck Institute, Institute of Molecular Biotechnology in Jena, and Applied Biosystems in Foster City, California, contributed to its sequence.

The sequencing showed that the number of base pairs on each of the 24 different chromosomes of the human genome ranged from 48 million to 250 million.

| HGP CHROMOSOME SEQUENCE | | | | |
|---|---|---|---|---|
| ***Chromosome*** | ***Date Sequence Published*** | ***Length (mm)*** | ***Number of Genes*** | ***Number of Base Pairs*** |
| 1 | 2006 May | 85 | 2,616 | 249,250,621 |
| 2 | 2005 February | 83 | 1,733 | 243,199,373 |
| 3 | 2006 April | 67 | 1,388 | 198,022,430 |
| 4 | 2005 February | 65 | 1,076 | 191,154,276 |
| 5 | 2004 September | 62 | 1,185 | 180,915260 |
| 6 | 2003 October | 58 | 1,328 | 171,115,067 |
| 7 | 2003 July | 54 | 1,250 | 159,138,663 |
| 8 | 2006 January | 50 | 920 | 146,364,022 |
| 9 | 2004 May | 48 | 1,101 | 141,213,431 |
| 10 | 2004 May | 46 | 1,001 | 135,534,747 |
| 11 | 2006 March | 46 | 1,618 | 135,006,516 |
| 12 | 2006 March | 45 | 1,338 | 133,851,895 |
| 13 | 2004 April | 39 | 466 | 115,169,878 |
| 14 | 2003 February | 36 | 890-1,050 | 107,349,540 |
| 15 | 2006 March | 35 | 916 | 102,531,392 |
| 16 | 2004 December | 31 | 1,075 | 90,354,753 |
| 17 | 2006 April | 28 | 1,462 | 81,195,210 |
| 18 | 2005 September | 27 | 401 | 78,077,248 |
| 19 | 2004 April | 20 | 1,598 | 59,128,983 |
| 20 | 2001 December | 21 | 727-733 | 63,025,520 |
| 21 | 2000 May | 16 | 325-355 | 48,129,895 |
| 22 | 1999 December | 17 | 601-679 | 51,304,566 |
| X | 2005 March | 53 | 1,089-1,129 | 155,270,560 |
| Y | 2003 June | 20 | 78-140 | 59,373,566 |

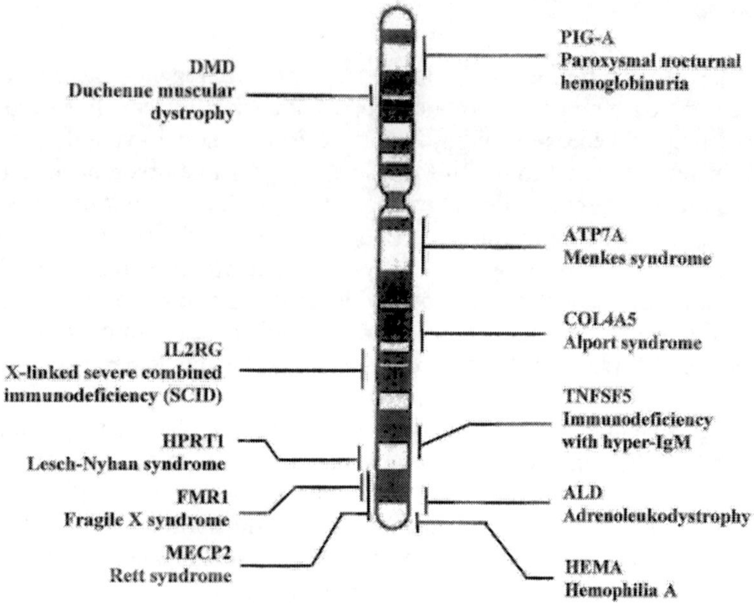


**Figure 22**
**X-Chromosome Map**
Courtesy of the National Institutes of Health

The excitement that completion of the human genome mapping and sequencing project generated almost overshadowed a surprising and totally unexpected consequence of the research. The IHGSC in the same papers published in *Nature* and *Science* in February 2001 questioned whether the human genome contained 100,000 genes. Its new studies lowered the number of protein-coding genes in the human genome to 30,000 or 35,000, which was less than half of the generally-accepted number. The function of the remaining 50 percent junk DNA is not clear (RNA and protein are also present), although geneticists believe it must in some way contribute to individual human differences. The new number is only about twice the number of genes in the worm or fly, whereas the percentage of junk DNA in the human genome is much greater than the worm's 7 percent and the fly's 3 percent.

The reason for the striking difference in the human gene number is a consequence of the different computational methods and gene-finding programs that geneticists have employed. Some programs estimate the number of genes by looking for patterns that establish the beginning and end of a gene. This tends to overestimate the number because it counts any segment that looks like a gene. A second program compares segments of a sequence with segments of known genes and proteins and tends to underestimate the number because it recognizes only those genes similar to already existing and known genes. The difficulty of detecting small genes, the same gene coding for several protein products, gene overlap, and some genes coding only for RNA are other factors complicating total gene number calculations, and this suggests that a more accurate human protein-coding gene total awaits further research. In fact the IHGSC's latest estimated number of human protein-coding genes published in October 2004 is 20,000-25,000 genes. Researchers have confirmed 19,599 protein-coding genes and another 2,188 DNA segments that they expect will code for genes.

The scientific community believes that the mapping and sequencing of the human genome will lead to an understanding of genetic diseases, including Alzheimer's disease, and to the discovery of new drugs to treat genetic diseases, such as schizophrenia and bipolar disorder. Progress in drug development remains yet elusive.

## VIII. DEVELOPMENTS IN CLONING AND STEM CELLS

### *What is cloning?*

Cloning is the artificial reproduction of a gene, or of several genes, or of an entire organism. Its main objective is to produce numerous identical copies of a specific gene. For simple organisms the process requires isolating the gene, inserting it into a bacterium, and allowing the bacterium to reproduce and thereby copy the inserted gene with the bacterium. Agriculturists have long used cloning to yield identical crop reproductions, following a sequence that requires cutting a tissue sample from a parent plant, exposing the sample to a mixture of nutrients and special hormones that allows it to grow roots, and then transferring the resulting clone to soil where it will grow normally. With this method a single plant can produce thousands of clones by dividing the tissue sample before it takes root. Scientists also have used these methods to produce thousands of artificial seeds from cells by inserting the cells into a specially prepared hormone-nutrient mix where they grow into genetically identical crops.

The advantage of cloning is that it gives many healthy plants that develop at the same rate for harvesting or for reproduction. The disadvantage is cloning's high cost and reduction in genetic diversity and therefore reduction of resistance to disease by the genetically-identical plants. Animal cloning is much more controversial than plant cloning because it suggests the next step, which is human cloning, or the creation of human life, and raises ethical and moral issues and questions God's role in the creative process and whether clones have souls.

The roots of animal cloning date from at least the late nineteenth-century embryo transfers of Walter Heape (1855-1929) at Cambridge who in 1890 carried out two angora rabbit embryo transfers. Sixty years later, in 1950, Elwin L. Willett (1913-58), director of the American Foundation for the Study of

Genetics, the research arm of the Wisconsin Scientific Breeding Institute in Madison and associated with the American Breeders Service (est. 1941), and Lester Casida (1904-86) at Wisconsin, transferred the first calf embryo.

More recently in the 1980s-90s researchers at Texas A&M, Wisconsin, and other universities, while continuing to perfect embryo transfer, cloned a number of different animals including frogs, mice, cows, and sheep from fetal cells. James Robl (b. 1956) then at Massachusetts (Amherst) and Steven Stice (b. 1961) then at Advanced Cell Technology in Worcester successfully cloned three Holstein calves using fetal cells. The calves, born at Texas A&M University's College of Veterinary Medicine on 20 January 1998 and named Charlie and George with the third unnamed, matured at nearby Camp Cooley Ranch, an 11,000 acre cattle operation near Franklin in central Texas. Ultimate Genetics, a biotechnology company at Camp Cooley assisted with the cloning.

In fetal cell cloning, Robl and Stice took a cell from a cow's fetus and into the cell inserted a gene to give the cell the genetic makeup they wanted to clone. They called it cell-B. Next, they removed the genetic material (nucleus) from the egg of another cow and placed the egg near the genetically-altered cell-B, then they fused the egg and cell-B with a 200-volt electrical discharge, and chemically stimulated the newly-formed cloned cell to induce normal cell division. They incubated the cloned cell for seven days in the laboratory, during which time it grew to about 100 cells, and then placed it in the uterus of a surrogate mother who nine months later gave birth to the genetic clone. They intended in this and similar experiments to produce disease-resistant farm animals and provide possible sources of organs and tissue for transplanting to humans. Researchers at Texas A&M University already had cloned disease-resistant mice that lack a specific protein, PrPs, and were working on cloning disease-resistant cows.

The most sensational reproduction was the cloning of Dolly (named for Dolly Parton), the sheep born on 5 July 1996 at the Roslin Institute in Edinburgh, Scotland, although the announcement came seven months later in February 1997. Ian Wilmut (b. 1944) and Keith Campbell (1954-2012) did the cloning. Dolly excited the scientific community because she was the first mammal cloned from an udder (mammary gland) cell of a 6-year old adult mammal, a female sheep (ewe) that Wilmut and Campbell killed and whose cell tissue they kept alive in their laboratory. Dolly was the first successful clone after 276 failures, and another 100 failures followed since Dolly's cloning. The cloning warranted attention because scientists previously believed that they could not reactivate dormant DNA in an adult cell and develop an embryo.

Plants, animals, and humans contain many different types of cells, such as nerve cells, muscle cells, and skin cells, and each type of cell has a special function. But every cell contains a complete set of chromosomes, or all the genetic information required for the production of any type of cell. As each type of cell develops, it deactivates or turns off the genetic information not required for the cell's specific function. Skin cells, for example, contain a deactivated type of DNA that dictates the development of hair cells. Similarly, liver cells contain the deactivated DNA of heart cells and do not become heart cells, and heart cells contain the deactivated DNA of brain cells and do not become brain cells. Wilmut and Campbell reversed the development of the genetic material contained in the nucleus of an adult cell and returned it to the state it was in when sperm first fertilized egg, that is, to an embryo. They then succeeded for the first time in reactivating all the dormant genetic material in an adult cell.

In this cloning method, called somatic-cell (body-cell) nuclear transfer, Wilmut and Campbell removed the nucleus containing the genetic material from unfertilized ewe egg cells of a Scottish Black Face sheep and placed the cells in a laboratory dish next to cultivated cells taken from a White Face Fin Dorset ewe's udder (mammary) tissue. They made the Fin Dorset's cells dormant by starving the cells of vital nutrients for five days, causing them to shut down. Wilmut and Campbell then applied a mild electric current to the cells, causing the second ewe's dormant cells to fuse with the first ewe's unfertilized egg cells. The cloned cell divided and developed into totipotent stem cells, which have the ability to form every type of sheep cell, and became an embryo, just as if a sperm cell had fertilized it. Finally, they placed the embryo in a third ewe, another Scottish Black Face sheep, to develop into a clone of the second sheep, the Fin Dorset. After 148 days, the normal gestation period for a Fin Dorset, the third ewe gave birth to Dolly.

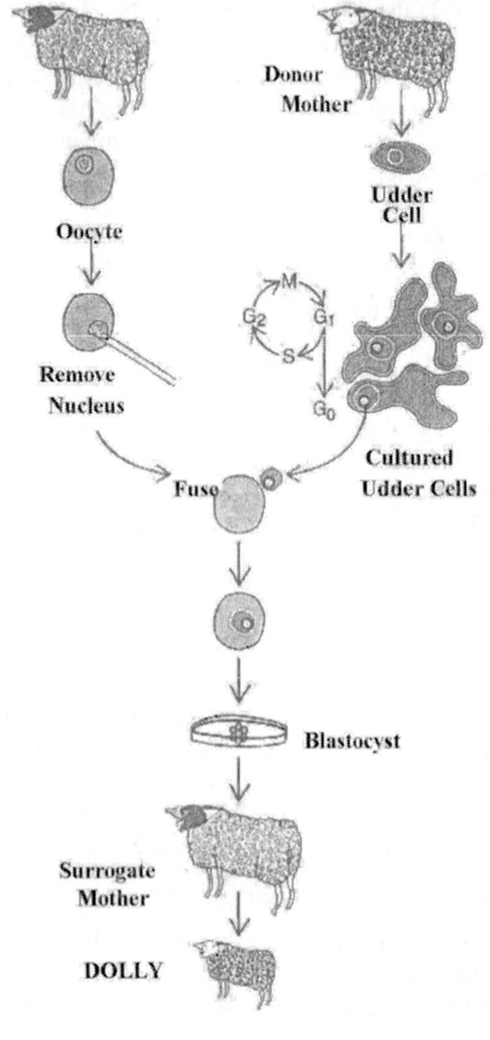

**Figure 23**
**Dolly's Cloning**
Courtesy of the National Institute of Health (left)
Courtesy of the Roslin Institute, The University of Edinburgh (right)

Two questions have arisen about Dolly: her age and her fertility. Scientists do not have a definitive answer to Dolly's age, whether at birth she was a newborn or already 6 years old. Dolly answered the second question by giving birth in April 1998 to a perfectly healthy lamb the Institute named Bonnie, whom Dolly conceived naturally or normally. Bonnie's birth helped to allay fears that Dolly might age prematurely and experience other age-related disorders. The most recent studies on Dolly indicated some shortening of the ends of her DNA, which is a well-established characteristic of older genes.

Roslin Institute researchers already had proved that cloned animals can reproduce in experiments with sheep that they cloned from fetal cells, but Dolly was the first clone produced from adult cells to give birth to an offspring. Unfortunately Dolly suffered from progressive lung disease and the Institute euthanized her on 14 February 2003, nearly seven years after her birth. Sheep have 11-12 year lifespans and commonly suffer from lung infections in their later years.

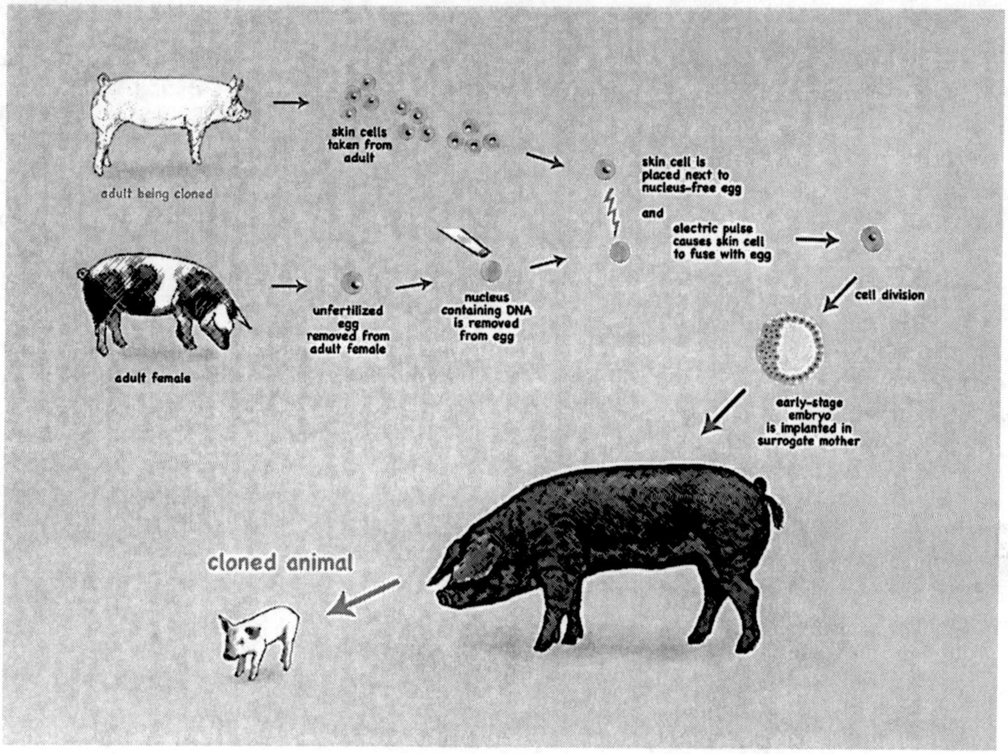


**Figure 24**
**Somatic-cell Nuclear Transfer Cloning of a Pig**
Courtesy of the National Human Genome Research Institute, National Institutes of Health

In other somatic-cell nuclear transfer cloning experiments, the biologist Ryuzo Yanagimachi (b. 1928) and his postdoctoral student Teruhiko Wakayama (b. 1967) at the University of Hawaii in Honolulu, reported in *Nature*, July 1998, that they had made more than 50 mouse clones, including dozens of adult mouse clones in about a month, and that of the first 22 clones, 7 were clones of clones (mouse gestation period is 19-21 days, 7-12 pups in a litter). They demonstrated with their *Honolulu technique*, as in Dolly's cloning, the reality of cloning from adult cells. They reversed the development of the genetic material (nucleus) of an adult cell, returning the nucleus to the state that existed when sperm first fertilized egg, its embryonic state, and found that when the cells grew they had produced a complete animal and not just one kind of tissue.

Yanagimachi and Wakayama used the nucleus from a mouse's cumulus cell, the spherical cell that surrounds an egg and nourishes it. They intended to use the nucleus from two other types of adult cells that they identified as Sertoli cells and brain cells, but concluded that cumulus cells were the easiest to clone and used them exclusively. They found that when they added the cumulus cell's nucleus to a mouse's egg cell from which they had removed the nucleus, allowed the altered egg cell to sit for six hours, and chemically stimulated the altered egg cell, the altered egg cell reactivated the cumulus cell's dormant cells, enabling it to direct the formation of a complete mouse genome. Their first mouse, a female named Cumulina, died of natural causes in May 2000, after a two-year seven-month lifetime, seven months longer than a mouse's average life span.

A few years later, in May 2003, Cesare Galli (b. 1961), an Italian researcher and his team at the Laboratory of Reproductive Technology near Cremona, after transplanting 841 embryos, succeeded in cloning the first horse, Primetea, a toffee-brown mare. Galli used somatic-cell nuclear transfer and a skin cell from Primetea's mother that he fused with a de-nucleated horse egg and implanted in the mother's uterus. Primetea remained in excellent health and in March 2008 gave birth to a healthy male foal.

Until November 2007 no scientist had cloned a primate. Using somatic-cell nuclear transfer, Shoukhrat Mitalipov (b. 1962) and his research team at the Oregon Health and Science University in Beaverton cloned embryos of rhesus macaques (short-tailed monkey). Mitalipov collected 304 eggs (oocytes) from 14 female macaques, removed the nucleus of each egg, and replaced it with the donor nucleus of skin cells taken from 9 adult male monkeys that resided at the Oregon National Primate Research Center in Hillsboro. The Oregon researchers used fresh eggs, stimulated the reaction with caffeine, allowed the cloned cells to develop, and from the 304 eggs, they obtained 35 early (5-day old) embryos, or blastocysts. A blastocyst is a hollow ball consisting of 30-150 cells. From the early embryos, they produced two lines of self-dividing embryonic stem cells. Stem cells are undifferentiated cells taken from the inside of an early embryo and have the potential to form any type of body cell, heart, skin, liver, or hair but not a complete organism. Their stem cells developed to give a heart and nerves but no complete monkey. A research team at Monash University in Melbourne, Australia, confirmed the Mitalipov team's results.

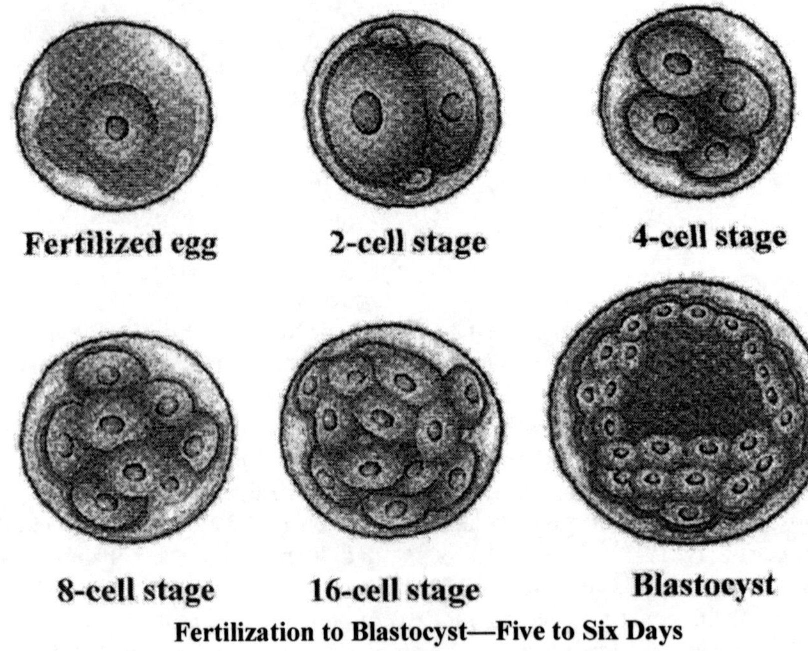


**Figure 25**
**Blastocyst**
BlueRingMedia/Shutterstock.com

Have scientists cloned a human? One of the first, if not the first claim of human cloning, occurred in 1978. David Rorvik (b. 1944), a respected author and medical writer for both *Time* and *The New York Times* claimed in his book, *In His Image*, published by J.B. Lippincott Company, a well-known publisher of medical texts, that in 1973 an aging, eccentric, unmarried American millionaire, code-named Max, approached him with an offer of one million dollars to manage a research project that would create a clone of Max. Rorvik agreed. He assembled a team of unnamed scientists on a tropical island somewhere beyond Hawaii in 1974, and in March 1976 after considerable experimentation the team led by a distinguished geneticist, code-named Darwin, succeeded in cloning a human egg cell containing Max's DNA. Darwin implanted the egg in a surrogate mother, an island resident code-named Sparrow. The cloned egg cell divided and developed into an embryo just as if a sperm cell had fertilized it. Nine months later in a California hospital Sparrow gave birth to the first human clone, a boy.

*In His Image* sold very well, nearly 700,000 copies, and earned $730,000 on book sales. It generated much controversy and skepticism. Most scientists questioned Rorvik's claim, arguing that the science of cloning was not sufficiently advanced to produce a human clone and that the procedure Rorvik described resembled an earlier frog cloning procedure. A $7 million defamation suit also resulted a few months after publication when J. Derek Bromhall, an Oxford geneticist whose 1975 cloning of a rabbit's genetic material to the advanced embryo stage, Rorvik mentioned in his book, sued Lippincott and Rorvik.

A United States District Court judge in Philadelphia in February 1981 ordered Rorvik to verify his claim by producing the boy clone, and when he did not, declared the book a fraud and a hoax. Lippincott settled the defamation suit out of court in April 1982, paying Bromhall about $100,000. Rorvik's motivation remains unclear. He admitted only that the clone died of a defect and has continued to defend his original claim.

Several more recent cloning claims have appeared in the news. Severino Antinori (b. 1948), an Italian physician in Rome, working with Panayiotis Zavos (b. 1944), a medical doctor at the Androgyny Institute in Lexington, Kentucky, announced in November 2003 that a woman participant in their cloning project would give birth shortly but never announced anything more. Zavos estimated a cost of about $50,000 for the cloning procedure.

At about the same time, Clonaid, a Bahamas-based organization founded in 1997 by a former French journalist Claude Vorilhon (b. 1946) and now known as Raël, announced it had cloned its first baby in February 2004. Clonaid claimed that it has since cloned a total of 13 babies in countries around the world, including Mexico, Spain, Italy, England, Australia, Hong Kong, and Brazil. No one has seen these clones, some of which required a $500,000 payment to Clonaid for their production. Scientists unanimously doubt Raël and the Raëlians who believe that the DNA of extraterrestrials was the source of human creation.

A more believable human cloning claim came from South Korea. Hwang Woo Suk (b. 1953), a professor in the Veterinary College of Seoul National University since 1986, announced in February 2004 that he had cloned a cell obtained from an adult woman. He used somatic-cell nuclear transfer. Hwang put the nucleus from the woman's cell into one of her own eggs from which he had removed the nucleus. The egg developed within five days into an early embryo, or blastocyst. He removed the embryo's inner cells which are the embryonic stem cells, cultured them in a medium containing cells from a mouse embryo to promote their survival and growth, and obtained a new human stem cell line before stopping the experiment and destroying the new stem cell line.

Hwang published his results in *Science*, 12 March 2004 and 17 June 2005. In the 2004 article he said he used 242 eggs and cumulus cells, which are somatic (body) cells Hwang obtained from voluntary female donors, to create a single embryonic stem cell line. In the 2005 article he claimed to have cloned embryos and obtained 11 different embryonic stem cell lines, each derived from a different donor that included young and old, male and female, using 185 eggs and somatic (skin) cells in a more efficient procedure.

Unfortunately for Hwang, triumph turned to tragedy. The university dismissed Hwang in March 2006 after a university panel on 10 January 2006 concluded he had faked the 11 embryonic stem cell lines. A Seoul National University committee investigating Hwang's 2004 and 2005 papers in *Science* and another 2005 paper in *Nature* in which he claimed to have cloned the first dog, Snuppy, an Afghan hound, declared Hwang's 2004 and 2005 *Science* papers fraudulent. Snuppy, it concluded, was a real clone. Hwang received a total of $31.8 million from the South Korean government for his research and $6.2 million in private funds.

The use of stem cells obtained from human embryos to derive stem cell lines did not originate with Hwang's discredited research. Stem cell research ignited the same debate following the 1998 path-breaking publication of James Thomson (b. 1958) announcing that he and his research group at Wisconsin for the first time had derived stem cell lines from a human embryo.

Thomson grew living cells outside the body, in this case human embryonic stem cells obtained from 5-day old fresh or frozen embryos left over after *in vitro* fertilization. He isolated 14 inner cell blastocysts and obtained five embryonic stem cell lines from five separate embryos. Thomson added a culture consisting of nutrients, hormones, growth factors, and blood serum to the container holding the stem cells. He also added cells from mouse embryos, a common procedure used to enable the stem cells to survive and grow in the culture, but their addition presented a new problem. The presence of animal cells increases the human stem cells' likelihood of acquiring animal viruses, proteins, and other potentially harmful organisms and thereby of infecting anyone treated with the human stem cells. The stem cell lines developed over a 6-8 month period to give muscle, bone, and gut, and Thomson believed that their study would benefit research in development biology, drug discovery, and transplantation medicine.

In January 2006 Thomson and his research group solved the problems resulting from the addition of animal cells. They succeeded for the first time in keeping alive two new human embryonic stem cell lines in their culture medium, an unconditional medium supplemented with 100 ng/ml basic fibroblast growth factor (FgF2), without the addition of animal cells. Both stem cell lines, derived from five blastocysts less than a week old, survived for up to 164 population doublings, for more than 7 months, without differentiating into specialized cells and still maintaining the developmental potential to form any type of body cell, although one line developed an abnormal chromosome after 4 months and the second line developed an abnormal chromosome after 7 months.

In November 2007 Thomson and independently Shinya Yamanaka (b. 1947) at Kyoto University and the Gladstone Institute for Cardiovascular Disease in San Francisco, made another significant advance in stem cell research. Yamanaka in the journal *Cell* and Thomson in *Science* announced that their research groups had changed human skin cells (fibroblasts) into cells that mimic embryonic stem cells. According to Thomson, they found a new way to make skin cells act like embryos, a new way to reprogram skin cells into multipurpose stem cells, or induced pluripotent stem cells (IPS), without using an embryo. Four genes added to the skin cells reprogrammed the skin cell's chromosomes, turning the skin cells into embryonic-like stem cells capable of developing into any of the human body's 220 cell types. Thomson and Yamanaka succeeded in producing stem cells that genetically matched the donor individual without having to clone an embryo of the individual and then removing stem cells from it, thereby destroying the embryo.

Thomson obtained his human cells from foreskin to which he added the genes Oct4, Nanog, Sox2, and Lin28. Yamanaka's four genes were Oct3/4, Sox2, c-Myc, and Klf4. The genes were among the 1,000 plus genes that the two researchers in 2003 determined were active in an embryonic stem cell but not in an adult cell. Because only four genes caused the reprogramming of an adult cell rather than a larger number, 8, 9, 10, or 12, the research moved relatively quickly. Yamanaka worked first with mouse cells. He succeeded in their reprogramming in a few months and then continued with human cells, using skin face cells from a 36-year old white woman and skin connective cell tissue from the joints of a 69-year old white man. Yamanaka grew the reprogrammed skin cells in a culture, obtaining brain tissue and heart muscle tissue that after twelve days started to beat. His team created eight new stem cell lines.

Although the reprogrammed human skin cells appear to meet every test of embryonic stem cells, Thomson and Yamanaka have not confirmed that reprogrammed cells are identical to stem cells obtained from embryos. They also have to resolve a second problem. To insert the four genes into the skin cells' chromosomes, Thomson and Yamanaka used retroviruses which enter the chromosomes randomly. This is the usual method of inserting genes into chromosomes. The randomness of the gene addition can cause genetic mutations, some of which convert normal cells into cancer cells and produce other complicating side effects. One of the genes Yamanaka inserted was a tumor-causing cancer gene. Nonetheless, Yamanaka believes that the cells will prove useful in understanding disease mechanisms and in producing safe and effective drugs.

***Ethical implications of cloning and stem cells***

Supporters of animal cloning, particularly scientists involved in cloning research, continue to stress cloning as a valuable source of organ and tissue transplants. These include a genetically altered cow's pancreas to produce human insulin and animal tissue for transplanting to people suffering from degenerative diseases such as Alzheimer's and Parkinson's; cows that produce albumin, a blood protein critical to people suffering from liver disease, malnutrition, and severe burns; and pharmaceuticals produced in the milk of cows and other livestock.

Critics of animal cloning contend that the cloning and transplanting of animal organs to humans could put new germs into the human population. The Massachusetts chapter of Earth First, an animal rights group, has protested that the cloning of living cells steps beyond the field of science and into the realm of creation. Jeremy Rifkin (b. 1945), president of *The Foundation on Economic Trends* in Bethesda, Maryland, and a well-known critic of genetic engineering, maintains that cloning and the mass production of a handful of commercial animal genotypes will seriously weaken the already dangerously narrow genetic diversity of existing animal species. Rifkin has questioned the value of children growing up in a society in which scientists have reduced much of the animal kingdom to patented inventions that they cloned and mass produced in part or whole on bio-industrial assembly lines.

Human cloning raises the most serious ethical issues because society at some point will have to answer the question of who should create life and whether a God any longer has a role in the creation process. Does the clone have a soul which most religious denominations profess a human has, or is it only a collection of atoms? This is the dilemma that human cloning faces in Western and other societies.

Another serious issue is the relation of the human clone to its parent. Is it an offspring or a twin of the parent that provided the cells for cloning? Even though a clone has the same genetic makeup and may appear physically the same as the donor, its psychological, moral, and ethical behavior, personality, intelligence, and physical skills may be entirely different and possibly a disappointment to the cloner. The clone is not an exact replica of the original donor but a time-delayed, much younger identical twin. Just as identical twins do not have identical fingerprints, neither do the clone and original donor have identical fingerprints.

Finally, are stem cells, those undifferentiated cells that undergo continued division to form other cells, such as skin cells or hair cells, a form of human life or a potential human life? If stem cells constitute human life, does the same argument hold that human egg and sperm cells are also potential human lives?

Lee Silver (b. 1952), a geneticist at Princeton, in *Remaking Eden* published in 1997, wrote optimistically that human cloning would occur in 10 years. Ten years have passed, but this view still remains prominent among many scientists. They believe that at this time no ethical scientist will attempt to clone because of possible malformations or diseases in the human clone.

## IX. CONCLUSION

Schleiden, Schwann, and Virchow recognized in the mid-1800s that the cell had in it the complete characteristics of life and was the structural unit of all living things. Miescher discovered DNA and RNA in 1869. Since that time numerous scientists have contributed significantly to the research that unraveled the structures of DNA and RNA, solved the base-triplet genetic code, and mapped and sequenced the human genome and the genomes of other organisms. These include the chromosome research of Flemming and Stasburger and that of the Americans Wilson, Morgan, Sutton, Muller, and Stevens in the 1880s-early 1900s. Fischer's amino acid and protein studies and the DNA-RNA analyses of Kossel, Levene, and later Todd, established the composition of these molecules and laid the groundwork for their structural determination.

The infusion of physical scientists, such as Wilkins and Crick, and the introduction of the mathematical-quantitative methods and the instruments of the physical sciences, such as the electron microscope, chromatography, radioactive tracers, and x-ray diffraction, particularly after World War II, greatly aided molecular biology's development. The invention of the electron microscope in1931 provided the first images of bacteriophages and later, images of DNA. Chargaff at Columbia in 1950 used paper chromatography and revealed the one-to-one ratio of the organic base pairs adenine-thymine and cytosine-guanine, A : T and C : G, in DNA. The radioactive phosphorus and sulfur tracers that Hershey and Chase at Cold Spring Harbor introduced in their crucial experiment in 1952 proved that DNA and not protein was the gene and the carrier of hereditary material. X-ray diffraction, as Franklin and Wilkins at King's College demonstrated, provided convincing evidence that DNA was a double helix with its four organic bases, A-T and C-G, paired inside the helixes. The cumulative evidence enabled James Watson and Crick at the Cavendish Laboratory in 1953 to solve DNA's molecular structure, a discovery many have proclaimed the greatest scientific achievement of the twentieth century.

Watson and Crick won the race for the double helix. As in most races for a scientific prize, such as the race for establishing longitudes in the 1700s or the twentieth-century race for the atomic bomb and the space race to the Moon, the stakes are high and sometimes bring out the worst in the participants, not the best. Watson and Crick deserved the Nobel Prize for their achievement, which ranks as probably the greatest discovery of the twentieth century or possibly of all time. The criticism they received did not come from envy of what they did, but for what they failed to do, that is, at least to acknowledge that they obtained important DNA x-ray diffraction data from Franklin without her permission and then never credited her, even in a footnote, as their source. Wilkins is guilty of the same ethical lapse. Watson and Crick briefly and in a very general way mentioned her unpublished work at the end of their 1953 paper. Watson also acknowledged her x-ray work fifteen years later in an apologetic Epilogue added to *The Double Helix*. To give credit, or acknowledge, someone who contributed significantly to solving DNA's structure in no way diminishes what Watson, Crick, and Wilkins accomplished. Nor would they have had to share the Nobel Prize with Franklin because by that time she was already dead.

Despite the Watson-Crick-Wilkins-Franklin issue, molecular biology achieved significant advancement in the last fifty years. Nirenberg and Gobind Khorana solved the genetic code in 1961. Smith, Boyer, and Stanley Cohen introduced gene sequencing in the 1970s. The Human Genome Project to map and sequence all the genes in the human body, whose roots go back to the 1980s, officially got underway in 1990. The first human chromosomes maps appeared in the 1990s and the completion of the mapping and sequencing of the human genome occurred in 2001.

Finally, if the human genome sequences apply to everyone because all humans share the same basic set of genes, why are humans so different in physical appearance? One answer may lie in the sequences of bases in a gene that do not code for an amino acid and in protein formation and which geneticists variously have called pseudogenes, junk DNA, or introns. Further research will provide the answer.

The successes in molecular biology have not been without controversy, particularly cloning and stem cell research, mainly because of the implications this research has regarding the fundamental question of what is life. What role should science have in creating life and where and how does a God become a part of the process? The social, moral, ethical, and theological issues that accompany the scientific advances in molecular biology remain the challenge of the future.

| BIOTECHNOLOGY MILESTONES | |
|---|---|
| 1953 | James Watson and Francis Crick discover the double helix structure of DNA. |
| 1961 | Marshall Nirenberg and Har Gobind Khorana break the genetic code carried by DNA. |
| 1967 | DNA synthesized. |
| 1970 | The first artificial gene synthesized. |
| 1973 | Stanley Cohen and Herbert Boyer perfect genetic engineering technique (recombinant DNA technology) to cut and paste sections of DNA. |
| 1975 | First artificial antibodies produced. |
| 1977 | Human gene transferred into bacteria for the first time. |
| 1979 | Human growth hormone synthesized. |
| 1980 | Supreme Court rules life forms are patentable. |
| 1982 | First transgenic animal, a mouse, produced. |
| 1983 | First artificial chromosome synthesized. |
| 1984 | First genetically-engineered vaccine produced. |
| 1986 | First field trial of a genetically-engineered plant (tobacco) carried out. |
| 1987 | First field trial of a genetically-altered bacterium completed. |
| 1988 | First United States animal patent, a transgenic mouse, awarded. |
| 1989 | First field trial of a genetically-altered virus carried out. |
| 1989 | First transplant of an alien gene into human beings carried out. |
| 1990 | Fifteen year, $3 billion Human Genome Project established. |
| 1992 | Low resolution genetic linkage map of entire human genome published. |
| 1993 | Francis Collins named director of NCHGR. |
| 1994 | Genetic-mapping five-year goal achieved one year ahead of schedule. |

| BIOTECHNOLOGY MILESTONES | |
|---|---|
| 1995 | High-resolution physical maps of chromosomes 16 and 19 obtained, moderate-resolution maps of chromosomes 3, 11, 12, and 22 published. |
| 1996 | International team completed yeast genome sequence. |
| 1997 | Dolly the sheep cloned. |
| 1998 | Fifty mice cloned, clones of the clones also produced. |
| 1999 | First human chromosome, chromosome 22, completely sequenced in December. |
| 2000 | Second human chromosome, chromosome 21, the smallest human chromosome, completely sequenced in May. Drafts of chromosome 5, 16, 19 completed. A 90 percent working draft of the human genome completed. |
| 2001 | Initial working draft sequence of the human genome published in *Nature* (15 February 2001) and in *Science* (16 February 2001). The third human chromosome, chromosome 20, completely sequenced in December. |
| 2002 | The National Academy of Sciences recommends a ban on human cloning. An international team of scientists completed the gene sequence of the mouse. |
| 2003 | Chromosome 14, the fourth human chromosome, completely sequenced; Human Genome Project completed in April. Y completely sequenced in June; 7 completely sequenced in July; 6 completely sequenced in October. |
| 2004 | Human chromosomes 13 and 19 completely sequenced in March; 10 and 9 completely sequenced in May; 5 completely sequenced in September. Human gene count estimates changed to 20,000 -25,000 in October. |
| 2005 | X chromosome completely sequenced in March. |
| 2008 | Duck-billed platypus genome sequenced, May |
| 2009 | Pig, HIV, parasitic flatworm, panda, corn, human rhinovirus strains, and horse genomes decoded. |
| 2011 | Orangutan genome sequenced, January 2011. |

## REVIEW QUESTIONS ON MOLECULAR BIOLOGY AND THE DOUBLE HELIX

1. What is the main theme(s) of this chapter?

2. What is the chemical composition of DNA? What is its structure? What is a gene?

3. Identify: William Bragg, Phoebus Levene, Oswald Avery, Alfred Hershey and Martha Chase, Alexander Todd, Erwin Chargaff, Rosalind Franklin, Maurice Wilkins, Linus Pauling.

4. What did Erwin Chargaff contribute to unraveling the structure of DNA?

5. What structural DNA model did Linus Pauling propose?

6. What did Rosalind Franklin's x-ray experiments contribute to solving the structure of DNA?

7. What approach or methodology did Watson and Crick follow in their search for the structure of DNA?

8. What is the difference between DNA and RNA?

9. Who proposed the successful triplet code of DNA?

10. What was Marshall Nirenberg's contribution to research on DNA?

11. What is a nucleotide?

12. How many genes are there in the human genome?

13. What is a mutation?

14. What is genetic engineering? What is cloning?

15. Are there ethical and moral implications resulting from the advances in molecular biology?

16. What is the Human Genome Project?

17. What is Watson doing in his retirement years?

# CHAPTER 9
# THE PHYSICAL SCIENCES IN THE NINETEENTH CENTURY

> Thus is established, I think for the first time a true direct relation and dependence between light and the magnetic and electric forces; and thus a great addition made to the facts and considerations which tend to prove that all natural forces are tied together and have one common origin. . . .
>
> Michael Faraday (1791-1867)
> *Philosophical Magazine*, November 1845

## I. INTRODUCTION

Physical science in the last decades of the nineteenth century appeared on the verge of completing the scientific revolution that began two centuries earlier. An electrodynamic interpretation of the universe and its laws arose from the electrical studies of Michael Faraday and James Clerk Maxwell in England and Joseph Henry in the United States. It required the existence of electromagnetic lines of force and electromagnetic fields permeating space. The new electrodynamic universe complimented the existing mechanical (machine) universe and largely completed the program of classical physical science, the term that defines or describes nearly all nineteenth-century physical science.

Several problems remained unsolved for most of the nineteenth century, but the scientific community did not see them as unsolvable. John Dalton's early nineteenth-century structureless indivisible atom, usually called the billiard ball chemical atom, failed to account for an atom's spectrum because atomic spectra seemed to require an atom with vibrating parts. Nor could the structureless atom account for the bonds that held together the atoms composing a molecule other than to rely on the affinity one atom had for another atom. The electron's discovery in 1897 solved these problems.

Henri Becquerel's discovery of radioactivity in 1896, resulting in an apparently endless source of energy emitted by structureless atoms of uranium, thorium, radium, and polonium, presented a second challenge because the energy emitted seemed to violate the recently proposed 1847 law of conservation of energy. Einstein's 1905 energy-mass relation, $E = mc^2$, accounted for the energy emitted.

Nineteenth-century scientists largely abandoned the mechanical, or particle, theory of light. Thomas Young's wave theory of light became the dominant model in the early 1800s. His revival of the wave theory also revived acceptance of the space-filling luminiferous ether. Physicists required this undetectable all-pervading medium to transmit light waves from the Sun and stars to Earth and other planets, and from their sources on Earth to any location on Earth. They also asked how fast these light waves traveled through space or through any medium. Beginning in 1849 French physicists invented the first terrestrial methods of calculating the speed of light. Roemer's 1676 inaccurate celestial method based on measuring the time interval between successive eclipses of one of Jupiter's moons had not provided very good values. Bradley's celestial method gave much better values but provided no way to vary the experimental conditions and improve the results.

Albert Michelson in the United States perfected the new and more accurate terrestrial method of calculating the speed of light. His measurements raised the question of whether the speed of the light waves moving through the ether had an absolute value. Was the speed independent of the direction in which the light moved with respect to the direction of Earth's motion? Or, did the speed depend on whether the light moved in a direction parallel to Earth's motion or perpendicular to Earth's motion, or in any other direction with respect to Earth's motion?

Michelson and most nineteenth-century physicists accepted the ether's reality, and in his well-known 1887 ether drift experiment he expected to detect a small but measurable direction-dependent difference in the speed of light. Michelson obtained no difference, casting doubt on the ether's presence. The issue remained unresolved until 1905 when Einstein introduced his special theory of relativity which dispensed with the ether's presence, not by disproving its existence, but by declaring and showing its irrelevance. How could he or anyone else disprove the existence of something that never existed. Einstein banished ether from physics just as Maxwell and Ludwig Boltzmann earlier in the 1860s-70s banished caloric, the fluid of heat, from thermodynamics.

Nineteenth-century chemical studies went from knowing only a chemical compound's composition to determining the arrangement or location of each atom in a molecule and thereby establishing the structure of a molecule, its structural formula. August Kekulé and Archibald Scott Couper in 1858 provided the first structural formulas for the carbon compounds of organic chemistry, particularly Kekulé's structure of benzene in 1865. Within a few years, in 1874, their planar, or flat, structures had acquired a spatial orientation.

About the same time, in 1869, Dimitri Mendeleev published his first Periodic Table of the chemical elements. He provided physical scientists with their long sought classification system for the steadily increasing number of elements. In the 1860s, chemists agreed on a standard for atomic weight determinations, accepting Jean Servais Stas's recommendation of oxygen, 0 = 16, as the standard weight with which to calculate the atomic weight of the other elements. Two new instruments, the Bunsen burner and especially the spectroscope, helped chemists to carry out their analyses of compounds, leading to advances in determining the composition and structure of matter. All of these developments occurred in Europe. Chemists in the United States began to make significant contributions to chemistry only in the twentieth century.

The study of heat, which had its origin in Joseph Black's 1760s research on specific heats and the heat content of different substances, became an important area of investigation known as thermodynamics. The first two laws of thermodynamics received a formal statement in the 1840s-1860s, the third law not until the 1920s. Thermodynamics itself became an integral component of a new branch of chemistry established in 1887, then called theoretical chemistry, and later physical chemistry. The American chemist-physicist J. Willard Gibbs at Yale made important contributions to thermodynamics with his investigations that clarified the meaning of entropy and his introduction of the all-important conceptions of free energy and the phase rule and phase diagram in 1875 and 1878.

Determinism dominated nineteenth-century physical and life sciences. Its central principle held that if we know precisely how a system functions at the present time we can predict its precise behavior at any time in the future. This principle followed directly from the belief in a Newtonian mechanical or machine-like universe. It remained largely unchallenged until the rapid series of discoveries and new theories that arose from 1895 to 1915. These were x-rays in 1895, radioactivity in 1896, the electron in 1897, quantum theory in 1900, special relativity in 1905, and general relativity in 1915. In the mid-1920s the physicists Werner Heisenberg and Erwin Schrödinger introduced the quantum mechanical or wave mechanical theory of matter, officially ending determinism in science. Because of these developments no completion ever occurred in the physical sciences. Classical physical science underwent a transformation that shattered the foundation of classical physical science and introduced a revolution in scientific thought that continues a century later.

In the nineteenth-century, science increasingly relied on the discovery of scientific principles and laws for its advancement. Application of these principles and laws led to many new scientific processes and inventions and thereby to science's increasing impact on society. The technological transformation that occurred in the second half of the nineteenth century occurred largely because of the new scientific outlook. Unfortunately, the technological transformation that brought the industrial world into existence and provided numerous benefits to society created serious environmental problems, particularly atmospheric and water pollution. Environmental pollution, either natural or human, had existed from the time of Earth's formation but increased significantly at this time. Nineteenth-century society, however, largely ignored the other side of the technological transformation. Recognition and acceptance that another side existed, the reality of human-caused environmental pollution, occurred only in the second half of the twentieth century.

## II. STUDY OF LIGHT

In the nineteenth century the scientific community abandoned the particle model of light and accepted the wave model of light. Thomas Young's double slit experiment of 1801 provided the crucial evidence. Adoption of the wave model required physicists to introduce a medium in which the light waves vibrated as they traveled through space. This was the all-pervading luminiferous ether that was rigid and elastic enough to allow the transverse light waves to vibrate in it and yet have an extremely low density that permitted the waves to pass through it unaffected. These were contradictory properties.

### *Characteristics of the wave theory of light*

From the time of the Greek natural philosophers, observers of nature recognized two kinds of wave motion, transverse and longitudinal.

A *transverse wave* results when the wave's direction of vibration is perpendicular to the direction in which the wave is traveling, as in a water wave.

A *longitudinal wave* results when the wave's direction of vibration is parallel to the direction in which the wave is traveling, as in a coiled spring or a sound wave.

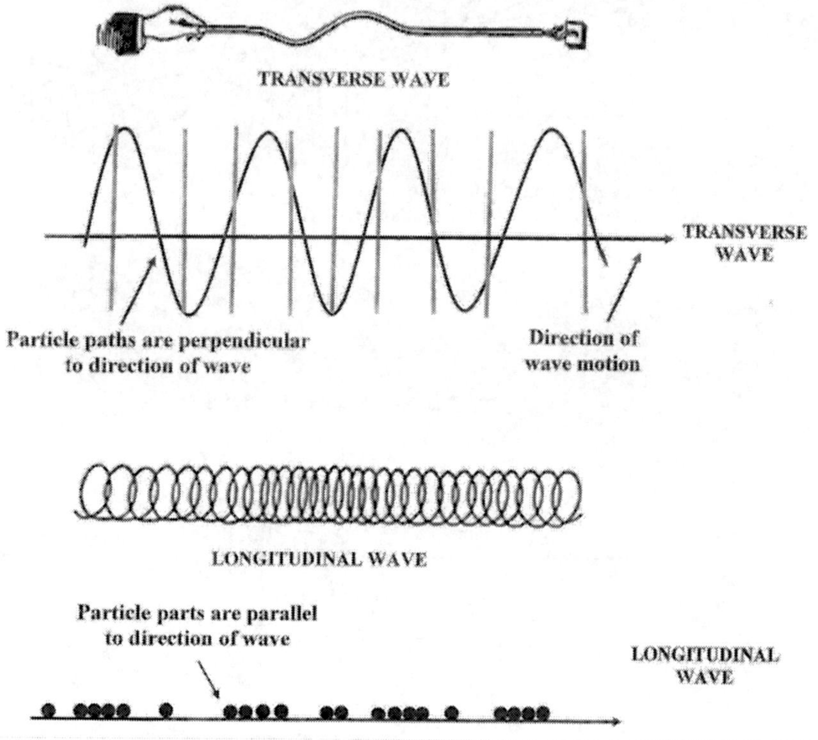


**Figure 1**
**Transverse and Longitudinal Waves**
Courtesy of the author
Source: Newton Henry Black and Harvey Nathaniel Davis. *New Practical Physics.* New York: Macmillan Company, 1913.

The refraction and diffraction of light were properties of light that scientists had discovered prior to the nineteenth century, although diffraction required a wave theory to account for it. Polarization of light was a nineteenth-century discovery that only a wave theory could explain.

*Refraction* is a bending or deflection of light waves when they pass from one medium to another of different density, as when light passes from air to water. The greater the refraction, the slower the velocity of the light wave in that medium. Snell in 1621 and Descartes in 1637 discovered the refraction of light.

*Diffraction* is the bending of light waves at a corner or edge of an opening. Grimaldi demonstrated this as a property of light waves in 1666.

*Polarization* occurs when light waves (which are transverse waves vibrating in all possible planes or planar surfaces in three-dimensional space), after passing through optically active substances such as calcium carbonate, quartz, and tourmaline, vibrate only in one or two planes, or planar surfaces, in three-dimensional space, and not in all possible planes. A later discovery, polaroid, which consists of two sheets of plastic with a thin layer of needle-like quinine iodosulfate crystals between them, produces the same effect.

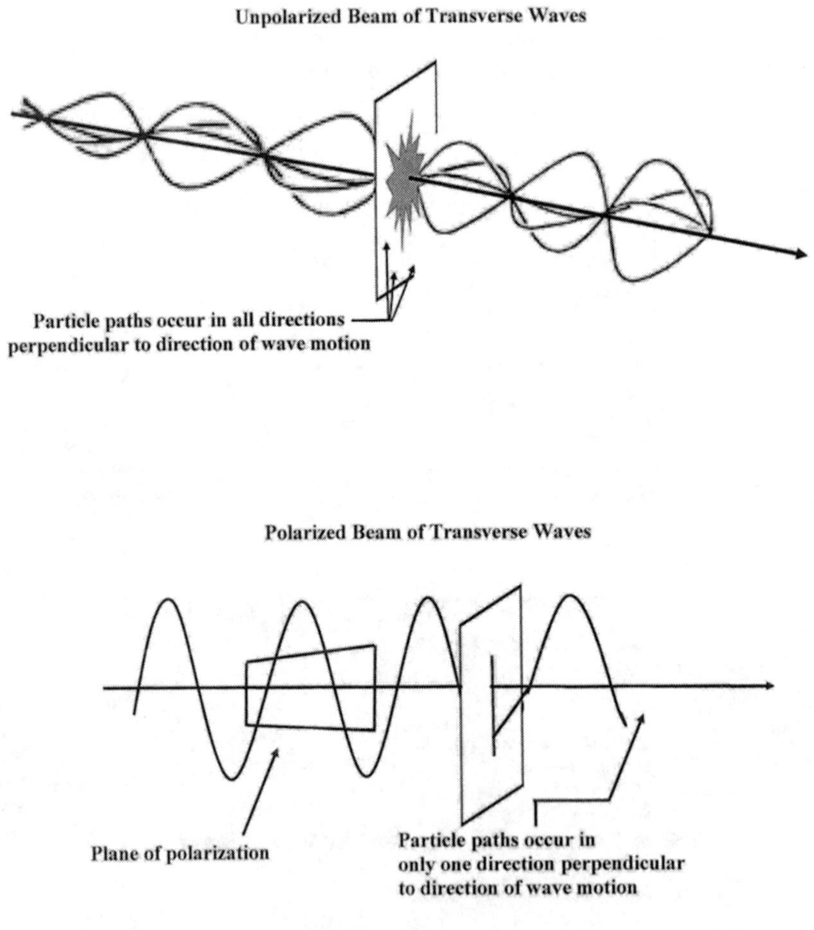


**Figure 2**
**Unpolarized and Polarized Beams of Transverse Waves**
Courtesy of the author

***Discovery of polarization***

Erasmus Bartholin (1625-98) at Copenhagen in 1669 observed double refraction in an Iceland spar crystal (calcium carbonate). The crystal refracted the single ray of unpolarized light waves traveling through it at two different angles causing two rays to emerge. When he rotated the crystal one ray remained fixed and the other ray rotated about the first ray. Bartholin had polarized the ray of light, but he drew no conclusions from his experiment.

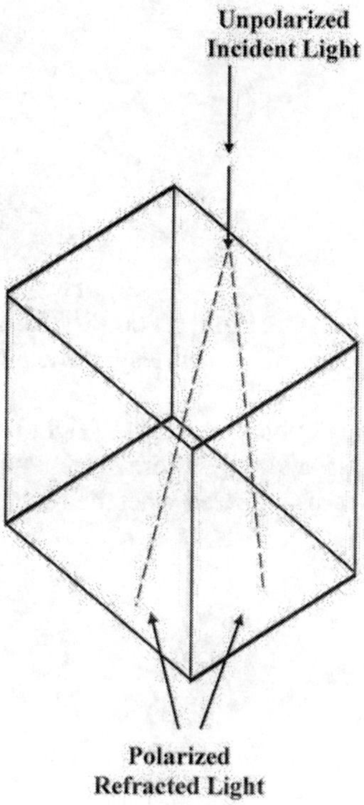


**Figure 3**
**The Behavior of the Extraordinary Rays.**
Courtesy of the author
Source: Erasmus Bartholin. *Experimenta crystalli islandici disdiaclastici quibus mira & insolita refractio detegitur.* Copenhagen (Hafniæ), Denmark: Daniel Paulli, 1669. trans. Werner Brandt. *Experiments with the double refracting Iceland crystal which led to the discovery of a marvelous and strange refraction.* Westtown, PA, 1959.

Etienne Malus (1775-1812), a French military engineer who supported the particle theory of light, in 1809 discovered two ways of polarizing light. His first method was by reflection of light off glass. Malus initially did not realize that the reflection polarized the light that struck the glass. When he held a calcium carbonate crystal in the path of the reflected polarized light, he observed a single ray emerging from the crystal.

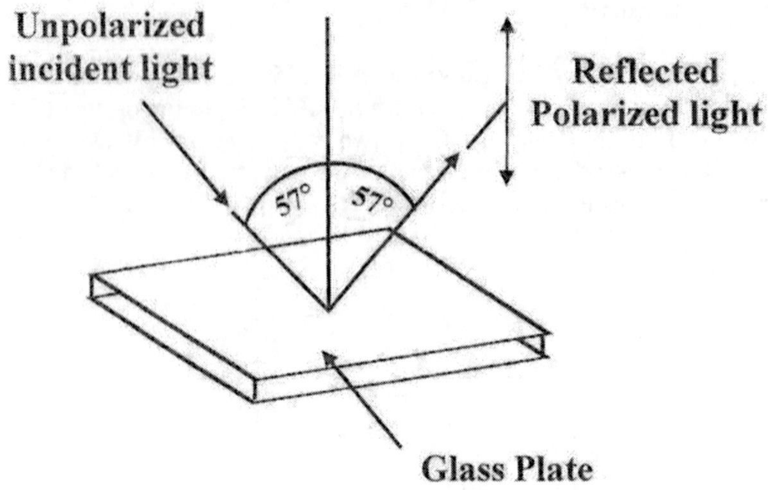


**Figure 4**
**Polarization by Reflection**
Courtesy of the author

His second method was by the double refraction of light passing through a calcium carbonate crystal. Malus reasoned that the two refracted rays were at right angles to each other (perpendicular), because one ray appeared and the other disappeared when he rotated the crystal 90°. This accounted for Bartholin's earlier observation.

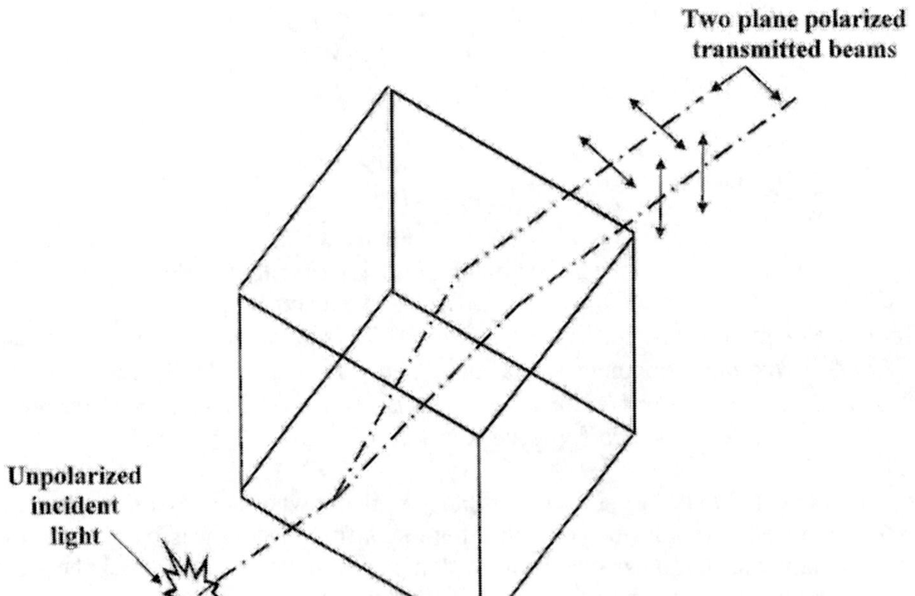


**Figure 5**
**Polarization Accompanying Double Refraction**
Courtesy of the author

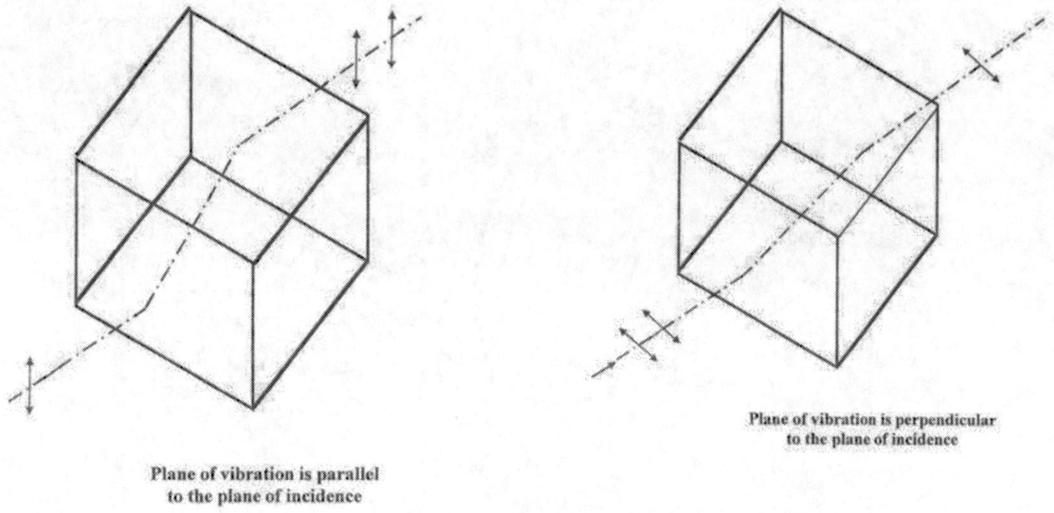


**Figure 6**
**Double Refraction with the Incident Light Polarized**
Courtesy of the author

David Brewster (1781-1868), a Scottish physicist in Edinburgh, showed in 1815 that at a surface, reflection occurs after refraction. The reflected ray splits from the refracted ray and the two rays are at right angles and polarized. Brewster played a leading role in establishing the British Association for the Advancement of Science in York in 1831. He held a positivist philosophy of science.

***Young's experiments demonstrating that light is a wave in 1801 and a transverse wave in 1817***

The experiments of Thomas Young (1773-1829) in London demonstrated that light was a wave motion and not a particle and that it was a transverse wave and not a longitudinal wave as some scientists believed at that time. Young's double slit diffraction experiment in 1801 provided convincing evidence for the wave theory because only waves, and not particles, could produce an interference pattern that consisted of repeating bands of light and dark when light rays struck a screen placed behind two other screens, through which they had passed.

The first of the three screens had a single slit through which light passed on to the second screen. The second screen had two slits made in it so that the light that traveled in a straight line through the first screen with one slit could not pass through the two slits of the second screen unless it was diffracted or bent at the edges of the slit of the first screen. Young used the interference pattern that resulted to calculate the wavelength of visible light. In *Lectures on Natural Philosophy* published in 1807, he made an analogy between the interference of light waves that produced bands of light and dark and the interference of sound waves that produced beats, an interference in which sound waves reinforced or cancelled each other.

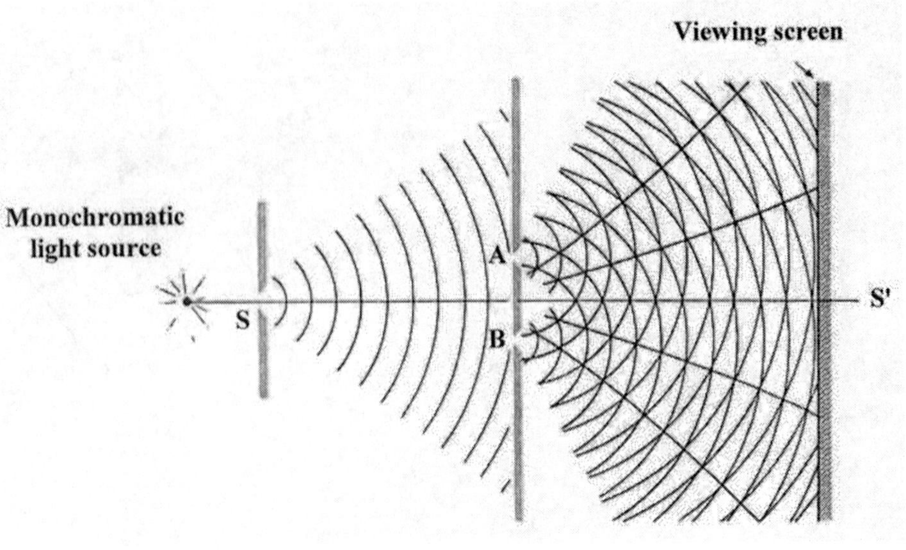


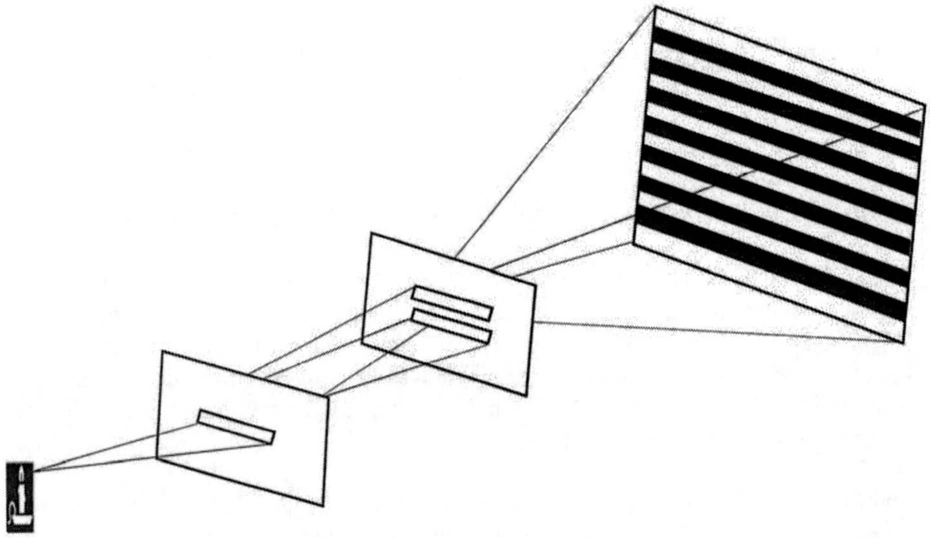

Plane of vibration is parallel to the plane of incidence.

Light travels in a straight line. Because of diffraction the screen's illumination is not uniform but consists of alternating bright and dark lines. If no diffraction occurs, no light reaches the viewing screen. The screen remains dark.

**Figure 7**
**Young's Single and Double Slit Experiment**
Courtesy of the author

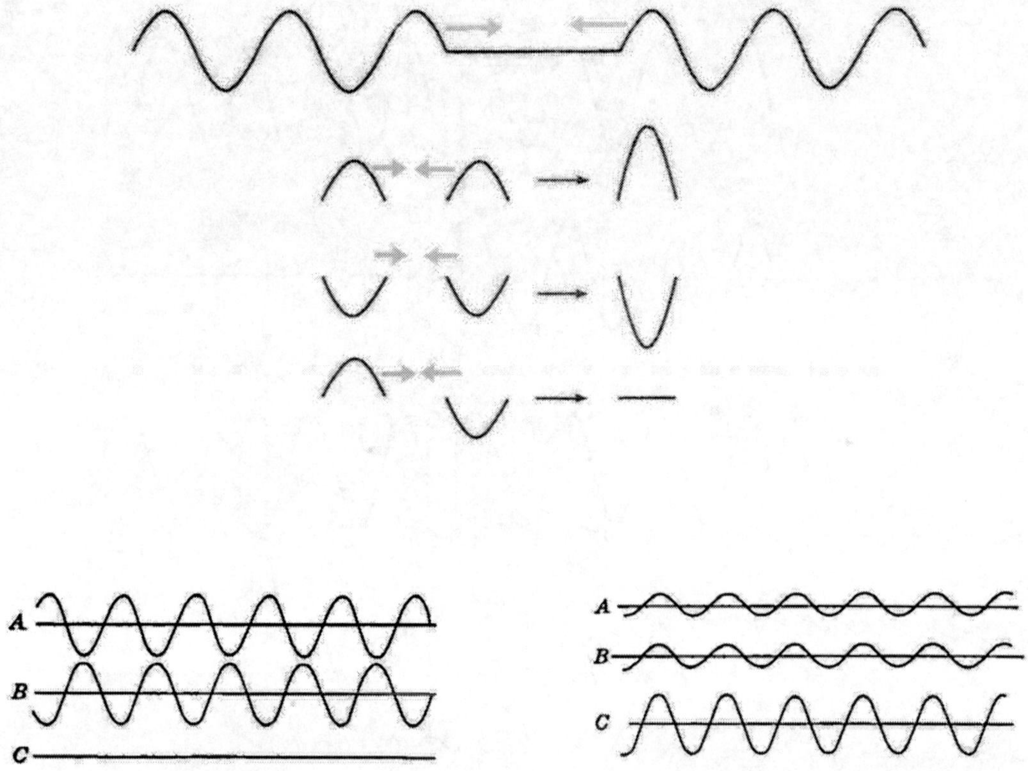


Two waves *A* and *B* of same period but opposite phase produce *interference C*.

Two waves *A* and *B* in step produce *reinforcement C*.

**Figure 8**
**The Principle of Superposition of Light Waves**
Courtesy of the author (top)
Source: Newton Henry Black and Harvey Nathaniel Davis. *New Practical Physics.* New York: Macmillan Company, 1913 (bottom).

Young proposed that light waves were transverse in 1817. In his crucial experiment he polarized light. This demonstrated that light was a transverse wave because he could polarize a transverse wave by restricting its vibrations, which are perpendicular to the direction in which the wave is traveling, to one plane and thereby reduce its intensity. He could not polarize a longitudinal wave, which has the wave's direction of vibration parallel to the direction in which the wave is traveling. To block the vibrations of a longitudinal wave is the same as blocking the wave's motion.

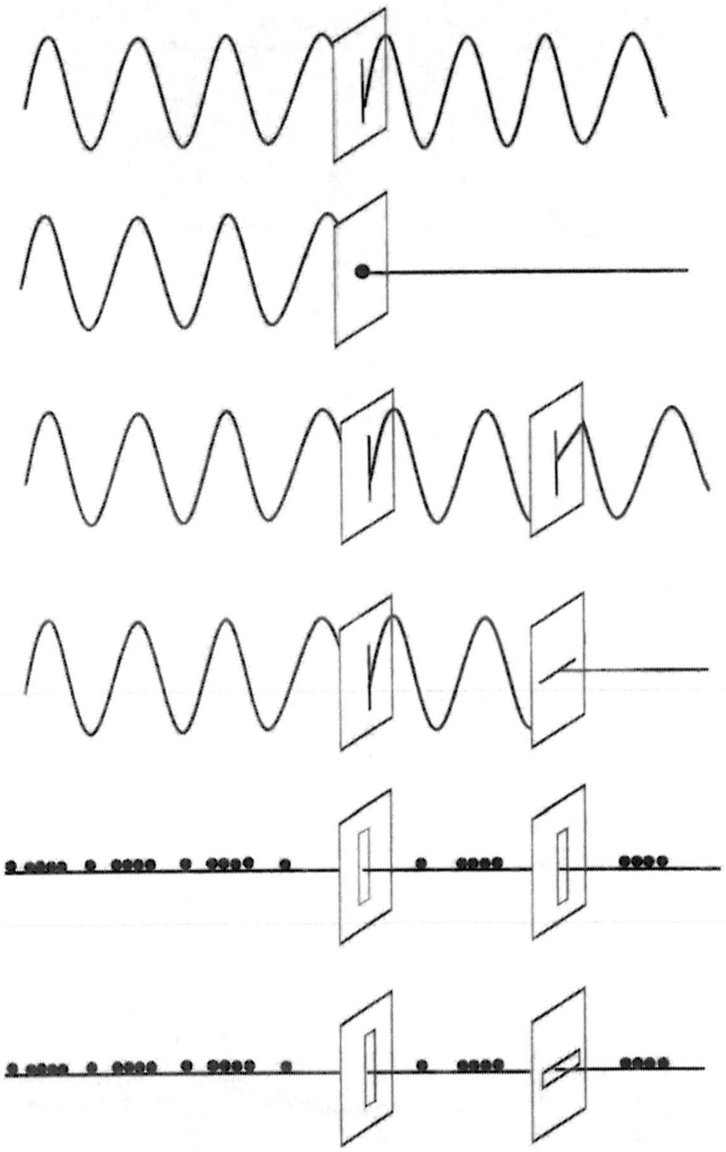

**Figure 9**
**Mechanical Analogies of Polarization**
Courtesy of the author

Augustin Fresnel (1788-1827), a physicist in Paris, explained Iceland spar crystal's double refraction and polarization of light using wave theory in 1814-15. As Malus had done, he showed that the two polarized rays were at right angles to one other. Fresnel believed ether was necessary to transmit waves. Louis Cauchy (1789-1857), a military engineer in Paris, developed a mathematical theory of the ether to account for the transmission of light waves. Particles did not need a medium for their transmission.

The astronomer Jean Arago (1786-1853) and the physicist Jean Biot (1774-1862) in Paris showed in 1815 that organic compounds rotated the plane of polarized light. This later provided a way to identify certain organic compounds by measuring the number of degrees the compound rotated the plane of polarized light to the right or to the left, dextro, *d*, right, or laevo, *l*, left. William Nicol (1768-1851), a physicist at Edinburgh, in 1828 constructed a calcium carbonate prism, known as the Nicol prism, to measure the degrees of rotation of polarized light.

### *Expanding the visible spectrum of light*

William Herschel (1738-1822), a German-born astronomer in London, discovered infrared light in 1800 when he detected a heating effect beyond the red end of the visible spectrum of light which has wavelengths in the range of 7,000 A to 4,000 A (1A $=10^{-8}$ cm), from red to violet. Infrared light has longer wavelengths than visible light.

Nearly 50 years later, Macedonio Melloni (1798-1854), a physicist in Perma, Italy, in his investigations of 1846-50 used prisms and lenses of sodium chloride to show that infrared light behaved the same as visible light. He used a thermopile (thermocouple) to detect infrared light beyond the red end of the spectrum.

In 1801, one year after Herschel discovered infrared light, Johann Ritter (1776-1810), an electrochemist at Jena, discovered ultraviolet light. He exposed silver chloride to invisible radiation beyond the violet end of the visible light spectrum (ultraviolet light) and saw the chloride darken. Ultraviolet light has shorter wavelengths than visible light.

### *Terrestrial methods of measuring the speed of light: Fizeau, Foucault, Cornu, and Michelson*

The celestial methods of Olaus Roemer in 1676 and James Bradley in 1729 were the only known method of measuring the speed of light prior to the development of the nineteenth-century terrestrial methods of Armand Fizeau, Jean Foucault, Maria Alfred Cornu in France and Albert Michelson in the United States.

The French experimenters were the first to develop accurate terrestrial methods of measuring the speed of light. In all these experiments, which culminated with Michelson's measurements in the United States, investigators reflected a ray of light back and forth between two mirrors a known distance apart. Knowing the distance the incoming and reflected ray of light traveled and the frequency at which a rotating toothed wheel (cogwheel) or stroboscope first blocked the reflected ray of light, or in Foucault's method the angle through which a mirror rotated, enabled them to calculate light's speed.

The first investigator, Armand Fizeau (1819-96) in Paris, in 1849 calculated $v$ = 313,000 km/sec (194,600 mi/sec) for the speed of light. Fizeau used a rapidly rotating toothed wheel with 720 teeth mounted on a system of wheels moved by weights, a counter to measure the toothed wheel's rate of rotation, and a carbon arc lamp to provide the light. The triangulation method gave an accurate measurement of the distance, d = 8.66 km (5.36 mi), between his reflecting mirror and toothed wheel.

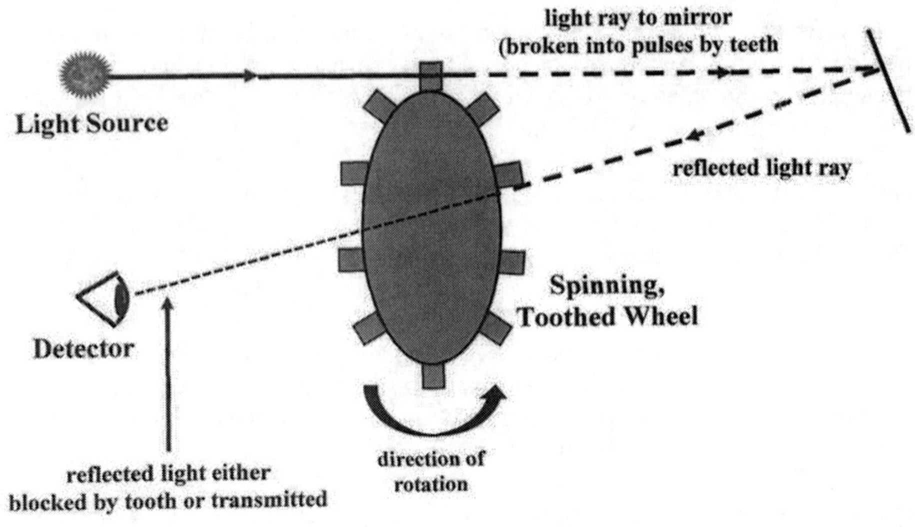


Fizeau used this apparatus to measure light's speed. The rotating toothed wheel (cogwheel) containing 720 cogs (teeth) and 720 openings (spaces) either permitted or blocked the reflected light beam from reaching the observer. Fizeau measured the time it took the light beam to travel from the toothed wheel to the mirror and back, a total distance of 10.72 miles.

**Figure 10**
**Fizeau's Method of Calculating the Speed of Light**
Courtesy of the author

Fizeau realized that if he rotated the toothed wheel from its initial stationary position to higher and higher frequencies and measured the frequency at which the first eclipse of the reflected light occurred, he could calculate the speed of light. He recorded a frequency of 12.6 revolutions per second (rev/sec) for the first eclipse. This meant that at that frequency the toothed wheel had rotated from the opening through which the light passed, to the first tooth that blocked the reflected light, or $\frac{1}{2}\ x\ \frac{1}{720} = \frac{1}{1440}$ of a revolution. When he doubled the frequency the reflected light again passed through an opening, when he tripled the frequency a second eclipse occurred, when he quadrupled the frequency the light again passed through an opening. Fizeau's main source of error lay in deciding exactly when (at which frequency of the rotating toothed wheel) the reflected light had reached the minimum (zero) brightness and its eclipse occurred. His calculation follows.

$$frequency = \frac{revolutions}{second} \text{ or}$$

$$frequency = \frac{revolutions}{time} \text{ and}$$

$$time = \frac{distance}{velocity} \text{ therefore}$$

$$frequency = \frac{revolutions}{\frac{distance}{velocity}} = \frac{\frac{1}{1440}}{\frac{10.72}{c}}$$

$$frequency = \frac{1}{1440} x \frac{c}{10.72}$$

$$12.6 = \frac{c}{1440x10.72}$$

$$c = 194{,}600\ mi/\ sec\ or\ 313{,}300\ km/sec$$

**Figure 11**
**Fizeau's Calculation of the Speed of Light**
Courtesy of the author

Prior to his calculation of the speed of light in air, Jean Foucault (1819-68) in Paris and his sometimes collaborator Fizeau, in 1850 calculated the speed of light in water and obtained a value about three-fourths its speed in air, or 223,000 km/sec. This evidence supported the wave theory of light rather than the particle theory which predicted a greater speed in water because water had a greater density than air. This meant water had more particles per unit volume and would attract the particles of light more strongly and therefore increase the speed of light.

Foucault calculated the speed of light in air in 1862. He used a rotating planar mirror *C*, rotating at a frequency $f$ = 500 rev/sec, which gave a measurable linear displacement of a reflected ray of light on a stationary semi-transparent planar mirror B, placed at distances from 4 to 20 meters, d = 4-20 meters from C (fig. 16). Foucault sent a ray of light to the rotating mirror C which when standing still reflected the light to B along the lower line. When C turned rapidly, it displaced the reflected light along the upper line, to a new position on B. Foucault could measure the amount of displacement on B. The displacement of the reflected ray on B represented the angle through which C had rotated (a small fraction of 360°) during the time light traveled from B to C and back. Knowing C's rotational frequency, the total distance (2d) the light ray traveled, and the angle through which C had rotated when it reflected the light ray (a fraction of 360°), Foucault in 1862 calculated the speed of light using the same equation as Fizeau, $c$ = 298,000 km/sec (185,150 mi/sec).

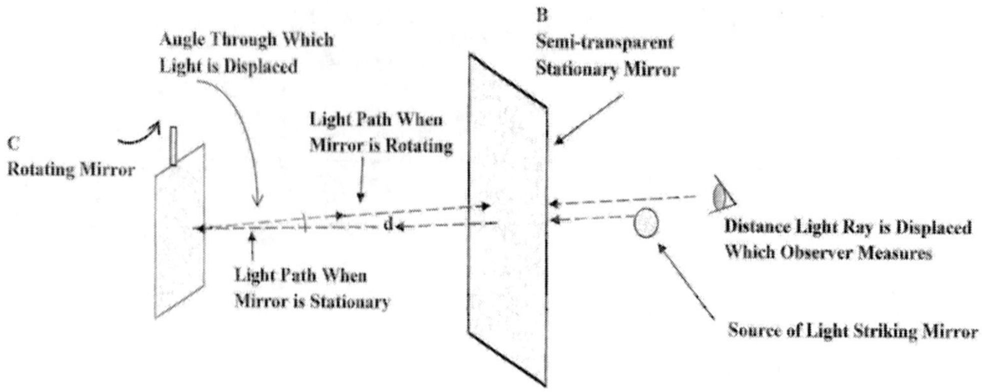


Foucault sent a ray of light that traveled from a stationary mirror, to a rotating mirror, and back to the stationary mirror for a total distance 2d. During this time the rotating mirror will have turned very slightly between the two times of light's arrival. This small rotation deflects the beam of light through a small angle from its original path, producing a measurable displacement proportional to the rotational speed of the mirror and inversely proportional to the speed of light. The greater the rotational speed of the mirror, the greater its displacement. The greater the speed of light, the smaller the displacement of the mirror.

**Figure 12**
**Foucault's Method of Calculating the Speed of Light**
Courtesy of the author

In addition to his work on the speed of light, Foucault in 1851 invented the Foucault pendulum with which he demonstrated mechanically Earth's daily rotation on its axis. He invented the gyroscope in 1852.

The third of the French investigators, Maria Alfred Cornu (1841-1902) in Paris, calculated $c$ = 298,500 km/sec in 1874 and $c$ = 300,400 km/sec (186,000 mi/sec) in 1878. Instead of Fizeau's toothed wheel, Cornu used a rotating stroboscope with 200 openings that rotated at 54,000 rev/min (900 rev/sec) when the first eclipse of light occurred. He placed the stroboscope and the reflecting mirror 23 kilometers apart.

The American physicist Albert Michelson (1852-1931), who spent a lifetime experimentally determining the speed of light, obtained values that surpassed those of all other experimenters. Born in Prussia, Michelson emigrated with his family to the United States in 1854. He graduated from the Naval Academy in 1873, taught at the Naval Academy, from 1883 at Case School of Applied Science in Cleveland (now Case-Western Reserve University), and from 1892 at the newly-established University of Chicago. Michelson began his experiments to measure the speed of light using a Foucault-rotating mirror-type experiment in 1877 while teaching at the Naval Academy. He published a very good value, 186,508 mi/sec (298,413 km/sec), in 1879, although three years later in 1882, Newcomb published a better value of 299,800 km/sec.

Nineteenth-century physicists believed that light consisted of transverse waves and required a medium for their transmission. The ether, which the Dutch scientist Christian Huygens reintroduced in 1678, was the medium. It filled all space, or was all-pervading. It therefore had a density value $d$ so small that it defied detection; yet, it had to have an elasticity $e$ of steel in order to transmit transverse waves traveling at the speed of light or according to the equation $c = \sqrt{e/d}$. These were clearly contradictory properties and presented a problem for the supporters of an ether medium. The particle model of light did not require a medium for its transmission and hence required no ether. Michelson never doubted ether's existence, and like most nineteenth-century physicists he believed Earth's orbital motion influenced the speed of light. Yet his interferometer experiments, which he began in 1880 and from which he expected to obtain an absolute value for Earth's orbital speed, cast serious doubt on the ether's existence.

Michelson began his interferometer experiments in 1880. Bell provided $500 in 1881, enabling Michelson to continue his research. Michelson's interferometer was a system of mirrors. When a beam of light struck a semi-transparent mirror, the mirror split the beam into two perpendicular paths. Each beam moving in its perpendicular path struck a mirror that reflected the light back along the same path to the semi-transparent mirror where they recombined. The two beams traveled an equal and known distance. In the experimental determinations carried out from 1880 to 1903, some in 1881 in Helmholtz's laboratory in Berlin where Michelson resided from 1880 to 1882, he found no difference in the speed of light traveling in the two perpendicular paths. They traveled an equal distance in an equal time. Speed did not depend on the direction in which light traveled.

In the most famous of his interferometer experiments, known as the Michelson-Morley ether drift experiment of 1887, Michelson and Edward Morley (1838-1923), a chemist at Case where Michelson had gone in 1883, constructed an extremely sensitive interferometer. It consisted of a stone block five feet square, one foot thick, which floated on a wooden frame set on a container filled with mercury to eliminate any vibrations. An Argand oil burner provided the light. The path length of each arm was 11 meters (36.1 feet), which they achieved by multiple reflections of the light. The interferometer rotated once in six minutes. Their interferometer measurements showed again that the speed of light did not depend on the direction in which the light was moving with respect to Earth's orbital motion.

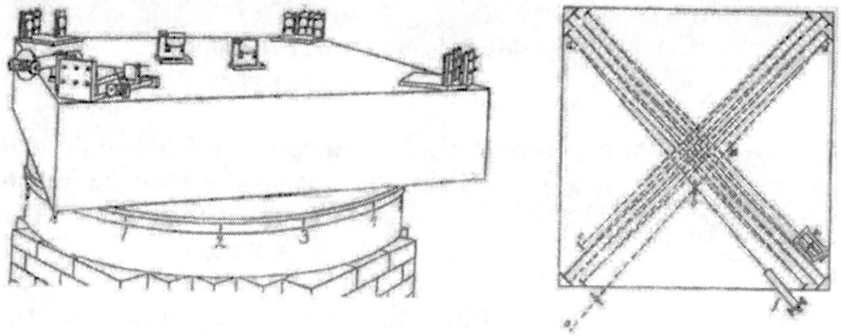

**Figure 13**
**Michelson-Morley Interferometer**
Source: A.A. Michelson. *Light Waves and Their Uses.*
Chicago: University of Chicago Press, 1903.

Michelson and most nineteenth-century physicists believed that light took less time to move across and back a known distance *perpendicular* to the direction in which Earth was moving through the ether than when it moved back and forth over the same distance *parallel* to the direction in which Earth was moving through the ether. This was analogous to rowing a boat in perpendicular and parallel directions over an equal distance in a river with a current. In this case, as shown mathematically using vector analysis, the boat took less time to move the same distance across the river and back perpendicularly than it took to move the same distance back and forth parallel to the current. Light, the physicists believed, behaved the same way.

Suppose two rowers row at the rate of 5 feet per second in still water, and the stream in which they are rowing flows at 4 feet per second. Rower one rows his boat 90 feet, upstream and back to the starting point. His upstream rate against the current is 5 feet per second minus 4 feet per second or 1 foot per second. His downstream rate with the current is 5 + 4 = 9 feet per second. His total time is the sum of the two times calculated from the formula $v = \frac{d}{t}$ or $t = \frac{d}{v}$. The upstream time is $t = \frac{90}{1}$ or 90 seconds, the downstream time $t = \frac{90}{9}$ or 10 seconds, and the total time is 100 seconds.

Rower two rows his boat across the stream and back the same distance of 90 feet. Using the Pythagorean theorem, the hypotenuse of a right-angle triangle represents his upstream rate of 5 feet per second, to compensate for the current's downstream rate, the vertical side represents the rate of the stream or 4 feet per second, and the horizontal side is his actual rate of travel which according to the Pythagorean theorem is $5^2 = 4^2 + x^2$ or 3 feet per second. The time each way is $t = \frac{d}{v}$ or $\frac{90}{3}$ or 30 seconds and the total time is 60 seconds.

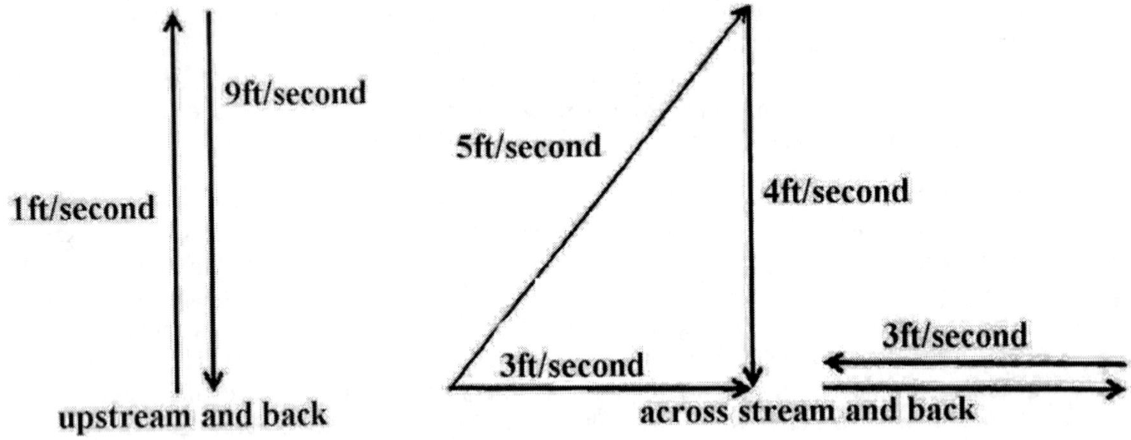


**Figure 14**
**Michelson's Rowboat Analogy**
Courtesy of the author

The null or zero result of the interferometer experiment cast doubt on the existence of an ether that transmitted light waves, or in which the light waves vibrated, and was the fixed reference against which to measure the speed of all objects, including the planets. Without the ether, all motion was relative to a chosen or arbitrary reference, no absolute or fixed reference against which to measure the speed of an object existed. Michelson and Morley expected to find a difference in the speed of light when they included Earth's motion in their calculations, and from the ratio of the two speeds or times they expected to calculate Earth's absolute orbital speed.

A question raised at that time asked whether Earth dragged the ether with it as it moved and masked the ether's presence. In 1893 Oliver Lodge(1851-1940) at Liverpool found that on passing light through an interferometer placed in a one-inch space between two steel disks, each 36 inches in diameter, rotating horizontally on a vertical axis, the rotation did not affect the speed of light. The disks, rotating at speeds up 3,000 revolutions per minute, apparently did not drag the ether along. Earth therefore did not drag the ether along as Earth moved in its orbit. This was contrary to what the physicist and wave theory of light proponent George Stokes (1819-1903) at Cambridge believed. Michelson's and Lodge's experiments indicated no apparent moving or stationary ether. Yet most physicists including Stokes, William Thomson, Lord Kelvin (1824-1907) at Glasgow, Hendrik Lorentz (1853-1928) at Leiden, and Lodge expected to obtain an interference fringe that depended on the speed of rotation and were unwilling to abandon belief in an ether.

Whirling machine for ether experiment, with pair of steel disks, 1 yard in diameter at top. From a photograph taken during preliminary tests before bolting the whirling machine to a stone pier.

**Figure 15**
**Lodge's Rotating Disks**
Source: Sir Oliver Lodge. *The Ether of Space.*
New York and London: Harper & Brothers, 1909.

How did Michelson and Morley's experiments enable them to measure the speed of light and at the same time determine whether the ether really existed? In the interferometer if the two light waves that the mirrors reflected at right angles to each other are traveling at the same speed they are perfectly in phase and do not interfere with each other. Instead, the waves add to one another and produce a bright band of light upon returning to the eyepiece detector. If the speed of the two reflected waves are sufficiently different so that the waves are out of phase by one-half wavelength, the waves cancel each other and give only a dark band, or no light. If the two reflected waves are out of phase by a small amount, a series of light and dark bands or fringes (narrow bands) will appear in the eyepiece detector, indicating some interference. Other out-of-phase reflected waves give similar interference patterns. The width of the band enables the experimenter to calculate the wavelength of light.

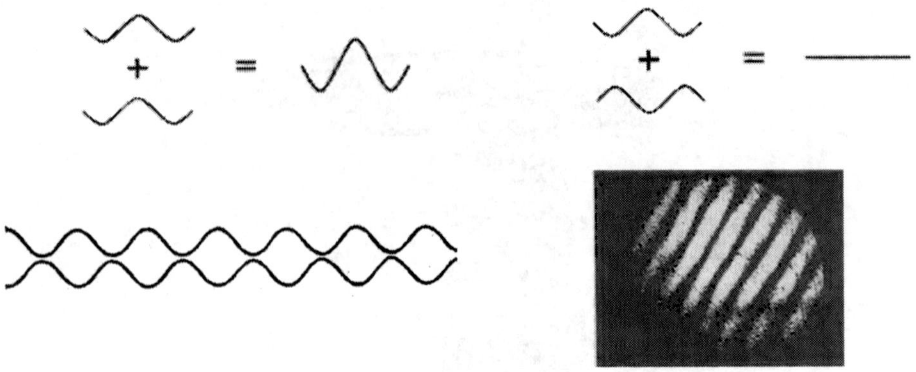

**Figure 16**
**Interference of Light Waves (above)**
**Interference Fringes (below)**
Top images courtesy of NASA
Source: A.A. Michelson. *Light Waves and Their Uses.*
Chicago: University of Chicago Press, 1903 (bottom left).
Source: US Department of Energy,
Office of Energy Research/Office of Fusion Energy (bottom right).

Michelson and Morley expected to see a narrow fringe. They expected the white light interference pattern to shift by 0.4 of a fringe as they rotated the interferometer through an angle of 90° in the horizontal plane, indicating a slight difference in the speed of light. They believed their interferometer could detect a shift of 0.4 of a fringe. When they performed the experiment from 8-12 July 1887 they concluded that if any displacement occurred because of the relative motion of Earth and the luminiferous ether, it had not exceeded 1/100 (0.01) of the distance between the fringes. Michelson and Morley obtained no evidence of interference.

The equation below relates the experimentally determined value of the ratio of the times of the two perpendicular reflected light waves with the speed of light *c*, and Earth's orbital speed *v*. The experimental value they obtained for the ratio was one, or $\frac{t_a}{t_b} = 1$.

**MICHELSON'S EQUATION RELATING *t, c,* and *v***

$$\frac{t_a}{t_b} = \sqrt{1 - \frac{v^2}{c^2}}$$

$v$ = 18.6 mi/sec, the known value of Earth's orbital speed

$c$ = 186,000 mi/sec, the speed of light

$t_a$ = time for light to travel perpendicularly to the direction in which Earth is moving

$t_b$ = time for light to travel parallel to the direction in which Earth is moving

Michelson and Morley observed no band or fringe width that an interference pattern should have produced. This indicated no difference in the two speeds despite the two different directions of the light waves. The value of *v*, Earth's orbital speed, appeared to have no effect on the speed of light.

Morley from 1902 to 1904 worked with the physicist Dayton Miller (1866-1941) at Case and once again confirmed the earlier results. Miller never accepted Einstein's later conclusion that the ether did not exist and continued into the 1920s to search for evidence of an ether drift that would prove the special theory of relativity incorrect. He thought he had evidence in 1921, but his claim fell short.

***Michelson's later measurements of the speed of light***

Michelson calculated a value of $c$ = 299,798 km/sec in 1927. In these experiments, which he did in Pasadena, Michelson reflected light between two mirrors, the first a rotating multi-sided plane mirror on top of Mount Wilson, where he also located a telescope for observation. The second mirror was actually an arrangement of a stationary concave and plane mirror on top of Mount San Antonio. He used rotating mirrors of glass and of steel, having 8, 12, and 16 sides, some of which his friend, the gyroscope inventor Elmer Sperry (1880-1930) in Washington, DC, provided. Michelson, who made 200 determinations with a given mirror, found the exact speed of the mirror's rotation by making a stroboscopic comparison with an electrically driven tuning fork that he calibrated with an invar pendulum. An air blast drove the mirror.

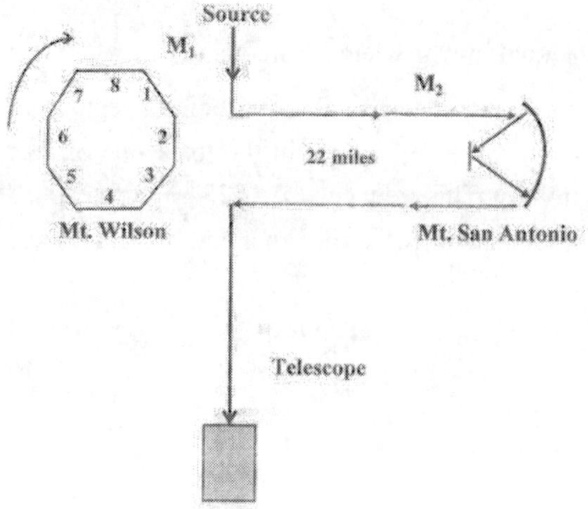


Many refinements of Michelson's original method of measuring the speed of light appear in this 1927 diagram of his apparatus, but the principle was essentially the same. The main change was in the length of light path.

**Figure 17**
**Michelson's Octagonal-Mirror Method of Measuring the Speed of Light 1927**
Courtesy of the author

Triangulation, a surveying technique in which the observer sets up a right-angle triangle relating known and unknown distances, enabled Michelson to measure very accurately the 35 km (22 miles) distance ($d$) that the light traveled between the two mirrors. The distance was accurate to 1 part in $10^6$, about 1/8 inch. In the triangulation method, the observer sights the object I, whose distance he wishes to determine, from two different positions $A$ and $B$. He knows the distance between $A$ and $B$, the base of the triangle and the line of position and arbitrarily chosen as 1 kilometer; and selects the two locations so that the distances $AB$ (horizontal side) and $AC$ (vertical side) form the sides of a right-angle triangle $CAB$. If $A$ represents Mount Wilson and $C$ Mount San Antonio and $ABC$ the angle of sight from $B$ to $C$ is 88° 22′, then knowing that the tangent of 88° 22′ (tan = opposite side over adjacent side), or $\frac{CA}{AB}$, is 35.070, Michelson calculated $CA$, the distance the light traveled, or tangent 88° 22′ = $\frac{CA}{AB}$, that is, $35.070 = \frac{CA}{1} = 35.070$ km.

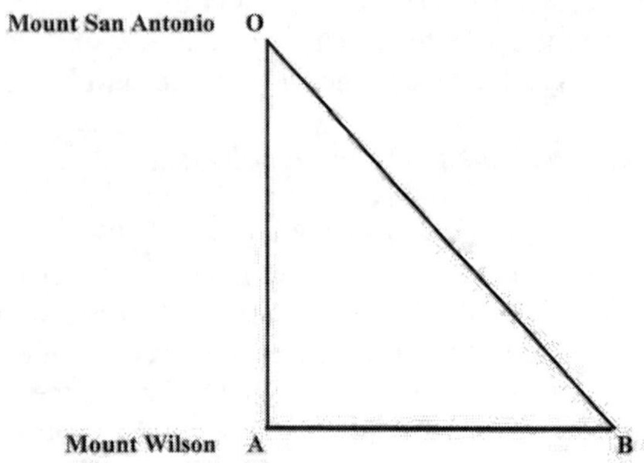


**Figure 18**
**Calculating Distances Using Triangulation**
Courtesy of the author

For the rotating octagonal mirror when its frequency of rotation reached $f = 528\ \frac{rev}{sec}$ the mirror had made $\frac{1}{8}$ of a revolution. The reflected light after being irregularly reflected at all lower frequencies of rotation, appeared again for the first time in the telescope exactly where it was when the octagonal mirror remained stationary. This frequency $f = 528\ \frac{rev}{sec}$ corresponds to a time $t = 0.00023$ seconds for the passage of light from one mirror to the other and back, and ⅛ revolution of the octagonal mirror.

$$\text{Frequency} = \frac{revolutions}{time}$$

$$\text{Time} = \frac{distance}{speed}$$

$$\text{Frequency} = \frac{revolutions}{\frac{distance}{speed}}$$

$$f = \frac{\frac{1}{8}}{\frac{2x35}{c}}$$

$$528 = \frac{\frac{1}{8}}{\frac{70}{c}}$$

$$c = 299.796 \pm 4 \text{ km/sec}$$

±

**Figure 19**
**Michelson's Calculation of the Speed of Light 1927**
Courtesy of the author

From 1930 to 1933, two years after Michelson's death from a cerebral hemorrhage, his research team measured the speed of light in an experiment in which they reflected light back and forth in an evacuated steel tube 1 mile long, 3 feet in diameter, built at a cost of $50,000 at the Irvine Ranch in Santa Ana, California. The air pressure inside the tube was 0.5 mm of mercury (760 mm is normal air pressure). The mirror placed inside the tube had 32 faces, and a compressed air turbine rotated the mirror on its axis. Michelson's group made 2,885 measurements, obtaining $c = 299{,}774\ \frac{km}{sec}$ ($186{,}164\ \frac{mi}{sec}$) in 1933.

Michelson received the Nobel Prize in physics in 1907 for his experiments on light. He was the first American scientist to win a Nobel Prize in science. During his career Michelson also used the interferometer to measure the angular diameter (width) of a heavenly body by comparing the light coming from the outside surface of each body and the interference pattern it produced. From the width of the fringe pattern for each body he calculated the distance between the points on the surface from which the light originated. He used this method to determine the angular diameter of Jupiter's largest satellites.

Michelson suggested using the wavelength of light as the standard of length, instead of the platinum-iridium meter bar kept in Paris. In 1893 he measured the meter using the wavelength of the red cadmium line as the unit length, defining the meter as 1,553,164 times the wavelength in air at 760 mm atmospheric pressure and 15° C. Michelson selected cadmium's red line because of the red line's clarity or monochromaticity.

Scientists adopted Michelson's method and his standard for the meter in 1925, although the standard wavelength today corresponds to the energy emitted for an electron transition in the gaseous element krypton. According to the International System of Units (SI) first adopted in 1960 at the Eleventh General Conference of Weights and Measures, the standard meter is a length 1,650,763.73 times the wavelength (0.605,780,211 $\mu m$) of light emitted by the unperturbed atomic energy level transition from a $5d_5$ orbital to a $2p_{10}$ orbital of krypton-86. The wavelength is the orange line in the krypton-86 spectrum, $\lambda = \frac{1}{1{,}650{,}763.73}\ m = 0.605{,}780{,}211\ um$. The last definition of the standard occurred in 1983, whereby the meter is the length light travels in a vacuum during a time interval of $\frac{1}{299{,}792{,}458}$ of a second. The speed of light is $c = 299{,}792{,}458\ \frac{m}{sec}$.

During World War I, Michelson served in Washington as a lieutenant commander in the United States Navy. He worked on several projects, particularly perfecting range finders for artillery. The National Academy of Sciences formally established the National Research Council (NRC) on 20 September 1916, during the presidency of Woodrow Wilson (1856-1924). The Academy directed the NRC to provide scientific assistance to the government on issues of national defense and to promote national security in wartime. Michelson and other American scientists including the chemist Arthur A. Noyes (1866-1936) at MIT who was chairman of the NRC, George Ellery Hale (1868-1938) director of the Mount Wilson Observatory, and the physicist Robert Millikan (1868-1935) at Chicago were members. Michelson worked on submarine detection. The NRC became a permanent organization on 11 May 1918.

After the war, Michelson resumed his interferometer measurements of stellar angular diameters. Hale invited Michelson and his assistant Francis Pease (1881-1938) to Mount Wilson. Pease already in September 1919 had used its 100-inch reflecting telescope, then the world's largest, to take the best pictures available of the Moon's surface. In 1920 they attached a 20-foot interferometer to Mount Wilson's 100-inch telescope and made the first measurements of the angular diameters of distant stars. They found that *Betelguese* had a diameter of 240 million miles (0.047 arc seconds), or 250 times the diameter of the Sun's 864,000 mile diameter.

***Other contributors to the study of light and astronomy***

Henry Rowland (1848-1901), a Rensselaer engineering graduate and in 1876 the first professor of physics at Johns Hopkins, made his greatest contributions in experimental physics. Rowland in 1875 demonstrated that a moving electric charge, just like an electric current flowing in a conductor, generates a magnetic field equal to the product of the quantity and velocity of the charge. He also measured accurately magnetic permeability in 1873-74, the ohm the unit of electrical resistance in 1878, and the mechanical equivalent of heat in 1879.

Rowland's most significant contribution was his 1882 invention of a concave diffraction grating that used a diamond screw to rule 14,000-20,000 closely-spaced lines on a concave glass or speculum surface. His gratings, which split white light into its component wavelengths, eliminated the need for prisms and lenses in the photography of spectra and enabled Rowland to map the solar spectrum more thoroughly than astronomers had done previously. His gratings made possible the direct photography and higher resolution of the heavenly spectra.

Hale, an 1890 graduate of MIT, accepted a position at the University of Chicago in 1892. In 1897 he established and directed Chicago's Yerkes Observatory at Williams Bay, Wisconsin, about 80 miles northwest of Chicago. Charles Yerkes (1837-1905), a Chicago street car magnate, provided the $300,000 required for its construction. The observatory had a 40-inch (2 meter) refracting telescope, then the world's largest.

Hale left Yerkes and Chicago for Pasadena, California in 1904. With financial support from the Carnegie Institution in Washington, DC, Hale completed the observatory at Mount Wilson in the four-year period 1904-08. Carnegie in December 1904 awarded Hale $150,000 a year for two years and assured future generous support. Mount Wilson had a 60-inch reflecting telescope. Hale's father earlier in 1896 had acquired the 60-inch mirror disk that Hale used in the telescope. In 1917 Hale constructed an even larger 100-inch reflecting telescope, the world's largest, this time with a $45,000 grant from John D. Hooker (1838-1911), a Los Angeles hardware millionaire, and $500,000 (final cost was $600,000) from the Carnegie Institution.

Hale believed that in most American observatories including Yerkes and Lick, the telescopes and instruments were built to solve problems already existing. Instead, the United States should construct new observatories, such as Mount Wilson and Mount Palomar, to solve the problems now in mind. Hale in 1928 secured financial support mainly from the Rockefeller Foundation to construct an even larger 200-inch reflecting telescope at Mount Palomar, 95 miles southeast of Mount Wilson. Hale wanted the telescope located away from the night lights of nearby cities and other airborne interference. The observatory, completed in 1948, ten years after Hale's death, and now named the Hale Observatory, contains one of the largest telescopes in the world, second only to the 236-inch Russian telescope at the Special Astrophysical Observatory at Zelenchukskaya in the Caucasas Mountains that began operating in 1976.

Hale's major invention was the spectroheliograph in 1889. It enabled him to photograph the light from a single spectral line of the Sun and to show the presence of hydrogen and calcium in the solar atmosphere. Hale, in 1908 was the first to detect the presence of strong magnetic fields in the sunspots. In 1919 working with Walter Adams (1876-1956) Hale discovered that for sunspots the polarity of their magnetic field reversed itself every 23 years. Beginning in 1921 Hale suffered from severe emotional problems and periods of insanity which led to his resignation from Mount Wilson and retirement in 1923. He died in 1938.

## III. DEVELOPMENTS IN CHEMISTRY

### *Molecular composition*

The first stage of a three-stage study of atoms and molecules that led to establishing the composition and structure of molecules began in the 1770s-80s with Lavoisier and his emphasis on the quantitative study of chemical composition. His quantitative approach required an empirical definition of an element and a compound. His definitions enabled chemists to determine whether a substance was simple and therefore an element incapable of breaking down into simpler components, or not simple and capable of breaking down into simple components and therefore a compound. Lavoisier provided the analytical methods with which to determine experimentally a substance's chemical composition and a system of nomenclature based on the substance's composition.

Chemistry entered the second stage of the three-stage study of atoms and molecules in the last half of the nineteenth century. The calculation of accurate atomic weights and chemical composition continued and led to the establishing of the structures of molecules and the classification of the chemical elements in the Periodic Table of 1869. All of the nineteenth-century breakthroughs in chemical composition, molecular structure, and classification occurred in Europe. American chemists made no major contributions to the study of chemical composition, molecular structure, and classification until the twentieth century. New instruments, particularly the spectroscope, that aided research on composition, structure, and classification also originated in Europe.

### *Laws of definite and multiple proportions*

Nineteenth-century chemists continued Lavoisier's quantitative approach. Joseph Proust (1754-1826), a French chemist who taught in Madrid, concluded as a result of his experiments on chemical combination carried out from 1794 to 1806 that elements combine in definite (fixed) proportions by weight, in small, whole-number ratios. This is the law of definite proportions. Proust's studies proved false Berthollet's belief that the composition of a chemical compound could vary over a wide range.

To account for the elements combining in fixed proportions by weight John Dalton (1776-1844), a Quaker school teacher in Manchester, introduced a new theory of atoms, known as the chemical atom. His atom was indivisible and structureless, and each kind of atom, oxygen, hydrogen, carbon, or any other, differed from every other kind only in weight. The atoms had different weights, and this was why the elements combined in definite proportions by weight. Dalton showed how his microscopic atoms accounted for the macroscopic whole-number combining ratios of elements observed in the molecular formulas of compounds. Dalton published his ideas in *A New System of Chemical Philosophy*, 3 volumes (1808, 1810, 1827).

John Dalton
1766-1844

Jöns Jacob Berzelius
1779-1848

**Figure 20**
**John Dalton and Jöns Jacob Berzelius**
Source: F.J. Moore. *A History of Chemistry.* New York: McGraw-Hill Book Company, 1918.

Dalton carried out numerous atomic weight determinations. He extended Proust's work on definite proportions and arrived at the law of multiple proportions in 1808. The law confirmed that when one element (atom) combines with another atom in more than one proportion by weight the two proportions will always be in a simple, whole-number ratio as shown for the oxides of nitrogen and carbon. The first compound was always a 1:1 proportion of weights and atoms, the second compound was in a 1:2 or 2:1 proportion. More complex compounds required further weight analyses.

| OXIDES OF NITROGEN AND CARBON |
|---|
| No, $N_2O$, $NO_2$<br>CO, $CO_2$ |

Dalton believed that the atoms making up water and ammonia combined in the simplest whole-number weight ratios according to his principle of simplicity. Therefore HO was the formula for water $H_2O$, and NH was the formula for ammonia $NH_3$, where Dalton assigned hydrogen an atomic weight of one, or H = 1. He established weights and formulas relative to this assumption starting with 1 to 1, 1 to 2, 2 to 1, etc., as the simplest and most probable combinations of different atoms. As a result of his erroneous starting assumption Dalton incorrectly concluded that oxygen had an atomic weight of O = 8, and nitrogen had an atomic weight of N = 7. Dalton calculated the correct atomic weight for carbon, C = 12.

Dalton introduced complicated chemical symbols and formulas for elements and compounds consisting mainly of a circle or circles with a dot or lines inside the circle. Oxygen was a circle, carbon was a black circle, hydrogen was a circle with a dot inside. The simplified Berzelius system replaced Dalton's picturesque but cumbersome symbols.

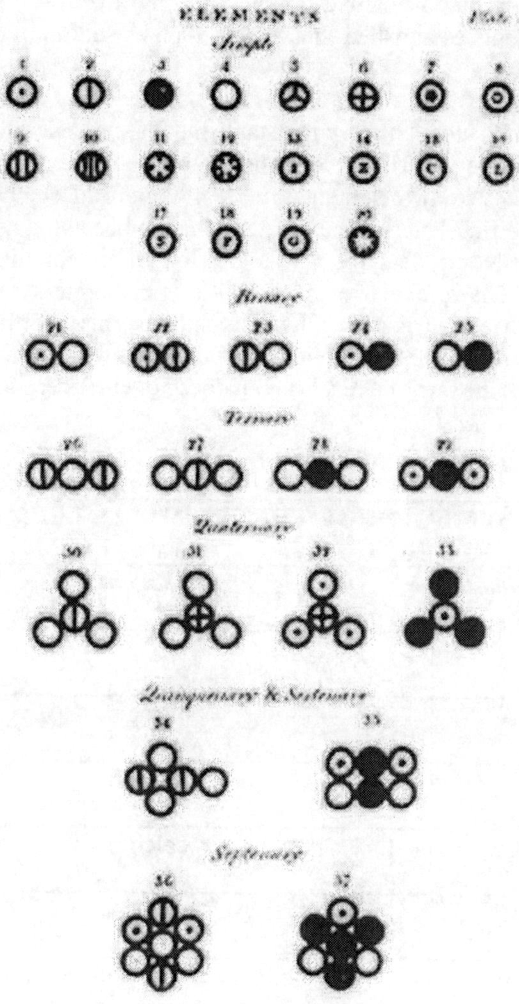

**Figure 21**
**Dalton's Chemical Formulas**
Source: John Dalton. *A New System of Chemical Philosophy.*
London: William Dawson & Sons, 1808.

***Reception of atomism and chemical symbols***

Davy and other positivistic natural philosophers, such as Berthollet, and William H. Wollaston (1766-1828) in London, opposed an atomic theory and the real existence of atoms. They said atoms were useful in understanding chemical combination and reactions, but their reality was beyond demonstration, or metaphysical.

Jons J. Berzelius (1779-1848) at Stockholm accepted Dalton's atomic theory, and in 1811 he proposed an electrical model of an atom to account for the chemists' long time problem of chemical affinity, or what kind of force held atoms in a chemical compound. His atom had one or more electric poles that had a positive end and a negative end that formed electrostatic bonds between atoms. It worked well for inorganic compounds that he could electrolyze, such as NaCl, HCl, KOH, but it could not account for the bonds in organic compounds that did not appear to be electrical, such as a linkage of carbon atoms C-C-C. Nor could it account for the replacement of H, which had a positive charge, by Cl, which had a negative charge, in the reaction between chlorine, $Cl_2$, and methane, $CH_4$, to give methyl chloride, $CH_3Cl$, or in electrolysis experiments involving hydrogen and chlorine. Electrolytic experiments showed that hydrogen had a positive charge and chlorine and the halogens had a negative charge. Electrolytic experiments established that metals such as sodium, potassium, calcium, and magnesium had positive charges.

Berzelius used Dalton's atomic theory to establish the atomic weights of the elements. He compiled his first atomic weight table in 1814 which when published in expanded form in1818 (1819), and 1826 (1828) contained the 49 known elements. In 1814 he introduced chemical symbols for the elements in which the capitalized first letter of the element's name became its symbol. For different elements that began with the same letter, the symbol was the first letter capitalized followed usually by the second letter in lower case. His symbols replaced Dalton's elaborate symbols and formulas. Berzelius used English, Latin, and German names. Chemists still use his shorthand notation for writing chemical formulas and equations. In 1835 Berzelius introduced the word *catalyst*, by which he meant a substance that alters (speeds up) the rate of a reaction but does not enter permanently into the reaction or become a part of the products.

| **SYMBOLS FOR THE CHEMICAL ELEMENTS** | |
|---|---|
| ***Symbol*** | ***Element*** |
| O | oxygen |
| N | nitrogen |
| C | carbon |
| Ca | calcium |
| Cl | chlorine |
| Ag | silver from *argentum* Latin |
| Au | gold from *aurum* Latin |
| K | potassium from *kalium* German |
| W | tungsten from *wolfram* German |

***Atomic weight calculations and atomic weight standards***

A main objective of nineteenth-century chemists was to establish the atomic weights of the elements. Berzelius, Joseph Gay-Lussac, Pierre Dulong and Alexis Petit, and Eilhardt Mitscherlich contributed to atomic weight calculations.

Berzelius calculated atomic weights and established chemical formulas using relations developed by his contemporaries, such as Joseph Gay-Lussac (1778-1850) in Paris, but he did not accept the 1811 hypothesis of Amedeo Avogadro (1776-1856) in Turin, Italy. Gay-Lussac in his 1808 law of combining volumes had shown that gases combine in small whole-number ratios. The ratio of hydrogen and oxygen in water is 2:1 the ratio of hydrogen and chlorine is 1:1 in hydrogen chloride, and the ratio of nitrogen and hydrogen is 1:3 in ammonia. These ratios represented the number of atoms of each element present in a molecule of each gas and provided a way to calculate the atomic weight of an atom and the formula of a compound.

Avogadro, as a result of his studies on gases, in 1811 proposed that equal volumes of gases at the same temperature and pressure contained the same number of molecules. His hypothesis offered a way to establish the molecular formula of a gas, thereby clarifying whether the gaseous elements hydrogen, nitrogen, oxygen, and the halogens (fluorine, chlorine, bromine, iodine) occurred naturally as monatomic or diatomic molecules, H or $H_2$, Cl or $Cl_2$, O or $O_2$, and to calculate the atomic weight of each element. It explained how one volume of hydrogen gas, $H_2$, combined with one volume of chlorine gas $Cl_2$, to give two volumes of HCl, and the whole number combinations of other reactions involving gaseous molecules, such as $O_2$ and $N_2$ and $H_2$ and $O_2$.

Avogadro's hypothesis later enabled chemists to define the atomic weight or volume of a chemical element. One mole of an atom or molecule contains $6.023 \times 10^{23}$ atoms or molecules (Avogadro's number), where 1 mole = 1 gram-molecular weight ($O_2$ = 32, C = 12) or 1 molecular volume (22.4 liters) of a substance.

Pierre Dulong (1785-1838) and Alexis Petit (1791-1820) in Alfort and later Paris established a relation in 1819 showing that for an element its specific heat multiplied by its atomic weight was a constant, or specific heat (per gram) x atomic weight = 6.2, which is the heat capacity of the element.

| DULONG AND PETIT'S LAW |
| --- |
| Specific heat x atomic weight of an element = 6.2 |

Eilhardt Mitscherlich (1794-1863) at Stockholm and later Berlin in his law of isomorphism in 1819 showed that compounds having the same crystalline structures such as sodium chloride and potassium chloride, NaCl and KCl, have the same atomic composition. His law enabled chemists to calculate a crystal's molecular or atomic weight.

By the 1850s Dalton's atomic theory had gained considerable acceptance, but chemists had no universal formulas for indicating a compound's chemical composition. Nor had they agreed on an element's atomic weight and therefore on the composition of compounds. The same compound could have different formulas.

To end the confusion on atomic weights, August Kékulé (1829-96) at Ghent, in 1860 called an international meeting of chemists at Karlsruhe, Baden, Germany. The 140 attendees represented a who's who of chemistry. At the conference Stanislao Cannizzaro (1826-1910) then at Genoa announced his rediscovery in 1858 of Avogadro's hypothesis. Chemists largely had ignored Avogadro's hypothesis mainly because he used the word molecule when referring to both an atom and a molecule, using them interchangeably, and he provided little experimental evidence to support his hypothesis. But as Cannizzaro recognized, and other chemists soon acknowledged, knowing the atomic weights of diatomic gaseous elements such as hydrogen, chlorine, and oxygen, and using Gay-Lussac's law of combining gaseous volumes they could determine the atomic weights of other elements these gases combined with, most of which were solids, and develop a complete and consistent set of atomic weights. Chemists still required an element to use as a standard with which to calculate, either directly or indirectly, the atomic weights of the elements that combined with their selected standard.

Jean Servais Stas (1813-91) at Brussels, who from 1860 to 1881 made excellent atomic weight determinations of the elements including carbon and oxygen, by 1865 had proposed oxygen O = 16 as the standard for atomic weight determinations. Because no single standard existed prior to Stas's proposal, some chemists adopted oxygen, mainly because it combined with many different elements, while others preferred hydrogen. Dalton assigned hydrogen a weight of H = 1, William Prout (1785-1850) in London assigned oxygen a weight of O = 10. Berzelius assigned oxygen a weight of

O = 100 as his standard atomic weight, although in 1826 he adopted O = 16 as his atomic weight. Francis Venable (1856-1934) at the University of North Carolina and its president from 1900 to 1914, wrote in *A Short History of Chemistry* (1894) that most chemists and chemical societies including Edward Morley, Josiah Cooke, and Theodore Richards in the United States adopted O = 16 as the atomic weight standard beginning in 1893-94.

Oxygen, O = 16, remained the standard until 1929 when William Giauque (1895-1982) and Herrick L. Johnston (1898-1965) at Berkeley from optical spectra discovered that oxygen gas consisted of three isotopes, O = 16 the most abundant isotope, O = 17, and O = 18. Because of this discovery, physicists adopted a new atomic weight standard in which the O = 16 isotope became the standard.

Chemists, as they had traditionally done, continued to use O = 16 as the standard even though they now knew that the true atomic weight of oxygen was not exactly 16 but the average of the three isotopes calculated according to the percentage of each isotope present and did not differ much from 16. Nevertheless a slight difference existed between the two standards and required a conversion between the physicists' and the chemists' standard which persisted until 1961. In that year the International Union of Pure and Applied Chemistry (IUPAC) and the International Commission on Atomic Weights adopted C = 12 as the new standard. The IUPAC chose C = 12 partly because of C = 12's use as a secondary standard in mass spectroscopy and because its adoption changed the atomic weight scale only by 42 parts per million.

***Molecular structure and chemical affinity***

The second stage of atomic and molecular study spanned the second half of the nineteenth century. It went beyond knowing only a compound's chemical composition to establishing the arrangement or location of each atom in a molecule, that is, to determining the molecule's structure. The most significant structural determination came in 1858 when first Kekulé and shortly after Archibald S. Couper (1831-92) then at Paris independently proposed a self-linking carbon atom (-C-C-C-). In the structural models that they proposed, based on the well-established analyses of carbon compounds, a carbon atom linked itself to other carbon atoms in a straight chain or a branched chain and to one or more hydrogen atoms. In these compounds, called hydrocarbon compounds, carbon had a valence of four, meaning that it formed four bonds with other atoms. Kekulé's and Couper's structures were planar not three-dimensional.

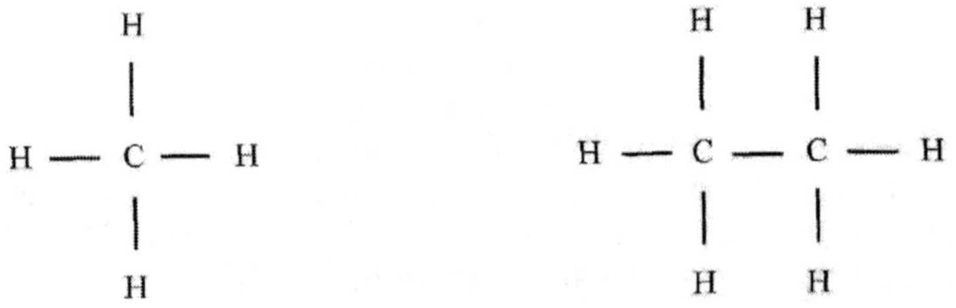


Methane Ethane

**Figure 22**
**Hydrocarbon Structural Formulas**
Courtesy of the author

The structure of benzene, known to have the molecular formula $C_6 H_6$, provided a serious challenge to structural chemistry because no straight chain or branch chain arrangement of self-linked carbon atoms, each forming four bonds, double or single, could account for benzene's molecular structure. Kekulé provided a satisfactory solution to benzene's structure in 1865. He proposed a hexagonal ring consisting of 6 carbon atoms with a hydrogen atom joined to each carbon atom. Alternating single and double bonds linked the carbon atoms and thereby maintained a valence of 4 for each carbon atom. Kekulé said the idea of a hexagonal ring structure came to him in a dream after considerable thinking about a possible structure. In his dream he saw six snakes, each of which bit another's tail thereby linking together to form a hexagon. Kekulé's structure never satisfied all chemists, and until the introduction of the molecular orbital theory in 1926-28 benzene's structure remained only partially resolved.

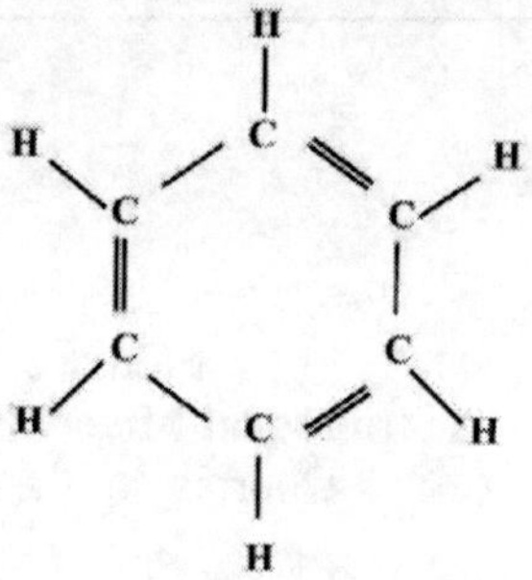

**Figure 23**
**Structure of Benzene, $C_6H_6$**
Courtesy of the author

Despite the disagreement on benzene's double bond arrangement, Kekulé's proposed benzene molecular structure marked the beginning of the coal tar industry and led to the structural determination and synthesis of dyes, such as mauve and indigo, and pharmaceuticals, such as aspirin. Previously only dyes and drugs obtained from natural sources were available. German organic chemists carried out numerous structural determinations that enabled them to synthesize these and many other valuable organic compounds, and to make Germany the leading global producer of organic chemicals.

The idea of a spatial orientation of the carbon atom's four bonds, the bonds directed in space rather than arranged in a plane (three dimensional rather than flat), appeared nearly a decade later. In September 1874 Jacobus van't Hoff (1852-1911) at Paris, and in November 1874 Joseph Le Bel (1847-1930) at Paris, independently observed that a carbon atom with four different atoms or atomic groups attached to it (an asymmetrical carbon atom, CHClBrF) had two different spatial arrangements and that the two arrangements were mirror images of each other, nonsuperimposable like a person's right and left hands. Le Bel did not propose a specific geometric arrangement, whereas van't Hoff proposed a tetrahedral (spatial) carbon atom in which the carbon atom's four bonds pointed to the corners of a tetrahedron and were at an angle of 109° 28′, as in methane $CH_4$. Van't Hoff showed that placing the carbon atom at the center of the tetrahedron and the four attached atoms at the corners made evident the two geometric arrangements.

The presence of an asymmetrical carbon atom (or any asymmetrical atom) in a compound or a solution of the compound accounted for the experimentally observed optical activity of certain organic compounds. That is, upon passing a ray of polarized light through a crystal of an optically active compound, the compound rotated the plane of polarized light to the right (dextro) or left (laevo) of a designated reference plane. Knowing the amount of rotation a compound produced, provided an analytical technique for identifying its presence in a substance or solution of unknown composition. Van't Hoff in 1901 received the first Nobel Prize in chemistry for his work on stereochemistry.

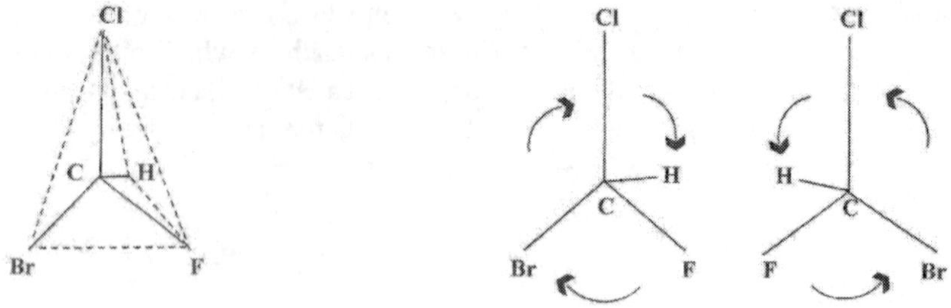


**Figure 24**
**Tetrahedral Molecular Structure**
Courtesy of the author

Nineteenth-century chemists did not know what held atoms in a molecule, other than to rely on the old standby of chemical affinity. J. J. Thomson (1856-1940) at the Cavendish Laboratory at Cambridge discovered the electron in 1897, and in his 1897 and 1899 publications he briefly discussed the electron's role in accounting for chemical affinity. Thomson went into greater detail in his 1904 publications, but to him and other physicists and chemists, chemical affinity was an electrostatic attraction that existed between oppositely-charged, positive and negative, atoms in a molecule. It failed, however, to account for the affinity or bonding of atoms in numerous compounds, especially the hydrocarbon compounds of organic chemistry that clearly contained no charged atoms. The solution to the non-electrostatic bonding of organic compounds and the transformation of the mysterious *affinity*, usually represented by a dash or dashes, into an electron theory of valence, came in 1916. Gilbert N. Lewis (1875-1946) at Berkeley, relying largely on his chemical intuition showed qualitatively how a pair of electrons accounted for the chemical bond in a compound's molecular and structural formulas.

Lewis's familiar electron dot formulas, where each pair of electrons represented the chemical (electron) bond in a molecule, provided a satisfying description of the bonds in a molecule, but they lacked a theoretical foundation. His work completed the third stage that Thomson had begun and laid the foundation for the fourth stage of the chemists' study of molecular structure. Beginning in the late 1920s and culminating in 1939 with the publication of *The Nature of the Chemical Bond*, Linus Pauling at Cal Tech showed the equivalency of Lewis's electron dot formulas with quantum mechanics and placed the chemists' molecular and structural formulas on a solid theoretical foundation.

***Classification of elements, the Periodic Table***

An early attempt to classify the elements came in 1815 when Prout proposed that the atomic weights of the elements were whole number multiples of the weight of hydrogen the lightest known element. Atomic weight determinations by Berzelius and other chemists showed the atomic weights of the elements were not exact multiples of 1, the implied atomic weight of hydrogen.

Another classification scheme appeared in 1829. Johann Döbereiner (1780-1849) at Jena proposed the law of triads. Döbereiner noted the similar chemical behavior of certain groups of three elements such as chlorine, bromine, iodine (Cl, Br, I); calcium, strontium, barium (Ca, Sr, Ba); copper, silver, gold (Cu, Ag, Au); and silicon, selenium, tellurium (Si, Se, Te). His classification helped but applied only to a small number of elements.

In the United States, Josiah Cooke (1827-94) a Harvard graduate who learned most of his chemistry in Europe, established Harvard's chemistry laboratory in 1850 with his own funds. Beginning in 1854 he carried out atomic weight measurements, and in an 1854 paper he introduced a periodic table based on his experimentally determined atomic weights. Cooke's belief that he could predict the properties of any undiscovered element and its compounds in any given series of elements foreshadowed the periodic table that the Russian chemist Dmitri Mendeleev published in 1869. Cooke's student, Theodore Richards (1868-1928) at Harvard, continued Cooke's atomic weight determinations.

John Newlands (1837-98) in London believed he had a solution for the classification problem and in 1865 he proposed the law of octaves. Newlands recognized a repeating pattern of chemical properties after every eighth element, as existed for chlorine, bromine, and iodine, when he listed the elements in order of increasing atomic weight. He published a table containing 60 elements in 1865. Newlands' octaves provided an insightful understanding of the elements' periodic behavior, but many chemists treated it with ridicule because of its relation to musical octaves. Yet, only four years later Dmitri Mendeleev's classification which contained some similarities to Newland's arrangement solved the classification problem.

Dmitri Mendeleev (1834-1907) in St. Petersburg (Leningrad), one of the Karlsruhe meeting's attendees, returned home impressed with Cannizzaro's emphasis on establishing molecular formulas and atomic weights. He began his investigation by arranging the 63 known elements in order of their atomic weights, as Newlands had done, and in doing so he discovered several recurring patterns. Mendeleev found that with few exceptions the valences of the elements, which for each element represented the number of atoms an element combined with, rose and fell, 1, 2, 3, 4 . . . then 3, 2, 1. When he placed the elements with the same valence in a vertical column of increasing atomic weight, the similarities in their chemical properties became apparent. The elements of atomic weights 19, 35, 80 and 127, fluorine, chlorine, bromine, and iodine (F, Cl, Br, I) had a valence of -1 and were corrosive. The metals lithium, sodium, and potassium (Li, Na, K, Rb) formed strongly alkaline (basic) solutions and had a valence of +1. The metals magnesium, calcium, and barium (Mg, Ca, Ba) formed weaker alkaline solutions and had a valence of +2. Mendeleev published his arrangement which he called the Periodic Table in 1869, and gave the first statement of the Periodic Law. According to Mendeleev's Periodic Law, the chemical properties of the elements are periodic functions of their atomic weights.

Mendeleev published a second table in 1871, a horizontal table with blank spaces in some of the columns and rows, in order to keep similar elements, such as fluorine, chlorine, and bromine, in the same group. From observing the chemical properties of elements above and below the gaps and left and right of the gaps he predicted the properties of the missing elements. He did not accept an electrochemical theory of the atom or believe that the electron, following its discovery in 1897, existed. He accepted ether as a medium.

| | | | | | |
|---|---|---|---|---|---|
| | | | Ti = 50 | Zr = 90 | ? = 180 |
| | | | V = 51 | Nb = 94 | Ta = 182 |
| | | | Cr = 52 | Mo = 96 | W = 186 |
| | | | Mn = 55 | Rh = 104,4 | Pt = 197,4 |
| | | | Fe = 56 | Ru = 104,4 | Ir = 198 |
| | | Ni = | Co = 59 | Pd = 106,6 | Os = 199 |
| H = 1 | | | Cu = 63,4 | Ag = 108 | Hg = 200 |
| | Be = 9,4 | Mg = 24 | Zn = 65,2 | Cd = 112 | |
| | B = 11 | Al = 27,4 | ? = 68 | Ur = 116 | Au = 197 ? |
| | C = 12 | Si = 28 | ? = 70 | Sn = 118 | |
| | N = 14 | P = 31 | As = 75 | Sb = 122 | Bi = 210 |
| | O = 16 | S = 32 | Se = 79,4 | Te = 128 ? | |
| | F = 19 | Cl = 35,5 | Br = 80 | J = 127 | |
| Li = 7 | Na = 23 | K = 39 | Rb = 85,4 | Cs = 133 | Tl = 204 |
| | | Ca = 40 | Sr = 87,6 | Ba = 137 | Pb = 207 |
| | | ? = 45 | Ce = 92 | | |
| | | ? Er = 56 | La = 94 | | |
| | | ? Yt = 60 | Di = 95 | | |
| | | ? In = 75,6 | Th = 118 ? | | |

| | I. Gruppe | II. Gruppe | III. Gruppe | IV. Gruppe | V. Gruppe | VI. Gruppe | VII. Gruppe | VIII. Gruppe zur I. Gruppe übergehend |
|---|---|---|---|---|---|---|---|---|
| Typische Elemente | H = 1<br>Li = 7 | Be = 9.4 | B = 11 | C = 12 | N = 14 | O = 16 | F = 19 | |
| Erste Periode { Reihe 1<br>» 2 | Na = 23<br>K = 39 | Mg = 24<br>Ca = 40 | Al = 27.3<br>— = 44 | Si = 28<br>Ti = 50? | P = 31<br>V = 51 | S = 32<br>Cr = 52 | Cl = 35.5<br>Mn = 55 | Fe = 56, Co = 59<br>Ni = 59, Cu = 63 |
| Zweite Periode { Reihe 3<br>» 4 | (Cu = 63)<br>Rb = 85 | Zn = 65<br>Sr = 87 | — = 68<br>(? Yt = 88?) | — = 72<br>Zr = 90 | As = 75<br>Nb = 94 | Se = 78<br>Mo = 96 | Br = 80<br>— = 100 | Ru = 104, Rh = 104<br>Pd = 104, Ag = 108 |
| Dritte Periode { Reihe 5<br>» 6 | (Ag = 108)<br>Cs = 133 | Cd = 112<br>Ba = 137 | In = 113<br>— = 137 | Sn = 118<br>Ce = 138? | Sb = 122<br>— | Te = 128?<br>— | J = 127<br>— | — — |
| Vierte Periode { Reihe 7<br>» 8 | —<br>— | —<br>— | —<br>— | —<br>— | —<br>Ta = 182 | —<br>W = 184 | —<br>— | Os = 199?, Ir = 198?<br>Pt = 197, Au 197 |
| Fünfte Periode { Reihe 9<br>» 10 | (Au = 197)<br>— | Hg = 200<br>— | Tl = 204<br>— | Pb = 207<br>Th = 232 | Bi = 208<br>— | —<br>Ur = 240 | —<br>— | |
| Grenzform der O-Verbindung | $R_2O$ | $R_2O_2$ oder RO | $R_2O_3$ | $R_2O_4$ od. $RO_2$ | $R_2O_5$ | $R_2O_6$ od. $RO_3$ | $R_2O_7$ | $R_2O_8$ oder $RO_4$ |
| Grenzform der H-Verbindung | | | ($RH_3$?) | $RH_4$ | $RH_3$ | $RH_2$ | RH | |

**Figure 25**

**Mendeleev's First and Second Periodic Tables**

Source: F.J. Moore. *A History of Chemistry.* New York: McGraw-Hill Book Company, 1918.

In a case of near simultaneous discovery, Lothar Meyer (1820-95) at Karlsruhe compared the atomic weights and recurring properties of the elements and recognized a periodicity. He published a periodic table in 1870, one year after publication of Mendeleev's first table.

Paul Emile Lecoq de Boisbaudran in Paris filled in one of Mendeleev's gaps in 1875 when he discovered, using spectroscopy, the element gallium and isolated it by electrolysis in 1876. It had the properties Mendeleev predicted. In 1879 Lars Fredrik Nilson at Uppsala isolated scandium by the analysis of rare-earth minerals, and in 1886 Clemens Winkler at Freiburg isolated germanium from a silver-sulfur-mercury ore. Each of the new elements had the properties Mendeleev predicted.

| DISCOVERY OF NEW ELEMENTS PREDICTED BY MENDELEEV'S PERIODIC TABLE | |
|---|---|
| ***Discoverer*** | ***Element*** |
| Paul Lecog Boisbaudran (1838-1912) at Paris | Discovered gallium, Z = 31, in 1875 |
| Lars Nilson (1840-99) at Uppsala | Discovered scandium, Z = 21, in 1879 |
| Clemens Winkler (1838-1904) at Freiburg | Discovered germanium, Z = 32, in 1886 |
| William Ramsay (1852-1916) and John William Strutt, Lord Rayleigh (1842-1919) at London | Discovered three of the rare gases, neon, Z = 10; argon, Z = 18; and krypton, Z = 36, in 1894 |

Mendeleev recognized that his law had several exceptions such as iodine (I) and tellurium (Te), cobalt (Co) and nickel (Ni), in which he had to place the heavier of the two adjacent elements before the lighter element in order to keep the elements with similar properties in the same group. Later, following the discovery of the rare gases in the 1890s, potassium (K) and argon (Ar) appeared in inverted order.

More accurate atomic weight determinations did not resolve the atomic weight inconsistencies. Elimination of the Periodic Table's atomic weight inconsistencies came in 1914, with the introduction of a new atomic property. As a result of his 1913-14 x-ray studies, the physicist Henry Gwyn-Jeffreys Moseley, working with Rutherford at Manchester, concluded that an element's identifying property was its atomic number Z, or number of protons, not its atomic weight as Mendeleev had proposed. The number of protons (positive charge) provided the correct interpretation of periodicity. Moseley restated Mendeleev's Periodic Law to indicate that the chemical properties of the elements are periodic functions of their atomic numbers. Niels Bohr at Copenhagen at about this same time realized that the configuration of the atom's electron shells gave atoms their chemical properties. Because the number of electrons always equals the number of protons in a neutral atom, Bohr's statement about electron configurations represented an alternative explanation of periodicity.

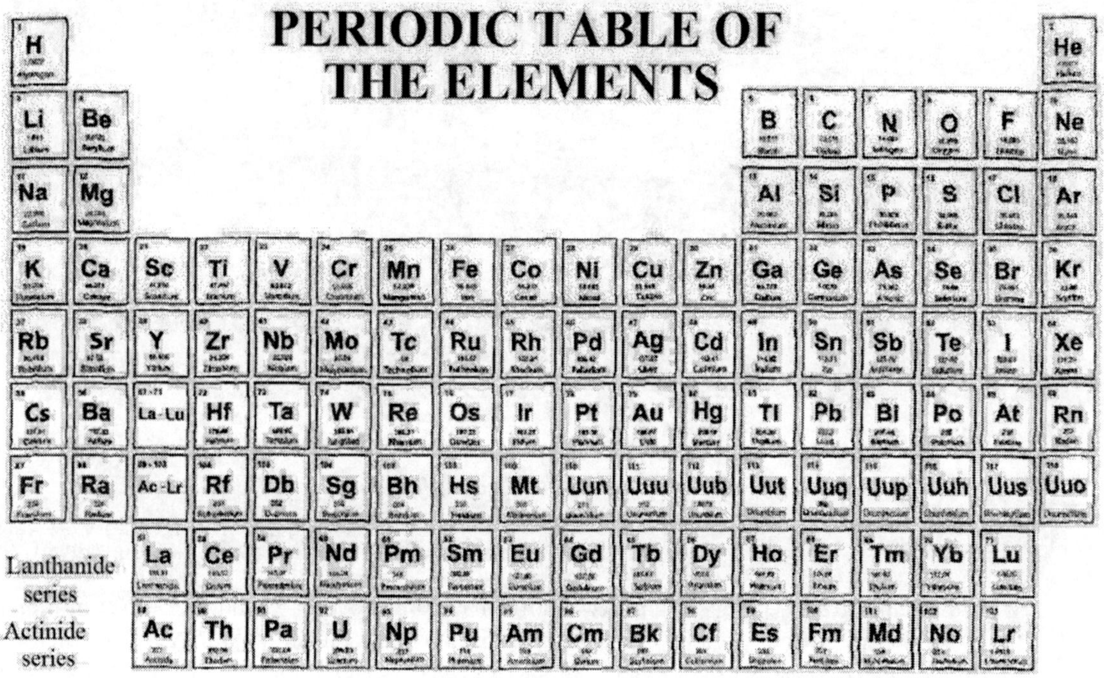


**Figure 26**
**Modern Periodic Table of the Elements**
charobnica/Shutterstock.com

***Chemical instrumentation***

As occurred in the research on molecular structure, most of the new instruments used in volumetric, gravimetric (weight), spectrographic, optical, and electrochemical investigations were European inventions. Americans once again had to wait until the twentieth century. A little known but noteworthy American invention was a special type of crucible to use with precipitates that became unstable when burning off the filter paper used to separate the precipitate. Frank Gouch (1825-1929), a student of Oliver Wolcott Gibbs and Josiah Cooke at Harvard, invented the Gouch crucible in 1878. Gouch was a government analyst before going to Yale in 1885.

Two much better known inventions of this time were the Bunsen burner and the spectroscope. Robert Bunsen at Heidelberg with his technician Peter Desaga (b. 1812) invented the Bunsen burner, a metal tube type heating device, in 1855. Their simple but very practical apparatus burned natural gas (methane, $CH_4$) or some other hydrocarbon gas and found widespread use in chemical laboratories.

Four years later, in 1859, Bunsen and Gustav Kirchhoff (1824-80) at Heidelberg made another important contribution to experimental science when they invented the prism spectroscope to analyze the light that a chemical element emitted when heated. Their spectroscope produced a pattern of bright lines called a bright line spectrum when the light passed through the prism and enabled them to identify a new element. The pattern of lines (spectrum) differed for each element, and a new or previously unidentified pattern of lines suggested the presence of a new element.

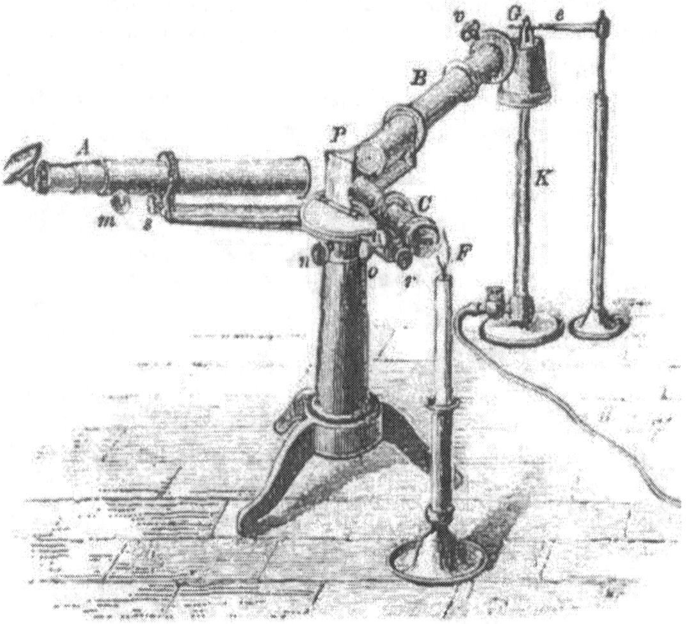


**Figure 27**
**Early Spectroscope**
Source: J. Dorman Steele. *Popular Physics.* New York: American Book Company, 1896.

With their spectroscope Bunsen and Kirchhoff in 1860 discovered cesium (Z = 53) in mineral water from Dürkheim, Germany. Other spectroscopic discoveries included helium (Z = 2) which the French astronomer Pierre Janssen (1824-1907) in Paris in 1868 discovered in the solar chromosphere during the solar eclipse that he observed in India, Boisbaudran's discovery of gallium (Z = 31), and argon (Z = 18) which William Ramsay (1852-1916) and John William Strutt, Lord Rayleigh (1842-1919), in London discovered independently in 1894. Ramsay received the 1904 Nobel Prize in chemistry and Rayleigh the 1904 prize in physics for their discovery and research on the noble gases (neon, argon, krypton).

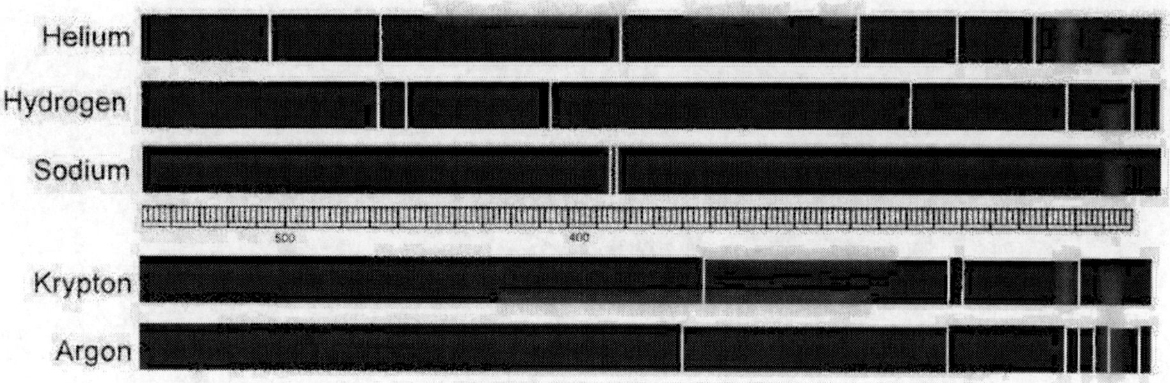


**Figure 28**
**Spectra of Elements**
Courtesy of NASA

## IV. ELECTRICAL THEORY AND APPLICATIONS

### *Electricity and magnetism*

Nineteenth-century scientists knew at this time that lightning was electrical and that lightning affected a compass needle, causing it to deflect. Nevertheless, the contributions of Hans Christian Oersted, André Ampere, Michael Faraday, James Clerk Maxwell, and others were necessary to show the relation between the two phenomena of electricity and magnetism. That lightning produced light also indicated a relation between light and electricity and therefore with magnetism.

### *Oersted's and Ampere's experiments*

Hans Christian Oersted (1777-1851) at Copenhagen belonged to the school of nature philosophers. Nature philosophers believed in the unity of the natural forces, the electric force, magnetic force, and any other force. All the natural forces were related to one another. Oersted's classroom demonstration in1820 showed the relation between the electric force and the magnetic force. He discovered that an electric current flowing in a wire conductor deflected a magnet (compass) that he held below the conductor. The electric force acted in a circular path, it produced circular motion. When Oersted reversed the current's direction, the direction of deflection also reversed itself. His experiment showed that a circular magnetic field surrounds a conductor and that it is perpendicular to the orientation of a conductor.

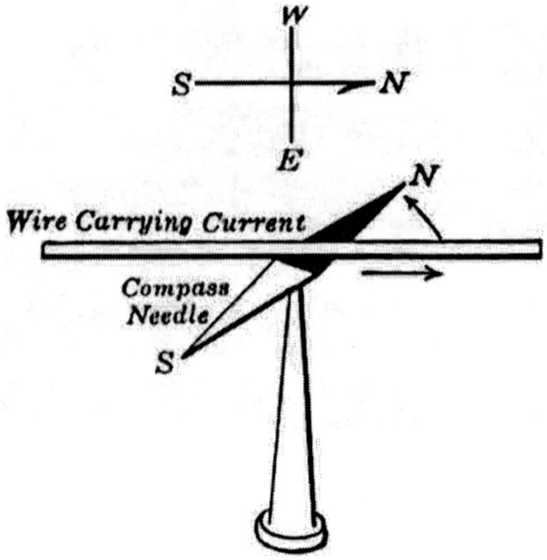


**Figure 29**
**Oersted's 1820 Experiment**
Source: Newton Henry Black and Harvey Nathaniel Davis. *New Practical Physics.*
New York: Macmillan Company, 1913.

Oersted's announcement of his discovery set off a flurry of activity. André Marie Ampere (1775-1836) in Paris began his experiments on electricity and magnetism immediately after hearing of Oersted's discovery. His right hand rule of 1820 showed the relation between the direction of the current, which for Ampere flowed from positive to negative, and the direction of deflection of a magnet's north pole. According to Ampere's rule if you grasped the wire in the palm, with the thumb pointing in the direction of the current, the fingers encircling the wire pointed in the direction of the north pole. Ampere had described the direction and circular shape of the magnetic field, and he determined the direction in which the current deflected a magnet's north pole.

You can demonstrate the circular force and the direction of the compass needle's deflection by passing a stiff wire conductor vertically through a small hole made in the center of a 1 ft by 1 ft horizontally held cardboard sheet, connecting the two ends of the conductor to a battery, and sprinkling iron filings on the cardboard. The filings will arrange themselves in a circular pattern. Placing several small compasses on the cardboard and around the conductor will show the direction in which the north pole of each compass points for a known direction of the current flowing in the conductor. Therefore knowing the current's direction enables you to predict the direction of the north pole's deflection and similarly to predict the direction of the current upon observing the direction of the north pole's deflection. This is the right hand rule: grasp the conductor in the palm of your right hand with your fingers encircling the conductor and your thumb pointing in the direction of the current. Your fingers will point in the direction in which the north pole of the compass is pointing.

In another experiment carried out in 1820 Ampere showed that two parallel wire conductors in each of which an electric current is flowing attract if the electric current flows in same direction and that they repel if the current is flowing in opposite directions (his right hand rule). One wire, if free to move rotates in a semicircle around the other. He constructed a solenoid (a coil of wire) in 1820 and with it showed that current flowing in a coil causes the coil to function as a magnet and to exhibit polarity. The magnetic effect increased with the number of coils.

Ampere in 1823 proposed an electrical theory of magnetism. He suggested that electric molecules circling in a substance produce a magnetic effect. He anticipated the electrical atom.

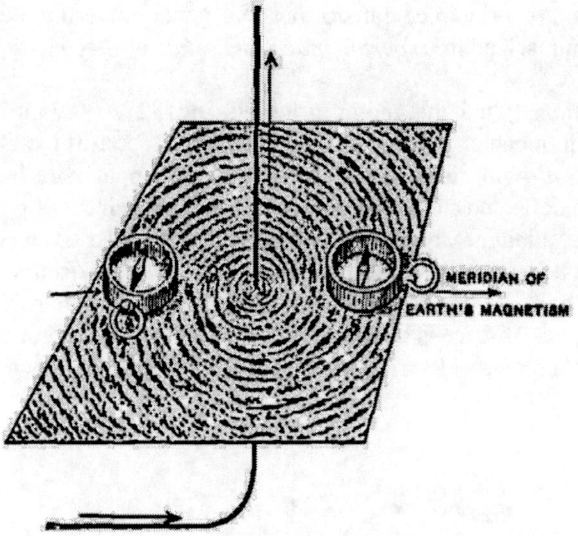


**Figure 30**
**The Magnetic Field Encircling a Current in a Straight Conductor**
Source: Newton Henry Black and Harvey Nathaniel Davis. *New Practical Physics.*
New York: Macmillan Company, 1913.

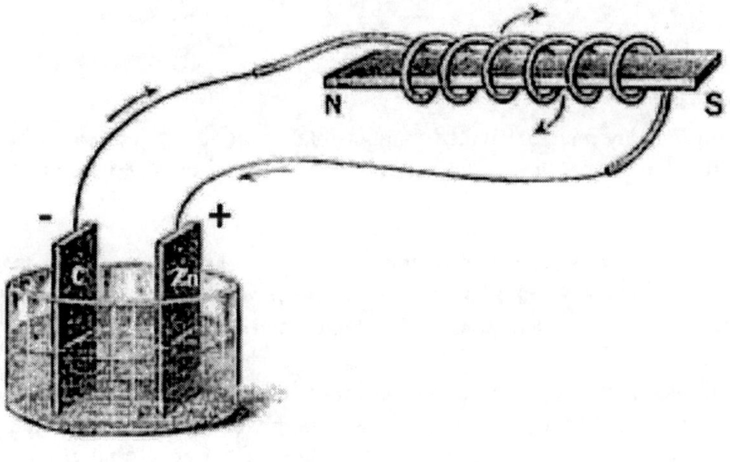


**Figure 31**
**An Electromagnet**
Source: Edwin J. Houston. *The Elements of Natural Philosophy.*
Philadelphia: Eldridge & Brother, 1900.

***Faraday's breakthrough: motor principle, electromagnetic induction, lines of force, generator, transformer***

Considerable electromagnetic research occurred in England, the result of Michael Faraday's experimental investigations. Faraday (1791-1867) at London's Royal Institution expanded the electromagnetic investigations of Oersted and Ampere. Faraday was an excellent experimental physicist, but not a great mathematician. His experiments established several important relations, specifically the motor principle, transformer and generator principle, and electromagnetic induction, which he later summarized in *Experimental Researches in Electricity*, 3 volumes, (1839, 1844, 1855).

Faraday first demonstrated the motor principle in 1821, showing that an electric motor converts electrical energy to mechanical motion. His experiments confirmed Oersted's 1820 demonstration that an electric current flowing in a conductor, such as a copper wire in an electrical circuit, produced a circular force that deflected a compass needle placed beneath the wire. Faraday went beyond Oersted in demonstrating that a magnet, if free to move, rotated around a fixed conductor, and a wire conductor if free to move rotated around a fixed magnet. The rotation occurs because the arrangement of the electromagnetic lines of force around a conductor is circular, the force acting to arrange the electromagnetic lines of force is circular not linear. This is the motor principle, the basic principle of an electric motor that converts electrical energy to mechanical energy.

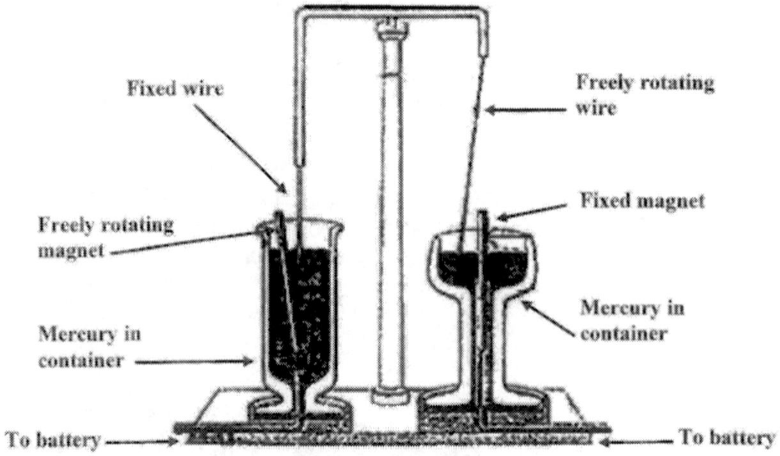


Left Faraday made a pivoted magnet rotate around a fixed wire when a current flowed through the apparatus. Right, a freely suspended wire rotated around a fixed magnet.

**Figure 32**
**Faraday's Motor Principle Electromagnetic Rotation**
Source: Michael Faraday. *Experimental Researches in Electricity.* Vol. 2. London: Richard and John Edward Taylor, 1844.

Faraday believed that if an electric current produced a magnetic force, then the reverse should occur, a magnetic force should produce an electric current. He brought a magnet near a coil of wire, each end of which he connected to one of the two terminals of a galvanometer, but nothing happened, to his disappointment. Ten years later, in 1831, upon returning to his electromagnetic experiments, he discovered that the *motion* of the magnet, moving a magnet back and forth inside a coil of wire whose two ends he connected to the terminals of a galvanometer, caused the galvanometer's needle to move back and forth, indicating the flow of an electric current in the coil, first in one direction and then in the opposite direction. Faraday had induced an alternating current in the coil, just as an electrically charged body induces a charge on a second body. To account for the induced alternating current, Faraday proposed that circular magnetic lines of force (a magnetic field) surrounded the magnet and that the coil of wire passed through (cut) the lines of force first in one direction and then in the opposite direction as he moved the magnet back and forth through the coil. The back and forth motion of the magnet produced, or induced, a current that moved back and forth in the coil, that is, an alternating current (ac) in the coil. This is the electrical inductive effect, or the principle of electromagnetic induction, and the first demonstration of the principle of the electrical generator which converts mechanical energy to electrical energy and of the transformer discussed below.

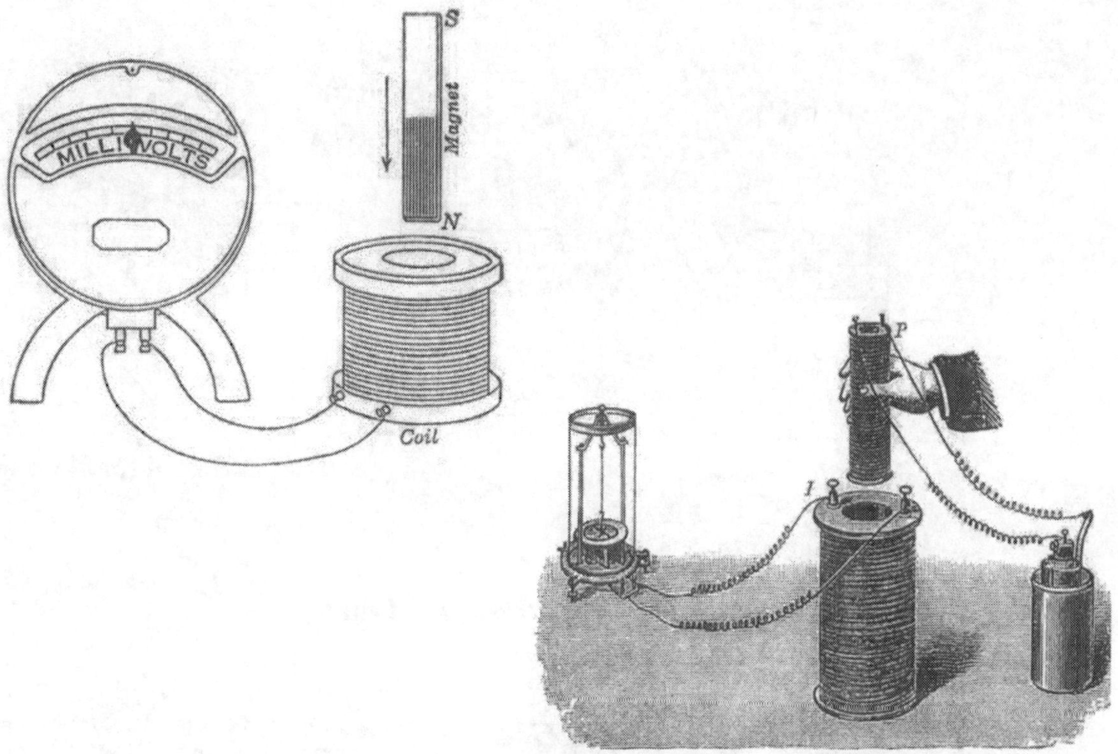


**Figure 33**
**Electromagnetic Induction**
Source: Newton Henry Black and Harvey Nathaniel Davis. *New Practical Physics.* New York: Macmillan Company, 1913 (left).
Source: J. Dorman Steele. *Popular Physics.* New York: American Book Company, 1896 (right).

That same year, 1831, Faraday demonstrated that the completing and breaking of an electrical circuit, or the starting and stopping of an electrical current flowing in one wire or coil of wire wound around an iron core, induced an alternating current (ac) in a second wire or coil of wire wound around the same iron core but in a separate location. Faraday recognized that just like the back and forth motion of a magnet in a coiled conductor (mechanical motion), the motion of an electric current in a wire or coil of wire induced the current in a second wire or coil of wire. Faraday again had demonstrated the principle of electromagnetic induction. The switch's closing and opening, which started and then stopped the flow of electricity, produced what Faraday called electromagnetic lines of force (an electromagnetic field) in the space surrounding the coil that alternately spread through the space upon completing the circuit and collapsed upon breaking the circuit. The lines or field moved in one direction (outward) as they expanded over the second coil and then in the opposite direction (inward) over the second coil as they collapsed, and thereby induced an alternating current in the second coil. The number of times the current reverses its direction in one second is the current's frequency, usually 60 cycles or 120 cycles. Faraday's electrical circuit consisting of coils and core enabled him to demonstrate the principle of the transformer.

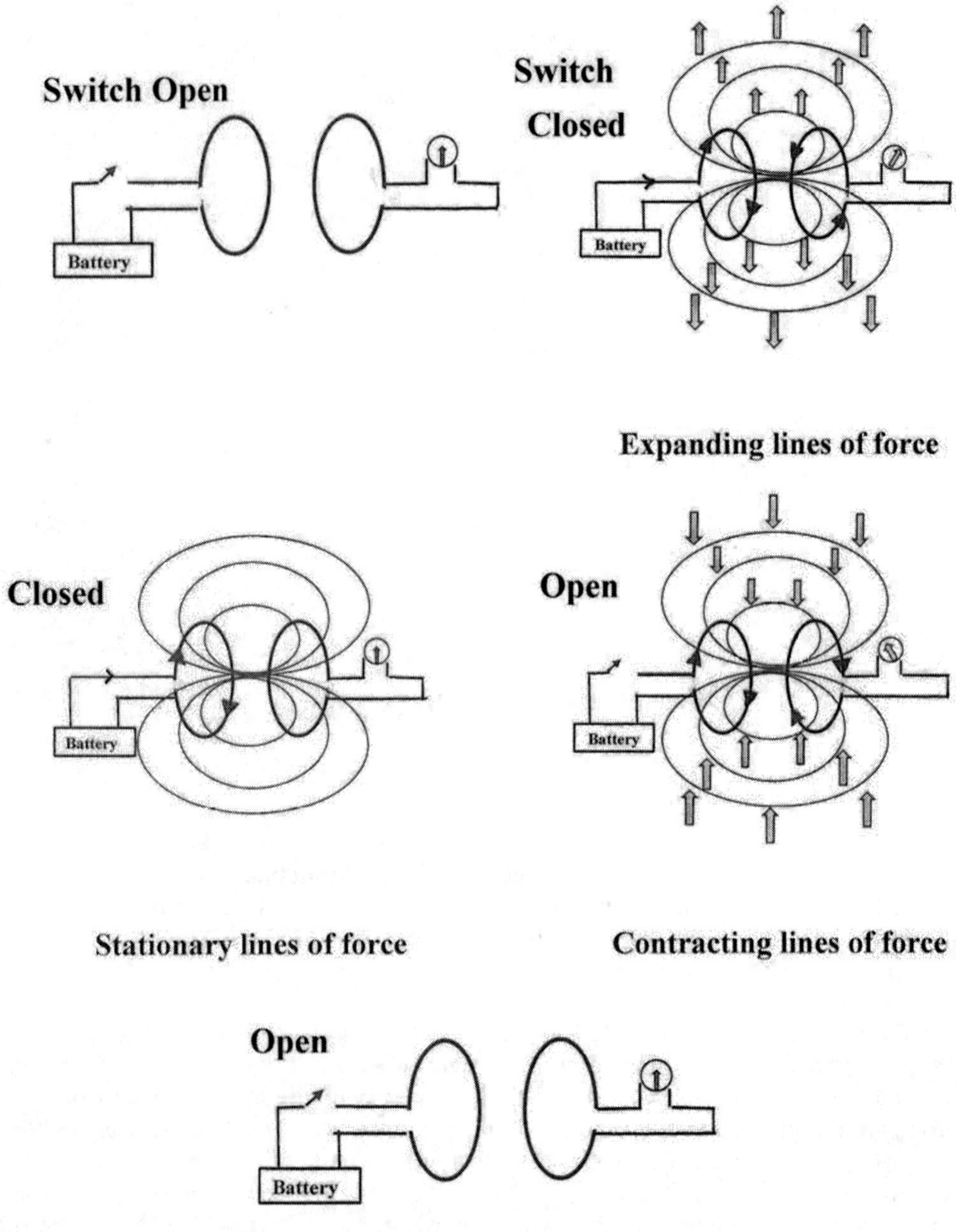


**Figure 34**
**Electromagnetic Induction, Lines of Force 1831**
Courtesy of the author

Faraday showed that in a transformer the voltage in each of two coils wound in separate locations on the same iron core depended directly on the number of windings (turns) in each coil. The greater the number of windings in a coil the higher the voltage induced in that coil. Doubling or tripling the windings in a coil doubled or tripled the voltage induced in that coil. A transformer therefore could increase the voltage of an electrical circuit (step up transformer) or decrease the voltage in an electrical circuit (step down) transformer.

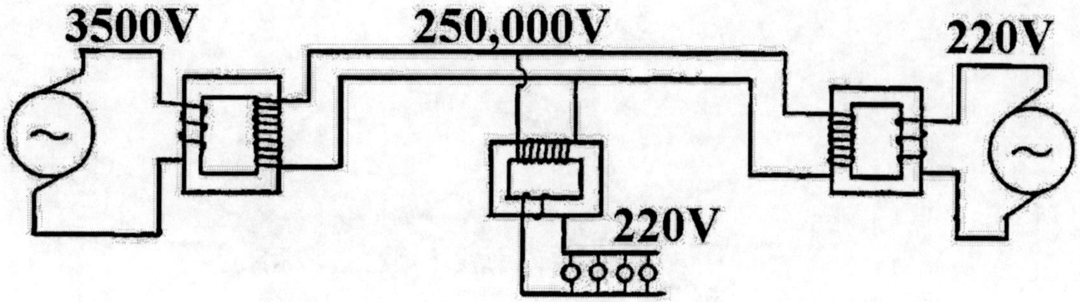


**Figure 35**
**A Simplified Diagram of a Typical AC Distribution System**
**Transformer (Electromagnetic Induction)**
Courtesy of the author

Faraday continued his experimental researches and that same year he constructed a direct current generator. It consisted of a copper disc having an axle passing through its center that he rotated between the poles of a horseshoe magnet. The rotating copper disc always moved in the same direction through the lines of force that Faraday believed surrounded the magnet and produced a small continuous direct (dc) current.

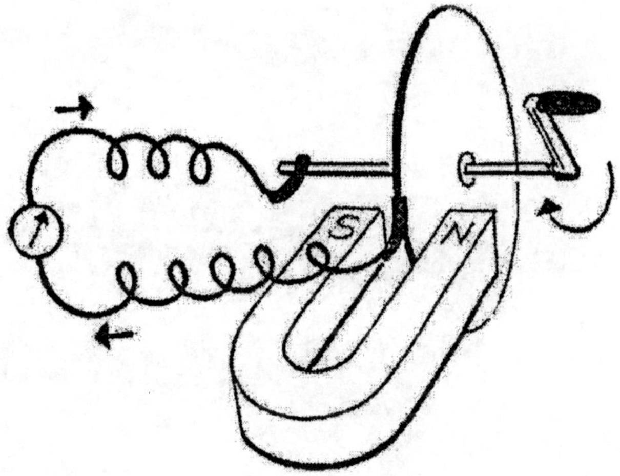


**Figure 36**
**Simple Version of Faraday's Disk DC Generator**
Source: John Tyndall. *Faraday as a Discoverer*. London: Longmans, Green, and Co., 1868.

Faraday in 1846 experimentally demonstrated the presence of lines of forces around a magnet by scattering iron filings on a piece of paper placed over the magnet. The iron filings oriented themselves in a pattern of lines around the magnet. Because a coil of wire through which a current was passing, behaved like a magnet, lines of force also surrounded the coil. Faraday used lines of force to supplement action at a distance. The electrical or magnetic effect moved along the lines of force. He did not abandon the mechanical philosophy. Instead he gave the electromagnetic theory a mechanical interpretation. James Clerk Maxwell (1831-79), a Scottish physicist then at King's College, London, first called the lines of force an *electromagnetic field* in 1864.

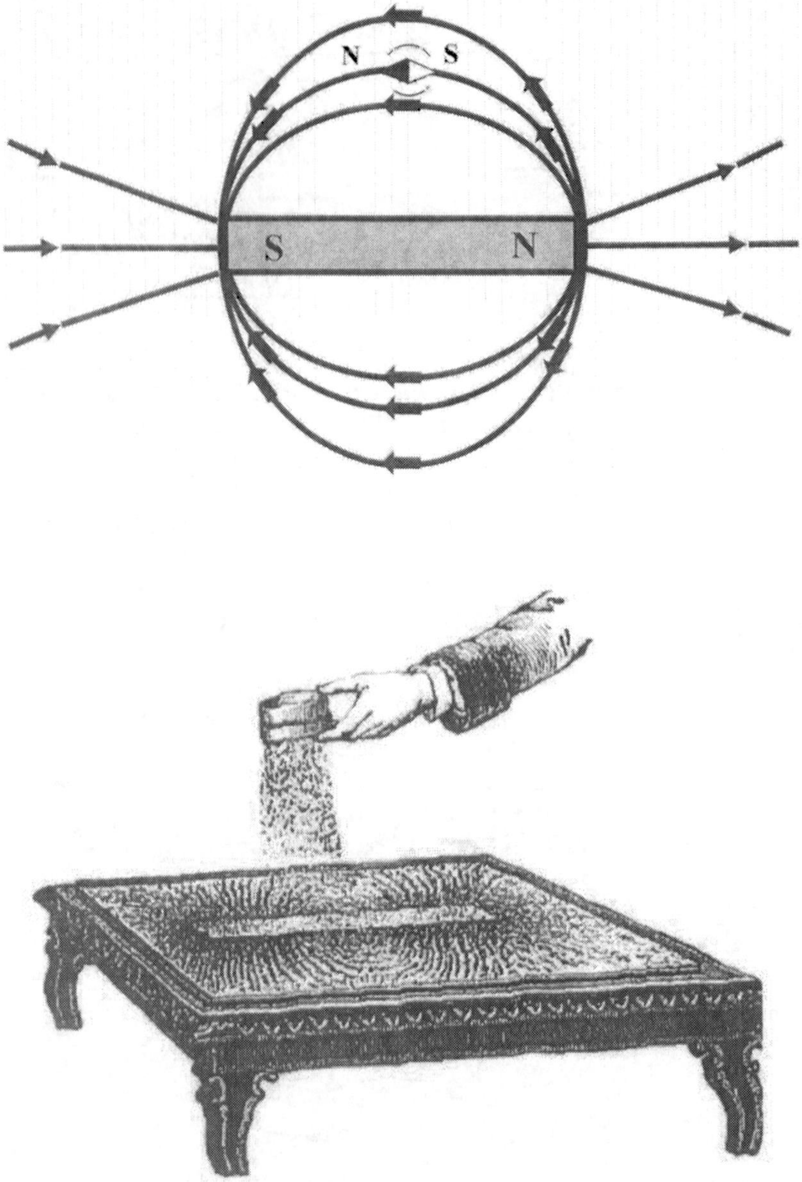


**Figure 37**
**Faraday's Lines of Force**
Courtesy of author (top)
Source: J. Dorman Steele. *Popular Physics.* New York: American Book Company, 1896 (bottom).

In an earlier related study Faraday showed in 1832-33 that in electrolytic reactions, such as the decomposition of water or salts or electroplating of metals, the reactions depended on the amount of electric charge ($q$), the electric current ($i$), and time ($t$). He established two laws of electrolysis, $q = it$, and $m = zit$ where $m$ = mass and $z$ = the electrochemical equivalent for a chemical element. Faraday introduced the words *cathode, anode, cation, anion, electrolyte, electrolysis* in 1833 to describe what was happening in an electrolytic cell. He demonstrated in 1846 that a magnetic field rotated the plane of polarized light. He made an analogy between electricity and gravity, proposing in 1849 that they both acted along lines of force.

| FARADAY'S LAWS OF ELECTROLYSIS | |
|---|---|
| $q = it$ | $m = zit$ |
| $q$ = | electric charge |
| $i$ = | electric current |
| $t$ = | time |
| $m$ = | mass of substance deposited |
| $z$ = | electrochemical equivalent for a chemical element |

| INVENTION OF ELECTRIC MOTORS AND GENERATORS | |
|---|---|
| ***Inventor*** | ***Invention*** |
| William Sturgeon (1783-1850) in London | Constructed the first electromagnet in 1823. His 0.45 pound horseshoe magnet lifted 9 pounds, 20 times its own weight. |
| Hippolyte Pixii (1805-35) in France | Built the first alternating current (ac) generator (dynamo) in 1832. |
| Zenobe Gramme (1826-1901) in France | Built the first direct current (dc) generator (dynamo) by introducing the commutator in 1869. The commutator converted alternating current to direct current. |
| Hippolyte Fontaine (1833-1917) in France | Built the first dc motor in 1873. |
| Galileo Ferraris (1847-97) in Italy | Built the first ac synchronous motor in 1885. |
| Nikola Tesla (1856-1943) in New York City | Built the first ac synchronous motor in 1887. |
| Lucien Gaulard (1850-88) in France | Built the first transformer in 1882. |
| George Westinghouse (1846-1914) in New York City | Began building transformers in 1885. |

***Measurement of electricity***

The discovery of one of electricity's most fundamental laws, the law relating voltage, current, and resistance, occurred in 1826. Georg Simon Ohm (1789-1854) at Cologne showed that the electric current in a conductor varied directly with the voltage applied and inversely with the resistance of the circuit, or $V = iR$.

| OHM'S LAW V = iR |
|---|
| $V$ = voltage (potential) |
| $i$ = current (amperes) |
| $R$ = resistance (ohms) |

Ohm showed that the electric current in a conductor depended on four factors: the type of conductor, the length of the conductor, the cross section (diameter) of the conductor, and the temperature. He used a thermocouple to measure the heating effect produced by a current in an electrical circuit in which he varied the voltage and resistance.

To honor Ohm, Ampere, and Volta for their fundamental electrical researches, the International Congress of Electrical Engineers in 1881 adopted the names volt, from Volta, as the unit of potential, ampere as the unit of current, and ohm as the unit of resistance.

***Henry's and Page's overlooked contributions***

Other than the study of electricity and magnetism, or electromagnetism, little original physics research occurred in the United States in the first half of the nineteenth century. Most of the electromagnetic research occurred in England, the result of Faraday's experimental investigations. Joseph Henry (1797-1878), the United States' foremost antebellum physicist, independently discovered electromagnetic induction and contributed significantly to the application of electromagnetic principles, particularly the invention of the telegraph. Henry taught at the Albany Academy and the College of New Jersey (Princeton) before becoming the Smithsonian Institution's first director at its establishment in 1846. Charles Page (1812-60), the patent examiner in Washington, DC, also did important research on electric motors and induction coils.

Henry, while teaching at Albany Academy in Albany, New York, discovered electromagnetic induction in 1830, before Faraday made his discovery, but Faraday had priority in publishing his experiments on electromagnetic induction and received credit as its discoverer. With the construction of electric motors, generators, and transformers the shift from steam power to electric power occurred. Henry never received credit for discovering electromagnetic induction, but his 1831 experiments with electromagnets led to another important discovery, the principle of the telegraph that Morse later exploited. Henry also showed in 1830 that in an electric motor an electric current produces a circular force that deflects a compass needle.

Page, who knew well the electromagnetic researches of Faraday and Henry, designed and constructed induction coils and electric motors. In 1837 he published a description of a prototype induction coil with a commutator to reverse an electric current's direction, thereby converting direct current (dc) into alternating current (ac). In 1838 Page modified an iron core transformer that he had previously constructed and built the first induction coil that had an automatic magnetic operating interrupter, or circuit breaker, known as Page's rocking magnetic interrupter.

Page worked on designing electric motors for producing locomotion and operating machinery, and in 1838 while in London he demonstrated a battery-powered electric train that moved around a circular railroad track. He continued working on electric motors, and in 1850 Congress awarded him $20,000 to build two electric motors for a locomotive. Page received one of Congress's first awards for scientific research, but Page's acceptance of the award made him a pariah among American scientists. They considered his or any scientist's acceptance of government funds an unethical act because the scientist received the award before having done anything to earn or merit the reward. The reward preceded the research required to earn the reward. Critics thought such awards represented patronage and favoritism, although their criticism seriously waned as the twentieth century dawned.

Page constructed and tested the battery-powered electric motors on 29 April 1851. He drove his electric train on a five-mile railroad track that ran from Washington, DC, to Blandensburg, Maryland. The electric train, carrying officials and dignitaries reached a speed of 19 miles per hour, traveling one mile in 39 minutes, before its batteries failed. Page never received any more funds from Congress to continue his electric motor and battery design, effectively ending his electric train research.

In his later years, after 1861, while serving as patent examiner, Page successfully petitioned Congress to acknowledge the originality of his patents on the induction coil. He convinced Congress to recognize him as the induction coil's inventor and not the German-born Parisian inventor Heinrich Rühmkorff (1803-77) who invented the Rühmkorff induction coil in 1851.

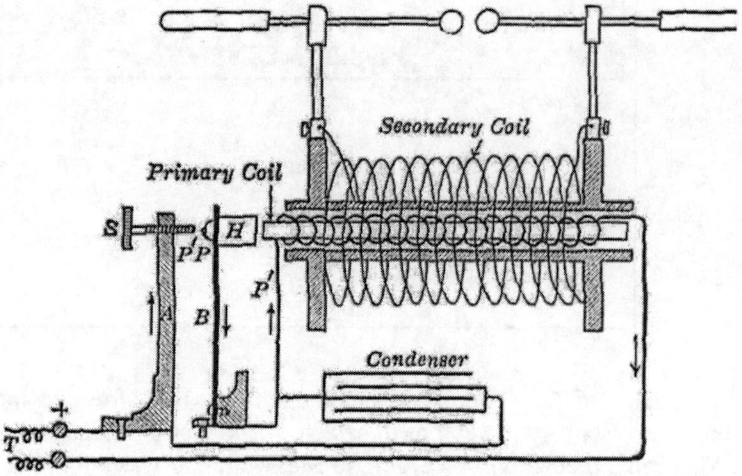


**Figure 38**
**Rühmkorff Induction Coil**
Source: Newton Henry Black and Harvey Nathaniel Davis. *New Practical Physics.* New York: Macmillan Company, 1913.

***Maxwell's development of the electromagnetic theory of radiation***

James Clerk Maxwell, at Aberdeen and from 1871 at the Cavendish Laboratory in Cambridge, developed the electromagnetic theory of radiation in which he proposed that the ether transmitted electromagnetic radiation as a wave through space at the speed of light. He discussed his electromagnetic theory in his papers on "Faraday's Lines of Force," in 1855-56; "On Physical Lines of Force," in 1861; in *A Dynamical Theory of the Electromagnetic Field* in 1865,and in his *Treatise on Electricity and Magnetism* in 1873. Maxwell provided the mathematical theoretical foundation for Faraday's experimental researches.

Maxwell, while at King's College, London, in 1861 and 1862 showed with his theory that the velocity of an electric charge (current) in a conductor such as a wire, the velocity of an alternating electric charge (current) produced by an ac generator and the electromagnetic field that radiated outward from it, the velocity of the displacement (electric) current in a dielectric (an insulator such as air), and the speed of light were the same and equal to the ratio of the electrostatic unit of charge and the electromagnetic unit of charge. Esu/emu = *c*, or 3 x $10^{10}$ cm/sec. The ratio of E/B therefore equaled *c*, or E = *c*B. His equations account for the generation of electrical and magnetic fields and provide the foundation of classical electrodynamics, optics, and electrical circuitry.

| **RATIO OF esu/emu** |
|---|
| $$\frac{esu}{emu} = c$$ $$\mathrm{c} = speed\ of\ light$$ $$c = 3\,x\,10^{10}\ cm/sec$$ |

| MAXWELL'S EQUATION FOR ELECTRIC FIELD INTENSITY |
|---|
| $E = c$B<br>$E$ = electric field intensity<br>$B$ = magnetic flux density<br>$c$ = speed of light |

Maxwell's equation E = $c$B, led him to conclude that light results from the motion of an electric charge (the electron). The discovery in 1897 of the electron and that the electrons in an atom radiated energy confirmed Maxwell's prediction. According to Maxwell light was a form of electromagnetic radiation and part of an electromagnetic spectrum that later included radio waves, infrared radiation, visible light, ultraviolet light, gamma rays, x-rays, and other kinds of electromagnetic radiation. His four mathematical equations published in *A Dynamical Theory* in 1865 established the following relations.

| CONCLUSIONS FROM MAXWELL'S EQUATIONS |
|---|
| An electric force acting on a dielectric (insulator) produces a displacement (electric) current in the dielectric proportional to the force.<br>An electrical current produces a magnetic field at right angles to the current's lines of force and proportional to the current's intensity.<br>A changing magnetic field produces an electric force proportional to the intensity of the magnetic field. This is what Faraday demonstrated with his electromagnetic induction experiments, particularly his coil and magnet experiments. |

Maxwell proved mathematically the correctness of Faraday's electromagnetic experiments. He believed that ether transmitted electromagnetic radiation as waves traveling at the speed of light. No action at a distance occurred, the transmission of radiation was a mechanical process acting along the lines of force or throughout the electromagnetic field.

In the United States, Henry Rowland (1848-1901), a Rensselaer engineering graduate and in 1876 the first professor of physics at Johns Hopkins, demonstrated in 1875 that a moving electric charge, just like an electric current flowing in a conductor, generated a magnetic field (B) equal to the product of the quantity (q) and velocity of the charge (v) for B=qv. He measured accurately the magnetic permeability in 1873-74, the ohm the unit of electrical resistance in 1878, and the mechanical equivalent of heat in 1879.

***Demonstrating the transmission of electromagnetic waves, wireless or radio transmission***

Heinrich Hertz (1857-94) at Karlsruhe, in his experiments in 1888 demonstrated the correctness of Maxwell's theory that the transmission of electromagnetic radiation was a wave motion. He used an induction coil to generate radio waves (microwaves) and detected the waves with a spark gap receiver 60 feet away. The waves produced a spark when they reached the receiver. They were real not imagined. Oliver Lodge at Birmingham in his tuning (syntony) experiments of 1894 and 1897 also demonstrated the reality of Maxwell's electromagnetic waves.

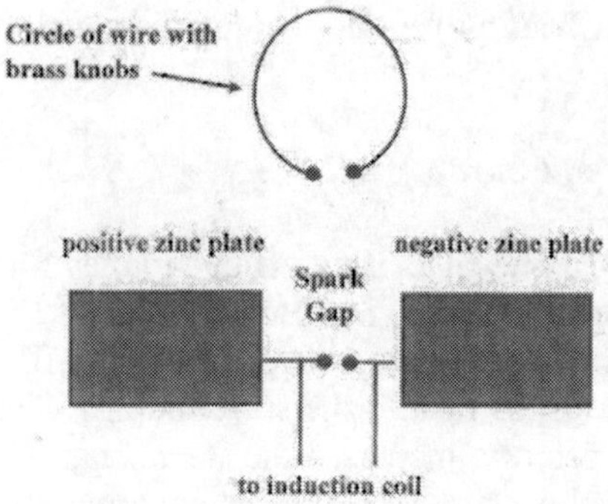


When the induction coil charges the zinc plates oppositely, a spark jumps across the gap. As the spark passes back and forth, it produces electromagnetic, or radio, waves. The figure shows the resonator that Hertz used as a receiver or detector.

**Figure 39**
**Hertz's Oscillator 1888**
Courtesy of the author

The first long distance transmissions of electromagnetic waves (radio waves) occurred in 1897-1901. Guglielmo Marconi (1874-1937), an Italian physicist-inventor experimenting in London, sent 3-10 mile wireless transmissions over land and water in 1897-98, across the 22-mile English Channel in 1899, and on 12 December 1901, from Cornwall, England, to St. John's, Newfoundland, a distance of 2,200 miles. Marconi sent the electromagnetic waves in pulses that imitated the dots and dashes of the Morse code. Marconi initiated radio transmission, and beginning in late 1902 he visited the United States intending to develop radio transmission commercially. He shared the 1909 Nobel Prize in physics.

Tesla in New York City in 1903-04 engaged in controversy with Marconi on the priority of the first successful wireless transmission. This led the United States Supreme Court much later in 1943 to declare some of Marconi's claims of patent infringement invalid, but not his patent of April 1900 (awarded April 901), and to acknowledge some of the patent claims of Tesla in 1900, Lodge in 1898, and the Boston physicist John Stone (1869-1943) in 1902-03 on wireless transmission.

Reginald Fessenden (1866-1932), a Canadian physicist then with the National Electric Signaling Company (NESCO) in Brant Rock, Massachusetts, who previously worked for both Edison and Westinghouse, demonstrated the first wireless voice transmission on 24 December 1906. Instead of transmitting electromagnetic waves in pulses that imitated the Morse Code's dots and dashes, Fessenden sent the waves as a continuous (radiowave) signal in which he varied (modulated) the amplitude, so that the amplitude modifications (am) matched the variations of the sound waves. The combination of radiowave frequency and the frequency of the sound waves (e.g. 500 vibrations per second, vps) being broadcast, generated a wave whose frequency increased and decreased (modulated) 500 times per second. The modulation causes the diaphragm of the receiver to vibrate at a frequency of 500 vps and reproduce the original sound. Fessenden transmitted Christmas music and Bible verses. The doomed ocean liner *Titanic* sent its distress signal by wireless on 14 April 1912.

## V. THERMODYNAMICS: THE STUDY OF HEAT

The three laws of thermodynamics emerged over the period 1842-1920 as a result of the experimental research of European and American investigators. The second law had its roots in Sadi Carnot's 1824 work on the efficiency of steam engines and preceded the 1842 and 1847 statements of the first law by nearly 20 years. Rudolf Clausius gave the generally accepted and clearest statement of the second law in 1850, and this 1850 date, along with Carnot's reliance on caloric in developing his earlier version of the second law, are the main reasons for the reversed numerical order of the first and second laws that became standard in twentieth-century textbooks on thermodynamics and physical chemistry. The third law in its three main forms was a twentieth-century conclusion first published in 1905 and in final form in 1920.

***Carnot, the heat engine and Carnot cycle***

Sadi Carnot (1796-1832), a one-time army officer and later a physicist in Paris, studied steam engines during the time of the industrial revolution. He was interested in improving the efficiency of Watt steam engines, which at that time were only about 5-7 percent efficient, with the other 90-95 percent of the energy input wasted and lost as heat to the surroundings. High-pressure steam, introduced in the first decade of the nineteenth century, increased the steam engine's efficiency to 17 percent, a considerable increase but still a low value. As a result of his experiments, Carnot in 1824 showed in a brief book *Réflexions sur la puissance motrice de feu (Reflections on the Motive Power of Fire)* that the efficiency depended on the temperature difference established in the steam engine, in which $T_1$ was the higher temperature of the hot (steam) cylinder and $T_2$ the lower temperature of the cold (condenser) cylinder. Carnot showed moreover that the efficiency did not depend on the liquid (water, alcohol, or any other liquid) that vaporized in the engine. Watt's separation of hot and cold cylinders in constructing his steam engine represented the first recognition of the temperature differential as a key factor in the steam engine's efficiency.

Carnot's 1824 publication did not include an equation to calculate the steam engine's efficiency. He gave a general description of an engine's operation in which he equated the work done by the fall in temperature and quantity of heat (caloric) to the work done by a quantity of water falling a known height. The waterfall analogy he derived from his father, the engineer and politician Lazare Carnot (1753-1823), who in 1784 studied the efficiency of water-powered machines and later died in political exile in Magdeburg.

Just as the falling water turned a paddle wheel and did work without consuming any water, Carnot believed that the fall of caloric (heat) would cause the piston in a steam engine to move and do work, by which he meant, lifting a weight through a height, with no loss of caloric. Caloric would flow from a higher temperature to a lower temperature to do work without any loss or conversion of caloric, therefore conserving caloric. Rudolf Clausius however showed in his 1850 statement of the second law of thermodynamics that when an engine converts heat to work, it converts some of the heat from the high-temperature boiler into work or mechanical energy and releases the rest of the heat at a lower temperature in the condenser. When the engine does work it does not conserve caloric, as Carnot believed.

Carnot was the first to interpret the operation of a steam engine, or any other engine, as a cyclic heat engine. Its operation was reversible, or what scientists call thermodynamic reversibility. Carnot concluded, moreover, that the construction of a perpetual motion machine was impossible and that all reversible engines operating between the same two temperatures had the same efficiency. The greater the temperature difference, the greater the efficiency. The equation below gives the thermodynamic efficiency of a reversible engine.

**CARNOT'S THERMODYNAMIC EFFICIENCY**

$$\text{Engine efficiency} = \frac{T_1 - T_2}{T_1}$$

$T_1$ = higher temperature (hot cylinder)
$T_2$ = lower temperature (condenser)

Carnot's work is significant because even though he accepted the caloric theory of heat, his thermal studies imply and contain as a consequence, a correct statement of the second law of thermodynamics, that heat always flows from a high temperature to a low temperature.

Sometime before his death from cholera during an 1832 outbreak, Carnot accepted that when an engine does work it converts some caloric to mechanical energy and therefore reduces the amount of caloric. At that time he abandoned caloric. Carnot never published these undated manuscripts (published posthumously in 1878), no other evidence indicates when he wrote these manuscripts.

Benoit Pierre Émile Clapeyron (1799-1864), a mining engineer who taught in St. Petersburg before returning to France in 1830 to teach at the École des Ponts et Chaussées in Paris in 1844, expanded on Carnot's ideas in an 1834 publication. Clapeyron accepted the caloric theory and presented Carnot's ideas in analytical form. He showed graphically, in which he represented the volume of a gas by the $x$ axis (horizontal, abscissa) and the pressure by the axis $y$ (vertical, ordinate), that the graphical area enclosed by the four reversible stages of an engine's cyclic operation (the Carnot cycle) represented the work the engine did.

For a typical twentieth-century steam engine operating between 450 K and 373 K, the efficiency is 17 percent, whereas the efficiency of an internal combustion engine is 39 percent, a diesel engine 56 percent, and a nuclear engine 33 percent.

Clapeyron in 1834 introduced a fundamental equation that Clausius in 1850 modified, establishing the relationship among the temperature at which a phase change occurs, as from a liquid to a vapor, the change in heat content, and the change in volume. The Clausius-Clapeyron equation appears below.

| CLAUSIUS-CLAPEYRON EQUATION |
|---|
| $\frac{\Delta H}{dT} = \frac{dp}{T\Delta V}$<br>p = pressure<br>T = temperature<br>ΔV = change in volume<br>ΔH = change in heat content |

***Work and the mechanical equivalent of heat***

Gaspard Coriolis (1792-1843) at the École Polytechnique in Paris in 1829 called energy *vis viva* and defined it as $\frac{1}{2}mv^2$. That same year he defined work as a force acting through a distance against a resistance.

| CORIOLIS'S DEFINITION OF WORK |
|---|
| Work = force x distance |

Germain Hess (1802-50), a Swiss-born chemist at St. Petersburg Technological Institute, applied thermodynamics to chemical reactions. He showed that the heat changes in chemical reactions were additive. The overall or net gain or loss of heat did not depend on the reaction pathway, whether the reaction occurred in a single step or in several steps, but only on the initial and final states of the reaction. This is Hess's Law of 1840.

James Joule (1818-89), an independently wealthy and privately educated English physicist, performed experiments on heat changes in his home laboratory near Manchester. Beginning in 1839 he established in a series of experiments a relation between the mechanical work a process did and the amount of heat produced. This relation is the mechanical equivalent of heat. In experiments done in 1845 he measured the quantity of heat produced by a known weight falling through a known distance and turning a paddle wheel immersed successively in water, oil, and mercury. He recorded the temperature increase and calculated the amount of heat H released using Black's equation, H= mass x temperature change x specific heat. He obtained a value of 819 ft-Lb/Btu. By 1847 Joule had obtained a good value, the equivalent of 781.8 ft-Lb for each British Thermal Unit (Btu) of heat produced.

In 1850 Joule calculated that the mechanical work required to raise the temperature of 1 pound of water 1 Fahrenheit degree was 772 ft-Lb. This relation became the definition of one British Thermal Unit, later revised to give 1 Btu = 778 ft-Lb, and is the mechanical equivalent of heat. Neither Joule, nor Rumford and Julius Robert Mayer (1814-78) before him, used the British Thermal Unit in their studies on heat, although by 1910 the British Thermal Unit had gained wide use among British engineers.

| JOULE'S MECHANICAL EQUIVALENT OF HEAT |
|---|
| weight x distance = mass x $\Delta t$ x specific heat |
| $w \times d = m \, \Delta t \, s$ |
| $wd = H$ |
| 772 ft-Lb = 1 Btu |

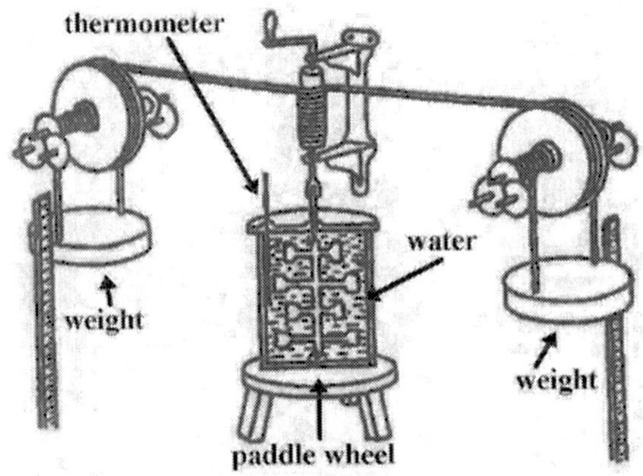


As the weight descended it turned the paddle wheel inside an insulated container of water. This converted the potential energy of the weight into the kinetic energy of the paddle wheel and then into heat energy of the water. Dividing the energy lost by the descending weight by the increase in the heat content of the water gave the mechanical equivalent of heat, 772 ft-Lb/Btu (4185 joules/kcal).

**Figure 40**
**Joule's Mechanical Equivalent of Heat 1845**
Source: Newton Henry Black and Harvey Nathaniel Davis. *New Practical Physics.* New York:Macmillan Company, 1913.

Joule's experiments on the expansion and contraction of gases, heating water electrically, and passing water through small holes to heat it by friction, showed that the expenditure of mechanical, electrical, or other forms of energy always resulted in a gain of heat energy. He realized that a conservation of energy resulted if he included the heat energy produced when doing mechanical work (or when doing any other form of work) as part of the total energy change involved in the process. This led Joule to conclude that the total energy evolved or consumed in a process is constant. Joule had given a statement of the first law of thermodynamics.

***Three laws of thermodynamics***

The first law of thermodynamics established experimentally that the energy of the universe is constant. Mayer, a medical doctor in Heilbronn, Germany, who in the 1850s suffered from periods of insanity and depression, in 1842 gave the earliest statement of the first law of thermodynamics. At this time the caloric theory, not the mechanical theory, of heat dominated scientific thought.

Mayer arrived at the first law from studying physiological processes while in Java in 1840 after serving as a ship's physician on a trip to the Dutch East Indies. From his studies of human body heat, he concluded that the maintenance of body heat in the tropics required less expenditure of energy than in a colder climate. The human body required less oxygen for combustion to maintain the normal body temperature, and the blood (because of oxygen's presence) appeared quite red even in the veins, where it normally has a darker color.

In his 1842 studies on energy, Mayer searched for a relation between heat and work, specifically how much work a certain amount of heat would produce. He measured the specific heats of gases to determine the quantity of heat gained and lost in a reaction, process, or change, and found that the specific heat of a gas at constant volume was smaller than when measured at constant pressure. Mayer said this occurred because at constant pressure the gas always expands against the atmosphere, and in so doing it used the extra heat absorbed to do work.

Mayer in 1845 applied the idea of conservation of energy (then called force) to electrical, magnetic, and chemical forces. Plants converted the Sun's energy, animals consumed this energy as food and then converted it to body heat and for use in their motions and movements. He calculated 1 Btu = 725 ft-Lb of mechanical work, a value lower than Joule's values. According to Mayer's *Dynamik des Himmels* published in 1848 the Sun's energy resulted from the conversion of kinetic to thermal energy during the Sun's contraction.

Hermann von Helmholtz (1821-94), a physician-physicist at Königsberg and later at Heidelberg and Berlin, established the first law in 1847 from physiological and physical experiments. Helmholtz's studies on animal (body) heat, which he believed was a mechanical process and not vitalistic, led him to investigate the conservation of energy. In his 1847 paper "On the Conservation of Force," Helmholtz proved that for a system of particles interacting through forces that depended only on the masses and separations of the particles, the energy of individual particles could change, but the total force (energy) of the system remained constant. Conservation of the total force (energy) occurred. Helmholtz applied the principle of conservation to other branches of science.

Joule in 1847 established the first law from experiments in which he converted potential energy to kinetic energy to thermal energy, such as his falling weight and paddle wheel experiment. He recognized that the gain of heat balanced the expenditure of mechanical energy through work done, friction, or air resistance and that in general a gain in heat energy always balanced the loss of another form of energy. This proved the conservation of energy.

Joule in 1840 established a relation for the heat generated by an electric current, $P = I^2R$ where $P$ = power, $I$ = current, and $R$ = resistance, and in 1862 with Kelvin showed that as a gas expanded freely its temperature fell. This is the Joule-Thomson effect. It provided evidence of the attraction among gaseous molecules. In expanding or moving apart, the gaseous molecules expended energy in overcoming this attraction, and the temperature fell. The Joule-Thomson effect found use in refrigeration with gases such as ammonia, sulfur dioxide, and later freon. The International Electrical Congress meeting in Paris in 1889 named the unit of work, the Joule, (1 calorie = 4.169 Joule) in his honor.

The second law of thermodynamics established experimentally that the entropy (disorder) of the universe strives to a maximum, or that heat always flows from a higher temperature to a lower temperature. The second law tells us the direction in which energy flows.

Rudolf Clausius (1822-88), a physicist at Zurich and later Wurzburg and Bonn, Germany, gave a statement of the second law in 1850. By that time he strongly favored the view that heat was not a substance but that heat "consists in a motion of the least parts of bodies," thereby rejecting caloric as the fluid of heat. Clausius reconciled Carnot's heat theory and Joule's mechanical equivalent of heat and concluded that heat always flowed from a body at a higher temperature to a body at a lower temperature and never spontaneously in the opposite direction.

Clausius recognized that the conservation of energy was not the only factor to consider in a physical or chemical process. He had to account for the energy that was lost or unavailable to do useful work as in a steam engine, or lost as heat through friction. He built on Carnot's recognition that the transfer of heat from a body at a higher temperature to one at a lower temperature does work, but he rejected the second part of Carnot's conclusion that the quantity of heat did not decrease and remained undiminished. He developed the second law and gave it a mathematical-quantitative interpretation (an equation) in 1854. At that time he used the expression, principle of equivalence of transformations, in discussing the energy changes taking place.

Clausius realized that the second law implied the dissipation of energy, and in 1865 he introduced the word *entropy* (Greek for transformation) and defined it as the ratio of the heat content to absolute temperature. Entropy was the energy unavailable to do useful work. It was unavailable thermal energy released in all physical and chemical processes because no process is 100 percent efficient. Clausius concluded that the entropy always increased in a closed system, one in which no heat entered from the outside and no heat escaped from the inside.

**SECOND LAW OF THERMODYNAMICS**

$$\Delta S = \frac{\Delta H}{T}$$

$\Delta S$ = change in entropy
T = absolute temperature
$\Delta H$ = change in heat content (enthalpy)

In a paper "On the Dynamical Theory of Heat" published in 1851, Kelvin gave a statement of the second law that reconciled Carnot's conclusions based on caloric theory with a mechanical theory of heat. He recognized prior to Clausius that the second law implied the dissipation of heat. Kelvin introduced the word *thermodynamics* in an 1854 paper published in the *Transactions of the Royal Society of Edinburgh.*

A statistical interpretation of the second law followed shortly. Maxwell in 1859 and 1866 publications and independently Ludwig Boltzmann (1844-1906) at Graz, later Vienna, in 1868 and 1872 publications gave a statistical interpretation of the second law. Maxwell and Boltzmann based their interpretation on the distribution of the velocities of gaseous molecules. They measured the viscosity of a gas to obtain the mean free paths of the gaseous molecules. The mean free path measured the average distance a molecule traveled between collisions and the frequency of collision among the gaseous molecules.

Maxwell and Boltzmann showed that at each temperature there was a distribution of velocities, some molecules had a low velocity, some had a high velocity, and the majority had velocities between these two values. As the temperature increased more and more molecules had a higher velocity, indicating that temperatures of the molecules depended directly on the molecules' motion. A graph plotting the percentage of molecules (y axis) against velocity of the molecules (x axis) gave a bell shape distribution for each temperature measured. They recognized that entropy measured a system's disorder.

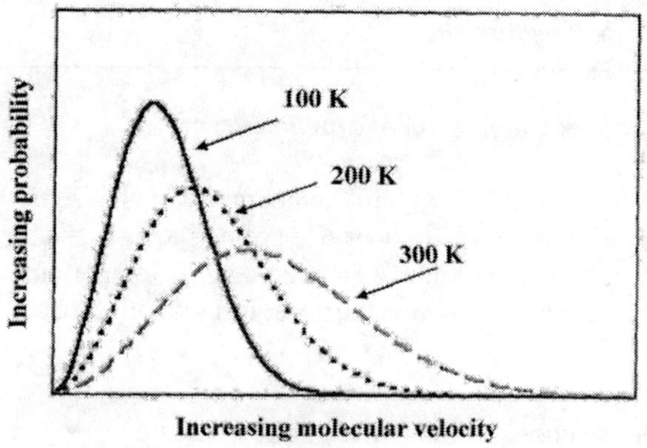


**Figure 41**
**Maxwell-Boltzmann**
**Distribution of Velocities**
Courtesy of the author

Recently creationists have claimed the incompatibility of Darwin's theory evolution and the second law of thermodynamics. The two scientific beliefs are not incompatible. The entropy of specific reactions taking place in our world can decrease as in snowflake formation, the development of organisms, or the organization of life. But the net entropy change for all processes taking place in a closed system as we consider our world and universe, is always positive. If we treat our universe as an open system, energy from outside our universe can enter the system and produce a positive net entropy change. If sufficient energy leaves the system the net entropy change could fall to zero or a negative value.

The third law of thermodynamics stipulates that the entropy of a perfect crystal is zero at absolute zero. Walther Nernst (1864-1941), a physical chemist at Göttingen and later Berlin, in 1906 introduced the heat theorem which said that in any reaction involving only solids and liquids (including solutions) the change in entropy approached zero as the temperature approached absolute zero. This was the first statement of the third law Nernst wrote in the multi-editioned classic *Theoretical Chemistry* first published in 1893. He received the 1920 Nobel Prize in chemistry for his work on the third law.

Five years later, Max Planck (1859-1947) at Berlin, in the third edition of his *Thermodynamik* (1911) argued that Nernst's heat theorem did not hold for solutions and required modification. Planck suggested assigning zero entropy to each element at absolute zero and interpreted the third law to mean that all pure solids and liquids had zero entropy at absolute zero.

Lewis and George E. Gibson (1884-1959) at Berkeley pointed out that entropy measured the randomness of a microscopic state, and even in a pure solid or liquid some randomness existed in its structure. Only a perfect crystal of a pure substance lost its entropy at absolute zero. In 1920 they gave the statement of third law that scientists use today: the entropy of a perfect crystal is zero at absolute zero. It is a logical extension of the second law.

| C.P. SNOW'S THREE LAWS OF THERMODYNAMICS |
|---|
| 1. You cannot win, you don't get something for nothing, no free lunch.<br>2. You cannot break even. You cannot return to the same energy state because the entropy or disorder always increases.<br>3. You cannot get out of the game, absolute zero is unattainable because no perfectly pure substance exists.<br>From *Encyclopedia Brittanica* online. |

***Kelvin: temperature scale, definition of kinetic energy***

Kelvin introduced the Absolute (Kelvin) temperature scale (represented by T) in 1848 and gave a definitive statement of it in his 1851 paper. He defined K or A = C° + 273.16. From his thermal studies on gases, in which he plotted volume V (*y* axis) against temperature T (*x* axis), Kelvin concluded that 273.16° C was the temperature at which the gaseous molecules lost all their energy and appeared to have no (zero) volume.

Kelvin showed that because a volume of a gas decreases by $\frac{1}{273.16}$ of its volume or 0.366 percent as it cools from 0° C to - 1° C, decreases by $\frac{2}{273.16}$ at -2° C, decreases by $\frac{10}{273.16}$ at -10° C, and similarly for other temperature reductions, the gas lost all of its volume at -273.16° C. He called this temperature absolute zero because any lower temperature would require the gas to have a negative volume, which was clearly an impossibility. Similarly, 1,000 cm³ of a gas at 0° C occupy the volumes given in the table and graph.

| KELVIN'S TEMPERATURE VOLUME RELATION | | |
|---|---|---|
| *Temperature* | | *Volume* |
| *C* | *K* | *(cm³)* |
| 10 | 283 | 1,636 |
| 0 | 273 | 1,000 |
| -10 | 263 | 963 |
| -23 | 250 | 915 |
| -48 | 225 | 824 |
| -73 | 200 | 783 |
| -273 | 0 | 0 |

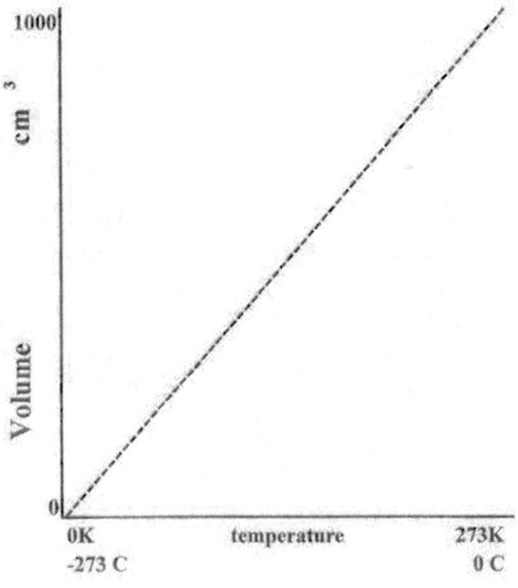


**Figure 42**
**The Volume of a Gas at Constant Pressure and Variable Temperatures**
Courtesy of the author

Kelvin introduced the relation KE = $\frac{1}{2}$ $mv^2$ for kinetic energy (energy of motion) in 1856. Previously scientists used force or *vis viva*, meaning force of life, or living force, and not energy in their discussions on the energy of motion. Leibniz in 1695 defined *vis viva* as $mv^2$ and said *vis viva* was conserved in motion. Thomas Young in 1807 used energy to represent quantity of work and set energy equal to $mv^2$. The derivation of the equation for kinetic energy follows.

| DERIVATION OF EQUATION FOR KINETIC ENERGY |
|---|
| Work = force x distance (height), or W = Fs<br>Potential energy = force x distance, or PE = Fs<br><br>From Newton's second law<br>F = weight = mass x gravitational acceleration, or F = mg.<br>Because s = h (height)<br>PE = mgh.<br><br>Galileo's two equations for falling bodies are v = gt and s = $\frac{1}{2}$ $gt^2$.<br><br>Solving the equation $s = \frac{1}{2}\ gt^2$ for $t$ and substituting for t in the equation v = gt gives<br><br>$v\ (velocity) = \sqrt{2gs}$ or $s = \frac{v^2}{2g}$.<br><br>Because h and s equal the distance the body has fallen $h = \frac{v^2}{2g}$.<br><br>A body which was at rest and with potential energy mgh and now moving with velocity v, therefore possesses kinetic energy KE.<br><br>$KE = mg\ x\ \frac{v^2}{2g}\ \ or\ KE = \frac{1}{2}mv^2$ |

***Contributions of J. Willard Gibbs***

J. Willard Gibbs (1839-1903) attended Yale University where his father was a philologist and professor of sacred theology. He graduated in 1858 and received his PhD from Yale in 1863 with a dissertation in engineering on the design and manufacture of gears. Gibbs then spent three years (1866-69) in Europe, one year in Paris, Berlin, and Heidelberg. In 1871 Yale appointed him to the first chair of mathematical physics in the United States. He received no salary with his appointment and lived off his inheritance. Gibbs declined an offer of $1,800 from Bowdoin College in 1873 and of $3,000 from Johns Hopkins in 1880 and remained at Yale for the rest of his career earning an unsubstantial salary of $2,000.

His classic papers on thermodynamics entitled "On the Equilibrium of Heterogeneous Substances" appeared in two parts totaling 321 pages in 1875 and 1878 in the *Transactions of the Connecticut Academy of Sciences*. He also sent copies to scientists, such as Maxwell, Kelvin, and Clausius, whom he believed would see the importance of his work. Gibbs adopted graphical (geometrical) solutions in solving chemical problems, but because his papers were highly mathematical and difficult to read, they did not have an immediate impact on many of his readers. The *phase rule* and *free energy* were the major ideas introduced in his papers on equilibrium.

Gibbs's *phase rule* was a mathematical expression that showed the conditions of equilibrium in a system by relating the number of phases, the number of components, and the degrees of freedom possible under given conditions. It enabled chemists to predict the stability of the states of matter, mixtures, or alloys, at different pressures, temperatures, and composition or concentration. In the early twentieth century chemists used the *phase rule* to determine the conditions at which gaseous reactions, such as the Haber synthesis of ammonia from its elements nitrogen and hydrogen and the Ostwald synthesis of nitric oxide from its elements nitrogen and oxygen, gave maximum yields.

Gibbs developed the following equation for a one component system, a system consisting of only one constituent, or substance, such as water or carbon dioxide.

**GIBBS PHASE RULE EQUATION**

$$F = C - P + 2$$

| | |
|---|---|
| F = | number of degrees of freedom or variables to specify (temperature, pressure, concentration). |
| C = | number of components (one pure substance, two pure substances). The components are the smallest number of pure constituents that make up separately each phase. Water is a one-component system, so is carbon dioxide. A mixture of water and oil, salt dissolved in water are two-component systems. |
| P = | number of phases (solid, liquid, gas). A gas or vapor has only phase because all gases are miscible. For solids and liquids the number of phases is indefinite. |

Gibbs developed a similar equation for systems with more than one component. The phase diagrams, which are a direct consequence of the phase rule, are graphs that show the equilibrium conditions that exist for the different phases of a substance. Phase diagrams are valuable in studying boiling points, freezing points, and other phase transitions.

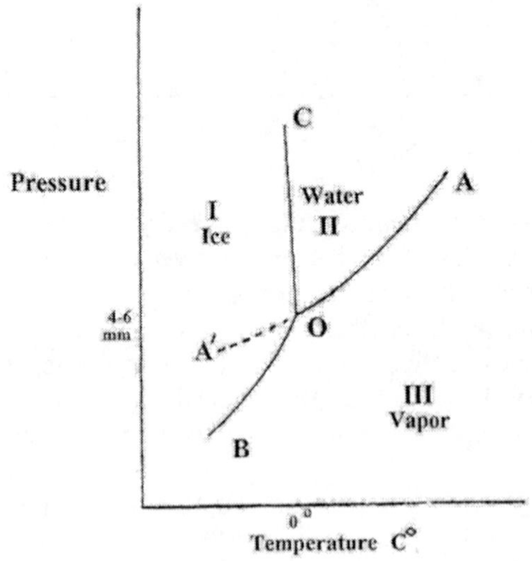


**Figure 43**
**Phase Diagram of the One-Component System Water**
Source: Alexander Findlay. *The Phase Rule and Its Applications.* London: Longmans, Green, and Co., 1906.

Gibbs defined *free energy* ($\Delta F$, since 1960s $\Delta G$) which related the heat content and the entropy of a reaction and showed the necessity of both in order to account for the energy change taking place in a chemical reaction. As a result of calorimetric studies that Julius Thomsen (1826-1909) at Copenhagen and Marcellin Berthelot (1827-1907) at Paris, carried out in 1852, most chemists at that time believed the quantity of heat evolved (released) in an exothermic reaction measured the reactants' affinity for each other. The terms exothermic reaction, a reaction that takes place with the emission of heat, and endothermic reaction, one that occurs upon adding heat, were definitions that Berthelot in 1865 introduced to the language of thermodynamics.

While this correlation between heat released and the reactants' affinity and therefore spontaneity or ease of reaction often seemed to hold, some reactions that were clearly endothermic occurred spontaneously. Confusion also surrounded the meaning of entropy after Clausius introduced the equation for entropy in 1854 and the term entropy in 1865, so that for a time even Maxwell thought entropy meant available energy rather that Clausius's meaning of entropy as energy lost or unavailable in a process or reaction. The confusion persisted until Gibbs showed in his 1875 and 1878 publications the necessity of determining experimentally both the quantity of heat (enthalpy) and entropy changes to establish a reaction's spontaneity at constant temperature and pressure. To measure spontaneity, Gibbs in 1875 defined a new function called a *potential* and expressed its dependence on enthalpy, entropy, and temperature with the equation below.

| GIBBS POTENTIAL |
| --- |
| $D\zeta = d\chi - td\eta$ |
| $\zeta$ = potential<br>$\chi$ = enthalpy<br>$\eta$ = entropy<br>$t$ = absolute temperature |

Helmholtz in 1882 derived a similar relation in which he introduced the term *free energy* synonymous to Gibbs's *potential.* For this reason, chemists and physicists today speak of the Gibbs-Helmholtz equation when describing *free energy* calculations.

| GIBBS - HELMHOLTZ EQUATION |
| --- |
| $\Delta G = \Delta H - T\Delta S$ |
| AG = change in Gibbs (free) energy<br>$\Delta H$ = enthalpy change<br>$T\Delta S$ = product of the entropy change and absolute temperature. |

Gibbs in 1876, Helmholtz in 1882, and van't Hoff in 1886, showed that experimentally determined equilibrium constant (*K*) and electromotive force (*E*) values for a chemical reaction provided additional ways of calculating standard Gibbs (free) energies. Knowing the free energy of a chemical reaction enabled chemists to predict whether a chemical reaction will occur. A negative $\Delta G$ is negative means the reaction will occur, a positive $\Delta G$ means it will not. The equations, using modern symbols, follow.

| GIBBS FREE ENERGY - EQUILIBRIUM CONSTANT | | |
|---|---|---|
| $\Delta G° =$ | | $-R\,T \ln K$ |
| $\Delta G° =$ | | free energy at standard conditions of T and P (pressure) |
| R | = | universal gas constant |
| $K$ | = | equilibrium constant (a dimensionless, constant, a function only of T) |
| T | = | absolute temperature |

| GIBBS FREE ENERGY-ELECTROMOTIVE FORCE | | |
|---|---|---|
| $\Delta G° =$ | | $-n\,F\Delta E°$ |
| $\Delta G° =$ | | free energy at standard conditions |
| N | = | number of moles (a quantity of mass) |
| F | = | Faraday (a quantity of electric charge) |
| $\Delta E_0$ | = | change in electromotive force for a reaction at standard conditions |

Physical scientists had difficulty understanding Gibbs's 1875 and 1878 papers on equilibrium because of their highly mathematical arguments with few examples to illustrate the meaning and significance of his ideas. Physical chemistry as a separate branch of chemistry did not yet exist, and the chemical industry had developed few processes in which to apply Gibbs's ideas. The German chemist Wilhelm Ostwald (1853-1932) at Leipzig translated Gibb's 1875 and 1878 papers into German in 1892, and in 1899 Henry Le Châtelier (1856-1936) at Paris had them translated into French.

Gibbs held a mechanical philosophy of matter, which accepted the reality of atoms and did not consider atoms simply a useful model or tool. He was not an energeticist as were the positivists Ostwald and the Austrian physicist Ernst Mach (1838-1916) at Vienna, both of whom denied the reality of atoms.

## VI. THE KINETIC-MOLECULAR THEORY OF GASES

The first quantitative study on gases was Robert Boyle's work on pressure-volume relations that resulted in Boyle's Law of 1662. Boyle showed that at constant temperature the volume of a gas varies inversely as its pressure, or $PV = k$.

About 75 years later, in 1738, the Swiss mathematician Daniel Bernoulli (1700-82) at Basel began a study of the motion and energies of gases, that is, a kinetic theory of gases. In this work, *Hydronamica* Bernoulli assumed that a gas consisted of atoms moving rapidly in all directions, an idea that goes back to Democritus. He showed that in a closed container the collisions of the atoms with the walls of the container produced a pressure that was inversely proportional to the volume of the container. The volume of the atoms in the container was negligible, and the volume of the container was essentially empty space. Bernoulli also found that the pressure of the gas varied directly with the energy or velocities of the atoms if the volume remained constant. The collisions of the gaseous atoms were elastic, like two billiard balls bouncing off one another.

The studies of Charles and Gay-Lussac in the 1780s-early 1800s led in 1787 to a relation between volume and temperature of a gas, V = kT, since known as Charles Law. The English physicist John Herapath (1790-1868) in Cranford, near London, in 1821 and John Waterston (1811-83) at Edinburgh in 1845 continued the kinetic study of gaseous molecules that involved elastic collisions of the molecules composing the gases. Waterston's most important contribution was his derivation of the equipartition of energy which showed that in a mixture of two or more gases at the same temperature the average kinetic energy of each kind of gaseous molecule was the same. A few years later in 1848 Joule made the first estimate of the velocity of gaseous molecules.

Beginning in the 1850s August Krönig (1822-79) at the Berlin Königliche Realschule in 1859, Clausius, Maxwell, and Boltzmann, working with a mechanical theory of heat, as opposed to the caloric theory of heat, made important contributions to the kinetic theory of gases that eventually put it on a firm foundation. Clausius in his 1857 paper, "The Nature of the Motion which we call Heat", used pressure measurements to calculate the mean speed of gaseous molecules. Clausius ignored molecular collisions, and in the statistical approach that he adopted he assumed that all the gaseous molecules had the same speed as they moved in all possible directions. In 1858 he introduced the idea of mean free path, which is the average distance traveled by molecules before collision, and the effective radius of a gaseous molecule. These ideas proved useful in explaining the small experimental values for gaseous diffusion rates and heat conductivities.

Maxwell, generally considered the founder of the kinetic theory of gases, also adopted a statistical approach, but unlike Bernoulli, Joule, Clausius, and all others who for simplicity had assumed that all of the gaseous molecules had the same speed, Maxwell believed that the collisions among the molecules gave them different speeds. The result of his 1859-60 mathematical statistical approach was the familiar bell-shaped curve showing the distribution or percentage of the molecules (as the y axis or ordinate) as a function of the velocities (x axis or abcissa). The formula that he derived from this work for the pressure of a gas was the same as the formula based on the belief that all the molecules had the same velocity, but Maxwell's equation resulted from the correct understanding of the molecular velocities.

A few years later in 1866-67 Maxwell used viscosity measurements (resistance to internal flow) of a gas to calculate the mean free path of a gas, that is, the average distance a gaseous molecule traveled between collisions. He now assumed that the collisions were not perfectly elastic and that the molecules having different velocities slid over one another, causing friction among the molecules and giving a measure of the viscosity of a gas. The greater the viscosity the longer the mean free path. Maxwell also showed that the viscosity of a gas did not depend on its density. His conclusions remain true today.

Boltzmann generalized Maxwell's conclusions in explaining the behavior of gases using statistical methods in his 1860s-70s publications on the kinetic theory of gases. He showed that Maxwell's distribution of velocities was the only possible equilibrium state of a gas, leading to the famous Maxwell-Boltzmann statistical interpretation of the second law of thermodynamics. Their work linked thermodynamics and mechanics. The equilibrium state represents the condition of maximum entropy or disorder according to Boltzmann's 1877 equation.

| BOLTZMANN EQUATION FOR ENTROPY | | |
|---|---|---|
| S | = | k log W |
| S | = | entropy |
| K | = | Boltzmann's constant, $1.38 \times 10^{-23}$ Joules/degrees Kelvin |
| W | = | probability or number of ways (microstates) of constructing a system |

## VII. SCIENTIFIC STUDIES ON ATMOSPHERIC AND WATER POLLUTION

### *Atmospheric pollution*

Atmospheric pollution has a long history, dating from the first natural disasters such as volcano eruptions, and the first burning of wood or roasting (smelting) of copper, lead, iron, aluminum, and other metallic ores, but it became a serious problem beginning with the industrial revolution in the late 1700s-early 1800s. The construction of the first industrial plants in England and their consumption of large amounts of coal for fuel, followed by a similar pattern of industrialization in Europe and the United States, brought considerable change to the environment. The rapid pace of industrialization in the last half of the nineteenth century, and the emergence of the automobile industry in the first decades of the twentieth century, added to the growing problem of atmospheric pollution. The automobile contributed to city sanitation because it put an end to the bodies of dead horses on the streets and to the tons of manure that horses dropped on the streets. A healthy horse deposited about 20 pounds of manure daily. Automobile exhaust, however, put large amounts of nitrogen oxides, sulfur dioxide, carbon monoxide, and other organic vapors into the atmosphere.

Nineteenth-century society and its emerging industrial entrepreneurs acknowledged the benefits the industrial revolution brought them, but they largely ignored contemporary writers' recognition and reporting of the disastrous effects of industrialization on air quality. Environmental pollution remained a neglected side of the industrial revolution. Moreover, few within the scientific community focused their research on atmospheric pollution. As a result of a widespread and lingering uninterest in atmospheric pollution and temperature fluctuations, significant studies on these environmental issues commenced only in the 1950s.

### *Acid rain*

Most of the nineteenth-century publications on industrialization's disastrous effects on air quality dealt with smoke pollution from burning soft bituminous coal. Carbon particles (soot) resulting from the combustion of coal and suspended in air constituted coal smoke, commonly called smoke. Fog is a suspension of water droplets in air. When smoke combines with fog, the combination of smoke and fog results in smog. The combustion of coal usually produces sulfur dioxide and nitric oxide, and their release and presence in the atmosphere accounts for the acidity of fog, smog, and rain. Few people recognized the problem of acid rain. The first and possibly most significant analysis appeared in an article that the Scottish chemist Robert Angus Smith (1817-84) published in 1852.

Smith, whom Queen Victoria (1819-1901) in 1863 appointed the first inspector in the Alkali Acts Administration, accidentally observed an unusual acidity in Manchester's rains during the winter of 1852. He believed the soot particles released into the atmosphere from burning bituminous soft coal, and the cause of the city's smog, were also the reason for the acidity of the precipitated fog. His 1852 article "On the Air of Towns" reported cases of stunted tree growth and an acidity of the air that was at times high enough to turn litmus paper red immediately during a rain. The acidity resulted from the conversion of sulfur in the coal to sulfur dioxide gas and its absorption by soot. Smith also estimated the amounts of carbonic acid and organic matter in the air and compiled figures showing a higher death rate in urban areas because of atmospheric pollution.

In a final set of experiments in which he bubbled polluted air through blood samples of inhabitants from different towns including Manchester, Smith concluded that the town's atmosphere had a peculiar effect on the blood's state. It caused a "remarkable reddening" "distinctly perceptible to the ordinary eye." Smith coined the phrase *acid rain* in his 1872 book *Air and Rain*.

Despite the significance of Smith's discovery of rain's acidity, his work failed to make an impact in its day. The explanations below answer the question: Why not? In the opening lines of the preface to *Air and Rain* Smith wrote, "This volume could not have been made interesting to readers without sacrificing that which I suppose to be its best point – namely the large number of facts. Another, more popular in character, might perhaps be made to follow with propriety. We have been wearied with speculations about the Air, and it seemed to me advisable to begin building a systematic foundation with experiments which may be repeated frequently, or they have already been repeated."

Smith consciously had chosen not to write *Air and Rain* as a popular account but as a compilation of the experimental evidence he had accumulated for years to highlight the damaging effects of pollution to air, land, and water. Sir Philip Joseph Hartog (1864-1947), an English chemist and educator in London, who early in his career taught briefly at Owens College at Manchester and wrote science articles for the *Manchester Guardian,* made this point in the sketch he published on Smith in 1897-98 in the *Dictionary of National Biography*.

Hartog added a second reason for Smith's neglect and his obscurity. He believed that the attention scientists paid to Pasteur's work on the germ theory diverted attention from the inorganic impurities in air, which included the acids, and hence from the significance of Smith's work. The English physicist, chemist, and historian of science, Sir Thomas Thorpe (1845-1925), who graduated from Owens College and probably knew Smith, arrived at the same conclusion writing that "as the chemist of sanitary science, Smith worked alone." Smith never married and left no direct heirs. His nephew, W. Anderson Smith, who in 1873 migrated to Melbourne, Australia, established a scholarship in sanitary science at Manchester in 1928 in honor of his uncle.

What Hartog, Thorpe, and other scientists of that time knew but left unsaid was that Smith had called attention to the other side of industrial development, the obvious pollution not only of the atmosphere but of land and water. The nineteenth century was a time of tremendous industrial development in the western world. Industrial development meant progress and progress meant a superior culture. The scientific community, with few exceptions, and most of society accepted the destructive effects of pollution as the price of progress. Those who questioned industrial progress were powerless to do anything anyway. Whatever and whenever protest emerged, the protests were usually and justifiably against the laborers' inhumane and brutal working and living conditions as described in several well-known nineteenth-century books, such as Charles Dickens' *Hard Times* (1854) and Henry George's *Progress and Poverty* (1879). Protest about the environment seldom occurred.

Another chemist, Lajos Ilosvay (1851-1836), at the Budapest Technical University, in an 1889 article on nitrous acid in the atmosphere established the presence of nitrogen compounds (nitrates) on the surface of grass, leaves, and branches and the presence of nitrous and nitric acid in the water of different kinds of soil. Ilosvay's article published in the *Bulletin de la Société Chimique de Paris* showed that the same soils in the presence of pure air contained no nitrogen compounds even after 10 days exposure. After exposing the soils to ordinary air for 26 hours he easily detected the presence of nitrous acid. Ilosvay did not discuss the formation of acid rain and its consequences, but his work clearly substantiated earlier work on acid rain.

In an important 1895 study on air quality and the acidity of the atmosphere because of coal burning, the chemist Charles F. Mabery (1850-1927) at the Case School of Applied Science in Cleveland detected the presence of excessive amounts of sulfurous-sulfuric acids and nitrous-nitric acids in Cleveland's air. Although the decay of plant and animal bodies provided ammonia and ultimately nitric acid, the main source of these acids was the quantity of coal consumed in Cleveland. It totaled 117,157 tons of clean burning (smokeless) anthracite and 924,602 tons of dirtier bituminous coal in 1889. Bituminous coal, according to Mabery, contained on average 1 percent sulfur, which although low, nevertheless released 9,246 tons of sulfur during combustion, or the equivalent of 80 tons daily or 28,305 tons of sulfuric acid annually.

The quantity of sulfur released varied throughout the city, but samples from one area totaled 11,851 grams (26 pounds, 1 pound = 454 grams) of sulfuric acid per 1,000,000 cubic meters of air and 12,512.6 grams of sulfurous acid per 1,000,000 cubic meters of air. An atmospheric volume of 1,000,000 cubic meters, a commonly-used volume, corresponded to the volume of air contained in an area of one square mile and 1.25 feet deep.

Mabery compared these figures to studies done in London and Manchester, England, and in Lille, France, which also had high concentrations of sulfuric acid because of the large quantities of smoke released during the combustion of bituminous coal. In London, the weight of sulfuric acid present in 1,000,000 cubic meters of air was 1,693.94 grams and for Manchester 2,443.74 grams. For Lille, the figures were 2 cubic centimeters of sulfurous acid per liter of air and 0.022 grams of sulfuric acid per liter of rain water or 2,200 grams per 1,000,000 cubic meters (1 liter = 10 cubic meters).

Mabery attributed the recent increase in the sooty condition of Cleveland's atmosphere to rapid growth in wealth and population, which resulted because of Cleveland's proximity to immense deposits of cheap fuel. The consequences, he said, were clearly detrimental. In the atmosphere of a sooty city, as everyone knows, the nails of a roof are nearly consumed after 10 or 15 years, whereas nails lasted 75 years in the country. Soot, which readily absorbed sulfuric acid, when deposited on trees, contributed to their decay and destruction. The soot that landed on the chemical laboratory's window sill when washed with water and tested with litmus paper gave a strong acid reaction and when analyzed showed the presence of sulfuric acid.

The devastating effect of sulfuric acid was most apparent in libraries. Mabery noted that in 1879 his colleague William Ripley Nichols (1847-86) at MIT had demonstrated the absorption of sulfuric acid in book bindings, causing the bindings to crumble after a time. Mabery concluded with the comment that a sooty atmosphere is not conducive to either personal comfort or to health.

*Global warming*

Except for Robert Angus Smith's and Mabery's publications on acid rain, only a few warnings, acts, or inventions that showed an awareness of atmospheric pollution problems occurred in the late nineteenth and early twentieth centuries. One of them came from Svante Arrhenius (1859-1927) the Swedish chemist at Stockholm who was one of the founders of physical chemistry. In a paper published in 1896 in the *Philosophical Magazine*, Arrhenius pointed out that doubling the amount of carbon dioxide in the atmosphere would produce an average temperature rise of 5-6°C. He was investigating the influence of heat-absorbing gases in the atmosphere (greenhouse effect) in response to questions that arose in the Physical Society of Stockholm regarding causes of an Ice Age. In his investigations Arrhenius used Samuel Langley's calculations of heat emissions in the lunar spectrum. Arrhenius did not comment on European and American industrial expansion that was putting more carbon dioxide in the atmosphere, but he obviously recognized that carbon dioxide's increased presence would lead to the atmosphere's trapping more of Earth's reflected heat like the glass of a greenhouse and produce a global warming.

Twelve years later in 1908, Charles Steinmetz (1865-1923), a German electrical engineer who came to the United States in 1889 and was a consulting engineer for General Electric Company in Schenectady, New York, issued another warning about the dangers of air pollution from burning coal in his book *The Future of Electricity*. In a 1910 speech he urged electrical engineers to promote the use of electricity to reduce its cost. This followed the establishment in Pittsburgh in 1907 of the International Association for the Prevention of Smoke, the first major anti-air pollution organization. The association later became the Air and Waste Management Association. As occurred with acid rain, significant studies on global warming began only in the 1950s.

## *Water pollution and the problem of waste disposal*

### *Water and disease*

For centuries physicians, scientists, engineers, and other concerned citizenry recognized the obvious connection between clean drinking water and good health. They boiled and filtered (strained) their drinking water to improve its appearance and taste and would not drink, or bathe, in dirty or visibly polluted water. The earliest known records on treating water to improve the aesthetic qualities of taste and smell date from 2000-1500 BC. Sanskrit writings record the use of filtration through sand, gravel, or charcoal, boiling, and aeration (exposure to sunlight) as recommended methods. The Egyptians recognized a link between the cloudiness of water and an unpleasant taste and appearance, and as early as 1500 BC they added alum, a known mordant or dye-fixer, to sink and settle the suspended particles. The ruins of Cretean and Assyrian cities dating from 1500-1400 BC indicate the existence of sanitary sewers. Hippocrates (460-377 BC) in Cos recommended boiling and straining rainwater to prevent it from having a bad smell and causing hoarseness. Aristotle in Athens cautioned Alexander the Great (356-23 BC) not to let his men drink from stagnant pools but to carry boiled water with them. Centuries later, in the 1200s, in Italy's medieval Ferrara, the city passed a law that limited the number of butcher shops in an area, in order to control the waste (blood and entrails) butchers discharged into the Po River and on the land.

During the 1700s filtration became an effective technique for removing particles from water. The Italian anatomist Luc Antonio Porzio (1639-1723) in Padua, in 1685 published the first known illustration of a filtration system that included sedimentation, straining, and sand filtration. It had three filtration units, each having two compartments, one with a downward flow filter and one with an upward flow filter. Porzio designed his system to provide soldiers with safe drinking water as a result of the sanitation problems he observed during the Austrian-Turkish war of 1685. In 1746 the Parisian scientist Joseph Amy (1697-1760) received a patent for a sponge, charcoal, and wood filtration system that he sold to homes, and in 1791 the English architect James Peacock (1738-1814) in London obtained the first British patent on a filtration system. His three-tank system filtered water by passing it through graded layers of sand and gravel.

Up to this time, no one connected unpleasant smelling and tasting water sources and the often-fatal intestinal diseases later identified as cholera, typhoid fever, and dysentery. Variations of the miasma, or bad air, theory of disease, particles emanating from rotting organic matter causing the bad air, and even divine intervention received greater acceptability until the mid-1800s' investigations of John Snow (1813-58). Snow, a vegetarian, teetotaling, practicing physician in London, survived three cholera epidemics during which time London's population grew from 1.95 million in 1841 to 3.2 million in 1860.

Snow investigated the cause of a cholera epidemic that in 1831 struck England, Europe, and North America for the first time. In 1849 after the 1848-49 cholera epidemic killed 14,600 Londoners, Snow in a short pamphlet questioned the prevailing miasmic theories and proposed instead a connection between cholera, a disease that killed with terrifying quickness in as little as eight hours, and a contaminated water supply. In his definitive study carried out during London's 1854 cholera epidemic in which another 10,675 people died, Snow used microbiological and statistical evidence to prove that drinking water pumped from the Thames River contaminated with animal and human feces had killed 600 people in only 2-3 days. Snow traced the source to polluted drinking water obtained from a pump on the corner of Broad Street and Cambridge Street in London's Soho district. The well lay only three feet from an old cesspit originally located under a nearby house.

Few people, including the London Board of Health, at first believed Snow. The Reverend Harry Whitehead (1825-96), vicar of St. Luke's Anglican Church, claimed divine intervention had caused the cholera outbreak. The medical establishment generally accepted the miasma theory that poisonous vapors released from decaying plant and animal matter into the air caused the disease. Ironically, in 1854 the Italian physician Filippo Pacini (1812-83) at the University of Florence, a believer in the germ theory of disease, discovered the short comma-shaped cholera bacterium (*Vibrio cholerae)* during the 1854-55 outbreak in Florence. He performed autopsies on cholera victims, and with his microscope observed the cholera bacterium in their intestinal mucosa (mucous membrane). The medical community largely ignored his work and attributed the disease to miasma. Pacini, who died poor, received belated recognition for his discovery, culminating in 1965 when the medical profession's International Committee on Nomenclature named the cholera bacterium *Vibrio cholerae Pacini 1854* in his honor.

Fortunately, advances in bacteriology enlightened medical opinion in the next decade. First, Pasteur in Paris beginning in 1861 introduced his germ theory of disease. Then in 1880 the German physician Carl Joseph Eberth (1835-1921) at Zurich discovered the rod-shaped bacterium responsible for typhoid fever (*Salmonella typhi bacillus*) in the spleen and mesenteric lymph nodes of victims. In 1883 another German physician, Robert Koch (1843-1910), at the Health Office in Berlin, rediscovered in the intestinal mucosa of cholera victims the bacterium responsible for cholera, thereby silencing Pacini's critics and skeptics. Koch's discovery was the latest of his significant medical breakthroughs. Already in 1881 in a 48-page paper Koch demonstrated the deadly effect of a calcium hypochlorite solution on bacterial cultures. In 1882 he isolated the bacterium that caused tuberculosis (*Tubercle bacillus*), likely his greatest discovery, and in 1884 his co-worker, the German bacteriologist George Gaffky (1850-1918), first isolated the rod-shaped typhoid fever bacterium from a victim's spleen, confirming Eberth's discovery.

Dysentery, the third member of the unholy trinity of waterborne diseases, is an infection of the large intestine. The cause is either a protozoan *Entamoeba histolytica* (amoebic dysentery) or bacteria of the genus *Shigella* (bacillary dysentery). Friedrich K. Lösch (1840-75) at St. Petersburg in 1875 clearly recognized that an intestinal protozoan (parasite) caused amoebic dysentery. Confirmation came in 1903 in the tropical disease studies of the German zoologist Fritz Schaudinn (1871-1906) at Berlin. Schaudinn in tests on himself demonstrated that *Entamoeba histolytica* was the pathogen of amoebic dysentery and not the harmless *Entamoeba coli*. He named both protozoa.

Discovery of the pathogen responsible for bacillary dysentery, or bloody diarrhea, occurred in 1897. The Japanese bacteriologist Kiyoshi Shiga (1871-1957), then at the Institute for Infectious Diseases in Tokyo, isolated the rod-shaped dysentery bacillus from tissue in the large intestine. Shiga named the bacterium *Bacillus dysenterie* (in 1930 renamed *Shigella dysenteriae*). Confirmation of Shiga's discovery came three years later when Simon Flexner (1863-1946), then at Pennsylvania, isolated the dysentery bacillus.

| THE THREE DEADLY WATERBORNE DISEASES | | | |
|---|---|---|---|
| *Disease* | *Discoverer* | *Year* | *Effect on Humans* |
| Cholera | John Snow in London established the link between contaminated water and cholera | 1854 | Abdominal pain and watery diarrhea, rapid dehydration and shock |
| | Filippo Pacini at Florence discovered the cholera bacterium (*Vibrio cholera*) | 1854-55 | |
| | Robert Koch in Berlin isolated the cholera bacterium | 1883 | |
| Typhoid Fever | Carl Joseph Eberth at Zurich discovered the bacterium *Salmonella typhi bacillus* | 1880 | Fever, rash, spleen and lymph node enlargement, gastrointestinal bleeding and ulceration |
| | George Gaffky in Berlin isolated the typhoid bacterium | 1884 | |
| Dysentery (amoebic) | Friedrich Lösch at St. Petersburg recognized that the parasite/protozoan *Entomoeba histolytica* caused dysentery | 1875 | Diarrhea and constipation accompanied by mucus and sometimes blood |
| | Fritz Schaudin at Berlin confirmed Lösch's work | 1903 | |
| Dysentery (bacillus) | Kyoshi Shiga in Tokyo isolated the bacterium *Bacillus dysenterie (Shigella dysenteriae)* | 1897 | Abdominal pain, diarrhea, and fever associated with poor hygiene |
| | Simon Flexner at Pennsylvania isolated the dysentery bacillus | 1900 | |

As a result of these advances in bacteriology and the increasing acceptance that water polluted with human and animal feces, animal parts, sewage, and other contaminants caused cholera, typhoid fever, and dysentery, cities in Europe and the United States in the nineteenth century took three important steps to purify their water supply systems.

*Three stages of water purification*

First came the construction of water distribution systems (waterworks) consisting of pumping stations and underground pipes to bring water from nearby rivers, lakes, or bays into the cities, and underground sewer systems to carry the raw sewage and other waste matter from the cities into the rivers, lakes, or bays far from the pumping stations. Air and sunlight (aeration) decomposed some of the undissolved solids in the polluted water. Sand filtration, and then coagulants with sand filtration units that removed undissolved solids and considerable bacteria, viruses, and protozoa (pathogens), followed. Finally in the first decades of the twentieth century, city after city began adding chlorine to its drinking water sources killing nearly all the pathogens and resulting in the elimination of waterborne diseases. Taking into account that a human on average drops 3.59 ounces of solid stool a day (possibly 91 pounds a year), that a horse can deposit as much as 20 pounds of manure per day and that some of the waste found its way into city water supplies, the public expected water purification.

- *Water distribution and sewer systems*

The rapid industrialization and urban population growth in the closing decades of the nineteenth century accelerated the construction of water distribution systems and use of piped water in the home. By that time the 16 largest cities had constructed waterworks. By the 1890s-early 1900s the addition of home bathrooms with sinks and bathtubs, and especially the significantly improved flush toilets, increased the need for sewer systems to deal with the problems of waste disposal and the water pollution it caused.

The earliest flush toilet in Europe dates from 1596 when Sir John Harrington (1561-1612), a London writer and Queen Elizabeth I's godson, built toilets for himself and for the queen. His toilet, named *Ajax*, consisted of a seat, a bowl, and a cistern (water tank) located behind the bowl and seat for removing the waste with running water.

Almost 200 years passed before Alexander Cummings (1733-1814), an Edinburgh-born London watchmaker, in 1775 patented an improved flush toilet. Cummings' toilet contained a sliding valve between the toilet bowl and the trap (a u-shaped joint), and pipes that carried waste to a sewer. Water contained in the lower part of the trap prevented bad odors from passing back through the toilet and into the surroundings. In later toilet designs the bad odors escaped through a pipe in the roof. Cummings' toilet cost 6 shillings 8 pence ($1.25 or £ 25.36 = $38.55 in 2002) to construct.

From 1861 Thomas Crapper (1836-1910), a London plumber, improved the flush toilet. He perfected the existing toilets' pull chain and cistern, which functioned to discharge sufficient water at high speed to cleanse the bowl. Contrary to popular legend, Crapper, who had nine toilet patents, did not invent the flush toilet. Until 1885, when Crapper's partner the London pottery maker Thomas Twyford (1849-1921) built the first one-piece all-porcelain trap-less toilet, toilets were of wood and metal.

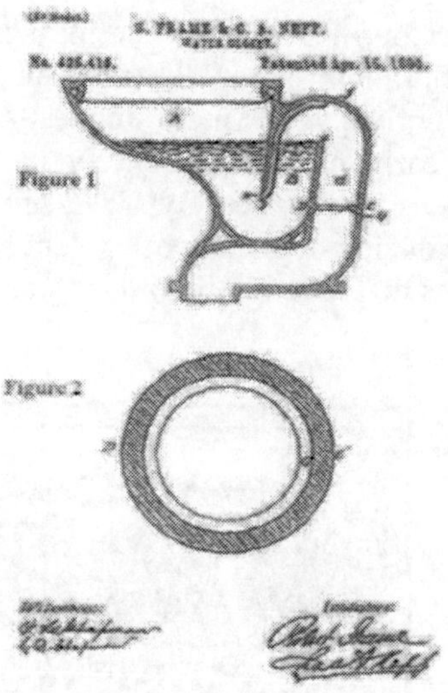


**Figure 44**
**Siphon Water Closet**
Source: US Patent 426, 416 of 15 April 1890

In the United States, several inventors contributed to the perfection of the flush toilet. They included James T. Henry and William Campbell in Philadelphia, who in 1857 received the first American patent for a plunger type flush toilet, and George Waring (1833-98) in Newport. Waring in 1882 transformed the jet siphon flush toilet that William Smith in San Francisco patented in 1877 into a full-size porcelain toilet. (Henry and Campbell, US Patent 18,972 of 29 December 1857; Smith, US Patent 190,919, filed 20 October 1876, patented 15 May 1877; Waring, US Patent 266,404 of 24 October 1882; Frame and Neff, US Patent 425,416 of 15 April 1890; Adee, US Patent 436,451 of 16 September 1890.)

The Dececo Company in Boston marketed the improved toilet as the Dececo siphon water closet (US Patent 425,416; 15 April 1890). Its features, such as its large area of standing water and self-siphoning action enabled Dececo to sell a significant number of toilets. In Washington, DC, most of the recently-constructed first class hotels, office buildings, including the Capitol Building, had Dececos. Because of these improvements, the number of toilets in American cities increased markedly. Boston in 1860 had 6,500 flush toilets, 100,000 in 1885. In 1880, 25 percent of households in Chicago, population 503,185, had toilets. The price of a toilet varied widely. A three-piece bathroom set consisting of a tub, lavatory (sink), and toilet that the J. L. Mott Iron Works in New York City listed in its 1898 *Trade Catalogue* cost anywhere from $200 to $1,200. Standard Sanitary Manufacturing Company in Pittsburgh gave a price of $69.75 for a bath, lavatory, and toilet. The Fred Adee Company in New York City sold its siphon toilet for $41.00 in 1898.

Toilet paper appeared at this time. Joseph Gayetty (b.1827) in New York City introduced Gayetty's medicated, or therapeutic, paper, in 1857. Each sheet had his name imprinted on it and contained aloe, a healing plant extract. A package of 1000 sheets sold for $1.00. Gayetty's toilet paper replaced newspapers, catalogue pages, and leaves. Twenty years later, in 1880, the British Perforated Paper Company patented and sold toilet paper in boxes containing small pre-cut squares. The brothers Thomas, Edward (1846-1931), and Clarence Scott, who in 1867 established the Scott Paper Company in Philadelphia, in 1890 introduced toilet paper on a roll. By that time Scott had become the country's leading toilet paper producer.

• *Slow and rapid sand filtration systems*

The second advance in water purification was the introduction of filtration systems in which water passed either slowly (slow filtration) or rapidly (rapid or mechanical filtration) through or over fine gravel or sand. Slow sand filtration, which in its simplest application dates from 2000-1500 BC and in improved design from the early nineteenth century in Europe, removed cloudiness, undissolved particles, and deadly pathogens from water. By 1900, scientists in Europe and the United States had proved its effectiveness in reducing outbreaks of cholera and typhoid fever in cities that filtered their water compared to those cities that failed to filter their water.

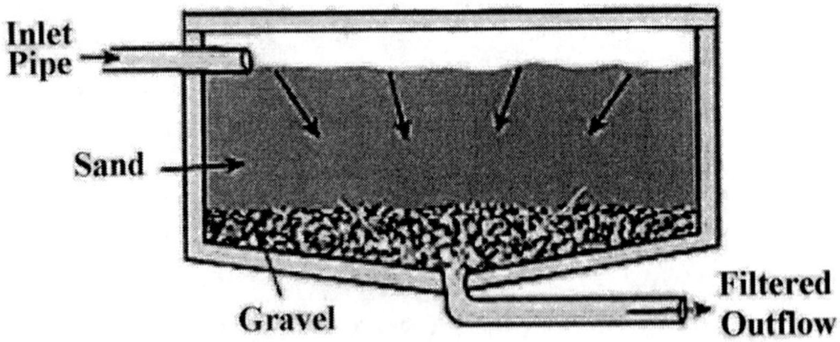


**Figure 45**
**Sand Filtration**
Source: © Kendall Hunt Publishing

In slow sand filtration, water passed down through a layer 1.0-1.5 meters or more of washed, uniform, fine sand at the rate of 100-200 liters per hour per square meter (3 to 6 gallons per hour per square foot) of sand area. It effectively treated low-turbidity small water sources such as well or spring water by removing clay and pathogens (polio virus, cholera, typhoid fever, and dysentery bacteria). The top 1 to 3 millimeters of sand, which after one or two days formed a sticky mat (schmutzdecke), contained biologically active microorganisms, such as predatory bacteria of the genus *pseudomonas* and trichoderma that came from the water. The microorganisms broke down organic matter and eliminated 91-99 percent of the pathogens. The sand filter bed eliminated, by straining, any suspended inorganic or organic matter. Plant workers regularly removed the top layer of sand to clean the filtration system, and after a few days a new sticky mat formed.

John Gibb, a bleacher in Paisley, Scotland, built a municipal water (slow sand) filtration system in 1804. Gibb supplied his bleachery and the entire town with filtered water he delivered by cart. By 1807 his sand-filtered system was pumping and piping filtered water in Glasgow. Robert Thom, an engineer in Paisley, improved Gibb's filtration system with his invention and patenting in 1827 of a water backlash process that reversed the direction of the water flow and much more effectively cleaned the sand filter. In Paris city officials in 1806 opened a large slow sand and charcoal filtration water treatment plant to purify water from the Seine River. The water settled for 12 hours before filtration. Every 6 hours, workers renewed the sticky surface mat that filtered the water. The plant operated for 50 years.

James Simpson (1799-1869), an English scientist and civil engineer, initiated the construction of slow sand filtration systems to treat municipal water supplies throughout the United Kingdom. In 1829 he constructed London's first slow sand filtration system, a one-acre filtration system that the Chelsea Water Works Company used to purify water from the Thames River. Yet two decades elapsed before London in 1852 passed the Metropolitan Water Act that required filtration of its water supplies.

The London cholera outbreak of 1854, which resulted from a contaminated public well, spurred more attention to polluted waterways. But in an 1855 letter to the *London Times* Faraday lamented the filth of the Thames, its bad smell and pale brown color, and the feculence that rolled up in dense clouds near the bridges. He feared it would become a fermenting sewer if the citizens of London continued to neglect its condition. Similarly, in 1867 a Royal Commission on River Pollution reported that the Aire and Calder Rivers were poisoned, corrupted, and clogged by refuse from mines, chemical works, slaughter-house garbage, and the sewage from houses.

**Figure 46**
**Faraday Cartoon**
Source: *Punch*, 21 July 1855

The Thames water still smelled in 1868, when Edward Frankland (1825-99), London's official water analyst from 1865 to 1876, carried out a four-month laboratory filtration study of London's raw sewage-laden water. Frankland passed the water through columns packed with substances such as coarse gravel and peaty soil. His study, which included chemical and bacterial analyses, left little doubt about the poor quality of the Thames water. Frankland issued monthly reports on London's water supply until his death in 1899. A second act, passed during the years of Frankland's studies, the River Pollution Control Act of 1876, tried to prevent the continuing deterioration of England's rivers by making it illegal to dump sewage into a waterway.

Frankland's filtration study preceded a major scientific and medical breakthrough in water purification that occurred in Germany. In 1892 Koch first clearly demonstrated the effectiveness of slow sand filtration in reducing cholera deaths in his comparative study of the drinking water in two German cities, Hamburg which did not filter its water and Altona which did. Both of these contiguous cities drew their water from the Elbe River, but Altona, which had to filter its water, now badly polluted because of its location downstream from Hamburg, experienced far fewer deaths from cholera than Hamburg: 230 per 100,000 population versus 1,344 per 100,000 population.

In the United States, Richmond, Virginia, on the James River operated the first slow sand filtration plant. Richmond began filtering its water in 1832 and had a centralized treatment facility serving 295 customers in 1833. Other cities with slow sand filtration systems included Elizabeth, New Jersey, 1855; Poughkeepsie, New York, 1872; Lawrence, Massachusetts, September 1893; Philadelphia 1902-11; Pittsburgh 1908; and Evanston, Illinois, 1914. Toronto, Ontario, Canada, had a slow sand filtration plant that filtered 180 million liters per day in 1909-11. The filtration increased to 275 million liters per day in 1914-17.

The slow sand filter installed at the Lawrence plant cost $31,000 per acre of filtering surface, whereas the 8 filters (filter beds) installed at the Albany, New York, plant in 1899 cost $45,600 per acre of filtering surface. The total cost of the Albany water treatment plant, with a daily capacity of 60 million liters (15 million gallons), was $499,890. The five filters at the Philadelphia plants each cost $145,000 and $187,000 per acre of filtering surface, including the cost of the 66 million liter (16.5 million gallons) reservoir. The cost of the largest plant, at Torresdale (1909), with a daily capacity of 960 million liters (240 million gallons) totaled $9,208,000. At the Pittsburgh plant, each of the 46 one-acre covered filters cost $78,000, the complete water treatment plant of 480 million liters (120 million gallons) daily capacity cost $5,970,971.

**TYPHOID FEVER DEATH RATE PER 100,000 POPULATION**

| *City* | *Several Years Before Installation of SSF* | *Several Years After Installation of SSF* | *Decrease Percentage* |
|---|---|---|---|
| Cincinnati | 54 | 10 | 81 |
| Columbus | 83 | 16 | 80 |
| Louisville | 58 | 17 | 71 |
| Lawrence (MA) operational 1893 | 122 | 17 | 86 |
| Minneapolis | 35 | 4 | 88 |
| New Orleans | 40 | 23 | 43 |
| Philadelphia operational 1906 | 63 | 20 | 68 |
| Pittsburgh operational 1908 | 120 | 18 | 85 |
| Hamburg (Germany) | 51 | 9 | 87 |

The circumstances that led to the construction of the slow sand filtration plant in Lawrence were similar to the events that precipitated Koch's 1892 study. The Massachusetts State Board of Health in 1887 had established an experiment station in Lawrence to study the treatment of drinking water obtained from the sewage-polluted Merrimack River. In the midst of the water tests, the 1890-91 typhoid fever epidemic struck Lawrence, an industrial city of 45,000 with a 45 percent immigrant population.

Hiram Mills (1836-1921), chief engineer, and Allen Hazen (1869-1930), chief chemist at the experiment station, in 1893 designed and constructed a slow sand filtration system to remove typhoid fever bacteria, and in post-1893 studies the biologist William Sedgwick (1855-1921) at MIT established its effectiveness in removing bacteria. The number of bacteria fell from 12,700 per cubic centimeter of water to 70 per cubic centimeter of water, Lawrence's typhoid fever deaths dropped by 86 percent following the installation of the slow sand filtration plant. Typhoid fever deaths in the United States dropped from 48 per 100,000 population in 1890 to 13 per 100,000 population in 1917 because of filtration and the introduction of chlorine early in the twentieth century.

Slow sand filtration remained widely used in the United States until about 1920. It proved uneconomical for large-volume water purification systems because slow sand filters require large land areas to supply sufficient amounts of filtered water. The average cost in 1917 was about $60,000 per acre of filtering surface, daily capacity per acre was 12 million liters (3 million gallons). It has limited application in the United States, mainly for small-scale water purification systems, but remains in use in Europe and Asia.

| **LAWRENCE, MASSACHUSETTS<br>TYPHOID FEVER DEATHS** | |
|---|---|
| ***Years*** | ***Deaths / 100,000*** |
| Years before SSF operational 1888-92 | 129 |
| Year SSF operational 1893 | 60 |
| Years after SSF operational 1894-98 | 17 |

Rapid sand filtration first appeared in the mid-1880s and gradually replaced many of the older slow sand filtration systems. Rapid sand filtration systems generally cost less to construct but more, sometimes twice as much, to operate and maintain as slow sand filtration systems. In rapid sand filtration, water passes through a coarse sand filter at the rate of 3,200-4,800 liters per hour per square meter (2 to 3 gallons per minute, or 120-180 gallons per hour per square foot) of sand area. For the same area, rapid sand filtration acts 20 to 40 times faster than slow sand filtration.

The introduction of coagulation hastened the development of rapid sand filtration. Coagulants are inexpensive chemical compounds such as calcium hydroxide also called milk of lime $Ca(OH)_2$ , iron (III) chloride $FeCl_3$, iron (II) sulfate $FeSO_4$, and alum $K_2SO_4 \cdot Al_2(SO_4)_3 \cdot 24H_2O$, a dye fixer. Alum in 1916 cost $20 per ton, iron (II) sulfate $11 per ton. When added to turbid water, they caused undissolved solids in the water to clump together (flocculation) because of electrostatic attraction. The clumping increased particle size and weight, enough to make the solids sink to the bottom of the container. Because of its rapidity, rapid sand filtration, with coagulants added, removes some bacteria, protozoa, and viruses, but it is not as effective as slow sand filtration in removing bacteria and protozoa, and removes few viruses.

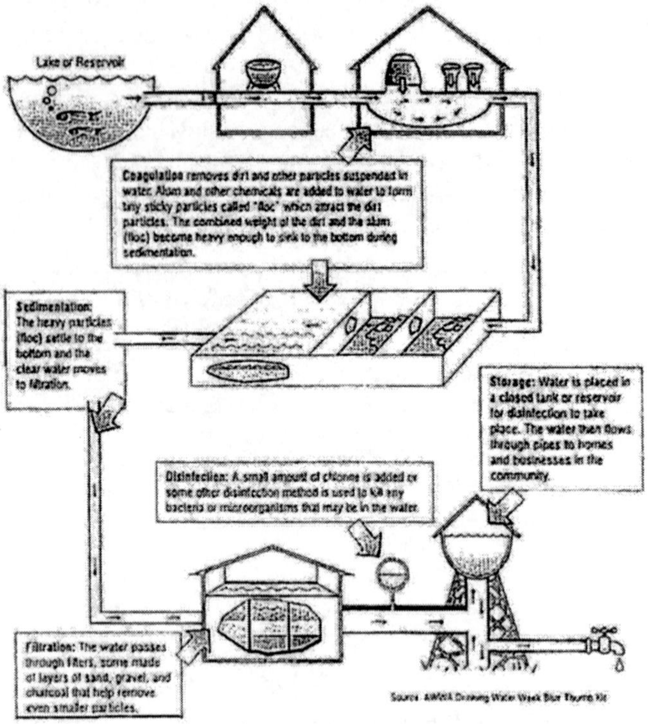


**Figure 48**
**Water Distribution and Purification System**
Courtesy of the EPA

Isaiah Smith Hyatt (1830-85) in Morristown, New Jersey, an associate of New Jersey's Newark Filtering Company and older brother of John Wesley Hyatt (1837-1920) celluloid's inventor, patented a coagulation-rapid filtration process in 1884. He used iron (III) sulfate $Fe_2$ $(SO4)_3$, iron (III) chloride $FeCl_3$, potassium permanganate $KMnO_4$, or any coagulating agent (a positively-charged chemical coagulant), to attract negatively-charged impurities in water (US Patent 293,740 of 19 February 1884). The coagulation or curdling produced a jelly-like substance, or flocculent (floc), that filtration easily removed. The Newark Company in 1885 installed the first Hyatt alum filter and coagulant system in Somerville, New Jersey. By the end of the nineteenth century, coagulation-rapid sand filtration systems outnumbered slow sand systems 10 to 1. Worcester, Massachusetts, in 1890, Louisville, Kentucky, in 1896, and New Milford, New Jersey, in 1903, were the first cities to apply coagulation and filtration. Worcester's chemical precipitation plant had six chemical precipitation settling basins, each 662 feet x 100 feet x 7 feet, for screening and treating raw sewage with calcium hydroxide. Each settling basin had the capacity to treat 350,000 gallons of sewage daily, although the plant's design enabled it to treat 3 million gallons of sewage daily. The treated sewage passed through a mixing channel into the six settling basins connected in series in about six hours. After two hours, when the top water layer became calm, workers drew it off for filtration and passed the bottom sludge to a centrifugal pump that discharged the sludge into nearby lagoons. The addition of ten more settling basins in 1893 increased the plant's capacity to 5.5 million gallons and the total cost of the precipitation plant to $200,000.

The combined coagulation-rapid sand filtration process passed its first crucial test in 1908. In 1884-85, the Louisville Water Company's chief engineer, Charles A. Hermany (1830-1908), who in 1861 succeeded Theodore R. Scowden (1815-81) now in Louisville, had tried settling the turbid water of the Ohio River in a reservoir, but failed mainly because the reservoir was too small to provide clear water.

George W. Fuller (1868-1934), a MIT educated chemist whom Hermany and Louisville's Board of Water Works hired in 1895, provided the solution. He pretreated the water with a combination of coagulation with alum and sedimentation before filtration. Fuller published a report of his successful combined treatment in 1898. Ten years later, in 1908, Louisville's Crescent Hill Water Filtration Plant tested its first filters and in 1909 began operation. Louisville's combined treatment eliminated turbidity and claimed 99 percent removal of bacteria from its water.

The success of Louisville's coagulation and sedimentation treatment contributed to Fuller's reputation as the father of American sanitary engineering. With the introduction of chlorine in 1914, Louisville's death rate from typhoid fever dropped 80 percent, from 71 per 100,000 in 1907 to 14 per 100,000 in 1915. Its coagulation and sedimentation treatment became the model and standard and paved the way for widespread adoption of rapid sand filtration in the United States.

- *Chlorine treatment*

The final advance in the water purification process was the addition of chlorine gas or hypochlorite compounds to kill deadly bacteria and disinfect water. The Swedish chemist Karl Scheele (1742-86) at Uppsala discovered chlorine gas in 1774 when he reacted hydrochloric acid with manganese dioxide. Establishing the chemical composition of substances posed a serious problem in Scheele's time, and decades passed before Humphry Davy at the Royal Institution in London in 1810 identified the gas as an element and named it chlorine, derived from the Greek *chloros* for pale green.

Chlorine's potential as a commercial product began in 1785 when Claude Louis Berthollet (1748-1822) in Paris first observed the bleaching properties of chlorine solutions on cloth. A few years later, in 1789, chlorine found limited use as a textile bleachery solution, named *Javelle water*, for the French town in which textile producers prepared the bleach by passing chlorine gas into an aqueous solution of potassium carbonate (potash). The reaction gave a potassium hypochlorite (KClO) solution. Using sodium carbonate (soda ash) gave an equally effective sodium hypochlorite (NaClO) bleaching solution.

Another 40 years elapsed before chlorine solution found use as a germicide and disinfectant. Antone Labarraque (1777-1850), a Paris chemist, in 1820 experimented with sodium chlorite solution as a disinfectant, and in 1827 Thomas Alcock (1784-1833), a London surgeon, published an essay in which he recommended using solutions of sodium hypochlorite or calcium hypochlorite, $Ca(ClO)_2$, to disinfect hospitals, reservoirs, and sewers.

The major breakthrough that supported Labarraque's and Alcock's research and led eventually to the widespread application of chlorine solution as a disinfectant, occurred in 1847. The Hungarian physician Ignaz Semmelweis (1818-65) at the General Hospital in Vienna ordered his medical students to wash their hands with bleaching powder (chlorinated lime or chloride of lime) solution, $CaCl(ClO).4H_2O$, each time they treated a new patient.

Semmelweis suspected the students were spreading deadly germs when they went directly from performing autopsies and dissecting corpses to the maternity ward and other wards for clinical instruction without washing their hands. The students examined pregnant women or attended to women in labor and, as Semmelweis proved, their germ-bearing hands caused the hospital's high female mortality rate from puerperal fever (child bed fever). The medical community largely opposed or ignored Semmelweis' results, which showed that the female mortality rate at the General Hospital dropped significantly from 18 to 2.4 percent. Physicians argued that they saved lives and did not contribute to their patients' deaths.

Snow's addition of chlorine to disinfect a pump water supply in London following the 1854 cholera outbreak, likely the first known application of chlorine gas to disinfect water, supported the earlier evidence demonstrating the effectiveness of chlorine and hypochlorites as disinfectants. Koch's 1881 demonstration of the deadly effect of calcium hypochlorite on bacterial cultures provided more evidence. Yet, chlorination for water purification began to gain general acceptance first in Europe and then in the United States only near the turn of the century. Perhaps the public feared the consequences of drinking a liquid containing the deadly gas chlorine or the irritant and toxic bleaching agents, the hypochlorite compounds. At that time public waterworks added either chlorine gas or the equally effective hypochlorites or chlorinated lime to purify their water supplies. These compounds were safer, provided the same protection against pathogens as chlorine gas, but more expensive.

One of the earliest commercial-scale applications of chlorine added to water as a disinfectant occurred in 1893, the year after Koch's cholera study, when the city of Hamburg added chlorinated lime to its water. A second application followed in1897 in Maidstone, Kent, England, when German Sims Woodhead (1855-1921), a Cambridge pathologist, used a sodium hypochlorite bleach solution as a temporary measure to sterilize potable water following a typhoid outbreak. The first municipal use of chlorine gas occurred in 1903 when the Belgium Ministry of Public Works added chlorine gas to disinfect drinking water in Middlekerke, Belgium. The next year, in 1904, London added sodium hypochlorite (10 percent available chlorine regulated to give a concentration of one part per million) to disinfect its drinking water, to stop a typhoid fever epidemic that struck the city. Alexander Houston (1865-1933), a medical doctor and a director of the London Metropolitan Water Board, and his associate Cecil McGowan, carried out the first systematic use of chlorine gas or a chlorine compound to treat a large water supply, namely London's.

Chlorine treatment began in the United States in 1908. The Jersey City Water Supply Company in Boonton, New Jersey, upon the recommendation of John Leal, a medical doctor and adviser to the Jersey City Water Supply Company, on 26 September 1908 added chlorine as sodium hypochlorite to its urban water supply, the Boonton Reservoir. The reservoir located 23 miles from Jersey City, a city of 200,000, began providing Jersey City with water in 1904. A Rockaway River tributary was polluting the reservoir, and Leal working with George W.Fuller of Hering and Fuller, Hydraulic Engineers and Sanitary Experts, designed the chlorine purification system.

Fuller (1868-1934 and his partner Rudolph Hering (1847-1923), father of the modern American municipal sewerage system, had excellent reputations in sanitary engineering. Their company established in New York City in 1901 manufactured water treatment equipment. Jersey City adopted their chlorination system because adding the cheap and abundant hypochlorite compound to the reservoir spared it the cost of having to build a more expensive sand filtration system. The chlorination cost 14¢ per million gallons of water, or $5.60 per day for Jersey City's 40 million gallon per day water consumption.

Jersey City provided the United States' first urban chlorination treatment. Other cities that introduced chlorination included Albany, New York, 1909; Minneapolis 1910, Cleveland and Pittsburgh 1911, Louisville 1914, Salt Lake City 1915, Redding, California, 1916, and Evanston Illinois, 1921. By 1918 over 1,000 American cities were treating 3 billion gallons of water per day.

Already in the 1890s European and American industries were producing the chlorine-containing chemical compounds so crucial to water purification. The Mathieson Alkali Works, which in 1892 constructed its Saltville, Virginia, plant, began in 1896 to produce bleaching powder (chlorinated lime) obtained from the electrolysis of brine. In 1897 Mathieson was producing chlorine-alkali products at its plant in Niagara Falls, New York. The Dow Chemical Company in Midland, Michigan, also began manufacturing bleaching powder in the 1890s. Later in 1924, James Cloyd Downs (1885-1957), a chemist at the Roessler and Hasslacher Chemical Company in Niagara Falls, New York, patented a widely used electrolytic process for the production of chlorine from molten sodium chloride. DuPont purchased Roessler and Hasslacher in 1930. The Solvay Process Company in Hopewell, Virginia, in 1936 introduced an alternative chlorine production process called the nitrosyl process.

Apparatus to deliver measured amounts of chlorine followed in 1913 when two unemployed engineers Charles Wallace (1885-1964) and Martin Tiernan (1883-1968) in New York City and later in Belleville, New Jersey, invented the first practical and effective apparatus for the controlled feeding of chlorine gas. Their $150 chlorinator, that six dry cells (6 x 1.5 volts) operated, piped the equivalent of a thimble-full (2-7 ml) of chlorine gas into 1 million parts of water (1 part per million = 1 milligram per liter, 1 ppm=1 mg/L). The Boonton Reservoir that provided Jersey City's drinking water installed the first Wallace-Tiernan chlorinator that same year.

Unquestionably the most significant advance in chlorination came in 1919 just after World War I ended. Abel Wolman (1892-1989), an engineer at the State of Maryland Department of Health in Baltimore who later taught at Johns Hopkins, and his coworker, Linn Enslow (1891-1957), a Health Department chemist, developed a safe and reliable method for determining the appropriate dose of chlorine for purification of any water source. Their work advanced the 1908 kinetic studies of Harriette Chick (1875-1977), a research student at the Lister Institute of Preventive Medicine in London, and Herbert E. Watson (1886-1980) at University College, London. Chick developed the first chemical rate of reaction equation that related a disinfectant's effectiveness in killing an organism and its contact time with the organism. That same year Watson modified Chick's equation to include a coefficient of specific lethality to give a rate equation known today as the Chick-Watson equation.

Wolman and Enslow noted that upon adding chlorine to water, the amount of chlorine absorbed varied significantly depending on the water quality, and they introduced the idea of chlorine demand to indicate the rates of chlorine absorption. The chlorine demand represented the difference between the amount of chlorine added and the amount remaining after a specified time interval. The higher the chlorine demand, the higher the bacterial count and presence of other organisms, and the more polluted the water. Over an eighteen-month period in 1917-18, they tested 350 water samples from the Potomac River and samples from other rivers, measuring the bacteria, acidity, and factors that possibly affected the taste, turbidity, and purity of a specific water source. They compared the relative chlorine demands (the amount of chlorine absorbed) of water sources containing different amounts of inorganic, organic, or biological matter, and determined the chlorine dose for safe water treatment.

Wolman and Enslow's initial chlorine concentration was 1 mg/L (1 ppm), high enough to disinfect most drinking water sources, the water temperature was 20°C (68°F), and the time intervals ranged from 5 to 60 minutes. They found that a 5 minute time interval at 20°C provided reliable results, with very little gain resulting from a 30 or 60 minute interval. They recommended adding another 0.2 mg/L of chlorine as a safety factor but opposed maintaining residual chlorine in the water distribution system. The residual chlorine, they said, would result in possible taste and odor problems. The Wolman-Enslow chlorine demand method remains the standard water pollution testing treatment used globally.

## VIII. CONCLUSION

The electromagnetic theory of nature had emerged to complement the older mechanical theory. Were they two distinct or different world views or were they two different descriptions and explanations of the same phenomena? Einstein's special theory of relativity settled the issue in 1905. Determinism dominated nineteenth-century physical and life sciences. Its central principle held that if we know precisely how a system functions at the present time we can predict its precise behavior at any time in the future. This principle followed directly from the belief in a Newtonian mechanical or machine-like universe. It remained largely unchallenged until the rapid series of discoveries and new theories that arose from 1895 to 1915. These were x-rays in 1895, radioactivity in 1896, the electron in 1897, quantum theory in 1900, special relativity in 1905, and general relativity in 1915. In 1925-27 the physicists Werner Heisenberg and Erwin Schrödinger introduced the new probabilistic quantum mechanical or wave mechanical theory of matter, officially ending determinism in science. Because of these developments no completion ever occurred in the physical sciences. Classical physical science underwent a transformation that shattered its foundation and introduced a revolution in scientific thought that continues a century later. Classical physics and determinism were out, relativity and quantum mechanics were in.

Thomas Young's revival of the wave theory of light in the early 1800s revived acceptance of the space-filling luminiferous ether. Physicists required this undetectable all-pervading medium to transmit light waves from the Sun and stars to Earth and other planets, and from their sources on Earth to any location on Earth. They also asked how fast these light waves traveled as they moved through space or through any medium. Beginning in 1849 the French physicists Fizeau, Foucault, and Cornu invented the first terrestrial methods of calculating the speed of light. They abandoned the celestial methods, Roemer's 1676 method of measuring the time interval between successive eclipses of a Jupiter moon and Bradley's star aberration method of 1729.

Michelson in the United States perfected the new and more accurate terrestrial method of calculating the speed of light. His measurements raised the question of whether the speed of the light waves moving through the ether had an absolute value. Was the speed independent of the direction in which the light moved with respect to the direction of Earth's motion? Or, did the speed depend on whether the light moved in a direction parallel to Earth's motion or perpendicular to Earth's motion, or in any other direction with respect to Earth's motion?

Michelson and most nineteenth-century physicists accepted the ether's reality, and in his well-known 1887 ether drift experiment he expected to detect a small but measurable direction-dependent difference in the speed of light. Michelson obtained no difference, casting doubt on the ether's presence. The issue remained unresolved until 1905 when Einstein introduced his special theory of relativity which dispensed with the ether's presence, not by disproving its existence, but by declaring and showing its irrelevance. Neither Einstein nor anyone else could disprove the existence of something that never existed. Einstein banished ether from physics, rendering it unnecessary, just as Maxwell and Boltzmann earlier banished caloric, the fluid of heat, from thermodynamics.

Physical science in the last decades of the nineteenth century had appeared on the verge of completing the scientific revolution that began two centuries earlier. An electromagnetic (electrodynamic) interpretation of the universe and its laws arose from the electrical studies of Faraday and Maxwell in England and Henry in the United States. Faraday, an excellent experimental physicist but not a great mathematician, expanded the electromagnetic investigations of Oersted and Ampere. His experiments established several important relations, specifically the motor principle, transformer and generator principle, and electromagnetic induction, which he later summarized in *Experimental Researches in Electricity*, 3 volumes, (1839, 1844, 1855). The electrodynamic interpretation of the universe, which required the existence of electromagnetic lines of force and electromagnetic fields permeating space underwent a quantitative mathematical advancement in the work of Maxwell and essentially completed the program of classical physical science.

Other problems remained unsolved, but the scientific community did not see them as unsolvable. Dalton's early nineteenth-century structureless indivisible atom, usually called the billiard ball chemical atom, failed to account for the spectrum of an atom because atomic spectra seemed to require an atom with vibrating parts. Nor could the structureless atom account for the bonds that held together the atoms composing a molecule other than to rely on the affinity one atom had for another atom. The electron's discovery in 1897 solved these problems.

In the 1860s, chemists agreed on a standard for atomic weight determinations, accepting Stas's recommendation of assigning oxygen an atomic weight of 16, 0 = 16, as the standard. Two new instruments, the Bunsen burner and especially the spectroscope, helped chemists to carry out their analyses of compounds, leading to advances in determining the composition and structure of matter.

Nineteenth-century chemical studies went from knowing only a chemical compound's composition to determining the arrangement or location of each atom in a molecule and thereby establishing the structure of a molecule, or its structural formula. Kekulé and Couper in 1858 were the first to provide structural formulas for the carbon compounds of organic chemistry, particularly Kekulé's structure of benzene in 1865. Within a few years, in 1874, their planar, or flat, structures had acquired a spatial orientation.

About the same time, in 1869, Mendeleev published his first Periodic Table of the chemical elements. He provided physical scientists with their long sought classification system for the steadily increasing number of elements.

The study of heat, which had its origin in Black's 1760s research on specific heats and the heat content of different substances, became an important area of investigation known as thermodynamics. In studies on thermodynamics, the caloric theory disappeared, replaced by the kinetic-molecular theory.

The first two laws of thermodynamics received a formal statement from Helmholtz and Clausius in the 1840s-1860s, the third law not until 1906 and in its final form in 1920. Thermodynamics itself became an integral component of a new branch of chemistry established in 1887, then called theoretical chemistry, and later physical chemistry. Gibbs made important contributions to thermodynamics with his investigations that clarified the meaning of entropy and his introduction of the all-important conceptions of free energy and the phase rule and phase diagram in 1875 and 1878.

In the nineteenth-century, science increasingly relied on the discovery of scientific principles and laws for its advancement. Application of these principles and laws led to many new scientific processes and inventions and thereby to science's increasing impact on society. The technological transformation that occurred in the second half of the nineteenth century occurred largely because of the new scientific outlook. Scientists increasingly understood the principles and laws of science. Unfortunately, the technological transformation that brought the industrial world into existence and provided numerous benefits to society created serious environmental problems, particularly atmospheric and water pollution. Environmental pollution, either natural or human, had existed from the time of Earth's formation but increased significantly at this time. Nineteenth-century society, however, largely ignored the other side of the technological transformation. Recognition and acceptance that another side existed, the reality of human-caused environmental pollution, occurred only in the second half of the twentieth century.

## REVIEW QUESTIONS ON THE PHYSICAL SCIENCES IN THE NINETEENTH CENTURY

1. What is the main theme of this chapter?

2. What was the guiding philosophy or belief that most nineteenth-century physical scientists accepted?

3. What was the accepted model of light in the nineteenth century? Why did scientists adopt a new model of light and who were its main contributors and supporters?

4. What similarities existed in the nineteenth-century methods used to measure the speed of light? How did they compare to the earlier methods of measuring the speed of light?

5. What was the significance of the Michelson-Morley 1887 ether drift experiment?

6. Who were the first two Americans to win a Nobel Prize in science?

7. Discuss the contributions of Joseph Proust, John Dalton, J.J. Berzelius, Dmitri Mendeleev, August Kekulé, Jacobus Van't Hoff, Jean Stas, and Joseph LaBel, to nineteenth-century chemistry.

8. Identify: Hans Christian Oersted, André Ampere.

9. What contributions did Michael Faraday make to nineteenth-century research on electricity and magnetism, electromagnetic induction, the motor principle, the generator principle, and transformers?

10. What new kind of energy or power production emerged in the first half of the nineteenth century?

11. What is Ohm's Law?

12. What theoretical contributions did James Clerk Maxwell make to nineteenth-century studies on electricity and magnetism?

13. What are the three laws of thermodynamics? Who was responsible for giving statements of these laws?

14. Identify: Sadi Carnot, Julius R. Mayer James Joule, Rudolf Clausius, Walther Nernst, G.N Lewis, and J. Willard Gibbs.

15. What scientific discovery in 1896 seemed to challenge the first law of thermodynamics (conservation of energy)?

16. Does the second law of thermodynamics negate Darwin's theory of evolution?

17. Who first identified acid rain?

18. Who first recognized the problem of global warming and the greenhouse effect?

# CHAPTER 10
# TWENTIETH-CENTURY PHYSICAL SCIENCE

Every great advance in science has
issued from a new audacity of imagination
John Dewey (1859-1952)
*The Quest for Certainty*, 1929

## I. INTRODUCTION

A second revolution occurred in the physical sciences in the first decades of the twentieth century, resulting in the rejection of determinism, the transformation of classical physics, and the acceptance of quantum theory, special and general relativity, and quantum mechanics. European scientists maintained their leadership in science, discovering x-rays, radioactivity, and the electron, and introducing quantum theory, the theories of special and general relativity, and quantum mechanics.

Max Planck and Albert Einstein laid the foundation of twentieth-century physical science. Planck, the father of quantum theory, in 1900 found that matter emitted radiant energy in discrete bundles, or quanta, according to the equation $E = hv$ and not in the continuum of classical physics. Einstein's 1905 equation for Brownian movement showed that matter was atomic. His $E = mc^2$ equivalency of 1905-7 indicated that matter had energy, just as the discovery of the electron in 1897, the proton in 1920, and the neutron in 1932 proved that matter contained electrically charged particles. The special theory of relativity that Einstein published in 1905 gave science a new theory of space, time, and matter. Starting with a few assumptions, that the speed of light was a constant in all frames of reference (a defined set of conditions) and independent of the direction light was moving relative to the direction its source was moving, that the laws of classical mechanics were valid in all frames of reference, and that the electromagnetic world obeyed the laws of classical mechanics, Einstein synthesized the mechanical and electromagnetic world views. His 1915-16 general theory of relativity described a new theory of gravity. Gravity was not Newton's force but a change that one body produced in the space-time geometry of another body.

Werner Heisenberg and Erwin Schrödinger found the existing electron models of the atom inadequate, and in 1925 and 1926 they proposed independently a new quantum mechanical model of the atom. Within a few years the application of quantum mechanics to chemistry enabled chemists to develop electron structures of molecules that accounted for a molecule's chemical behavior.

In the United States, Albert Michelson, Robert Millikan, Gilbert N. Lewis, Linus Pauling, and Robert Mulliken made important contributions to the physical sciences. Michelson continued to obtain excellent values for the speed of light, Millikan experimentally determined the charge on the electron, while Lewis, Pauling and Mulliken showed how the electron accounted for the chemical bond in molecules. Particle physics, which arose in the 1930s with the discovery of neutrons, neutrinos, and mesons, advanced rapidly in the late 1940s after World War II. In the 1950s, 1960s, and 1970s American physicists, such as Murray Gell-Mann, Steven Weinberg, and Sheldon Glashow, significantly contributed to particle physics with their studies on quarks and the four fundamental forces, and began to dominate its study.

World War I (1914-18) and the Great Depression (1929-41) were the major political, social, and economic events of the twentieth century prior to the outbreak of World War II in 1939. In the United States the 1920s represented a period of changing values in religious outlook and occupation. Financial good times prevailed for some Americans until the stock market crash of October 1929, whereas disorder best describes Europe's social, political, and economic situation. Franklin Roosevelt's New Deal did much to improve the lives of unfortunate and unemployed Americans during the 1930s depression years prior to the United States' entrance into World War II in December

1941. In Europe and Asia the political events of the 1930s foretold the disaster that would follow. The war introduced society to atomic energy and changed forever the role of the scientist in society.

## II. FUNDAMENTAL BREAKTHROUGHS IN PHYSICS AND CHEMISTRY

### *Discovery of cathode rays*

Heinrich Geissler (1814-79), a skilled glass blower in Bonn, Germany, in 1854 built the first mercury vacuum pump and with it produced the first vacuum discharge tube. It consisted of a large diameter glass tube with two copper terminal wires, or metal (copper, aluminum, iron) electrodes, sealed inside the tube, one terminal wire or electrode at each end of the tube. He connected the other end of one electrode to the positive terminal of a Rühmkorff coil and the end of the other electrode to the coil's negative terminal. The Rühmkorff coil was an induction coil, or high voltage source, that the German mechanic Heinrich Rühmkorff in Paris invented in 1851.

Geissler's pump removed air from a glass tube in the same way that a force pump raised water from a well. It moved a column of mercury in a thin glass tube up and down as in a barometer. The vacuum above the mercury column allowed him to remove the air from an enclosed container gradually until the air pressure in the container approached the air pressure above the mercury column. Geissler's vacuum pumps were of excellent quality and enabled physicists, such as Julius Plücker (1801-68) at Bonn, Eugene Goldstein (1850-1930) at Berlin, and William Crookes (1832-1919), editor of the *Chemical News* in London, to study high voltage (200-400 volts) electrical discharges in gases at low pressure. Their investigations with vacuum discharge tubes led to the discovery of cathode rays and to an understanding of their properties.

Plücker in 1858 detected a greenish glow around the cathode when current was flowing in his vacuum discharge tube. A strong magnet placed close to the tube deflected the glow. The magnetic deflection indicated that an electric charge likely produced the glow and provided evidence that later supported the existence of electrically-charged subatomic particles. Goldstein, who studied under Helmholtz, in 1876 introduced the name cathode ray to describe the rays moving in straight lines from the cathode (negative electrode) to the anode (positive electrode) of the discharge tube. By this time, scientists had renamed a gas discharge tube a cathode ray tube, or oscillograph. He believed the cathode rays were waves like light.

In his 1876 experiments, Goldstein used a perforated metal cathode and observed straight yellow beams passing through the holes in a direction opposite the cathode rays. He called the self-luminous beams canal rays. Unlike cathode rays, which had no glow or color and were apparent only when they collided with substances that fluoresced, such as air or other gases and glass, the luminous color canal rays produced depended on the residual gas in the tube, yellow in air, rose in hydrogen, yellow-rose in oxygen, and green-gray in carbon dioxide. Goldstein failed to deflect the canal rays in a magnetic field, but in 1898 Wilhelm Wien (1864-1928) then at the Technical High School in Aachen deflected the rays in a strong magnetic field. This showed that they carried a positive charge and were much more massive than cathode rays.

Crookes also found that cathode rays streamed in straight lines from cathode to anode. In 1885 he demonstrated that cathode rays cast shadows, a magnet brought close to the tube deflected them in a direction indicating that the rays carried a negative charge, they heated objects, and had mass because they turned the vanes of a small paddle wheel placed inside the tube. J.J. Thomson (1856-1940) at the Cavendish Laboratory, Cambridge, later showed that the vane-turning was a heating effect caused by gas molecules striking one side of the vanes and not a result of the nearly massless cathode rays striking the vanes. Crookes and most British and French physicists believed that cathode rays were tiny corpuscles, whereas German physicists, such as Goldstein, Eilhard Wiedemann (1852-1928) at Erlangen, and Hertz at Bonn thought cathode rays were waves moving through the ether. Helmholtz disagreed with the German wave school and considered the beams particles.

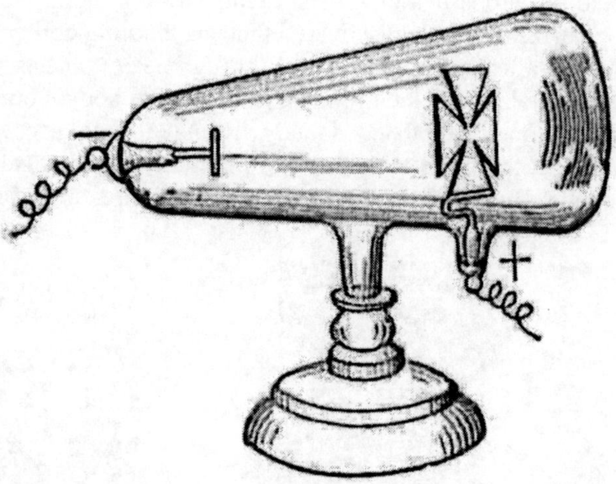


**Figure 1**
**Crookes' Tube**
Source: Newton Henry Black and Harvey Nathaniel Davis. *New Practical Physics.*
New York: Macmillan Company, 1913.

Crookes' discharge tube design, known later as a Crookes tube, became the technological foundation of television. Alan Campbell-Swinton (1863-1930), a Scottish electrical engineer in London, recognized the cathode ray tube's television principle in 1908. He proposed that a moving image would result upon scanning or sweeping the electron stream in a cathode ray tube back and forth in rows from top to bottom and varying the electron stream's intensity. Another Scot, John Logie Baird (1888-1946), using scanning discs and photoelectronics, in 1926 produced the first television transmission when he sent a picture between two attic rooms in his Soho, London, home laboratory. Fifty scientists from the Royal Institution witnessed the transmission. In 1927 Baird sent pictures by telephone line 438 miles from London to Glasgow. In 1928 he sent pictures from London to an amateur radio operator in Hartsdale, a small town 20 miles north of New York City.

***Serendipitous discovery of x-rays***

The cathode ray experiments of Plücker, Goldstein, and Crookes caught the physicists' attention. So did an 1892 experiment of Hertz and Philipp Lenard (1862-1947) at Bonn. They discovered the cathode rays' ability to pass through gold or aluminum foil placed in the rays' path and cause the air to fluoresce. Lenard received the 1905 Nobel Prize in physics for his cathode ray investigations. He also claimed the discovery of x-rays in 1893.

Wilhelm Röntgen (1845-1923) at Würzburg continued the cathode ray experiments. In November 1895, while studying the ultraviolet light a highly-evacuated cathode ray tube emitted when the rays struck the glass, he noted the fluorescence the ultraviolet light produced in nearby crystals of sodium chloride or zinc sulfide. To observe better the crystals' fluorescence, Röntgen tightly wrapped his glass tube in black paper to mask the fluorescence produced when cathode rays struck the glass tube and then darkened his laboratory. In one of his experiments, Röntgen saw that a cardboard screen coated with the yellow-green fluorescent compound barium platinum cyanide $BaPt(CN)_4.4\ H_20$ and located two meters from his tube, was fluorescing. Ultraviolet light, which traveled only 2-3 centimeters in air and which the black paper blocked from leaving the glass tube, could not have caused the fluorescence. The screen's location led Röntgen to conclude that an unknown ray or radiation had passed out the tube at the anode end and struck the coated screen. Because Röntgen could not identify the emission, he called it an x-ray, x for unknown.

Röntgen established the properties of x-rays, demonstrating that a magnet did not deflect them and that he could not polarize them. He discovered their penetrating ability, showing that the fluorescence on the screen persisted after passing the x-rays through heavy objects, such as books and metals, and demonstrated their value in medicine in viewing the human body's bones and objects, such as bullets and metals, that sometimes penetrated a body. Röntgen announced his discovery in a lecture on 23 January 1896 and at that time took the first public x-ray photograph, the hand of his Würzburg colleague, the biologist Rudolf Kölliker (1817-1905). He already had taken a 30-minute x-ray of his wife's hand on 22 December.

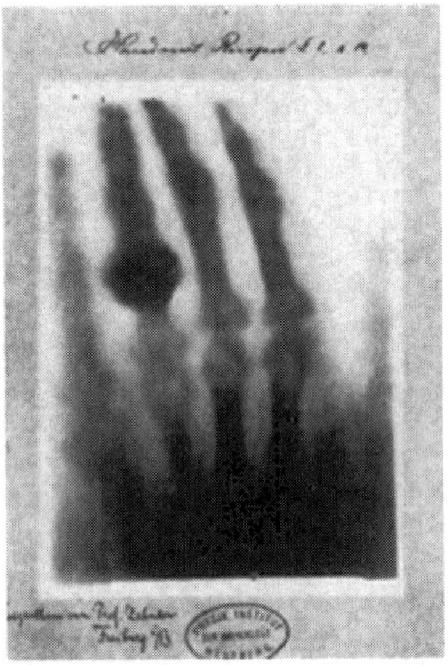

**Figure 2**
**Röntgen and X-Ray Photo of Hand**
Courtesy of the National Library of Medicine (left)
Courtesy of NASA (right)

Unlike the American dentist William Morton (1819-68) then in Boston, Röntgen did not patent his discovery of x-rays. Morton in 1846 patented the use of ether as anesthesia only to have the US Patent Office invalidate his patent in December 1862 because Morton claimed a monopoly on a law of nature. Röntgen's unwillingness to patent his discovery permitted the x-rays' immediate application in medicine, which in the United States occurred on 3 February 1896.

Edwin B. Frost (1866-1935), a physicist-astronomer, and his brother, the physician Gilman Frost (1864-1942), at Dartmouth took a 20-minute x-ray of Gilman's patient, fourteen-year old Eddie McCarthy, who fell while ice skating and fractured his left wrist. A photographer took a picture of the Frosts giving McCarthy the x-rays, and a report of their results appeared in the 14 February issue of *Science*. They used the same cathode ray tube that Frank Austin, an assistant in the Dartmouth physics laboratory, had used a week earlier on 27 January to x-ray a wooden box containing a coin, knife, pair of scissors, and keys. They also x-rayed Gilman Frost's hand and his wife's hand. Austin learned of Röntgen's discovery from a *New York Sun* article dated 26 January 1896, and after obtaining his x-ray photographs reported the results to Edwin Frost.

That same year, 1896, two other Americans, the physicist Arthur Goodspeed (1860-1943) at Pennsylvania and the well-known Philadelphia architecture-science photographer William Jennings (1860-1946), claimed that on 22 February 1890 they accidentally had produced an x-ray photograph of two 25¢ coins Jennings placed on a photographic plate in the university's physical lecture room. Goodspeed gave an account of the event at a meeting of the American Philosophical Society in Philadelphia on 21 February 1896 and published a two-page article in *Science*, 13 March 1896.

Goodspeed said Röntgen's publication prompted him to reexamine the serendipitous event of 1890. The Yale physicist Arthur Wright (1836-1915) had an equally credible claim to the first x-ray photograph taken in the United States. Wright's laboratory notes of 27-28 January 1896 describe experiments in which he placed a lead pencil, a pair of scissors, and a 25-cent coin on a photographic plate and x-rayed them for 15 minutes. Wright published his results in the *Engineering and Mining Journal* 1 February 1896, the *New York Journal* 2 February 1896, and in two more journals by March 1896.

By this time American physicians were asserting the importance of x-rays in surgery. They needed x-rays to protect their patients from unnecessary medical treatments and themselves from malpractice. New x-ray medical uses included some of the first x-ray photographs of a living fetus in utero that Edward Parker Davis (1856-1937), professor of obstetrics at Jefferson Medical College in Philadelphia, obtained in February 1896. Two months later, in April 1896, Francis H. Williams (1852-1936), a physician at Boston City Hospital, used x-ray photographs to compare pathological tissue with normal tissues, and correctly diagnosed tuberculosis, a serious and contagious disease. That same month Michael Pupin (1858-1935) at Columbia introduced a new application of x-rays. Pupin used x-rays to treat a woman in Chicago who suffered from breast cancer.

The application of x-ray photography to medicine had limitations. Radiologists could not study soft tissues, such as the digestive tract and the circulatory system, because of their relative transparency compared to the human skeleton. The different opacities of air and water, water being more opaque, provided the first clue to solving the problem. If the complete transparency of air spaces to x-rays indicated the presence of an air space, the increased opacity of an organ (soft tissues) filled with water or other opaque substances to the passage of x-rays should indicate the presence of the organ's soft tissue. Experimental work in Germany in 1896 confirmed this hypothesis. It demonstrated the absorption of x-rays by the three halogens, chlorine, bromine, and iodine, and by organs filled with solutions of these elements, but the big breakthrough occurred the following year.

George Pfahler (1874-1957), a physician at the Philadelphia General Hospital and later at the Medico-Chirurgical College of Philadelphia and the University of Pennsylvania Medical School, in 1897 accidentally discovered that bismuth compounds were opaque to x-rays. Pfahler was working with bismuth compounds because physicians in the 1890s commonly prescribed bismuth compounds in treating gastric ulcers to relieve pain and promote healing. When Pfahler examined plates he had made of a patient who had taken bismuth for a stomach ulcer, he saw the visible outline of the patient's stomach. Pfahler did not immediately follow through with his discovery, but he had demonstrated bismuth's potential as a contrast medium. Bismuth compounds had another advantage in that they did not alter the chemical reactions of the gastric liquids, which was a potential problem when using halogens. From that time on, radiologists, such as Pfahler's colleagues Goodspeed and Henry Pancoast (1875-1939) at Pennsylvania, regularly employed bismuth compounds instead of the halogens in the x-ray photography of the gastro-intestinal tract.

The pharmacopial preparation, bismuth subnitrate [$4BiNO_3\ (OH)_2BiO\ (OH)$], a white, water insoluble, non-toxic compound used in cosmetics, became the choice for early research in digestion. Walter B. Cannon (1871-1945), professor of physiology at Harvard Medical School, used it in his pioneering studies of 1897-98 to elucidate the physiology of the intestinal tract. Cannon found that he could administer the subnitrate orally, and in this way he monitored its movement through the entire digestive system. On 23 April 1893 he fed bread mixed with the subnitrate to a cat and noted a series of wavelike contractions of the cat's stomach.

That autumn he tested the subnitrate on a human patient for the first time. Though it worked well, Cannon soon found that the subnitrate caused uncomfortable and sometimes serious side effects in patients. He experimented with barium sulfate ($BaSO_4$), a white or yellowish non-toxic tasteless powder, and discovered that it produced the same results without any harmful effects. Cannon reported his findings in the *American Journal of Physiology* in 1904. Radiologists continued to use bismuth subnitrate extensively until 1910 when barium sulfate replaced it on a significant scale. Barium sulfate's low cost also increased its attractiveness as a desirable substitute.

The years from 1896 to 1913 were x-ray photography's formative years, or gas-filled tube years, during which radiology emerged as an independent medical field. Samuel H. Monell (1857-1918) in New York City, one of a growing number of physicians proficient in both gas tube physics and medicine, who called themselves radiologists, in 1896 proposed the establishment of an x-ray society. Monell found no support, but four years later John Rudis-Jicinsky (1862-1901), a Czech-born physician in Cedar Rapids, Iowa, revived the idea in a letter to Hebar Robarts (1852-1922), a St. Louis physician. On 26 March 1900 the two physicians, together with representatives from nine states, held a meeting in Robarts' office and established the Röntgen Ray Society of the United States. The representatives elected Robarts president and Rudis-Jicinsky secretary. The society in 1901 changed its name to American Röntgen Ray Society and has remained a functioning organization.

In the formative years of x-ray investigations, several experimenters recognized or acknowledged dangers, such as skin burns and loss of hair, from prolonged exposure to x-rays. Pupin, Thomas Edison (1847-1931), and Elihu Thompson (1853-1937), scientist-inventor with the General Electric Company in Lynn, Massachusetts, referred to the burns their assistants experienced in working with x-rays. Thomson, in late November 1896, deliberately burned his left little finger in demonstrating the effects of x-rays on humans. Clarence Dally (1865-1904), one of Edison's chief assistants at the West Orange, New Jersey, Laboratory, was not as fortunate. He died in October 1904 as a result of radiation burns to his hands and arms that over eight years led to a fatal skin cancer. Dally's death, the first x-ray fatality in the United States, convinced Edison, who suffered radiation damage to his eyes, to halt x-ray investigations at the West Orange laboratory.

In 1899, three years after Dally began his fateful experiments, Frederick Cottrell (1877-1948) at Berkeley and inventor of the Cottrell electrostatic precipitator in 1907, while at an Oakland, California, industrial fair, used x-rays to confirm a glassblower's claim of eating glass. The glassblower received a one and one- half hour exposure to the rays, and Cottrell earned twenty dollars for his service. One week later the glassblower became dangerously ill and threatened a lawsuit. He dropped the lawsuit only after Cottrell's two collaborating physicians agreed to provide the glassblower with free treatment. Understandably, Cottrell's interest in x-rays waned.

X-ray testing claimed another victim in 1910. Mihran Kassabian (1870-1910), the skiagrapher (radiographer) and instructor in electrotherapeutics at Philadelphia's Medico-Chirurgical Hospital and in 1903, director of Philadelphia's Röntgen Ray Hospital, examined 3,000 patients, and exposed to x-rays (fogged) more than 800 negatives in his brief career. Kassabian described his work in a well-known 540-page book *Röntgen Rays and Therapeutics* published in 1907. He experienced x-ray burns on his hands beginning in April 1900 as a result of his extensive experiments, and in 1908 developed a malignancy on his left hand. Two years later Kassabian died of radiation induced cancer.

In addition to the medical application and wartime application in the Boer War in South Africa in 1899-1902, customs officials, police detectives, and the military quickly put x-rays to use. Police in London and Paris in 1896 used x-rays to inspect the contents of suspicious packages. French customs in 1897 used x-rays to inspect luggage, and in 1896-97 the War Ministries in Germany and England began using x-rays to inspect for cannon flaws and metal weld quality in weaponry. The Wright brothers proposed the same use of x-rays in the United States. Andrew Carnegie (1835-1919) in Pittsburgh already in 1896 had adopted x-rays to test the quality of some steels.

The scientific community found Röntgen's discovery of such significance that in 1901 he received the first Nobel Prize awarded in physics. Any scientist living in 1901 who had made an important contribution to physics remained an eligible candidate for the prize. Röntgen's selection from such a prestigious group of competitors was truly impressive. He received 150,782 Swedish Crowns, (3.76 Swedish Crowns = $1.00 US), equivalent in 2010 to 7 million Swedish Crowns or $936,000 ö(7.48 Swedish Crowns = $1 US), all of which he donated to the University of Würzburg. The currency inflation that arose in Germany after World War I left Röntgen, like many Germans, in dire financial straits. He died in 1923 from carcinoma of the intestines.

World War I played a significant role in x-rays' development as a diagnostic tool. William Coolidge (1873-1975) at the General Electric Company in Schenectady, New York, in 1913 had invented the Coolidge tube, an improved x-ray tube. In the Coolidge tube, a heated tungsten filament instead of a rarified gas released electrons that struck the tungsten anode and produced x-rays. No gas was present in the tube. This provided clearer pictures and more certainty in determining the amount of x-radiation a patient received during examination.

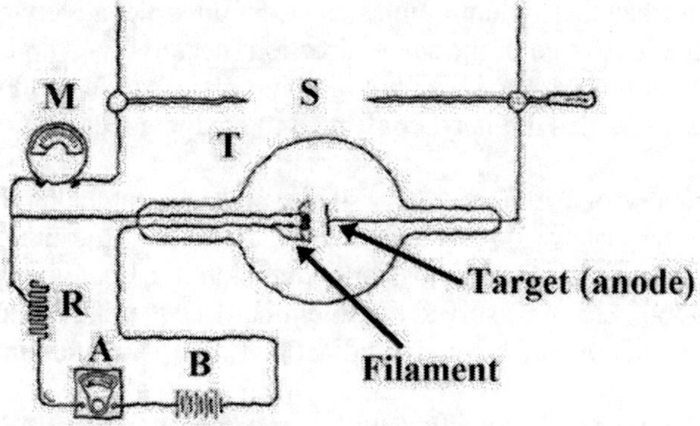


**Figure 3**
**Coolidge X-ray Tube**
Source: William D. Coolidge. "A Powerful Rontgen Ray Tube with a Pure Electron Discharge." *Physical Review* 2, no. 6 (1913).

Using the Coolidge tube, Marie Curie constructed portable x-ray units that French physicians used in battlefield conditions to determine the locations of fractures and to pinpoint shrapnel and bullets in the wounded. Coolidge's tube, which ended radiology's gas tube era, greatly enhanced the performance of x-rays in medicine and initiated modern radiology.

Studies that began in 1912 with J.J. Thomson and particularly in 1913 with Niels Bohr (1885-1962) at Copenhagen and other physicists led to the recognition of quantized energy levels in an atom, and to a scientific explanation of how x-rays originate in an atom. X-rays, which are high energy and high frequency rays, result from cathode rays' (electrons) striking an atom or molecule and displacing the atom's or molecule's electrons from their lowest energy level to a higher energy level. Upon stopping the cathode ray bombardment, the displaced electrons fall to a lower energy level closer to the nucleus, or from energy level $n_2$ or higher to the lowest energy level $n_1$. The energy released by the electrons falling in the atom or molecule is an x-ray.

***Serendipitous discovery of radioactivity***

A second serendipitous discovery followed from Röntgen's discovery of x-rays. Antoine Henri Becquerel (1852-1908) at the Ècole Polytechnique in Paris in February 1896, investigated whether the yellow crystals of potassium uranyl sulfate ($K_2SO_4.UO_2SO_4.2H_2O$), the fluorescent compound he was studying, might produce x-rays. Becquerel knew from Röntgen's work that x-rays emanated from the fluorescent spot on the cathode ray tube, and he suspected a connection between the fluorescence and x-radiation. To test his hypothesis, Becquerel wrapped a photographic plate in black paper, put a thin sheet of the uranium crystals on top of it, and placed the plate and crystals in sunlight for a few hours to initiate the fluorescence. He believed that if the fluorescence the uranium compound emitted contained x-rays, the rays would pass through the paper and expose (fog) the plate.

The uranium compound did expose the plate, but it caused the same plate exposure when Becquerel, during the few cloudy days that followed, did not place the plate and uranium compound in sunlight. Sunlight had nothing to do with the exposure. Becquerel, who had inherited a collection of uranium compounds from his father, the physicist Alexandre Becquerel (1820-91), found that all his uranium compounds exposed a photographic plate, establishing that uranium produced the effect. Becquerel noted that uranium, the element that Martin Klaproth (1743-1817) at Berlin discovered in 1789, was about 3.5 times as active as the uranium sulfate compound he first tested. In his 1899 investigations Becquerel showed that the uranium emissions contained negatively-charged particles, or a negative emission. A magnet deflected in the same direction, negatively-charged cathode rays and the uranium emissions that struck and exposed the photographic plate. Marie and Pierre Curie and Friedrich Giesel (1852-1927) in Brunswick, Germany, confirmed the magnetic deflection that same year.

Becquerel's discovery of the uranium emissions presented a challenge to the law of conservation of energy because physicists could not account for the energy the uranium compounds released while experiencing no apparent equivalent loss or conversion of some other form of energy. Einstein's 1905 energy-mass relation ($E = mc^2$) solved the dilemma. Becquerel also noted that a small radium sample he carried in his pocket burned his body, indicating that it, like uranium, released energy.

The scientific community's fascination with x-rays and comparative lack of interest in Becquerel's uranium emissions, along with Becquerel's diminishing commitment to his uranium studies, led Marie Curie (1867-1934) to renew uranium's investigation. Maria Sklodowska came from Warsaw in Russian-dominated Poland to Paris in 1891 to attend the Sorbonne (University of Paris). She met the physicist Pierre Curie (1859-1906) in 1894, they married in a civil ceremony the next year. Pierre since 1882 was teaching at the Municipal School of Industrial Physics and Chemistry (EPCI) in Paris, and in 1895 received his PhD in physics, for what scientists since 1911 have called the Curie temperature, the temperature at which a magnetic substance such as iron loses its magnetism. The Curies were non-conformists, likely religious agnostics, although both were born and raised in predominately Catholic countries. Marie was highly motivated and ambitious. Pierre, a dreamer and an idealist, dabbled for a time with spiritualism. Both were independent thinkers. Their scientific philosophy placed them in the empiricist-positivist school of thought.

With Becquerel's support, Marie, beginning in 1897, studied uranium's emissions as the subject of her doctoral research at the Sorbonne. Working in poor conditions in a shed at the EPCI, she showed that the intensity of the emissions depended only on the amount of uranium present, and not on the temperature, pressure, or any other physical property of uranium. Radioactivity, she concluded, was a property of the atom and did not depend on external factors.

Marie investigated all the other known chemical elements, about 80, testing them for the emission of ionizing radiation with the piezoelectric quartz electrometer that Pierre and his brother Jacques (1856-1941) invented in 1880. Only thorium compounds emitted radiation. Marie then analyzed pitchblende, a dull black acid-soluble substance containing 55-100 percent uranium oxides ($UO_2$ and $UO_3$), a brownish-black compound mixture the glass industry used as an additive to produce fluorescent glass. Pigments and ceramic glazes were other products. To her surprise, she found pitchblende about two and a half times more radioactive than either uranium (atomic number $Z = 92$) or thorium ($Z = 90$). Suspecting something of scientific importance, Marie, joined by Pierre in 1898, found that pitchblende's activity resulted from the presence of two new metallic elements, polonium ($Z = 84$) discovered in July and radium ($Z = 88$) discovered in December. The Curies in 1898 introduced the term *radioactivity* to describe the energy the atoms of uranium, thorium, polonium, and radium released.

To provide enough polonium and radium for their research, the Curies obtained at no cost about seven tons of discarded pitchblende residue from a government mine in Joachimsthal, Bohemia. Beginning in 1900 they processed the pitchblende by dissolving it in sulfuric or nitric acid and following the radioactivity's path with Pierre's electrometer. After several months of hard labor in a primitive facility, they obtained from the seven tons of pitchblende a 0.2-0.3 gram sample containing radium and polonium. Using fractional crystallization, a separation process Marie essentially invented to separate compounds of different solubility, and which in their work required several thousand crystallizations, they separated a polonium-bismuth fraction as a sulfide and showed it was about 400 times as radioactive as uranium. They attributed the radioactivity to the new element polonium. Bismuth was not a radioactive element. In 1902 they separated enough of a polonium-bismuth fraction to identify polonium chemically. They crystallized and separated a tiny amount (0.1 gram) of the less soluble yellow-white radium chloride ($RaCl_2$ ) from the colorless barium chloride ($BaCl_2.2H_20$) present, and used the spectrum of radium that Eugene Demarçay (1852-1904) at Paris obtained from their sample to confirm that they had discovered a second new element. Their radium sample was 900 times more active than uranium. Marie isolated radium as the pure metal in 1910.

The Curies in 1901 calculated that 1 gram of radium underwent 500 disintegrations per hour and released 140 calories of heat per hour. Radium gave off 4 billion times more heat than uranium, which Ernest Rutherford (1871-1937) then at McGill University in Montreal showed in 1901, released $1x10^{-11}$ calories per second ($36 \times 10^{-9}$ calories per hour). Their experiments demonstrated for the first time the large amount of energy released in radioactive decay, as well as the dangers of radioactive decay, particularly gamma rays, to the human body.

In 1907 Marie defined the unit of radiation (the curie, c) as the number of disintegrations that one gram of radium emitted. The curie is equal to 37 billion disintegrations emitted by a radioactive substance in one second. The Becquerel (Bq), defined as one radioactive disintegration per second, is the standard international (SI) unit of radioactivity. Another radiation unit is the rem (röntgen equivalent man) first introduced in 1945, defined in 1947, and accepted as the non-SI in 1971. The rem measures the actual biological effects of radiation absorbed in human tissue, or the absorbed dose equivalent, and accounts for the quantity of radiation, or dose, and the quality, or kind, of radiation causing biological damage. Both US and international commissions on radiation and on standards since 1976 have recommended the sievert (Sv) as the SI unit replacing the rem, where 1 rem = 0.01 sievert (1J/kg).

Marie's exposure to radiation likely reached 1 rem per week, 0.03 rem per week is dangerous. A chest x-ray is equivalent to an exposure of 0.007 rem. The national average for occupationally-exposed persons is 0.5 rem per year (the limit is 5 rem per year) and 0.17 rem per year for the general public. The average dose a person receives per year of exposure to natural radiation is about 0.3 rem.

Both Curies permanently damaged their fingertips, experienced weakness, and lost weight. Marie in 1903 suffered a miscarriage five months into her second pregnancy. Nevertheless, she kept a glass vial containing radium salts on a bedside stand because she liked the soft blue glow radium produced in the dark. Pierre deliberately burned himself on the arm to test the amount and intensity of the heat radium emitted. A 10-hour exposure to a low activity sample produced immediate redness and a wound that required four months to heal.

In the three-year period from 1903 to 1906 the Curies' professional careers improved considerably. Marie completed her doctoral research and obtained her PhD degree in June 1903, the first woman in France to receive a PhD in science. Six months later, in December 1903, the Curies and Becquerel received the Nobel Prize in physics for their researches on radioactivity. Because of weakness (radiation sickness) they experienced while doing their research, neither of the Curies traveled to Stockholm to accept their Nobel Prize. Pierre delivered their Nobel address two years later. He spoke on the possible dangers of radioactivity.

Their half share of the prize, 70,679 Swedish Crowns or $18,750 (3.77 Swedish Crowns = $1.00 US), for the first time provided them funds for research. The following year the Curies received prestigious appointments. Pierre in 1904 became professor of physics at the Sorbonne, and Marie, head of a new physics laboratory scheduled for construction at the Sorbonne. The French Academy of Sciences elected Pierre a member in 1905.

Their good fortune ended tragically in 1906 when Pierre died as a result of a heavily-loaded horse-drawn carriage running over him as he walked from his laboratory. Marie, now left to raise two young daughters, Irene (1897-1956) and Eve (1904-2007), aged nine and two, received Pierre's professorship in 1906, becoming the first woman professor at the Sorbonne. She submitted her name for membership in the French Academy of Sciences in 1910, but the all-male Academy denied her. She lost by two votes and never again applied for membership. The French Academy, founded in 1666, did not elect its first female member until 1962. Marguerite Perey (1909-75), a chemist-physicist at the Radium Institute in Paris, won election for her discovery in 1939 of the radioactive element francium. She died of cancer. The Royal Society in England treated women scientists no better. It elected it first female members in 1945.

In 1911 Marie received a second Nobel Prize, this time the sole recipient in chemistry for the discovery of polonium and radium and for her 1910 isolating of radium as the pure metal. But controversy arose when the news broke of her 1911-12 affair with their family friend and Pierre's former student, the physicist Paul Langevin (1872-1946) at the Collège de France in Paris. Langevin achieved fame for his quantitative work in 1905 on paramagnetism and diamagnetism in atoms and molecules. His wife threatened to kill Marie, and because of the scandal the Royal Swedish Academy of Sciences told Marie that she was unwelcome in Stockholm and should decline the prize until she removed the embarrassment of her affair. Marie refused to accept the Academy's terms and received the Nobel Prize in Stockholm. She wrote to one of the Nobel committee's members, "I believe there is no connection between my scientific work and the facts of my private life."

In the years following her receiving a second Nobel Prize, Marie devoted her time to establishing the Radium Institute, serving as a major fundraiser and as a senior scientist following its completion in late August 1914. At this time, exposing a patient to the emissions from radium, radon, and other radioactive elements became an accepted medical procedure for treating many types of cancer.

World War I had just begun on 1 August, and during the war, Marie assisted by her daughter, Irene, operated 20 portable x-ray stations to take photographs of wounded soldiers. Later, they provided training for women in the use of radiological techniques with radium and radon. After the war in 1921 and 1929, Marie, on invitation from Mrs. William Brown Melony (Marie Mattingly Melony, 1877-1943) in New York City, toured the United States on a fund raising mission. Melony was a leading journalist and editor of the women's magazine the *Delineator.* The president Warren G. Harding (1865-1923) hosted Marie at a White House reception in 1921 and presented her with a gram of radium then worth $100,000. A committee of wealthy American women and American scientists that Melony organized and named the Marie Curie Radium Campaign raised the money to buy the radium when they learned her institute had only a single gram of radium for research. Laboratories in the United States possessed 50 grams of radium.

While in the United States in 1921, Marie received an honorary ScD from Columbia, Chicago, Pennsylvania, Smith, Wellesley, and Yale, the first that Yale awarded. Several physicists and chemists at Yale objected to awarding her an honorary degree, among them Bertram Boltwood, a leading radiochemist, although he relented and met with her during her visit. The Physics Department at Harvard blocked Harvard's awarding Marie an honorary degree, and the National Academy of Sciences never selected her for membership. They believed her scientific work did not merit their prestigious awards.

Marie and Pierre Curie

**Marie with President Warren G. Harding on the steps of the White House, 1921**

**Figure 4**
© Shutterstock, Inc. (left)
Courtesy of the National Institute of Standards and Technology, US Department of Commerce (right)

A connection may not have existed between her private life and scientific work but did exist between her work and her health. For years both Curies suffered burns, weakness, and periods of illness from the effects of radiation poisoning, which they largely ignored. Two workers at the Radium Institute in 1925 died of radiation poisoning, one from aplastic anemia (low red, white, platelet cell counts), the other from leukemia (cancer of the blood). Marie attributed their deaths to work-related exhaustion. Marie died of aplastic anemia, no doubt a result of her work on radioactivity. Even today her laboratory notebooks, many of which she personally destroyed because of her secretive personality, and some of her other personal items have remained radioactive. In 1995 the French government moved the Curies from their burial place in the cemetery in Sceaux, a commune in the southern suburb of Paris, to the Pantheon, France's national mausoleum, in Paris.

**Figure 5**
**Tombs of Pierre and Marie Curie, Pantheon, Paris**
Courtesy of the author

Following Becquerel's discovery that his uranium sulfate compound emitted a negatively-charged particle or a negative emission, Paul Villard (1860-1934), a physicist at the École Normal Supérieure in Paris, used photographic plate fogging to show in 1900 that radioactive substances gave off a neutral emission undeflected by a magnet and that easily penetrated or passed through thin metal foils. In 1903 Rutherford and Frederick Soddy (1877-1956) at McGill discovered a third emission that had little penetrating ability, and from the direction of its defection in a strong electric and magnetic field showed that it had a positive charge.

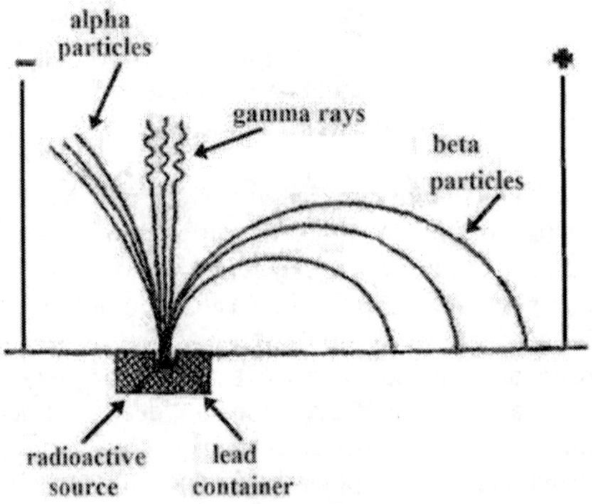


**Figure 6**
**The Three Types of Radioactive Emissions**
Courtesy of the author

Rutherford named the three types of emissions, beta for the negative emission, gamma for the neutral emission, and alpha for the positive emission. He classified the three types by their penetrating powers and the conductivity they produced in air, and like Becquerel and Villard, investigated their magnetic defection and photographic effects. Rutherford in 1906 proved that the alpha particle was a helium nucleus by passing alpha particles through a very thin glass wall of a tube into an evacuated container and identifying helium as the gas in the container. For his work on the chemistry of radioactive substances Rutherford received the 1908 Nobel Prize in chemistry.

| THE THREE RADIOACTIVE EMISSIONS | |
|---|---|
| ***Particle or Emission*** | ***Discoverer*** |
| Beta: a negatively-charged particle | Henri Becquerel 1899 |
| Gamma: a ray having no charge | Paul Villard 1900 |
| Alpha: a positively-charged particle | Ernest Rutherford 1903 |

The discovery and identification of the three kinds of radioactive emissions led Boltwood in 1905 to propose that radioactive elements decayed over time and that uranium eventually decayed into lead. Soddy, now at Aberdeen and Kasimir Fajans (1887-1975), a Polish chemist at Karlsruhe, in 1913 independently developed Boltwood's idea of radioactive decay. They established the radioactive disintegration series and the presence of isotopes, which are atoms of elements that have the same atomic number (Z), or number of positive charges (protons) in the nucleus, but a different atomic weight (same number of protons but a different number of neutrons in the nucleus). They showed that a radioactive atom decayed into a new atom by emitting an alpha particle, which was a helium nucleus $^{4}_{2}He^{2+}$, or by emitting a beta particle, which was a negatively-charged electron.

In alpha emission, the element undergoing radioactive decay lost two positive charges and occupied a new position, two places to its left in the Periodic Table, as occurred when uranium (Z=92) decayed to thorium (Z=90). Alexander Fleck (1889-1968), who assisted Soddy, showed in 1913 that in beta emission the element undergoing radioactive decay gained one positive charge and occupied a new position one place to the right in the Periodic Table, as occurred when uranium (Z=92) changed to neptunium (Z=93). The decay series that started with uranium or thorium ended with the formation of lead (Z=82), a stable element.

Soddy and Fajans recognized the existence of isotopes, a Greek word that Soddy introduced in December 1913, meaning in the same place, because not enough unoccupied places remained in the Periodic Table in which to place all the newly discovered radioactive atoms. They realized that although the atomic weight of an element's radioactive isotopes differed slightly, the chemical properties of each element's radioactive isotopes remained the same, and the radioactive isotopes were therefore atoms of the same element. Soddy won the 1921 Nobel Prize in chemistry for his discovery of isotopes and his contributions to the chemistry of radioactive decay.

***Divisible atom***

Early nineteenth-century atomic models, particularly Dalton's 1808 chemical atom, implied that an atom had no parts and was indivisible. Chemists also recognized that to account for atoms to combine and form a molecule, some kind of attractive force acted on the atoms. Berzelius proposed an electrical force, and in 1812 he published an electrochemical dualistic theory in which positive and negative charges resided in opposite parts of every atom like the poles of a magnet. Faraday's electrolytic experiments of 1814 provided additional evidence for the existence of electrically-charged atoms.

The literature of the next 25 years revealed that the electrical atom and atoms of electricity (the atomicity of electricity) were popular ideas even before the electron's discovery in 1897. Wilhelm Weber (1804-91) at Göttingen, in 1872 proposed that every ponderable (weighable) atom had an electric atom attached to it. Two years later, in 1874, the Irish physicist George Johnstone Stoney (1826-1911) at Queen's University in Dublin, in his electrolytic studies introduced a fundamental or unit quantity of electricity that he called an *electrine*. Stoney related the unit quantity to the breaking of a chemical bond in a molecule. Helmholtz in his celebrated 1881 Faraday Lecture before the

Chemical Society at the Royal Institution in London and Svante Arrhenius (1859-1927) at Stockholm in his 1887 paper on his electrolytic dissociation theory provided strong experimental evidence supporting the idea of electrical atoms present in atoms of matter. Nine years later, in 1896, the physicists Pieter Zeeman (1865-1928) at Leiden and Hendrik Lorentz (1853-1928) under whom Zeeman studied at Leiden, recognized that the spectrum of an atom of an element indicated that something on or in an atom had to move or vibrate in order to produce the atom's spectrum.

All of these studies provided sufficient evidence for the existence of electrical atoms or atoms of electricity, but they had not proved that the electric atom existed independently of the atoms of matter. Proof came in 1897, when J. J. Thomson experimentally demonstrated the independent existence of the electric charge, the electrical atom or electron. Lorentz and Zeeman in physics in 1902, Arrhenius in chemistry in 1903, and Thomson in physics in 1906 received Nobel Prizes for their research on the electron and its behavior in atoms.

Thomson discovered the electron, which he called a corpuscle, while studying electrical discharges in a highly evacuated cathode ray tube. He showed that the cathode ray was a negatively-charged particle by deflecting it with an applied magnetic field in the same direction as the deflection of a known negatively-charged body. He demonstrated that the corpuscles behaved the same way regardless of the kind of gas present in the discharge tube or the elements used for the electrodes.

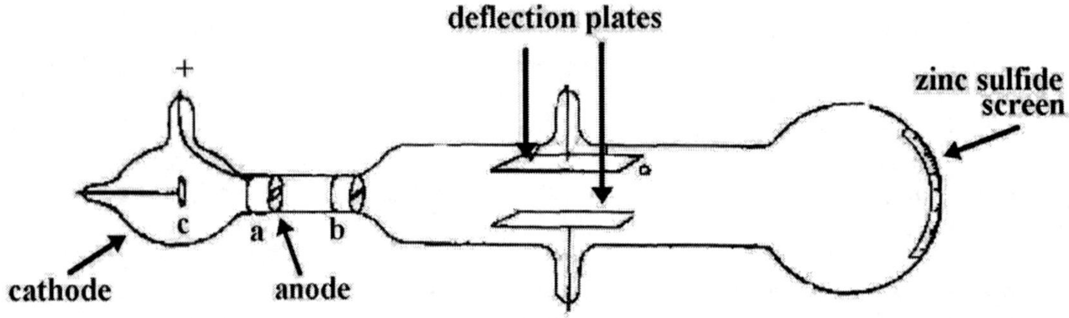


**Figure 7**
**J. J. Thomson's Cathode Ray Tube**
Source: J. J. Thomson. "Cathode Rays." *Philosophical Magazine* 44 (1897).

Prior to his work on electric discharges in cathode ray tubes containing different gases, Thomson in 1883 believed that an atom was likely a vortex ring in the electromagnetic ether, analogous to a smoke ring in air. His studies on the atomic structure of matter and his theory of the vortex atom led to his cathode ray experiments and to his discovery of the electron. Thomson proved the atom's divisibility and the existence of a negatively-charged particle within the atom. In 1897 he proposed a new atomic model in which he arranged the electrons in concentric rings in a neutralizing sphere of positive electricity. Each kind of atom, hydrogen, helium, lithium, neon, had a sufficient number of electrons, sometimes thousands of them, to account for the atom's mass. This was Thomson's positive-sphere atomic model, also called the plum-in-pudding atom.

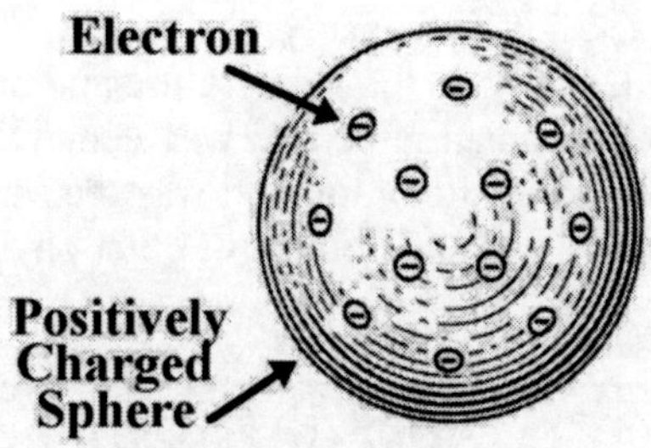


**Figure 8**
**Thomson's Positive Sphere Atom**
Source: United States Energy Research and Development Administration, Office of Public Affairs

Thomson experimentally determined the ratio of charge to mass for the electron, e/m, by balancing the electric and magnetic fields applied to the cathode rays produced in the cathode ray tube. He obtained the value e/m = $3 \times 10^{17}$ electrostatic units/gram (esu/g). Previous attempts by Hertz and other physicists to deflect the cathode rays with an electric field failed because of a poor vacuum in the cathode ray tube. This allowed the particles of the residual gas to acquire an electric charge that weakened or neutralized the electric field.

Thomson in 1898 measured the electron's charge by sending electrons at 400 volts into a dust-free air-filled container, in which the electrons collided with and attached themselves to water droplets sprayed into the container. He observed the negatively-charged droplets' rate of fall upon applying an electric field to the container, and from these experimental measurements calculated the electron's charge. He obtained mean values of $6.5 \times 10^{-10}$ esu when air filled the container and $6.7 \times 10^{-10}$ esu when he substituted hydrogen gas for air.

The next year, in 1899, Thomson demonstrated convincingly that the electron's charge was the same as the smallest known charge, the +1 charge on the hydrogen atom calculated from the electrolysis of water, or e = $6.3 \times 10^{-10}$esu. Knowing the e/m value for the hydrogen atom (ion) from the electrolysis of water and the e/m value for the electron, Thomson concluded that if the electron's charge had the same value as the charge on a hydrogen atom (ion), but differed only in sign, -1 and +1, the electron's mass was 1/1000 (more accurately 1/1836) the mass of a hydrogen atom, or m = $2.1 \times 10^{-27}$ gram.

| THOMSON'S VALUES FOR THE ELECTRON | |
|---|---|
| e/m | $3 \times 10^{17}$ esu/g |
| e | $6.5 \times 10^{-10}$ esu |
| m | $2.1 \times 10^{-27}$ g |

Thomson turned the Cavendish Laboratory, established in 1874, into the foremost center of research in experimental physics, particularly in the first half of the twentieth century. He received the 1906 Nobel Prize in physics for his discoveries with high voltage gas discharge tubes. Seven of his assistants or successors at the laboratory, among them Rutherford in chemistry in 1908 and his son George Paget Thomson (1892-1975) in physics in 1937, also won Nobel Prizes.

***Quantum theory 1900***

In 1859 Gustav Kirchoff (1824-87), a physicist at Heidelberg, was investigating the radiation emitted by a body, particularly the radiation from a black body. The physicists' typical black body was a dull black cast iron block of any size with a small hole that absorbed all the energy (all wavelengths or frequencies) that fell on the body. It appeared black to an observer because it reflected no light.

Kirchoff argued that the black body was also a perfect emitter, or radiator, of energy and showed that the energy it emitted depended only on the temperature and not on the composition of the black body's walls or on anything else. For radiation of any wavelength (frequency), the ratio of the wavelength (frequency) of the absorbed and emitted radiation was the same at a given temperature. Kirchoff gave the equation below for the energy E and pointed out that physicists had to determine the value of J, an unknown function.

| **KIRCHOFF'S EQUATION** |
|---|
| $$E = J\,(T, v)$$ |
| E = energy |
| J = an unknown function |
| T = absolute temperature |
| $v$ = frequency of the emitted energy |

Josef Stefan (1835-93) at Vienna, in 1879 expanded Kirchoff's ideas on radiation. He argued from experimental evidence on the absorption and emission of radiation from hot bodies that he and other physicists, such as John Tyndall (1820-93) at the Royal Institution, had obtained.

Tyndall showed in his 1860s studies on the heat losses of an incandescent platinum wire that the total energy all hot bodies emitted was proportional to the fourth power of the absolute temperature. A blackened thermocouple, capable of measuring the infrared radiation emitted, provided Stefan with accurate temperature readings. He calculated the number of calories that a one-square centimeter surface of a black body radiated in one second, and to relate the emitted radiation and temperature, he proposed the equation $E \approx T^4$, or $E = \theta\, T^4$, where $\theta$ is the Stefan-Boltzmann constant. Stefan estimated for the first time that the temperature of the Sun was 6,000° C.

Five years later, in 1884, Stefan's former student Ludwig Boltzmann, then at Graz, using Maxwell's 1864 results on electromagnetic radiation and Clausius's work on the second law of thermodynamics, derived Stefan's experimentally-established equation, thereby placing it on a solid theoretical foundation.

| **STEFAN-BOLTZMANN EQUATION** |
|---|
| $$E = \theta\, T^4$$ |
| E = energy |
| $\theta$ = an unknown function, the Stefan-Boltzmann constant |
| T = absolute temperature |

By that time, Boltzmann had accepted an atomic theory of matter. His argument rested on the kinetic studies of Maxwell and Clausius, in which their application of the laws of dynamics and probability to gaseous molecules supported an atomic theory. In an 1877 paper and in other papers published in the 1870s Boltzmann pointed out the twofold significance of their kinetic studies. Maxwell and Clausius gave a probabilistic interpretation of entropy as an increasing state of disorder, and a mathematical derivation for a discontinuous or quantized distribution of the energy of gaseous molecules. The observed macroscopic thermodynamic behavior of matter, he said, resulted from its microscopic atomic behavior, or from atomism. The second law of thermodynamics interpreted statistically provided evidence for the movement of the microscopic components, or the atoms, present in the system.

Boltzmann suffered from depression brought on partially by the attack of several prominent positivist physicists on his atomism. Ernst Mach (1838-1916) then at Prague, Pierre Duhem (1861-1916) at Bordeaux, and energeticists, such as Wilhelm Ostwald at Leipzig, rejected the reality of atoms and the atomistic foundation of Boltzmann's work. Boltzmann committed suicide in September 1906 by hanging himself while on vacation with his wife and one of his daughters near Trieste on the Adriatic Sea.

The question remained whether the Stefan-Boltzmann equation held for all wavelengths. In 1896 Wien derived an equation (Wein's distribution law) for a black body that showed that the energy of the oscillators present in the black body, for a given wavelength, varied inversely as the fifth power of the wavelength, or E $(\lambda_{max}T) \propto \lambda^{-5}$ T. Wien's equation fit experimental observations within one percent at high frequencies (shorter wavelengths) and low temperatures (T = Kelvin degrees), where $\lambda$ was less than 3,100 microns. His equation gave energy values that were too low at longer wavelengths. Wien's discoveries on the laws of heat radiation earned him the 1911 Nobel Prize in physics.

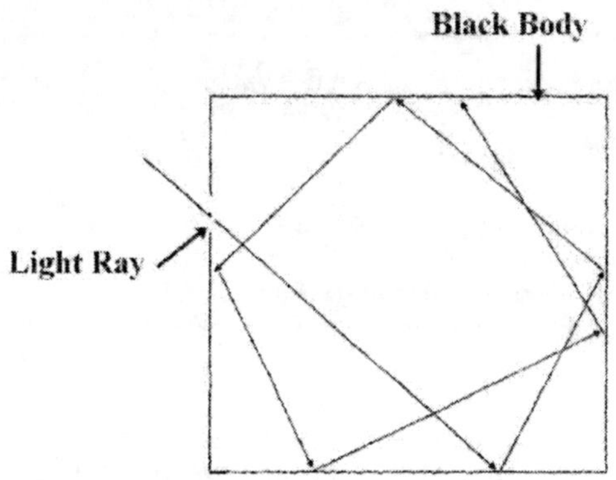


Wien proposed that a container with a small hole in its walls would simulate a black body. The inner walls of the black body absorb light entering the container through the hole. Any light initially reflected from the wall eventually will strike another part of the wall and be absorbed, because the chances of its leaving via the hole are negligibly small. Upon heating the container, blackbody radiation escapes from the hole, permitting measurement of its frequency and energy at different temperatures.

**Figure 9**
**Black Body Model**
Courtesy of the author

Lord Rayleigh (1840-1919), working in his private laboratory in Essex, and James Jeans (1877-1946) at Cambridge in 1900-5, derived a similar equation, E $(\lambda T) \propto \lambda^{-4}T$, to account for black body radiation. The Rayleigh-Jeans law matched the experimental data to within one percent at low frequencies (long wavelengths) and high temperatures, but diverged as the wavelengths approached zero and caused the energy to have an infinite value. This set the stage for the black body radiation equation that Max Planck at Berlin derived in 1900.

Using classical electromagnetic theory, Planck assumed in a black body, the presence of simple electric oscillators with frequencies that would produce an exchange of energy by emission and absorption and account for the thermal equilibrium of a black body. His assumption was incorrect, and Planck turned to thermodynamics and treated the internal energy as a function of entropy. By interpolating Wien's radiation law that held at high frequencies and low temperatures and the Rayleigh-Jeans equation that held at low frequencies and high temperatures, Planck derived a new radiation law that agreed with the experimental results obtained for black body radiation at all wavelengths. His law explained what was happening physically in black body radiation.

Planck, however, remained unsatisfied because his empirically-derived equation lacked a theoretical foundation. In searching for an explanation, he applied Boltzmann's probabilistic interpretation of entropy, and then made the radical assumption that a black body emitted its energy only in discrete, or quantized, frequency values, and not in a continuum. These two insights enabled Planck to arrive at his equation and a theoretical explanation for the relation between energy and frequency.

Planck concluded as a result of his black body studies that the atoms or molecules in every substance emit or absorb radiation only in integral multiples of the substance's fundamental frequency, or lowest energy value. They did not emit or absorb energy in a continuum that allowed every possible energy value. If the substance's lowest energy value was 1, then the permitted quantized energy values were 2, 3, 4, 5 and all other integral values. He derived the equation below.

| **PLANCK'S EQUATION** |
|---|
| $E = hv$ |
| E = energy radiated or absorbed |
| h = Planck's constant or quantum of action, $6.624 \times 10^{-27}$ erg-seconds |
| $v$ = frequency of the energy radiated or absorbed |

In 1905 Einstein applied Planck's equation to explain the photoelectric effect. Robert Millikan (1868-1953) at Chicago, in 1906 confirmed Planck's equation by calculating accurate values of $h$ upon varying E and $v$. For this work and his measurement of the electron's charge, Millikan received the 1923 Nobel Prize in physics. Planck received the Nobel Prize in physics in 1918. It came at a time Planck experienced a series of personal tragedies. He lost one son in World War I, in 1916 at the Battle of Verdun. A twin daughter died in 1917 during childbirth, the surviving twin died during childbirth in 1919. Later, in January 1945, the Nazis hanged another son, Erwin, for his role in a failed July 1944 plot to assassinate Adolf Hitler. Planck strongly opposed the Nazi regime, but because of Planck's scientific esteem and reputation in Germany and internationally, he escaped Hitler's vengeance. Planck was a religious person, a Lutheran, but his God permeated the universe and resembled more the pantheist's immanent God rather than the transcendent God of traditional Christianity.

**Figure 10**
**Postage Stamp Commemorating Max Planck**


***Albert Einstein and the new scientific world***

Einstein was born in Ulm, Germany, on 14 March 1879. He acquired an aptitude for mathematics from his father Hermann, an unsuccessful businessman, and a love for music and reading from his mother Pauline. Although Jewish, he attended a Catholic elementary school, where several teachers, failing to understand his unorthodox approach to problem solving, declared he would never succeed in anything. His parents moved to Milan, Italy, in 1894, leaving Einstein behind to finish his education at the Luitpold Gymnasium, but he soon dropped out of school and followed them to Italy, probably to avoid military service in Germany. He renounced his German citizenship in 1896 and that same year entered the Federal Institute of Technology (ETH) in Zurich, Switzerland.

After graduating and becoming a Swiss citizen in 1901, Einstein failed to obtain a high school teaching position, and in 1902 he accepted a position as a junior official in the Swiss patent office in Bern. He earned 3,500 Swiss francs (about $700) annually, considered a fairly good salary. During his years at the patent office Einstein developed and refined his ideas on light, matter, space, and time. In 1905 he received a PhD from the University of Zurich after publication of four major papers on the photoelectric effect, Brownian movement which was his dissertation, special relativity, and mass-energy equivalence.

In 1909 the University of Zurich appointed him associate professor of theoretical physics. From there Einstein went to Prague, to the ETH, and in 1914 to Berlin as director of the Kaiser Wihelm Institute of Physics. He remained in Berlin, except for brief stints as visiting professor at the California Institute of Technology, and already had left Germany for the United States in December 1932, shortly before Hitler became Germany's chancellor in January 1933. He traveled in Europe in 1933 before leaving for Belgium and then Dover and Oxford, England, arriving in the United States in October 1933 to accept an offer to join the newly-established Institute of Advanced Study in Princeton. Louis Bamberger (1858-1944) and his sister Caroline Bamberger Fuld (1864-1944), owners of the R. H. Macy and Company department store in Newark, New Jersey, at the urging of Abraham Flexner, in 1930 donated $5 million to fund the Institute as a place for refugee scientists, many of whom were Jews. When asked to give an annual salary figure for his position, Einstein replied $3,000. The Institute paid him $16,000 per year.

Einstein met Mileva Maric (1875-1948), a classmate from Novi Sad, Serbia, while both were students at the ETH in 1896. They married in Bern in January 1903, one year after the birth of their daughter Lieserl on 26 or 27 January 1902, whom they gave up for adoption that same year, most likely to her parents in Novi Sad. He never saw his daughter, who seems to have vanished and may have died from scarlet fever at the age of two. Later, they had two sons, Hans Albert (1904-73) a hydraulic engineer at Berkeley and Eduard (1910-65) a schizophrenic who spent the last 33 years of his life in his mother's care and in the Burghölzli Institute at the University of Zurich. Einstein divorced Mileva in 1919, the same year he married his first cousin Elsa Löwenthal (1876-1936), the widowed mother of two daughters. Einstein admitted his failure as a father and husband in both marriages and engaged in several extramarital affairs.

Though an ardent pacifist, Einstein signed a letter to Franklin Roosevelt in 1939, urging the president to consider developing a nuclear bomb. Because of ill health, personal misgivings, and the Federal Bureau of Investigation's (FBI) labeling him an extreme security risk for his pacifist and socialist beliefs, Einstein never worked on the Manhattan Project. His FBI file covering the years 1932-55, and released in 1983 in accordance with the Freedom of Information Act (1966 and 1974), contains wild, unsubstantiated statements. The file statements accused Einstein of running an espionage ring, of being the *brain* behind an alleged communist plot to take over Hollywood, of inventing a miracle ray, of orchestrating the 1932 kidnapping and murder of Charles Lindbergh's 20-month old son, and of devising a robot that read human minds and exercised mind control.

Much later, in 1980, a rumor circulated that Einstein had contributed to the Repository of Germinal Choice, a sperm bank that Robert Graham (1907-97), the inventor of the plastic eyeglass lens in 1947, established in Escondido, California, in 1978. Unfortunately for the rumor, Einstein died 25 years earlier. The Repository, to which Graham intended Nobel Prize winners to donate, received a donation from William Shockley (1910-89) who shared the 1956 Nobel Prize in physics for inventing the transistor. Shockley ventured into the field of genetics, and beginning in 1965 claimed that blacks were genetically inferior to whites.

Einstein became a United States citizen in 1940, retired from the Advanced Institute in 1945, and devoted his remaining years to private studies and the search for a unified theory that would unite gravity with the three other natural forces, the strong and weak nuclear forces and the electromagnetic force. Israel offered him the presidency in 1952, but he declined.

Einstein died of a cerebral hemorrhage in April 1955. Seven hours after his death, Thomas Harvey (1912-2007), the pathologist at the Princeton Hospital where Einstein died, removed Einstein's brain for his own research before cremation of Einstein's body. Harvey received permission for the removal from Einstein's son Hans Albert only after informing the family that he had done the removal. About three months later the hospital fired Harvey for his refusal to return Einstein's brain. Steven Levy (b.1951), a New York City journalist and senior writer for *Wired*, rediscovered Einstein's brain in 1978 while interviewing Harvey in Harvey's office in Wichita, Kansas. Levy had gone to Wichita to ask Harvey about Einstein's brain. During the interview he learned that Harvey had divided Einstein's brain into 240 pieces, each one $cm^3$, and had preserved the pieces in two jars of formalin that he kept in his office. Harvey returned to Princeton in 1996, and in 1998 he returned Einstein's brain to the Princeton Medical Center. He acknowledged that during the forty-three years he had Einstein's brain, he had given a few pieces to brain researchers.

The few studies, other than Harvey's cursory examination, that took place since 1955, include those of Marian Diamond, a neuroanatomist at Berkeley, in 1985, and Britt Anderson, a neurologist at Alabama, in 1996. More recently, a 1999 study by Sandra Witelson, a neuroscientist at McMaster University in Hamilton, Ontario, found that in Einstein's brain, the parietal lobes which account for visual thinking and spatial reasoning were quantitatively different from the control group. Each hemisphere of Einstein's brain was 1 cm wider (15 percent) than the control group brains of 35 men and 56 women. The parietal lobes were relatively wider and structured as one unit rather than the usual two units separated by the Sylvan fissure, and the brain was more spherical. Einstein's brain weighed 1,230 grams, which placed it within the average range, but below the mean (1,400 grams) for men of his age.

A later study by Dahlia Zaidel, a neuroscientist at UCLA in 2001, reported that the neurons on the left side of Einstein's hippocampus, the region of the brain responsible for memory and word associations, were consistently larger than the neurons on the right side. The brains of the 10 people of ordinary intelligence that she studied showed only minimal and inconsistent asymmetry. Zaidel suggested that the larger neurons in the left hippocampus may imply stronger nerve connections between the hippocampus and the neocortex in the left side than in the right side. The neocortex is that part of the brain where detailed, logical, analytical, and innovative thinking occur, but Zaidel said she did not know yet whether the asymmetry originated at birth, during Einstein's brain's development, or resulted from an abnormality. The relation between this asymmetry and Einstein's genius, she said, remains unknown.

A different perspective on Einstein's genius comes from Walter Isaacson (b.1952), an American author in Washington, DC, in his 2007 biography *Einstein: His Life and Universe*. Isaacson, and other biographers before him, believed that Einstein's contempt for authority and his rebellion against rote learning helped make Einstein the most creative scientific genius of our time. Einstein's slowness in learning to speak led him to think in pictures, and according to Isaacson, to become an imaginative and creative scientist.

How did Einstein evaluate his success? Einstein spoke of his persistence and perseverance, his reluctance to abandon a problem as an important strength he possessed. Imagination, he said, was always more important than knowledge. Einstein believed intuition and deduction provided the key to understanding nature. Ideas are free creations of the mind.

Einstein said he lost his religion at age 12. He acknowledged a God, but his God was nature or was in nature. His God was not a personal deity but more a pantheistic entity, a cosmic force. Einstein considered himself an agnostic not an atheist, and to believe in a personal God to whom you prayed was naive. Life after death was wishful thinking. In a letter of January 1954, he wrote, the Bible was "a collection of honorable but still primitive legends which are nevertheless pretty childish." Judaism was "an incarnation of the most childish superstitions" and "Jews were no better than other human groups." In Einstein's thinking, "science without religion is lame, religion without science is blind." Einstein always maintained an "unbounded admiration for the structure of the world so far as science can reveal it." He never accepted the indeterminism and probability of quantum mechanics, arguing that beneath its veneer lay the certainty of physical law. This is the origin of his famous statement, "God does not play dice."

In the first decades of the twentieth century Einstein proposed two new theories of the universe. The first, the special theory of relativity published in 1905, was a new theory of space, time, and matter. Classical physics' absolute space, or space filled with the luminiferous ether, against which to measure the speed of light, of the planets, and of any other objects moving through it, did not exist. There was no absolute frame of reference, and all motion was relative to a chosen frame of reference. Absolute time did not exist, time and the passage of time were relative and depended on the observer's frame of reference. Time was therefore a fourth variable, or dimension, and was necessary along with an object's x, y, and z spatial dimensions to specify the location of an object in space. Time was the fourth dimension in space. Matter was no longer absolute, but like time, relative or dependent on its frame of reference. Einstein proved that matter and energy were interconvertible, so that a loss of matter by an object undergoing a reaction occurred because of the conversion of matter to energy.

The second, the general theory of relativity, proposed in 1915 and published in 1916, was a new theory of gravity. Gravity was not an attractive force as Newton showed but a distortion of one body's geometrical space by a second body.

Einstein published three other significant papers in 1905 on the photoelectric effect, Brownian movement, and mass-energy equivalence.

*Photoelectric effect March 1905*

A photoelectric effect occurs when light of sufficient energy shines on a metallic surface, such as zinc, and causes the metallic surface to emit electrons. When the electrons flow from the metal into a conducting circuit, they produce an electric current in the circuit. Hertz first observed this light-induced electric current, the photoelectric effect, in 1887, ten years before J. J. Thomson discovered the electron. His one-time assistant, Lenard at Kiel, Germany, observed the photoelectric effect in 1889, but neither Hertz, Lorentz, nor classical physics' electromagnetic theory of light waves (radiation) could account satisfactorily for the electric current the electrons released from the metal produced.

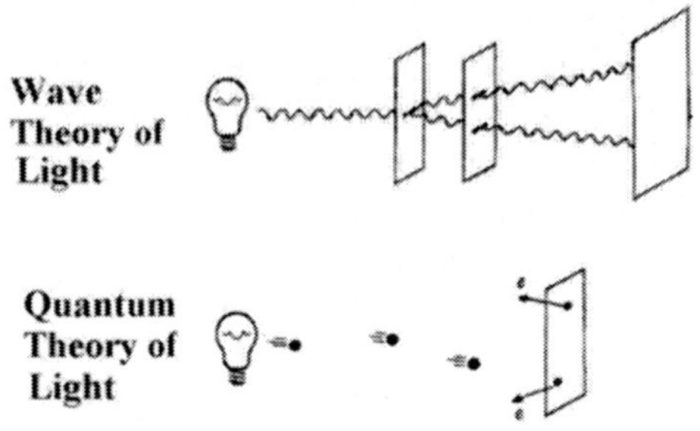


The wave theory of light explains diffraction and interference phenomena, which the quantum theory cannot explain. The quantum theory of light explains the photoelectric effect, which the wave theory cannot explain.

**Figure 11**
**Wave and Quantum Theory of Light**
Courtesy of the author

Einstein proposed a quantum theory of light to solve the dilemma, and thereby established the validity of Planck's relation, E = $h\nu$, that light consists of particles (photons). Einstein showed that a photon's energy depends on its frequency and not its intensity, or brightness, as classical physics suggested. When photons of sufficient energy struck the electrons on the metallic surface, they imparted enough energy to the electrons, enabling the electrons to escape from the surface and produce an electric current. Einstein received the 1921 Nobel Prize in physics for discovering the law of the photoelectric effect.

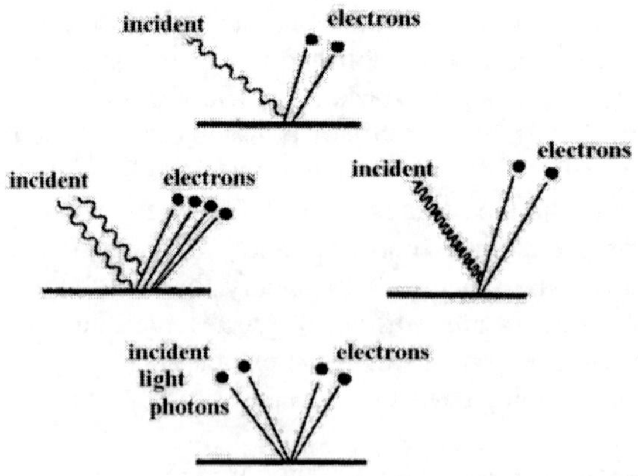


**Figure 12**
**The Photoelectric Effect**
Courtesy of the author

*Brownian movement May 1905*

Brownian movement is the random motion of small particles immersed in a liquid, such as pollen grains in a container of water. Robert Brown in London, in 1827 first observed with a microscope the random motion in a liquid of different kinds of particles less than one millimeter long. As with the photoelectric effect, Einstein was not the first to observe the random motion. He went beyond Brown's physical description and provided a mathematical explanation based on an atomic theory of matter. Einstein argued convincingly that the irregular visible motion or pathway of a large particle when randomly struck by tiny particles, represented the irregular motion or pathway of the invisible atoms or molecules that produced the Brownian movement. This occurred when the invisible water molecules struck and bounced off the larger visible pollen grains added to a container of water.

The equation Einstein derived for Brownian movement enabled Jean Perrin (1870-1940) at Paris, in 1909 to calculate the sizes of molecules and atoms and to determine Avogadro's number from his experimental measurements of the particles' rates of diffusion in a liquid. Avogadro's number is the number of molecules (or atoms) present in one molecular volume of a gas (defined as 22.4 liters) or one-molecular weight (mole) of a substance. Its value is $6.022 \times 10^{23}$ molecules per mole. Einstein's theoretical argument and Perrin's experimental evidence provided convincing evidence for the existence of atoms, and satisfied Oswald and most of the remaining skeptics with the exception of Mach. Perrin in 1926 received the Nobel Prize in physics for resolving Brownian movement.

**EINSTEIN'S EQUATION FOR THE BROWNIAN MOVEMENT**

$$D = \frac{RT}{6\pi Nvr}$$

| | | |
|---|---|---|
| D | = | diffusion coefficient |
| R | = | universal gas constant |
| T | = | absolute temperature |
| N | = | number of moles of a substance |
| v | = | viscosity of fluid |
| r | = | radius of sphere |

*Special theory of relativity June 1905*

Einstein undertook his work on the special theory of relativity to unify the mechanical and electrodynamic (electromagnetic) interpretations of nature, as indicated by the title of his 1905 paper "On the Electrodynamics of Moving Bodies." His new theory of space, time, and matter dealt with the special case of uniform, non-accelerated motion. It explained why the laws of nature appeared the same to all observers moving with constant speed relative to one another.

According to the classical mechanics of Galileo and Newton, laboratory measurements showed that the laws governing the motion of a body were the same regardless of whether the frame of reference (location) in which an observer established the laws, such as Earth, was at rest or moving at constant speed in a straight line. The same laws governed the motion of a body regardless of whether the motion took place on a ship, train, or airplane that was at rest or moving at a constant speed.

The electromagnetic theory of radiation developed by Maxwell and Lorentz stipulated, however, that light did not obey this rule, and measurements of light's resultant speed as it moved in different directions through the all-pervading ether and with respect to Earth's motion should show the effect of Earth's motion. Yet, none of Michelson's interferometer experiments, especially the Michelson-Morley 1887 interferometer experiment, had shown any difference in the speed of light because of Earth's motion. The speed of light did not vary.

Attempts to save the ether and to use the ether to account for the negative result of the 1887 Michelson-Morley interferometer experiment, led to the investigations of George Fitzgerald (1851-1901) at Trinity College, Dublin, in 1889; Henri Poincaré (1854-1912) at Paris in the 1890s; and Lorentz in 1892. Lorentz and Fitzgerald independently developed an equation known as the Lorentz-Fitzgerald contraction that showed mathematically that a body decreased in length when the body moved against the ether. The length decreased as the speed of the body increased, and masked the difference in the speed of light when it traveled in the interferometer's two perpendicular paths.

According to the Lorentz-Fitzgerald contraction, in Michelson's interferometer experiments, the length of the interferometer's arm that was parallel to Earth, as Earth moved against the ether, decreased just enough to make the speed of light equal to the speed measured when the light was traveling perpendicular to the direction that Earth was moving in space. The speed of light only appeared to be constant because of the decrease in a body's length (the interferometer arm) as it moved against the ether. Therefore, the distance the light traveled when a body moved against the ether was less than the distance it traveled when a body moved perpendicular to the ether and did not undergo any shortening. The decrease in length and the decrease in time taken to travel the shorter distance were just enough to increase the measured speed of light in the parallel direction, and thereby make the speed of light appear constant, regardless of the light's direction, just as in Michelson-Morley's 1887 interferometer experiment.

Lorentz, in his publications on the electron in 1899 and 1904 (Lorentz transformations), also developed an equation that showed mathematically how the mass of a body increased with its speed. The implication that a body's mass increased with its speed dates from Maxwell's 1861 paper "On Physical Lines of Force," his 1865 paper "A Dynamical Theory of the Electromagnetic Field," and from J. J. Thomson, who showed in an 1881 paper that to put an electrically-charged body in motion required more energy than to put an uncharged body in motion. This observation led Thomson to conclude that electrostatic energy possessed a new kind of mass called electromagnetic mass. A body's electromagnetic mass increased with increasing speed, and increased a body's inertial (mechanical) mass. Other physicists, including Wien in 1900 and Max Abraham (1875-1922) at Göttingen, an opponent of special relativity, who in 1902 examined Thomson's idea, concluded that a body's electromagnetic mass was its inertial mass.

Lorentz derived a similar equation for the passage of time, showing that time dilated, or slowed down, as the speed of a particular frame of reference increased.

The Lorentz and Fitzgerald equations were the same as the equations Einstein derived for length, mass, and time in special relativity, but they derived them to save the ether, not disprove it. Lorentz, Fitzgerald, and other classical physicists did not abandon the all-pervading stationary ether that provided the medium required for the transmission of light waves and the fixed reference against which to measure the speed of light and the speed of Earth. Absolute space and time remained the fixed references against which to measure all motions. They developed *ad hoc* explanations, or after the fact hypotheses, to account for ether, absolute space and time. They did not deduce their explanations from a theory.

Einstein believed that the principle of relative motion in classical mechanics applied to all phenomena, including electromagnetic, and therefore to light, which Maxwell in his 1873 *A Treatise on Electricity and Magnetism* defined as the motion or vibration of an electric charge in the ether. The speed of light should not depend on Earth's motion or on the uniform motion of any body. Einstein's special theory of relativity and its new analysis and understanding of space, time and matter, led to his abandoning the ether that classical physicists believed filled space and provided the medium that transmitted light waves.

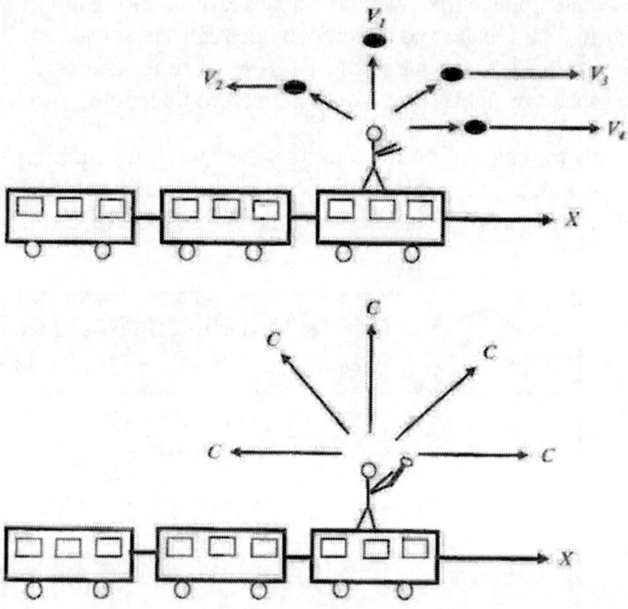


**Figure 13**
**The Speed of an Object, Classical Theory, Variable (top)**
**The Speed of Light, Constant (below)**
Courtesy of the author

Einstein made the following assumptions in arriving at his special theory of relativity. The speed of light was constant for all frames of reference, that is, the speed of light remained the same regardless of whether the light source remained at rest or moved at any (uniform) non-accelerating speed. The faster a body moved, the less the speed of its source (Earth) affected the final (resultant) speed, and for a body moving at the speed of light, the speed of the source (Earth) had no effect on the resultant speed.

The ether does not exist, nor does an absolute reference (the ether-filled space or absolute space) against which to measure the speed of a moving body. Therefore, all motion is relative to a chosen reference. The laws of physics are valid for all chosen frames of reference.

Four dimensions are necessary to define a body's location in space. In addition to the three dimensions of space, x, y, and z, Einstein added time, t, as a variable whose value depended on the speed of the source in its chosen frame of reference, or location. Time no longer had an absolute value, only relative values. Hermann Minkowski (1864-1909) at Göttingen, from 1902 provided a geometrical interpretation of space-time, giving it a formal mathematical treatment in 1907 and in his 1908 book *Time and Space*. Minkowski died the following year following an operation for appendicitis.

*Consequences of the special theory*

Special relativity showed that the values of length, mass, and time duration depended on the relative speeds of the frames of reference. As the speed of a body in one frame of reference increased relative to an observer in another frame of reference, the length $l$ of the body contracted, the mass $m$ of the body increased, and the time $t$ dilated, expanded, or slowed down, relative to the observer. A body, such as a person, present in a frame of reference (on a different planet) that moved uniformly at a speed greater than the speed of Earth orbiting the Sun, would have a shorter length, a greater mass, and experience time passing more slowly, relative to its length, mass, and time measured on Earth.

Anyone in this new frame of reference would not know that length, mass, and time duration were different, they would appear that way to an observer looking in from another frame of reference, perhaps from Earth. The greater the difference between the speeds of the two frames of reference, the greater the difference in length, mass, and passage of time. If a body existed in a frame of reference that moved at the speed of light, what would happen to its length, its mass, and to the passage of time?

Einstein's equations for calculating the length, mass, and time duration were the same as those Lorentz and Fitzgerald developed, but instead of being *ad hoc* and introduced to save the ether, they resulted directly from Einstein's starting assumption that the speed of light was constant for all frames of reference.

**RELATIVISTIC CHANGE IN LENGTH, MASS, AND TIME**

$$l = l_0 \sqrt{1 - \frac{v^2}{c^2}}$$

$$m = \frac{m_0}{\sqrt{1 - \frac{v^2}{c^2}}}$$

$$t = \frac{t_0}{\sqrt{1 - \frac{v^2}{c^2}}}$$

v = speed of object
c = speed of light
$l_o$ = rest length of object or body
$m_o$ = rest mass of object or body
$t_o$ = time at rest

*Mass-energy equivalence September1905*

Einstein derived the mass-energy equivalency using evidence obtained from several sources. Kelvin's 1856 equation for kinetic energy showed a relation among energy, mass, and the speed of a body squared, $KE = \frac{1}{2}mv^2$. Maxwell's four equations that described the interconnectedness of the electric field, magnetic field, electric charge, and electric current, and which he first published in his 1861 paper "On Physical Lines of Force," showed that the mass of a body provided a direct measure of a body's energy. According to Maxwell, light had momentum (mass × velocity, mc), and therefore indicated a relation between its mass and energy (velocity). Einstein's own studies on the photoelectric effect and his conviction that a photon transfers mass, and that the energy of a photon was equivalent to a quantity of mass; on an atom's radioactive decay, in which one atom decayed into another, and in so doing released energy, which very likely came from the decaying atom; his equation from special relativity that showed the relation between relativistic mass and the speed of light squared (the transformation of energy of motion into the increasing mass of the body) led to the same mass-energy connectedness.

A series of investigations that the physicist Walter Kaufmann (1871-1947) at Göttingen began in 1901-02 and continued into 1906, provided additional crucial experimental evidence of a mass-energy relation. Kaufmann found that the electrons that radium emitted, and which he accelerated to very high speeds in an electric or a magnetic field, experienced a change in mass that depended on the electron's speed. The electron's charge did not change as its speed increased, but the ratio of charge to mass, e/m, decreased. Kaufmann concluded that the electron's mass increased as its speed increased, and in 1903 he proposed, as had J.J. Thomson, Lorentz, and Abraham at about the same time, that the electron's mass was purely electromagnetic mass, a manifestation of electric charge. Inertia (inertial mass) was only its apparent mass.

All of this evidence convinced Einstein of a mass-energy relation and to conclude in his brief 1905 paper that the amount of energy released, or produced, in a reaction resulted from the reactants' loss of mass. The mass of a body was a measure of its energy content, and if the energy changed by an amount L, Einstein believed that the mass (m) changed by $L/9\,x\,10^{20}$, or $\Delta m = L/c^2$. In 1907 he published the equation that related energy, mass, and the speed of light, using for the first time the symbols E, m, and c in the equation $E = mc^2$. This equation showed that energy traveled at the speed of light, established the convertibility of mass and energy, and linked the laws of conservation of mass and conservation of energy. The Einstein mass-energy conversion showed that the complete conversion of a one gram mass released 90 trillion joules of energy, equivalent to 20,000 tons of TNT.

*Verification of the special theory*

Four different kinds of experiments have established the validity of special relativity beyond reasonable doubt: relativistic mass increase and half-life increase in particle accelerators, the regularity of pulsar emissions, and neutrino detection.

In a particle accelerator, such as a cyclotron, the mass of a positively-charged particle, usually a proton or a deuteron, increases as the particle accelerates in a circular path from one dee (positive electrode) to the accelerator's other dee (negative electrode). Each dee electrode, so-called because of its D shape, is a flat semicircular-shaped box placed between the poles of an electromagnet and connected to an alternating current high voltage source, or oscillator. The oscillator rapidly reverses the charge on each dee causing the positively-charged particle to get a push and a pull (repulsion and attraction) as it moves in a circular path from one dee to the other. Because of the particle's relativistic mass increase as it accelerates in the cyclotron, the cyclotron's oscillating frequency, which is the frequency at which the dees reverse their charges, requires continuous synchronization (adjustment) so that the charged particle receives its push and pull exactly at the time it moves from one dee to the other.

The cyclotrons constructed in the mid-1940s and 1950s at the University of California Radiation Laboratory at Berkeley (1945-46), the University of Chicago (1951), Brookhaven National Laboratory on Long Island, New York (1953), the European Center for Nuclear Research (CERN) in Geneva, Switzerland (1954), and Cornell in Ithaca, New York (1956) are frequency-adjusted cyclotrons, called a synchrocyclotron or synchrotron. Vladimir Veksler (1907-66) at the Lebedev Physics Institute (Soviet Academy of Sciences) in Moscow in 1944, and Edwin McMillan (1907-91) at Berkeley in 1945 independently invented the synchrocyclotron. Its successful operation provided clear proof of the relativistic mass increase.

Studies of the mu meson's (muon) radioactive decay have shown that its half-life increases with acceleration in a particle accelerator. In a 1977 experiment that an international team of 12 physicists carried out at the CERN Muon Storage Ring, the muon's normal lifetime of about two-millionths of a second (2.2 μs) increased 29.33 times when moving in a closed circular orbit at speeds near that of light. The muon decays into electrons, neutrinos, and anti-neutrinos. Detection of the decay particles provided scientific evidence of the time dilation that special relativity predicted.

In 1967 Antony Hewish (b. 1924) *A* =and his graduate student Jocelyn Bell (b. 1943) at Cambridge, while surveying the sky with a newly constructed giant phase-arrayed antenna that detected all scintillating sources in a large region, accidentally discovered the first pulsar PSR 1919+21, or PSR B1919+21. The numbers indicate the pulsar's position in the sky, in this case the constellation Vulpecula. A pulsar (pulsating radio star) is a rapidly rotating, highly magnetized neutron star of very high density, that because of its rotational energy generates an electrical field and emits a regular sequence of rapid radio pulses (electromagnetic radiation). Each pulse lasts a few hundredths of a second, with one second or less between pulses, and enables a pulsar to function as an extremely accurate clock. PSR 1937+21, which Donald Backer (1943 - 2010) at Berkeley and his four associates discovered in 1982, emits as many as 641 pulses per second.

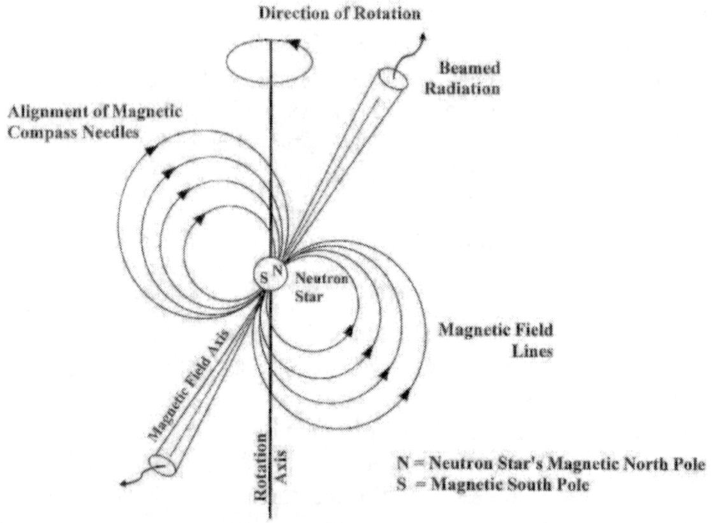


**Figure 14**
**Pulsar**
Courtesy of the author

As the pulsar moves in its orbit, it approaches and recedes from Earth. If the speed of its radio pulses, which travel at the speed of light, depended on the pulsar's direction of motion, then the pulses emitted as the pulsar approaches Earth would travel faster and reach Earth sooner than those emitted when the pulsar recedes from Earth.

Over distances of thousands of light years, the speed and time it takes the pulses to reach Earth would vary continuously. Instead of seeing the pulsar or receiving radio pulses at the same time intervals, astronomers would receive a continuum of pulses if the motion of the pulsar (the source of the pulses) affected the speed and time taken for the pulses to reach Earth. The experiments of Backer and other physicists beginning in 1983, showed that light reaching Earth from pulsars arrived at the same time regardless of whether the pulsar in its orbit was moving towards or away from Earth. Hewish, but not Bell, shared the 1974 Nobel Prize in physics for the discovery of the pulsar.

Frederick Reines (1918-98) and Clyde Cowan (1919-74) in 1956 discovered the neutrino while working at the US government's Savannah River Plant nuclear reactor in South Carolina. Reines belatedly in 1995 received the Nobel Prize in physics for discovering the neutrino. Cowan, its co-discoverer, died years earlier and ineligible to share the award. Beginning in the 1970s, neutrino detectors, which are large tanks holding hundreds or thousands of tons of ice, water, or an organic liquid scintillator, such as b-PDB,2-(4-tert-butyphenyl)-5-(4-biphenyl)-1,3,4-oxadiazole dissolved in a large volume of mineral oil, have provided evidence of neutrinos arriving on Earth from a nearby supernova and from deep space. The collisions of neutrinos with neutrons and protons of ice, water, or organic liquid scintillator molecules in the tank, create muons that emit blue light and ultraviolet light. Hundreds or thousands of photomultipliers located at the tanks detect the neutrinos. Their nearly simultaneous detection at different geographic locations shows that the neutrinos arrive on Earth at the same time, regardless of whether Earth is moving towards or away from the incoming neutrinos.

Supernova 1987 (SN 1987A), located in the large Magellanic Cloud, which is itself an irregular dwarf galaxy that orbits the Milky Way galaxy and lies in the constellation Dorado, provided the neutrinos. SN 1987A's discovery in February 1987 by observers at the Las Campanas Observatory in Chile and in Nelson, New Zealand, showed that it collapsed at a distance of 51.4 kiloparsecs (168,000 light-years) from Earth about 168,000 years ago. Because of the supernova's closeness to Earth, the light reaching Earth was visible to the naked eye from the southern hemisphere. SN 1987A was the first naked-eye visible supernova since Kepler's 1604 discovery. About three hours before the light arrived on Earth, three neutrino observatories located in three different locations simultaneously detected bursts of neutrinos.

One of the observatories, Japan's Kamioka Observatory at the Institute for Cosmic Ray Research, detected a burst of 11 neutrinos in its 3,000-ton ultra-pure water detector. Its detector, constructed in 1983, is 7,000 meters underground in the Mozumi Mine of the Kamioka Mines and Smelting Company near the city of Hida, Gifu.

The second detection occurred at the larger Irvine-Michigan-Brookhaven (IMB) 8,000-ton water detector. The University of California-Irvine, University of Michigan, and Brookhaven National Laboratory in 1982 constructed IMB 2,000 feet (633 meters) below ground, in a Morton Company salt mine near Cleveland, Ohio. It detected a burst of 8 neutrinos. Reines directed the detection experiments.

The third detection occurred at the 300-meter deep Baksan Scintillation Telescope constructed in 1977 at the Baksan Neutrino Observatory at Prielrusye in the Baksan Gorge in the Caucasus. The telescope consists of 3,200 individual container-detectors equipped with photomultipliers, each holding 150 liters of an organic liquid scintillator (a target mass of 330 tons of liquid scintillator). The Baksan telescope detected a burst of 5 neutrinos.

Although the three observatories detected a total of only 24 neutrinos, the bursts were the first that astronomers had observed directly. The actual measured time for the burst was 13 seconds at Kamioka, 6 seconds at IMB, and 9.3 seconds at Baksan. The simultaneous arrival and detection of the neutrinos at three widely-separated detecting locations, in which each of the detectors was moving either toward or away from the path of the incoming neutrinos provided evidence that Earth's speed had no effect on the speed of the neutrinos. The large detectors at the three observatories into which neutrinos passed, confirmed the neutrinos' presence.

*General theory of relativity 1915, 1916*

Einstein's general theory of relativity was a new theory of gravity. It dealt with accelerated motion. Einstein's starting point was the well-known observation that all bodies falling from the same height fall with the same acceleration in the absence of air resistance. This was the invariable factor that led him to search for a link between the force of gravity and accelerated motion, and resulted in his equation for general relativity. Einstein believed inertial mass, which is the mass that determines the difficulty of changing an object's motion and calculated using Newton's second law,

$$F = ma,$$

was equivalent to gravitational mass, the mass that determines how strongly two bodies such as Earth $m_e$ and any body $m_b$ on Earth's surface attract each other. Newton's law of universal gravitation provides the equation for the calculation of gravitational mass.

$$F = \frac{Gm_1m_2}{R^2}$$

$$F = \frac{Gm_em_b}{R^2}$$

Their equivalency resulted in the uniformity of gravitational acceleration, g, as shown by the relation for an object's falling at Earth's surface and other kinds of acceleration, or

$$F = ma = m_bg$$

$$m_bg = \frac{Gm_em_b}{R^2}$$

Canceling $m_b$ on each side of the equation gives

$$g = \frac{Gm_e}{R^2}$$

$$g = \frac{6.67 \times 10^{-11} m^3 sec^{-2} kg^{-1} \times 5.972 \times 10^{24}\, kg}{(6.37 \times 10^6\, m)^2}$$

$$g = 9.8 \text{ meters/second}^2 \text{ (32 ft/sec}^2\text{)}$$

In other words, gravitational acceleration is indistinguishable from other kinds of acceleration, and gravitational mass is indistinguishable from inertial mass. This is Einstein's equivalence principle. Einstein believed that an accelerating force caused a ray of light to follow a curved path. He argued that a ray of light passing through a gravitational field experienced a similar bending and followed a curved path. Space-time, according to Einstein was therefore curved.

In the equation which appears below in its simplest form, the left side represents the space-time curvature, that is, how space is curved. The right side represents the matter-energy content of space-time. It contains all of the required information about the location and motion of matter.

**EINSTEIN'S EQUATION FOR GENERAL RELATIVITY**

$$G_{ab} = \kappa T_{ab}$$

| | | |
|---|---|---|
| Gab | = | Einstein tensor (a two dimensional quantity) |
| $T_{ab}$ | = | stress-energy tensor |
| K | = | a constant |
| K | = | $\frac{8\pi G}{c^4}$ at slow speeds relative to light and in a weak gravitational field |

An alternative form of the equation is $G_{ab} = 8\pi\, T_{ab}$ where c = G = 1.

Newton provided a mathematical equation for the calculation of gravity. He showed how gravity works but not why it obeys an inverse square law. Einstein went beyond Newton, and with his general theory offered an explanation of why gravity acts the way it does and what it is.

In Einstein's theory space is geometric. Gravity is no longer an attractive force. It is the influence of one body on the geometry of the space-time in which a second body is moving. Inertial mass and gravitational mass are equivalent. You experience the same force when you are accelerating in a rocket in space at 9.8 m/sec$^2$ (32 ft/sec$^2$) as gravity exerts on you when you are at rest on Earth's surface.

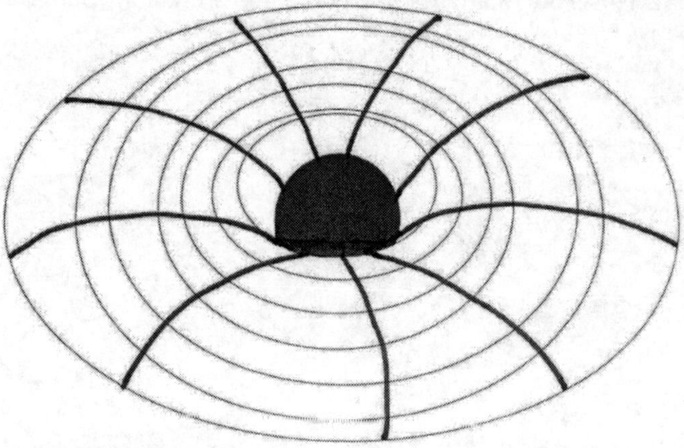

Einstein believed gravity was the result of matter influencing the curvature of space-time, rather than a force of attraction. An analogy is that of a heavy metal sphere in the middle of an elastic sheet, causing the sheet to curve in its vicinity. A small spherical object rolled across the sheet would fall towards the sphere because of the sheet's curvature. In a similar way, objects *fall* to Earth under the influence of Earth's gravity.

**Figure 15**
**Gravity as space-time curvature**
Courtesy of the author

Einstein's universe was also static, not expanding. In a 1917 paper he introduced a constant called the cosmological constant, or cosmological member, a measure of the outward repulsive force that had to exist to counter the inward pull of gravity, otherwise the universe would collapse. Because the gravitational force bent the rays of light, light eventually curved and reached its starting point resulting in curved space. The radius of curvature was constant and the universe did not expand.

Willem de Sitter (1872-1934), a Dutch astronomer at Leiden, in 1917 interpreted Einstein's general theory as showing the curvature of space increasing and the universe expanding like a giant bubble. The 1920s spectral studies on galaxies by Edwin Hubble (1889-1953) at the Mount Wilson Observatory in California led to the same conclusion. Einstein accepted Sitter's interpretation and abandoned the cosmological constant after he read Hubble's 1929 paper on an expanding universe. He later referred to his addition of the cosmological constant to his original theory as his greatest scientific blunder.

*Verification of the general theory*

Experimental confirmation of the bending or curving of light when it passed near an intense gravitational field, such as the Sun, was the first proof of general relativity. Sir Arthur Eddington (1882-1944) at the Cambridge Observatory observed the bending of light from stars during a 1919 solar eclipse. Eddington's expedition traveled to Principe Island in the Gulf of Guinea off the coast of

West Africa. World War I delayed any earlier solar eclipse test. The solar eclipse blocked the Sun's light and allowed astronomers to see that the Sun's gravitational field bent the starlight passing through it. Eddington sent Einstein a telegram announcing his experimental confirmation of Einstein's general theory.

A second proof was Einstein's prediction of Mercury's perihelion. This is the place or point on Mercury's elliptical orbit when Mercury is closest to the Sun. The perihelion does not occur at the same point on Mercury's orbit, but is slowly shifting, or advancing, closer to the Sun as Mercury's orbit spirals inward. This slow rotation of the perihelion around the Sun is a planet's precession.

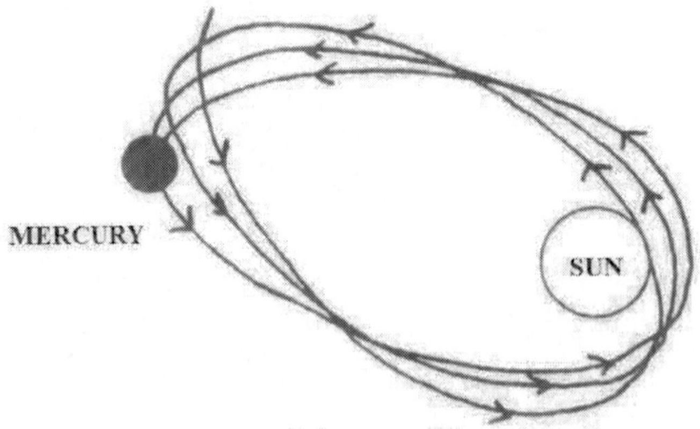


When Mercury undergoes a revolution around the Sun, it precesses. It does not complete a closed ellipse. Classical physics failed to explain this phenomenon, but Einstein's general theory of relativity gave a precise explanation for Mercury's precession.

**Figure 16**
**Precession of Mercury**
Courtesy of the author

All planets undergo precession, which results from the pull of the planets on one another. Newton's gravitational theory and force law could account for all the planetary precessions, except for Mercury the planet closest to the Sun. Mercury's experimentally measured precession was 5,600 seconds of arc per century, where one second of arc equals 1/3600 of a degree. Newton's calculated value was 5,557 seconds of arc per century, a difference of 43 seconds of arc per century. Einstein's 1916 paper predicted correctly that the precession, the advance or shift in the position of Mercury's perihelion because of the Sun's intense gravitational field, was an extra 43 seconds of arc per century. Einstein's prediction agreed perfectly with the experimentally measured value.

The red shift was a third proof. Hubble, a lawyer who switched to astronomy because of Hale's and Millikan's influence on him while at the Mount Wilson Observatory in California, provided the evidence. He studied the radial speeds of galaxies that Vesto Slipher (1875-1969) at the Lowell Observatory in Flagstaff, Arizona, beginning in 1912 had determined to be recessional speeds with respect to Earth, and resulted in a shift of the absorption lines of the galaxies' spectra. The absorption lines shifted to the red end of the spectra. The Doppler-like red shift of light from the stars was a displacement of the spectral lines from their normal positions in their spectrum to the red or longer wavelength positions in the spectrum. The red shift occurred when the light traveled through an intense gravitational field, like that of the Sun, just as Einstein had predicted, and it increased with increasing distance of the stars from Earth. The farther the stars or galaxy, the greater the red shift. The increased gravitational field caused the frequency of the light to decrease, thereby increasing the wavelength.

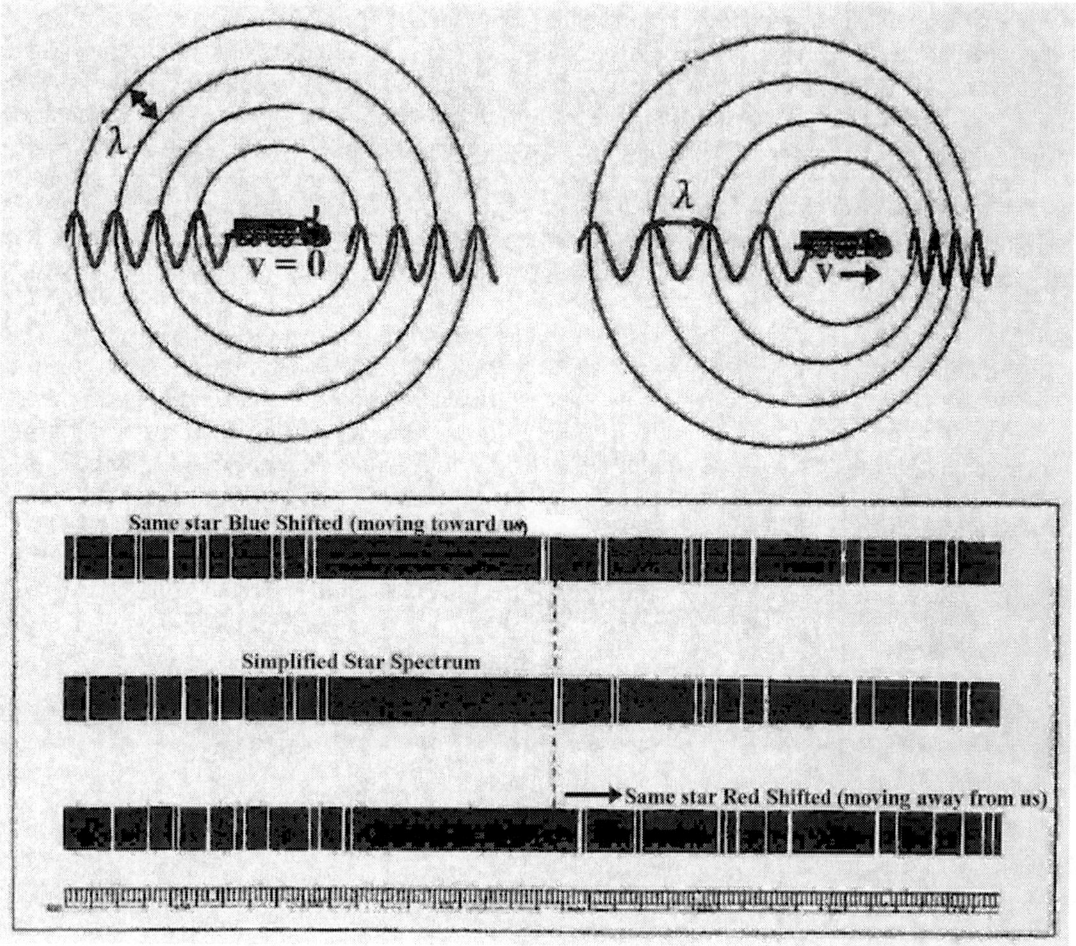


**Figure 17**
**The Red Shift**
Courtesy of NASA

Hubble's law of 1929 showed that the recessional speed of a star was proportional to the star's distance from Earth. As the distance increased, the recessional speed increased, and eventually reached the speed of light. This suggested that the universe was steadily expanding. This distance represents the Hubble radius of the universe, estimated at 13 billion light years, and implies a spherical universe with a diameter of 26 billion light years.

| HUBBLE'S LAW |
|---|
| $V = H_0 d$ |
| V =recessional (radial) speed of a star galaxy as measured by the Doppler red shift |
| $H_0$ =Hubble's constant measures the rate of expansion of the universe |
| d =distance of the star |

Astronomers calculated an initial $H_0$ value of 500 kilometers per second per megaparsec (km/s/Mpc), where 1 megaparsec = 1 million parsecs or 3.0857 x $10^{19}$ kilometers. Later $H_0$ values calculated prior to the 1990s ranged from $H_0$ = 50-100 km/s/Mpc and created some uncertainty. The inverse of $H_0$, $1/H_0$, has the units of time and gave $1/H_0$ values of 10-20 billion years as the estimated age of the universe. Since the 2000s, $H_0$ values have ranged from 67 to 72. NASA in 2013 calculated $H_0$ =70.4 and 71 and the age of the universe, $1/H_0$ =13.8 billion years. NASA used the Hubble telescope to make measurements from observations of 800 cepheid variables in 18 galaxies, some 65 million light years from Earth and most recently its Wilkinson Microwave Anisotropy Probe (WMAP) to measure cosmic microwave background, which is a constant background temperature in the universe also called its afterglow.

The discovery of the first binary pulsar PSR 1913+16 also supported general theory's prediction that two masses moving relative to each other emit energy in the form of gravity waves. Joseph Taylor (b. 1941) and his graduate student Russell Hulse (b. 1950) at the University of Massachusetts in Amherst, using the 300-meter radiotelescope at Arecibo, Puerto Rico, to search for pulsars, discovered the first binary pulsar, PSR 1913+16, in 1974. A binary pulsar is a pulsar orbiting another star, often a neutron star or white dwarf that may or may not be a pulsar. The two stars are rapidly rotating in elliptical orbits around their common center of mass. Each of the binary pulsar's two neutron stars, located 16,000 light years from Earth, has a radius of 10 kilometers and a mass 1.4 times the Sun's mass of $1.98 \times 10^{30}$ kilograms. Only one of PSR 1913+16's neutron stars is a pulsar.

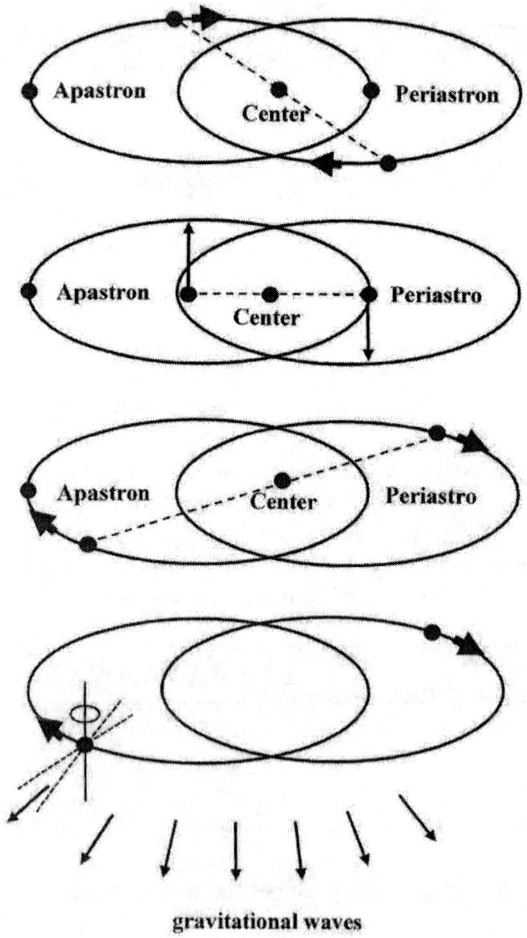


**Figure 18**
**Binary Pulsar**
Courtesy of the author

Taylor and Hulse timed the pulses, and discovered a systematic variation in the pulses arrival time. Some of the pulses arrived earlier than expected, some later. The variations became repetitive and had a period of 7.75 hours. Taylor and Hulse concluded that the observed behavior of PSR 1913+16 matched the predicted behavior for a pulsar in a binary orbit with another star.

Gravity waves result when extremely massive bodies distort the space around them. They originate in space-time regions where gravity is relativistic and the speeds of large masses approach the speed of light. Binary pulsars provide a stellar system that enables scientists to measure the gravitational effect of one large body on another closely located body in a gravitational field. Taylor and Hulse found that the two pulsar orbits were shrinking as they radiated orbital energy, and that the periastron (the rotation of the pulsar's elliptic orbit) of PSR 1913+16 advanced by 4 degrees per year. Taylor in 1978 showed that the observed orbital path values agreed with the theoretically calculated value of the binary pulsar's orbital path within 0.5 percent, and provided indirect proof of gravitational waves. Taylor and Hulse, then at Princeton, shared the 1993 Nobel Prize in physics for discovery of the first binary pulsar and their later research on pulsars.

Kip Thorne (b. 1940), until 2009 a theoretical physicist at the California Institute of Technology (Caltech), in the 1980s predicted that gravity wave detectors would enable scientists to observe gravity waves before 2000. Proof of their existence would verify general relativity's prediction that the waves are transverse, traceless, and travel with the speed of light. They are the force carriers of massless gravitons with a spin value s = 2 (the only massless spin-2 particles), where spin *s* is a particle's intrinsic angular momentum that depends only on the type of particle, and does not change. The graviton, which is a boson (a force carrier), has no mass because the gravitational field (force) acts over an infinite (unlimited) range and weakens as the inverse square of the distance.

**Figure 19**
**Gravity Waves from Two Orbiting Black Holes**
Courtesy of NASA

Tests to detect gravity waves began in 1992. At that time Thorne, Ronald Drever at Caltech, and Ranier Weiss (b.1932) at MIT, with $365 million in funding from the National Science Foundation, began construction of a Laser Interferometer Gravitational-Wave Observatory (LIGO), operating one interferometer in Livingston, Louisiana, and a second interferometer in Hanford, Washington. Each of the completed interferometers consists of two evacuated concrete tubes (arms) lined with stainless steel, each four feet in diameter and 2.5 miles long that intersect in a L-shape. Two mirrors (test masses) hang like pendulums inside each tube, one near the vertex of the L and one at the end of each tube. Precision laser beams sent down each of the perpendicularly-arranged tubes sense any small motions of the mirrors resulting from gravitational waves entering the tube from space. Gravitational waves interact with matter by compressing an object in one direction and stretching it in the perpendicular direction. This is the reason for LIGO's L-shape.

From 2005 to 2007, LIGO scientists sent a laser beam down each tube to a laser beam splitter located at the L's vertex that divides the laser beam in half and sends each half through one of the arms. The laser beam in each arm bounces back and forth between the mirrors, and then returns to the intersection where the two beams interfere with each other. If the two beams traveled the same distance in the two arms, they will overlap and cancel each other exactly when they reunite.

Because gravitational waves stretch and compress a body in perpendicular directions, a gravitational wave passing through a detector will expand the distance between the mirrors in one arm of the L and compress it in the other. By bouncing a high-energy laser beam back and forth between the mirrors in each arm, and then allowing the two beams to interfere with each other, an interference pattern will result. Any change in the interference pattern resulting from a change in the distance between the mirrors will provide strong evidence of gravitational waves. LIGO's evacuated tubes, which prevent air molecules and outside vibrations from affecting the mirrors, enable LIGO's scientists to detect changes as small as $10^{-16}$ cm occurring in the distance between the mirrors, and therefore the existence of gravitational waves.

The two observatories, in Livingston and Hanford, which are about 2,000 miles apart, provide another check on the actual detection of gravitational waves. The occurrence of an event simulating the presence of a gravitational wave, such as a falling tree or a laser fluctuation, at one observatory is highly unlikely to occur nearly simultaneously at the other observatory. Gravitational waves travel at the speed of light. This gives a detection delay of about 10 milliseconds between the two observatories. The delay will enable scientists to reject a possible detection (noise) observed at one observatory but not at the other, within the 10 millisecond time difference for the gravitation waves to reach each observatory.

The European Gravitational Observatory's French-Italian VIRGO detector, completed in 2003 at Cascina near Pisa, Italy, and having the same design as the American detectors, has sent laser beams down its tunnels. Other gravity wave detecting observatories in Hannover, Germany (GEO); Tokyo, Japan (TAMA); and Perth, Australia (AIGO), are developing space-based observatories. No laser interferometer has yet detected gravity waves.

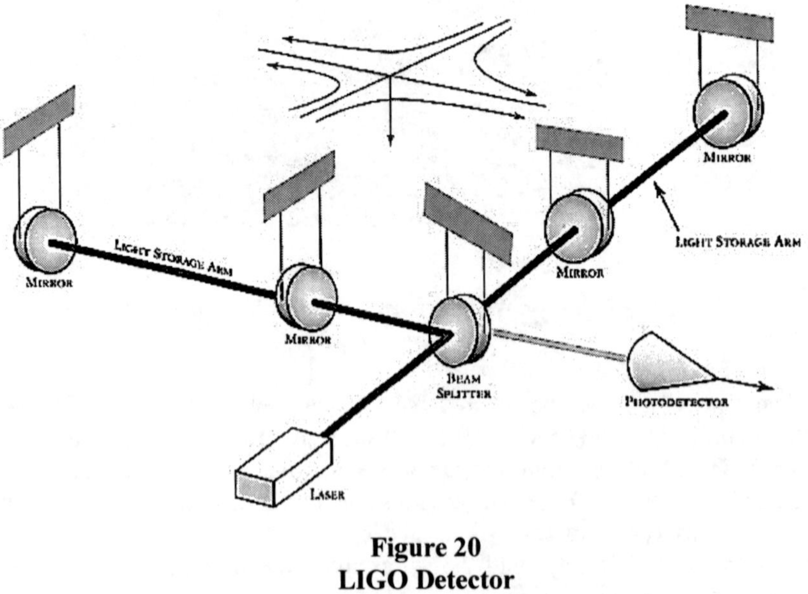


**Figure 20**
**LIGO Detector**
Courtesy of NASA

*String and superstring theory of gravity*

Steven Weinberg (b. 1933) at Texas, in the 1960s noted that when pushed to short distances and high energies, Einstein's general theory of relativity gave *mathematical nonsense*. General theory predicted a value of infinity for the contribution to the force between two particles that were emitting and reabsorbing short wavelength quanta, such as those present in the atom and its nucleus.

The breakdown of Einstein's general theory at short wavelengths, and its failure to account for the strong nuclear force led, beginning in 1968, to a string theory that included only bosons (photons, gluons, and the undiscovered graviton). The all-inclusive superstring theory that included bosons and fermions (quarks, electrons, protons, and neutrons) followed in 1971. The two theories had their origin in Gabriele Veneziano's 1968 paper on the strong nuclear force, and listed as founders Veneziano (b.1942) then at MIT and CERN, Leonard Susskind (b.1940) at Stanford, Holger Nielsen (b.1941) at Copenhagen, the 2008 physics Nobel Prize winner Yoichiro Nambu (b.1921) at Chicago, Edward Witten (b.1951) at Princeton, Michio Kaku (b. 1947) at CCNY in New York City, John Schwarz (b. 1941) at Caltech, Joël Scherk (1946-80) a French theoretical physicist then at Caltech, and Alexander Vílenkin (b.1949) a Ukrainian-born physicist at Tufts. They expected the two string theories would provide for the first time a mathematically consistent theory of gravity and the three other fundamental forces, the strong nuclear, weak nuclear, and electromagnetic force. The two theories would reconcile quantum mechanics and general relativity.

Following the publication of Veneziano's 1968 paper and Nambu's, Nielsen's, and Susskind's independent interpretation in 1968 and 1969 of Veneziano's particle model as a quantum mechanical theory of one-dimensional bosonic strings, Schwarz, then at Berkeley, and other physicists developed one-dimensional string and superstring theories to account for the strong nuclear force. Their publications preceded the introduction in 1973-74 of quantum chromodynamics (QCD), a rival theory that treats particles as point-like rather than string-like, and became the accepted theory of the strong nuclear force. The initial string and superstring theories failed, however, to give a realistic model or theory of the strong nuclear force, a theory of hadrons, which are subatomic particles of two kinds, baryons (protons and neutrons) and mesons, and consist of quarks bound together by the strong nuclear force. The failure led Schwarz and Scherk to propose in 1974 that superstring theory, instead, provided a basis for unifying gravity and the three other fundamental forces, rather than a theory of hadrons. String and superstring theories contained the graviton, a massless boson with a spin value of 2 and the carrier of gravity waves, and led Schwarz and Scherk to propose a unified theory of quantum mechanics and general relativity.

Schwarz found most remarkable that all consistent string theories *required* the existence of gravity. You cannot leave out gravity, whereas point-particle theories did not require gravity's existence. According to Schwarz, nature consisted of one-dimensional vibrating strings (energy), not particles, zero-dimensional points, or waves. The strings were either open, with free ends (hadrons), or closed loops connected at both ends (gravitons), and were continually opening and breaking apart. Each string had no width and was of the Planck length $10^{-35}$ m. The strings vibrated in many modes, and the vibrations stood in one-to-one correspondence with an elementary particle, that is, each mode of vibration represented an elementary particle, such as an electron, photon, meson, or quark. Susskind, Nielsen, and Nambu independently proposed that the elementary particles were in fact states of excitation of a relativistic string. Scherk, Schwarz's one-time collaborator, continued his research on string theory and quantum gravity but became despondent over his marriage's breakup and committed suicide in 1980.

Vĩlenkin, in publications beginning in 1981, dealt with the gravitational properties of cosmic strings. He reviewed their dynamics in flat space-time, gave some exact solutions of Einstein's equation for static straight strings, and discussed the gravitational field and radiation produced by oscillating closed loops.

Vilenkin and Schwarz remain convinced that further refinement of string and superstring theory will lead to the long sought and elusive *theory of everything*. Other proponents include Brian Greene (b.1963) at Columbia, author of the best seller *The Elegant Universe* (1999), and Michael Green (b.1946), Schwarz's longtime collaborator and Stephen Hawking's successor as Lucasian professor at Cambridge. Green and Schwarz initiated the 1984 superstring revolution in which they formulated superstring theory in ten dimensions and thereby cancelled the anomalies in the theory. Critics, such as Sheldon Glashow (b. 1932) at Harvard, have continued to argue, as they have since string and superstring theory emerged in the 1970s, that string and superstring theories lack testable predictions. If we can't test it, it's not science. That is the message that the authors of two recent publications have conveyed, Peter Woit (b. 1957), a Columbia mathematician, in *Not Even Wrong* (2006) and Lee Smolin (b.1955) at the Perimeter Institute for Theoretical Physics, University of Waterloo (Ontario), in *The Trouble with Physics* (2006).

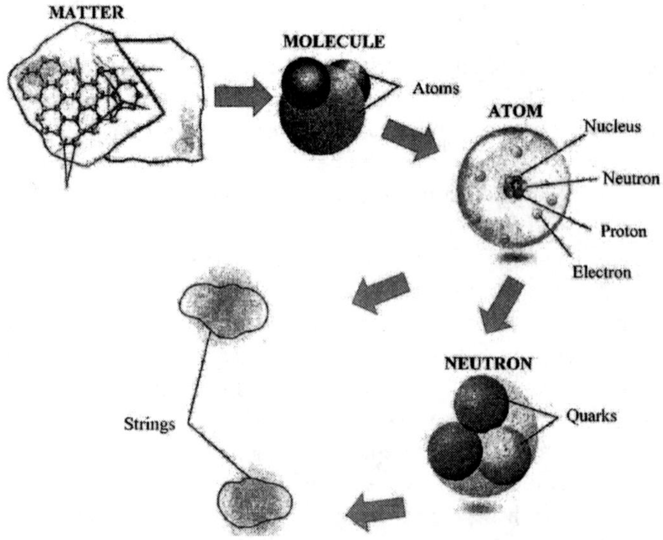


**Figure 21**
**String Theory**
© Designua, 2014. Used under license from Shutterstock, Inc.

## III. SEARCH FOR PRECISE ATOMIC WEIGHTS

The quest for precise atomic weights became the lifelong pursuit of Theodore Richards (1868-1928). As a graduate student at Harvard, Richards worked with Cooke on determining the atomic weight of oxygen. Later, he studied with Ostwald at Leipzig and Nernst at Göttingen. Beginning in 1894 and continuing in the 1900s, Richards and his students at Harvard determined accurate atomic weights of 28 elements, among them sodium, zinc, magnesium, iron, and cobalt. In 1913 Richards determined the atomic weight of lead present in different minerals, and found that lead's atomic weight in uranium minerals was less than the atomic weight of lead in other lead minerals and compounds. His work verified Soddy's 1913 prediction that isotopes of the chemical elements, such as lead, really existed. Richards was the second American to win a Nobel Prize in science, receiving the 1914 prize in chemistry for his work on atomic weights.

The oxygen isotope, O-16, remained the standard against which to compare the atomic weights of other elements until 1929. In that year, Giauque and Johnston at Berkeley, in their study of oxygen's spectrum discovered that oxygen gas consisted of three isotopes, O-16 the most abundant isotope, O-17, and O-18. Because of this discovery, physicists adopted a new atomic weight standard in which the O-16 isotope became the standard.

Chemists, as they had traditionally done, continued to use the average atomic weight of the three isotopes as the standard, setting it equal to O-16. They knew that the true atomic weight of oxygen was not exactly 16, but the average of the three isotopes calculated according to the percentage of each isotope present. The chemists' atomic weight did not differ much from 16. Nevertheless a slight difference existed between the two atomic weight standards and required a conversion between the physicists' and the chemists' standard, which persisted until 1961. In that year the International Union of Pure and Applied Chemistry (IUPAC) and the International Commission on Atomic Weights adopted the carbon isotope, C-12, as the new standard. The IUPAC chose C-12 partly because of C-12's use as a secondary standard in mass spectroscopy and because its adoption changed the existing oxygen atomic weight scale only by 42 parts per million.

## IV. ELECTRON: THE FUNDAMENTAL ELECTRIC CHARGE

Crookes and Arthur Schuster (1851-1934) at Manchester, in the late 1870s-80s observed that a magnetic field deflected cathode rays produced in a vacuum discharge (cathode ray) tube in a direction that indicated the rays had a negative charge. But neither Crookes, Schuster, nor any of the many physicists studying cathode rays succeeded in deflecting the rays upon applying an electric field, which should have occurred if the rays carried a negative charge. Improved cathode ray tubes that had a better vacuum resolved the issue.

In 1897 J.J. Thomson, who used an improved highly evacuated cathode ray tube, first demonstrated that an electric field deflected cathode rays. Thomson's demonstration provided irrefutable evidence that the rays were really negatively-charged particles. It enabled him to measure the ratio of the particle's charge to its mass (e/m) and to calculate the particle's mass, m = 1/1000 of a hydrogen atom's mass. The particle, which Thomson called a corpuscle, was much smaller than an atom and suggested that other subatomic particles might exist. Years earlier, in 1874, Stoney had called the fundamental unit of electricity (Thomson's corpuscle) the *electrine*, changing it in 1891 to the *electron* from the Greek word for amber, a resin that when rubbed produced a static electric charge.

J.J. Thomson in 1897 made the first measurements of the electron's charge. Millikan, at Chicago and from 1921 at Caltech, obtained more accurate values with his oil drop experiments. Millikan conducted these experiments from about 1906 to 1917, switching from water droplets to oil droplets in 1909 to minimize any effects of evaporation.

In his many experiments, Millikan studied the rate of fall of electrically-charged water and oil droplets in a closed container. He allowed the droplets to fall gravitationally through a hole in a horizontally-arranged positively-charged plate towards a second horizontally-arranged negatively-charged plate located a few centimeters below the first plate. The two plates had a potential difference of several thousand volts, creating a uniform electric field between them. Millikan then ionized the air molecules in the container by sending x-rays into the container. This released electrons from the air molecules and caused an oil droplet to attach itself to an electron, and sometimes fall more slowly towards the lower negatively-charged plate, or even rise. He increased the voltage and recorded the increase in voltage required to stop a particular drop from falling and remain suspended in the air. At that voltage the electric force on the drop exactly balanced the gravitational force acting on the drop.

Millikan believed that the smallest increase in the voltage required to stop the drop from falling (the minimum change in the drop's motion) resulted from the additional electron charge on the oil drop, and that by balancing the electrical and gravitational motions before and after the addition of the electric charge, he had measured the value of the electron's charge. Using known or calculated values of *m, g,* and E, he found that the values of the charges he measured in his experiments were always whole number multiples of the lowest charge. This confirmed Faraday's 1834 electrolytic studies and that electricity, like matter, was atomic in composition. Millikan obtained an average value of $e = 4.8 \times 10^{-10}$ emu or $1.6 \times 10^{-19}$ coulomb.

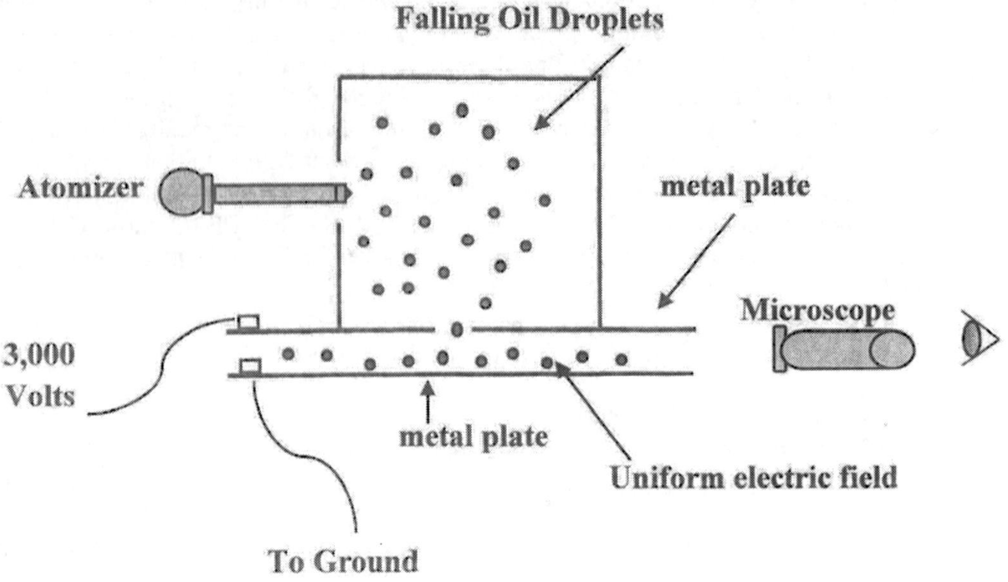


**Figure 22**
**Millikan's Oil Drop Experiment**
Courtesy of the author

| **MILLIKAN'S CALCULATION OF THE ELECTRON'S CHARGE** |
|---|
| $Ee = mg$<br><br>$e = \dfrac{mg}{E}$<br><br>$e$ = electron charge<br>E = electric field strength<br>$m$ = mass of electron<br>$g$ = gravitational force |

While at Caltech, Millikan investigated a type of radiation that Victor Hess (1883-1964) at the Institute of Radium Research of the Vienna Academy of Sciences, beginning in 1911, had discovered during his balloon ascensions. Hess made ten balloon ascensions to heights of 1,000-5,000 meters and used the rate of discharge of a gold leaf or quartz fiber electroscope to detect atmospheric radiation levels that were much higher (four times) than expected. Hess believed the radiation originated in the cosmos or outer space.

Millikan agreed with Hess on the radiation's origin. He named the radiation cosmic rays in 1925 and proposed that cosmic rays were high-energy electromagnetic waves. Investigations by Walther Bothe (1891-1957) in 1929 at the Max Planck Institute for Medical Research in Berlin and Arthur Holly Compton (1892-1962) at Chicago in the 1930s, using a pair of Geiger counters to detect the ray's charge proved that the rays, because of their deflection by a magnetic field, were charged elementary particles. Later studies identified the rays as mainly protons and a smaller number of helium nuclei. These studies showed that some of the cosmic rays originated in the cosmos, as Hess and Millikan originally proposed, but that most of the rays that reached Earth originated in the Sun's chromosphere and were most intense during solar flares.

Millikan, the son of a congregational minister, studied the classics while an undergraduate at Oberlin College, but his interests shifted to physics, and in 1895 he obtained his PhD in physics at Columbia, working under Pupin. Millikan received the first PhD in physics from Columbia. He actively promoted harmony between science and religion, maintained strong religious beliefs, and remained a committed conservative Republican. He received the 1923 Nobel Prize in physics for his experimental determination of the electron's charge and verification of the photoelectric effect.

Hess shared the 1936 Nobel Prize in physics for his discovery of cosmic rays. Carl Anderson (1905-91) at Caltech, who discovered the positron in 1932, was the other recipient. Hitler's annexation of Austria in 1938 forced Hess, a Catholic with a Jewish wife, to leave. He became a physics professor at Fordham University in New York City. After World War II ended, Hess spoke strongly against nuclear testing.

## V. THEORIES OF ATOMIC STRUCTURE

### *Multi-electron atom*

In constructing atomic models following the electron's discovery in 1897, physicists and chemists had no idea how many electrons an atom contained. Because an electron had so little mass, they believed an atom had to contain many electrons to account for the atom's mass. J.J. Thomson proposed the first multi-electron atomic model in 1899. His hydrogen atom, the lightest and therefore the simplest atom, contained 1,000 electrons because, as he had demonstrated, the electron was about 1/1000 times lighter than a hydrogen atom. The heavier atoms contained multiples of 1,000, although by 1906 he believed the number of electrons in an atom equaled an atom's weight. Thomson arranged the electrons in concentric rings inside a neutralizing spherical volume of positive electricity that spread throughout and filled the atom, and accounted for the remaining atomic weight. This was Thomson's *plum in pudding* atom. At this time the more massive proton and neutron were unknown.

Other scientists introduced multi-electron atoms. James Jeans then at Cambridge proposed in 1901 that a hydrogen atom had 700 electrons, and each kind of atom had a number of electrons in proportion to its weight. Kelvin in 1902 believed that an atom contained thousands of millions of electrons, and Lodge at Birmingham suggested in 1903 that hydrogen had 1,000 electrons, sodium 10,000-30,000 electrons, and mercury 100,000 electrons. Thomson's and Kelvin's atoms were positive sphere models, whereas Lodge in 1902 adopted an interleaved atomic model composed of positive and negative electrons interleaved together.

Lenard at Kiel, with his dynamic theory of 1903, and Lewis then at MIT, with his cubic atom in 1902, adopted an interleaved model. Lewis believed that the positive charge consisted of discrete particles, and that its localization determined the localization of the electrons. The positively-charged particle and the electron were spatially related to one another, perhaps coupled at the corners of the cubes.

A third atomic model that Hantaro Nagaoka (1865-1950) at Tokyo proposed in 1904, had negative electrons encircling a positively-charged center of an atom in the same way the planets circled the Sun in our solar system. Nagaoka's Saturnian atom resembled the nuclear atom that Rutherford, then at Manchester, introduced in 1911.

***Ernest Rutherford and the nuclear atom***

Rutherford's alpha particle (ray) and beta particle scattering experiment, in which he sent alpha particles (helium nuclei) against a thin foil of gold, platinum, or other metals, about 1/50,000th of an inch, or 2,000 atoms thick, provided the evidence for a nuclear atom. Most of the alpha particles passed through the foil unaffected, but some deflected at various angles, and others bounced straight back. This suggested that most of the atom was empty space, but that it also had a dense positively-charged component that sent the positively-charged alpha particles straight back in a 180° reflection. Rutherford's atom appeared to have its mass concentrated at the center with the negative electrons moving in circles around the positive center.

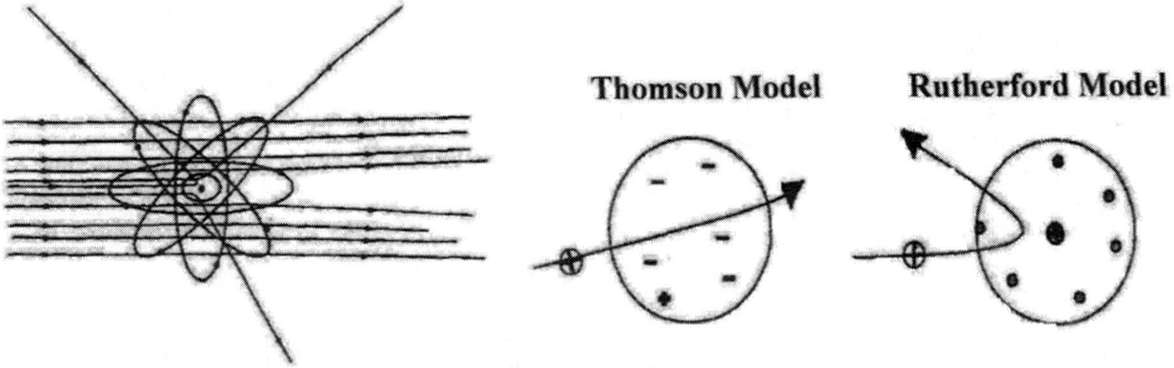


**Figure 23**
**Rutherford's Alpha Particle Scattering Experiment with Gold Foil (left)**
**Thomson and Rutherford Atomic Models, Alpha-Particle Scattering Patterns (center-right)**
Courtesy of the United States Energy Research and Development Administration, Office of Public Affairs (left)
Courtesy of the author (center-right)

His 1911 experiment disproved Thomson's atomic model which had the electrons arranged in rings in a sphere of uniformly distributed positive charge and would not have produced such large angular deflections. The scattering Rutherford observed when using thin foils of different metals, supported by the 1913 alpha scattering experiment of Hans Geiger (1881-1945) and Ernest Marsden (1889-1970) working in his laboratory, enabled Rutherford to conclude that an atom's nucleus carried a positive charge with a value approximately equal to one-half its atomic weight.

Rutherford in 1920, and now at the Cavendish Laboratory, first used the word proton to indicate the unit positive charge in the nucleus. The two fundamental charged particles known at this time and believed to make up each kind of atom were the electron, with a charge of -1, and the proton, about 1,836 times as heavy as the electron, with a charge of +1.

***Establishment of atomic numbers***

The acceptance of an atom with a positively-charged center immediately raised a new question. How many positive charges did the nucleus of each kind of atom contain? Rutherford's alpha scattering experiments provided the first clue, the answer came a few years later. Moseley in 1913-14 proposed that the experimentally determined x-ray spectrum of each kind of atom indicated a uniform increase in the number of positive charges in atom's nucleus. His new periodic law of 1914, replacing Mendeleev's 1869 periodic law based on the increasing atomic weights of the elements, established experimentally that the properties of the elements were periodic functions of their nuclear charge, (atomic number Z), which Moseley represented with the letter N.

Moseley came from a family distinguished in the sciences. At Eton College, his interest in the physical sciences began while studying with Thomas C. Porter (1860-1933), one of the first English scientists to experiment with x-rays. He entered Trinity College, Oxford, in 1906 on a science scholarship and received a master's degree in 1910. From Oxford, Moseley went to Manchester, where for two years he served as lecturer and demonstrator in physics while conducting research with Rutherford. A fellowship in 1912 enabled him to relinquish teaching duties and devote full time to research. When Great Britain entered World War I in August 1914, Moseley, now back at Oxford, immediately entered the military and secured a commission as an army signal officer in the Royal Engineers. He did not survive the war. During the disastrous Battle of Suvla Bay on the Aegean Coast of Gallipoli Peninsula in European Turkey, while telephoning a military order to Gallipoli on 10 August, 1915, a Turkish bullet passed through his head, killing him instantly. Moseley, having never married, left his estate to the Royal Society. His untimely death was for science the most costly single loss of life in World War I.

Under Rutherford, Moseley's initial research problem was to determine the number of beta particles emitted in the disintegration of one atom of radium B ($^{214}_{82}Pb$). Although no one doubted the emission of a single particle per atom, Moseley's research confirmed this assumption and introduced him to high vacuum techniques and radioactivity.

By this time, in 1911-13, two important breakthroughs had occurred in atomic structure and x-ray research. First, Charles Barkla (1877-1944) at Edinburgh showed from the scattering of x-rays by gaseous molecules that each element emits a characteristic x-ray spectrum, and that for the lighter elements the number of electrons in an atom was equal to one-half the element's atomic weight. Barkla's experiments strongly supported the evidence from Rutherford's, and Geiger and Marsden's alpha scattering experiments that an atom's nucleus carried a positive charge approximately equal to one-half its atomic weight. He received the 1917 Nobel Prize in physics for his experiments on x-ray scattering. Second, in 1912, the Braggs at Leeds extended Max von Laue's 1912 discovery of x-ray crystalline diffraction to establish that same year the fundamental equation, $n\lambda = 2d \sin \theta$. It related the x-ray's wavelength and the distance between the atomic or ionic planes in a crystal.

Starting with these new developments, Moseley began experiments in which he photographed the x-ray spectra of elements, using every element from aluminum to gold as a target (anticathode). He found that the wavelength, and accordingly the frequency, of the emitted radiation called the K line varied regularly from one element to the next when it followed the order of increasing atomic weights given in the Periodic Table. Except for the pairs cobalt-nickel, argon-potassium, and tellurium-iodine, his sequence was identical with Mendeleev's order based on increasing atomic weight. When he plotted the number of the place each element occupied in the Periodic Table (its atomic number) against the square root of the element's corresponding K frequency, he obtained a nearly perfectly straight line. From the evidence of x-ray spectra alone, Moseley concluded that an atom contained a fundamental quantity that increased in a regular manner upon passing from one element to the next, and that this quantity was the charge on the central positive nucleus, or its atomic number.

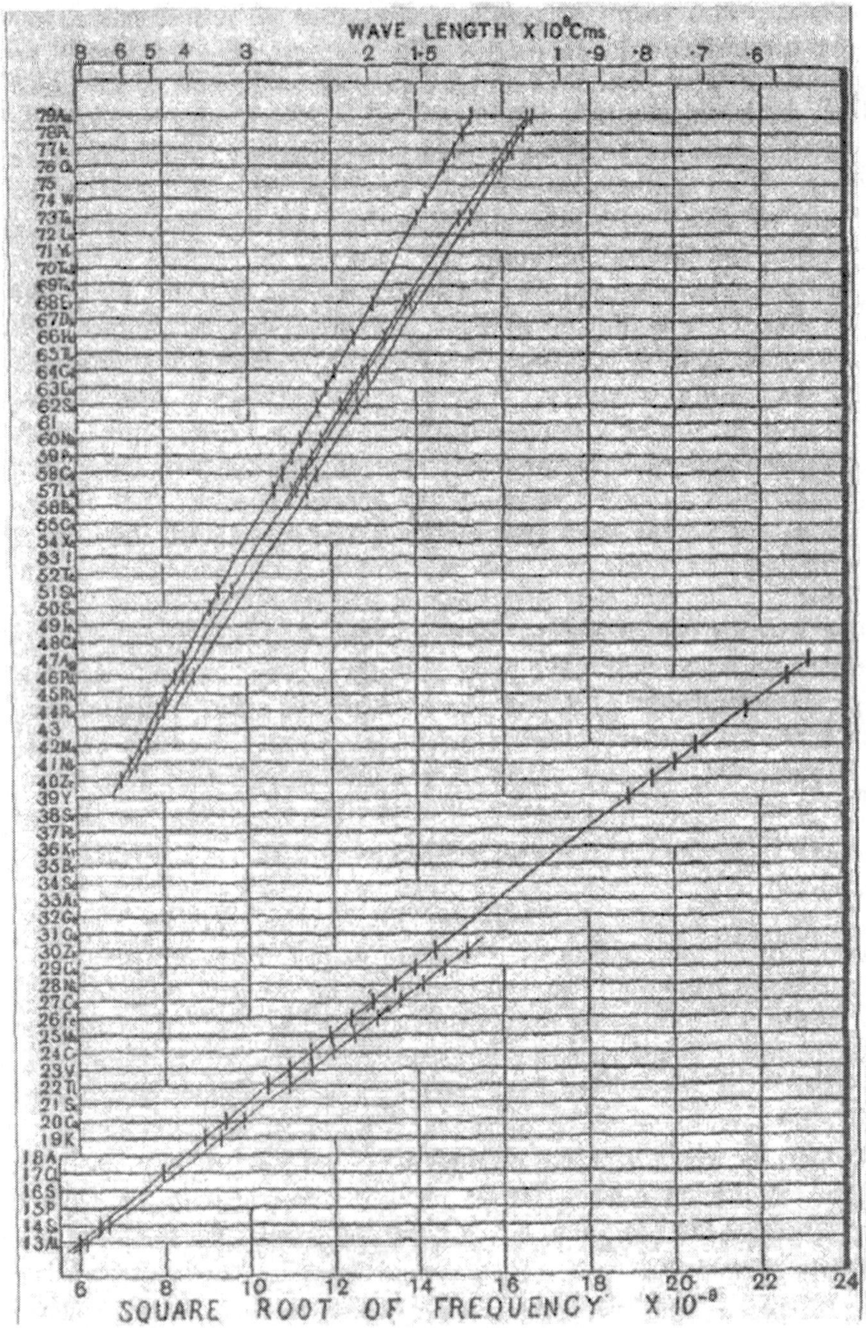


**Figure 24**
**Moseley Plot of Characteristic X-Rays**
Source: Henry Moseley,"The High-Frequency Spectra of the Elements," Part II. *Philosophical Magazine* 27(1914).

Because of the high improbability that two different stable elements would have the same x-ray spectrum, Moseley's work settled definitely the total number of elements lying between hydrogen (Z = 1) and uranium (Z = 92), the two extremes in the Periodic Table. His findings, which appeared in the *Philosophical Magazine* in 1913, ended the debate surrounding the number and order of the rare earth elements. When Georges Urbain (1872-1938) at Paris, took his samples of rare earth elements to

Oxford for examination, Moseley showed him the characteristic spectral lines, confirming twenty years of Urbain's research, in which he used magnetic susceptibilities to identify the different rare earth elements.

***Bohr's theory of the hydrogen atom: Sommerfeld's and Pauli's contributions***

Bohr, a member of a well-educated family, studied physics at the University of Copenhagen, where his dissertation on the electron theory of metals earned him a doctorate in 1911. Immediately afterwards, he left for England, spending a short time with J. J. Thomson at the Cavendish Laboratory before joining Rutherford at Manchester. Upon returning to Denmark in 1916, Bohr became professor of theoretical physics at Copenhagen. In 1920, when the university established the Institute for Theoretical Physics, he became its first director. Bohr remained at the Institute, later renamed the Neils Bohr Institute, until his death, leaving only in 1943-45 to avoid imprisonment during the Nazi occupation of Denmark.

Bohr fled to Sweden, then to Great Britain, and finally to the United States where he worked on the Manhattan Project in Los Alamos. He returned to Copenhagen in 1945 to become chairman of the Danish Atomic Energy Commission. Bohr had serious concerns about the military applications of atomic energy, and to promote its peacetime use he organized the first Atoms for Peace Conference in Geneva in 1955.

In 1911 Rutherford introduced his atomic model, a dense positively-charged nucleus surrounded by a revolving negatively-charged electron cloud. According to the laws of classical electrodynamics, his atom was unstable because the accelerating electrons continuously emitted energy and eventually spiraled into the nucleus. The electrons' spectral frequencies, therefore, were changing continuously, but theory failed to agree with the experimentally observed, sharply defined frequencies of an atom's spectral lines. Bohr resolved these issues by abandoning classical electrodynamics and applying Planck's quantum theory to Rutherford's electron cloud. He proposed two postulates. First, the electrons revolved about the nucleus in stationary states (fixed orbits) of different energy value that he designated by $n$ values $n = 1, 2, 3, 4 \ldots$ for the first, second, third, fourth, and higher orbits. He called $n$ his principal quantum number. An electron did not radiate energy as it revolved in its orbit, it absorbed or emitted energy only when it moved between the orbits. Second, the spectral frequency of the emitted energy multiplied by Planck's constant $E = h\nu$ equaled the energy difference between two orbits, such as $n_2 - n_1$ or $n_3 - n_2$, where the lower $n$ values were closer to the nucleus, more stable, and by arbitrary convention, of lower energy.

Bohr's quantized atom, which he published in 1913 and in two essays in 1920 and 1921, accounted reasonably well for the spectra of the one-electron hydrogen atom and singly-ionized helium, but failed when applied to atoms with two or more electrons. For multi-electron atoms, Bohr initially arranged all the electrons having a particular energy value in the same orbit in a single plane. His principal quantum number $n$, which determined the size of the orbit, held only for circular orbits, and his equation for an atom's energy levels neglected the influence of a magnetic field that oriented the orbits in different spatial planes. Despite his model's shortcomings, Bohr's successful linking of atomic structure and quantum theory and his atom's success in accounting for hydrogen's spectrum earned him the Nobel Prize in physics in 1922.

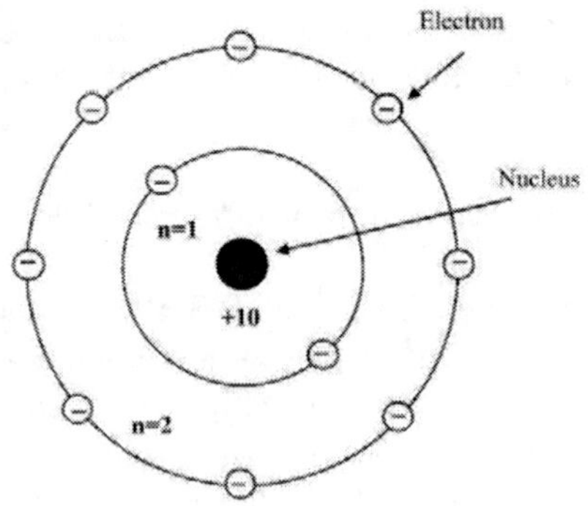


**Figure 25**
**Bohr's Atom**
Courtesy of the author

The German physicist Arnold Sommerfeld (1869-1951) at Munich, in 1915 and 1917 remedied both shortcomings of the one-electron hydrogen atom and helium ion. In 1915 Sommerfeld proposed that an atom had elliptical orbits as well as Bohr's circular orbits, and that the elliptical orbits accounted for the closely-spaced (fine) spectral lines observed in hydrogen's spectrum. The electron's revolutions were not exactly periodic, and because of the relativistic change in the electron's mass with the change in velocity, the orbiting electron experienced a very slow precession, causing it to move in an elliptical orbit. The 1916 spectral studies of Friedrich Paschen (1865-1947) at Tübingen experimentally confirmed Sommerfeld's explanation of the closed-spaced lines in the infrared region of hydrogen's spectrum.

Sommerfeld introduced the azimuthal quantum number $l$ that related the major and minor axes and hence the shape of each ellipse. Particles revolving around one another have an angular momentum, and the quantum number $l$ representing angular momentum, gave the shape of the electron orbit in an atom. For $n = 1$, $l = 0$ and indicated a circular orbit. For $n = 2$, $l = 0$ and 1, $l = 0$ indicated a circular orbit, and $l = 1$ an elliptical orbit. For $n = 3$, $l = 0, 1, 2$, $l = 0$ was circular, and $l = 1$ and $l = 2$ represented two different shaped ellipses.

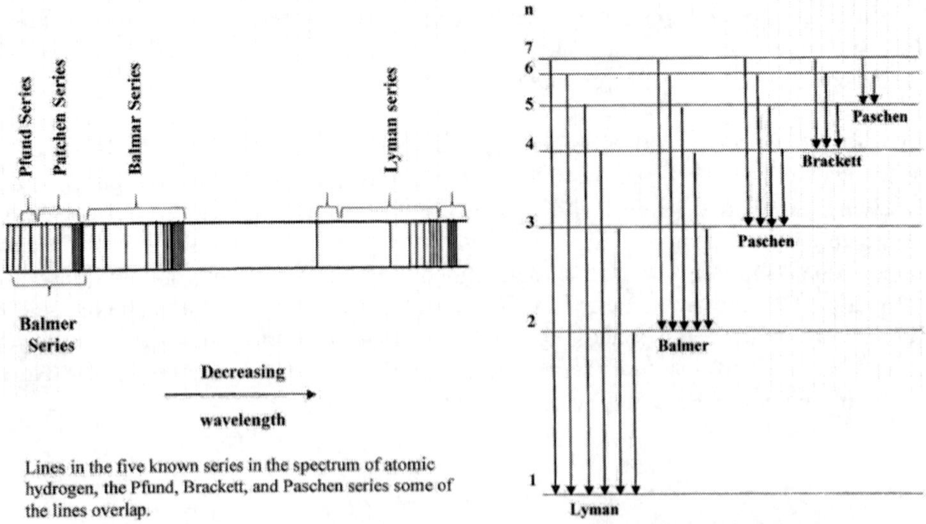


Lines in the five known series in the spectrum of atomic hydrogen, the Pfund, Brackett, and Paschen series some of the lines overlap.

Electronic transitions associated with the five series in the spectrum of atomic hydrogen.

**Figure 26**
**Line Spectrum of Hydrogen**
Courtesy of the United States Energy Research and Development Administration, Office of Public Affairs

Sommerfeld in 1917 proposed a third quantum number, the magnetic quantum number *m*, to account for the Zeeman effect, the splitting of an atom's spectral lines under the influence of a magnetic field. The physicist Pieter Zeeman (1865-1943) at Leiden, discovered the splitting in 1897. Sommerfeld's magnetic quantum number gave the orbit's magnetic orientation in space with respect to an applied magnetic field. Only certain magnetic orientations existed and the number depended on the *l* values (the orbit's shape). For each *l* value, Sommerfeld showed that the *m* values were $m = 2l + 1$. If $l = 0$, *m* has one value, $m = 0$. If $l = 1$, $m = 2l + 1$ and has three values -1, 0, +1. Each *m* value represented a different orientation of an electron orbit in space, such as along the x, y, and z, axes, and a different energy value.

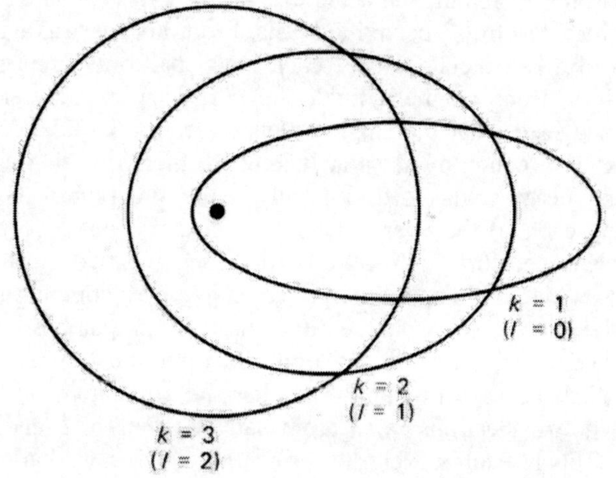


**Figure 27**
**Bohr-Sommerfeld Atom**
Courtesy of the United States Energy Research and Development Administration, Office of Public Affairs

The Zeeman magnetic effect led Johannes Stark (1874-1957) at Aachen to search for a splitting of an atom's spectral lines under the influence of an electric field. Stark observed a splitting of hydrogen's spectral lines in 1913. Three years later, in 1916, Karl Schwarzchild (1873-1916) at Potsdam and independently Paul Epstein (1883-1966) at Munich, showed how quantum theory and the electron's quantum numbers accounted for the splitting observed in hydrogen's fine line spectrum when placed in an electric field. Stark received the 1919 Nobel Prize in physics for his discovery of the electric splitting known as Stark effect. Stark in 1930 joined the Nazi Party and with Lenard, a party member since 1924, fervently supported the removal of Jewish scientists from academic and industrial positions. Stark in 1947 received a four-year prison sentence for his Nazi activities. Lenard, who faced no charges, died in 1947.

Sommerfield in 1920 discussed a fourth quantum number that he called the inner quantum number to account for a different kind of Zeeman spectral line splitting observed in some atoms. A normal Zeeman effect, which occurred in atoms such as the rare gases and zinc, gave an equally-spaced triplet when placed in a magnetic field. (It occurs in all atoms that have no net electron spin). In atoms that have an odd number of valence electrons, such as the alkali metals, the magnetic field split the triplet into 4, 6, or more lines with wider spacings than the normal triplet. This is the anomalous Zeeman effect. (It occurs in all atoms that have unpaired electrons and in which the net electron spin is therefore not zero.)

Alfred Cornu at Paris, Thomas Preston (1860-1900) at Dublin, and Michelson at Chicago, several years earlier had observed the anomalous splitting and discussed it in presentations in

October-December in 1897 and in papers in 1898. Waldemar Voight (1850-1919) at Göttingen, in 1912 used classical electromagnetic theory in an unsuccessful attempt to account for the anomalous splitting using the D-lines of sodium's optical spectrum. But physicists had no explanation for the anomalous Zeeman effect until the 1920s. Heisenberg, in a 1921 paper and more completely in a 1925 paper, and ultimately Wolfgang Pauli's introduction in 1925 of a fourth quantum number, the electron spin quantum number *s*, provided the definitive account for the increased number, or splitting of the spectral lines in a magnetic field, or for the anomalous Zeeman effect.

Sommerfeld's former student Wolfgang Pauli (1900-58), then at Hamburg, continued Sommerfeld's work on an electron's quantum numbers, examining experimental evidence on an atom's valence (outer) electrons obtained from spectral analysis. From his spectral analysis, Pauli in 1925 found that for the spectrum of alkali metals, where each metal had only one valence electron, each spectrum when examined in a strong magnetic field, showed twice as many energy states as quantum theory predicted. Pauli suggested that the single valence electron could have two different energy values, differing only in the direction of its spin. It implied, therefore, that every electron in an atom could have two different energy values, differing only in the direction of its spin. *Two* electrons, instead of one, could occupy each of the energy levels in an atom, not just the outermost energy level that contained the valence electron or electrons, provided the two electrons had different spins. Pauli concluded that if an electron's spin, interpreted as a sphere spinning on an arbitrary axis through its center, was parallel (clockwise), or in the direction of the *m* quantum number, it had a different energy value from an electron spinning in the opposite direction (counterclockwise) of the *m* quantum number. Pauli called the fourth quantum number the *s* quantum number, and assigned it values of $+1/2s$ and $-1/2s$. If two electrons in an atom had the same *n, l,* and *m* values, each had to have a different *s* value. This is Pauli's exclusion principle, which says that no electron in an atom can have the same numerical value for each of the four quantum numbers used to describe the energy of an electron in an atom.

Pauli's exclusion principle gave an atom a stable electron structure and allowed physicists and chemists to establish the complete electron configuration for every atom in the Periodic Table. His crucial deduction of the spin quantum number received additional experimental confirmation in 1927 from the spectral studies of Samuel Goudsmit (1902-78) and George Uhlenbeck (1900-88) at Leiden. Pauli received a belated Nobel Prize in physics in 1945.

The investigations of Sommerfeld, Pauli, Edmond C. Stoner (1899-1968) at Leeds, and others led Bohr to assign each electron a separate orbit, and to orient the orbits in space, not in a single plane. The first orbit had one energy level and held 2 electrons, the second orbit had two energy levels and held a maximum of 8 electrons, the third orbit had three energy levels and held a maximum of 18 electrons, and the fourth orbit had 4 energy levels and held a maximum of 32 electrons.

Despite these modifications, Bohr's atom never satisfactorily accounted for the physical and chemical behavior of multi-electron atoms. By 1925-26 the new quantum theory of Heisenberg and Schrödinger had completely replaced Bohr's atom.

## VI. CHEMICAL BOND

### *Electrostatic theories of the chemical bond*

Chemists who belonged to a new school of thought, first called theoretical chemistry and later in 1887 physical chemistry, began a systematic study of the atom's electrical behavior in the first decades of the twentieth century. They believed that their science required more than preparing new compounds and cataloging their behavior. Like the physicists, they placed before themselves the much higher ideal of gaining a clear insight into the electrical atom and the forces acting within it, and to developing a mathematical quantitative theory of the chemical bond.

William A Noyes (1857-1941), then at the National Bureau of Standards in Washington, DC, expressed the theoretical or physical chemists' conviction in his 1904 address to the International Congress of Arts and Sciences in St. Louis. Chemists, he said, should not be content with rounding out organic chemistry as a descriptive science, nor even with adding to the number of empirical rules which enable us to predict certain classes of phenomena. We must, instead, place before ourselves the much higher ideal of gaining a clear understanding of atoms and molecules, and of the forces or motions that are the real reason for the phenomena we study.

Until 1916 and the publication of Lewis's valence theory that accounted for both polar and nonpolar bonds, most of the electron theories of chemical bonding or valence were electrostatic, or polar, theories that required the transfer of an electron or electrons from one atom to another resulting in oppositely-charged ions whose positive and negative charges held the ions together. The electrostatic attraction replaced the affinity that chemists had used since the nineteenth century to account for the attraction holding atoms in a molecule. The electrostatic theory worked quite well for many inorganic compounds, but failed almost completely when applied to organic or carbon compounds whose reactions revealed little evidence that they contained oppositely-charged ions.

Stark then at Hanover and Bohr provided exceptions. From 1908 to 1915, Stark published a valence theory in which one or two electrons shared between a pair of atoms constituted the chemical bond holding the atoms together. Bohr in 1913 represented the H-H chemical bond in hydrogen ($H_2$) and each of the four C-H bonds in methane ($CH_4$) by a pair of electrons but did not define the bond in all molecules as a shared electron pair.

Chemists and physicists who developed electrostatic theories included J.J. Thomson and Richard Abegg (1869-1910) at Breslau. Thomson introduced his theory in 1899 and discussed it in his Silliman lectures at Yale University in 1903. Abegg, who formerly served as an assistant to Nernst and studied under Arrhenius at Stockholm, in 1904 proposed an electron atomic model and an electrostatic bond. Abegg's *rule of eight* gave some indication of the maximum number of electrons involved in the union of atoms. Abegg observed that the same atom in its different compounds had either a positive or a negative charge and that the absolute sum of the two charges was often eight. Chlorine had a -1 charge in sodium chloride ($NaCl$) and a +7 charge in perchloric acid ($HClO_4$), nitrogen's charge was -3 in ammonia ($NH_3$), and +5 in nitric acid ($HNO_3$). According to Abegg, most atoms had available for bonding anywhere from one to eight electrons. His rule was the first publication dealing with the number of bonding (valence) electrons and enjoyed great popularity among physical chemists. Abegg, a balloon enthusiast since 1900 after some balloon flights while a lieutenant in the German Army reserves, perished six years later from injuries he sustained in a balloon accident in Silesia (Poland).

American chemists who developed electrostatic bond theories included Julius Stieglitz (1862-1937) at Chicago; Harry Shipley Fry (1878-1949) and Stieglitz's former pupil Lauder S. Jones (1869-1960) at Cincinnati; K. George Falk (1880-1953), John M. Nelson (1876-1925), and Hal T. Beans (1876-1960) at Columbia; and William Noyes at Illinois. They published widely on the electron theory of valence in the first two decades of the twentieth century. Lewis, then at MIT, in 1902 discussed an electrostatic theory of valence based on his cubic atomic model but did not publish his theory. He conceived of the cubic atom while attempting to explain to an elementary chemistry class the periodic repetition of the elements' chemical properties (Periodic Law), and related an atom's valence to the numbers of electrons an atom either gained or lost to give an outermost completed cube of eight electrons. A polar (ionic) compound resulted whenever two or more atoms completed their outermost cubes by gaining or losing electrons, and each formed a stable octet. The electrostatic attraction between the positively- and negatively-charged cubes produced the polar bond between them.

***G.N. Lewis's shared electron pair chemical bond: a valence theory for polar and nonpolar bonds***

*Biographic*

Born in Weymouth, near Boston in 1875, Lewis received his primary education at home from his father, an independent-minded lawyer, and a mother educated in a Massachusetts seminary. His family moved to Lincoln, Nebraska, in 1884, and in 1889 Lewis began his formal education at the University of Nebraska preparatory school. Lewis transferred to Harvard in 1893 and graduated in 1896 with a BS degree. After teaching for a year at Phillips Academy at Andover, he returned to Harvard for a MA degree in 1898 and a PhD in chemistry under Richards in 1899. He taught chemistry at Harvard for one year before going abroad on a traveling fellowship to study under Oswald at Liepzig and Nernst at Göttingen. Lewis returned to Harvard in 1901 and taught thermodynamics and electrochemistry until 1904. Several of his longtime interests date from this early period, the most important of which were thermodynamics, valence theory, and photochemistry.

Lewis spent 1904 as Superintendent of Weights and Measures in the Philippine Islands and Chemist in the Bureau of Science in Manila. He returned to the United States in 1905 and joined the group of physical chemists that Arthur A. Noyes (1866-1938) assembled at MIT. Noyes's laboratory became the first center for physical chemists in the United States. Lewis remained at MIT for seven years, publishing over thirty papers including "Outlines of a New System of Thermodynamic Chemistry" and "The Free Energy of Chemical Substances." They were among the most important in a long series of papers on the experimental determination of free energy, and led in 1923 to the publication of Lewis's monumental *Thermodynamics and the Free Energy of Chemical Substances*.

In 1912 Lewis left MIT to become dean of the College of Chemistry and chairman of the Chemistry Department at the University of California, Berkeley. His recruitment of exceptional faculty and curricular reforms made Berkeley the model for American chemical education. Berkeley's production of first-rate chemists soon rivaled that of leading German universities.

Lewis's work on thermodynamics became more intense at Berkeley as he transformed an essentially theoretical science into a practical tool for chemists. His other interests, such as atomic structure and valence, continued to develop, and led to his 1916 paper "The Atom and the Molecule." His most extensive treatment of valence appeared in 1923 as *Valence and the Structure of Atoms and Molecules*.

World War I and his commission as a major in the Chemical Warfare Service interrupted Lewis's work at Berkeley. As chief of the service's Defense Division, he instructed American army officers on the effects of poison gas and how to defend against poison gas attacks. Lewis received the Distinguished Service Medal in 1922 for his valuable contribution. He returned to Berkeley after World War I and concluded his work in thermodynamics and chemical valence with his major publications on both topics in 1923.

**Figure 28**
**Gilbert N. Lewis**
Public Domain

*Cubic atom and shared electron pair chemical bond*

Lewis made significant contributions to atomic structure and valence with his theories of the cubic atom and shared electron pair bond. The cubic atom was a static model in which the electrons sat motionless at the corners of a cube, unlike Bohr's dynamic model that had the electrons moving in rings around a central nucleus. Lewis introduced a chemist's atom that accounted for the structure of chemical compounds. He further developed his electron pair bond but abandoned the cubic atom in the same 1916 paper in which he first published it.

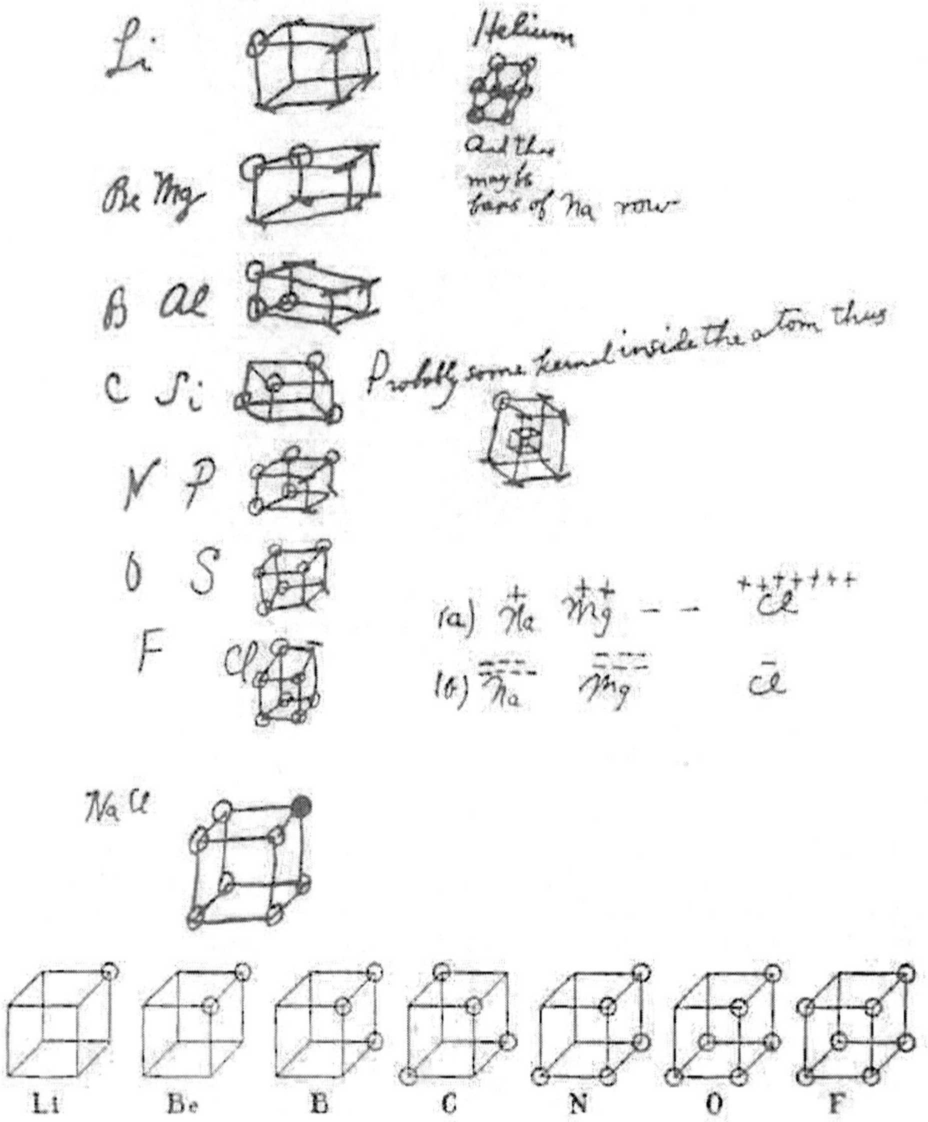


**Figure 29**
**Origin of the Lewis Cubic Atom**
Source: Gilbert Newton Lewis. *Valence and the Structure of Atoms and Molecules.* New York: Chemical Catalog Company, 1923 (top).
Source: Gilbert Newton Lewis. "The Atom and the Molecule." *Journal of the American Chemical Society* 38, no. 4 (1916) (bottom).

The idea of electrons in an atom arranged at the corners of concentric cubes led Lewis to conclude incorrectly that helium, because of its well-known unreactiveness and lightness (low density), consisted of a single completed cube of eight electrons and eight neutralizing positive charges. Evidence from the radioactive decay series, Rutherford's identification in 1906, and definitely in 1909, of the alpha particle as a positively-charged helium atom, $He^{2+}$, with a relative mass of four units strongly suggested, however, that the helium atom had two, rather than eight electrons. No other element lay between hydrogen and lithium. Helium was inert and resembled the members of the rare gas or Group 0 family.

The evidence from radioactive decay supported the conclusions of Rutherford's alpha and beta scattering experiment of 1911. Rutherford's scattering experiment showed that an atom's total positive charge occupied a very small and dense volume of radius $10^{-12}$ - $10^{-13}$ cm at the atom's center, called its nucleus, and not a surface of radius $10^{-8}$ cm as in Thomson's positive sphere model. This startling discovery resulted in Rutherford's theory of the nuclear atom. His experiment demonstrated that the number of electrons surrounding an atom's nucleus was nearly equal to one-half the atom's weight. Because Rutherford proved that the alpha particle was really an ionized helium atom with a +2 charge and a mass of four, helium obviously contained only two electrons.

As a direct consequence of Rutherford's work, a Dutch lawyer and amateur physicist Antonius van den Broek (1870-1926), then in Noordwijk–Zee, Holland, in 1911 proposed that the integral number with which physicists represented the positive charge of an atom's nucleus was the same as the number (the ordinal number) chemists used to indicate the sequence of elements in the Periodic Table. He called this number the atomic number and defined it as either the number of positive charges in an atom's nucleus or the number of electrons surrounding the nucleus of a neutral atom.

Van den Broek's speculations, likewise, supported the laws of disintegration proposed in 1913 chiefly by Soddy and Fajans. An atom undergoing alpha decay lost two positive charges and four units of mass, and it agreed in chemical behavior with the atoms in the group two places to its left in the Periodic Table. Radium in Group II decayed to radon in Group 0, which in turn decayed to polonium in Group VI. By 1914 Moseley's x-ray spectral analysis established the atomic number of each element and the total number of electrons present in an atom of an element.

Because of the advances that occurred in atomic structure from 1902 to 1916, Lewis now recognized that the electron pair played the same role in helium as the group of eight in the heavier elements and accounted for the stability of atoms and molecules. In row one of the Periodic Table, consisting of hydrogen and helium, the rule of two replaced the rule of eight. For heavier elements, such as oxygen, sodium, magnesium, and chlorine, knowing the correct atomic number enabled Lewis to construct cubic atom structures that agreed with the known atomic numbers.

With the cubic atom, Lewis accounted for the formulas of inorganic chemistry's simple (binary) polar compounds, but it left unexplained the vast number of nonpolar or organic compounds that did not consist of ions. Lewis solved this problem in 1916 by postulating that the cubes or atomic shells were interpenetrable. Because of shell interpenetrability, an electron(s) from one atom could form part of another atom's shell but not belong exclusively to either atom. Hence, neither atom lost or gained the electron(s). This was the mechanism for electron pair sharing, or chemical bond formation. Only in purely polar compounds was the electron transfer complete.

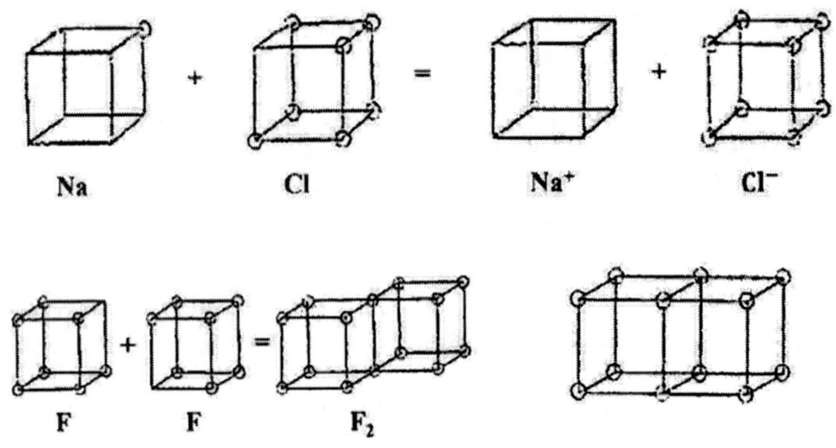


**Figure 30**
**Ionic (Polar) Bond, Single and Double Bonds**
Source: Gilbert Newton Lewis. "The Atom and the Molecule."
*Journal of the American Chemical Society* 38, no.4 (1916).

*Introduction of electron dot or Lewis formulas*

With the cubic atom now a thing of the past, Lewis needed a new way to express conveniently and clearly his shared electron pair bond. The new notation appeared in his 1916 paper in which Lewis introduced the now well-known electron dot or Lewis formulas. The use of dots to represent bonds in chemical formulas was not an entirely new idea. Organic chemists, long before the discovery of the electron and its application to valence theory, had represented single, double, and triple bonds with one, two, and three dots, respectively. In the Lewis theory, the important innovation was that each dot no longer represented a valence bond of unknown constitution but an electron, and each pair of dots, or colon (:), symbolized the electron pair shared by two atoms and constituted the chemical bond. For the single bond in the nonpolar chlorine molecule Lewis wrote Cl:C1. To indicate polarity in a molecule, he placed the pair of dots closer to the more negative element, for example, the chlorine atom in iodine monochloride, I :C1.

Lewis offered an electron theory of valence capable of accounting for both polar and nonpolar bonds. His fundamental conception, the shared electron pair bond, satisfied the long-established rule of eight, and its representation by the simple and pictorially satisfying electron dot formulas made his theory operable. Lewis readily wrote formulas and easily resolved many of the difficulties that had plagued theoretical chemistry for so many years, particularly the formulas for ammonia $NH_3$ and the ammonium compounds, the diatomic gases $H_2$, $O_2$, and $N_2$, and the oxyacids.

In spite of these achievements, specifically that the shared electron pair bond and the electron dot formulas became permanent parts of chemical tradition, Lewis's 1916 paper was a transitional one, containing residues of the past and unsolved problems for the future. The cubic atom had no place in the new theory, its limitations were too fundamental and too obvious. Yet, Lewis used it in the paper, as he must have used it in his own theory's development, as a lead-in to the shared electron pair bond. The cubic atom was a geometric representation that he easily adapted to the transfer of electrons and to the sharing of one or two pairs (but not three pairs) of electrons while preserving eight electrons as the complete atomic shell. The cubic atom became obsolete, and in Lewis's 1916 paper he replaced the eight electrons at the cube's corners with four electron pairs, one pair at each corner of a tetrahedron. Lewis's shared electron pair bond became the starting point for the new quantum chemistry.

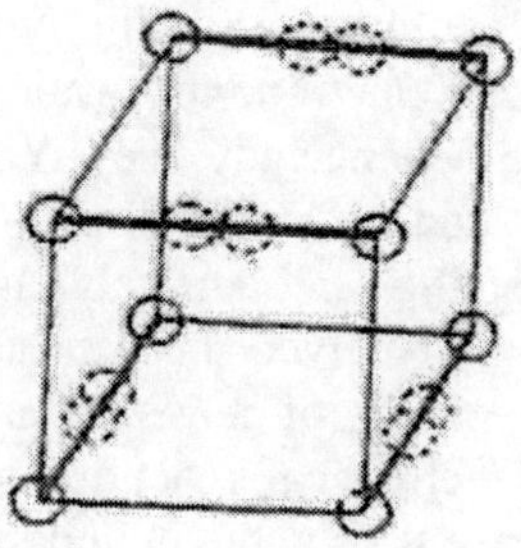

**Figure 31**
**Lewis Electron Pairs**
Source: Gilbert Newton Lewis. "The Atom and the Molecule."
*Journal of the American Chemical Society* 38, no. 4 (1916).

But electrons are not simply dots on paper. They have an experimental reality, consisting of real negative charge, and the question of how two negative charges could form a stable nonpolar bond was a question Lewis really could not answer in 1916. Lewis seemed to accept an idea that Alfred L. Parson (1889-1970) had developed earlier but published only in 1915 while studying in Lewis's laboratory in Berkeley. Parson, a British-born (in India) Harvard graduate student, spent the years 1913-15 as a visiting student at Berkeley.

In Parson's magneton theory, an idea that had its roots in the electrical studies of André Marie Ampére (1775-1836) and Langevin in Paris, each electron was not only an electric charge, but also a small magnet. A magnetic force of attraction, therefore, enabled the two negatively-charged electrons to pair. Lewis pointed out that molecules containing an odd number of electrons were very few in number and were less stable than even-numbered molecules. They almost always reacted to form a structure with an even number of electrons. This last fact, the scarcity of molecules containing an odd number of electrons, was most important, because by 1916 Lewis had concluded that the nonpolar valence bond consisted of an electron pair shared between two atoms.

A magnetic electron was of necessity a moving electron, whereas in the Lewis atom the electrons were essentially motionless. Lewis reconciled these opposing views to allow for the motion of electrons and yet retain his idea of electron pair sharing in his 1917 and 1923 papers and in his book *Valence*, but his novel idea of two negative electrons attracting one another and hence pairing, lacked theoretical support until Pauli enunciated the exclusion principle in 1925. According to this principle, no two electrons in an atom could have the same numerical value for the four quantum numbers used to describe an electron's energy state. A pair of electrons having identical values for the *n, l,* and *m* quantum numbers, had to have each electron's spin axis, designated by the *s* quantum number, oriented in opposite directions to one another. In other words, the two electrons were paired.

Pauli's principle limited the total number of electrons possible in an atom and permitted their correct distribution within each atom. Applied to the hypothesis of electron pairing in a chemical bond, it appeared equivalent to what Lewis in 1924 had suggested in his magneto-chemical theory. If the electrons in a nonbonded atom were paired magnetically, Lewis wrote, then a magnetic pairing also occurred between two electrons in different atoms upon forming a nonpolar bond. Indeed, Lewis's shared electron pair consisted of two electrons in identical energy states except for their opposing or paired spins. Its saturation followed directly from Pauli's principle.

### *Langmuir's contribution to valence theory*

Following the introduction of the shared electron pair bond in 1916 and the publication on the static atom in 1917, Lewis did little to popularize his new valence theory. Service in World War I and interests in other areas of physical chemistry, chiefly thermodynamics, consumed most of his time. Chemists' acceptance of the shared electron pair resulted partly from the efforts of Irving Langmuir at the General Electric Company in Schenectady, New York. In the two-year period 1919- 21, Langmuir wrote extensively and lectured widely in the United States and abroad on the Lewis theory. His interesting and convincing personality and admirable methods of presentation clearly gave the theory considerable popularity. Langmuir received the America Chemical Society's Nichols Medal in 1920 for his development of the Lewis theory of atoms and molecules. He won the Nobel Prize in chemistry in 1932 for his research on the chemical reactions between molecules adsorbed on surfaces. His molecular adsorption research proved valuable in understanding catalysis.

In a 1919 paper Langmuir showed that atoms, molecules, or ions with the same number of electrons had the same electron structure and usually similar physical properties, especially if each atom, molecule, or ion had the same electrical charge. Langmuir called these structures *isosteres*, and was probably the first chemist to support them with crystallographic data.

Carbon dioxide $CO_2$, and nitrous oxide $N_2O$, each had 22 electrons arranged identically in the molecules. Their physical properties, such as density, critical temperature, and pressure, refractive index, and dielectric constant, were nearly equal. They were, therefore, isosteres. Carbon monoxide CO, the nitrogen molecule $N_2$, and the cyanide ion CN, were also isosteres, each containing 14 electrons in identical electron structures. But in this case, only the neutral molecules, CO and $N_2$, had closely related physical properties: namely their freezing and boiling point, critical temperature and pressure, viscosity and density.

Langmuir's 1919 paper had numerous cubic models scattered throughout, but he never once employed the Lewis dot notation, which would have illustrated the identity of the electron structures and the sharing of the electron pair bonds.

| **LEWIS STRUCTURES OF $CO_2$ AND $N_20$** |
|---|
| :Ö:: C:: Ö: and :N::N:: Ö: |

Langmuir also attempted to account for the behavior of the transition elements. Lewis had omitted them in his 1916 paper, and Langmuir hoped to include their electron structures and the electron structures of all atoms and molecules, within his theoretical framework. In 1919 Langmuir coined the term *covalent bond* to replace the rather cumbersome shared electron pair bond.

The Lewis theory, particularly in England after Langmuir introduced it to British chemists, often became the octet theory of valence, and even Langmuir's theory of valence. Lewis, of course, found this obvious neglect of his 1916 paper inexcusable. It led to a growing sense of embarrassment and to the beginning of a long lasting grudge between the two men. Lewis believed that Langmuir had never observed the established rules of priority regarding the origin of the cubic atom and the shared electron pair. Langmuir's octet theory, Lewis pointed out, was identical to what he had concisely, yet completely, developed in his 1916 paper on valence. To refer to the Lewis-Langmuir theory of valence, as chemists often did, seemed to imply some sort of collaboration between them, and to Lewis, this was out of the question.

Lewis received nominations for the Nobel Prize in chemistry in every year from 1922 to 1935, except 1923. The 34 nominations from Americans and Europeans including Langmuir, Richards, Haber, and Planck, covered all aspects of his work, valence theory, thermodynamics, deuterium, and acid-base chemistry, but he never received the prize. The Nobel committees even thought his contributions merited a Nobel Prize. From 1901 to 1935, only three Americans won in chemistry, Richards in 1914, Langmuir in 1932 for surface chemistry, not valence theory; and Harold Urey (1843-1981), Lewis' student, in 1934 for his discovery of deuterium ($^2_1 H$), which Lewis first isolated and determined its properties.

In the 1940s-60s, several of Lewis's Berkeley colleagues received the Nobel Prize in chemistry, among them William Giauque (1895-1982) in 1949, Glenn Seaborg (1912-99) in 1951, Willard Libby (1908-80) in 1960, and Melvin Calvin (1911-97) in 1961. The disappointment of not winning a Nobel Prize, along with his failing health, likely contributed to Lewis's despair and death on 23 March 1946. On that day, Lewis had gone to his laboratory where he was experimenting with hydrogen cyanide. Later that afternoon laboratory assistants found him dead near a broken ampule of hydrogen cyanide.

## VII. COMPOSITION OF THE NUCLEUS

### *Proton and proton-electron theory*

The experiments of J.J. Thomson, Rutherford, and others on the ionization of hydrogen atoms in a high voltage discharge tube provided protons (hydrogen ions) for investigations on the structure of an atom. These studies showed that the proton's radius, $10^{-13}$ cm, was very small compared to the radius of a hydrogen atom, $10^{-8}$ cm; its mass, $1.67 \times 10^{-27}$ kg, was 1836 times the rest mass of an electron; and its specific charge was equal to the electron's charge, $1.602 \times 10^{-19}$ coulomb. The methods that measured the e/m ratio for an electron, such as the deflection of an electron beam in an electric and a magnetic field, and methods that measured the electron's charge, such as Millikan's 1909 oil drop experiment, also measured these values for the proton.

J.J. Thomson first suggested in 1906 that an atom had a positively-charged component that accounted for most of the atom's mass. His atomic model, lacking a dense central region, failed to explain the results Rutherford obtained in his 1911 alpha and beta particle scattering experiment. In Rutherford's atomic model, electrons occupied the regions around a tiny central nucleus that contained all the positive charge and most of the atom's mass. By 1913 Rutherford had determined that a hydrogen nucleus, or *H-particle*, had one positive charge and a helium nucleus (alpha particle) had two. That same year Moseley used an element's x-ray spectrum to establish the number of positive charges in the nucleus of each known element. To emphasize the fundamental position of the hydrogen nucleus with respect to other elements, Rutherford in 1920 named the H-particle *proton* from the Greek *proteios* meaning of *first importance.*

Later that same year, Rutherford proposed a proton-electron theory of the nucleus. In Rutherford's theory, an atom's nucleus contained only protons (+) and electrons (-), and it always had an excess of protons to account for the nucleus's positive charge and the atom's mass. Oxygen had 16 protons and 8 electrons in its nucleus, giving a net charge of +8; nitrogen had 14 protons and 7 electrons, giving a net charge of +7; and similarly for carbon, which had 12 protons and 6 electrons, and hydrogen, which had one proton and no electrons in its nucleus. The electron, being an extremely light particle, did not affect an atom's weight. It helped balance the repulsive force that existed between the positively-charged protons and stabilized the nucleus.

Rutherford in 1920, following the earlier comments of Nernst and William L. Bragg, suggested the possible existence of a neutral particle formed by the union of the two fundamental particles, the proton and electron. He thought the particle would have very novel properties, particularly a near-zero electric field that would enable it to move freely through matter. Its neutrality, he believed, made its existence almost necessary to explain the formation and stability of the nuclei of the heavy elements, which contained large numbers of protons.

Continued studies on the proton led Pauli in 1924 to propose that the nucleus of an atom had an intrinsic spin angular momentum, or spin, and therefore a nuclear magnetic moment. In classical physics, the magnetic moment measured the force a magnet exerted on an electric current, and thereby the energy required to turn a magnet from a position parallel to a magnetic field to a position perpendicular to a magnetic field. To account for the spin angular momentum of the nucleus, Pauli envisioned each nuclear particle, a proton and later a neutron, as a particle spinning on its axis and behaving like a tiny bar magnet with a north and south pole. Each particle's spin produced a nuclear magnetic moment $\mu$, in which $\mu \propto I$, the nuclear spin (total angular momentum or total nuclear spin). When a strong external magnetic field acted on the nucleus, each nuclear particle's magnetic moment aligned itself with or against the field, in the same way bar magnets tended to align themselves in a magnetic field.

A few years later, experiments by numerous investigators confirmed Pauli's hypothesis. They included the experiments of Friedrich Hund (1896-1997) at Rostock and later Leipzig on band spectra, Samuel Goudsmit at Leiden and later Michigan on the fine structure of spectral lines, and Otto Stern (1888-1969) at Hamburg on the magnetic moment of atoms and sub-atomic particles. Stern's experiments with Otto Frisch (1904-79) in 1933, in which they deflected a beam of protons in a magnetic field, gave values for the proton's magnetic moment and spin angular momentum. Their measurement of its magnetic moment clearly suggested that the proton was not a fundamental particle.

Isidor Rabi (1898-1988) at Columbia, in 1934 and 1936 and more precisely in 1939, experimentally determined the proton's magnetic moment. Rabi in 1938 introduced a new molecular beam method called nuclear magnetic resonance (NMR) that enabled him to measure directly and to very high precision a particle's magnetic moment. He applied a small oscillating magnetic field at right angles to a much larger constant field which produced a re-orientation of the nuclear spin and magnetic moment, and provided a value for the proton's magnetic moment. The present value of the proton's magnetic moment obtained from NMR measurements is +2.7928 nuclear magnetons, for the neutron -1.9130 nuclear magnetons, and for the deuteron +0.8574 nuclear magnetons. Stern, a Jew who left Hamburg in 1933 and accepted a position at the Carnegie Institute of Technology in Pittsburgh, won the Nobel Prize in physics in 1943. Rabi, who in the late 1920s worked with Stern, won the Nobel Prize in physics in 1944.

***Neutron and proton-neutron theory***

In the two year period 1930-32, four physicists almost discovered a second particle inside an atom's nucleus. The fifth physicist succeeded. In 1930 Walter Bothe (1891-1959) and Hans Becker at Berlin, and in 1932 Frédéric (1900-58) and Iréne Joliot-Curie (1897-1956) in Paris, bombarded beryllium, boron, and lithium with alpha particles and detected a penetrating radiation that ejected protons from paraffin and other hydrogen-containing substances. They believed that the high energy radiation was gamma radiation.

James Chadwick (1891-1974), Rutherford's pupil at Cambridge, repeated these experiments in 1932, but reached a different conclusion. Upon bombarding beryllium, boron, and lithium with alpha particles, he concluded that the radiation that ejected protons from paraffin and other hydrogen-containing substances was a neutral particle present in the nucleus, with a mass almost equal to the proton's mass. What experimental evidence convinced Chadwick?

Two experimental results: The maximum range of protons ejected from paraffin indicated that the ejecting source (a particle) had a mass nearly equal to a proton's mass. The ranges of the recoil atoms (the paths of the recoiled nitrogen atoms) in an expansion chamber filled with nitrogen gas, and into which the beryllium radiation passed, indicated the presence in the beryllium radiation of a particle whose mass nearly equaled a proton's mass. The beryllium radiation did not consist of gamma radiation.

To show that the particle did not have an electric charge, Chadwick demonstrated that his particle easily passed through a 10-20 cm thickness of lead, whereas a piece of lead 0.25 mm thick stopped a positively-charged proton having the same mass and velocity as the particle. Because the penetrating power of particles having the same mass and speed depends only on the particle's charge, Chadwick concluded that the particle present in the beryllium radiation had a very small charge compared to the proton's charge or no charge at all. It left no ionization in its path and therefore no trail in a cloud chamber. Chadwick called the neutral particle a neutron. He received the Nobel Prize in physics in 1935 for his discovery. Bothe shared the 1954 Nobel Prize in physics with Max Born for his work on light quanta and photoelectrons, which laid the foundation for nuclear spectroscopy.

The same year that Chadwick discovered the neutron, 1932, Werner Heisenberg (1901-76) at Berlin, who studied under Sommerfeld at Munich and assisted Bohr at Copenhagen, proposed a proton-neutron model of the nucleus to replace the proton-electron nuclear model. His model accounted for both the atom's charge and mass. Oxygen O-16 had 8 protons and 8 neutrons in its nucleus, nitrogen N-14 had 7 protons and neutrons, carbon C-12 had 6 protons and neutrons. Exchange forces held the protons and neutrons within the tiny confine, $10^{-13}$ cm, of the nucleus.

Alfred Landé (1888-1976), Sommerfeld's student who came to Ohio State in 1930 and remained there, and Hans Bethe (1906-2005), who left Nazi Germany in 1933 and accepted a position at Cornell in 1935, using evidence from a Stern-Gerlach experiment, showed that the neutron, like the proton, had a magnetic moment The neutron beam which passed through an inhomogeneous magnetic field experienced a deflection. Landé published his evidence for the neutron's magnetic moment in 1934, Bethe in 1936. Its presence suggested that the neutron was not a fundamental particle and likely consisted of electrically-charged particles. Later studies from the 1950s to the 2000s, such as those in 1967, 1972, and 2013, established the neutron's stability within the nucleus, but outside of the nucleus it had a half-life of 10.4 minutes. It underwent beta decay, emitting an electron and an antineutrino to give a proton.

### *Later developments*

In 1954, two decades after Landé's and Bethe's neutron studies, Maurice Goldhaber (1911-2011) at Brookhaven National Laboratories and Cowan and Reines at Los Alamos, calculated that the proton had a maximum half-life of $10^{28}$ years. Yet, the long held belief in the proton's indivisibility did not change until 1964. By that time the scientific community had recognized that the existing quantum theory of fundamental particles failed to account for properties such as spin and charge. The quantum theory required modification, and it came that year in the quark theory of the theoretical physicist Murray Gell-Mann (b. 1929) at Caltech. New predictions of the proton's half-life also followed. Sheldon Glashow (b. 1932) at Harvard, in 1974 predicted $10^{31}$ years, and in the 1980s physicists accepted a half-life of $10^{32}$ years.

## VIII. DEVELOPMENT OF QUANTUM MECHANICS: MATRIX MECHANICS AND WAVE MECHANICS

### *Background*

Max Born (1872-1970) at Göttingen introduced the term *quantum mechanics* in the title and introduction of his 1924 paper "Über Quantummechanik" published in the *Zeitschrift für Physik.* Physicists already had incorporated wave particle duality, special and general relativity, unified field theory, and quantum theory into their discipline. These investigations preceded Heisenberg's new kinematics, or matrix mechanics, of 1925, the efforts of Paul Dirac (1902-92) at Cambridge, and other physicists to integrate their advances in quantum theory with Heisenberg's matrix mechanics, and Schrödinger's 1926 wave mechanics. The resulting synthesis of the old and new quantum theory led in 1927 to a statistical interpretation of quantum mechanics and to Heisenberg's uncertainty principle, the two foundations on which quantum mechanics rested.

Heisenberg's matrix mechanics revolutionized physics because it rejected classical physics' visualization of atoms and their particles that in the 1920s became a part of the search for the underlying physical conceptions of quantum mechanics. Pauli opposed and challenged Bohr's conviction that the ideas of classical physics had their roots in visualization and were irreplaceable. He provided the starting point for Heisenberg's new matrix mechanics of 1925, and its origin and development resulted in Pauli's and Heisenberg's close collaboration. Ultimately, quantum mechanics emerged from the synthesis of Bohr's Copenhagen interpretation that included complementarity, Heisenberg's uncertainty principle, and the statistical interpretation of Schrödinger's wave equation.

***Electron duality: wave or particle***

Bohr's 1913 atom with its quantized electron orbits satisfactorily accounted for the spectrum of the one-electron hydrogen atom and the helium ion $He^{1+}$. Why the electron resided only in the restricted or quantized orbits and in no others received a partial answer in 1923, when Arthur Holly Compton (1892-1962) at Chicago provided new evidence supporting the duality of light. When a particle (photon) of light struck an electron, the collision scattered, or deflected, the electron and caused the photon's wavelength to increase, indicating the photon's wave nature, and that it had lost energy. In the collision with an electron, the photon clearly behaved like a particle, and its behavior led Compton to suggest the name *photon* for a quantum of light, or light particle. His discovery, called the Compton effect, strengthened the evidence for the duality of light and earned Compton a share of the 1927 Nobel Prize in physics with C.T.R. Wilson. Wilson received the prize for his invention of the cloud chamber.

That same year, 1923, Louis de Broglie (1892-1987) at the Sorbonne argued that the electron, like the photon, possessed duality. If a wave can behave like a particle, a particle can behave like a wave. By equating Planck's $E = h\nu$ and Einstein's $E = mc^2$, and substituting $\nu = c / \lambda$, where $\nu$ = frequency, $c$ = speed of light, and $\lambda$ = wavelength, de Broglie revealed the relation between particles and waves.

| WAVE-PARTICLE DUALITY |
|---|
| $\frac{hc}{\lambda} = mc^2$ <br> $\lambda = \frac{h}{mc}$ |

According to de Broglie, each of the Bohr atom's stable quantized orbits corresponded to an integral number of standing waves. One, two, three, or more wavelengths occupied an orbit. Only those orbits that contained an integral number of wavelengths were stable, and corresponded to Bohr's orbits, n =1, 2, 3 . . . .

Experimental evidence supporting the electron's wave property appeared four years later in 1927. Clinton Davisson (1881-1958) and Lester Germer (1896-1971) at Bell Laboratories in New York City, and independently George Thomson, J.J. Thomson's son, at Aberdeen, successfully diffracted electrons, thereby establishing the electron's wave behavior. Davisson and Germer used a single crystal of nickel, Thomson used thin gold foil (crystals of gold) in his diffraction experiments. De Broglie received the 1929 Nobel Prize in physics for his discovery of the electron's wave property. Davisson and Thomson shared the 1937 Nobel Prize in physics for their experimental discovery of the diffraction of electrons by crystals.

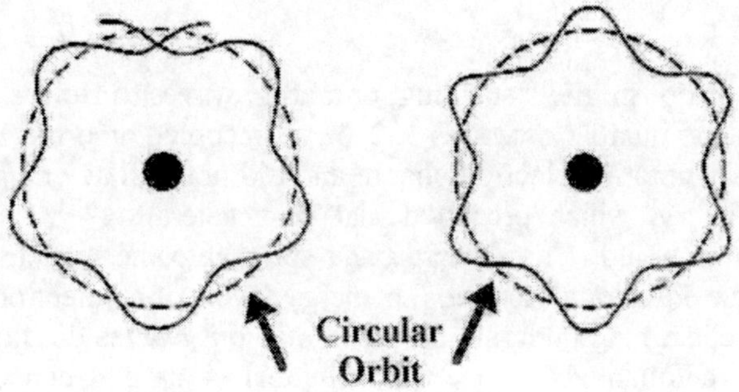


**Figure 32**
**De Broglie's Standing Waves**
Courtesy of the United States Energy Research and Development Administration, Office of Public Affairs

***Matrix mechanics***

The limited applicability of Bohr's quantized atom to a one-electron atom or ion initiated the search for a more comprehensive atomic model. The new quantum mechanical model appeared in 1925 in Heisenberg's paper on matrix mechanics, and in 1926 in Schrödinger's four papers on wave mechanics.

Bohr proposed his quantized atom to account for hydrogen's spectral lines. His model and its modifications by Sommerfeld and de Broglie provided an easily visualized picture of the electron arrangement in an atom and the permitted electron energy level changes, or transitions from one orbit to another orbit, in an atom. Heisenberg, who studied under Sommerfeld at Munich and worked with Born at Göttingen and Bohr at Cophenhagen, abandoned these visual models. Like Bohr, he started with observable phenomena, hydrogen's spectrum. But instead of introducing another visual model, with Born's help he developed a matrix algebra, called matrix mechanics, that accounted mathematically for the wavelengths of the lines in hydrogen's spectrum and thereby for the electron's energy levels (orbits or locations) and its transitions between energy levels.

Heisenberg's 1925 paper contained a second important idea that he called *exchange energy.* It provided the starting point for the application of quantum mechanics to the electron theory of the chemical bond. Heisenberg believed that as two hydrogen atoms approached each other, the electron from the first hydrogen atom found itself increasingly attracted to the positively-charged nucleus of the second hydrogen atom. Likewise, the electron of the second atom found itself increasingly attracted to the positively-charged nucleus of the first hydrogen atom. At a certain distance, the two electrons moved back and forth between the two nuclei and established a rapid electron exchange, possibly billions of exchanges per second. According to Heisenberg, the two electrons lost their identities and no longer belonged to the first or the second hydrogen atom. Heisenberg's exchange energy led later to the idea of the electron pair bond that Walther Heitler (1904-81), Sommerfeld's former student, and Fritz London (1900-54) at Zurich, and independently, Linus Pauling (1901-94) at Caltech developed.

Despite its success in accounting for the electron configuration of the hydrogen atom and for the approximate configurations of heavier atoms, Heisenberg's non-visual matrix mechanics proved unsatisfying to many physicists and chemists. Schrödinger's wave mechanics, developed a few months later, provided a more acceptable visual model of an atom's electron configuration. Later, in 1944, John von Neuman (1903-57), at the Institute for Advanced Study in Princeton, showed the mathematical equivalence of Schrödinger's wave mechanics and Heisenberg's matrix mechanics.

*Wave mechanics*

Schrödinger applied de Broglie's standing, or matter, waves to Bohr's quantized electron orbits. In his atom, only an integral number of waves 1, 2, 3 . . . occupied an orbit. Each electron standing wave did not represent an accelerating electron charge and did not radiate energy. It did not violate Maxwell's electromagnetic theory, which predicted that an accelerating electric charge, such as an electron orbiting in an atom, would radiate energy and spiral into the nucleus, causing the atom to collapse. Schrödinger's wave equation described the energy levels of an electron in an atom. It showed that any orbit that did not contain an integral number of standing waves did not exist. An atom with a distinct number of orbits, or quantized atom, was a logical or natural consequence of the electron's wave properties, and not an arbitrary imposition on the atom. Schrodinger's equation for the electron's orbits in hydrogen, the simplest atom, and in atoms with higher atomic numbers, appears below.

| SCHRÖDINGER WAVE EQUATION |
|---|
| $\frac{-h}{8\pi^2 m}\nabla^2\psi + \mathcal{V}\psi = E\psi$ |

In this equation *E* represents the total energy of the hydrogen atom, and is the sum of the potential energy $\mathcal{V}$ and the kinetic energy represented by the first term on the left side of the equation. $\nabla^2$ (del squared, a differential operator called a Laplacian) relates the electron's spatial location. The ψ term (psi), an *eigenfunction* or wave function, represents the unknown in the equation. The equation has many unknowns that give acceptable solutions, namely, the *n*, *l*, and *m* quantum numbers. The ψ wave functions depend on the electrons' coordinates. They do not give the electron's exact position in the atom, only the probability of finding the electron in the atom. The higher the value of the square of ψ for a particular location in the atom, the more likely is the electron's presence in that location. This is the electron density. The electron is *smeared,* or spread, in space, and represented as a dense cloud having a particular shape called an atomic orbital, rather than a discrete particle. The atomic orbital may or may not contain an electron, and its shape and size depend on the wave function ψ.

Upon solving the Schrödinger equation for an atom's ψ values, each ψ solution gives an *n*, *l*, and *m* value, or quantum number, for an electron. The quantum numbers emerge naturally as solutions to the Schrödinger equation, and not arbitrarily as in the Bohr equation, and have essentially the same meaning they already had. The spin quantum number *s* did not arise as a solution to Schrödinger's equation, but from Goudsmit and Uhlenbeck's spectroscopic observations. They accounted for their experimental observation by proposing a spherical electron spinning in two directions, either clockwise or counterclockwise, around an orbiting axis that passed through its center.

The solutions to the Schrödinger wave equation, which are the *n*, *l*, and *m* quantum numbers, appear below.

The principal quantum number *n* gives the size of the orbital and the permitted orbitals and has only integral values, 1, 2, 3, 4. . . .

The *l* quantum number determines the orbital's shape, and for a given value of *n* it has the values 0, 1, 2, 3. . . . , or *n* - 1 different values. If *n* is 2, its *l* values are 0 and 1. If *n* is 3, its *l* values are 0, 1, and 2.

The *m* quantum number indicates the orbital's spatial orientation. For each *l* value there are 2 *l* + 1 *m* values (*m*=2 *l* + 1) and therefore 2 *l* + 1 different atomic orbitals, or spatial orientations, that have the same *n* and *l* quantum numbers.

To name an atomic orbital, the quantum number $n$ appears as an integer in front of the $l$ value, and the $l$ values 0, 1, 2, 3, . . . are assigned the letters *s, p, d, f,* where $s = 0, p = 1, d = 2$, and $f = 3$. The letters *s, p, d,* and *f* represent the names sharp, principal, diffuse, and fundamental, already used in spectroscopic notation to define the different series of lines making up an atomic spectrum. This is the origin of the notation 1*s*, 2*s*, 2*p*, 3*d*, or 4*f* adopted in quantum mechanics.

For all $n$ values, when $l = 0$, the *s* orbital has the shape of a fuzzy spherical shell. When $l = 1$, each of the three *p* orbitals has the shape of a fuzzy dumbbell, each oriented in space on the x, y, and z axes ($p_x$, $p_y$, $p_z$ ). When $l = 2$, the five *d* orbitals ($d_{xy}$ ,$d_{xz}$ , $d_{yz}$ , $d_x{}^2 - {}_y{}^2$ , $d_z{}^2$) have more complicated shapes oriented in space, such as two pairs of dumbbells. When $l$ =3, the seven f orbitals have different shapes and spatial orientations.

In accordance with Pauli's principle, each of the atomic orbitals contains 0, 1, or 2 electrons. The single *s* orbital contains a maximum of two electrons or $s^2$, the three *p* orbitals contain a maximum of six electrons, or $p^6$ ($p^2{}_x$, $p^2{}_y$, $p^2{}_z$), the five *d* orbitals contain a maximum of 10 electrons or $d^{10}$ ($d^2{}_{xy}$, $d^2{}_{xz}$, $d^2{}_{yz}$, $d^2{}_x{}^2 - {}_y{}^2$, $d^2{}_z{}^2$ ), and the seven *f* orbitals, a maximum of 14 electrons or $f^{14}$.

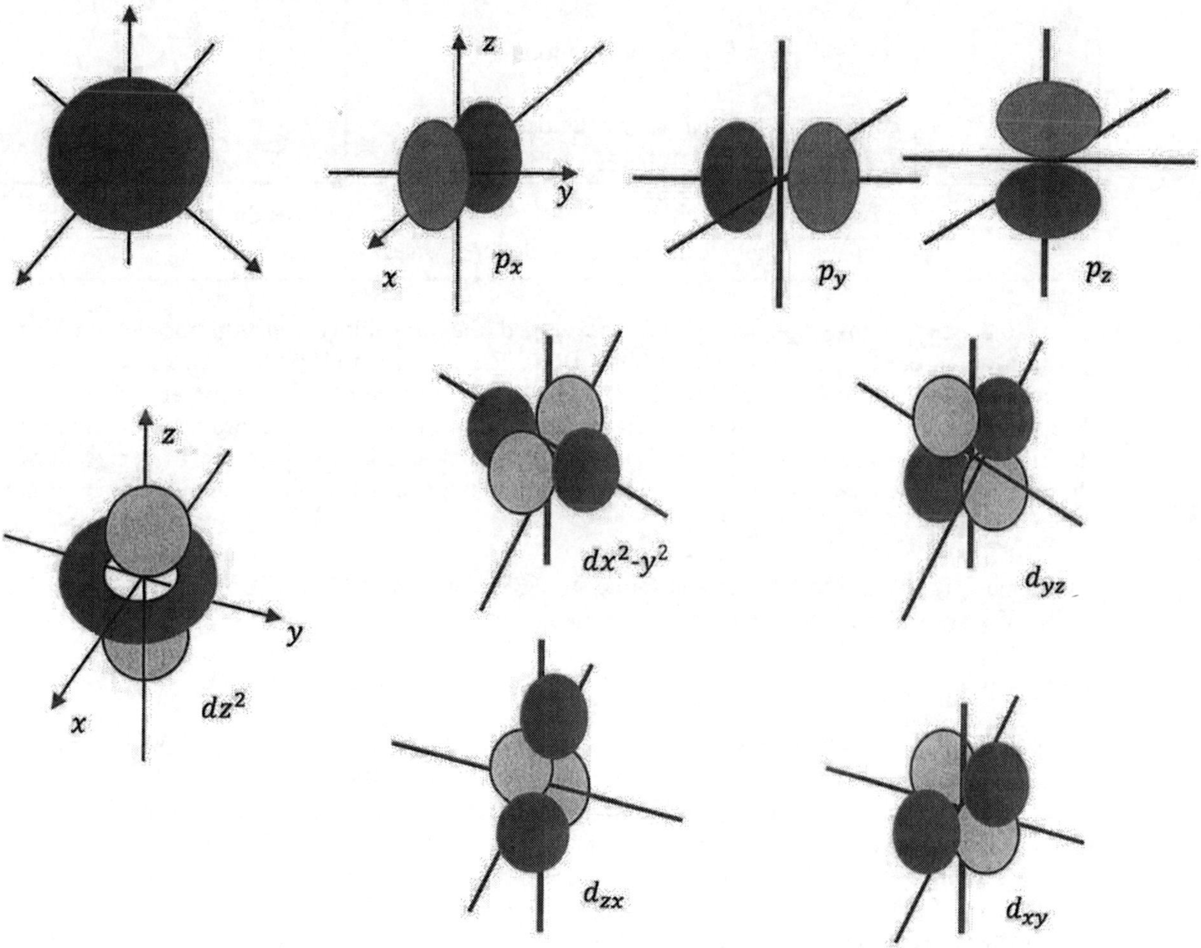


**Figure 33**
**Arrangement of Electron Energy Levels in an Atom, *s, p, d,* Orbitals**
Courtesy of the author

| QUANTUM NUMBERS OF AN ATOM | | |
|---|---|---|
| ***Quantum Number*** | ***Possible Values*** | ***Description of Orbital*** |
| *n* | Only integral values<br>1, 2, 3, 4, 5 . . . | Gives the size of an orbital |
| *l* | *l* has *n*-1 values<br><br>For $n = 1$   $l = 0$<br>For $n = 2$   $l = 0, 1$<br>For $n = 3$   $l = 0, 1, 2$ | Gives the shape of an orbital |
| *m* | *m* has $2\,l + 1$ values<br><br>For $l = 0$   $m = 1$ spatial orientation<br>(0)<br>For $l = 1$   $m = 3$ spatial orientations<br>(-1, 0, +1)<br>For $l = 2$   $m = 5$ spatial orientations<br>(-2, -1, 0,+1, +2) | Gives the spatial orientation of an orbital |
| *s* | *s* for an electron is either<br>½ or -½ , either clockwise or counterclockwise | Gives the spin direction of an electron |

Three independent investigators in 1926 developed the probability function for determining the position of an electron in its atomic orbital. Born, Pascal Jordan (1902-80) at Rostock, and Dirac proposed that the electron waves described the probability of finding the electron at a given location. The wave function squared, $\psi^2$, represented the probability function of the coordinate for an electron's position. This interpretation was necessary in order to deal with the problem of representing the discrete value of a quantum, such as an electron or any other particle, with a continuously propagating wave.

Heisenberg in 1932, Schrödinger and Dirac in 1933, Pauli in 1945, and Born in 1954 won Nobel Prizes in physics for the contributions to quantum mechanics.

***Hund's rule***

Hund at Leipzig, in 1925 proposed an order of electron orbital filling, arguing that the electrons in an atom filled the orbitals in order of increasing energy starting with $n = 1$ followed by $n = 2$, $n = 3$, $n = 4$..., and singly before pairing. According to Hund, the electrons occupying the orbitals singly, one electron in each of the three *p* orbitals for example, had parallel spins resulting in greater total spin and greater stability. This is Hund's rule of maximum multiplicity, or more commonly called Hund's rule. Hydrogen (Z = 1) had the electron configuration H $1s^1$, helium (Z = 2) had the configuration He $1s^2$, lithium (Z = 3) Li $1s^2 2s^1$, Be (Z = 4) Be $1s^2\ 2s^2$, and boron (Z = 5) B $1s^2\ 2s^2\ 2p^1$. The electron configurations for some of the higher atomic numbers appear below.

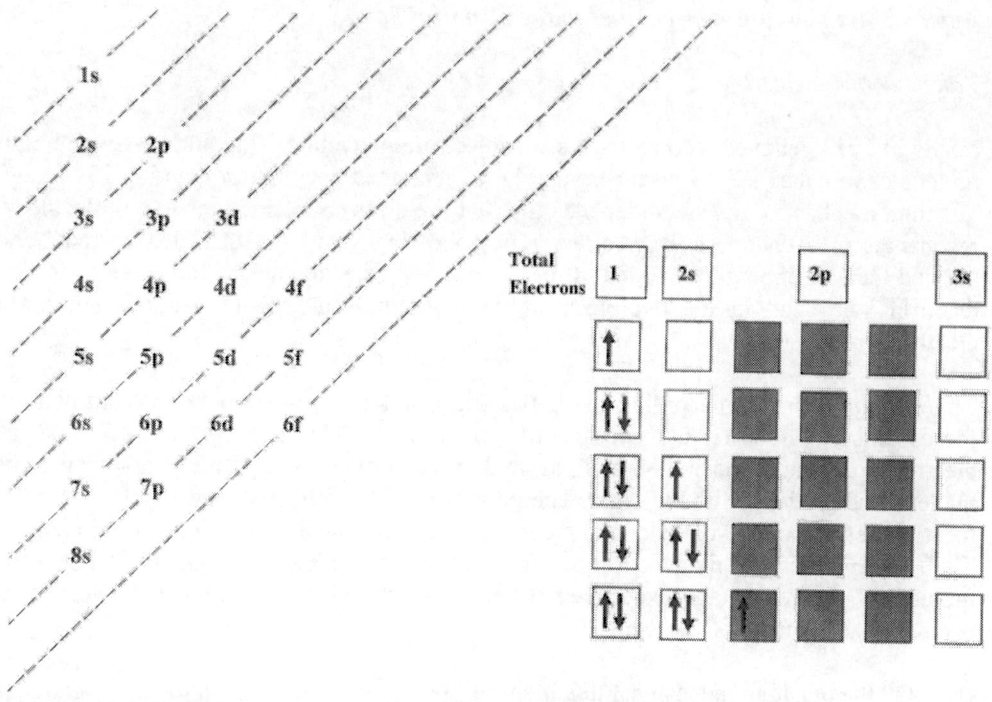


**Figure 34**
**Electron Orbital Configurations Illustrating Hund's Rule**
**Order of energy levels starting from right to left with 1s, 2s, 2p, 3s, 3p, 4s (left) and electron addition (right)**
Courtesy of the author

***Heisenberg's uncertainty principle***

In 1927, Heisenberg introduced his uncertainty principle to deal with another problem of quantum mechanics, that of trying to locate an electron in its atomic orbital. According to Heisenberg, an observer cannot determine the position and the momentum p (defined as p = mass × velocity) of an electron, or of any particle, at the same time. The greater the certainty in measuring one of them, such as the electron's position, the less becomes the certainty in measuring the other, the electron's momentum. The act of observing an electron's position using photons (light) of short wavelength, and therefore of high energy, produces a change in the electron's momentum. Using longer wave, or lower energy, photons to reduce the change in momentum increases the uncertainty in the electron's position. The product of the uncertainty of position Δx and momentum Δp is approximately equal to Planck's constant, h = 6.624 × $10^{-27}$ erg-sec, or Δx Δp ≃ *h/4π*. Heisenberg's uncertainty principle removed causality (cause and effect) and determinacy from physics and replaced it with statistical probability.

$$\Delta p\, \Delta x \geq \frac{1}{2}\hbar$$

$$\Delta E\, \Delta t \geq \frac{1}{2}\hbar$$

$$where\ \hbar = \frac{h}{2\pi}$$

### *Application of quantum mechanics to the chemical bond*

*Valence bond method*

Lewis in 1923 believed incorrectly that the electron interactions (the bond energy) holding atoms in a molecule were magnetic. Heisenberg and Dirac remained unconvinced, and in 1926 they used the new quantum mechanics to show independently that the electron interactions in a molecule resulted from a resonance, or exchange energy, effect. The following year, in 1927, Heitler and London at Zürich applied the Heisenberg-Dirac idea of resonance, or exchange energy, to account for the energy of the electron pair bond in the two-electron hydrogen molecule, H··H, where each dot represents an electron.

The exchange energy, according to Heitler and London, was the sum of the attractive and repulsive forces existing between the two positive hydrogen nuclei and two negative electrons. The two electrons exchanged positions, and resulted in each electron's partially associating itself with one nucleus and partially with the other. The balancing of the attractive and repulsive forces occurred at a location between the two nuclei and created a stable chemical bond with a certain length and strength. Their interpretation of the electron pair bond provided approximate agreement between the hydrogen molecule's theoretical and experimental values for the heat of dissociation, moment of inertia, and oscillational frequency.

By 1928 the quantum mechanical treatment of the electron pair bond in the hydrogen molecule seemed beyond doubt. A paper that John Van Vleck (1899-1980) at Wisconsin published in 1928 provided convincing evidence that the exchange energy accounted for the energy of the chemical bond in the hydrogen molecule. Van Vleck calculated that the magnetic attraction was almost entirely negligible in comparison to the electrical attraction, amounting to less than 0.1 percent of the electrical attraction in the hydrogen molecule. That same year, London suggested that the exchange energy of two electrons, one from each of the atoms undergoing combination, was *in every case* the energy of the nonpolar bond. The two electrons forming the bond had paired their spins so that the spins of the two electrons were in opposite directions and were, therefore, incapable of further combination.

Pauling, who met Heitler and London while in Germany in 1926-27, agreed with London's interpretation, and pointed out that London's theory was in simple cases entirely equivalent to the Lewis theory of the shared electron pair bond. Quantum mechanics, he wrote in a brief 1928 paper, more a note, provided a theoretical justification for the rule of eight, or octet theory, at least for the first row elements.

To quote Pauling, "the shared electron structures assigned by Lewis to molecules such as $H_2$, $F_2$, $Cl_2$, $CH_4$, etc., are also found for them by London. The quantum mechanics explanation of valence is, moreover, more detailed and correspondingly more powerful than the old picture. For example, it leads to the result that the number of shared bonds possible for an atom of the first row is not greater than four, and for hydrogen not greater than one, for neglecting spin, there are only four quantum states in the L-shell ($n = 2$) and one in the K-shell ($n = 1$)."

In the next part of his paper, Pauling went beyond Heitler and London's breakthrough. Using his chemical intuition and knowledge of valence and chemical structure, Pauling discussed qualitatively how Heitler and London's exchange energy provided a quantum mechanical explanation for the four identical bonds associated with a carbon atom in a hydrocarbon molecule. The physicists' electron configuration for carbon (Z = 6), C $1s^2\,2s^2\,2p^1\,2p^1$, indicated a carbon atom that had only two unpaired electrons and formed two bonds instead of four. Yet, chemists experimentally had established beyond doubt carbon's tetrahedral bonding arrangement in numerous compounds, in which the carbon atom formed four equivalent bonds directed in space to the corners of a tetrahedron.

According to Pauling, the formation of a chemical bond required an electron exchange energy in which an atom's electrons acquired a more stable electron configuration. In the tetrahedral carbon atom, the electron exchange energy resulting from forming four bonds provided enough energy to rearrange the carbon atom's normal electron configuration with two unpaired electrons that formed two bonds into a new configuration with four unpaired electrons. Each electron occupied an identical orbital, called a hybrid orbital, permitting the carbon atom to form four equivalent bonds.

Pauling knew from methane's ($CH_4$) chemical properties and reactions that the carbon atom's four carbon-hydrogen bombs were equivalent. In order to have the carbon atom's four valence electrons, $2s^2\ 2p^1,\ 2p^1,\ 2p^0$ (the third *p* orbital is empty), distributed equally, he developed mathematically in December 1930 the idea of hybridization. In Pauling's bonding arrangement, the single pair of *s* electrons and the two unpaired *p* electrons ($2s^2\ 2p^1\ 2p^1\ 2p^0$) rearranged to give four identical $sp^3$ orbitals ($sp^3\ sp^3\ sp^3\ sp^3$ where the 3 refers to the hybrid orbital not the number of electrons) each with a single unpaired electron available to form a carbon-hydrogen bond. The one spherical *s* orbital merged with the three dumbbell-shaped *p* orbitals to give four $sp^3$ orbitals in which one-half of each dumbbell occupied a much greater space than the other half. The bond angle, the angle between any two orbitals, was 109° 28′. This angle allowed the orbitals to arrange themselves as far from each other as possible and thereby minimize electron-electron repulsion. For hydrocarbon molecules (carbon-hydrogen compounds) such as methane, each of carbon's four unpaired electrons paired with an unpaired electron from another atom, such as the unpaired electron in a hydrogen atom, and formed four bonds directed to the corners of a tetrahedron, the structure chemists experimentally established for methane. Other s-p hybrids, *sp* and $sp^2$, had similar dumbbell hybridized shapes. Compounds having these hybridizations had different spatial arrangements of their bonds.

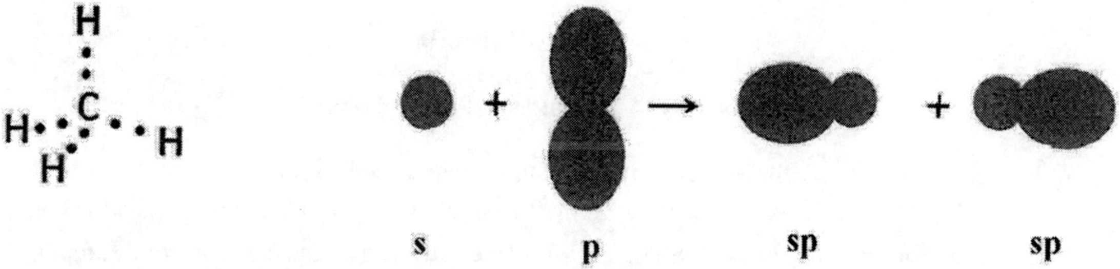


**Figure 35**
**Pauling's Hybrid Orbitals Showing Methane $sp^3$ (left) and sp Hybridization (right)**
Courtesy of the author

Pauling spent three years working out the mathematics to support his idea that hybrid orbitals and hybridization accounted for the tetrahedral carbon atom. His quantum mechanical proof, which verified Lewis's 1916 idea that a pair of electrons constituted the nonpolar bond, appeared in an April 1931 article entitled "The Nature of the Chemical Bond." Pauling used crystalline structures to derive other hybrid bond structures, and at that time showed that a $dsp^2$ hybridization (four equivalent bonds) corresponded to a square planar structure, and a $d^2sp^3$ hybridization (six equivalent bonds) to an octahedral structure in molecules. He tested his deductions from wave mechanics against the experimental evidence he obtained from x-ray crystallography. Like Heitler, London, and John C. Slater (1900-76) at MIT, whose 1931 paper discussed many of the same topics as Pauling's paper, Pauling considered each atom in a molecule to retain to some degree its electron configuration even when chemically bonded. This interpretation of the electron pair bond is the valence bond (VB) method. Pauling continued his investigations on chemical bonding in the 1930s and summarized them in his classic 1939 book *The Nature of the Chemical Bond.*

*Molecular orbital theory*

In the years 1932-35, shortly after the introduction of the valence bond method, Hund, John E. Lennard-Jones (1894-1954) at Bristol and later Cambridge, and Robert Mulliken (1896-1986) at Chicago, proposed an alternative interpretation of the hydrogen molecule's electron structure. Their proposed structure was similar to the structure that Edward Condon (1902-74) published in 1927, when he was a National Research Fellow at Göttingen and Münich with Born and Sommerfeld. Like Condon, they assumed that the electrons of the combining atoms belonged to the molecule as a whole. The atomic orbitals combined to form new molecular orbitals, designated as $\sigma$ and $\pi$, that held the electron pairs that constituted the chemical bonds in a molecule. This is the molecular orbital (MO) theory. The * represents an antibonding σ or π orbital.

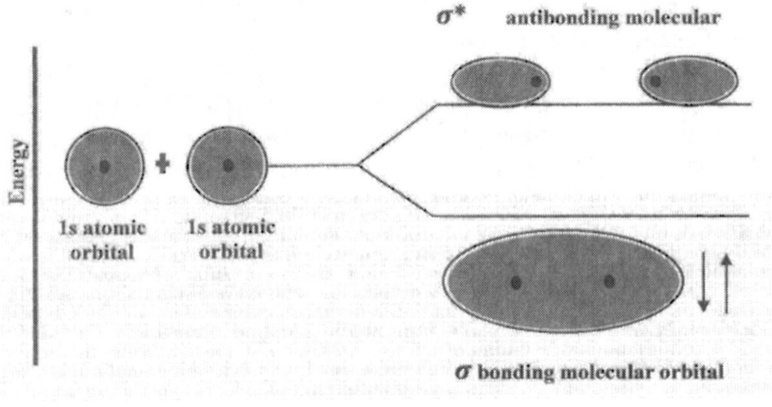


**Figure 36**
**Molecular Orbitals**
Courtesy of the author

For simple molecules, the energy of the orbitals increased in the order $\sigma_{1s}\,\sigma^*_{1s}\,\sigma_{2s}\,\sigma^*_{2s}\,\sigma_{2p}\,\pi_y\,\pi_x\,\pi^*_y\,\pi^*_x\,\sigma^*_{2p}$. The hydrogen molecule, with 2 electrons, had the electron configuration $\sigma^2{}_{1s}$. The oxygen molecule, with 16 electrons, had the electron configuration $\sigma^2_{1s}\,\sigma^{*2}_{1s}\,\sigma^2_{2s}\,\sigma^{*2}_{2s}\,\sigma^2_{2p}\,\pi^2_y\,\pi^2_x\,\pi^{*1}_y\,\pi^{*1}_x\,\sigma^*_{2p}$.

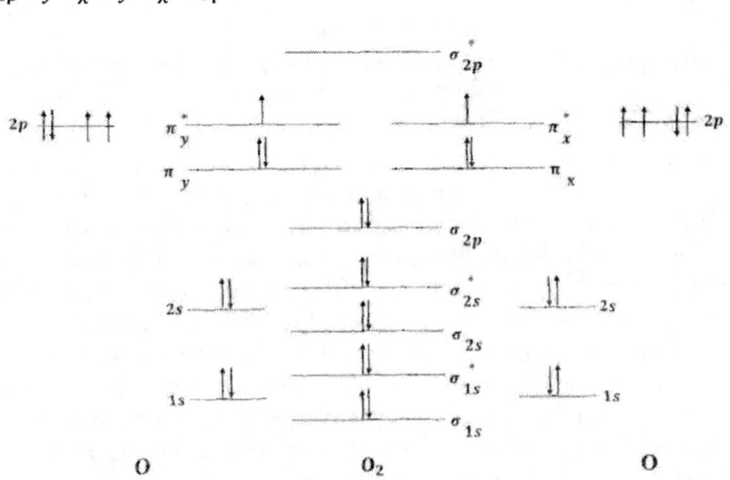


**Figure 37**
**Molecular Orbitals for Oxygen**
Courtesy of the author

Lennard-Jones's model, which he first proposed in 1929 while at Bristol, included features of both the valence bond method and the molecular orbital theory. His model, known as the linear combination of atomic orbitals (LCAO), required an atom to use some of its orbitals to construct a molecular orbital. In method and approach, LCAO resembled the MO theory.

In their application of the two theories to account for the electron structures of molecules, physicists and chemists found the MO theory of the chemical bond more useful than the VB method in understanding the bonding in carbon's aromatic (ring) compounds, such as benzene and naphthalene and their many derivatives. MO theory was broader in approach, accounted for the polarity of molecules, explained oxygen's paramagnetism, simplified calculations of electron probabilities, and provided a good understanding of molecular spectra to which the VB method paid little attention. The two theories agreed in most of their conclusions, although the MO theory was conceptually simpler.

Many of the physicists and chemists who contributed to the development of quantum mechanics and its application to the chemical bond received a Nobel Prize in physics or chemistry, although sometimes recognition came slowly. Heisenberg, Schrödinger, and Dirac, received Nobel Prizes in physics only a few years after publishing their papers on quantum mechanics. Heisenberg won in 1932, Schrödinger and Dirac in 1933. Pauli had a longer wait, not until 1945. Born in physics and Pauling in chemistry received their Nobel Prizes in 1954.

For Mulliken, it seemed like recognition would never come. Over the three-year period, 1932-35, Mulliken published a series of fourteen papers that gave the most complete explanation and description of the MO theory of chemical bonding. He not only contributed significantly to a much greater understanding of the chemical bond, but his work represented the origin and inspiration of many other developments in molecular spectroscopy and structure. Mulliken received the Nobel Prize in 1966 in chemistry for his herculean effort. Other issues may have contributed to Mulliken's belated recognition, but Pauling's skill and success in popularizing and promoting the rival VB method clearly played a major role. Undergraduate chemistry textbooks In the 1940s, 1950s, and into the 1960s discussed mainly VB not MO valence theory.

In his posthumously published autobiography, *Life of a Scientist* (1988), in which he reflected on his lifetime in science, Mulliken discussed the rival theories of chemical bonding and the scientific community in which they emerged. He modestly assessed his role in developing MO theory, generously acknowledged the contributions of others, and never wavered in his belief in MO theory's superiority. Hund's 1928 paper, he said, introduced the symbols $\sigma$, $\pi$, $\delta$ to classify an electron according to its angular momentum. That same year, Mulliken first assigned quantum numbers to each electron occupying a molecule's experimentally-determined energy levels, and in 1932 he proposed the idea of molecular orbitals to designate the energy levels. Mulliken called the wave function for an electron in an atom a *molecular orbital*, which was an abbreviation for a one-electron orbital wave function, and he used Hund's symbols $\sigma$ and $\pi$ to name the molecular orbitals.

*Life of a Scientist* provides several other revelations about Mulliken's scientific world. Mulliken wrote that early in his scientific career, like many chemists, he was very much the inductive, experimental scientist, and found theory fruitless when used in interpreting erroneous models. Outside of his scientific endeavors, he adopted a deterministic, behaviorist philosophy of life, which put a damper on his feelings, and only later did he learn to empathize and sympathize with others. After World War I, Mulliken, like many of the disillusioned, went through an existentialist period in which life seemed pointless and futile until he decided to follow a path of scientific hedonism.

During his long career, spent mostly at Chicago, Mulliken traveled widely and met just about everyone working on quantum mechanics, spectra, and chemical bonds. They included Heisenberg, Schrödinger, Born, Heitler, London, Dirac, and Lennard-Jones. Slater was a close friend. Hund became a close and longtime friend, and received considerable praise from Mulliken for his contribution to MO theory. Lewis, as he was for Pauling, became Mulliken's guide in the study of electrons. Mulliken had good relations with his rival Pauling, but he believed Pauling's masterful salesmanship and showmanship persuaded chemists to adopt the VB method. Pauling, he said, had presented the VB method in *The Nature of the Chemical Bond* as the only theory of bonding.

## IX. SPECTROSCOPY AND CHEMICAL BONDS

The numerous advances in spectroscopy that contributed so much to understanding the structures of atoms and molecules and the chemical bond led to a new research area that revolutionized chemistry. Ahmed Zewail (b. 1946) at Caltech, in 1985 invented a new branch of spectroscopy called femtosecond spectroscopy to study atoms and molecules in slow motion and observe what happens when chemical bonds break and form. Zewail received the 1999 Nobel Prize in chemistry for his invention of femtochemistry.

Femtochemistry represented the latest and most significant advance in the study of reaction rates that began over a century ago when Arrhenius in 1889 developed a simple equation showing reaction rate as a function of temperature. Arrhenius's equation, which considered microscopic systems and relatively long reaction times, required the existence of an intermediate state in the change from reactants to products. Decades later, in 1935, Henry Eyring (1901-81) at Princeton and independently Michael Polanyi (1891-1976) then at Manchester, introduced the term *transition state* and assumed that the reactants passed through the intermediate transition state very rapidly, on the same time scale as molecular vibrations. Eyring in his 1935 paper developed a theory of absolute reaction rates, or transition state theory, for the reactions of individual molecules in a microscopic system. But until Zewail's invention of femtochemistry and publication of his results in 1985, experiments on such a small time scale remained a dream. Femtosecond (fs) spectroscopy, uses an ultrafast laser that emits light flashes in intervals as brief as 10-100 fs, which is the time it takes for the atoms in a molecule to perform one vibration. This high-speed camera records the spectra of molecules, the intermediates in the transition state that exist in the course of a chemical reaction.

In his femtosecond spectroscopic experiments, Zewail mixed the original reactants as beams of molecules in a vacuum chamber, and with the ultrafast laser injected two pulses: first, a powerful pump pulse that struck the molecule and excited it to a higher energy state, and then a weaker probe pulse at a wavelength chosen to detect the original molecule or its altered form. The pump pulse signaled the start of the reaction, while the probe pulse examined the resulting changes. Varying the time interval between the two pulses by causing the probe pulse to make a detour via mirrors (light travels 0.03 mm in 100 fs) enabled Zewail to see how quickly the original molecule changed. The new shapes the molecule takes upon its excitement and passage through one or more transition states have spectra that serve as fingerprints. Comparing the spectra fingerprints and the time elapsing with theoretical simulations based on results of quantum chemical calculations of spectra and energies for the molecules in their various states, enabled Zewail to understand what happened.

With Zewail's femtosecond spectroscopy, chemists reached the time scale at which reactions occur. No chemical reactions occur faster than this time scale. Chemists can study atoms and molecules in slow motion, see what happens when chemical bonds break and form, and understand why some reactions occur and others do not, and explain why the speed and yield of reactions depend on temperature. Femtosecond spectroscopy provides a better understanding of catalytic behavior, the mechanisms of life processes and the synthesis of future medicines, and the design of molecular electronic components.

## X. QUANTUM MECHANICS AND THE STRUCTURE OF THE NUCLEUS

Since the 1920s-30s and the introduction and development of quantum mechanics, physicists have proposed about 37 different structural theories or models of the nucleus, but no one theory or model has proved consistent and satisfying. The search has followed two different approaches. The first approach entailed a study of the properties and interactions of the elementary particles to obtain an understanding of the nuclear forces, and from them deduce the nucleus's structure. The second and more experimental approach required physicists to obtain enough data to show regularities and correlations essential to understanding the nucleus's structure.

One of the proposed models, the liquid-drop model, had its roots in an analogy that Gamow made in 1930, two years before the neutron's discovery. Gamow compared the behavior of the nucleus to the drop of a liquid. He pointed out that the binding energies and volumes of atomic nuclei were proportional to the number of nuclear particles (nucleons), and that this proportionality indicated a saturation of the nuclear forces. Bohr developed the liquid-drop model in 1936, and following the discovery of nuclear fission in December 1938, he used the model to account for nuclear binding energies and nuclear fission. Bohr believed the nucleons (protons and neutrons) strongly interacted with each other, just as the molecules in a drop of liquid reacted with each other. The interactions enabled each nucleon to share any excess energy it had with other nucleons. The resulting energy states or levels were the quantum states of the nucleons as a whole and not of a single nucleon.

A second model, the nuclear shell model, or independent-particle model, differed from the liquid-drop model in assuming that a weak interaction existed among the nucleons. It held that a certain number of protons, neutrons, or protons plus neutrons called magic numbers, or shell numbers, gave a nucleus unusual stability regardless of the number of other nucleons. The magic numbers 2, 8, 20, 28, 50, 82, 126 were analogous to the number of electron pairs in an atom's valence orbital that determined its chemical stability or reactivity, such as 2 and 8 electrons in the rare gases.

Several physicists in the United States, France, and Germany proposed the idea of a magic number of nucleons occupying quantized energy levels in the nucleus. James H. Bartlett (1904-2004) at Illinois proposed a nuclear shell model with quantized energy levels in 1932. The German refugee physicists in Paris, Walter Elsasser (1904-91) at the Institute Henri Poincaré-Centre National de la Recherche Scientifique in 1933 (later at the Scripps Institute of Oceanography in La Jolla, California) and Karl Guggenheimer at the Collège de France in 1934, suggested a nuclear shell model based on the magic number of nucleons, particularly 50 and 82. They noted that 50 and 82 were the atomic numbers of tin and lead, which were very stable elements and had a large number of isotopes, and that many other elements also had 50 or 82 neutrons. Three years later in separate 1937 publications, Hermann Schüler at the Astrophysical Observatory in Potsdam, and Theodor Schmidt (1908-86), a mathematical physicist and Shüler's former friend and assistant at the Observatory, showed a correlation between a nucleus with quantized energy levels (nuclear shells) and nuclear magnetic moments of the nucleons.

The nuclear shell model accounted for the energy levels in the nucleus using essentially the same *s*, *p*, *d*, *f* quantum notation as the electron configuration of the elements. The principal quantum number $n$ for its energy levels is $n = 1, 2, 3, 4 \ldots$, and the model requires that each nucleon couples its orbital angular momentum (quantum number $l$) with its spin (quantum number $s$ *or* $m_s = 1/2$) to give the nucleus's spin-angular momentum. An element's spin quantum number is either integer or half integer. The helium nucleus ${}^{4}_{2}He$ had the configuration $1s^2$, oxygen's nucleus ${}^{16}_{8}O$ had the configuration $1s^2\ 1p^6$, and calcium ${}^{40}_{20}Ca$, $1s^2\ 1p^6\ 1d^{10}\ 2s^2$.

The nuclear shell model also accounted for an element's parity, which is either even or odd. The reactions of the nucleons determine parity. If a nucleon undergoes a reaction, such as beta decay, then undergoes a reflective symmetry transformation that produces a mirror image of the nucleon and its properties, such as nuclear spin reversal, but leaves unchanged the results of the reaction, the parity is even and parity is preserved. The experimental evidence, such as the spin of the beta particle, is the same for the nucleon and its mirror image, making it physically impossible to distinguish between the two nucleons, and does not violate any natural laws. If the mirror image is inverted, thereby restoring the property to its original condition, but the nuclear reaction now produces a result contrary to experimental evidence, the spin of the beta particle is opposite, the parity is odd and parity is not conserved. The shell model's main weakness: it lacked a theoretical foundation and failed to provide agreement between some of the nucleus's calculated and measured properties, such as the nuclear magnetic moment for elements with an odd proton and their total angular momentum.

Until 1948-49, the nuclear shell model attracted little attention mainly because it lacked a theoretical foundation and failed for higher atomic numbers, and because of Bohr's more successful liquid-drop model. At that time, Maria Goeppert-Mayer (1906-72) at the Argonne National Laboratory and the University of Chicago, provided convincing evidence that supported the peculiar stability of elements having 50 and 82 protons or 50, 82, and 126 neutrons. Elements with these numbers were in addition to the stability of elements having magic numbers 2, 8, 20, and 28. Goeppert-Mayer pointed out that the element tin, Z = 50, had 10 stable isotopes, the largest number of any element. Six stable nuclei had 50 neutrons (Kr-86, Rb-87, Sr-88, Y-89, Zr-90, Mo-91), and seven stable nuclei had 82 neutrons (Xe-136, Ba-138, L a-139, Ce-140, Pr-141, Nd-142, Sm-144). Nuclei with the magical number of neutrons far exceeded the normal two or three nuclei with the same number of neutrons. Helium, one of the most abundant and stable elements in the universe, had 2 protons and 2 neutrons, oxygen 8 protons and 8 neutrons, and calcium (Z = 20) had six isotopes. Goeppert-Mayer noted that the nuclei of some of the universe's abundant elements, such as O-16, Ca-40, Sn-118, and Pb-208, had nuclei that contained the magic number of protons, neutrons, or both.

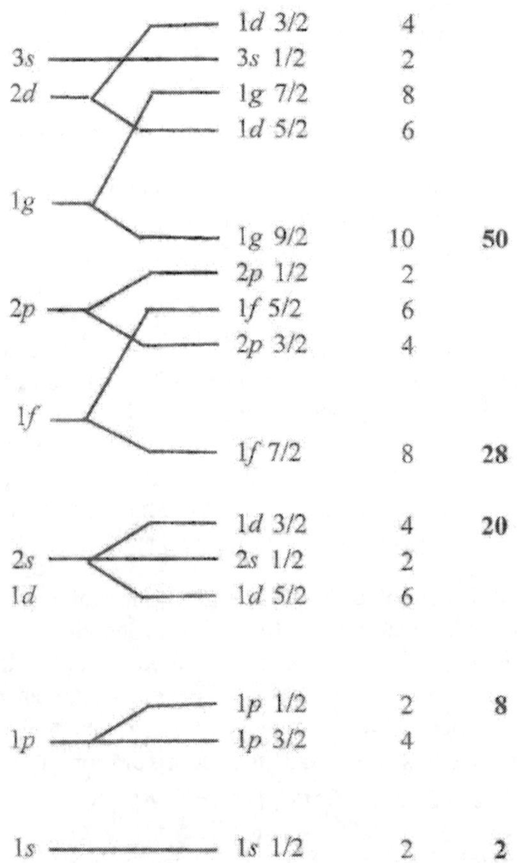


**Figure 38**
**Nuclear Shell Model**
Courtesy of the author

For her contributions to understanding the nucleus's shell structure, calculating its energy levels and magic numbers, Goeppert-Mayer shared the 1963 Nobel Prize in physics with Hans Jensen (1907-73) and Eugene Wigner (1902-95). Jensen, then at Hanover and Heidelberg, in 1949 independently calculated the nuclear energy levels of the nuclear shell model and the magic numbers. Wigner, a Hungarian refuge physicist who came to Princeton in 1934, contributed to nuclear theory and understanding of the fundamental particles particularly the law of conservation of parity in 1937. He adopted the liquid-drop model of the nucleus, and coined the term 'magic numbers' used to describe the nucleus in the nuclear shell model. Conservation of parity remains valid for the strong (nuclear) and electromagnetic forces, but beta decay experiments in 1956, and in 1957 with cobalt nuclei giving beta particles with a counterclockwise spin or left-handedness, showed that parity did not hold for the weak (nuclear) force.

A third nuclear model, the collective or unified model, appeared in 1951. Aage Bohr (1922-2009), Niels Bohr's son, and Ben Roy Motielson (b. 1926) at the Niels Bohr Institute and Nordita (Nordic Institute for Theoretical Atomic Physics (est. 1957) in Copenhagen, were major contributors to the collective model. As its name implies, it retains the analogy between the nucleus and the liquid-drop and combines it with the main ideas of the nuclear shell model. Bohr, Mottelson, and Leo Rainwater (1917-86) at Columbia shared the 1975 Nobel Prize in physics for their research on the link between particle motion and the structure of an atom's nucleus.

## XI. SEARCH FOR FUNDAMENTAL PARTICLES

### *Fundamental particle physics*

Fundamental (elementary) particle physics and nuclear physics represent two branches of modern physics, although belief in the reality of elementary or fundamental particles dates from the time of the atomists Leucippus (490 BC) in Miletus (Asia Minor) and Democritus (470-380 BC) in Abdera (ancient Greece). Their theories were intuitive, or speculative, and never tested experimentally. Like the particles of the earlier atomists, the chemical billiard ball atoms that Dalton introduced in the first decade of the nineteenth century were indivisible, but unlike the earlier atomic theories, each kind of atom differed from every other kind of atom only in weight, not shape. Dalton's profound insight, his theory of the chemical atom, enabled chemists to determine the atomic weights of the elements and to establish chemical formulas.

Throughout most of the nineteenth century, the atom remained indivisible and, therefore, fundamental. Studies on the kinetic theory of gases and spectroscopy suggested otherwise, but the first demonstration of an atom's divisibility came only in 1897, when J.J. Thomson discovered the electron and quickly confirmed its presence in all matter.

By this time, a revolution in the physical sciences already had begun. Röntgen discovered x-rays in 1895, Henri Bacquerel and the Curies observed the radioactivity of atoms in 1896, and within a few years, Planck introduced the quantum theory (1900) and Einstein his special theory of relativity (1905). In 1911 Rutherford discovered a second sub-atomic particle, officially naming it the proton in 1920, and in 1932 Chadwick found a third, the neutron. In the intervening years, Bohr in 1913 applied the quantum theory to the electron arrangement in the atom. At this time, physicists still operated under the principle of determinism. They believed that their knowledge of a system at a given time enabled them to predict the state of the system at any given time in the future.

All of this changed by the mid-1920s. The publication of Heisenberg's matrix mechanics in 1925 and uncertainty principle in 1927, Schrödinger's wave mechanics in 1926, and Dirac's and Born's papers in 1926 gave quantum mechanics a probabilistic interpretation that most physicists rather quickly accepted. The electron now appeared in a new guise. Physicists had treated it as a particle with a definite location in an orbit, but the diffraction experiments of George Thomson and independently those of Davisson and Germer in 1927 confirmed de Broglie's proposal that the electron also functioned as a wave with a far less defined location.

The following year, in 1928, Dirac's relativistic wave equation developed from general principles and without any presumptions, accounted for the properties of electron spin ($s = 1/2$). His equation introduced special relativity into Schrödinger's wave equation, and provided a possible reconciliation between the two distinct and opposing theories of special relativity and quantum mechanics. Dirac's work received further refinement in 1947 in the theory of quantum electrodynamics (QED) that linked special relativity with quantum mechanics.

With the discovery of the neutron in 1932, the structure of the atomic world seemed quite complete, consisting of four fundamental particles: electrons, protons, neutrons, and photons. But several problems lingered and were central to physics research in the 1930s-40s. The first was antimatter. Dirac's 1928 wave equation permitted four different electron-energy states rather than the observed two. Two of the four states described an electron with negative energy and a negative charge, and led Dirac in 1928 to predict the existence of a particle similar to the electron, but existing in a positive energy state and having a positive charge. Dirac's particle was the positron, the first of the antiparticles.

Carl Anderson (1905-91) at Caltech, who worked with Millikan on the study of cosmic rays, discovered the positron in 1932, four years after Dirac's prediction of antiparticles. He inserted a lead plate in his cloud chamber to slow down the high speed particles entering the cloud chamber and observed a particle that had a path identical to an electron's path except that it curved in the opposite direction. Anderson named the new particle the positron. For the positron's discovery, Anderson shared the 1936 Nobel Prize in physics with Victor Hess.

A second problem dealt with beta decay, the emission of an electron from the nucleus. This led Enrico Fermi (1901-54) then at the University of Rome to announce in 1933 that in addition to a much stronger force that held the nucleons (protons and neutrons) in the nucleus, another weaker force existed. This weak force enabled a nucleon to change from a neutron to a proton with the emission of a beta particle (electron) and a new particle of zero mass and charge that Fermi in 1933 called the neutrino. The neutrino carried off any energy that the beta particle did not acquire and thereby conserved the total amount of energy exchanged in the reaction (conservation of energy).

A third problem centered on the force holding the protons and neutrons in the nucleus. The gravitational force was too weak, and a zero electrical force existed between the proton and neutron. Clearly, another force acted within the nucleus. Hideki Yukawa (1907-81) at Kyoto, in 1935 proposed a theory of the strong force in which the force acting within the tiny nucleus had enough strength to overcome the repulsion between positively-charged protons. He predicted the existence of a particle he called a meson (middle) that carried or exchanged this force, and from Heisenberg's uncertainty principle concluded that it had a mass about 200 times the electron's mass, but less than the proton's mass.

The following year, in 1936, Anderson, in cloud chamber experiments with cosmic rays on Pike's Peak in Colorado, discovered a new particle that physicists at first thought was Yukawa's particle. The cloud chamber path of Anderson's particle curved less than an electron's path and more than a proton's path, which suggested a particle of mass intermediate between the electron and proton masses.

Anderson initially, in 1938, called the new particle a mesotron (intermediate particle) but shortened its name to meson. Cesare Lattes (1924-2005), a Brazilian physicist at Bristol, called it the mu-meson in`1947, and in 1951 Fermi correctly named it the muon. Anderson's muon failed to react with the proton and neutron, and contrary to initial expectations, it did not have the properties of Yukawa's particle. As shown in 1961, Anderson's muon was a heavier electron. It had a negative charge and a mass 207 times the electron's mass, but decayed in $1.52 \times 10^6$ seconds to give an electron, a neutrino, and an anti-neutrino. Its positively-charged anti-particle decayed to give a positron, a neutrino, and an anti-neutrino.

The muon is a fundamental particle, like the electron, and has no internal (quark) structure. It belongs to the lepton (Greek, meaning small mass) family, which consists of six fundamental particles: electron, muon, lepton (tau lepton) that Martin Perl (b. 1927) at Stanford discovered in 1975, and three kinds of neutrinos. Mesons, and baryons, belong to a different group of particles, called hadrons, and consist of two quarks, a quark and an antiquark, whereas a baryon consists of three quarks.

The discovery of Yukawa's particle occurred in 1947 when Cecil Powell (1903-69), Lattes and Giuseppe Occhialini (1907-93) at Bristol, discovered a new particle in photographic plate experiments they conducted high up (5,500 meters) in the Bolivian Andes at Chacaltaya. Powell, who worked with Rutherford at Cambridge, abandoned the cloud chamber, which required that water droplets condense around ions to form a visible path or track, and then photographing the path. Instead, Powell had the particles he was trying to detect and identify strike a photographic emulsion on which they produced a pattern of small dark spots. Powell's emulsion process replaced the two-stage production and photographing of cloud chamber paths and photographed the paths directly.

While testing his photographic emulsion to see how cosmic rays affected it, Powell observed curved paths that corresponded to a particle with a mass 273 times the electron's mass and heavier than Anderson's muon. Powell's particle had all the properties of Yukawa's predicted particle. Lattes named the new particle the pi-meson to distinguish it from Anderson's muon. Yukawa received the Nobel Prize in physics in 1949 for his prediction that mesons existed. He was the first Japanese scientist to receive a Nobel Prize. Powell received the Nobel Prize in physics in 1950 for his discovery of the pi-meson, since 1951 called a pion.

Exciting as these discoveries were, they were not without controversy. Ernest Stuckelberg (1905-84) at Zurich, in 1935 developed a pion theory identical to Yukawa's, but withheld it from publication because of Pauli's ridicule. Seven years later, in 1942, the Japanese physicists Shoichi Sakata (1911-70) who studied under Yukawa, Seitaro Nakamura, Yasutaka Tanikawa, and Takesi Inoue at Nagoya University proposed a two-meson theory that resolved the problem of the strong force. Their theory remained unknown in the west until its publication in 1946, shortly before Robert Marshak (1916-92) at Rochester and Hans Bethe at Cornell in 1947 published a two-meson theory. According to the two-meson theory the heavier pi-meson (mass, $m = 273\ m_e$) interacted strongly with nucleons and accounted for part of the nuclear force. The lighter μ-meson (mass, $m = 207\ m_e$), which is the main constituent of the hard component of cosmic rays at sea level, interacts weakly with nucleons and contributes nothing to the nuclear force.

***Discovery and classification of quarks***

Research on the muon and pion, on other particles and anti-particles present in the nucleus, on the basic forces found in the universe, and on establishing the conservation laws of particle physics continued in the immediate post-World War II years. By 1947 physicists had identified and suspected the existence of about eleven particles that included the four pioneer particles (photon, electron, proton, neutron); two of their three antiparticles; a neutrino-antineutrino pair; a pion and two muons. Within the next ten years, the number doubled with the discovery of five more mesons (the k-mesons, or kaons) and another group of seven particles called hyperons. George Rochester (1908-2001) and Clifford Butler (1922-99) in 1947 discovered the first of the five k-mesons, a kaon or k-meson, in cosmic ray experiments in a Wilson cloud chamber in the Cosmic Ray Laboratory at Manchester and in the French Pyrenees.

Murray Gell-Mann, and independently Kazuhiko Nishijima (1926-2009) at Osaka City University, noted that the five mesons and seven hyperons, although easily created in particle collisions, decayed much more slowly than expected for their large masses and cross sections. They had life times longer than the expected $10^{-10}$ seconds, and the collisions always produced pairs of the particles. Because the collisions always produced a k-mesons pair and a hyperon pair, Gell-Mann in 1953 named the 12 k-mesons and hyperons strange particles, and postulated that the particles had a new conserved property that he called *strangeness.* He represented strangeness with the quantum number *s*.

By the late 1950s, particle physics had expanded to include new types of short-lived particles ($10^{-25}$ seconds) that physicists initially called resonances, and recognized as particles with definite masses and spins. With the existence of so many new particles, physicists began to search for some kind of classification scheme. One of the first came in 1961, when Gell-Mann, and independently Yuval Ne'eman (1925-2006) at Tel Aviv University, proposed the *eightfold way* of classifying the group of particles called hadrons, which then consisted of mesons, such as Powell's pion, and a second group of particles called baryons, that at that time had only two known members, the proton and neutron. Gell-Mann and Ne'eman then united groups of particles (8, 10, or 27 members) that had the same values for the three quantum numbers, baryon number, spin, and parity, into a larger set or supermultiplet. Particles with spin = 0, the pions and kaons, formed a group of eight mesons (octet); a second group with spin = 1 formed another octet; a third group with $s = 1/2$ formed an octet of baryons; and a group of ten baryons with $s = 3/2$ formed a decimet. The name eightfold way came from the eightfold path of Buddhism and referred to the eight-member families to which many sets of particles belonged.

A more complete and detailed model for the hadrons appeared in 1964, in the quark (pronounced kwark) theory proposed independently by Gell-Mann and George Zweig (b. 1937), his Russian-born Caltech colleague. Gell-Mann and Zweig argued that not even the proton and neutron were fundamental particles. The neutron had zero charge, yet had a sizeable magnetic moment, whereas a fundamental particle had no magnetic moment. They believed that the neutron, as well as the proton, consisted of a more fundamental particle called a quark, a name Gell-Mann borrowed from a line in James Joyce's 1939 comic fiction *Finnegan's Wake*, "Three quarks for Muster Mark, sure hasn't got much of a bark." The quest for quarks was, therefore, the search for the building blocks of matter, the origin of the universe.

According to the theory, matter consisted of three fundamental particles (the three kinds of flavors), and their anti-particles, called quarks. Gell-Mann designated them *u*, *d*, and *s*, with *u* for up and *d* for down, indicating the direction of the quark's spin, and *s* for strange. The strange quark *s* accounted for the properties of the strange particles. It had a strangeness of -1. Gell-Mann assigned *strangeness* to any particle produced by the strong (nuclear) force, but which decayed by the weak or electromagnetic force. The difference between the number of strange quarks and strange antiquarks equaled the *strangeness* of a particle. The theory's most revolutionary feature was the charge Gell-Mann assigned to each quark. Whereas every known particle contained an integral charge, the quarks had fractional charges: 2/3 for *u*, -1/3 for *d* and -1/3 for *s*.

Two simple rules accounted for all the known hadrons (mesons and baryons): (1) each meson was a quark-antiquark pair or quark doublet and (2) each baryon consisted of three quarks, each antibaryon of three antiquarks, and each was a quark triplet. A proton was p = *uud*, a neutron, n = *udd*. Ne'eman and independently Gell-Mann in 1962 predicted the existence and properties of the omega-minus baryon Ω, a heavy strange particle that had half the proton's mass. Investigators at Brookhaven discovered it in 1964. Gell-Mann received the 1969 Nobel Prize in physics for his contributions to the classification of elementary particles.

| QUARKS (FUNDAMENTAL PARTICLES, FERMIONS) | | | | | | |
|---|---|---|---|---|---|---|
| *Name* | *Symbol* | *Mass (Million Electron Volts MeV)* | *Charge* | *Spin* | *Baryon Number* | *Date Proposed or Confirmed* |
| up | u | 360 | + 2/3 | 1/2 | 1/3 | 1964 proposed<br>1970 confirmed |
| down | d | 360 | - 1/3 | 1/2 | 1/3 | 1964 proposed<br>1970 confirmed |
| strange | s | 540 | - 1/3 | 1/2 | 1/3 | 1964 proposed<br>1970 confirmed |
| charm | c | 1,500 | + 2/3 | 1/2 | 1/3 | 1974 confirmed |
| bottom (beauty) | b | 5 GeV | - 1/3 | 1/2 | 1/3 | 1977 confirmed |
| top (truth) | t | 174 GeV | + 2/3 | 1/2 | 1/3 | 1987 proposed<br>1995 confirmed |

***Baryon number:*** A quantum number that is conserved in nearly all interactions. B = 1/3 (number of quarks – number of antiquarks).
***Fermions:*** Obey Fermi-Dirac statistics and Pauli's exclusion principle, no two fermions can have the same four quantum numbers. Fermions have a spin of ½ or multiples of 1/2, 3/2, 5/2 . . . .

In the 1960s, matter therefore appeared to consist of two groups of fundamental particles: quarks and a second group of four fundamental particles call leptons. The Belgian physicist Leon Rosenfeld (1904-74), then at Manchester, coined the term lepton, meaning small or slight, in 1948. The four known leptons were the negatively-charged electron and negatively-charged muon and the electron neutrino and muon neutrino that Leon Lederman (b. 1922), Jack Steinberger (b. 1921), and Melvin Schwartz (1932-2006) at Columbia Brookhaven National Laboratory, discovered in 1962. They are fundamental particles containing no quarks, and therefore have a zero baryon number.

Physicists accepted the reality of leptons, such as the electron and neutrino, but until the late 1960s questioned the existence of fractionally-charged quarks, and whether a quark existed as a free particle not combined with other quarks, and in violation of QED theory. At that time, from 1967 to 1973, inelastic electron-nucleon scattering experiments with high-speed electrons (20 billion electron volts, 20 GeV) at the Stanford Linear Accelerator (SLAC) and at MIT showed that the proton had three point-like scattering centers, each with the fractional electrical charge predicted by theory. The Stanford and MIT groups measured each charge center's spin, and they too agreed with prediction. In fact, the particle accelerators experiments not only justified quark theory, they extended it. Jerome I. Friedman (b. 1930) and Henry Kendall (1926-1999) at MIT and Richard Taylor (b. 1929) at Stanford received the 1990 Nobel Prize in physics for their discovery.

Experimental proof of the existence of free quarks proved more elusive. Brian McCusker (b. 1919), at Manchester in 1940 and from 1961 to 1985 at the University of Sydney in Australia, conducted some of the early cosmic radiation experiments searching for evidence of free quarks. Unlike the majority of physicists, McCusker believed that his 1969 experiments and the joint Japanese-Brazilian Centauro Experiment (1971) provided very good evidence for the existence of free quarks in high-energy radiation. According to McCusker, if free quarks existed in normal matter, their presence could account for the fractional electron charge that Millikan reported in 1910 in one of his oil drop experiments. William Fairbank (1917-89) and his team at Stanford in 1970 also obtained a fractional electron charge with its Millikan-type low-temperature experiments using niobium spheres. Experiments to detect free quarks have continued to the present, but none has provided evidence requiring a second look.

Because two or three quarks in a hadron had the same quantum numbers and therefore violated the Pauli Exclusion Principle, the quark theory required modification. This led Oscar Greenberg (b. 1932) at Maryland and independently Yoichiro Nambu (b. 1921) at Chicago and Moo-Young Han (b. 1934) at Osaka City University (now at Duke), in their 1964 quantum chromodynamics theory (QCD) to propose a new quark quantum number called color charge. Color determined the combining of quarks to form the elementary particles and accounted for the strong force between quarks in a hadron.

Greenberg and Nambu and Han considered color analogous to the positive and negative charge of electrically-charged particles, and arbitrarily assigned color three varieties, red, green, and blue to a quark, and three anticolor varieties antired, antigreen, and antiblue to an antiquark. The three quark colors and the three antiquark colors combined to form particles with no net color, just as positive and negative particles combined to form electrically-neutral atoms. Quarks interacted by emitting and absorbing a neutral particle of zero mass (a force carrier) called a gluon, each of which carried a color-anticolor pair. Eight gluons exchange and transmit the strong force between quarks. A red quark can interact with a blue quark by exchanging a red-antiblue gluon.

The color quantum number was at first arbitrary. Confirmation of quark color came in 1979 when physicists at the German National Laboratory in Hamburg showed that high-energy particle collisions emitted gluons. The colored gluons transmitted the chromodynamic force that binds the quarks. Other studies, such as those of Glashow and Weinberg, showed the mathematical similarity among QCD, QED, and the unified theory of electromagnetic and weak interactions.

Arguments for a fourth quark appeared at this time. Initially, the arguments for its existence were purely esthetic. Physicists wanted a fourth quark, four quarks to match the four known leptons. But in 1970 a fourth quark seemed necessary to explain the decay of a neutral k-meson or kaon. Experimental confirmation of a heavy meson containing a fourth quark, the charm quark *c*, came late in 1974 in independent experiments by Burton Richter (b. 1931) at Stanford and Samuel Ting (b. 1936) at Brookhaven. They shared the 1974 Nobel Prize in physics for their discovery.

By the summer of 1977, Lederman and his group at Fermilab had announced the discovery of another even heavier meson containing a fifth quark, bottom (beauty) *b*, three times heavier than charm. The Stanford group and a team at Cornell's Electron Storage Ring in 1980 provided evidence for the fifth quark. The finding of bottom inaugurated a search for its partner, and in 1985 evidence of a meson containing a sixth quark, top (truth) *t* appeared at CERN in Geneva. The Fermilab confirmed the top quark in 1995. Lederman, Steinberger, and Schwartz shared the 1988 Nobel Prize in physics for their discovery of the muon neutrino and other contributions to particle physics. Nambu shared the 2008 Nobel Prize in physics for his contributions to subatomic physics.

At present, quark theory requires the existence of six quarks, each of which comes in three colors for a total of eighteen types (plus their eighteen antiquarks and eight gluons that provide the strong force that holds together the quarks.

| EXAMPLES OF HADRONS (MESONS AND BARYONS) | | | | | |
|---|---|---|---|---|---|
| **Mesons $q\,\bar{q}$ ( 19 known)** | | | | | |
| ***Name*** | ***Symbol*** | ***Quark Content*** | ***Electric Charge*** | ***Mass GeV/c²*** | ***Spin*** |
| pion | $\pi^o$ | $u\,\bar{u}$ | 0 | 0.135 | 0 |
| pion | $\pi^+$ | $u\,\bar{d}$ | +1 | 0.140 | 0 |
| kaon | $K^+$ | $u\,\bar{s}$ | +1 | 0.494 | 0 |
| kaon | $K^0$ | $d\,\bar{s}$ | 0 | 0.498 | 0 |
| rho | $\rho^+$ | $u\,\bar{d}$ | +1 | 0.770 | 1 |
| B-zero | $B^0$ | $d\,\bar{b}$ | 0 | 5.279 | 0 |
| eta-c | $\eta_c$ | $c\,\bar{c}$ | 0 | 2.980 | 0 |
| D+ | D+ | $c\,\bar{d}$ | +1 | 1.8689 | 0 |
| **Baryons (fermions, 14 known)** | | | | | |
| ***Name*** | ***Symbol*** | ***Quark Content*** | ***Electric Charge*** | ***Mass GeV/c²*** | ***Spin*** |
| proton | p | *uud* | +1 | 0.938 | 1/2 |
| anti-proton | $\bar{p}$ | $\bar{u}\,\bar{u}\,\bar{d}$ | -1 | 0.938 | 1/2 |
| neutron | n | *udd* | 0 | 0.940 | 1/2 |
| sigma-plus | $\Sigma^+$ | *uus* | +1 | 1.189 | 1/2 |
| omega-minus | $\Omega^-$ | *sss* | -1 | 1.672 | 3/2 |
| lambda | $\Lambda$ | *uds* | 0 | 1.116 | 1/2 |
| lambda (charm) | $\Lambda^+c$ | *udc* | +1 | 2.281 | 1/2 |
| A hyperon (Greek, above or beyond) is any baryon that contains one or more strange quarks. All hyperons have short lifetimes. | | | | | |

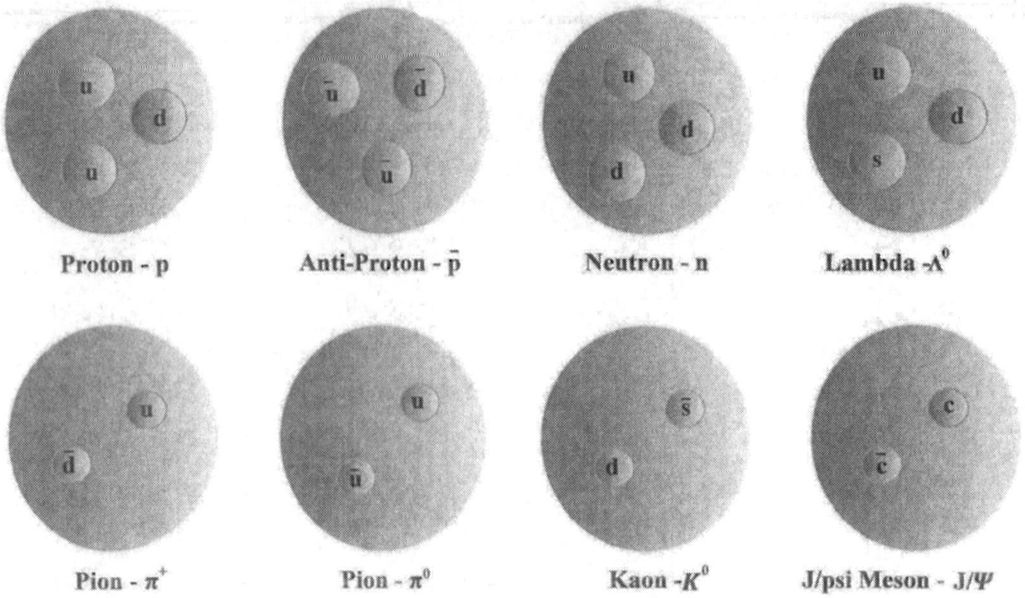


**Figure 39**
**Baryon and Meson Quark Makeup**
Courtesy of the author

The lepton family had remained unchanged until 1975 when Martin Perl at Stanford, using the two-year old electron-positron collider called SPEAR, discovered tau, a charged lepton with mass about twice that of the proton. Discovery of the tau neutrino in 2000 by the DONUT collaboration (Direct Observation of the Nu Tau) at Fermilab brought the number of leptons to three families (generations) of two leptons each, six in total, and matching the three families or six known quarks. Perl shared with Reines the 1995 Nobel Prize in physics for his discovery of the tau lepton.

The generations of particles in the presently accepted model of particle physics, called the standard model, differ only in mass. Their quantum numbers and other properties and interactions remain the same. The standard model has three generations with increasing mass, in moving from a lower generation to the next higher generation. In the lepton family, the mass increases in moving from the first generation electron, to the second generation muon, to the third generation tau particle. The model of the atom usually depicted shows it consisting of three first generation particles, the proton, neutron, and electron.

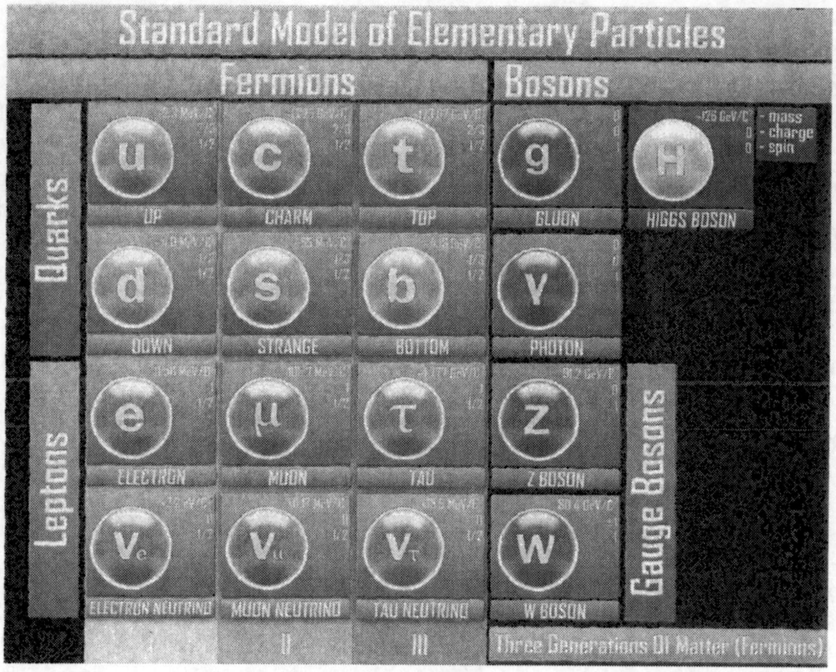


**Figure 40**
**The Standard Model**


| LEPTONS (FUNDAMENTAL PARTICLES, FERMIONS) | | | | |
|---|---|---|---|---|
| ***Name*** | ***Symbol*** | ***Charge*** | ***Mass MeV/c²*** | ***Spin*** |
| electron | e | -1 | 0.511 | ½ |
| neutrino (electron) | $\nu_e$ | 0 | 0 ($2.2x10^{-6}$) | ½ |
| muon | $\mu^-$ | -1 | 105.7 (207e) | ½ |
| neutrino (muon) | $\nu_u$ | 0 | 0 (<0.17) | ½ |
| tau | $\tau^-$ | -1 | 1,777 (3,477e) | ½ |
| neutrino (tau) | $\nu_\tau$ | 0 | 0 (<15.5) | ½ |
| ***Lepton number*** is the difference between the number of leptons (+1) and antileptons (-1). Its sum is conserved in all interactions. | | | | |

*The four forces*

The recognition that four forces governed the universe evolved over a period of about 350 years. Newton in 1687 provided the first quantitative theory of the gravitational force. Maxwell's 1864 theory of electromagnetic radiation accounted for the electromagnetic force that acted on charged particles and electric currents. The discovery of the neutron in 1932 led Heisenberg and other physicists to propose the (nuclear) strong force that enabled them to understand the properties of atomic nuclei. Two years later, in 1934, Fermi added the (nuclear) weak force to deal with nuclear beta decay.

The four basic forces, strong, electromagnetic, weak, and gravitational, have relative strengths that decrease in the order 1: $10^{-2}$: $10^{-13}$: $10^{-39}$. The range over which each force acts varies. The strong and weak forces act within the dimensions of the nucleus. The strong force acts only on hadrons, whose two classes are baryons (proton, neutron, and other particles), which consist of three quarks, and mesons (pions, kaons, and other particles), which consist of a quark and its antiquark. The weak force acts on hadrons and leptons (electron, muon, and other fundamental particles). The electromagnetic force is of infinite range but acts only on hadrons, leptons, and photons. Gravity acts over an infinite range and affects all particles.

Each force acts through an intermediary particle that belongs to the family of force carriers called a boson. Bosons have zero baryon number and whole-number spins of 0, 1, 2, 3. They do not obey Pauli's exclusion principle. Yukawa's pi meson (pion) accounts for the strong force holding the nucleons in the nucleus, although physicists believe this force really resides in the gluons that bind the quarks in the nucleons. $W^+$, $W^-$, and $Z^0$ bosons carry Fermi's weak force. An exchange of photons causes the electromagnetic force according to QED theory. QED's principal contributors, Sin-Itiro Tomonaga (1906-79) at Tokyo University, and independently Richard Feynman (1918-88) at Caltech, and Julian Schwinger (1908-94) at Harvard, showed that high-speed subatomic particles, such as electrons and positrons, interacted by exchanging photons (electromagnetic energy). The force acted over an infinite range because the exchange particle, the photon, was massless. Tomonaga studied with Heisenberg at Leipzig, Feynman with John Wheeler (1911-2008) at Princeton, and Schwinger under Rabi at Columbia and Oppenheimer at Berkeley. They shared the 1965 Nobel Prize in physics and would have received the prize much earlier if not for Bohr's opposition to their conclusions. Most physicists, Thorne, Drever, Weiss, to name a few, believe the massless and still undetected graviton produces the gravitation effect.

Until 1960 physicists believed that the four basic forces accounted for all interactions that occurred among the fundamental particles and for gravity. In that year, however, the Harvard theoretical physicist Sheldon Glashow developed mathematically a theoretical model that combined the electromagnetic and weak forces into a single force called the electroweak force. A few years later, in 1967, Steven Weinberg at Harvard and independently Abdus Salam (1926-96) at the International Centre of Theoretical Physics in Trieste, incorporated an additional component, the Higgs mechanism, into Glashow's theoretical model. The new model showed how the masses of all the fundamental particles, the quarks, leptons, and other predicted but not yet discovered particles arose. It laid the foundation for the *standard model* of fundamental particles, a theory that proposes that all particles of matter consist of quarks and leptons (6 quarks, 6 leptons, and the anti-particle of each), and that the exchange of the force carriers of the electromagnetic, weak, and strong forces account for the interactions of the fundamental particles.

The standard model achieved maturity in 1973, when the linear accelerator experiments at Stanford in the mid-1970s demonstrated that protons consisted of fractionally-charged quarks. It received experimental confirmation in 1982-83 at CERN, when Carlo Rubbia (b.1934) detected the first weak force carriers, the $W^+$, $W^-$, and $Z^0$ bosons, and showed that the bosons had the masses that the standard model predicted. Additional confirmation came with the discovery of bottom quark in 1977, the top quark in 1995, and tau neutrino in 2000. Weinberg, Salam, and Glashow shared the 1979 Nobel Prize in physics for their contributions to the electroweak force. Rubbia with Simon van der Meer (b.1925-2011) at CERN shared the 1984 Nobel Prize in physics.

The most recent evidence supporting the standard model occurred in July 2012 with the discovery of the long sought Higgs boson ($H^o$) at the $10 billion Large Hadron Collider at CERN in Geneva. Named for the physicist Peter Higgs (b.1929) at Edinburgh, who with five other physicists, in 1964 proposed a mechanism that explained why the elementary particles such as the W and Z bosons, quarks, and leptons had mass, and required the existence of a particle since called the Higgs boson. The Higgs mechanism interpreted mass as a particle's resistance to movement in a Higgs field that permeated space, and its interaction with the field. The particles which were massless at the beginning of the universe, immediately acquired mass upon reacting with the theoretical space-permeating field. Physicists often call the Higgs boson the *god particle,* a term Leon Lederman allegedly coined in his 1993 book *The God Particle,* although the term actually came from his publisher.

| BOSONS (FORCE CARRIERS) | | | | | | | |
|---|---|---|---|---|---|---|---|
| ***Name*** | ***Symbol*** | ***Mass GeV/c²*** | ***Electric Charge*** | ***Force*** | ***Spin*** | ***Range (Meters)*** | ***Strength*** |
| photon | $\gamma$ | 0 | 0 | unified electroweak | 1 | $10^{-18}$ - infinite | $10^{-2}$ - $10^{-6}$ |
| W boson | $W^-$ | 80.4 | -1 | unified electroweak | 1 | $10^{-18}$ - infinite | $10^{-2}$ - $10^{-6}$ |
| W boson | $W^+$ | 80.4 | +1 | unified electroweak | 1 | $10^{-18}$ - infinite | $10^{-2}$ - $10^{-6}$ |
| Z boson | $Z^o$ | 91.187 | 0 | unified electroweak | 1 | $10^{-18}$ - infinite | $10^{-2}$ - $10^{-6}$ |
| gluon | g | 0 | 0 | strong (color) | 1 | $10^{-15}$ | $6x10^{-39}$ |
| graviton | g, G | 0 | 0 | gravity | 2 | infinite | $6x10^{-39}$ |
| Higgs boson | $H^o$ | 125.3 | 0 | | 0 | | |

The conservation laws of physics have their source in experimentation, they rule modern physics and determine what cannot occur. In addition to the classical conservation laws of energy, momentum, mass, and charge, physicists have established the conservation of baryon number in most reactions, and lepton number, strangeness, and parity in particle physics. This means that the sum of the quantum numbers of all incoming particles equals the sum of the quantum numbers resulting from the reactions. Conservation of baryon number accounted for the stability of the proton, which being the lightest baryon cannot decay to other baryons or to mesons or leptons. Conservation of lepton number governs the reactions of electrons, muons, neutrinos, and tau particles, whereas studies in the 1950s confirmed that conservation of neither strangeness nor parity quantum numbers occurred in all the reactions. Parity is the spatial inversion, or change in sign, of the three spatial coordinates x, y, z, as from +1 to -1. It affects the symmetry of a particle such that a particle and its anti-particle may have opposite intrinsic parity. When parity is conserved, the laws of nature, the experimental results, remain the same for a particle and its mirror image. Reflective symmetry exists for the particle and its mirror image.

## XII. CONCLUSION

The years 1890-1935 represent one of the most prolific periods in the history of science. Historians of science often call it the second scientific revolution. A major shift in scientific thinking occurred, resulting in the demise of classical physics and determinism, and the genesis of relativity, quantum theory, and probability in science. A rapid series of significant breakthroughs put an end to classical science and challenged some of the accepted doctrines of the physical sciences.

The collapse started in 1895, with Röntgen's discovering x-rays, a powerful radiation whose source scientists could not account for until the introduction in 1913 of Bohr's quantum model of electron orbits. The next year, in 1896, Becquerel's discovery of radioactivity in uranium compounds and the Curies' demonstration that other elements in addition to uranium were radioactive challenged the recently-proposed law of conservation of energy. The explanation came with Einstein's discovery of a mass-energy relation $E = mc^2$ in 1905 and Soddy's and Fajan's establishment of the first nuclear decay series and isotopes in 1913.

Planck's quantum theory $E = h\nu$ in 1900 continued the transformation. It showed that the emission of energy from an atom occurred in discrete quanta and not in a continuum that included all possible energy values. It received confirmation in Einstein's explanation of the photoelectric effect and Bohr's account of the hydrogen atom's spectrum.

Einstein's theory of special relativity in 1905 was even more revolutionary than quantum theory. Special relativity presented a new way of understanding space, time, and matter-energy and did away with absolute space and time. Einstein started with the assumption that the speed of light was constant, ether was unnecessary, and introduced a new relativistic, rather than an absolute interpretation of the universe. Einstein continued with his relativistic ideas, and in 1915-16 in general relativity, he proposed a new understanding of gravity in which a geometrical curved space replaced Newton's gravitational force.

The atom also underwent major transformation, going from Dalton's indivisible structureless billiard ball model to an atom with three particles, the electron in 1897, the proton ($H^+$) in 1911, and the neutron in 1932. The arrangements of the particles within an atom changed from Thomson's 1899 positive sphere model, to Rutherford's nuclear atom in 1911, to Bohr's quantum atom in 1913, and finally, to the two quantum mechanical models introduced in the mid-1920s: Heisenberg's matrix mechanics in 1925 and Schrödinger's wave mechanics in 1926. Schrödinger's model incorporated the wave-particle duality that emerged from de Broglie's combining of Einstein's $E = mc^2$ and Planck's $E = h\nu$ equations. Born, Dirac, and Jordan in 1926 gave the quantum theory a probabilistic interpretation that dominates atomic structure today. Von Neuman in 1944 showed the equivalence of the two models.

Another important breakthrough occurred in the period 1928-39 when Pauling, and independently, Mulliken applied quantum mechanics to Lewis's qualitative electron pair theory of the chemical bond. Pauling's valence bond method and Mulliken's molecular orbital theory solved the problem of how atoms combine, which is the foundation of modern chemistry.

In the decades that followed, Gell-Mann's quark theory and its development by physicists, such as Greenberg, Richter, and Ting, contributed significantly to elucidating the nucleus' structure. Their research enabled scientists to classify the fundamental particles in the standard model and to understand the forces that held the particles within the tiny confines of the nucleus.

What is the future of particle physics? What will the new particle accelerators expose? Have physicists reached the end of new particle discoveries with the discovery of three generations of quarks and leptons? Why do quarks and leptons have the masses they do? Having moved from molecule to atom, to the nucleus and electrons, and now to quarks, should physicists look for constituents of quarks and leptons?

In the 1970s Glashow and other physicists investigating grand unification theories proposed that the quarks present in all protons and neutrons will eventually decay into electrons and muons, both of which are leptons. Glashow in 1974 calculated the proton's half-life as $10^{31}$ years. Since that time, physicists in the United States, Russia, and Japan have suggested a lifetime of $10^{32}$-$10^{33}$ years. Other studies have shown that the proton is not a perfectly round particle but slightly flattened at the poles.

Physicists have asked whether the higher generations of quarks and leptons are only exalted states of the first generation, and whether astronomical observations set a limit on the total number of quarks and lepton generations, a number not larger than four. Experimental measurements of the hydrogen/helium ratio in the universe have shown that the number of types of neutrinos affects the abundance of helium. The observed helium abundance agrees with the standard model's three types of neutrinos and the improbability of more leptons.

Finally, will physicists unite the standard model of particle physics with the string theory of gravity? The union would give a single theory that accounts for the quantized microscopic world of the atom and the macroscopic world in which the gravitational force governs the behavior of matter.

## REVIEW QUESTIONS
## ON TWENTIETH-CENTURY PHYSICAL SCIENCE

1. What is the main theme of this chapter?

2. What did Wilhelm Röntgen, Henri Becquerel, and Marie and Pierre Curie contribute to our understanding of matter and energy?

3. Whose experimental work in 1897 provided the first clear evidence that the atom was not a structureless indivisible sphere?

4. What did Max Planck's 1900 quantum theory of energy tell scientists?

5. Why did Albert Einstein undertake his scientific work?

6. What was Einstein's philosophy of science?

7. What were Einstein's publications of 1905? Explain what Einstein claimed in these publications.

8. Why and how do lengths, masses, and times change according to special relativity?

9. What is Einstein's equation showing the relation between mass and energy? How does this equation account for the energy released in nuclear fission and fusion?

10. How does Einstein's theory of general relativity account for gravity?

11. Compare the atomic models of J.J. Thomson, Ernest Rutherford, Niels Bohr, G.N. Lewis, Erwin Schrödinger, and Werner Heisenberg.

12. What was the significance of Robert Millikan's oil drop experiment?

13. Compare the electron theories of the chemical bond proposed by J.J. Thomson, Richard Abegg, G.N. Lewis, Linus Pauling, Friedrich Hund, and Robert Mulliken.

14. What is femtochemistry and who invented it?

15. What is the nuclear shell model?

16. What is a quark, and quark theory?

17. What are the four fundamental forces present in the natural world?

18. What is the physicists' standard model?

19. Identify: Steven Weinberg, Sheldon Glashow, Murray Gell-Mann.

# CHAPTER 11
# ENERGY AND THE ENVIRONMENT: CONFLICT OR COMPATABILITY

The Italian navigator has landed in the new world.

Arthur Holly Compton to
James B. Conant, 2 December 1942

## I. INTRODUCTION

A significantly increased understanding of energy production, energy conversion, and energy conservation occurred in the last half of the nineteenth century and early twentieth century. The acquisition of this new knowledge enabled scientists to summarize their discoveries in three laws of energy, or laws of thermodynamics, and to understand more fully the significance of Planck's quantum theory $E=h\nu$, Einstein's mass-energy equivalency $E=mc^2$, and the Heisenberg-Schrödinger quantized atom.

Beginning in the late 1930s, the discovery of nuclear energy, a new form of energy, dramatically and permanently changed the scientific world. Scientists in Western Europe, the United States, the Soviet Union, and Japan were part of the new nuclear age. Nuclear fission and nuclear fusion, whether in the form of a nuclear bomb or a nuclear reactor, introduced an energy source capable of providing one thousand to one million times as much energy as conventional fossil fuel sources. The construction of the first atomic (fission) bombs in 1945, the first hydrogen (fusion) bomb in 1952, the first nuclear fission reactors and first experimental fusion reactors in the 1950s generated controversy that continues to this day.

Solar energy was a second *new* form of energy that although as old as the universe experienced a revival in the 1950s-60s. Solar energy's roots reach back to ancient times, but it began to achieve a degree of general acceptability only in the 1970s, thanks largely to the pioneering work of Farrington Daniels and a few other solar prophets.

The slow recognition of environmental issues resulting from the increased domestic and industrial consumption of fossil fuels paralleled the slow recognition of alternative energy sources such as solar energy. Like solar energy, environmental issues had deep roots, but only since the 1960s did society begin to acknowledge the damage environmental pollution caused to Earth's water resources and its atmosphere. Earlier efforts to deal with environmental pollution occurred in the United States in the nineteenth century. The pollution prevention programs, however, took aim mainly at water pollution, because no one would drink or bathe in visibly polluted water. These programs led to improved water quality.

The earlier prevention programs that dealt with atmospheric pollution focused on smoke pollution because society saw and experienced its consequences. For the sometimes less-visible consequences of air pollution, such as acid rain, global warming, and the presence of lead and mercury in the atmosphere, which were the direct result of an automobile-dominated and fossil fuel burning society, decades elapsed before any action occurred. Eugene Likens documented the serious consequences of acid rain. Roger Revelle, Charles Keeling, James Hansen, and others experimentally demonstrated the reality of global warming. Clair Patterson significantly increased society's awareness of the lead-in-air pollution problem, showing that much of the lead eventually settled in water or on land. Despite the scientific evidence indicating the extent and seriousness of environmental pollution, progress in improving air quality has continued to lag, fueled largely by unsubstantiated claims of financially-, politically-, and industry-supported doubters.

## II. EARLY DEVELOPMENTS IN NUCLEAR FISSION REACTIONS

### *First transmutations*

Ernest Rutherford at Cambridge first produced transmutation, the conversion of one element into another element, in 1919. Using a Wilson cloud chamber to detect the presence of an electrically-charged particle, or ion, he bombarded nitrogen gas with an alpha particle (helium nuclei) obtained from the nuclear decay of thorium to give oxygen gas and a hydrogen atom (proton).

$$^{14}_{7}N + ^{4}_{2}He \rightarrow ^{17}_{8}O + ^{1}_{1}H$$

J.J. Thomson had already, in 1897, discovered the electron, the first subatomic particle, and used a cloud chamber to determine its charge. Rutherford in 1911 discovered the hydrogen nucleus, the first nuclear particle, naming it the proton in 1920. Rutherford detected the proton and alpha particle because, when present as charged particles in the atmosphere, they ionized the oxygen and nitrogen molecules making up the atmosphere.

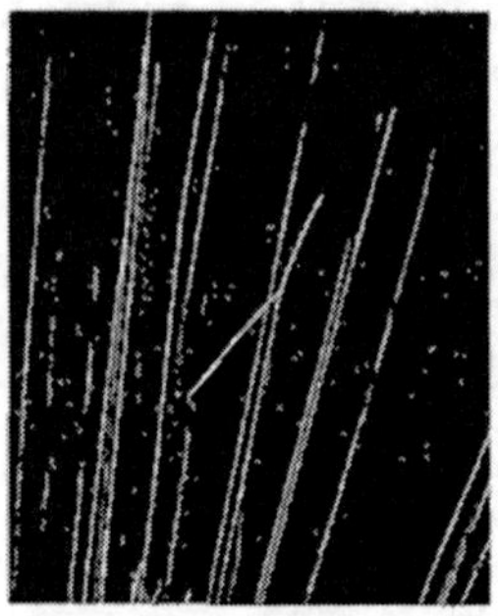

**Figure 1**
**Photograph of the Disintegration of a Nitrogen Nucleus by an Alpha Particle, 1932**
Courtesy of The Royal Society London

The Scottish physicist Charles T. Wilson (1869-1959), at Cambridge, developed the cloud chamber in the years 1895 to 1910. Wilson used x-rays to ionize dust-free air saturated with water vapor to form a fog consisting of tiny water droplets. Each of the electrically-charged (ionized) particles in the cloud chamber acted as a nucleus for a water droplet. The visible pathway of the water droplets made evident the path (tracks) of the electrically-charged particles. William Harkins (1873-1951), at Chicago, used a cloud chamber to obtain photographs that confirmed Rutherford's work on proton formation. Harkins' experiments led him to consider the presence of some other particle in the nucleus besides protons and electrons. In a 1920 cloud chamber experiment Harkins observed an alpha particle reacting with a nitrogen atom to produce a fluorine atom and a particle with a mass equivalent to a proton. Cloud chamber tracks indicated that the particle had no charge and that the fluorine atom decayed to give an oxygen atom and a proton.

$$^{4}_{2}He + ^{14}_{7}N \rightarrow ^{17}_{9}F + ^{1}_{0}n$$

$$^{17}_{9}F \rightarrow ^{16}_{8}O + ^{1}_{1}H$$

This sequence of reactions led Harkins in 1920 to suggest the existence of a neutron-like particle in the nucleus and to introduce the word neutron in a 1921 article published in the *Philosophical Magazine*.

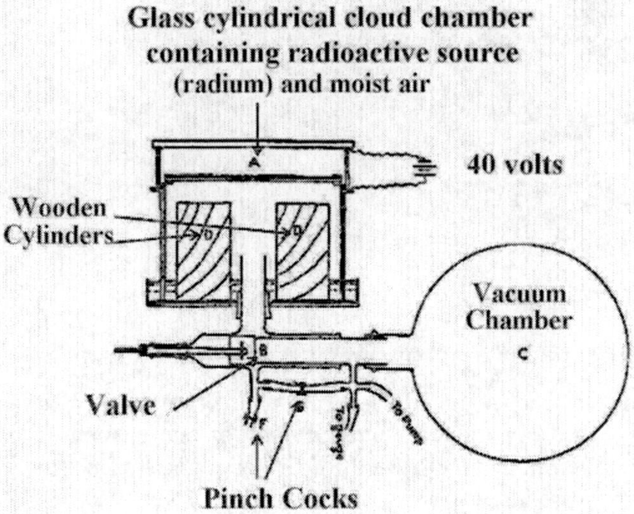

**Figure 2**
**Wilson Cloud Chamber**
Photograph shows alpha particle tracks from radium (right)
Source: C.T.R.Wilson. "On an Expansion Apparatus for making Visible the Tracks of Ionizing Particles in Gases and some Results obtained by its Use." *Proceedings of the Royal Society* (London) A, 87 (1911).

***Discovery of the neutron***

James Chadwick, who studied under Rutherford at Manchester, continued his studies in Germany in 1913, but the outbreak of World War I led to his wartime internment. Upon his release he worked with Rutherford at Cambridge on the alpha particle cloud chamber experiments. Chadwick continued the alpha particle bombardments and in February 1932 discovered the neutron. He repeated the 1930 experiments of Walther Bothe (1891-1957) and Heinrich Becker (1911-42) at Giessen and the 1932 experiments of Frédéric Joliot-Curie (1900-58) and Iréne Joliot-Curie (1897-1956) in Paris. Both groups had failed to detect the neutron's presence.

Bothe, who obtained his PhD in physics under Planck, spent the years 1914-19 as a Russian prisoner in Siberia. He shared the 1954 Nobel Prize in physics with Born. He and Becker bombarded separately beryllium (Be-9), boron (B-11), and lithium (Li-7) with alpha particles derived from polonium and detected a penetrating radiation that passed through lead. They believed the radiation was gamma radiation. The Joliot-Curies repeated Bothe and Becker's experiments early in 1932 before Chadwick announced his discovery. They also observed a penetrating radiation that passed through lead and thought it was gamma radiation. The Joliet-Curies even placed paraffin in front of the emitted radiation (neutrons), knocking out protons, but never doubted that the radiation was gamma.

In separate experiments Chadwick bombarded beryllium (Be), boron (B), and lithium (Li) with alpha particles derived from polonium. He equated the neutron's presence with the appearance of a penetrating radiation that ejected protons (hydrogen nuclei) from paraffin. The production of protons led Chadwick to conclude that the ejecting particle (neutron) had the same mass as a proton but no charge, and because the neutron was an uncharged subatomic particle he could not test its ability to ionize the oxygen and nitrogen molecules in the atmosphere. Aston's mass spectrograph measurements of nitrogen's (N-14) and boron's (Be-11) atomic mass enabled Chadwick to calculate the neutron's mass.

**NEUTRON PRODUCTION FROM ALPHA BOMBARDMENT**

$$^{4}_{2}He + ^{9}_{4}Be \rightarrow ^{12}_{6}C + ^{1}_{0}n$$

$$^{4}_{2}He + ^{11}_{5}B \rightarrow ^{14}_{7}N + ^{1}_{0}n$$

$$^{4}_{2}He + ^{7}_{3}Li \rightarrow ^{10}_{5}B + ^{1}_{0}n$$

A few months later, in April 1932, John D. Cockcroft (1897-1967) and Ernest T.S.Walton (1903-95) at Cambridge carried out the first artificially induced transmutation of one element into another element. They carried out the transmutation in a three-stage, 36-inch long, glass accelerator, using as a projectile the proton, a particle not produced by radioactive decay. They electrically accelerated the protons to an energy of 400,000 electron volts (eV) and bombarded lithium with the protons to give two alpha particles. The brightness of the scintillations the reaction emitted and the density of the tracks produced in the cloud chamber in which the reaction occurred convinced Cockcroft and Walton that they had produced two alpha particles. Just as significant was the fact that the transmutation reaction confirmed Einstein's mass-energy equivalence, $E=mc^2$.

**COCKCROFT AND WALTON'S MASS-ENERGY CONVERSION**

$$^{7}_{3}Li + ^{1}_{1}H \rightarrow ^{4}_{2}He + ^{4}_{2}He$$

$$7.0182\ amu + 1.0081\ amu \rightarrow 4.0039\ amu + 4.0039\ amu$$
$$Loss\ in\ Mass\ \Delta m = Mass\ Reactants - Mass\ Products$$
$$\Delta m = 8.0263\ amu - 8.0078\ amu$$
$$\Delta m = 0.0185\ amu$$

$$E = 0.0185\ amu \times 931 \frac{MeV}{amu}$$
$$E = 17.22\ MeV$$

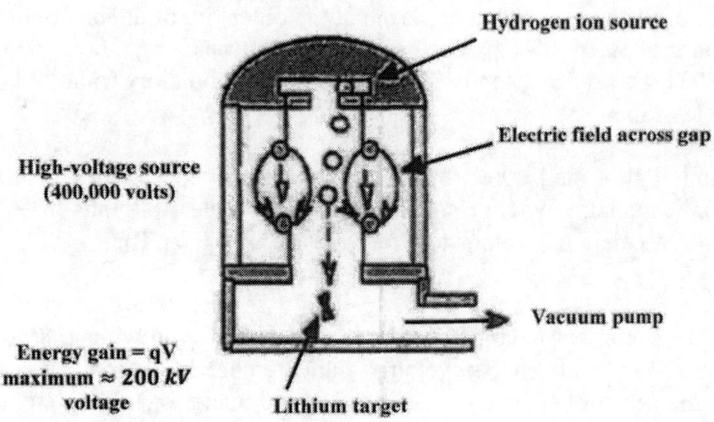


**Figure 3**
**Cockcroft and Walton's Direct Current Particle Accelerator**
Courtesy of the Department of Energy

Cockcroft left Cambridge in 1946 to become director of the Atomic Energy Research Establishment at Harwell. He and Walton shared the 1951 Nobel Prize in physics for their pioneering work on the transmutation of atomic nuclei by artificially-accelerated atomic particles.

***Cyclotron and artificial radioactivity***

Two years before Cockcroft and Walton accelerated protons in a linear particle accelerator, Ernest O. Lawrence (1901-58) at Berkeley, in 1930, began constructing his first circular particle accelerator. Lawrence believed his circular particle accelerator would produce higher voltages than a comparable-sized linear accelerator and thereby provide a much better understanding of the atom's nucleus.

Lawrence initially accelerated positive ions in a linear particle accelerator. He conceived the idea of multiple-acceleration of positively-charged particles to great speeds in 1929 while examining an article by the Norwegian engineer Rolf Wideröe (1902-96), then at the Rhenish-Westphalian Technical University in Aachen, Germany. Wideröe accelerated potassium and sodium ions to an energy of 50,000 electron volts in a linear particle accelerator. Unable to read Wideröe's German language article, Lawrence studied the diagrams, recognizing that he could reduce the accelerator's size by applying a magnetic field to make the charged particles move in a spiral path. Working with his first graduate students, Niels E. Edlefsen (1893-1971) and M. Stanley Livingston (1905-86), in an old building near the home of the physics department, Lawrence in spring 1930 constructed a test model of a 4-inch diameter circular particle accelerator using copper wire, a glass plate, and red sealing wax. He initially called his accelerator a magnetic resonance accelerator and later, in 1935, a cyclotron.

In January 1931 Lawrence constructed an all-metal 4.5 inch diameter cyclotron, about the same size as his test model. The oscillator generated a voltage of 1,800 volts and accelerated protons to an energy of 80,000 eV. His second cyclotron consisted of a 4-inch diameter cylindrical box set between the north and south poles of an electromagnet. Inside the box he placed two hollow D-shaped insulated electrodes (dees); connected them to an oscillator, a high voltage alternating current source; and evacuated the box. When the cyclotron was operating, the oscillator rapidly reversed the electric charge on each dee, changing the positive dee to negative and the negative dee to positive. The particles (protons) that the cyclotron accelerated entered at the center of the box. The perpendicular electromagnetic field forced the accelerated particles to move in a spiral path, while the high voltage alternating positive or negative charge on the dees accelerated the particles on each trip around the dees to higher speeds. The particles' speed, and therefore their energy, increased as they moved closer to the outside of the box. A deflector placed at the outer rim of the box deflected the beam of particles toward its target. Each of Lawrence's early cyclotrons cost about $25.00. Later that year, on 26 August 1931, Lawrence established the Radiation Laboratory in an old civil engineering building on the Berkeley campus.

Cockcroft and Walton and other researchers had achieved energies of 400,000 eV with which to bombard atomic nuclei. Lawrence's goal was 1 million electron volts (MeV), and in August 1931 he and Livingston accelerated protons to an energy of 1.1 MeV. Their cyclotron, 11 inches in diameter, cost $1,000.

Lawrence in 1932 constructed a much larger 27-inch cyclotron with an 80-ton magnet. Located in the Radiation Laboratory, Lawrence's cyclotron in September 1932 accelerated protons to 3.6 MeV, about five percent the speed of light, and disintegrated the nuclei of lithium atoms. In 1935 he accelerated and sent deuterons (H-2) at an energy of 5 MeV and alpha particles at an energy of 10 MeV against targets such as lithium, and in 1936 against platinum, converting platinum to gold. Lawrence, following G.N. Lewis's suggestion, initially called the deuteron that Harold Urey in 1931 discovered by spectrum analysis of liquid hydrogen, a *deuton*. In 1933 he favored *deuteron*. Rutherford preferred the name *diplon*.

The use of larger cyclotrons at Berkeley continued with the construction in 1937 of a 37-inch cyclotron and a 60-inch cyclotron in 1939. The 60-inch cyclotron in 1940 enabled Edwin McMillan (1907-91) and Philip Abelson (1913-2004) to discover the first transuranium element, neptunium (Z=93), and Glenn Seaborg (1912-99) and McMillan to discover a second transuranium element, plutonium (Z=94). A 184-inch synchrocyclotron that produced 340 MeV protons followed in 1946. Between 1931 and 1940 Lawrence obtained $550,000 for apparatus and operation of the Radiation Laboratory and during the same period (1932-39) increased the staff from ten to sixty. In only a few years the cyclotron had become a powerful investigative instrument that renewed interest in radioactivity and led to the discovery of many new isotopes.

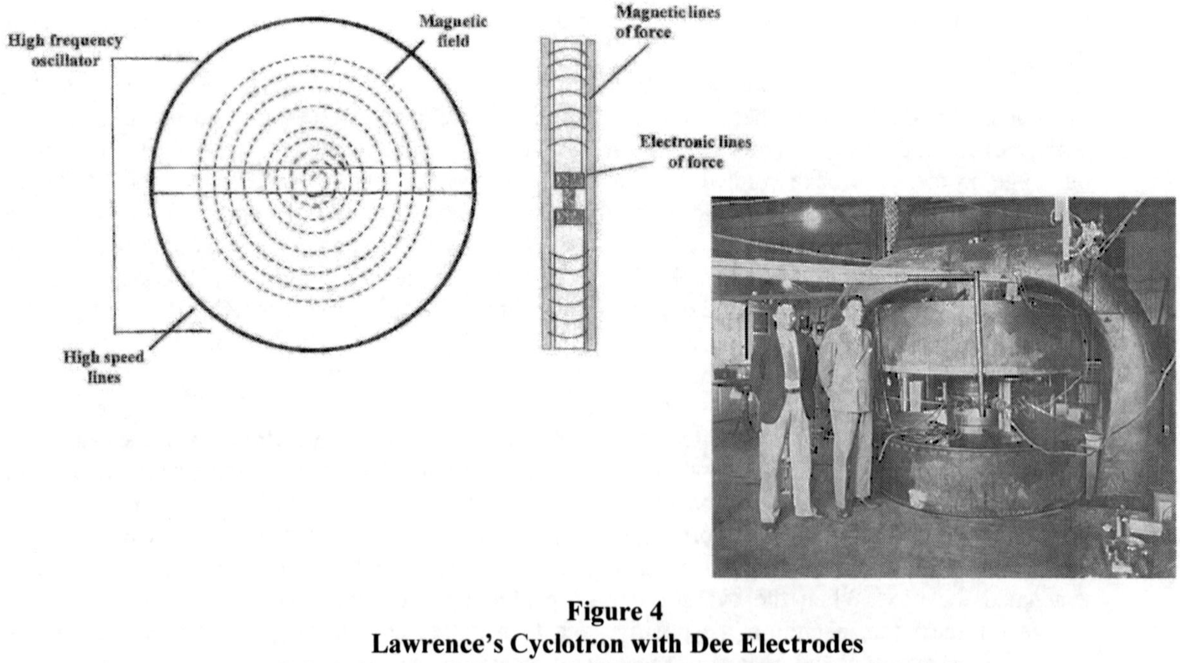


**Figure 4**
**Lawrence's Cyclotron with Dee Electrodes**
E.O. Lawrence, US Patent 1,948,384 of 20 February 1934 (left)
Courtesy of the Department of Energy (right)

The Berkeley Radiation Laboratory became an official department at Berkeley, with Lawrence its director in 1936. At the laboratory he and his close friend J. Robert Oppenheimer (1904-67) helped devise an electromagnetic method for separating the fissionable uranium U-235 isotope used in the atomic bomb dropped on Hiroshima on 6 August 1945. In 1952 Edward Teller (1908-2003), at the Los Alamos Laboratory in New Mexico, and Lawrence, both of whom supported the development of the thermonuclear-fusion, or hydrogen, bomb, played a prominent role in establishing a second laboratory for research on nuclear weapons. This was the Lawrence Radiation Laboratory at Livermore in northern California. Later Lawrence and Teller spoke against the proliferation of nuclear weapons.

Lawrence received the 1939 Nobel Prize in physics for his invention of the cyclotron and its impact on particle physics. While participating in a nuclear test ban conference in Geneva, Switzerland, in 1958, Lawrence became ill and died after ulcerative colitis surgery. Albert Ghiorso (1915-2010) and the three other Berkeley scientists, who in 1961 discovered element number 103, named the new element lawrencium (Lr) in his honor.

The Joliot-Curies in 1932 missed the discovery of the neutron. They continued their experimentation on nuclear bombardment, and in January-February 1934 they discovered a new type of radioactive decay. The Joliot-Curies sent alpha particles derived from polonium against the lighter elements boron, magnesium, and aluminum. Each reaction, as suspected, produced a proton or a neutron, and the newly discovered neutrino. What happened next astonished them. When they stopped the alpha particle bombardment, the emission of radiation did not stop immediately but continued for a time. The other product, in addition to the proton, neutron, and neutrino, was an unstable radioactive isotope of another light element, such as P-30, that underwent additional nuclear decay as shown below in the equations for aluminum bombardment.

**ARTIFICIAL RADIOACTIVITY**

$$^{4}_{2}He + ^{27}_{13}Al \rightarrow ^{30}_{15}P + ^{1}_{0}n$$

$$^{30}_{15}P \rightarrow ^{30}_{14}Si + ^{0}_{+1}e$$

The Joliot-Curies used alpha particles to convert boron (B-10) into radioactive nitrogen (N-13), and magnesium (Mg-24) into radioactive aluminum (Al-28) and silicon (Si-27). They had prepared the first radioactive isotopes, radioisotopes that did not occur in nature but were artificial, or man-made. The Joliot-Curies had discovered artificial radioactivity, a process that other laboratories duplicated, and over a twenty-year period, by 1956, had prepared at least 872 artificial radioactive isotopes (radioisotopes). Their discovery opened up new fields of research in medicine and chemistry, particularly the use of radioiodine (I-131) and radiophosphorus (P-32) in thyroid gland studies, and as tracers (P-32, S-35) that revealed the pathways of chemical reactions, and the use of radiocarbon in the C-14 dating method.

The Joliot-Curies received the 1935 Nobel Prize in chemistry. After World War II, in 1946, they held important positions in the French Atomic Energy Commission. Both had socialist-communist sympathies and paid a high price because of their sympathies. Frédéric lost his position as high commissioner of the French Atomic Energy Commission in 1950, following his admission of membership since 1942 in the French Communist Party. The American Chemical Society in 1953 rejected Irène's membership application because of her socialist beliefs. Ten years later both were dead. Irene died of leukemia, Frédéric died from internal hemorrhaging following surgery for liver disease, probably related to radiation.

***Fermi and neutron bombardment***

Enrico Fermi represents one of the rare cases of a scientist who excelled as both theorist and experimentalist. He also possessed a remarkable memory for scientific information. With the rise of the new physics, which the discovery of x-rays, radioactivity, and especially quantum mechanics exemplified, Fermi aimed to revolutionize physics in Italy. From the time of his appointment at the University of Rome in 1926, he vowed to make the university a center for theoretical physics. The study of nuclear reactions, particularly the inducing of artificial radioactivity by neutron bombardment of atoms rather than Joliot-Curie's alpha particle or proton bombardment, became his starting point.

**Figure 5**
**Enrico Fermi in 1948, with an incorrect equation he wrote for α and printed as a US 34 cent stamp in 2001. The correct equation is**

$$\alpha = \frac{e^2}{\hbar c}$$

Courtesy of the Library of Congress

The neutron was as heavy as the proton, but because the neutron had no charge, a particle accelerator could not accelerate it. Fermi did not find this a problem. Instead, he believed that the neutron might more easily react with or penetrate the positive nucleus of an atom, because unlike the proton and alpha particle, the neutron had no repulsive force to overcome. Radon's decay to give an alpha particle, and the reaction of an alpha particle with beryllium provided the neutrons Fermi needed for the bombardment.

$$^{222}_{86}Rn \rightarrow\ ^{218}_{84}Po +\ ^{4}_{2}He$$

$$^{9}_{4}Be +\ ^{4}_{2}He \rightarrow\ ^{12}_{6}C +\ ^{1}_{0}n$$

Fermi's four-man experimental group, in early 1934, began bombarding hydrogen (Z=1) with neutrons but detected no evidence of artificial radioactivity as a result of neutron absorption until fluorine (Z=9). A Geiger counter detected alpha particles emitted by the radioactive nitrogen isotope.

$$^{1}_{0}n +\ ^{19}_{9}F \rightarrow\ ^{16}_{7}N +\ ^{4}_{2}He$$

An extremely important discovery about neutron bombardment emerged from these experiments. Fermi found that slowing down the neutrons by passing them through paraffin wax or water, both of which were hydrogen-rich compounds containing other light elements, gave a higher conversion to irradiated products. Fermi concluded that this occurred because the neutron, when it struck a hydrogen atom in wax or water, gave up much of its energy to the hydrogen atom which weighed about the same as the neutron. When the neutron struck a heavy nucleus it often bounced off with little loss of energy or speed resulting in a lower conversion to irradiated products.

The experiments continued throughout 1934, and by the end of the year Fermi's group published 25 papers on neutron bombardment. In February 1935 they published a comprehensive survey of their work and showed that although lead did not slow down neutrons, a thin layer of boron or of cadmium, in addition to wax and water, effectively slowed the neutron. As a result, neutron absorption and the production of radioactive isotopes of stable elements increased.

Fermi's group in 1935 continued the neutron bombardment of heavier elements including uranium, the heaviest naturally-occurring element (Z=92). In conducting the uranium experiments, they covered the uranium target with a thick aluminum foil that blocked the reaction products from reaching their detector. As a result, Fermi failed to recognize that a fission, or splitting, of the uranium atom had occurred. Fermi believed instead that the neutron bombardment produced new elements having atomic numbers higher than uranium, the transuranium elements Z=93 or Z=94, although he had not identified any product of the bombardment. He did not suspect that the radiation their group detected resulted from the fission of uranium by slow neutrons. The Italian patent that Fermi's group obtained in 1935 was consequently on inducing radioactivity with slow neutrons and not on nuclear fission. (US Patent 2,206,634 awarded 2 July 1940. The Atomic Energy Commission awarded Fermi's team $300,000 in 1953 for use of the patent).

All of these experimental developments set the stage for Fermi's reception in December 1938 of the Nobel Prize for physics. Even before the Nobel committee informed Fermi of his award, he made plans to leave Benito Mussolini's fascist Italy. Fermi's wife was Jewish, and when Mussolini (1883-1945) imposed a serious crackdown on Italian Jews in November 1938, Fermi decided the time had come for departure. He notified the Italian authorities that he was departing Italy for six months. After traveling to Stockholm for the Nobel award, Fermi left Sweden for London and then New York City, where he arrived in January 1939 to accept a position at Columbia University. Although pleased that Fermi had received the Nobel Prize, the Mussolini government strongly criticized Fermi for not wearing a fascist uniform and not giving the fascist salute upon receiving the prize.

***Nuclear fission comes to the United States***

After four years of neutron bombardment experiments and no confirming evidence of transuranium elements, Otto Hahn (1879-1968) and Fritz Strassman (1902-80) at Berlin questioned the results of Fermi's neutron experiments. On 18 December 1938 they bombarded uranium (U-235) with neutrons, and instead of producing a heavier element such as Fermi's hoped-for element 93, they suspected that barium (Ba-141), a second much lighter element, and radiation (neutrons) were likely products. Hahn and Strassman published their results in January 1939 but did not interpret them as evidence of a fission reaction.

Lise Meitner (1878-1968), Hahn's collaborator at Berlin for thirty years and a Jew-turned-Protestant in 1903, already in late 1938 had fled Hitler's Germany for Copenhagen. There, through her nephew Otto Frisch (1904-79), who worked in Niels Bohr's laboratory at Copenhagen from 1934 to summer 1939, she met Bohr and with Bohr's help moved to Stockholm. Bohr since 1933 had helped German scientists escape Germany and seek positions in the west. Hahn mailed Meitner the results of the Hahn-Strassman uranium bombardment experiments, and during the 1938 Christmas holidays when Frisch visited Meitner in Stockholm, they discussed Hahn's letter. Whereas Hahn had hesitated to publish his suspicion that the neutron bombardment of uranium produced the lighter element barium (Ba-141), Meitner, who had worked with Hahn on the uranium-neutron bombardments, and Frisch had no doubt about the result. The neutrons striking the uranium nucleus split the uranium nucleus into two lighter atoms.

Frisch called the splitting reaction nuclear fission, thereby coining the word fission. He and Meitner concluded that the fission produced the two lighter elements barium (Ba-141) and krypton (Kr-92), more neutrons, and released about 200 MeV of energy.

$$ {}^{235}_{92}U + {}^{1}_{0}n \rightarrow {}^{236}_{92}U \rightarrow {}^{141}_{56}Ba + {}^{92}_{36}Kr + 3{}^{1}_{0}n + 200\ MeV\ or\ 2 \times 10^{10}\ Kj/mol $$

They believed that Bohr's liquid drop model provided a theoretical model for the nuclear fission of an element. According to the theory that Bohr developed from 1936 to 1939, the nucleus resembled a liquid drop. The nucleons (protons and neutrons) in the nucleus, just as the molecules in a drop of liquid, interacted very strongly with each other. As a result, all the nucleons shared any excess energy associated with an individual nucleon. The energy levels of the nucleus, according to Bohr, were the quantized energy levels of the nucleus and not of a single nucleon. When a neutron struck the nucleus and reacted with it, the nucleus became elongated and divided, producing lighter elements and more neutrons. These neutrons struck more nuclei, resulting in more fission and more neutrons, which in turn produced more fission and more neutrons and the very real possibility of a chain reaction.

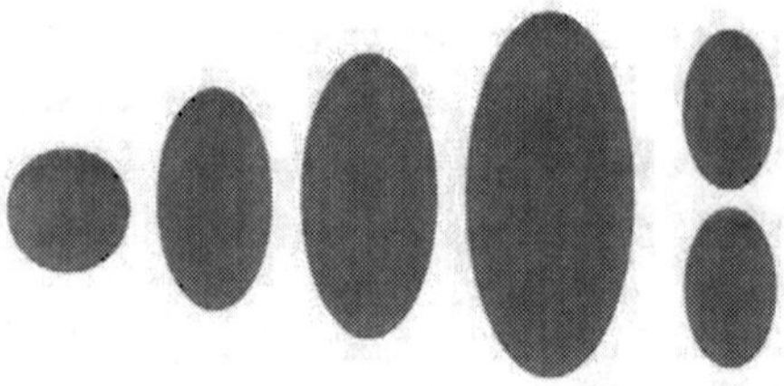

**Figure 6**
**Bohr's Liquid Drop Model of Nuclear Fission**
Courtesy of the author

After meeting with Meitner in Stockholm, Frisch returned to Copenhagen and on 3 January 1939 relayed their conclusion to Bohr. Bohr was leaving on the liner *MS Drottningholm* for New York City on 7 January to visit Einstein in Princeton, Fermi at Columbia, and on 26-28 January attend the Fifth Washington Conference on Theoretical Physics at George Washington University in Washington, DC. On 26 January while at the conference, he informed the attendees of Meitner and Frisch's report that uranium (U-235) underwent fission when bombarded with neutrons. One week after his announcement, Bohr received a letter that Frisch in Copenhagen mailed on 22 January. Frisch wrote that in his uranium bombardment experiments he detected fission fragments in an ionization chamber, thereby confirming his splitting of the uranium atom with neutrons. A few days later, on 11 February 1939, Meitner and Frisch published in *Nature* for the first time that Hahn and Strassman in their 1938 experiment had split the uranium atom. Their paper introduced the word fission to describe the splitting.

Bohr's announcement of nuclear fission at the conference caught nearly everyone by surprise, and as with all announcements of scientific breakthroughs it experienced some skepticism. Oppenheimer at Berkeley remained a doubter for about an hour once he saw the experimental evidence. Fermi re-thought his belief that uranium's fission was remote, about 10 percent. Bohr enlisted the cooperation of John Wheeler (1911-2008) at Princeton, and on 1 September 1939, the same day that Hitler invaded Poland, they published a paper in the *Physical Review* explaining why the rare U-235 isotope was more fissionable than the common U-238 isotope. U-235 with an odd number of neutrons, 235-92 = 143 neutrons, versus U-238 with an even number of neutrons 238-92 = 146 neutrons and only a slightly lower neutron/proton (n/p) ratio for U-235 than for U-238, 143/92 to 146/92, was more unstable and therefore more likely to undergo fission.

| COMPARISON OF URANIUM ISOTOPES | |
|---|---|
| *U-235* | *U-238* |
| 92 protons<br>143 neutrons<br>1.55 n/p ratio | 92 protons<br>146 neutrons<br>1.58 n/p ratio |

The race to sustain a nuclear fission reaction had begun. Fermi in January 1939 led a group at Columbia; Frisch at Copenhagen and Frèdèric Joliet-Curie in Paris led separate groups. They verified that bombarding uranium with slow moving neutrons released a tremendous amount of energy and with the rest of the scientific community quickly foresaw the impact of fission in war. The key to a nuclear fission bomb was to sustain a nuclear chain reaction which required bombarding uranium with slow neutrons. Joliot-Curie used heavy water (hydrogen-2 oxide, deuterium oxide) to slow down the neutrons because it absorbed fewer neutrons. Fermi believed carbon, graphite specifically, was the most efficient way to slow the neutrons. His initial results were disappointing, and he suspected impurities in the graphite. Leo Szilard (1898-1964), a Hungarian physicist at Berlin, who left Germany arriving in London in 1933 and at Columbia in 1938, helped Fermi prepare a pure form of graphite.

Szilard, sometimes described as a dreamer, volatile, immodest, and a scientific misfit, had envisioned the idea of a nuclear chain reaction in September 1933, after reading in *The Times* (London) of Rutherford's dismissing industrial-scale matter-energy conversion as moonshine. Six months later on 12 March 1934, Szilard filed a patent with the British Patent Office (British Patent 440,023 awarded 12 December 1935) *Improvements in or relating to the Transmutation of Chemical Elements*. His patent discussed neutron generation and the likelihood of a sustainable chain reaction, isotope separation, and energy released in nuclear transmutation. A second patent that he filed on 28 June 1934 with the same title (British Patent 630,726 accepted 30 March 1936) described a mechanism for a sustainable nuclear chain reaction. By this time, 1936, Szilard feared the consequences of his chain reaction mechanism, its possibly releasing a large amount of energy, and to secure its secrecy he assigned patent number 630,976 to the British Admiralty. It remained unpublished until 28 September 1949.

Szilard's second patent described a process in which a neutron struck a beryllium atom, converting it to helium and releasing two neutrons. The two neutrons struck two more beryllium atoms, releasing two more neutrons, thereby initiating a chain reaction. The chain reaction was linear not exponential, it also generated *heavy neutrons*, *2n,* that were crucial to sustaining the reaction. Szilard's nuclear fuel consisted of indium and beryllium, bromine or uranium. He introduced the idea of a critical mass, and included a diagram of a nuclear reactor with cooling pipes that removed and used the heat from the reaction for the generation of electricity and that produced radioisotopes. Szilard correctly envisioned the idea of a self-sustained nuclear chain reaction and critical mass, but the reaction he proposed does not occur. The University of California-San Diego's archives have the original copies of Szilard's two nuclear fission patents.

Hahn, who received the 1944 Nobel Prize in chemistry for his work on nuclear fission, had no role in Germany's World War II atomic bomb project. He considered suicide in 1939, once he realized his discovery of fission would lead to the construction of atomic bombs. He thought again about suicide in 1945 during his brief internment after the war, but instead he actively promoted nuclear disarmament. Meitner declined to participate in the British and American atomic bomb projects, whereas Frisch worked on the American atomic bomb project at Los Alamos. Szilard played a key role in the American project's success.

## III. MANHATTAN PROJECT: CONTROLLING NUCLEAR FISSION AND RACING TO BUILD AN ATOMIC BOMB

### *Organization and early research*

Fermi in 1934 had a tiger by the tail but never knew it. Controlling a nuclear reaction came only after understanding the science and engineering of nuclear fission, particularly knowing how to separate isotopes, produce slow neutrons, and sustain a chain reaction. Technology did not lead the science of fission.

At the beginning of World War II on 1 September 1939, the fundamentals of nuclear fission were circulating in western journals. The advent of war quickly raised the question of how to apply nuclear fission to the war effort. Scientists envisioned a bomb device having thousands of times the destructive power of a conventional explosive. Reality suggested enormous technical problems. Such a device would require tremendous resources applied over years and perhaps make the United States at that time the only nation capable of pursuing such a project.

The American military and congress did not rush to develop nuclear fission because they did not understand it fully. With a war imminent, they focused on short-term projects, just as the military and politicians did in Germany. Once nuclear research began, however, fear became the great motivator. American scientists knew of Germany's theoretical fission research before communications ended and believed they were playing catch-up. This fear led Szilard, with the help of two other Hungarian-refugee physicists, Eugene Wigner (1902-95) at Princeton, and Teller, then at George Washington University, to write a letter expressing their concern to President Franklin D. Roosevelt. The letter of 2 August 1939, which Szilard and Einstein signed, urged Roosevelt to initiate research on an atomic bomb. The possibility that Hitler's scientists might develop nuclear weapons horrified Einstein, a lifelong pacifist.

Despite its urgency, the Szilard letter fell to the bottom of Roosevelt's priorities, and its message conveyed only when Alexander Sachs (1893-1973), a New Deal official and Roosevelt economic advisor, on 11 October 1939 personally delivered the letter to Roosevelt. The letter led Roosevelt to authorize the next day a government examination of the three physicists' concerns. He also appointed Lyman Briggs (1874-1963), a physicist-engineer and director of the National Bureau of Standards, to organize a secret advisory committee on uranium. The Briggs Advisory Committee's nine civilian and military participants met for the first time in Washington, DC, on 21 October and agreed on a $6,000 grant to Columbia to support Fermi and Szilard's research on developing an atomic bomb. The Briggs committee awarded the grant on 20 January 1940 and provided a second grant of $40,000 in November 1940.

The Briggs committee, largely because of Briggs' inattention to nuclear developments outside the United States, particularly in Britain, acted slowly. About 18 months passed before Vannevar Bush (1890-1974), an engineer and former president of the Carnegie Institution in Washington, DC, and head of the National Defense Research Council (NDRC), on 28 June 1941 after consultation with Roosevelt, established the Office of Scientific Research and Development (OSRD) with headquarters in Washington, DC. As its director, Bush assumed all responsibility for fission research including the Briggs Advisory Committee, which became the S-1 Project of the NDRC with Briggs reporting to Bush.

The next year, on 13 August 1942, the United States government established the Manhattan Project in New York City for the construction of atomic bombs. General Leslie Groves (1896-1970), the army engineer who in 1940 oversaw construction of the Pentagon in Washington, DC, assumed command in September 1942. The Manhattan Project, officially the Manhattan Engineer District, in its early years had at least ten secret Manhattan, New York City sites, including the project's first headquarters, and employed 5,000 people. The Pupin Physics Laboratories at Columbia at its peak employed 700 people. The Baker and Williams Warehouses on West 20th Street in 1942-3 held 173.5 tons of green-yellow-red sodium uranate ($Na_2 UO_4$), yellow sodium uranyl carbonate ($2 Na_2 CO_3 \cdot UO_2 CO_3$), and black uranium dioxide ($UO_2$). The Woolworth Building at 233 Broadway Avenue had 3,700 employees, many of whom worked on U-235 concentration, and the Army Corps of Engineers had its North Atlantic Division and later its headquarters, on the eighteenth floor of a skyscraper located at 270 Broadway Avenue. The Manhattan Project eventually employed 125,000 people and cost $2 billion ($22 billion in 2010).

Although the organizational side of the American nuclear fission program initially moved slowly, the science of nuclear fission had made and continued to make notable progress. Arthur Dempster (1866-1950) at Chicago, using a mass spectrometer, discovered the U-235 isotope in 1935. His investigations and other mass spectrometer investigations showed that in naturally-occurring uranium, the ratio of U-238 atoms to U-235 atoms was 140 to 1. In March 1940 John Dunning (1907-75), a member of Fermi's research group at Columbia, demonstrated that fission occurred in the rare U-235 isotope (0.7 percent natural uranium) and not in the more plentiful U-238 isotope (99.3 percent natural uranium).

A few months later, in June 1940, McMillan and Abelson at Berkeley identified a new element, neptunium (Z=93) in uranium fission fragments. The following year, on 24 February 1941, Seaborg and his research team at Berkeley showed that the much more abundant but non-fissionable U-238 isotope when struck by a neutron absorbed the neutron to give U-239. The U-239 isotope underwent beta (electron) decay to give neptunium, Np-239 (Z=93), which decayed to give another new element that Seaborg named plutonium, Pu-239 (Z=94). On 28 March 1941 Seaborg's team showed that Pu-239 was fissionable, determining in May 1941 that Pu-239 was 1.7 times more fissionable than U-235. He obtained a weighable sample of Pu-239 in August 1942.

$$^{238}_{92}U + ^{1}_{0}n \rightarrow ^{239}_{92}U \rightarrow ^{239}_{93}Np + ^{0}_{-1}e$$

$$^{239}_{93}Np \rightarrow ^{239}_{94}Pu + ^{0}_{-1}e$$

***Construction of the first atomic piles***

Fermi's group at Columbia succeeded in solving two fundamental problems of reactor design. Tests in 1940 showed that highly purified graphite was a better moderator than water because it absorbed fewer neutrons. In a nuclear reactor, which Fermi first called a *pile* because his group constructed its first test pile by placing graphite layer upon layer, the moderator slowed down the neutrons sufficiently for fission to occur. The moderator's neutron multiplication factor (k) had to exceed 1, k>1, to initiate a chain reaction, that is, the number of neutrons emitted by the fission had to exceed the number of neutrons that initiated the fission.

Investigations on uranium indicated that U-238 had a greater nuclear cross section than U-235 and would combine with neutrons released in the fission reaction. Its presence even in small amounts would reduce the number of neutrons available for the fission of U-235. To eliminate this problem, Fermi and Szilard, beginning in July 1941, constructed a graphite-uranium lattice reactor in which they placed the uranium (uranium metal; black uranium dioxide $UO_2$; and brown uranium dioxide $UO_2$, $U_3O_8$, and $UO_3$) in small lumps (slugs) surrounded by graphite blocks. They used uranium oxides because uranium metal of high enough purity did not exist. The bottom of the pile contained the radon-beryllium neutron source. Fermi and Szilard followed the neutron intensity as it spread up the pile by measuring the radioactivity induced in rhodium foil put in the pile. Their small test pile at Columbia, with which they studied graphite's absorption of neutrons, had a three-foot square base and an eight-foot height, built in the pattern shown below.

## URANIUM PILE BUILDUP

uranium-uranium oxides + graphite + uranium-uranium oxides + graphite + uranium-uranium oxides + graphite

When, in early 1942, the OSRD decided to concentrate the nuclear pile research at the University of Chicago Metallurgical Laboratory, the Columbia pile program shut down and construction began on the first Chicago pile. In March 1942 Fermi's team moved to Chicago.

The Chicago pile, called CP1, located in a squash court under the west stands at the Alonzo Stagg Field (the university's athletic stadium) was much bigger than the earlier test piles at Columbia. CP1 resembled a slightly squashed sphere 25 feet wide, 20 feet tall, constructed of 400 tons of graphite as the moderator and 6 tons of uranium metal and 40 tons of black $UO_2$ and brown $UO_2$-$U_3O_8$-$UO_3$ uranium oxides as the fuel. The uranium oxides were not present as a 40-ton block, but dispersed in 22,000 egg-like lumps within the reactor, similar to the Columbia reactor.

CP1's graphite moderator slowed down neutrons for effective fission. It had control rods to stop the fission if Fermi's team could not regulate the reaction rate. Wooden boards covered with neutron-absorbing cadmium, that the team could lower into the pile and block the neutrons' path to the fuel, served as control rods. At a height of 57 layers the pile contained sufficient uranium fuel to go critical. On 2 December 1942 CP1 produced the world's first controlled nuclear chain reaction. CP1 initially delivered 0.5 watt of power which increased to a maximum of 200 watts.

**Figure 7**
**Chicago Pile 1 (CPI), World's First Reactor**
Courtesy of Argonne National Laboratory

A second pile, CP2, constructed at nearby Argonne, Illinois, with materials from CP1, operated from March 1943 until 1954 and delivered 100,000 watts, 500 times as much as CP1. CP2 investigated neutron capture probability, control systems, and instrumentation. In November 1943 another pile, the plutonium producing air-cooled graphite X-10 demonstration plant, was operating at the Clinton Engineer Works in Oak Ridge, in the mountains of east Tennessee. In September 1944 a 250 million watt water-cooled uranium reactor went critical in Hanford, Washington.

Most of the uranium ore (mainly a mix of uranium oxides) used in the early pile research, and later in the atomic bombs, came from Belgium. Edgar Sengier (1879-1963), director of the Belgian Union Minière du Haut Katanga in the Belgian Congo (now Zaire), west central Africa, provided 1,200-1,250 tons of high-grade uranium ore (up to 65 percent uranium) from its Shinkolobwe mine. The ore arrived in New York City in September-October 1940, where Sengier secretly stored it in 2,000 steel barrels in a vegetable oil warehouse on Staten Island near the Bayonne Bridge.

Sengier came to New York City in October 1939 to manage Union Minière's office and remained there for the duration of the war. He acted to prevent Germany from obtaining the ore after the British chemist Henry Tizard (1885-1959) at the Department of Scientific and Industrial Research in London told him that German scientists might make an atomic bomb from uranium. General Leslie Groves bought the uranium ore for a dollar a pound, about $2.5 million. From 1942 to 1944 Groves bought 30,000 tons of uranium ore from Sengier.

Evidence also exists that in May 1945 the United States obtained 560 kilograms of low-grade uranium dioxide ($UO_2$) when it captured the German submarine *U-234* in the Atlantic Ocean south of the Grand Banks of Newfoundland. The German submarine with 56 passengers on board was transporting uranium dioxide, aircraft parts, two crated (disassembled) Messerschmitt 262 jets, and other military equipment to Japan. When Germany surrendered on 7 May 1945 Admiral Karl Dönitz (1891-1980), Germany's new president succeeding Hitler, ordered all German submarines including *U-234* to surrender. The US destroyer *Sutton* intercepted *U-234* south of the Grand Banks. *Sutton's* crew members boarded the German submarine and brought it to the Portsmouth Naval Shipyard located near Portsmouth, New Hampshire. At that time, on 14 May, the German uranium ore fell into American hands.

***Construction of the first atomic bombs***

Following Fermi's demonstration of a chain reaction in the CP1 reactor, the center of fission research in March 1943 moved from Chicago to Los Alamos, New Mexico, to a location called Site Y. There, work began on the construction of two kinds of fission bombs, one using U-235 as the fission material and the other using Pu-239.

Groves selected Oppenheimer to direct the bomb's construction. Oppenheimer since early 1942 had worked on fast neutron calculations and neutron diffusion at Berkeley, when Groves appointed him the Manhattan Project's scientific director. Groves based his selection on Oppenheimer's reputation as the United States' leading theoretical physicist, well known for his contributions to relativistic quantum mechanics and his prediction of the positron, which Carl Anderson discovered in 1932. Groves made the decision despite Oppenheimer's known association with American communist party (Communist Party USA) members or fellow travelers. Oppenheimer's former girlfriend, his wife, his brother Frank (1912-85), also a physicist, and Frank's wife, were or had been members of the Communist Party USA. Despite claims of his membership Oppenheimer never joined the Communist Party USA.

In constructing the first atomic bombs at Los Alamos, the purified U-235 and a small amount of Pu-239 came from three uranium isotope separation plants (Y-12, K-25, and S-50) and a Pu-239 demonstration plant (X-10) at Oak Ridge. The Y-12 plant used electromagnetic separation; K-25 was a gaseous diffusion separation plant, and S-50 a thermal diffusion separation plant. Electromagnetic separation and gaseous diffusion provided most of the U-235, a small amount came from Lawrence's cyclotron at Berkeley. Most of the Pu-239 came from Hanford. The other problems to solve included (1) determining the mass of uranium or plutonium, called the critical mass of the nuclear fuel, necessary for a successful bomb and (2) designing a mechanism to detonate or initiate the fission. Herbert Anderson (1914-88), Fermi's long-time associate at Columbia and Chicago, and Joseph McKibben (b. 1914) and his graduate student David Frisch (1918-91) from the University of

Wisconsin, in 1943-44, contributed significantly to establishing the critical mass for fission. Critical mass was a relatively new term that the French physicist Francis Perrin (1901-92), Joliot-Curie's colleague in Paris, introduced in May 1939.

Using a cyclotron from Harvard, two Van de Graaff accelerators from Wisconsin, and a third accelerator from Illinois, McKibben and Frisch calculated a critical mass of 15 kilograms for a U-235 fission bomb and 5-6 kilograms for a Pu-239 fission bomb. Neither element was in plentiful supply. McKibben and Frisch worked with a plutonium speck estimated to weigh only 142 micrograms, but still sufficient to demonstrate the feasibility of plutonium's fission.

The detonation of the fission bomb proved formidable because the uranium and plutonium bombs required different detonating methods. The simpler method, the gun method, which Oppenheimer in July 1942 took charge of, worked for U-235 but was too slow for Pu-239. It combined two sub-critical masses of U-235 in a long steel barrel (tube), using a chemical explosive either to shoot together two sub-critical hemispheres to give a sphere of critical mass or to fire a sub-critical hollow bullet against a sub-critical spike, thereby giving a critical mass of U-235 sufficient to produce a chain reaction. Oppenheimer and the Los Alamos physicists had considerable confidence in the gun method. They knew very well the science of projectiles fired from guns and never tested the gun method prior to dropping *Little Boy*, the U-235 bomb, on Hiroshima.

The gun method failed to provide a suitable mechanism for Pu-239's fission because some of Pu-239's fission-released neutrons reacted with Pu-239 to form Pu-240. Pu-240 underwent a very rapid fission and released too many neutrons before the gun method could produce a critical mass of Pu-239. This slow formation of the critical mass from two sub-critical masses prevented a chain reaction from occurring and instead caused the Pu-239 fission reaction to fizzle.

The second method, the implosion method that Seth Neddermeyer (1907-88), previously at the National Bureau of Standards, developed, followed earlier suggestions by Richard Tolman (1881-1948), a Manhattan Project advisor at Caltech, and Robert Serber (1909-97), previously at Illinois and now at Los Alamos. It required rapidly compressing a large diameter plutonium sphere to a smaller sphere to achieve a critical mass. The Los Alamos scientists tested the implosion method at the Trinity site, an Army Air Force bombing range near Alamogordo 200 miles south of Los Alamos, on 16 July 1945. Serber in 1942 introduced the names, *Thin Man*, *Little Boy,* and *Fat Man*, indicating the shape of the first three atomic bombs.

The Trinity bomb, 1.5 meters (59 inches) in diameter, was a 6.2 kilogram Pu-239 bomb, that Serber in 1942 originally named *Thin Man*, because it employed an early unsuccessful plutonium gun design that the Manhattan Project leaders abandoned on 17 July 1944. They renamed the plutonium bomb *Gadget* and tested the implosion method. Its components included an initiator, a tamper (casing) and chemical explosives. The initiator, polonium (Po-210 and Be-9), provided the neutrons as shown in the equations below. Every alpha particle released produced 30 neutrons.

$${}^{210}_{84}Po \rightarrow {}^{206}_{82}Pb + {}^{4}_{2}He$$

$${}^{9}_{4}Be + {}^{4}_{2}He \rightarrow {}^{8}_{4}Be + {}^{1}_{0}n\ (neutrons) + {}^{4}_{2}He$$

A sub-critical concentration of Pu-239's critical mass of 6-8 kilograms, arranged in the shape of a solid ball, surrounded the initiator, and a 108 kilogram tamper of natural uranium surrounded the Pu-239. The Hanford plant produced the Pu-239. The uranium tamper directed neutrons back into the plutonium once the chain reaction began. It also held the bomb together for the fraction of a second before fission occurred. A ring consisting of 2,400 kilograms (5,400 pounds) of chemical explosives surrounded the tamper. The explosives produced the spherical shock, the shock waves, that imploded the tamper and compressed the plutonium from a grapefruit size to a tennis ball size.

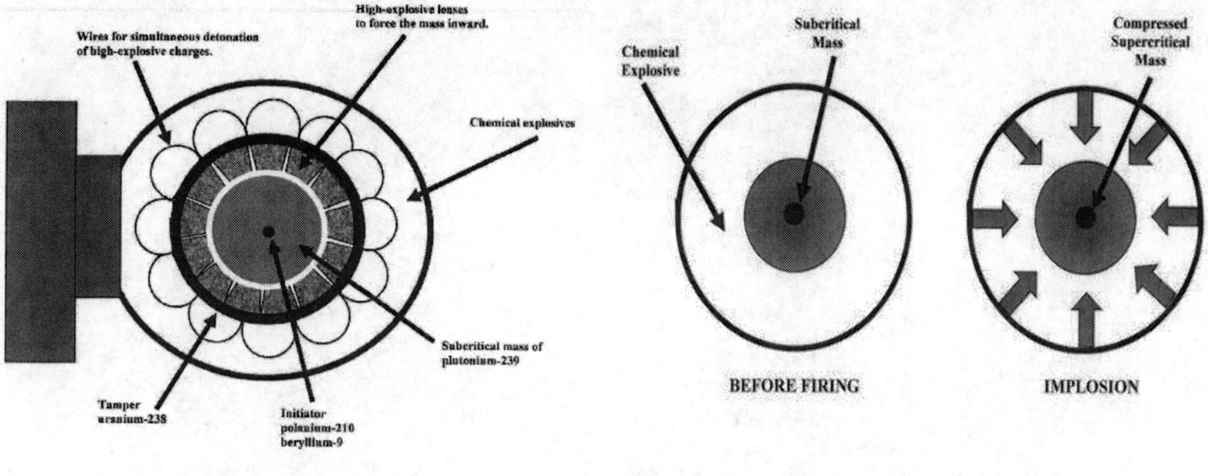


**Figure 8**
**Implosion Method**
Courtesy of the author (left)
Courtesy of the Department of Energy (right)

Neddermeyer led the initial work on the implosion method, but when the program slowed, Oppenheimer in 1944 brought in George Kistiakowsky (1900-82) to replace Neddermeyer. Kistiakowsky, a Harvard explosives expert, built the high explosive lenses that focused or directed inward the shock waves resulting from the detonation of the explosives. The shock waves compressed the plutonium sphere to the center where it achieved critical mass and underwent fission. Other scientists who made significant contributions to the implosion method, while at Los Alamos, included John von Neumann, a Hungarian-born Princeton physicist; Hans Bethe, a German physicist who came to Cornell in 1935; Louis T.E. Thompson (1891-1978), a navy ballistics expert; George Chadwick, (1910-94) a navy ordnance engineer; William Parsons (1901-53), a navy ordnance specialist; and Charles Critchfield (d. 1994) and Edwin L. Rose, ordnance experts.

The implosion method was much more efficient than the gun method, 17 percent compared to 1.4 percent for the gun method. For this reason, implosion became the method used in all later fission bombs. *Gadget* sat atop a 100-foot steel tower, and when detonated it produced a blast equivalent to about 20,000 tons of TNT. The blast completely vaporized the 100-foot tower, melted the sand around it to give radioactive glass, and left a crater 3 meters deep and 330 meters wide. People 100 miles away felt the shock wave, the mushroom cloud ascended 7.5 miles.

The tower holding Gadget. Oppenheimer and Groves examining the tower's remains following the explosion.

**Figure 9**
**Gadget at Trinity Test Site**
Courtesy of the Department of Energy

Three weeks after the Trinity test, on 6 August 1945, the United States dropped the first atomic bomb on Hiroshima, a Japanese military base and industrial center with a population of 320,000. The blast killed 80,000 Japanese and injured another 40,000. Eleven imprisoned American crewmen from two B-24 bombers that the Japanese army shot down during a raid on 28 July 1945 also died in the blast.

The Hiroshima atomic bomb named *Little Boy* was 120 inches long, 28 inches in diameter (315 cm x 71 cm), and weighed 4.5 tons. U-235 underwent fission. Because of U-235's low 1.4 percent efficiency, the gun method of detonation projected a hollow 38-kilogram U-235 bullet down the gun barrel of interior diameter 6.5 inches and length 71 inches (16.5 cm x 180 cm), at a final speed of 984 feet per second (300 meters per second). It struck the 26-kilogram U-235 spike, and tightly packed together enough U–235, 64 kilograms, to produce the 15-kilogram critical mass of U-235 required for fission to occur. The low 1.4 percent efficient fission reaction converted only about 0.6 grams of mass into energy ($E=mc^2$), giving an explosion equivalent to 15,000 tons of TNT.

**Tinian Island 1945, Little Boy loading onto Enola Gay**

Little Boy's design consisted of a gun that fired one mass of uranium-235 at another mass of uranium-235, creating a supercritical mass. The two masses came together in a time shorter than the time between spontaneous fissions. When the two masses are together, the initiator introduces a burst of neutrons; the chain reaction begins and continues until the energy released becomes great enough to blow the bomb apart.

| | |
|---|---|
| **Length:** | 120.0 inches (10 feet / 3.0 meters) |
| **Diameter:** | 28.0 inches (71.1 cm) |
| **Weight:** | 9,700 lbs (4,400 kg) |
| **Yield:** | 15 kilotons (+/- 20%) |

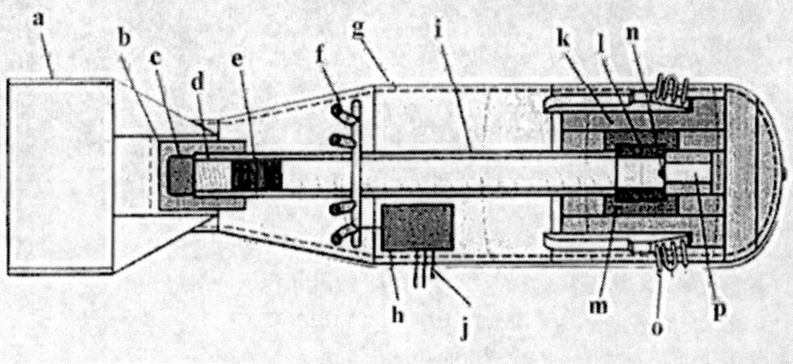


| LITTLE BOY INTERNAL COMPONENTS | | | |
|---|---|---|---|
| a. Box tail fins | b. Steel gun breech assembly | c. Detonator | d. Cordite (conventional) explosives |
| e. U-235 projectile, six rings (26 kg) in a thin steel can | f. Baro sensing ports and manifold | g. Bomb casing wall | h. Arming and fusing equipment |
| i. Gun barrel, steel, 10 cm diameter, 200 cm length | j. Arming wires | k. Tamper assembly, steel | l. U-235 target, two rings (38 kg) |
| m. Tamper/reflector assembly, tungsten carbide | n. Neutron initiator | o. *Archie* fusing radar antennas | p. Recess to eject the boron safety plug (not shown) |

**Figure 10**
**Little Boy**
Courtesy of the US National Archives (left)
Courtesy of Los Alamos National Library (right)
Courtesy of the author (bottom)

Paul Tibbetts (1915-2007), an experienced Army Air Corps colonel, commanded the 12-member crew of the *Enola Gay*, a modified B-29 Superfortress that dropped the atomic bomb on Hiroshima. The B-29, the first intercontinental bomber, flew its initial combat mission in June 1944. Tibbetts named his B-29 after his mother. Two other B-29s went on the mission, one observational and one with instruments to measure the blast's effects. The three B-29s departed from Tinian Island in the Northern Mariana Islands in the early morning and arrived over Hiroshima 6 ½ hours later, at 8:15 am local time. The Japanese did not attempt to intercept the three B-29s because they wanted to conserve their dwindling fuel supplies and aircraft and because the three B-29s did not indicate a major air raid but likely a weather reconnaissance mission. *Little Boy*, dropped from an altitude of 31,000 feet, exploded 43 seconds later at a height of 1,890 feet above its target, the Aloi Bridge in central Hiroshima.

Weight: 10,800 pounds
Length: 128 inches
Diameter: 60 inches
Yield: 21,000 tons of TNT
Fissile materal: Pu-239

**Figure 11**
**Fat Man Implosion-Triggered Fission Bomb**
Courtesy of the US National Archives

Three days later, on 9 August, the United States dropped a plutonium bomb called *Fat Man* on Nagasaki, a large industrial center of 444,000 people. The industrial city of Kokura was the designated target, but clouds and smog prevented its bombing. Kyoto and Nigata were the other targets initially considered. Groves selected the bomb sites and chose cities not previously bombed so that he could assess the destructive effects of the atomic bombs. Henry Stimson (1867-1950), secretary of war, vetoed Kyoto, Japan's ancient capital, cultural-religious center, and location of numerous shrines. To bomb Kyoto would lead to resentment and unnecessarily strained long-term relations between the United States and Japan.

*Fat Man,* 10.75 feet long and 5 feet in diameter (325 cm x 152.4 cm), weighed 5 tons (4,545 kg). It required 5 kilograms of pure plutonium (Pu-239) for detonation by implosion and was equivalent to about 21,000 tons of TNT. The Hanford plant provided the Pu-239. Detonated at an altitude of 1,800 feet like the Hiroshima atomic bomb, *Fat Man* killed an estimated 45,000 Japanese and injured another 40,000. The lower number of deaths occurred because of Nagasaki's hilly terrain compared to Hiroshima's flat terrain. Korean and Chinese forced laborers and approximately 3,200 Japanese-American citizens were other casualties of the two bombings.

Charles Sweeney (1919-2004), an Army Air Corps major, commanded the modified B-29 named *Bockscar (Bock's Car)* that bombed Nagasaki. Sweeney commanded *The Great Artiste*, a B-29 that was the blast measurement instrumentation support plane for the Hiroshima bombing. Frederick Bock (1918-2004), the commander of *Bockscar*, switched places with Sweeney the day of the bombing. Groves had moved the Nagasaki bombing ahead two days because of a poor weather forecast, and because of the change Sweeney's B-29, originally designated to bomb Nagasaki, had not completed its conversion from an instrumentation aircraft back to a bomber capable of transporting an atomic bomb. As occurred at Hiroshima, the Japanese did not intercept the three B-29s because they thought the aircraft were on a reconnaissance mission, not a bombing mission.

After the war ended on 14 August 1945 about 4,000 military personnel went to Hiroshima and Nagasaki to clean up the devastated cities and provide assistance to the survivors. Within a year of the bombings, 40,000 US troops occupied Hiroshima, 27,000 occupied Nagasaki. Many experienced radiation exposure, mainly from gamma rays, because of the dangerous levels of atmospheric radiation.

The Japanese government named those who suffered from radiation exposure but survived the bombings *hibakusha*, meaning *explosion-affected* people. It initially shunned the *hibakusha,* but the Atomic Bomb Survivors Relief Law of 1957 effectively defined the *hibakusha* as those who were within about 2 kilometers of the epicenters of the bombs. A small number, about one percent, also received a special medical allowance and two yearly medical examinations. A second law in 1968 granted some *hibakusha* a small monthly monetary allowance. Then, after more than 25 years of inaction, the Japanese government in 1994 combined the two earlier laws into a new law called the Hibakusha Aid Law that provided a 100,000 yen monthly compensation, about $1,000. The government acknowledged only 2,000 *hibakusha*, whereas most Japanese authoritative sources say the number is 300,000.

Robert Miller (1922-2006), later at the National Cancer Institute in Bethesda, Maryland, in 1954 went to Japan to study the effects of the two bombings. He was chief of pediatrics for the Atomic Bomb Casualty Commission (ABCC) that President Harry Truman (1884-1972) and the US National Academy of Sciences established in November 1946. At the time of its establishment, members of the ABCC had gone to Japan and worked with the Japanese National Research Council which was studying the immediate and long-term effects of radiation on the bombings' survivors. Miller was one of the 150 American scientists, and 1,061 Japanese scientists, who were ABCC members in the 1950s. He and his colleagues, in an important 1972 study and other studies on pregnant Japanese women and their unborn children, found that the increased exposure to radiation led to an increase in mental retardation and smaller head circumference in the children. The effects were evident at four to 17 weeks of gestational age and at exposures of 50 rad in Hiroshima and 150 rad in Nagasaki. The Commission's study ended in 1975, nearly 30 years after its inception. At that time, on 1 April 1975, the Radiation Effects Research Foundation, which the US (DOE) and Japanese governments support financially, succeeded the ABCC. Its main laboratories are in Hiroshima and Nagasaki.

Both Tibbetts and Sweeney believed that dropping the atomic bombs on Hiroshima and Nagasaki, which by 2009 caused 410,000 deaths, saved more lives than they destroyed because the atomic bombs eliminated the need for an invasion of Japan. The invasion, Operation Downfall, was to begin on 1 November 1945.

***Controversy over dropping the atomic bomb***

A major issue that arose at the time of the atomic bomb's construction was the moral justification and necessity of dropping an atomic bomb on Japan once the war in Europe ended in May 1945.

Truman, president only since April 1945 following Roosevelt's death, and his advisors both military and civilian, particularly Henry Stimson, supported dropping the bomb. An eight-member special committee known as the Interim Committee, established in May 1945 and which met for the first time on 9 May 1945, also supported the decision. Its members included Stimson and the scientists Karl Compton (1887-1954), president of MIT; Vannevar Bush, director of the OSRD; and James Conant, (1893-1978) president of Harvard. They argued that the bombing would eliminate the need to invade Japan and thereby save millions of American and Japanese lives. Most of the scientists who worked on the bomb's construction, including Oppenheimer and Fermi, supported its use.

Grove's 1962 memoir, *Now It Can Be Told: The Story of the Manhattan Project*, summed up his reasons for supporting the decision to drop the atomic bomb. Groves attended one or two of the Interim Committee meetings and knew that the Committee recommended using the bomb against a Japanese military target as soon as possible and without prior warning. His most important reason, like the committee's, was that the bombing would save lives by shortening the war. Groves also believed the project's momentum and size made it too big to end. Justification of the $2 billion spent on the project, the enhancement of personal careers and other organizations involved in the project, revenge for the Japanese bombing of Pearl Harbor, and the bomb's use as leverage against the Soviet Union were other reasons. The Soviets had not yet constructed an atomic bomb, but nevertheless, the Allies considered them a serious military threat in Europe and Asia.

Not all the Manhattan scientists supported dropping the atomic bomb on Japanese cities. A poll that Arthur Compton took at the Metallurgical Laboratory in June 1945 indicated that of the 150 scientists 72 percent favored demonstration on a military objective or an uninhabited location. The remaining 28 percent chose either no use of the bomb or its direct use on Japan.

Szilard opposed dropping the atomic bomb for moral reasons. He initiated a petition against dropping it without warning the Japanese of its consequences and offering them the chance to surrender. Sixty-seven scientists at the Clinton Laboratories in Oak Ridge signed the petition before the army stopped its circulation. Szilard's petition of July 1945 remained classified until September 1958.

Despite the long-promoted view that dropping the atomic bomb was essential to ending the war and preventing the deaths resulting from an invasion of Japan, the question remains whether an invasion was even necessary. The Japanese had a 2,000,000 man army; about 1,000 aircraft, many obsolete, and a shortage of fuel and pilots; but no navy. The Allies therefore could have imposed a naval blockade to force the Japanese to surrender. The Allies' air force controlled the skies, enabling them to bomb Japan at will using conventional explosives. Finally, if Truman had guaranteed not to try Emperor Hirohito (Showa, 1901-89) as a war criminal, and the Japanese had a Truman guarantee while negotiating an end to the war with the Soviets in June and July 1945, Japan may have surrendered. Instead, Truman delayed in announcing a decision on the Emperor, the United States bombed Hiroshima, and the Soviets on 8 August declared war on Japan, ending the Japanese-Soviet negotiations. The Nagasaki bombing occurred the next day on 9 August, followed by Japan's unconditional surrender on 14 August. Truman, the US Congress and military never seriously considered these alternatives, other than not executing Hirohito as a war criminal, thereby leaving the necessity of the atomic bomb's dropping a never-debated question, and the controversy unresolved.

The fiftieth anniversary of the Hiroshima and Nagasaki bombings reignited the controversy. The Smithsonian's National Air and Space Museum in 1994 planned to exhibit the *Enola Gay*, highlighting its role in the use of strategic bombing and in ending World War II. The exhibition's accompanying text included debates on the controversy surrounding the bomb's dropping. Several groups, including the American Legion, the Air Force Association, some members of Congress and the media, criticized the exhibit, arguing that the exhibit condemned the United States' wartime conduct. They claimed the exhibit failed to emphasize the atomic bomb's role in ending the Pacific war and focused too much on the horrors of the atomic bomb and the Japanese victims.

Historians countered that the bomb debates were essential to provide context and to indicate the ongoing controversy surrounding the bomb's use. Because of the dispute, the Smithsonian on 30 January 1995 cancelled the exhibit and from 1995 to 1998 displayed only the *Enola Gay's* forward fuselage, vertical stabilizer, two engines, an aileron, and propellers with a brief accompanying text. The *Enola Gay,* now restored completely, went on display in December 2003 at the National Air and Space Museum's Steven F. Udvar-Hazy Center, near Dulles International Airport, in Washington, DC. The exhibit's text contains no controversial information.

*Bockscar* had a less controversial history. The Army Air Force in August 1946 sent it for storage to the Davis-Monthan Army Air Field near Tucson, Arizona. *Bockscar's* title in September 1946 passed to the Air Force Museum (National Museum of the United States Air Force) at the Wright-Patterson Air Force Base near Dayton, Ohio. After its flight to Dayton in September 1961, *Bockscar* has remained on display at the National Museum.

## IV. GERMAN, SOVIET, BRITISH, AND JAPANESE ATOMIC BOMB PROJECTS

### *Germany's atomic bomb project*

In 1937 Hitler wrote that he would not revoke or modify national policies, even for scientists. "If the dismissal of Jewish scientists means the annihilation of contemporary German science, then we shall do without science for a few years." His policies forced many scientists to leave Germany and other German-occupied countries. About 40 of those who remained formed the nucleus of Germany's atomic bomb project. Erich Schumann (1898-1985) and Robert Esau (1884-1955) at Berlin, Carl von Weizsäcker (1912-2007) at Berlin and later Strasbourg, Kurt Diebner (1905-64) director of the Army Ordnance Office Nuclear Testing Station in Gottow near Berlin, Walther Bothe at Heidelberg, Walter Gerlach (1889-1979) at Munich, Wilhelm Groth (1904-77) and Paul Harteck (1908-85) at Hamburg contributed significantly to the German project.

Several of these scientists had met on 29 April 1939 at the Reich Ministry of Education in Berlin to discuss the possibility of a nuclear chain reaction and its military application. Some early work on nuclear fission began at that time at Göttingen when two of the participants, Georg Joos (1894-1959) and Wilhelm Hanle (1901-93), with Reinhold Mannkopff (1894-1978), organized the first Uranium Club. The research ended in August 1939 when they left for military training.

On 1 September 1939 the Army Ordnance Office established the second Uranium Club, and at a meeting that Diebner held in Berlin on 16 September 1939 the German military formally took control of the atomic bomb project. From that time until June 1942 the German atomic bomb project led all other atomic bomb projects. Heisenberg, Diebner's rival, attended meetings of the second Uranium Club. He directed German reactor research first at Leipzig in 1941 and from 1942 at the *Virus Haus* Laboratory at the Kaiser Wilhelm Institute (KWI) in Berlin-Dahlem.

The United States launched the Manhattan Project because it believed the German scientists were racing to build the first atomic bomb. No one in the scientific community doubted the capability of Germany's scientists, hence the great urgency. Germany in 1938-39 had acquired thousands of tons of uranium ore contained in Czechoslovakia's uranium mines, 3,500 tons of Belgian uranium ore in May 1940, the world's only heavy water plant at Vemork near Rjukan in the Telemark region of Norway in April 1940, and the Joliot-Curie's 7 MeV cyclotron at the Collége de France in Paris in June 1940.

German scientists also constructed several experimental reactors. Heisenberg in 1941-42 constructed four reactors while at Leipzig, Diebner constructed three reactors at Gottow, and Heisenberg after moving to the KWI constructed seven reactors through 1944. Allied bombing destroyed the Leipzig nuclear facility in 1943 and the KWI in 1944. Only towards the end of the European war did the United States learn that the race was never close and that the German atomic bomb project began to lose its momentum about the time the American project achieved its initial successes.

Five reasons led to the German failure. First, the German military based its strategy on blitzkrieg and quick victories. Beginning in the summer of 1942, it gave long-term projects, anything requiring more than six months, low priority and directed its attention to other research areas such as the V-1 and V-2 rocket programs. Heisenberg kept reiterating that an atomic bomb remained a long-term project, too expensive, and too uncertain. Germany lacked resources and factories and did not have any large-scale process to separate the uranium isotopes because their thermal diffusion separation process had failed. Manfred von Ardenne (1907-97), in 1943, carried out an electromagnetic separation of uranium's isotopes at his private laboratory in Berlin, but his 60-ton cyclotron provided too little U-235 for the construction of an atomic bomb.

Germany also lacked scientists. Hitler had expelled or run off many essential scientific minds because they were either Jewish or sympathetic to the Jewish cause. Some of these key scientists such as Szilard, Wigner, and Teller, came to the United Sates and worked on the Manhattan Project.

Second, the German anti-Nazi author, Robert Jungk (1913-94), in *Brighter Than a Thousand Suns* (1950), believed that German nuclear scientists did not want to give Hitler an atomic bomb and possible world domination. They deliberately dragged their feet, manipulated data for developing fission, and steered the fission research away from bomb construction. Having seen what had happened to their Jewish colleagues, Jungk argued that the German nuclear scientists decided not to support the Nazi regime. At the same time they knew that refusal to participate in a wartime project guaranteed a ticket to the Russian front, a firing squad, or prison camp. Heisenberg visited his friend and colleague Bohr at Copenhagen in September 1941 and kept stressing the difficulty and uncertainty in building an atomic bomb. He wanted Bohr to take his subtle hint and encourage scientists outside of Germany not to work very enthusiastically on an atomic bomb project. Bohr, however, thought Heisenberg was acting defensively and protecting Germany from future incrimination.

During their 1941 meeting, Heisenberg gave Bohr a drawing of a nuclear reactor which Bohr later brought to the United States and showed to Bethe who recognized immediately that it was a reactor. Bethe, a major contributor to the Manhattan Project in Los Alamos from 1942 to 1945, was an expert on critical mass determinations and implosion. Heisenberg also allowed Bohr to bring to the United States a letter dated 14 April 1941 in which he described the large size and expense of the German atomic bomb project. Fear of the German potential grew in the United States as a result of the Bohr-Heisenberg meeting. Yet, according to Jungk, the German scientists working on the project were more interested in building a nuclear reactor than an atomic bomb and, in 1945, near the end of the war, they built a sub-critical reactor near Haigerloch in Württemburg, a region in southwest Germany between Baden and Bavaria.

Gerlach, who was familiar with the Haigerloch region, suggested building a laboratory in the region's limestone river valley to avoid the Allied bombings that destroyed the Leipzig and KWI facilities. Upon inspecting the Haigerloch region the German scientists discovered a cave dug in the side of an 80-foot cliff that served as the beer storage cellar of an inn (the Schwanen Inn). The beer cellar cave at the bottom of the cliff was originally a railroad tunnel, the Castle Church sat at the top of the cliff. The cave, which the German scientists rented on 29 July 1944, became the location of the Haigerloch nuclear reactor. From August 1944 to the end of February 1945 they moved reactor components, uranium, and heavy water by truck from Berlin to Haigerloch.

The reactor constructed between February and the first week of March 1945, consisted of an aluminum cylinder, 210 cm high and 210 cm in diameter. The aluminum cylinder surrounded a magnesium cylinder. Graphite bricks (5 x 10 x 50 $cm^3$) weighing 10 tons filled the space between the two concentric cylinders and prevented neutrons from escaping. A concrete cylinder about 305 cm in diameter surrounded the aluminum cylinder, and ordinary water for cooling filled the space between the concrete and aluminum. The uranium fuel consisted of 664 uranium metal cubes each 5 cm long attached by 78 chains to a lid that lowered the cubes into the magnesium cylinder. About 1.5 tons of heavy water, poured slowly into the magnesium cylinder containing the uranium cubes, moderated (controlled) the fission reaction. The Haigerloch reactor (B-VIII experiments), which was about 50 percent of the size required to become critical, operated only a few days from late March to mid-April before shutdown.

A 50-member *Alsos* (Army Logistics Support to Other Services) team led by Colonel Boris Pash (1900-95), the overall military leader of *Alsos*, reached Haigerloch on 23 April 1945, only a few weeks after shutdown. *Alsos,* an Allied team and branch of the Manhattan Project, consisted mainly of American and British scientists. Its members went to Germany and Europe to interrogate German scientists and bring them to the United States to continue their research. Pash captured the 80 estimated nuclear scientists at their Haigerloch offices or at their homes in nearby Hechingen where the Kaiser Wilhelm Institute for Physics had relocated in 1943-44. The captured included six of the nine leading nuclear scientists: Hahn, Harteck, Weizsäcker, Heisenberg's former student Erich Bagge (1912-96), Horst Korsching (1912-98), and Karl Wirtz (1910-94). The other three, Heisenberg, Diebner, and Gerlach, left Haigerloch two weeks earlier and surrendered to *Alsos* by the first week of May. Pash's team dismantled the reactor and after coaxing Wirtz recovered the 1.5-2 tons of metallic uranium cubes the German scientists had buried in a nearby plowed field. Pash shipped the reactor and uranium to the United States. After the war the Haigerloch region became part of the French Zone of Germany.

Most Allied authorities and investigators, including Groves and those who went to Germany in 1944-45, doubted Jungk's argument for the German failure. Samuel Goudsmit (1902-78), the scientific leader of *Alsos*, discussed his rejection of Heisenberg's explanation in his 1947 book *Alsos*.

A third reason contributing to Germany's failed atomic bomb project and at variance with Jungk's hypothesis was the German scientists' mistake in choosing a moderator for their reactor. Bothe's alleged critical error in determining the cross section of the carbon atom ended the plan for carbon in the form of graphite as a moderator to slow the neutrons. Bothe tested a graphite sphere of 1.1 meter diameter submerged in water and concluded that a controlled fission reaction required an extraordinarily large quantity of graphite and uranium. Impure graphite containing too much boron probably caused the error. The German scientists, thinking a graphite moderator would not work, turned their attention to heavy water (deuterium oxide) and to another problem.

Fourth, the German scientists lost their only heavy water source. The first attack on the Norsk Hydro heavy water plant in Vemork, a nighttime British glider attack in Nazi-occupied Norway in November 1942, failed. Both gliders and one of the two bombers towing each glider crashed. The 34 paratroopers on the two gliders and the seven-member bomber crew either died in the crashes or the Nazis captured and executed them. In a second successful attack in February 1943, Norwegian paratroopers blew up the Vemork heavy water plant. The Norsk Hydro heavy water plant constructed in 1934 (closed in 1971) had an annual production capacity of 12 tons of heavy water. This was the only German source of the 500 kilograms of heavy water required for the neutron moderator in their experimental reactor then located at the Kaiser Wilhelm Institute at Berlin. The Germans restarted the plant in April, but a November 1943 raid by 143 American B-17 bombers forced the plant's shut down in early 1944. At that time the Germans arranged to ship by ferry to Germany their total supply of heavy water, about 600 kilograms. Norwegian commandos learned of the shipment, and on 20 February 1944 they attached 8.5 kilograms of plastic explosives to the ferry's keel and sank the ferry on Lake Tinnsjø, resulting in the loss of the entire supply. At least 14 Norwegian civilians perished in the sinking.

Finally, Heisenberg after initially estimating in June 1942, that a uranium fission chain reaction required a quantity of enriched U-235 the size of a football or pineapple and telling Hahn this meant a mass of 50 kilograms, revised his mass to two tons of enriched U-235. The uranium enrichment required a large-scale process to separate the uranium isotopes and obtain enough U-235, but no German separation process existed.

When the war in Europe ended with Germany's unconditional surrender on 7 May 1945 (V-E Day, 8 May), Heisenberg and eight other German nuclear scientists, Diebner, Bagge, Gerlach, Wirtz, Korsching, Hahn, Harteck, and Weizsäcker already in April-early May, had surrendered to *Alsos* who sent them to Farm Hall, a large country house in Godmanchester near Cambridge, England. *Alsos* also incarcerated Max von Lau who did not work on the fission process. The ten scientists remained at Farm Hall for six months (3 July 1945 - 3 January 1946) during which time the Allies provided them with good food, sleeping quarters, a tennis court, and a piano for Heisenberg. Unknown to them, a British intelligence team called *Operation Epsilon* in late 1945 taped the German scientists' conversations. These tapes, entitled *The Farm Hall Transcripts,* 270 pages declassified only in February 1992, revealed the astonishment of Heisenberg and the other German scientists upon hearing on 6 August 1945 the BBC's 6:00 pm radio report of the American atomic bomb dropped on Hiroshima.

Heisenberg at first thought the report was a hoax, but when he and the other scientists heard more details they soon accepted its truth. In fact within hours after hearing the BBC report Heisenberg recreated the atomic bomb's probable operation. Eight days later Heisenberg in a lecture to his colleagues described its operation in detail. Whereas earlier he had given various estimates of 50 or 500 or 5,000 kg for the critical mass of enriched U-235 undergoing a fission chain reaction, Heisenberg now estimated a critical mass of 16-210 kg. What Heisenberg failed to accomplish in his several years of research, he produced only days *after* the news of the American success and presented his corrected results to his research team.

Following their release from Farm Hall the German scientists returned to the Allied occupied sectors of Germany and Berlin. The question remains whether Heisenberg truly understood the problem of nuclear fission and decided to delay the German effort, or did his miscalculations mislead him? Heisenberg previously had faced the Nazis' wrath, and they called him a *white Jew*. He knew a refusal to work with the Nazis likely amounted to a death camp sentence.

Heisenberg is, nevertheless, a good example of how the culture and politics of the time affected and influenced science. Because science does not occur in a vacuum, did the culture and politics of the time distort Heisenberg's moral sense? Was he just a good actor, or both? Those who dedicate themselves to scientific careers usually cannot avoid or escape the culture and politics in which they find themselves. They are people of their times and subject to its fortunes or misfortunes.

***Britain's atomic bomb project: tube alloys project***

Rudolf Peierls (1907-95) and Frisch, two German physicists who left Nazi-controlled Germany in the 1930s and were now at Birmingham, in February 1940 made the first valid critical mass calculations for the fission of U-235 in an atomic bomb. Their report confirmed Bohr's hypothesis that U-235 was more likely than U-238 to give a chain reaction and calculated that 10 kg (22 pounds) rather than tons of U-235 would sustain the reaction.

Peierls who came to Cambridge in 1933 received government permission to remain in England at Manchester and in 1937 was a professor at Birmingham when Frisch arrived in 1939. The Frisch-Peierls report led to the establishing of Britain's six-member MAUD committee in April 1940. MAUD, a code name officially adopted in June 1940, was the first name of Bohr's childrens' governess. The MAUD committee proved the feasibility of constructing an atomic bomb. Henry Tizard, at the Department of Scientific and Industrial Research, led MAUD. Other members included Cockcroft and William Penney (1909-91).

Frisch already in January 1939 directly observed fission upon detecting fission fragments in an ionization chamber, and with Meitner coined the word *fission* in their *Nature* article published the next month. In summer 1939 he moved to Birmingham. From 1943 to 1945 Frisch was a member of the British Mission, the team of British and Canadian scientists who worked for the Manhattan Project mainly at Los Alamos and Berkeley in accordance with the 19 August 1943 Quebec Agreement that Churchill and Roosevelt secretly signed in Quebec City.

The cooperative Quebec Agreement required Britain to suspend its fission research until after the war's end. With its signing, the large number of British scientists working on the British atomic bomb project joined the Manhattan Project, 35 at Berkeley, 19 at Los Alamos. Two Canadian scientists, J. Carson Mark (1913-97) a weapons designer and Louis Slotin (1910-46) who died accidentally from a lethal dose of radiation, were at Los Alamos. The Los Alamos group included the British Mission's leader James Chadwick, Peierls, Marcus Oliphant (1901-2000), James Tuck (1910-80) an expert on explosive lenses for implosion, Penney, and Klaus Fuchs (1911-88) the German-born Soviet atomic spy. Frisch headed the critical assembly group. After the war he returned to England to work for Britain's Atomic Energy Research Establishment at Harwell.

Cockcroft, appointed director of the Canadian Atomic Energy Project, in April 1944 converted the factious, low-morale 300-member Anglo-Canadian-French, heavy water research laboratory at the University of Montreal into a harmonious productive team. At the same time, Cockcroft directed the design of the Canadian NRX heavy water reactor at Chalk River, Ontario, that operated from 1947 to 1993. When the Montreal Laboratory closed in 1946 he returned to England to direct the Harwell Atomic Energy Research Establishment.

Penney, an expert on the construction and operation of the atomic bomb, went to Los Alamos in 1944. He observed the Trinity plutonium bomb explosion site, flew in one of the B-29s over Nagasaki on 9 August, and then traveled to Nagasaki to study the damaging effect of the atomic bomb blast wave on the city. Penney was one of the five members of the Los Alamos brain trust with Oppenheimer, Neumann, Parsons, and Norman Ramsey (1915-2011) the Harvard physicist and 1989 Nobel Physics Prize winner. After the war, Penney returned to England to the High Explosive Research Project at Woolwich Arsenal where he led the team responsible for the overall design, development, and fabrication of the implosion method used in Britain's first atomic bomb.

***Japan's atomic bomb project***

During World War II American scientists correctly recognized the possibility of Japanese physicists producing an atomic bomb. Yoshio Nishina (1890-1951), an internationally respected high-energy physicist at the Riken Institute for Physical and Chemical Research, a private research foundation in Tokyo since 1931, constructed a small 25-inch (63.5 cm) cyclotron in 1936 to study the acceleration of charged particles. With Lawrence's assistance, Nishina in 1937 purchased in the United States a 200-ton magnet, then the world's largest, and in 1939 completed construction of a 60-inch (152.4 cm) cyclotron. The magnet cost 110,000 yen (1 yen=50¢), the construction cost, which included the electromagnet (prepared by wiring the magnet), added another 180,000 yen. After some initial problems the cyclotron became operational in 1941. Bunsaku Arakatsu (1890-1973) at Kyoto Imperial University constructed a third cyclotron, a 41-inch (105 cm) cyclotron that accelerated deuterons to an energy of 15 MeV, beginning in June 1941. Masashi Kikuchi built Japan's fourth cyclotron at Osaka Imperial University in 1935-37.

Nishina, by October 1940, convinced Lieutenant-General Takeo Yasuda (1889-1964) and Imperial Japanese Army authorities of the possibility of nuclear fission, and in July 1941 the Japanese army began its atomic bomb project (Ni-Go) with Nishina as its director and Lieutenant Colonel (Major) Tatsusaburo Suzuki (b. 1912), a physicist at Osaka, serving as the army's coordinator. The Imperial Japanese Navy launched a second fission project in 1942. Arakatsu at Kyoto led the project. His team included Hideki Yukawa (1907-81) who in 1949 became the first Japanese to win a Nobel Prize in physics, Sakae Shimizu (1915-2003), Yoshiaki Uemura (d. 1974), and Ryokei Nakane (b. 1920).

Nishina investigated the different methods of separating (enriching) fissionable U-235 from U-238 and concluded that gaseous diffusion of uranium hexafluoride ($UF_6$) provided enough uranium to build a 20-kiloton bomb. Despite Nishina's optimism, Nakane, who worked on the enrichment at Kyoto, believed the Japanese scientists proceeded at a leisurely pace and did not expect to complete their research before the war ended. Recently discovered papers, 23 pages of Imperial Army papers, have confirmed Nakane's long-held position. (Paul) Kazuo Kuroda (1917-2001), a former Imperial University of Tokyo chemistry professor who may have worked on the fission project, smuggled the papers out of Japan in 1949 when he came to the United States. He was at Minnesota and from 1952 in the chemistry department at Arkansas. Shortly after his death in 2001, his Dutch-born widow sent the papers to the Riken Institute.

The Imperial Army papers indicated that Nishina and the Riken physicists underestimated the amount of U-235 required for an atomic bomb and that they did not understand the mechanics of the bomb's construction and operation. Although the Imperial Army supported the bomb's construction, it spent only $1 million on the project, an amount hardly comparable to the United States' $2.5 billion Manhattan Project. The Riken project ended in April 1945 when B-29 bombers destroyed Riken's gaseous diffusion research laboratories. Several unsubstantiated reports dating from 1946 claimed that after the bombing the Japanese moved their nuclear facility to Hungnam (Konan) in North Korea and exploded an atomic device on 12 August 1945. Arakatsu and Kuroda stressed that the Japanese nuclear program remained at the laboratory scale and have denied unequivocally any Konan explosion. No creditable evidence supports the 1946 reports.

Arakatsu's Kyoto bomb program, called F-Go, also received little funding, about $300,000 from the Imperial Japanese Navy by August 1940. It made little progress. Two more recently recovered notebooks (2006) of Shimizu and Uemura support this opinion. Shimizu's 220-page *Laboratory Notes 2* (ca. 1941) describes the development of the cyclotron on which he worked from 1942. Uemura's 75-page *Research Diary* dates from 1941 and contains his research on metals exposed to gamma radiation. The General Headquarters of the Allied Forces seized the two notebooks and other documents during the 1945 to 1952 postwar American occupation of Japan. They are part of the 10,000 uncatalogued Japanese wartime documents stored in the United States Library of Congress in Washington, DC.

Shimizu was a member of the Kyoto investigating team who went to Hiroshima only days after the bombing and confirmed the United States' dropping of an atomic bomb. Later, he confirmed the detonation of a hydrogen bomb (Castle Bravo) at the Bikini Atoll in the Marshall Islands on 1 March 1954. Shimizu analyzed ash collected from a Japanese fishing boat with 23 contaminated fishermen operating near the bomb site.

The Japanese had no domestic sources of uranium dioxide ($UO_2$) and tried to obtain the dioxide from Korea and from other nations, including Czechoslovakia and Germany, its World War II ally. The best evidence of this is the United States' capture of the German submarine *U-234* which had aboard 560 kilograms of low grade uranium dioxide $UO_2$. Later that year in November 1945, after the Japanese surrender, the American occupying army destroyed the four Japanese cyclotrons and dumped them into Tokyo Harbor.

The Japanese scientists failed to reach the practical stage of fission development for three reasons. The Japanese lacked a plentiful supply of uranium, the Japanese government never established a national program that brought together all its nuclear scientists as the United States did with the Manhattan Project, and the Japanese military, while enthusiastically supporting the building of an atomic bomb, never provided sufficient financial support.

***Soviet Russia's atomic bomb project***

Soviet scientists knew of Hahn's discovery of fission and recognized that a fission bomb would release a tremendous amount of energy. Peter Kapitza (1894-1984) in Moscow, the Soviets' leading physicist and 1978 Nobel Prize winner for his work in low-temperature physics, believed that a single atomic bomb could destroy a large city. The Soviets, however, lacked sufficient scientific manpower. They had stretched their natural resources and industrial capacity for the war effort and could not support such a draining project. As a result, their fission project began after those in Germany, Britain, and the United States.

Soviet fission research was largely the product of one man, Igor Kurchatov (1903-60), head of the Nuclear Physics Laboratory at the Physico-Technical Institute in Leningrad (St. Petersburg). Kurchatov left Leningrad for Moscow to become director of fission research in 1943. Even with a country in disarray and shortages of everything, the Soviets on Christmas Eve 1946 activated Europe's first self-sustaining nuclear reactor located at the Kurchatov Institute in Moscow. They detonated their first atomic bomb, a plutonium implosion bomb named *First Lightning* (*Joe-1* in the West), on 29 August 1949, at their Semipalatinsk test site in the northeastern province of East Kazakhstan, Kazakhstan (south central Russia). The first public news of the Soviets' detonation of an atomic bomb came on 23 September 1949.

The Soviets copied exactly the design of *Fat Man,* the Nagasaki implosion bomb, using the information that Fuchs, a member of the German Communist Party since 1932, provided in 1945. Fuchs sketched an accurate drawing of *Fat Man*, giving its measurements and showing the blocks of heavy explosives and heavy uranium tamper. The Soviet implosion bomb had a 22-kiloton yield. Kurchatov remained unknown in the West until 1956, leaving western intelligence agencies, especially the United States, questioning how the *backward* Soviets made such tremendous gains in so little time with so few resources and allegedly inferior science and engineering expertise.

Manhattan Project scientists suspected, however, that Soviet scientists had the expertise to design and construct an atomic bomb and never believed the US atomic bomb monopoly would last for long, only until mid-1953, when they expected the Soviets to detonate an atomic bomb. Nuclear fission information that Soviet spies obtained, probably confirmed what Soviet scientists knew, and they had available uranium from the German atomic bomb project. This made essential the need for the United States to detect and monitor the Soviet project, and any other nation's atomic bomb project, although the Soviets were the only nation the United States perceived as a threat. For this reason, General Dwight Eisenhower (1890-1969) in September 1947 directed the Army Air Force to monitor air samples over the North Pacific Ocean for evidence of radioactive particles and therefore of Soviet atomic bomb tests.

Eisenhower's directive led on 16 September 1947 to the establishment of the Army Air Force's Technical Applications Center (TAC) and its important subdivision, the Long Range Detection Program (LRDP). With the establishment of the Air Force as a separate service two days later, on 18 September 1947, the TAC and the LRDP became part of the Air Force.

The LRDP began monitoring air samples over the North Pacific Ocean, during the next year it detected no enhanced radiation in the atmosphere. Then on 3 September 1949 an LRDP B-29 weather reconnaissance airplane (WB-29), flying for 3 hours at 18,000 feet over Alaska in the North Atlantic Ocean east of Soviet Russia's Kamchatka Peninsula, trapped a very high number of airborne radioactive particles. The WB-29's traps were 9 inch x 22 inch filter papers (bagged filter pairs) attached to the underside of the fuselage. On landing at Eielson Air Force Base in Fairbanks, Alaska, the TAC checked the filters for evidence of enhanced radioactivity.

When necessary, a more thorough analysis took place at Tracerlab Incorporated, a detection laboratory at Berkeley. The scientists at Tracerlab, established in Boston in 1946, included former Manhattan Project scientists who developed the instruments for the analysis. The analysis required dissolving the filters and separating the fission products that included radioactive isotopes of barium, cerium, lead, molybdenum, and zirconium. Measuring each isotope's rate of decay established each isotope's time of creation. If the tests showed that the creation of each isotope occurred at the same time, and no natural explanation existed, a nuclear explosion provided the only explanation.

Photo from the 1 Jan-31 March 1948 History of the 375th Reconnaissance Squadron (Very Long Range) Weather, taken at Fairfield-Suisun Air Force Base, California. Provided by Bernie Barris, Lt. Col. USAF (Ret), Historian and Archivist for the Air Weather Reconnaissance Association. Official US Air Force photograph, courtesy of Air Force Weather History Office.

**Figure 12**
**Bagged Filter Pairs Attached to Fuselage**
Official US Air Force photograph, courtesy of Air Force Weather History Office

The air samples had a normal or expected decay rate count of 50 counts per minute. Tracerlab's first measured count was 85 counts per minute, its second measured count increased to 153 counts per minute. Over the next week American and British military airplanes tracked the radioactivity as high altitude winds blew it over North America across the Atlantic Ocean. Tracerlab's reading on 5 September 1949 reached 1,000 decay ray counts per minute, and its analysis established that the Soviets had detonated an atomic bomb at 12:00 am Greenwich time on 29 August 1949. The detection of both U-238 and Pu-239 in the samples indicated that the Soviets likely had detonated a plutonium bomb with a uranium tamper (shell) and identical to the Nagasaki bomb. The amount of debris indicated a yield of 20 kilotons. Nearly a month later, on 23 September 1949, President Truman announced to the public that the Soviets had detonated an atomic bomb. This ended the US's four-year monopoly on atomic bomb production.

Several factors contributed to the Soviets' quick catch-up. First, Joseph Stalin (1879-1953), the Soviet dictatorial leader, had great motivating skills. After receiving information on the atomic bomb from Fuchs and hearing of the American bombs over Hiroshima and Nagasaki, he ordered his scientists to develop an atomic bomb by 1948. Laventry Beria (1899-1953), the ruthless former chief of the KGB, the Soviet secret police, headed the project. Stalin secretly placed the scientists in Sarov, a town in the Gorki oblast on the Volga River, designated it Arzamas-16, and gave the project top priority in all resource allocations. Second, espionage played a role. British and US intelligence looked for leaks and found several. The best known Soviet spies were the American husband and wife Julius (1918-53) and Ethel Greenglass Rosenberg (1915-53) in New York City and Fuchs at Los Alamos.

## V. ATOMIC SPY SCANDALS

### *Soviet spy scandals of the 1950s*

Fuchs worked on the British atomic bomb project at the Harwell Atomic Research Establishment in 1942 and from late-1943 on the Manhattan Project first at Columbia and from August 1944 at Los Alamos under Bethe before returning to Harwell in 1949. At Los Alamos he investigated plutonium implosion (*Fat Man*) and fusion and reported his results to the Soviets. In 1946-47 he gave the Soviets information on Teller's first proposed hydrogen bomb explosion mechanism that proved unworkable. Teller and Ulam discovered the correct mechanism in 1951, the year after Fuchs's outing and conviction. Fuchs also informed the Soviets about American monthly production of U-235 (100 kg) and plutonium (20 kg). This information enabled the Soviets to determine the number of atomic bombs the United States possessed.

Fuchs's espionage activities remained unknown until early 1950. The Soviet KGB intelligence officer Igor Gouzenko (1919-82) at the Soviet Embassy in Ottawa, Canada, who defected to the west in September 1945, informed the British that Fuchs established contact with the Soviets in August 1941 and spied for them (until late 1949). Fuchs, now back at Harwell, confessed to espionage in January 1950. His testimony implicated other Soviet espionage agents. After a brief 90-minute trial, a British court, in March 1950, sentenced Fuchs to prison for 14 years, the maximum sentence it could impose. Upon release in June 1959, Fuchs left for Dresden, East Germany, where he resumed his scientific work and remained until his death in 1988.

The Army Security Agency established in 1945 and located in Farmington Hall in Farmington, Virginia, broke the Soviet code in 1946, which led to the government's charge that the Rosenbergs were Soviet spies. Julius, a 1939 City College of New York (CCNY) graduate in electrical engineering, worked on radar equipment for the Army Signal Corps at Fort Monmouth, New Jersey, from 1940 to 1945. A member of the Young Communist League since 1936, he confessed in 1950 to membership since 1942 in the Communist Party USA, as were most members of his espionage ring, but not to espionage. A second account that undermined Julius's denial of espionage eventually unfolded in the 1995 declassified *Venona Project*, a World War II American-British espionage operation established in 1943, that decoded Soviet intelligence reports. The Soviets learned of *Venona* in 1945. The declassified information revealed the details of Julius's espionage and the perjured testimony of Ethel's brother David Greenglass (1922-2014) and David's wife Ruth (1925-2008). Their testimony victimized Ethel. Both Greenglasses had joined the Young Communist League in early 1943.

The KBG agent Semyon Semenov (1911-86), assigned to the Soviet Embassy in New York City recruited Julius to spy for the Soviets in September 1942, after Julius in autumn 1941 approached Communist Party USA officials about passing classified documents to the Soviets. Bernard Schuster (b. 1904), the organizational secretary in New York City and a high-ranking member of the Communist Party USA, arranged the 1942 meeting in New York City's Central Park. When Semenov returned to Moscow in 1944, Alexandre Feklisov (1914-2007) at the embassy became Julius's liaison. By this time Julius had passed documents on a proximity fuse for guided missiles and Lockheed's new jet fighter plane, the P-80 Shooting Star.

Julius recruited David and Ruth Greenglass and Morton Sobell (b. 1917). When the army in 1943 drafted David Greenglass, a good machinist and draftsman, and sent him to Oak Ridge and then to Los Alamos in 1944, Julius already knew from Soviet infiltration of the atomic bomb work in Los Alamos. He had Ruth Greenglass speak to her husband about sending atomic bomb information and diagrams. David Greenglass agreed, and for 16 months, from November 1944 to March 1946 when he left the army, Greenglass spied for the Soviets, passing notes and sketches of the implosion type atomic bomb's high explosive lens (a lens mold, lens arrangement, and an implosion experiment) to Julius and to another Soviet spy, Harry Gold.

Sobell, a 1939 CCNY electrical engineering graduate, Julius's classmate, and member of the Communist Party USA, also passed information on radar and artillery (an aircraft gun) to the Soviets. Sobell worked for General Electric in Schenectady, New York, and Reeves Electronics in New York City and in 1939 moved to Washington, DC, to work at the Navy Bureau of Ordnance while spying for the Soviets. When accused of espionage Sobell, his wife, and two children, in 1950 fled to Mexico City intending to go to Europe. Instead Mexican government agents turned Sobell over to the FBI. The United States government charged and in 1951 convicted Sobell and the Rosenbergs with conspiracy to commit espionage. Sobell maintained his innocence but received a 30-year sentence of which he served 17 years and 9 months in Alcatraz and other federal prisons. Following release in 1969, Sobell who currently resides in the Bronx, New York City, continued to claim innocence until September 2008 when he acknowledged passing information to the Soviets. His admission also confirmed Julius Rosenberg's espionage.

Harry Gold (1910-72) in Brooklyn, who studied chemical engineering for two years at Drexel Institute in Philadelphia but likely knew little about nuclear fission, began spying for the Soviets in 1934. Gold was probably a member of the Communist Party USA. He served as a courier, bringing information on the atomic bomb from Fuchs, Gold, and Julius Rosenberg to the Soviet General Consul Anatoli Yakovlev (1913-93) in New York City. Gold, who never married, received a 30-year sentence in 1951. He served 15 years in the Lewisburg, Pennsylvania, Federal Prison and gained his release in 1966. He died in Philadelphia.

When American and British intelligence in early 1950 learned that Fuchs had spied for the Soviets, Fuchs's confession led to Gold's arrest, Gold's arrest led to Greenglass, and Greenglass' arrest led to the Rosenbergs' arrests that same year. Gouzenko at the Soviet Embassy in Ottawa, Canada, provided the initial information that implicated Fuchs, then Gold, and then David Greenglass who acknowledged that Julius was a spy. In order to spare his wife from espionage charges, Greenglass changed his grand jury testimony during the Rosenbergs' 1951 trial and testified that Ethel actively participated in the spy ring. Ruth Greenglass also changed her grand jury testimony to implicate Ethel.

The prosecution had only weak evidence of espionage to use against Ethel, a one-time aspiring singer, actress, secretary, and member of the Young Communist League and in 1942 the Communist Party USA. The Greenglass' perjured testimony, the government's principal evidence, nevertheless enabled the prosecution to obtain a guilty verdict on the charge of conspiracy to commit espionage. These were the Cold War years and Cold War fears. The push for a hydrogen bomb, the rise of Joe McCarthy (1908-57) Republican senator from Wisconsin (1947-57), the Red Scare, the beginning of the Korean War (1950-53), and the atomic spy scandals led to an exaggerated fear of an international and national communist threat that destroyed the lives and careers of many Americans.

The prosecution had a much stronger case against Julius, and he received a guilty verdict on the same charge of conspiracy to commit espionage. Two years later on 19 June 1953 the Rosenbergs died in the electric chair at Sing Sing Correctional Facility in Ossining, New York. They were the only American civilians executed for espionage during the Cold War. Irving Kaufman (1910-92), a Truman-appointed District Court Judge for the Southern District of New York, presided at the Rosenbergs and Sobell trial. Shortly after the Rosenberg's execution, Abel Meeropol (Lewis Allan 1903-86) a well-known Manhattan songwriter and his wife adopted the Rosenberg's two sons, Michael Meeropol (b. 1943) and Robert Meeropol (b. 1947). In 1975, 22 years after their parents' execution, Michael later an economics professor in Springfield, Massachusetts, and Robert later a Springfield, Massachusetts, lawyer, published a book *We Are Your Sons* in which they defended their parents' innocence. The recently released *Venona Papers* (1995), however, undermined Julius's claim of innocence, while failing to present any convincing evidence of Ethel's guilt. The Meeropols have since accepted the *Venona Papers* findings and Sobell's testimony of 2008.

David Greenglass received a 15-year sentence and served 10 years before his release in 1961 from the Federal House of Detention in New York City. His wife received immunity from prosecution in return for his testimony. Upon his release from prison Greenglass and his wife lived in Queens, New York, under an assumed name. He admitted in 1996 to Sam Roberts, urban affairs correspondent for the *New York Times* that he gave false testimony about Ethel at the Rosenbergs' 1951 trial. Universal agreement exists that the atomic bomb information that David Greenglass provided Julius Rosenberg was of little value. The Soviets already had the information Julius obtained and used it to verify their own results or what they acquired from other spies.

Two other defectors attracted much less attention than the Rosenbergs and Fuchs. Bruno Pontecorvo (1910-93), a Fermi protégé, allegedly spied for the Soviets while working at the Harwell Atomic Energy Research Establishment. The British government never charged Pontecorvo with espionage. He defected to the Soviet Union in 1950 continuing his studies on neutrinos and muons at the Institute for Nuclear Research in Dubna where he remained until his death. Theodore (Ted) Hall, born Hotzberg (1925-99), an American physicist, began spying for the Soviets in 1944, the same year he went to Los Alamos. Hall provided the Soviets information on *Fat Man*, the plutonium bomb, independently confirming Fuchs's information. After the war Hall moved to Chicago and completed his graduate studies at the University of Chicago. He left the United States for Cambridge, England, in 1962. The FBI in 1951 questioned Hall about espionage activities, using information from the *Venona Project*, but the US government never prosecuted Hall. The West learned of Hall's career as a spy only in 1997.

### *The Griffin: a little-known Allied spy*

Paul Rosbaud (1896-1963), an Austrian-born metallurgist, was possibly Britain's most valuable World War II spy. As a distinguished science editor for the German publisher Springer Verlag in Berlin, Rosbaud was in constant communication with Europe's leading scientists. As editor, he had the freedom he needed to collect and transmit scientific information to his contacts in Britain. He provided information on Germany's successful rocket research at Peenemünde, established in 1937 in northern Germany on the Baltic coast, and on Germany's unsuccessful atomic bomb research in Berlin-Dahlem, Hamburg, Munich, Haigerloch, and other locations.

Rosbaud spied not for wealth, he left only £500 (about $2,000) at his death, nor for fame, he remains virtually unknown today, but because he had a strong dislike of everything Nazi. Rosbaud became an anglophile in 1915 while a British prisoner of war. Twenty years later, as *The Griffin*, he would establish contact with the British Secret Service and work actively to undermine the Nazi regime.

Rosbaud undertook his first mission in November 1939, a few months before Germany occupied Norway. He sent the British Embassy in Oslo, a copy of Adolf Beck's *Magnesium und seine Legierungen*, 1939 (*Technology of Magnesium and Its Alloys,* English translation, 1940). In this definitive 512-page study, eighteen experts wrote on magnesium, including its use in aerial incendiary bombs, aircraft and rocket construction. Not only did Beck's book leave Germany, but hidden inside its covers was a second valuable communication. The hidden seven-page manuscript, or *Oslo Report*, contained a list of new German weapons such as an aircraft-launched, remote-controlled, rocket-propelled glider, an electronic triggering device that activated a proximity fuse on anti-aircraft shells (with a sample), radar, and the first mention in an intelligence report of the Peenemünde experimental rocket site. No one signed the *Oslo Report*, and British intelligence, fearing it might contain planted information, reacted cautiously to its contents. The *Oslo Report* nevertheless contributed to the British war effort, particularly radar and aerial warfare. In 1961 Rosbaud revealed that he had sent Beck's book to Oslo. His admission received confirmation upon disclosure of the *Oslo Report's* full text in 1985.

Convincing evidence exists, however, that Hans Ferdinand Mayer (1895-1980), a physicist and director of the Siemens Research Laboratory in Berlin, wrote the *Oslo Report* during a two-day business trip to Oslo, on 1 and 2 November 1939, and mailed it to the British Embassy in Oslo. Although the *Oslo Report* became more widely known in 1947, Mayer stipulated that his authorship remain unknown until after his and his wife's death. Mayer came to the United States as part of Operation Paperclip. He taught from 1947 to 1950 at Cornell before returning to Siemens in Munich in 1950. Acknowledgement of his authorship occurred much later in 1989.

One of Rosbaud's reports described the Peenemünde area where, since 1937, Walter Dornberger (1895-1980) and Wernher von Braun (1912-77) tested rockets in violation of the Versailles Treaty's ban on Germany's production of heavy artillery, although the treaty did not specifically mention rockets. In the summer of 1941 Rosbaud learned of increased activity at Peenemünde through his publishing contacts and from colleagues, and he quickly arranged a visit. By August-September 1941 he sent the first reports that included rocket testing schedules, a rough description of the V-2's shape and dimensions, and Peenemünde's coordinates with building locations. British intelligence remained skeptical because Rosbaud's main priority was the atomic bomb and not missiles. It took three more reports, on 22 December 1942, 1 January 1943, and 31 March 1943, all of which followed the V-2's first successful launch on 3 October 1942, before Rosbaud finally succeeded in sparking interest in Peenemünde and led to its nighttime bombing by the Royal Air Force in August 1943 and by the US's Eighth Air Force in July-August 1944. Both raids were largely ineffective and slowed V-2 production by only two months.

The list of Rosbaud's intelligence activities on German nuclear research is long and impressive, some even pre-date the beginning of Germany's atomic bomb project on 29 April 1939. In the period 1942-45 he sent 10-15 messages that were so accurate and comprehensive that the British at times wondered whether Rosbaud was a double agent. Rosbaud helped Lise Meitner escape from Berlin to Stockholm in July 1938 and then had her assist him in passing *book code* messages to the British. Book code messages are selected words in a book that both the sender and receiver possess and which code for the words used in the message. He arranged for the rapid publication of Hahn's paper on fission in *Naturwissenshaften* in January 1939. By the summer of 1943 Rosbaud had informed the

British that the German atomic bomb project was effectively over. In early 1944, after the American bombing of the heavy water plant at Vemork, Norway, he disclosed the Germans' secret plan to ship the entire 600-kilogram heavy water supply to Germany and thereby set in motion the ferryboat sinking with loss of the entire supply on 20 February 1944.

After the war when Heisenberg claimed the German scientists did not build an atomic bomb because they did not want Hitler to have it and implied that they were on a higher moral plane than other scientists, Rosbaud called their claim a myth. The German scientists, he said, never built the bomb because they never knew how. The Heisenberg version appeared in Jungk's *Brighter Than a Thousand Suns* in which Jungk's critics claim he dehumanized the British and American scientists but credited the German scientists with moral fortitude for not building an atomic bomb. Rosbaud left Berlin in late 1945 and resumed his publishing activities in London where he cofounded the Pergamon Press in 1948.

## VI. POST-WAR NUCLEAR DEVELOPMENTS

### *Controlling the nuclear program*

With the end of World War II and the construction of atomic bombs and nuclear reactors a reality, the future control of the American nuclear program emerged as the main issue. Should military control continue or should a new nuclear agency assume control? In 1946, a new civilian agency, the Atomic Energy Commission (AEC), since 1975 the Nuclear Regulatory Commission, assumed control of nuclear energy development.

The first congressional act on nuclear control maintained the status quo and left the military mainly in control. This act, the May-Johnson bill sponsored by two Democrats, Representative Andrew May (1875-1959) from Kentucky and Senator Edwin Johnson (1884-1970) from Colorado, became law on 3 October 1945. Violation of its terms, such as revealing classified nuclear energy documents, called for a 10-year prison sentence and a $100,000 fine. Oppenheimer, Lawrence, and Fermi accepted May-Johnson mainly because its passage would not likely delay nuclear funding and research. Groves supported continuing military (army) control.

A significant number of scientists at the Metallurgical Laboratory and Oak Ridge, members of the Federation of Atomic Scientists established on 1 November 1945, and congressional members disagreed with continuing military control. Harold Urey called May-Johnson "The first totalitarian bill ever written by congress. You can call it a communist bill or a Nazi bill, whichever you think is worse." Because of the considerable opposition, support waned for May-Johnson, and in late 1945 Truman withdrew his initial advocacy of the bill's swift passage. One of the critics, Brien McMahon (1903-52), Democratic senator from Connecticut, in December 1945-January 1946 held hearings on a bill that stipulated civilian control of atomic energy. Groves adamantly opposed McMahon's bill, but the Democratic-controlled Senate on 1 June 1946 and the Democratic-controlled House of Representatives on 20 July approved the bill. Truman signed the McMahon Act, better known as the Atomic Energy Act, on 1 August 1946. The act established a five-member Atomic Energy Commission (AEC) with David Lilienthal (1899-1981) its first director and resulted in the abolishing of the Manhattan Engineer District (Project) on 15 August 1947.

The AEC established a General Advisory Committee (GAC) with Oppenheimer as its first chairman from 1947 to 1952. When Truman on 23 September 1949 announced that the Soviets exploded their first atomic bomb on 29 August 1949, the GAC debated the construction of the more powerful hydrogen bomb. The former Manhattan Project scientists held divided opinions. Oppenheimer, Fermi, and Rabi didn't want a crash program to develop a hydrogen bomb, whereas Teller, the initial and continuing driving force behind the hydrogen bomb, favored a crash program. The Cold War environment enabled Teller to win the debate, and on 31 January 1950 Truman ordered the construction of a hydrogen bomb. The successful development of the Teller-Ulam hydrogen bomb design in 1951, demonstrated convincingly the technological possibility of constructing a hydrogen bomb. The new design convinced the previously skeptical Oppenheimer.

The Cold War fears, of which the push for the hydrogen bomb was only one example, victimized many Americans. The victims included Oppenheimer, who only a decade earlier had achieved fame within the scientific community and the public as scientific leader of the Manhattan Project and father of the atomic bomb. Because of his past associations with communists and his initial opposition in 1949 to the hydrogen bomb's development, Oppenheimer's loyalty to the United States became an issue.

At the month-long series of inquiries on Oppenheimer's conduct, held from 12 April to 6 May 1954, the AEC investigated Oppenheimer's loyalty and his associations with communists. Oppenheimer stumbled badly at the hearings, failing to clarify his viewpoint, and contradicting himself on a number of issues including his past associations. He was his own worst enemy, in addition to letting the AEC stack the deck against him by listening to his supposedly private telephone and other conversations. As a result, the chairman of the AEC from 1953 to 1958, Lewis Strauss (1896-1974), a strong supporter of the hydrogen bomb who disliked Oppenheimer intensely, upon the two-to-one negative vote of the AEC's Personnel Security Board, suspended Oppenheimer's security clearance until resolution of Oppenheimer's loyalty issue. He blamed Oppenheimer for the security lapse that enabled Fuchs to avoid detection at Los Alamos.

Most of the scientific community, including Fermi, Bethe, and Serber as did Bush, Conant, and Groves, supported Oppenheimer. Teller and Lawrence were two of the few scientists who did not. Teller thought Oppenheimer might be a security risk, but he never believed Oppenheimer was a spy. Lawrence said Oppenheimer was trying to wreck research on new weapons.

Roger Robb (1907-85), a Washington, DC, lawyer and counsel for the newly-established AEC, established in 1946, and William Borden (ca. 1920), former executive of Congress's Joint Atomic Energy Committee, believed Oppenheimer had spied for the Soviets from 1939 to1942. McCarthy and his good friend, William Jenner (1908-85), Republican senator from Indiana and well-known communist hunter, also believed Oppenheimer was a Soviet spy. Oppenheimer fell dramatically and remained an outsider to the AEC for nearly ten years. Redemption came in November 1963, when the president Lyndon Johnson awarded Oppenheimer the Fermi medal, the AEC's highest honor, for his contributions to nuclear science and technology. Oppenheimer, a heavy smoker, died from lung cancer four years later.

***Proposal for the first nuclear reactor 1945-47***

On 13 August 1942, the US government formally established the Manhattan Project and began recruiting scientists to construct an atomic bomb. The need for Manhattan scientists continued during the war years, and in the autumn of 1944 Farrington Daniels (1889-1972), a physical chemist at Wisconsin, received a call to report to the Manhattan Project's Metallurgical Laboratory at Chicago. He served as associate director of the chemistry division and as director from July 1945 to mid-1946, replacing Arthur Holly Compton (1892-1962). Daniels had no direct involvement in bomb research. His research focused on reactor principles and the design and construction of a small, helium gas-cooled reactor, 6 feet in diameter x 5 ½ feet high, that operated at a high-temperature of 760° C and generated 5-50 megawatts of power. Daniels hoped his low cost, easily transported reactor would contribute to peacetime energy production in both industrial and nonindustrial countries. A 5-megawatt reactor provided enough energy to power about 2,000 homes (1 megawatt ≡ 500 homes normal energy demand, 200 homes peak energy demand).

In April 1945, with the war's end in sight, Daniels obtained $20,000 from the Manhattan Project to continue his nuclear research. That August Groves arranged for Daniels to present his reactor proposal at a meeting of the Manhattan Project's seven-man Advisory Committee on Research and Development scheduled for Washington on 8-9 March 1946. Daniels sold the committee on his reactor as a new and promising alternative source of energy, and he received $3 million to design, construct, and operate as rapidly as possible an experimental reactor at Oak Ridge, Tennessee.

Daniels and the Oak Ridge Power Pile staff of forty scientists and engineers made a good start on designing a reactor that minimized the danger of overheating and featured underground installation of uranium dicarbide ($UC_2$) fuel rods inserted in graphite blocks as the moderator, and molybdenum-boron control rods. But the Cold War had just begun, and in 1947, only months after the go-ahead decision, the AEC canceled Daniels' reactor proposal. The AEC told Daniels he could continue the research, but it had decided to halt all construction until it studied alternative reactor designs and was ready to proceed with a prototype of the final design. In the end Daniels neither completely designed nor constructed the high-temperature helium gas-cooled reactor. With the Cold War emphasis on the military's nuclear research, his peacetime nuclear reactor gave way to the navy's nuclear program.

The AEC studied three reactor designs, rejecting Daniels' reactor because it was too big for the navy's proposed nuclear powered submarines. The AEC approved the other two designs, both submitted by the navy for submarines: the light water-cooled, highly-enriched U -235 fueled reactor, used in the 320-foot long *Nautilus*, the first nuclear powered submarine (commissioned 30 September 1954) and the sodium-cooled, highly- enriched U-235 fueled reactor used initially in the 353-foot long *Sea Wolf*, the second nuclear- powered submarine (commissioned 30 March 1957). Daniels never intended to put his gas-cooled reactor in a submarine, and so the AEC's decision disappointed but did not surprise him. He correctly surmised that the AEC wanted reactors for the military, not for the production of kilowatts. What happened, Daniels said, was that "Admiral (Hyman) Rickover saw the power vacuum, sailed in with the navy, went off with the atomic submarine and did a grand job. But we let [Soviet] Russia get the credit for building the first peacetime atomic power plant--milking cows and lighting a village--in 1954. We could have had a peacetime atomic power plant in 1948 or '49 at the latest. We were all ready to start in detail and design in '47."

By the time word of the cancellation reached Daniels, he had resigned as director of the Metallurgical Laboratory. He had been commuting between Oak Ridge and Madison, and now, completely frustrated by the AEC's decision, he returned to full-time teaching and research at Wisconsin. The one positive result of this unhappy experience, Daniels said, was the new social awareness he saw arise in the Chicago and Oak Ridge scientists after the United States dropped two atomic bombs on Japan in August 1945. They accepted responsibility for the far-reaching impact they expected their secretive wartime nuclear research to have, and they welcomed the challenge of explaining this new and powerful kind of energy to postwar society.

More than ten years passed before underground installation, uranium carbide fuel rods, graphite moderator, and helium gas coolant, design features Daniels had included in his proposed reactor, became characteristic of the advanced design high-temperature gas-cooled reactors constructed in the 1960s-70s. They included the 40-megawatt Peach Bottom Reactor 1 in Delta, Pennsylvania, that the Philadelphia Electric Company operated from 1966 to 1974, and the 3-megawatt ultra-high temperature reactor experiment (UHTREX) at the Los Alamos Scientific Laboratory that operated from 1959 to about 1971. Britain's 20-megawatt Dragon reactor in Winfrith, England, that the United Kingdom Atomic Energy Authority operated from 1965 to 1976 and West Germany's THTR 300-megawatt reactor that the North Rhine Westphalia State Government and Hochtemperatur-Kernkraftwerk GmbH (HKG) operated from 1985 to 1988, were other high-temperature gas-cooled reactors.

### *First nuclear reactors*

The world's first peacetime nuclear power plant (station) to which Daniels referred was the Soviet Union's state-owned nuclear plant at the science city of Obninsk about 110 km southwest of Moscow. Constructed under Kurchatov's direction in the three-year period January 1951-June 1954, the plant consisted of an enriched uranium fueled (uranium dioxide, $UO_2$) single water-cooled stainless steel reactor with a graphite moderator. It had an electrical generating capacity of 6 megawatts but actually produced 5 megawatts, enough to power 2,000 homes, and for 10 years remained the Soviet Union's only nuclear power plant. The Obninsk plant ceased operation in 2002.

The United Kingdom's first nuclear power plant, the Calder Hall nuclear power station in Sellafield, England, began operation in 1956. It was the first commercial-size nuclear power plant to generate electricity. Each of its pressurized gas-cooled ($CO_2$), graphite-moderated four reactors consumed natural uranium (unenriched either as the metal or $UO_2$) and had an initial capacity of 60 megawatts reduced to 50 megawatts in 1973. The British government decommissioned Calder Hall in 2003.

The US government-developed Shippingport nuclear power plant in Shippingport, Pennsylvania, 25 miles north of Pittsburgh, was the United States' first commercial-size nuclear power plant. The 60-megawatt light water-moderated-and-cooled reactor (LWR) whose design Rickover significantly influenced, began operation in December 1957, following a 32-month construction period. The AEC provided $55 million, Duquesne Light Company provided the remaining $5 million of the $60 million (NRC gave a $72.5 million) total cost. The reactor initially used 4 tons of enriched natural uranium metal and natural $UO_2$. Shippingport replaced its 58-ton original reactor core in 1965, converted to a breeder reactor using a thorium-U-233 core in 1977, and ceased operation in 1982.

Within five years after Shippingport's construction, three more US commercial-size nuclear power plants were in operation. Commonwealth Edison Company in June 1960 (since 2003 Exelon in Chicago) began operation of its Dresden Nuclear Power Plant, a $45 million, 210-megawatt of electricity, boiling water reactor power plant in Dresden, Illinois. The Dresden plant shut down in October 1978. The Yankee Atomic Electric Company in Massachusetts completed construction of a $39 million, 175-megawatt, pressurized water power plant in November 1960 in Rowe, Massachusetts, and began commercial operation in 1961. It shut down in February 1992 with full decommission in August 2007. Consolidated Edison Company's Indian Point Station started operation of its 265-megawatt plant in 1962, using thorium (Th-232) at first and then $UO_2$ as its fuel. Plant operation ceased in October 1974, because of an unsatisfactory core-cooling system. Located on the Hudson River 35 miles north of Manhattan, New York City, the $121 million nuclear plant provided electricity to about one million people (1 megawatt ≡ 3,700 people). Beginning in 1966 and continuing for 20 years the generating capacity of nuclear power plants in the United States increased dramatically from a single plant capacity of about 450 megawatts to 1,200 megawatts.

Most American reactors were of two types: (1) uranium fueled (enriched U-235 as the metal or $UO_2$), light water moderated and cooled (LWR) and (2) plutonium fueled Pu-239 breeder reactor with a liquid sodium coolant. The reactor vessel design and the power plant outlay were similar for both kinds of reactors. The nuclear fuel, uranium or plutonium, underwent several stages of processing and refinement prior to its use in a reactor.

For the uranium fueled light water reactor, the processing and refinement required crushing and grinding the uranium ore which is mainly pitchblende [70-90 percent $UO_3.U_2O_5$ ($U_3O_8$)] or uraninite (largely $UO_2$ with some $UO_3$) in a mill to give a product, called yellowcake. Yellowcake consists naturally of 99.3 percent U-238 and 0.7 percent U-235. U-235 undergoes fission readily, but because of its low concentration in yellowcake and all naturally-occurring uranium compounds, it required enrichment. The enrichment by reaction with nitric acid and hydrogen fluoride gas or fluorine gas converts solid yellowcake to gaseous uranium hexafluoride $UF_6$ and separates by gaseous diffusion the U-235 and U-238 isotopes present in $UF_6$. In gaseous diffusion the $UF_6$ passes through a cascade (series) of diffusion barriers made of nickel powder and electro-deposited nickel mesh that increasingly separate the lighter U-235 isotope from the heavier U-238. Enrichment to 3 percent U-235 is sufficient for operation of a nuclear reactor.

Reaction with hydrogen and steam converts $UF_6$ to uranium dioxide ($UO_2$). Conversion of $UF_6$ to $UF_4$ by reduction with hydrogen [later, hydrolysis using tin (II) chloride $SnCl_2$ as the reducing agent] and reaction with calcium or magnesium gives uranium metal.

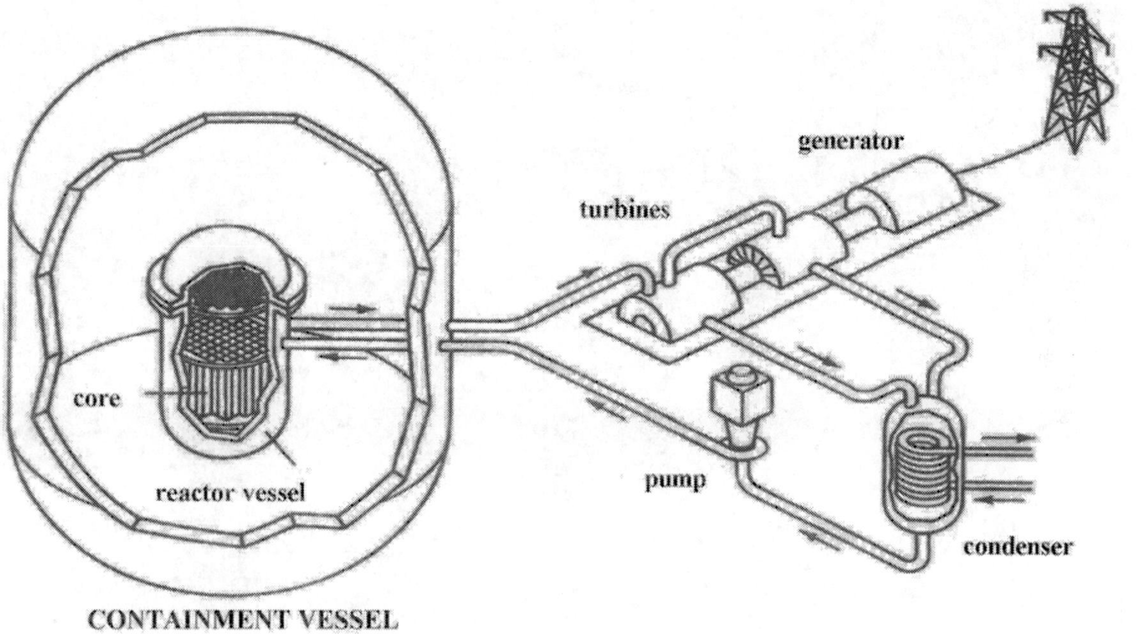


**Figure 13**
**Light Water Reactor**
Courtesy of the US Department of Energy

A breeder reactor converts the U-238 separated by gaseous diffusion to long-lived fissionable Pu-239. U-238, when struck by a neutron released by the polonium-beryllium $^{210}_{84}Po - ^{8}_{4}Be$ initiator, reacts with the neutron to form U-239 which undergoes two rapid beta decay reactions to give Pu-239. The design of a breeder reactor enhances the fission process, so that each neutron striking a Pu-239 atom produces a fission reaction releasing three neutrons, each of which then strikes another Pu-239 atom, releasing nine neutrons, initiating the chain reaction. At the same time, within the breeder reactor some of the neutrons released in the fission of Pu-239 strike U-238 atoms, converting them to Pu-239 which then undergoes fission. In the breeder reactors constructed at Hanford, neutrons emitted by a controlled fission of a U-235 core and absorbed by a blanket of U-238 produced fissionable Pu-239 within the blanket.

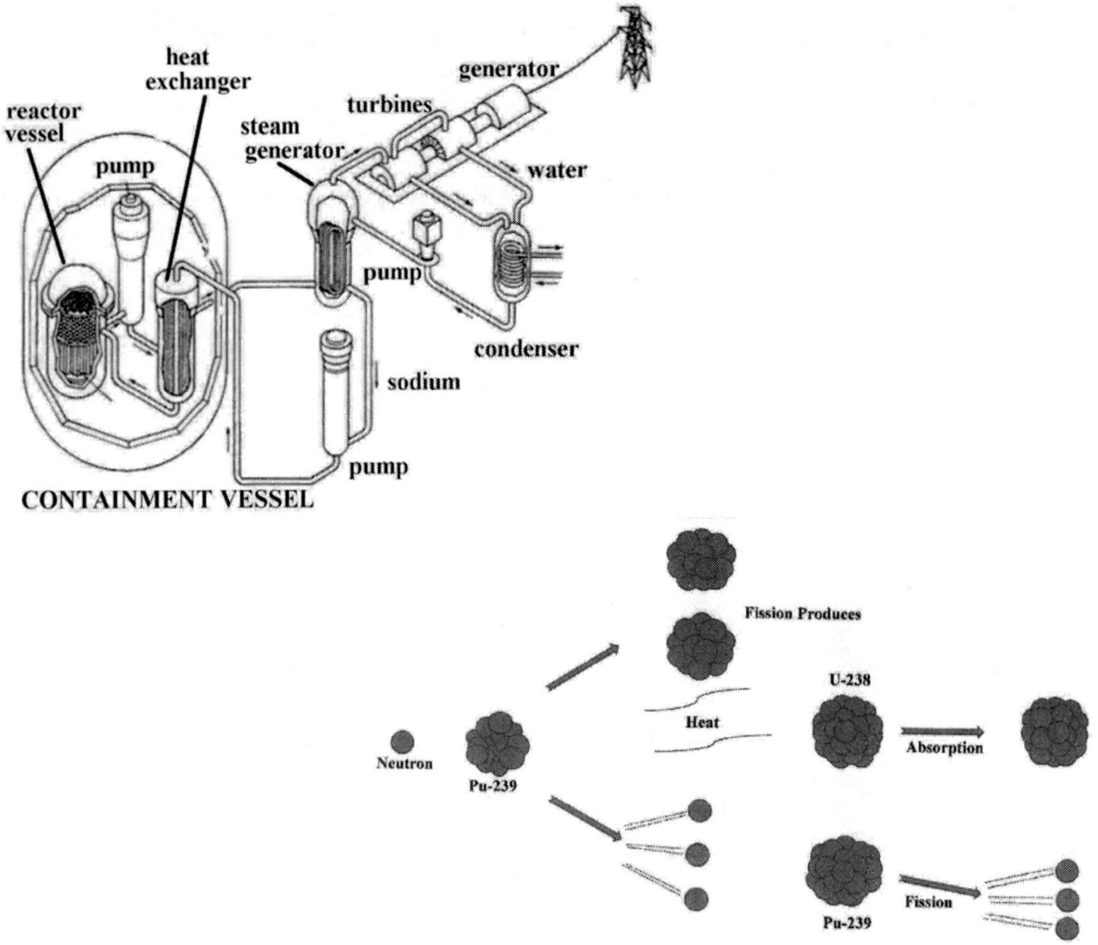


**Figure 14**
**Liquid Metal Fast Breeder Reactor**
Courtesy of the US Department of Energy

| REEDER REACTOR – PLUTONIUM FISSION |
|---|
| $^{238}_{92}U + ^{1}_{0}n \rightarrow ^{239}_{92}U$ |
| $^{239}_{92}U \xrightarrow{half\ life\ =\ 23\ minutes} ^{239}_{93}Np + ^{0}_{-1}e\ (breeding\ reaction)$ |
| $^{239}_{93}Np \xrightarrow{half\ life\ =\ 2.3\ days} ^{239}_{94}Pu + ^{0}_{-1}e\ (breeding\ reaction)$ |
| $^{239}_{94}Pu + ^{1}_{0}n \rightarrow ^{90}_{38}Sr + ^{146}_{56}Ba + 3\ ^{1}_{0}n\ (fission\ reaction)$ |

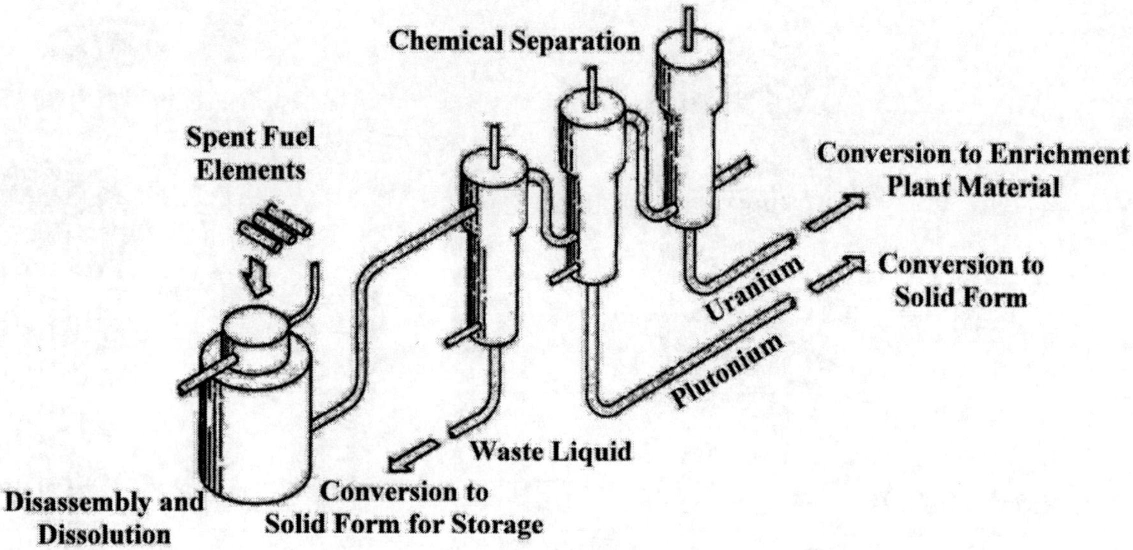


**Figure 15**
**Reactor Fuel Reprocessing**
Courtesy of the US Department of Energy

Because U-235 light water reactors produce some plutonium during their operation, reprocessing and recovering the plutonium is a source for the plutonium fuel used in a breeder reactor. Reprocessing the fuel discharged from a breeder reactor provides a second source of plutonium. This required shaping the plutonium or uranium fuel into small cylinders, each about ½ inch diameter and ¼ inch high, called a fuel pellet, and then stacking and sealing the fuel pellets in 12-foot long hollow tubes. These tubes were the reactor's fuel pins or fuel rods which when bundled in groups of 60 or more gave the fuel assembly. Spacing the fuel rods in an assembly allowed the light water or liquid sodium coolant to flow between them. A stainless steel vessel, called the reactor vessel, enclosed the reactor core (the fuel rods and the control rods). A concrete structure surrounded the reactor vessel.

A group of about 200 fuel assemblies constituted the reactor's core. The core produced the heat in the reactor. Moveable control rods constructed of a substance that absorbed neutrons, such as cadmium steel, and inserted vertically throughout the core, controlled the nuclear chain reaction. Inserting the rods into the core slowed the reaction by absorbing neutrons and produced less heat. Removing the rods from the core accelerated the reaction and produced more heat. The required coolant, usually water in most US reactors, pumped continuously through the core from bottom to top, absorbs the heat produced by the fission. The heat converts some of the water coolant at the top of the reactor vessel to steam which passes out of the top of the reactor vessel to a turbine(s). The steam rotates the turbine blades and their rotation produces an electric current in a generator. The current flows through an electrical grid and provides electricity for a city or cities.

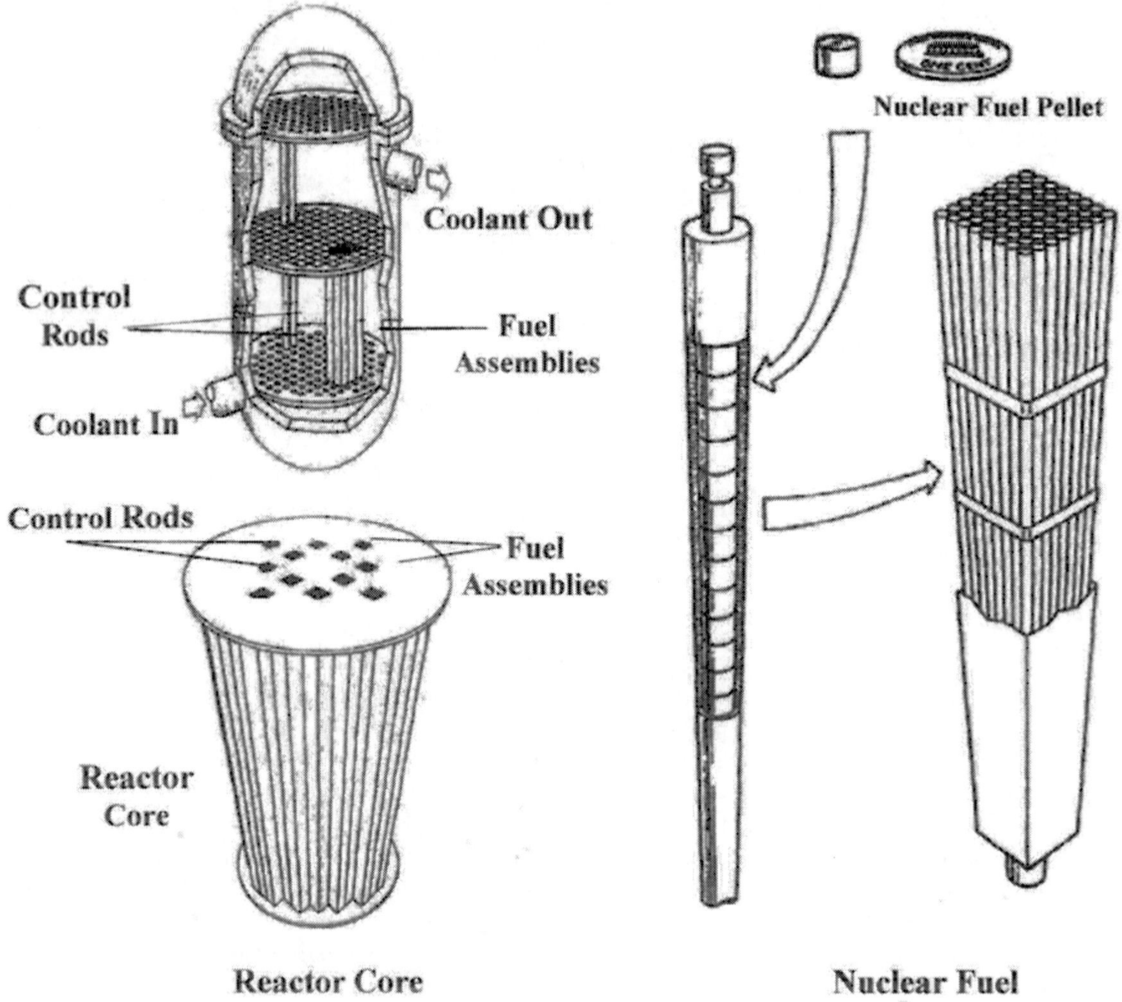


**Figure 16**
**Reactor Vessel, Core, and Fuel**
Courtesy of the US Department of Energy

***Later reactor developments, waste disposal***

From the year of Obninsk's startup in 1954 the number of nuclear power plants increased globally until Chernobyl's nuclear reactor number 4 in Ukraine exploded in April 1986. By 1984 about 340 nuclear plants in 26 countries were producing electricity. In France in 1984 37 nuclear plants generated 59 percent of France's electricity. In Japan 28 nuclear plants generated 23 percent of its electricity. The Soviet Union had 50 nuclear plants operating at the end of 1985. These included 18 large graphite reactors, each with a capacity of 1,000 megawatts or higher. They provided 11 percent of the Soviets' electricity. In the United Kingdom nuclear power plants produced 17 percent of its electricity.

The United States' 94 nuclear power plants generated 14 percent of its electricity, although new plant construction ceased in 1978. New plant orders that totaled 231 through 1974, fell to 15 between 1974 and 1978 and to none after that. Utility companies cancelled orders for more than 100 nuclear plants during the period 1974-82. Two factors proved too much for the nuclear industry to overcome and led to the cessation of new plant construction. Plant costs escalated from about $560 million for a 1,300 megawatt plant in 1971 to $2.4 billion in 1987. The partial meltdown of the core in reactor number 2 (TMI-2) at the Three Mile Island power plant near Middleton, Pennsylvania, in March 1979 permanently shut down the number two reactor. Three Mile Island's reactor number one (TMI-1) commissioned in 1974 with a generating capacity of 802 megawatts has continued to operate. TMI-2 commissioned in 1978 had a generating capacity of 906 megawatts.

In 2010 the 436 commercial nuclear plants operating in 30 countries provided about 15 percent of electricity produced globally. This is 5 percent less than the amount of electricity generated by natural gas 20 percent, hydropower 16.4 percent, and far less than coal-generated electricity 41 percent. The nuclear plants have a total capacity of 372,000 megawatts of electricity.

The United States obtains about 20 percent of its electricity from its 104 nuclear power plants. Several other countries obtain more. Japan (54 nuclear plants), Germany (17 nuclear plants), and Finland (4 nuclear plants) obtain about 25 percent, whereas France with 58 nuclear plants leads all countries with about 75 percent of its electricity from nuclear power plants. In addition to the commercial nuclear power plants, 250 research reactors are operating in 56 countries and another 220 reactors power ships and submarines.

With recent increasing concern over greenhouse gas emissions and global warming, the nuclear power industry has gained increasing support as an alternative to fossil fuel consumption. The three fossil fuels, coal, oil, and natural gas, are the main source of greenhouse gases such as carbon dioxide and methane. Clearly a nuclear power plant will not emit greenhouse gases, but its operation introduced a different set of problems, particularly the problem of nuclear waste disposal. The storage of nuclear waste, the long-lived high-level radioactive isotopes from the reactor core or from nuclear weapons, has remained controversial for decades because some of the isotopes have half-lives ranging from 20,000 to 100,000 years.

Plutonium-239 is the most dangerous of these isotopes because of its extremely high toxicity. Uranium-235 is the second most dangerous isotope present in the reactor core's spent fuel. Nuclear scientists can reprocess and reuse both elements as fission fuel and ideally move their highly radioactive residue along with any other non-reprocessed spent fuel or high-level radioactive waste to a permanent disposal site. The wastes include U-234, Np-237, Pu-238, and Am-241.

The US Nuclear Regulatory Commission, the Department of Energy (DOE), and private industry have proposed several nuclear waste disposal processes, but no agreement exists on which is safest. At Hanford, concrete tanks lined with steel and buried under one foot of concrete and six feet of soil hold the high-level waste. The National Academy of Sciences in 1957, the Congressional Office of Technology Assessment in 1982, and DOE have recommended a deep geologic multiple barrier method: seal the high-level waste in metal or ceramic corrosion-resistant canisters lowered 2,500 feet underground and stored in basalt or salt formations. These natural formations are self-sealing, stable, and have remained water free for millions of years. The deep geologic method remains the most favored disposal process in the United States and other countries.

High-level waste storage tanks. The 30 high-level radioactive waste storage tanks at the Atomic Energy Commission's Savannah River Plant in South Carolina were built of carbon steel, surrounded by concrete encasements two to three feet thick, set about 40 feet in the ground and back-covered with dirt. Each had a capacity of from 750,000 to 1,300,000 gallons. Most tanks were equipped with cooling coil assemblies to remove heat produced by radioactive decay of fission products in the waste. Pictured here are two steel tanks before concrete encasement. They consist of a tank within a tank. The outer liner on the tank in the background is finished. The rounded, inner primary steel liner is visible at the top of the tank in the foreground.

**Figure 17**
**Nuclear Waste Disposal**
Courtesy of the US Department of Energy

Short-lived, low-level radioactive waste, such as plant workers' clothing, tools, filters, cold water pipes and other components of a nuclear power plant, is not a serious problem because the isotopes in the waste have half-lives of only 10-50 years during which they lose most or all of their radioactivity. Nevertheless, some short-lived isotopes are dangerous. These are the fission-produced isotopes that escape into the environment and sometimes pass into the human body. Cesium-137, which has a half-life of 30.07 years, is water soluble and settles in muscle tissue and bones. Strontium-90, half-life of 28.8 years, passes readily into bones and bone marrow. Iodine-131, half-life of 8.02 days, lodges in the thyroid gland. These and other highly radioactive isotopes emit beta particles and gamma rays as they decay. If the emissions strike a DNA molecule in a cell they may cause a mutation in the cell, destroying the cell's ability to function properly. The result is usually a serious disease, most often cancer.

Nuclear power plants routinely, and accidentally, release tritium (H-3 or T, radioactive hydrogen) into the atmosphere or water sources (lakes, rivers) as HT gas or as water HTO. Tritium has a half-life of 12.3 years and emits beta particles as it decays. Like most radioactive substances it can give off harmful radiation for up to 10 half-lives. When it gets into the human body either as the gas or as liquid water, its beta emissions produce harmful cell mutations resulting in cancer. No filtration process exists to remove tritium from a nuclear power plant's liquid and gaseous environmental emissions, nor does the US Nuclear Regulatory Commission require the elimination or cessation of tritium emissions.

## VII. NUCLEAR FUSION

### *First fusion reaction*

In nuclear fusion small nuclei join to form larger nuclei and release energy. In 1920 as a consequence of his mass spectrograph studies, Aston showed that four hydrogen nuclei weighed more than a helium atom, 4 x1.0073 = 4.092 compared to 4.0015. Aston's discovery led Eddington in his 1920 address to the British Association for the Advancement of Science meeting in Cardiff to propose that the hydrogen to helium conversion could account for the Sun's energy. The Sun consists mainly of hydrogen and some helium, and the mass loss occurring in the fusion would result in an energy gain according to Einstein's $E = mc^2$ equation. The fusion reaction had enabled the Sun to shine for about 100 million years.

$$4\left({}_{1}^{1}H\right) \rightarrow {}_{2}^{4}He + 2\left({}_{1}^{0}e\right) + 28.8\,MeV$$

Fourteen years later, in 1934, Rutherford, Harteck, and Oliphant at the Cavendish Laboratory carried out the first fusion reaction. They used a particle accelerator to shoot a deuterium nucleus (H-2 or D) against a deuterium nucleus target to produce helium gas. Hydrogen gas contains only about one part deuterium in 6,500 parts of hydrogen, but the small $1.00 per gram cost of separating deuterium either by the electrolysis of high purity heavy water or fractional distillation of liquid hydrogen compensated for having to work with such a low concentration source of deuterium.

$${}_{1}^{2}H + {}_{1}^{2}H \rightarrow {}_{2}^{4}He + 24\,MeV$$

These earlier developments led to a key advance in understanding fusion. Weiszäcker at Berlin, introduced his carbon-nitrogen-oxygen cycle (CNO) in 1938, and independently in 1939 Bethe at Cornell, introduced his proton-proton (p-p) chain mechanism. Each sequence of reactions showed how fusion of the lighter elements, especially hydrogen but also helium, beryllium, and lithium, produced the high temperatures of the Sun and stars. Bethe calculated the Sun's temperature to within 20 percent of its presently accepted value of 16 million degrees Kelvin.

### *Construction of the first fusion or hydrogen bombs*

The understanding that the fusion reactions occurring in the Sun and stars produced the high temperatures essential to sustaining fusion led to the search for methods of duplicating fusion reactions on Earth. The initial research in the United States began in the Cold War years and focused on constructing a fusion (hydrogen) bomb, or *Super Bomb* for military purposes. Work on fusion reactors for the peacetime production of energy followed within a few years.

Teller, who refused to work on the atomic bomb while at Los Alamos, was the foremost advocate of the hydrogen bomb. His early fusion bomb design done from 1946 to 1950-51, used the high temperature produced by an atomic (fission) bomb to compress deuterium atoms and initiate a fusion of deuterium atoms. His idea was not original. Fermi in autumn 1941 had suggested to Teller that the energy from a fission reaction could initiate a fusion reaction.

$${}_{1}^{2}D + {}_{1}^{2}D \rightarrow {}_{2}^{3}He + {}_{0}^{1}n + 3.25\,MeV$$

In his super bomb design Teller placed the deuterium around a sphere of fissionable material, U-238. He proposed adding a small amount of tritium gas (H-3 or T) to initiate the reaction because the deuterium-tritium fusion had a lower ignition temperature and occurred much faster than the deuterium-deuterium fusion. Only traces of tritium, which decays to helium in about 12.3 years, occur in nature. Teller therefore used the reaction between neutrons released in the deuterium-tritium fusion and lithium-6 deuteride, a colorless - gray solid, to produce a sufficient amount of tritium, with helium the other product.

$$ {}_{1}^{2}D + {}_{1}^{3}T \rightarrow {}_{2}^{4}He + {}_{0}^{1}n + 17.6\ MeV $$

$$ {}_{3}^{6}Li\ (from\ Li-6\ deuteride) + {}_{0}^{1}n \rightarrow {}_{2}^{4}He + {}_{1}^{3}T $$

Stanislaw Ulam (1909-84), a Polish-born physicist at Wisconsin, who since 1943 worked under Teller at Los Alamos, carefully examined Teller's early bomb design. He showed in 1951 that the design would not produce a fusion reaction mainly because Teller underestimated the amount of tritium required. Fuchs, who was at Los Alamos from 1944 to 1946 and participated in the early fusion bomb research, passed this information to the Soviets.

The successful two-stage fusion bomb design came in January 1951. Ulam separated the fusion reaction from the fission reaction. He proposed putting a fission bomb at one end of a fusion bomb casing, the nuclear material (deuterium atoms) for a fusion bomb at the other end, and using the neutron flux, or hydrodynamic shock wave of the exploding and rapidly expanding fission bomb core, to compress the deuterium atoms. A lithium-6 deuteride fuel capsule provided the deuterium atoms.

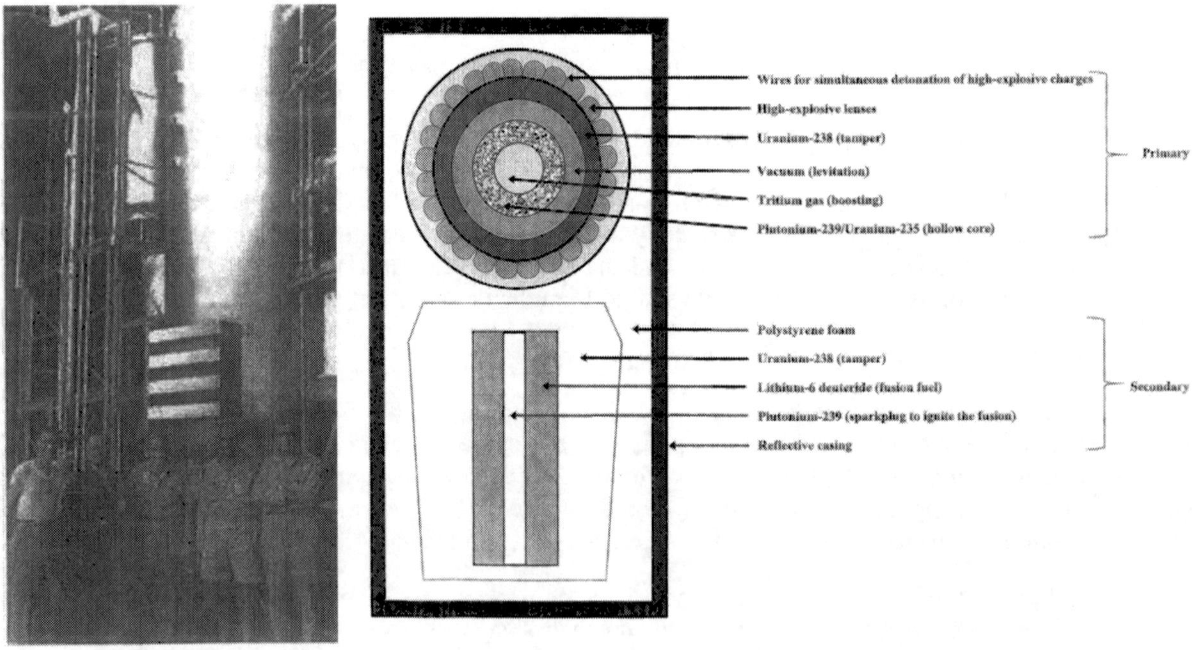


**Figure 18**
**Left: Mike, First Fusion Explosive Device**
**Right: Teller-Ulam Configuration**
Courtesy of Department of Energy/Los Alamos National Laboratory (left)
Courtesy of the author (right)

Teller in February 1951 modified Ulam's hydrodynamic shock mechanism showing that the radiation (x-rays) emitted by the fission more efficiently compressed deuterium. Radiation shock (radiation implosion), from March-April 1951 called the Teller-Ulam configuration after its two rivals, worked. Teller, however, increasingly downplayed Ulam's contribution and claimed himself sole inventor of the hydrogen bomb design. The United States used the Teller-Ulam configuration in *Mike* (*Ivy Mike*, the *Sausage*), its first hydrogen bomb, detonated on 1 November 1952 at Eniwetok in the Marshall Islands 3,000 miles west of Hawaii. *Mike* weighed 82 tons and had a 20-foot high, 10-foot diameter steel casing. Because *Mike* used liquefied deuterium, about 100-200 liters, supplemented with a few grams of a tritium-deuterium gas mixture, it remained in a freezing apparatus at a low temperature (-417° F). *Mike* had an explosive force of 10.4 megatons (10.4 million tons) of conventional explosives (TNT). It was not really a portable bomb but a fusion device.

Sixteen months later the United States detonated a true hydrogen bomb, a dry bomb called *Shrimp* that had a lighter aluminum casing. *Shrimp* measured 179.5 inches long, 53.9 inches wide, weighed 23,500 pounds, and used about 220 kilograms of solid lithium deuteride enriched to 30-40 percent Li-6 deuteride as its source of deuterium atoms. The presence of considerable Li-7 deuteride, which when neutrons struck it, released tritium and significantly boosted the explosive force. The test, *Castle Bravo* that occurred on 1 March 1954 on the Bikini Atoll in the Marshall Islands, had an explosive force of 15 megatons of TNT.

***Soviet fusion bomb***

The Soviets detonated RDS-6, a 400-kiloton yield portable nuclear bomb that the US named *Joe-4*, on 12 August 1953 at their Semipalatinsk test site in northern Kazakhstan. High explosives compressed alternating layers of U-235 as the oxide and the metal, lithium-6 deuteride and lithium-6 tritide provided the deuterium. Analysis of the bomb's debris told American scientists that *Joe-4* was a single-stage boosted atomic bomb and not a two-stage hydrogen bomb like *Mike*. It achieved compression using high explosives rather than radiation. Bethe and the Manhattan scientists considered *Joe-4* a dead end approach. Only 15-20 percent of its yield came from fusion and no scaling up its yield was possible.

Detonation of the Soviets' first hydrogen bomb (RDS-37) followed on 22 November 1955, at their testing site in Kazakhstan. It used lithium-6 deuteride and had an explosive force of 1-1.6 megatons. Kurchatov worked on the Soviets' hydrogen bomb program with the physicists Andrei Sakharov (1921-89) and Vitaly Ginzburg (1916-2009). Sakharov's research led to his independent discovery of the Teller-Ulam fusion bomb configuration which the Soviets called *Sakharov's Third Idea*.

The Soviets' analysis of the radioactive fallout from *Mike* and *Castle Bravo,* rather than the erroneous information they obtained from Fuchs, which predated the 1951 Teller-Ulam configuration, likely aided their construction of a fusion bomb. The large compression that occurred inside *Mike* led to nuclear reactions that produced a high neutron density. Heavy nuclei in the compressed material absorbed these neutrons, resulting in the production of a large number of elements with high atomic numbers, such as einsteinium-255 ($Z = 99$) and fermium-255 ($Z = 100$). The Soviets detected these elements and suspected correctly that extreme compression as in a fusion bomb had produced the high atomic number elements. Their analysis indicated that the reaction in *Mike* occurred in two stages and that the first stage fission produced the compression. Bethe noted that British scientists had conducted an analysis of radioactive fallout from the *Castle Bravo* and Soviet nuclear tests and that their analysis aided the British in developing their own hydrogen bomb design. Ghiorso at Berkeley first detected einsteinium and fermium in the radioactive fallout from *Mike*.

After the Soviets detonated their first hydrogen bomb in 1955, Kurchatov and Sakharov spoke against nuclear bomb testing and for peaceful development of nuclear energy. Kurchatov, however, in 1960 died unexpectedly of a blood clot in his brain while sitting on a park bench in Moscow discussing nuclear research with a colleague. Sakharov, at the Lebedev Physics Institute in Moscow and since 1948 the scientist most responsible for developing the Soviets' hydrogen bomb, became a political activist in the 1970s. He won the Nobel Peace Prize in 1975, which the Soviet government forced him to decline, and publicly opposed the 1979 Soviet invasion of Afghanistan. This led in 1980 to his exile to the closed city of Gorky. Mikhail Gorbachev (b. 1931), the Soviet Union's General Secretary from 1985 to 1991, allowed Sakharov to return to Moscow in 1986.

Ginzburg, a Russian-born Jew, lost his position on the Soviet hydrogen bomb project when Stalin initiated his anti-semitic and anti-intellectual campaign in the early 1950s. Ginzburg spoke openly on the social, religious, and scientific issues in the Soviet Union and, after its breakup in December 1991, in Russia. He was a member of the US National Academy of Sciences and Britain's Royal Society. He shared the 2003 Nobel Prize in physics for his work on superconductors.

Since the development of the American and Soviet hydrogen bombs, Britain in 1957, China in 1967, and France in 1968 detonated hydrogen bombs. India and Pakistan detonated low-yield hydrogen bombs in 1998, and for years Israel has remained deliberately ambiguous about its nuclear arsenal including atomic bombs since the late 1960s and a hydrogen bomb.

***Post World War II developments and cooperation in nuclear fusion reactor design***

Nuclear energy, fission-fusion bombs and reactors, during the Cold War years provoked much, sometimes heated, discussion about its future development internationally. Scientists and statesmen expressed different opinions about the promise of nuclear energy. In the United States, President Eisenhower, in December 1953, gave his famous *Atoms for Peace* talk before the General Assembly of the United Nations in New York City. His talk promoted the peaceful use of nuclear energy and the establishment of an international atomic energy agency. Pauling led protests against nuclear weapons testing in the late 1950s-early 1960s and lost his position at Caltech as a result. Urey strongly condemned the proliferation of nuclear weapons. Oppenheimer in 1954 became a government outcast and security risk because of the nuclear weapons issue, while Teller ardently proposed the US development of nuclear weapons.

Despite the intense Soviet-West competition on nuclear weapons, the first sign of international cooperation on nuclear energy appeared during the Cold War years. The International Atomic Energy Agency, established in Vienna in July 1957, declared as its mission the peaceful use of nuclear energy. Its statute, which 81 nations including the United States and the Soviet Union unanimously approved in October 1956, promoted nuclear verification and security, safety and technology transfer.

Cooperation in the nuclear fusion community began in 1958, at the second Atoms for Peace Conference in Geneva. At the conference the Soviets, British, and Americans agreed to no longer classify fusion research, excluding laser fusion research on weapons which still remains classified. The conference revealed two things. The Soviets did not lead the West in fusion development, and all government-financed fusion programs had followed practically the same path.

*Soviet, European, American, and Japanese nuclear fusion programs*

Four members constituted the emerging nuclear fusion community, and they dominated the development of fusion reactors. They were the Soviet tokamak program at the Kurchatov Institute of Atomic Energy in Moscow; the Joint European Torus program (JET) at Culham Laboratory near Oxford; the American program at Princeton, Lawrence Livermore National Laboratory at Livermore, California, and Los Alamos; and the Japanese program at the Japan Atomic Energy Research Institute (JAERI, since 2005 the Japan Atomic Energy Agency) in Tokai (Naka), 120 km north of Tokyo.

Kurchatov started the Soviet reactor program in 1950. The following year Kurchatov, Sakharov, and the 1958 Nobel Prize winning physicist Igor Tamm (1885-1971) at the institute, invented the tokamak type reactor (Russian acronym for toroidal chamber and magnetic coil). Consisting of a toroidal chamber and magnetic coil that looked like a fat wired-up steel doughnut, the tokamak confined in a near vacuum (pressure =$10^{-3}$ torricelli or less) the ions of gaseous deuterium and tritium, or plasma, with both an external magnetic field and a self-induced magnetic field. Lev Artsimovich (1909-73), director of the institute and a fourth father of the tokamak, with other Soviet physicists continued to improve the design, modifying the geometry of the magnetic field alignment to reduce the presence of heat-stealing impurities. In 1968 Soviet T-3 and TM-3 tokamaks produced temperatures of 10 million degrees Celsius and confined plasmas for 10/1000 second. This was 50 percent better than predicted. One year later plasma confinement time doubled to 20/1000 second. Maximum confinement times in the 2000s increased to a few seconds but still remained too low for power production.

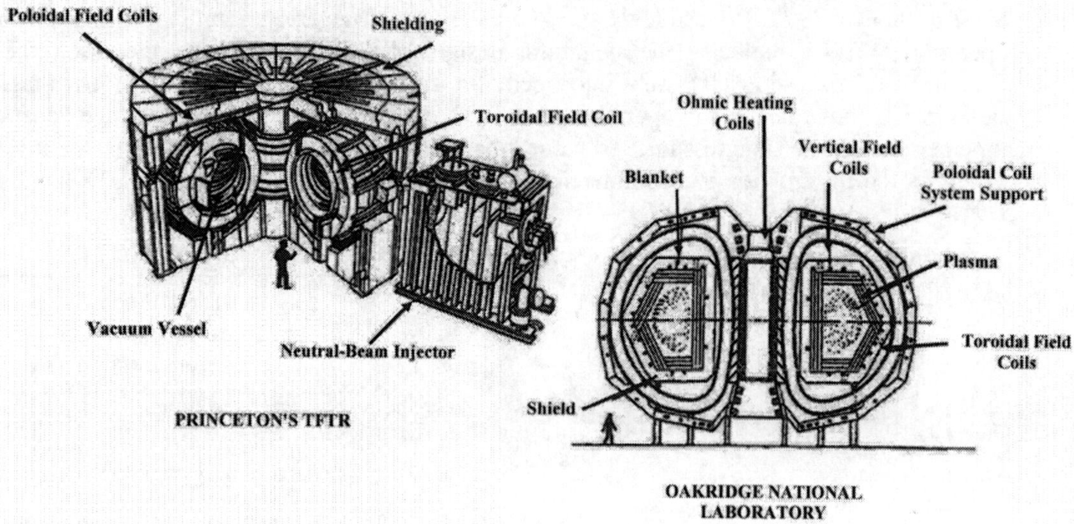


**Figure 19**
**Tokamak Fusion Reactors**
Courtesy of the US Department of Energy

The JET program built on British fusion research that began in 1946, when George Thomson at the Imperial College of Science in London filed a patent on a *pinch* fusion reactor that was essentially the same as the tokamak. The similarities may have resulted from the early 1950s' Soviet spy scandals involving Fuchs and Pontecorvo, both of whom were at Britain's Harwell Atomic Energy Research Establishment. Lack of funding shut down Harwell's Zeta pinch reactor in 1969, leaving Britain without a fusion program until 1977, when Culham Laboratory near Oxford became JET's home. Though located in England, JET had a German director and a French design team director. It became Europe's showpiece of big science, particularly rivaling the American program.

With 5 megamperes current capability, JET's \$1 billion D-shaped tokamak achieved its first plasma by mid-1983. It ignited a mixture of 86 percent deuterium and 14 percent tritium to produce nearly 2 megawatts of power for 2 seconds. In 1997 JET produced 16.1 megawatts of power and sustained 10 megawatts for 0.5 seconds. This remains the highest energy output from any deuterium-tritium fusion reactor. In fusion, tritium releases high-energy (14 MeV) neutrons that cause radiation damage to the reactor's magnets and other components. Because of tritium's radioactivity and its rarity on Earth (and therefore produced), Seaborg in 1991 proposed instead that fusion researchers concentrate on using the much cleaner reaction of helium-3.

$$ {}^{3}_{2}He + {}^{2}_{1}H \rightarrow {}^{4}_{2}He + {}^{1}_{1}H + 18.3\ MeV $$

The United States started its fusion program about the same time as the Soviets. In 1951 Lyman Spitzer (1914-97), at Princeton, invented the stellarator, a figure-eight shaped plasma confinement reactor. Los Alamos researchers in 1952 investigated a circular pinch reactor, and Richard Post (b. 1918) at Livermore in 1952 constructed a meter-long, 5 cm diameter, pyrex glass, straight-line mirror type reactor. As the research continued on achieving higher plasma temperatures and longer confinement times and on the relation between them, the reactors got bigger and costlier. In the early 1950s Spitzer was working with a 3.66 meter (12-foot) long stellarator and expected to spend $1 million on fusion research over the next few years. The AEC spent $1 million in 1951-53, $1.7 million in 1954, $18 million in 1958, and $36 million in the 1960s to construct Princeton's C-Stellarator. Fusion research totaled $24.2 million in 1963, increased to $50 million in 1970, and $300 million in 1977. Princeton's Large Torus (PLT), the world's largest tokamak when it started operation in 1975, replacing the stellarator design, cost $15 million. Its successor, the Tokamak Fusion Test Reactor (TFTR), went into operation in late 1982 with a $1 billion price tag, about the same as JET, and remained in operation until 1997. Fusion research received a huge boost during this time period because Congress in 1980 passed the $20 billion Magnetic Fusion Engineering Act, which called for the construction of a commercial-size demonstration fusion reactor by 2000.

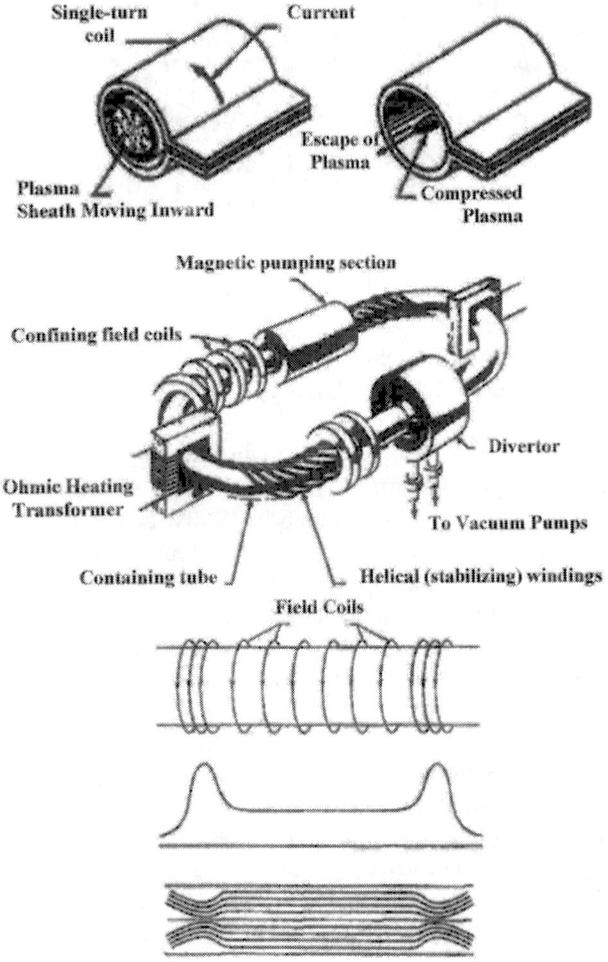


Theta Pinch
(a) shock-heating phase,
(b) quiescent compression phase

Model C Stellarator
The confining field coils and helical windings surround the entire tube.They produce an axial magnetic field and the helical windings provide the twist required to inhibit drift of plasma to the walls. The helical field has a stabilizing effect. The magnetic pumping section heats the plasma and the divertor removes impurities.

Simple Magnetic Mirror System
(a) windings of the field coils closer at ends than in middle
(b) how the strength of the magnetic field varies along the tube
(c) the form of the lines of force, closer and curved at the ends

**Figure 20**
**Pinch, Stellarator, and Mirror Fusion Reactors**
Courtesy of the US Department of Energy

In their nearly 50 years of fusion research, Princeton's investigators saw confinement times and temperatures increase from 1/1000 second and 1 million degrees Celsius to 5/10 second and 300 million degrees Celsius. At these high temperatures the plasma was producing about 1/3 as much energy as it consumed. In December 1993, TFTR became the first fusion reactor to experiment extensively with a 50-50 mix of deuterium and tritium. Its production of 10.7 megawatts in 1994 set a world record and provided sufficient energy for 3,000 homes. In 1995 TFTR achieved a plasma temperature of 510 million degrees Celsius, the highest laboratory temperature ever reached.

Financial support for Post's mirror type reactor at Livermore continued throughout the 1960s-70s because the AEC wanted to see which type of reactor would reach the energy breakeven point first. But in 1984, only four years after Congress passed the $20 billion Magnetic Fusion Engineering Act of 1980, the Reagan administration greatly reduced support for fusion research. DOE funding of fusion research fell to $375 million in 1995 and to less than $250 million in 1996, resulting in the shutdown of the Livermore mirror reactor and closing of Princeton's TFTR. A later stellarator, the National Compact Stellarator Experiment, which began construction in 2003, ended in 2008 because of a funding reduction. Princeton's remaining project is its National Spherical Torus Experiment (NSTX), a tokamak type fusion reactor.

Despite the reduction in funding, fusion research has continued to advance in the United States. A particularly promising project is the laser-stimulated fusion reactor (inertial confinement fusion) at the National Ignition Facility (NIF) constructed from 1997 to 2009 at the Livermore Laboratory. Laser fusion research at Livermore dates from 1962, two years after the laser's invention in 1960. The $4 billion laser-stimulated fusion reactor consists of 192 lasers that focus and compress on a thimble-size target of deuterium and tritium. The resulting heat (100 million degrees Celsius and higher) and pressure produce the fusion. The project, whose formal dedication occurred in June 2009, began ignition experiments in late 2010. Tests on NIF's fusion reactor in September 2013 for the first time produced more energy than the amount of energy the deuterium- tritium fuel absorbed. NIF's shift in focus from nuclear weapons research to fusion tests have suggested the reality of ignition (energy in ≡ energy out) and self-sustaining fusion and a place for nuclear fusion in society's energy future.

Japanese researchers were latecomers to the international fusion community, having established the JAERI semi-government corporation in 1956. In 1961, like many American and European researchers, they began the study of basic plasma physics using small reactors such as the JAERI's JFT-2 in 1972 and the JFT-2a in 1974 which contained the world's first *divertor*. The divertor allowed for the controlled passage of the excess heat and unwanted particles (helium He-4) from the deuterium-tritium plasma. The Arab oil embargo occurred in 1973-74, and because of it and possible future embargos the Japanese in 1976 decided to build a large tokamak. Completed in 1985 at the Naka Fusion Research Establishment, their JT-60 was the last of the giant tokamaks, measuring 15 meters wide, 13 meters high, and weighing 5,000 tons. JT-60 was slightly larger than Princeton's TFTR and had a *divertor.* Political factors have prevented the Japanese from using radioactive tritium as a fuel. Instead JT-60 uses deuterium and in 1998 surpassed energy breakeven with a Q-factor of 1.25, where Q =1 when energy out equals energy in. Despite its late start, in the 1980s Japan had 57 research reactors of the 300 reactors operating globally. As a sign of its increasing prestige in controlled fusion, Kyoto was the site of the 1986 meeting of the International Atomic Energy Agency. Japan's long-range national energy plan includes a commercial fusion reactor by the 2030s.

*Cooperation and competition in fusion*

The 1958 Atoms for Peace conference gave birth to the global fusion community. Within months, fusion researchers were exchanging their results at conferences and traveling to one another's laboratories. Cockcroft visited Kurchatov's institute in November 1958, visits among British, Soviet, and American scientists followed, and in 1961 the British and Soviets signed a five-year pact to ooperate in nuclear fission and fusion research. In 1969, after the Soviets claimed that their tokamaks had achieved a higher fusion temperature and longer plasma confinement time, a British team went to Moscow and confirmed the correctness of their claim.

The cooperative spirit continued, and in late 1986 the governments of the United States, the Soviet Union (in December 1991 Russia), the European Economic Community, and Japan agreed to design a fusion test reactor called the International Thermonuclear Experimental Reactor (ITER). They hoped ITER would prove the engineering feasibility of fusion as an electric power source. Based in Munich, the ITER design team had a $170 million budget and an ambitious three-year time allotment extended to six years and then to 20 years to complete the design. In June 2005 ITER participants agreed to construct the fusion reactor at Cadarache, near Aix-en Provence in southern France. This is the location of France's nuclear research center, Commissariat á l'Energie Atomique. Site preparation at Cadarache began in January 2007.

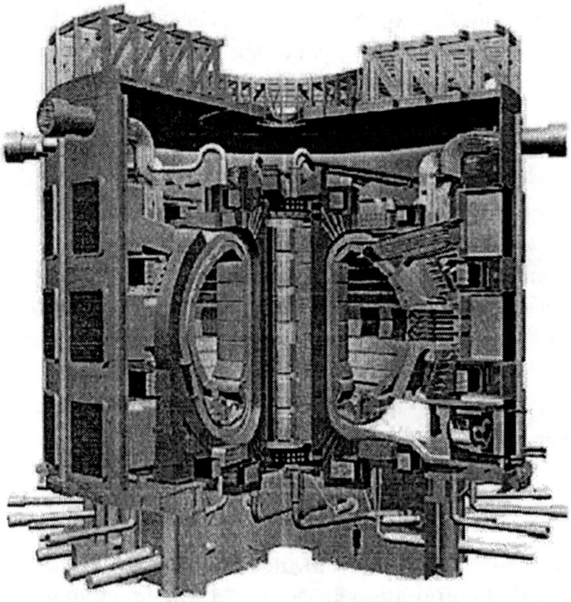

**Figure 21**
**ITER Tokamak Fusion Reactor**
Courtesy of the US Department of Energy

Formal establishment of ITER occurred on 24 October 2007. The estimated total construction and operation cost of the reactor was 10 million euros over 30 years, which ITER's expanded seven members (Russia, China, European Union, United States, India, South Korea, and Japan) agreed to share. The reactor has a designed energy output of 500 megawatts for an energy input of 50 megawatts, a ten-to-one ratio, Q = 10.

The competition in fusion research and development consisted of several kinds. One was national in scope and at first focused on the Cold War competition between the West and the Soviets. After lifting the Soviet-West ban on certain classified research in 1958, competition rose among the Europeans including the British and the Americans, and then later with the Japanese.

Competition also existed between theorists, mainly physicists such as Spitzer and Post who sought a better understanding of the basic plasma physics, and experimentalists such as as Britain's Sebastian Pease (1922-2004), former director of the Culham Laboratory, Paul-Henri Rebut (b.1936), JET's director and in 1992 director of ITER, and Evgeny Velikhov (b.1935) at the Kurchatov Institute. They favored a more empirical approach and wanted to construct bigger reactors hoping to achieve energy breakeven, regardless of whether they understood why. The technology would lead the science. Even here competition surfaced because experimentalists disagreed on the type of fusion reactor to build: magnetic (tokamak or straight line) or laser type. Competition also existed between fission and fusion researchers or what the public perceived as a struggle between dirty and clean types of nuclear energy.

In *Fusion*: *The Search for Endless Energy* published in 1990, Robin Herman, a former health and medical writer for the *Washington Post* and now an assistant dean at the Harvard School of Public Health, asked the question that, for every fusion follower, has remained unanswered more than twenty years later. Why has controlled fusion taken so long to develop and would more money and support have made a difference? Herman argued that the technical problem was and remains complex and that theory has offered little help. If, in Herman's opinion, both theory and experimental application have been and remain the problems, on which should fusion researchers focus? Herman sided with the experimentalists and said bigger reactors such as ITER are necessary for fusion to succeed. She concluded that success seemed inevitable if the fusion researchers build the reactors to explore the plasma. Prominent fusion researchers Pease, Rebut, and Velikhov supported this position.

Not all fusion researchers agreed with the experimentalists on the direction of future research. Spitzer placed the emphasis on the basic physics and said that fusion has not yet demonstrated its technical feasibility. Post, who pioneered the development of the mirror-type straight line fusion reactor, also criticized the increasing emphasis on experiment, as well as the United States' virtual abandonment of all fusion designs other than the tokamak and the weapons-related laser or particle beam programs. Post believes a return to more balanced fusion research is the true path to sustained fusion. Nevertheless, commercial fusion reactors and energy from fusion always seem to be 50 years away.

## VIII. SOLAR ENERGY AS AN APPROPRIATE ALTERNATIVE TECHNOLOGY

### *Introductory*

The United States emerged as an independent nation during the enlightenment, a time period in which the idea of progress dominated the thoughts and actions of many western European societies. Belief in progress quickly took root in the United States, and for about 200 years progress, bolstered by an unfailing faith in the country's abundant natural resources, guided the American outlook. Supplies of cheap energy seemed almost limitless and allowed the United States to shift from wood to coal to petroleum almost without a pause. Several signs warning of possible energy shortages appeared during the 200 years of growth particularly in the 1920s and immediately after World War II. But with the continuing abundance of domestic coal deposits, the discovery of domestic and easily accessible foreign petroleum sources in the 1930s-40s, and the emergence of nuclear energy and its promotion as the solution to everyone's energy problems in the 1950s-60s, the fear of future energy shortages disappeared, or so it seemed. Hardly anyone noticed that by 1962 the United States had become a net petroleum importer. Its imports increased to about one-half its total petroleum consumption at the time of the 1973-74 Arab embargo on petroleum exports to the United States and continued through the Iranian hostage crisis of 1979-81. The era of cheap petroleum, despite some wild price fluctuations in the past thirty or forty years, had ended.

A combination of high costs and environmental concerns about radioactive pollution and waste disposal also precipitated a crisis in the nuclear industry that virtually ended plant construction by the mid-1980s. The nuclear industry experienced severe problems, but the petroleum industry's instability, more than anything else, triggered the energy crisis and initiated the search for alternative energies that included solar, wind, and geothermal energy and synthetic liquid fuels. Neither government nor industry invested heavily in alternative energy programs. Nevertheless, their limited commitment represented public awareness of conventional fossil fuel limits and the potential of alternative energy sources as appropriate technologies that could eliminate or reduce the impact of future energy crises. The public's concern over air pollution resulting from fossil fuel combustion was not yet driving the search for appropriate alternative technologies.

Ernst Friedrich Schumacher (1911-77), an influential German-born economist living in London, England, founded the appropriate technology movement. Schumacher coined the term *intermediate technology* in 1963, and in 1966 he founded the Intermediate Technology Development Group (ITDG). The term *appropriate technology* appeared for the first time in 1968 at the ITDG's Conference on the Further Development in the United Kingdom of Appropriate Technologies for, and their Communication to, Developing Countries, held at St. Cross College, Oxford University. In 1973 Schumacher published his successful book promoting appropriate technology *Small is Beautiful: A Study of Economics as if People Mattered.*

Schumacher in 1974 defined intermediate (appropriate) technology as "the know-how and the equipment designed to help the poor help themselves." He believed it should meet four criteria: (1) create jobs in rural areas, thereby reducing the urban migration characteristic of developing countries; (2) be labor-intensive rather than capital intensive, create jobs with a minimum of investment to reduce high unemployment in nonindustrial countries; (3) be simple with a minimum demand for skilled labor and management; (4) aim its production at local use. This was the scientific outlook that guided many of the post-World War II solar researchers, who focused initially on the development of solar cookers and solar stills, as appropriate alternative technology and later on large-scale solar development and production.

***History of solar energy as appropriate technology***

An early application of solar energy as appropriate alternative technology appeared in the solar-heated homes of fifth-century BC Hellenic Greece. By that time, entire areas of the Grecian landscape had lost their trees, an image Plato dramatically conveyed with his remark that his native Attica resembled the bones of a wasted body. With little wood to produce the charcoal they needed for heating and cooking, Greek city dwellers often oriented their homes toward the southern horizon and sometimes planned entire cities such as Priene (population 4,000), Olynthus (population 2,500), and the island of Delos (est. population 25,000), to give the inhabitants equal access to the winter Sun. They knew that in winter the Sun moved across the southern sky in a low arc, whereas its summer path followed a high arc. By orienting their homes to allow winter's sunlight easy entrance through a southerly portico (a covered porch), Greek city dwellers reduced the use of charcoal-burning braziers, saving both fuel and money.

The Romans and Chinese constructed similar solar-oriented homes and buildings. Particularly noteworthy in Roman architecture was the use of glass or other transparent materials to cover windows and openings. This permitted an essentially one-way passage of heat and light into a room, creating a heat trap or what the Romans called a heliocaminus solar furnace. Solar mirrors and lenses were another early application. Two proposals that described the construction of giant solar mirrors to reflect sunlight on a designated site appeared in the writings of the Oxford Franciscan priest Roger Bacon (1219-92) in the Middle Ages and Leonardo da Vinci (1452-1519) during the Renaissance. Bacon intended his mirrors for the military, da Vinci's mirrors were for industrial and recreational use. Neither actually constructed a mirror.

Several hundred years elapsed before scientific laboratories in the mid-1700s acquired convex lenses large enough (12 inches or more in diameter) to concentrate sufficient sunlight and thereby provided a new and powerful method of producing the high temperatures required in many experiments. Priestley in England and Lavoisier in France used lenses in combustion studies that led to the discovery of oxygen and its role in combustion.

The solar distillation plant that the Swedish engineer Charles (Carlos) Scott Wilson (1831-1901) constructed in 1872 near the harbor of Antofagasta, in Las Salinas, northern Chile, was very likely the world's first large-scale application of solar energy as appropriate technology. Wilson constructed the plant at a large saltpeter (potassium nitrate) mine in northern Chile to supply miners and mules (and miners at a nearby silver mine) with fresh water. The plant occupied 4,722 square meters (51,200 square feet) and contained 64 solar distillation assemblies, each 61 meters long and 1.22 meters wide (200 feet x 4 feet). About 100 solar stills, each consisting of a tent-like glass roof covering a wooden distillation tray made watertight with asphalt, constituted a solar assembly. A windmill pump elevated saline water from nearby wells to a basin on high ground for distribution to the distillation trays. On bright days the Sun's rays passed through the glass roofs, which trapped some of the infrared radiation reflected by the water's surface, and vaporized the saline water that filled the trays. The vapor condensed as fresh water on the air-cooled underside of each roof and ran down into channels on the sides of each still.

The plant was a technological and economic success, producing daily about one pound per square foot or 6,000 gallons (22,600 liters) of fresh water at an efficiency of 35 percent. Despite considerable opposition from water carriers who earned their living by transporting fresh water from the Andes Mountains by mule, the plant continued to operate until construction of the first pipeline that brought water from the Andes to Antofagasta and the mine's exhaustion of saltpeter and closing in 1912. Traces of the foundation, ditches, and many small pieces of glass remain at the plant's site.

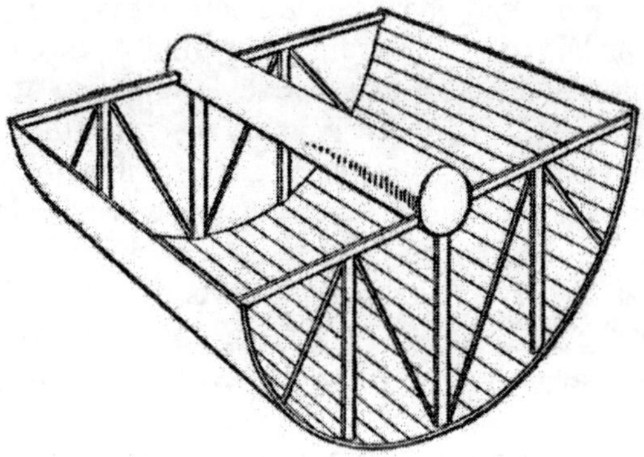

**Figure 22**
**John Ericsson Solar Engine 1872**
Source: Svante A.Arrhenius. *Chemistry in Modern Life*. New York: D. Van Nostrand Company. 1925.

Solar-powered steam engines constructed in the late nineteenth and early twentieth centuries represented another application of solar energy as appropriate technology. They include the 1860s-70s engines of Augustin Mouchot (1825-1912), a mathematics teacher at the Lyceé de Tours, in France; the 1872 engine of John Ericsson (1803-89), engineer, inventor, and warship designer, in New York City; and the 1890s-1910s engines of Frank Shuman (1862-1918), a self-taught engineer and inventor, in Philadelphia. Mouchot's engine consisted of a large copper funnel silvered on the inside. When he turned the funnel in the direction of the Sun, the Sun's rays struck the funnel's inner wall. The funnel reflected the rays against the axis which served as the steam boiler. Mouchot's engine produced 0.11 horsepower at a 3 percent efficiency. Ericsson improved Mouchot's engine by replacing its reflecting funnel with an assembly of small glass ribs silvered on the back and the silver coating then varnished. He mounted the ribs in an iron frame so that they formed a curved reflecting surface (a parabolic arc) and placed a small cylindrical steam boiler, of the same length as the glass ribs, in the focus of the parabola. Ericsson arranged the parabolic reflector perpendicular to the Sun's rays. It produced 0.1 horsepower for each square meter of reflecting surface. Shuman constructed his solar engine in Maadi, a few miles south of Cairo, Egypt, in 1913. It consisted of five horizontally-placed, long, parabolic (cylindrical) mirrors that occupied an area 60 meters long x 4 meters wide. Each mirror focused the Sun's reflected rays on a cast iron pipe (the boiler) of the same length as the mirror, and fixed above the mirror. It looked like a long version of Ericsson's engine pictured above. Shuman's engine generated enough steam to operate a 75-horsepower steam engine at about a 50 percent efficiency. It pumped about 6,000 gallons of water per minute from the Nile River to irrigate nearby cotton fields. With the exception of the small 0.33-0.50 horsepower engines that Charles G. Abbott (1872-1973) at the Smithsonian Institution built in the 1930s, solar engine research made little headway.

Investigations in other areas of appropriate solar technology, such as home heating, were also quite limited with the first significant financial commitment coming from Godfrey Lowell Cabot (1861-1962), an oil, gas, and mineral industrialist, in Boston. His $650,000 grant to MIT in 1938 enabled Hoyt Hottel (1903-98) and MIT engineers to begin two decades of research on solar collectors for home heating.

***A solar pioneer's circuitous route to solar energy applications: the contributions of Farrington Daniels***

Daniels came late to the study of solar energy. His early scientific research interest was chemical kinetics, the study of the speed of a chemical reaction, some of which investigated light-induced or photochemical reactions. His interest in solar energy as appropriate technology began in 1947. With his nuclear reactor program going nowhere, Daniels abandoned nuclear research and, building upon his earlier photochemical kinetics research, began a twenty-five year crusade as the foremost American advocate of solar energy as appropriate technology. At a time when both nuclear and solar energy were in their infancy and when most of the scientific community favored nuclear energy development, Daniels challenged mainstream scientific thinking to argue solar energy's potential in nonindustrial countries and geographic regions that received ample sunlight. Daniels calculated ample sunlight to equal 500 kilocalories per square foot per day. Mexico, India, and many of the Pacific Ocean islands met Daniels's criteria; and so did several industrialized countries, such as Australia and the Sunbelt region of the United States. The success of solar researchers already working around the world with inexpensive equipment convinced Daniels of solar energy's bright future.

Both the theory and practical application of solar science were part of Daniels' early program. Later, in the 1960s, after more than a decade of promoting, organizing, and fund-raising in support of solar energy as appropriate technology, Daniels dealt increasingly with practical applications. He directed his interest to two areas that he believed clearly needed considerable research and development. They were the design and operation of relatively inexpensive, easily constructed, solar cookers and solar stills for use in nonindustrial countries. Each depended on direct use of the Sun's energy, used readily obtained, assembled, or constructed components, and had comparatively low research and development costs.

In developing his solar apparatus, Daniels showed a sensitivity to the traditions and cultures of different societies and recognized that acceptance and use of solar cookers and stills or other solar apparatus required a society to perceive them as appropriate to solving its problems or improving its lifestyle. On-site testing and user construction from locally obtained materials were therefore crucial to successful long-term adoption. Both user and inventor had to value the technology as appropriate, which sometimes was a problem because of unforeseen technological difficulties and unsuspected sociological misunderstandings. Daniels' goal of finding the technology that best fit the situation without causing environmental and social disruptions and problems, reflected the spirit of Schumacher's later, and independently developed, appropriate technology movement that arose in England in 1963.

Because of their rather easy construction and operation, Daniels considered solar cookers a promising technological application in nonindustrial countries where many families had annual incomes of less than $150. Their widespread adoption, he said, was also important environmentally in promoting soil conservation by reducing the destruction of valuable grass, trees, and shrubs for firewood. Researchers around the world had experimented with solar cookers during the 1950s, and in 1955 after having received a $250,000 Rockefeller Foundation grant to establish a solar energy laboratory at Wisconsin, Daniels initiated a study of solar cookers. The Solar Energy Laboratory with its emphasis on research and applications beneficial to nonindustrial countries, received $635,000 over 10 years from the Rockefeller Foundation.

The solar cookers developed during the late 1950s-early 1960s consisted of a reflector constructed by covering a four-foot diameter lightweight circular steel rim with three-inch wide strips of aluminized mylar of 70 percent reflectivity, a rotating parabolic or U-frame that held the reflector and whose two ends were attached to a horizontal rod, and a circular grill that sat at the rod's midpoint. When field tested in Wisconsin, the cooker delivered 500-600 watts to an eight-inch diameter, four-inch high metal cooking kettle and had a water-boiling time of about ten minutes. Construction material, parts, and hand labor costs totaled $25, although Daniels estimated a reduction to $16 if mass produced in lots of 10,000. An electric or gas kitchen range in the United States cost $150-200.

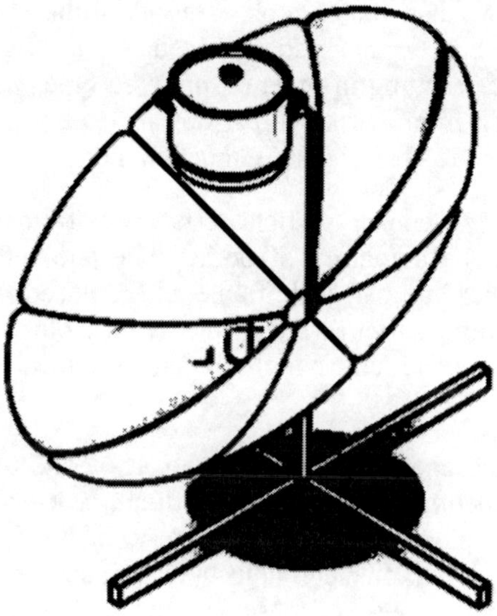

**Figure 23**
**Illustration of a Daniels' Wisconsin Solar Cooker**


To test the physical durability and social acceptance of their cookers over an extended period of time required considerable on-site field work, which the Rockefeller grant made possible. From 1957 through 1963, Solar Laboratory researchers tested over 200 cookers, most of them at the Colorado River Indian Reservation in Arizona and in rural villages near Torreon in northern Mexico and near Oaxaca in southern Mexico. Each of the sites selected had plenty of sunshine but little wood for fuel, and the local inhabitants had low incomes.

The first field tests on forty-one cookers in Arizona in 1958 largely failed for both technological and sociological reasons. The cooker's lightweight rim and framework proved too flimsy to withstand the pounding it took from the roaming farmyard pigs and strong swarming desert winds. Its aluminized mylar reflector produced a blinding glare that forced the Indians to wear sunglasses while cooking and deteriorated easily when exposed to strong sunlight and non-ideal weather conditions. The cookers were inconvenient because they cooked only one item at a time. The construction problems were serious but not insurmountable. A far more difficult challenge came from the reluctance of many Indians to adopt the cookers. If white people did not use them, why should they? The Indians wanted modern ovens not cheap solar cookers.

The tests in Mexico were not appreciably better. Only half the cookers remained in use one year after their introduction. Solar Laboratory researchers had gone to Mexico following the receipt of a letter from a Catholic priest who wrote: "The women of our village go to the hills twelve miles away and bring back wood on their backs and yet we have sunshine all around us. Can't you do something to help us?" The Solar Laboratory researchers complied. Beginning in the winter of 1961, with the assistance of university anthropologists who lived in the village near Torreon, they carried out a two-year test of twenty-six cookers.

Their cookers experienced many of the same problems as in Arizona, namely, weak frames that vibrated in the wind and aluminized mylar reflectors that failed to withstand continuous exposure to the Sun. Many of the villagers abandoned the cookers because kerosene (32¢ per liter) and sometimes free wood were available. An aversion to cook outdoors in the Sun, cloudiness or the appearance of clouds once the cooking had begun also impeded usage. Daniels encouraged complete local construction of the cookers, resulting in several improved types having clay or earthenware cooking containers that did not crack from thermal expansion as expected because they diffused the focused sunlight as well as factory-made steel or aluminum containers.

The failure of the lightweight cooker in field tests away from the Solar Energy Laboratory was disappointing. To improve the durability of both frame and reflector and the reflector's lifetime, Daniels constructed a sturdier iron parabolic frame and replaced the aluminized mylar reflector with about 1,800 one-inch square glass mirrors cemented to a cloth layer stretched over the frame. The design was more complicated, but still simple enough for local construction from easily available materials.

To ensure that the improved cooker worked as well at on-site tests as it did in a more controlled environment, Daniels travelled to the village of Teotitlan, near Oaxaca, in southern Mexico, in April 1962. There he joined two university anthropologists to begin field construction and testing of two solar cookers, each with a 48-inch diameter adjustable parabolic reflector. The cookers Daniels tested cost $9-$16, about equal to two weeks of a Mexican worker's salary, and generated the equivalent of 600 watts of electric heat.

The redesigned cooker worked well, boiling a quart of water in ten minutes, but once again the villagers stopped using the cookers after only a few months because no one wanted to cook outdoors in the Sun.

The years of field testing mainly in Mexico, indicated that the acceptance of solar cookers depended considerably on the users' cooking and eating habits. The cookers worked well for boiling and stewing in covered containers, satisfactorily for frying (tortillas, meat, eggs), and poorly for baking. Because solar cooking was not possible in the early morning or in the evening Daniels believed the cookers would find their greatest acceptance in cultures that customarily ate a heavy noon meal, which typically occurred in most of Central and South America.

Widespread adoption of solar cooking in Mexican villages or elsewhere, according to Daniels, awaited the design of a small portable heat storage cooker that combined the best features of the reflector and hot box types. He proposed a three-stage cooker in which the operator heated the cooker (a reflector and a one-liter cylinder of iron or aluminum) in the Sun for two or three hours to capture sufficient heat energy, transferred the cylinder to an insulated box indoors to retain the energy, and cooked with the one kilowatt-hour of stored energy a few hours later. Daniels acknowledged the difficulty, not the impossibility, of constructing a heat storage cooker that delivered at least one kilowatt-hour at 200° C (392° F). Yet ten years after the field tests in Mexico, no dramatic advances had occurred in solar cooking.

Compared to solar cooking, solar distillation experienced considerable progress in the 1960s. About twenty large solar stills, each having a basin (area) exceeding 1,000 square feet, such as those at Coober Pedy in Australia, Mahdia in Tunisia, Patmos and Kimolos in Greece, Las Marinas in Spain, and Daytona Beach in Florida, were producing daily about 20,000 gallons of fresh water from salt water.

By this time Daniels had begun testing inexpensive and easily constructed solar stills mainly at the Solar Energy Laboratory. Each still had only three or four components. One of his early designs consisted of a wooden rectangular tray 12 feet long, 2.5 feet wide, 1-2 inches deep, arranged on the ground with its long side directed east-west and fitted on each long side with a sloping wooden trough about 2 inches deep. A large, thin (1/32 inch) sheet of black butyl rubber draped over the tray and troughs and provided a watertight floor. A gently sloping A-shaped roof, constructed by covering a wooden frame and ridgepole with sun-resistant, water-wettable, transparent or clear plastic (mechanically scratched tedlar or mylar), sat on the tray. Black polyethylene bags filled with sand secured the bottom edges of the plastic roof covering. Solar radiation passing through the roof heated the one inch of seawater or brackish water added to the tray to temperatures greater than 65° C (150° F). A rubber tube led the distilled water from the troughs into bottles set below ground level.

In a second type of A-shaped roof still, Daniels replaced the rubber-draped wooden tray and troughs with a two-inch thick concrete platform and concrete troughs one inch deep. A third type eliminated the wooden ridgepole and had a single gently sloping (tilted) plastic roof.

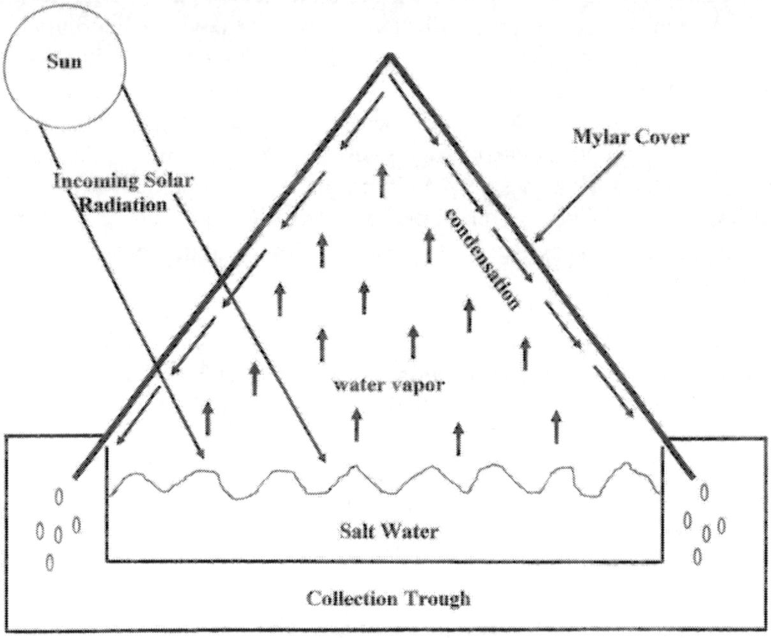


**Figure 24**
**Illustration of a Daniels' Solar Still**
Courtesy of the author

All the stills required less than a day to construct and were sturdy enough to resist heavy wind damage. Their plastic roofs had an estimated five-year lifetime. On a bright summer day, each type of still with a 25-square foot (2.33 square meters) basin produced over two gallons of distilled water. Its efficiency was 30-40 percent and compared favorably with the efficiencies of most solar stills. On such a day about 500 calories of solar radiation struck each square centimeter of a horizontal surface (500 langleys per day) and theoretically (100 percent) distilled a layer of water one centimeter thick, or a volume of one cubic centimeter. Each still had a construction cost of $20 or 80¢ per square foot, and even if the stills operated at 100 percent efficiency, the production of fresh water for agricultural or other purposes clearly required a large number of stills.

Because the construction materials cost little and the assembly and operation quite simple, as they were for solar cookers, Daniels believed that the best places to introduce solar stills were the nonindustrial countries or locations where fresh water was difficult to obtain and often expensive. These included the many South Pacific islands that received little rain during the summer months.

Up to this time Daniels had operated his stills mainly in Wisconsin, but in the early 1960s two opportunities enabled him to test them in the South Pacific at Rangiroa, an atoll situated 20 degrees south of the equator in the Tuamotu Archipelago 200 miles north by northeast of Tahiti, and in the Galápagos Islands. These regions often experienced fresh water shortages and were ideal sites to show that seawater distillation with simple solar stills provided sufficient amounts of fresh water.

The two-month South Pacific research trip to Rangiroa, beginning in late January 1963, was part of a National Academy of Sciences' program that sent American scientists abroad to share their expertise. In Rangiroa, Daniels constructed four stills, three with wooden frames, the other with concrete. Each measured 3 feet x 12 feet x 4 inches and had a floor of dark plastic or of clear plastic (mylar or tedlar) laid over black cloth and an A-shaped roof consisting of a wooden frame supported by a bamboo ridgepole and covered with water-wettable clear mylar or tedlar.

Because of Rangiroa's highly favorable location, Daniels expected significant improvement on yields obtained in Wisconsin and even in Florida. Only two days in February provided sufficient solar radiation of 546 langleys (Ly where 1Ly = 41,840 joules/$m^2$ or 11.622 watt-hours/$m^2$) to give good results, about one gallon per day per still, for an 18.7 percent efficiency. His twenty-two-page report of 8 March, submitted to the Pacific Science Association, described his distillation experiments and affirmed his belief that inexpensive and easily constructed solar stills economically could provide fresh water to water-short islands such as Rangiroa, even if operated only for brief periods of the year. Solar distillation clearly was an appropriate technology with which to convert seawater into fresh water, although Daniels recognized the necessity of much higher efficiencies before the process would find widespread use.

The second test of the solar stills, in the Galápagos Islands, about 650 miles off the coast of Ecuador, followed one year later in 1964. Scientists had done fieldwork in the arid, sunny, lowlands of the fourteen larger and many smaller uninhabited islands for decades, and for decades they had coped with the islands' meager 3-4 inch annual rainfall and resultant shortages of drinking water. Consequently in 1964, when the Galápagos International Scientific Project invited Daniels and fifty-two scientists from different countries to participate in its six-week field program, Daniels welcomed the opportunity to test his solar stills and perhaps improve the Galápagos's drinking water problem.

On the Galápagos, Daniels calculated the solar radiation hourly each day, but of the three highest values, 285, 345, and 500 langleys, only the last provided enough radiation to produce sufficient distillation. He constructed three stills, but ironically sun-sickness forced Daniels to leave before obtaining any measurements. Galápagos was Daniels's last field trip abroad, although he continued his solar research in Wisconsin, summarizing his results in 1964 with the publication of *Direct Use of the Sun's Energy*, his classic work on solar energy.

*Promotion of solar energy as appropriate technology: funds, stills, cookers*

Daniels, like many of the solar pioneers, was an idealist in promoting solar energy. At the same time, he was a realist and risk taker and in the 1960s reluctantly accepted that widespread use of solar energy remained far in the future. His twenty-five years of solar research had witnessed some success in cooking and increasing success in water heating and distillation and in home heating and cooling. The solar heated systems that Maria Telkes (1900-95) at MIT built in 1948 for an experimental solar-heated home in Dover, Massachusetts, and the 3,200 square foot home that George Löf (1903-2009) at Colorado State University built in 1957 in Denver, were clearly the result of increasingly promising research on solar heating and cooling. Yet the small group of solar researchers, such as Daniels, Telkes, and Löf, most of whom knew each other quite well, acknowledged practical success in only one field, the development of solar cells (photovoltaics) for use in space vehicles. This was a significant advance, but it represented progress in solar research and development quite different from their solar interests.

Daniels in 1971 conceded the obvious. Government had recognized the significance of solar energy in its space program, but neither government nor industry envisioned a large market for solar energy as either an alternative or an appropriate technology on Earth and had dismissed solar energy as too expensive to develop, or impractical to use, or both. The universities' lack of interest in solar research, other than photovoltaics, mainly because solar research cut across so many branches of science and engineering, also troubled him. Solar research, Daniels remarked, seemed to fit everywhere, but unfortunately it belonged nowhere. Where, he asked, were the departments of solar science and engineering? Who was teaching the solar science and engineering courses so crucial to the success of a nationwide solar energy program and to the future development of solar energy? Any long-term and large-scale commitment to solar energy research, Daniels said, probably had to wait another twenty-five years.

Solar scientists in the 1950s-early 1970s accepted that neither government, industry, or university had made a long-term commitment to solar development, other than photovoltaics for the space program. The lack of commitment has continued in the 2000s. As a result, solar energy enjoyed brief revivals during times of crises only to face a long period of neglect once the crisis ended. Nevertheless, the invention of improved solar apparatus and development of solar science and technology has remained the motivating force behind solar programs and the commitment of scientists to the future of solar energy.

In the United States total spending on solar research and development in 1976 was $110 million, far more than in any nation, but solar energy provided only one-half of one percent (0.50 percent) of the United States' total energy requirements. The two most active areas of research were solar heating and cooling of homes and buildings and photovoltaic conversion. By the 1980s research spending had increased significantly, and during the 1990s the federal government poured about $1 billion and private industry several more billion dollars into solar research. Most of the spending, $3 billion, went into photovoltaic research, which remains the area of greatest success. The operation in early 1995 of NASA's fully integrated solar dynamic space power system at its Lewis Research Center in Cleveland, Ohio, demonstrated photovoltaics' progress.

Solar energy programs on photovoltaics, solar heating, and solar distillation also progressed in Germany, France, Japan, Russia, and Australia in the mid-1970s. In the 1980s-90s significant research occurred in the development of appropriate solar technology in nonindustrial countries such as India, mainly through the efforts of small nonprofit and charitable organizations that received support from the United Nations, United States federal and state government agencies, and their own governments.

In India early interest in solar cookers dating from the 1950s began to wane in the 1960s-70s but experienced a revival in the 1980s. At that time, in 1982, the Indian government established the

Ministry of Non-Conventional Energy Sources, and as one of its objectives the ministry promoted the use of a family- size solar box cooker having a single reflecting mirror. The cooker had a 15-20 year lifetime and cost 1,500-2,000 rupees (1,000 rupees = $17.84). The ministry provided a subsidy of 30 percent to reduce its cost (later 15 percent), and in the twelve-year period 1982-94, when the program ended, Indian families had purchased 340,000 cookers. Only China used more solar cookers. Solar researchers in India claimed they had overcome conceptual difficulties in solar refrigeration, solar pump design, solar desalination, and other areas, but low technological input, heavy initial costs, and sociological factors delayed significant solar energy use.

Solar Cookers International (SCI) is An American organization that actively promoted solar cookers as appropriate technology. Seventeen educators, public health administrators, and engineers established SCI at a meeting at the University of the Pacific in Stockton, California, in 1987. In 1996 SCI moved its headquarters in Sacramento. At that time of its establishment, global population had just reached 5 billion. Half the world was cooking with wood, and the United Nations Food and Agriculture Organization predicted correctly that by the year 2000, 2.4 billion people would experience wood shortages.

SCI worked through UNESCO (United Nations Educational Scientific and Cultural Organization) as solar pioneers had done thirty years earlier, with the United Nations High Commission for Refugees (UNHCR) and CARE. It introduced thousands of inexpensive but efficient solar cookers costing $4-$10 ($20 in 2010) to societies in Central and South America (Costa Rica, Bolivia, Guatemala), Asia (India, Nepal refugee camps), and Africa (Zimbabwe, Tanzania, and especially the refugee camps in eastern Ethiopia, and at Kakuma in north Kenya and until 1997, at Dadaab in eastern Kenya).

SCI promoted solar cooker usage primarily to reduce the rapidly escalating global consumption of wood for cooking, where wood remained available, and to lessen the problem of spending days searching for wood in those areas having little wood available. SCI also aimed with solar cookers to reduce the use of animal (camel, cow) dung as fuel in regions that traditionally burned dung, as in India, Nepal, and parts of Africa, but not in Latin America. The application of solar cookers eliminated the irritating dung smoke that damages the lungs as well as the environment, and the hygienic problems associated with handling dung and food at the same time. Burning the dung also reduced the amount that went back into the ground as a fertilizer.

D***evelopment of photovoltaic cells: in space, on Earth***

In 1980 Ken Butte and John Perlin published *A Golden Thread*, a comprehensive and authoritative history of solar energy that remains one of the best books on the subject. A later book, Perlin's *From Space to Earth, the Story of Solar Electricity* published in 1999, is an equally impressive history of the twentieth-century solar photovoltaic development. He describes how a diverse group that included scientists, engineers, and other advocates used the chemical elements selenium and later silicon to convert the Sun's thermal energy into electricity. The group includes researchers, such as the chemist Elliot Berman (b. 1944), founder of Solar Power Corporation incorporated in Delaware in 1967, and which in 1973 established an office in Braintree, Massachusetts; the Swiss engineer Markus Real in Zurich; the petroleum industry; the United States military and Coast Guard; governments; and Bernard Versperien (1924-2003), a French Catholic missionary priest who came to Mali, West Africa, in 1951.

The history of solar cells, or photovoltaics, dates from the selenium cells the British scientists William Grylls Adams (1836-1915), a professor of mathematics at King's College, London, and his pupil Richard Evans Day constructed in 1876 and the American inventor Charles Fritts (1850-1903) in New York City constructed in 1883. After decades of neglect, Gerald Pearson (1905-87), a Bell Laboratories scientist, in 1953 constructed a much more efficient silicon cell. One year later in 1954 two other Bell scientists, Darryl Chapin (1906-95) and Calvin Fuller (1902-94), worked with Pearson to develop an improved silicon cell.

Silicon solar cells were not cheap in 1953, but the US air force and army immediately saw them as the ideal source of power for their top-secret satellite project. The navy, however, won the ensuing competition to launch the United States' first satellite, and it rejected photovoltaic cells as untried technology. The chance of adopting solar cells seemed futile until Hans Ziegler (1911-99) waged a relentless campaign and convinced the navy to use a dual power system of chemical batteries and solar cells on its troubled Vanguard satellite. Ziegler, a German-born scientist who came to the United States in 1947 with Operation Paperclip, worked for the United States Signal Corps in Fort Monmouth, New Jersey. His relentlessness paid off. The chemical batteries failed within weeks, whereas the silicon solar cells worked for about eight years after Vanguard I's successful launch on 17 March 1958.

Solar cells successfully powered both American and Soviet satellites, such as Sputnik III, during the late 1950s-60s, but they were expensive and considered a temporary solution in the United States. Nuclear-powered satellites were to provide a permanent solution, but nuclear power in space never fulfilled its expectation, and by the late 1960s solar cells became the accepted power source in space satellites.

Economics continued to delay terrestrial solar cell use until the early 1970s. At that time significant improvements in solar cell design that Berman, with financial support from Exxon introduced, reduced the price from about $100 per watt in 1970 to $10-20 per watt in 1973. Berman recognized that silicon used in solar cells, unlike other semiconductors, functioned effectively with lower quality silicon and that solar cells therefore did not require expensive semiconductor-grade crystalline silicon. Cheaper silicon wafers that the space industry and rapidly growing semiconductor industry rejected proved suitable for generating solar power.

Berman's breakthrough resulted in the first mass purchase of solar cells beyond those of the space industry. The US government required petroleum companies operating in the Gulf of Mexico to provide navigation lights and warning horns on their oil platforms, and for years petroleum companies, such as Exxon, Amoco, and Royal Dutch Shell used heavy expensive chemical batteries as their electrical source. Solar cells proved so superior that by 1980 they completely replaced the chemical batteries. A similar adoption occurred in the gas and oil fields which required electricity to prevent corrosion problems in wellheads and piping and cathodic protection for natural gas production but lacked power lines for transmission. The petroleum industry remains a mass consumer of solar cells.

Solar cell advocates such as Berman, expected the US Coast Guard to make the first mass purchase of solar cells to power its many isolated buoys and lighthouses such as those in Alaska. The Coast Guard, like the navy earlier, resisted, and only a crusading effort by Lloyd Lomer, a Coast Guard lieutenant-commander, forced the Coast Guard in the late 1970s-early 1980s to develop a photovoltaic program. Solar cells remain today the main power source of navigational aids globally.

Similar successful conversions to solar cells occurred in the United States and globally in the 1970s-80s. In the railroad industry solar cells powered the industry's microwave communication system and in remote telecommunication systems worldwide. Australia, a large country with a small widely dispersed population, in 1978 began developing a photovoltaic-powered communication network that exemplified the feasibility of solar cells.

During these years, new applications of solar cells as appropriate technology arose in the world's nonindustrialized regions, such as Mali and Kenya in West Africa, Tahiti, the Dominican Republic, and Central America. Father Versperien, who in 1967 established a West African experimental farm-village called Terriya-Bugou to study solar and wind irrigation methods, developed a photovoltaic water pumping program to tap the deep aquifers that ran beneath the semiarid Sahel region, south of the Sahara Desert. His pumping program, the Versperian Mali Aqua Viva, established in 1974 after the 1968-73 Sahelian drought that killed at least 100,000,became the model for the nonindustrial world. It resulted in the number of operational photovoltaic pumps increasing from ten in the 1970s to tens of thousands in the 1990s.

The recently developed small solar-power systems (SSPS) represent the continuing application of solar energy as appropriate technology that Daniels and other solar scientists had promoted with the development of solar cookers and stills. By 2000 SSPSs had provided electric lighting and communication for about 200,000 families in Central America, Asia, and Africa. At a time when seventy percent of the nonindustrialized world's population remained without electricity and had to rely on nineteenth-century light sources, such as smoky kerosene lamps or wood fires, a photovoltaic SSPS provided a clean and inexhaustible source of electricity to individual households.

Three groups, beginning in the 1980s-90s, promoted SSPSs in the rural regions of nonindustrialized countries. They included small private companies-government energy agencies in countries such as India, China, Sri Lanka, Nigeria, Uganda, and South Africa; the US Department of Energy; and US nonprofit organizations such as the Solar Electric Light Fund (SELF) that Neville Williams, a journalist, author, former DOE consultant, and solar advocate, founded in 1990 in Washington, DC, and Enersol Associates that Richard Hansen, a former Westinghouse Electric mechanical engineer, founded in North Chelmsford, Massachusetts, in 1984. After dissolving Enersol in 2009, Hansen continued his solar program with Solez Inc. in North Chelmsford, a for-profit business and technology development company he established in 1990. Solez commercializes the distribution of photovoltaic energy in regions or countries that lack electrical distribution systems, particularly those in the Caribbean and Latin America, such as Honduras and the Dominican Republic.

Working with local government officials and the Power & Sun Company in Sri Lanka, and Gansu PV Company in China, Enersol and SELF brought electricity to rural regions that lay beyond existing electrical grids and where dam construction for hydroelectric power generation presented serious problems. These regions existed because they did not offer commercial utilities big industrial loads or the favorable economics of urban areas. Nor did government authorities see any economic justification for constructing dams that required flooding of fertile valleys and population displacement and the acquiring of large debts all in the name of increased productivity in these regions. By 1997 SELF established 11 self-sustaining solar projects in 11 countries in Asia, Africa, and South America in Brazil's Xixuaú Xipariná Ecological Reserve. Since that time, SELF in 1997 established the Solar Electric Light Company (SELCO) in Bangalore, India, to sell solar home systems; installed in 2000, a 1.5 kilowatt solar system that produced sufficient electricity to operate 20 computers at Myeka High School in Durban, South Africa; and in 2003 with support from DOE equipped three villages in northern Nigeria's Jigawa State with solar power systems for water pumping, street lights, and a health clinic. In 2006 SELF and the International Crops Research Institute for the Semi-Arid Tropics (ICRISAT), installed solar water pumping and drip irrigation systems in two West African villages of Benin's Kalale district.

Each SSPS system consisted of an electrical circuit in which a photovoltaic solar module produced the current that charged a lead storage (automobile) battery which in turn operated light bulbs; radio cassettes; small, 12 volt, black-and-white televisions; or other small appliances placed in the circuit. A standard single-household roof or movable pole-mounted solar module generally contained thirty-six, 10-cm-diameter solar cells each composed of a single-crystal or polycrystalline silicon wafer. Such a module operating in the tropics in direct sunlight at noon produced 40 watts of electricity, enough to power three 9-watt florescent lights for three hours, three hours of radio, and three hours of television. The charged lead storage battery provided a five-day power supply during overcast periods.

Because of the modules' mass production and low cost, villages or communities that purchased SSPSs either imported the modules completely assembled or assembled them locally. Costs for a complete SSPS system including shipping, tariffs, installation, and all equipment, which has a twenty-year lifetime, range from $350 to $750. The solar lantern, an even cheaper system, cost less than $100. It consists of a smaller solar module and a built-in lead storage battery that provides enough electricity for one light bulb and a radio.

Williams, Hansen, and others who worked with SSPS systems in nonindustrial countries recognized that the electrification process worked best by installing decentralized photovoltaic units. They did not repeat the West's power-generating history of constructing huge centralized photovoltaic systems and distributing the electricity by wire, and thereby advanced photovoltaic development. Their decentralized systems represented a technological transfer from the nonindustrial world to the industrialized world. Real in 1987 became the first to apply the decentralized photovoltaic application in the west. He installed three-kilowatt solar cells on 333 homeowners' roof tops in Zurich. His successful experiment provided an effective alternative to the necessity of constructing large centralized photovoltaic power plants.

In the early 1970s (1973), photovoltaic-generated electricity per kilowatt hour cost $2.50-7.50, about 10-15 times as much as it cost in the early 1990s (10-15 times 25-50¢ in 1995). The cost per kilowatt hour dropped drastically in that 20-year period and has remained in the 25-50¢ range for the last ten years: 15-20¢ in 2005, 30-60¢ in 2005, 30¢ in 2006, 18-23¢ in 2009, 20-30¢ in 2010. As Perlin noted in *From Space to Earth,* at 25¢ per kilowatt hour (1999 cost), photovoltaic-generated electricity cost less than using batteries or diesel generators, less than laying underground transmission lines anywhere in the world, and less than constructing overhead transmission lines over a 250-yard distance. Solar photovoltaic cells remain more expensive than consuming electricity from existing power lines, but the production of thin film photovoltaic cells rather than the present thick crystalline cells has reduced the final cost of solar cells.

Progress also occurred in the design and development of solar stills that use photovoltaics to provide safe drinking water to the many in society without a safe supply. This was the objective of Sunseed Desert Technology, an English charitable organization, and of the Texas State Energy Conservation Office. The Energy Conservation Office in the early 2000s completed solar demonstration projects to purify water for those in Webb County in south Texas and El Paso County in central west Texas who reside in colonias, which are unincorporated settlements on the United States-Mexico border. The colonias lack a safe water supply and waste water treatment and have experienced severe water-borne diseases such as cholera, typhoid fever, and dysentery. The 8.1 kilowatt solar-powered reverse osmosis system produces 14,000 gallons of clean water in a 24-hour period.

Despite the advances made in photovoltaics and other branches of solar research in the last half century, an indifference to the lack of solar programs persisted within the larger energy community. Solar research, nevertheless, has continued at two of the US government's 17 national laboratories: the National Renewable Energy Laboratory (NREL) in Golden, Colorado, and Sandia National Laboratories (SNL) in Albuquerque, New Mexico; in a few industrial laboratories; in joint government–industry projects; and in universities.

Solar World AG, an international oil company established in 1988 in Bonn, Germany; and Showa Shell Sekiyu KK, Royal Dutch Shell's Japanese subsidiary established in1985 in Tokyo, have ongoing solar programs. BP Solar (formerly Solarex Corporation established in 1973, a subsidiary of Amoco Oil Company) in San Francisco and Frederick, Maryland, ended its long-running solar program in December 2012.

In California, Photovoltaics for Utility Systems Applications (PV USA), a 650- kilowatt solar plant near Davis, which DOE, the California Energy Commission, and nine utility companies have funded since 1987, and the DOE-state electric utilities-private industry $150 million Solar One (1982-88) and $48.5 million Solar Two (1996-98) in the Mojave Desert near Barstow, about 100 miles northeast of Los Angeles, are or were among a handful of joint government-industry projects. The recently-dedicated (February 2014) Ivanpah Solar Energy Generating System in Ivanpah Dry Lake, southern California, is another joint government-industry solar project. Its participants include NRG, Google, and BrightSource Energy with Bechtel the engineering, procurement, and construction contractor. DOE's Loan Programs Office provided a $1.6 billion loan guarantee. Ivanpah is the world's largest concentrating solar plant (CSP), having the capacity to generate 392 megawatts of electricity, enough to power 94,400 average homes. Ivanpah and four other CSP projects, when completed, will have a combined capacity of 1.26 gigawatts of electricity. Large solar power plants also are operating in Arizona, Nevada, Colorado, Florida, New York, and other states.

Within the university only a small number of solar programs exist today. They include Delaware, Stanford, Princeton, Chicago, Georgia Institute of Technology, Pennsylvania State, Purdue, South Florida, Northern Colorado, Colorado School of Mines, Wisconsin, and MIT.

Improved solar energy storage is a recent promising result of the university solar programs. Daniel Nocera and his group at MIT in 2008 prepared a new catalyst that led to the development of a simple, efficient, and inexpensive process for storing solar energy. Their solar storage process has provided a solution to a long-standing and formidable problem that for decades hindered solar energy's development and application. The catalyst (cobalt metal-phosphate-electrode) greatly increased the yield of oxygen in a solar-initiated electrolysis of water. Another inexpensive catalyst they developed in 2011 produced hydrogen gas at the other electrode.

In the first stage of Nocera's process, solar energy striking a photovoltaic cell generates an electric current which, when passed through the electrolytic cell containing the water and the electrodes, produces oxygen and hydrogen gas. In the second stage, combining the two gases in a fuel cell, or storing them separately for later combining in a fuel cell, produces water as the only product and generates carbon-free electricity for a home or building consumption. Nocera envisions facilities equipped with solar panels, photovoltaic cells, electrolytic cells, and fuel cells, enabling them to use any excess solar-generated electricity to produce oxygen and hydrogen in an electrolytic cell, store the gases, and then later combine them in a fuel cell to produce electricity when photovoltaic electricity is unavailable. His solar storage process would eliminate the need for a central electrical generating system that requires the consumption of fossil fuels.

***Conclusion***

Solar energy's development as either an alternative or an appropriate technology has remained a somewhat unfulfilled promise. A breakthrough appeared close at hand in the mid- and later nineteenth century with the solar engine researches of scientists and engineers that included Augustin Mouchot and Charles Tellier (1828-1913) in France and John Ericsson, Aubrey Eneas (fl. 1890-1920), and Frank Shuman in the United States. Rapid consumption of the conventional fuels, wood and coal, in the industrializing countries, contrasted with the scarcity and expense of these fuels in many nonindustrializing countries, had seriously concerned and strongly motivated these nineteenth-century solar researchers.

But two events quickly changed the energy picture and put to rest the most promising solar projects. First, the improvement in coal mining techniques and better railroad transportation globally led to increased coal production and decreased price. Second, the large petroleum discoveries at Titusville, Pennsylvania, in 1859 and then at Spindletop, Texas, in 1901, ushered in the era of abundant and cheap liquid fuel. Another solar breakthrough, to which Daniels, Löf, Telkes, and others made major contributions with their research in solar heating and cooling, appeared imminent shortly after World

War II ended, but the excitement that nuclear energy generated and the increasing availability of oil from the Middle East once again sidetracked solar research.

In the 1970s, the Arab oil embargo of 1973-74 and the Iranian hostage crisis of 1979-81 sparked another renewal of interest in solar energy. The US federal budget in 1974 included $13 million for solar research, compared to the 1960s nearly zero spending other than for NASA's solar cell research. Funding peaked at $286 million in 1980 before falling to $42 million in 1990, increasing again in the early 1990s and then decreasing beginning in 1996. The 2010 US federal budget for solar research was $235.9 million, increased to $263.1 million in 2011 but down from $302.4 million proposed for 2011, and decreased even more to $176 million in 2012. About 50 percent is for photovoltaics. The table below gives government spending for solar research in recent years.

| US GOVERNMENT SPENDING ON SOLAR RESEARCH | |
|---|---|
| ***Year*** | ***Amount in Millions of Dollars*** |
| 1999 | 87.4 |
| 2000 | 81.1 |
| 2001 | 81.0 (approximately) |
| 2002 | 81.0 (approximately) |
| 2003 | 82,300 |
| 2004 | 83,393 |
| 2005 | 85,074 |
| 2006 | 83,113 |
| 2007 | 157,028 |
| 2008 | 166,300 |
| 2009 | 172,414 |
| 2010 | 243,396 |
| 2011 | 259,556 |
| 2012 | 288,951 |
| 2013 | 290,719 |
| 2014 | 356,500 |

Source: US DOE, FY BUDGET

## IX. ENVIRONMENT AND POLLUTION

### *Introductory*

The availability of naturally-occurring sources of energy, mainly wood, coal, and petroleum, significantly contributed to the emergence of our modern world. Science and engineering aided the discovery, development, and application of these energy sources making them more available by continuing to discover their locations, reducing their development and application costs, and thereby increasing their availability to society. Science and engineering also led to the discovery of additional energy sources such as nuclear and photovoltaic energy. All of these energy sources clearly brought vast benefits, but until the 1960s society largely ignored the unpleasant consequences of their application. Awareness seldom led to action. As a result, atmospheric pollution problems such as acid rain and global warming and the disposal of nuclear waste went unchecked for decades. The environmental polluters dismissed the deadly air and nuclear pollution disasters that didn't require laboratory confirmation and should have sparked environmental reform.

One of these disasters occurred in the Meuse Valley in eastern Belgium. From 1-5 December 1930 industrial smog (smoke plus fog) containing sulfur dioxide gas ($SO_2$) and fluorine compounds killed 63 people. In Donora, Pennsylvania, from 25-31 October 1948, sulfur dioxide, carbon monoxide, and particulate matter released by a sulfuric acid plant and a steel plant killed 20 people in 3 days. The greatest disaster of that time was the London smog of 5-8 December 1952, in which particulate matter containing soot and lead killed 4,000 people in 5 days and 8,800 in the following months.

More recently, in 1984, in a disaster that occurred in Bhopal, India, 42 metric tons of methyl isocyanate leaking from a tank in the Union Carbide pesticide plant killed 2,000 people immediately, another 20,000 by 2002, and left another 120,000 chronically ill. In 1986, the Chernobyl nuclear reactor located near Pripyat, Ukraine, exploded, killing 28 people within four months, another 28 by 2004 for a total of 56 deaths, and another 4,000 expected deaths. The explosion released 400 times more radiation than the Hiroshima atomic bomb.

Because we live in a scientifically-advanced world, science, sometimes to the exclusion of common sense and the inclusion of economics, has become both the reason for effective action, as in the case of water pollution in the late-nineteenth and early-twentieth centuries, or for limited or a lack of action, as in the case of air pollution in the twentieth century. Society constructed effective water treatment systems to deal with water pollution, whereas it passed and continues to pass ineffective regulations to deal with air pollution.

Environmental awareness is a relatively recent development, beginning for the most part in the 1960s. Yet even today a significant segment of society continues to deny that an air pollution problem even exists and that even if a pollution problem does exist, the damage or danger is not serious enough to warrant any real change in our linking, equating, or identifying industrialization with progress and progress with cultural superiority and advancement. The remaining sections of this chapter focus on the major issues in environmental history, particularly regulation of water pollution, acid rain, global warming (climate change) and the serious and neglected problem of lead in the environment.

**ORIGIN OF THE WORD SMOG**

The word *smog* dates from 1905 when Henry Antoine des Voeux, a scientist (medical doctor) and Honorary Treasurer of the Coal Smoke Abatement Society, used it in a paper he presented at a meeting of the Public Health Congress in London. In his paper "Fog and Smoke" he discussed the combination of smoke and sulfur dioxide resulting from the burning of large quantities of coal in English homes and factories.

### *Water purification*

In 1914 the Department of Treasury enacted the first set of standards required for disinfecting drinking water. It set a maximum of 2 coliforms (2 detected pathogens) per 100 milliliters (mL) in 1914, lowered the standard to 1 coliform per 100 mL and included standards for copper, lead, and zinc in 1925. The standard applied only to interstate water carriers, such as trains, buses, and boats and to contaminants capable of causing contagious diseases. Later revisions of drinking water standards occurred in 1946, 1962, 1974, 1986, and 1996. The 2000s standard for total coliforms is zero.

The 1974 Safe Drinking Water Act for the first time required all public water systems to comply with the government's health standards, exempting only non-community water systems. Congress passed the 1974 act after a Federal study showed the presence of carcinogens, lead, and harmful bacteria in large and small public drinking water systems, such as those of New Orleans, Boston, and Pittsburgh. The 1974 act required local systems to monitor only 20 substances which by 2000 grew to 91 where it has remained. These include arsenic, chromium (hexavalent), perchloroethylene (dry cleaning solvent), and uranium, all of which are carcinogens, and lead, but they represent a tiny fraction of the 830 contaminants that scientists at the EPA and the US Geological Survey have found harmful even at small concentrations.

.Nevertheless, the National Public Water System Compliance Report for 2007-2008 concluded that the majority (73 percent) of the 155,000 public water systems in the United States were in compliance with drinking water standards. A significant number of systems still remain in noncompliance. In 2008, 8,438 systems were not in compliance, the vast majority of them small systems, which by definition serve 501 to 3,300 users (13 percent) of total systems. Medium systems (3 percent noncompliance) serve 3,301 to 10,000 users (10 percent of total systems), large systems serve 10,001 to 100,000 users, and very large systems (3 percent noncompliance) serve more than 100,000 users (78 percent of total systems). In the United States, the Environmental Protection Agency (EPA) in 1998 set the current minimum-maximum chlorine dosage for drinking water at 0.2-4.0 milligrams per liter (mg/L). It followed the World Health Organization (WHO) which in 1993 recommended a maximum chlorine concentration of 0.2-5 mg/L.

***Acid Rain***

*What is acid rain?*

Acid rain is an atmospheric mixture of wet and dry particles that contains higher than normal concentrations of nitric and sulfuric acids. When deposited from the atmosphere in wet form (rain, snow, and fog), and dry form (gases) the deposition forms an acidic rain. Acid rain (acidic deposition) commonly means the wet deposition.

The United States and other countries, such as those of the European Union, since the 1970s have made serious efforts to reduce the emissions of the main pollutants that form acid rain. Over the twelve-year period 1990-2002, this has resulted in a 37 percent decrease in human-caused sulfur dioxide emissions and a 17 percent reduction in human-caused nitrogen oxides. Among the benefits of a reduction in acidic deposits is a reduced acidity of soils and waterways and regional haze. Sulfate particles formed from the reaction of sulfur dioxide with atmospheric oxygen and water vapor can reduce visibility by as much as 50-70 percent in some regions of the United States. Reduction of nitrogen oxides emissions ($NO$, $NO_2$) results in a reduction of ozone ($O_3$) formation, a known lung irritant, and of lung diseases.

*Measuring acid rain*

Soren Sorensen (1868-1939), a Danish chemist at the Carlsberg Laboratory in Copenhagen, in 1909 introduced the idea of pH to indicate the hydrogen ion concentration of an acidic solution and therefore the strength of an acidic solution. He defined pH as the negative logarithm of the hydrogen ion concentration so that the pH readings would have a positive value and range from 0 to 14.

Electrolytic experiments that measured the conductivity of aqueous solutions and of pure water showed that some of the aqueous solutions of acids, bases, and salts were good conductors, whereas pure water had a very slight conductivity. This indicated that water HOH only slightly dissociated into hydrogen $H^+$ and hydroxyl $OH^-$ ions and that the concentrations of hydrogen ion $[H^+]$ and hydroxyl ion $[OH^-]$ in pure water were equal and very low, $10^{-7}$ moles/liter. Pure water had a pH of 7 (and a pOH of 7) and was neutral, neither acidic nor basic (alkaline). Acidic solutions had pH values less than 7, the lower the more acidic, basic solutions had pH values greater than 7, the higher the more basic. The ion product of water equaled $10^{-14}$ moles/liter.

| THE DISSOCIATION OF WATER AND pH |
|---|
| $HOH \rightleftarrows H^+ + OH^-$ <br> $pH = -\log [H^+]$ |

| ION PRODUCT OF WATER |
|---|
| $[H^+] \times [OH^-] = HOH$ <br> $1 \times 10^{-7} \times 1 \times 10^{-7} = 1 \times 10^{-14}$ |

Other experimental evidence had shown that mixing equivalent solutions of any strong acid with any strong base produced a neutral solution and that the heat of neutralization was equal for each acid-base mixing. This evidence led Svante Arrhenius (1859-1927) at Stockholm, in 1884, to propose his acid-base theory of hydrogen and hydroxyl ions in an aqueous solution. Arrhenius concluded that an acid solution contained hydrogen ions and that the strength of the acidic solution depended on the acid's degree of dissociation into hydrogen ions and therefore on the hydrogen ion concentration. Similarly, in a basic solution, the base dissociated into hydroxyl ions and the strength of the basic solution depended on the degree of dissociation. In a neutralization reaction between a strong acid and a strong base, the hydrogen and hydroxyl ions combined to form neutral water.

**pH VALUES OF DILUTE SOLUTIONS**

| Solution | pH |
|---|---|
| HCl | 1.1 |
| $H_2SO_4$ | 1.2 |
| $CH_3COOH$ | 2.9 |
| $H_3BO_3$ | 5.2 |
| Rainwater | 5.6 |
| $H_2O$ pure | 7.0 |
| $NaHCO_3$ | 8.4 |
| $NH_4OH$ | 11.1 |
| NaOH, KOH | 13.0 |

| OTHER pH FACTS |
| --- |
| A pH of 7 indicates a neutral solution, such as pure water.<br>Normal rain water has a pH of 5.6 and is slightly acidic because carbon dioxide ($CO_2$) present in the atmosphere reacts with the rain (moisture, water vapor) to form carbonic acid, a weak acid (slightly dissociated acid).<br><br>$CO_{2(g)} + H_2O_{(l)} \rightarrow H_2CO_{3\,(aq)}$<br>$H_2CO_3 \rightarrow H^+ + HCO_3^-$<br>$HCO_3^- \rightarrow H^+ + CO_3^{-2}$<br><br>An acidic solution (HCl, $H_2SO$, $HNO_3$) has a pH less than 7, the lower the number the more acidic.<br><br>An alkaline (basic) solution (NaOH, KOH) has a pH greater than 7. |

*History of acid rain*

In the years preceding the studies of Arrhenius and Sorensen on the acidity and alkalinity of aqueous solutions and the measurement of their conductivity, Robert Angus Smith (1817-74) in his 1852 paper "On the Air in Towns" provided the first experimental evidence of rainfall's acidity and its harmful consequences. Later in his 1872 book *Air and Rain,* Smith coined the expression *acid rain.*

During the next century the industrializing nations such as the United States and Britain did little to limit or control the passage of sulfur dioxide ($SO_2$) and nitrogen oxides, particularly nitrogen dioxide ($NO_2$) into the atmosphere. In the United States the earliest regulations, the Air Pollution Control Act of 1955 and the Clean Air Act of 1963, placed no restrictions on sulfur dioxide emissions, the main cause of acid rain, or on nitrogen oxides emissions. The year, 1963, however, marked a turning point in the history of acid rain.

Noye M. Johnson (1922-2012) and Eugene Likens (b.1935), ecologists at Dartmouth, with two other scientists F. Hubbard Bormann and Robert S. Pierce of the US Forest Service, in 1963 began the long-term Hubbard Brook Ecosystem Study at the twelve-square mile Hubbard Brook Experimental Forest in the White Mountains of New Hampshire. Hubbard Brook was the first study since Smith's linking air pollutants such as sulfur dioxide to the acidity of rain (acid rain). It became the model for the study of ecological science. In the Hubbard Brook study, Bormann and Likens experimentally varied an entire ecosystem in order to establish its biological, chemical, and physical patterns and characteristics and the effects of acid rain on natural ecosystems. The results enabled them to evaluate human influence on the chemistry of soil, plants, trees, precipitation and clouds, and the ecology and biogeochemistry of aquatic ecosystems. Their work, particularly the 1974 paper in *Nature*, "Acid rain: A Serious Regional Environmental Problem," initiated the later passage of national legislation that dealt with the acid rain problem.

During this same period another ecologist, Harold Harvey at the University of Toronto, began research on what ecologists called a *dead lake*, one in which acid rain had destroyed the lakes' living systems. Harvey investigated fish deaths in lakes in Killarney Park, near Sudbury, Ontario. He published his results in 1971.

*Causes of acid rain*

The research of the Likens-Bormann's group at Dartmouth and the Forest Service, Harvey, and other ecologists triggered a debate on the cause of acid rain. Few scientists, namely the physicist S. (Siegfried) Fred Singer (b.1924) an Austrian-born former professor of environmental science at Virginia and now a research fellow at the conservative Independent Institute in Washington, DC; and William Nierenberg (1919-2000) former director at the Scripps Institute of Oceanography in La Jolla, California, from 1965 to 1986, doubted that acid rain had a natural and a human cause. Both were well known scientists, but neither had done any actual research on acid rain. They aimed to cast doubt among the public about the cause of acid rain, always claiming the insufficiency of evidence supporting human-caused acid rain. The doubters received financial support from conservative think tanks and industrial groups such as the National Association of Manufacturers, a Washington, DC, conservative advocacy group that in a 1987 statement challenged human- caused acid rain. Manufacturers, especially the major consumers of fossil fuels, opposed any regulations to reduce acid rain, mainly for economic reasons. They did not want to install any plant equipment designed to reduce and even eliminate the gaseous emissions responsible for acid rain. Political ideology, their opposition to *big activist* government, not scientific evidence directed their thinking.

Most ecologists agreed that acid rain resulted primarily from human causes, particularly the combustion of fossil fuels by utility companies. The main fossil fuels, coal, petroleum, and natural gas, contain varying amounts of sulfur and nitrogen and toxic metals such as mercury and lead that they release during their combustion. In the United States, about 66 percent of sulfur dioxide and 25-40 percent of the nitrogen oxides (NO, $NO_2$) came from the utilities companies' combustion of fossil fuels for electrical generation. Automobile exhaust contributes most of the remaining nitrogen oxides, about 50 percent in the atmosphere.

The main natural causes of acid rain are volcano emissions, lightning strikes, cow flatulence, and decaying vegetation. Volcanoes emit, mainly in decreasing quantities, water vapor $H_2O_{(g)}$, $CO_{2(g)}$, $SO_{2(g)}$, and smaller amounts of hydrogen $H_{2(g)}$, carbon monoxide $CO_{(g)}$, hydrogen sulfide $H_2S_{(g)}$, hydrogen chloride $HCl_{(g)}$, hydrogen fluoride $HF_{(g)}$, methane $CH_{4(g)}$, ammonia $NH_{3(g)}$, and silicon (IV) fluoride $SiF_{4(g)}$. The harmful concentrations of volcanic emissions remain close or only about 10 kilometers, from a volcano. Remote sensing instruments identify and determine the quantity of each gas. Lightning discharges in the atmosphere are the main natural causes of nitrogen oxides. Cow flatulence and decaying vegetation produce most of the methane emissions.

*Consequences of acid rain*

The investigations of Likens, Bormann, Harvey, and others beginning in the 1960s and extensive studies by the EPA and the Canadian government showed clearly that acid rain affected natural systems, such as waterways, soil, plants, and forests; human-constructed objects such as buildings and monuments; and the health of humans. Some of their conclusions follow.

A lower pH of a body of water indicates increasing acidity, and when the pH of the water falls below 5 many kinds of fish eggs do not hatch. At an even lower pH some species of adult fish cannot survive as happened in the well-known ecological tragedy that occurred in the 1950s-70s in the waterways of the Adirondack Mountains in northeast New York State. Bass cannot survive below pH 5.5, trout below pH 5, and perch below pH 4.5.

Acid rain flowing through soils in a watershed also released aluminum from the soils into the stream and lakes located in the watershed. Aluminum is the most abundant metal in Earth, making up 7 percent of Earth's crust. Increased aluminum levels proved toxic to fish, and when combined with increased acidity of the water, subjected fish to considerable stress. The combined effect greatly reduced their ability to survive in the new hostile environment.

EPA studies in the 1980s-90s on lakes and streams in the United States demonstrated the acidity of 580 streams in the Mid-Atlantic Coastal Plain. An even higher number, 90 percent, in the New Jersey Pine Barrens were acidic, making its acidity rate the highest in the United States. More than 1,350 streams in the Mid-Atlantic Highlands were acidic. An 808-page study that the Canadian government's Environment Canada published, *State of Canada's Environment, 1996,* confirmed the acidity of 14,000 lakes in eastern Canada.

The EPA's ecological studies have shown how acid rain contributed to soil and forest degradation particularly in the highly-elevated forest regions of the eastern United States, such as the Appalachian Mountains which stretch along the eastern seaboard. Fortunately, certain soils because of their alkalinity have the ability to neutralize (buffer) some of the acidity. The carbonates present (limestone $CaCO_3$, dolomite $CaCO_3.MgCO_3$) react with the sulfates and nitrates of the acid rain and in a typical acid-base reaction reduce the acidity of the soil. Nebraska and Indiana in the Midwest have soils with a high buffering capacity whereas mountainous regions in the northeast, such as the Adirondack and Catskill Mountains in New York State, do not.

In addition to acid rain's dissolving a soil's helpful nutrients and minerals and washing them away, the aluminum that acid rain releases from the soil is toxic to trees and plants. Aluminum reduces the availability of the essential nutrient phosphorus by reacting with it to form aluminum-phosphorus compounds. The combination of aluminum with sulfur to form aluminum-sulfur compounds leads to stunted roots.

Acid rain and the dry deposition of acidic particles has resulted in the corrosion (oxidation) of metals and alloys such as copper and bronze and the deterioration of objects constructed of limestone and marble. These include buildings, bridges, monuments, statues, and tombstones. The sulfuric acid in the rain reacts with calcium carbonate in limestone and marble to form calcium sulfate which flakes off the object. Acid rain also attacks the paints used on houses and automobiles.

Human health has suffered because of acid rain. The gaseous pollutants, sulfur dioxide and the nitrogen oxides (NO, $NO_2$), react with oxygen and water vapor in the atmosphere to form the sulfate and nitrate particles present in acid rain. When inhaled, these particles lead to an increase in the number and severity of heart and lung diseases such as asthma and bronchitis. The nitrogen oxides lead to a second serious medical problem. Oxygen ($O_2$) reacts with atomic oxygen (O) produced by the photodecomposition (sunlight) of $NO_2$ in the presence of volatile organic compounds to produce ozone ($O_3$). Ozone causes lung inflammation including asthma and emphysema.

$$\underset{oxygen}{O_2} + \underset{nitrogen\ oxides}{O(NO, NO_2)} + \underset{volatile\ organic\ compounds}{VOC} \rightarrow \underset{ozone}{O_3}$$

| SMOG TIMELINE DEFINING OZONE | |
|---|---|
| *Year* | *Standard* |
| 1971 | EPA sets one-hour ozone standard at 80 parts per billion (ppb) |
| 1979 | One-hour standard revised to 120 ppb. |
| 1991 | Number of counties nationwide violating one-hour standard reaches 371. |
| 1997 | EPA replaces 1979 standard with an eight-hour standard set at 84 ppb. |
| 2002 | After legal challenges, EPA starts to put 1997 standard into effect. |
| 2003 | The number of counties out of attainment with new eight-hour standard hits 506. |
| 2008 | EPA replaces 1997 standard with an eight-hour standard of 75 ppb. |
| 2009 | EPA withdraws the year-old standard over concerns that it is too lax. |
| 2010 | EPA replaces the 1997 standard and proposes a new eight-hour standard set between 60 and 70 ppb. |

*Remedies for reducing acid rain*

Scientists have developed four successful solutions that have reduced the damaging effects of acid rain on the environment. They are (1) liming, the addition of an alkaline (basic) substance, limestone (calcium carbonate) or lime (calcium oxide, $CaO$), to lakes, (2) coal washing and the installation of scrubbers to smokestacks and selective catalytic reduction units to power plants, (3) the use of electrical precipitators such as the Cottrell precipitator, and (4) conservation and the use of alternative energies.

The liming of lakes is a short-term, expensive process that Sweden and Norway have used extensively since 1969, but which the United States seldom has used. Svante Odén (1924-86) at the Swedish University of Agriculture Sciences in Uppsala in 1968 analyzed precipitation data from a European network of atmospheric chemistry stations established in 1954-55. The area he studied included Scandinavia's southern regions and showed clear evidence of acidified lava and water. The principal acidifying agent was sulfuric acid ($H_2SO_4$) that resulted from the increasing consumption of fossil fuels, such as high-sulfur content coal. Its increasing concentration increased the hydrogen and sulfate ions concentrations in the atmosphere. The lime or limestone added reacts with the acidified water to neutralize or reduce the acidity which increases the pH of the acidified water. Liming does not prevent pollution by high-sulfur fossil fuels, but it does enable fish and other lake inhabitants to survive until a reduction in harmful acidic emissions occurs.

Washing coal with water before combustion to give clean coal reduces the sulfur content by 2.5 percent and thereby the sulfur dioxide emitted when burning high-sulfur coal. A scrubber installed in a smokestack (spray tower scrubbers) removes sulfur dioxide present in the hot exhaust gases by injecting a lime or limestone water slurry into the smokestack. The limestone reacts with the sulfur dioxide to produce neutral calcium sulfate, a solid that is then removed from the scrubber. Scrubbers remove 70 to more than 90 percent of the sulfur dioxide emissions from coal burning plants.

A British company, the London Power Company, installed the first scrubber (flue gas desulfurization unit, FGD) in 1931 at its Battersea Station. British industrialists installed a second scrubber at the Swansea Power Station in 1935 and a third in 1938 at the Fulham Power Station. All three ceased operation during the World War II years. In the United States, the Tennessee Valley Authority (TVA) in the 1950s, Wisconsin Electric in 1965, and Detroit Edison in 1966 carried out pilot plant studies of scrubber effectiveness. The TVA in 1971 built a 1-megawatt full-scale demonstration test unit at its Colbert Fossil Plant in Alabama, and in 1972 the EPA funded the construction cost of three 10-megawatt prototype scrubbers at TVA's Shawnee Fossil Plant located 10 miles north of Paducah, Kentucky. The Shawnee plant initiated the FGD technology used globally today. Union Electric Company in St. Louis, Missouri, and Kansas Power and Light Company in Lawrence, Kansas, in 1967 and 1968 installed full-scale limestone scrubbers at their power stations. Both of these stations had steam electric-generating capacities greater than 100 megawatts.

By June 1973, the number of scrubbers at coal burning plants globally had increased to 42, mainly Japan's 36 scrubbers, such as those at the Mitsui Aluminum Plant near Omuta and the Japan Synthetic Rubber Plant near Chiba, and in the United States at Boston's Edison's Mystic Plant. The plants ranged in capacity from 5 megawatts to 250 megawatts. By 2000. the number of scrubbers had increased to 678 in 27 countries, mainly in the United States, Germany, and Japan. About 80 percent use lime or limestone scrubbing, the remainder use a spray-dry (an activated alumina compound) scrubber or a magnesium oxide scrubber.

Scrubbers are expensive, $100 per kilowatt, or $100-200 million per unit, but have proved very effective. The EPA showed in the 2000s that the installation of scrubbers reduced current sulfur dioxide emissions to 40 percent of their 1990 levels. The FGD scrubbers that the Dayton Power and Light Company installed at its Killen, Ohio, plant in May 2007 are the largest in North America. They remove 97 percent of the sulfur dioxide emissions. A $170 million scrubber installed in 2009 at the 650-megawatt South Carolina Electric and Gas (SCE&G) Fossil Plant near Charleston, South Carolina, removes 95 percent of the sulfur dioxide emissions.

The combustion of any fossil fuel in a power plant, in addition to sulfur dioxide produces nitrogen oxides, although automobile exhaust accounts for about 50 percent of the nitrogen oxides. Scrubbers and precipitators dealt effectively with sulfur dioxide and some of the other pollutants emitted by industrial plants, but until the 1970s no operational anti-pollution device removed nitrogen oxides emitted by power plants, about 40 percent, and the 50 percent or more in automobile exhaust.

The Englehard Corporation in Iselin, New Jersey, patented a selective catalytic reduction (SCR) system for industrial power plants in 1957. Aqueous ammonia or urea passes into and reacts with (reduces) nitrogen oxides using a vanadium oxide or tungsten oxide catalyst in a catalytic reduction unit (catalyst bed). The reaction converts the nitrogen oxides to nitrogen gas and water. SCR reduces nitrogen oxides emissions by 70 to 95 percent. The IHI Corporation, a Japanese company in Tokyo that manufactures heavy industrial and aerospace equipment, installed the first large-scale SCR unit in 1978. SCE&G installed a SCR unit at its 650-megawatt Charleston plant in 2004, at a cost of $65 million. The unit reduced nitrogen oxides emissions by 70 percent.

Carl Keith (1920-2008) and John Mooney (b. 1929) at the Englehard Corporation solved the problem of automobile nitrogen oxides emissions in 1973 when they invented the catalytic converter. The mandatory installation in 1975 of a catalytic converter on all automobiles manufactured in the United States, Canada, and most other countries significantly reduced nitrogen oxides emissions, and the hydrocarbon and carbon monoxide emissions, found in automobile exhaust.

Electrostatic precipitation, the best known of which is the Cottrell precipitator, uses high voltages, 30,000 to 80,000 volts, to remove sulfur dioxide and other pollutants emitted from smokestacks. The Cottrell precipitator consists of a high-voltage negatively-charged wire installed vertically in the smokestack. The particles of the gaseous pollutants passing out of the smokestack acquire a negative electric charge by contact, and when they collide with the positively-charged walls of the smokestack they are neutralized and fall to the bottom of the smokestack for removal.

A reduction in the amount of fossil fuels consumed provides the least expensive solution to the acid rain problem. Alternative energies such as solar, wind, and nuclear energy provide other ways of reducing the amount of fossil fuels consumed and therefore of the emissions that are the primary source of acid rain.

*Government regulations to control acid rain*

The EPA in 1971 set a maximum limit on sulfur dioxide in the air, and in 1977, revisions to the 1970 Clean Air Act required scrubbers to remove sulfur dioxide, installed in all new power plants. Scrubbers today remove about 90 percent of sulfur dioxide emissions.

The combustion of fossil fuels, particularly unwashed coal, also releases mercury, an element that does not cause acid rain but produces neurological disorders in humans. Mercury emissions remained unregulated until the Clean Air Act as amended in 1990 gave the EPA the authority to restrict its emissions. Nevertheless, in 1995 industrial plants in the United States released 50 tons of mercury into the atmosphere.

Most of the mercury emissions originate outside the United States. Of the 5,500-8,800 tons of global annual emissions, in 2013, 50 percent came from Asia, mainly China. In the US about 1 percent came from coal plants, giving a total of 3 percent from all US sources. The Bush administration's Clear Skies Act of 2003 called for a reduction of US emissions to 34 tons in 2010 and 10 tons in 2013, but both objectives failed. The EPA, under considerable pressure from the Bush administration, allowed power plant lawyers to write significant parts of the legislation regulating mercury emissions. The pro-industry regulations greatly weakened the restrictions on mercury emissions and led in February 2008 to a court hearing in Washington, DC. At the hearing, the US Court of Appeals for the DC Circuit found that the EPA had illegally violated the Clean Air Act's requirements by not enforcing regulations to control mercury emissions. Moreover, the problem of mercury emissions is international and any serious reduction in mercury emissions requires an international solution, which remains unfulfilled.

Congress, in 1980, passed a second act to deal with the acid rain problem. The Acid Deposition Act of 1980 initiated a 10-year research program under the direction of the National Acidic Precipitation Assessment Program (NAPAP). NAPAP provided its first evaluation of acid rain in the United States in 1991. It showed that 5 percent of New England's lakes were acidic, mainly because of sulfates; 2 percent of the lakes could not support trout, and 6 percent were unsuitable for some minnow species. NAPAP noted that the amounts of nutrients in the soil had decreased and historical monuments had suffered damage.

In 1990 Congress passed amendments to the 1970 Clean Air Act. Title IV of the amendments established the Acid Rain Program. The Acid Rain Program introduced a Cap and Trade Program to control sulfur dioxide emissions. Title IV had two stages, the first beginning in 1995, the second in 2000, and proposed to reduce permanent power plant emissions of sulfur dioxide from their 1980 total of 18 million tons by 10 million tons, about a 50 percent reduction. Stage One limited sulfur dioxide emissions from the 110 largest power plants to 8.7 million tons of sulfur dioxide per year. Stage Two began in 2000 and included all US power plants larger than 25-megawatts capacity. Total sulfur dioxide emissions in 2007 were 8.9 million tons sulfur dioxide, which met the 1990 requirement.

In the Cap and Trade Program that Congress established to deal with the problem of sulfur dioxide emissions, each polluter receives or can purchase an emission allowance for each ton (or for any designated unit) of a specific pollutant it emits. If the plant emits fewer than its allotted tons of the pollutant, it can sell its extra tons and thereby reduce its operating cost or the cost of equipment it may have purchased to reduce its polluting emissions. Since establishing the Cap and Trade Program in 1990, sulfur dioxide emissions in the United States had decreased 40 percent by 2010 and 65 percent since 1976. During this same period the European Union achieved a 70 percent reduction in sulfur dioxide emissions.

The 1990 Clean Air Act amendments placed no nationwide cap on nitrogen oxides emissions. The Act aimed to reduce nitrogen oxides emissions by 400,000 tons between 1996 and 1999 and allowed each plant to establish its own maximum. Total emissions, which were 375,000 tons in 1975, rose to about 600,000 tons in 1995 but since that time have fallen to 300,000 tons in 2010.

In 2005 the EPA announced the Clean Air Interstate Rule (CAIR) to deal with the problem of interstate pollution from power plant emissions of sulfur dioxide and nitrogen oxides. CAIR plans to reduce by 70 percent sulfur dioxides emissions in 28 eastern states and DC from their 2003 level and the nitrogen oxides emissions by 60 percent.

| **FOG, SMOKE, SMOG, PHOTOCHEMICAL SMOG** | |
|---|---|
| *Fog* | is a suspension of water vapor droplets in the air. |
| *Smoke* | is a suspension of carbon particles in the air produced by the combustion of coal and has an irritating odor. Electrostatic precipitation, such as a Cottrell precipitator, can remove larger smoke particles. |
| *Smog* | is a combination of smoke from the combustion of coal and fog. It occurs in heavy industrial areas. Sulfur dioxide gas ($SO_2$) produced by the combustion of sulfur usually present in the coal, is often a component of smog. |
| *Photochemical Smog* | results from the reaction of sunlight (photons) with the pollutants emitted by the combustion of coal, such as sulfur dioxide ($SO_2$) and the nitrogen oxides, nitric oxide (NO) and nitrogen dioxide ($NO_2$), and with volatile organic compounds such as gasoline, paints, and organic solvents. At least 50 percent of the nitrogen oxides come from the reaction between atmospheric nitrogen and oxygen at the high temperatures present in automobile internal combustion engine exhaust. Power plant emissions produce another 40 percent. The photochemical reactions with the nitrogen oxides in automobile exhaust and with the volatile organic compounds produce ozone ($O_3$) which is a highly irritating and toxic gas present in photochemical smog. Aldehydes, peroxybenzoyl nitrate, nitrogen oxides, and the volatile organic compounds are other noxious pollutants found in photochemical smog. |

***Climate change (global warming): its three-stage history***

*Introductory*

While Smith, Mabery, and Cottrell were investigating how the increasing combustion of fossil fuels, particularly coal, contributed to the acidity of rain, other nineteenth-century scientists began to investigate a different fossil fuel problem. They suspected that the increasing fossil fuel combustion was increasing the concentration of atmospheric carbon dioxide (anthropogenic carbon dioxide, $CO_2$), changing the atmosphere's composition, and that the changing atmosphere could change Earth's climate. The small number of nineteenth-century scientists who accepted a changing atmosphere-changing climate theory initially focused on accounting for the ice ages and glacier formation and to a lesser extent on global warming. Despite the tremendous technological transformation that was occurring in the western world in the last half of the nineteenth century and the recognition that the transformation was spewing huge amounts of heat-absorbing gases such as carbon dioxide into the atmosphere, much of the scientific community remained convinced that little evidence supported a warming of Earth's surface temperature. If any increase of Earth's surface air temperature had occurred, the warming would be slow and insignificant. Belief in a slow and insignificant warming continued into the 1940s. The nineteenth-century believers in a changing atmosphere-changing climate theory, a minority in the scientific community, represent the first stage in the history of climate change.

As the twentieth century passed through its early decades, from 1900 through the 1940s, and the evidence of any significant change in climate still lacking, research on the changing atmosphere-changing climate theory gradually lost the scientific community's interest and faded from discussion. The successive tragedies of World War I, the Great Depression, and World War II pushed climate science theories further into the background. During this same time period, from 1900 through the 1940s, the causes of ice ages and glacier formation, the first applications of numerical methods to analyze weather patterns, and other climate issues remained relevant research interests of atmospheric scientists. When linked to new evidence from infrared absorption spectra and gas chromatography published in the 1950s and to global atmospheric surface temperature measurements for the same time period, the scientific community looked anew at the forgotten nineteenth-century changing atmosphere-climate change theory. Absorption spectra and gas chromatography showed an increasing concentration of atmospheric carbon dioxide compared to pre-1800s-50s concentrations, the atmospheric temperature measurements revealed a corresponding global surface temperature rise. The decades from 1900 through the 1940s represent the second stage in the history of climate change.

The research that began in the 1950s and continues in the 2000s represents the third stage in the history of climate change. The introduction of high speed computers and computer programs that performed the numerous, difficult, and tedious calculations that the laws of physics and chemistry required, such as the differential equations of fluid dynamics and thermodynamics, resulted in numerical models that much more comprehensively and accurately simulated the consequences of climate-related variables (temperature, pressure, concentration) on the atmosphere. At that time Wallace Broecker (b. 1931), a geoscientist at Columbia, upon evaluating the recent evidence, introduced in a scientific context the expression *global warming*. (Broecker "Are We on the Brink of a Pronounced Global Warming?" *Science,* 8 August 1975). The scientific evidence supporting Earth's warming, obtained from surface air temperature measurements at thousands of land-based meteorological (monitoring) stations and sea surface measurements from ships, moored buoys, and since 1978 from satellites with remote sensing instruments; from weather balloons that provided vertical profiles of temperature, humidity, and winds; and instruments that measured subsurface ocean temperature, salinity, and currents, continued to accumulate, so that by the first decade of the twenty-first century, climate science researchers nearly unanimously accepted the reality of the globe's warming. The combined evidence showed that the average global surface temperature, which for the period 1951-80 was 14° C (57° F), gradually had warmed over the twentieth century. It increased by more than 0.8 ° C (1.4° F) over the past century, much of the increase occurring in the last 35 years. For comparison, a decrease of 5° C (9° F) represents the difference between today's global temperature and an ice age. Little disagreement exists within the scientific community. The challenges to the scientific evidence have come from the political side and the doubters in the coal and petroleum industries, who have an economic interest in denying the scientific evidence. The scientific debate, if ever one existed, has ended but unfortunately the political debate has not. Climate politics unfortunately has replaced climate science.

*First stage 1800-1900: establishing a changing atmosphere-changing climate theory to account for ice ages and other climate changes*

The geologist James Hutton (1726-97) in Edinburgh, the father of uniformitarianism, and other geologists and observers by the late 1700s had begun to suspect a connection between ice ages and climate change. They found evidence of earlier glaciers, such as randomly located boulders that they called *erratics* and scoured rocks, in regions that in the 1700s were too warm for glaciers to exist and believed that thousands of years ago glaciers covered most of northern Europe.

One observer, Jean Pierre Perraudin (1767-1858), a chamois hunter in Loutier, France, upon finding huge granite boulders in the Val de Bagnes region of Switzerland's western Alpine valleys, in 1815 proposed that glaciers moving down the mountains had deposited the boulders in the valley. Ignaz Venetz (1788-1859), a Swiss naturalist-engineer in Valais, as a result of his own observations in the Val de Bagnes which he published in 1833, agreed with Perraudin. So did Venetz's friend, the early disbeliever Jean de Charpentier (1786-1855), a German-Swiss geologist in Bex, Switzerland, who discussed his Swiss glacier studies in 1835, six years before their publication in 1841. Louis Agassiz, the Swiss scientist, who in 1846 came and remained at Harvard, also concluded from his 1837 studies of European and North American glaciers that ice ages had occurred on Earth. His analysis of glacial activity was the first scientific study of glaciers and clearly supported Perraudin's glacier theory. Agassiz converted William Buckland (1784-1856), the former Oxford catastrophist who previously tried to link the randomly deposited boulders and scoured rocks to the biblical flood.

By the 1870s the presence of huge boulders far removed from existing glaciers convinced most scientists that ice ages had occurred and that Earth's climate had changed. A natural disaster such as volcanic activity leading to smog formation and a blocking of the Sun's rays could have produced a cooling of Earth's lower atmosphere, and if Earth's temperature decreased enough an ice age could occur. A change in solar activity also could produce a change in Earth's atmosphere. To many scientists, the atmosphere likely held the answer to what caused ice ages and Earth's climate to change.

The earliest investigations linking atmospheric change with climate change occurred in 1824 and 1827. Joseph Fourier (1768-1830), a leader of the French analytical school of mathematics at the École Polytechnique in Paris, concluded from the 1767 glass-covered solar cooker experiments of Horace Bénédict de Saussure (1740-99), a physicist in Geneva, that Earth's atmosphere retained some of the heat energy it received from the Sun. Fourier, like Saussure, believed that Earth's surface, upon receiving solar radiation of visible light wavelengths (4,000-7,000 angstroms), absorbed some of it, reflected some of it into space as lower energy, longer wavelength, infrared radiation (greater than 7,000 angstroms), and that Earth's atmosphere trapped some of the reflected infrared radiation, preventing its escape into space and thereby warming Earth. The reflected radiation fell in the low energy infrared region of the electromagnetic spectrum because Earth's surface and surrounding atmosphere had relatively low temperatures and therefore reflected low energy radiation. The trapped radiation produced a warming effect instead of a cooling and led Fourier to compare Earth with its surrounding atmosphere to a box covered with glass (a greenhouse). Fourier's analysis showed that because of its atmosphere Earth had a temperature higher (about 15° C or 59° F) than it would have had without a heat-absorbing atmosphere.

An explanation of how Earth's atmosphere prevented infrared radiation from escaping and produced Earth's warming came in 1859. John Tyndall (1820-93), at the Royal Institution in London, while studying possible causes of recently-discovered ice ages that may have occurred 7,000-10,000 years ago, found in his investigations on the transparency of gases that, contrary to the prevailing scientific opinion, not all gases were transparent to radiation. Oxygen ($O_2$) and nitrogen ($N_2$), which made up nearly 99 percent of the atmosphere, were transparent to infrared radiation, but $CO_2$ (0.03 percent), water vapor ($H_2O$), and coal gas (a mixture of hydrocarbon gases including methane $CH_4$ produced by the distillation of coal) were not and prevented the radiation of heat (infrared radiation) into space. Tyndall believed water vapor absorbed the most infrared radiation and produced the greatest warming of Earth's surface.

Scientific interest in the cause of ice ages, a changing atmosphere, the carbon cycle, and the effect of heat-absorbing atmospheric gases such as $CO_2$ continued throughout the nineteenth century. Arvid Högbom (1857-1940) and Arrhenius at Stockholm shared an interest in these climate science issues, and aimed to resolve them. Arrhenius, in response to lively discussion that arose at meetings of the Physical Society of Stockholm on the causes of an ice age, investigated the influence of heat-absorbing gases in the atmosphere, and in April 1896 he published his results in the *Philosophical Magazine.* Arrhenius used Samuel Langley's 1890 measurements of the Moon's infrared absorption spectrum and Högbom's 1894 estimate of human-produced (anthropogenic) $CO_2$ and showed that doubling the amount of $CO_2$ in the atmosphere produced a global average temperature rise of 5-6° C. The mean global temperature calculated from measurements recorded at meteorological stations in the mid-to-late 1800s was about 13.5° C (56.3° F). Twentieth-century measurements of air bubbles trapped in ice cores gave a concentration of 290 ppm by volume $CO_2$ in the atmosphere in the 1890s.

Langley, until 1891 at the Allegheny Observatory in Pittsburgh, was trying to calculate the Moon's surface temperature by measuring the amount of infrared radiation leaving the Moon and reaching Earth. He used a bolometer, an electrical instrument that he invented in 1878. It consisted of a rock salt (NaCl) prism that produced an infrared spectrum, two thin strips of metal on which infrared radiation fell and produced an electric current, a battery, and a galvanometer to measure the electric current. Langley examined the infrared spectrum when the Moon was in different locations in the sky and its rays reached Earth's surface at 35 to 40 degree angles. He found that when the rays struck Earth at a low angle and therefore passed through more $CO_2$ and water vapor, a greater absorption of the infrared radiation occurred than when the radiation struck Earth at a higher angle. Langley calculated that 38 percent of the heat from the Moon passed through the atmosphere and that the Moon had a surface temperature a little above 0° C and a *mean effective temperature* of 45° C (113° F).

Högbom's study of the carbon cycle led him to calculate the amount of $CO_2$ natural sources emitted and to compare that amount with the amount of $CO_2$ emitted mainly because of the increased industrial combustion of coal. He showed that at the current 1890s low rate of increasing $CO_2$ emissions, the amount of atmospheric $CO_2$ would double over a very long time, likely thousands of years.

Neither did Arrhenius seriously worry about European and American industrial expansion that was putting more $CO_2$ into the atmosphere. He recognized that $CO_2$'s increased presence in the atmosphere would trap more of Earth's reflected heat like the glass of a greenhouse and produce a global warming. But he believed that at the current pace of industrial expansion, a 5-6° C temperature rise from doubling $CO_2$'s concentration would require a 3,000-year time span. Arrhenius discussed global warming again in his 1908 book, *Worlds in the Making: The Evolution of the Universe.*

At about the time Arrhenius was examining the $CO_2$, ice ages, and climate change connection, two Americans, the geologists Thomas Chamberlin (1843-1928) then at Chicago and Cyrus Tolman (1873-1942) at Stanford, in 1899 undertook similar investigations. Chamberlin theorized that a changing $CO_2$ atmospheric concentration might cause significant global climate changes, including glacier formation and melting. Tolman did some of the earliest studies on the effect of oceans on $CO_2$'s global distribution.

*Second stage 1900-1940s: loss of interest in a changing atmosphere-changing climate theory*

The decline of interest in a changing atmosphere-changing climate theory in the decades from 1900 through the1940s did not lead to a complete abandonment of atmospheric studies, some of them continued to investigate atmospheric $CO_2$ concentrations, climate change and ice ages, and weather forecasting. Climate change studies in the 1920s-40s included those of Charles Schuchert (1856-1942) at Yale and Alfred Lotka (1880-1949) a mathematician and chemist at Johns Hopkins. Schuchert wrote in 1919 on the volcanic origin of $CO_2$. Lotka in 1925, in *Elements of Physical Biology* calculated that continued consumption of coal at the present rate would lead to a doubling of carbon dioxide emissions in 500 years. "Economically we are living on our capital; biologically we are changing radically the complexion of our share in the carbon cycle by throwing into the atmosphere, from coal fires and metallurgical furnaces ten times as much $CO_2$ as in the natural biological process of breathing. . . . (T)hese human agencies alone would . . . double the amount of carbon dioxide in the entire atmosphere. . . ."

One year later in his 1926 book, *The Biosphere,* Vladimir Vernadsky (1863-1945), a Russian-Ukrainian geochemist at St. Petersburg (Leningrad), examined the connection between Earth's atmosphere and living processes. A second Russian, the geochemist Vladimir A. Kostitzin (1883-1963) in 1935 in *Evolution de l'atmosphere, circulation organique, époques glaciares,* discussed carbon, oxygen, and nitrogen concentrations and their abundance in the atmosphere and soil and the use of atmospheric models to deal with their cycles. Kostitzin spent his later years in Paris.

The ongoing studies on the causes of climate change and ice ages led in 1941 to their first comprehensive evaluation. Milutin Milankovitch (1879-1958), a Serbian astrophysicist at Belgrade, proposed an explanation of ice ages based on irregularities in Earth's three motions. Milankovitch noted that as Earth traveled in its orbit around the Sun, cyclical variations in Earth's orbital eccentricity, obliquity, and precession affected Earth-Sun geometry and therefore the amount of solar energy that reached Earth.

The variations in Earth's orbital eccentricity changed the orbit's shape from a nearly perfect circle to an oval, as Earth when viewed from above, or looking down from the north pole, revolved counter–clockwise from west to east around the Sun. The variations changed Earth's distance from the Sun as Earth moved through a 90,000-100,000-year cycle and produced a maximum 20-30 percent difference in the amount of solar energy reaching Earth.

The variations in obliquity (axial tilt) changed the angle that Earth's axis made with the plane of Earth's orbit, as Earth rotated counter-clockwise on its axis from west to east. Earth's axial tilt, which in the 2000s is 23½°, varies from 22.1° to 24.5° in a 41,000-year cycle. The variations in obliquity determine the strength of the seasons. As the axial tilt increased, the seasonal contrasts increased, resulting in colder winters and hotter summers. When the axial tilt decreased the seasonal contrasts diminished.

The third motion, precession, varied the times at which the seasons begin. Precession occurs because as Earth rotates (spins) on its axis, its axis slowly wobbles (rotates) westward through the 360 degrees of the constellations, in the 2000s from Pisces to Aquarius. The 360-degree rotation takes about 26,000 years and results in the seasons' beginning earlier and earlier in the year. Because Earth is not a perfect sphere and bulges at the equator, the gravitational attraction of the Sun produces a torque or pull on the bulge that causes the slow rotation called precession.

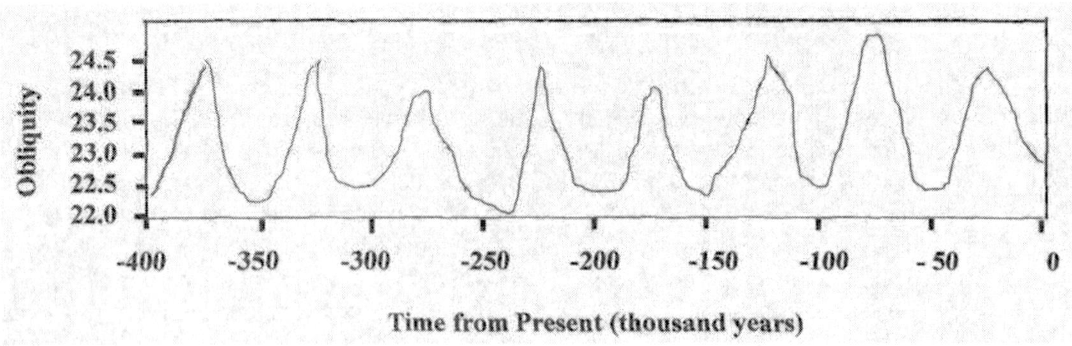


**Figure 25**
**Milankovitch Cycles, Orbital Obliquity**
Courtesy of NOAA

The Milankovitch cycles represented the time periods during which the irregularities in the orbital motions occurred. As he and others pointed out, during these long time periods occasions arose when the sunlight reaching certain regions of Earth decreased and failed to melt the snow in the higher latitudes. The snow reflected more of the sunlight reaching it, causing further cooling, and over a long enough time produced ice sheets and glaciers. When the region again received more sunlight, possibly thousands of years later, the ice age would end and a warming period began. Milankovitch published his papers on climate change, some dating from 1912, in 1941 in German, in a single volume entitled *Canon of Insolation of the Earth and Its Application to the Problem of the Ice Ages.* An English translation, *Canon of Insolation of the Ice-Age Problem*, appeared in 1969.

Later studies in the 1950s-60s that applied numerical methods and computer models to climate research provided convincing evidence and support for Milankovitch's theory of orbital cycles, climate change, and weather patterns. The theoretical meteorologist Norman A. Phillips (b. 1923) at MIT and from 1974 at the National Weather Service, National Meteorological Center, and Joseph Smagorinsky (1924-2005), the founding director of the National Oceanic and Atmospheric Administration (NOAA)-Geophysical Fluid Dynamics Laboratory at Princeton, beginning in the 1950s introduced numerical methods that led to the first computer models (equations, programs) of weather and climate. Phillips' simulation-based approach and six basic equations he introduced represented the physics of hydrodynamics and for the first time reproduced the global flow patterns of Earth's atmosphere. Phillips' simulation-based method became the accepted method that climate scientists used to develop circulation models that enabled them to understand and determine how energy moved through the atmosphere. His computer model allowed scientists to reproduce the entire atmosphere's wind and pressure patterns. Applied to a two-layered atmosphere, it advanced the scientific world's understanding of the atmosphere's general circulation, particularly heat and moisture transfer. Phillips demonstrated for the first time the ability of a simple numerical model, since called a General Circulation model, to predict weather conditions, and in 1956 he began using computer models to predict weather patterns and disturbances. His models transformed weather forecasting from an individual effort to a group of experts using computer programs to make weather predictions. Smagorinsky played a major role in establishing a national weather monitoring system.

The International Geophysical Year (IGY), which started on 1 July 1957, also directed scientists to Milankovich's cycles and climate change. It launched a collaborative effort by scientists from 64 countries to study and unravel the complexities of Earth, its atmosphere, and climate. One important study, the *butterfly effect*, that Edward Lorenz (1917-2008), a meteorologist at MIT proposed in 1961 and named in 1972, showed the extreme difficulty if not the impossibility of predicting weather forecasts too far into the future. Lorenz's *butterfly effect* demonstrated the unpredictability, or chaotic state, of climate systems and how a minor change in the initial conditions randomly could trigger a major change in the future climate. It served as a prelude to Lorenz's pioneering studies on chaos theory that he published in 1963 and discussed at the 1965 conference on Causes of Climate Change held in Boulder, Colorado. His studies supported Milankovitch and other scientists who believed that small changes in the eccentricity of Earth's orbit and axial tilt caused past ice ages.

Beginning in 1966, evidence of short-term Pleistocene Epoch (The Great Ice Age 2-3 million to10,000 years ago) and Holocene Epoch (the time since the Pleistocene, the last ice age about 10,000 years ago) sea-level changes provided additional support for Milankovitch's cycles and for the existence of earlier glaciation and deglaciation periods. An analysis of the patterns of deposit on uplifted ancient coral reefs in Barbados and along the shore of Papua, New Guinea, provided the evidence. The pattern of deposits, visible as steps, indicated that when an ice age began, the oceans' waters froze into polar ice caps, causing the ocean level to fall and exposing the coral reefs. When the ice age ended, the ocean level rose and warmed during the interglacial period, and new coral reefs formed. The motion of Earth's shifting tectonic plates pushed up the plates, revealing the pattern of visible steps. Measurements of the uranium in the reefs indicated the years between the ice ages and the duration of each age. As early as 1966 and continuing into the 1970s-80s, cores drilled from the seabed supported the existence of earlier periods of glaciation and deglaciation and Milankovitch's theory of ice ages.

*Third stage 1950s-present: revival of the changing atmosphere-changing climate theory, the role of $CO_2$*

*Callendar and Plass*

Guy Callendar (1898-1964), a Canadian-born steam engineer and amateur meteorologist in West Sussex, England, continued and extended the nineteenth-century climate change studies of Fourier, Tyndall, and Arrhenius. His records, consisting of 95 notebooks (8 boxes) in the archives of the University of East Anglia in Norwich, contain weather charts and tables, global temperature measurements from meteorological stations and locations around the world, and atmospheric $CO_2$ concentrations taken from 1898 to 1901 at the Jodrell Laboratory at Kew Gardens, London, and from the eastern United States in 1930-38. Callendar recognized that Earth's temperature had risen during the years 1900-40. He linked the rise to the retreat of glaciers, the increasing concentration of atmospheric $CO_2$ compared to pre-industrial concentrations, and as recent evidence from infrared absorption band spectra revealed, the presence of methane and water vapor in the atmosphere.

Beginning in 1938 with a paper "The Artificial Production of Carbon Dioxide and its Influence on Climate," and in publications through 1964, Callendar evaluated temperature measurements from 200 meteorological stations around the globe, in Edinburgh, Copenhagen, Oxford, New York City, Honolulu, relying primarily on the Smithsonian Institution's *World Weather Records* (WWR) and atmospheric $CO_2$ concentrations taken at Kew. He concluded that during the past half century, humans had added about 150 billion ($150 \times 10^9$) tons of $CO_2$ to the atmosphere and that global temperatures had increased at the rate of 0.005° C per year. About 75 percent of the $CO_2$ remained and according to Callendar would cause an estimated future mean temperature increase of 0.003° C per year.

Callendar thought the warming trend particularly significant north of the 45th parallel. He closely connected increased coal burning with increasing atmospheric $CO_2$ concentrations and believed that the increased $CO_2$ concentrations, which had risen 10 percent during the last century, could account for the increased global warming that scientists had observed. He estimated that doubling the atmospheric $CO_2$ concentration would produce a gradual 2° C rise over the next few centuries and saw the rise as a *negative environmental effect.* One of Callendar's primary interests in studying atmospheric temperature change was, like other climate scientists, to discover the cause of ice ages. His publications received some attention in the 1940s-50s climatology textbooks, but generated little enthusiasm within the scientific community, because from about 1900 to the 1950s, scientists had largely dismissed $CO_2$'s significance in affecting climate change.

The physicist A. Nelson Dingle (1916-2011) at Michigan in 1954, Giles Slocum at the US Weather Bureau in Washington, DC, in 1955, and Stig Fonselius (1921-2003), a hydrographer at Stockholm in 1956, claimed that nineteenth- and twentieth-century measurements did not show a significant increase in $CO_2$ concentrations. Dingle said that the $CO_2$ increase might have resulted from a change in the general atmospheric circulation and not from a global increase in $CO_2$ concentrations. Most critics believed incorrectly that the oceans' waters would absorb completely the excess $CO_2$ and that even if a few percent increase in $CO_2$'s concentration had occurred, the increase likely produced no increase in the air temperature near Earth's surface. The few scientists who did not reject Callendar's $CO_2$-global warming theory included Gilbert Plass, Roger Revelle, Hans Suess, and Charles Keeling in the United States. Revelle and Suess in 1957 renamed Callendar's $CO_2$-global warming theory the Callendar effect.

Gilbert Plass (1920-2004), another Canadian-born physicist, worked in the academic and industrial worlds. He spent the war years 1942-45 at the Manhattan Project's Chicago Metallurgical Laboratory; 1950s-62 with Research Laboratory, Aeronutronic Division of Ford Motor Company in Newport Beach, California; until 1968 at the Southwest Center for Advanced Studies in Dallas, Texas; and since 1968 until his death at Texas A&M University. He provided infrared spectral evidence that supported the connection between climate change and increasing atmospheric $CO_2$ concentrations. In papers published from 1953 to 1959, Plass argued that none of the other climate change theories, such as variations in solar energy that Earth received or the amount of volcanic dust in the atmosphere, accounted for most of the facts scientists had discovered about climate change. In his 1956 paper published in *Tellus*, a European meteorology and oceanography journal, Plass calculated that doubling Earth's $CO_2$ concentration would produce a 3.6°C increase in Earth's surface temperature whereas a 50 percent reduction of $CO_2$ would decrease the surface temperature by 3.8° C. He predicted a 30 percent increase in $CO_2$'s atmospheric concentration and an Earth surface temperature 1° C higher in 2000 than in 1900. Fifty years later, the United Nations Intergovernmental Panel on Climate Change in its fourth report published in 2007 estimated a 2-4.5° C temperature increase for a doubling of the $CO_2$ concentration, a 37 percent $CO_2$ increase since the pre-industrial period; and a 0.7° C rise for the 1900-2000 interval.

Plass made his calculations and remarkable predictions in his 1956 paper, without having available Revelle and Suess's work on $CO_2$ dissolution in the oceans, Keeling's studies on $CO_2$'s rate of increase in the atmosphere, or the research of Willi Dansgaard (1922-2011) at Copenhagen who, beginning in 1952 and 1954 used Greenland's ice cores to study past climate conditions.

Dansgaard originated the procedure of drilling into a thick sheet of ice, removing a core of ice usually 500-1,500 grams, and analyzing the air molecules carried down and trapped within the snowflakes that fell to Earth. The falling snow over the centuries formed the thick layers of ice that covered the Arctic and Antarctic. He made his first analyses of the air bubbles trapped in the ice cores in 1964, extracting cores from a depth of about 1,360 meters, at Camp Century, a nuclear-powered research station that the US Army Corps of Engineers from 1959 to 1966 operated in Northwestern Greenland. Dansgaard determined the O-18/O-16 ratios of the trapped air bubbles and showed that a decrease of 0.7 ppm in the O-18 isotope corresponded to a 1° C drop in air temperature (a 1 ppm decrease in O-18 corresponded to a 1.5° C decrease). He concluded that 80-180 year cycles of warming and cooling dominated the last 800 years. Dangaard's analyses, especially his 1979-81 analyses of air bubbles in ice cores located at the Dye Radar Station in South Greenland, established ice core science as a fundamental experimental tool in discovering the climate of the past. Drills can reach depths of 3,500 meters and extract cores 4-6 meters long.

Plass, an excellent spectroscopist, relied primarily on the distinct infrared absorption band spectra of water vapor, $CO_2$, and ozone he obtained and used computers in climate modeling. His infrared spectra and the computer models he developed enabled him to show that the wavelength range of the water vapor and $CO_2$ absorption bands were independent and that their spectral lines did not overlap. This disproved the prevailing belief that $CO_2$'s absorption of infrared radiation was insignificant and negligible compared to that of water vapor. Plass pointed out that the concentration of water vapor diminished very rapidly with height, whereas $CO_2$ had a uniform distribution throughout the atmosphere.

In Plass's time, scientists estimated an atmospheric $CO_2$ concentration of about 300 ppm (by volume). Plass and most physicists believed a natural balance existed between $CO_2$ removed from the atmosphere, about $60 \times 10^9$ (60 billion) tons per year mainly by photosynthesis, and $CO_2$ returned by the respiration and decay of plants and animals. Volcano release of $CO_2$ also balanced $CO_2$ removal in the formation of carbonates that occurred in the weathering of igneous rocks. He feared, however, that humanity had upset the balance, and as Callendar had noted, human consumption of fossil fuels would emit large amounts, $6 \times 10^9$ tons, of $CO_2$ per year. Plass maintained correctly, contrary to prevailing scientific opinion, that the oceans would not absorb all the $CO_2$ and that this extra $CO_2$, mainly human-generated, had caused most of Earth's warming.

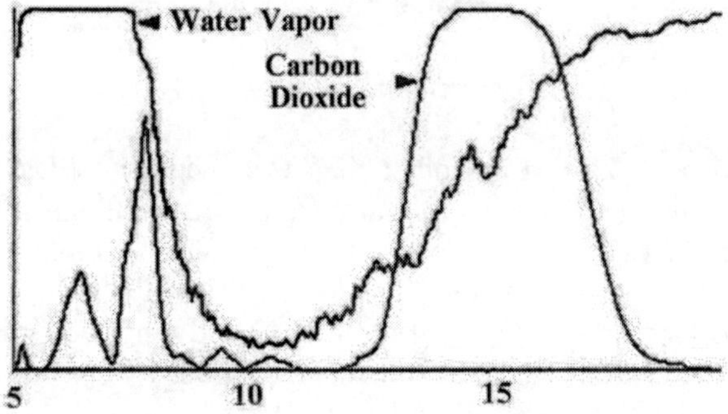


Carbon dioxide is the most frequently mentioned greenhouse gas, but water vapor absorbs infrared (heat) radiation much more strongly. Carbon dioxide is significant because it closes a "window" that would otherwise allow certain infrared wavelengths to escape Earth's water vapor blanket.

**Figure 26**
**Absorption Spectra for Water Vapor and Carbon Dioxide in Earth's Atmosphere**
Courtesy of NASA

Lewis Kaplan (b. 1917), a theoretical physicist then at MIT, criticized Plass's 1956 research and published his comments in *Tellus* in 1960 and had an exchange of letters with Plass in *Tellus* in 1961. He made some of the first criticisms of global warming based on his own scientific research and not only on a perceived weakness in the data of climate change science proponents. Despite using more recent infrared spectra than Plass, he concluded incorrectly (1) that Plass's estimated 3.8° C temperature drop if the $CO_2$ concentration fell by half, was 2-3 times too high, and (2) that $CO_2$ variations therefore could not play a role in the ice age cycle unless the variations were 10 times as large (an order of magnitude).

Plass was not totally correct in all his calculations because he focused solely on $CO_2$ and in 1956 had no knowledge of the impact of 1970s research on greenhouse gas emissions such as methane ($CH_4$), nitrous oxide ($N_2O$), the chlorofluorocarbons (CFC, in 1974), and on aerosol emissions. But contrary to Kaplan's charges, the scientific evidence has confirmed Plass's recognizing the significance of rapidly rising atmospheric concentrations of $CO_2$ on global warming.

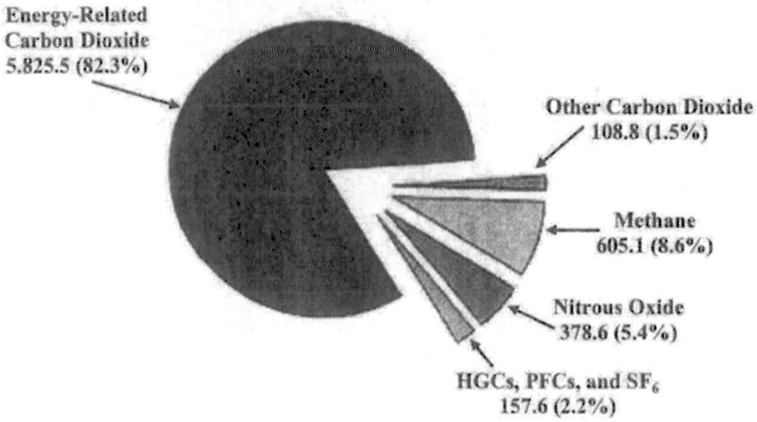


**Figure 27**
**US Anthropogenic Greenhouse Gas Emissions by Gas, 2006**
**(Million metric tons of carbon dioxide equivalent)**
Source: Energy Information Administration. Emissions of Greenhouse Gases in the United Sates 2006 Washington, DC, November 2007

*Roger Revelle, $CO_2$ absorption, oceans, and climate change*

Well into the 1950s, few scientists other than Callendar and Plass, challenged the dogma that the oceans dissolved enough $CO_2$ to prevent any significant resultant global warming. The research that shattered this near-infallible dogma came from the oceanographer Roger Revelle (1909-91) at the Scripps Institute of Oceanography established in 1903 in La Jolla, California. Scripps established an affiliation with the University of California in 1912 and with the University of California, San Diego, upon its founding in 1960. Revelle, a believer in the ocean-$CO_2$ absorption model, came to Scripps in 1931, and served as its director from 1951 to 1964.

Revelle's 1950s pioneering studies of the ocean waters' turnover rate and their buffering effect showed that, contrary to prevailing scientific opinion, the world's oceans could not absorb unlimited quantities of atmospheric carbon dioxide ($CO_2$). His ocean studies and other studies that showed a steadily increasing concentration of anthropogenic (human caused) carbon dioxide in the atmosphere left no doubt about Earth's future warming.

• *$CO_2$ absorption, oceans' turnover rate and buffering*

Serious scientific study of the oceans effectively began with the British Treasury-sponsored *HMS Challenger* expedition of December 1872 - May 1876. Only a few small scientific expeditions in the North Sea and Scandinavia preceded the *Challenger* expedition, and none rivaled *Challenger's* scope and significance. The $16 million global expedition of 69,000 nautical miles gathered and analyzed for the first time extensive data on the biology, geology, and chemistry of the oceans and ocean floors. Its scientists measured the oceans' temperatures, depths, currents, and salinity. They found that at depths greater than 500 meters, life existed despite the darkness, high pressure, and extreme cold. *Challenger* established the oceans' alkalinity and that they contained about 50 times as much dissolved $CO_2$ as the atmosphere.

This last and clearly significant discovery led global warming proponents from Arrhenius to Callendar, Plass, and especially Revelle to question why the oceans did not absorb more or even all of the atmospheric $CO_2$ and thereby eliminate or significantly reduce any global warming. They suspected that two factors determined the $CO_2$ absorption rate and therefore the quantity of $CO_2$

absorbed over long time periods: (1) the rate at which the oceans *turned over*, that is, the time required for the oceans' surface water to sink to the oceans' depths and for the deep ocean water to rise to the surface, and (2) the oceans' buffering effect, which indicated their resistance to dissolving unlimited quantities of atmospheric $CO_2$.

Revelle's research on the oceans' turnover rate originated in the 1954 Bikini Atoll-Pacific Ocean radioactive fallout studies that scientists from US government and academic institutions, such as the Naval Medical Research Institute in Bethesda, Maryland; Los Alamos National Laboratory in New Mexico; Scripps; and the University of Washington in Seattle, provided the Office of Naval Research in Arlington, Virginia. The Bikini tests of March-May 1954 were one of 66 atmospheric and surface nuclear bomb tests conducted from 1946 to 1958 on the Bikini and Enewetak atolls, which were two of the 29 atolls and 5 islands that make up the Marshall Islands. The fallout from the Castle Bravo 15-megaton hydrogen bomb, the first of four hydrogen bombs detonated on the Bikini Atoll's 23 islands, consisted of metallic ions of 50-100 micrometer size, such as ions of strontium - 90, 91, 92, and cesium -137 that chemically bonded to water molecules, and radioactive calcium oxide particles from ocean-bottom corals. The radioactive fallout killed one and severely poisoned the other 22 crewmen of the Japanese tuna fishing vessel, *Daiko Fukuryu Maru* (*Lucky Dragon 5*), located 80 miles east of Bikini and contributed to the global protests condemning atomic and hydrogen bomb construction. The 13 years of Bikini tests also revived the twin horrors of Hiroshima and Nagasaki and brought new Japanese fears of long lasting or permanent radioactive pollution of their Pacific fishing grounds.

The Bikini tests enabled the Office of Naval Research to determine how quickly the oceans turned over and carried the radioactive fallout to their depths. From measurements of the concentration of radioactive carbon present in the fallout, initially short-lived C-11 and then C-14, as it moved through the ocean water, Revelle estimated that the turnover took about 300 years. Later studies in the 1960s-70s by the physicists Minze Stuiver (b. ca. 1935) at Yale and now at Washington and Hans Suess (1909-93) at Scripps and the geochemist Wallace Broecker at Columbia, supported a much longer turnover rate of about 500-1,000 years. They considered the longer turnover rate rapid enough for the oceans to dissolve sufficient atmospheric $CO_2$ and thereby prevent the temperature above the surface water from rising too quickly. In addition to the oceans' slow mixing, Christopher Sabine at the NOOA Pacific Marine Environmental Laboratory in Seattle, showed in a 2004 publication that about 50 percent of the $CO_2$ in the oceans remained in the upper 10 percent of the oceans' depth, at a depth of less than 400 meters, and that the average depth to which the $CO_2$ sank was about 1,000 meters. In this upper region of the ocean the temperature changed rapidly with depth. Only 7 percent of total anthropogenic $CO_2$ sank to a depth greater than 1,500 meters. The circulation, which depends on the ocean water's temperature and density (salinity) differences, prevents the dissolved $CO_2$ from sinking farther into the oceans' interior, whose deepest regions vary from about 5 to 7 miles, or from 8,200 to 11,500 meters.

The second factor, the buffering effect, resulted from the interaction of atmospheric $CO_2$ and the oceans' surface water. The high percentage of dissolved $CO_2$ that remained at or near the surface produced the buffer and thereby prevented more atmospheric $CO_2$ from dissolving. Revelle's long time interest and concern about ocean waters' $CO_2$ (g) buffering began in May 1955, when Scripps participated in the Office of Naval Research's one-day nuclear project called Operation Wigwam. In the Wigwam Operation of 14 May, the Navy and the Atomic Energy Commission detonated a 30-kiloton atomic bomb 660 meters (2,000 feet) underwater at a site about 500 miles southwest of San Diego. The Navy's objective was to determine the vulnerability of submarines to deeply detonated nuclear bombs and the effectiveness of deep underwater nuclear bombs in warfare. While studying the radioactive fallout Revelle observed that the fallout spread itself in a thin layer over a 100-kilometer area but only to a depth of one meter. The fallout in the upper layer of the water appeared to remain there indefinitely and occupied a volume of water only one-fiftieth to one-hundredth the volume of the fallout area. Revelle surmised that atmospheric $CO_2$, in contact with the oceans, possibly behaved the same way, remaining dissolved in a thin surface layer of water and preventing more atmospheric $CO_2$ from dissolving. Because of ocean water's buffering effect, the $CO_2$ either had to remain in the atmosphere or pass back into the atmosphere, thereby causing Earth's temperature to rise.

Revelle did not discover the oceans' buffering effect, both Callendar and Plass foresaw the oceans' buffering effect, but once again he and Suess in 1957 provided one of the first clear explanations of its mechanism and consequences. Buffering occurred at the interface between atmospheric $CO_2$ and the oceans' surface water because some of the $CO_2$ dissolved and reacted with the ocean water to form carbonic acid and negatively-charged bicarbonate ion, and carbonate ion, collectively called dissolved inorganic carbon (DIC). The dissolving of atmospheric $CO_2$ reduced the $CO_2$ concentration in the atmosphere and the total alkalinity of the oceans. The chemical equations below show the reactions responsible for the production of DIC and the hydrogen ions that increase the oceans' acidity (lowers the pH value), and the resulting buffering of the oceans' surface waters.

$$CO_2(g) + H_2O(l) \leftrightarrow \underset{\text{Carbonic acid}}{H_2CO_3}$$

$$H_2CO_3 \leftrightarrow \underset{\text{Hydrogen ion}}{H^+} + \underset{\text{Bicarbonate ion}}{HCO_3^-}$$

$$HCO_3^- \leftrightarrow H^+ + \underset{\text{Carbonate ion}}{CO_3^{2-}}$$

The total alkalinity measures the total concentration of negatively–charged ions in an ocean water sample. Most are bicarbonate and carbonate ions, only a few other well–known negative ions such as borate which compete for hydrogen ions, are present and in small concentrations, so that measuring the bicarbonate and carbonate ion concentrations effectively measures the sample's total alkalinity.

• *Processes to measure DIC and total alkalinity*

Revelle, Suess, and Andrew Dickson at Scripps, and other scientists beginning in the 1960s developed separate processes to measure the DIC and the total alkalinity of the oceans' surface water. The first process, the DIC-determining process, measured the $CO_2$ (g) pressure and occurred in two stages. Stage one converted all the DIC to $CO_2$ (g) by reacting an ocean water sample contained in an evacuated extraction flask with dilute phosphoric acid. In stage two, the $CO_2$ (g) and the other gases present in the extraction flask, mainly water vapor, nitrogen, and oxygen, passed through a series of cryogenic (low-temperature) traps. The traps selectively condensed each gas at a different temperature and permitted its extraction or collection in an ampule. An electronic constant-volume manometer (ECM) measured the $CO_2$ pressure in the $CO_2$ ampule and therefore its concentration (its number of moles).

The second process measured the total alkalinity of an ocean water sample by making a potentiometric titration of a weighed sample in a container, sealed to the atmosphere and kept at constant temperature. The process required adding small volumes of dilute hydrochloric acid (0.1 molar HCl) to the container and measuring the change in the electromotive potential (emf, or voltage) of the ocean water with a standard combination type pH electrode. A computer recorded the volume of acid added, the emf, and temperature, and produced a titration curve with two inflections. The first inflection point on the titration curve indicates the conversion of carbonate ions to bicarbonate ions upon adding hydrogen ions (HCl) to the solution. Adding more acid converts the bicarbonate ions to $CO_2$ and produces the second inflection point on the titration curve. The second inflection point measures the total alkalinity of the ocean water sample.

*Revelle Factor (R): measuring the buffering effect*

Calculations of $CO_2$'s buffering effect published into the 1950s showed that oceanic and land (photosynthesis) sinks absorbed only about 50 percent of the $CO_2$ (g) emitted during the industrial period of the last 200–250 years. They led Revelle to search for a mathematical equation that related the change in the concentration, or partial pressure, of the $CO_2$ ($\Delta$ $pCO_2$) in the atmosphere and the change in concentration of dissolved inorganic carbon ($\Delta$ DIC) in the ocean. The equation he introduced to measure quantitatively the oceans' surface layer's resistance to absorbing $CO_2$ (g) and what he called " the peculiar buffer mechanism of sea water" appeared in a preliminary form in a 1957 paper he and Suess published in *Tellus*. Broecker renamed Revelle's "peculiar buffer mechanism" the Revelle factor (R) in 1979.

$$\text{R (Revelle Factor)} = \frac{\Delta pCO_2/pCO_2}{\Delta[\text{DIC}]/[\text{DIC}]}$$

According to the equation, a low R factor meant that the ocean surface layer offered little resistance to atmospheric $CO_2$'s dissolving in the ocean, whereas a high R factor offered high resistance to more $CO_2$'s dissolving in the ocean. Revelle and Suess used C-14 to follow the exchange of $CO_2$ between the atmosphere and the oceans and calculated an R factor of about 10. The R factor in the oceans in the early 2000s varies from 8-9 in low latitude subtropical and tropical waters to 13-15 in cold, high north and south latitude waters such as the Southern Ocean off Antarctica. R is also one unit higher than in preindustrial times because of the higher concentration of atmospheric $CO_2$ mainly anthropogenic in origin.

*Suess carbon isotope determinations*

Suess, Revelle's collaborator in these studies, joined Scripps in 1955 when Revelle, then serving as director, invited him. Suess previously worked on hydrogen-deuterium (hydrogen-2) exchange and heavy water chemistry for Germany's World War II atomic bomb project while at Hamburg and served as a scientific advisor to Norse Hydro, the Norwegian heavy water plant in Vemork. In 1948-49, he and the physicist Hans Jensen (1907-73) then at Hamburg, studied the magic numbers of the nuclear shell model. Suess came to the United States to the University of Chicago in 1950 at the invitation of Harrison Brown (1917-86). At Chicago and then with the US Geological Survey in Washington, DC, Suess continued to perfect the C-14 radioactive dating method. His measurement of radioactive C-14 in atmospheric $CO_2$ enabled him to show in 1955, while at Scripps, that the increase in atmospheric $CO_2$ concentrations resulted from the combustion of fossil fuels and was anthropogenic.

Suess proved that fossil fuels were so old that they contained no measurable amount of C-14, which has a half-life of 5,568 years. When burned, the $CO_2$ a fossil fuel released, diluted the concentration of C-14 already present in the atmosphere (and in plants) and thereby indicated that the $CO_2$ added resulted from the combustion of a fossil fuel. The C-14 dilution is the *Suess effect*. Suess also measured the distribution of C-14 and tritium (hydrogen-3) in the oceans to determine the age of an ocean sample. In 1956 he used C-14 dating to shorten the chronology of the last of the seven, possibly 11, glacier periods (ice ages) that occurred 110,000 –10,000 years ago in the northern hemisphere. An isotope ratio mass spectrometer compared values of the C-14/C-12 ratio in the marine carbonate of mid-Atlantic and Caribbean deep sea cores, with the C-14/C-12 ratio in modern wood and from the lower C-14/C-12 ratio in modern wood showed that the last glacier period reached its maximum southward extent 17,000-18,000 years ago. Earth's temperature, which began falling about 80,000 years ago, reached its minimum temperature about 15,000 years ago.

• *Average lifetime for a $CO_2$ molecule in the atmosphere*

By this time Revelle had become aware of Callendar's and Plass's studies on global warming, and he began to speak publicly about the long-term consequences of global warming. He increasingly believed that Earth's temperature would rise in the next 100 years because of greenhouse gases but acknowledged his uncertainty about the number of degrees, possibly 2° C or 10° C. The uncertainty led Revelle and Suess, beginning in 1956, to continue making mass spectrometric measurements to determine how much $CO_2$ the oceans could absorb from the atmosphere and how long the absorption would take. An isotope mass ratio spectrometer compared C-14/C-12 and C-13/C-12 ratios in marine carbonate (shells) and in a wood (plant) standard. Suess's 1954-55 mass spectrometric measurements of the C-13/C-12 ratios showed that the ratio was 2.5 percent higher in marine carbonate than in land plants. Suess had calculated an apparent age of about 400 years for marine carbonate, and from the slight decrease that they found in the C-14 concentration of terrestrial plants in the last 50 years concluded that on average a molecule of $CO_2$ remained in the atmosphere for about 10 years before it dissolved in the ocean. Their 10-year lifetime corresponded to a net exchange rate of about $10^7$ mole of $CO_2$ per second (1 mole $CO_2$ = 44 grams) for a square meter of ocean surface. They assumed that except for a small increase in atmospheric $CO_2$ resulting from an addition of $CO_2$, containing no C-14, no other changes had occurred in the oceanic and atmospheric reservoirs and that the small $CO_2$ increase would not affect the exchange rate.

Revelle and Suess noted that their 10-year lifetime was 100 times longer than Plass's proposed time. Their calculations suggested that the oceans had absorbed most of the $CO_2$ released by fossil fuel combustion since the beginning of the industrial revolution and that at the present rate of industrial emissions the oceans could absorb an excess of $CO_2$. They published these results in their 1957 *Tellus* paper.

The equation in which they proposed that t (ocean), the average lifetime of a carbon atom as a member of a marine reservoir, equaled the average apparent C-14 age of the marine carbonate appears below.

$$\frac{t\,(\text{ocean})}{t\,(\text{atm})} = \frac{k_1}{k_2} = \frac{S_O}{A_O}$$

| | | |
|---|---|---|
| t (ocean) | = | average lifetime of carbon in the ocean (marine carbonate) before it becomes atmospheric $CO_2$ |
| t (atom) | = | average lifetime of a $CO_2$ molecule in the atmosphere before it dissolved in the ocean |
| $k_1$ | = | 1/t (atm) = rate of $CO_2$ transfer per year from atmosphere to ocean |
| $k_2$ | = | 1/t (ocean) = rate of $CO_2$ transfer per year from ocean to atmosphere |
| $S_O$ | = | total carbon of the marine carbon reservoir at equilibrium condition, at time zero |
| $A_O$ | = | atmospheric $CO_2$ carbon at time zero |

Measurements of atmospheric $CO_2$ concentrations obtained from observatories such as the Mauna Loa Observatory in Hawaii beginning in 1960 and continuing to the 2000s have confirmed Revelle and Suess's short 10-year atmospheric residence (exchange) time of $CO_2$. They show that for a total atmospheric concentration of 720-750 gigatons (1gigaton = 1 billion tons = $10^9$ tons or 360-375 ppm) and an annual removal of 230-200 gigatons (115-100 ppm) of atmospheric $CO_2$ by land and mainly ocean sinks, a $CO_2$ molecule resides in the atmosphere for about 3.1 years before returning to the ocean. Harmon Craig (1926-2003), Revelle's colleague at Scripps, using two different methods, in 1957 and 1958 calculated residence times of 4-10 years or about 5 years average. Studies done in the 1980s-2000s also give short residence times of 5-15 years.

Revelle and Suess recognized however, that the real significance of their ocean studies was not the short atmospheric residence time of a $CO_2$ molecule. What really mattered was the increasingly large amounts of anthropogenic $CO_2$ entering the atmosphere and upsetting the natural carbon cycle.

• *How much $CO_2$ remains in the atmosphere*

In the carbon cycle most of the $CO_2$ entering the atmosphere, about 200 gigatons per year, traded places with an equal amount leaving the oceans. But beginning with the industrial revolution more $CO_2$ from fossil fuel combustion, about 3.3 gigatons per year, has entered than left the atmosphere. Although $CO_2$ rather quickly dissolved in the oceans most of it remained in the surface waters which were becoming increasing saturated. The oceans' slow circulation and turnover rate which ranged from 500 years to thousands of years prevented the transfer of $CO_2$ from the surface water to deep water. Moreover the oceans' buffering effect prevented the surface water from absorbing unlimited quantities of $CO_2$.

Revelle and Suess's 1957 paper showed a clear connection between fossil fuel combustion and increased atmospheric $CO_2$ concentrations and rising atmospheric temperatures. It contains one of the most prophetic and quoted statements made about global warming.

*Human beings are now carrying out a large scale geophysical experiment of a kind that could not have happened in the past nor be reproduced in the future. Within a few centuries we are returning to the atmosphere and oceans the concentrated organic carbon stored in sedimentary rocks over hundreds of millions of years. This experiment, if adequately documented, may yield a far-reaching insight into the processes determining weather and climate. It therefore becomes of prime importance to attempt to determine the way in which carbon dioxide is partitioned between the atmosphere, the oceans, the biosphere* (the thin layer around Earth in which life exists) *and the lithosphere* (the solid part of Earth).

Revelle initially calculated that 80 percent of the $CO_2$ passing into the atmosphere, most of it coming from industrial emissions, would remain in the atmosphere. He always feared that if industrial $CO_2$ emissions rose exponentially in future decades, the oceans would not remove the emissions and the increased atmospheric $CO_2$ concentrations would contribute to making Earth's temperature rise a reality. The scientific evidence has validated his greatest fear.

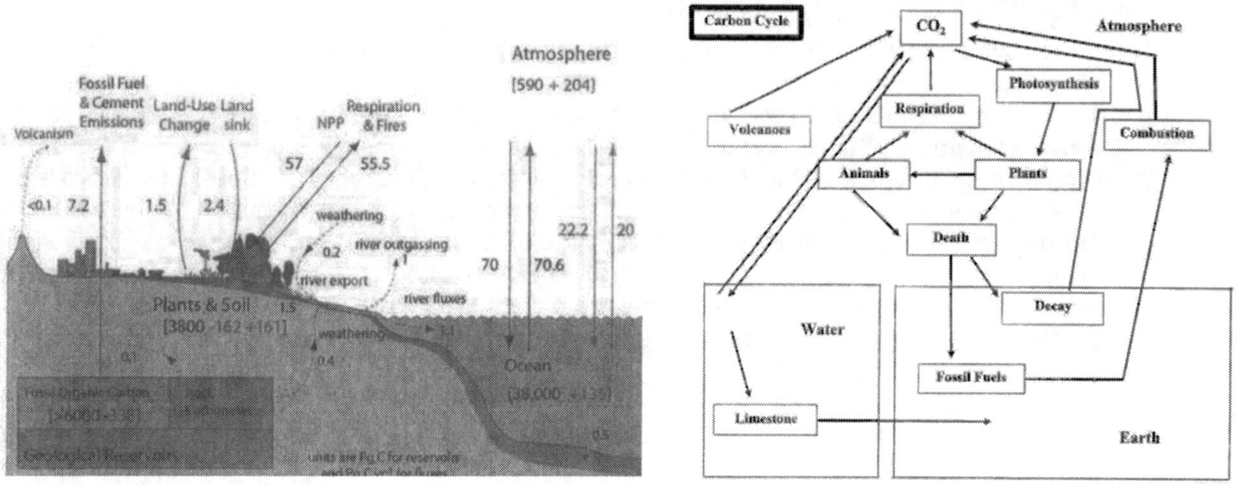


**Figure 28**
**The Carbon Cycle**
Courtesy of NOAA (left)
Courtesy of the author (right)

In 1971 Broecker calculated that the oceans would remove 35±10 percent of the $CO_2$ produced by fossil fuel combustion, leaving 55 to 75 percent in the atmosphere. In 1975 Broecker, using the results the United Nations published in *World Energy Supply Statistical Papers*, Series J, 1950-59, assumed that 50 percent of the $CO_2$ remained in the atmosphere. That same year, he became one of the first scientists, along with Plass, to use models to predict a future global warming. Later, in 1977, Revelle, with new data on projected $CO_2$ emissions, concluded that a 40 percent rise in atmospheric $CO_2$ concentration would occur over the next centuries. A more recent 2004 study by Sabine showed that the oceans removed about 30 percent of the anthropogenic $CO_2$ in the atmosphere leaving about 70 percent. He found that the oceanic uptake reduced net atmospheric $CO_2$ concentration by 55 ppm compared to an atmospheric $CO_2$ concentration without an oceanic sink.

In other publications, Revelle expanded the work of previous researchers and investigated the reaction mechanisms in the carbon cycle by which dissolved $CO_2$ returned to the atmosphere in the form of volatile organic compounds that dissociated easily to give free $CO_2$. He recognized that the $CO_2$ reactions were part of an exchange cycle between the oceans and the atmosphere and that transfer of $CO_2$ from the oceans to the atmosphere rapidly returned large amounts of $CO_2$ to the atmosphere. He identified some of the many pathways, noting that the pathways depended on factors that included ocean water temperature differences, and variations in the oceans' chemical composition and organisms present in the surface waters. The faster the $CO_2$ evaporation rate the less acidic the oceans' waters. The faster the evaporation rate, however, the more quickly the atmospheric $CO_2$ concentrations increased which resulted in more atmospheric warming and hence more rapid evaporation of $CO_2$ present in the oceans. In an experiment in which Revelle increased the atmospheric $CO_2$ concentration above a tank of sea water from 350 ppm to 400 ppm and used radioactive C-11 to follow the reaction pathway, more $CO_2$ returned to the atmosphere than remained dissolved in the sea water. Recent studies in the 2000s have shown that each 1° C rise in the ocean's surface temperature produces a 7 percent increase in the rate at which $CO_2$ returns to the atmosphere.

• *Other contributions and significance*

Revelle was a navy reservist, and in 1941, six months before Pearl Harbor, the navy called Revelle to active duty. He remained in the navy until 1948, serving as a commander in the Bureau of Ships and as a branch head in the Office of Naval Research. He maintained his connections with the navy and during his career he received considerable financial support from the navy. Later, in 1957-58, Revelle served as chairman of the US National Committee on the International Geophysical Year (IGY). The Scripps Institute participated in the IGY and became the principal center for the Atmospheric Carbon Dioxide Program. Revelle was one of the fourteen panelists who in November 1965 published the 317-page "Restoring the Quality of Our Environment," *Report of the Environmental Pollution Panel, President's Science Advisory Committee.* The panel's study was the first authoritative government report to recognize that $CO_2$ emissions from fossil fuels posed a potential global problem. He chaired the National Academy of Sciences Energy and Climate Panel which in 1977 warned of the serious consequences of human-caused $CO_2$ emissions. Of the 40 percent of human-caused $CO_2$ that stayed in the atmosphere, the Panel noted that two-thirds of it came from fossil fuels, the remaining one-third from forest clearing.

Revelle's 1950s pioneering studies of the ocean waters' turnover rate and their buffering effect showed that contrary to prevailing scientific opinion the world's oceans could not absorb unlimited quantities of atmospheric carbon dioxide ($CO_2$). His studies left no doubt about Earth's future warming resulting from the steadily increasing concentration of anthropogenic (human caused) carbon dioxide in the atmosphere. The Revelle factor (R) provided the first quantitative measure of the oceans' ability to dissolve atmospheric $CO_2$.

*Confirming the link between a changing atmosphere and a changing climate: Charles Keeling and the Keeling Curve*

Revelle invited Suess to the Scripps Institute in 1955. In 1956, he brought the chemist Charles Keeling (1928-2005) to Scripps to head the Atmospheric Carbon Dioxide Program, that in 1957-58 established meteorological monitoring stations in Antarctica and Mauna Loa, Hawaii. At that time, from 1953 to 1956, Keeling was Harrison Brown's postdoctoral student at Caltech and had investigated the equilibrium between carbonate ion ($CO_3^{2-}$) in surface waters and limestone ($CaCO_3$) and $CO_2$ in the atmosphere. A gas manometer that Keeling constructed to measure $CO_2$'s partial pressure, enabled him to calculate the concentrations of $CO_2$ removed from the atmosphere and the acidified water.

Keeling's early investigations included daytime measurements of atmospheric $CO_2$ concentrations in Pasadena. He found considerable variation in the concentrations and attributed the variation to industrial pollution. To test his hypothesis Keeling in 1955-56 took additional measurements in areas that had no urban pollution and no vegetation (forests and grasslands) and found that the $CO_2$ concentrations remained fairly constant over time. Keeling had discovered what he called the *atmospheric background concentration* of $CO_2$, and it led him, beginning in 1957-58, to consider measuring atmospheric $CO_2$ concentrations around the world to see if a global pattern existed. Later, scientists adopted the idea of making global measurements and determining the atmospheric background concentration of other greenhouse gases, such as methane ($CH_4$) and nitrous oxide ($N_2O$), and of the chlorofluorocarbons (CFCls) that were destroying the ozone layer in the stratosphere.

Another set of measurements taken in a vegetated region at Big Sur near Monterey, California, during the day and at night revealed a second discovery, an unsuspected daily pattern of $CO_2$ concentrations. The atmosphere contained more $CO_2$ at night than during the day and when corrected for the presence of water vapor had a nearly constant $CO_2$ afternoon concentration of 310 ppm. To measure the $CO_2$ concentrations Keeling collected the air samples in evacuated 5-liter flasks. He isolated the $CO_2$ by freezing it in a liquid nitrogen trap (bp -321° F) and distilling at dry ice temperature (sublimation temperature -109.3° F) to separate the $CO_2$ from condensed water vapor. A gas manometer measured the $CO_2$ concentration. (Dry ice is solid $CO_2$. Sublimation temperature is the temperature at which a gas changed directly to a solid or the solid to a gas).

Keeling also determined the daytime and nighttime ratios of the carbon isotopes C-13/C-12 in his $CO_2$ samples and discovered that they too differed, the C-13/C-12 ratio in his $CO_2$ samples was smaller at night than during the day. An isotope ratio mass spectrometer measured the C-13/C-12 isotope ratio. Elemental carbon consists mainly of two stable nonradioactive isotopes, C-12, 98.9 percent; C-13, 1.1 percent; and a trace of radioactive C-14. An isotope ratio mass spectrometer can measure the ratio to 1 part in 100,000 and determine the C-13/C-12 ratios in atmospheric $CO_2$ from different sources. The Big Sur measurements showed that $CO_2$ produced by the combustion of fossil fuels (old organic matter) had a smaller C-13/C-12 ratio than that extracted from carbonates in the oceans and in mountain regions, or emitted by volcanoes and other natural processes.

Keeling made additional measurements in 1955-56 in several California forests, the Olympic Rain Forest in Washington, and Arizona's high mountain forests, and obtained the same day/night pattern of $CO_2$ concentrations and C-13/C-12 isotope ratios as previously. From these studies Keeling concluded that his lower afternoon measurements of $CO_2$ in vegetated regions, the lowest of which he obtained in the early morning near sunrise, likely represented the stable background concentrations of $CO_2$ found in the northern hemisphere. Respiration in which plants and animals took in oxygen and emitted $CO_2$ largely accounted for $CO_2$'s higher night-time concentrations.

His measurements also established that the C-13/C-12 ratio in $CO_2$ resulting from respiration in vegetated regions was lower than the C-13/C-12 ratio in $CO_2$ in regions largely free of vegetation. This occurred because in photosynthesis, which took place in the daytime in sunlight, vegetation removed $CO_2$ and water from the atmosphere. Vegetation more easily assimilated C-12 than C-13 and thereby increased the C-13/C-12 ratio. During respiration, which was the opposite of photosynthesis, vegetation took in oxygen and during the night released $CO_2$ which had a higher concentration of C-12 and decreased the C-13/C-12 ratio. This remarkable discovery of day-night variation of $CO_2$ concentrations is the Keeling Plot.

In 1956, and now at Scripps, Keeling began his expanded program to measure global atmospheric $CO_2$ concentrations in the background air. He received IGY funds from Harry Wexler (1911-62) at the US Weather Service in Silver Spring, Missouri, and used them to purchase four infrared gas analyzers, each equipped with a strip chart to measure continuously the $CO_2$ concentration. The analyzers directly compared the partial pressure of atmospheric $CO_2$ collected in the flasks with that of prepared mixtures of $CO_2$ in nitrogen gas. A gas manometer measured the $CO_{2\ partial}$ pressure (concentration) in the prepared reference mixtures. All the stations reported their measurements as the dry air mole fraction in ppm of $CO_2$.

$$\text{Dry air mole fraction} = \frac{\text{number of molecules of } CO_2}{\text{total number of molecules in air including } CO_2 - \text{water vapor}} \times 1000$$

$$= \text{ppm of } CO_2$$

Three of the analyzers continuously measured $CO_2$ concentrations at the meteorological stations established in 1957-58 at the South Pole in Antarctica, Mauna Loa (the Mauna Loa Observatory, MLO) in Hawaii, and at Scripps in La Jolla, California. The fourth analyzed air samples collected in the glass flasks at other stations and by other methods beginning in 1958. These included Little America in Antarctica, on ice floes in the Arctic, and by aircraft of the 55th Weather Reconnaissance Squadron of the US Air Force flying at five to six kilometers above the Pacific and Arctic Oceans. Beginning in 1960 Keeling obtained $CO_2$ concentrations from oceanographic expeditions.

Continuous measurements began at the South Pole in September 1957. The MLO, which is part of the NOAA, provided continuous records beginning in March 1958. NOAA established the observatory on the island of Hawaii on the large active (4,600-meter) volcano Mauna Loa (long mountain) because of its undisturbed air, remote location, and minimal influence of vegetation and human activity. It provides readings obtained at an altitude of 3,400 meters.

At this time, uncertainty existed about the buffering effect of the oceans on atmospheric $CO_2$ concentrations. The daily $CO_2$ concentration record from the MLO and other stations over a few years and the graph obtained from the record, called the Keeling Curve, ended the uncertainty. The global increase in atmospheric $CO_2$ concentration was a fact.

On its first day of operation in March 1958, the Mauna Loa infrared gas analyzer recorded an atmospheric $CO_2$ concentration of 313 ppm. The concentration rose 1 ppm in April, reached a maximum in May, and then began to decline, falling to its lowest value in October before increasing again. The concentration readings repeated the same seasonal pattern in 1959 and 1960. Keeling noted that humanity was seeing for the first time nature's withdrawal of $CO_2$ from the atmosphere for plant growth (photosynthesis) during the summer and returning it to the atmosphere during the winter and following spring. Of equal significance, the annual average atmospheric $CO_2$ concentration increased from 1958 to 1959 and even more in 1960. Keeling published his results in his 1960 *Tellus* article "The Concentration and Isotopic Abundances of Carbon Dioxide in the Atmosphere."

**Figure 29**
**Mauna Loa Observatory**
Courtesy of NOAA

The Mauna Loa measurements, which Keeling recorded until April 1974 and NOAA has continued to the present, showed that the increasing atmospheric $CO_2$ concentrations matched the increasing industrial and domestic consumption of fossil fuels. Keeling calculated that if 57 percent of airborne fossil fuel emissions remained airborne, the increasing fossil fuel consumption matched very well the experimentally established $CO_2$ concentration curve over time. The slope of the curve was gradual in the first years and steeper in the ensuing years.

Keeling later compared the Mauna Loa record with the variations in atmospheric $CO_2$ concentrations over the past 420,000 years. Polar ice cores beginning in the late 1960s provided air samples for the very old $CO_2$ concentrations which measured about 200 ppm during the ice ages and about 280 ppm during the warmer interglacial periods. Ice core samples taken from the Law Dome glacier in the Antarctic showed that atmospheric $CO_2$ concentrations remained steady at 270-280 ppm for the previous 1,000 years before beginning to rise sharply in the 1800s.

The ice core $CO_2$ concentrations contrasted sharply with the Mauna Loa continuous records that began in 1958 and showed a 35 percent increase in $CO_2$ concentrations for the years after the industrial revolution and a 6 percent increase in the last 20 years. Keeling cautioned, moreover, that if the fossil fuel consumption rate continued at the same accelerated rate, the projected $CO_2$ concentrations could rise to 1,500 ppm, pushing Earth into a new period of geologic history having a climate significantly different from what existed at the present time. Recent Mauna Loa readings were 378 ppm in 2005, 381 ppm in 2006, 383 ppm in 2007, 389 ppm in 2010, 396.78 ppm in May 2012, 398.35 ppm in May 2013, and 401.320 ppm in May 2014.

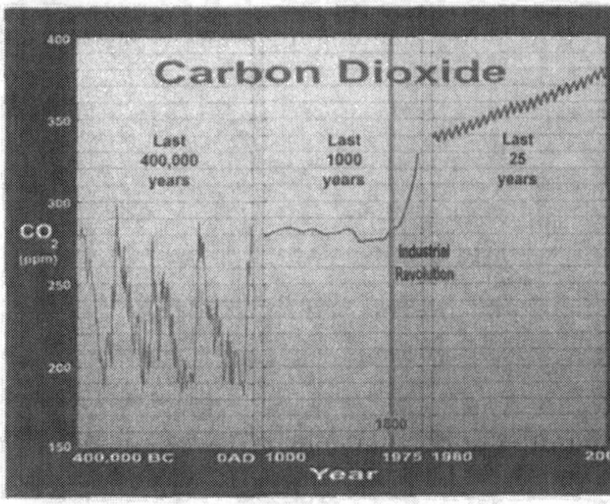


**Figure 30**
**Atmospheric Carbon Dioxide from Law Dome Ice Cores**
Courtesy of NASA

Keeling, and before him Callendar, Plass and Revelle, believed that the increasing fossil fuel combustion contributed significantly to the increasing concentration of atmospheric $CO_2$ and to Earth's warming. The figures that various industries, such as metallurgical and electrical (power plant) compiled, documenting the increasing industrial consumption of fossil fuels, supported their belief and enabled researchers at Scripps and elsewhere to calculate the amount of $CO_2$ released into the atmosphere. They reaffirmed that over the last 50 years the rate of $CO_2$ increase was about 57 percent of the rate expected if all (100 percent) of the $CO_2$ emissions had remained in the atmosphere.

The fossil fuel consumption figures provided three important outcomes. First, the environment, land and water, had absorbed and removed the other 43 percent of the $CO_2$ emissions. Second, if the $CO_2$ emissions continued at the rapidly increasing rate, the possibility existed that the environment could not continue to absorb higher concentrations of $CO_2$, resulting in continued global warming. Third, because the $CO_2$ increases closely paralleled the increased combustion of fossil fuels, and no other known natural sources of $CO_2$ emissions were significant enough to account for the recent increases of atmospheric $CO_2$, human-caused emissions provided the most logical explanation.

Keeling summed up the evidence linking fossil fuel consumption with increasing atmospheric $CO_2$ concentrations and global warming with these conclusions. No known sinks of $CO_2$ sufficient to absorb all the $CO_2$ emissions from fossil fuel combustion exist. For more than 10,000 years prior to the industrial revolution, atmospheric $CO_2$ concentrations changed little, indicating that the recent $CO_2$ increase did not result from natural causes. The decrease in the C-13/C-12 isotope ratio found in atmospheric $CO_2$ indicates that the $CO_2$ increase resulted from the combustion of fossil fuels, which consist of old organic matter and have a lower C-13/C-12 ratio than $CO_2$ from respiration and other natural processes. The Keeling Curve that Keeling first published in 1960 has provided, at the present time, the most widely recognized experimentally-established historical representation of the link between global warming-climate change and the atmospheric concentration of $CO_2$.

**Atmospheric CO at Mauna Los Observatory**

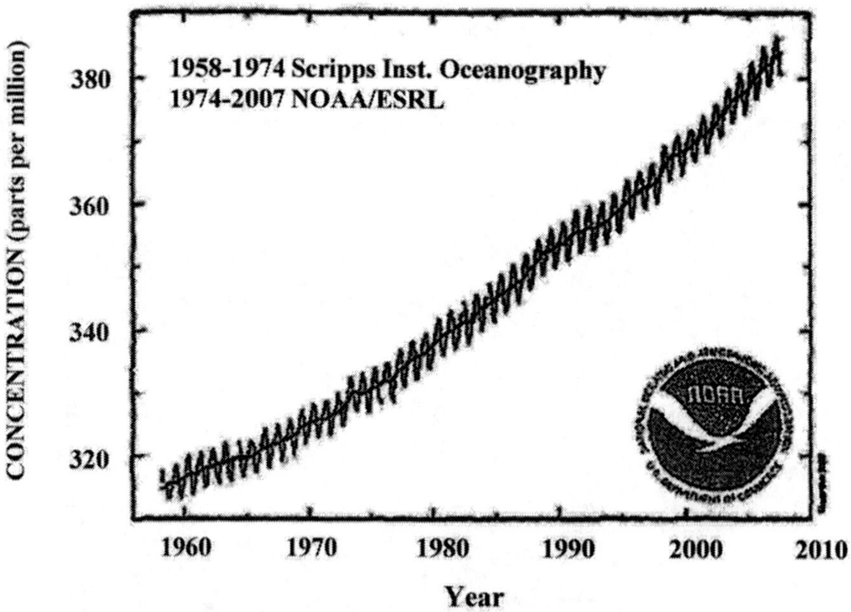


**Figure 31**
**Keeling Curve Mauna Loa Record**
Courtesy of NOAA

*Is the debate over: James Hansen's contributions to the changing atmosphere-changing climate theory*

James Hansen (b. 1941), a physicist at NASA's Goddard Institute for Space Studies (GISS) in New York City since 1967, the institute's head 1981-2013, and since 2013 a professor at Columbia's Earth Institute, has become the leading spokesman for climate scientists who support the $CO_2$ changing atmosphere-changing climate theory. Following his earlier 1960s-70s planetary studies of Venus's atmosphere, Hansen and the Goddard research group, developed their first climate change model, a global atmospheric circulation model, in 1974 and have continued to improve their models' reliability. The GISS model incorporated six fundamental equations of physics and chemistry that accounted for atmospheric conditions: an equation of motion, an equation of continuity (conservation of mass), an equation of state for a global grid, the first law of thermodynamics (conservation of energy), a hydrostatic equation, and a conservation equation for water vapor. They used the model to observe atmospheric system simulations and for experimental extended-range forecasting (five 48-hour periods and one 30-day period), not for long-term climate simulation. The model enabled them to calculate cloud and water vapor fields and from these results determine energy transfer by solar and terrestrial radiation. Their model's atmosphere reproduced an energy cycle in good agreement with climatological observations, including (the computed zonal mean fields of) winds, temperature, humidity, transports, diabatic heating, evaporation, precipitation, and cloud cover. GISS's first model successfully simulated the major climate features at North American sea-level pressure, about 1,013 millibar (mb), and at an elevation with a pressure reading of 500 mb, about 5.5 km. This is the steering level, the height at which weather systems beneath move in the same direction as winds at the 500 mb level. Their model provided a realistic simulation of continents, oceans, and topography.

A 1981 one-dimensional radiative-convective model calculated temperatures as a function of height. The calculated temperatures led Hansen to conclude that actual measured atmospheric $CO_2$ concentrations would produce an earlier-than-predicted warming. Hansen noted that atmospheric $CO_2$ concentrations increased from 280-300 ppm in 1880 to 335-340 ppm in 1980 mainly because of fossil fuel combustion. From the mid-1960s to 1980 the global temperature rose by 0.2° C, resulting in a 0.4° C warming in the past century. The temperature increase agreed with the greenhouse effect calculations based on measured increases of atmospheric $CO_2$. Hansen's one-dimensional model and a three-dimensional model he developed to include seasonal solar insolation (radiation received) variation, a 70-meter mixed layer ocean that accounted for heat capacity but no horizontal heat transfer, and a realistic global geography, predicted a mean global warming of 2-3.5° C for a doubling of $CO_2$'s concentration from 300 to 600 ppm. Possible twenty-first century consequences of the warming included drought-prone regions in North America and central Asia because of climate zones' shifting, West Antarctic ice sheet erosion and a global sea level rise, and the opening of the Northwest Passage.

In 1983 Hansen introduced a three-dimensional general circulation model that permitted long-range climate experiments lasting from three months to five years. A high speed computer (IBM 360/95 or Amdahl V-6) solved simultaneous equations for conservation of mass, energy, and momentum, an equation of state for a global grid, and the ideal gas law. The model included variables such as air convection schemes that mixed moisture, heat, and momentum; computed snow depths, cloud cover, and ground temperature; and estimated the change in mean global surface temperatures resulting from future greenhouse gas emissions.

A few years later in 1987 Hansen and Sergej Lebedeff (1928-90), at Sigma Data Services Corporation in New York City, undertook an analysis of surface air temperature records from 1,700 continental and island meteorological stations on every continent for the years 1880 to 1985. The World Weather Records (WWR) that the Smithsonian Institution and later NOAA published provided the temperature records. A three-dimensional general circulation model that they developed created an artificial global temperature history and estimated the error resulting from incomplete spatial coverage of Earth.

The results, which were similar in both hemispheres, indicated a warming of 0.5–0.7° C in the last century and revealed a strong warming trend between 1865 and 1880 that raised the global mean temperature in 1980 and 1981 to the highest-ever instrumentally-recorded level. Hansen noted that a 1940 warming period occurred mainly in the northern latitudes whereas the 1980-81 warming was more global. When extended to 1988 and then to 1999 the same warming pattern continued and showed that the four warmest recorded years were in the 1980s with 1981 and 1987 the highest, and that 1998 was the warmest year since 1880. The rate of temperature change in the 1980s-90s exceeded all rates of change recorded since 1880.

By 2001 Hansen had improved the validity of his temperature measurements by adjusting for the time of day he made a measurement, balancing the urban-suburban measurements, and including satellite measurements of urban night life intensity. He further refined his measurements in 2002-2005 by accounting for the influence of black carbon (carbon black, soot) on regional climate. A photometer, an instrument that measures the amount of light passing through it, detects the black carbon produced by the incomplete combustion of fossil fuels, biofuels or biomass. The particles, which are black because they absorb sunlight, reduce the amount of light passing through the photometer and instead heat the atmosphere and produce a warming effect. The black carbon effect is usually regional and short-lived because the carbon particles eventually fall to the ground.

The general circulation models that Hansen and other scientists used in climatology studies are mathematical equations of physics and chemistry that account for factors that influence atmospheric conditions, such as variations in composition and temperature, and aim to simulate (reproduce) and predict accurately atmospheric conditions. Their adoption began early in the twentieth century. The physicist Vilhelm Bjerknes (1862-1951) then at Stockholm, in 1904 introduced the first mathematical model in climatology and weather forecasting. His method of numerical forecasting that required his performing numerous and sometimes difficult mathematical calculations and solving of differential equations made his model's application impractical. The English mathematician and scientist Lewis Fry Richardson (1881-1953), who held positions at the National Physical Laboratory in London, in industry and in the academic world, continued Bjerknes's mathematical applications in climatology studies and published *Weather Prediction by Numerical Process* in 1922. Bjerknes's and Richardson's numerical approach and the methods based on mathematical functions that followed in the 1930s-40s proved inadequate because they provided only partial mathematical solutions. The resulting climate models reproduced some of the variables of the atmospheric layers but did not include all of the hydrodynamic and thermodynamic equations required to account for the general circulation of Earth's atmosphere.

Improved climate models appeared in the 1950s. With the introduction of computers such as ENIAC which made the first weather forecast in 1950 and computer programs such as IBM's Fortran in 1956-67, computer programs have performed the mathematical calculations that checked the validity of the variables included in the climate models. Climate models that reproduced and predicted the influence of weather variables began to appear in 1956 when Norman Phillips at MIT and the National Weather Service introduced his numerical methods and produced the first general circulation model of Earth's atmosphere.

These earlier breakthroughs were significant and laid the foundation for Hansen's improved and refined general circulation models. The agreement between his models' projections and actual observations, including temperature readings from satellites in the 1990s, prompted Hansen to argue that the debate had changed from whether global warming was occurring to discovering the rate of global warming and what society should do about the warming.

Hansen's climate science research led to numerous awards including election to the US National Academy of Sciences in 1996 and awards from the AAAS in 2006 and the American Meteorological Society in 2009. In recent years he advocated repeatedly and outspokenly for action on global warming. His public statements have drawn criticism from NASA during the George W. Bush years, from CEOs of coal and oil companies, and from his friend, the theoretical physicist Freeman Dyson (b. 1923) at the Advanced Institute for Study in Princeton, New Jersey. Dyson, a global warming skeptic, said Hansen had turned his science into ideology. The scientific community thinks otherwise.

***Deniers of the changing atmosphere-changing climate theory: Lomborg, Lindzen, and others, their six usual criticisms***

Critics and criticism of the $CO_2$ changing atmosphere-changing climate theory have existed almost from the time scientists began publishing articles indicating Earth's warming. Much of the criticism came in the 1980s when the scientific evidence showed a definite rise in the global annual mean temperature curve. Two events in particular both in 1988 triggered the renewed criticism. First, Hansen in June summarized before the US Senate Committee on Energy and Natural Resources his years of research that showed Earth's warming. Hansen testified that Earth had been warmer in the months from January to June 1988 than in any five-month period since the beginning of temperature recordings in 1858. He believed with 99 percent certainty that the anticipated warming trend did not result from a natural variation or cycle of Earth's temperature. The increasing concentration of $CO_2$ and other greenhouse gases caused the warming.

Second, the United Nations Environment Program (UNEP) and the World Meteorological Organization (WMO) in December established the Intergovernmental Panel on Climate Change (IPCC). One month earlier, in November 1988, 35 nations including the United States met in Geneva at the first plenary session of the IPCC to examine climate change and the consequences of global warming on humanity. The session led in December to the IPCC's formal establishment. The WMO and UNEP, organizers of the session and IPPC's co-founders, fund the IPCCs Secretariat; the WMO hosts the Secretariat in Geneva. They have opened membership to any nation belonging to the United Nations or WMO. The IPCC, numbering about 2,500 scientific experts in climatology, ecology, economics, medicine, and oceanography from 60 countries, published its first climate assessment report in 1990, stressing the necessity of reducing greenhouse gas emissions. Other IPCC reports followed in 1992, 1995, 2001, and 2007. Their reports, with other climate change reports, indicated that the ten warmest years globally have occurred since 1990, with 1998 the warmest, followed by 2002, 2003, 2001, and 1997.

The critics seized on Hansen's research, the IPCC reports, and numerous other reports supporting a rise in the global annual mean temperature, directing their attacks on the perceived errors, shortcomings, and weaknesses in the reports. The critics range from outright deniers of global warming to those who acknowledge the reality of global warming but deny its human origin and significance. They argue that most of the warming resulted from natural causes and focus their criticisms on six issues.

The first and the most common criticism is that the computer-generated climate models are unreliable because they cannot predict abrupt weather changes. This is correct, but abrupt weather changes affect weather forecasting and not longer-term regional and global climate. Some of the models have underestimated the extent of climate change, and the information obtained is not as inclusive of variables and broadly applicable as that obtained from actual meteorological stations. Nevertheless, since the 1980s, three-dimensional general circulation models have incorporated with good results, variables such as deep ocean circulations that might affect the dissolution of $CO_2$ in water and match data from meteorological stations about 75 percent.

A second criticism claims that scientists have manipulated data to give results that support global warming as allegedly revealed in November 2009 in the hundreds of e-mails, some from 1999, illegally obtained from the accounts of climate researchers at the University of East Anglia. The critics denounced and pronounced guilty of fraud Philip Jones (b. 1952), head of East Anglia's Hadley

Climate Research Unit, and Michael Mann (b. 1965), director of the Earth System Science Center at Penn State and author of the 1998 *Hockey Stick Graph of Temperature Change*. Data manipulation is a serious charge and threatens to undermine all science not just climate science. Fortunately science has the time-tested procedure of duplication, verification, and external review that has exposed most but not all fraud in science.

This criticism results largely from the failure of some climate change opponents, usually newspaper and magazine writers, radio and television commentators, and politicians, to understand or accept scientific methodology or the scientific method. Scientists in every discipline always have evaluated and adjusted their models, theories, or laws as they obtained more and better data. As the number of meteorological reporting stations increased and better models evolved, the climate science community examined continuously and modified regularly its published results. The objective is to describe, explain, and predict with as much certainty as possible Earth's changing climate, or in other words, to understand Earth's climate as it appears to be, and never claim absolute certitude. Affirmation since 2001 from national academies of sciences of many countries including Canada, the United Kingdom, France, China, India, and Germany; a 2005 statement by the US National Academy of Sciences, another by the AAAS in 2007, a 2009 American Geophysical Union survey, a 2010 report published in the *Proceedings of the National Academy of Sciences*, made clear that more than 90 percent of scientific community's active researchers on climate change have accepted Earth's long-term warming. They agree that the increasing atmospheric concentration of industry-produced $CO_2$ has contributed significantly to the warming. Their affirmation of the scientific evidence confirming climate change has led to considerable certainty about the reality of climate change.

A third criticism maintains that solar activity is responsible for most of the global warming and not the increasing anthropogenic $CO_2$ concentrations in the atmosphere. The international scientific community long agreed that solar activity is a factor and has shown that the combined effect of solar activity and greenhouse gases matched the global warming curve for the last 130 years. Other evidence demonstrated that volcanic activity over the last 1,000 years contributed in a minor way to global warming, and when combined with solar activity the two activities still failed to account for the warming that occurred since the 1800s. The evidence instead demonstrated the consequences of recent $CO_2$ additions to the atmosphere.

A fourth criticism charges that global warming proponents have neglected the effect of cloud coverage, namely that low-in- the- sky dense clouds can reflect sunlight into space and thereby reduce the impact of global warming. Alternatively, scientific evidence has shown that high-in-the-sky thin clouds produce the opposite effect. They allow sunlight to pass through them but effectively block infrared radiation from escaping, thereby warming Earth just like greenhouse gases. Depending on their location and properties, clouds can either produce a warming or cooling effect. Despite the global warming deniers' claims, atmospheric scientists have not ignored the effect of cloud coverage, they have not yet developed sufficiently reliable climate models that incorporate cloud coverage. Their experiments indicate only that cloud influence will not reduce the warming significantly.

The fifth criticism deals with the presence of water vapor in the atmosphere. Water vapor is a known greenhouse gas and accounts for the greatest greenhouse effect. It accounts for 36-66 percent of the greenhouse effect for a clear sky and 66-85 percent on a cloudy day. Its atmospheric concentration varies with temperature, ranging from 0.01 percent in cold climates to 2 percent in humid climates. Human activity has little effect on atmospheric water vapor concentration.

Water vapor affects global warming through what scientists call *feedback*, in this case water vapor feedback. Compared to other greenhouse gases such as $CO_2$ and methane that spend decades or centuries in the atmosphere, the water vapor molecule has an average 9-10 day lifetime in the atmosphere. During this short time it magnifies the greenhouse effect of other gases. It causes *feedback* because of the Clausius-Clapeyron equation, a well-known nineteenth-century (post-ca. 1834) thermodynamic equation which shows that as the temperature of a gas such as the atmosphere increases, it holds more water vapor per unit volume. Because of this effect, Earth's warming that results from increasing concentrations of greenhouse gases causes an increase in atmospheric water vapor concentrations. The increase in water vapor concentrations in turn produces further warming, and gives a positive *feedback*. Other atmospheric and climate changes, such as temperature or absorption, can counter a positive *feedback* and prevent excessive atmospheric temperature increases from occurring.

The sixth and final criticism draws its support from data obtained from weather satellites. Weather satellites began with the launching of the US Navy's Vanguard II in February 1959 and NASA's TIROS-1 (Television Infrared Observation Satellite) in April 1960. Major breakthroughs in satellite imagery followed in 1966 when Verner Suomi (1915-95) at Wisconsin invented the instrumentation, such as the spin scan camera, that made possible satellite-produced images. Suomi in 1971 improved his camera and monitors for displaying satellite-produced images and in 1972 developed a computer system, Man-Computer Interactive Data Access System (McIDAS), to store the considerable new meteorological data and imagery his instruments provided. In 1980 he introduced the visible infrared spin scan radiometric atmospheric sounder (VAS). Launched in 1980 aboard a geostationary satellite, the Geostationary Operational Environmental Satellite-4 (GOES-4), 138 inches high, 35 inches in diameter, weighing 874 pounds, VAS detected and recorded the atmospheric temperature and water vapor content.

A geostationary satellite, such as GOES-4, revolves in the same direction as Earth rotates on its axis and completes one orbit in 24 hours, the same time period for one Earth rotation. The satellite therefore appears stationary to an observer on Earth. GOES-4, which orbited Earth at an altitude of 35,800 km (22,300 miles) directly over the equator, ended its operational period in November 1988. At that time NASA sent it into a new orbit a few hundred kilometers above its operating orbit called a graveyard orbit in order to minimize collisions with operational spacecraft and reduce space debris. Weather satellites launched after 1980, GOES-9 and GOES-10 continue to operate.

The new satellite data (NASA has about 20 satellites and sensors), according to the critics, instead of supporting global warming indicated a slight cooling of Earth's temperature and challenged measurements from land-based stations that, since the 1950s-60s, revealed a warming. Hansen's correction of the satellite temperatures in the 1990s showed that no disagreement existed between the two sets of temperature measurements.

Many, perhaps most, of the critics contesting the reality or the extent and consequences of global warming, such as the economists Hans Labohm (b. 1941) and Bjorn Lomborg (b. 1965), the English politician-journalist, Christopher Monckton (b. 1952), and Lawrence Solomon (b. 1948), come from outside the scientific community. Critics from within the scientific community, mainly physicists, represent a small group of less than 10 percent, and include Lewis Kaplan, Richard Lindzen (b. 1940), S. Fred Singer, Patrick Michaels (b. 1950), Frederick Seitz (1911-2008), Robert Jastrow (1925-2008), and William Nierenberg.

Labohm, a professor at the Dutch Institute of International Relations Clingendael and the Netherlands Defense College in The Hague, and his co-authors, Simon Rozendaal, a writer, and Dick Thoenes, a chemical engineer, in *Man-Made Global Warming: Unravelling a Dogma* (2004) doubt the reality of anthropogenic global warming. They believe that $CO_2$ and other greenhouse gases produce the greenhouse effect, but that anthropogenic $CO_2$ represents only 5 percent of the total $CO_2$ emissions. They cite the lack of long-term temperature records because the thermometer (Fahrenheit) dates only from 1724, doubt the value of climate models because the models do not account for the cloud effect on atmospheric temperatures, and suspect that scientists have manipulated scientific data to further their own interests.

Lomborg, an economist at the Copenhagen Business School, former member of Greenpeace, and author of *The Skeptical Environmentalist* (1998), made the same arguments. He questioned the ice core and coral records, the tree ring records that Mann at Penn State used to determine temperatures for the last 1,000 years, the influence of sulfur aerosols to reflect sunlight and thereby counter the greenhouse gases, the effect of evaporation of water vapor, a known greenhouse gas, and cloud effect on global warming. Lomborg asked whether the economic cost of dealing with the consequences of greenhouse gas emissions, which he believed did not contribute as much as suspected to global warming, was a good investment. Lomborg insisted that the scientific community deal with these issues, which it did apparently to his satisfaction. In 2011 the skeptic Lomborg agreed that the scientific evidence supported anthropogenic global warming, and he ended his opposition.

Solomon, a Romanian-born Canadian, founder in 1980 of the Energy Probe Research Foundation in Toronto and author of *The Deniers* (2008), owes his lofty status among the skeptics and doubters largely to his misrepresentation of a statement that Revelle made in a 14 July 1988 letter to James Bates (b. 1941), former San Diego-area Democratic congressman (1983-91), and repeated in a 1991 article. Revelle expressed his caution about the greenhouse effect, saying "My own personal belief is that we should wait another 10 or 20 years to really be convinced that the greenhouse effect is going to be important for human beings, in both positive and negative ways." Solomon transformed Revelle's caution into a denial of global warming. Twenty years have passed since Revelle's death in 1991, and the evidence of global warming has continued to grow.

The leading global warming skeptic in the 1990s-2000s is Lindzen, an atmospheric physicist at MIT, author of more than 230 articles and books on the atmosphere, a contributor to the IPCC's 1995 report, and member of the US National Academy of Sciences. His skepticism that began in 1989, and as he said caused him to switch political parties from Democratic to Republican, resulted largely from concerns about the validity of computer models used to predict climate change. Lindzen wrote in a 1992 article that the current climate models, which were then in their infancy, overestimated the amount of global warming because they overestimated the effect of water vapor *feedback*, the temperature rise resulting from the increasing concentration of water vapor in the atmosphere. Reducing the feedback factor would reduce or even eliminate future global warming. Nor do the climate models deal adequately with the ability of clouds to reduce global warming. Lindzen's position on water vapor feedback and cloud coverage has evolved since the 1990s, a result of the recent data analyses and models of atmospheric scientists such as Andrew Dessler (b. 1964) at Texas A&M. Their 2000s studies have provided alternative interpretations of the impact water vapor and cloud coverage have on global warming.

Lindzen agrees that the $CO_2$ levels today are about 30 percent higher than in pre-industrial years, but contrary to the IPCC's 2007 report, he believes that a doubling of $CO_2$ concentrations would result in a 0.6° C (1.08° F) increase and not IPCC's 2-4.5° C increase. He argues that Earth's climate sensitivity (how much Earth's climate will warm because of the increased greenhouse effect upon doubling the atmospheric concentration of $CO_2$) is less than 1° C (low climate sensitivity) based on volcanic eruptions and satellite data. Hansen and other proponents using the two methods of determining Earth's climate sensitivity, paleoclimate history (ice core analysis, sea level and deep ocean temperature changes) and climate models, maintain Earth's climate sensitivity is 3-4° C (high climate sensitivity). Hansen and Makiko Sato's 2012 study of the oxygen O-18 record in ocean sediments to estimate past deep sea level and ocean temperature changes indicated a climate sensitivity of 2.5-3.5° C. Dessler's 2014 analysis of atmospheric concentrations of ozone and aerosols also supports Earth's high climate sensitivity. He found that ozone and aerosols, although they have short atmospheric lifetimes, are highly localized. They remain concentrated where emitted and have a greater impact on climate change than twentieth-century warming records indicated. His data analysis agrees with climate models and paleoclimate records which suggest a climate sensitivity of 3.3° C (6° F).

Singer and Seitz, a former president of Rockefeller University in New York City and of the National Academy of Sciences (1962-69) are the other well-known longtime skeptics from the scientific community, Singer from at least 1989 and Seitz from at least 1990. Both were prominent scientists but neither ever did or has done any significant research on climate change in twenty years.

The Independent Institute in Oakland, California, and Washington, DC, the home of global warming deniers, published Singer's attack on global warming *Hot Talk, Cold Science: Global Warming's Unfinished Debate,* in 1997 (second edition 1999). Seitz wrote the foreword. Seitz's credibility further eroded in 1998. The National Academy disassociated itself from Seitz in 1998 after he circulated a one-page letter asking scientists to sign a petition called the Oregon Petition (Petition Project). The Oregon Institute of Science and Medicine (OISM), a nonprofit public foundation that Arthur Robinson (b.1942) founded in 1980 near Cave Junction, Oregon, and currently directs, organized the petition. The petition advocated that the US government reject the Kyoto protocol of December 1997. Seitz claimed 15,000 scientists signed the petition and that the number increased to 31,000 in 2007-8. He also heavily promoted a 12-page article that Robinson and three other global warming critics at the OISM published in 1998, adopting a format and style deceptively designed to look like an article published in the peer-reviewed National Academy's *Proceedings*. The article contained serious errors and concluded that $CO_2$ did not contribute to global warming. The National Academy disavowed the article emphasizing that neither it nor any peer- reviewed journal had published it.

In their 2010 publication, *Merchants of Doubt,* Naomi Oreskes, a historian of science at the University of California, San Diego, and Erik Conway at NASA's Jet Propulsion Laboratory identified Singer, Seitz, and the physicists Michaels, Jastrow, and Nierenberg as the leaders or former leaders of the climate change skeptics. Michaels is at the conservative Cato Institute in Washington, DC, Jastrow was at NASA and Columbia, and Nierenberg formerly directed Scripps (1965-86).

Seitz, Jastrow, and Nierenberg in 1984 founded the George C. Marshall Institute, a conservative institute in Washington, DC. The other critics are founders or affiliates of conservative institutes, such as the Competitive Enterprise Institute (founded 1984) and the Institute for Energy Research (founded 1989) in Washington, DC; the Heartland Institute (founded 1984) in Chicago; the Independent Institute (founded 1986) in Oakland, California, and Washington, DC; and the American Tradition Institute (founded 2009) in Washington, DC, and Denver. The critics' main objective is to cast doubt on the reality of climate change.

This is the central theme of Oreskes and Conway's book, so clearly reflected in its title, *Merchants of Doubt.* Oreskes and Conway note that many of the doubters really are not climate science experts. Their strategy of confusing the public and doubting the evidence of climate science researchers has worked very well to discredit all scientific research that the doubters consider erroneous and threatening to their values of complete freedom from government regulations, a totally free market, and an unchecked capitalism. Only their values create wealth, power, and strength, nothing should deter from pursuing them. The eternal debate about creationism and evolution, especially in the United States, is the classic example of how casting doubt on the results of scientific research and scientific theories can influence a society in which too many are misinformed or uninformed about science and its impact on society to make intelligent decisions.

*Kyoto Protocol (1997) and Copenhagen Accord (2009)*

The 1957-58 International Geophysical Year, which lasted from 1 July 1957 to 31 December 1958, effectively marked the beginning of an international cooperative effort in which scientists from 64 nations pledged to study Earth's atmosphere and climate. It occurred during the Cold War years at the time of another scientific milestone, the Soviet Union's launching of Sputnik, the first artificial satellite, in October 1957.

During the next 30 years scientists from these nations published a number of reports that examined the reality of global warming. One of the reports, a 1979 National Academy of Sciences report, indicated that doubling the $CO_2$ concentration would lead to an atmospheric temperature increase of 1.5-4.5° C. A second report, a Toronto, Ontario, conference report published in 1989 following the Toronto Conference on the Changing Atmosphere in June 1988, called for a 20 percent reduction in $CO_2$ global emissions by 2005 with the ultimate objective of reducing emissions by 50 percent.

The reports appeared to have some impact on alerting the public about the long-term consequences of human-caused, essentially industrial, $CO_2$ emissions. They received additional credibility in June 1988 when Hansen, the global warming advocate, appeared before the US Senate's Energy and Natural Resources Committee and from the IPCC's first published report in 1990, stressing the necessity of reducing greenhouse gas emissions, and from the IPCC's reports of 1992, 1995, 2001, and 2007.

All the IPCC reports concluded that Earth's temperature was increasing and that human activity, particularly industrial emissions of $CO_2$, was the main cause of the warming. Unfortunately, none of the reports proposed specific measures on which the United States and the other nations concerned about global warming agreed. The first report with specific measures to regulate $CO_2$ emissions came in December 1997 with the Kyoto Protocol. The protocol, for the first time, legally obligated the industrial nations that ratified it, to reduce the industrial gas emissions, particularly $CO_2$, that were warming Earth. By the time of the protocol's formal ratification in February 2005, 141 nations had ratified it, increasing to 191 nations in 2010.

The United States, which produced about 15 percent (2012) of the industrial $CO_2$ emissions responsible for global warming, was the only major industrial nation not to ratify the protocol. Germany, Japan, the United Kingdom, Canada; China and India, both rapidly growing industrial and polluting nations, ratified the protocol. But because the Kyoto ratifiers considered China, now the world's largest emitter at 29 percent (2012 for $CO_2$) and India (6 percent 2012), developing nations, China and India did not have to begin making emission reductions until 2012. This was the year in which Kyoto's first commitment period ended.

The Kyoto Protocol aimed by 2012 for a 5.2 percent reduction in greenhouse gas emissions from the 1990 global emissions. Since its adoption in December 1997 and ratification in February 2005, several membership changes suggested that Kyoto would not meet its commitments and fail. The United States, which had aimed for a 7 percent reduction, withdrew its support in early 2001. The US President George W. Bush (1946) concluded that the Kyoto Protocol would hurt the US economy and did not agree to China's and India's exemption from the same provisions required of other industrialized nations. Canada, which ratified the protocol in 2002 and targeted a 6 percent reduction, abandoned its emissions reduction target after the January 2006 election of the Conservative Party. Russia, which accounts for 17 percent of global greenhouse emissions, committed to the protocol in September 2004.

To enforce the protocol, making it a legally binding document and not a statement of intent, the combined greenhouse gas emissions of the developed nations that ratified the protocol had to total at least 55 percent of global greenhouse emissions. The United States' refusal to ratify the protocol made crucial Russia's commitment to Kyoto's ratification and to Kyoto's 2012 goal of a 5.2 percent average reduction of greenhouse gases from their 1990 total.

The nations signing the protocol believed that the production of more fuel efficient automobiles and fewer coal-fired plants would reduce greenhouse gas emissions. They introduced three specific measures to help achieve the reductions: (1) a nation unable to meet its emission reductions could purchase emission credits from other nations whose emission totals remained below their targeted total, (2) a nation could earn emission credits by carrying out a joint implementation project that resulted in an emissions reduction in an industrialized nation, such as a reforestation project that removes $CO_2$ from the atmosphere, and (3) a nation could participate in a *clean development mechanism* through which wealthy nations promoted the transfer of environmentally-friendly technologies such as solar and wind to developing nations and thereby encouraged international investment in developing nations.

Failure of a nation to meet its targeted emission reduction by the end of the first commitment period in 2012 required that nation to make up the remainder by the end of the second commitment period in 2016. It will receive a 30 percent penalty on its remaining commitment, requiring it to reduce emissions further, and lose the right to sell emissions credits.

Studies that the United Nations Framework Convention on Climate Change (UNFCC) and the Netherlands Environmental Assessment Agency (NEAA) showed that during Kyoto's first commitment period 1990-2012 more successes than failures occurred and that the sum of emissions from countries with targets fell, whereas emissions in other countries, particularly China and India, increased significantly.

| **PERCENT CHANGES IN GREENHOUSE GAS EMISSIONS EXCLUDING LAND USE* 1990-2011** | | |
|---|---|---|
| ***Country*** | ***Increase*** | ***Decrease*** |
| Canada | 18.7 | |
| United States | 8.0 | |
| Japan | 3.2 | |
| France | | -12.2 |
| Germany | | -02.6 |
| United Kingdom | | -27.8 |
| Russian Federation | | -30.8 |
| *land use: total emissions and removals relating to land use changes and forestry (greenlands, wetlands, settlements) | | |

| **CARBON DIOXIDE EMISSIONS 34.5 BILLION METRIC TONS (2012)** | |
|---|---|
| ***Country*** | ***Percent*** |
| China | 29 |
| United States | 15 |
| European Union | 11 |
| India | 6 |
| Russian Federation | 5 |
| Japan | 4 |

The NEAA reported a gradual increase of greenhouse gas emissions from 1990 to 2012 and that Kyoto had failed to slow global carbon dioxide emissions which comprise the greatest percentage of greenhouse gases (60 percent). Global emissions of $CO_2$ decreased in 2008, increased by 5 percent in 2010, increased by 3 percent in 2011, reaching an all-time high of 34 billion metric tons in 2011 and only a slight increase to 34.5 million billion metric tons in 2012. $CO_2$ emissions for the decade had an average annual increase of 2.7 percent.

## CLIMATE CHANGE CHRONOLOGY

| ***Report, Event, or Conference*** | ***Year and Location*** | ***Result*** |
|---|---|---|
| Causes of Climate Change, Conference | August 1965 Boulder, Colorado | Trivial astronomical shifts of Earth's orbit might have triggered past ice ages. |
| National Oceanic and Atmospheric Administration (NOAA) | Established 10 July 1970 | Combined air and sea research, becomes world's leading financier of climate research. |
| International Council of Scientific Unions, Report | 13 October 1976 | Increasing industrial activity over next two centuries expected to raise atmospheric $CO_2$ content fourfold, possibly eightfold. |
| US National Academy of Sciences, Report | November 1979 | Doubling $CO_2$ concentration could increase temperature 1.5-4.5° C. |
| Environmental Protection Agency (EPA), Report | 18 October 1983 | Called for immediate action to deal with expected changes in global temperature and climate. |
| Montreal Protocol on Substances that Deplete the Ozone Layer, Conference | September 1987, Montreal | An accord to protect Earth's ozone layer. |
| Our Changing Atmosphere: Implications for Global Security, Conference | June 1988, Toronto | Called for a 20 percent reduction in $CO_2$ emissions by 2005 with an ultimate reduction of 50 percent. |
| Intergovernmental Panel on Climate Change (IPCC) | Established 6 December 1988 | International effort by 35 countries to evaluate global warming trends and their impact. |
| First IPCC Report | 26 May 1990 | Cut greenhouse gas emissions by 60 percent over next century to prevent a sharp rise in global temperatures. |
| Earth Summit Conference sponsored by United Nations Framework Convention on Climate Change (UNCCC) | June 1992 Rio de Janeiro | Representatives from 178 nations met to discuss fundamental environmental problems. |
| Second IPCC Report | September 1995 | Earth has entered a period of climate change, important to reduce greenhouse gases over the next century. |
| Kyoto Protocol | December 1997 | Industrial countries for the first time made legally binding their pledge to reduce emissions of waste industrial gases ($CO_2$) that were warming the atmosphere. |
| Third IPCC Report | February 2001 | Presented evidence that human activities produced most of Earth's warming over the last 50 years. |
| US National Academy of Sciences Report | June 2001 | Eleven leading atmospheric scientists reported that human activities largely caused Earth's warming. |

| CLIMATE CHANGE CHRONOLOGY | | |
|---|---|---|
| ***Report, Event, or Conference*** | ***Year and Location*** | ***Result*** |
| Kyoto Protocol | July 2001 | Representatives of 178 countries reach compromise agreement that required industrialized countries to cut greenhouse gas emissions. United States did not sign the compromise. |
| Fourth IPCC Report | February, 2007 | Global warming was unequivocal and human activity very likely causing most of the temperature rise since 1950. |
| Copenhagen Accord | December 2009 | The United States, China, India, and other countries (193 total) that account for 96 percent of the global emissions of greenhouse gases voluntarily agreed to curb emissions. In June 2010, 133 nations committed to the Accord. |
| United Nations Climate Change Conference | 29 November - 10 December 2010 Cancun, Mexico | Called for a $100 billion a year green climate fund by 2020 to assist poorer countries in financing emissions reductions and the establishment of a climate technology center and network. |

The most recent of the climate change agreements is the Copenhagen Accord. Representatives of 193 nations meeting for two weeks in December 2009, at a session of the Conference of Parties to the United Nations' Framework Convention on Climate Change, produced a three-page document that they did not formally adopt. They voted instead to *take note* of it. The United States, China, India, Canada, Brazil, and the European Union were among the 133 nations who signed the accord and voluntarily agreed to reduce greenhouse gas emissions. Those signing the accord produced about 86 percent of the emissions. The United States and Canada pledged a 17 percent reduction, China 40-45 percent, India 20-25 percent, and Japan 20-25 percent reductions.

By fulfilling the emission reductions the Copenhagen Accord projects that emissions will peak in 2020 at 44 gigatons and not exceed 50 gigatons and that the global temperature increase will not reach 2° C. The Copenhagen Accord, though seen as endorsing the Kyoto Protocol, is voluntary and not legally binding and therefore sets no real targets or timelines to reduce gaseous emissions. Wealthy nations, including the United States and the European Union, also have pledged to provide billions of dollars, $30 billion a year until 2012 and $100 billion a year by 2020, to the poorer nations in Africa and small island states like Haiti, Fiji, Suriname, and Tonga, to develop clean energy sources, such as wind, hydro, and solar, but their pledge to deliver the funds to the nations requiring them has no guarantee.

*Summary and concluding thoughts on climate change*

Recent studies in the 1990s-2000s have indicated that global warming resulting from a changing atmosphere has contributed to the sea level rise of the last 1,000-2,000 years and to the 0.3-0.6° C (1° F) increase in Earth's surface temperature in the past 150 years. Environmental analyses of marsh deposits from known depths of marshes along the United States' eastern coast in the Long Sound and Delaware Bay area have used micro-faunal indicators, sediment chemistry, lithologic parameters, and a time frame based on Pb-210 and C-14 dating and anthropological data to show that the sea level rise has increased since 1650 with the greatest increase occurring about 1900 and rising 4-10 inches over the last century, or 0.06-0.08 inches per year (1.6-1.9 mm per year).

The analyses also suggested that the twentieth-century warming has exceeded the warming of the last 400-600 years and possibly of the last 1,200-1,500 years. Glaciers, which are present on every continent except Australia and are therefore excellent global indicators of regional change, have shrunk and receded, giving additional support to a warming trend since the early 1800s. Analysis of gas bubbles trapped in glaciers-ice cores in fact have shown good correlation with climate. Low concentrations of carbon dioxide and methane are characteristic of glacial periods, whereas high concentrations of the two gases characterize interglacial periods.

The reason for the warming is that infrared radiation absorbing gases, or greenhouse gases, namely water vapor ($H_2O$), carbon dioxide ($CO_2$), nitrous oxide ($N_2O$), methane ($CH_4$) and ozone ($O_3$), which are normally present in Earth's atmosphere, absorb infrared radiation (heat) emitted by Earth's surface and atmosphere. The heat absorbed warms the atmosphere so that the atmosphere emits more infrared radiation, most of which acts to warm Earth's surface and lower atmosphere. Oxygen ($O_2$) and nitrogen ($N_2$) do not absorb infrared radiation.

Earth's average surface air temperature is about 12° C (54° F) because of greenhouse gases present in the atmosphere, and the rapid increase in the concentration of greenhouse gases that has occurred since the industrial era began about 200 years ago clearly suggests future climate changes. Average surface air temperature has increased 0.78° C (1.4° F) since the beginning of the twentieth century, most of it, 0.61° C (1.1° F), occurring in the last 30 years. The atmospheric concentration of $CO_2$, increased 30 percent from about 280 parts per million (ppm) in the pre-industrial era to about 364.28 ppm in December 1997, 396.78 ppm in May 2012,and 398.86 ppm in December 2014, mainly as a result of fossil fuel combustion and cement production. These two polluting sources contributed about 6.5x$10^{15}$ grams (1.41 x $10^{13}$ or 14.1 trillion pounds) of carbon in 1996, 33.4 billion metric tons (91 percent) in 2010.

The greenhouse gases differ in their heat-absorbing ability. The non-naturally occurring chlorofluorocarbon gases (freons) CFC-11 ($CCl_3F$) and CFC-12 ($CCl_2F_2$) along with the other fluorocarbons generated in industrial processes, such as hydrofluorocarbons (HFCs) and perfluorocarbons (PFCs), are the most heat absorbing. Nitrous oxide absorbs 270 times more heat than $CO_2$, and methane absorbs 21 times more heat than $CO_2$. The contribution of each gas to global warming also depends on its atmospheric concentration. Consequently, the chlorofluorohydrocarbons, while absorbing the most infrared radiation, contribute far less to global warming because of their low concentration compared to water vapor and $CO_2$.

The main contributors in descending order are water vapor 36-72 percent, $CO_2$ 9-26 percent, methane 4-9 percent, and ozone 3-7 percent. The concentration of water vapor is highly variable because warm air holds more water vapor than cold air, and unlike the other greenhouse gases, water vapor frequently changes its state and exists in the liquid and solid states. The table below summarizes the increase in concentration of all the greenhouse gases over the last 200 years, or the industrial era.

| **CHANGE IN CONCENTRATION OF GREENHOUSE GASES** | | |
|---|---|---|
| ***Greenhouse Gas*** | ***Main Sources of Gas Pre-Industrial Period to 1996 200-Year period*** | ***200-Year Increase in Atmospheric Concentration*** |
| $CO_2$ | Fossil fuel combustion and cement production contributed 6.5 x $10^{15}$ grams of carbon in 1996, 33.4 billion metric tons (91percent) in 2010. Changes in land use contributed 1.6 ± 1.0 x $10^{15}$ grams of carbon in 1996, 3.3 billion metric tons (9 percent) in 2010. | 280-364 ppm in 1997<br>379 ppm in 2005<br>396.78 ppm in 2012 |
| $CH_4$ | Fossil fuel sources contributed 70-120 x $10^{12}$ grams of $CH_4$ in 1996. Rice cultivation, animal husbandry, biomass burning, and landfills contributed 200-350 x $10^{12}$ grams of $CH_4$ in 1996. | 700-1,721 ppb in 1994<br>1,774 ppb in 2005 |
| $N_2O$ | A total of 3-8 x $10^{12}$ grams of nitrogen per year for the 1980s came mainly from agricultural and industrial sources, such as adipic acid and nitric acid production. | 275-312 ppb in 1994<br>319 ppb in 2005 |
| Chlorofluorocarbons | Manmade compounds used as refrigerants and in spray propellants and foam blowing are the main contributors. | not present significantly prior to 1850 |

Climate scientists have reconstructed northern hemisphere temperature anomaly trends from 1000 AD to the present, using figures obtained from dendroclimatic (tree-climate), coral, and ice-core proxy records calibrated by instrumental measurements. The measurements show that, except for a brief period in the mid-1400s, the temperature seldom increased or decreased by more than about 0.25° C from a warm temperature until the last 150 years. Thorium isotope measurements in fossils present in coral reefs in the Bahamas and Mexico's Yucatan's Peninsula tell the same story. The United Nations IPCC in 1995 and the UNEP Convention on Climate Change, Conference of Parties 5 in Bonn in October-November 1999, predicted that unless society reduces greenhouse gases emissions, Earth's average surface temperature by 2100 will increase about 1.1-3.3° C (2-6° F), or a best estimate of 2° C (3.5° F), compared with an increase of 2.8-5° C (5-9° F) since the last Ice Age 18,000-20,000 years ago.

No catastrophic volcanic emissions of $CO_2$ have occurred since the end of the Precambrian eon about 550 million years ago. The volcanic activity at that time raised atmospheric $CO_2$ concentration about 12 percent, or about 350 times today's values. Global warming occurred, leading to limestone formation as a result of $CO_2$ absorption by calcium oxide (lime) and to the emergence of higher (multicellular) life forms through photosynthesis. Volcanic activity occurring in the 200-300 previous years represents about one percent of human caused gaseous emissions.

Adding to the warming problem is the recent increasing recognition of soot (black carbon or carbon black) as a contributor to global warming. Physicians long have recognized soot as a health hazard particularly in lesser developed countries, but only recently have climate scientists investigated its contribution to global warming. The IPCC did not mention soot as a warming agent as recently as in its 2007 summary report. Yet leading climate scientists, such as Veerabhadran Ramanathan at Scripps and Mark Jacobson at Stanford, see soot as having a serious impact on climate change.

$CO_2$ is the major contributor to rising global temperatures, accounting for as much as 30 percent, but recent studies suggest that soot may be the second greatest contributor at 18 percent. It does not travel as far as $CO_2$ which moves around the globe, and unlike $CO_2$ which remains in the atmosphere for years, soot settles after a few weeks.

What is the origin of the soot particles that are contributing to global warming? Most of the soot particles come from the inefficient wood and dung burning cooking stoves used in the poorer villages, towns, and cities of Africa and Asia. The soot particles travel through the atmosphere, absorbing the Sun's energy and warming the air. Some of soot settles on glaciers, such as the Himalayan glaciers, melting the ice and increasing the likelihood of severe flooding in mountain communities. Soot from India has reached the Maldive Islands and the Tibetan Plateau.

An immediate effective and economical solution is to replace the inefficient cookers with cleaner burning efficient cookers that pulverize the fuel before burning and have fans to promote combustion. Solar cookers, which do not burn wood, provide another clean and efficient choice.

The industrialized nations of the West, Europe and the United States, and in the East, in Japan and increasingly in China, also contribute soot to the atmosphere with emissions from diesel engines and coal plants. Scrubbers and filters when installed remove most of soot particles and prevent their escape into the atmosphere.

***Lead and the environment***

Today's global community largely has accepted the reality of environmental pollution. This has improved significantly from the situation that prevailed 40 or 50 years ago. The global community concerns itself in the 2000s with the impact and consequences, or the *now* of environmental pollution, which is what it should be, although the focus on the *now* often results in our ignoring or forgetting those who first made us aware of environmental pollution issues. An environmental issue has an origin, or beginning; a recognition of; and it always involves a *who, what, when, where*, and *why*.

What we often have, however, is an ending without a beginning. We know the consequences of the scientific or technological event on the environment but not its origin, which we need to complete the historical record and to provide some perspective on current environmental pollution issues. Everyone has heard of acid rain, but few have heard of Robert Angus Smith, the Scottish chemist who first brought the problem to our attention in 1852 and coined the term *acid rain* in his 1872 book *Air and Rain*. Similarly we know about global warming and greenhouse gases but seldom associate either issue with Arrhenius, the Swedish chemist and 1903 Nobel Prize winner who discussed global warming in his 1896 article published in the *Philosophical Magazine*.

The same lack of public awareness existed for many other scientists, such as Sir Richard Doll (1912-2005) and Arie Jan Hagen Smit (1900-77), scientists who were well known within the scientific community but virtually unknown to the public. Doll was an English epidemiologist in London and later at Oxford. His research in 1950 and 1955 established the link between lung cancer and workers in the nickel, asbestos, and coal tar industries in particular and between cigarette smokers and lung cancer in general. Smit, a Dutch chemist, joined the faculty at Caltech in 1937. Beginning in 1948, Smit's investigations demonstrated the link between the smog (smoke plus fog) polluting southern California's air and the combustion products present in automobile exhaust and industrial plant emissions. Smit identified ozone ($O_3$), a strong irritant and a toxin if inhaled, in automobile exhaust and that photochemical smog consisted mainly of ozone and industrial plant and automobile emissions, especially nitric oxide (NO) and sulfur dioxide ($SO_2$). Sunlight acting on the airborne combustion products increased the smog's intensity.

Lead pollution was another well-known twentieth-century environmental pollution issue, about which we know the impact and consequences, but know little about the scientist and the ground-breaking research he did to make the public aware of the risk posed to society because of lead's presence in the environment. Clair Patterson (1922-95), an American geochemist at Chicago and Caltech, was the scientist responsible for ending the debate over lead as a dangerous pollutant. Patterson has remained one of those scientists well known within his scientific specialty but largely unknown to the public. His conclusions put an end to the 40-year government-industry claim that the presence of lead in the environment posed little or no risk to society.

Patterson's research on airborne lead and lead pollution, dating from 1963, is important for a second reason. It dramatically demonstrated the issues of scientific objectivity, intimidation of scientists, denial of science, scientific fraud, and the politicization of science, which since the 1980s has become too common in science. The scientist's goal is to provide an objective description and explanation of the world as it appears to be. How then could reputable scientists, such as Patterson and Robert Kehoe (1893-1992) at the University of Cincinnati Medical School and chief medical consultant of Ethyl Gasoline Corporation in New York City, totally disagree on lead concentrations in the environment and in the human body? Specifically, they disagreed on what constituted natural versus normal (typical) lead concentrations in the environment and human body. More broadly they debated the important questions of what scientific truth is and how a scientist, the scientific community, and society determine scientific truth. The debate revealed the enormous political and economic clout of DuPont, Standard Oil (Exxon), and General Motors in defining scientific truth and not scientific evidence. Patterson's conclusions on lead pollution, not Kehoe's, have withstood all challenges to their validity. He struggled long before the scientific community eventually accepted his conclusions. Patterson remains one of the many dedicated individuals who contributed to improving our environment and our health, and we never knew he did it.

*Recognition of lead poisoning goes back to ancient times*

Lead was the heaviest of the seven metals known in antiquity. For this reason, astrologers and alchemists associated it with a planet, just as they did with each of the other six metals. They associated lead with Saturn, the slowest-moving planet, which they also believed was the heaviest planet. The alchemists obtained elemental lead by roasting (smelting) its naturally-occurring compound, lead sulfide (PbS).

Ancient and later civilizations recognized the uses as well as the dangers of lead. The earliest known lead mine dates from 6500 BC at the Hittite city of Catal Huyuk in Turkey, and the earliest known lead artifact is a lead bead necklace with an estimated age of 6,000-8,000 years, found in Anatolia, the peninsula (Anatolian Peninsula, Asia Minor) that links Europe and Asia. The ancients very likely observed the signs of lead poisoning associated with lead miners, such as colic (acute abdominal pain), palsy (tremors), and gout (arthritis or painful inflammation of the joints). Egyptian papyrus scrolls mention lead compounds (likely lead acetate or nitrate) used for homicidal purposes.

The first documented accounts of lead-related illnesses appear much later with the Greeks. Hippocrates (460-370 BC) of Cos observed symptoms of colic in a metal worker and related gout to food and wine but did not connect gout with lead. The first report on lead poisoning as the cause of colic and anemia appeared in the *Alexipharmaca* of the Greek physician and poet Nicander (fl. 130) of Colophon. Nicander's codex, first printed in 1499 and often jointly with his *Therica*, described white lead (basic lead carbonate or ceruse) poisoning as constriction of the palate and gums, asperity (roughness) of the tongue, hiccups, dry cough, nausea, heaviness of the head, and dullness or lethargy (torpor).

Early civilizations never mined and smelted large quantities of lead. They obtained lead as a by-product in the refining of silver and gold and used it mainly in glazes, alloys, vessels, kitchenware, and decoration, and in medicine and cosmetics. Large-scale lead production began in 50 BC with the Romans who operated several large mines and smelters in Spain and Greece. Their strongly atmospheric-polluting metallurgical enterprise produced an average 60,000 long tons of lead per year for 400 years, an amount not surpassed until the nineteenth-century industrial revolution. While continuing the older established uses of lead metal and its compounds, the Romans extensively used lead metal in plumbing (drain pipes for their baths) and aqueducts because it resisted corrosion and was malleable and ductile. The Latin word for lead, *plumbum*, is the origin of the English word plumber. They added lead (II) oxide, litharge PbO a yellow to red oxide made by heating lead in air, to wine to sweeten its taste and enhance its color. Because of lead's widespread use throughout the Roman Empire, lead intake varied from 35 milligrams per day to 250 milligrams per day, compared to an intake in the United States in the 1980s of 0.3 milligrams per day. The high Roman intake has led to speculation that lead-poisoning contributed to the decline of the Roman Empire.

A few Roman writers, such as Marcus Vitruvius Pollio (d. 25 BC) and Horace (65-8 BC), acknowledged the dangers of smelting lead and of using it in plumbing and aqueducts. Pollio, a Roman engineer and architect, wrote in *On Architecture* that lead workers had pale gray complexions. In the casting process the lead receives the currents of air, and the fumes from it occupy the members of the body, and burning them thereon, rob the limbs of the blood's virtues. He and Horace in *The Epistulae* (22-8 BC) commented on the purity of fresh water in the countryside compared to the town water.

The widespread use of elemental lead and its compounds continued throughout the Middle Ages, Renaissance, and into the modern centuries. So did observations of lead's poisoning but with little change or action by lead producers, users, and government officials at any level. In the mid-1400s the printing industry used lead alloys. Johann Gutenberg (1398-1468) in Mainz, in printing the first Bible in 1454-55 used a tin-antimony-lead alloy for his type metal and a copper-lead slurry for his ink. Lead compounds, many of which are brightly colored such as yellow lead (II) oxide and red lead oxide $Pb_3O_4$, found wide use in glass manufacturing. In the 1500s Agricola (1494-1556), a physician and commentator on mining, wrote in *De Re Metallica* (1556) of the corrosive qualities of mine dust that ate away the lungs. He recommended ventilation of mine shafts and veils for miners to protect themselves from dust. Adopting a tactic widely used today, Agricola said that such problems rarely happen and only insofar as workmen are careless.

In Italy, in the 1700s, Bernardino Ramazzini (1633-1714), a physician at the University of Modena and considered the father of occupational medicine, wrote on the diseases and problems of 52 occupations. In his book *De Morbis Artificum Diatriba* published in 1700, Ramazzini stressed that physicians should ask all the questions Hippocrates recommended and also ask, what is your occupation? In his description of occupational diseases found in lead mine workers, Ramazzini observed their skin "is apt to bear the same color of the metal. Demons and ghosts are found to disturb the miners. At first tremors appear in the hands, soon they are paralyzed."

In England, physicians traced the great gout epidemics of the 1700s to the large quantities of heavily-leaded port wines imported from Portugal. A century later Charles Dickens (1812-70) in his essay "The Small Star in the East," a description of patients in the East London Hospital published in December 1868, wrote of the horrible effects of lead poisoning on women who worked for 18 pence a day in London's white lead mills. Describing one such woman he wrote, "Her brain is coming out at her ear, and it hurts her dreadful."

Across the Atlantic, Benjamin Franklin in Philadelphia in 1783 discussed the "dry gripes" (colic) and "dangles," the wrist drop that affected tinkers, painters, and typesetters and those who had drunk rum distilled in lead vessels.

Lead found additional uses in the lead chamber process to manufacture sulfuric acid beginning in 1746, lead storage batteries in 1859 and 1881, and in 1895 as a shield from x-rays and other forms of radiation. Still to come in the twentieth century, were new dangers from lead poisoning following the introduction and mass production of lead-based paints and the colorless liquid antiknock additive tetraethyllead.

Throughout the centuries of lead usage, many commentators observed and wrote of the lead-related health problems, but no one took any or much action either to reduce the risk or to avoid the use of lead and its compounds.

*Lead in paint issue*

Despite the convincing evidence, beginning in Australia in 1892 and linking children's illnesses and deaths to paint containing lead compounds, such as litharge PbO, yellow lead (II) oxide; lead carbonate $2PbCO_3$. $Pb(OH)_2$, white lead; lead sulfate $PbSO_4$, white lead; and lead oxide $Pb_3O_4$, red lead, the US Congress did nothing to control or ban lead in paint. Outside the United States numerous countries began banning white lead interior paint: France, Belgium, Canada and Austria in 1904; Australia in 1914; Tunisia and Greece in 1922; Czechoslovakia in 1924; Britain and Sweden in 1926; Spain and Yugoslavia in 1931; and Cuba in 1934. The Third International Labor Conference of the League of Nations in Geneva in 1922, at which 400 delegates from 40 nations discussed lead trade regulations, also recommended a ban on lead in indoor paint.

In the United States, action on lead's removal from paint moved in the opposite direction. When medical researchers in 1892 and again in 1904, 1908, and 1914 linked child poisoning to white lead paint used inside homes and on furniture and toys, National Lead Company, the leading American producer of lead-based paints, with offices in New York City and other American cities, in 1907 began promoting its *Dutch Boy White Lead Paint*. It named its paint after the original 1600s manufacturing process, and put an image of a child on the paint can's label. National Lead continued to counter negative publicity by targeting its marketing and advertising to children with ads such as its *Cater to the Children* in 1918 and *Don't Forget the Children* in 1920.

The promotion of lead-based paints continued throughout the 1920s and 1930s, even though National Lead acknowledged in 1921 that lead-based paint, which contained anywhere from 15 percent to 40-50 percent lead, was poisonous. Two years later, in 1923, National Lead, while still aiming its ads at children and promoting the use of lead-based paints in homes, hospitals, schools, buildings, and on children's toys, added a new potentially life-threatening dimension to its advertising, claiming that lead was healthy and helped to guard your health. In 1928 the lead manufacturers established the Lead Industries Association to silence the critics and give lead manufacturers a positive image.

The warnings against the use of lead-based paints continued, nevertheless. One of the most serious charges came in 1933 from within the lead industry, from Robert Kehoe, Ethyl Gasoline Corporation's chief medical consultant. Kehoe, an authority on the environmental effects of airborne lead on humans and the foremost promoter of tetraethyllead, strongly urged the elimination of lead-based paint from children's environment. Ten years later, in 1943, another report, this one in *Time* magazine, linked lead poisoning to children's permanently-reduced IQs, learning disabilities, and behavioral problems. The Lead Industries Association challenged the *Time* article, arguing that the connection between lead poisoning in infancy and later mental retardation remained unproved. Kehoe criticized the lead industry's position, dismissing it as inconsistent with the scientific evidence.

The charges and countercharges continued until 1954, when the American Medical Association (AMA) recommended labeling lead paint a poison and its use discontinued in home and building interiors and around children. Following the AMA's statement, cities such as New York in 1954 and Baltimore in 1957 limited the amount of lead in paint to 1 percent and screened homes for lead-based paint. But Congress remained immobile until 1970, when it passed the Lead Poisoning Prevention Act that called for the phase-out of lead-based paints. The prohibition of lead-based paint came in February 1978, only after the United States from 1910 to 1977 used 4,000 tons of lead compounds in painting homes and other products. Lead-based paints manufactured as of 1992 can contain no more than 600 ppm (0.06 percent) lead to meet the criterion of a non-hazardous product. Nine years after banning lead-based paint, Congress in 1987 banned the organomnetallic compound tetraethyllead as an antiknock additive in gasoline.

*First preparation of tetraethyllead*

The British chemist Edward Frankland (1825-99) in 1849, while at the University of Marburg in Germany, prepared the first organometallic compounds, dimethyl zinc and diethyl zinc. His work spurred chemists to prepare other organometallic compounds, including tetraethyllead, the first organolead compound that Carl J. Löwig (1803-90), a German chemist at Zurich, synthesized in 1853. Löwig added a 6:1 lead: sodium alloy to a solution of the colorless liquid, ethyl iodide, dissolved in ether, to give an insoluble lead product and an ether solution that contained tetraethyllead, as his main product. Löwig discarded the ether solution without determining its composition, but his description of its preparation, which was one of the methods used 70 years later to produce tetraethyllead commercially, makes clear that Löwig had prepared tetraethyllead.

Tetraethyllead's isolation and identification as a pure compound came six years later in 1859. George B. Buckton (1818-1905), an English chemist in Haslemere (Surrey), reacted diethylzinc, an explosive colorless organometallic liquid, with an excess of crystalline lead (II) chloride and separated the products by distillation. Auguste Cahours (1813-91), a French chemist at the Sorbonne, prepared tetramethyllead in 1861, and in 1879 Frankland and Aubrey Lawrence at the Royal School of Mines in London improved Buckton's preparation by distilling the crude tetraethyllead with steam to obtain a much purer product, a pleasant-smelling, colorless, oily liquid.

*Tetraethyllead and the antiknock problem*

Charles Kettering (1876-1958) initiated the research that eventually transformed tetraethyllead from an interesting laboratory liquid compound to the gasoline additive that solved the internal combustion engine's knock (rough combustion) problem. Kettering, an electrical engineer who graduated from Ohio State University in 1904, in 1909 founded Dayton Engineering Laboratories (Delco) in Dayton, Ohio, to work on developments in the automotive industry. His first major invention in February 1911, an electric automobile ignition and self-starter system that eliminated hand cranking, appeared in the 1912 Cadillac. Cadillac was one of the divisions of the General Motors Company that William C. Durant (1861-1947) organized in Flint, Michigan, in September 1908 and reorganized as the General Motors Corporation (GMC) in October 1916. General Motors Corporation moved to Detroit in 1923.

At that time no one in the automobile and fuel industries knew what caused engine knock, and they included Kettering's ignition and starter system as a possible cause. Engine knock was not a serious problem in the low-compression internal combustion engine of that day, but Kettering wanted to build a more powerful and efficient high-compression engine and put it in an upscale automobile to challenge Henry Ford's dominance of the automobile industry. Unfortunately engine knock increased with increasing compression ratio.

The engine knock problem remained neglected until 1916 when Kettering hired Thomas Midgely (1889-1944), a Cornell University mechanical engineering graduate, and assigned him the problem. That same year in May, Kettering sold Delco for $9 million to United Motors Corporation, an automotive parts and equipment company established in 1916 in Detroit, that became part of recently-organized General Motors Corporation in 1918 (renamed United Delco in 1960 and AC-Delco in 1974). Kettering formed a new company, Dayton Metal Products Company, that in January 1919 merged-with the General Motors Corporation becoming the General Motors Research Division in Detroit. Kettering and Midgely remained with General Motors with Kettering becoming vice-president for research in 1920 and Midgely resuming the antiknock research that he had begun in 1916 but had set aside during the World War I years. Kettering, determined to show General Motors that it had acquired an efficient research organization, gave Midgely two weeks, until 30 January 1919, to find a gasoline antiknock additive.

When Kettering and Midgely first sought an antiknock additive in December 1916, Midgely already had established that knocking was not, as widely believed, the result of pre-ignition (ignition occurring before the maximum compression of the gasoline in the cylinder) but resulted from a sudden pressure increase after ignition. They also decided to color the gasoline red because they knew from the physics of light that the color red absorbed more heat energy. The heat absorption should cause the gasoline to evaporate more completely and thereby eliminate the gasoline's rough combustion or knock.

The cold weather survival of the trailing arbutus, a small plant with pinkish flowers and a leaf with a red back that the two Ohioans, Kettering and Midgely, had seen grow and bloom in Ohio's spring snows, seemed to support their red-color heat absorption hypothesis. They decided to test their hypothesis using a red-colored gasoline additive. Not having the gasoline-soluble red aniline ($C_6H_5NH_2$) dye available in their laboratory, Kettering and Midgely experimented with iodine crystals dissolved in kerosene in a small Delco engine. The brownish-black solution eliminated the knock, but iodine added one dollar to the cost of a gallon of gasoline and was too reactive, converting all the metal parts it contacted into metallic iodides. Other tests with kerosene-soluble red dyes, failed, although colorless, explosive ethyl iodide inhibited engine knock. This last test disproved the red color theory but indicated that iodine although impractical was an effective antiknock additive.

World War I temporarily had suspended the search for an antiknock additive, but once Midgely resumed the research in early 1919, he and his co-researcher Thomas A. Boyd decided to test aniline and on 30 January 1919, the day of Kettering's deadline, found that aniline worked. They also knew of aniline's several undesirable properties: its very bad odor, likely toxicity, oxidation in air, and its attacks on metals. For two years, Midgely and his research team had little to show for their efforts until spring 1921, when Kettering directed them to a patent on a newly synthesized heavy metal compound, the colorless-to-yellow liquid selenium oxychloride ($SeOCl_2$) that Victor Lenher (1873-1927) at Wisconsin had obtained. Midgely's tests showed that selenium oxychloride had good antiknock properties but was corrosive and, as later evidence showed, deadly. Lenher died in 1927 from selenium poisoning.

Their shift to testing heavy metal compounds proved crucial to solving the antiknock problem. Midgely's long-time interest and understanding of the Periodic Table and knowledge that BASF in Ludwigshafen was testing the straw-colored, toxic liquid, heavy metal compound iron carbonyl [pentacarbonyl iron, $Fe(CO_5)$] as an antiknock compound in gasoline engines directed and drove their search. Diethylselenide, $Se(C_2H_5)_2$, a toxic liquid which when decomposed gave a yellowish heavy oil with an unpleasant odor, prevented knocking. Diethyltelluride, $Te(C_2H_5)_2$, a red or yellow toxic liquid, worked better but gave off a strong garlic odor that never seemed to fade. Midgely discussed his research on antiknock additives with Robert Wilson (1893-1964), director of the Research Laboratory of Applied Chemistry at MIT whose work with General Motors included research on heavy metal antiknock compounds. To guide his research Wilson had devised a Periodic Table that highlighted the properties of the heavier metals and obtained encouraging results with some of them. Midgely consequently continued the research path he already had started, testing and obtaining good results with tetraethyltin, $Sn(C_2H_5)_4$ a toxic colorless liquid, which suggested that lead, the next heavier element in the tin group, should prove effective. It was.

When Midgely on 9 December 1921 tested in a Delco-Light engine a 1 percent solution of tetraethyllead (TEL), $Pb(C_2H_5)_4$, a colorless, oily, pleasant-smelling toxic liquid of specific gravity 1.65, in kerosene, he found that tetraethyllead provided better antiknock protection than any other organic or organometallic compound previously tested. More dilute solutions were equally impressive, a one-fortieth of one percent tetraethyllead-kerosene solution (0.125 percent) tested as effectively as a 1.3 percent aniline-kerosene solution.

One problem remained. The combustion of a gasoline-tetraethyllead solution left yellow lead (II) oxide deposits on the combustion chamber's interior, spark plugs, exhaust valves, and anything else it contacted, thereby diminishing both engine performance and life. Midgely needed another gasoline additive that could convert any solid lead products formed in the engine to volatile (gaseous) lead products that passed from the engine into the exhaust system, and into the atmosphere. After considerable testing Boyd found that 1,2-dibromoethane, a compound that Löwig had synthesized in 1853, converted the solid lead products formed in the engine to lead (II) bromide, which was volatile at the engine's temperature.

*Commercial production of tetraethyllead, fatal consequences*

Following the 1921 discovery that tetraethyllead was an effective antiknock agent General Motors, DuPont Chemical Company in Wilmington, Delaware, and Standard Oil in Bayway, New Jersey, began production of *Ethyl Gasoline*, the name that Kettering decided to call the new leaded gasoline and which noticeably did not contain the word lead. Pierre du Pont (1870-1954), president of General Motors, initiated production. He contacted his cousin Irénée du Pont (1876-1963), president of DuPont, for the purpose of having DuPont construct a tetraethyllead plant. DuPont at the time owned 38 percent of General Motors stock. Midgely, in April 1922, had improved Löwig's synthesis of tetraethyllead by replacing explosive ethyl iodide with cheaper but corrosive ethyl bromide, and by September 1922, DuPont had constructed a small unit that produced 1 gallon per day of tetraethyllead at an 85 percent yield.

Several months later, in March 1923, Midgely and Kettering provided a rosy economic outlook for tetraethyllead's commercial production. They believed that ethyl gasoline could gain 20 percent of the gasoline market. With a production and distribution cost of one cent per gallon and a profit of three cents per gallon, and with total gasoline sales of eight billion gallons (190 million barrels) annually, ethyl gasoline sales would gross $48 million and net a profit of $32 million.

Construction of the first commercial and semi-commercial plants at Dayton; Deepwater, New Jersey; and Bayway, New Jersey, followed later that year and in 1924, General Motors Chemical Company founded in 1923, began producing tetraethyllead using Standard's new ethyl chloride process in its

7 gallon per day Dayton plant. In Standard's ethyl chloride process, cheaper ethyl chloride replaced ethyl bromide. Charles Kraus (1875-1967), a Standard consultant at Clark University, in Worcester, Massachusetts, in 1922 developed the more economic ethyl chloride process to manufacture tetraethyllead using ethyl chloride and a 1:1 sodium/lead alloy that gave 70-85 percent yields of tetraethyllead. DuPont in August 1923 completed a 100 gallon per day plant in Deepwater, initially using Midgely's ethyl bromide process. Production increased to 700 gallons per day in summer 1924, and in January 1925 DuPont, having adopted Standard's ethyl chloride process, began operation of a 1,000 gallon per day plant. In September 1924, Standard began operating a 100 gallon per day pilot plant at its Bayway location.

During this two-year period of plant construction, General Motors and Standard in August 1924 merged their production units to form a new company called Ethyl Gasoline Corporation (renamed Ethyl Corporation in 1942) in New York City with Kettering as president and Midgely as a vice president. A patent issue forced the merger. General Motors had patents for the use of tetraethyllead in gasoline as an antiknock additive, whereas Standard owned Kraus's more economic ethyl chloride process for the preparation of tetraethyllead.

But almost as soon as the three tetraethyllead plants began operation, lead's toxicity once again revealed itself, resulting in worker deaths and illnesses. Midgely and three of his coworkers already in late 1922 had experienced digestive problems, reduced body temperature and blood pressure as a result of their work on tetraethyllead. At DuPont's Deepwater plants 300 workers suffered from lead poisoning. One worker died in September 1923, three in 1924, and four in 1925. At Standard's Bayway plant five workers died in October 1924 and the remaining 44 hospitalized after experiencing palsy and hallucinations. At the General Motors Dayton plant two more deaths occurred and 40 more workers hospitalized in April 1924. Some of the hospitalized wore straightjackets.

Because of the illnesses and death, the Bureau of Mines in Pittsburgh, at the urging of Hugh Cumming (1869-1948), surgeon general of the United States from 1920 to 1936, and Kettering of General Motors, conducted an eight-month investigation, beginning in September 1923 and published its report in October 1924. The Bureau, in 188 trials lasting for 3 and 6 hours, subjected dogs, monkeys, rabbits, guinea pigs, and pigeons to the exhaust from a small Delco engine that burned ethyl gasoline and concluded that human intake of lead released in automobile exhaust was negligible. The Bureau's sanitized report, and even the public apathy that prevailed following the news of the first workers' deaths, failed to silence the concerns that arose following the deaths at Standard. General Motors on 1 May, 1925, suspended production of tetraethyllead at the three manufacturing sites, and after Kettering met with Cumming, Cumming on 20 May, convened at the Butler Building, a Renaissance mansion and former embassy, in Washington, DC, a one-day conference of 100 (87) industrialists, public health officials and academic physicians to determine the future of tetraethyllead. The United States Public Health Service published a 100-plus-page conference report entitled "US Public Health Service, Proceedings of a Conference to Determine Whether or Not There is a Public Health Question in the Manufacture, Distribution or use of Tetraethyl Lead Gasoline," *PHS Bulletin No. 158*, (Washington, DC: US Treasury Dept., August 1925)

In attendance at the conference was Frank A. Howard (1890-1964), a patent lawyer and vice-president at Standard, who called tetraethyllead *a gift of God* that would enable the United States to conserve oil. Alice Hamilton (1869-1970), a Harvard medical doctor, refused to accept that tetraethyllead was the only gasoline additive that could eliminate engine knock. Midgely's own research in 1920-21 had shown that a 30 percent ethyl alcohol- gasoline blend eliminated engine knock. Others, such as the physiologist Yandell Henderson (1873-1944) at Yale and David Edsall (1869-1945) at Harvard, spoke against the use of tetraethyllead and warned of the unhealthy consequences of inhaling lead particles in automobile exhaust emissions. Edsall in 1918 established Harvard's first industrial hygiene program and focused it on lead poisoning. The conference's standout performer was Robert Kehoe, the clinical pathology professor at the University of Cincinnati. Largely because of his testimony the ban on tetraethyllead lasted only a year, until 26 May 1926.

*Robert Kehoe industrial apologist, the rise of tetraethyllead*

Kehoe received his undergraduate and medical degrees from the University of Cincinnati in Columbus, Ohio, and was a medical school faculty member in 1924 when Kettering of General Motors hired him as a consultant and asked him to investigate the hazards of tetraethyllead production at the Dayton ethyl plant. His 1925 report certified tetraethyllead's non-toxicity and led Kettering, Ethyl Gasoline Corporation, and DuPont to complete and open in December 1930 the $130,000 Kettering Laboratory in the University of Cincinnati's medical school with Kehoe as its director. Kettering, Ethyl, and DuPont already in 1925 had made Kehoe the Ethyl Gasoline Corporation's Medical Director and a corporate officer at General Motors. Totally lacking modesty and professing infallibility, Kehoe became the unchallenged spokesman and final authority for the tetraethyllead industry. His operational philosophy or rule *show me the evidence* became the tetraethyllead industry's defense and set the pattern for all other industries when faced with any challenge to the safety and impact of their products on humans and their environment. His near monopoly on airborne lead research, dictatorial pronouncements, and the General Motors-Standard Oil-DuPont political clout and marketing-advertising skill enabled tetraethyllead to remain a toxic pollutant for 50 years. Kehoe's rule helps explain why the science of occupational disease in the United States fared poorly in comparison to European research, particularly in England and Germany.

What did Kehoe claim in his 1925 report, and in later publications on tetraethyllead's toxicity and workers' health at General Motors' Dayton Laboratory? When Kehoe found that his control group, consisting of workers who had not handled tetraethyllead, had lead in their blood, he concluded that lead was a natural constituent of the human body and therefore presented no danger to humans. When some of those who had clashed with Kehoe at the Washington conference, such as Hamilton, Henderson, and Edsall, continued to argue that lead was a cumulative poison and that Kehoe's control group was likely contaminated, he looked for a new control group in an area unaffected by industrial lead pollution. Kehoe in 1933 selected the inhabitants in rural villages around Ixtlahuaca outside of Mexico City as his new control group. Tests for lead in the villagers' soil, air, water, and eating ware detected lead only in their eating ware (cooking ware, tableware, pots and pans). Tests on their blood, stool, and urine indicated the presence of lead.

Kehoe's test results convinced him that environmental lead -- lead in the soil, air, and water -- did not pass into the body and lead was, therefore, a natural constituent of the body. He published his conclusions in 1933, despite his own tests and common knowledge that Mexican lead - glazed ceramic cooking ware, tableware, pots and pans contained lead and were the likely and obvious source of the lead he found in the villagers. A concentration of 20 micrograms/deciliter (ug/dL) of lead in the blood was natural and therefore normal to Kehoe. Lead poisoning did not occur until the concentration reached 80 ug/dL of lead in the blood. No widespread lead poisoning from airborne lead existed, it affected only factory workers in the lead industry where industry officials could impose safety policies. Lead added to gasoline and released in automobile exhaust did not threaten public health. Kehoe's conclusion: you either had or did not have lead poisoning.

Kehoe and the lead industry insisted that the victims of exposure to airborne lead particles had to prove lead caused their illnesses. The US's first air pollution act of 1955 (Air Pollution Control Act) that granted $5 million annually for five years for pollution research and its 1960 amendment that extended research funding for four more years did nothing to prevent airborne lead pollution. Kehoe's rule remained the law of the land without any challenge until the 1960s, first with the amendment in 1962 that called for an examination of automobile exhaust, followed by the Clean Air Act of 1963 and its amendments in 1965, 1966, 1967, and 1969. But Congress, and later the Environmental Protection

Agency (EPA) established in July 1970, moved slowly despite the obvious consequences of lead's escaping into the environment from both leaded paint and tetraethyllead.

In the 1966 (7-9, 14-15 June) Senate hearings on the Clean Air Act, Kehoe testified before Senator Edmund Muskie (1914-96) and the Subcommittee on Air and Water Pollution and once again strongly defended the adding of tetraethyllead to gasoline. A summary of his testimony follows.

Kehoe first informed Chairman Muskie that he knew so much about lead's toxicity that he would have to leave out much information, otherwise the Subcommittee hearings would last all week. Scientists knew enough about tetraethyllead that increasing its concentration in gasoline was without risk. We have investigated "no other hygienic problem in the field of air pollution . . . over such a prolonged period of time and with such definitive results." When Muskie said that the Public Health Service disagreed with some of his statements and that unanswered questions remained, Kehoe replied that he had a greater responsibility than anyone else in the country to ensure the non-toxicity and safety of tetraethyllead. The evidence that tetraethyllead presented no hazard has never been better. When first sold in Dayton in February 1923 for 25¢ a gallon, one gallon of ethyl gasoline contained 3 milliliters of tetraethyllead. In 1959 the Surgeon General Leroy Burney [1906-88] approved an increase to 4 milliliters per gallon, or 1.0 gram per liter (1mL per liter=1 gram per liter). The EPA decreased the existing concentration of 0.52 grams per liter to 0.28 grams per liter only in 1982 and to 0.026 grams per liter in 1986). The high concentration of tetralethyllead permitted in gasoline for decades allowed Kehoe to make his unsubstantiated claim.

Muskie next asked whether scientists could find a substitute for tetraethyllead. Kehoe reiterated that tetraethyllead introduced no danger to the field of public health. The work at the Kettering Laboratory showed that lead is an inevitable element in Earth's surface, in its vegetation and in animal life. Man cannot avoid absorbing lead while living on this planet, but the amounts present in the body had accumulated over several thousand years and were *natural*, meaning safe.

Kehoe claimed that during the last thirty years, from 1937 to 1966, atmospheric lead levels had decreased in Cincinnati despite the tremendous growth in automobiles and amount of traffic. He neglected to say that during the World War II years and tight coal supplies Cincinnati, and other cities, had to burn dirty, lead-containing coal, whereas once the war ended and the greater availability of cleaner coals, the rate of growth of atmospheric lead concentrations decreased significantly but not the total atmospheric lead concentration. Kehoe blamed stationary sources, such as industrial plants, as the problem, not automobiles. He failed to mention that in the earlier studies on atmospheric lead in Cincinnati investigators took more samples from industrial sites than they took in later studies, which accounted for the apparent present downward trend.

Kehoe concluded his testimony with the totally confident remarks that not the slightest evidence existed that anyone had ingested harmful amounts of atmospheric lead and suffered any harmful effects from lead intake below the level of lead poisoning. In response to Muskie's question, whether the medical community agreed with his assessment, Kehoe replied that many people had less certainty than he had, but he had more experience in the field of lead toxicity than any living person.

Kehoe's argument that lead was a natural constituent in the body and that no harmful effects occurred from ingesting amounts of lead below the level of lead poisoning, rested more on his reputation than on solid scientific evidence. His total confidence in his conclusions, and the failure of most of the scientific establishment to challenge his self-proclaimed expertise, such as his erroneous interpretation of the Cincinnati air pollution data that his own laboratory later refuted, allowed him to dictate policy on lead's toxicity for 50 years.

*Clair Patterson and a new evaluation of lead's toxicity: the fall of tetraethyllead and other airborne lead sources*

Forty years earlier, at the 1925 meeting the Surgeon General Hugh Cumming had called on tetraethyllead's toxicity, Hamilton, Henderson, and Edsall, had criticized Kehoe's conclusions with little success. In 1966, at the Muskie Subcommittee hearings, Patterson challenged his senior Kehoe, this time with much more success. With his slogan, show me the evidence, Kehoe had convinced the scientific community and others who followed the lead controversy that lead was a *natural* constituent of the human body and that over time humans had adapted to the presence of environmental lead. The small amount of lead in the body represented the average amount found in humans and therefore not harmful. You either had or didn't have lead poisoning.

Patterson argued that Kehoe's claim of a natural concentration of lead in humans was actually the typical concentration and existed in the human body only because of human contamination of the environment. Natural should indicate the body's lead concentration before humans caused contamination, not the body's lead concentration after humans caused contamination. Typical did not mean harmless. Patterson also rejected Kehoe's argument that the several thousand years that humans had contaminated the environment with lead provided enough time for the body to have adapted to the environmental change. Not enough time had elapsed, and the accumulation of lead concentrations at levels below the accepted value for lead poisoning presented a real danger to an individual's health.

Patterson was not a medical doctor, but a geochemist who graduated from Grinnell College in Grinnell, Iowa. He began graduate studies in chemistry at Iowa in the early 1940s, but in 1944 the Manhattan Project first in Chicago and then in Oak Ridge, Tennessee, summoned him because of his expertise on molecular spectroscopy. He worked with mass spectrometers on the electromagnetic separation of uranium-235 (U-235) from its other isotopes. After the war, in 1951, Patterson received his PhD from the University of Chicago.

Harrison Brown, a geochemist at Chicago best known for his World War II work on isolating plutonium used in the first and third US atomic bombs, in the 1940s had become interested in iron meteorites and the age of Earth. With the war now over, he started Chicago's geochemistry program to measure trace element abundance using the new analytical technique of mass spectrometry developed during the war. He believed that lead isotope measurements from iron meteorites would reveal the isotopic composition of lead when the solar system first formed. Brown suspected that the uranium concentration in an iron meteorite was negligible, a correct assumption, so that the lead isotope concentrations in an iron meteorite at the present time were mainly naturally-occurring lead (Pb-204) and not lead formed from uranium's radioactive decay. These lead isotope concentrations would be the same as the lead isotope concentrations at the time of the meteorite's formation and the same for Earth at the time of its formation.

Patterson began this research for his dissertation in 1948, completing it in 1951. Working with samples from a billion-year-old Precambrian granite and other rocks, he used alpha counting to measure uranium's and thorium's rate of decay to radioactive lead and emission spectroscopy to measure the amount of lead produced by the decay. He obtained accurate ages of his samples. Patterson did not study meteorites at this time, their study began the next year, in 1952, when he moved with Brown to the California Institute of Technology. There he built the laboratory facilities and perfected the techniques for microgram and sub-microgram measurements of lead isotopic concentrations.

Working with the troilite sulfide phase of the Canyon Diablo iron meteorite, Patterson in 1953 measured its lead isotopic composition, showing that it contained no lead isotopes from uranium decay. All its lead was naturally-occurring lead (Pb-204), and its concentration equal to the concentration of Pb-204 on Earth, assuming that the Diablo Meteorite and Earth had originated at the same time from the Big Bang. Knowing the amount of naturally-occurring lead on Earth, Patterson now could calculate how much of the total amount of lead resulted from the radioactive decay of uranium, thorium, and actinium (present only in trace amounts as Ac-227 in uranium ores) which gave respectively the three lead isotopes Pb-206, Pb-208, and Pb-207. From uranium's and thorium's long half-life (Ac-227 had a half- life of only 21.8 years) he could determine how many years had to pass to produce the total amount of lead found on Earth. The number of years was Earth's age, which Patterson calculated at 4.5 billion years (currently 4.55 billion years), improving on the previously accepted value of 3.3 billion years determined from terrestrial lead sulfide (galena) deposits.

Patterson's 1950s studies on lead isotopes enabled him to isolate microgram quantities of lead in rocks and other geological samples. He expanded his isotopic measurements to ocean sediments, and in a 1962 paper on the balance between lead escaping into the environment (input) and lead's removal by depositing in the oceans as sediment, he estimated that the amount of anthropogenic (human-generated) lead dispersed into the environment each year was about 80 times the rate of deposit into ocean sediments. This study strongly indicated a serious imbalance in lead's geochemical cycle and became the starting point for Patterson's investigations on airborne lead pollution.

The first of these investigations published in 1963 showed that deep ocean water contained 3-10 times less lead than surface water. Most elements behaved the opposite way and supported his argument that anthropogenic lead input had distorted lead's natural geochemical cycle. In a highly critical and strongly criticized 16-page paper "Contaminated and Natural Lead Environments of Man," published in 1965 in *Archives of Environmental Health*, Patterson challenged the accepted wisdom that industrial lead only doubled the amount of environmental lead. He argued that lead contamination of samples and incorrect analyses of lead concentrations contributed to this erroneous view. His compilations showed that large amounts of industrial lead were passing into the environment from gasoline, solder (a tin-lead alloy), paint, and pesticides compared to the natural lead geochemical cycle. His measurements of lead concentration in human blood samples indicated that the lead concentrations were 100 times that of pretechnological societies, 0.2 ppm (0.2 mg/L) versus 0.002 ppm (0.002 mg/L), which he considered the natural concentration of lead. If the lead concentrations he measured in 1965 doubled, Patterson believed they almost would have reached the scientifically-accepted concentration of lead poisoning.

In carrying out these measurements and experiments, Patterson continued the extraordinary techniques he had introduced in his earlier research on lead's isotopes and Earth's age to avoid lead contamination and establish the true lead concentrations of his samples. He filtered the laboratory's air, gowned and washed the experimenters, purified the reagents, such as chloroform, and water supply of trace amounts of lead as low as 0.0008 grams of reagent, washed the polyethylene (teflon) laboratory ware with nitric acid (4M), and substituted pure quartz ware whenever possible. Patterson used a high resolution thermal ionization source mass spectrometer to make stable isotope dilution measurements.

Kehoe was one of the *Archives* reviewers, and he read Patterson's paper at the time of its submission. He agreed to its publication, writing in his review to the journal's editor that "I should let the man, with his obvious faults, speak in such a way as to display these faults. . . . The inferences are remarkably naive. . . . It is an example of how wrong one can be in his biological postulates and conclusions, when he steps into this field, of which he is so woefully ignorant and so lacking in any concept of the depth of his ignorance, that he is not even cautious in drawing sweeping conclusions."

Most toxicologists and health specialists sided with Kehoe, and not Patterson. Herbert Stockinger (d. 1998) at the National Institute for Occupational Safety and Health called Patterson a rabble rouser. The World Health Organization at its March 1965 meeting in Rome argued that environmental lead contributions to the human body had not changed in any significant way in 20 years. In short, the critics said that Patterson, a geochemist, had strayed out of his field and should stick to rocks instead of people.

In addition to the *Archives* article's critics, Patterson had to deal with the Ethyl Corporation. Its representatives met with Patterson in 1965 and tried to buy him off by offering "research support that would yield results favorable to their cause." Following the Ethyl meeting Patterson lost his long-standing contract with the Public Health Service and a well-paying contract with the American Petroleum Institute. Some members of the Board of Trustees at Caltech tried to have Patterson fired. This occurred only two years after Caltech's administrative officials all but forced Linus Pauling to resign shortly after he won his second Nobel Prize, the 1963 Nobel Peace Prize, for his opposition to the proliferation of nuclear weapons. Pauling's sin was his alleged lack of patriotism and communist sympathies. Patterson's sin was challenging the establishment.

Patterson did not resign, nor did he resume his rock-dating research. Instead he launched a parallel state and national attack on lead pollution that led to the start of California's air control standards and to his appearance before Muskie's 1966 Subcommittee on Air and Water Pollution. Patterson had sent a letter to Muskie offering to appear and received an invitation to testify on 15 June, one week after Kehoe's testimony. His testimony focused on several issues that he believed required greater clarification and public understanding. The first issue was the mistaken belief that the natural lead cycle and human activity contributed about the same amount of lead to the environment. The amounts were significantly different. Natural recycling contributed 10,000 tons of lead annually whereas industrial emissions contributed millions of tons. Moreover, public health officials and agencies by collaborating with lead industry officials were misleading the public about the real danger from industrial lead pollution. They failed to distinguish clearly between *natural* and *normal* (*typical*) concentrations of lead in the body, in food, and in the environment. To Patterson, normal concentrations were really typical concentrations of lead found in the body, food, and environment and the results of lead contamination, and definitely not natural concentrations.

Regarding lead concentrations in the blood, Patterson pointed out the significant difference of 0.11-0.17 ppm lead in the blood level concentrations of people living in the cities and those living outside the cities, which contradicted the American petroleum industry's position of "no evidence of an increase in the past since sometime in the 1920s." As for Kehoe's claim that lead levels in Cincinnati in the last 30 years had decreased, Patterson quoted data from the National Air Sampling Network that showed lead concentrations in Cincinnati and other cities actually were increasing and that Kehoe's data were invalid. Patterson also reported the results of his recent 1963 work on lead concentrations in the snow of Greenland showing large increases in the surface snow versus deeply buried snow.

For the next two decades, Patterson continued to present evidence supporting the important distinction between natural and normal concentrations of lead. He focused his attention on atmospheric lead concentrations obtained from three sources: lead in snow strata trapped in 200-300 meter deep ice cores of Greenland and Antarctica, lead in mummies and ancient skeletons, and lead in the food chain and canned food.

Patterson's studies on the Greenland and Antarctica snow strata and ice cores that began in 1969 showed that from the time of the industrial revolution, the concentration of atmospheric lead had increased in both regions. In Greenland, the surface snow and ice contained greater than 100 times more lead than the ice cores obtained from deeper strata. Lead concentrations in the ice cores had risen fourfold during the 1800s industrial revolution, and threefold from the 1920s to the 1960s after the introduction of tetraethlyllead. In Antarctica, Patterson's initial studies showed the` atmospheric lead pollution was 1/10 that of Greenland, whereas his later more accurate studies showed the effect was 1/2 of Greenland's. The slices he obtained from the ice cores enabled him to date each sample and thereby show the historical increase of lead in the atmosphere and its subsequent fall to Earth.

Beginning in 1975, Patterson's measurements on lead, barium, and calcium concentrations in the bones of 1,600 year old Peruvian skeletons showed a 700-1,200 (0.2 ppm versus 0.002 ppm) times increase of lead concentrations in modern humans but no change in the concentrations of two metals, barium and calcium, not present in industrial pollutants. He provided the same evidence of increasing lead concentrations in the food chain with his studies in 1979-81 on tuna fish muscle, demonstrating that the Environmental Protection Agency (EPA), like Kehoe earlier, was using initial lead concentrations (laboratory blanks) that were too high to permit accurate analyses of lead concentrations below 1 ppm. He recommended that the EPA and the Bureau of Foods use the Bureau of Standards' mass spectrometers which had the capability to do ppm analyses and that the EPA obtain better analytical equipment, such as mass spectrometers.

In the 1980 paper on tuna, Patterson examined the amounts of lead entering the food chain because of lead solder present in canned goods. A 1978 report the National Marine Fisheries Services Laboratory in Baltimore published, estimated that canned albacore tuna muscle contained only about twice as much lead as fresh albacore tuna muscle. The fresh tuna contains the smallest concentration of lead yet measured in any biological tissue, or 700 nanograms lead ($700x10^{-9}$) per gram canned tuna versus 400 nanograms lead ($400x10^{-9}$) per gram fresh tuna. Patterson found in his 100-300 mg samples, 1,400 nanograms lead per gram canned tuna and 0.3 nanogram lead per gram fresh tuna. A comparison of the barium concentration in the two samples showed it had doubled. A later National Marine Fisheries Services test indicated a value of 20 nanograms lead per gram fresh tuna, which was much higher than Patterson's value but still 70 times less than Patterson's determination of the lead content of canned tuna. The evidence presented in these studies convinced the scientific community, and led in 1993 to the EPA's removal of lead solder from all food containers.

*Patterson's vindication*

Because of the public's increasing awareness that airborne lead, most of which came from leaded gasoline, produced serious health effects, the EPA in 1970 commissioned the National Research Council of the National Academy of Sciences (NAS) to investigate the relation between airborne lead and health. The report, *Lead: Airborne Lead in Perspective*, published in 1972, was a sham, actually an embarrassment to the Academy's objective scientists. As soon as the NAS announced the committee's membership and its consultants, the results were a foregone conclusion. No member of the committee, nor most of the committee's consultants, had worked on airborne lead, and most of them came from industry. The consultants included Kehoe and Lloyd Tepper from the Kettering Laboratory; Gordon Stopps, Kamran Habibi, and John Perrard from DuPont, and. Gary Ter Haar from the Ethyl Corporation.

The NAS passed on Patterson and other scientists who had done considerable research on airborne lead, considering them either disruptive or biased and not objective. The bypassed included Patterson; John Goldsmith, head of California' Epidemiology division in the Health Department; Henry Schroeder at Dartmouth Medical School whose research focused on low dose lead transgeneration; and Tsaihwa J. Chow, Patterson's Caltech colleague who wrote the 1973 article *Our Daily Lead* and had studied atmospheric lead deposits in Greenland's ice cores, although the committee appointed him as a consultant.

Industry scientists wrote much of the report, which focused heavily on discussions of lead in plants and animals instead of humans. Stopps who had long defended tetraethyllead was harmless, which was the very issue the committee was to resolve, wrote the important section on adult epidemiology and lead alkyl. Chow wrote about the committee membership's obvious conflicts of interest. In the end, the report specified that no conclusive data showed that atmospheric lead concentrations of less than 2 micrograms per cubic meter ($2x10^{-6}$ g/m$^3$ or 2 ug/m$^3$) of air contributed to blood lead levels and no evidence of toxicity at low levels of lead existed. The EPA as recent as 1981 allowed a tolerance of 0.15 mg lead as dust, fumes, and inorganic compounds per cubic meter of air (1.5 mg/m$^3$) and

1.5 micrograms lead per cubic meter (1.5 $ug/m^3$ ) of ambient (external surrounding) air. The lead industry clearly had received an exoneration to continue the status quo, but the report's obvious bias gave it a short lifetime that lasted only until 1980 when NAS published a second more balanced report, *Lead in the Human Environment.*

The reasons for the greater balance in the 1980 report were Patterson's appointment as a member of the 12-member panel and the panel's decision to publish majority and minority statements. The majority report, which Patterson found insufficiently critical, nevertheless recommended several important changes and re-evaluations of current thinking on airborne lead. Briefly the report recommended reducing lead hazards for urban children and adopting a better definition of the margin between toxic and typical (normal) lead levels for adults. It finally acknowledged that typical atmospheric lead concentrations were 10-100 times the natural lead backgrounds for average populations and 1,000-10,000 times greater than the natural lead backgrounds for urban populations. The majority report called for more research on the connection between lead ingestion and intellectual ability.

In his 78-page minority report, Patterson argued against additional studies on airborne lead. The scientific community had convincing evidence, and he wanted immediate action on reducing or eliminating completely the sources that put lead into the atmosphere. This meant leaded gasoline, soldered food containers, paints, and glazes. He pointed out that in the period 1850-1923 scientists believed that Americans had virtually no lead in their bodies, they had a natural level near zero. After Kehoe published his flawed studies beginning in 1925 the new natural zero lead level became the much higher typical lead level found in present-day Americans. Kehoe had raised the natural lead level to a concentration that was much closer to the lead level of a lead- poisoned person. Patterson, as he had done earlier, again corrected this misrepresentation in his minority report. Before 1970 the US government's Centers for Disease Control had set an acceptable upper limit of lead in children's blood at 60 ug/dL. It lowered this to 25 ug/dL after 1970, and to 10 ug/dL in 1991. Kehoe's value for lead poisoning was 80 ug/dL, later studies have confirmed that 70 ug/dL causes severe lead poisoning.

In the end, Patterson argued that it was "intrinsically wrong to mine and smelt a highly toxic substance such as lead on a scale of millions of tons per year and then disperse it within human environments. The majority of this Committee on Lead . . . fails to explicitly acknowledge this obvious truth." Looking ahead to the future of science and engineering, Patterson wanted scientists and engineers to adopt humanitarian principles in their work on new engineering developments because the presence or absence of vital kinds of knowledge obtained from these developments may well be the factor that tips the scale for or against the survival of our hominid species in the face of challenges presented by impending developments in genetic engineering.

*Conclusion*

What is Patterson's legacy? Patterson's 1965 paper "Contaminated and Natural Lead Environments of Man" introduced and made clear the crucial difference between natural and typical (normal) concentrations of lead in humans. He repeatedly stressed the difference until the scientific and medical communities finally accepted its validity. In so doing, he forced the scientific and medical communities to go beyond the chemical study of lead and its compounds. The two communities had to look at lead's effects on humanity at concentrations below the level of lead toxicity and examine lead's effects on learning, particularly in young children. Patterson exposed the Kehoe rule that permitted the 70-year acceptance of tetraethyllead as an antiknock additive in gasoline and the subsequent release of lead in automobile exhaust into the environment and humans occupying the environment. Evidence suggests that tetraethyllead contributed 98 percent of environmental lead pollution.

As a result of tetraethyllead's removal that began in 1973 with complete removal in 1987, nine years after Congress in 1978 prohibited lead in paint, lead concentrations in the blood of adults and children fell by nearly 80 percent, decreasing in children from 15 ug/dL in 1978, to 3.6 ug/dL in 1998. During that time, the Centers for Disease Control lowered acceptable lead blood concentrations from 60 ug/dL before 1970 to 25 ug/dL after 1970 and to 10 ug/dL in 1991.

In 1991, John Sinfelt (1931-2011) at Standard Oil in Westfield, New Jersey, in a follow-up to previous patents in 1969 and 1971, in September 1974 patented a bimetallic catalytic process in which a platinum-iridium catalyst enabled petroleum refiners to produce cheaply a high-octane lead-free gasoline. His invention contributed significantly to the improvement of air quality in the United States and abroad.

Removal of lead from other sources followed, such as plumbing fixtures and lead solder from food containers in 1988, and lead foil from bottled wine in 1992. From 1971 to 1991 the lead content of Greenland snow decreased 7.5 times. Outside of the United States, the Rio de Janeiro Environmental Summit in 1992 called for a worldwide lead phase out, and in 2000 the European Union banned leaded gasoline, as have several Asian countries including China, Japan, Korea, and India.

Patterson's last writings sum up his outlook on science, engineering, and society. He saw the scientific mind working to discover the secrets of the world and the engineering mind seeking to control the natural world. The scientist would unravel the science of lead and its effects on society whereas the engineer would develop the engineering applications that, if wisely utilized, would eliminate the large amounts of lead that pollute the environment.

| LEAD FACTS |
|---|
| ***Testing for lead*** |
| Anthropogenic lead (human-distributed) lead is Earth's largest and most widespread heavy metal contaminant. It is not an essential trace element for life, as are many other metals such as cobalt, zinc, selenium, chromium, and iron. A simple test for lead's presence is to digest lead-containing soil in 4-molar (4M) nitric acid and to examine the recovered lead using a graphite furnace atomic absorption spectrophotometer. Studies from the Box Springs Mountains near Riverside in southern California have shown that its soil contained almost 3.3 milligrams of native lead per kilogram of soil compared with 19 milligrams anthropogenic lead (from air pollution) per kilogram of undisturbed soil and 10 milligrams anthropogenic lead per kilogram of disturbed (by rodents) soil. The numbers represent 20-38 kilograms of anthropogenic lead per hectare of soil (10,000 square meters or 2.47 acres) deposited from air pollution. |
| ***How lead affects humans*** |
| Lead is a neurotoxin and accumulates in the body. Exposure to lead lowers childrens' IQ test scores. Symptoms of lead poisoning include fatigue, discolored (blue-gray) teeth and gums, jaundice, colic, numbness, palsy (trembling and lack of motor control), muscle cramps and contractions, pain in the extremities, paralysis and loss of sensation, hallucinations, kidney failure, seizures and coma, brain damage, and death. |
| ***How tetraethyllead works*** |
| Tetraethyllead must undergo thermal decomposition in the automobile engine to give ethyl radicals and gaseous lead atoms. The atmosphere oxidizes lead atoms to give lead monoxide, which either in gaseous form or as a fog of fine particles inactivates intermediates of the hydrocarbon oxidation, such as alkyl hydroperoxides, and thereby inhibits the chain-branching reactions that could result in detonation or engine knock. Tetraethyllead is therefore a gasoline-soluble and volatile source of lead atoms at the operating temperature of the engine. The intact tetraethyllead molecule is not the effective antiknock agent, it is merely a lead carrier. |

## X. CONCLUSION

Energy and the environment always have interested and sometimes concerned society. They acquired even greater interest and concern following the discovery and development of nuclear fission and fusion in the 1940s-50s and a heightened awareness of environmental issues in the late 1950s-early 1960s.

The discovery of the subatomic particles, electrons, protons, and neutrons, led to an investigation of their properties and to their use in reactions with the chemical elements. Rutherford, Chadwick, Fermi, Hahn, and others carried out many of the investigations that resulted in the discovery of the subatomic particles and a new branch of physical science called nuclear science. Rutherford's identification of the proton as a hydrogen ion, $H^+$, the alpha particle as a helium ion, $He^{2+}$, and Fermi's and Hahn's use of the neutron in nuclear reactions led to the discovery of nuclear fission in 1938 and later to the discovery of nuclear fusion, and to production of new and heavier elements beyond uranium (Z=92) in the Periodic Table.

Nuclear fission became the driving force in nuclear studies because its discovery coincided with the rise of Hitler in Germany and the beginning of World War II. The long-term consequence of these studies was the race to build the first fission bomb, an atomic bomb, which the United States'

Manhattan Project accomplished in July 1945. The Soviet Union followed with its first atomic bomb in 1949, other nations such as Britain, and China joined the nuclear club within a few years. Work on nuclear fusion also began during the World War II years and resulted in the United States' constructing the first fusion (hydrogen) bomb in 1952 followed again by the Soviets in 1955. The nuclear arms race to construct more and bigger nuclear bombs, which began after World War II, would continue until the collapse of the Soviet Union in December 1991.

Research on nuclear reactors for the production of energy for domestic consumption, rather than energy for weapons, began in the United States, the Soviet Union, England, France, and other countries after World War II. Production of nuclear reactors remained secondary to the production of nuclear weapons during the Cold War years, not only because of the US government's claim of national security but because of the safety issues associated with nuclear reactors especially the disposal of spent nuclear fuel and radioactive emissions. The nuclear accidents at Three Mile Island in Pennsylvania, in 1979, Chernobyl in Ukraine in 1986 and most recently in March 2011 at Japan's Fukushima Daiichi nuclear power plant, which experienced an overheating and partial meltdown of three of its six boiling water reactors, have contributed to a loss of confidence in nuclear energy as a domestic energy source. Fusion reactors continue to remain an unfulfilled promise and unlikely to provide a dependable energy source in the near future.

Alternative and renewable sources of energy, particularly solar energy, have increased total energy output globally but still remain a minor part of overall energy production. The United Nations Environment Program's (UNEP), *Global Trends in Renewable Energy Investment 2012* (fifth edition) reported that in 2010 renewable energy accounted for only 16.7 per cent of global energy consumption with solar and wind contributing 8.2 per cent and biomass 8.5 per cent. Global investment in renewable energy nevertheless increased 37 per cent in 2010, another 17 per cent in 2011, and reached a record high of $257 billion invested in 2011. China spent the largest amount $52 billion, the United States followed with $51 billion (57 per cent increase). India $12 billion (62 per cent increase mainly on solar), Brazil $7 billion (8 per cent increase), Germany, and Italy were the other main investors. The 2011 investment represents a 600 per cent increase over the 2004 amount and a 94 per cent increase on the 2007 amount.

Solar energy investment accounted for more than half of the 2011 total renewable energy investment, or $147 billion, a 52 per cent increase resulting mainly from strong demand for rooftop photovoltaic installations in Italy and Germany, small-scale photovoltaics in China and the United Kingdom, and large-scale concentrating solar power (CSP) installations in Spain and the United States. Investment in wind power totaled $84 billion in 2011.

Only with continued government incentives and industry's willingness to develop and promote their widespread application, will alternative and renewable energies become an increasingly important part of energy production. All of this leads back to a continued dependence on fossil fuels, particularly coal, oil, and natural gas, as the major energy sources, accounting for more than 80 percent of global electrical consumption in 2011.

The continued and growing dependence on fossil fuels has produced serious environmental problems, such as acid rain and global warming. Scientists have discovered the causes and cures for both problems, and the overwhelming majority of the scientific community no longer disputes their reality. The resistance to reducing and possibly eliminating their climate threatening effects is unfortunately political and economic, not scientific. This has resulted in the denial that one or both are real and has delayed the implementation of practical remedies within the United States and globally.

## REVIEW QUESTIONS ON ENERGY AND THE ENVIRONMENT

1. What is the main theme of this chapter?

2. Who carried out the first nuclear fission reaction?

3. Discuss the discovery of the neutron in 1932 to include the work of the Bothe and Curie proton experiments.

4. Who constructed the first cyclotron?

5. Discuss Frédéric and Irene Joliot-Curie's discovery of artificial radioactivity.

6. What did Enrico Fermi in Italy contribute to the study of nuclear fission, and also Otto Hahn and Lise Meitner in Germany?

7. Discuss the early work on controlled nuclear fission. Who was responsible for the first controlled fission reaction, when, and where?

8. Identify: Leslie Groves, J. Robert Oppenheimer, Los Alamos, Hiroshima, Nagasaki.

9. Discuss the German, Soviet, and Japanese fission programs.

10. What is the difference between an atomic bomb and a nuclear reactor, fission, and fusion?

11. Discuss the atomic spy scandals of the 1950s, the Rosenbergs and other spies.

12. Which nations constructed the first nuclear fission and fusion reactors?

13. What are the main problems confronting the use of nuclear fission reactors?

14. Discuss Farrington Daniels' solar energy research, appropriate technology?

15. What is acid rain, its cause, and how has society (scientists and government) dealt with it?

16. Who is Robert Angus Smith, Gene Likens, S. Fred Singer?

17. What is global warming? Is it real or scientific fraud?

18. Identify: Svante Arrenhius, Guy Callendar, Gilbert Plass, Roger Revelle, David Keeling, James Hansen, Bjorn Lomborg, William Nierenberg?

19. Who are the leading climate change-global warming deniers?

20. What is the Kyoto Protocol, Copenhagen Accord?

21. Discuss Clair Patterson and the environment – lead issue.

22. Is there conflict between energy production and environmental well-being?

23. Does nuclear energy have a future in the United States, and in other countries?

# EPILOGUE

*Science Changed the World* has provided an integrated examination of the origin, growth, and impact of science in different times and in different cultures from ancient to modern, from East to West. Its eleven chapter themes show the transition of science from a record keeping period that began with the Babylonians, Egyptians, and Chinese about 2000 BC to a theoretical-philosophical period that arose in Greece in 600-400 BC, and to a mathematical-quantitative period that defined the sixteenth- and seventeenth-centuries (1600-1700) scientific revolution that occurred in Europe.

The transitions did not occur simultaneously in the physical and life sciences and in Eastern and Western societies. The physical sciences completed the three-stage transition in the sixteenth and seventeenth centuries whereas the life sciences moved from record keeping to theoretical-philosophical science in the later eighteenth and nineteenth centuries and to mathematical-quantitative science in the twentieth century. The transitions occurred first in Western societies. As a result of east-west contacts by western missionaries and other travelers to the East, Western science spread to the East beginning in the late 1700s in Japan and in China in the late 1800s, although Jesuit missionaries brought Western astronomy to the Chinese court as early as1583.

*Science Changed the World* also has provided an understanding of what science is and how science differs from engineering and technology and from the humanities, Of equal importance, it has shown how scientists acquire scientific knowledge and develop scientific theories using both an inductive and a deductive process, and the scientific method, that serendipity sometimes plays a role in scientific discovery and invention, and that scientists have an epistemology and an ontology of science. These are important issues and concerns for historians of science and for which *Science Changed the World* has provided answers.

Building on the foundation provided by earlier natural philosophers, scientists, and medical practitioners of ancient times, such as Hippocrates, Aristotle, Hipparchus, Ptolemy, Archimedes, and Galen, while at the same time challenging their expertise, scientists in the Renaissance (1400-1600) brought about a revival, a rebirth, of science through the studies of Copernicus, Vesalius, and others. Their studies ushered in the scientific revolution of the sixteenth and seventeenth centuries (1600-1700) with its emphasis on the search for the mathematical laws of nature and acceptance of the mechanical philosophy, the study of matter in motion. The contributions of Boyle, Harvey, Galileo, Kepler, and Newton define the scientific revolution.

In the eighteenth century (1700s) the study of composition, the classification of matter, and its electrical behavior directed the studies of physical scientists such as Cavendish, Priestley, and Lavoisier, whereas classification of plants and animals directed the studies of naturalists such as Linnaeus. Beginning in the nineteenth century (1800s) the electrodynamic worldview emerged to complement the existing mechanical worldview. Faraday and Maxwell made major contributions to its development, as did Michelson with his studies on the speed of light. Evolutionary theories, particularly Darwin's theory of natural selection, represented the shift from classification in the life sciences to the search for theories and laws, as had happened two centuries earlier in the physical sciences. Mendel's studies on plant breeding solved Darwin's problem of inheritance and established the new science of genetics. The life sciences in the United States were more notable for the debates about polygenesis and evolution rather than for American contributions, but this would change in the twentieth century.

The twentieth century (1900s) saw the end of classical physical science and determinisn and the beginning of a second scientific revolution initiated by several late nineteenth-century discoveries such as x-rays, radioactivity, and the electron, by Planck's quantum theory and Einstein's theories of special and general relativity, and other twentieth-century developments such as quantum mechanics and the structure of the atom an its nucleus. The discovery of nuclear fission and fusion followed in the 1930s-40s.

The major breakthroughs in the physical sciences in the twentieth century, namely quantum theory, special and general relativity, nuclear structure, quantum mechanics, and the electronic structure of the atom came from European scientists, such as Planck, Einstein, Rutherford and Chadwick, Heisenberg and Schrödinger. American scientists developed many of these ideas, particularly the application of quantum mechanics to the electron theory of valence (Pauling and Mulliken) and the quark theory (Gell-Mann, Greenberg, Richter, Ting).

A similar pattern emerged later in the century with the discovery of nuclear fission and fusion. European scientists discovered fission and fusion, whereas American scientists or European scientists in the United States, such as Oppenheimer, Fermi, Bethe, and Teller, were the first to demonstrate the impact of fission and fusion with the construction of the first fission (atomic) and fusion (hydrogen) bombs in the years 1945-52. The Soviet Union, England, and United States were the first nations to construct nuclear reactors for the peaceful production of energy after World War II ended.

Solar energy was a third *new* form of energy that, although as old as the universe, experienced a revival in the 1950s-60s. Solar energy's roots reach back to ancient times, but it began to achieve a degree of general acceptability only in the 1970s, thanks largely to the pioneering work of Daniels and a few other solar prophets.

The life sciences underwent their own revolution. In the first decades of the twentieth century, important breakthroughs occurred, particularly in the study of genetics, chromosome mapping, and the characteristics and role of the chromosomes and the genes located on them. The contributions of Wilson, Sutton, Stevens, and other geneticists led to understanding the role of the sex chromosomes in creating life and the genes linked to diseases.

Watson and Crick in 1953 discovered the structure of DNA, the molecule responsible for existence of life and in 1961 Nirenberg, and independently Khorana, solved the genetic code. These discoveries paved the way for the later developments in cloning and stem cell research.

At about the same time society began to recognize its damaging of the environment with its pollution of air and water, resulting largely from the increased consumption of fossil fuels. The slow recognition of environmental issues resulting from the increased domestic and industrial consumption of fossil fuels paralleled the slow recognition of alternative energy sources such as solar energy. Like solar energy, environmental issues had deep roots, but only since the 1960s did society begin to acknowledge the damage that environmental pollution caused to Earth's water resources and its atmosphere. Earlier efforts to deal with environmental pollution occurred in Europe and in the United States in the nineteenth century. The pollution prevention programs, however, aimed mainly at water pollution, because no one would drink or bathe in visibly polluted water. These programs led to improved water quality.

The earlier prevention programs that dealt with atmospheric pollution focused on smoke pollution, because society saw and experienced its consequences. For the sometimes less-visible consequences of air pollution, such as acid rain, global warming, and the presence of lead and mercury in the atmosphere, which were the direct result of an automobile-dominated and fossil fuel burning society, decades elapsed before any action occurred. Likens documented the serious consequences of acid rain. Plass, Revelle, Keeling, Hansen, and others experimentally demonstrated the reality of global warming. Patterson significantly increased society's awareness of the lead-in-air pollution problem, showing that much of the lead eventually settled in water or on land.

Despite the scientific evidence indicating the extent and seriousness of environmental pollution, progress in improving air quality has continued to lag, fueled largely by unsubstantiated claims of financially, politically, and industry-supported doubters. Scientists have discovered the causes and cures for acid rain, global warming, and other forms of environmental pollution, and the overwhelming majority of the scientific community no longer disputes their reality. The resistance to reducing and possibly eliminating their climate-threatening effects is unfortunately political and economic, not scientific. This has resulted in the denial that one or both are real and has delayed the implementation of practical remedies within the United States and globally.

We have studied the science of the past, are experiencing the science of the present, and can only contemplate and look forward eagerly to the science of the future.

The table below outlines the eleven themes that have dominated the history of science and which *Science Changed the World* has discussed.

| *Chronology of the History of Science* | | |
|---|---|---|
| *Time Period* | *Theme* | *Representative Topics* |
| Introduction to the History of Science: Its Language, Objectives, and Scope | The language, objectives, and scope of the history of science | History of science.<br>Science, engineering, and technology<br>Science and the humanities: science and religion.<br>Three approaches used in studying the history of science.<br>Philosophies of science. |
| Ancient Science: from Earliest Times to the Fall of Rome 476 AD | The transition from an oral tradition to record keeping 2000 BC to the theoretical-philosophical science of the Greeks (600-400 BC) | Universe of Plato, Aristotle, Hipparchus and Ptolemy; theories on structure of matter, Democritus, Aristotle and Alchemy; theories of medicine, Hippocrates, Galen; science in Pre-Columbia America: Mayans, Aztecs, Incas; science in the East: Chinese, Japanese, Hindu |
| Science and its Application in the Middle Ages 400-1400 | An age of faith, a lack of scientific incentives, technology led science | Alchemy; impetus theory of motion; Alhazen: light; invention of the compass and gunpowder |
| Science in the Renaissance 1400-1600 | A rebirth of learning, accepting and challenging the authority of the ancients | Leonardo da Vinci; Copernicus: heliocentric model of the universe; Vesalius: anatomy; Paracelsus: alchemy |
| Scientific Revolution of the Sixteenth and Seventeenth Centuries<br>1600-1700 | The mechanical philosophy, the study of matter in motion and the search for the mathematical laws to account for the motion dominated the physical sciences. Determinism directed the scientific outlook. | Bacon and Empiricism; Gilbert and magnetism; Harvey and the circulation of the blood; Boyle and the gas laws, van Helmont and the composition of matter; Kepler's three laws of planetary motion; Galileo's quantification and laws of falling bodies, Newton's three laws of motion, law of universal gravitation |
| Physical and Life Sciences in the Eighteenth Century: the Enlightenment<br>1700-1800 | The study of composition and the classification of matter, theories of heat, and the electrical behavior of matter directed the physical sciences; classification occupied the naturalists. Impact of science and mechanics on society. | Lavoisier's revolution in chemistry; law of conservation of mass, definition of an element; theories of heat; development of steam power; advances in electrical theory and applications; classification in nature |

| *Chronology of the History of Science* | | |
|---|---|---|
| *Time Period* | *Theme* | *Representative Topics* |
| Darwin's Century | A transition from classification (record keeping) to theories and laws occurs in geology and natural history. Mendel's new science of genetics solved Darwin's problem of inheritance. | Age of Earth, theories of evolution: Buffon, Lamarck, Erasmus Darwin; geological developments: catastrophists and uniformitarians; Charles Darwin, natural selection; Mendel and genetics; evidence of evolution; science and religion; Scopes trial, creationists |
| Molecular Biology and the Double Helix | A new field of study emerges in the life sciences, building on previous advances and the moving of physical scientists and their mathematical and experimental methods and instruments to the life sciences. | Earlier developments: cell theory, chromosomes, nucleic acids; Watson and Crick's structure of DNA; Genetic code; genetic engineering; human genome project; ethical issues; cloning and stem cells |
| Physical Sciences in the Nineteenth Century 1800-1900 | The electrodynamic worldview complements the existing mechanical worldview. Classical physics and determinism dominate but begin to collapse at the century's end. | Light: properties, Michelson and speed of light; developments in chemistry; Maxwell: electricity and magnetism; three laws of thermodynamics |
| Twentieth-Century Physical Science 1900-2000 | Theoretical science, quantum theory, special and general theories of relativity, and quantum mechanics become the foundation of physical science. They replace classical physics and determinism. | Fundamental breakthroughs: x-rays, radioactivity, electron; quantum theory: Planck; Einstein's theory of special relativity and general relativity; atomic structure; quantum mechanics: Heisenberg and Schrödinger; elementary particles |
| Energy and the Environment: Conflict or Compatibility | Scientists search for new sources of energy, such as nuclear energy, solar energy. Environmental concerns emerge. | Nuclear fission: Manhattan Project, spy scandals; German, British, Japanese, and Soviet fission research; nuclear fusion; postwar developments in nuclear energy; alternative energies: solar energy; environmental concerns: acid rain, climate change, lead in the environment |

# BIBLIOGRAPHY


***Chapter 1: Introduction to the History of Science: its Language, Objectives, and Scope***

Fuller, Steve. *Thomas Kuhn: A Philosophical History of Our Time.* Chicago: University of Chicago Press, 2002.

Kauffman, Stuart. *At Home in the Universe*. New York: Oxford University Press, 1995.

Kuklick, Bruce. *A History of Philosophy in America, 1720-2000.* New York: Oxford University Press, 2001.

Kuhn, Thomas S. *The Structure of Scientific Revolution*. Chicago: University of Chicago Press, 1962, 3d ed. 1996.

Nickles, Thomas, ed. *Thomas Kuhn: Contemporary Philosophy in Focus*. Cambridge: Cambridge University Press, 2003.

Pine, Ronald C. *Science and the Human Prospect.* Belmont, CA: Wadsworth Publishing Company, 1989.

Roberts, Royston. *Serendipity.* New York: John Wiley & Sons, 1987.

Sardar, Ziauddin. *Thomas Kuhn and the Science Wars: Postmodern Encounters*. Cambridge: Icon Books, 2000.

Sharrock, Wes N. and Rupert J. Read. *Kuhn: Philosopher of Scientific Revolutions.* Malden, MA: Polity, 2002.

Tipler, Frank. *The Physics of Immortality.* New York: Doubleday, 1994.

***Chapter 2: Ancient Science: From Earliest Times to the Fall of Rome 476AD***

Africa, Thomas W. Science and the State in Greece and Rome. New York: John Wiley & Sons, 1968.

Aveni, Anthony. *Native American Astronomy.* Austin: University of Texas Press, 1977.

Aveni, Anthony. *Skywatchers of Ancient Mexico.* Austin: University of Texas Press, 1980.

Baker, Robert H. *Astronomy*. 7th ed. New York: D. Van Nostrand Company, 1959.

Berry, Arthur. *A Short History of Astronomy, From Earliest Times through the Nineteen Century.* London: John Murray, 1898.

Burnet, John. *Early Greek Philosophy.* London: A&C Black, 1920.

Childe, V. Gordon. *Man Makes Himself.* London: Watts and Co., 1936.

Cornford, F.M. *Plato's Cosmology: The Timaeus of Plato.* London: Kegan Paul, 1937.

Clagett, Marshall. *Greek Science in Antiquity.* New York: Collier Books, 1963.

Dodds, E. R. *The Greeks and the Irrational.* Berkeley: University of California Press, 1951.

Dreyer, J. L. E. *A History of Astronomy from Thales to Kepler.* New York: Dover Publications, 1953 (reprint of 1906 edition).

Durham. Frank. *Frame of the Universe: A History of Physical Cosmology.* New York: Columbia University Press, 1983.

Farrington, Benjamin. *Greek Science.* Baltimore: Penguin Books, 1966.

Farrington, Benjamin. *Science in Antiquity.* New York: Oxford University Press, 1969.

Frankfort, Henri, et al. *Before Philosophy, the Intellectual Adventure of Ancient Man.* New York: Penguin Books, 1946.

Galen. *On the Natural Faculties.* ed. and trans. Arthur John Brock. London: G.P. Putnam's Sons, 1916. (The Loeb Classical Library).

Gill, Jerry H. *Native American Worldviews: An Introduction.* Amherst, NY: Humanity Books, an Imprint of Prometheus Books, 2002.

Grob, Gerald N. *The Deadly Truth: A History of Disease in America.* Cambridge: Harvard University Press, 2002.

Guerrini, Anita. *Natural History and the New World 1524-1770.* Philadelphia: American Philosophical Society Library, 1986.

Hall, Alfred Rupert and Marie Boas Hall. *A Brief History of Science.* New York: New American Library, 1964.

Hawkins, Gerald S. *Stonehenge Decoded.* New York: Barnes & Noble Books, 1993.

Heath, Thomas L. *Aristarchus of Samos. The Ancient Copernicus. A History of Greek Astronomy to Aristarchus Together with Aristarchus's Treatise on the Sizes and Distances of the Sun and Moon. A New Greek Text with Translation and Notes.* Oxford: Clarendon Press, 1913. .

Heath, Thomas L. *Greek Astronomy.* New York: E.P. Dutton and Company, 1932.

Hippocrates. *The Genuine Works of Hippocrates.* trans. Francis Adams. London: Sydenham Society, 1849.

Hodges, Henry. *Technology in the Ancient World.* New York: Barnes & Noble Books, 1992.

Johnson, Steven. *The Invention of Air.* New York: Riverhead Books, 2008.

Kahn, Charles. *Anaximander and the Origins of Greek Cosmology.* New York: Columbia University Press, 1960.

Krupp, E. C. *Echoes of the Ancient Skies: The Astronomy of Lost Civilizations.* New York: Harper & Row, 1983.

Landels, J. G. *Engineering in the Ancient World.* Berkeley: University of California Press, 1981.

Lloyd, G. E. R. *Aristotle: the Growth and Structure of his Thought.* Cambridge: Cambridge University Press, 1968.

Lloyd, G. E. R. *Early Greek Science: Thales to Aristotle.* New York: W.W. Norton & Company, 1970.

Lloyd, G. E. R. *Greek Science after Aristotle.* New York: W.W. Norton & Company, 1973.

Mason, Stephen F. *A History of the Sciences.* New York: Collier Books, 1971.

Multhauf, Robert P. *The Origins of Chemistry.* New York: Franklin Watts, 1967.

Needham, Joseph. *Science and Civilization in China.* Cambridge: Cambridge University Press, 2000.

Neugebauer, Otto. *The Exact Sciences in Antiquity.* New York: Dover Publications, 1969.

Newman, William R. and Lawrence M. Principe. *Alchemy Tried in the Fire: Starkey, Boyle, and the Fate of Helmontian Chymistry.* Chicago: University of Chicago Press, 2002.

Principe, Lawrence. *The Secrets of Alchemy.* Chicago: University of Chicago Press, 2013.

Ronan, Colin A. *The Shorter Science and Civilization in China.* Cambridge: Cambridge University Press, 1986.

Ross, William D. *Aristotle.* 6th ed. London: Routledge, 1995.

Sambursky, S. *The Physical World of the Greeks.* New York: Collier Books, 1962.

Sarton, George. A History of Science. 2 vols. Cambridge: Harvard University Press, 1952, 1959.

Sarton, George. *Ancient Science and Modern Civilization.* New York: Harper and Brothers, Publishers, 1959.

Santillana, Giorgio de. *The Origins of Scientific Thought.* New York: Mentor Books, 1961.

Singer, Charles. *The Evolution of Anatomy.* London: Kegan, Paul, Trench, and Trübner 1925.

Singer, Charles. *Greek Biology and Greek Medicine.* Oxford: Clarendon Press, 1922.

Sivin, Nathan, ed. *Science and Technology in East Asia.* New York: Science History Publications, 1977.

Solomon, Char. *Tatiana Proskouriakoff: Interpreting the Ancient Maya.* Norman: University of Oklahoma Press, 2002.

Toulmin, Stephen & June Goodfield. *The Architecture of Matter.* Chicago: University of. Chicago Press, 1976.

Toulmin, Stephen & June Goodfield. *The Fabric of the Heavens: The Development of Astronomy and Dynamics.* New York: Harper & Row, Publishers, 1961.

Van Helden, Albert. *Measuring the Universe: Cosmic Dimensions from Aristarchus to Halley.* Chicago: University of Chicago Press, 1985

Van Melsen, Andrew G. *From Atomos to Atom: The History of the Concept of Atom.* New York: Harper and Row, Publishers, 1960.

Van der Waerden, B. L. *Science Awakening*. New York: John Wiley & Sons, 1963.

Whyte, Lancelot, Law. *Essay on Atomism: from Democritus to 1960*. New York: Harper & Row Publishers, 1963.

***Chapter 3: Science and its Application in the Middle Ages 400-1400***

Agricola, Georgius. *De re metallica libri.* trans. Herbert D. and Lou H. Hoover. New York: Dover Publications, 1950.

Al-Khalili, Jim. *The House of Wisdom, How Arabic Science Saved Ancient Knowledge and Gave us the Renaissance.* New York: Penguin Books, 2010.

Bacon, Roger. *Secretum Secretorum, Opera hactenus inedita.* ed. Robert Steele. Oxford: Clarendon Press, 1909.

Bacon, Roger. *The Art and Science of Logic*. (Medieval Sources in Translation, 47). Toronto: Pontifical Institute of Medieval Studies, 2009.

Bacon, Roger. *On Signs* (Medieval Sources in Translation, 54). Toronto: Pontifical Institute of Medieval Studies, 2013.

Bacon, Roger. *The 'Opus Maius.'* ed. John Henry Bridges. 3 vols. Oxford: Clarendon Press, 1897 - 1900.

Crombie, Alistair C. *Medieval and Early Modern Science*. 2 vols. Garden City, NY: Doubleday Anchor Books, 1959.

Cutler, Alan. *The Seashell on the Mountain Top.* New York: Penguin Books, 2004.

Demaitre, Luke E. *Medieval Medicine: The Art of Healing from Head to Toe.* Santa Barbara, CA: Praeger Publishers, 2013.

Easton, Stewart C. *Roger Bacon and His Search for a Universal Science: A Reconstruction of the Life and Work of Roger Bacon in the Light of His Own Stated Purposes.* New York: Russell and Russell, 1952 (reprinted 1971).

Eisenstein, Elizabeth L. *The Printing Revolution in Early Modern Europe.* 2d ed. Cambridge: Cambridge University Press, 2005.

Forbes, R. J. and E. J. Dijksterhuis. *A History of Science and Technology: Ancient Times to the Seventeenth Century.* Vol. 1. New York: Penguin Books, 1963.

Gilson, Etienne. *Reason and Revelation in the Middle Ages.* New York: Charles Scribner's Sons, 1938.

Gimpel, Jean. *The Cathedral Builders*. New York: Harper & Row, Publishers, 1984.

Gimpel, Jean. *The Medieval Machine*. New York: Penguin Books, 1976.

Glacken, Clarence J. *Traces on the Rhodian Shore*. Berkeley: University of California Press, 1976.

Grant, Edward. *Physical Science in the Middle Ages*. New York: Cambridge University Press, 1978.

Grant, Edward. *Science and Religion, 400 BC- AD 1550: From Aristotle to Copernicus*. Baltimore: Johns Hopkins University Press, 2006.

Hackett, Jeremiah, ed. *Roger Bacon and the Sciences: Commemorative Essays*. Leiden: Brill Publishers, 1997.

Hall, A. Rupert and Maris Boas Hall. *A Brief History of Science*. New York: New American Library, 1964.

Haskins, Charles Homer. *The Rise of Universities*. Ithaca, NY: Cornell University Press, 1967.

Hughes, Jonathan. *The Rise of Alchemy in Fourteenth-Century England: Plantagenet Kings and the Search for the Philosopher's Stone*. London: Continuum Publishing, 2012.

King, David, ed. *Islamic Astronomy and Geography*. London: Ashgate Variorum, 2012.

Klemm, Friedrich. *A History of Western Technology*. trans. Dorothea W. Singer. New York: Charles Scribner's Sons, 1959.

Knowles, David. *The Evolution of Medieval Thought*. New York: Random House, 1962.

Lindberg, David C., trans. *Roger Bacon and the Origins of Perspectiva in the Middle Ages*. Oxford: Clarendon Press, 1996.

Lindberg, David C. *The Beginnings of Western Science: 600 BC to AD* 1450. Chicago: University of Chicago Press, 1992.

Lindberg, David C. *Studies in the History of Medieval Optics*. London: Variorum Reprints, 1983.

Lindberg, David C., ed. *Science in the Middle Ages*. Chicago: University of Chicago Press, 1978.

Mason, Stephen. *A History of the Sciences*. New York: Collier Books, 1971.

Parker, R. A. *The Calendars of Ancient Egypt*. Chicago: University of Chicago Press, 1950.

Russell, Jeffrey Burton. *Inventing the Flat Earth*: *Columbus and Modern Historians*. Westport, CT: Praeger Publishers, 1997.

Sobel, Dava. *Longitude*. New York: Penguin Books, 1995

Taylor, F. Sherwood. *A Short History of Science and Scientific Thought*. New York: W.W. Norton & Company, 1963.

Turner, Howard R. *Science in Medieval Islam*. Austin: University of Texas Press, 1995.

***Chapter 4: Science in the Renaissance 1400-1600***


Armitage, Angus. *Copernicus.* New York: A. S. Barnes and Company, 1962.

Boas, Marie. *The Scientific Renaissance, 1450-1630.* New York: Harper & Row, Publishers, 1962.

Buckhardt, Titus. *Alchemy.* Baltimore: Penguin Books , 1971.

Clark, Kenneth. *Leonardo da Vinci.* New York: Penguin Books, 1988.

Connell, Evan S. *The Alchemist's Journal.* San Francisco: North Point Press, 1991.

Debus, Allen G. *Man and Nature in the Renaissance.* Cambridge: Cambridge University Press, 1978.

Debus, Allen G. *The Chemical Dream of the Renaissance.* Cambridge, England: W. Heffer & Sons Ltd., 1968.

Debus, Allen G. English Paracelsians. Chicago: University of Chicago Press, 1968.

Dibner, Bern. *Leonardo da Vinci: Machines and Weaponry.* Norwalk, CT: Burndy Library, 1976.

Drake, Stillman and I. E. Drabkin. *Mechanics in Sixteenth-Century Italy.* Madison: University of Wisconsin Press, 1969.

Dreyer, J. L. E. *Tycho Brahe: A Picture of Scientific Life & Work in the Sixteenth Century.* Gloucester, MA: Peter Smith, 1977 reprint. Original edition published by Adam & Charles Black, Edinburgh, 1890.

Eliade, Mircea. *Myth and Reality.* New York: Harper & Row Publishers, 1968.

Eliade, Mircea. *The Forge and the Crucible.* New York: Harper & Row Publishers, 1971.

Garin, Eugenio. *Science and Civic Life in the Italian Renaissance.* Garden City, NY: Doubleday & Company, 1969.

Gibbs-Smith, Charles. *The Inventions of Leonardo da Vinci.* New York: Charles Scribner's Sons, 1978.

Gingerich, Owen. *The Book Nobody Read.* New York: Penguin Books, 2004.

Gingerich, Owen and James MacLachlan. *Nicolaus Copernicus.* New York: Oxford University Press, 2005.

Goldstein, Thomas. *Dawn of Modern Science.* Boston: Houghton Mifflin Company, 1980.

Grafton, Anthony. *Cardano's Cosmos: The Worlds and Works of a Renaissance Astrologer.* Cambridge: Harvard University Press, 2001.

Hesse, Mary. *Forces and Fields: The Concept of Action at a Distance in the History of Physics.* . New York: Philosophical Library, 2008.

Holmyard, E. J. *Alchemy.* Baltimore: Penguin Books, 1957.

Hooykaas, R. *Religion and the Rise of Modern Science*. Vancouver, BC: Regent College Publishing, 2000.

Koestler, Arthur. *The Sleepwalkers: A History of Man's Changing Vision of the Universe.* New York: Penguin Books, 1989.

Kristeller, Paul Oskar. *Eight Philosophers of the Italian Renaissance.* Stanford: Stanford University Press, 1964.

Kuhn, Thomas. *The Copernican Revolution.* New York: Random House, 1959.

Lind, L. R. *The Epitome of Andreas Vesalius.* Cambridge: MIT Press, 1949.

Lindberg, David C. *Theories of Vision from Al Kindi to Kepler.* Chicago: University of Chicago Press, 1976.

Long, Pamela. *Artisan/Practitioners and the Rise of the New Sciences, 1400-1600.* Corvallis, OR: Oregon State University Press, 2011.

Merton, Robert. *Science, Technology, and Society in Seventeenth-Century England.* New York: Harper & Row, Publishers, 1970.

Murray, Linda. *Michelangelo*. London: Thames and Hudson Ltd., 1985.

Murray, Peter and Linda. *The Art of the Renaissance*. London: Thames and Hudson Ltd., 1963.

Nuland, Sherwin B. *Leonardo da Vinci*. New York: Penguin Books, 2000.

O'Malley, C. D. *Andreas Vesalius of Brussels 1514-1564*. Berkeley: University of California Pr1964.

O'Malley, C. D. and J. B. de C. M. Saunders. *Leonardo da Vinci on the Human Body*. New York: Schuman Publishers, 1952.

Ornstein, Martha. *The Role of Scientific Societies in the Seventeenth Century*. Chicago: University of Chicago Press, 1938.

Pagel, Walter. *Paracelsus: An Introduction to Philosophical Medicine in the Era of the Renaissance.* Basel: Karger Publishers, 1958.

Popper, Nicholas. *Walter Raleigh's History of the World and the Historical Culture of the Late Renaissance.* Chicago: University of Chicago Press, 2012.

Read, John. *Prelude to Chemistry*. Cambridge: MIT Press, 1966.

Rosen, Edward. *Copernicus and the Scientific Revolution.* Malabar, FL: Krieger Publishing, 1984.

Rosen, Edward. *Three Copernican Treatises*. New York: Dover Publications, 1959.

Rossi, Paolo. *Francis Bacon: From Magic to Science*. Chicago: University of Chicago Press, 1968.

Sarton, George. *Six Wings: Men of Science in the Renaissance.* Bloomington, IN: Indiana University Press, 1957.

Singer, Dorothea Waley. *Giordano Bruno: His Life and Thought*. New York: Henry Schuman, 1950.

Singleton, Charles S., ed. *Art, Science, and History in the Renaissance.* Baltimore: Johns Hopkins Press, 1968.

Taylor, Pamela. *The Notebooks of Leonardo da Vinci.* New York: New American Library, 1962.

Vesalius, Andreas. *The Epitome of Andreas Vesalius*. trans. with preface and intro. L.R. Lind. Cambridge: MIT Press, 1969.

Vesalius, Andreas. *The Illustrations from the Works of Andreas Vesalius of Brussels.* trans. and notes J. B. de C.M. Saunders and C. D. O'Malley. New York: Dover Publications, 1973.

Vitruvius. *The Ten Books on Architecture.* New York: Dover Publications, 1960.

Vollmann, William T. *Uncentering the Earth: Copernicus and the Revolutions of the Heavenly Spheres*. New York: W. W. Norton & Company, 2006.

Webster, Charles. *From Paracelsus to Newton.* New York: Cambridge University Press, 1982.

Wiener, Philip P. and Aaron Noland, eds. *Roots of Scientific Thought.* New York: Basic Books, 1957.

Wightman, W. P. D. Science and the Renaissance. New York: Hafner Publishing, 2 vols., 1962.

Yates, Frances A. *The Art of Memory.* Chicago: University of Chicago Press, 1966.

Zammottio, Carlo, Augusto Marinoni, and Anna M. Brizio. *Leonardo The Scientist.* New York: McGraw-Hill Book Company, 1980.

Zophy, Jonathan W. *A Short History of Renaissance and Reformation Europe*. 2d ed. Upper Saddle River, New Jersey: Prentice Hall, 1999.

### *Chapter 5: The Scientific Revolution of the Sixteenth and Seventeenth Centuries*

Andrade, E. N. da C. *Sir Isaac Newton*. Garden City, NY: Doubleday & Company, 1954.

Ashby, Eric. *Technology and the Academics*. London: Macmillan, 1963.

Basalla, George. *The Rise of Modern Science.* Lexington, MA: D. C. Heath and Company, 1968.

Boyle, Robert. *The Sceptical Chymist*. London: J. M. Dent & Sons, 1967.

Brown, Harcourt. *Scientific Organizations in 17th Century France, 1620-1680*. Baltimore: Johns Hopkins Press, 1934.

Brooke, John Hedley. *Science and Religion: Some Historical Perspectives.* New York: Cambridge University Press, 1991.

Bullough, Vern L., ed. *The Scientific Revolution*. New York: Holt, Rinehart and Winston, 1970.

Burns, William E. *The Scientific Revolution in Global Perspective*. New York: Oxford University Press, 2016.

Burtt, Edwin. A. *The Metaphysical Foundations of Modern Physical Science.* London: Kegan Paul, Trench, Trubner, 1923.

Butterfield, Herbert. *The Origins of Modern Science.* New York: Free Press, 1965.

Caspar, Max. *Kepler.* trans. C. Doris Hellman. London: Abelard-Schuman, 1959.

Christianson, Gale E. *In the Presence of the Creator: Isaac Newton and His Times.* New York: Free Press, 1984.

Christianson, Gale E. *Isaac Newton.* New York: Oxford University Press, 1996.

Christianson, Gale E. *The Wild Abyss: the Story of the Men Who Made Modern Astronomy.* New York: Collier Macmillan Publishers, 1978.

Cohen, I. Bernard. *The Birth of a New Physics.* rev. ed. New York: W.W. Norton & Company, 1985.

Dear, Peter. *Revolutionizing the Sciences: European Knowledge and its Ambitions, 1500-1700.* 2nd ed. Princeton: Princeton University Press, 2009.

Descartes, René. *Discourse on Method and Meditations.* Indianapolis, IN: Bobbs-Merrill Company, 1960.

Descartes, René. *Discourse on Method, Optics, Geometry, and Meteorology.* trans. Paul J. Olscamp. Indianapolis, IN: Bobbs-Merrill Company, 1965.

Descartes, René. *Meditations on First Philosophy.* trans. Laurence J. Lafleur. Indianapolis, IN: The Bobs-Merrill Company, 1980.

Descartes, René. *The Philosophical Writings of Descartes.* trans. John Cottingham, Robert Soothoff, Dugald Murdoch. 2 vols. Cambridge: Cambridge University Press, 1985.

Descartes, René. *Le Monde, ou Traite de la lumiere.* trans. and intro. Michael S. Mahoney. New York: Arabis Books, 1979.

Descartes, René. *Principles of Philosophy.* trans. with notes V. R. Miller and R. P. Miller. Dordrecht, Holland & Boston, MA: Reidel Publishing, 1983.

Dijksterhius, E. J. *The Mechanization of the World Picture.* New York; Oxford University Press, 1969.

Dobbs, Betty Jo Teeter. *The Foundations of Newton's Alchemy or 'The Hunting of the Greene Lyon'.* Cambridge: Cambridge University Press, 1976.

Drake, Stillman. *Discoveries and Opinions of Galileo.* Garden City, NY: Doubleday & Company, 1957.

Drake, Stillman. *Galileo at Work: His Scientific Biography.* Chicago: University of Chicago Press, 1978.

Drake, Stillman. *Galileo Studies, Personality, Tradition, and Revolution.* Ann Arbor: University of Michigan Press, 1970.

Durham, Frank and Robert D. Purrington. *Frame of the Universe*. New York: Columbia University Press, 1983.

'Espinasse, Margaret. *Robert Hooke*. Berkeley: University of California Press, 1962

Finocchiaro, Maurice, ed. *The Galileo Affair, A Documentary History.* Berkeley: University of California Press, 1989.

Finocchiaro, Maurice. *Retrying Galileo, 1633-1992.* Berkeley: University of California Press, 2007.

Galilei, Galileo. *Dialogue Concerning the Two Chief World Systems- Ptolemaic & Copernican.* trans. Stillman Drake. Berkeley: University of California Press. 1967. Original Italian edition, Florence, 1632.

Galilei, Galileo. *Dialogue Concerning Two New Sciences.* trans Henry Crew and Alfonso de Salvio. New York: Dover publications, 1954. Original Italian edition, Leiden, 1638.

Galilei, Galileo. *Discoveries & Opinions of Galileo.* trans. Stillman Drake. New York: Doubleday Anchor Books, 1957.

Geymonat, Ludovico. *Galileo Galilei.* New York: McGraw-Hill Book Company, 1965.

Gilbert, William. *De Magnete magneticisque corporibus et de magno magnete Tellure physiologia nova*. Londini: 1600.

Gribbin, John. *The Fellowship.* New York: The Overlook Press, 2008.

Hall, Alfred Rupert. *From Galileo to Newton: 1630-172*0. New York: Harper & Row, 1963.

Hall, Alfred Rupert. *Philosophers at War: The Quarrel Between Newton and Leibniz.* Cambridge: Cambridge University Press, 2002.

Hall, Alfred Rupert. *The Revolution in Science, 1500-1750.* 3d ed. New York: Longman Group Ltd., 1983.

Hall, Marie Boas., ed. *Nature and Nature's Laws*. New York: Harper& Row, Publishers, 1970.

Harvey, William. *De Motu Cordis: Anatomical Studies on the Motion of the Heart and Blood.* 4th ed. trans. Chauncey D. Leake. Springfield, IL: Charles C. Thomas, 1958.

Heilbron, John L. *Electricity in the Seventeenth and Eighteenth Centuries: A Study of Early Modern Physics.* Berkeley: University of California Press, 1979.

Heilbron, John. *The Sun in the Church.* Cambridge: Harvard University Press, 2001.

Henry, John. *The Scientific Revolution and the Origins of Modern Science*. 3rd ed. New York: Palgrave Macmillan, 2008.

Hofstadter, Dan. *The Earth Moves: Galileo and the Roman Inquisition.* New York: W.W. Norton & Company, 2009.

Hooykaas, R. *Religion and the Rise of Modern Science.* Edinburgh: Scottish Academic Press, 1972.

Jacob, Margaret C. *The Cultural Meaning of the Scientific Revolution.* New York: Alfred A. Knopf, 1988.

Jacob, Margaret C. *The Scientific Revolution: A Brief History with Documents*. Boston: Bedford/St. Martin's, 2010.

Jardine, Lisa. *The Curious Life of Robert Hooke: The Man Who Measured London*. New York: Harper Collins Publishers, 2004.

Jones, Richard Foster. *Ancients and Moderns*. Berkeley: University of California Press, 1965.

Kaplan, Morton F., ed. *Homage to Galileo*. Cambridge: MIT Press, 1965.

Kargon, Robert H. *Atomism in England from Hariot to Newton.* Oxford: Oxford University Press, 1966.

Kearney, Hugh. *Science and Change 1500-1700.* New York: McGraw-Hill Book Company, 1971.

King, Henry C. *The History of the Telescope.* London: Charles Griffin, 1955.

Koestler, Arthur. *The Sleepwalkers: A History of Man's Changing Vision of the Universe*. New York: Grosset & Dunlap, 1963.

Koestler, Arthur. *The Watershed.* New York: Anchor, 1960.

Koyré, Alexandre. *From the Closed World to the Infinite Universe.* Baltimore: Johns Hopkins University Press, 1968.

Koyré, Alexandre. *Newtonian Studies*. Chicago: University of Chicago Press, 1968.

Kuhn, Thomas S. *The Copernican Revolution.* Cambridge: Harvard University Press, 1957.

Kuhn, Thomas. *The Structure of Scientific Revolutions*. 2d ed. Chicago: University of Chicago Press, 1970.

Langford, J. J. *Galileo, Science, and the Church.* Ann Arbor: University of Michigan Press, 1966.

Lodge, Oliver. *Pioneers of Science and the Development of their Scientific Theories*. New York: Dover Publications, 1960.

MacLachlan, James. *Galileo Galilei*. New York: Oxford University Press, 1997.

Manuel, Frank E. *A Portrait of Isaac Newton.* Cambridge: Harvard University Press, 1968.

Manuel, Frank E. *Isaac Newton Historian*. Cambridge: Harvard University Press, 1963.

Manuel, Frank E. *The Religion of Isaac Newton*: The Fremantle Lectures 1973. Oxford: Clarendon Press, 1974.

McClellan, James E., III. *Science Reorganized: Scientific Societies in the 18th Century*. New York: Columbia University Press, 1985.

Merton, Robert. *Science, Technology, and Society in Seventeenth-Century England.* New York: Harper & Row, Publishers, 1970.

Middleton, W. E. Knowles. *The Scientific Revolution.* Cambridge, MA: Schenkman Publishing Company, 1963.

Newman, William R. and Lawrence M. Principe. *Alchemy Tried in the Fire: Starkey, Boyle, and the Fate of Helmontian Chymistry.* Chicago: University of Chicago Press, 2002.

Newton, Isaac. *Opticks*. New York: Dover Publications, 1952. Based on the Fourth Edition London, 1730.

Newton, Isaac. *The Mathematical Principles of Natural Philosophy*. New York: Philosophical Library, 1964.

Ornstein, Martha. *The Role of Scientific Societies in the Seventeenth Century.* New York: Arno Press, 1975.

Panek, Richard. *Seeing and Believing, How the Telescope Opened Our Eyes and Minds to the Heavens.* New York: Penguin Books, 1998.

Purnell, Jr., Frederick, Michael S. Pettersen, and Mark C. Carnes. *The Trial of Galileo: Aristotelianism, the "New Cosmology," and the Catholic Church*, 1616-33. New York: Pearson Education, 2008.

Purver, Margery. *The Royal Society: Concept and Creation.* Cambridge: MIT Press, 1967.

Redondi, Pietro. *Galileo: Heretic.* trans. Raymond Rosenthal. Princeton: Princeton University Press, reprint ed. 1989.

Sabra, A. I. *Theories of Light from Descartes to Newton.* London: Oldbourne, 1967.

Santillana, Giorgio de. *The Crime of Galileo.* Chicago: University of Chicago Press, 1955.

Shackelford, Jole. *William Harvey*. New York: Oxford University Press, 2003.

Shea, William R. *Galileo's Intellectual Revolution: Middle Period, 1610-1632.* New York: Neale Watson Academic Publications, 1977.

Smith, Alan G. R. *Science and Society*. New York: Harcourt Brace Jovanovich, 1972.

Sobel, Dava. *Galileo's Daughter*. New York: Penguin Books, 2000.

Sobel. Dava, trans. *Letters to Father: Suor Maria Celeste to Galileo, 1623-1633*. New York: Penguin Books, 2001.

Sullivan, J. W. N. *Isaac Newton.* New York: MacMillan, 1938.

Thayer, H. S., ed. *Newton's Philosophy of Nature.* New York: Hafner Publishing Company, 1953.

Toulmin, Stephen and June Goodfield. *The Fabric of the Heavens*. New York: Harper & Row, Publishers, 1961.

Voelkel, James R. *Johannes Kepler*. New York: Oxford University Press, 1999.

Westfall, Richard S. *The Construction of Modern Science: Mechanisms and Mechanics.* New York: John Wiley & Sons, 1971.

Westfall, Richard S. *Never at Rest: A Biography of Isaac Newton*. Cambridge: Cambridge University. Press, 1980.

Westfall, Richard. *Science and Religion in Seventeenth-Century England.* New Haven, CT: Yale University Press, 1958.

Willey, Basil. *The Seventeenth Century Background.* Garden City, NY: Doubleday & Company, 1953.

Wolf, Abraham. *A History of Science, Technology, and Philosophy in the 16th and 17th Centuries.* London: G. Allen & Unwin, 1935.

Yates, Frances. *Giordano Bruno & the Hermetic Tradition.* New York: Random House, 1969.

Zinsser, Judith P. *La Dame d'Esprit: A Biography of the Marquise Du Châtelet*. New York: Penguin Books, 2006.

***Chapter 6: The Physical and Life Sciences in the Eighteenth Century (the Enlightenment)***

Bedini, Silvio A. *Jefferson and Science*. (Monticello Monograph Series, Charlottesville, Virginia: Thomas Jefferson Foundation, Inc.; distributed by Chapel Hill: University of North Carolina Press, 2002.

Bell, Madison Smartt. *Lavoisier in the Year One.* New York: W. W. Norton & Company, 2005.

Berry, A. J. *Henry Cavendish: His Life and Scientific Work.* London: Hutchinson, 1960.

Brown, Sanborn. *Count Rumford.* Garden City, NY: Doubleday & Company, 1962.

Burns, William E. *Science and Technology in Colonial America.* Westport, CT: Greenwood, 2005.

Bury, J. B. *The Idea of Progress.* New York: Dover Publications, 1955.

Chaplin, Joyce E. *The First Scientific American: Benjamin Franklin and the Pursuit of Genius*. New York: Basic, 2006.

Cohen, Bernard I. *Benjamin Franklin.* New York: Charles Scribner's Sons, 1975.

Cohen, Bernard I. *Franklin and Newton. An inquiry into speculative Newtonian experimental science and Franklin's work in electricity as an example thereof.* Philadelphia, PA: University of Pennsylvania Press, 1976.

Cohen, Bernard I. *Science and the Founding Fathers.* New York: W.W. Norton & Company, 1997.

Conant, James B. *On Understanding Science: An Historical Approach.* New Haven: Yale University Press, 1947.

Crackel, Theodore J. *West Point: A Bicentennial History.* Lawrence: University Press of Kansas, 2002.

Crosby, Jr., Alfred W. *The Columbian Exchange*. Westport, CT: Greenwood Press, 1972.

Delbourgo, James. *A Most Amazing Scene of Wonders: Electricity and Enlightenment in Early America.* Cambridge: Harvard University Press, 2006.

Donovan, Arthur. *Antoine Lavoisier: Science, Administration and Revolution.* New York: Cambridge University Press, 1996.

Dray, Philip. *Stealing God's Thunder: Benjamin Franklin's Lightning Rod and the Invention of America*. New York: Random House, 2005.

Evans, Howard Ensign. *Pioneer Naturalists*. New York: Henry Holt & Company, 1993.

Fara, Patricia. *An Entertainment for Angels: Electricity in the Enlightenment*. New York: Columbia University Press, 2002.

Fox, Robert. *The Caloric Theory of Gases: From Lavoisier to Regnault*. New York: Oxford University Press, 1971.

Franklin, Benjamin. *Experiments and Observations on Electricity*. Commentary by I. Bernard Cohen. (Octavo Digital Editions). Oakland, CA: Octavo, 2003.

Gaustad, Edwin S. *Benjamin Franklin.* New York: Oxford University Press, 2006.

Gay, Peter. *The Enlightenment: an Interpretation*. New York: Random House, 1966.

Green, John C. *American Science in the Age of Jefferson.* Ames: Iowa State University Press, 1984.

Gronim, Sara S. *Everyday Nature: Knowledge of the Natural World in Colonial New York*. New Brunswick, NJ: Rutgers University Press, 2007.

Guerlac, Henry. *Antoine-Laurent Lavoisier, Chemist and Revolutionary*. New York: Charles Scribner's Sons, 1975.

Guerlac, Henry. *The Crucial Year*. Ithaca, NY: Cornell University Press, 1961.

Hankins, Thomas L. *Science and the Enlightenment*. Cambridge: Cambridge University Press, 1985.

Heilbron, John L. *Electricity in the 17th & 18th Centuries: A Study of Early Modern Physics.* Berkeley: University of California Press, 1979.

Hiebert, Erwin N., Aaron J. Ihde, and Robert Schofield. *Joseph Priestley: Scientist, Theologian, and Metaphysician.* Cranbury, NJ: Bucknell University Press, 1980.

Hindle, Brook. *The Pursuit of Science in Revolutionary America.* Chapel Hill: Published for the Institute of Early American History and Culture, Williamsburg, VA: University of North Carolina Press, 1956.

Holmes, Frederick L. *Claude Bernard & Animal Chemistry: The Emergence of a Scientist*. Cambridge: Harvard University Press, 1974.

Holmes, Frederick L. *Lavoisier and the Chemistry of Life: An Exploration of Scientific Creativity*. Madison, WI: University of Wisconsin Press, 1985.

Hoermann, Alfred R. *Cadwallader Colden: A Figure of the American Enlightenment.* Westport, CT: Greenwood Press, 2002.

Hoffmann, Nancy El. And John C. Van Horne, eds. *America's Curious Botanist: A Tercentennial Reappraisal of John Bartram, 1699-1777.* (Memoirs of the American Philosophical Society, 249). Philadelphia: American Philosophical Society, 2004.

Isaacson, Walter. *Benjamin Franklin: An American Life.* New York: Simon & Schuster, 2004.

Jackson, Joe. *A World on Fire: A Heretic, and Aristocrat, and the Race to Discover Oxygen.* New York: Penguin Books, 2005.

Jefferson, Thomas. *Notes on the State of Virginia.* New York: Penguin Books, 1999.

Johnson, Steven. *The Invention of Air*. New York: Riverhead Books, 2008.

Kaplan, Lawrence S. *Thomas Jefferson: Westward the Course of Empire.* Wilmington, DE: Scholarly Resources Inc., 1999.

Lavoisier, Antoine. *Elements of Chemistry*. New York: Dover Publications, 1965. trans. from original French edition 1789 by Robert Kerr 1790.

Manuel, Frank E., ed. *The Enlightenment*. Englewood Cliffs, NJ: Prentiss-Hall, 1965.

McKie, Douglas. *Antoine Lavoisier, Scientist • Economist • Social Reformer.* New York: Collier Books, 1962.

Mendelsohn, Everett. *Heat and Life: the Development of the Theory of Animal Heat.* Cambridge: Harvard University Press, 1964.

Mitchill, Samuel L. *A Discourse on the Character and Services of Thomas Jefferson, More Especially as a Promoter of Natural and Physical Science*. (11 October, 1826). Charlottesville: University of Virginia, 1982.

Morgan, Edmund S. *Benjamin Franklin.* New Haven: Yale University Press, 2002.

Parrish, Susan Scott. *American Curiosity: Cultures of Natural History in the Colonial British Atlantic World.* Chapel Hill: University of North Carolina Press, 2006.

Priestley, Joseph. *Memoirs and Correspondence of Joseph Priestley.* (Rare biographical sources). 2 vols. Bristol: Thoemmes, 2002. (Reprint of 1817-1822 edition)

Schiffer, Michael Brian. *Draw the Lightning Down: Benjamin Franklin and Electrical technology in the Age of Enlightenment.* Berkeley: University of California Press, 2003.

Schofield, Robert S. *The Lunar Society of Birmingham: A Social History of Provincial Society and Industry in 18th Century England.* Oxford: Clarendon Press, 1963.

Schofield, Robert S. *Mechanism & Materialism: British Natural Philosophy in the Age of Reason.* Princeton: Princeton University Press, 1969.

Schofield, Robert E., ed. *A Scientific Autobiography of Joseph Priestley: Selected Scientific Correspondence.* Cambridge: MIT Press, 1966.

Siegfried, Robert. *From Elements to Atoms: A History of Chemical Composition.* Philadelphia: American Philosophical Society, 2002.

Tucker, Tom. *Bolt of Fate: Benjamin Franklin and His Electric Kite Hoax.* New York: Public Affairs Books, 2003.

Van Doren, Carl. *Benjamin Franklin.* New York: Viking Press, 1938.

Whyte, Lancelot Law. *Roger Joseph Boscovich.* New York: Fordham University Press, 1961.

Wiley, Basil. *The Eighteenth Century Background: Studies on the Idea of Nature in the Thought of the Period.* Boston: Beacon Press, 1961.

Zall, Paul M., ed. *Franklin on Franklin.* Lexington: University Press of Kentucky, 2000.

***Chapter 7: Darwin's Century***

Agassiz, Elizabeth Cary, ed. *Louis Agassiz: His life and Correspondence.* 2 vols. Bristol,UK: Thoemmes Press, 2003. Originally published 1885.

Agassiz, Louis. *Agassiz on Evolution.* 2 vols. Bristol, UK: Thoemmes Press, 2003. Reprint of works by Agassiz.

Agassiz, Louis. *Essay on Classification.* intro. Edward Lurie. Mineola, NY: Dover, 2004. (originally published 1857)

Alters, Brian J. and Sandra M. Alters. *Defending Evolution in the Classroom: A Guide to the Creation/Evolution Controversy.* Boston: Jones and Bartlett Publishers, 2001.

Appleman, Philip, ed. *Darwin.* New York: W.W. Norton & Company, 1970.

Asimov, Isaac. *The Human Body: Its Structure and Operation.* New York: Signet, 1963.

Asimov, Isaac. *The Wellsprings of Life.* New York: Signet, 1960.

de Beer, Gavin, Sir. *Charles Darwin: Evolution by Natural Selection.* Garden City, NY: Doubleday, 1964.

Bernasconi, Robert, ed. *American Theories of Polygenesis.* 7 vols. Bristol, UK: Thoemmes Press, 2002.

Bowler, Peter J. *Evolution: the History of an Idea.* Berkeley: University of California Press, 1984.

Bowler, Peter. *The Earth Encompassed.* New York: W.W. Norton & Company, 1993.

Brace, C. Loring. *"Race" Is a Four Letter Word: The Genesis of the Concept.* New York: Oxford University Press, 2005.

Brent, Peter. *Charles Darwin: A Man of Enlarged Curiosity.* New York: W.W. Norton & Company, 1981.

Brooks, John L. *Just Before 'The Origin': Alfred Russel Wallace's Theory of Evolution.* New York: Columbia University Press, 1984.

Browne, Janet. *Darwin's Origin of Species: A Biography.* New York: Atlantic Monthly Press, 2007.

Burchfield, Joe D. *Lord Kelvin and the Age of the Earth.* New York: Science History Publications, 1975.

Burrow, J. W. *Evolution and Society: A Study in Victorian Social Theory.* London: Cambridge University Press, 1966.

Clark, Ronald W. *The Survival of Charles Darwin: A Biography of a Man and an Idea.* New York: Random House 1984.

Coleman, Simon and Leslie Carlin, eds. *The Cultures of Creationism: Anti-Evolutionism in English-Speaking Countries.* Farnham Surrey: Ashgate Publishing, 2004.

Coleman, William. *Biology in the Nineteenth Century: Problems of Form, Function, and Transformation.* Cambridge: Cambridge University Press, 1977.

Coyne, Jerry A. *Why Evolution is True.* New York: Penguin Books, 2009.

Cuddy, Lois A. and Claire M. Roche, eds. *Evolution and Eugenics in American Literature and Culture, 1880-1940: Essays on Ideological Conflict and Complicity.* Lewisburg: Bucknell University Press, 2003.

Darwin, Charles. *The Autobiography of Charles Darwin: 1809-1882.* New York: W.W. Norton & Company, 1958.

Darwin, Charles. *The Descent of Man.* New York: Penguin Classics, 2004.

Darwin, Charles. *The Expression of Emotions in Man and Animals.* New York: Penguin Classics, 2009.

Darwin, Charles. *The Origin of Species.* New York: Mentor, 1958.

Darwin, Charles. *The Voyage of the Beagle.* Garden City, NY: Doubleday & Company, 1962.

Darwin, Charles and T. H. Huxley. *Autobiographies.* Oxford: Oxford University Press, 1974.

Depew, David J. and Bruce H. Weber. *Darwinism Evolving: Systems Dynamics and the Genealogy of Natural Selection.* Cambridge: MIT Press, 1995.

Dowd, Michael. *Thank God for Evolution.* New York: Penguin Books, 2009.

Driscoll, Marsha, Elizabeth E. Dunn, Dann Siems, and B. Kamran Swanson. *Charles Darwin, The Copley Medal, and the Rise of Naturalism: 1862-1864.* Saddle River, NJ: Longman, 2010.

Dunbar, Carl O. *Historical Geology.* New York: John Wiley & Sons, 1949.

Dupree, A. Hunter. *Asa Gray, 1810-1888.* New York: Atheneum, 1968.

Dyson, George B. *Darwin Among the Machines: The Evolution of Global Intelligence.* Reading, MA: Perseus Books, 1997.

Eiseley, Loren. *Darwin's Century: Evolution and the Men Who Discovered It.* Garden City, NY: Doubleday & Company, 1961.

Edelson, Edward. *Gregor Mendel and Roots of Genetics*. New York: Oxford University Press, 1999.

Eldredge, Niles and Ian Tattersall. *The Myths of Human Evolution.* New York: Columbia University Press, 1982.

Farber, Paul Lawrence. *Mixing Races: From Scientific Racism to Modern Evolutionary Ideas.* Baltimore: Johns Hopkins University Press, 2011.

Futuyma, Douglas J. *Science on Trial: The Case for Evolution.* New York: Pantheon Books, 1982.

Gates, Alexander E. *A to Z of Earth Scientists: Notable Scientists.* New York: Facts on File, 2003.

Gillispie, Charles Coulston. *Genesis and Geology.* New York: Harper & Row Publishers, 1959.

Ginger, Ray. *Six Days or Forever? Tennessee v. John Thomas Scopes.* Oxford: Oxford University Press, 1958.

Gladfelter, Elizabeth Higgins. *Agassiz's Legacy: Scientist's Reflections on the Value of Field Experience.* Oxford: Oxford University Press, 2002.

Glass, Bentley, Owsei Temkin and William L. Straus, Jr., eds. *Forerunners of Darwin: 1745-1859.* Baltimore: Johns Hopkins Press, 1959.

Godfrey, Laurie R, ed. *Scientists Confront Creationism.* New York: W.W. Norton & Company, 1984.

Gould, Stephen Jay. *Ever Since Darwin.* New York/ London: W.W. Norton & Company, 1977.

Gould, Stephen Jay. *The Mismeasure of Man.* New York: W.W. Norton & Company, 1891.

Haeckel, Ernst. *The Evolution of Man.* 2 vols. New York: D. Appleton and Company, 1897.

Hazen, Robert M. *The Story of Earth: The First 4.5 Billion Years, From Stardust to Living Planet.* New York: Penguin Books, 2012.

Helferich, Gerard. *Humboldt's Cosmos.* New York: Penguin Books, 2004.

Herbert, Sandra. *Charles Darwin and the Question of Evolution.* New York: Bedford/St. Martin's, 2011.

Himmelfarb, Gertrude. *Darwin and the Darwinian Revolution.* New York: W.W. Norton & Company, 1962.

Hofstadter, Richard. *Social Darwinism in American Thought.* Boston: Beacon Press, 1955.

Horner, Jack and James Gorman. *How to Build a Dinosaur: The New Science of Reverse Evolution.* New York: Penguin Books, 2010.

Huxley, Thomas H. *Man's Place in Nature.* Ann Arbor, MI: University of Michigan Press, 1959.

Irvine, William. *Apes, Angels, and Victorians.* New York: McGraw-Hill Book Company, 1955.

Israel, Charles. *Before Scopes: Evangelicalism, Education, and Evolution in Tennessee, 1870-1925.* Athens: University of Georgia Press, 2004.

Jackson, John P., Jr. and Nadine M. Weidman. *Race, Racism, and Science: Social Impact and Interaction.* New Brunswick: Rutgers University Press. 2006.

Johnson, Phillip E. *Darwin on Trial.* Downers Grove, IL: InterVarsity Press, 1993.

Jones, Steve. *Darwin's Ghost: The Origin of Species Updated.* New York: Random House, 1999.

Keynes, Randal. *Darwin, His Daughter, and Human Evolution.* New York: Riverhead Books, 2001.

Koestler, Arthur. *The Case of the Midwife Toad.* New York: Vintage Books, 1971.

Larson, Edward J. *Summer for the Gods: The Scopes Trial and America's Continuing Debate over Science and Religion.* Cambridge: Harvard University Press, 1997.

Larson, Edward J. *Trial and Error: The American Controversy over Creation and Evolution.* New York: Oxford University Press, 1985.

Lurie, Edward. *Louis Agassiz: A Life in Science.* Chicago: University of Chicago Press, 1960.

Lyell, Charles. *Principles of Geology.* New York: Penguin Books, 1997.

Mayr, Ernst. *This Is Biology: The Science of the Living World.* Cambridge: Harvard University Press, 1997.

McIver, Tom. *Anti-Evolution, A Reader's Guide to Writings before and after Darwin.* Baltimore: Johns Hopkins University Press, 1992.

Magner, Lois. *A History of the Life Sciences.* New York: M. Dekker, 1994.

Malthus, Thomas Robert. *On Population.* New York: Random House, 1960.

McDonald, Roger. *Mr. Darwin's Shooter: A Novel.* New York: Atlantic Monthly Press, 1998.

Midgley, Mary. *Evolution as a Religion.* London/ New York: Methuen, 1985.

Miller, Kenneth R. *Only a Theory.* New York: Penguin Books, 2008.

Moran, Jeffrey P. *The Scopes Trial: A Brief History with Documents.* Boston, New York: Bedford/St. Martin's, 2002.

Numbers, Ronald L. *Darwinism Comes to America.* Cambridge: Harvard University Press, 1998.

Numbers, Ronald L. *The Creationists: From Scientific Creationism to Intelligent Design.* Expanded edition. Cambridge: Harvard University Press, 2006.

Olby, Robert C. *Origins of Mendelism.* New York: Schocken Books, 1966.

Palumbi, Stephen R. *The Evolution Explosion: How Humans Caused Rapid Evolutionary Change.* New York: W.W. Norton & Company, 2001.

Pigliucci, Massimo. *Denying Evolution: Creationism, Scientism, and the Nature of Science.* Sunderland, MA: Sinauer Associates, Publishers, 2002.

Pyne, Lydia V. and Stephen J. Pyne. *The Last World: Ice Ages, Human Origins, and the Invention of the Pleistocene.* New York: Penguin Books, 2013.

Quammen, David. *The Reluctant Mr. Darwin.* New York: W. W. Norton & Company, 2007.

Rea, Tom. *Bone Wars: The Excavation and Celebrity of Andrew Carnegie's Dinosaur.* Pittsburgh: University of Pittsburgh Press, 2001.

Regal, Brian. *Henry Fairfield Osborn: Race and the Search for the Origins of Man.* Burlington, VT: Ashgate Publishing, 2002.

Rudolph, John L. *Scientists in the Classroom: The Cold War Reconstruction of American Scientific Education.* New York: Palgrave, 2002.

Russell, Robert John, ed. *Fifty Years in Science and Religion: Ian G. Barbour and His Legacy.* Farnham Surrey: Ashgate Publishing, Ashgate Science and Religion Series, 2004.

Ryan, Frank X., ed. *Darwinism and Theology in America, 1850-1930.* 4 vols. Bristol, UK: Thoemmes Press, 2002.

Serafini, Anthony. *The Epic History of Biology.* Cambridge, MA: Perseus Publishing, 1993.

Simpson, George Gaylord. *This View of Life: The World of and Evolutionist.* New York: Harcourt, Brace & World, 1964.

Stanton, William. *The Leopard's Spots: Scientific Attitudes Toward Race in America, 1815-1859.* Chicago: University of Chicago Press, 1960.

Stauffer, Robert C., ed. *Charles Darwin's Natural Selection, Being the Second part of his Big Species Book Written from 1856 to 1858.* New York: Cambridge University Press, 1975.

Stefoff, Rebecca. *Charles Darwin and the Evolution Revolution.* New York: Oxford University Press, 1996.

Swisher, Carl Celso, III, Garniss H. Curtis, and Roger Lewin. *Java Man: How Two Geologists Changed Our Understanding of Human Evolution.* Chicago: University of Chicago Press, 2001.

Toulmin, Stephen and June Goodfield. *The Discovery of Time.* New York: Harper & Row Publishers, 1962.

Wade, Nicholas. *Before the Dawn.* New York: Penguin Books, 2006.

White, Michael, and John Gribbin. *Darwin: A Life in Science.* New York: Dutton, 1995.

Wilson, R. J., ed. *Darwinism and the American Intellectual.* Homewood, IL: Dorsey Press, 1967.

Witham, Larry A. *Where Darwin Meets the Bible: Creationists and Evolutionists in America.* Oxford : Oxford University Press, 2002.

Woodward, Thomas. *Doubts about Darwin: A History of Intelligent Design*. Grand Rapids, MI: Baker Books, 2003.

Young, David. *The Discovery of Evolution.* 2d ed. New York: Cambridge University Press, 2007.

***Chapter 8: Molecular Biology and the Double Helix***

Allen, Garland E. *Life Science in the Twentieth Century*. Cambridge: Cambridge University Press, 1975.

Angier, Natalie. *The Canon: A Whirligig Tour of the Beautiful Basics of Science.* Boston: Houghton Mifflin Company, 2007.

Asimov, Isaac. *The Genetic Code.* New York: Orion Press, 1962.

Creager, Angela N. H. *The Life of a Virus: Tobacco Mosaic Virus as an Experimental Model, 1930-1965.* Chicago: University of Chicago Press, 2002.

Crick, Francis. *Of Molecules and Men.* Seattle: University of Washington Press, 1966.

Crick, Francis. *What Mad Pursuit: A Personal View of Scientific Discovery.* New York: Basic Books, 1988.

Davies, Kevin. *Cracking the Genome: Inside the Race to Unlock Human DNA.* Baltimore: Johns Hopkins University Press, 2002.

Edelson, Edward. *Francis Crick and James Watson and the Building Blocks of Life*. New York: Oxford University Press, 1998.

Fredrickson, Donald S. *The Recombinant DNA Controversy: A Memoir: Science, Politics, and the Public Interest, 1974-1981.* Washington: ASM Press, 2001.

Friedberg, Errol C. *The Writing Life of James D. Watson*. Cold Spring Harbor, NY: Cold Spring Harbor Laboratory Press, 2005.

Holmes, Frederic Lawrence. *Meselson, Stahl, and the Replication of DNA: A History of "The Most Beautiful Experiment in Biology."* New Haven: Yale University Press, 2001.

Inglis, John R., Joseph Sambrook, and Jan. A. Witkowski, eds. *Inspiring Science: Jim Watson and the Age of DNA.* Cold Spring Harbor, NY: Cold Spring Harbor Laboratory Press, 2003.

Jacob, Francois. *The Logic of Life: A History of Heredity.* New York: Vintage Books, 1976.

Judson, Horace. *The Eighth Day of Creation: Makers of the Revolution in Biology.* New York: Simon and Schuster, 1979.

Kevles, Daniel J. *The Baltimore Case*. New York: W.W. Norton & Company, 1998.

Lewontin, Richard. *It Ain't Necessarily So: The Dream of the Human Genome and Other Illusions.* New York: New York Review of Books, 2000.

Luria, S. E. *A Slot Machine, a Broken Test Tube: An Autobiography.* New York: Harper & Row, 1984.

Maddox, Brenda. *Rosalind Franklin: The Dark Lady of DNA.* New York: Harper Collins, 2003.

McElheny, Victor K. *Watson and DNA: Making a Scientific Revolution.* Cambridge, MA: Perseus, 2003.

Mead, Clifford and Thomas Hager, eds. *Linus Pauling: Scientist and Peacemaker.* Corvallis: Oregon State University Press, 2001.

Olby, Robert. *The Path to the Double Helix.* Seattle: University of Washington Press, 1974.

Ridley, Matt. *Francis Crick: Discoverer of the Genetic Code.* New York: Harper Perennial, 2006.

Sayre, Anne. *Rosalind Franklin and DNA.* New York: W.W. Norton & Company, 1975.

Schrödinger, Erwin. *What is Life? & Mind and Matter.* Cambridge: Cambridge University Press, 1967.

Shine, Ian and Sylvia Wrobel. *Thomas Hunt Morgan: Pioneer of Genetics.* Lexington: University Press of Kentucky, 1976.

Smith, Jane S. *The Garden of Invention: Luther Burbank and the Business of Reading Plants.* New York: Penguin Books, 2010.

Stahl, Frank W., ed. *We Can Sleep Later: Alfred D. Hershey and the Origins of Molecular Biology.* Cold Spring Harbor, NY: Cold Spring Harbor Laboratory Press, 2000.

Stent, Günther S. *The Coming of the Golden Age: A View of the End of Progress.* Garden City, NY: Published for the American Museum of Natural History by the Natural History Press, 1969.

Stern, Alexandra Minna. *Eugenic Nation: Faults and Frontiers of Better Breeding in Modern America.* Berkeley: University of California Press, 2005.

Stern, Alexandra Minna. *Telling Genes: The Story of Genetic Counseling in America.* Baltimore: Johns Hopkins University Press, 2012.

Thackray, Arnold, ed. *Private Science: Biotechnology and the Rise of the Molecular Sciences.* (The Chemical Sciences in Society Series). Philadelphia: University of Pennsylvania Press, 1998.

Tone, Andrea, ed. *Controlling Reproduction: An American History.* Wilmington, DE: Scholarly Resources, 1997.

Venter, J. Craig. *A Life at the Speed of Light: From the Double Helix to the Dawn of Digital Life.* New York: Penguin Books, 2014.

Venter, J. Craig. *A Life Decoded: My Genome: My Life.* New York: Penguin Books, 2007.

Watson, James. *The Double Helix.* New York: Mentor, 1968.

Watson, James D. *Genes, Girls, and Gamow: After the Double Helix.* First American edition. New York: Alfred A. Knopf, 2002. Previously Published as *A Passion for DNA: Genes, Genomes, and Society.* Cold Spring Harbor, NY: Cold Spring Harbor Laboratory Press, 2000.

Wilkins, Maurice. *The Third Man of the Double Helix: The Autobiography of Maurice Wilkins.* New York: Oxford University Press, 2003.

Wohlsen, Marcus. *Biopunk.* New York: Current, 2011.

***Chapter 9: The Physical Sciences in the Nineteenth Century***

Basalla, George, William Coleman, and Robert H. Kargon, eds. *Victorian Science*. Garden City, NY: Doubleday & Company, 1970.

Bednarek, Janet R. Daly and Roger D. Launius, eds. *Reconsidering a Century of Flight.* Chapel Hill: University of North Carolina Press, 2003.

Bonner, Thomas Neville. *Iconoclast: Abraham Flexner and a Life in Learning*. Baltimore: Johns Hopkins University Press, 2002.

Brady, Tim, ed. *American Aviation History.* Carbondale: Southern Illinois University Press, 2000.

Bradley, Duane. *Count Rumford.* Princeton, NJ: D. Van Nostrand Company, 1967.

Brown, Sanborn C. *Count Rumford: Physicist Extraordinary.* Garden City, NY: Doubleday & Company, 1962.

Browne, Charles, *A Half-Century of Chemistry in America, 1876-1926.* Philadelphia: American Chemical Society, 1926.

Bruce, Robert. *The Launching of Modern American Science 1846-1876.* New York: Knopf, 1987.

Buttmann, Gunther. *The Shadow of the Telescope: A Biography of John Herschel.* New York: Charles Scribner's Sons, 1970.

Carnot, Sadi. *Reflections on the Motive Power of Fire*. New York: Dover Publications, 1960.

Carter, Bill and Merri Sue Carter. *Simon Newcomb: America's Unofficial Astronomer Royal*. St. Augustine, FL: Mantanzas Publishing, 2006.

Cheney, Margaret. *Tesla: Man Out of Time.* New York: Laurel, 1981.

Christensen, Dan Charly. *Hans Christian Oersted: Reading Nature's Mind.* Oxford: Oxford University Press, 2013.

Comstock, J. L. *Elements of Chemistry*, 13th ed. New York: Pratt, Woodford & Co., 1847.

Corn, Joseph J. *The Winged Gospel: America's Romance with Aviation.* Baltimore: Johns Hopkins University Press, 2002.

Crouch, Tom D. *Wright Brothers and the Invention of the Aerial Age.* Washington: National Geographic Society, 2003.

Crouch, Tom D. *A Dream of Wings: Americans and the Airplane, 1875-1905.* London: W.W. Norton & Company, 2002.

Crowther, J. G. *British Scientists of the Nineteenth Century*. London: Routledge & Kegan Paul Ltd., 1962.

Daniels, George H., ed. *Nineteenth-Century American Science*. Evanston, IL: Northwestern University Press, 1972.

Dibner, Bern. *Oersted and the Discovery of Electromagnetism*. New York: Random House, 1962.

Dubos, René J. *Louis Pasteur: Free Lance of Science*. Boston: Little, Brown and Company, 1950.

Dubos, René. *Pasteur and Modern Science*. Garden City, NY: Doubleday & Company, 1960.

Eckert, Michael. *Heinrich Hertz.* Hamburg: Ellert & Richter, 2010.

Everitt, C. W. F. *James Clerk Maxwell: Physicist and Natural Philosopher*. New York: Charles Scribner's Sons, 1975.

Faraday, Michael. *The Chemical History of a Candle*. New York: The Viking Press, 1960.

Forbes, Nancy and Basil Mahon. *Faraday, Maxwell, and the Electromagnetic Field.* Amherst, NY: Prometheus Books, 2014.

Freund, Ida. *The Study of Chemical Composition: an Account of its Method and Historical Development*. New York: Dover Publications, 1968.

Furter, William F. *A Century of Chemical Engineering.* New York: Plenum Press, 1982.

Gillispie, Charles, Coulston and Rafaele Pisano. *Lazare and Sadi Carnot: A Scientific and Filial Relationship*. Dordrecht: Springer, 2013.

Goldberg, Stanley and Roger H. Stuewer, eds. *The Michelson Era in American Science 1870-1930.* New York: American Institute of Physics, 1988.

Gray, Alonzo. *Elements of Chemistry*. New York: Gould, Newman & Saxton, 1840.

Greenaway, Frank. *John Dalton and the Atom.* London: Heinemann Press, 1966.

Harman, P. M. *Energy, Force, and Matter: A Conceptual Development of Nineteenth-Century Physics*. Cambridge: Cambridge University Press, 1983.

Hiebert, Erwin N. *Historical Roots of the Principle of Conservation.* Madison, WI: State Historical Society of Wisconsin, 1962.

Heibert, Erwin N. *The Uses and Abuses of Thermodynamics in Religion.* Cambridge: American Academy of Arts and Sciences, 1966.

Helmholtz, Hermann von. *Popular Scientific Lectures*. New York: Dover Publications, 1962.

Hesse, Mary B. *Forces & Fields*. Totowa, NJ: Littlefield, Adams & Co., 1965.

Hill, Libby. *The Chicago River: A Natural and Unnatural History*. Chicago: Lake Claremont Press, 2000.

Hughes, Thomas Parke. *Elmer Sperry: Inventor and Engineer.* Baltimore: Johns Hopkins Press, 1971.

Hunt, Bruce J. *The Maxwellians.* Ithaca, NY: Cornell University Press, 1994.

Hunt, Bruce J. *Pursuing Power and Light; Technology and Physics from James Watt to Albert Einstein.* Baltimore: Johns Hopkins University Press, 2010.

Jaffe, Bernard. *Michelson and the Speed of Light.* Garden City, NY: Anchor Books, 1960.

Jensen, William B., ed. *Mendeleev and the Periodic Law: Selected Writing 1869-1905.* New York: Dover Publications, 2005.

Johnson, George. *The Purification of Public Water Supplies*, Department of the Interior, United States Geological Survey, Water-Supply Paper 315, Washington: Government Printing Office, 1913.

Koenigsberger, Leo. *Hermann von Helmholtz*. New York: Dover Publications, 1965.

Lemon, Harvey Brace. *From Galileo to the Nuclear Age*. Chicago: University of Chicago Press, 1958.

MacDonald, D. K. C. *Faraday, Maxwell, and Kelvin*. Garden City, NY: Doubleday & Company, 1964.

Markowitz, Gerald, and David Rosner. *Deceit and Denial: The Deadly Politics of Industrial Pollution.* (California / Milbank series on health and the public, 6). Berkeley: University of California Press, 2002

Maxwell, James Clerk. *Treatise on Electricity and Magnetism*. Oxford: Clarendon Press, 1873.

Mayewski, Paul Andrew and Frank White. *The Ice Chronicles: The Quest to Understand Global Climate Change.* Hanover, NH: University Press of New England, 2002.

Mazurs, Edward G. *Graphic Representations of the Periodic System During One Hundred Years.* Tuscaloosa, AL: University of Alabama Press, 1957.

Melosi, Martin V. *Effluent Americas: Cities, Industry, Energy, and the Environment*. Pittsburgh: University of Pittsburgh Press, 2001.

Merchant, Carolyn. *The Columbia Guide to American Environmental History.* New York: Columbia University Press, 2002.

Merz, John Theodore. *A History of European Scientific Thought in the Nineteenth Century.* 2 vols. New York: Dover Publications, 1965.

Morgan, H. Wayne. *Industrial America: The Environment and Social Problems, 1865-1920.* Chicago: Rand McNally College Publishing Company, 1974.

Morrell, Jack and Arnold Thackray. *Gentlemen of Science: Early Years of the British Association for the Advancement of Science*. Oxford: Clarendon Press, 1982.

Nye, Mary Jo. *Before Big Science: The Pursuit of Modern Chemistry and Physics, 1800-1940.* Cambridge: Harvard University Press, 1996.

Patterson, Elizabeth C. *John Dalton and the Atomic Theory*. Garden City, NY: Doubleday & Company, 1970.

Pisarzhevsky, O.N. *Dimitry Ivanovich Mendeleyev: His Life and Work.* Moscow: Foreign Languages Publishing House, 1954.

Posin, Daniel Q. *Mendeleyev: The Story of a Great Scientist.* New York: McGraw-Hill Book Company, 1948.

Ramberg, Peter J. *Chemical Structure, Spatial Arrangement: The Early History of Stereochemistry,* 1874-1914. Farnham Surrey: Aldershot-Ashgate, 2003.

Reingold, Nathan, ed. *Science in America since 1820.* New York: Science History Publications, 1976.

Rocke, Alan J. *Chemical Atomism in the Nineteenth Century: From Dalton to Cannizzaro*. Columbus: Ohio State University Press, 1984.

Rocke, Alan J. *Image and Reality: Kekule, Kopp, and the Scientific Imagination*. Chicago: University of Chicago Press, 2010.

Rocke, Alan J. *Natiuonalizing Science: Adolphe Wurtz and the Battle for French Science*. Cambridge: MIT Press, 2001.

Rocke, Alan J. *The Quiet Revolution: Hermann Kolbe and the Science of Organic Chemistry.* Berkeley: University of California Press, 1993.

Rukeyser, Muriel. *Willard Gibbs*. Garden City, NY: Doubleday, Doran & Company, 1942.

Russell, Colin A. *Michael Faraday*. New York: Oxford University Press, 2000.

Scerri, Eric. *A Tale of Seven Elements*. Oxford: Oxford University Press, 2013.

Scerri, Eric. *The Story of the Periodic System: Its Development and Significance*. New York: Oxford University Press, 2007.

Seitz, Frederick. *The Cosmic Inventor: Reginald Aubrey Fessenden (1866-1932).* (Transactions of the American Philosophical Society, 89 part 6). Philadelphia: American Philosophical Society, 1999.

Sharlin, Harold I. *The Convergent Century: the Unification of Science in the Nineteenth Century.* New York: Abelard-Schuman, 1966.

Siegel, Daniel M. *Innovation in Maxwell's Electromagnetic Theory: Molecular Vortices, Displacement Current, and Light.* New York: Cambridge University Press, 1991.

Smith, Robert Angus. *Air and Rain.* London: Longmans, Green, and Company, 1872.

Strathern, Paul. *Mendeleyev's Dream*. New York: Berkley Books, 2000.

Turner, Gerard L'E. *Nineteenth-Century Scientific Instruments.* Berkeley: University of California Press, 1983.

Tyndall, John. *Faraday as a Discovery*. New York: Thomas Y. Crowell Company, 1961.

Vallery-Radot, René. *The Life of Pasteur*. Garden City, NY: Garden City Publishing Co., 1923.

Van't Hoff, Jacobus H. *The Arrangement of Atoms in Space*. New York: Longmans, Green, and Co., 1898.

Weber, A. S., ed. *Nineteenth-Century Science: An Anthology*. Peterborough, Ontario: Broadview Press, 1999.

Wheeler, Lynde Phelps. *Josiah Willard Gibbs*. New Haven: Yale University Press, 1962.

Williams, Howard R. *Edward Williams Morley: His Influence on Science in America*. Easton, PA: Chemical Education Publishing Company, 1957.

Wise, Tad. *Tesla*. Atlanta, GA: Turner Publishing, 1994.

Wright, Orville. *How We Invented the Airplane*. New York: Dover Publications, 1988.

***Chapter 10: Twentieth-Century Physical Science***

Anderson, David L. *The Discovery of the Electron*. Princeton, NJ: D. Van Nostrand, 1964.

Andrade, E.N. da C. *Rutherford and the Nature of the Atom*. Garden City, NY: Doubleday, 1964.

Asimov, Isaac. *Worlds Within Worlds: The Story of Nuclear Energy, Vol. 1. Atomic Weights-Energy, Electricity*. Washington, DC: United States Energy Research and Development Administration, Office of Public Affairs, 1972.

Barnett, Lincoln. *The Universe and Dr. Einstein*. New York: Mentor Books, 1962.

Bergmann, Peter G. *The Riddle of Gravitation*. New York: Charles Scribner's Sons, 1968.

Bernstein, Jeremy. *Cranks, Quarks, and the Cosmos*. New York: Basic Books, 1993.

Bernstein, Jeremy. *Einstein*. New York: Penguin Books, 1976.

Biquard, Pierre. *Frédéric Joliot-Curie: The Man and His Theories*. New York: Paul S. Eriksson, 1966.

Birks, J.B., ed. *Rutherford at Manchester*. New York: W. A. Benjamin, 1963.

Bleich, Alan Ralph. *The Story of X-Rays: From Röntgen to Isotopes*. New York: Dover Publications, 1960.

Bohr, Niels. *Atomic Theory and the Description of Nature*. Cambridge: Cambridge University Press, 1961.

Boorse, Henry A. and Lloyd Motz, eds. *The World of the Atom*. 2 vols. New York: Basic Books1966.

Born, Max. *Experiment and Theory in Physics*. New York: Dover Publications, 1956.

Born, Max. *My Life & My Views*. New York: Charles Scribner's Sons, 1968.

Born, Max. *Physics in My Generation*. London: Pergamon Press, 1956.

Born, Max. *The Restless Universe.* New York: Dover Publications, 1951.

Cajori, F. *History of Physics In Its Elementary Branches: Including the Evolution of Physical Laboratories.* New York: Macmillan, 1929.

Carroll, Sean. *The Particle at the End of the Universe: How the Hunt for the Higgs-Boson Leads Us to the Edge of a New World.* New York: Penguin Books, 2012.

Cassidy, David. *Uncertainty: The Life and Science of Werner Heisenberg.* New York: Henry Holt and Company, 1993.

Chadwick, James. *Radioactivity and Radioactive Substances.* London: Sir Isaac Pitman & Sons, 1932.

Chernick, Cedric. L. *The Chemistry of the Noble Gases.* Oakridge, TN: United States Atomic Energy Commission/Division of Technical Information, 1967.

Clark, Ronald W. *Einstein: The Life and Times.* New York: Avon Books, 1971.

Cline, Barbara L. *Men Who Made a New Physics.* Chicago: University of Chicago Press, 1987.

Close, Frank. *The Infinity Puzzle: Quantum Field Theory and the Hunt for an Orderly Universe.* New York: Basic Books, 2011.

Conn, G. K. T. and H. D. Turner. *The Evolution of the Nuclear Atom.* London: Iliffe Books, 1965.

Crowther, J. G. *British Scientists of the Twentieth Century.* London: Routledge & Kegan Paul, 1952.

Crowther, J. G. *The Cavendish Laboratory: 1874-1974.* New York: Science History Publications, 1974.

Cuny, Hilaire. *Albert Einstein: The Man and His Theories.* Greenwich, CT: Fawcett Publications, 1962.

Curie, Eve. *Madame Curie.* Garden City, NY: Garden City Publishing Co., 1940.

Curie, Marie. *Pierre Curie.* New York: Dover Publications, 1963.

Curie, Marie. *Radioactive Substances.* New York: Philosophical Library, 1961.

D'Abro, A. *The Rise of the New Physics: Its Mathematical and Physical Theories.* Vol. 2. New York: Dover Publications, 1951.

De Broglie, Louis. *Matter and Light the New Physics.* New York: W.W. Norton & Company, 1939.

Dowie, Mark. *American Foundations: An Investigative History.* Cambridge: MIT Press, 2001.

Einstein, Albert. *Essays in Science.* New York: Philosophical Library, 1934.

Einstein, Albert. *Ideas and Opinions.* New York: Crown Publishers, 1954.

Einstein, Albert. *Out of My Later Years.* New York: Philosophical Library, 1950.

Einstein, Albert. *Relativity: The Special and General Theory.* New York: Crown Publishers, 1961.

Einstein, Albert and Leopold Infeld. *The Evolution of Physics: From Early Concepts to Relativity and Quanta.* New York: Simon and Schuster, 1938.

Feynman, Richard P. *"Surely You're Joking, Mr. Feynman!" Adventures of a Curious Character.* New York: W. W. Norton & Company, 1985.

Frank, Philipp. *Einstein: His Life and Times.* New York: Alfred A. Knopf, 1947.

Gabor, Andrea. *Einstein's Wife.* New York: Penguin Books, 1996.

Galison, Peter. *Einstein's Clocks, Poincaré's Maps: Empires of Time.* New York: W.W. Norton & Company 2003.

Galison, Peter. *How Experiments End.* Chicago: University of Chicago Press, 1987.

Galison, Peter. *Image and Logic: A Material culture of Microphysics.* Chicago: University of Chicago Press, 1997.

Ghent, Percy. *Roentgen: A Brief Biography.* Toronto: The Hunter-Rose Company, 1929.

Glasstone, Samuel. *Inner Space: The Structure of the Atom.* Washington, DC: United States Energy Research and Development Administration, Office of Public Affairs, 1975.

Gleick, James. *Chaos: Making a New Science.* New York: Penguin Books, 1987.

Goertzel, Ted and Ben Goertzel. *Linus Pauling: A Life in Science and Politics.* New York: Basic Books, 1995.

Goldberg, Dave. *The Universe in the Rearview Mirror: How Hidden Symmetries Shape Reality.* New York: Penguin Books, 2013.

Goldsmith, Barbara. *The Inner World of Marie Curie.* New York: W.W. Norton & Company, 2005.

Gribbin, John. *In the Beginning: The Birth of the Living Universe.* New York: Little, Brown and Company, 1993.

Gribbin, John and Mary Gribbin. *Richard Feynman: A Life in Science.* New York: Penguin Books, 1999.

Hager, Thomas. *Linus Pauling and the Chemistry of Life.* New York: Oxford University Press, 1998.

Hecht, Selig. *Explaining the Atom.* New York: Viking Press, 1954.

Heilbron, J. L. *Ernest Rutherford and the Explosion of Atoms.* New York: Oxford University Press, 2003.

Heilbron, J. L. *The Dilemmas of an Upright Man: Max Planck as Spokesman for German Science.* Berkeley: University of California Press, 1986.

Heim, Karl. *Christian Faith and Natural Science.* New York: Harper & Row, Publishers, 1953.

Hoffmann, Banesh. *Albert Einstein: Creator and Rebel.* New York: New American Library, 1972.

Hiebert, Erwin N. *The Conception of Thermodynamics in the Scientific Thought of Mach and Planck.* Freiburg: Ernst Mach Institute, 1968.

Holton, Gerald. *Einstein, History, and Other Passions: The Rebellion against Science at the End of the Twentieth Century.* Cambridge: Harvard University Press, 1996.

Hughes, Donald J. *The Neutron Story.* Garden City, NY: Doubleday & Company, 1959.

Illy, József, ed. *Albert Meets America: How Journalists Treated Genius During Einstein's 1921 Travels.* Baltimore: Johns Hopkins University Press, 2006.

Isaacson, Walter. *Einstein: His Life and Universe.* New York: Simon & Schuster, 2007.

Jeans, James. *Physics and Philosophy.* Ann Arbor: University of Michigan Press, 1958.

Jensen, William B. *The Lewis Acid-Base Concepts: An Overview.* New York: John Wiley & Sons, 1980.

Jerome, Fred. *The Einstein File: J. Edgar Hoover's Secret War against the World's Most Famous Scientist.* New York: St. Martin's Press, 2002.

Jolly, William L. *From Retorts to Lasers: The Story of Chemistry at Berkeley.* Distributed to the College of Chemistry, University of California, Berkeley, CA, 1978.

Kakalios, James. *The Amazing Story of Quantum Mechanics.* New York: Penguin Books, 2011.

Kaku, Michio. *Einstein's Cosmos.* New York: W. W. Norton & Company, 2004.

Kargon, Robert H. *The Rise of Robert Millikan.* Ithaca, NY: Cornell University Press, 1982.

Kauffman, George B. *Alfred Werner: Founder of Coordination Chemistry.* Berlin: Springer-Verlag, 1966.

Kauffman, George B., ed. *Classics in Coordination Chemistry: Part 1, The Selected Papers of Alfred Werner.* New York: Dover Publications, 1976.

Kevles, Daniel. *The Physicists: The History of a Scientific Community in Modern America.* Cambridge: Harvard University Press, 1995.

Kragh, Helge. *Quantum Generations: A History of Physics in the Twentieth Century.* Princeton: Princeton University Press, 1999.

Laughlin, Robert B. *A Different Universe: Reinventing Physics from the Bottom Down.* New York: Basic Books, 2005.

Lewis, Gilbert Newton. *Valence and the Structure of Atoms and Molecules.* New York: Chemical Catalog Company, 1923.

McCusker, Brian. *The Quest for Quarks.* Cambridge: Cambridge University Press, 1983.

Miller, Arthur I. *Imagery in Scientific Thought: Creating Twentieth-Century Physics*. Boston: Birkhäuser, 1984.

Millikan, Robert. *The Electron, Its Isolation and Measurement and the Determination of Some of Its Properties*. Chicago: University of Chicago Press, 1924.

Mlodinow, Leonard. *Feynman's Rainbow*. New York: Warner Books, 2003.

Mulliken, Robert. *Life of a Scientist*. Berlin: Springer-Verlag, 1989.

Navarro, Jaume. *A History of the Electron: J.J. and G.P. Thomson*. Cambridge: Cambridge University Press, 2012.

Newman, Louise Michele. ed. *Men's Ideas/Woman's Realities: Popular Science, 1870-1915*. New York: Pergamon Press, 1985.

Nitschke, W. Robert. *The Life of Wilhelm Conrad Röntgen: Discoverer of the X Ray*. Tucson, AZ: University of Arizona Press, 1971.

Nye, Mary Jo. *From Chemical Philosophy to Theoretical Chemistry: Dynamics of Matter* and *Dynamics of Disciplines, 1800-1950*. Berkeley: University of California Press, 1993.

Ouellette, Jennifer. *Black Bodies and Quantum Cats: Tales from the Annals of Physics*. New York: Penguin Books, 2005.

Overbye, Dennis. *Einstein in Love*. New York: Penguin Books, 2001.

Pais, Abraham. *'Subtle is the Lord...' The Science and the Life of Albert Einstein*. New York: Oxford University Press, 1982.

Palmer, W. G. *Valency: Classical and Modern*. Cambridge: Cambridge University Press, 1946.

Panek, Richard. *The Invisible Century: Einstein, Freud, and the Search for Hidden Universes*. New York: Penguin Books, 2004.

Paradowski, Robert J. "The Structural Chemistry of Linus Pauling." PhD diss., University of Wisconsin, 1972.

Pasachoff, Naomi. *Marie Curie and the Science of Radioactivity*. New York: Oxford University Press, 1996.

Penrose, Roger. *The Emperor's New Mind: Concerning Computers, Minds, and the Laws of Physics*. New York: Penguin Books, 1991.

Quinn, Susan. *Marie Curie: A Life*. New York: Addison-Wesley Publishing Company, 1995.

Redei, Miklos and Michael Stöltzner, eds. *John von Neumann and the Foundations of Quantum Physics*. (Vienna Circle Institute Yearbook, 8). Dordrecht: Kluwer Academic Press, 2001.

Reeves, Richard. *A Force of Nature: The Frontier Genius of Ernest Rutherford*. New York: W.W. Norton & Company, 2008.

Romer, Alfred, ed. *The Discovery of Radioactivity and Transmutation.* New York: Dover Publications, 1964.

Romer, Alfred, ed. *Radiochemistry and the Discovery of Isotopes*. New York: Dover Publications, 1970.

Schilpp, Paul. *Albert Einstein: Philosopher-Scientist*. New York: Tudor Publishing Company, 1951.

Seife, Charles. *Decoding the Universe*. New York: Penguin Books, 2006.

Servos, John. *Physical Chemistry from Ostwald to Pauling: the Making of a Science in America.* Princeton: Princeton University Press, 1990

Silverberg, Robert. *Niels Bohr: The Man Who Mapped the Atom*. Philadelphia: Macrae Smith Company, 1965.

Stranges, Anthony N. *Electrons and Valence: Development of the Theory*. College Station: Texas A&M University Press, 1982.

Thomson, George. *J. J. Thomson: Discoverer of the Electron*. Garden City, NY: Doubleday & Company, 1966.

Thomson, J. J. *Electricity and Matter*. New York: Charles Scribner's Sons, 1904.

True, Webster P., ed. *The Structure of the Universe*. Vol. 1. New York: Simon and Schuster, 1960.

Tyson, Neil deGrasse and Donald Goldsmith. *Origins: Fourteen Billion Years of Cosmic Evolution.* New York: W.W. Norton & Company, 2005.

Weinberg, Steven. *The First Three Minutes: A Modern View of the Origin of the Universe*. New York: Basic Books, 1977.

White, Michael and John Gribbin. *Einstein: A Life in Science.* New York: Dutton, 1993.

White, Michael and John Gribbin. *Stephen Hawking: A Life in Science.* New York: Dutton, 1992.

Wigner, Eugene P. *Symmetries and Reflections: Scientific Essays*. Cambridge: MIT Press, 1970.

Wright, Stephen, ed. *Classical Scientific Papers, Physics.* London: Mills & Boon, Limited. 1964.

Zackheim, Michele. *Einstein's Daughter" The Search for Lieserl*. New York: Riverhead Books, 1999.

Zewail, Ahmed H. *Voyage Through Time: Walks of Life to the Nobel Prize.* Cairo/New York: American University in Cairo Press, 2003.

***Chapter 11: Energy and the Environment: Conflict or Compatibility***

Aczel, Amir D. *Present at the Creation: Discovering the Higgs Boson*. New York: Crown Publishers, 2010.

*Atomic Power and Space: A History*. Washington, DC: US Department of Energy, 1987.

Barkey, Michael B., ed. *Environmental Stewardship in the Judeo-Christian Tradition: Jewish, Catholic, and Protestant Wisdom on the Environment*. Washington, DC: Interfaith Council for Environmental Stewardship, 2000.

Baxter, James Phinney. *Scientists Against Time*. Cambridge: MIT Press, 1952.

Bell, Michael Mayerfield. *An Invitation to Environmental Sociology*. Thousand Oaks, CA: Pine Forge Press, 1998.

Bird, Kal and Martin J. Sherwin. *American Prometheus: The Triumph and Tragedy of J. Robert Oppenheimer*. New York: A. A. Knopf, 2005.

Blumberg, Stanley A. and Gwinn Owens. *Energy and Conflict: The Life and Times of Edward Teller.* New York: G. P. Putnam's Sons, 1976.

Bowen, Mark. *Censoring Science: Inside the Political Attack on Dr. James Hansen and the Truth of Global Warming*. New York: Plume, 2009.

Brand, Stewart. *Whole Earth Discipline*. New York: Penguin Books, 2010.

Brodsky, Robert F. *On the Cutting Edge: Tales of a Cold War Engineer at the Dawn of the Nuclear, Guided Missile, Computer, and Space Ages*. New York: Gordian Knot, 2006.

Brown, Gerald E. and Chang-Hwan Lee, eds. *Hans Bethe and His Physics.* Singapore: World Scientific Publishing Company, 2006.

Bush, Vannevar. *Modern Arms and Free Men: A Discussion of the Role of Science in Preserving Democracy.* New York: Simon and Schuster, 1949.

Buti, Ken and John Perlin. *A Golden Thread: 2500 Years of Solar Architecture and Technology.* Palo Alto: Cheshire Books, 1980.

Canaday, John. *The Nuclear Muse: Literature, Physics, and the First Atomic Bomb.* Madison: University of Wisconsin Press, 2000.

Carson, Cathryn and David Hollinger, eds. *Reappraising Oppenheimer: Centennial Studies and Reflections.* (Berkeley Papers in History of Science, 21). Berkeley: Office for History of Science and Technology, University of California, Berkeley, 2005.

Childs, Herbert. *An American Genius: The Life of Ernest Orlando Lawrence, Father of the Cyclotron.* New York: E.P. Dutton & Co., 1968.

Christianson, Gale E. *Greenhouse: The 200-Year Story of Global Warming.* New York: Penguin Books, 1999.

Compton, H. *Atomic Quest: A Personal Narrative.* New York: Oxford University Press, 1956.

Conant, Jennet. *109 East Palace: Robert Oppenheimer and the Secret City of Los Alamos*. New York: Simon & Schuster, 2005.

Conant, Jennet. *Tuxedo Park: A Wall Street Tycoon and the Secret Palace of Science that Changed the Course of World War II.* New York: Simon & Schuster, 2002.

Cooper, Dan. *Enrico Fermi and the Revolutions of Modern Physics*. New York: Oxford University Press, 1999.

Cornwell, John. *Hitler's Scientists: Science, War, and the Devil's Pact*. New York: Penguin Books, 2003.

Cronin, James. *Fermi Remembered.* Chicago: University of Chicago Press, 2004.

Cronon, William, ed. *Uncommon Ground, Rethinking the Human Place in Nature*. New York: W.W. Norton & Company, 1996.

Cunningham, William P., Mary Ann Cunningham, and Barbara Saigo. *Environmental Science: A Global Concern*. 8th ed. New York: McGraw-Hill, 2005.

Daniels, Farrington. *Direct Use of the Sun's Energy*. New Haven: Yale University Press, 1964.

Davis, N. Pharr. *Lawrence and Oppenheimer.* New York: Fawcett World Library, 1968.

Dessler, Andrew. *Introduction to Modern Climate Change.* New York: Cambridge University Press, 2012.

Dessler, Andrew and Edward A. Parson. *The Science and Politics of Global Climate Change: A Guide to the Debate,* 2d ed. New York: Cambridge University Press, 2010.

Dudley, William, ed. *The Environment: Opposing Viewpoints*. San Diego: Greenhaven Press, 2001.

Feinberg, J. G. *The Story of Atomic Theory and Atomic Energy*. New York: Dover Publications, 1960.

Fermi, Laura. *Atoms in the Family: My Life with Enrico Fermi.* Chicago: University of Chicago Press, 1954.

*First Reactor*, *The*. US Energy Research and Development Administration, Office of Public Affairs, Washington, DC, 1967.

Fleming, James Rodger. *Fixing the Sky: the Checkered History of Weather and Climate Control.* New York: Columbia University Press, 2010.

Fleming, James Rodger. *Historical Perspectives on Climate Change*. New York: Oxford University Press, 1998.

Ford, Kenneth W. *Building the H-Bomb: A Personal History.* Singapore: World Scientific Publishing Company, 2015.

Freese, Barbara. Coal: *A Human History*. New York: Penguin Books, 2004.

Funk, McKenzie. *Windfall: The Booming Business of Global Warming*. New York: Penguin Books, 2014.

Gerber, Michele Stenehjem. *On the Home Front: The Cold War Legacy of the Hanford Nuclear Site*, 2d ed. Lincoln: University of Nebraska Press, 2002.

Gerasimov I. P., ed. *Man, Society, and the Environment*. Moscow: Progress Publishers, 1975.

Glasstone, Samuel. *Controlled Nuclear Fusion.* US Atomic Energy Commission/Division of Technical Information. Oak Ridge, TN, June 1964.

Glasstone, Samuel. *Fusion Energy.* US Department of Energy. Office of Energy Research/Office of Fusion Energy. US Department of Energy Technical Information Center. Oak Ridge, TN, 1980.

Gordin, Michael D. *Red Cloud at Dawn: Truman, Stalin, and the End of the Atomic Monopoly*. New York: Picador, 2009.

Gosling, F. G. *The Manhattan Project: Science in the Second World War*. Washington, DC: US Department of Energy, 1990.

Greenbaum, Leonard. *A Special Interest*. Ann Arbor: University of Michigan Press, 1971.

Grodzins, Morton and Eugene Rabinowitch, eds. *The Atomic Age: Scientists in National and World Affairs*. New York: Simon and Schuster, 1965.

Guha, Ramachandra. *Environmentalism: A Global History*. New York: Longman, 2000.

Guillerme, André E. *The Age of Water: The Urban Environment in the North of France, AD 300-1800*. College Station: Texas A&M University Press, 1988.

Haley, James, ed. *Pollution*. Farmington Hills, MI: Greenhaven Press, 2003.

Hargittai, István. *Martians of Science: Five Physicists Who Changed the Twentieth Century*. New York: Oxford University Press, 2006.

Heisenberg, Werner. *Philosophic Problems of Nuclear Science*. New York: Pantheon Books, 1952.

Herken, Gregg. *Brotherhood of the Bomb: The Tangled Lives and Loyalties of Robert Oppenheimer, Ernest Lawrence, and Edward Teller.* New York: Henry Holt and Company, 2002.

Herman, Robin. *Fusion: The Search for Endless Energy*. Cambridge: Cambridge University Press, 1990.

Hoddeson, Lillian, and Vicki Daitch. *True Genius: The Life and Science of John Bardeen: The Only Winner of Two Nobel Prizes in Physics.* Washington: Joseph Henry Press, 2002.

Hoyle, Fred. *The Nature of the Universe*. New York: Harper & Brothers, 1950.

Hughes, Jeff. *The Manhattan Project: Big Science and the Atomic Bomb.* New York: Columbia University Press, 2003.

Hull, McAllister H. and Amy Bianco. *Rider of the Pale Horse: A Memoir of Los Alamos and Beyond.* Albuquerque: University of New Mexico Press, 2005.

Hunner, Jon. *Inventing Los Alamos: The Growth of an Atomic Community.* Norman: University of Oklahoma Press, 2004.

Hydrick, Carter P. *Critical Mass: How Nazi Germany Surrendered Enriched Uranium for the United States' Atomic Bomb*. Bloomington, IN: Authorhouse, 2003.

Jungk, Robert. *Brighter Than a Thousand Suns: A Personal History of the Atomic Scientists.* New York: Harcourt Brace, 1958.

Katz, Eric, ed. *Death by Design: Science, Technology, and Engineering in Nazi Germany.* New York: Longman, 2006.

Keeling, Charles D. "Rewards and Penalties of Monitoring the Earth," *Annual Review of Energy and the Environment*, 23 (1998): 25-82.

Kelly, Cynthia, ed. *Oppenheimer and the Manhattan Project: Insights into J. Robert Oppenheimer, "Father of the Atomic Bomb."* Hackensack, NJ: World Scientific Publishing Company, 2006.

Kennett, Lee. *A History of Strategic Bombing.* New York: Charles Scribner's Sons, 1982.

Kipphardt, Heinar. *In the Matter of J. Robert Oppenheimer.* New York: Hill and Wang, 1968.

Lasby, Clarence. *Project Paperclip.* New York: Atheneum, 1971.

Laurence, William L. *The Story of the Atomic Bomb.* New York: Alfred Knopf, 1946.

Lewis, Richard S. and Jane Wilson, eds. *Alamagordo Plus Twenty-Five Years.* New York: Viking Press, 1971.

Mann, Michael M. *The Hockey Stick and the Climate Wars.* New York: Columbia University Press, 2012.

Masco, Joseph. *The Nuclear Borderlands: The Manhattan Project in Post-Cold War New Mexico.* Princeton: Princeton University Press, 2006.

Mason, Richard. *Oppenheimer's Choice: Reflections from Moral Philosophy.* Albany: State University of New York Press, 2006.

McKibben, Bill, ed. *The Global Warming Reader: A Century of Writing about Climate Change.* New York: Penguin Books, 2011.

McMillan, Priscilla Johnson. *The Ruin of J. Robert Oppenheimer and the Birth of the Modern Arms Race.* New York: Viking, 2005.

Michnovicz Gibson, Toni and Jon Michnovicz. *Images of America: Los Alamos 1944-1947.* Charleston, SC: Arcadia Publishing, 2005.

Midgley, Thomas, IV. *From the Periodic Table to Production: The Biography of Thomas Midgley, Jr., the Inventor of Ethyl Gasoline and Freon Refrigerants.* Corona, CA: Stargazer Publishing Co., 2001.

Molella, Arthur, and Joyce Bedi, eds. *Inventing for the Environment.* Cambridge: MIT Press, 2003.

Norris, Robert S. *Racing for the Bomb: General Leslie R. Groves, the Manhattan Project's Indispensable Man.* South Royalton, VT: Steerforth Press, 2002.

Oppenheimer, J. Robert. *The Open Mind.* New York: Simon and Schuster, 1955.

Oreskes, Naomi and Erik M. Conway. *Merchants of Doubt.* New York: Bloomsbury Press, 2010.

Patterson, Walter C. *The Plutonium Business: And the Spread of the Bomb*. San Francisco: Sierra Club Books, 1984.

Perlin, John. *From Space to Earth: The Story of Solar Electricity*. Ann Arbor, MI: Aatec Publications, 1999.

Perlin, John. *Let it Shine-The 6,000-Year Story of Solar Energy*. Novato, CA: New World Library, 2013.

Polenberg, Richard, ed. *In the Matter of J. Robert Oppenheimer: The Security Clearance Hearings*. Ithaca, NY: Cornell University Press, 2002.

Pollack, Henry. *A World Without Ice*. New York: Penguin Books, 2010.

Powers, Thomas. *Heisenberg's War*. New York: Alfred A. Knopf, 1993.

Quaratiello, Arlene Rodda. *Rachel Carson: A Biography*. Westport, CT: Greenwood Press, 2004.

Rabi, I.I., Robert Serber, Victor F. Weisskopf, Abraham Pais, Glenn T. Seaborg. *Oppenheimer*. New York: Charles Scribner's Sons, 1969.

Rhodes, Richard. *Dark Sun: The Making of the Hydrogen Bomb*. New York: Simon & Schuster, 1995.

Rhodes, Richard. *The Making of the Atomic Bomb*. New York: Simon & Schuster, 1986.

Rifkin, Jeremy. *The Hydrogen Economy*. New York: Tarcher/Putnam, 2002.

Roleff, Tamara L., ed. *Pollution: Opposing Viewpoints*. San Diego: Greenhaven Press, 2000.

Rouzé, Michel. *Robert Oppenheimer: The Man and His Theories*. New York: Fawcett World Library, 1965.

Russell, Colin A. *The Earth, Humanity and God*. London: University College London Press, 1994.

Seaborg, Glenn Theodore, with Eric Seaborg. *Adventures in the Atomic Age: From Watts to Washington*. New York: Farrar Strauss & Giroux, 2001.

Segrè, Emilio. *Enrico Fermi, Physicist*. Chicago: University of Chicago Press, 1970.

Seife, Charles. *Sun in a Bottle: The Strange History of Fusion and the Science of Wishful Thinking*. New York: Penguin Books, 2008.

Shapiro, Jerome F. *Atomic Bomb Cinema: The Apocalyptic Imagination on Film*. New York: Routledge, 2001.

Smith, Alice Kimball. *A Peril and a Hope: The Scientists' Movement in America 1945-47*. Cambridge: MIT Press, 1970.

Smyth, Henry D. *Atomic Energy: For Military Purposes*. Princeton: Princeton University Press, 1945.

Stoff, Michael B., Jonathan F. Fanton, and R. Hal Williams, eds. *The Manhattan Project: A Documentary Introduction to the Atomic Age*. New York: McGraw-Hill, 1991.

Stoll, Steven. *U.S. Environmentalism Since 1945: A Brief History with Documents*. Boston: Bedford/ St. Martin's, 2007.

Swimme, Brian. *The Universe is a Green Dragon: A Cosmic Creation Story*. Santa Fe, NM: Bear & Company, 1984.

Thorpe, Charles. *Oppenheimer: The Tragic Intellect.* Chicago: University of Chicago Press, 2006.

Van De Mark, Brian. *Pandora's Keepers: Nine Men and the Atomic Bomb.* Boston: Little, Brown and Co., 2003.

Ward, Bob. *Dr. Space: The Life of Werner von Braun.* Annapolis: Naval Institute Press, 2005.

Warren, Louis S., ed. *American Environmental History.* Malden, MA: Blackwell, 2003.

Weart, Spencer. *The Discovery of Global Warming*, 2d ed. Cambridge: Harvard University Press, 2008.

Westwick, Peter J. *The National Labs: Science in an American System, 1947-1974.* Cambridge: Harvard University Press, 2003.

Yergin, Daniel. *The Prize*. New York: Simon & Schuster, 1991.

Yergin, Daniel. *The Quest*. New York: Penguin Press, 2011.

### ***Biographical Collections***

Asimov, Isaac. *Asimov's Biographical Encyclopedia of Science and Technology*. Garden City, NY: Doubleday, 1972.

*Concise Dictionary of American Biography*, 3d ed. New York: Charles Scribner's Sons, 1980.

*Concise Dictionary of Scientific Biography*. New York: Charles Scribner's Sons, 1981.

*Dictionary of American Biography*. New York: Charles Scribner's Sons, 20 volumes 1928-1936. Successor is the American National Biography, Oxford University Press.

Elliott, Clark A. *Biographical Dictionary of American Science: The Seventeenth through the Nineteenth Centuries*. Westport, CT: Greenwood Press, 1979.

Farber, Eduard. *Great Chemists*. New York: Interscience Publishers, 1961.

Gillispie, Charles Coulston, ed. *Dictionary of Scientific Biography* and *New Dictionary of Scientific Biography.* New York: Charles Scribner's Sons, 1981. Electronic version combines both publications as *Complete Dictionary of Scientific Biography.*

Greene, Jay E., ed. *McGraw-Hill Modern Men of Science*. New York: McGraw-Hill, 1966.

Jaffe, Bernard. *Men of Science in America*. New York: Simon and Schuster, 1958.

McNeil, Ian and Lance Day, eds. *Biographical Dictionary of the History of Technology*. New York: Routledge, 1996

Roysdon, Christine and Linda Kharti. *American Engineers of the Nineteenth Century*. New York: Garland Publishing, 1978.

### Dictionaries and Encyclopedias

Bynum, W.F., E.J. Browne, and Roy Porter, eds. *Dictionary of the History of Science*. Princeton, NJ: Princeton University Press, 1981.

Bynum, W.F. and Helen Bynum, eds., revised edition. *Dictionary of Medical Biography*. 5 vols. Westport, CT: Greenwood Press, 2007.

Dunbar, Gary S. *The History of Modern Geography*. New York: Garland Publishing, 1985.

Hetherington, Norriss S., ed. *Encyclopedia of Cosmology*. New York: Garland Publishing, 1993.

Macey, Samuel L., ed. *Encyclopedia of Time*. New York: Garland Publishing, 1994.

Magill, Frank N., ed. *Magill's Survey of Science: Earth Science Survey*. Pasadena, CA: Salem Press, 1990.

Merchant, Carolyn. *The Columbia Guide to American Environmental History*. New York: Columbia University Press, 2002.

Millar, David, Ian, John & Margaret. *Cambridge Dictionary of Scientists*. New York: Cambridge University Press, 1996.

Oakes, Elizabeth H., ed. *International Encyclopedia of Women Scientists*. New York: Facts on File, 2002.

Ogilvie, Marilyn Bailey. *Women in Science: Antiquity through the Nineteenth Century*. Cambridge: MIT Press, 1986.

Paehlke, Robert C., ed. *Conservation and Environmentalism*. New York: Garland Publishing, 1995.

Rothenberg, Marc, ed. *The History of Science in the United States*. New York: Garland Publishing, 2001.

Stannard, Eric, ed. *Oxford Encyclopedia of the History of American Science, Medicine, and Technology*. New York: Oxford University Press, 2014.

Tattersall, Ian, Eric Delson, and John A. Van Couvering, eds. *Encyclopedia of Human Evolution and Prehistory*. New York: Garland Publishing, 1988.

### General Histories

Alioto, Anthony. *A History of Western Science*. 2d ed. Englewood Cliffs, NJ: Prentice Hall, 1993.

Allen, Garland. *Life Science in the Twentieth Century*. New York: John Wiley & Sons, 1975.

American Society of Civil Engineers, Committee on History and Heritage of American Civil Engineers. *American Civil Engineering History: ASCE 150 Years: Proceedings of the Fought National Congress on Civil Engineering History and Heritage, November 2-6, 2002, Washington, DC.* Reston, VA: American Society of Civil Engineers, 2003.

Anton, John Peter. *American Naturalism and Greek Philosophy*. Amherst, NY: Humanity Books, 2005.

Appleby, Joyce. *Shores of Knowledge: New World Discoveries and the Scientific Imagination*. New York: W. W. Norton & Company, 2013.

Asimov, Isaac. *A Short History of Biology*. Garden City, NY: Natural History Press, 1964.

Baigrie, Brian S. *Scientific Revolutions: Primary Texts in the History of Science*. Upper Saddle River, NJ: Pearson Education, 2004.

Bala, Arun. *The Dialogue of Civilizations in the Birth of Modern Science*. New York: Palgrave Macmillan, 2006.

Barth, Frederik, ed. *One Discipline, Four Ways: British, German, French, and American Anthropology*. Chicago: University of Chicago Press, 2005.

Bernal, John D. *Science in History*. vol. 1. New York: Cameron Associates, 1954.

Bochner, Salomon. *The Role of Mathematics in the Rise of Science*. Princeton: Princeton University Press, 1981.

Boorse, Henry A., Lloyd Motz, and Jefferson Hane Weaver. *The Atomic Scientists: A Biographical History*. New York: John Wiley & Sons, 1989.

Boorstin, Daniel. *The Discoverers*. New York: Random House, 1983.

Bowler, Peter J. *The Earth Encompassed*. New York: W.W. Norton & Company, 1993.

Bowler, Peter J. *Evolution: The History of an Idea*. Berkeley: University of California Press, 2003.

Bowler, Peter J. and Iwan Rhys Morus. *Making Modern Science*. Chicago: University of Chicago Press, 2005.

Bradford, Marlene. *Scanning the Skies: A History of Tornado Forecasting*. Norman: University of Oklahoma Press, 2001.

Bridson, Gavin D. R., James J. White, and Lugene B. Bruno. *American Botanical Prints of Two Centuries: Catalogue of an Exhibition: 27 April-31 July 2003*. Pittsburgh, PA: Hunt Institute for Botanical Documentation, Carnegie Mellon University, 2003.

Brock, William H., *Chemistry*. New York: W.W. Norton & Company, 1993.

Brooks, Michael. *Free Radicals: The Secret Anarchy of Science*. New York: Overlook Press, 2012.

Bryson, Michael A. *Visions of the Land: Science, Literature, and the American Environment from the Era of Exploration to the Age of Ecology.* (Under the Sign of Nature: Explorations in Ecocriticism). Charlottesville: University Press of Virginia, 2002.

Bud, Robert and Deborah Jean Warner, eds. *Instruments of Science: An Historical Encyclopedia.* New York: Garland Publishing, 1998.

Burke, James. *Connections.* Boston: Little, Brown, 1978.

Burke, James. *The Day the Universe Changed.* Boston: Little, Brown, 1995.

Burns, William E. *Knowledge and Power: Science in World History.* New York: Pearson, 2011.

Burstall, Aubrey. *A History of Mechanical Engineering.* Cambridge: MIT Press, 1965.

Bynum, W. F. and Helen Bynum, eds. *Dictionary of Medical Biography.* 5 vols. Westport, CT: Greenwood Press, 2007.

Carey, John. *Eyewitness to Science.* Cambridge: Harvard University Press, 1995.

Clagett, Marshall, ed. *Critical Problems in the History of Science.* Madison: University of Wisconsin Press, 1969.

Crellin, John K. *A Social History of Medicines in the Twentieth Century: To Be Taken Three Times a Day.* New York: Pharmaceutical Products Press, 2004.

D'Abro, A. *The Evolution of Scientific Thought: From Newton to Einstein.* 2d ed. New York: Dover Publications, 1949.

Daniels, George, ed. *Nineteenth Century American Science.* Evanston, IL: Northwestern University Press, 1972.

De Groot, Jean, ed. *Nature in American Philosophy.* Studies in Philosophy and History of Philosophy, 42. Washington, DC: Catholic University of America Press, 2004.

De Solla Price, Derek J. *Little Science, Big Science.* New York: Columbia University Press, 1963.

Deutsch, David. *The Beginning of Infinity: Explanations that Transform the World.* New York: Penguin Books, 2011.

Dowie, Mark. *American Foundations: An Investigative History.* Cambridge: MIT Press, 2001.

Duffin, Jacalyn. *History of Medicine.* Toronto: University of Toronto Press, 1999.

Dunbar, Robin. *The Trouble with Science.* Cambridge: Harvard University Press, 1995.

Ede, Andrew and Lesley B. Cormack. *A History of Science in Society.* Peterborough, Ontario, Canada: Broadview Press, 2004.

England, J. Merton. *A Patron for Pure Science: The National Science Foundation's Formative Years, 1945-57.* Washington, DC: National Science Foundation, 1982.

Evans, Harold, Gail Buckland, and David Lefer. *They Made America: From the Steam Engine to the Search Engine, Two Centuries of Innovators*. New York: Little, Brown, 2004.

Ferngren, Gary B., ed. *The History of Science and Religion in the Western Tradition: An Encyclopedia*. New York: Garland Publishing, 2000.

Forrest, Derek. *Hypnotism: A History*. New York: Penguin Books, 1999.

Garfield, Simon. *On the Map: A Mind-Expanding Exploration of the Way the World Looks*. New York: Penguin Books, 2013.

Gjertsen, Derek. *The Classics of Science: A Study of Twelve Enduring Scientific Works*. New York: Lilian Barber Press, 1984.

Greenberg, Arthur. *Chemical History Tour: Picturing Chemistry from Alchemy to Modern Molecular Science*. New York: John Wiley & Sons, 2000.

Gross, Alan G. *The Rhetoric of Science*. Cambridge: Harvard University Press, 1996.

Hall, Courtney Robert. *History of American Industrial Science.* New York: Library Publishers, 1954.

Hatton, John and Paul B. Plouffe. *Science and its Ways of Knowing*. Upper Saddle River, NJ: Prentice-Hall, 1997.

Hargittai, István. *Candid Science: Conversations with Famous Chemists.* Ed. Magdolna Hargittai. London: Imperial College Press, 2000 Distributed by World Scientific Publishing Company, Hackensack, NJ.

Hargittai, István. *Candid Science II: Conversations with Famous Biomedical Scientists.* ed. Magdolna Hargittai. London: Imperial College Press, 2002. Distributed by World Scientific Publishing Company, Hackensack, NJ.

Hargittai, István. *Candid Science III: Conversations with More Famous Chemists.* ed. Magdolna Hargittai. London: Imperial College Press, 2003. Distributed by World Scientific Publishing Company, Hackensack, NJ.

Harrison, Edward. *Masks of the Universe*. New York: Macmillan, 1985.

Hellman, Hal. *Great Feuds in Technology: Ten of the Liveliest Disputes Ever*. New York: John Wiley and Sons, 1998.

Hempstead, Colin A. and William E. Worthington, eds. *Encyclopedia of Twentieth Century Technology*. 2 vols. New York: Routledge, 2005.

Henry, John. *A Short History of Scientific Thought*. New York: Palgrave Macmillan, 2012.

Hilts, Victor L. *Statist and Statistician: The Development of Science.* North Stratford, NH: Ayer Publishing Company, 1981.

Holton, Gerald. *The Scientific Imagination: Case Studies.* Cambridge: Cambridge University Press, 1978.

Holton, Gerald and Stephen G. Brush. *Physics, the Human Adventure: From Copernicus to Einstein and Beyond*. New Brunswick, NJ: Rutgers University Press, 2001.

Hughes, Thomas Parke. *American Genesis*. New York: Penguin Books, 1989.

Ifrah, Georges. *From One to Zero: A Universal History of Numbers*. New York: Penguin Books, 1985.

Ihde, Aaron J. *The Development of Modern Chemistry*. New York: Harper & Row, 1964.

Kaku, Michio. *Physics of the Future: How Science Will Shape Human Destiny and Our Daily Lives by the Year 2100*. New York: Doubleday, 2011.

Kargon, Robert H., ed. *The Maturing of American Science*. Washington, DC: American Association for the Advancement of Science, 1974.

Kasson, John F. *Civilizing the Machine*. New York: Penguin Books, 1976.

Kevles, Daniel, *The Physicists: The History of a Scientific Community in Modern America*. Cambridge: Harvard University Press, 1995.

Knight, David. *Ideas in Chemistry: A History of the Science*. New Brunswick, NJ: Rutgers University Press, 1992.

Kramer, Edna E. *The Nature and Growth of Modern Mathematics*. Princeton: Princeton University Press, 1981.

Kranzberg, Melvin, and Carroll Pursell, eds. *Technology in Western Civilization*. New York: Oxford University Press, 1967.

Laidler, Keith J. *To Light Such a Candle: Chapters in the History of Science and Technology*. Oxford: Oxford University Press, 1998.

Layton, Jr., Edwin T., ed. *Technology and Social Change in America*. New York: Harper & Row, Publishers, 1973.

Lederman, Leon and Judith Scheppler. *Portraits of Great American Scientists*. Amherst, NY: Prometheus Books, 2001.

Lemon, Harvey Brace. *From Galileo to the Nuclear Age*. Chicago: University of Chicago Press, 1958.

Levere, Trevor and Frederic L. Holmes, eds. *Instruments and Experimentation in the History of Chemistry*. Cambridge: MIT Press, 2000.

MacLachlan, James. *Children of Prometheus: A History of Science and Technology*. Toronto: Wall & Thompson, 1989.

Marks, John. *Science and the Making of the Modern World*. London: Heinemann Educational Books, 1983.

Marsh, J. E. *The Origins and the Growth of Chemical Science*. London: John Murray, 1929.

Mason, Stephen. *A History of the Sciences*. New York: Macmillan, 1962.

Medawar, P.B. and J.S. Medawar. *Aristotle to Zoos: A Philosophical Dictionary of Biology.* Cambridge: Harvard University Press, 1983.

Melosi, Martin V. *Coping with Abundance: Energy and Environment in Industrial America.* New York: Alfred A. Knopf, 1985.

Meyer, Ernst Von. *A History of Chemistry: From Earliest Times to the Present Day.* London: MacMillan, 1898.

Meyers, Morton A. *Prize Fight: The Race and the Rivalry to be the First in Science.* New York: Palgrave Macmillan, 2012.

Millar, David, Ian, John & Margaret. *Cambridge Dictionary of Scientists.* New York: Cambridge University Press, 1996.

Nadis, Fred. *Wonder Shows: Performing Science, Magic, and Religion in America.* New Brunswick, NJ: Rutgers University Press, 2005.

Nordenskiold, Erik. *The History of Biology: A Survey.* New York: Knopf, 1928.

Nye, David E. *America as Second Creation: Technology and Narratives of New Beginnings.* Cambridge: MIT Press, 2003.

Oakes, Elizabeth H. *A to Z of Chemists. (Notable Scientists).* New York: Facts on File, 2002.

Olah, George A. *Life of Magic Chemistry.* New York: John Wiley-Interscience, 2001.

Olson, Richard G. *Science and Religion, 1450-1900: From Copernicus to Darwin.* Baltimore: Johns Hopkins University Press, 2004.

Osler, Margaret J. *Reconfiguring the World.* Baltimore: Johns Hopkins University Press, 2010.

Pantin, C. F. A. *The Relations Between the Sciences.* London: Cambridge University Press, 1968.

Parascandola, John and James C. Whorton, eds. *Chemistry and Modern Society: Historical Essays in Honor of Aaron J. Ihde.* Washington, DC: American Chemical Society, 1983.

Partington, James R. *A History of Chemistry.* 4 vols. New York: St. Martin's Press, 1962.

Pullman, Bernard. *The Atom in the History of Human Thought.* New York: Oxford University Press, 1998.

Pursell, Carroll W., ed. *Technology in America: a History of Individuals and Ideas.* Cambridge, MA: MIT Press, 1981.

Pursell, Carroll W., ed. *The Machine in America: A Social History of Technology*, 2d ed. Baltimore: Johns Hopkins University Press, 2007.

Reingold, Nathan, ed. *Science in America Since 1820.* New York: Science History Publications, 1976.

Ronan, Colin A. *Science: Its History and Development Among the World's Cultures.* New York: Facts On File Publications, 1982.

Rosner, Lisa M., ed. *Chronology of Science: From Stonehenge to the Human Genome Project.* Santa Barbara, CA: ABC-CLIO, 2002.

Rossi, Paulo. *The Birth of Modern Science*. Oxford: Blackwell Publishers, 2001.

Rossiter, Margaret W. *Women Scientists in America: Before Affirmative Action, 1940-1972*. Vol. 2. Baltimore: Johns Hopkins University Press, 1998.

Rossiter, Margaret W. *Woman Scientists in America: Forging a New World Since 1972*. Vol. 3. Baltimore: Johns Hopkins University Press, 2012.

Rossiter, Margaret W. *Women Scientists in America: Struggles and Strategies to 1940.* Vol 1. Baltimore: Johns Hopkins University Press, 1982.

Rothenberg, Marc, ed. *The History of Science in the United States.* New York: Garland Publishing, 2001.

Shapin, Steven. *Never Pure*. Baltimore: Johns Hopkins University Press, 2010.

Silver, Brian L. *The Ascent of Science*. New York: Oxford University Press, 1998.

Szpiro, George G. *Poincaré's Prize: The Hundred-Year Quest to Solve One of Math's Greatest Puzzles*. New York: Penguin Books, 2008.

Taylor, Sherwood F. *A Short History of Science and Scientific Thought.* New York: W. W. Norton & Company, 1963.

Thackray, Arnold. *Atoms & Powers: An Essay on Newtonian Matter-Theory and the Development of Chemistry*. Cambridge: Harvard University Press, 1970.

Toulmin, Stephen and June Goodfield. *The Architecture of Matter.* Chicago: University of. Chicago Press, 1976.

Toulmin, Stephen and June Goodfield. *The Fabric of the Heavens: The Development of Astronomy and Dynamics.* New York: Harper & Row, 1961.

Trigg, George L. *Crucial Experiments in Modern Physics*. New York: Crane, Russak & Company, 1975.

Turnbull, Herbert Westren. *The Great Mathematicians*. New York: Simon and Schuster, 1962.

Van Helden, Albert. *Measuring the Universe: Cosmic Dimensions from Aristarchus to Halley.* Chicago: University of Chicago Press, 1985.

Van Melsen, Andrew G. *From Atomos to Atom: The History of the Concept of Atom.* Duquesne University Press, 1952.

Warren, Louis S., ed. *American Environmental History.* Malden, MA: Blackwell, 2003.

Weinberg, Steven. *To Explain the World: The Discovery of Modern Science*. New York: Harper Collins Publishers, 2015.

Williams, Trevor. *A Short History of Twentieth-Century Technology.* New York: Oxford University Press, 1982.

Wright, Russell O. *Chronology of Energy in the United States.* Jefferson, NC: McFarland, 2003.

### *Internet Sources*

Historical and scientific societies, foundations, professional organizations, universities, government agencies, and private researchers have placed on-line a large number of scientific papers, books, apparatus, and other sources chronicling past and current developments in the sciences. These groups include the History of Science Society (HSS), the Society for the History of Technology (SHOT), the American Chemical Society (ACS), the American Institute of Physics (AIP), the American Association for the Advancement of Science (AAAS), the Intergovernmental Panel on Climate Change (IPCC), the National Aeronautical and Space Agency (NASA), the US Department of Energy (DOE), Los Alamos National Laboratory (LANL), Library of Congress (LOC), National Academy of Sciences (NAS), the Chemical Heritage Foundation, Carmen Giunta Classic Chemistry, the Royal Society, The History of Science Museum in Florence, the Deutsches Museum in Munich, Museé des Arts et Métiers, and many others, too numerous to list. A simple internet search makes them readily available. For example, papers published in *Science*, the official journal of the AAAS, are available from the early 1900s. Articles, reports, and diagrams on energy, the atomic bomb (fission), the hydrogen bomb (fusion), and nuclear reactors are at several government sites such as DOE, LANL, and Argonne National Laboratory (ANL). Sources on climate change are available at DOE, NASA, National Oceanic and Atmospheric Association (NOAA), and IPCC. University sites such as Oregon State University contain Linus Pauling's papers. The papers of Joseph Henry, Thomas Edison, Alexander Graham Bell, Charles Darwin, and Albert Einstein are also on-line. The list given here represents only a small sample of available on-line sources.

### *Journals*

*Agricultural History*: official journal of the Agricultural History Society, published quarterly
*Annals of Science*: published quarterly
*British Journal for the History of Science*: official journal for the British Society for the History of Science, published quarterly
*Bulletin for the History of Chemistry*: official journal for the History of Chemistry Division, American Chemical Society, published semi-annually
*Bulletin of the History of Medicine*: official journal for the American Association for the History of Medicine, published quarterly
*Environmental HISTORY*: official journal for the American Society for Environmental History, published quarterly
*Historical Studies in the Physical and Biological Sciences*, published semi-annually
*History of Science*, published quarterly
*History and Technology,* published quarterly
*ICON*: official journal for the International Committee for the History of Technology, published annually
*Isis*: official journal of the History of Science Society, published quarterly
*Isis Current Bibliography:* indexes articles, books, and dissertations in the history of science. It covers all disciplines and time periods, published annually
*Journal of Chemical Education*: published monthly
*Journal of the History of Biology*: published quarterly
*Osiris*: a thematic journal, published annually by the History of Science Society
*Physics Teacher*: published by the American Association of Physics Teachers, nine issues per year
*Physics Today*: published by the American Association of Physics Teachers, monthly
*Social Studies of Science*: official journal of the Society for the Social Studies of Science, published bi-monthly

*Technology and Culture*: official journal for the Society of the History of Technology, published quarterly

### *Nobel Lectures*

Biographies and Nobel Lectures of Nobel Prize recipients in chemistry, physics, and medicine (physiology). Elsevier Publishing Company published the biographies and lectures in each discipline from 1901 to 1970, when it discontinued the series. Beginning in 1991 World Scientific Publishing Company in Singapore has published the volumes including the 1901-70 volumes and continuing to 2005.

### *Other Collections*

*Landmarks of science* and *landmarks II.* New York: Readex Microprint, 1967- present. A microprint collection of journals and monographs on the history of science, covering years from 1482 to 1812. A major reference collection which cuts across all fields of physical, biological, mathematical and earth sciences.

*Energy and agriculture*. Sanford, NC: Microfilming Corporation of America. Provides primary source materials on the development of energy from agricultural products as well as alternative energy sources such as solar, wind, geothermal and nuclear. Also stressed is the importance of energy for the production, processing and distribution of agricultural products.

*Nuclear Weapons. Arms Control and the Threat of Thermonuclear War: Special Studies, 1969-1981.* Frederick, MD: University Publications of America, 1982. The studies in this collection consider nearly all of the crucial questions on the nuclear arms race. They offer comprehensive analyses of the relative strengths, weaknesses and cost effectiveness of the key weapons systems. The studies explore the international political climate which has given rise to the renewed arms race and review all aspects of the SALT negotiations and the substantial political, ideological and technical barriers which have stood in the way of bona fide progress in limiting nuclear weapons.

Bindocci, Cynthia Gay. *Women and Technology: An Annotated Bibliography and Research Guide.* New York: Garland Publishing, 1993.

***Source Books (Primary Documents)***


Leicester, Henry M., ed. *A Source Book in Chemistry, 1900-1950.* Cambridge: Harvard University Press, 1968.

Leicester, Henry M. and Herbert M. Klickstein, eds. *A Source Book in Chemistry, 1400-1900.* New York: McGraw-Hill, 1952.

Magie, William F., ed. *A Source Book in Physics.* Cambridge: Harvard University Press, 1935.

Mather, Kirtley F., ed. *A Source Book in Geology, 1900-1950.* New York: McGraw-Hill, 1939.

Shapley, Harlow, ed. *A Source Book in Astronomy, 1900-1950.* Cambridge: Harvard University Press, 1960.

Shapley, Harlow and H. E. Howarth, eds. *A Source Book in Astronomy.* New York: McGraw-Hill, 1929.

# INDEX

## A

# B

# C

# D

# F

# G

# H

# I



# J



# K

# M

# N

## O

# Q



# R

## S

# T

## U



## V



## W

# X



# Y



# Z

CPSIA information can be obtained at www.ICGtesting.com
Printed in the USA
LVOW09s0129010815

448471LV00001B/3/P

9 781465 284488